Edited by
Julio Alvarez-Builla,
Juan Jose Vaquero,
and José Barluenga

Modern Heterocyclic Chemistry

Related Titles

Majumdar, K. C., Chattopadhyay, S. K. (eds.)

Heterocycles in Natural Product Synthesis

2011

ISBN: 978-3-527-32706-5

Eicher, T., Hauptmann, S., Speicher, A.

The Chemistry of Heterocycles

Structure, Reactions, Synthesis, and Applications

Third Edition

2011

ISBN: 978-3-527-32868-0

Yudin, A. K. (ed.)

Catalyzed Carbon-Heteroatom Bond Formation

2010

ISBN: 978-3-527-32428-6

Ma, S. (ed.)

Handbook of Cyclization Reactions

2009

ISBN: 978-3-527-32088-2

L. Ackermann (Ed.)

Modern Arylation Methods

2009

ISBN: 978-3-527-31937-4

H. Yamamoto, K. Ishihara (Eds.)

Acid Catalysis in Modern Organic Chemistry

2 Volumes

2008

ISBN: 978-3-527-31724-0

J. Royer (Ed.)

Asymmetric Synthesis of Nitrogen Heterocycles

2009

ISBN: 978-3-527-32036-9

Edited by
Julio Alvarez-Builla, Juan Jose Vaquero, and José Barluenga

Modern Heterocyclic Chemistry

Volume 2

WILEY-VCH Verlag GmbH & Co. KGaA

The Editors

Prof. Dr. Julio Alvarez-Builla
Universidad de Alcalá
Facultad de Farmacia
Dpto. de Química Organíca
Campus Universitario s.n.
Alcalá de Henares
28871 Madrid
Spain

Dr. Juan Jose Vaquero
Universidad de Alcalá
Dpto. de Química Organíca
Ctra. Madrid-Barcelona km 33
Alcalá de Henares
28871 Madrid
Spain

Prof. Dr. José Barluenga
Universidad de Oviedo
Instituto Universitario de
Química Organometálica "Enrique Moles"
33071 Oviedo
Spain

Library of Congress Card No.: applied for

British Library Cataloguing-in-Publication Data
A catalogue record for this book is available from the British Library.

Bibliographic information published by the Deutsche Nationalbibliothek
The Deutsche Nationalbibliothek lists this publication in the Deutsche Nationalbibliografie; detailed bibliographic data are available on the Internet at http://dnb.d-nb.de.

Typesetting Thomson Digital, Noida, India
Printing and Binding betz-druck GmbH, Darmstadt
Cover Design Schulz Grafik-Design, Fußgönheim

Printed in the Federal Republic of Germany
Printed on acid-free paper

Print ISBN: 978-3-527-33201-4
oBook ISBN: 978-3-527-63406-4

Contents

List of Contributors

Ramón Alajarín
Universidad de Alcalá
Departamento de Química Orgánica
Alcalá de Henares
28871 Madrid
Spain

Benito Alcaide
Universidad Complutense de Madrid
Facultad de Química
Departamento de Química Orgánica I
28040 Madrid
Spain

Pedro Almendros
Instituto de Química Orgánica General
(CSIC)
Juan de la Cierva, 3
28006 Madrid
Spain

Julio Alvárez-Builla
Universidad de Alcalá
Facultad de Farmacia
Departamento de Química Orgánica
Alcalá de Henares
28871 Madrid
Spain

Venkataramanarao G. Anand
Regional Research Laboratory (CSIR)
Chemical Sciences and Technology
Division
Photosciences and Photonics Section
Trivandrum 695 019
India

José Barluenga
Universidad de Oviedo
Instituto Universitario de Química
Organometálica "Enrique Moles"
Julián Clavería 8
33006 Oviedo
Spain

Miguel Bayod
Asturpharma S.A.
Peña Brava 23
Polígono Industrial Silvota
33192 Llanera, Asturias
Spain

Jan Bergman
Karolinska Institute
Department of Biosciences and
Nutrition
Unit of Organic Chemistry
Novum Research Park
141 57 Huddinge
Sweden

Ulhas Bhatt
Albany Molecular Research, Inc.
Albany, NY 12212
USA

Carolina Burgos
Universidad de Alcalá
Departamento de Química Orgánica
Alcalá de Henares
28871 Madrid
Spain

María-Paz Cabal
Universidad de Oviedo
Instituto Universitario de Química
Organometálica "Enrique Moles"
Julián Clavería 8
33006 Oviedo
Spain

Luis Castedo
Universidad de Santiago de Compostela
Facultad de Química
Departamento de Química Orgánica
15782 Santiago de Compostela
Spain

Tavarekere K. Chandrashekar
Regional Research Laboratory (CSIR)
Chemical Sciences and Technology
Division
Photosciences and Photonics Section
Trivandrum 695 019
India

and

Indian Institute of Technology
Department of Chemistry
Kanpur 208 016
India

Ugo Chiacchio
Università di Catania
Dipartimento di Scienze Chimiche
Viale Andrea Doria 6
95125 Catania
Italy

Stefano Cicchi
Università degli Studi di Firenze
Dipartimento di Chimica "Ugo Schiff"
via della Lastruccia 13
50019 Sesto Fiorentino-Firenze
Italy

Ana M. Cuadro
Universidad de Alcala
Departamento de Química Orgánica
Alcalá de Henares
28871 Madrid
Spain

José Elguero
University of Aveiro
Instituto de Química Médica (CSIC)
Department of Chemistry
Juan de la Cierva, 3
28006 Madrid
Spain

José L. García-lvarez
Universidad de Oviedo
Instituto Universitario de Química
Organometálica "Enrique Moles"
Departamento de Química Orgánica e
Inorgánica
Unidad asociada al CSIC
Julian Claveria 8
33006 Oviedo
Spain

Cristina Gómez de la Oliva
Instituto de Química Médica (CSIC)
Juan de la Cierva, 3
28006 Madrid
Spain

Concepción González-Bello
Universidad de Santiago de Compostela
Facultad de Ciencias
Departamento de Química Orgánica
27002 Lugo
Spain

Pilar Goya Laza
Instituto de Química Médica (CSIC)
Juan de la Cierva, 3
28006 Madrid
Spain

Bernardo Herradón
Instituto de Química Orgánica (CSIC)
Juan de la Cierva, 3
28006 Madrid
Spain

Xue-Long Hou
The Chinese Academy of Sciences
Shanghai Institute of Organic Chemistry
Shanghai-Hong Kong Joint Laboratory in Chemical Synthesis and State Key Laboratory of Organometallic Chemistry
354 Feng Lin Road
Shanghai 200032
China

Hui Huang
The Chinese Academy of Sciences
Shanghai Institute of Organic Chemistry
Shanghai-Hong Kong Joint Laboratory in Chemical Synthesis
354 Feng Lin Road
Shanghai 200032
China

Tomasz Janosik
Karolinska Institute
Department of Biosciences and Nutrition
Unit of Organic Chemistry
Novum Research Park
141 57 Huddinge
Sweden

Amparo Luna
Universidad Complutense de Madrid
Facultad de Química
Departamento de Química Orgánica I
28040 Madrid
Spain

Fabrizio Machetti
Instituto Chimica dei Composti Organometallica del CNR c/o
Dipartimento di Chimica Organica "Ugo Schiff"
Via Madonna del Piano 10
50019 Sesto Fiorentino (Firenze)
Italy

François Mathey
Nanyang Technical University
School of Physical and Mathematical Sciences
Division of Chemistry and Biological Chemistry
21 Nanyang Link
637371 Singapore
Singapore

S. Shaun Murphree
Allegheny College
Department of Chemistry
520 N, Main Street
Meadville, PA 16335
USA

Carmen Ochoa de Ocariz
Instituto de Química Médica (CSIC)
Juan de la Cierva, 3
28006 Madrid
Spain

Teresa M.V.D. Pinho e Melo
Universidade de Coimbra
Departamento de Química
3004-535 Coimbra
Portugal

Julia Revuelta
Instituto de Quimica Organica General (CSIC)
Grupo de Quimica Organica Biologica
C/Juan de la Cierva, 3
28006 Madrid
Spain

Sylvie Robin
Université de Paris-Sud
ICMMO
Laboratoire ed Synthèse Organique et Méthodologie
Université de Paris-Sud
91405 Orsay
France

Giovanni Romeo
Università di Messina
Dipartimento Farmaco-Chimico
Via SS Annunziata
98168 Messina
Italy

Gérard Rousseau
Université de Paris-Sud
ICMMO
Laboratoire ed Synthèse Organique et Méthodologie
Université de Paris-Sud
91405 Orsay
France

Javier Santamaría
Universidad de Oviedo
Instituto Universitario de Química Organometálica "Enrique Moles"
Departamento de Química Orgánica e Inorgánica
Unidad asociada al CSIC
Julian Claveria 8
33006 Oviedo
Spain

Artur M.S. Silva
University of Aveiro
Department of Chemistry
3810-193 Aveiro
Portugal

Alagar Srinivasan
Regional Research Laboratory (CSIR)
Chemical Sciences and Technology Division
Photosciences and Photonics Section
Trivandrum 695 019
India

Augusto C. Tomé
University of Aveiro
Department of Chemistry
3810-193 Aveiro
Portugal

Carlos Valdés
Universidad de Oviedo
Instituto Universitario de Química Organometálica "Enrique Moles"
Julián Clavería 8
33006 Oviedo
Spain

Juan J. Vaquero
Universidad de Alcala
Departamento de Química Orgánica
Alcalá de Henares
28871 Madrid
Spain

José M. Villalgordo
Villalpharma S.L.
Polígono Industrial Oeste
C/Paraguay, Parcela 7/5-A, Módulo A-1
30169 Murcia
Spain

David J. Wilkins
Key Organics Ltd.
Highfield Industrial State
Camelford
Cornwall PL32 9QZ
UK

Henry N.C. Wong
The Chinese University of Hong Kong
Institute of Chinese Medicine, and
Central Laboratory of the Institute of
Molecular Technology for Drug
Discovery and Synthesis
Department of Chemistry
Center of Novel Functional Molecules
Shatin, New Territories
Hong Kong SAR, China

and

The Chinese Academy of Sciences
Shanghai Institute of Organic
Chemistry
Shanghai-Hong Kong Joint Laboratory
in Chemical Synthesis
354 Feng Lin Road
Shanghai 200032
China

Jie Wu
Fudan University
Department of Chemistry
220 Handan Road
Shanghai 200433
China

Larry Yet
299 Georgetown Ct
Albany, NY 12203
USA

Kap-Sun Yeung
Bristol-Myers Squibb Pharmaceutical
Research Institute
5 Research Parkway
P.O. Box 5100
Wallingford, CT 06492
USA

8
Five-Membered Heterocycles: 1,2-Azoles. Part 1. Pyrazoles

José Elguero, Artur M.S. Silva, and Augusto C. Tomé

8.1
Introduction

Pyrazoles belong to the family of azoles, which for some authors include pyrroles and indoles while for others it contains only by imidazoles, benzimidazoles, pyrazoles, indazoles, 1,2,3-triazoles, benzotriazoles, 1,2,4-triazoles, tetrazoles and pentazoles. The dubious position of pyrroles is due to their very different reactivity and less aromatic stability.

We have carried out a search in the *Chemical Abstracts* "on line" (1987–2004) using the following truncated words: pyrazol*, indazol*, imidazol*, benzimidazol*, triazol* (thus treating together 1,2,3 and 1,2,4-triazoles), benzotriazol*, tetrazol* and pentazol*. In this way, fused compounds are included although they will not be discussed in detail in this chapter. Table 8.1 gives the number of references and the percentages in each case.

The evolution of the number of publications between 1999 and 2004 show an increase of all of them, with the most cited, imidazoles and pyrazoles, growing the fastest.

From Figure 8.1 is can be concluded that benzazoles are much less studied than the corresponding azoles and that an increase in the number of nitrogen atoms diminishes the importance of the azoles, with pentazoles being only a curiosity.

In the *Chemical Abstracts* 2004, the word pyrazol* appears 1931 times. Regarding these 1931, it is possible to classify them (at the price of some simplifications) into six different fields (Table 8.2).

As expected, the medicinal chemistry aspects of pyrazoles dominates largely the 2004 production, but note their great importance as ligands in coordination chemistry. As materials, the main applications are as dyes, inks and luminescent devices.

Modern Heterocyclic Chemistry, First Edition.
Edited by Julio Alvarez-Builla, Juan Jose Vaquero, and José Barluenga.

Table 8.1 Relative importance of azoles in the literature.

Azol*	Total	%
Pyrazol*	20701	22.12
Indazol*	1451	1.55
Imidazol*	31118	33.25
Benzimidazol*	10158	10.85
Triazol*	15688	16.76
Benzotriazol*	6429	6.87
Tetrazol*	8008	8.56
Pentazol*	42	0.05

Another way to classify pyrazoles is to made a class of 1,2-azoles formed by pyrazoles, isoxazoles and isothiazoles. This is not entirely satisfactory, because there is an important difference between pyrazoles on the one hand and isoxazoles and isothiazoles (Chapter 9) on the other: the presence of an NH (NR) in pyrazoles (**1–3**)

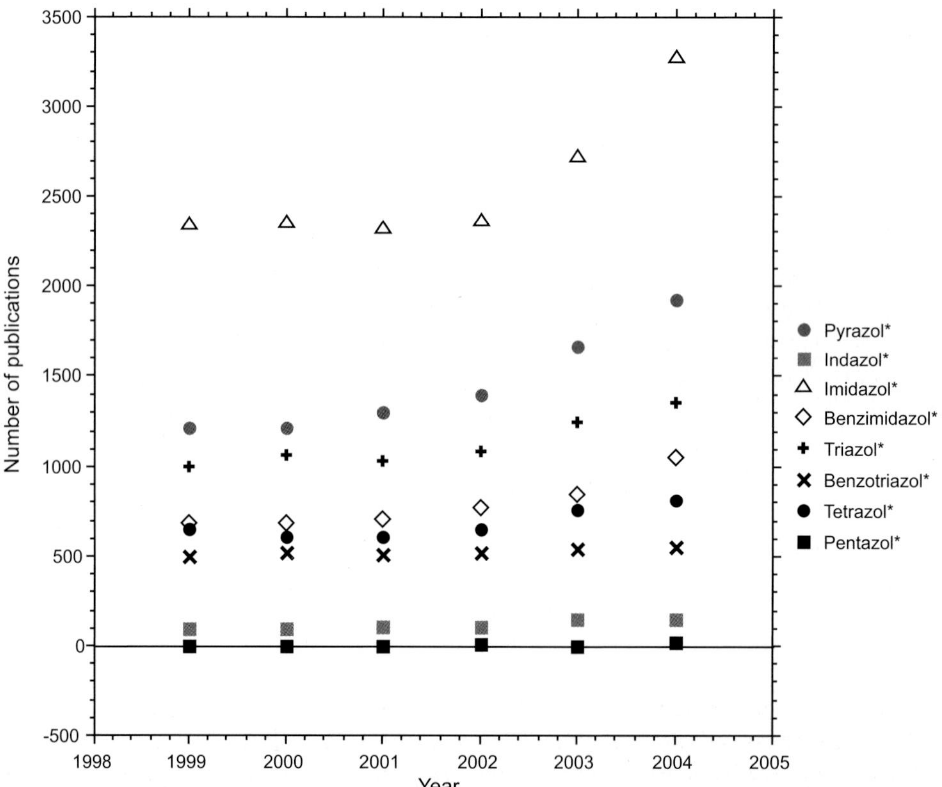

Figure 8.1 Evolution of the number of publications devoted to azoles between 1999 and 2004.

Table 8.2 Use of pyrazoles according to the number of citations.

Pyrazol*	References	%	% Without biol. appl.
Biological applications	785	40.65	—
Coordination chemistry	423	21.91	36.91
Synthesis	375	19.42	32.72
Properties	125	6.47	10.91
Application as materials	118	6.11	10.30
Reactivity	105	5.44	9.16

that confers to this heterocycle a specific behavior to the point that we have treated it separately from the two other heterocycles.

1*H*-pyrazole (**1**) 1*H*-indazole (**2**) 2*H*-indazole (**3**)

8.1.1 Nomenclature

We have represented above the three main heterocycles of this chapter. Notably, the migration of the proton of a pyrazole from N1 to N2 changes the numbering of the ring, that is 3-methyl-1*H*-pyrazole (**4a**) becomes 5-methyl-1*H*-pyrazole (**4b**), whereas in indazoles the same migration transforms 3-methyl-1*H*-**5a** into 3-methyl-2*H*-indazole (**5b**).

4a **4b** **5a** **5b**

A series of books or chapters in books have been devoted to this heterocycle, including its benzo derivative, indazole, with exclusion of the very large topic of pyrazoles fused with other heterocycles. In 1966, Kost and Grandberg published the first systematic approach to pyrazole chemistry [1]. It is still a valuable source of information because it summarizes in a short and clear way the knowledge accumulated in Russia after the seminal contribution of Karl Friedrich von Auwers (1863–1939) to the chemistry of these compounds. A year later, a book appeared that contains tables describing many pyrazoles, indazoles and their reduced derivatives, one of the authors, Fusco, was at that time very active in the chemistry of pyrazoles [2]. A book on all azoles is very useful since it represents a more structural and comparative study with much spectroscopic data [3]. In 1984 (updated in 1996) *Comprehensive Heterocyclic Chemistry* provided an extensive treatment of the structure and reactivity of pyrazoles and indazoles but with a concise report on synthetic aspects [4, 5]. The synthesis is well developed in *Houben-Weyl, Methoden der Organischen Chemie* and its continuation, *Science of Synthesis*, which contains synthetic recipes [6–9].

8.2 General Reactivity

Depending on the oxidation degree, pyrazoles can be classified into different families (Scheme 8.1). Indazoles saturated in the benzene ring (4,5,6,7-tetrahydroindazoles) are better considered as 3(5),4-tetramethylenepyrazoles. The 3*H*-indazoles, a third isomer of indazoles, are unstable if one (or both) of the substituents at position 3 is an H atom, that is, there are no 3*H*-indazole tautomers.

8.2.1 Relevant Physicochemical Data, Computational Chemistry and NMR Data

Figure 8.2 shows the structures of the three parent compounds, giving an image of their compact character. Table 8.3 summarizes the properties of pyrazole (**1**) and 1*H*-indazole (**2**) (the other tautomer, **3**, is unstable – see under tautomerism).

In general, pyrazoles unsubstituted at position 1, NH-pyrazoles, and NH-indazoles are solids, the exception being some pyrazoles substituted at position 4, like 4-methylpyrazole [28]. The N-substitution, especially the N-alkylation, is accompanied by a large decrease in the melting point.

Figure 8.3 presents, in a simplified manner, the geometry of **1** in the gas phase (both from MW spectroscopy and from high-level theoretical calculations). Pyrazole is planar and to a first approximation has a regular pentagonal geometry. Closer examination reveals alternating single (C3-C4)/(C5-N1) and double (C4-C5)/(N2-C3) bonds. Particularly relevant to determining the position of the NH, when it is not observed in X-ray crystallography due to disorder, is the fact that the angle on N1 is always larger than that on N2 (the same happens at C3 and C5).

Structural assignment of pyrazoles and indazoles is usually carried out by NMR (Schemes 8.2–8.4), at one time by using tables of reference compounds and/or coupling constants [29] but now more often by 2D spectroscopy, such as NOESY, which detects the proximity of substituents [30].

The identification of pyrazole isomers is easy by NMR when there is only one substituent at positions 3(5). The ^{1}H NMR chemical shifts are dependent on the

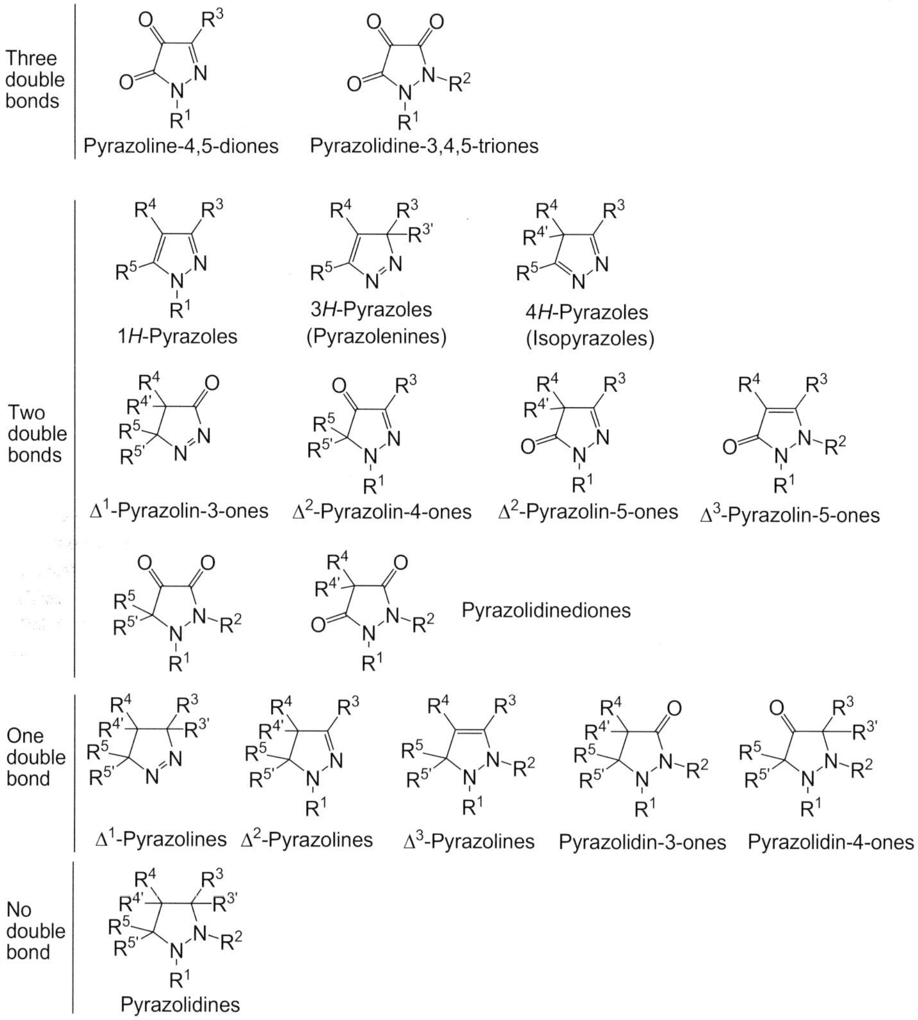

Three double bonds: Pyrazoline-4,5-diones; Pyrazolidine-3,4,5-triones

Two double bonds: 1H-Pyrazoles; 3H-Pyrazoles (Pyrazolenines); 4H-Pyrazoles (Isopyrazoles); Δ^1-Pyrazolin-3-ones; Δ^2-Pyrazolin-4-ones; Δ^2-Pyrazolin-5-ones; Δ^3-Pyrazolin-5-ones; Pyrazolidinediones

One double bond: Δ^1-Pyrazolines; Δ^2-Pyrazolines; Δ^3-Pyrazolines; Pyrazolidin-3-ones; Pyrazolidin-4-ones

No double bond: Pyrazolidines

Scheme 8.1 Different structures of pyrazole derivatives depending on their oxidation degree

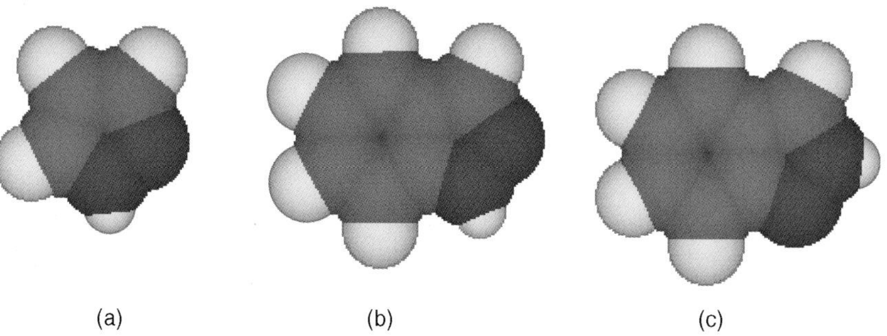

Figure 8.2 Space filling models of (a) pyrazole (**1**), (b) 1H-indazole (**2**) and (c) 2H-indazole (**3**).

Table 8.3 Properties of the parent compounds.

Property	Pyrazole (1)	Indazole (2)
Mp (°C)	69–70	145–149
Bp (°C)	186–188 (758 mmHg)	269–270 (743 mmHg)
Log *P*	0.13–0.26	1.82
Dipole moment (μ, D)	1.92 (benzene)	1.60 (benzene)
pK_a (proton addition)	2.52	1.31
pK_a (proton loss)	14.21	13.80
Enthalpies of formation (kJ mol^{-1})	179.4 [10]	243.0 [10]
UV (λ_{max} nm, log ε)	211 (3.49) [4]	250 (3.65), 284, 296 (3.52) [4]
IR (cm^{-1})	3524 (ν_{NH} gas) [4]	Source: [11]
X-ray (CSD) [12]	PYRZOL	INDAZL
^{1}H NMR	Source: [13]	Source: [13, 14]
^{13}C NMR	Source: [15]	Source: [16]
^{15}N NMR	Source: [17]	Source: [17]
MS	Source: [18]	Source: [19]
PES	Source: [20]	Source: [21]
MW	Source: [22]	Source: [23]
Best theoretical calculations	Source: [24]	Source: [25]
Aromaticity (benzene = 0.991)	0.900: [26]	0.808: [26][a)]

a) Naphthalene: 0.811, 2*H*-indazole: 0.792 (using as criteria the HOMA) [27].

solvent, but the $^3J_{34}$ vs. $^3J_{45}$ coupling constants are not, so the fact that $J_{45} > J_{34}$ is always a useful test. The ration J_{45}/J_{34} depends on the substituent on the nitrogen, with EWG (like tosyl) the difference is large and the criteria easy to apply. With EDG (like amino) the difference tends to blur and $J_{34} \approx J_{45}$ [31].

For isomeric pyrazoles bearing different substituents at positions 3 and 5, the ^{13}C chemical shifts of carbons C3 and C5 is a better method of assignment if both isomers are available. If only one is obtained, one must rely on a comparison with the large amount of data available [15]. In contrast, indazoles are easily identified by ^{13}C NMR spectroscopy [16].

^{15}N NMR spectroscopic data can be routinely obtained from unlabelled samples (natural abundance). But their utility as a diagnostic tool is limited. Note that the ^{15}N chemical shifts are very sensitive to hydrogen bonds and, obviously, to protonation.

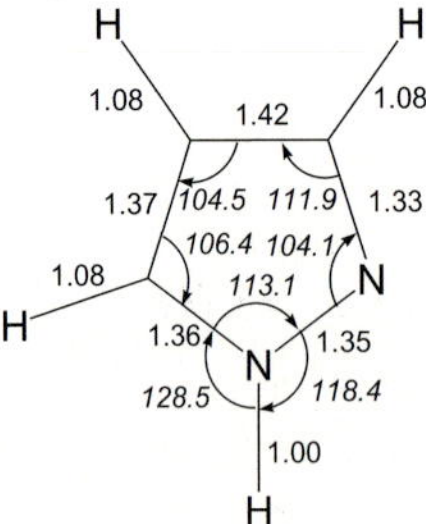

Figure 8.3 Geometry of pyrazole.

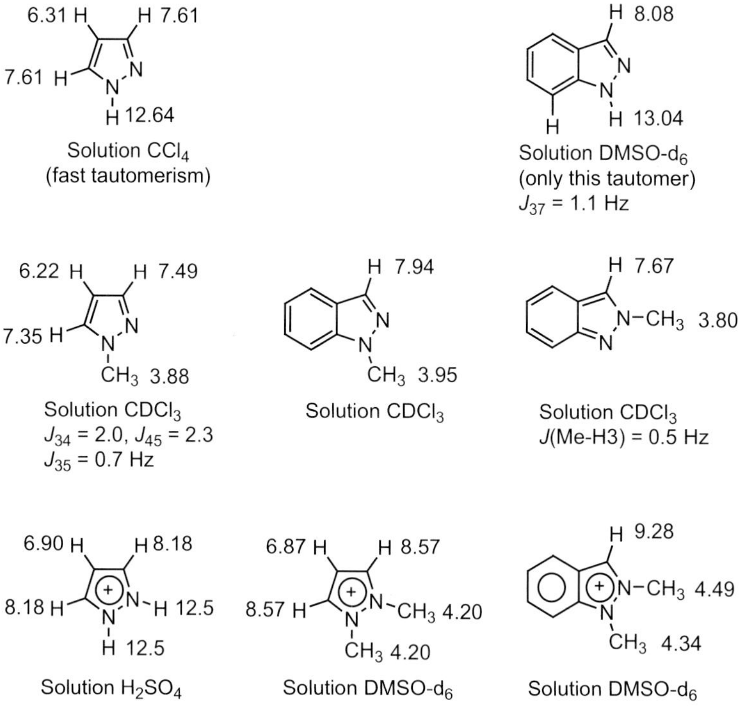

Solution CCl_4 (fast tautomerism)

Solution DMSO-d_6 (only this tautomer) J_{37} = 1.1 Hz

Solution $CDCl_3$ J_{34} = 2.0, J_{45} = 2.3 J_{35} = 0.7 Hz

Solution $CDCl_3$

Solution $CDCl_3$ J(Me-H3) = 0.5 Hz

Solution H_2SO_4

Solution DMSO-d_6

Solution DMSO-d_6

Scheme 8.2 ^{1}H NMR spectra [δ (ppm) and J (Hz)] of some representative compounds

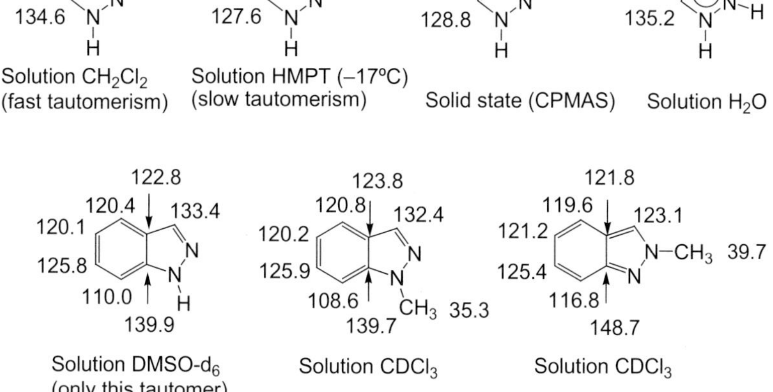

Scheme 8.3 ^{13}C NMR spectra [δ, (ppm)] of some representative compounds

NMR information on the non-aromatic derivatives of pyrazoles is very abundant, particularly on Δ^2-pyrazolines and on pyrazolones [4, 7]. Scheme 8.5 contains some multinuclear NMR data on Δ^2-pyrazolines.

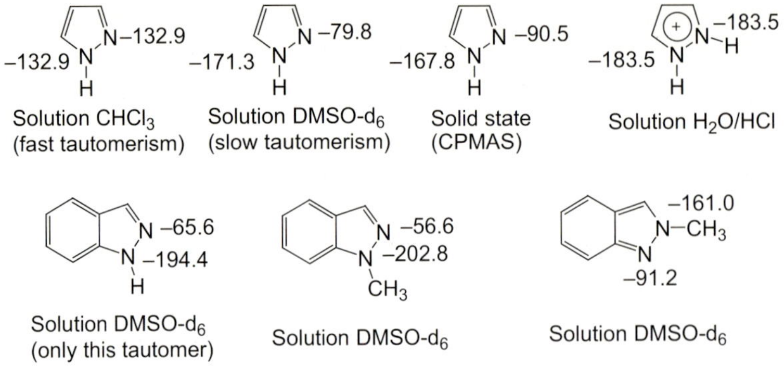

Scheme 8.4 ^{15}N NMR spectra [δ, (ppm)] of some representative compounds

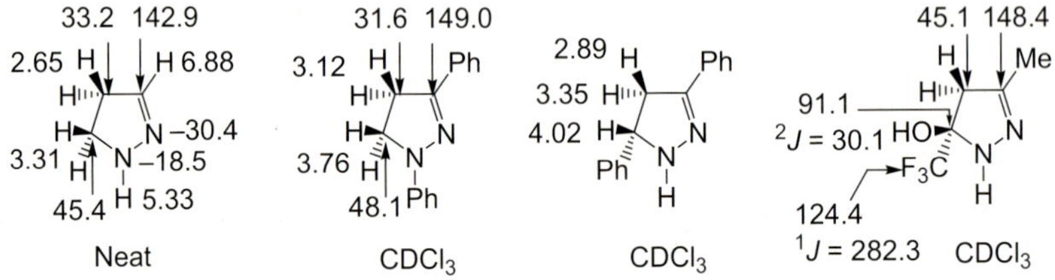

Scheme 8.5 NMR data on some Δ^2-pyrazolines

8.2.2
Tautomerism

The study of the tautomerism of NH-pyrazoles and indazoles (annular tautomerism) and that of functional compounds (e.g., pyrazolones) has been of considerable importance for the development of structural chemistry [32–34]. Some protonated cations also shown tautomerism. This huge amount of data is difficult to summarize, but the principal conclusions are as follows:

1) Tautomerism occupies a multidimensional space (Scheme 8.6): the three states of the matter, the thermodynamic and kinetic aspects, and the proton (prototropy versus elementotropy). All these aspects have been observed in pyrazoles although those situations marked with a gray circle are very rare. Theoretical studies of the tautomerism of pyrazole and its derivatives mainly concern the gas phase.
2) Concerning thermodynamic aspects, the main conclusion about annular tautomerism of pyrazoles (**6a** ⇌ **6b**) is that the equilibrium tautomeric constant K_T is never far from 1 [35–37]. The preference of **6a** versus **6b** has been analyzed using the Taft–Topson model [35, 36]. BH_2, COF, CO_2H, and CHO substituents stabilize the **6a** tautomer, while the **6b** tautomer is stabilized by OH, F, NH_2, Cl, $CONH_2$, CN, NO_2, and CH_3 groups (this paper contains a wealth of information about the IR of NH-pyrazoles) [37]. A more detailed analysis of the annular tautomerism of 3(5)-aminopyrazoles (R=NH_2) shows that both tautomers are present in solution and even in the solid state [38].

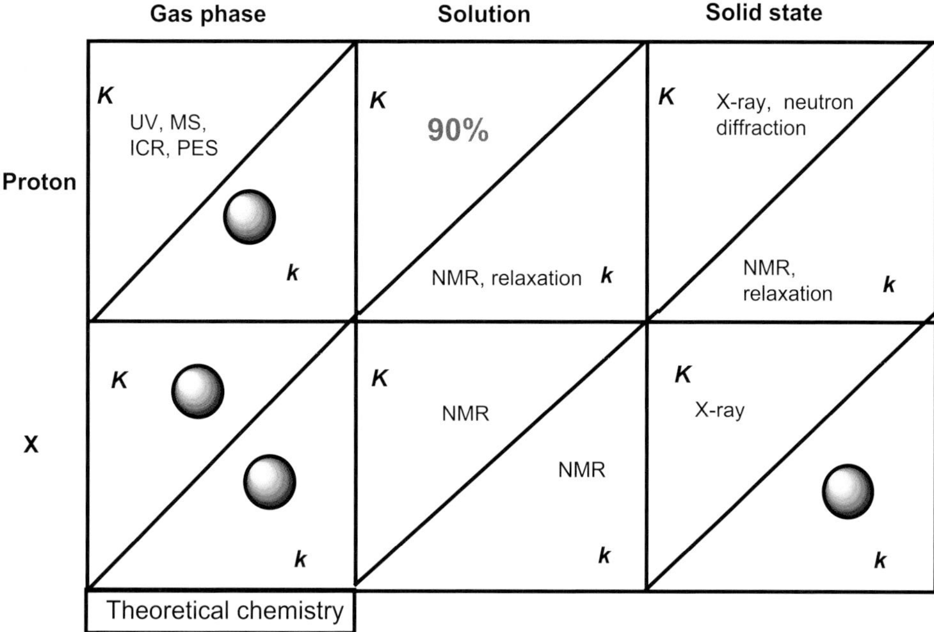

Scheme 8.6 The tautomerism twelve regions

For indazole, the 1*H*-tautomer (**2**) is more stable than the 2*H*-tautomer (**3**) by about 20 kJ mol^{-1}, and this tendency cannot be reversed by phase effects; consequently, in solution and in the solid state the only tautomer present is **2** [39]. To displace the equilibrium towards one of the tautomers it is necessary to use a combination of substituents, aza and annelation effects [39]. For instance, substituents like NO_2 and CO_2Me at position 3 favor the 2*H*-tautomer, the replacement of a CH by an N atom at position 7 favors the 1*H*-tautomer (by lone-pair/lone-pair repulsion in **7b**), and a fused benzo group at positions [*f*] and [*g*] favors, respectively, the 1*H*- (**8a**) and 2*H*-tautomers (**9b**).

6a ⇌ **6b** **8a** ⇌ **8b**

7a ⇌ **7b** **9a** ⇌ **9b**

3) In terms of kinetic aspects, the transfer of group R between both nitrogen atoms is an intermolecular process in the case of prototropy [40]. This requires other

molecules (another pyrazole, water, a solvent, etc.) or a surface, for instance that of the measuring instrument [41]. Other groups on the nitrogen migrate either intermolecularly (COR, CH_3, etc.) or intramolecularly (SiR_3, GaR_2, GeR_3, SnR_3, HgR) with barriers that can be very low in some cases [34, 41].

4) The tautomerism of functional derivatives, such as pyrazolones, is well understood and is no longer a research subject [32, 34]. However, there remains always some interesting cases, for instance that of 4-nitroso-5-aminopyrazoles **10** [42]. The compound exists as a mixture of rotamers **10a/10c** of the amino-nitroso tautomer, rather than a mixture of amino-nitroso/imino/oxime tautomers **10a/10b**.

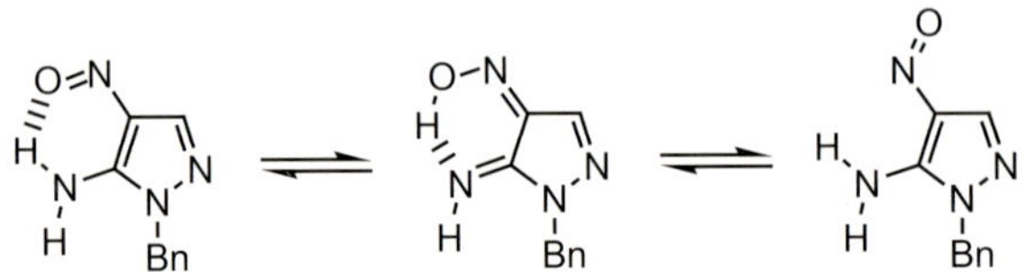

10a **10b** **10c**

8.3
Relevant Natural and/or Useful Compounds

The common belief is that heterocycles are natural if they can be found in DNA (nucleosides and nucleotides) or in proteins. That is why indole (tryptophan), pyrrole (proline, porphyrins), imidazole (histidine, biotine) and benzimidazole (vitamin B12) are considered "naturals" while pyrazole is not. One of the rare natural products that contains a pyrazole ring is Withasomnine [5,6-dihydro-3-phenyl-4*H*-pyrrolo[1,2-*b*]pyrazole (**11**) an alkaloid isolated from the roots of an Indian medicinal plant, *Withania somnifera*]: its simplicity means that several synthesis are known, with the most recent described in 2002 [43, 44].

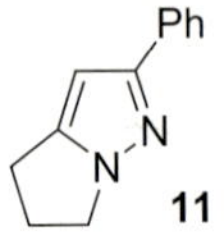

11

Other pyrazoles found in nature are pyrazofurin or pyrazomycin (an antibiotic isolated from the fermentation broth of *Streptomyces candidus*), formycin (a naturally occurring isomer of adenosine) and L-β-pyrazolylalanine (found in the seeds of many species of Cucurbitaceae).

From this it must not be concluded that the pyrazole skeleton is not a good scaffold for making drugs. A recent review examines the topic "Pyrazoles as Drugs: Facts and Fantasies" [45]. This review describes the past and the present of pyrazole derivatives in medicinal chemistry. From an important past, exemplified in the analgesic and anti-inflammatory pyrazolones and pyrazolinediones, not devoid of severe complications, to a glorious present with some of the most important drugs of recent times (sildenafil, celecoxib) being pyrazole derivatives.

Analgesics were, by far, the main area of biological activity of pyrazoles, often associated with antipyretic activity. They belong to three main classes: pyrazolin-5-ones [antipyrine –phenazone – (**12**), pyramidon (**13**), dipyrone (**14**)], pyrazolin-3,5-diones [phenylbutazone (**15**)] and pyrazoles [actually, an acetic acid, lonazolac (**16**)] [46].

Indazole proved to be an interesting nucleus in this field. Structural modifications of the anti-inflammatory agent bendazac (**17**), an indazole derivative, have been realized and, in some cases, the synthesized compounds showed analgesic effects along with anti-inflammatory properties [47].

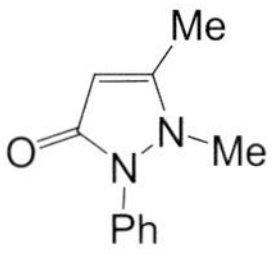

12 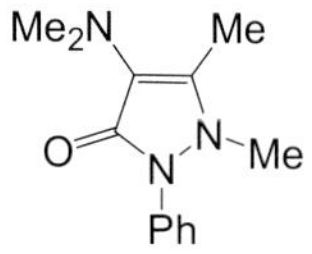 **13** 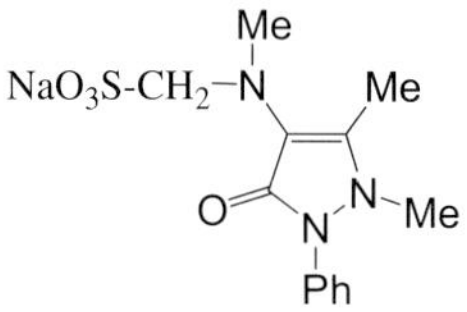 **14**

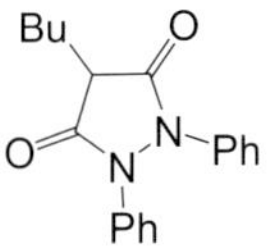

15 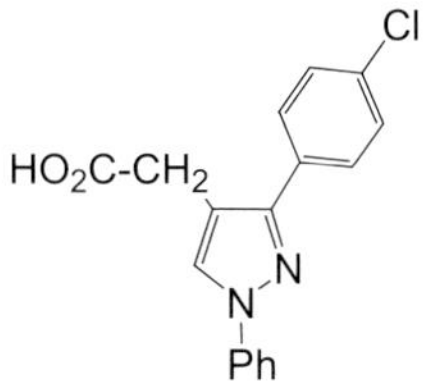 **16** 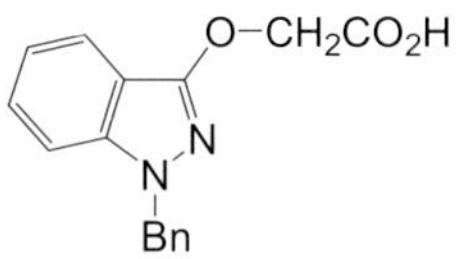 **17**

Lesopitron (**18**) (E-4424), a pyrimidylpiperazine substituted by 1-butyl-4-chloropyrazole, was introduced as a new non-benzodiazepine anxiolytic acting on 5-HT_{1A} receptors. It differed from other 5-HT_{1A} receptors ligands mainly because of its greater anxiolytic potency, its lack of sedative effects, its sustained activity even on long-term treatments and its lack of withdrawal problems [48]. Lesopitron, currently in advanced clinical trials (phase III), and has been shown to be efficient and safe in patients with generalized anxiety disorder. Of particular interest is zaleplon, *N*-[3-(3-cyanopyrazolo[1,5-*a*]pyrimidin-7-yl)phenyl]-*N*-ethylacetamide (**19**), a BZ_1-receptor selective ligand [49]. Zaleplon (Sonata, Wyeth-Pharma, MI **10165**) is a non-benzodiazepine sedative hypnotic that has been recently introduced for clinical use. It is indicated for short-term treatment of insomnia and presents the advantages of showing weak anxiolytic activity and reduced risk of tolerance.

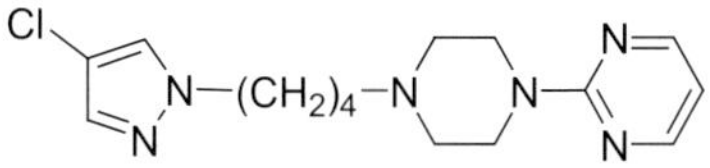

18 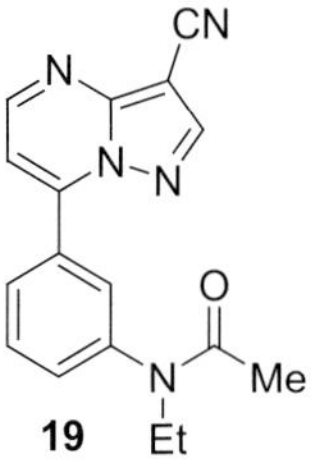 **19**

The implication of glutamate receptors in memory processes has initiated great interest in anti-Alzheimer therapy. In an attempt to obtain heterocyclic analogs of glutamic acid, a synthetic strategy has been developed to prepare pyrazoles that show biological activity at central glutamate receptors [50].

The discovery by Sanofi [51] in 1994 of the pyrazole SR141716A (**20**) has been of great interest in the field of cannabinoids because this study reported the first cannabinoid antagonist possessing nanomolar affinity. This selective and orally active CB_1 receptor subtype antagonist has become an experimental tool for insights into CB_1 subtype recognition and activation and for clinical applications such as treatment of psychosis, eating disorders or memory deficits [52]. Recent studies of analogs retaining the central pyrazole structure of **20**, tested for CB_1 binding affinity and in a battery of *in vivo* tests, suggest that the structural properties of 1- and 5-substituents are primarily responsible for the antagonist activity of SR141716A.

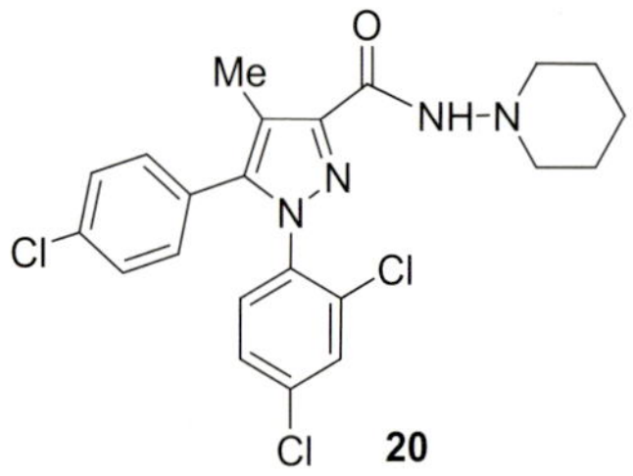

20

The availability in 1997 of the highly specific antagonist SR144528 (**21**) [53] for the CB_2 receptor has allowed the investigation of both the architecture of ligand binding sites, an approach that is difficult due to the structural disparity of cannabinoid agonists, and the respective contribution of cannabinoid receptor subtypes in functional cannabinoid effects *in vivo*. Its potential therapeutic applications include immune disorders such as rheumatoid arthritis, multiple sclerosis, psoriasis, infections and asthma.

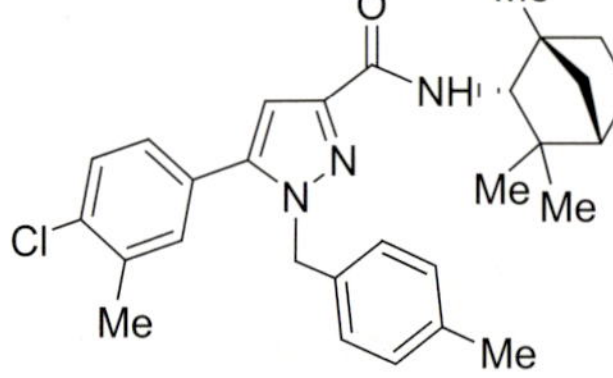

21

Various derivatives of SR141716A have been synthesized [54] and Makriyannis *et al.* [55] have reported a study of structure–activity relationships of pyrazole derivatives as cannabinoid receptor antagonists and have proposed structural requirements for CB_1 antagonistic activity.

6-Aryl-tetrahydropyridazin-3-ones are prototypes of cardiotonic agents, having both inotropic and vasodilator activities. Imazodan and bemoradan are two representatives of this family, which are supposed to exert their actions by selective inhibition of phosphodiesterase type 3 enzymes (PDE3). Many different structural variations have been devised in this series and, related to pyrazoles, the most interesting compound from a series of heterocyclic benzimidazolyl-pyridazinones is meribendan (**22**). It inhibited myocardial PDE3, showed an interesting calcium sensitizing effect and was selected for development as a positive inotrope [56].

The synthesis and inhibitory effects on cyclooxygenase, lipoxygenase and thromboxane synthetase of 3-amino-4,5-dihydro-1*H*-pyrazoles and related compounds have been reported. Among these, the trifluoromethylphenyl derivative **23** is the most interesting [57].

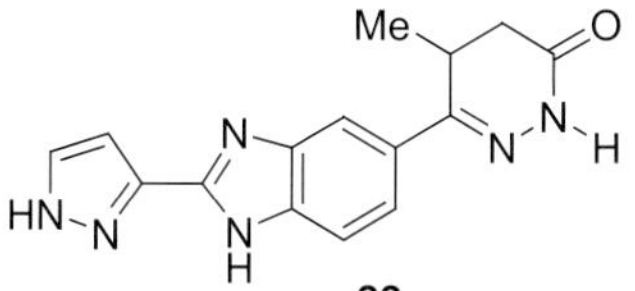

22 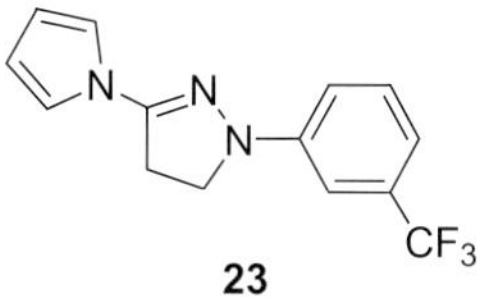 **23**

Progress in understanding inflammatory processes has led to the search of inhibitors of both the cyclooxygenase (COX) and lipoxygenase (LOX) pathways of the arachidonic acid cascade. In this way tepoxalin (**24**) has been prepared and found to be a potent anti-inflammatory agent [58].

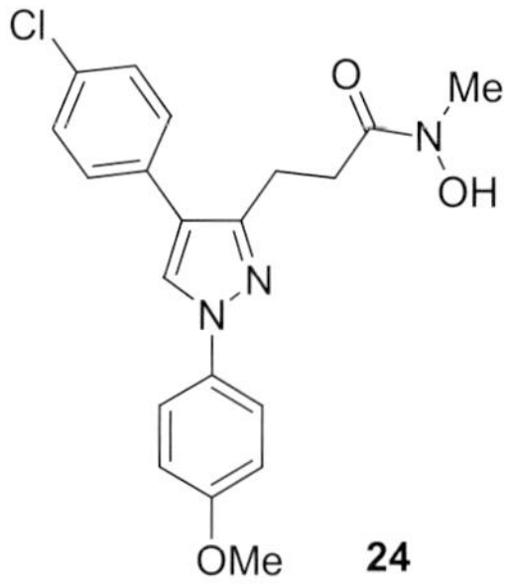

24

The discovery of a second, inducible form of cyclooxygenase (COX-2) that exists along with the constitutive form (COX-1) led to the hypothesis that selective inhibitors of COX-2 would be anti-inflammatory without causing the side effects associated with inhibition of COX-1 in the gastrointestinal tract and kidney. This is the moment most promising approach at present and has ultimately led Searle to SC-58125 (**25**) and then to *celecoxib* SC-58635 (**26**) (**MI** 1968), which is useful for the treatment of rheumatoid arthritis and osteoarthritis [59]. Other pharmaceutical companies have explored this avenue; for instance Fujisawa has developed **27** [60].

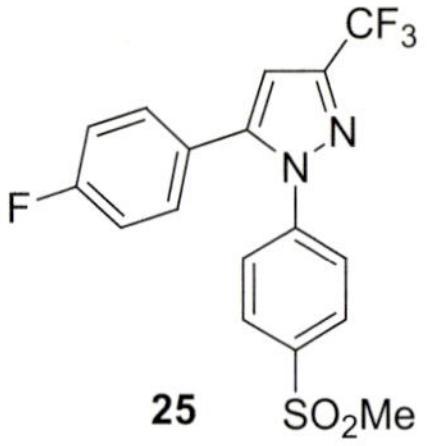

25

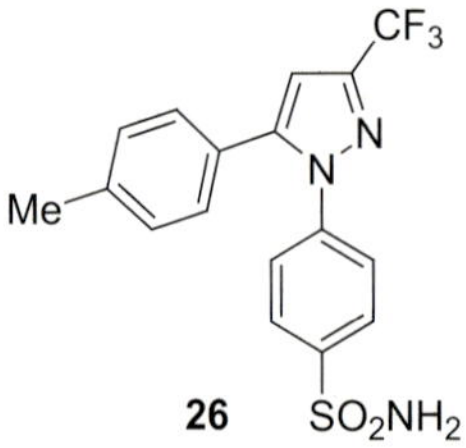

26

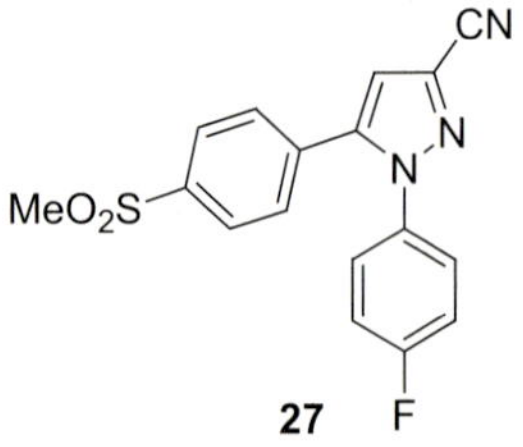

27

Other groups, like ASTA, have approached the problem by inhibiting the enzyme 5-LOX. By analogy with zileuton, one of the first launched 5-LOX inhibitors for the treatment of asthma, they prepared a series of 1,5-disubstituted indazol-3-ones, the most potent being **28** [61]. Mosti *et al.* have also reported indazoles related to angelicin, like compound **29a**, which shows good anti-inflammatory and antipyretic properties, while **29b** shows significant local anesthetic activity [62].

For the treatment of diabetes, a series of hypoglycemic agents derived from pyrazoles have been prepared and tested. The most interesting antidiabetic in this field is WAY-123783 (**30**) – obtained after extensive SAR studies; it acts by blocking SGTL (sodium-glucose co-transporter) in the kidney [63].

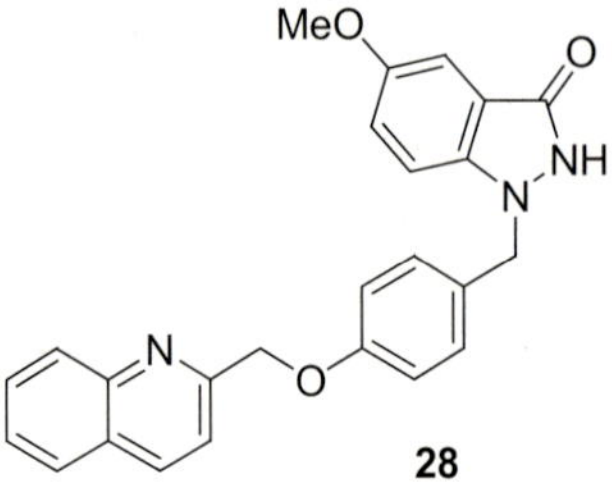

28

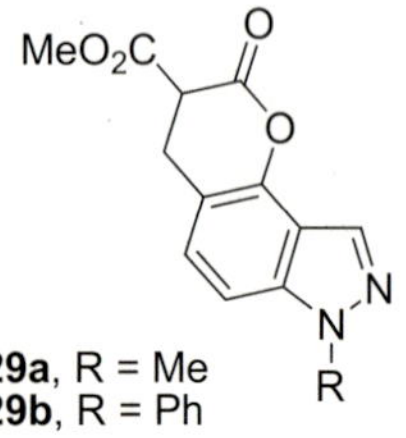

29a, R = Me
29b, R = Ph

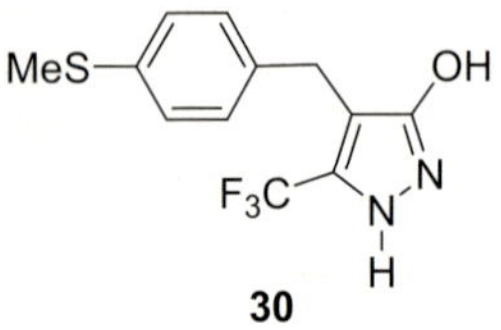

30

For the treatment of obesity, Henke from Glaxo Wellcome has optimized a series of 3-(1*H*-indazol-3-ylmethyl)-1,5-benzodiazepines – potent and orally active CCK-A agonists derived from **31** [64].

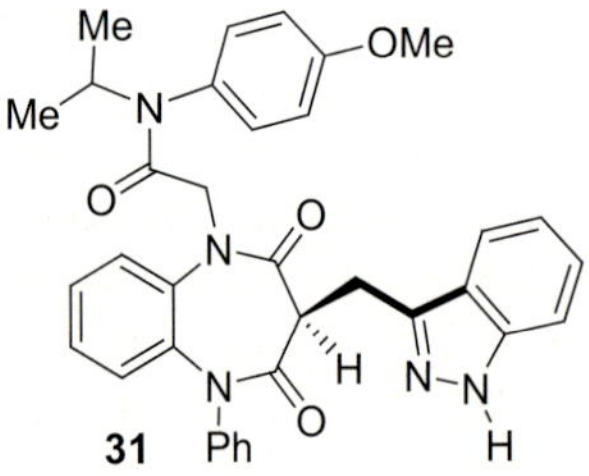

31

Steroidal pyrazoles have long been known. Kirschke [6] has reported several of these compounds, like cortivazol (**32**) (X-ray structure) [65] and stanozolol (**33**), both important and commonly used drugs. Cortivazol (**32**) is an anti-inflammatory glucocorticoid while stanozolol (**33**) is an anabolic steroid used as androgen. Nivazol (**34**) also belongs to the glucocorticoid class [66].

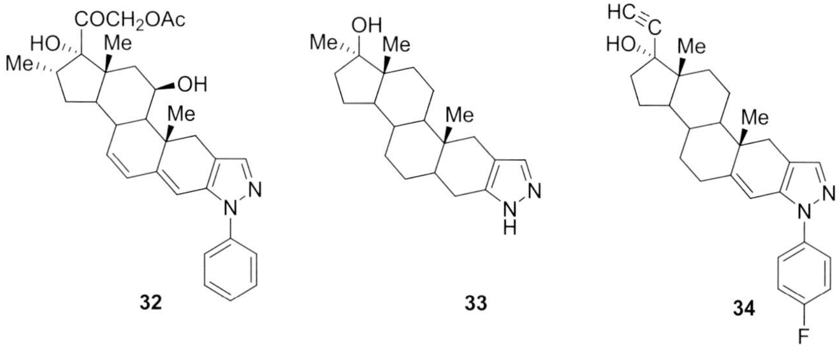

Liver alcohol dehydrogenase (EC 1.1.1.1) catalyzes the first step in alcohol metabolism and is a rational target for inhibiting alcohol metabolism. Prevention of poisoning by methanol and damaging effects of ethanol metabolism are potential applications of inhibitors of alcohol dehydrogenase. From the pioneering work of Theorell [67] it is known that pyrazole and some of its 4-substituted derivatives (4-methyl, 4-iodo and 4-bromo) are potent inhibitors of ethanol metabolism *in vivo*. Pyrazoles have been proposed as therapeutic agents for treatment of alcohol intoxication. Unfortunately, pyrazole is itself toxic and may not be useful for long-term treatment of humans.

Although some interesting efforts have been made, including X-ray studies and molecular modeling [68], 4-methylpyrazole (fomepizol) continues to be the most efficient and less toxic of all the liver alcohol dehydrogenase inhibitors and inactivators. Note also that pyrazole itself, an alcohol dehydrogenase inhibitor, has dual effects on *N*-methyl-D-aspartate (NMDA) receptors of hippocampal pyramidal cells, agonist and noncompetitive antagonist [69].

One of the most prominent of the anticancer agents with a pyrazole skeleton is pyrazoloacridine (PZA, **35**), 2-(*N*,*N*-dimethylaminopropyl)-9-methoxy-5-nitro-(6*H*)-pyrazolo[3,4,5-*kl*]acridine (NSC 366140). PZA is the first of a new class of rationally synthesized acridine derivatives to undergo clinical testing as an anticancer agent. Recent studies suggest that PZA might be a dual inhibitor of DNA topoisomerases I and II and exerts its effects by diminishing the formation of topoisomerase-DNA adducts. Consistent with this unique mechanism of action, PZA exhibits broad-spectrum antitumor activity in pre-clinical models *in vivo*. In addition, this agent displays several remarkable properties, including solid tumor selectivity, activity against hypoxic cells, and cytotoxicity in noncycling cells. PZA has been studied in phase I trials in adults and children, and is currently undergoing broad phase II trials in several tumor types. No significant anti-tumor activity has been seen in gastrointestinal malignancies and prostate cancer. Owing to its unique properties, combination studies with other antineoplastic agents are in progress [70].

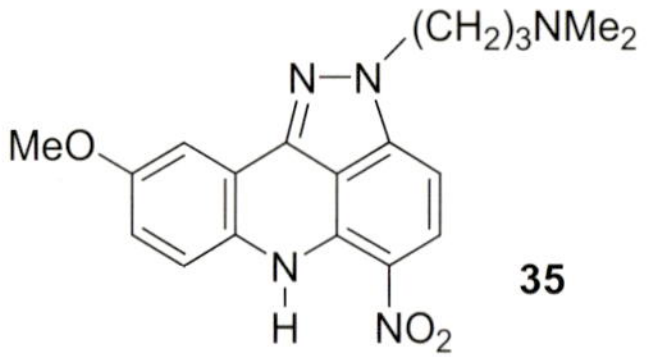

$(CH_2)_3NMe_2$ N–N MeO N H NO_2 **35**

We end this section by summarizing the most promising fields of application of pyrazoles and indazoles with four compounds. First, edaravone –norphenazone–, 3-methyl-1-phenyl-2-pyrazolin-5-one (**36**), is a very simple compound, known from old, that it is a free radical scavenger and a very potent antioxidant agent against lipid peroxidation. *In vivo* studies have revealed that edaravone shows brain-protective activity in a transient ischemia model [71]. The second one is tolfenpyrad, 4-chloro-3-ethyl-1-methyl-*N*-[4-(*p*-tolyloxy)benzyl]pyrazole-5-carboxamide (**37**) one of the most promising insecticides recently discovered in Japan [72].

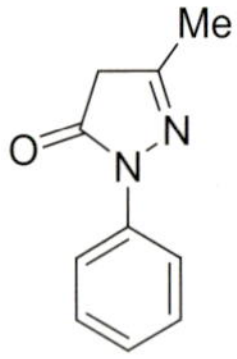

Me O N N **36**

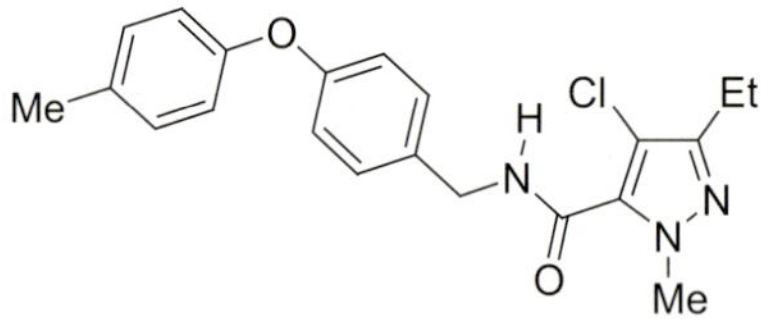

Me O H N O Cl Et N N Me **37**

The normal biological functions regulated by nitric oxide are attributed to three nitric oxide synthase (NOS) isoforms (neuronal, endothelial or inducible macrophage). A dysfunction of these enzymes is implicated in various diseases such as Alzheimer's disease, septic shock, inflammatory arthritis, schizophrenia, impotence and susceptibility to infection. 1*H*-Pyrazole-1-carboxamidines are competitive inhibitors of all three isoforms: the most selective compound, 1*H*-pyrazole-*N*-(3-aminomethylanilino)-1-carboxamidine, is 100-fold selective for neuronal NOS over endothelial NOS [73]. In the field of neuroprotective activity, studies carried out by Wolff and Gribin [74] on inhibition of nitric oxide synthase by indazole agents confirmed the proposal that 5-nitro-, 6-nitro- and 7-nitroindazoles exert inhibitory actions by interaction of nitric oxide synthase such that oxygen does not bind. 7-Nitroindazole (**38**), a selective inhibitor of neuronal nitric oxide synthase, has been studied for neuroprotective activity and has been used to investigate the role of nitric oxide [75–77].

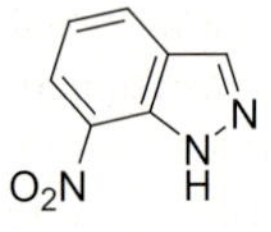

N N H O_2N **38**

On 27 March 1998 the US Food and Drug Administration approved sildenafil citrate (Viagra) (**39**) for treating male erectile dysfunction (MED). The drug works by inhibiting cyclic guanosine monophosphate (cGMP) phosphodiesterase Type 5 (PDE5). Further structural manipulations have included α-thiagra, the thiophene bioisostere [78], and Monagra, a chiral 5-(2-methyl-2,3-dihydro-7-benzofuryl)-pyrazolopyrimidone analog [79]. At Bristol-Myers Squibb a PDE5 screening of a series of pyrazolopyridines identified a lead compound with modest potency. Based on this template, and using parallel synthesis, from a large and diverse library emerged a new pyrazolopyridine showing comparable *in vitro* functional PDE5 inhibition when compared to sildenafil and improved PDE isozyme selectivity. Thus, due to its pharmacokinetic profile, it is expected to have fewer PDE-related side effects than sildenafil [80].

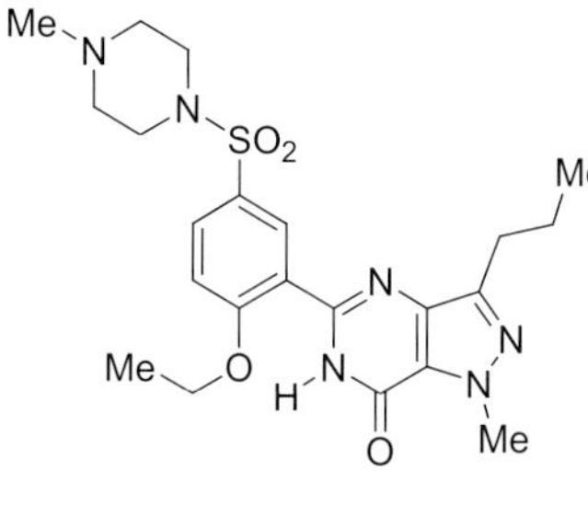

39

8.4 Synthesis of Pyrazoles and Indazoles

Since the synthesis of pyrazoles and indazoles have very few items in common it is better to treat them separately. Ring transformations, when a heterocycle is transformed into a pyrazole (or indazole), are discussed in the synthesis section. On the other hand, cases where a pyrazole (or indazole) is transformed into another heterocycle are reported under reactivity (Section 8.5). For the same reasons, oxido-reduction reactions, for example, transformation of pyrazolines into pyrazoles or vice versa, will also be covered under reactivity.

8.4.1 Synthesis of Pyrazoles

There are many methods for the synthesis of the pyrazole ring; it can be formed both by cyclization and by cycloaddition reactions. The reaction of hydrazines with 1,3-dicarbonyl compounds (or their equivalents) is probably the most general and versatile method; however, a disadvantage, in some cases, is the formation of mixtures of isomeric pyrazoles from unsymmetrical dicarbonyl compounds. The 1,3-dipolar cycloaddition reaction of diazo compounds, nitrilimines and azomethine imines with alkynes is also a general route to pyrazoles.

Scheme 8.7 shows the different possibilities for the creation of the pyrazole ring according to the bonds formed.

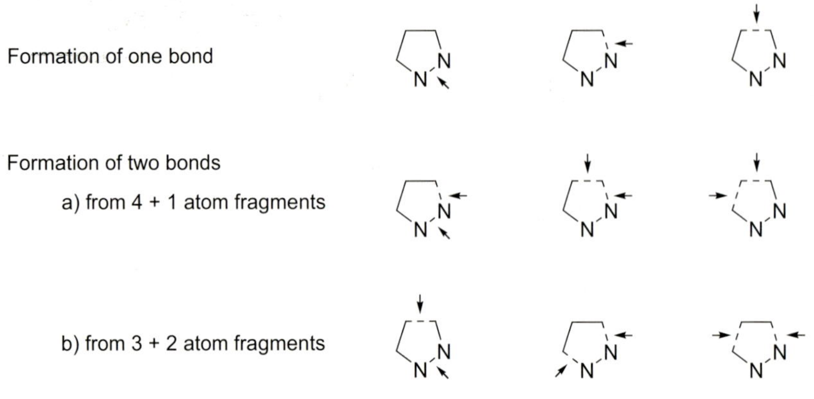

Formation of one bond

Formation of two bonds

a) from 4 + 1 atom fragments

b) from 3 + 2 atom fragments

Scheme 8.7

The synthesis and chemistry of pyrazoles has been the subject of recent reviews [6, 8, 81–84].

8.4.1.1 Formation of One N–N Bond

Dioximes of 1,3-dicarbonyl compounds give oxidative cyclization to 4*H*-pyrazole-1,2-dioxides **42** (Scheme 8.8). Various oxidants have been used, namely lead(IV) acetate [85, 86] and *N*-bromoacetamide [87–89]. Dehydration of dioximes **41** with thionyl chloride leads to the monooxides **43** [88, 89]. Base-induced cyclization of 1,3-diketone dioximes affords 1-hydroxypyrazoles, but in low yield (10–30%) [90].

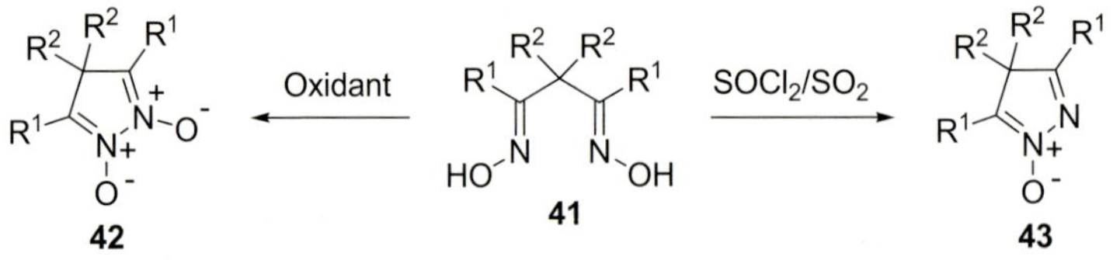

Scheme 8.8

Imines **44** react with thionyl chloride, at room temperature, to furnish 1,2,6-thiadiazine 1-oxides **45**. Thermal extrusion of sulfur monoxide leads to pyrazoles **46** (Scheme 8.9). Alternatively, reaction of imines **44** with thionyl chloride in pyridine at 90 °C leads directly to pyrazoles **46** [91, 92].

Iminohydrazones **47** give NH-pyrazoles **48** when treated with an aqueous solution of sulfuric acid (Scheme 8.10). In contrast, they are converted into the *N*-alkenyl derivatives **49** when treated with an equimolar amount of anhydrous trifluoroacetic acid in dry THF [93].

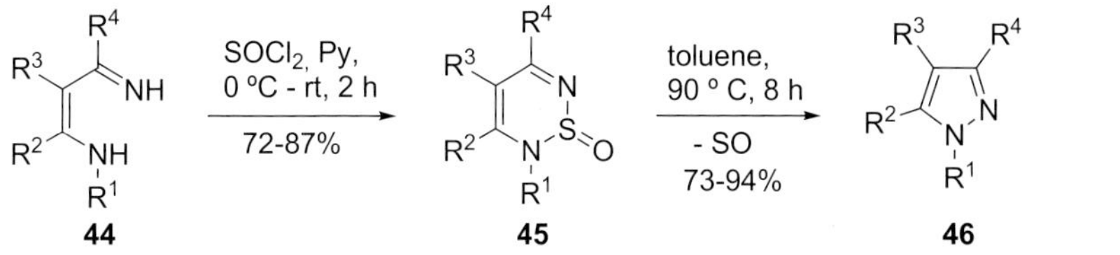

R^1 = Ph, *o*-Tolyl, *p*-Tolyl; R^2 = Ph, 4-ClC_6H_4; R^3 = H, Me, Cl, Br; R^4 = Ph, *p*-Tolyl, cyclohexyl

Scheme 8.9

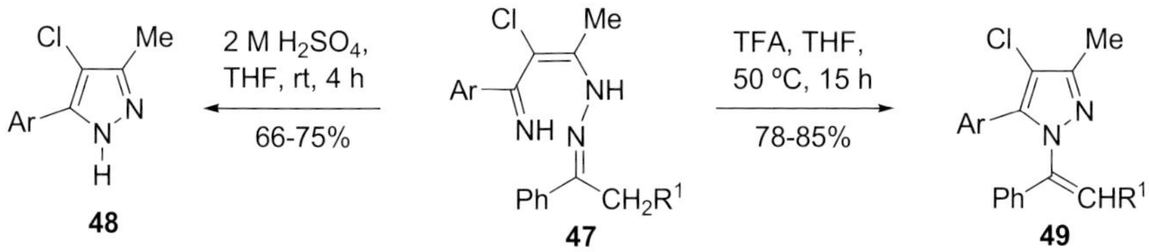

Ar = Ph, 4-ClC_6H_4, 4-MeC_6H_4; R^1 = H, Me

Scheme 8.10

8.4.1.2 Formation of One N–C Bond

Oxidative cyclization of arylhydrazones **50** leads to 1,3,5-trisubstituted pyrazoles **51** in high yields (Scheme 8.11). Lead tetraacetate [94], manganese dioxide [95] and thianthrene cation radical perchlorate [96] have been used as the oxidants in these transformations.

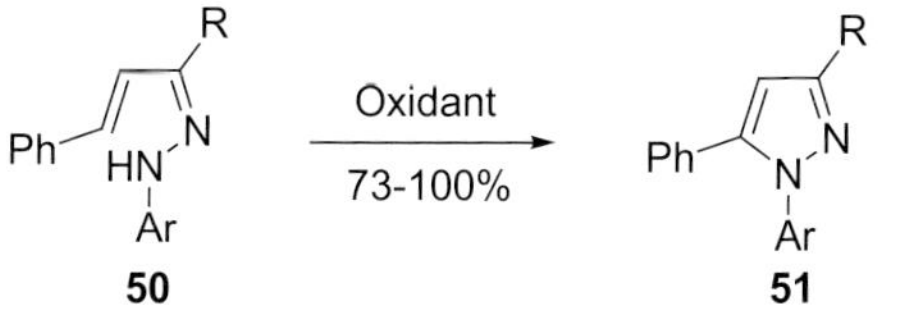

R = Me, Ph

Scheme 8.11

Diazoalkenes **52** give 1,5-electrocyclizations to 3*H*-pyrazoles **53**, which isomerize spontaneously to 1*H*-pyrazoles **54** (Scheme 8.12). These reactive intermediates can be generated by alkaline decomposition of ethyl alkenylnitrosocarbamates [97], tosylhydrazones of α,β-unsaturated carbonyl compounds [98] or *N*-methoxypyrida-

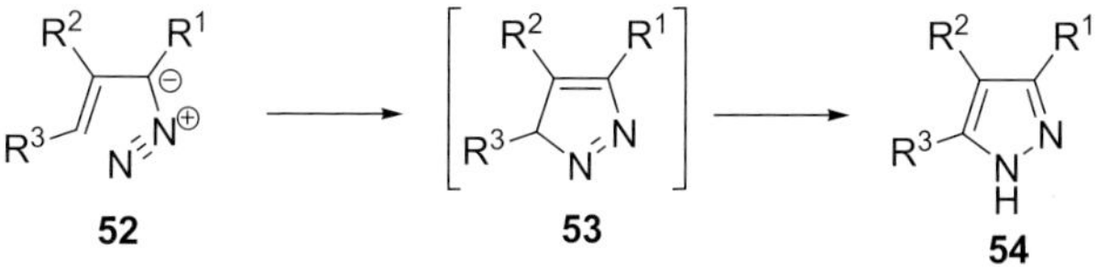

Scheme 8.12

zinium salts [99] or from the reaction of carbonyl compounds with ethyl lithiodiazoacetate [100].

8.4.1.3 Formation of One C–C Bond

Dieckmann cyclization of hydrazones **55** leads to 4-hydroxypyrazole-5-carboxylic acid derivatives **56** in moderate to good yields (Scheme 8.13) [101, 102]. Trifluoroacetic anhydride-pyridine induced cyclization of hydrazones **57** affords 1-alkyl-4-trifluoromethylpyrazoles **58** [103].

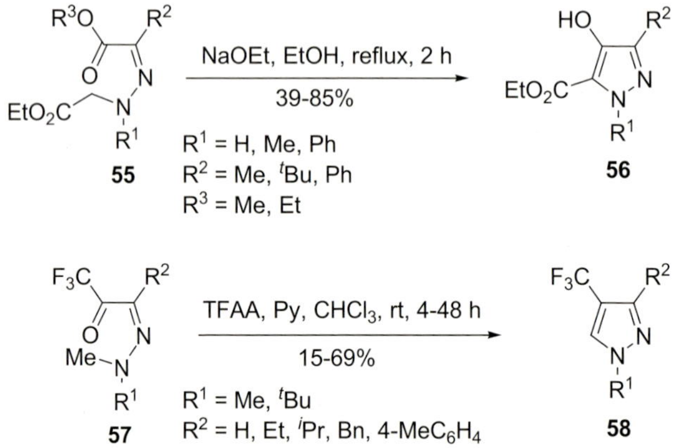

NaOEt, EtOH, reflux, 2 h; 39-85%
R^1 = H, Me, Ph
R^2 = Me, ^{t}Bu, Ph
R^3 = Me, Et
55 → **56**

TFAA, Py, $CHCl_3$, rt, 4-48 h; 15-69%
R^1 = Me, ^{t}Bu
R^2 = H, Et, ^{i}Pr, Bn, 4-MeC_6H_4
57 → **58**

Scheme 8.13

When heated with trace amounts of concentrated hydrochloric acid at 140 °C for 100 h, azines of methyl ketones **59** are converted into 3,5-dialkyl-5-methyl-2-pyrazolines **60** in 65–79% yields (Scheme 8.14) [104]. Heating these azines with nickel or cobalt(II) halides at 200 °C also affords pyrazolines **60** in good yields [105]. Under these conditions, other azines give pyrrole derivatives [105]. When acetone azine is heated at 100 °C in the presence of trace amounts of $TiCl_3$, 90% conversion into pyrazoline **60** (R^1=Me) is effected after 20 h [106].

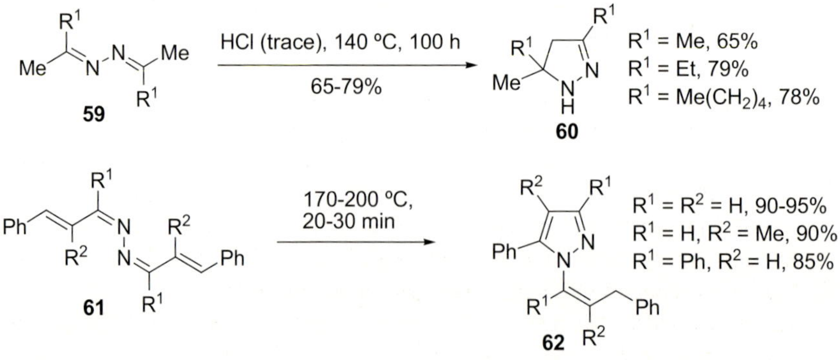

HCl (trace), 140 °C, 100 h; 65-79%
59 → **60**
R^1 = Me, 65%
R^1 = Et, 79%
R^1 = $Me(CH_2)_4$, 78%

170-200 °C, 20-30 min
61 → **62**
R^1 = R^2 = H, 90-95%
R^1 = H, R^2 = Me, 90%
R^1 = Ph, R^2 = H, 85%

Scheme 8.14

Cinnamaldehyde azine and derivatives **61** are converted into pyrazoles **62** in high yields when heated at about 200 °C (with or without solvent) (Scheme 8.14) [107]. Other α,β-unsaturated azines behave similarly [108, 109]. The thermal decomposition of polyhalogenated propenal azines can be used for the synthesis of mono- and dihalogenated *N*-unsubstituted pyrazoles [110].

Treatment of benzyl phenyl ketazine with two equivalents of LDA generates a dianion that, when heated at 65 °C for 1 h in THF-HMPA, provides 3,4,5-triphenylpyrazole [111].

8.4.1.4 Formation of Two Bonds

8.4.1.4.1 From 4 + 1 Atom Fragments

Formation of One C–N and One N–N Bond [N-C-C-C + N] Iminoposphoranes **63** react with 2-azido-cyclopent-1-enecarbaldehyde **64** to afford azidoimines **65**. When heated in refluxing toluene, these compounds extrude nitrogen and cyclize to pyrazoles **66** (Scheme 8.15) [112].

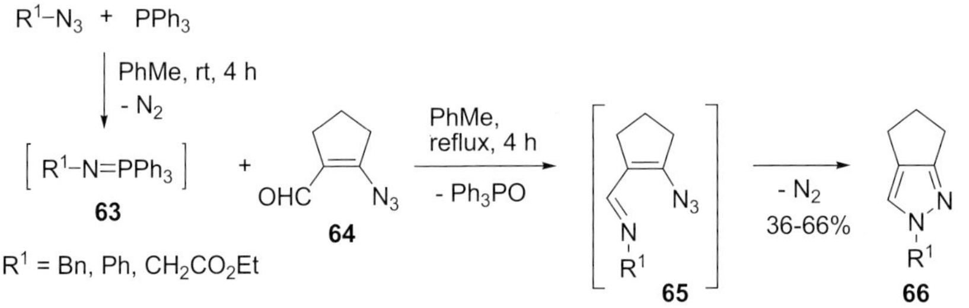

R^1–N_3 + PPh_3 → (PhMe, rt, 4 h; - N_2) [R^1–N=PPh_3] **63** + OHC / N_3 **64** → (PhMe, reflux, 4 h; - Ph_3PO) [**65**] → (- N_2; 36-66%) **66**

R^1 = Bn, Ph, CH_2CO_2Et

Scheme 8.15

Nitrosation of α,β-unsaturated oximes unsubstituted in the α-position with sodium nitrite and acetic acid leads to pyrazole 1,2-dioxides **68** and **69** while the α-substituted ones afford 1-hydroxypyrazole 2-oxides **70** (Scheme 8.16) [113–115]. Nitrosation of oximes **67** with butyl nitrite in the presence of pyridine and copper sulfate affords copper complexes **71**, which can be conveniently converted into the 1-hydroxypyrazole 2-oxides **70** [116].

Formation of One C–N and One C–C bond [N-N-C-C + C] Hydrazones of methyl ketones react with the Vilsmeier–Haack reagent to afford 1*H*-pyrazole-4-carbaldehydes **74** (method A, Scheme 8.17) [117]. The reaction time can be reduced from 4–5 h to 35–50 s if microwave irradiation is used instead of conventional heating [118]. An interesting variation of method A consists in the substitution of phosphorus oxychloride by cyanuric chloride (2,4,6-trichloro[1,3,5]triazine, TCT) (method B) [119]. This variation requires milder reaction conditions.

Condensation of the Vilsmeier–Haack reagent with arylhydrazones of β-ketoesters and γ-ketoesters yields, respectively, pyrazole-4-carboxylic acid esters [120] **75** and 4-pyrazoleacetic acid esters [121] **76** (Scheme 8.18).

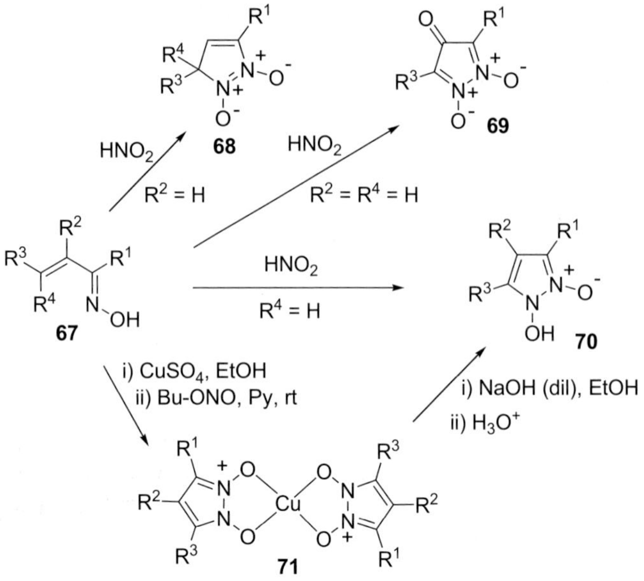

HNO2
R2 = H
68
HNO2
R2 = R4 = H
69
67
HNO2
R4 = H
70
i) CuSO4, EtOH
ii) Bu-ONO, Py, rt
i) NaOH (dil), EtOH
ii) H3O+
71

Scheme 8.16

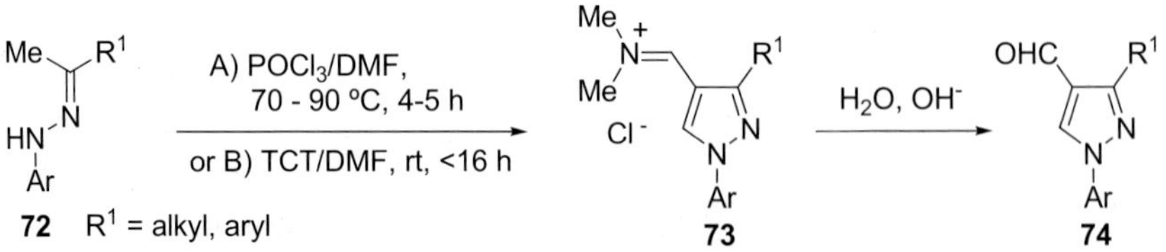

A) POCl3/DMF, 70 - 90 °C, 4-5 h
or B) TCT/DMF, rt, <16 h
H2O, OH-
72 R1 = alkyl, aryl
73
74

Scheme 8.17

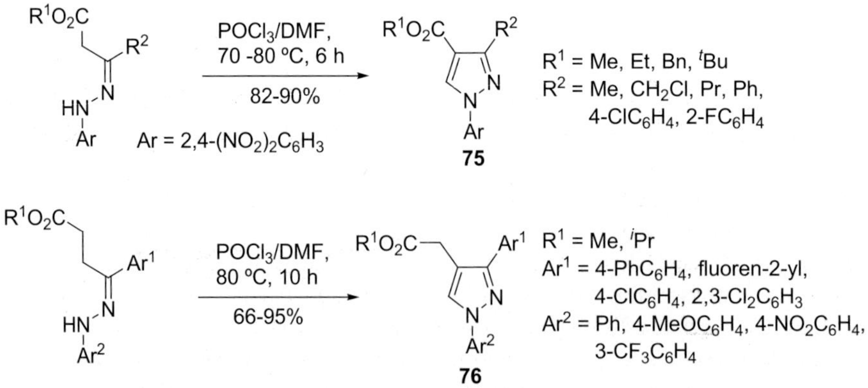

POCl3/DMF, 70 -80 °C, 6 h
82-90%
Ar = 2,4-(NO2)2C6H3
75
R1 = Me, Et, Bn, tBu
R2 = Me, CH2Cl, Pr, Ph, 4-ClC6H4, 2-FC6H4

POCl3/DMF, 80 °C, 10 h
66-95%
76
R1 = Me, iPr
Ar1 = 4-PhC6H4, fluoren-2-yl, 4-ClC6H4, 2,3-Cl2C6H3
Ar2 = Ph, 4-MeOC6H4, 4-NO2C6H4, 3-CF3C6H4

Scheme 8.18

Aminocarbonylhydrazones of methyl ketones [122] or of β-ketoesters [120] react with the Vilsmeier–Haack reagent to afford, respectively, *N*-unsubstituted pyrazole-4-carbaldehydes **77** and *N*-unsubstituted pyrazole-4-carboxylic acid esters **78** (Scheme 8.19).

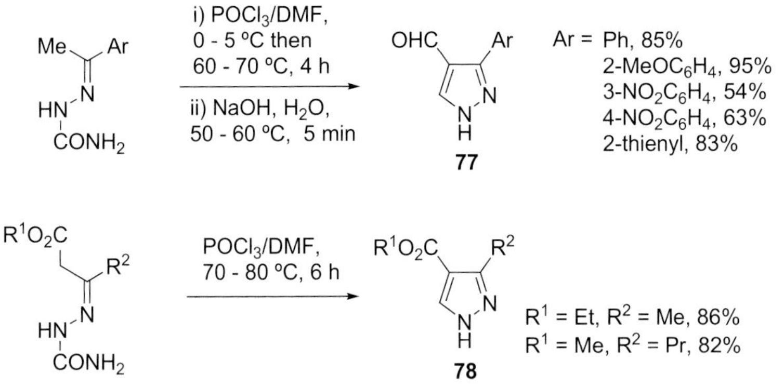

i) $POCl_3$/DMF, 0 - 5 °C then 60 - 70 °C, 4 h; ii) NaOH, H_2O, 50 - 60 °C, 5 min

77 Ar = Ph, 85%; 2-$MeOC_6H_4$, 95%; 3-$NO_2C_6H_4$, 54%; 4-$NO_2C_6H_4$, 63%; 2-thienyl, 83%

$POCl_3$/DMF, 70 - 80 °C, 6 h

78 R^1 = Et, R^2 = Me, 86%; R^1 = Me, R^2 = Pr, 82%

Scheme 8.19

Acetophenone azines **79** also react with the Vilsmeier–Haack reagent to afford 1*H*-pyrazole-4-carbaldehydes **80** in excellent yields (Scheme 8.20) [123].

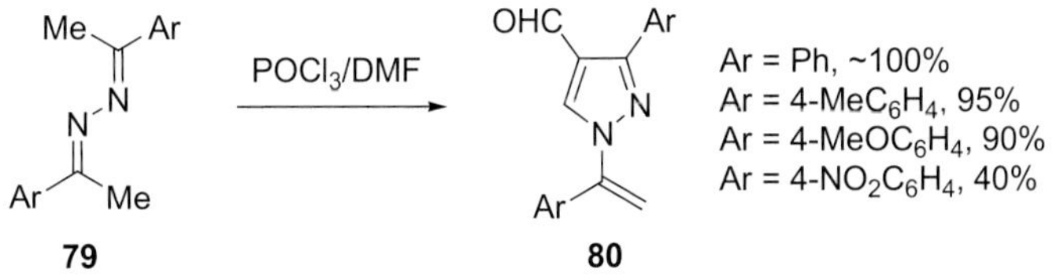

79 → $POCl_3$/DMF → **80**

Ar = Ph, ~100%
Ar = 4-MeC_6H_4, 95%
Ar = 4-$MeOC_6H_4$, 90%
Ar = 4-$NO_2C_6H_4$, 40%

Scheme 8.20

Anions generated from hydrazones with an α-hydrogen undergo a series of reactions affording N-heterocycles, namely pyrazoles [124]. The reaction of dilithiated hydrazone anions with electrophiles (esters [125–129], acyl chlorides [128, 130], nitriles [131], amides [128], α-haloketones [128], aldehydes [132] or diethyl carbonate [133]) followed by acid-catalyzed ring closure furnishes adequately functionalized pyrazoles (Scheme 8.21).

The anion derived from the *N*-phenyl-α-phosphinylhydrazone **81** reacts with aromatic aldehydes to afford hydrazones **82**. Heating at 100 °C in toluene leads to pyrazoles **83** in excellent yields (Scheme 8.22) [134]. In a similar way, deprotonation of the *N*,*N*-dimethylhydrazone **84**, addition of isocyanates (X=O) or isothiocyanates (X=S) and cyclization leads to pyrazoles **86** [135].

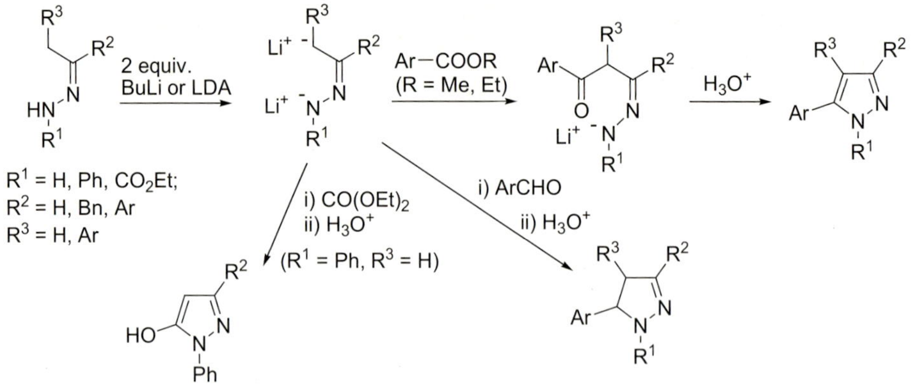

Scheme 8.21

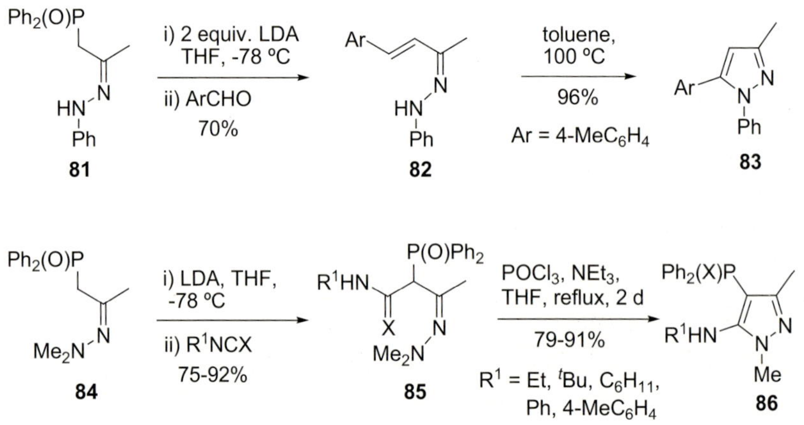

Scheme 8.22

Treatment of a solution of diazabutadiene **87** in toluene with 1–10 mol% of Pd(0) catalyst under an atmosphere of CO (1–2 atm) at 100 °C for 15 min affords 1,3,4-triphenylpyrazol-5-one (**88**) in excellent yield (Scheme 8.23) [136].

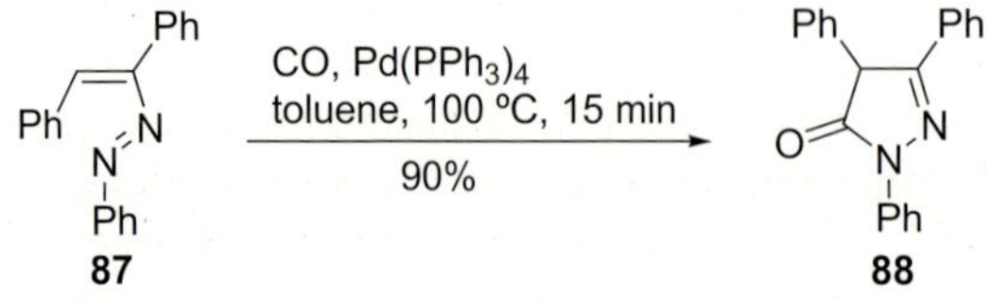

Scheme 8.23

8.4.1.4.2 From 3 + 2 Atom Fragments

Formation of Two C–N Bonds [C-C-C + N-N] The addition of hydrazines (double nucleophiles) to three-carbon units featuring two electrophilic carbons in a 1,3-relationship is one of the most versatile routes to pyrazoles. The condensation of hydrazines with 1,3-diketones, for instance, is perhaps the most common route to the construction of the pyrazole ring (Scheme 8.24) [137–139]. Several other 1,3-bis(electrophilic) compounds react with hydrazines to yield pyrazoles, namely β-ketoaldehydes, β-ketoesters, β-ketoamides, β-ketonitriles, vinyl ketones, alkynyl ketones, and so on.

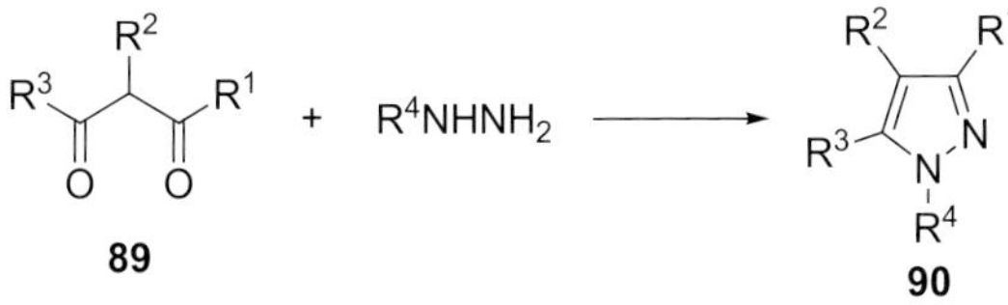

Scheme 8.24

Substituted 5-alkylamino and 5-(arylamino)pyrazoles **93** can be prepared in one-pot synthesis from a β-ketoamide **91**, an aryl or alkyl hydrazine and Lawesson's reagent (Scheme 8.25) [140]. Hydrazones **92** are probable intermediates. This method has also been applied for the solid-supported synthesis of 5-(*N*-monosubstituted-amino)pyrazoles [141].

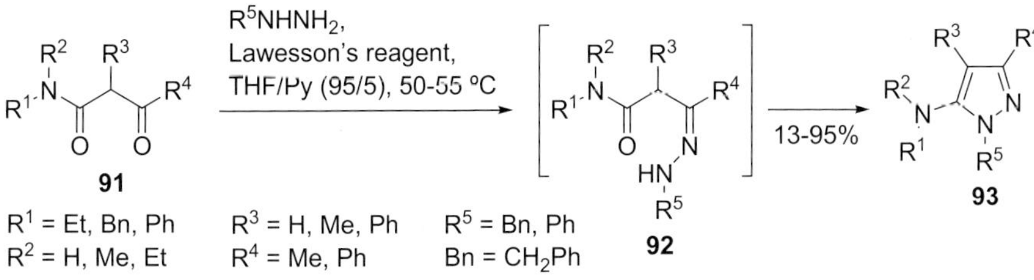

Scheme 8.25

Diketo oximes **95** (prepared from 1,3-diketones **94**) react with hydrazines to yield 4-nitrosopyrazoles **96** (Scheme 8.26) [142, 143]. However, if a large excess of hydrazine is used the isolated products are the corresponding 4-amino-3,5-disubstituted pyrazoles **97** [144].

Enamines of general type **98** react with hydrazine derivatives to afford pyrazoles **99** (Scheme 8.27) [145]. For instance, reaction of (+)-camphor derivative **100** with hydrazine or benzylhydrazine leads to pyrazoles **101** [146]. Also, compounds **102** react with phenylhydrazine to afford 1,4,5-trisubstituted pyrazoles **103** in moderate to excellent yields [147]. Similarly, ethyl 4-dimethylamino-2-oxo-3-butenoate **104** and its diester analogue **105** react with a range of hydrazine derivatives to afford pyrazoles **106** and **107**, respectively [148].

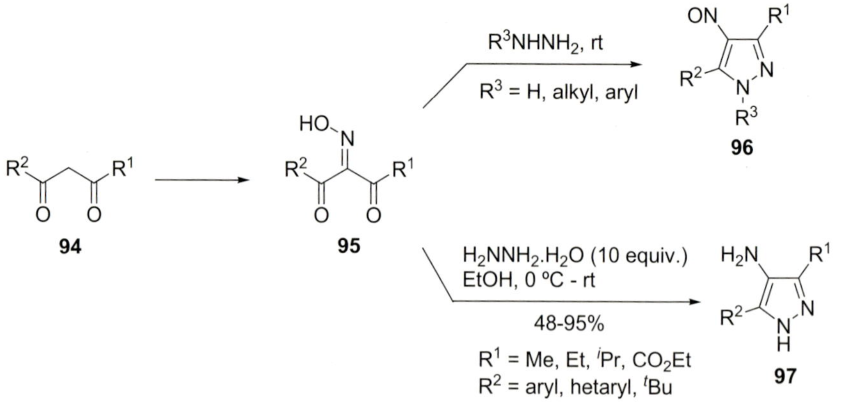

Scheme 8.26

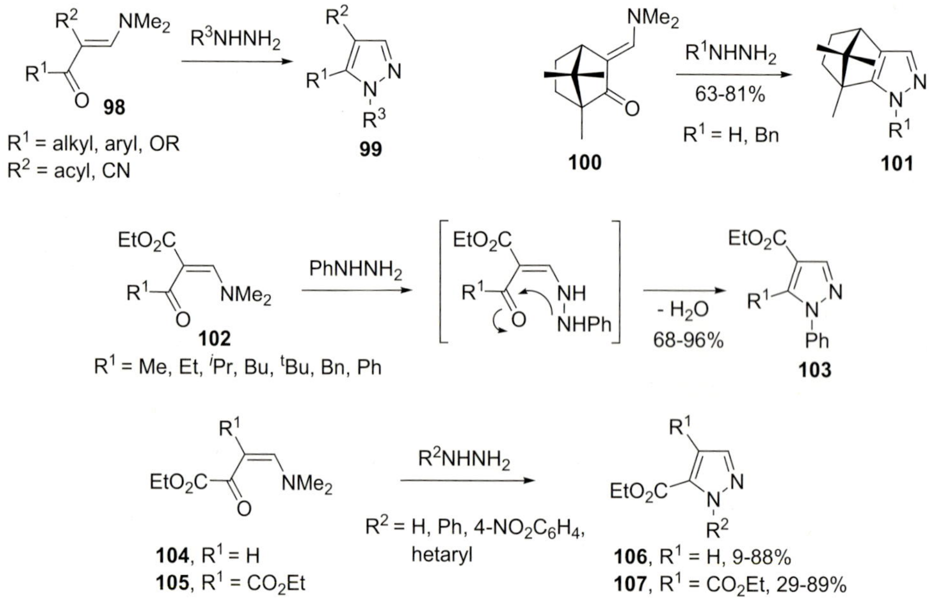

Scheme 8.27

The use of microwave irradiation in this type of chemistry allows the formation of pyrazole derivatives in a few minutes while it requires several hours under conventional heating [149]. The formation of pyrazoles from polymer-bound 2-acyl-3-aminopropenoates has also been described [150, 151].

Cyclocondensation of α-oxoketene *N,S*-acetals **108** with phenylhydrazine gives regioselectively 3- or 5-(*N*-cycloamino)pyrazoles just by variation of the reaction conditions (Scheme 8.28) [152]. Reaction of 2-cyanoketene *N,S*-acetals with substi-

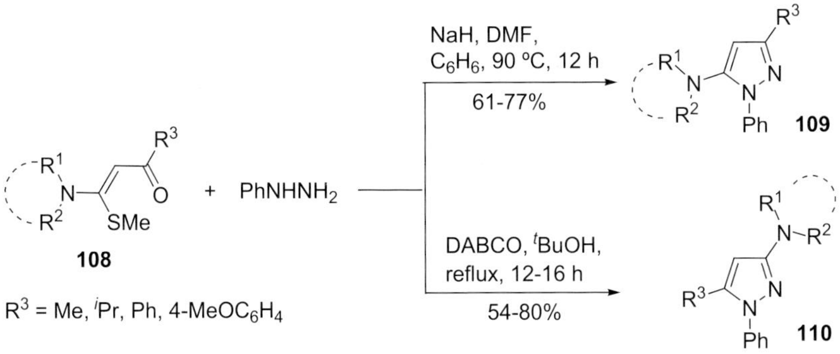

Scheme 8.28

tuted hydrazines in refluxing ethanol containing a catalytic amount of piperidine gives the corresponding 5-amino-3-anilino-1*H*-pyrazole-4-carboxamides [153].

α-Oxoketene dithioacetals **111** react with phenylhydrazine to give selectively 5-methylthio-1-phenyl-3,4-substituted/annulated pyrazoles **112** (Scheme 8.29) [154]. Regioisomeric 3-methylthio-1-phenyl-4,5-substituted/annulated pyrazoles **114** are obtained selectively from the reaction of β-oxodithioesters **113** with phenylhydrazine [154].

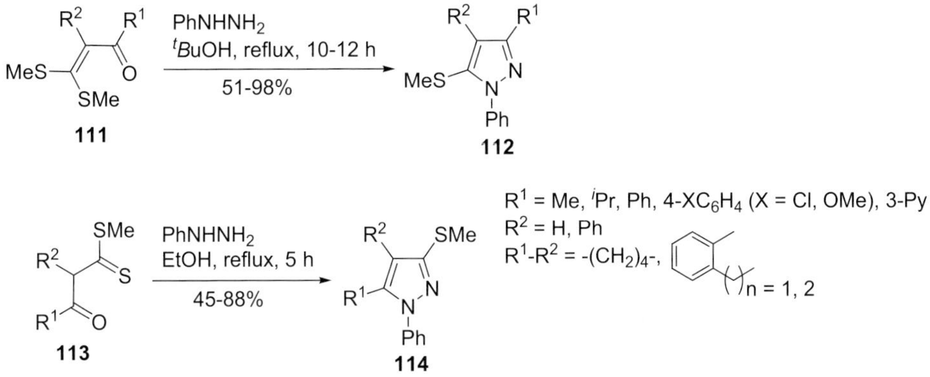

Scheme 8.29

Cyclocondensation of 1-bis(methoxy)-4-bis(methylthio)-3-buten-2-one (**115**) with hydrazine hydrate gives pyrazole **116** with a masked aldehyde functionality (Scheme 8.30) [155]. The corresponding 5(3)-cycloaminopyrazole derivatives **118** can also be synthesized in a one-pot sequence by prior displacement of one of the methylthio groups of **115** by the respective amine. Hydrolysis of the dimethylacetal moiety of **116** with aqueous acetic acid (50%) affords the corresponding pyrazole-3 (5)-carbaldehyde in 95% yield.

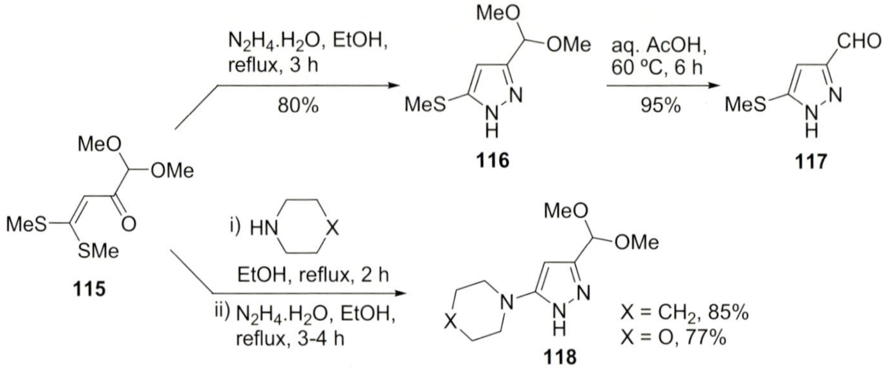

Scheme 8.30

Baylis–Hillman adducts **119** and **121** react with hydrazine hydrochlorides to afford regioselectively 1,3,4,5-tetrasubstituted pyrazoles **120** and **122** respectively (Scheme 8.31) [156].

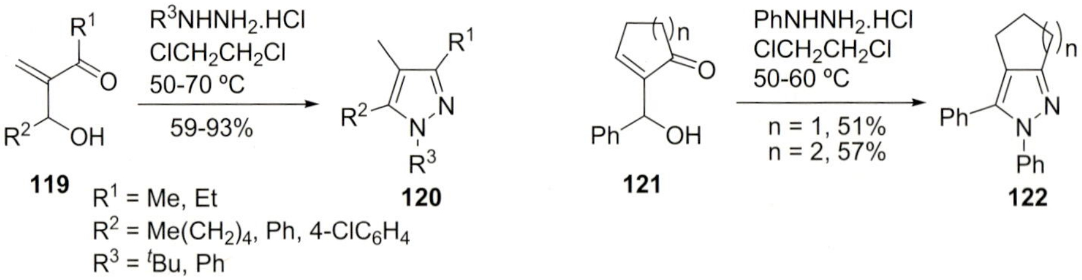

Scheme 8.31

5-Trichloromethyl-1-phenyl-1*H*-pyrazoles **124** and 5-trichloromethyl-1,2-dimethylpyrazolium chlorides **125** can be synthesized in 80–98% yield by the cyclocondensation of β-alkoxyvinyl trichloromethyl ketones **123** with phenylhydrazine and 1,2-dimethylhydrazine dihydrochloride, respectively, under microwave irradiation and using toluene as solvent (Scheme 8.32) [157]. While the use of microwave and classical methods are comparable for making pyrazoles, the pyr-

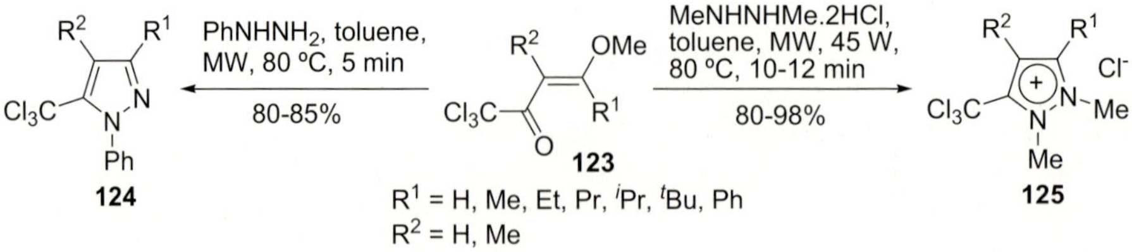

Scheme 8.32

azolium chlorides can be obtained in a significantly shorter time and in some cases better yield. The trifluoromethyl analogues of **123** react with 7-chloro-4-hydrazinoquinoline to afford 1*H*-pyrazol-1-yl quinolines in high yields [158].

Treatment of α-benzotriazolyl-α,β-unsaturated ketones **127** with monosubstituted hydrazines leads to the regioselective synthesis of benzotriazolylpyrazolines **128**, which, by treatment with a base, can be converted into the trisubstituted pyrazoles **129** (Scheme 8.33) [159]. Alkylation of pyrazolines **128** at the 4-position of the pyrazoline ring is a versatile route to unsymmetrical 1,3,4,5-tetrasubstituted pyrazolines **130** and -pyrazoles **131** [159]. The α-benzotriazolyl-β-ethoxy-α,β-unsaturated ketones **132** react with hydrazines to afford directly the corresponding 4-benzotriazolylpyrazoles **133** [160].

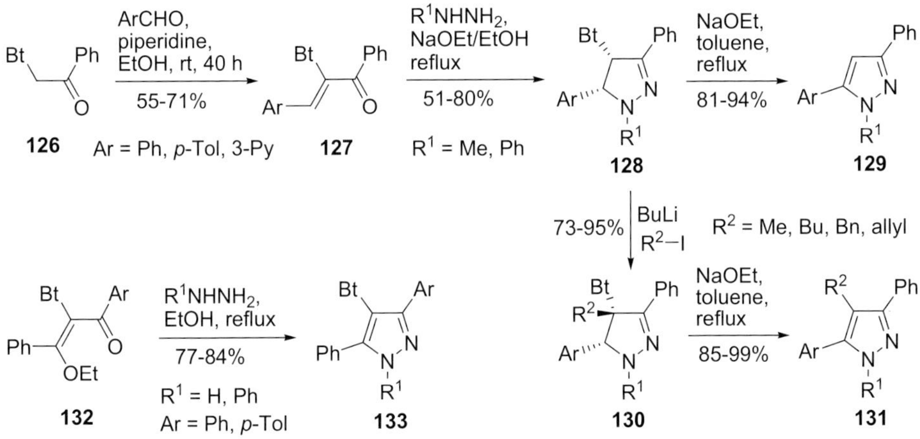

ArCHO, piperidine, EtOH, rt, 40 h; 55-71%
R^1NHNH_2, NaOEt/EtOH reflux; 51-80%
NaOEt, toluene, reflux; 81-94%
126 Ar = Ph, p-Tol, 3-Py 127 R^1 = Me, Ph 128 129
73-95% BuLi R^2–I R^2 = Me, Bu, Bn, allyl
R^1NHNH_2, EtOH, reflux; 77-84%; R^1 = H, Ph; Ar = Ph, p-Tol
NaOEt, toluene, reflux; 85-99%
132 133 130 131

Scheme 8.33

Chalcones **134** react with substituted hydrazines to yield 1-substituted-3,5-diaryl-4,5-dihydro-1*H*-pyrazoles **135** (Scheme 8.34) [84]. With hydrazine hydrate, in acetic

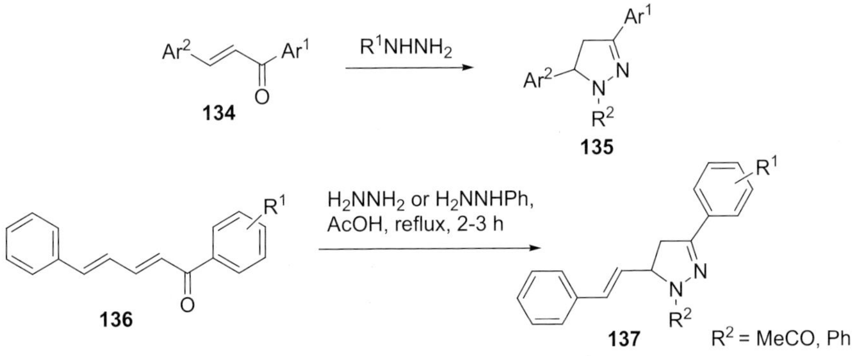

R^1NHNH_2
134 135
H_2NNH_2 or H_2NNHPh, AcOH, reflux, 2-3 h
136 137 R^2 = MeCO, Ph

Scheme 8.34

acid, they afford the 1-acetyl derivatives; some of them are potent and selective inhibitors of monoamine oxidase [161]. Chalcones **134** undergo a rapid cyclization with phenylhydrazine under solvent-free and silica-supported conditions using microwave irradiation to afford 4,5-dihydro-1*H*-pyrazoles **135** in good yields in 2–3 min [162]. Chalcone-epoxides react with hydrazine hydrate, in refluxing ethanol, to afford 3,5-diaryl-1*H*-pyrazoles in high yield [163].

Cinnamylideneacetophenones **136** react with hydrazine hydrate or phenylhydrazine to afford the dihydropyrazoles **137** (Scheme 8.34) [164]. In a similar process, chalcones, bis(chalcones) and oligo(chalcones) react with hydrazine hydrate or hydrazine derivatives to yield pyrazolines, bis(pyrazolines) and oligo(pyrazolines), respectively [165, 166].

Addition of hydrazine to alkynyl ketones is a simple and regioselective route to pyrazoles. Some examples are shown in Scheme 8.35 [167–170].

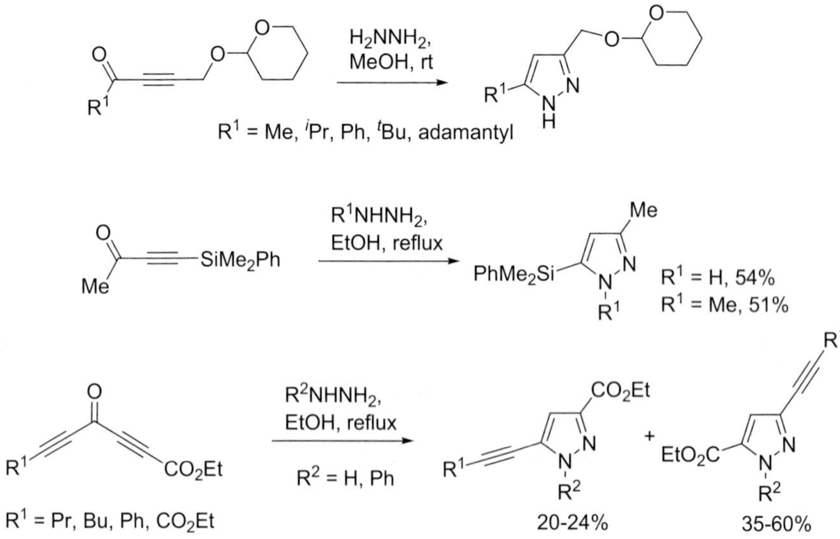

Scheme 8.35

Alkynylcarbene complexes **138** react with hydrazines **139** to form selectively 1,4-disubstituted pyrazoles **140** (Scheme 8.36) [171]. With methylhydrazine it leads to the cyclic aminocarbene complex **141**, which can be demetallated to 1-methyl-5-phenylpyrazole (**142**).

Aryl benzophenone hydrazones **143** are convenient substitutes of arylhydrazines in the synthesis of pyrazoles. These hydrazones react with various 1,3-bifunctional substrates under acidic conditions to afford adequately functionalized pyrazoles in good yields (Scheme 8.37) [172, 173]. The obtained regioselectivity is consistent with transhydrazonation followed by subsequent cyclization. This synthetic route is especially attractive for the synthesis of 1-hetaryl-pyrazoles, since *N*-hetaryl benzo-

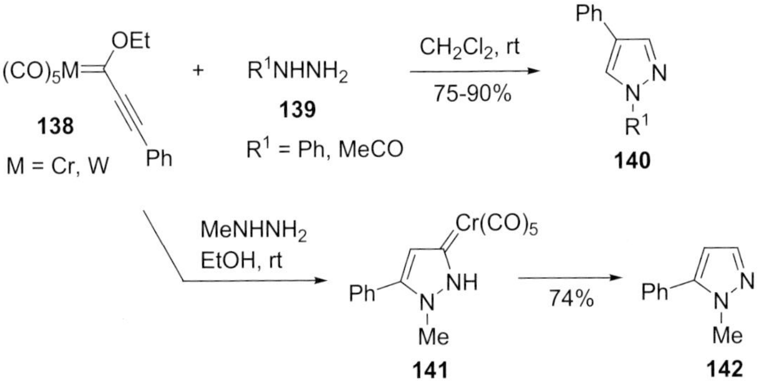

Scheme 8.36

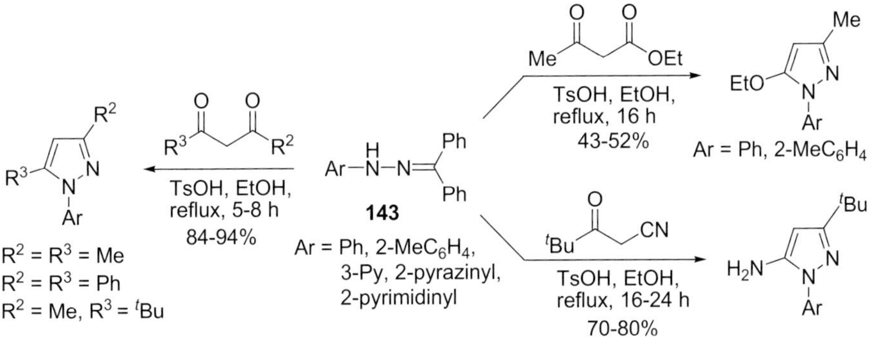

Scheme 8.37

phenone hydrazones can easily be prepared following Buchwald–Hartwig procedures [173].

The cyclocondensation of hydrazines with 1,3-dihalopropanes **144** or propane-1,3-diol ditosylate (**145**) in aqueous alkaline medium and under microwave irradiation affords 4,5-dihydropyrazoles **146** as major products instead of the anticipated pyrazolidines **147** (Scheme 8.38) [174].

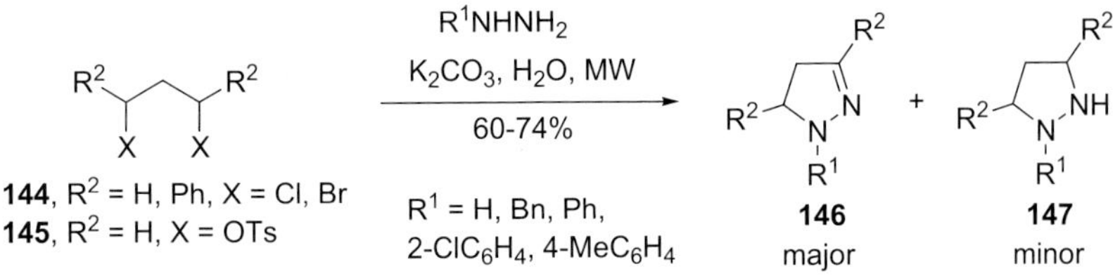

Scheme 8.38

Formation of One C–N Bond and One C–C Bond [N-N-C + C-C] The 1,3-dipolar cycloaddition of diazoalkanes, nitrilimines or azomethine imines to alkynes, alkenes or to functionalized alkenes (enamines or enol ethers, for instance) is a versatile method for the synthesis of pyrazoles. The dipole is the source of the [CNN] fragment while the dipolarophile contributes with the [CC] fragment. This method suffers from the disadvantage that a mixture of two isomeric pyrazoles may be formed when unsymmetrical dipolarophiles are used.

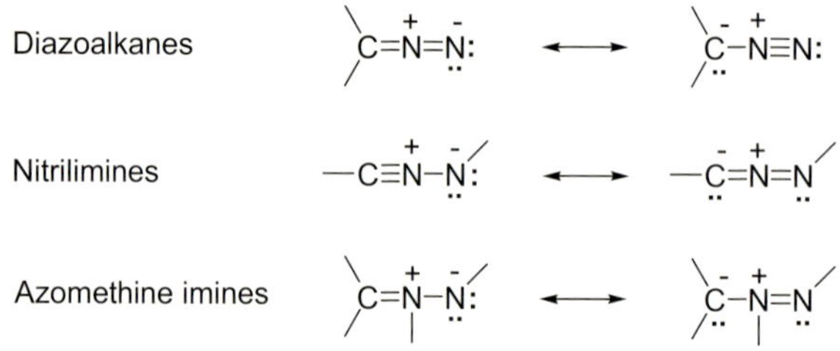

Diazoalkanes $\quad >C=\overset{+}{N}=\overset{-}{N}: \longleftrightarrow >\overset{-}{C}-\overset{+}{N}\equiv N:$

Nitrilimines $\quad -C\equiv\overset{+}{N}-\overset{-}{N}: \longleftrightarrow -\overset{-}{C}=\overset{+}{N}=N$

Azomethine imines $\quad >C=\overset{+}{N}-\overset{-}{N}: \longleftrightarrow >\overset{-}{C}-\overset{+}{N}=N$

Diazoalkanes react with alkenes and alkynes to give, respectively, Δ^1-pyrazolines and 3*H*-pyrazoles [175–177]. However tautomerization may take place, especially if diazomethane or a monosubstituted diazoalkane is used.

The orientation of the dipole–dipolarophile interaction is mainly governed by electronic effects, as described by the FMO theory. However, as indicated in Scheme 8.39 [178], the regioisomer ratios are also strongly dependent on steric effects, which are more pronounced in the case of alkynes.

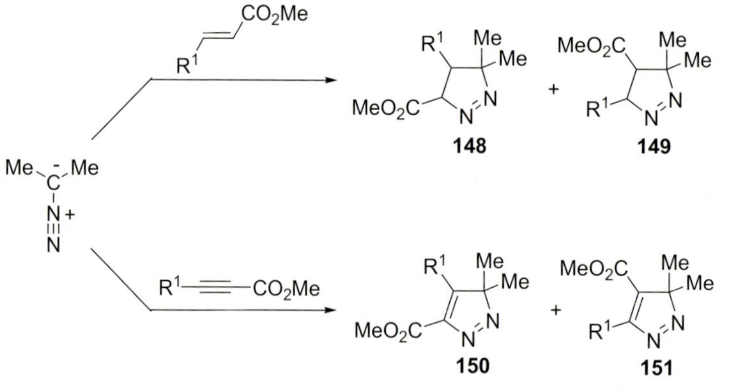

$Me_2C^-{-}N^+{\equiv}N$ + $R^1CH{=}CHCO_2Me$ → **148** + **149**

R^1	148 : 149
H	100 : 0
Me	91 : 9
Et	80 : 20
^{i}Pr	47 : 53
^{t}Bu	0 : 100

$Me_2C^-{-}N^+{\equiv}N$ + $R^1{-}{\equiv}{-}CO_2Me$ → **150** + **151**

R^1	150 : 151
H	100 : 0
Me	63 : 37
Et	39 : 61
^{i}Pr	19 : 81
^{t}Bu	0 : 100

Scheme 8.39

Diazoalkanes react with arylalkynes and acylalkynes to afford, with high regioselectivity, pyrazoles **152** and **153**, respectively (Scheme 8.40). In both cases, the polarization of the triple bond is such that the nucleophilic carbon atom of the dipole attacks the terminal position of the alkyne. With acylarylalkynes, a mixture of the two possible regioisomers **154** and **155** is frequently obtained. For each alkyne, the ratio **154** : **155** is strongly dependent on the diazoalkane used [177, 179].

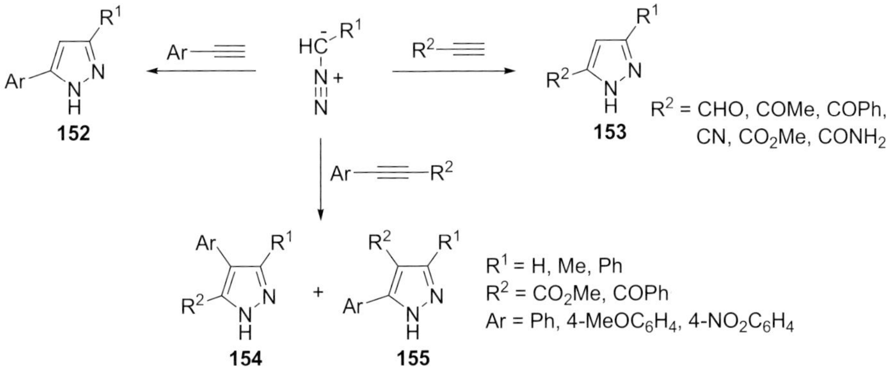

R^2 = CHO, COMe, COPh, CN, CO_2Me, $CONH_2$

R^1 = H, Me, Ph
R^2 = CO_2Me, COPh
Ar = Ph, 4-$MeOC_6H_4$, 4-$NO_2C_6H_4$

Scheme 8.40

Pyrazole itself can be synthesized from the reaction of diazomethane with acetylene. When this reaction is carried out under pressure, almost quantitative yields are obtained [180]. Under identical conditions, addition of diazoethane, ethyl diazoacetate or α,ω-bis(diazo)alkanes **157** to acetylene gives rise to the corresponding pyrazoles **156** or bispyrazoles **158** (Scheme 8.41) [180].

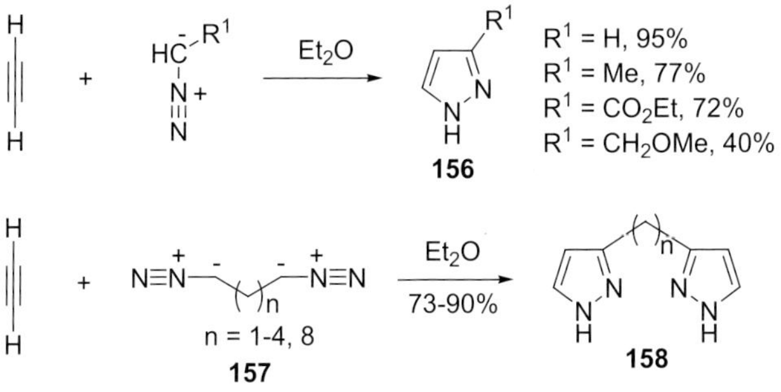

R^1 = H, 95%
R^1 = Me, 77%
R^1 = CO_2Et, 72%
R^1 = CH_2OMe, 40%

n = 1-4, 8

Scheme 8.41

Addition of an ethereal diazomethane solution to fluoro(tributylstannyl)acetylene (**159a**), at −30 °C, gives the corresponding 5-tributylstannyl-4-fluoropyrazole (**160a**) in 98% yield (Scheme 8.42) [181]. The trifluoromethyl analogue **160b** can be obtained in the same way [182].

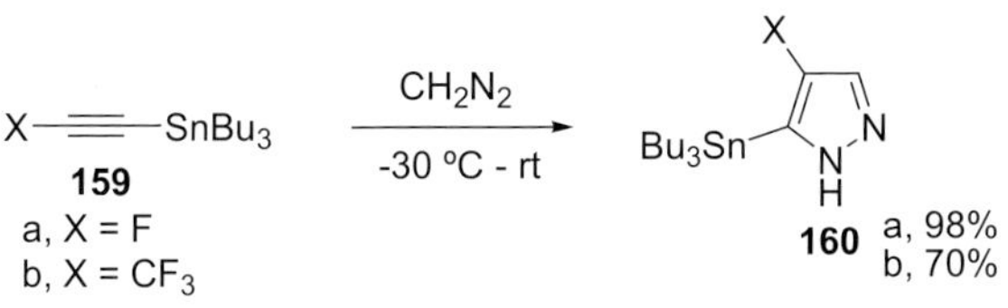

$$X{-}{\equiv}{-}SnBu_3 \xrightarrow[-30\,°C\text{ - rt}]{CH_2N_2} \mathbf{160}$$

159 a, X = F; b, X = CF_3

160 a, 98%; b, 70%

Scheme 8.42

A one-pot procedure for the preparation of 1*H*-pyrazoles involving aryldiazomethanes generated *in situ* has been reported [183]. The 1,3-dipoles are generated *in situ* from aldehydes, via tosylhydrazone sodium salts, and then react with arylacetylenes to furnish regioselectively 3,5-diaryl-1*H*-pyrazoles **161** (Scheme 8.43). When they are generated in the presence of *N*-vinylimidazole, an acetylene equivalent, 3-substituted-1*H*-pyrazoles **162** are obtained.

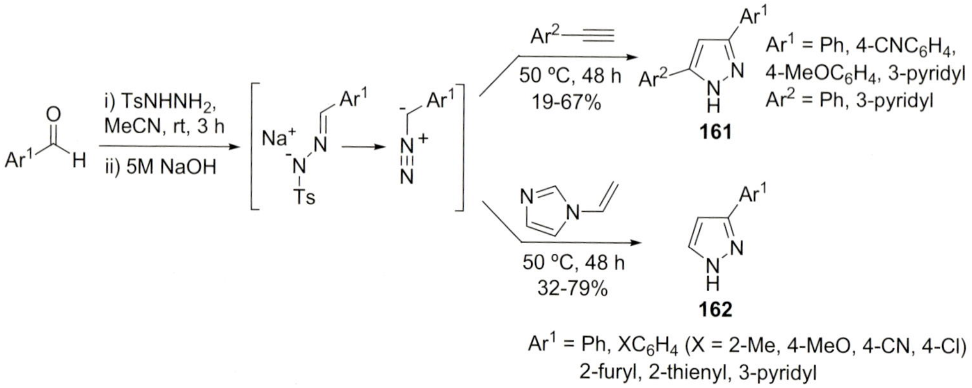

i) $TsNHNH_2$, MeCN, rt, 3 h
ii) 5M NaOH

50 °C, 48 h
19-67%

161

Ar^1 = Ph, 4-CNC_6H_4, 4-$MeOC_6H_4$, 3-pyridyl
Ar^2 = Ph, 3-pyridyl

50 °C, 48 h
32-79%

162

Ar^1 = Ph, XC_6H_4 (X = 2-Me, 4-MeO, 4-CN, 4-Cl)
2-furyl, 2-thienyl, 3-pyridyl

Scheme 8.43

The generation of aryldiazomethanes **164** from bromovinyl tosylhydrazones **163** leads to benzopyrano[1*H*]pyrazoles **165** in high yields (Scheme 8.44) [184].

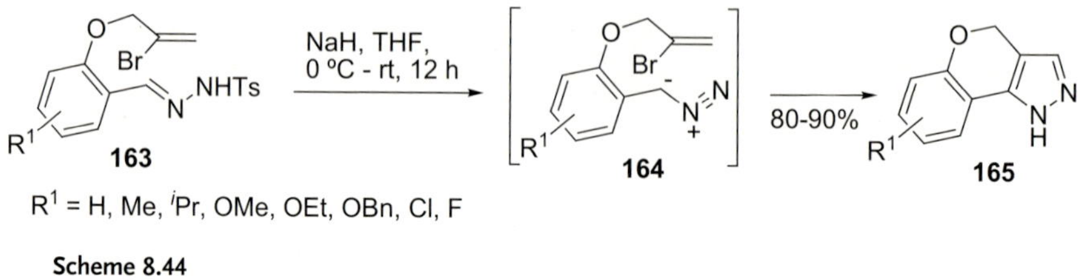

NaH, THF,
0 °C - rt, 12 h

163 **164** **165**

80-90%

R^1 = H, Me, ^{i}Pr, OMe, OEt, OBn, Cl, F

Scheme 8.44

An efficient $InCl_3$-catalyzed 1,3-dipolar cycloaddition of diazocarbonyl compounds and alkynes to synthesize pyrazoles has been reported recently [185]. The reaction is carried out in water, at room temperature, and the catalyst, which stays in the aqueous phase after the work-up, can be reused without loss of catalytic activity. The reaction is applicable to various α-diazocarbonyl compounds and alkynes with a carbonyl group at the neighboring position. As an example, methyl α-diazoarylacetates **166** react with methyl propiolate to afford pyrazoles **167** and **168** in excellent combined yields (Scheme 8.45).

Nitrilimines are also interesting intermediates in the synthesis of pyrazoles and Δ^2-pyrazolines [186, 187]. These 1,3-dipoles are typically generated by dehydroha-

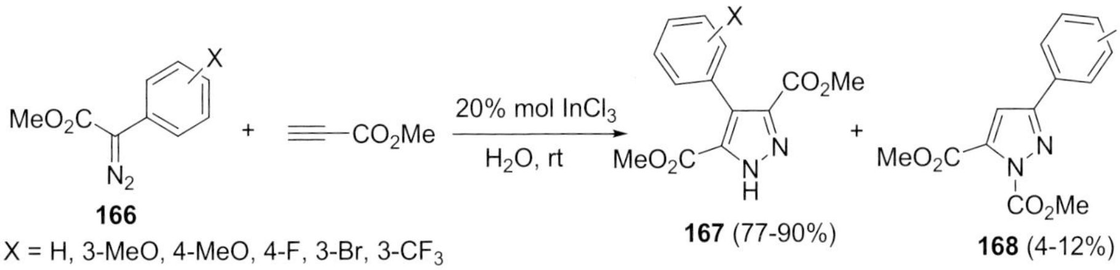

Scheme 8.45

logenation of *N*-arylhydrazonoyl halides **169** with triethylamine in an inert solvent in the presence of the dipolarophile (Scheme 8.46) [188]. Useful alternatives for the generation of nitrilimines are the cycloreversion [189] of tetrazoles **170**, oxathiadiazolinones **171** and 1,3,4-oxadiazolin-2-ones **172** and the oxidation of aldehyde hydrazones with chloramine T, lead tetraacetate [190] or with (diacetoxy)iodobenzene [191, 192].

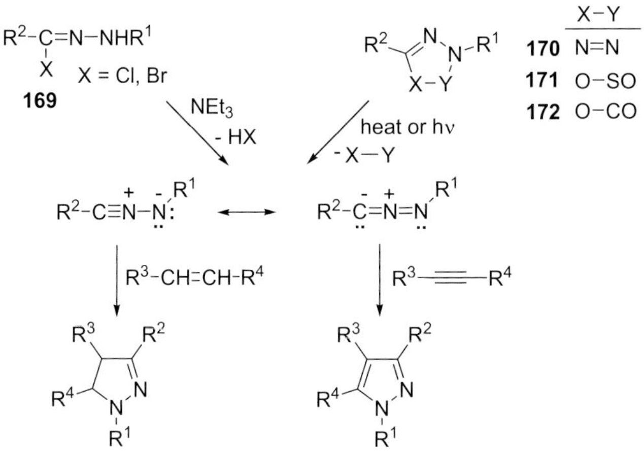

Scheme 8.46

3-Diethylaminoacrylonitrile (**173**) reacts readily with nitrilimines generated from hydrazonoyl chlorides **174** and triethylamine to yield selectively 1,3-disubstituted pyrazole-4-carbonitriles **175** (Scheme 8.47) [193]. Bis-pyrazolophanes have been

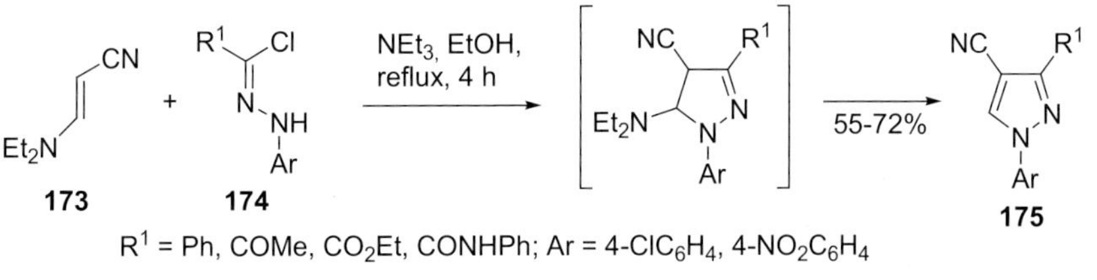

Scheme 8.47

prepared via double cycloadditive macrocyclization of bis-hydrazonoyl chlorides with bis-allyl ethers, a bis-vinyl ether and a bis-propargyl ether [194].

Diaryl azines react with maleic anhydride, maleimide or N-substituted maleimides to give pyrazolo[1,2-*a*]pyrazole derivatives **177** by (1,3 : 2,4)-dipolar cycloadditions (also known as *crisscross* cycloadditions) (Scheme 8.48) [195–197]. Azomethine imines **176** are probable intermediates in these transformations.

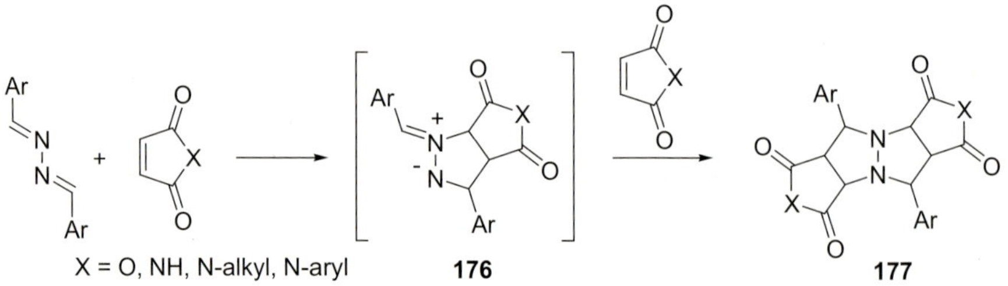

X = O, NH, N-alkyl, N-aryl **176** **177**

Scheme 8.48

Azomethine imines of types **178** and **180** have been used in the synthesis of aza-β- and aza-γ-lactams (Scheme 8.49) [198, 199].

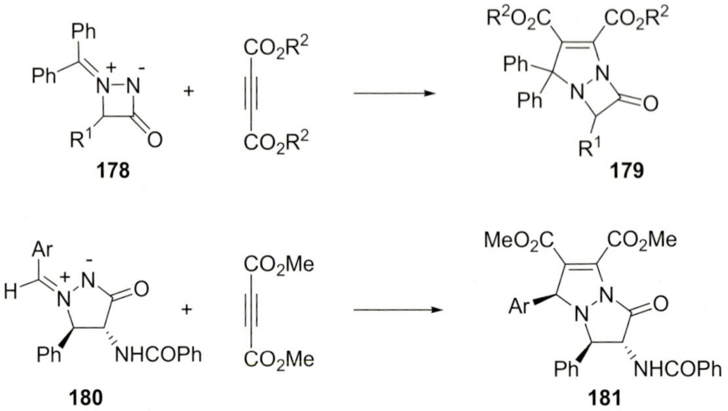

178 **179** **180** **181**

Scheme 8.49

A highly enantioselective catalytic intermolecular [3 + 2] cycloaddition of acylhydrazones **182** to electron-rich alkenes has been reported [200]. Tetrahydropyrazoles **183** are obtained in high yield and with high ee (95–98%) (Scheme 8.50). The reaction proceeds in the presence of a chiral zirconium catalyst prepared from zirconium propoxide ($Zr(OPr)_4$), (*R*)-3,3′-I_2BINOL (**184**) and propanol. A concerted mechanism has been proposed for this reaction.

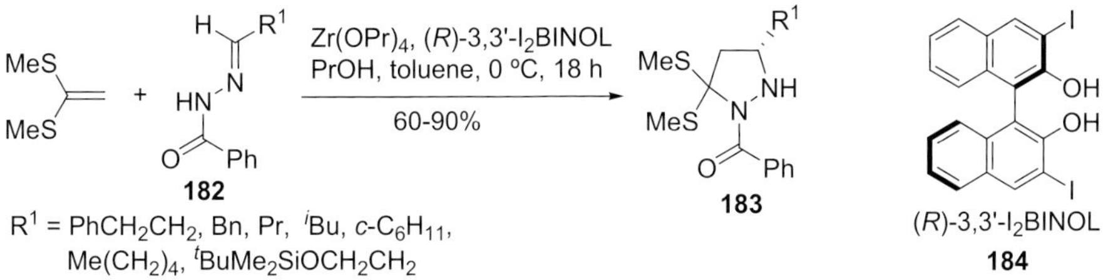

MeS, MeS ketene dithioacetal + **182** (H, R^1, HN–N, O, Ph) → Zr(OPr)$_4$, (R)-3,3'-I$_2$BINOL, PrOH, toluene, 0 °C, 18 h; 60-90% → **183** (MeS, MeS, R^1, NH, N, O, Ph)

R^1 = PhCH$_2$CH$_2$, Bn, Pr, ^{i}Bu, c-C$_6$H$_{11}$, Me(CH$_2$)$_4$, ^{t}BuMe$_2$SiOCH$_2$CH$_2$

(R)-3,3'-I$_2$BINOL **184**

Scheme 8.50

Formation of Two C–N Bonds and One C–C Bond [N-N + C-C + C] A four-component one-pot construction of pyrazoles via a palladium-catalyzed coupling of terminal alkynes, hydrazine (or methylhydrazine), carbon monoxide and aryl iodides has been described recently (Scheme 8.51) [201]. The reaction proceeds at room temperature and an ambient pressure of carbon monoxide in an aqueous solvent system. The reaction is completely regioselective and the yields are excellent. Under similar conditions, the reaction with phenylhydrazine does not afford the corresponding pyrazole.

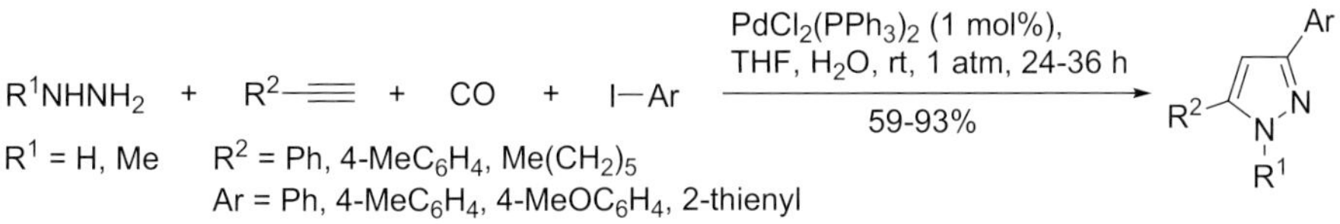

R^1NHNH$_2$ + R^2–≡ + CO + I–Ar → PdCl$_2$(PPh$_3$)$_2$ (1 mol%), THF, H$_2$O, rt, 1 atm, 24-36 h; 59-93% → pyrazole (R^2, Ar, N, N–R^1)

R^1 = H, Me R^2 = Ph, 4-MeC$_6$H$_4$, Me(CH$_2$)$_5$
Ar = Ph, 4-MeC$_6$H$_4$, 4-MeOC$_6$H$_4$, 2-thienyl

Scheme 8.51

8.4.1.5 From Other Heterocycles

Many heterocyclic compounds can be converted into pyrazoles. However, despite some synthetically interesting exceptions, in most cases the starting heterocycles are not readily available and the routes are unlikely to be general. A few examples of ring enlargement, ring modification and ring contraction reactions leading to pyrazoles are shown in the following schemes.

Ring enlargement of diaziridinone **185** by reaction with bifunctional carbanions leads to pyrazolinone **186** or to spiroheterocycles **187** (Scheme 8.52) [202].

Sydnones give 1,3-dipolar cycloadditions with a range of dienophiles to yield pyrazoles. For instance, the reaction of 3-phenylsydnone (**188**) with *meta-* and *para*-diethynylbenzene in refluxing xylene gives *meta-* and *para*-phenylenedipyrazoles **190**, respectively (Scheme 8.53) [203]. Similarly, the reaction of *para*-phenylene-3,3′-disydnone (**191**) with phenylacetylene provides 3,3′-diphenyl-1,1′-*para*-phenylenedipyrazole (**192**). A range of polysubstituted pyrazoles have been obtained in high yield and with elevated regioselectivity from the reaction of 3,4-disubstituted sydnones with 1-aryl-3,3,3-trifluoromethylpropynes [204]. The 1,3-dipolar cycloaddition reactions of nitrilimines with a great variety of heterocyclic compounds, many of them leading to pyrazoles, have been reviewed [187].

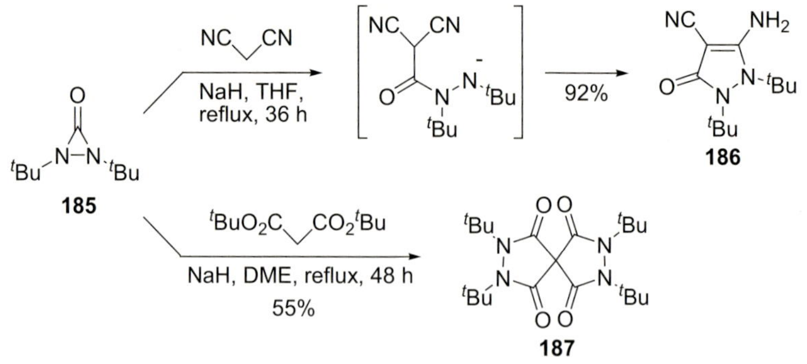

Scheme 8.52

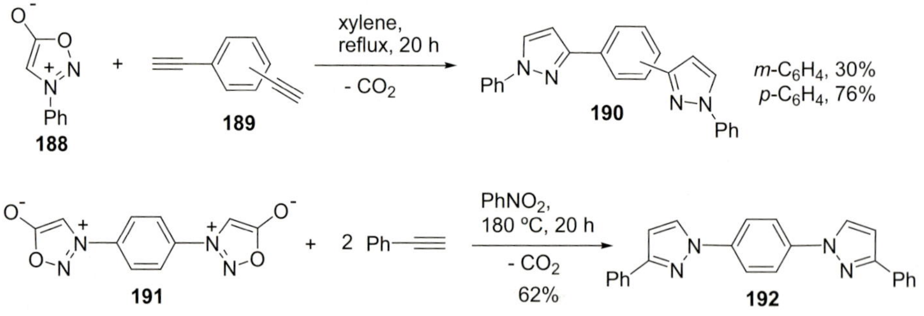

Scheme 8.53

Pd(0)-catalyzed carbonylation of 1,2-diaza-1,3-butadienes **196**, generated *in situ* by thermal extrusion of SO_2, CO_2 or COS from heterocyclic precursors **193–195**, respectively, under 1–2 atm of CO affords pyrazol-5-ones **197** in good yields (Scheme 8.54) [136].

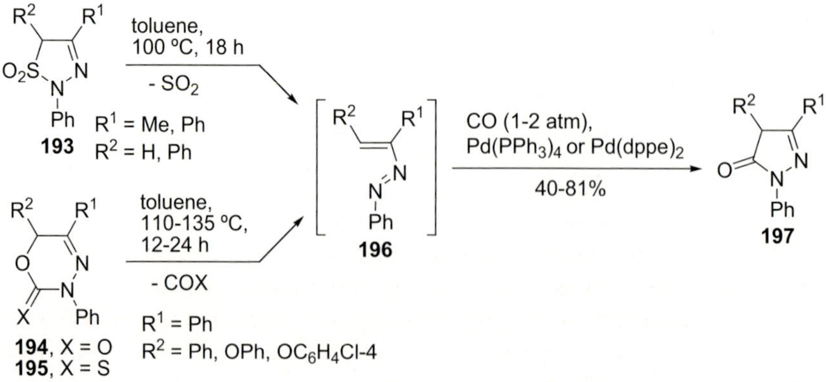

Scheme 8.54

Addition of hydrazine derivatives to (*S*)-1-acylpyrrolidin-2-ones **198** in refluxing acetic acid leads to (*S*)-*N*-acyl-3-(1-substituted-5-hydroxy-1*H*-pyrazol-4-yl)alanine methyl esters **199** (Scheme 8.55) [205].

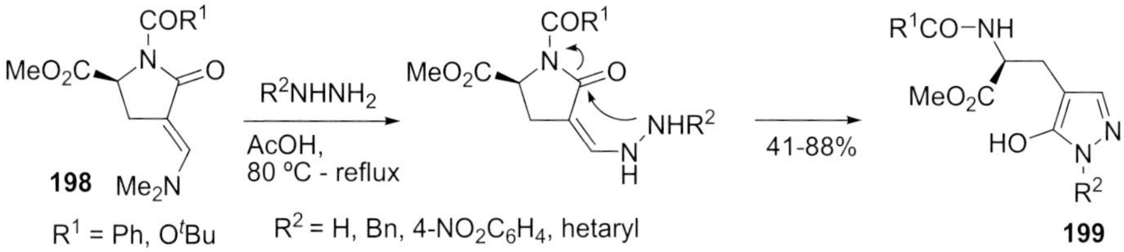

R^2NHNH_2, AcOH, 80 °C - reflux; 41-88%

198 R^1 = Ph, O*t*Bu; R^2 = H, Bn, 4-$NO_2C_6H_4$, hetaryl; **199**

Scheme 8.55

3-Benzoyl-2-substituted-5-phenylfurans **200** react with hydrazine to afford the corresponding 4-benzoylmethyl-3(5)-phenylpyrazoles **201** (Scheme 8.56) [206].

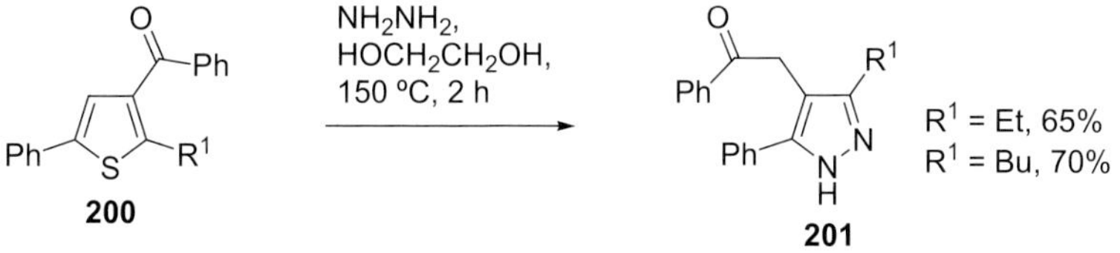

NH_2NH_2, $HOCH_2CH_2OH$, 150 °C, 2 h

200 → **201**: R^1 = Et, 65%; R^1 = Bu, 70%

Scheme 8.56

Furan-2,3-diones **202** react with hydrazines under different conditions to yield pyrazole-3-carboxylic acid hydrazides **203** (Scheme 8.57) [207].

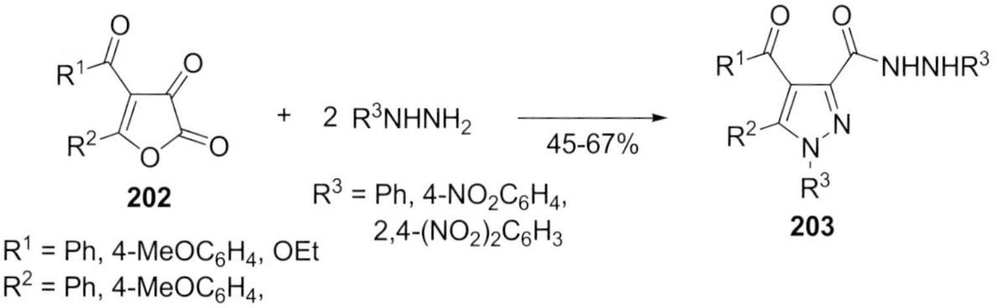

202 + 2 R^3NHNH_2 → (45-67%) **203**

R^1 = Ph, 4-$MeOC_6H_4$, OEt; R^2 = Ph, 4-$MeOC_6H_4$; R^3 = Ph, 4-$NO_2C_6H_4$, 2,4-$(NO_2)_2C_6H_3$

Scheme 8.57

4,5-Dihydro-5-(hydroxyimino)-4-oxothiophene-3-carboxylic esters **204** react with hydrazines to yield pyrazole-3- or -5-thiohydroxamic acids (**205** and **206**, respectively), depending on the substituents R^1 and R^2 (Scheme 8.58) [208].

2,3-Dihydro-4*H*-pyran-4-ones **207** undergo rapid condensation with arylhydrazines in the presence of montmorillonite KSF clay to afford enantiomerically pure 5-substituted pyrazoles **208** in good yields (Scheme 8.59) [209]. In the absence of the clay, no reaction is observed between the pyranones and arylhydrazines. The catalyst

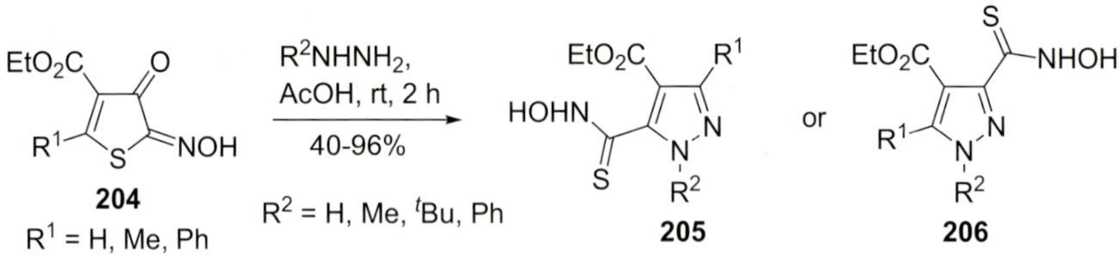

Scheme 8.58

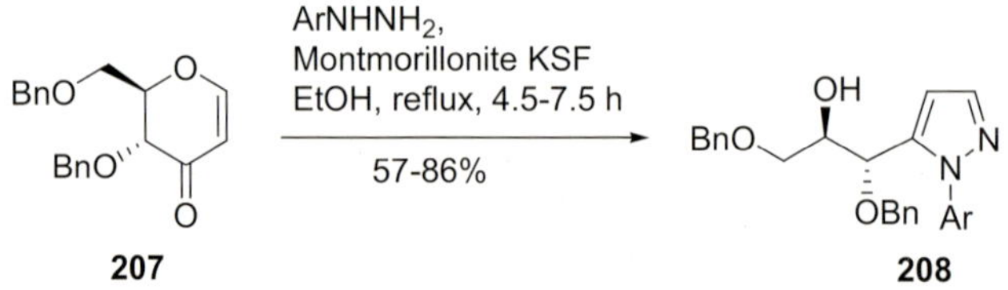

Scheme 8.59

can be recovered by simple filtration and reused three times without any significant decrease in activity after being washed with methanol and activated at 120 °C.

2-Formyl glycals **209** undergo rapid condensation with arylhydrazines under solvent-free conditions to give the corresponding optically pure 4-substituted pyrazoles **210** in good yields with high selectivity (Scheme 8.60) [210]. Like arylhydrazines, hydrazine hydrate itself also affords the respective pyrazoles in good yields.

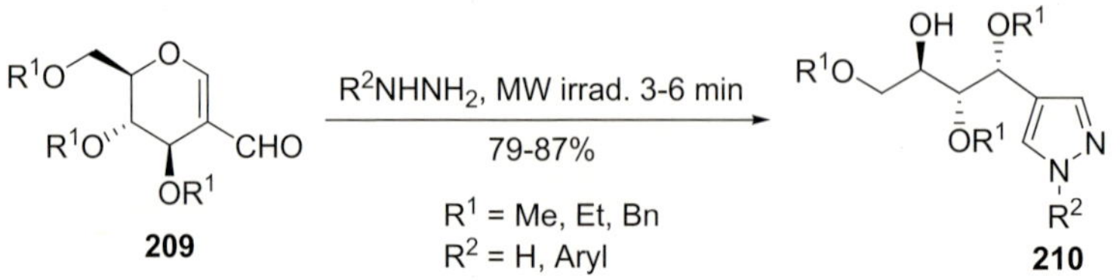

Scheme 8.60

2-(Methyl, phenyl, or styryl)chromones **211** react with methylhydrazine to afford the corresponding 3-(2-benzyloxy-6-hydroxyphenyl)pyrazole derivatives **212** (Scheme 8.61) [211]. Similarly, treatment of 3-aroylflavones **213** with hydrazine gives a mixture of the two aroylpyrazoles **214** and **215** [212].

Treatment of the 3-(3-aryl-3-oxopropenyl)chromen-4-ones **216** with hydrazine hydrate in hot acetic acid afforded the 1-acetyl-3-aryl-5-[3-(2-hydroxyphenyl)pyrazol-4-yl]-2-pyrazolines **217** in good yields (Scheme 8.62) [213]. Oxidation of the 2-pyrazoline ring with DDQ gave the bispyrazoles **218**. *N*-Deacylation occurred during the oxidation.

3-Acyl-2*H*-pyran-2,4-diones **219** react with one equivalent of phenylhydrazine to give hydrazones **220a–d** (R^3=Ph) (Scheme 8.63) [214]. Reaction of **219a** with

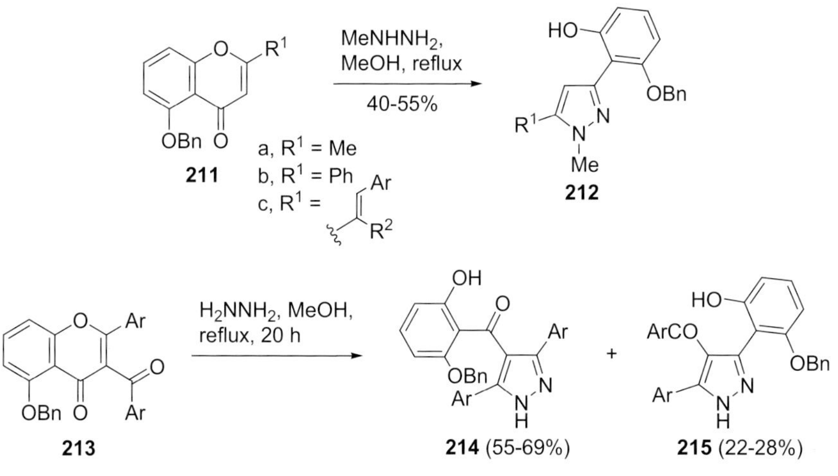

MeNHNH$_2$, MeOH, reflux
40-55%

211
a, R^1 = Me
b, R^1 = Ph
c, R^1 = (Ar, R^2 substituted vinyl)

212

H$_2$NNH$_2$, MeOH, reflux, 20 h

213

214 (55-69%) + 215 (22-28%)

Scheme 8.61

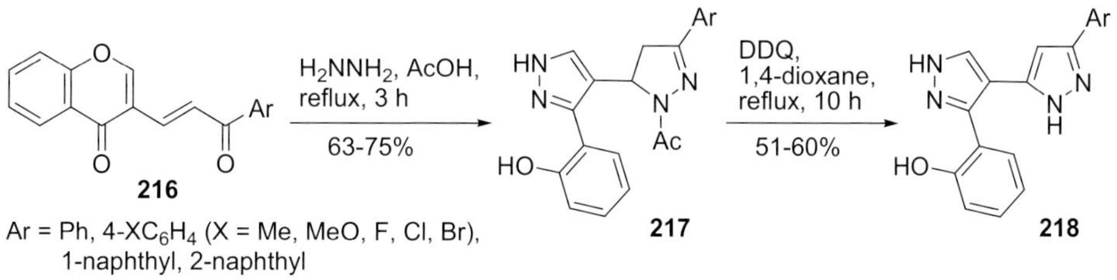

H$_2$NNH$_2$, AcOH, reflux, 3 h
63-75%

DDQ, 1,4-dioxane, reflux, 10 h
51-60%

216

Ar = Ph, 4-XC$_6$H$_4$ (X = Me, MeO, F, Cl, Br), 1-naphthyl, 2-naphthyl

217

218

Scheme 8.62

R^{3}NHNH$_2$, C$_6$H$_6$, reflux, 5 min
75-92%
R^3 = H, Ph

220

AcOH, reflux, 1 h
65-84%

221

219

a, $R^1 = R^2$ = Me
b, R^1 = Me, R^2 = Et
c, R^1 = Me, R^2 =Ph
d, R^1 = Ph, R^2 = Me
e, $R^1 = R^2$ = Ph

PhNHNH$_2$ (2 equiv)
($R^1 = R^2$ = Me)

222

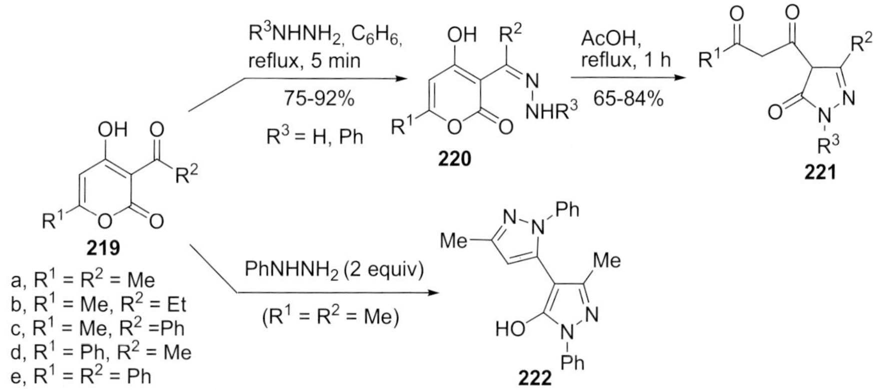

Scheme 8.63

hydrazine hydrate affords **221a** (R^3=H) [215]. Hydrazones **220** can be converted into pyrazolin-5-ones **221** in good yields [214]. 3-Acetyl-2*H*-pyran-2,4-dione **219a** reacts with two equivalents of phenylhydrazine to give the pyrazolylpyrazole **222** [214].

Surprisingly, 2*H*-pyran-2,4-dione **223** reacts with phenylhydrazine to afford dihydropyrazole **224** while with hydrazine it gives pyrazole **225** (Scheme 8.64) [216]. The formation of **225** involves a decarboxylation process, not observed in the reaction of compound **219a** with hydrazine [215]. 3-Aryl-1-(3-coumarinyl)propen-1-ones also react with hydrazines to afford 1-substituted 5-aryl-3-(3-coumarinyl)-2-pyrazolines [217].

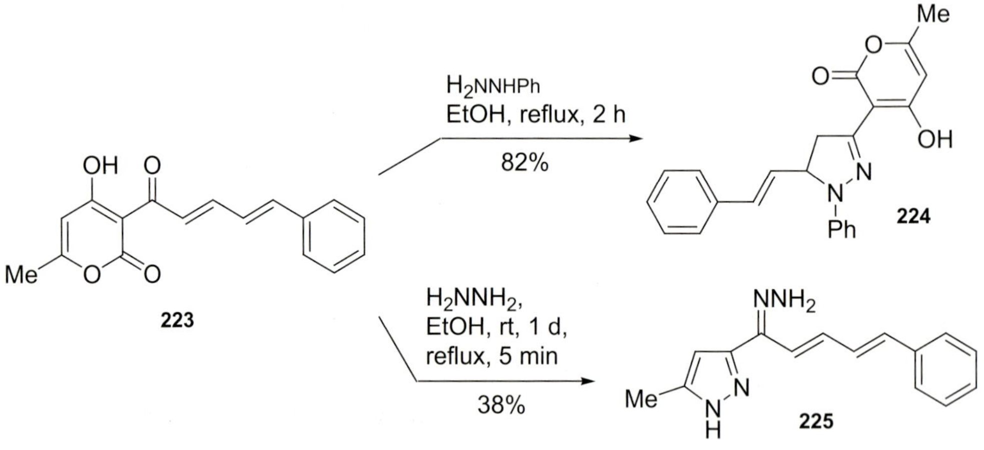

Scheme 8.64

N-Methoxypyridazinium salts **226** react with hydroxide ion to give vinyl diazomethanes **227**, which cyclize to 3(5)-acyl-1*H*-pyrazoles **228** when heated in benzene (Scheme 8.65) [99].

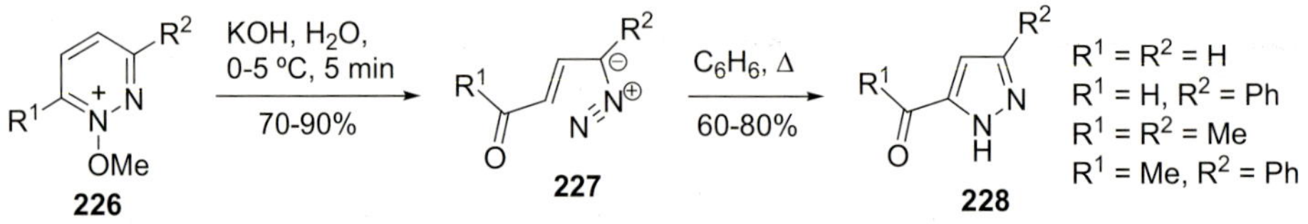

Scheme 8.65

Under the action of hydrazines, pyrimidines give ring contraction transformations to pyrazoles. For instance, 5-nitropyrimidine reacts with hydrazine hydrate at room temperature for 2 days, or at 100 °C for half an hour, to yield 4-nitropyrazole in nearly quantitative yield [218]. More recent work has shown that treatment of 4-(dimethylamino)-6-chloro (or 6-methoxy)-5-nitropyrimidine with hydrazine hydrate or methylhydrazine (2 equiv.) leads to 3,5-diamino-4-nitropyrazole and 3,5-diamino-1-methyl-4-nitropyrazole, respectively, in moderate yields (Scheme 8.66) [219]. The conversion of 5-acylpyrimidines and 5-acyl-uracils into a range of pyrazole derivatives has been reviewed [220].

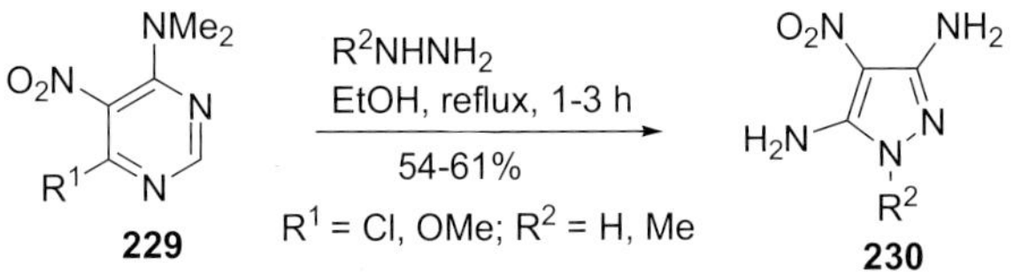

R^2NHNH_2
EtOH, reflux, 1-3 h
54-61%
R^1 = Cl, OMe; R^2 = H, Me
229 → 230

Scheme 8.66

A novel one-step synthesis route to fully substituted pyrazol-4-ols was reported recently. It is based on the reaction of thietanone **231** with 1,2,4,5-tetrazines **232** (Scheme 8.67) [221]. All of the elements of the thietanone, except its sulfur, are incorporated in the products. Quoting the authors of that work, this is a simple yet non-obvious method for the construction of pyrazol-4-ols.

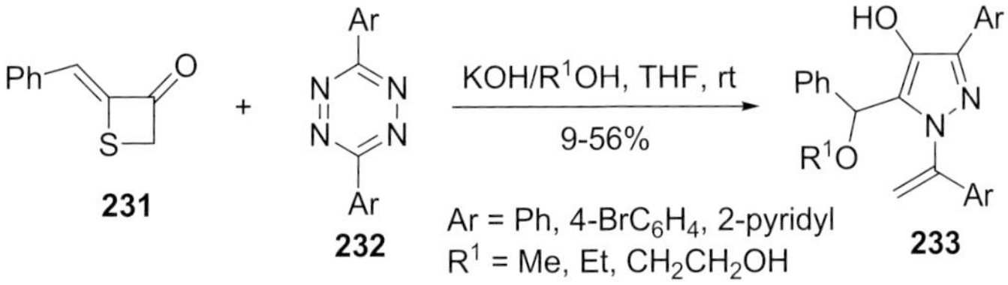

231 + 232 → 233
KOH/R^1OH, THF, rt
9-56%
Ar = Ph, 4-BrC_6H_4, 2-pyridyl
R^1 = Me, Et, CH_2CH_2OH

Scheme 8.67

Thermolysis of the tetrazolo[1,5-*a*]pyrimidines **234** and tetrazolo[1,5-*b*]pyridazine **237** gives 1-cyanopyrazoles in good yields (Scheme 8.68) [222]. When these heterocyclic compounds are heated at about 10–20 °C over their melting temperature evolution of nitrogen is observed. The reactions are particularly fast (few minutes)

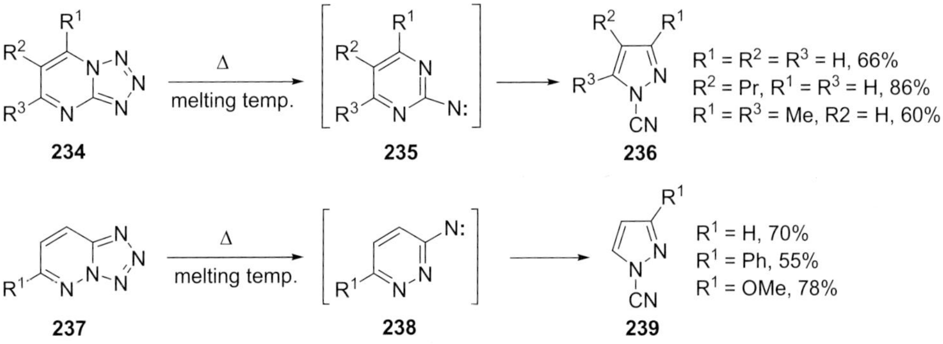

234 → [235] → 236
Δ
melting temp.
$R^1 = R^2 = R^3$ = H, 66%
R^2 = Pr, $R^1 = R^3$ = H, 86%
$R^1 = R^3$ = Me, R2 = H, 60%

237 → [238] → 239
Δ
melting temp.
R^1 = H, 70%
R^1 = Ph, 55%
R^1 = OMe, 78%

Scheme 8.68

and the pyrazole derivatives **236** and **239** are the only detectable products. Nitrenes **235** and **238** are probable intermediates in these ring contraction reactions.

8.4.2 Synthesis of Indazoles

There are several methods for the synthesis of indazoles [4, 5, 7, 9, 223, 224]. Most start from benzene derivatives, where the pyrazole ring is formed by ring closure. However, a few examples start from pyrazoles. The different possibilities for the construction of the indazole ring can be regarded according to the bonds formed, as described for the pyrazoles, and their synthesis will be organized in that way.

The major part of indazole ring-closure procedures involves creating a bond between the two nitrogen atoms (N–N) as the last step; nevertheless, ring closure by creation of a N–C bond through the formation of N2–C3 or N1–C7a bond is also common. A few examples involving a C3–C3a ring closure are also reported.

8.4.2.1 Formation of One N–N Bond

One of the simplest syntheses of indazoles by an N–N ring closure involves the reduction of *N-ortho*-nitrobenzaldimines **240** with triethyl phosphite, to afford 2-aryl-2*H*-indazoles **241** (Scheme 8.69) [225]. The same type of imines (**242**) can be converted into 2*H*-indazoles **244** by reductive *N*-heterocyclization of *N*-(2-nitrobenzylidene)amines **242** with the catalyst system dichlorobis(triphenylphosphine)palladium(II)–tin(II) chloride at 100 °C for 16 h under 20 kg cm^{-2} of initial carbon monoxide pressure (Scheme 8.70) [226]. Carbon monoxide operates as a reducing agent of the nitro substituent into a nitrene intermediate **243**, which strongly coordinates palladium, along with the generation of carbon dioxide. The electrophilic nitrene can then attack the nitrogen atom of the imino group to give the corresponding 2-substituted 2*H*-indazoles **244**.

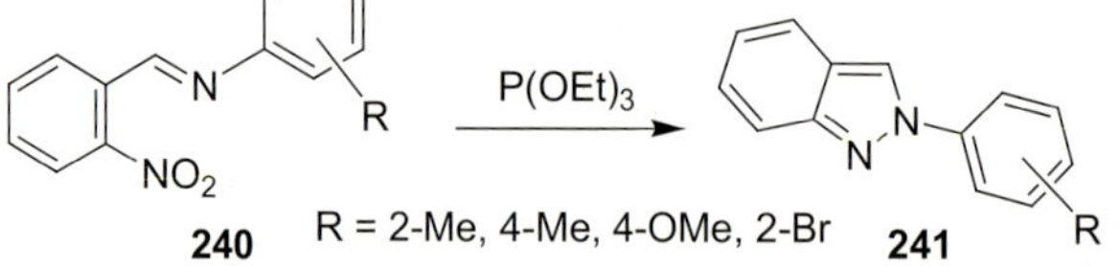

Scheme 8.69

2-Aryl-3-chloro-2*H*-indazoles **247** are obtained by the treatment of *ortho*-azidobenzanilides **245** with thionyl chloride at reflux (Scheme 8.71). The mechanism proposed for this transformation probably involves the initial formation of *ortho*-azidobenzimidoyl chlorides **246**, which cyclize into **247** by a concerted pericyclic process with loss of nitrogen [227]. The required anilides **245** can be prepared in high

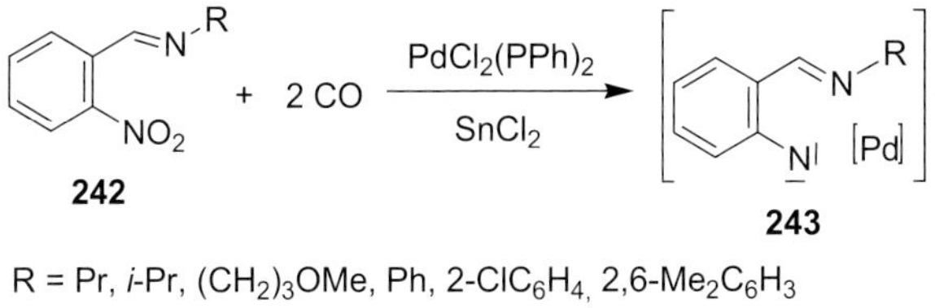
PdCl2(PPh)2
SnCl2
+ 2 CO
[Pd]
242
243
244
R = Pr, i-Pr, (CH2)3OMe, Ph, 2-ClC6H4, 2,6-Me2C6H3

Scheme 8.70

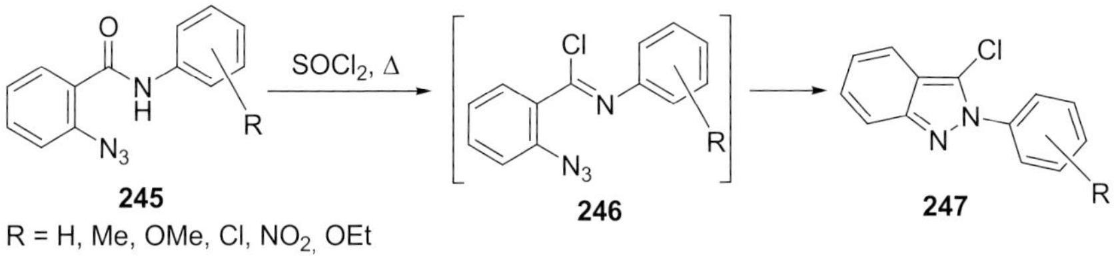
SOCl2, Δ
245
246
247
R = H, Me, OMe, Cl, NO2, OEt

Scheme 8.71

yield from the reaction of the appropriate arylamine with *ortho*-azidobenzoyl chloride in pyridine solutions. 2-Substituted-2*H*-indazoles **250** are obtained from *N*-(2,4,6-trinitrobenzylidene)anilines and hydrazones **248** by a similar synthetic process (Scheme 8.72). Treatment of **248** with sodium azide leads to the regiospecific substitution of the *ortho*-nitro group by the azido group, and the thermolysis of the obtained **249** give 4,6-dinitro-2-substituted-2*H*-indazoles **250**. In some cases, compounds **248** are converted into 2*H*-indazoles **250** even in the process of the azide formation [228]. The thermal decomposition of other 2-azidobenzylideneamine derivatives into 2-substituted-2*H*-indazoles is also described [229].

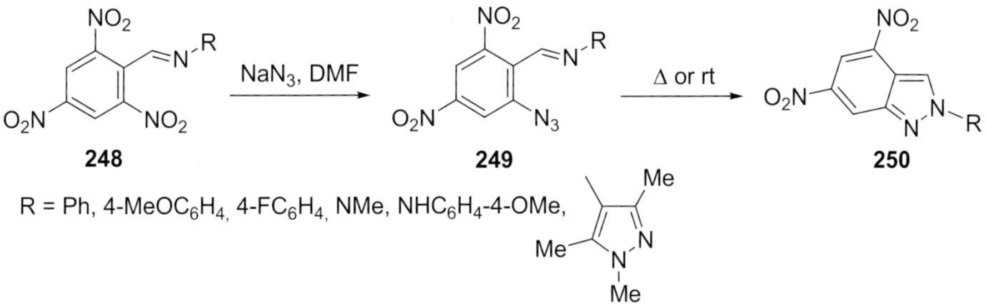
NaN3, DMF
Δ or rt
248
249
250
R = Ph, 4-MeOC6H4, 4-FC6H4, NMe, NHC6H4-4-OMe,

Scheme 8.72

2-Amino-3-(alkyl or aryl)amino-2*H*-indazoles **253** are also prepared from the reaction of *ortho*-azidobenzaldimines **251** with tertiary phosphines followed by acid hydrolysis of the obtained iminophosphoranes **252** (Scheme 8.73) [230].

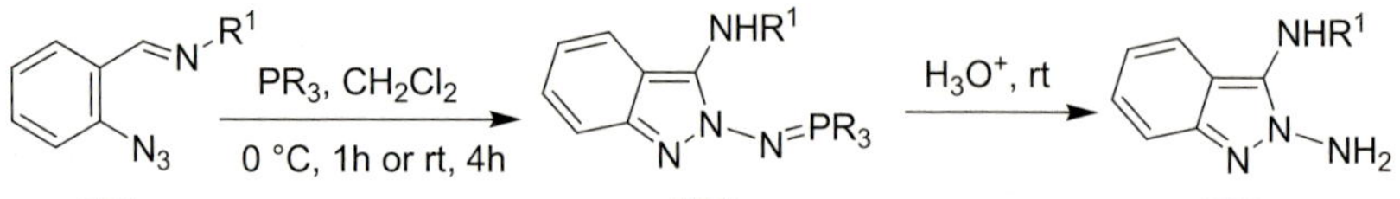

Scheme 8.73

2-Aryl-2H-indazoles are also prepared by a base-catalyzed reaction of *ortho*-nitrobenzyl triphenylphosphonium bromide **254** and aryl isocyanates (Scheme 8.74) [231]. Deprotonation of the triphenylphosphonium bromide **254** with DBU gives a purple ylide **255**. When sodium hydride is used, the expected indazoles are accompanied by small amounts of 2-nitrotoluene as a by-product. Treatment of the ylide with aryl isocyanates, bearing electron-withdrawing or electron-donating substituents, affords 2-aryl-2H-indazoles **256** in moderate to good yields; the nitrogen of the nitro group being transformed into the indazole N1 atom.

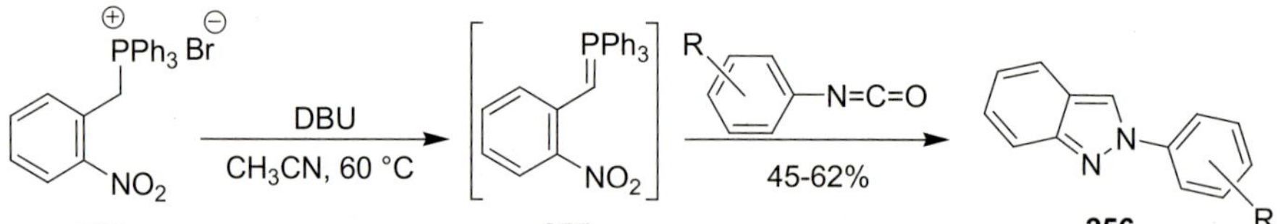

R = 4-OMe, 2-OMe, 4-OPh, 4-CO_2Et, 4-F, 3-CN, 3,4,5-$(OMe)_3$

Scheme 8.74

2H-Indazole-2-oxides **260** are obtained via the 1,7-electrocyclization of non-stabilized azomethine ylides **257**, formed from *ortho*-nitrobenzaldehydes and sarcosine, onto the nitro group to give the unstable benz-1,2,6-oxadiazepine **258**, which undergoes a ring contraction, resulting in the elimination of formaldehyde and the formation of 2-methyl-2H-indazole-1-oxides **260** in moderate yield (32–40%) (Scheme 8.75) [232, 233]. In these reactions, 3-methyl-5-aryl-oxazolidines **259** are also formed, resulting from the reaction of the starting *ortho*-nitrobenzaldehydes with an azomethine ylide obtained from formaldehyde generated *in situ* and the excess of sarcosine.

2H-Indazole-1-oxides **260** are deoxygenated in the presence of Pd–C to afford 2-methyl-2H-indazoles **261** [232, 233].

A N−N bond in the synthesis of 1-substituted-1H-indazoles **263** and **265** can be created by dehydration of oximes **262** with acetic anhydride [223] or by heating the oxime acetate **263** in the melt at 170 °C, under vacuum [234], respectively (Scheme 8.76).

8.4.2.2 Formation of One N2–C3 Bond

One of the most common methods for the synthesis of indazoles involving a N2−C3 ring closure starts from an *ortho*-toluidine (Scheme 8.77) [235]. Acetylation of the

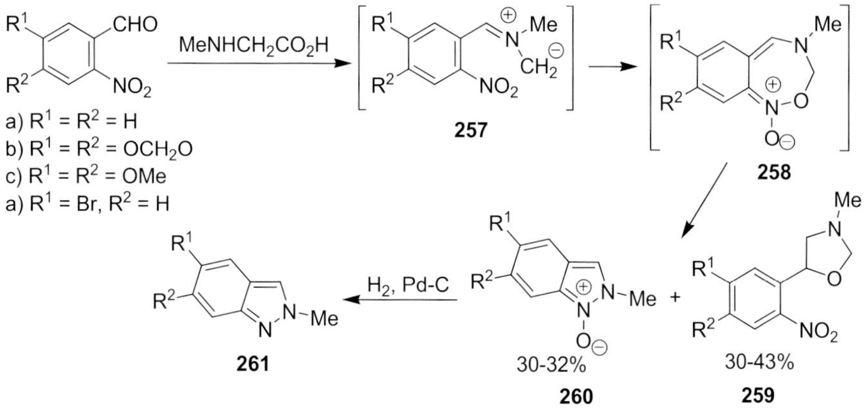

Scheme 8.75

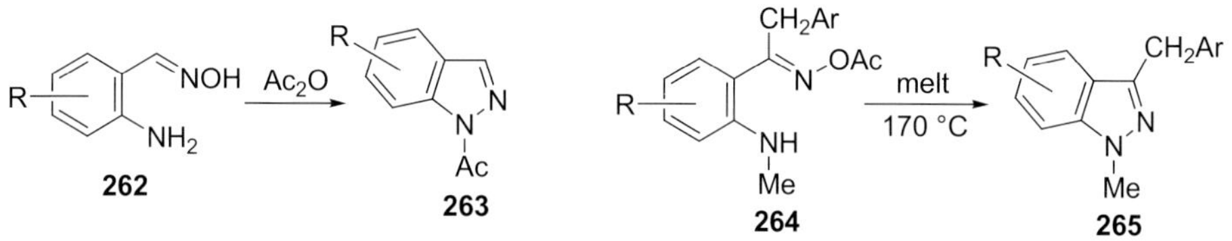

Scheme 8.76

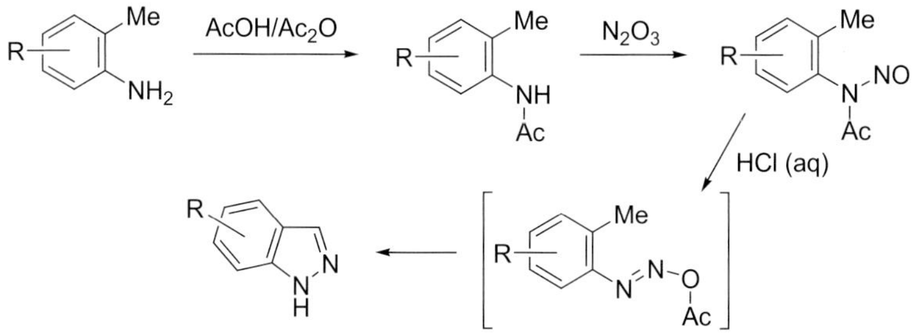

Scheme 8.77

aniline, followed by nitrosation with nitrous gases, formed by the action of nitric acid on sodium nitrite, and subsequent intramolecular azo coupling, with an initial acyl migration, leads to the 1*H*-indazoles. This classic protocol is also employed in the synthesis of several potential biologically active *N*-substituted-1*H*-indazoles, although with small changes in the experimental procedure [236–238].

The diazotization of *ortho*-toluidines in acid or neutral aqueous solution is also a well-know procedure to prepare indazole rings [239, 240]. However, the reactions are successful only for *ortho*-methylbenzenediazonium salts bearing an electron-withdrawing nitro or halogen group on the aromatic ring and involve the isolation of an explosive *ortho*-methylbenzenediazonium chloride. The improvement of this synthetic procedure allows the preparation of 1*H*-indazoles **268** bearing electron-withdrawing or electron-donating substituents from the reaction of non-explosive *ortho*-alkylbenzenediazonium tetrafluoroborates **267** (Scheme 8.78) [241, 242]. *ortho*-Alkylbenzenediazonium tetrafluoroborates **267** are obtained from the reaction of *ortho*-alkylanilines **266** with sodium nitrite in fluoroboric acid or sodium nitrite in hydrochloric acid followed by the addition of sodium tetrafluoroborate. Treatment of **267** with two equivalents of potassium acetate, 5 mol% of 18-crown-6 in ethanol-free chloroform at room temperature affords 1*H*-indazoles **268** in moderate to good yields. The presence of the phase transfer catalyst 18-crown-6 is essential; in its absence no indazole is formed.

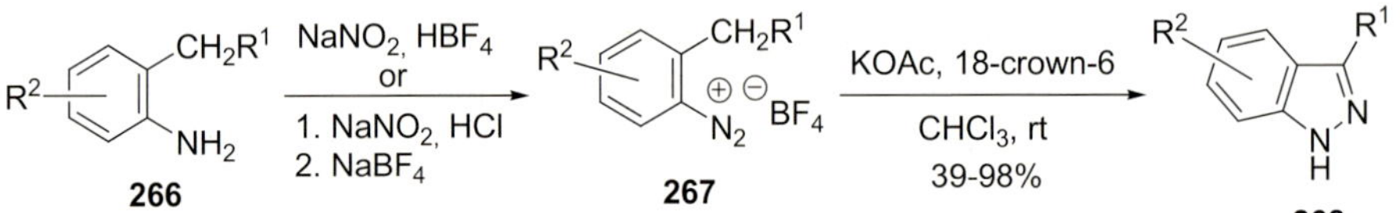

266 → 267: $NaNO_2$, HBF_4 or 1. $NaNO_2$, HCl 2. $NaBF_4$
267 → 268: KOAc, 18-crown-6, $CHCl_3$, rt, 39-98%

R^1 = H, Me
R^2 = H, 4-NO_2, 5-NO_2, 6-NO_2, 4-Cl, 5-Cl, 3-Me, 4-Me, 5-Me, 6-Me, 7-Me, 4-OMe

Scheme 8.78

3-(2-Fluorophenyl)-1*H*-indazole **271** is prepared from the diazotization of the corresponding 2-aminobenzophenone **269** under strongly acidic conditions (HBF_4), followed by reduction with sodium dithionite (Scheme 8.79) [243]. Zhang *et al.* describe a similar synthesis, save the reduction of the intermediate **270**, which is made with SO_2 [244].

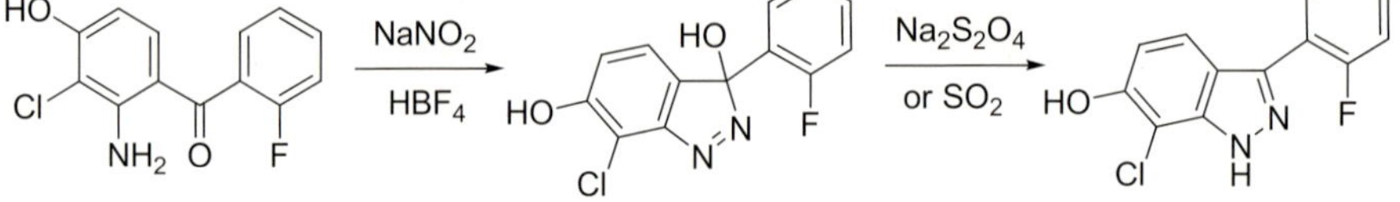

269 → 270: $NaNO_2$, HBF_4
270 → 271: $Na_2S_2O_4$ or SO_2

Scheme 8.79

The most common synthesis of 1,2-dihydro-3*H*-indazol-3-ones **274** (or their enolic forms 3-hydroxy-1*H*-indazoles) starts from diazonium salts of anthranilic acid **272**, which are reduced with sulfites or sulfur dioxide to the corresponding *ortho*-hydrazinobenzoic acid derivatives **273** (Scheme 8.80) [224]. Cyclization to **274** may be carried out with phosphoryl chloride, by boiling in nitrobenzene or refluxing in an

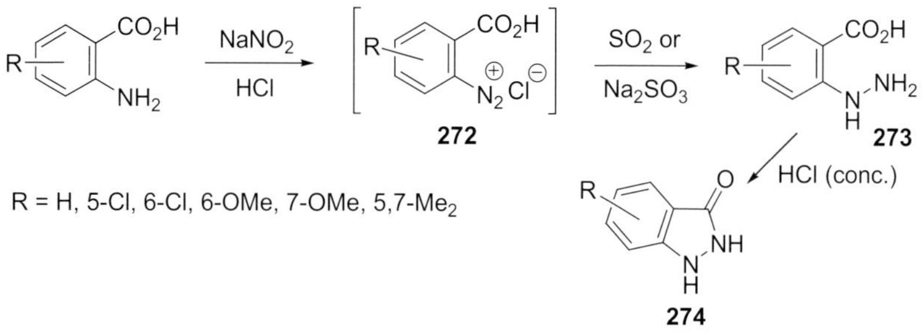

R = H, 5-Cl, 6-Cl, 6-OMe, 7-OMe, 5,7-Me_2

Scheme 8.80

aqueous solution of either acidic or buffered with sodium acetate. A variation of this method consists in the nitrosation of *N*-substituted anthranilic acids or esters, with nitrous acid, followed by reduction of the formed nitroso derivatives. This reduction must be done with sodium hydrosulfide when the *N*-substituent is a hydrogen or halogen atom or an alkyl group; for compounds with an *N*-aryl group as substituent the reduction must be carried out with zinc in acetic acid or lithium aluminium hydride [224].

8.4.2.3 **Formation of One N1—C7a Bond**

Aromatic hydrazides are converted into indazol-3-ones when treated with an excess of butyllithium (Scheme 8.81) [245]. Benzoylhydrazines **275** afford 2*H*-indazol-3-ones **276** in 61–80% yield. The same transformation has been carried out replacing butyllithium by sodium or potassium hydride, but the corresponding 2*H*-indazol-3-one was obtained in lower yields (49 and 56%, respectively, when R=H). 2*H*-Indazol-3-ones **277–279** are obtained by the same procedure from appropriate aromatic hydrazides, while aliphatic and heterocyclic hydrazides afford the corresponding aldehydes [245].

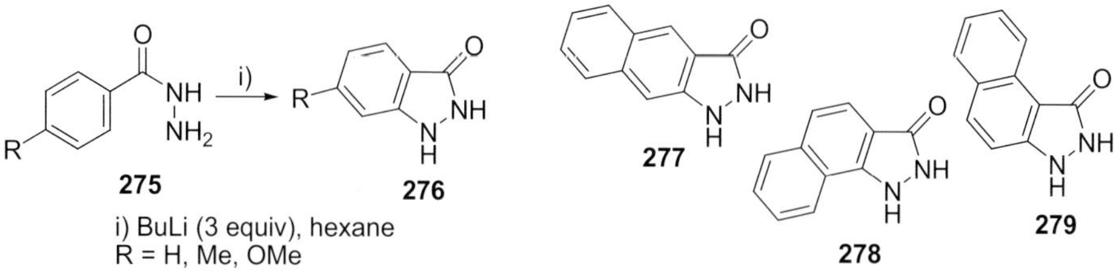

i) BuLi (3 equiv), hexane
R = H, Me, OMe

Scheme 8.81

3-Substituted-1-tosyl-6-fluoro-1*H*-indazoles **283** are obtained from 2-chlorobenzoylhydrazides **280** (Scheme 8.82) [246]. Treatment of **280** with thionyl chloride affords the corresponding imidoyl chloride **281**, which reacts with piperazine derivatives, in the presence of DABCO, to yield imidates **282**. This is a trick reaction, the best yields being obtained when 1.1 molar equivalents of piperazine derivatives

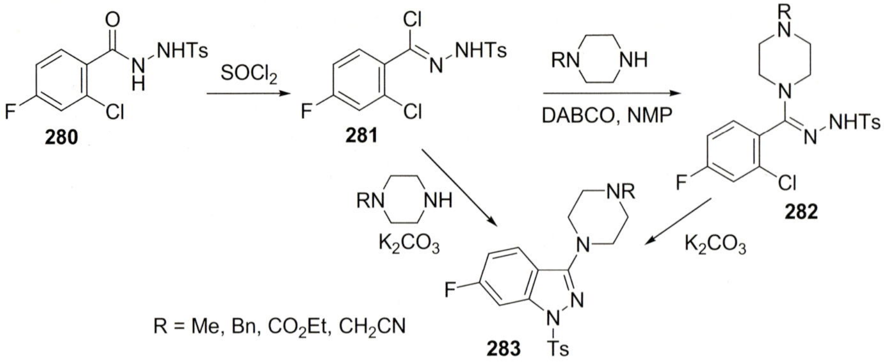

Scheme 8.82

and 0.7 molar equivalents of DABCO are used; the presence of an amine more basic than piperazine is required. However, imidoyl chlorides **281** are also transformed into 1*H*-indazoles **283** in one-pot transformation; after the *in situ* formation of imidate **282**, milled potassium carbonate is added to the mixture and the 3-substituted-1-tosyl-6-fluoro-1*H*-indazoles **283** are obtained.

Another general method involving the formation of the N1–C7a bond consists in the cyclization of phenylhydrazone derivatives. Lead tetraacetate oxidation of phenyl ketone phenylhydrazones **284** leads to the formation of azoacetates **285**, which furnish 1-phenyl-1*H*-indazoles **286** when treated with Lewis acids (e.g., $AlCl_3$, BF_3.Et_2O) (Scheme 8.83) [247–252]. 1-Benzyl-3-(5-hydroxymethyl-2-furyl)indazole **YC-1**, an indazole possessing important biological applications, is obtained from the hydrazone of ketone **287**, although in a modest yield (Scheme 8.84) [253]. **YC-1** is obtained in higher yield starting from the unsubstituted-2*H*-indazol-3-one [254].

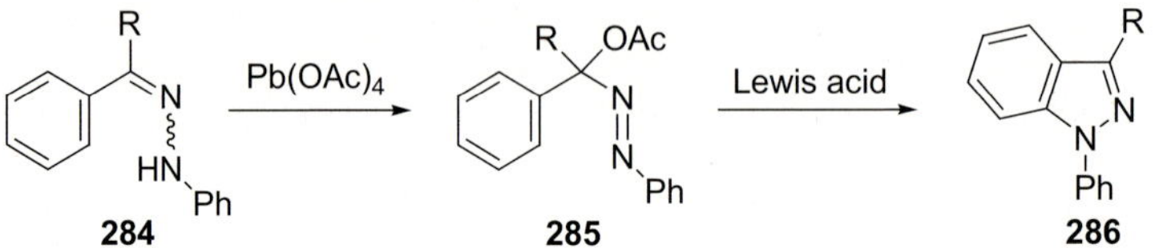

Scheme 8.83

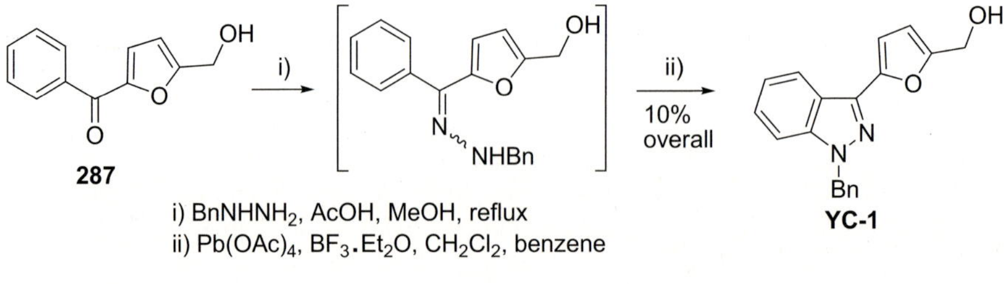

Scheme 8.84

A variation of this method consists in the cyclization of *para*-nitrophenylhydrazones of several acetophenones, benzophenones and benzaldehyde by reaction with polyphosphoric acid at high temperature (150–165 °C), affording 1-(4-nitrophenyl)-1*H*-indazoles [255, 256]. Another variation involves the cyclization of arylhydrazines possessing a leaving group in the *ortho* position (F, NO_2, OH or OR) [243, 257–261]. These methods also require harsh experimental conditions, such as very high temperatures (200–270 °C), although there are references to the synthesis of indazoles by the reaction of *ortho*-fluorobenzophenone or pentafluoroacetophenone with hydrazine in refluxing ethanol or toluene, respectively [243, 262]. Treatment of 2′,6′-(dialkoxy- or dihydroxy)acetophenone or benzophenone hydrazones **288** with polyphosphoric acid gives 4-(alkoxy- or hydroxy)-3-substituted-1*H*-indazoles **289** in moderate to good yields (Scheme 8.85) [263, 264].

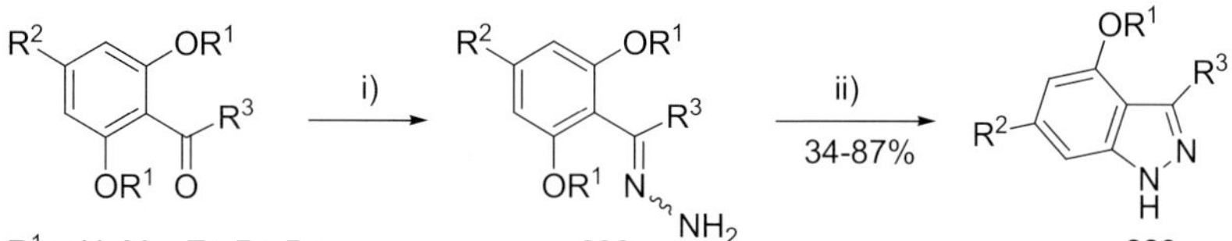

R^1 = H, Me, Et, Pr, Bu
R^2 = H, Me, OMe
R_3 = Me, Ph, 4-ClC_6H_4, 4-BrC_6H_4, 4-OMeC_6H_4, 4-$NO_2C_6H_4$

i) NH_2NH_2.H_2O, AcOH, 110-120 °C
ii) PPA, 115-135 °C (34-87%)

Scheme 8.85

The palladium-catalyzed cyclization of arylhydrazones of 2-bromobenzaldehydes and 2-bromoacetophenones **290** constitutes an easy, efficient method for the synthesis of 1-aryl-1*H*-indazoles **291** (Scheme 8.86) [265–267]. *N*-Arylindazole derivatives are synthesized in a one-pot reaction of 2-bromobenzaldehydes with arylhydrazines in the presence of a catalytic amount of a palladium catalyst, a phosphorous chelating ligand and sodium *tert*-butoxide [265]. The use of preformed hydrazones is the key to obtaining higher yields, milder reaction conditions and to extend the scope of this method to heterocyclic substrates and to 2-chlorobenzaldehyde [267]. In addition, better yields can be obtained using $Pd(dba)_2$ and chelating phosphines

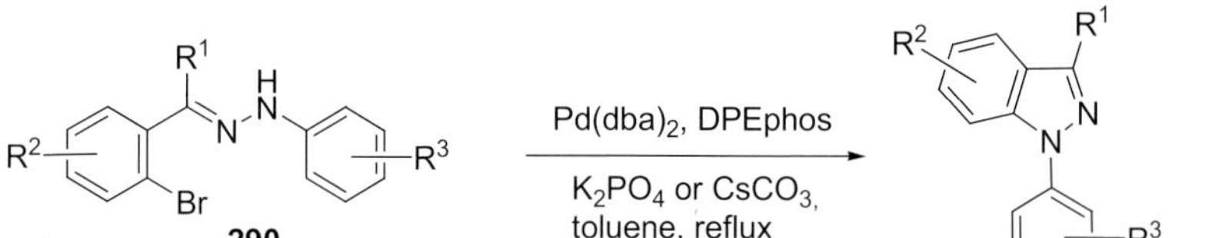

Pd$(dba)_2$, DPEphos
K_2PO_4 or $CsCO_3$, toluene, reflux

R^1 = H, Me
R^2 = H, NO_2, Me, Br
R^3 = H, NO_2, CF_3, Me, OMe, Br

Scheme 8.86

(*rac*-BINAP, DPEphos and dppf) in the presence of a base, such as caesium carbonate or potassium phosphate. This method is applicable for the synthesis of a wide range of 1-aryl-1*H*-indazoles bearing electron-donating and electron-withdrawing substituents.

Various 2-aryl-2*H*-indazoles **293** and 1-aryl-1*H*-indazoles **296** can be prepared by the palladium catalyzed intramolecular amination of the corresponding *N*-aryl-*N*-(2-bromobenzyl)hydrazines **292** and *N*-aryl-*N*′-(2-bromobenzyl)hydrazines **295**, followed by spontaneous aromatization of the formed aryl-substituted-2,3-dihydro-1*H*-indazoles (Schemes 8.87 and 8.88, respectively) [268, 269]. [*N*-Aryl-*N*′-(2-bromobenzyl)hydrazinato]triphenylphosphonium bromides **294**, intermediates in the synthesis of **295**, also underwent cyclization under suitable conditions to afford 1-aryl-1*H*-indazoles **296** (Scheme 8.88). Toluene is used as solvent when starting with *N*-aryl-*N*′-(2-bromobenzyl)hydrazines **295** and 1,4-dioxane with [*N*-aryl-*N*′-(2-bromobenzyl)hydrazinato]triphenylphosphonium bromides **294** due to the insolubility of the latter in toluene.

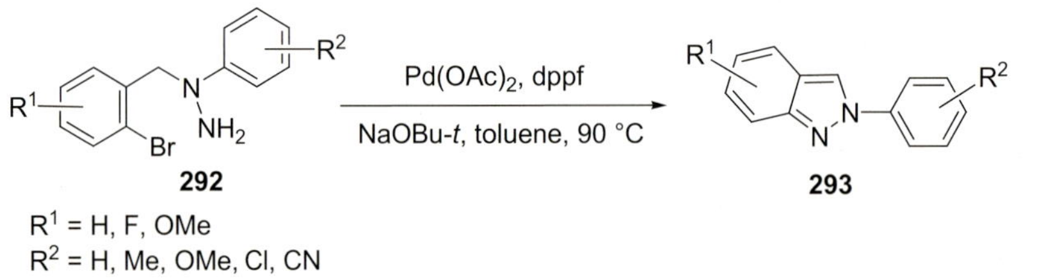

R^1 = H, F, OMe
R^2 = H, Me, OMe, Cl, CN

Pd(OAc)$_2$, dppf; NaOBu-*t*, toluene, 90 °C: **292** → **293**

Scheme 8.87

R^1 = H, F, OMe
Ar = Ph, MeC$_6$H$_4$

294 → NaOH, 60 °C → **295**
295 → Pd(OAc)$_2$, dppf, NaO^{t}Bu, 1,4-dioxane, 90 °C → **296**
294 → Pd(OAc)$_2$, dppf, NaO^{t}Bu, toluene, 90 °C → **296**

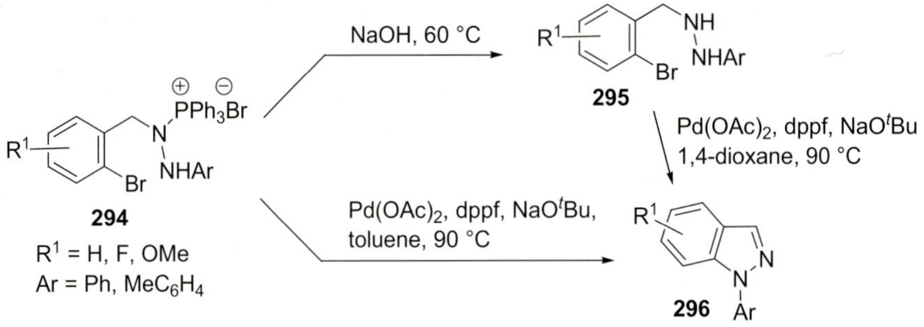

Scheme 8.88

1-Aza-2-azonia allene salts **298**, formed by oxidation of hydrazones **297** with *tert*-butyl hypochloride followed by treatment with $SbCl_5$ in dichloromethane at −50 °C, can be used as starting materials for the preparation of 1-substituted-1*H*-indazoles **300** (Scheme 8.89) [270]. Treatment of **298** with $SbCl_5$ induces an intramolecular cyclization to afford 3-pyridyl-1*H*-indazolium hexachloroantimonates **299**. It is worth

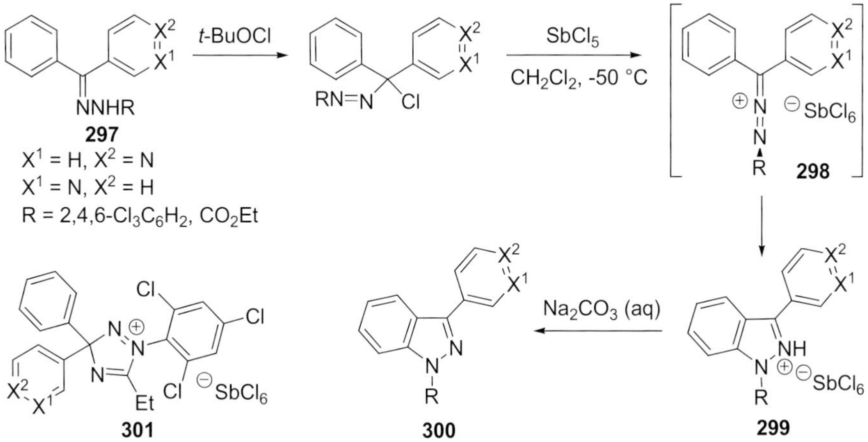

Scheme 8.89

noting the complete regioselectivity of the cycloaddition. The reaction of 1-aza-2-azonia allene salts **298**, bearing a trichloropyridyl substituent, with propionitrile gives 3*H*-1,2,4-triazolium hexachloroantimonates **301** instead of **300**.

8.4.2.4 **Formation of One C3−C3a Bond**

The cyclization of *para*-nitrophenylhydrazones of ketones and aldehydes with polyphosphoric acid gives 1-arylindazoles through a C3–C3a ring closure [271], while *N′*, *N′*-diphenylhydrazides are converted into 1-phenylindazoles by means of trifluoromethanesulfonic anhydride [yields from 2% (R=H) to 50% (R=Ph)] [272].

1*H*-Indazoles **303** can also be obtained in moderate yield from the reaction of nitroarenes with aromatic hydrazones in alkaline medium (Scheme 8.90) [273]. The presence of electron-withdrawing groups in both reagents is required for the formation of 1*H*-indazoles **303**. Other substituents originate displacement of the chlorine atom or hydrogen atom in the 4-position of the nitroarene by the hydrazone anion. Although the mechanism of this reaction is not clear, a possible reaction sequence could be described as an initial attack of the hydrazone anion to

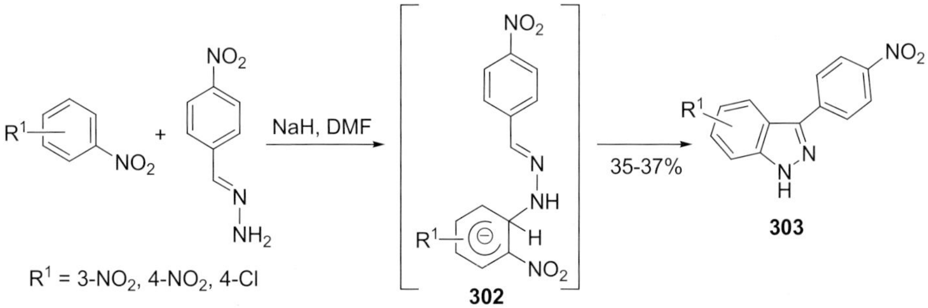

Scheme 8.90

the *ortho*-position of the nitroarene, to give a Meisenheimer intermediate (**302**). The presence of an electron-withdrawing substituent on the nitroarene facilitates the attack of the hydrazone anion and stabilizes the anion **302**. Displacement of the nitro group leads to indazoles **303**.

1,2-Dihydro-3*H*-indazol-3-ones **308** are also obtained from anilines (Scheme 8.91). Treatment of substituted anilines with phosgene in toluene affords carbamoyl chlorides **304**, which react with sodium azide in methanol to give carbamoyl azides **305**. Thermal cyclization of **305** gives indazol-3-ones **308** [224, 274, 275]. This cyclization must proceed via intermediates **306**, which are converted into isocyanates **307**, which then cyclize to indazol-3-ones **308**. The existence of intermediate **306** is confirmed by the concomitant formation of benzimidazolinones **309** [224, 274–276].

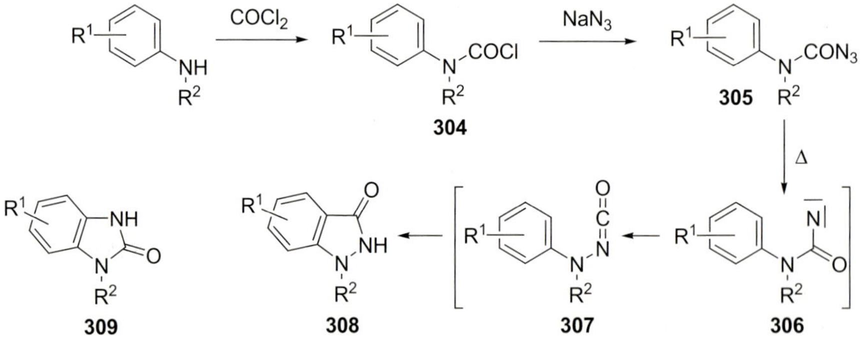

R^1 = H, Cl, NO_2
R^2 = H, 2-ClC_6H_4, 3-ClC_6H_4, 4-ClC_6H_4, 4-$OMeC_6H_4$, Bu

Scheme 8.91

8.4.2.5 Formation of Two Bonds

8.4.2.5.1 From 4 + 1 Atom Fragments

Formation of One C–N and One C–C Bond [C-C-N-N + C] 3-Dimethylamino-1-phenyl-1*H*-indazoles are obtained from the reaction of *N,N*-diphenylhydrazine with phosgene iminium chloride (Scheme 8.92) [277].

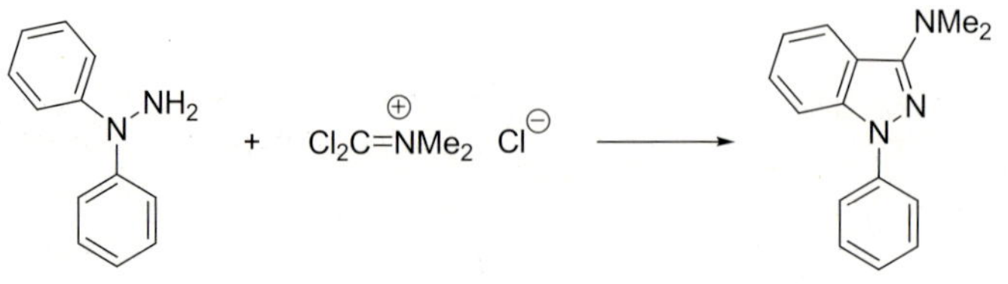

Scheme 8.92

Formation of One C–N and One N–N Bond [C-C-C-N + N] Several 1*H*-indazol-3-carboxamide or carboxylate derivatives are prepared from 2-nitrophenylacetic acid derivatives (Scheme 8.93) [278, 279]. The nitro group is reduced to the amino derivative, then *tert*-butyl nitrite or sodium nitrite in acetic acid promotes the cyclization to the corresponding 1*H*-indazole. The hydrolysis of the ester and amide groups with a great excess of sodium hydroxide in water affords the corresponding 1*H*-indazol-3-carboxylic acids.

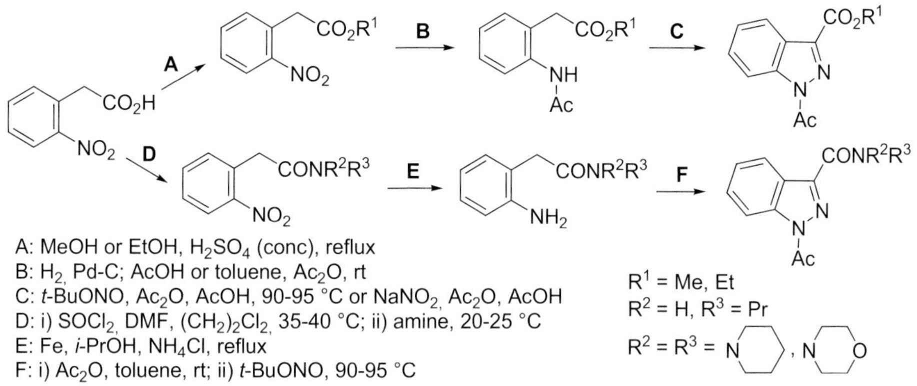

A: MeOH or EtOH, H_2SO_4 (conc), reflux
B: H_2, Pd-C; AcOH or toluene, Ac_2O, rt
C: *t*-BuONO, Ac_2O, AcOH, 90-95 °C or $NaNO_2$, Ac_2O, AcOH
D: i) $SOCl_2$, DMF, $(CH_2)_2Cl_2$, 35-40 °C; ii) amine, 20-25 °C
E: Fe, *i*-PrOH, NH_4Cl, reflux
F: i) Ac_2O, toluene, rt; ii) *t*-BuONO, 90-95 °C

R^1 = Me, Et
R^2 = H, R^3 = Pr
R^2 = R^3 = N(piperidine), N(morpholine)

Scheme 8.93

8.4.2.5.2 **From 3 + 2 Atom Fragments**

Formation of Two C–N Bonds [C-C-C + N-N] Like pyrazoles, indazoles can be prepared by the [C-C-C + N-N] approach. The standard method consists of the condensation of a 1,3-difunctional compound with hydrazine or hydrazine derivatives. 1*H*-indazoles **311** are synthesized by cyclization of hydrazones of 2-mesyloxyphenyl ketones **310** (Scheme 8.94) [280]. Conversion of the appropriate mesylate **310** into the desired 1*H*-indazoles **311**, by the reaction with hydrazine derivatives, proceeds through the hydrazone intermediate, which cyclizes to indazoles by nucleophilic aromatic substitution of the mesylate group. The reaction needs a slightly acidic medium to catalyze the conversion of the ketone into the hydrazone and it is imperative the use of a Dean-Stark apparatus to eliminate the water, to avoid mesylate hydrolysis to the corresponding phenol, which does not cyclize to indazoles under these reaction conditions.

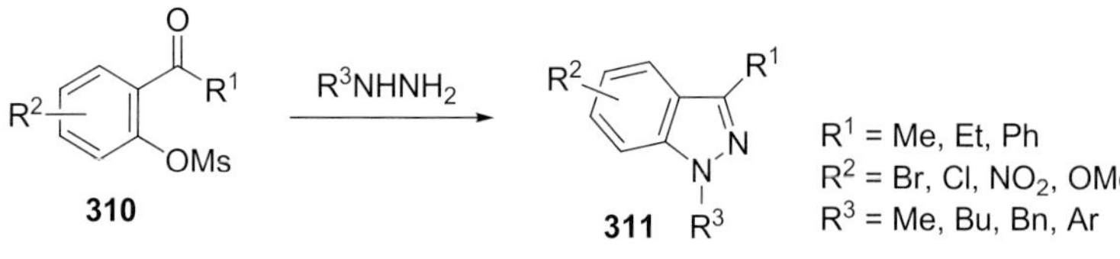

310 → (R^3NHNH_2) → **311**

R^1 = Me, Et, Ph
R^2 = Br, Cl, NO_2, OMe
R^3 = Me, Bu, Bn, Ar

Scheme 8.94

A related method involves the reaction of benzoic acid derivatives bearing a leaving group in the *ortho* position with hydrazine. For instance, treatment of methyl 2-fluorobenzoate **312** with hydrazine in refluxing butanol gives the corresponding 1*H*-indazole-3-ones **314** (Scheme 8.95) [261]. Under identical conditions, *ortho*-fluorobenzonitriles **313** afford 3-amino-1*H*-indazoles **315** (Scheme 8.95) [237, 281, 282].

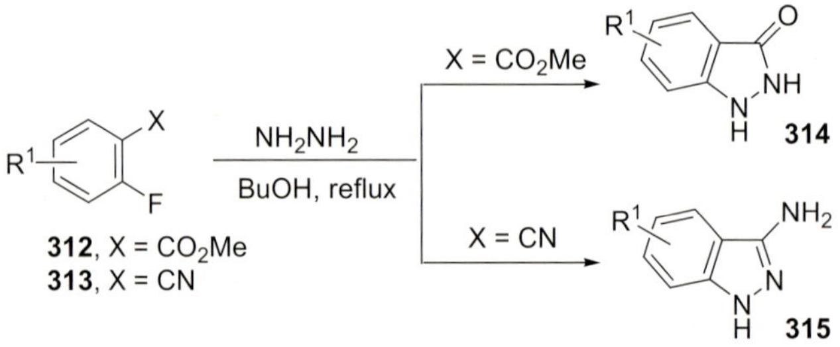

312, X = CO_2Me
313, X = CN

NH_2NH_2, BuOH, reflux; X = CO_2Me → **314**; X = CN → **315**

Scheme 8.95

4,5,6,7-Tetrahydro-2*H*-indazoles **317** are obtained from the reaction of the Baylis–Hillman adducts of cyclohexen-1-ones (**316**) with hydrazine derivatives. These tetrahydroindazoles (**317**) are oxidized to 2-substituted-2*H*-indazoles **318** by treatment with DDQ (Scheme 8.96) [283]. 3-Substituted-1-aryl-4,5,6,7-tetrahydro-1*H*-indazoles **320** are obtained by the reaction of diketone **319** with arylhydrazines (Scheme 8.97) [284]. Various 4,5-dihydro-1*H*-benzo[g]indazole-based ligands for cannabinoid receptors have been prepared by a similar procedure [285]. Tetrahydro-

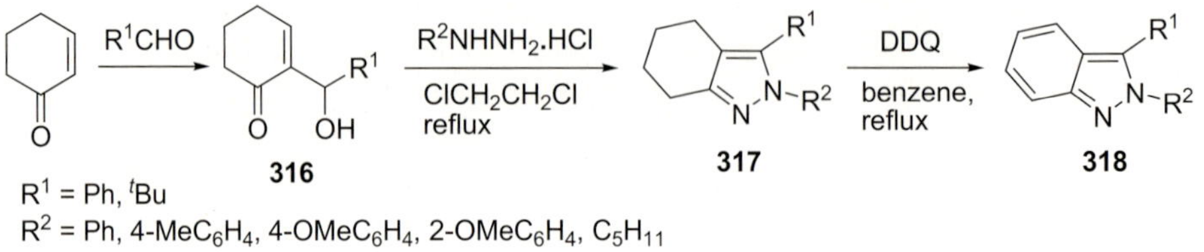

R^1CHO → **316**; $R^2NHNH_2.HCl$, $ClCH_2CH_2Cl$, reflux → **317**; DDQ, benzene, reflux → **318**

R^1 = Ph, ^tBu
R^2 = Ph, 4-MeC_6H_4, 4-$OMeC_6H_4$, 2-$OMeC_6H_4$, C_5H_{11}

Scheme 8.96

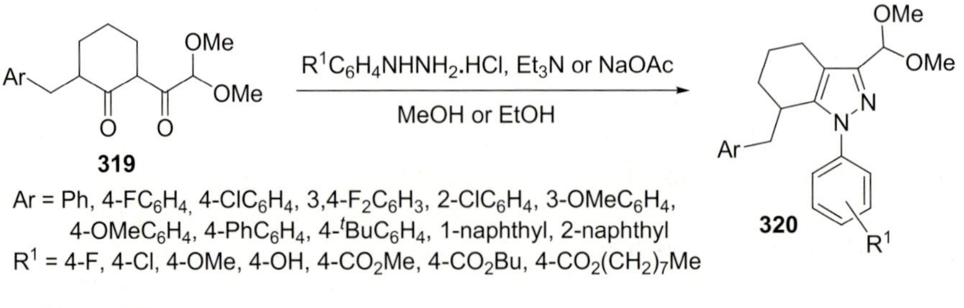

319 → $R^1C_6H_4NHNH_2.HCl$, Et_3N or NaOAc, MeOH or EtOH → **320**

Ar = Ph, 4-FC_6H_4, 4-ClC_6H_4, 3,4-$F_2C_6H_3$, 2-ClC_6H_4, 3-$OMeC_6H_4$, 4-$OMeC_6H_4$, 4-PhC_6H_4, 4-$^tBuC_6H_4$, 1-naphthyl, 2-naphthyl
R^1 = 4-F, 4-Cl, 4-OMe, 4-OH, 4-CO_2Me, 4-CO_2Bu, 4-$CO_2(CH_2)_7Me$

Scheme 8.97

and hexahydro-1H- and 2H-indazole derivatives are also obtained from the reaction of appropriate chalcone-type compounds or β-diketones with hydrazine derivatives [283, 284, 286–288].

Formation of One C–C and One C–N Bond [C-N-N + C-C] The 1,3-dipolar cycloaddition reactions of diazo compounds with benzyne is a versatile method for the synthesis of indazoles (Scheme 8.98) [289–291]. Benzyne is conveniently generated *in situ* from the reaction of anthranilic acid with alkyl nitrite in an aprotic media (mixtures of dichloromethane : acetone or acetonitrile) [292]. Aoyama *et al.* describe the synthesis of 3-trimethylsilyl-1H-indazoles **322** from the [3 + 2] cycloaddition reactions of lithium trimethylsilyldiazomethane with benzyne, generated from halobenzenes and lithium 2,2,6,6-tetramethylpyperidine (LTMP) (Scheme 8.99) [293]. LDA, a less hindered base, can also be used, but a significant decrease in the yield of indazole is observed. The reaction mechanism involves a nucleophilic attack of $(CH_3)_3SiC(Li)N_2$ to the benzyne, subsequent cyclization of the indazole intermediate **321** and reaction with water to afford 3-trimethylsilyl-1H-indazoles **322** (Scheme 8.99).

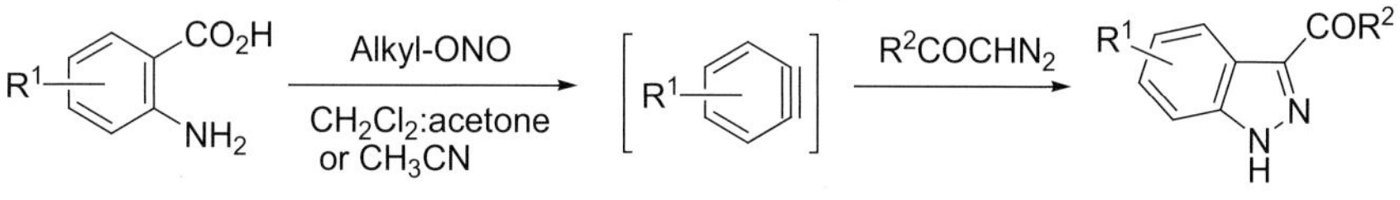

Scheme 8.98

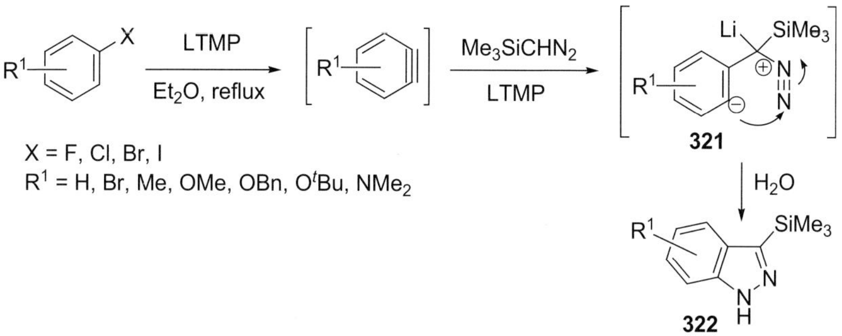

Scheme 8.99

1,3-Dipolar cycloaddition reactions of diazomethane to quinones originate the corresponding 1H-indazole-4,7-diones (Scheme 8.100) [294, 295].

8.4.2.6 **Other Synthetic Methods**

Dehydrogenation of tetrahydro-1H-indazoles in boiling decalin with 5% palladium on activated carbon gives the corresponding 1H-indazoles (Scheme 8.101) [296].

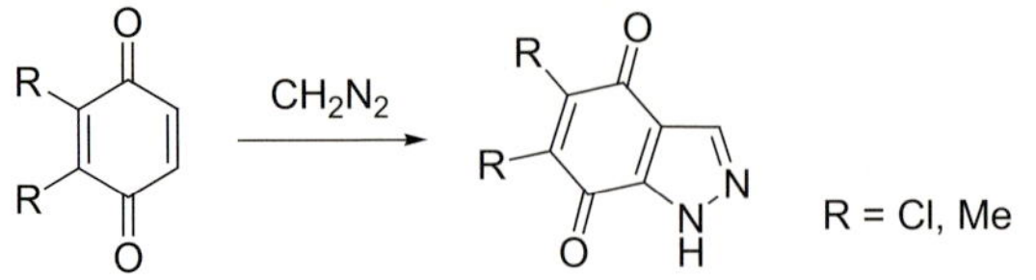

CH_2N_2

R = Cl, Me

Scheme 8.100

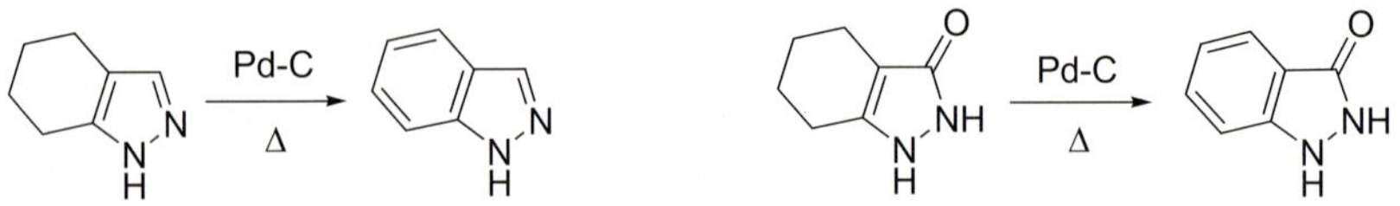

Pd-C Δ

Pd-C Δ

Scheme 8.101

Most syntheses of indazoles proceed from benzene derivatives, where the pyrazole ring is generated by ring closure. However, a few examples of the synthesis of indazoles starting from pyrazoles are known [297–299]. One of them involves the condensation of 3-substituted pyrazole-4-carbaldehydes **323** with diethyl succinate in the presence of potassium *tert*-butoxide, affording esters **324** (mixture of geometric isomers), which undergo cyclization to 1*H*-indazoles **325** by reaction with sodium acetate in acetic anhydride (Scheme 8.102) [297]. The ring closure step probably takes place by an electrocyclization process after the enolization of the mixed anhydride **326**.

$(CH_2)_2(CO_2Et)_2$, KO^tBu, ^tBuOH, reflux; Ac_2O, NaOAc, reflux

R= Me, ^{i}Bu, 4-$MeOC_6H_4$, 2-ClC_6H_4, CO_2Et

Scheme 8.102

Another example of synthesis of indazoles from pyrazoles consists in the reaction of stable chromium and tungsten Fischer dienyl carbenes **328**, formed from a [3 + 2] cycloaddition reaction of alkenylethynyl carbene **327** with trimethylsilyl diazomethane, with isocyanides to give highly functionalized 1*H*-indazoles **329** (Scheme 8.103) [298].

1,3-Diphenyl-1*H*-indazoles **333** are obtained from base induced addition–elimination of 5-cyanomethyl-1,3-diphenylpyrazoles **331** to various α-oxoketene

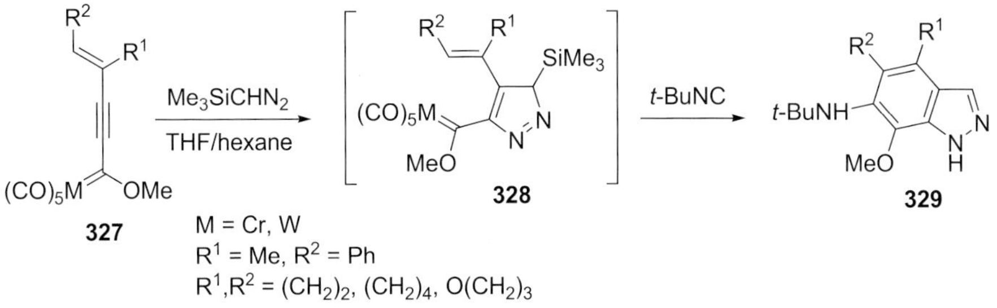

M = Cr, W
R^1 = Me, R^2 = Ph
R^1,R^2 = $(CH_2)_2$, $(CH_2)_4$, $O(CH_2)_3$

Scheme 8.103

dimethylthio acetals **330**, followed by acid-assisted cycloaromatization of the resulting adducts **332** (Scheme 8.104) [299]. The cycloaromatization of adducts **332** is more efficient (in terms of yield and work-up) in the presence of *para*-toluenesulfonic acid (PTSA) than with various protic and Lewis acids. By using cyclic of α-oxoketene dimethylthio acetals it is possible to obtain annulated indazoles.

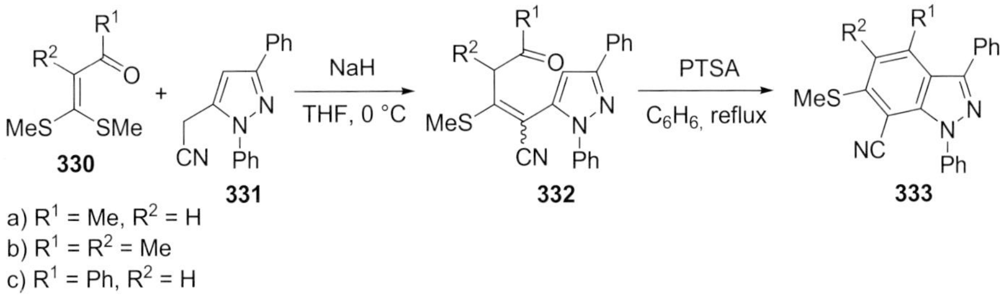

a) R^1 = Me, R^2 = H
b) R^1 = R^2 = Me
c) R^1 = Ph, R^2 = H

Scheme 8.104

1-Phenyl-5-vinyl-1*H*-pyrazole (**334**) gives cycloaddition reactions with several dienophiles to furnish 1-phenyl-1*H*-indazole derivatives **335–337** (Scheme 8.105) [300]. The reaction of **334** with methyl propiolate affords 1-phenyl-1*H*-indazole **336** as a result of a double Diels–Alder cycloaddition with extrusion of ethylene. 1-Phenyl-1*H*-indazole **338** is obtained by oxidation of its dihydro derivative **335** by treatment with DDQ.

Highly substituted-1*H*-indazoles **341** have been obtained from the cycloaddition reaction of 1-aryl-3-phenyl-1,6-dihydropyrano[2,3-*c*]pyrazoles **339** with dialkyl acetylenedicarboxylates in refluxing DMF (Scheme 8.106) [301]. The resulting cycloadducts **340** spontaneously eliminate acetone to give 1-aryl-6,7-dialkoxycarbonyl-4-methyl-3-phenyl-1*H*-indazoles **341**.

N-Unsubstituted pyrazole *ortho*-quinodimethanes **343**, generated by thermal extrusion of sulfur dioxide from *NH*-pyrazole-fused 3-sulfolenes **342**, react with dienophiles to give 4,5,6,7-tetrahydro-1*H*-indazoles **344–346** (Scheme 8.107) [302]. Thermolysis of sulfones **342a,b** in refluxing toluene or chlorobenzene in the presence

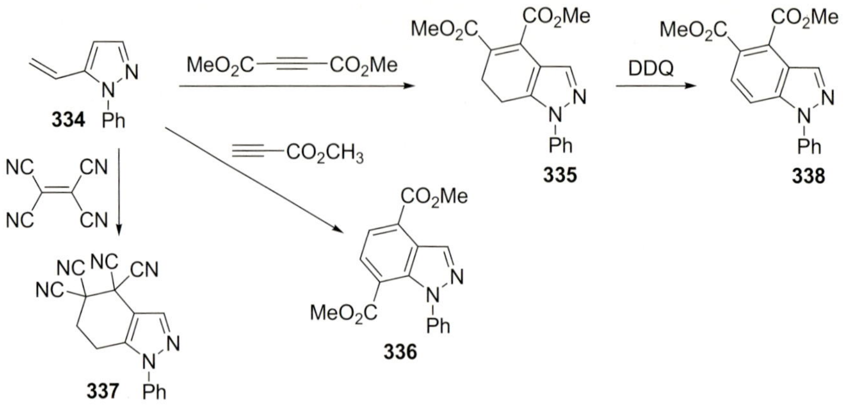

Scheme 8.105

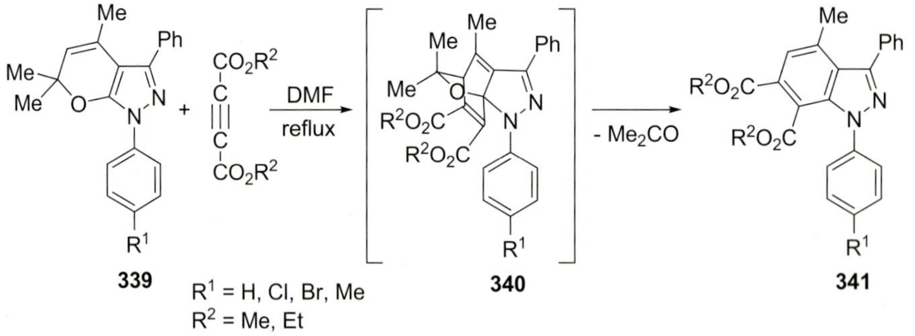

Scheme 8.106

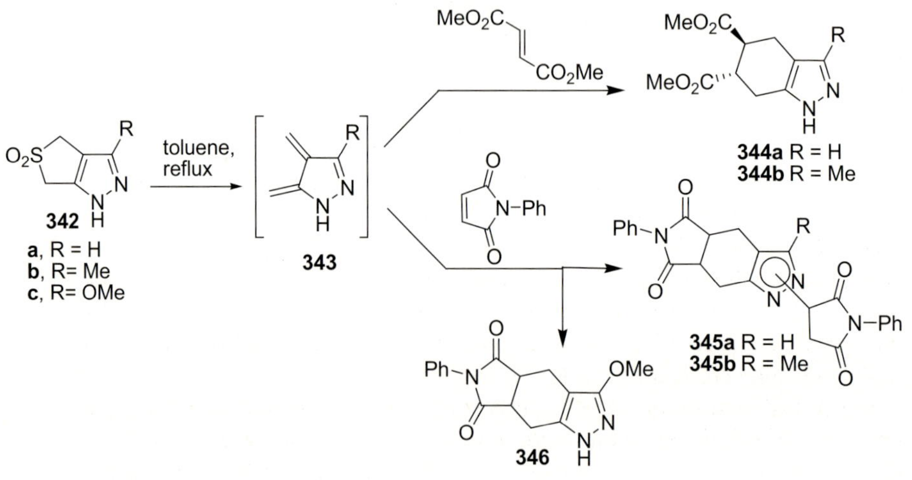

R^1 = H, Cl, Br, Me
R^2 = Me, Et

Scheme 8.107

of dimethyl fumarate gives the 1 : 1 Diels–Alder cycloadducts **344a,b** without the competition of the Michael addition. However, thermolysis of the same sulfones in the presence of *N*-phenylmaleimide gives the 2 : 1 cycloadducts **345a,b**. Adducts **345a, b** were obtained as diastereomeric mixtures of the 1- and 2-substituted indazole isomers. Thermolysis of sulfone **342c** in the presence of *N*-phenylmaleimide gives the tetrahydro-1*H*-indazole **346** as the only product.

8.4.2.7 **Ring Synthesis from Heterocycles**

Flash vacuum pyrolysis of 2,5-diaryltetrazoles **347** at 400–500 °C gives almost quantitative yields of 3-aryl-1*H*-indazoles **350** (Scheme 8.108) [303]. Under these conditions, tetrazoles **347** eliminate nitrogen, leading to nitrilimines **348**, which undergo cyclization onto the remote aromatic ring to afford 3-aryl-3*H*-indazoles **349**, which spontaneously isomerize to 3-aryl-1*H*-indazoles **350**. 3-Substituted-1*H*-indazoles are also prepared by flash thermolysis (400–500 °C) of 2-substituted-4-phenyl-1,3,4-oxadiazolin-5-ones [303, 304]. In this case, the same type of nitrilimines is also involved, being generated by elimination of carbon dioxide.

347
FVP
450-500 °C
- N_2
348
$R^1 = R^2$ = H, Me
350
349

Scheme 8.108

Treatment of 1*H*-[1-3]triazolo[1,5-*a*]benzimidazole **351** with dimethyl sulfate at elevated temperature results in the formation of benzimidazolyl-1*H*-indazole **354** [305]. Instead of the simple methylation to yield **352**, a ring opening reaction take place and, via formation of a nitrenium cation, indazole **353** is formed (Scheme 8.109). This intermediate is then alkylated to the dimethyl salt **354**. Similar transformations occur when **351** is treated with trifluoroacetic acid at reflux, resulting in the formation of 1*H*-indazole **353** (R=H) [306].

The photochemistry of some 3,5-disubstituted-1,2,4-oxadiazoles **355** bearing a nitrogen nucleophilic group, such as an *ortho*-aminophenyl moiety, at C3 of the oxadiazole ring leads to the concomitant formation of indazoles **357** and benzimidazoles **359** (Scheme 8.110) [307]. The photochemistry of **355** is characterized by

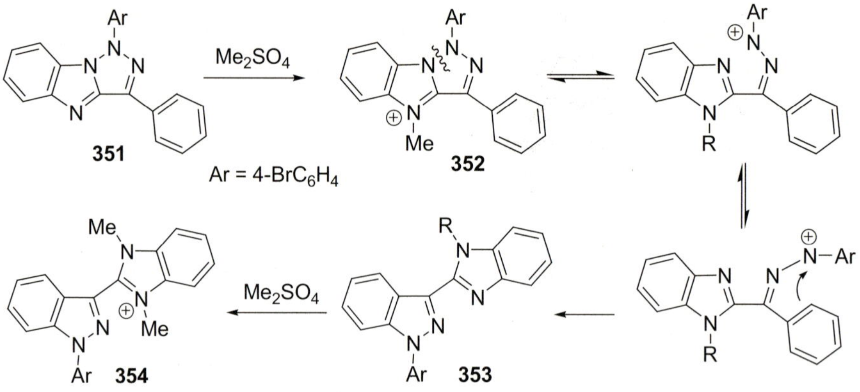

Scheme 8.109

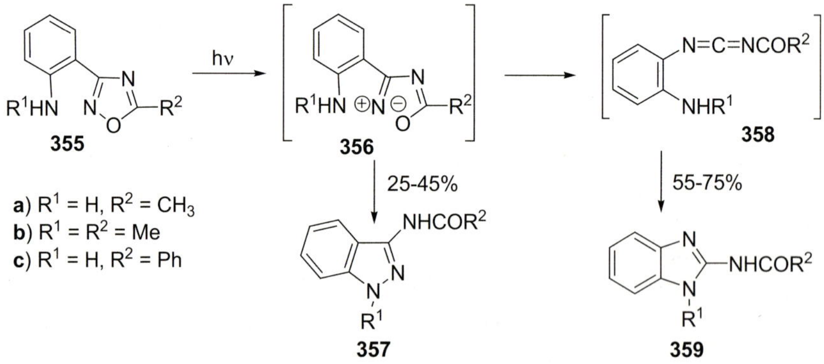

Scheme 8.110

photolysis of the O–N bond to give open compounds **356**, which rearrange to 1*H*-indazoles **357** through the N–N bond closure. The photolytic species **356** can also be converted into carbodiimides **358**, precursors of the benzimidazoles **359**.

1*H*-Indazoles **357** can also be obtained from 3,5-disubstituted 1,2,4-oxadiazoles **355** in almost quantitative yield, by heating these compounds, without solvent, at a temperature much higher than their melting points [308].

8.5 Reactivity

The reactivity of pyrazoles is related to the tautomerism of the neutral forms **360** and to their acid (conjugated anion **361**) and basic properties (conjugated cation **362**) (Scheme 8.111). It is very important when discussing the reactivity of pyrazoles and indazoles to determine the form that reacts. The "pyridinic" N2 atom is susceptible to

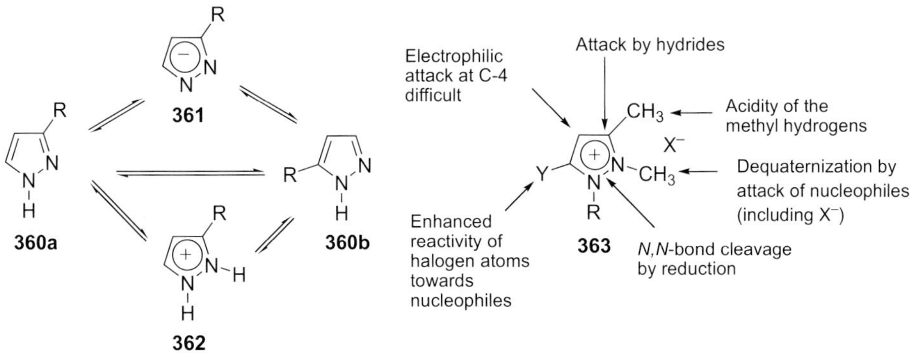

Scheme 8.111

electrophilic attack and the "pyrrolic" N1 is unreactive, but the N1 proton can be removed by bases to afford the anion **361**. Electrophilic attack on C4 is generally preferred. Since indazoles have the pyrazolic C4 position substituted, electrophilic attack on indazoles takes place in the 3-position and in the benzene ring (positions 5 and 7).

We have also represented in Scheme 8.111 the effect of a positive charge (pyrazolium salts) on pyrazole reactivity. In contrast, the anions (pyrazolates) show the expected inversion of reactivity when compared with the cations. Note that in general *N*-alkyl pyrazoles and indazoles are prepared from the corresponding anions.

8.5.1
Reactions with Electrophilic Reagents

8.5.1.1 Electrophilic Attack at Nitrogen

This is the most characteristic reaction of pyrazoles. The reactivity of the nitrogen atom in neutral pyrazoles and indazoles corresponds to that of pyridine N atom. Notably, the apparent rate of formation of an *N*-substituted derivative depends more on the rate of reaction of a given tautomer than on the tautomeric equilibrium constant.

8.5.1.1.1 Basicity of Azoles Pyrazoles are medium to weak bases, much weaker than for instance imidazole ($pK_a = 6.95$). Some significant values are reported in Table 8.4 from a large collection described in Reference [309]. In general, the effect of the substituents is additive and the acidity and basicity pK_as are linearly related [309].

8.5.1.1.2 Acidity of Azoles Pyrazole and indazole are very weak acids, unless they bear a strong EWG such as NO_2 (Table 8.4).

8.5.1.1.3 Metal Ions (see also Section 8.6.4.3) Pyrazoles and indazoles form sodium, potassium and silver salts that are hydrolyzed to a large extent by water. The resulting anions react very readily with electrophiles. Pyrazoles, pyrazolate anions

Table 8.4 pK_a for proton addition (basicity) and proton loss (acidity).

Azole	pK_a (basicity)	pK_a (acidity)
Pyrazole	2.52	14.21
3(5)-Methyl	3.32	—
4-Methyl	3.09	—
3,4,5-Trimethyl	4.63	—
4-Nitro	−1.96	9.05
3,5-Dinitro	—	3.14
1-Methyl	2.09	—
1-Phenyl	0.44	—
Indazole	1.31	13.80
1-Methyl	0.42	—
2-Methyl	2.02	—

and polypyrazolylborates (scorpionates) [310] are much used ligands in coordination chemistry.

8.5.1.1.4 ***N*-Alkylation and *N*-Arylation** *N*-Alkylation is one of the most important and most studied reactions of pyrazoles and indazoles. Alkylations have been carried out using alkyl halides (usually iodides and bromides), dialkyl sulfates, arenesulfonates, diazomethane and dialkyl phosphates. Microwave irradiation has proved very effective for carrying out this reaction [311, 312]. With neutral pyrazoles and alkyl halides, the alkylation yields salts of the corresponding acids.

The orientation of the entering group strongly depends on the substituents on the pyrazole ring, on the nature of the alkylating agent and on the experimental conditions. Phase transfer catalysis has been used with success to prepare *N*-substituted pyrazoles [4]. When polyhalogenoalkanes are used as alkylating agents, poly-*N*-pyrazolylalkanes (useful ligands in coordination chemistry) are obtained.

Only activated halogenated benzenes (*para*-fluoronitrobenzene, 1-fluoro-2,4-dinitrobenzene, picryl chloride) [313] and halogeno substituted heterocycles (such as 3,6-dichloropyridazine, cyanuric chloride and brominated derivatives) [314] react with pyrazoles and indazoles [4]. Pyrazolate anions react with hexafluorobenzene to yield hexapyrazolylbenzenes (propellenes, Section 8.6.4.3) [5, 315]. An efficient synthesis of β-hydroxyethyl-pyrazoles **364** and **365** from propylene and styrene oxide using Cs_2CO_3 has been reported [316].

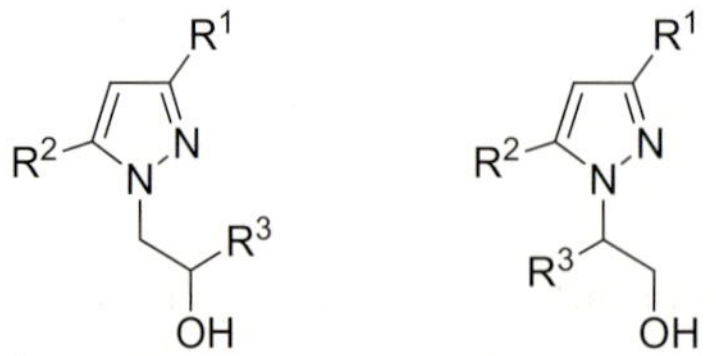

R1 R2 N N R3 OH
364

R1 R2 N N R3 OH
365

Through the use of Bi, B and Pb derivatives and copper-catalyzed cross-coupling reactions it is possible to make N–C bonds between pyrazoles and unsubstituted phenyl rings [317–320], or between indazoles and aryl rings [321].

8.5.1.1.5 **N-Acylation (see also Section 8.6.4.2)** *N*-Acetylated pyrazoles are obtained from *N*-unsubstituted pyrazoles by treatment with acetyl chloride (alone or in the presence of pyridine) or acetic anhydride. The fact that the isomeric structure of azolides is thermodynamically controlled has been used to prepare the less accessible 1-alkylpyrazoles regioselectively (Scheme 8.112) [4].

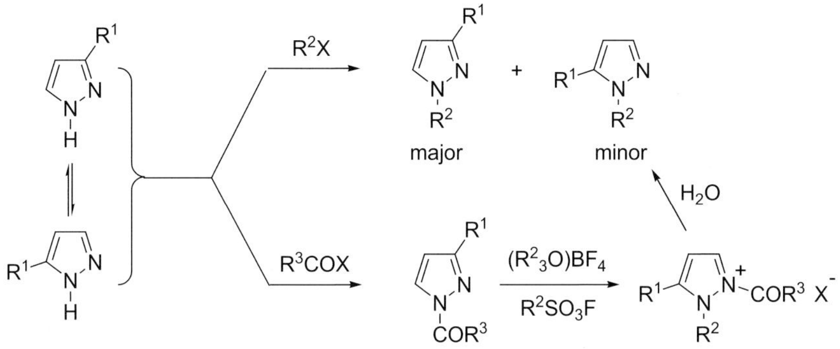

R^1 R^2X R^1 R^1 + major minor H_2O R^3COX $(R^2{}_3O)BF_4$ R^2SO_3F COR^3 R^2 X^-

Scheme 8.112

Acylation of 3(5)-aminopyrazole **366** with chloroacetyl chloride affords a mixture of **367** and **368**, both of which rearrange in the solid state to 3-(chloroacetamido)pyrazole **369** (Scheme 8.113) [322].

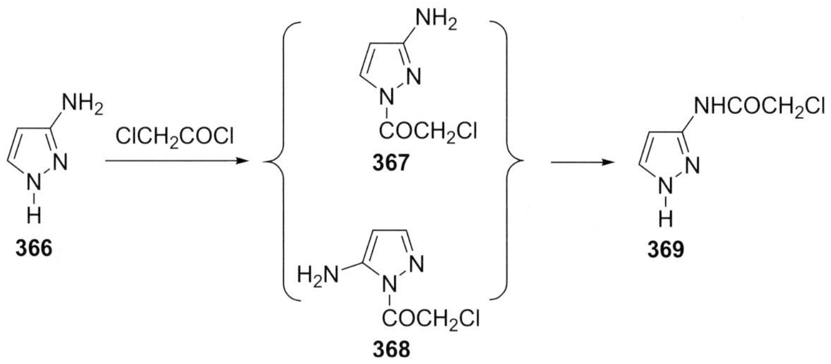

NH_2 $ClCH_2COCl$ $COCH_2Cl$ **366** **367** **368** $NHCOCH_2Cl$ **369**

Scheme 8.113

8.5.1.1.6 **Michael Addition** *N*-Unsubstituted pyrazoles and indazoles add to compounds containing activated double and triple bonds [1–5]. Amongst C–C double and triple bonds, maleic anhydride, acrylic acid esters and nitriles, acetylenecarboxylic and -dicarboxylic esters, quinones, and some α,β-unsaturated ketones have been used with success. With an activated C–C triple bond two successive additions can occur if the intermediate alkene is reactive enough; this occurs, for instance, with DMAD. For additions on C–O double bonds see Section 8.6.4.2.

8.5.1.1.7 ***N*-Halogenation** *N*-Halogenated pyrazoles are unstable compounds (Cl > Br > I) that are seldom isolated. 1-Bromopyrazole resembles NBS and can act as a source of the electrophilic brominium ion [4].

8.5.1.1.8 ***N*-Amination and *N*-Nitration** The powerful aminating agents hydroxylamino-*O*-sulfonic acid and *O*-mesitylenesulfonylhydroxylamine have been used to aminate pyrazole and indazole (60% of 1-amino and 40% of 2-aminoindazole) [4]. Amination of *C*-aminopyrazole **370** affords both diamino isomers **371** and **372** (Scheme 8.114) [323].

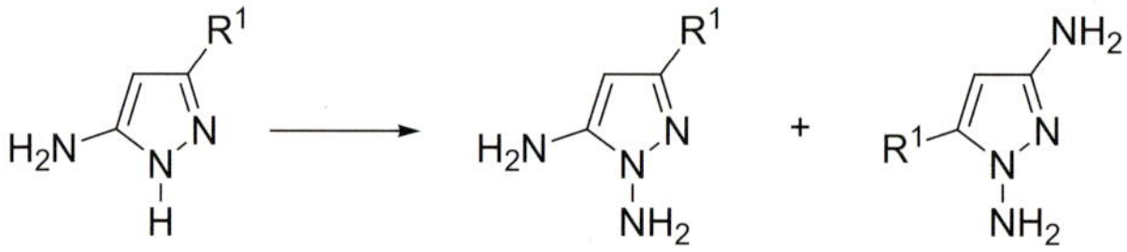

Scheme 8.114

8.5.1.2 Electrophilic Attack at Carbon

Pyrazole is less reactive towards electrophiles than pyrrole. As a neutral molecule it reacts as readily as benzene and, as an anion, as readily as phenol. Pyrazole cations (pyrazolium ions), formed in strong acid media, show a pronounced deactivation (nitration, sulfonation, Friedel–Crafts reactions). Electrophilic attack on pyrazoles takes place at C4.

8.5.1.2.1 **Nitration** Pyrazole is very stable in acid media and even under rather vigorous conditions neither ring opening nor ring oxidation was observed. Nitration occurs at the 4-position; in the case of 4-R substituted pyrazoles, mono- and dinitration at positions 3 and 5 are observed [4].

8.5.1.2.2 **Sulfonation** Direct sulfonation of the pyrazole ring is rather difficult due to cation formation and takes place at position 4 only on prolonged heating with 20% oleum [4, 5].

8.5.1.2.3 **H/D Exchange** Qualitatively it was observed that in D_2SO_4 exchange of 1-methylpyrazole occurs initially at C4 and then simultaneously at C3 and C5, while in 1,2-dimethylpyrazolium it occurs only at C4 [4, 5].

8.5.1.2.4 **Halogenation** Halogenation is one of the most studied electrophilic substitutions in the pyrazole series [4]. Many reagents can chlorinate pyrazoles:

chlorine-water, chlorine in carbon tetrachloride, hypochlorous acid and chlorine in acetic acid (one of the best experimental procedures). Bromine in chloroform and bromine in acetic acid are the reagents used most often to brominate pyrazoles. To effect polybromination of pyrazoles the use of iron as catalyst is necessary. Pyrazole does not react with iodine although pyrazolsilver is converted into 4-iodopyrazole.

Ultrasound irradiation using *N*-halosuccinimides [324], *N*-iodosuccinimide [325] and a mild and efficient method for the regioselective iodination of pyrazoles have been reported [326].

8.5.1.2.5 **Acylation: Vilsmeier–Haack and Friedel–Crafts reactions** 1-Substituted pyrazoles are formylated and acetylated at C4. *C*-Alkylation of pyrazoles is rather uncommon and only groups like benzyl or adamantyl can be introduced directly on the 4-position.

8.5.1.2.6 **Diazo Coupling and Nitrosation** Generally, pyrazoles do not react with diazonium salts. However, when an activating group (hydroxy, alkoxy, amino) is present at position 3 or 5, the reaction proceeds easily at position 4. The reaction is very common in pyrazolone chemistry; pyrazolone diazo coupling is an important industrial reaction since the resulting azo derivatives are important dyestuffs, like tartrazine. The behavior of pyrazoles towards nitrosation is similar to that towards diazo coupling.

8.5.2 Reactions with Oxidizing Agents

Pyrazoles are resistant to oxidation but with agents like potassium permanganate the indazole ring is completely destroyed. Side chains can be oxidized; for instance, methyl groups into carboxylic acids.

8.5.2.1 Oxidation

The most important of all oxidation reactions in the chemistry of pyrazoles is the aromatization of pyrazolines into pyrazoles. Various oxidizing agents transform pyrazolines into pyrazoles: sulfur, bromine, chloranil, potassium permanganate, lead dioxide and mercury(II) acetate are all effective.

The problem is not totally solved, as shown by the numerous papers devoted to this topic. Oxidations using chlorine in CCl_4 [327], chloranil [164, 328], trichloroisocyanuric acid [329], Pd/C in acetic acid [330], DDQ [331] and $Bi(NO_3)_3$ with MW irradiation [332] have been reported.

8.5.3 Reactions with Nucleophilic Reagents

Little is known about nucleophilic attack on an unsubstituted carbon atom of pyrazoles. Some nucleophiles do not attack the heterocyclic ring carbon atoms but instead the substituent linked to nitrogen with subsequent *N*-deprotection, for instance dequaternization (pyrazolium salts) [4] or debenzylation (pyrazoles and indazoles) [333].

8.5.3.1 Reduction by Complex Hydrides

When an electron (mass spectrometry, electrochemistry) attacks a pyrazolium salt, two different radicals are formed, one leading to a pyrazoline (reduction of a C−C double bond) and the other to an open-ring diamine (N−N bond cleavage) [334]. With lithium aluminium hydride, Δ^3-pyrazolines and pyrazolidines are obtained from quaternary pyrazolium salts [4].

8.5.4 Reactions with Bases

This section corresponds to topics like metallation at ring carbon atoms (mainly lithiation) [335], hydrogen/deuterium exchange in neutral pyrazoles and pyrazolium cations, ring cleavage via *C*-deprotonation (opening to β-amino acrylonitriles or *ortho*-cyano anilines in the case of indazoles) – all of them of secondary importance.

8.5.5 Reactions of *N*-Metallated Pyrazoles

The main use of *N*-metallated pyrazoles and indazoles (sodium or silver salts) is for the preparation of *N*-substituted derivatives.

8.5.6 Reactions of *C*-Metallated Pyrazoles

This is a field of growing importance related to Suzuki, Miyaura, Sonogashira [325] and other cross-coupling reactions. Examples of C-arylation [52, 54, 154, 336], C-alkylation [154, 159] and C-alkynylation [325, 337, 338] have been reported.

8.5.7 Reactions with Radicals

Expansion of pyrazoles into pyridazines by the action of dichlorocarbenes has been reported [4]. Similarly, indazole and chloroform at 555 °C yields 2-chloroquinazoline. Little interest has been shown in the radical reactions (methyl and phenyl radicals) of pyrazoles because they afford mixtures of *C*-substituted derivatives.

A complete study of the chemistry of ground and excited state of *ortho*-pyrazolylphenyl nitrenes **373** (Scheme 8.115) has been carried out by Carra, Bally and Albini: different kinds of heterocycles have been isolated, amongst them **374–376** [339].

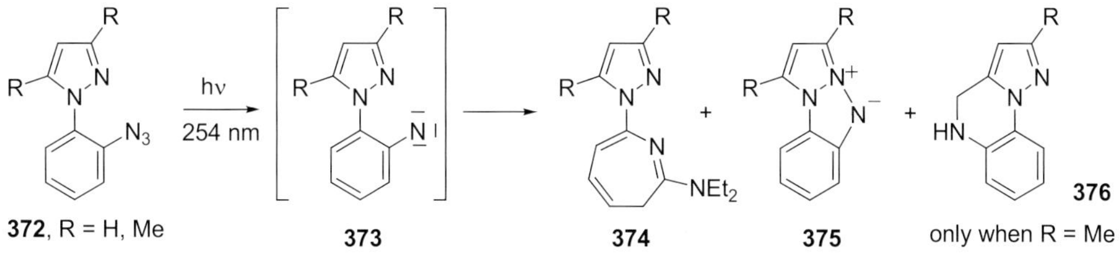

R R R R R
R N N hν 254 nm R N N R N N + R N N+ N− + HN N N
N3 N N N NEt2 376
372, R = H, Me 373 374 375 only when R = Me

Scheme 8.115

8.5.8
Reactions with Reducing Agents

The reduction of pyrazole and 1-phenylpyrazole with H_2, in the presence of palladium on active charcoal, gives the pyrazolines or, under more drastic conditions, the pyrazolidines [4].

8.5.9
Ring Transformations

8.5.9.1 Ring Opening without Fragmentation

Scheme 8.116 shows two of the most illustrative examples of this kind of reaction: the opening of 1-substituted indazoles **377** into 2-alkylaminobenzonitriles and the rearrangement of 1-(*ortho*-nitrophenyl)pyrazoles **379** into benzotriazole 1-oxides (*cis* and *trans* **381**) through intermediate azo compound **380** [4].

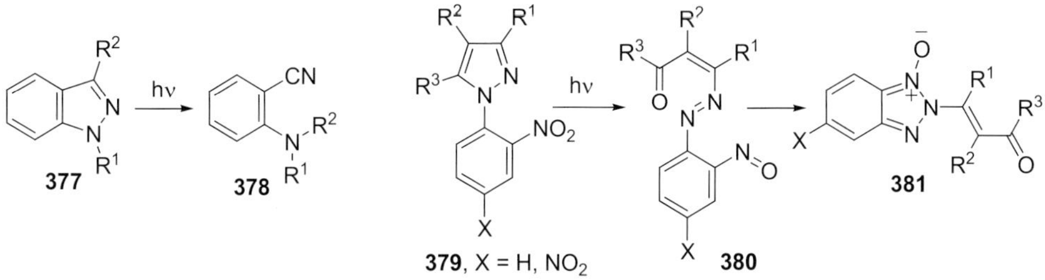

R2 hν CN R2 R1 R3 N N hν NO2 R2 R3 R1 O N=N N O O− N+ N N R1 R3 X R2 O
377 R1 378 R1 379, X = H, NO2 X 380 381

Scheme 8.116

8.5.9.2 Ring Isomerization

The most studied reaction is the transformation of pyrazoles into imidazoles and of 2-substituted indazoles into benzimidazoles.

8.5.9.3 Ring Enlargement

In dilute sulfuric acid (pH 2–4) rearrangement of indazoles **382** (Scheme 8.117) into benzimidazoles is suppressed and dihydroazepinones **383** and **384** are formed [4].

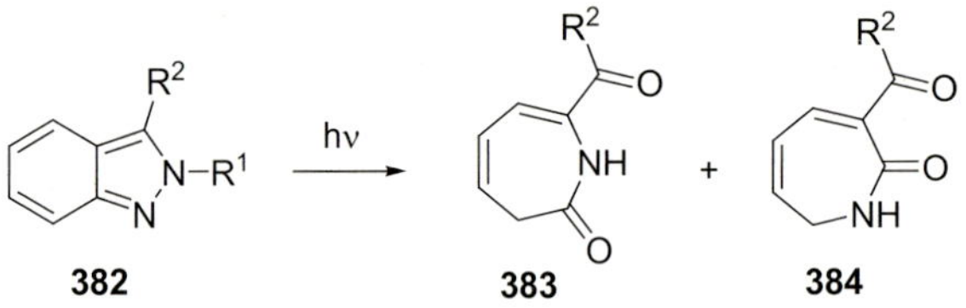

Scheme 8.117

8.5.10
Electrocyclic Reactions

Concerning Diels–Alder and 1,3-dipolar cycloadditions, pyrazole never reacts as an alkene towards a diene or a 1,3-dipole nor as an azadiene towards a dienophile. There are very few examples of reactions related to this topic: some are reported in Schemes 8.105–8.107 and 8.118 [4, 5].

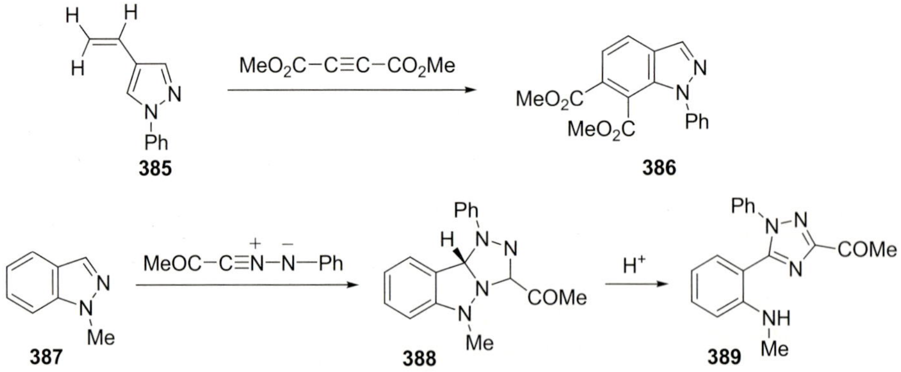

Scheme 8.118

Stephanidou-Stephanatou *et al.* have described the sequence of reactions reported in Scheme 8.119. From the *N*-benzoyl derivative **390**, through bromination and dehydrobromination, pyrazole *ortho*-quinodimethane **392** was generated and trapped with several dienophiles to afford the Diels–Alder adducts **393** [340].

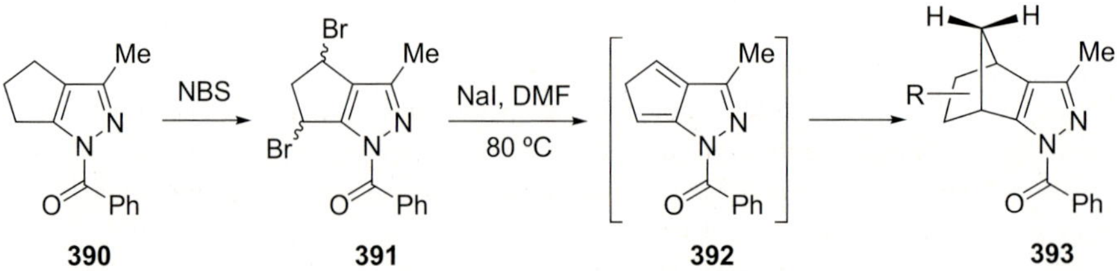

Scheme 8.119

8.6
Derivatives

8.6.1
C-Substituted Pyrazoles and Indazoles

In pyrazoles the simplest way to characterize the carbon atoms is to consider that C3 is similar to the pyridine α-position, C4 to the pyrrole β-position and C5 to both the γ-pyridine and the α-pyrrole positions.

8.6.2
Oxy- and Aminopyrazoles and Indazoles

Probably the most studied pyrazole derivatives are the oxy (pyrazolones and indazolones) and the amino derivatives in the order 5-substituted > 3-substituted ≫ 4-substituted. For a long time, the only comprehensive source on pyrazolones was the book from Wiley and Wiley of 1964 [341]. Fortunately, Varvounis *et al.* have updated it recently (they use 3-ones for the 3- and 5-ones, a logical but uncommon decision) [342].

Pyrazolin-4-ones exist as 4-hydroxypyrazoles and are much less common [4–6, 8], although deserving attention for their biological potentialities. Indazolones exist as such and not as 3-hydroxyindazoles [343]. The chemistry of pyrazolidinones has been reviewed [344].

3-, 4- and 5-Aminopyrazoles are interesting in themselves and also as precursors of many pyrazoles with fused five- [345], and six-membered rings [323, 346–349], as well as larger heterocyclic rings [350].

8.6.3
Other Substituents

The synthesis and reactivity of many different *C*-substituted pyrazoles have been described in the literature. Some of the most important (because of much studied or because very rare) are: CHO [351], C≡CR [352], NO_2 [353–355], ^{11}B [356], ^{19}F [357–360], ^{31}P [361, 362], ^{32}S [363], ^{28}Si (for instance TMS) [364] and ^{127}I [326, 337, 365, 367].

8.6.4
N-Substituted Pyrazoles and Indazoles

8.6.4.1 Quaternary Pyrazolium and Indazolium Salts

Pyrazolium **394** and indazolium salts **395** main importance was as precursors of Δ^3-pyrazolines and indazolines [4, 5]. But in recent years the extraordinary development of ionic liquids has promoted the study of pyrazolium salts as an alternative to imidazolium salts [366, 367]. Pyrazolium salts have been used to generate carbenes [368], and several papers dealing with their mass spectrometry have been published [369].

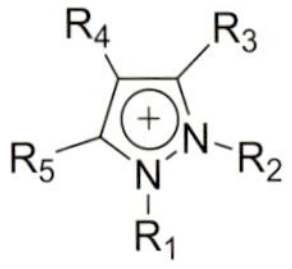

394

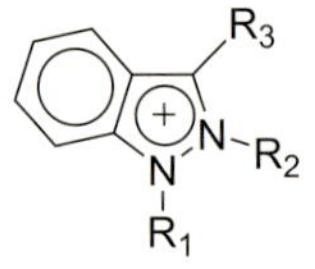

395

8.6.4.2 Azolides

N-Acyl azoles (azolides) are much studied compounds both for their use as synthons and for their structural properties [370]. Pyrazolides **396** and related compounds (like the addition products to aldehydes **397** and **398**) are in an equilibrium similar to prototropy and only the most stable isomers are usually isolated [371].

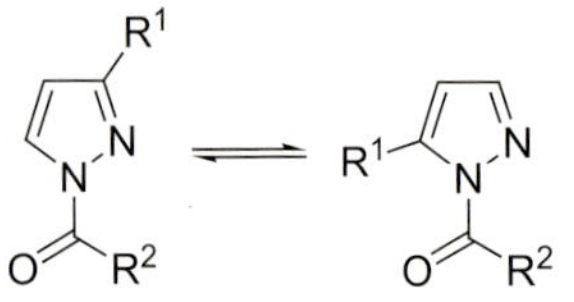

396

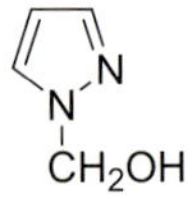

397

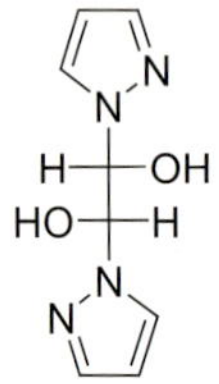

398

8.6.4.3 Coordination Chemistry of Pyrazoles

As we recalled in the introduction, the use of pyrazoles as ligands is one of the most prominent in their chemistry. Not considering biological properties, which include most patents, this aspect represent 36% of all 2004 references, according to the *Chemical Abstracts* (Table 8.2).

Since this topic has been covered in several reviews, only a classification of the pyrazoles as ligands is described here (see Refs [4, 5, 372–375]).

- Simple pyrazoles: pyrazolate anions. Both examples of pyrazoles acting as exobidentate ligands (**399**) and, much less common, as endobidentate – chelating – agent (**400**).

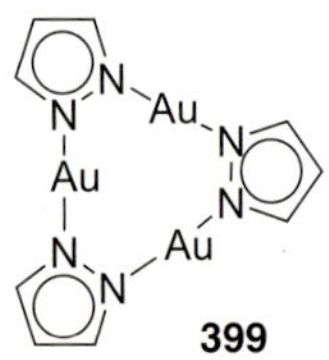

399

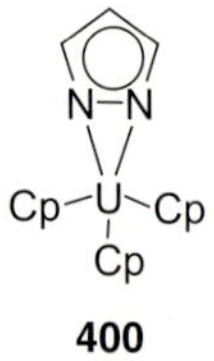

400

- Simple pyrazoles: neutral pyrazoles. These compounds act as 2-monohapto-pyrazoles.
- Bis-, tris- and poly-pyrazolyl derivatives: Ligands such as **401** (polypyrazolyl-methanes) [376], **402** [poly(pyrazolylmethyl)benzenes] [377], **403** (polypyrazolyl-benzenes, propellenes) [315], and **404** (polypyrazolylazines) [378] often form chelate complexes with different metals.

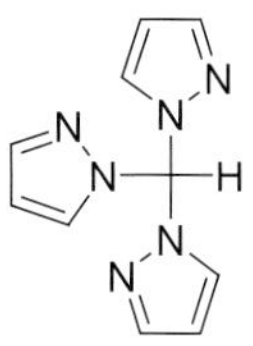

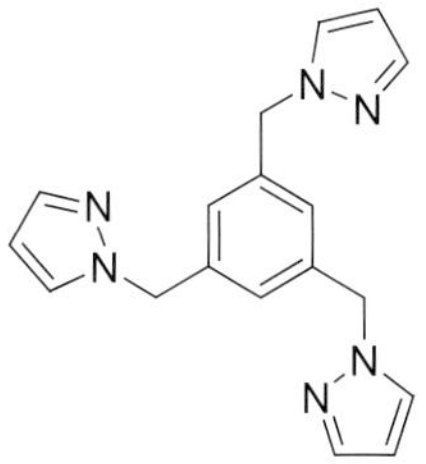

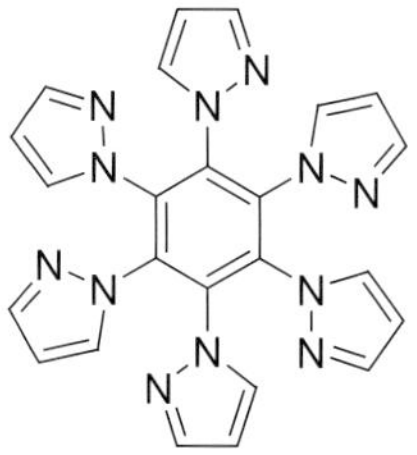

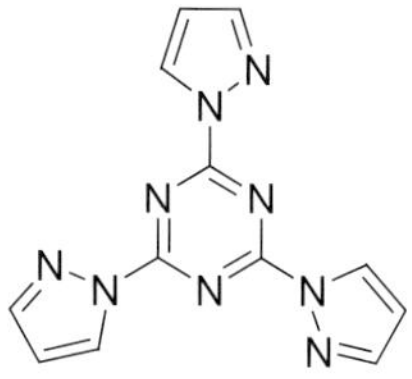

401 402 403 404

- Tris(pyrazolyl)borates (scorpionates) [310, 379]: The synthesis by Trofimenko of the pyrazolylborates (bis, tris and tetrakis, the tris **405** being the most interesting) is one of the major discoveries of pyrazole coordination chemistry. The anionic nature of **405**, which can be isolated with the Cp anion, confers to **405** and related compounds very rich coordination chemistry.

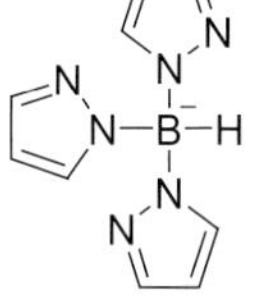

405

- Ferrocenylpyrazoles: The possibility that the pyrazole has a *C*-substituent with interesting properties has enriched the usefulness of pyrazoles in organometallic chemistry. Amongst these substituents, ferrocene is one of the most promising. Two representative structures are **406** [380] and **407** [381]. Other ferrocene-based pyrazolylborates (similar to **407**) have been described by Wagner *et al.* [382] Multidentate ferrocenyl-pyrazoles have been described by Thiel *et al.* [383].

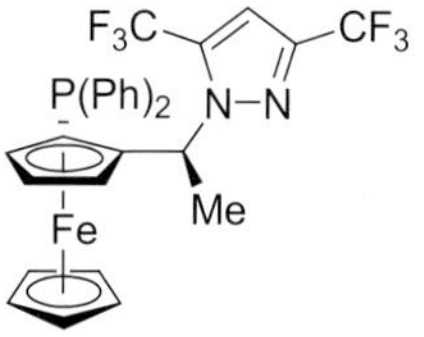

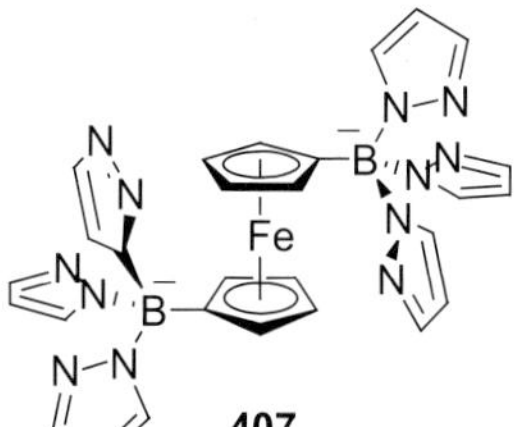

(*S*)-(*R*)-**406** **407**

8.6.5
Chiral Derivatives

The increasing interest in chiral compounds comes from the pharmaceutical industry and from the catalytic properties of coordination complexes (e.g., asymmetric hydrogenation). The chiral pyrazoles can be classified into three groups: (i) the chirality is in the substituent; (ii) the chirality is in a fused ring, tetrahydroindazoles from natural chiral ketones; (iii) the chirality pertains to a ring that is partially – pyrazolines – or totally – pyrazolidines – saturated.

8.6.5.1 Pyrazoles bearing *N*- or *C*-Chiral Substituents

Pyrazoles derived from steroids, carbohydrates and vitamins share the chiral properties of the starting materials. Thus compound **408**, an analogue of vitamin D, has been prepared [384]. Others, like **409**, a CNS agent, have been obtained by resolution [385]. Pyrazoles substituted at position 3 by (2*R*)-bornane-10,2-sultam **410** have been described and their annular tautomerism determined by X-ray and NMR spectroscopy [386]. Compounds related to Tröger's bases, such **411**, retain the inherent chirality of this family of compounds [387]. The synthesis of enantiomerically pure 5-substituted pyrazoles from 2,3-dihydro-4*H*-pyran-4-ones and from 2-formyl glycals has been reported (Schemes 8.59 and 8.60) [209, 210].

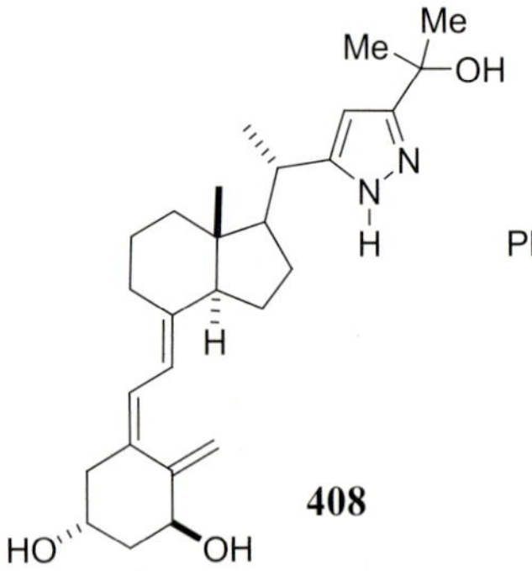

408 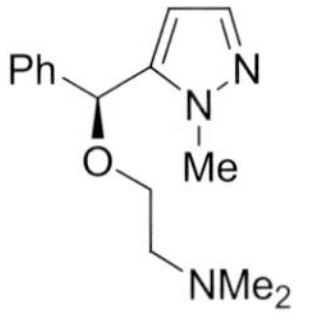 **409** 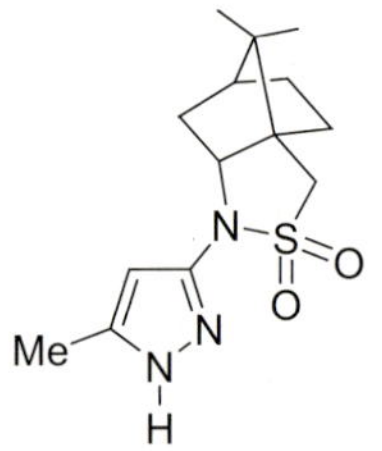 **410** 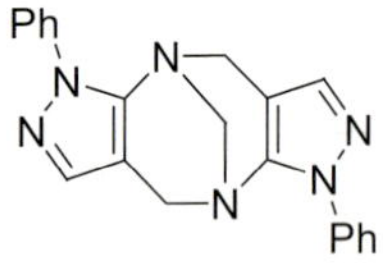 **411**

8.6.5.2 Tetrahydroindazoles from the Chiral Pool

Here, the most important compound by far is camphopyrazole – (4*S*,7*R*)-7,8,8-trimethyl-4,5,6,7-tetrahydro-4,7-methano-2*H*-indazole – **412**; this compound and related ones have been the subject of many studies: synthetic [146, 388–390], structural [391], reactivity [392, 393] and coordination chemistry [394, 395]. Other ketones found in natural sources have also been used to prepare compounds such as **413** [396–399].

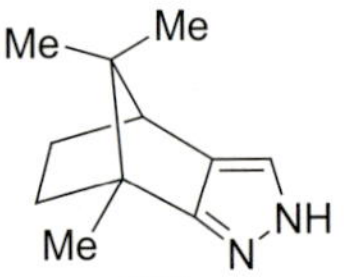

412 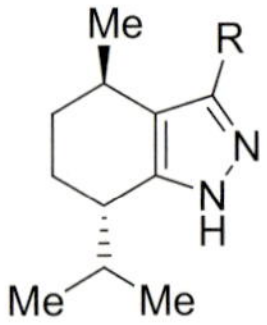 **413**

8.6.5.3 Pyrazolines and Pyrazolidines

These compounds have several stereogenic centers: carbons C4 and C5 in Δ^2-pyrazolines **414** and carbons C3, C4 and C5 in pyrazolidines **415**. Even taking into account the nitrogen inversion, stereogenic nitrogen atoms should be considered. Chiral pyrazolines have been prepared by resolution [400], and by inclusion in a chiral host [401]. The formation of a ruthenium complex from a racemic pyrazoline lead to diastereomeric structures **416** that have been separated [402].

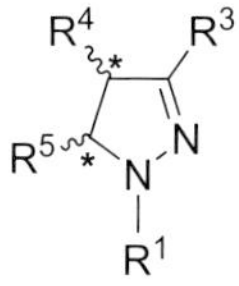

414 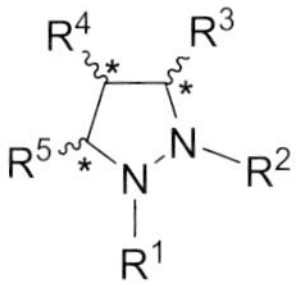 **415** 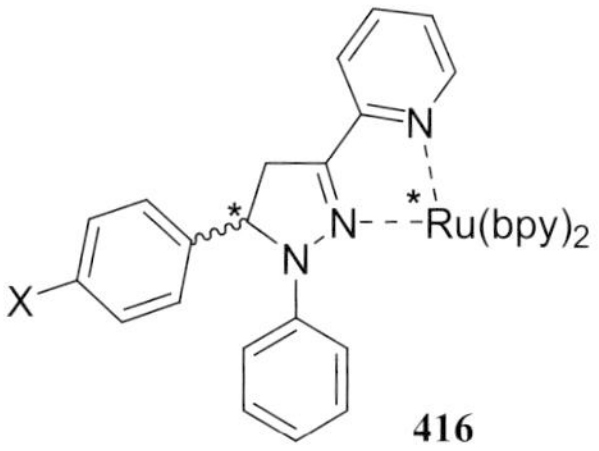 **416**

Using cycloaddition methodologies a series of pyrazolines and pyrazolidines enantiomerically pure have been prepared by Barluenga (**417**, and then reduced to pyrazolidines) [403], **418** [404], Molteni (**419**) [405], Ortuño (**420**) [406] and Kobayashi (**421**) (Scheme 8.50) [200]. A highly enantioselective [3 + 2] acylhydrazone-enol ether cycloaddition has been reported for the preparation of mono-benzoyl (similar to **421**) and dibenzoylpyrazolidines [407]. This is a very interesting field that it is expected to become very important in the near future.

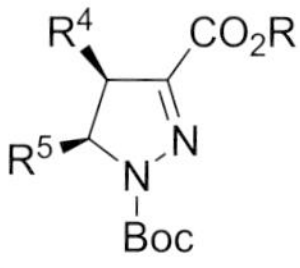

417 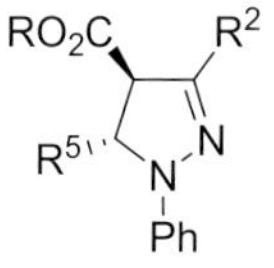 **418** 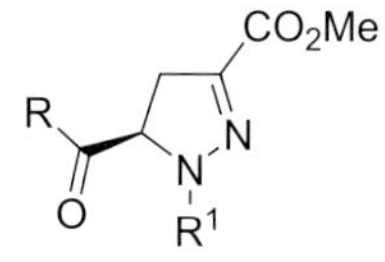 **419** 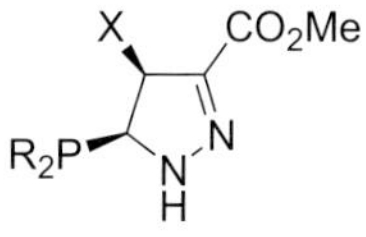 **420** 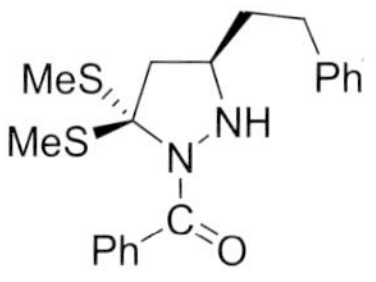 **421**

A series of compounds by Sibi are worth mentioning in this context: **422** (used for enantioselective intermolecular free radical conjugate additions) [408], **423** (a new ligand system) [409] and **424** (a pyrazolidinone template used with chiral Lewis acids) [410].

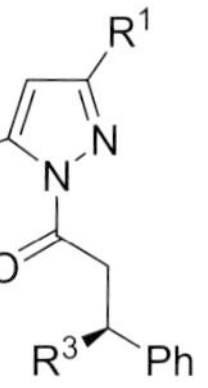

422 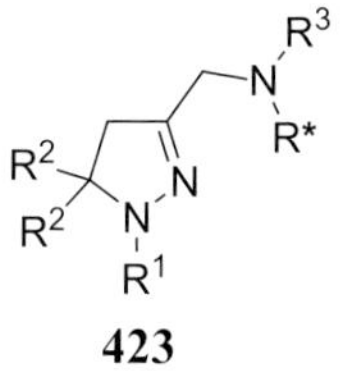 **423** 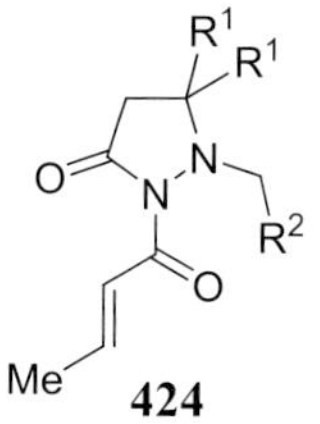 **424**

Finally, the resolution of compounds **425** and **426**, whose chirality resides exclusively in the nitrogen atoms, has been reported by Kostyanovsky [411]. For previous studies on chiral nitrogen atoms in pyrazole reduced derivatives due to hindered inversion see Reference [412].

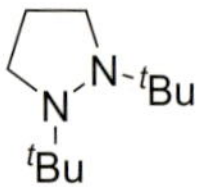

425 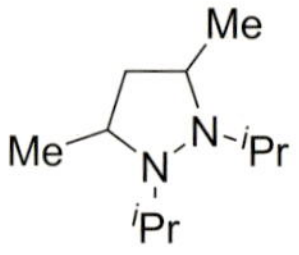 426

8.6.6 Macrocyclic Pyrazoles

This is an important field that has been treated in detail in other monographs [4, 5]. Particularly relevant are the results of Navarro *et al.* [413–415] (structures **427** and **428**) and of Kohnke *et al.* (**429**) [416].

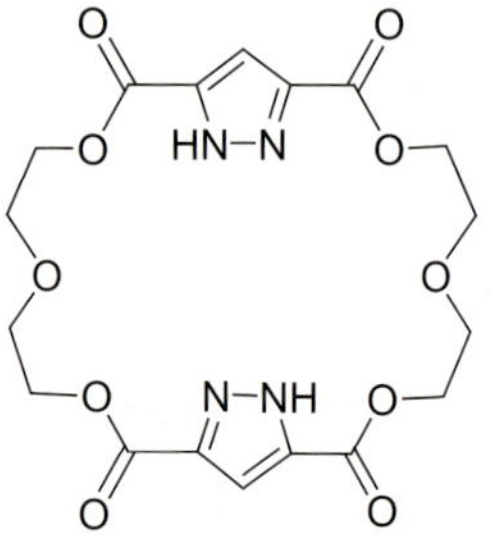

427 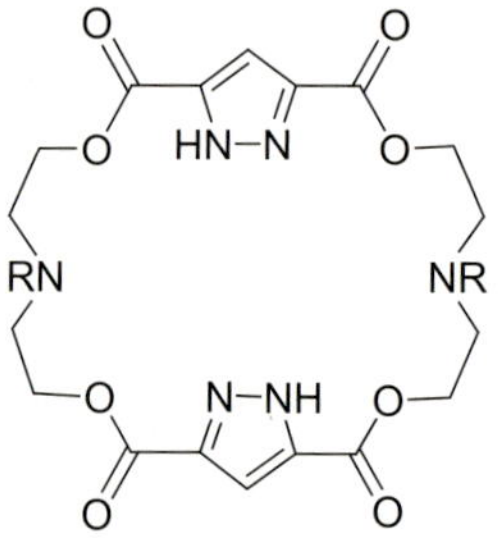 428 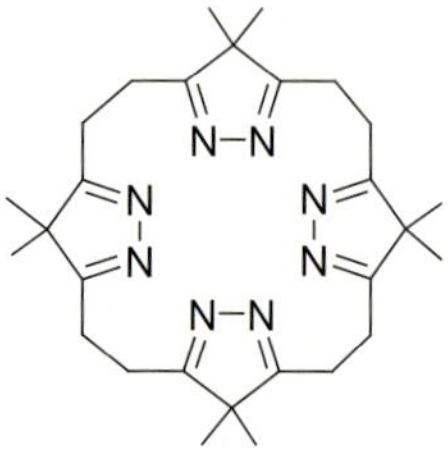 429

8.6.7 Labeled Derivatives

For biological purposes or by spectroscopic necessities (microwave, IR, NMR, etc.) many labeled pyrazoles have been prepared, mainly deuterium and ^{15}N derivatives for spectroscopy [4, 5, 417], and radioligands for biological studies: tritium [418, 419], ^{11}C [420], ^{18}F [421], and ^{125}I [422].

References

1 Kost, A.N. and Grandberg, I.I. (1996) *Progress in Pyrazole Chemistry – Advanced Heterocyclic Chemistry*, **6**, 347.

2 Behr, L.C., Fusco, R., and Jarboe, C.H. (1967) in *Pyrazoles, Pyrazolines, Pyrazolidines, Indazoles and Condensed Rings* (ed. R.H. Wiley), Interscience, New York.

3 Schofield, K., Grimmett, M.R., and Keene, B.R.T. (1976) *The Azoles*, Cambridge University Press, Cambridge.

4 Elguero, J. (1984) *Pyrazoles and their Benzo Derivatives* (ed. K.T. Potts), Comprehensive Heterocyclic Chemistry (series eds, A.R. Katritzky and C.W. Rees), Pergamon Press, Oxford.

5 Elguero, J. (1996) *Pyrazoles* (ed. I Shinkai), Comprehensive

Heterocyclic Chemistry II (series eds A.R. Katritzky, C.W. Rees, and E.F.V. Scriven), Pergamon, Oxford.

6 Kirschke, K. (1994) 1H-Pyrazole, in *Houben-Weyl, Methoden der Organischen Chemie, Volume E8b, Hetarenes III/2* (ed. E. Schauman), Georg Thieme Verlag, Stuttgart, p. 400.

7 Stadlbauer, W. (1994) Indazole (Benzopyrazole), in *Houben-Weyl, Methoden der Organischen Chemie, Volume E8b, Hetarenes III/2,* Georg Thieme Verlag, Stuttgart, p. 764.

8 Stanovnik, B. and Svete, J. (2004) *Pyrazoles,* in *Science of Synthesis,* Volume 12 (ed. Neier), *Five-Membered Hetarenes with Two Nitrogen or Phosphorus Atoms.* Electronic Edition, 15.

9 Stadlbauer, W. (2004) 1H- and 2H-indazoles, in Science of Synthesis, Volume 12 (ed. Neier), *Five-Membered Hetarenes with Two Nitrogen or Phosphorus Atoms.* Electronic Edition, 227.

10 http://webbook.nist.gov/chemistry/ NIST Chemistry WebBook.

11 Bigotto, A. and Zerbo, C. (1990) *Spectroscopy Letters,* **23**, 65.

12 http://www.ccdc.cam.ac.uk/products/csd/ Cambridge Structural Database.

13 Batterham, T.J. (1973) *NMR Spectra of Simple Heterocycles,* Interscience, New York.

14 Fruchier, A., Pellegrin, V., Schimpf, R., and Elguero, J. (1982) *Organic Magnetic Resonance,* **18**, 10.

15 Begtrup, M., Boyer, G., Cabildo, P., Cativiela, C., Claramunt, R.M., Elguero, J., Garcia, J.I., Toiron, C., and Vedsø, P. (1993) *Magnetic Resonance in Chemistry,* **31**, 107.

16 Elguero, J., Fruchier, A., Tjiou, E.M., and Trofimenko, S. (1995) *Chemistry of Heterocyclic Compounds,* 1006.

17 Claramunt, R.M., Sanz, D., López, C., Jiménez, J.A., Jimeno, M.L., Elguero, J., and Fruchier, A. (1997) *Magnetic Resonance in Chemistry,* **35**, 35.

18 (a) van Thuijl, J., Klebe, K.J., and van Houte, J.J. (1970) *Organic Mass Spectrometry,* **3**, 1549; (b) Luitjen, W.C.M.M. and van Thuijl, J. (1979) *Organic Mass Spectrometry,* **14**, 577.

19 Maquestiau, A., Van Haverbeke, Y., Flammang, R., Pardo, M.C., and Elguero, J. (1975) *Organic Mass Spectrometry,* **10**, 558.

20 (a) Cradock, R.H., Findlay, R.H., and Palmer, M.H. (1973) *Tetrahedron,* **29**, 2173; (b) Palmer, M.H. and Beveridge, A.J. (1987) *Chemical Physics,* **111**, 249.

21 Palmer, M.H. and Kennedy, S.M.F. (1978) *Journal of Molecular Structure,* **43**, 33.

22 Hargittai, I., Brunvoll, J., Foces-Foces, C., Llamas-Saiz, A.L., and Elguero, J. (1993) *Journal of Molecular Structure,* **291**, 211.

23 Velino, B., Cane, E., Trombetti, A., Corbelli, G., Zerbetto, F., and Caminati, W. (1992) *Journal of Molecular Structure,* **155**, 1.

24 Llamas-Saiz, A.L., Foces-Foces, C., Mó, O., Yáñez, M., Elguero, E., and Elguero, J. (1995) *Journal of Computational Chemistry,* **16**, 263.

25 Catalán, J., De Paz, J.L.G., and Elguero, J. (1996) *Journal of the Chemical Society-Perkin Transactions 2,* 57.

26 Krygowski, T.M. and Cyranski, M.K. (2001) *Chemical Reviews,* **101**, 1385.

27 Krygowski, T.M., Cyranski, M.K., Czarnocki, Z., Häfelinger, G., and Katritzky, A.R. (2000) *Tetrahedron,* **56**, 1783.

28 Goddard, R., Claramunt, R.M., Escolastico, C., and Elguero, J. (1999) *New Journal of Chemistry,* **23**, 237.

29 Holzer, W., Kautsch, C., Laggner, C., Claramunt, R.M., Pérez-Torralba, M., Alkorta, I., and Elguero, J. (2004) *Tetrahedron,* **60**, 6791.

30 Holzer, W. and Seiringer, G. (1993) *Journal of Heterocyclic Chemistry,* **30**, 865.

31 Claramunt, R.M., Sanz, D., Alkorta, I., and Elguero, J. (2005) *Magnetic Resonance in Chemistry,* **43**, 985.

32 Elguero, J., Marzin, C., Katritzky, A.R., and Linda, P. (1976) *The Tautomerism of Heterocycles,* Academic Press, New York.

33 Elguero, J., Katritzky, A.R., and Denisko, O.V. (2000) *Advances in Heterocyclic Chemistry,* **76**, 1.

34 Minkin, V.I., Garnovskii, A.D., Elguero, J., Katritzky, A.R.,

and Denisko, O.V. (2000) *Advances in Heterocyclic Chemistry*, **76**, 157.

35 Abboud, J.L.M., Cabildo, P., Cañada, T., Catalán, J., Claramunt, R.M., De Paz, J.L.G., Elguero, J., Homan, H., Notario, R., Toiron, C., and Yranzo, G.I. (1992) *The Journal of Organic Chemistry*, **57**, 3938.

36 Hammadi, A.E. and Mouhtadi, M.E. (2000) *THEOCHEM*, **497**, 241.

37 Jaronczyk, M., Dobrowolski, J.C., and Mazurek, A.P. (2004) *THEOCHEM*, **673**, 17.

38 Quiroga Puello, J., Insuasty Obando, B., Foces-Foces, C., Infantes, L., Claramunt, R.M., Cabildo, P., Jiménez, J.A., and Elguero, J. (1997) *Tetrahedron*, **53**, 10783.

39 Alkorta, I. and Elguero, J. *Journal of Physical Organic Chemistry*, **18**, 719.

40 (a) de Paz, J.L., Elguero, J., Foces-Foces, C., Llamas-Saiz, A., Aguilar-Parrilla, F., Klein, O., and Limbach, H.-H. (1997) *Journal of the Chemical Society-Perkin Transactions 2*, 101; (b) Schweiger, S. and Rauhut, G. (2003) *Journal of Physical Chemistry A*, **107**, 9668.

41 Alkorta, I. and Elguero, J. (2005) *Heteroatom Chemistry*, **16**, 628.

42 Holschbach, M.H., Sanz, D., Claramunt, R.M., Infantes, L., Motherwell, S., Raithby, P.R., Jimeno, M.L., Herrero, D., Alkorta, I., Jagerovic, N., and Elguero, J. (2003) *The Journal of Organic Chemistry*, **68**, 8831.

43 Allin, S.M., Barton, W.R.S., Bowman, W.R., and McInally, T. (2002) *Tetrahedron Letters*, **43**, 4191.

44 (a) Mikhalenok, S.G., Kuz'menok, N.M., and Zvonok, A.M."Alkaloids of the pyrrolo[1,2-*b*]pyrazole series: Synthesis of withasomnine and its analogs", in Selected Methods for Synthesis and Modification of Heterocycles, 1, 369–392; (b) Kartsev V.G. (ed.), InterBioScreen Monograph Series Press, Moscow, Russia.

45 Elguero, J., Goya, P., Jagerovic, N., and Silva, A.M.S. (2002) *Targets in Heterocyclic Systems, Vol. 6*, Italian Society of Chemistry, Roma, p. 52.

46 Merck Index, 11th Edition, (2001).

47 Mosti, L., Menozzi, G., Fossa, P., and Schenone, P. (1992) *Farmaco (Societa Chimica Italiana: 1989)*, **47**, 567.

48 Farré, A. and Frigola, J. (1994) *Drugs Future*, **19**, 651.

49 Heydorn, W.E. (2000) *Expert Opinion on Investigational Drugs*, **9**, 841.

50 Bowler, A.N., Dinsmore, A., Doyle, P.M., and Young, D.W. (1997) *Journal of the Chemical Society-Perkin Transactions 1*, 1297.

51 Rinaldi-Carmona, M., Barth, F., Heaulme, M., Shire, D., Calandra, B., Congy, C., Martinez, S., Maruani, J., and Neliat, G. (1994) *FEBS Letters*, **350**, 240.

52 Pertwee, R.G. (2000) *Exp Opin Invest Drugs*, **9**, 1.

53 Rinaldi-Carmona, M., Barth, F., Millan, J., Derocq, J.-M., Casellas, P., Congy, C., Oustric, D., Sarran, M., Bouaboula, M., Calandra, B., Portier, M., Shire, D., Breliere, J.-C., and Le Fur, G. (1998) *The Journal of Pharmacology and Experimental Therapeutics*, **284**, 644.

54 (a) Barth, F. (1998) *Expert Opinion on Therapeutic Patents*, **8**, 301; (b) Goya, P. and Jagerovic, N. (2000) *Expert Opinion on Therapeutic Patents*, **10**, 1529.

55 Lan, R., Liu, Q., Fan, P., Lin, S., Fernando, S.R., McCallion, D., Pertwee, R., and Makriyannis, A. (1999) *Journal of Medicinal Chemistry*, **42**, 769.

56 Jonas, R., Klockow, M., Lues, I., Prücher, H., Schliep, H.J., and Wurziger, H. (1993) *European Journal of Medicinal Chemistry*, **28**, 129.

57 Frigola, J., Colombo, A., Parés, J., Martínez, L., Sagarra, R., and Roser, R. (1989) *European Journal of Medicinal Chemistry*, **24**, 435.

58 (a) Murray, W., Wachter, M., Barton, D., and Forero-Kelly, Y. (1991) *Synthesis*, 18; (b) Murray, W.V. (1993) *Tetrahedron Letters*, **34**, 1863.

59 (a) Penning, T.D., Kramer, S.W., Lee, L.F., Collins, P.W., Koboldt, C.M., Seibert, K., Veenhuizen, A.M., Zhang, Y.Y., and Isakson, P.C. (1997) *Bioorganic & Medicinal Chemistry Letters*, **7**, 2121;

(b) Penning, T.D., Talley, J.J., Bertenshaw, S.R., Carter, J.S., Collins, P.W., Docter, S., Graneto, M.J., Lee, L.F., Malecha, J.W., Miyashiro, J.M., Rogers, R.S., Rogier, D.J., Yu, S.S., Anderson, G.D., Burton, E.G., Cogburn, J.N., Gregory, S.A., Koboldt, C.M., Perkins, W.E., Seibert, K., Veenhuizen, A.W., Zhang, Y.Y., and Isakson, P.C. (1997) *Journal of Medicinal Chemistry*, **40**, 1347; (c) Habeeb, A.G., Rao, P.N.P., and Knaus, E.E. (2001) *Journal of Medicinal Chemistry*, **44**, 3039.

60 Tsuji, K., Nakamura, K., Konishi, N., Tojo, T., Ochi, T., Senoh, H., and Matsuo, M. (1997) *Chemical & Pharmaceutical Bulletin*, **45**, 987.

61 Schindler, R., Fleischhauer, I., Höfgen, N., Sauer, W., Egerland, U., Poppe, H., Heer, S., Szelenyi, I., Kutscher, B., and Engel, J. (1998) *Archiv Der Pharmazie*, **331**, 13.

62 Mosti, L., Lo Presti, E., Menozzi, G., Marzano, C., Baccichetti, G., Falcone, G., Filipelli, W., and Piucci, B. (1998) *Farmaco (Societa Chimica Italiana: 1989)*, **53**, 602.

63 Kees, K.L., Fitzgerald, J.J., Jr, Steiner, K.E., Mattes, J.F., Mihan, B., Tosi, T., Mondoro, D., and McCaleb, M.L. (1996) *Journal of Medicinal Chemistry*, **39**, 3920.

64 Henke, B.R., Aquino, C.J., Birkcmo, L.S., Croom, D.K., Dougherty, R.W., Ervin, G.N., Grizzle, M.K., Hirst, G.C., James, M.K., Johnson, M.F., Queen, K.L., Sherrill, R.G., Sugg, E.E., Suh, E.M., Swewczyk, J.W., Unwalla, R.J., Yingling, J., and Willson, T.M. (1997) *Journal of Medicinal Chemistry*, **40**, 2706.

65 Czerwinski, E.W. (1991) *Acta Crystallographica Section C-Crystal Structure Communications*, **47**, 2598.

66 Spence, C.D., Coghlan, J.P., Denton, D.A., Mills, E.H., Whitworth, J.A., and Scoggins, B.A. (1986) *Journal of Steroid Biochemistry*, **25**, 411.

67 Theorell, H., Yonetani, T., and Sjöberg, B. (1969) *Acta Chemica Scandinavica (Copenhagen, Denmark: 1989)*, **23**, 255.

68 (a) Fries, R.W., Bohlken, D.P., and Plapp, B.V. (1979) *Journal of Medicinal Chemistry*, **22**, 356; (b) Horjales, E., Eklund, H., and Braenden, C.I. (1987) *Journal of Molecular Biology*, **197**, 685; (c) Rozas, I., Arteca, G.A., and Mezey, P.G. (1991) *International Journal of Quantum Chemistry. Quantum Biology Symposium*, **18**, 269; (d) International Journal of Quantum Chemistry Echevarria, A., Martin, M., Pérez, C., and Rozas, I. (1994) *Archiv Der Pharmazie*, **327**, 303.

69 Pereira, E.F.R., Aracava, Y., Aronstam, R.S., Barreiro, E.J., and Albuquerque, E.X. (1992) *The Journal of Pharmacology and Experimental Therapeutics*, **261**, 331.

70 (a) Horwitz, J.P., Massova, I., Wiese, T.E., Wozniak, A.J., Corbett, T.H., Sebolt-Leopold, J.S., Capps, D.B., and Leopold, W.R. (1993) *Journal of Medicinal Chemistry*, **36**, 3511; (b) Adjei, A.A. (1999) *Investigational New Drugs*, **17**, 43.

71 Watanabe, K., Morinaka, Y., Iseki, K., Watanabe, T., Yuki, S., and Nishi, H. (2003) *Redox Report: Communications in Free Radical Research*, **8**, 151.

72 Nonaka, N. (2003) *Agrochemicals Japan*, **83**, 17.

73 Lee, Y., Martasek, P., Roman, L.J., and Silverman, R.B. (2000) *Bioorganic & Medicinal Chemistry Letters*, **10**, 2771.

74 Wolff, D.J. and Gribin, B.J. (1994) *Archives of Biochemistry and Biophysics*, **311**, 300.

75 Raman, C.S., Li, H., Martásek, P., Southan, G., Masters, B.S.S., and Poulos, T.L. (2001) *Biochemistry*, **40**, 13448.

76 Sopková-de Oliveira Santos, J., Collot, V., and Rault, S. (2002) *Acta Crystallographica Section C-Crystal Structure Communications*, **C58**, 0688.

77 Claramunt, R.M., Sanz, D., López, C., Pinilla, E., Torres, M.R., Elguero, J., Nioche, P., and Raman, C.S. (2009) *Helvetica Chimica Acta*, **92**, 1952.

78 El-Abadelah, M.M., Sabri, S.S., Khanfar, M.A., Voelter, W., Abdel-Jalil, R.J., Maichle-Mössmer, C., and Al-Abed, Y. (2000) *Heterocycles*, **53**, 2643.

79 Al-bojuk, N.R., Eñ-Abadelah, M.M., Sabri, S.S., Michel, A., Voelter, W., M.

Mössmer, C., and Al-abed, Y. (2001) *Heterocycles*, **55**, 1789.

80 Yu, G., Mason, H.J., Wu, X., Wang, J., Chong, S., Dorough, G., Henwood, A., Pongrac, R., Seliger, L., He, B., Normandin, D., Adam, L., Krupinski, J., and Macor, J.E. (2001) *Journal of Medicinal Chemistry*, **44**, 1025.

81 Makino, K., Kim, H.S., and Kurasawa, Y. (1998) *Journal of Heterocyclic Chemistry*, **35**, 489.

82 Makino, K., Kim, H.S., and Kurasawa, Y. (1999) *Journal of Heterocyclic Chemistry*, **36**, 321.

83 Elmaati, T.M.A. and El-Taweel, F.M. (2004) *Journal of Heterocyclic Chemistry*, **41**, 109.

84 Lévai, A. (2002) *Journal of Heterocyclic Chemistry*, **39**, 1.

85 Stephanidon-Stephanaton, J. (1985) *Journal of Heterocyclic Chemistry*, **22**, 293.

86 Kotaly, A. and Papageorgiou, V.P. (1985) *Journal of the Chemical Society-Perkin Transactions 1*, 2083.

87 Hansen, J.F., Kim, Y.I., Griswold, L.J., Hoelle, G.W., Taylor, D.L., and Vietti, D.E. (1980) *The Journal of Organic Chemistry*, **45**, 76.

88 Gnichtel, H. and Schonherr, H.-J. (1966) *Chemische Berichte*, **99**, 618.

89 Gnichtel, H. and Boehringer, U. (1980) *Chemische Berichte*, **113**, 1507.

90 Fitton, A.O., Patel, R.N., and Millar, R.W. (1986) *Journal of Chemical Research-S*, 124, (M), 1101.

91 Barluenga, J., Lópes-Ortiz, J.F., Tomás, M., and Gotor, V. (1981) *Journal of the Chemical Society-Perkin Transactions 1*, 1891.

92 Barluenga, J., Tomás, M., Lópes-Ortiz, J.F., and Gotor, V. (1983) *Journal of the Chemical Society-Perkin Transactions 1*, 2273.

93 Barluenga, J., Iglesias, M.J., Muñiz, L., and Gotor, V. (1986) *Journal of Heterocyclic Chemistry*, **23**, 459.

94 Gladstone, W.A.F. and Norman, R.O.C. (1966) *Journal of the Chemical Society C*, 1536.

95 Bhatnagar, I. and George, M.V. (1968) *Tetrahedron*, **24**, 1293.

96 Kovelesky, A.C. and Shine, H.J. (1988) *The Journal of Organic Chemistry*, **53**, 1973.

97 Brewbaker, J.L. and Hart, H. (1969) *Journal of the American Chemical Society*, **91**, 711.

98 Grandi, R., Messerotti, W., Pagnoni, U.M., and Trave, R. (1977) *The Journal of Organic Chemistry*, **42**, 1352.

99 Tsuchiya, T., Kaneko, C., and Igeta, H. (1975) *Journal of the Chemical Society. Chemical Communications*, 528.

100 Padwa, A., Kulkarni, Y.S., and Zhang, Z. (1990) *The Journal of Organic Chemistry*, **55**, 4144.

101 Farkas, J. and Flegelová, Z. (1971) *Tetrahedron Letters*, **12**, 1591.

102 Just, G. and Kim, S. (1977) *Canadian Journal of Chemistry*, **55**, 427.

103 Kamitori, Y., Hojo, M., Masuda, R., Ohara, S., Kawasaki, K., and Yoshikawa, N. (1988) *Tetrahedron Letters*, **29**, 5281.

104 Sloan, K.B. and Rabjohn, N. (1970) *Journal of Heterocyclic Chemistry*, **7**, 1273.

105 Stapfer, C.H. and D'Andrea, R.W. (1970) *Journal of Heterocyclic Chemistry*, **7**, 651.

106 King, F. and Nicholls, D. (1978) *Inorganica Chimica Acta*, **28**, 55.

107 Stern, R. and Krause, J.G. (1968) *The Journal of Organic Chemistry*, **33**, 212.

108 Albright, T.A., Evans, S., Kim, C.S., Labaw, C.S., Russiello, A.B., and Schweizer, E.E. (1977) *The Journal of Organic Chemistry*, **42**, 3691.

109 Schweizer, E.E. and Hirwe, S.N. (1982) *The Journal of Organic Chemistry*, **47**, 1652.

110 Frêche, P., Gorgues, A., and Levas, E. (1976) *Tetrahedron Letters*, **17**, 1495.

111 Tamaru, Y., Harada, T., and Yoshida, Z. (1978) *The Journal of Organic Chemistry*, **43**, 3370.

112 Aubert, T., Tabyaoui, B., Farnier, M., and Guilard, R. (1988) *Synthesis*, 742.

113 Freeman, J.P. (1962) *The Journal of Organic Chemistry*, **27**, 1309.

114 Freeman, J.P., Gannon, J.J., and Surbey, D.L. (1969) *The Journal of Organic Chemistry*, **34**, 187.

115 Freeman, J.P. and Gannon, J.J. (1969) *The Journal of Organic Chemistry*, **34**, 194.

116 Hansen, J.F. and Luther, M.L. (1993) *Journal of Heterocyclic Chemistry*, **30**, 1163.

117 Kira, M.A., Abdel-Rahman, M.O., and Gadalla, K.Z. (1969) *Tetrahedron Letters*, **10**, 109.

118 Selvi, S. and Perumal, P.T. (2002) *Journal of Heterocyclic Chemistry*, **39**, 1129.

119 De Luca, L., Giacomelli, G., Masala, S., and Porcheddu, A. (2004) *Synlett*, 2299.

120 Sridhar, R., Sivaprasad, G., and Perumal, P.T. (2004) *Journal of Heterocyclic Chemistry*, **41**, 405.

121 Xu, D.D., Lee, G.T., Jiang, X., Prasad, K., Repic, O., and Blacklock, T.J. (2005) *Journal of Heterocyclic Chemistry*, **42**, 131.

122 Kira, M.A., Aboul-Enein, M.N., and Korkor, M.I. (1970) *Journal of Heterocyclic Chemistry*, **7**, 25.

123 Kira, M.A., Nofal, Z.M., and Gadalla, K.Z. (1970) *Tetrahedron Letters*, **11**, 4215.

124 Mangelinckx, S., Giubellina, N., and Kimpe, N. (2004) *Chemical Reviews*, **104**, 2353.

125 Foote, R.S., Beam, C.F., and Hauser, C.R. (1970) *Journal of Heterocyclic Chemistry*, **7**, 589.

126 Duncan, D.C., Trumbo, T.A., Almquist, C.D., Lentz, T.A., and Beam, C.F. (1987) *Journal of Heterocyclic Chemistry*, **24**, 555.

127 Huff, A.M., Hall, H.L., Smith, M.J., O'Grady, S.A., Waters, F.C., Fengl, R.W., Welsh, J.A., and Beam, C.F. (1985) *Journal of Heterocyclic Chemistry*, **22**, 501.

128 Matsumura, N., Kunugihara, A., and Yoneda, S. (1985) *Journal of Heterocyclic Chemistry*, **22**, 1169.

129 Meierhoefer, M.A., Dunn, S.P., Hajiaghamohseni, L.M., Walters, M.J., Embree, M.C., Grant, S.P., Downs, J.R., Townsend, J.D., Metz, C.R., Beam, C.F., Pennington, W.T., VanDerveer, D.G., and Camper, N.D. (2005) *Journal of Heterocyclic Chemistry*, **42**, 1095.

130 Beam, C.F., Reames, D.C., Harris, C.E., Dasher, L.W., Hollinger, W.M., Shealy, N.L., Sandifer, R.M., Perkins, M., and Hauser, C.R. (1975) *The Journal of Organic Chemistry*, **40**, 514.

131 Beam, C.F., Foote, R.S., and Hauser, C.R. (1972) *Journal of Heterocyclic Chemistry*, **9**, 183.

132 Reames, D.C., Harris, C.E., Dasher, L.W., Sandifer, R.M., Hollinger, W.M., and Beam, C.F. (1975) *Journal of Heterocyclic Chemistry*, **12**, 779.

133 Wilson, J.D., Fulmer, T.D., Dasher, L.P., and Beam, C.F. (1980) *Journal of Heterocyclic Chemistry*, **17**, 389.

134 Palacios, F., Aparicio, D., and Santos, J.M. (1994) *Tetrahedron*, **50**, 12727.

135 Palacios, F., Aparicio, D., and Santos, J.M. (1996) *Tetrahedron*, **52**, 4123.

136 Boeckman, R.K., Jr, Reed, J.E., and Ge, P. (2001) *Organic Letters*, **3**, 3651.

137 Dastrup, D.M., Yap, A.H., Weinreb, S.M., Henryb, J.R., and Lechleiter, A.J. (2004) *Tetrahedron*, **60**, 901.

138 Wang, Z.-X. and Qin, H.-L. (2004) *Green Chemistry*, **6**, 90.

139 Giuntini, F., Faustino, M.A.F., Neves, M.G.P.M.S., Tomé, A.C., Silva, A.M.S., and Cavaleiro, J.A.S. (2005) *Tetrahedron*, **61**, 10454.

140 Dodd, D.S. and Martinez, R.L. (2004) *Tetrahedron Letters*, **45**, 4265.

141 Dodd, D.S., Martinez, R.L., Kamau, M., Ruan, Z., Van Kirk, K., Cooper, C.B., Hermsmeier, M.A., Traeger, S.C., and Poss, M.A. (2005) *Journal of Combinatorial Chemistry*, **7**, 584.

142 Habraken, C.L., Beenakker, C.I.M., and Brussee, J. (1972) *Journal of Heterocyclic Chemistry*, **9**, 939.

143 Aiello, E., Aiello, S., Mingoia, F., Bacchi, A., Pelizzi, G., Musiu, C., Setzu, M.G., Pani, A., La Colla, P., and Marongiu, M.E. (2000) *Bioorganic and Medicinal Chemistry*, **8**, 2719.

144 Majid, T., Hopkins, C.R., Pedgrift, B., and Collar, N. (2004) *Tetrahedron Letters*, **45**, 2137.

145 Stanovnik, B. and Svete, J. (2004) *Chemical Reviews*, **104**, 2433.

146 Groselj, U., Bevk, D., Jakse, R., Recnik, S., Meden, A., Stanovnik, B., and Svete, J. (2005) *Tetrahedron*, **61**, 3991.

147 Menozzi, G., Mosti, L., and Schenone, P. (1987) *Journal of Heterocyclic Chemistry*, **24**, 1669.

148 Hanzlowsky, A., Jelencic, B., Recnik, S., Svete, J., Golobic, A., and Stanovnik, B. (2003) *Journal of Heterocyclic Chemistry*, **40**, 487.

149 Giacomelli, G., Porcheddu, A., Salaris, M., and Taddei, M. (2003) *European Journal of Organic Chemistry*, 537.

150 Westman, J. and Lundin, R. (2003) *Synthesis*, 1025.

151 De Luca, L., Giacomelli, G., Porcheddu, A., Salaris, M., and Taddei, M. (2003) *Journal of Combinatorial Chemistry*, **5**, 465.

152 Peruncheralathan, S., Yadav, A.K., Ila, H., and Junjappa, H. (2005) *The Journal of Organic Chemistry*, **70**, 9644.

153 Elgemeie, G.H., Elghandour, A.H., and Elaziz, G.W.A. (2004) *Synthetic Communications*, **34**, 3281.

154 Peruncheralathan, S., Khan, T.A., Ila, H., and Junjappa, H. (2005) *The Journal of Organic Chemistry*, **70**, 10030.

155 Mahata, P.K., Kumar, U.K.S., Sriram, V., Ila, H., and Junjappa, H. (2003) *Tetrahedron*, **59**, 2631.

156 Lee, K.Y., Kim, J.M., and Kim, J.N. (2003) *Tetrahedron Letters*, **44**, 6737.

157 Martins, M.A.P., Pereira, C.M.P., Beck, P., Machado, P., Moura, S., Teixeira, M.V.M., Bonacorso, H.G., and Zanatta, N. (2003) *Tetrahedron Letters*, **44**, 6669.

158 Bonacorso, H.G., Cechinel, C.A., Oliveira, M.R., Costa, M.B., Martins, M.A.P., Zanatta, N., and Flores, A.F.C. (2005) *Journal of Heterocyclic Chemistry*, **42**, 1055.

159 Katritzky, A.R., Wang, M., Zhang, S., Voronkov, M.V., and Steel, P.J. (2001) *The Journal of Organic Chemistry*, **66**, 6787.

160 Abdel-Fattah, A.A.A. (2005) *Synthesis*, 245.

161 Chimenti, F., Bolasco, A., Manna, F., Secci, D., Chimenti, P., Befani, O., Turini, P., Giovannini, V., Mondovì, B., Cirilli, R., and La Torre, F. (2004) *Journal of Medicinal Chemistry*, **47**, 2071.

162 Azarifar, D. and Maleki, B. (2005) *Journal of Heterocyclic Chemistry*, **42**, 157.

163 Bhat, B.A., Puri, S.C., Qurishi, M.A., Dhar, K.L., and Qazi, N.G. (2005) *Synthetic Communications*, **35**, 1135.

164 Lévai, A., Patonay, T., Silva, A.M.S., Pinto, D.C.G.A., and Cavaleiro, J.A.S. (2002) *Journal of Heterocyclic Chemistry*, **39**, 751.

165 Pinto, D.C.G.A., Silva, A.M.S., Cavaleiro, J.A.S., and Elguero, J. (2003) *European Journal of Organic Chemistry*, 747.

166 Meier, H. and Hormaza, A. (2003) *European Journal of Organic Chemistry*, 3372.

167 Cuadrado, P., González-Nogal, A.M., and Valero, R. (2002) *Tetrahedron*, **58**, 4975.

168 Grotjahn, D.B., Van, S., Combs, D., Lev, D.A., Schneider, C., Rideout, M., Meyer, C., Hernandez, G., and Mejorado, L. (2002) *The Journal of Organic Chemistry*, **67**, 9200.

169 Adamo, M.F.A., Adlington, R.M., Baldwin, J.E., Pritchard, G.J., and Rathmell, R.E. (2003) *Tetrahedron*, **59**, 2197.

170 Bishop, B.C., Brands, K.M.J., Gibb, A.D., and Kennedy, D.J. (2004) *Synthesis*, 43.

171 Aumann, R., Jasper, B., and Fröhlich, R. (1995) *Organometallics*, **14**, 2447.

172 Haddad, N. and Baron, J. (2002) *Tetrahedron Letters*, **43**, 2171.

173 Haddad, N., Salvagno, A., and Busacca, C. (2004) *Tetrahedron Letters*, **45**, 5935.

174 Ju, Y. and Varma, R.S. (2005) *Tetrahedron Letters*, **46**, 6011.

175 Regitz, M. and Heydt, H. (1984) in *1, 3- Dipolar Cycloaddition Chemistry, Vol 1* (ed. A. Padwa), Ch 4, Wiley, New York.

176 Di, M. and Rein, K.S. (2004) *Tetrahedron Letters*, **45**, 4703.

177 Bastide, J., Henri-Rousseau, O., and Aspart-Pascot, L. (1974) *Tetrahedron*, **30**, 3355.

178 Koszinowski, J. (1984) Ph.D. Thesis, University of Munich, 1980. Cited by Huisgen, R. In 1,3-Dipolar Cycloaddition Chemistry, Vol. 1 (ed. A. Padwa), Ch. 1, Wiley, New York.

179 Sasaki, T. and Kanematsu, K. (1971) *Journal of the Chemical Society C*, 2147.

180 Reimlinger, H. (1959) *Chemische Berichte*, **92**, 970.

181 Hanamoto, T., Koga, Y., Kido, E., Kawanami, T., Furuno, H., and Inanaga, J. (2005) *Chemical Communications*, 2041.

182 Hanamoto, T., Hakoshima, Y., and Egashira, M. (2004) *Tetrahedron Letters*, **45**, 7573.

183 Aggarwal, V.K., Vicente, J., and Bonnert, R.V. (2003) *The Journal of Organic Chemistry*, **68**, 5381.

184 Chandrasekhar, S., Rajaiah, G., and Srihari, P. (2001) *Tetrahedron Letters*, **42**, 6599.

185 Jiang, N. and Li, C.-J. (2004) *Chemical Communications*, 394.

186 Caramella, P. and Grünanger, P. (1984) in *1,3-Dipolar Cycloaddition Chemistry, Vol. 1* (ed. A. Padwa), Ch. 3, Wiley, New York.

187 Shawali, A.S. (1993) *Chemical Reviews*, **93**, 2731.

188 Shawali, A.S. and Párkányi, C. (1980) *Journal of Heterocyclic Chemistry*, **17**, 833.

189 Bianchi, G., De Michelli, C., and Gandolfi, R. (1979) *Angewandte Chemie (International Edition in English)*, **18**, 721.

190 Roy, A., Sahabuddin, S., Achari, B., and Mandal, S.B. (2005) *Tetrahedron*, **61**, 365.

191 Chen, D.W. and Chen, Z.C. (1995) *Synthetic Communications*, **25**, 1617.

192 Xia, M. and Pan, X.-J. (2004) *Synthetic Communications*, **34**, 3521.

193 Ghozlan, S.A.S., Abdelhamid, I.A., Gaber, H.M., and Elnagdi, M.H. (2005) *Journal of Heterocyclic Chemistry*, **42**, 1185.

194 Molteni, G., Pilati, T., and Ponti, A. (2003) *Tetrahedron*, **59**, 9315.

195 Wagner-Jauregg, T. (1976) *Synthesis*, 349.

196 Rádl, S. (1997) *Aldrichimica Acta*, **30**, 97.

197 Tomé, A.C., Cavaleiro, J.A.S., Domingues, F.M.J., and Cremlyn, R.J. (2005) *Phosphorus, Sulfur, Silicon*, **180**, 2617.

198 Taylor, E.C. and Sobieray, D.M. (1991) *Tetrahedron*, **47**, 9599.

199 Pezdirc, L., Jovanovski, V., Bevk, D., Jakse, R., Pirc, S., Meden, A., Stanovnik, B., and Svete, J. (2005) *Tetrahedron*, **61**, 3977.

200 Yamashita, Y. and Kobayashi, S. (2004) *Journal of the American Chemical Society*, **126**, 11279.

201 Ahmed, M.S.M., Kobayashi, K., and Mori, A. (2005) *Organic Letters*, **7**, 4487.

202 Komatau, M., Kajihara, Y., Kobayashi, M., Itoh, S., and Ohshiro, Y. (1992) *The Journal of Organic Chemistry*, **57**, 7359.

203 Stille, J.K., Harris, F.W., and Bedford, M.A. (1966) *Journal of Heterocyclic Chemistry*, **3**, 155.

204 Meazza, G. and Zanardi, G. (1993) *Journal of Heterocyclic Chemistry*, **30**, 365.

205 Skof, M., Svete, J., and Stanovnik, B. (2000) *Heterocycles*, **53**, 339.

206 Okuro, K., Furuune, M., Miura, M., and Nomura, M. (1992) *The Journal of Organic Chemistry*, **57**, 4754.

207 Ilhan, I.O., Saripinar, E., and Akçamur, Y. (2005) *Journal of Heterocyclic Chemistry*, **42**, 117.

208 Robey, R.L., Alt, C.A., and Van Meter, E.E. (1997) *Journal of Heterocyclic Chemistry*, **34**, 413.

209 Yadav, J.S., Reddy, B.V.S., Srinivas, M., Prabhakar, A., and Jagadeesh, B. (2004) *Tetrahedron Letters*, **45**, 6033.

210 Yadav, J.S., Reddy, B.V.S., Satheesh, G., Lakshmi, P.N., Kumar, S.K., and Kunwar, A.C. (2004) *Tetrahedron Letters*, **45**, 8587.

211 Pinto, D.C.G.A., Silva, A.M.S., and Cavaleiro, J.A.S. (2000) *Journal of Heterocyclic Chemistry*, **37**, 1629.

212 Pinto, D.C.G.A., Silva, A.M.S., Almeida, L.M.P.M., Cavaleiro, J.A.S., and Elguero, J. (2002) *European Journal of Organic Chemistry*, 3807.

213 Lévai, A., Silva, A.M.S., Pinto, D.C.G.A., Cavaleiro, J.A.S., Alkorta, I., Elguero, J., and Jekö, J. (2004) *European Journal of Organic Chemistry*, 4672.

214 Gelin, S., Chantegrel, B., and Nadi, A.I. (1983) *The Journal of Organic Chemistry*, **48**, 4078.

215 Bendaas, A., Hamdi, M., and Sellier, N. (1999) *Journal of Heterocyclic Chemistry*, **36**, 1291.

216 Ait-Baziz, N., Rachedi, Y., Hamdi, M., Silva, A.M.S., Balegroune, F., Thierry, R., and Sellier, N. (2004) *Journal of Heterocyclic Chemistry*, **41**, 587.

217 Lévai, A., Jeko, J., and Brahmbhatt, D.I. (2005) *Journal of Heterocyclic Chemistry*, **42**, 1231.

218 Van der Plas, H.C., Jongejan, H., and Koudijs, A. (1978) *Journal of Heterocyclic Chemistry*, **15**, 485.

219 Guillard, J., Goujon, F., Badol, P., and Poullain, D. (2003) *Tetrahedron Letters*, **44**, 5943.

220 Takagi, K. and Hubert-Habart, M. (1996) *Journal of Heterocyclic Chemistry*, **33**, 1003.

221 Suen, Y.F., Hope, H., Nantz, M.H., Haddadin, M.J., and Kurth, M.J. (2005) *The Journal of Organic Chemistry*, **70**, 8468.

222 Simoni, D., Rondanin, R., Furnò, G., Aiello, E., and Invidiata, F.P. (2000) *Tetrahedron Letters*, **41**, 2699.

223 Behr, L.C. (1967) in *The Chemistry of Heterocyclic Compounds, Vol. 22* (ed. A. Weissberger), Interscience, New York, p. 289.

224 Baiocchi, L., Corsi, G., and Palazzo, G. (1978) *Synthesis*, 633.

225 Cadogan, J.I.G., Cameron-Wood, M., Mackie, R.K., and Searle, R.J.G. (1965) *Journal of the Chemical Society*, 4831.

226 Akazome, M., Kondo, T., and Watanabe, Y.J. (1994) *Organic Chemistry*, **59**, 3375.

227 Ardakani, M.A., Smalley, R.K., and Smith, R.H. (1979) *Synthesis*, 308–309.

228 Kuvshinov, A.M., Gulevskaya, V.I., Rozhkov, V.V., and Shevelev, S.A. (2000) *Synthesis*, 1474.

229 Krbechek, L. and Takimoto, H. (1964) *The Journal of Organic Chemistry*, **29**, 1150.

230 Molina, P., Arques, A., and Vinader, M.V. (1990) *The Journal of Organic Chemistry*, **55**, 4724.

231 Taher, A., Ladwa, S., Rajan, S.T., and Weaver, G.W. (2000) *Tetrahedron Letters*, **41**, 9893.

232 Nyerges, M., Fejes, I., Virányi, A., Groundwater, P.W., and Tőke, L. (2001) *Tetrahedron Letters*, **42**, 5081.

233 Nyerges, M., Fejes, I., Virányi, A., Zhang, W., Groundwater, P.W., Blaskó, G., and Tőke, L. (2004) *Tetrahedron*, **60**, 9937.

234 Matassa, V.G., Maduskuie, T.P., Jr, Shapiro, H.S., Hesp, B., Snyder, D.W., Aharony, D., Krell, R.D., and Keith, R.A. (1990) *Journal of Medicinal Chemistry*, **33**, 1781.

235 Huisgen, R. and Bast, K. (1973) *Org. Syntheses, Vol. V*, Wiley, New York, p. 650.

236 Sun, J.H., Teleha, C.A., Yan, J.-S., Rodgers, J.D., and Nugiel, D.A. (1997) *The Journal of Organic Chemistry*, **62**, 5627.

237 Cui, J.J., Araldi, G.-L., Reiner, J.E., Reddy, K.M., Kemp, S.J., Ho, J.Z., Siev, D.V., Mamedova, L., Gibson, T.S., Gaudette, J.A., Minami, N.K., Anderson, S.M., Bradbury, A.E., Nolan, T.G., and Semple, A.E. (2002) *Bioorganic & Medicinal Chemistry Letters*, **12**, 2925.

238 Souers, A.J., Gao, J., Wodka, D., Judd, A.S., Mulhern, M.M., Napier, J.J., Brune, M.E., Bush, E.N., Brodjian, S.J., Dayton, B.D., Shapiro, R., Hernandez, L.E., Marsh, K.C., Sham, H.L., Collins, C.A., and Kym, P.R. (2005) *Bioorganic & Medicinal Chemistry Letters*, **15**, 2752.

239 Huisgen, R. and Nakaten, H. (1951) *Justus Leigbigs Annalen der Chemie*, **573**, 181.

240 Behr, L.C. (1967) in *Pyrazoles, Pyrazolines, Pyrazolidines, Indazoles and Condensed Rings* (ed. R.H. Wiley), John Wiley & Sons, Inc., New York, p. 295.

241 Bartsch, R.A. and Yang, I.-W. (1984) *Journal of Heterocyclic Chemistry*, **21**, 1063.

242 Schumann, P., Collot, V., Hommet, Y., Gsell, W., Dauphin, F., Sopkova, J., MacKenzie, E.T., Duval, D., Boulouard, M., and Rault, S. (2001) *Bioorganic & Medicinal Chemistry Letters*, **11**, 1153.

243 Shutske, G.M., Allen, R.C., Försch, M.F., Setescak, L.L., and Wilker, J.C. (1983) *Journal of Medicinal Chemistry*, **26**, 1307.

244 Zhang, D., Kohlman, D., Krushinski, J., Liang, S., Ying, B.-P., Reilly, J.E., Dinn, S.R., Wainscott, D.B., Nutter, S., Gough, W., Nelson, D.L.G., Schaus, J.M., and Xu, Y.-C. (2004) *Bioorganic & Medicinal Chemistry Letters*, **14**, 6011.

245 Barton, D.R., Lukacs, G., and Wagle, D. (1982) *Journal of the Chemical Society. Chemical Communications*, 450–452.

246 Leroy, V., Lee, G.E., Lin, J., Herman, S.H., and Lee, T.B. (2001) *Organic Process Research & Development*, **5**, 179.

247 Gladstone, W.A.F. and Norman, R.O.C. (1966) *Journal of the Chemical Society C*, 1527.

248 Butler, R.N. (1968) *Chemistry & Industry*, 437.

249 Kaushik, M.P., Lal, B., Raghuveeran, C.D., and Vaidyanathaswamy, (1982) *The Journal of Organic Chemistry*, **47**, 3503.

250 Vivona, N., Frenna, V., Buscemi, S., and Condò, M. (1985) *Journal of Heterocyclic Chemistry*, **22**, 29.

251 Yan, B. and Hubert, G. (1996) *Tetrahedron Letters*, **46**, 8325.

252 Lee, F.-Y., Lien, J.-C., Huang, L.-J., Huang, T.-M., Tsai, S.-C., Teng, C.-M., Wu, C.-C., Cheng, F.-C., and Kuo, S.-C. (2001) *Journal of Medicinal Chemistry*, **44**, 3746.

253 Kuo, S.-C., Lee, F.-Y., and Teng, C.-M. (1995) European Patent EP667345; *Chem. Abstr.* (1995) 89, 43221.

254 Gordon, D.W. (1998) *Synlett*, 1065.

255 Frasca, A.R. (1962) *Tetrahedron Letters*, **24**, 1115.

256 Dennler, E.B. and Frasca, A.R. (1966) *Tetrahedron*, **22**, 3131.

257 Meyer, V. (1889) *Berichte der Deutschen Chemischen Gesellschaft*, **22**, 319.

258 Pummerer, R., Buchta, E., and Deimler, E. (1951) *Chemische Berichte*, **84**, 583.

259 Krishnan, R., Lang, S.A., Jr, Lin, Y.-I., and Wilkinson, R.G. (1988) *Journal of Heterocyclic Chemistry*, **25**, 447.

260 Halley, F. and Sava, X. (1997) *Synthetic Communications*, **27**, 1199.

261 Patel, M., Rodgers, J.D., McHugh, R.J., Jr, Johnson, B.L., Cordova, B.C., Klabe, R.M., Bacheler, L.T., Erickson-Viitanen, S., and Ko, S.S. (1999) *Bioorganic & Medicinal Chemistry Letters*, **9**, 3217.

262 Hathaway, B.A., Day, G., Lewis, M., and Glaser, R. (1998) *Journal of the Chemical Society-Perkin Transactions 2*, 2713.

263 Zhenqi, Z., Tongsheng, X., Xiaonai, C., Yuzhu, Q., and Zheng, Z. (1993) *Journal of the Chemical Society-Perkin Transactions 1*, 1279.

264 Guofu, Q., Jiangtao, S., Xichum, F., Lamei, W., Wenjin, X., and Xianming, H. (2004) *Journal of Heterocyclic Chemistry*, **41**, 601.

265 Cho, C.S., Lim, D.K., Heo, N.H., Kim, T.-J., and Shim, S.C. (2004) *Chemical Communications*, 104.

266 Inamoto, K., Katsuno, M., Yoshino, T., Suzuki, I., Hiroya, K., and Sakamoto, T. (2004) *Chemistry Letters*, **33**, 1026.

267 Lebedev, A.Y., Khartulyari, A.S., and Voskoboynikov, A.Z. (2005) *The Journal of Organic Chemistry*, **70**, 596.

268 Song, J.J. and Yee, N.K. (2000) *Organic Letters*, **2**, 519.

269 Song, J.J. and Yee, N.K. (2001) *Tetrahedron Letters*, **42**, 2937.

270 Amer, A.M. (1998) *Monatshefte Fur Chemie*, **129**, 1293.

271 Dennler, E.B. and Frasca, A.R. (1967) *Canadian Journal of Chemistry*, **45**, 697.

272 Barone, R., Camps, P., and Elguero, J. (1979) *Anales De Quimica*, **75**, 736.

273 Uehata, K., Kawakami, T., and Suzuki, H. (2002) *Journal of the Chemical Society-Perkin Transactions 1*, 696.

274 Palazzo, G., Corsi, G., Baiocchi, L., and Silvestrini, B. (1966) *Journal of Medicinal Chemistry*, **9**, 38.

275 Selwood, D.L., Brummell, D.G., Budworth, J., Burtin, G.E., Campbell, R.O., Chana, S.S., Charles, I.G., Fernandez, P.A., Glen, R.C., Goggin, M.C., Hobbs, A.J., Kling, M.R., Liu, Q., Madge, D.J., Meillerais, S., Powell, K.L., Reynolds, K., Spacey, G.D., Stables, J.N., Tatlock, M.A., Wheeler, K.A., Wishart, G., and Woo, C.-K. (2001) *Journal of Medicinal Chemistry*, **44**, 78.

276 Koga, N., Koga, G., and Anselme, J.-P. (1972) *Tetrahedron*, **28**, 4515.

277 Hervens, F. and Viehe, H.G. (1973) *Angewandte Chemie-International Edition*, **12**, 405.

278 Yoshida, T., Matsuura, N., Yamamoto, K., Doi, M., Shimada, K., Morie, T., and Kato, S. (1996) *Heterocycles*, **43**, 2701.

279 Wrobleski, S.T., Chen, P., Hynes, J., Jr, Lin, S., Norris, D.J., Pandit, C.R., Spergel, S., Wu, H., Tokarski, J.S., Chen, X., Gillooly, K.M., Kiener, P.A., McIntyre, K.W., Patil-Koota, V., Shuster, D.J., Turk, L.A., Yang, G., and Leftheris, K. (2003) *Journal of Medicinal Chemistry*, **46**, 2110.

280 Caron, S. and Vazquez, E. (1999) *Synthesis*, 588–592.

281 Kaltenbach, R.F., Patel, M., Waltermine, R.E., Harris, G.D., Stone, B.R.P., Klabe, R.M., Garber, S., Bacheler, L.T., Cordova, B.C., Logue, K., Wright, M.R., Erickson-Viitanen, S., and Trainor, G.L. (2003) *Bioorganic & Medicinal Chemistry Letters*, **13**, 605.

282 Stocks, M.J., Barber, S., Ford, R., Leroux, F., St-Gallay, S., Teague, S., and Xue, Y. (2005) *Bioorganic & Medicinal Chemistry Letters*, **15**, 3459.

283 Lee, K.Y., Gowrisanar, S. and Kim, J.N. (2005) *Tetrahedron Letters*, **46**, 5387.

284 Connolly, P.J., Wetter, S.K., Beers, K.N., Hamel, S.C., Haynes-Johnson, D., Kiddoe, M., Kraft, P., Lai, M.T., Campen, C., Palmer, S., and Phillips, A. (1997) *Bioorganic & Medicinal Chemistry Letters*, **7**, 2551.

285 Murineddu, G., Ruiu, S., Mussinu, J.-M., Loriga, G., Grella, G.E., Carai, M.A.M., Lazzari, P., Pani, L., and Pinna, G.A. (2005) *Bioorganic and Medicinal Chemistry*, **13**, 3309.

286 El-Rayyes, N.R. and Al-Jawhary, A. (1986) *Journal of Heterocyclic Chemistry*, **23**, 135.

287 Lóránd, T., Kocsis, B., Emôdy, L., and Sohár, P. (1999) *European Journal of Medicinal Chemistry*, **34**, 1009.

288 Stadlbauer, W. and Hojas, G. (2003) *Journal of Heterocyclic Chemistry*, **40**, 753.

289 Huisgen, R. and Knorr, R. (1961) *Die Naturwissenschaften*, **48**, 716.

290 Ried, W. and Schön, M. (1965) *Justus Leigbigs Annalen der Chemie*, **689**, 141.

291 Garcia-Abbad, E., Garcia-López, M.T., Garcia-Muñoz, G., and Stud, M. (1976) *Journal of Heterocyclic Chemistry*, **13**, 1241.

292 Friedman, L. and Logullo, F.M. (1969) *The Journal of Organic Chemistry*, **34**, 3089.

293 Shoji, Y., Hari, Y., and Aoyama, T. (2004) *Tetrahedron Letters*, **45**, 1769.

294 Conway, G.A., Loeffer, L.J., and Hall, I.H. (1983) *Journal of Medicinal Chemistry*, **26**, 876.

295 Tapia, R.A., Carrasco, C., Ojeda, S., Salas, C., Valderrama, J.A., Morello, A., and Repetto, Y. (2002) *Journal of Heterocyclic Chemistry*, **39**, 1093.

296 Ainsworth, C. (1957) *Journal of the American Chemical Society*, **79**, 5242.

297 Baraldi, P.G., Cacciari, B., Spalluto, G., Romagnoli, R., Braccioli, G., Zaid, A.N., Pineda, M.J., and de las Infantas, J.P. (1997) *Synthesis*, 1140.

298 Barluenga, J., Aznar, F., and Palomero, M.A. (2001) *Chemistry – A European Journal*, **7**, 5317.

299 Peruncheralathan, S., Khan, T.A., Ila, H., and Junjappa, H. (2004) *Tetrahedron*, **60**, 3457.

300 Medio-Simón, M., Laviada, M.J.A., and Sepúlveda-Arques, J. (1990) *Journal of the Chemical Society-Perkin Transactions 1*, 2749.

301 Matsugo, S. and Takamizawa, A. (1984) *Synthesis*, 852.

302 Tomé, A.C., Cavaleiro, J.A.S., and Storr, R.C. (1996) *Synlett*, 531.

303 Wentrup, C., Damerius, A., and Reichen, W. (1978) *The Journal of Organic Chemistry*, **43**, 2037.

304 Reichen, W. (1976) *Helvetica Chimica Acta*, **59**, 1636.

305 Hajós, G., Riedl, Z., Mátyus, P., Maes, B.U.W., and Lemière, G.L.F. (2005) *Journal of Heterocyclic Chemistry*, **42**, 421.

306 Soós, T., Hajós, G., and Messmer, A. (1997) *The Journal of Organic Chemistry*, **62**, 1136.

307 Buscemi, S., Vivona, N., and Caronna, T. (1996) *The Journal of Organic Chemistry*, **61**, 8397.

308 Vivona, N., Cusmano, G., Macaluso, G., Frenna, N., and Ruccia, M. (1979) *Journal of Heterocyclic Chemistry*, **16**, 783.
309 Catalán, J., Abboud, J.-L.M., and Elguero, J. (1987) *Advances in Heterocyclic Chemistry*, **41**, 187.
310 Trofimenko, J. (1999) *Scorpionates: The Coordination Chemistry of Polypyrazolylborate Ligands*, World Scientific Pub Co Inc, New Jersey.
311 Pérez, E.R., Loupy, A., Liagre, M., de Guzzi Plepis, A.M., and Cordeiro, P.J. (2003) *Tetrahedron*, **59**, 865.
312 de la Hoz, A., Díaz-Ortíz, A., and Moreno, A. (2005) *Chemical Society Reviews*, **34**, 164.
313 Wang, X.-J., Tan, J., Grozinger, K., Betageri, R., Kirrane, T., and Proudfoot, J.R. (2000) *Tetrahedron Letters*, **41**, 5321.
314 Zoppellaro, G. and Baumgarten, M. (2005) *European Journal of Organic Chemistry*, 2888.
315 Claramunt, R.M., Elguero, J., Escolástico, C., Fernández-Castaño, C., Foces-Foces, A., Llamas-Saiz, A.L., and Santa, Maria M.D. (1997) *Targets in Heterocyclic Systems*, **1**, 1.
316 Duprez, V. and Heumann, A. (2004) *Tetrahedron Letters*, **45**, 5697.
317 López-Alvarado, P., Avendaño, C., and Menéndez, J.C. (1995) *The Journal of Organic Chemistry*, **60**, 5678.
318 Lam, P.Y.S., Clarck, C.G., Saubern, S., Adams, J., Winters, M.P., Chan, D.M.T., and Combs, A. (1998) *Tetrahedron Letters*, **39**, 2941.
319 Antilla, J.C., Baskin, J.M., Barder, T.E., and Buchwald, S.L. (2004) *The Journal of Organic Chemistry*, **69**, 5578.
320 Cristau, H.J., Cellier, P.P., Spindler, J.F., and Taillefer, M. (2004) *European Journal of Organic Chemistry*, 695.
321 Collot, V., Bovy, P.R., and Rault, S. (2000) *Tetrahedron Letters*, **41**, 9053.
322 Clarke, D., Mares, R.W., McNab, H., and Riddell, F.G. (1994) *Magnetic Resonance in Chemistry*, **32**, 255.
323 Blake, A.J., Clarke, D., Mares, R.W., and McNab, H. (2003) *Organic and Biomolecular Chemistry*, **1**, 4268.
324 Stefani, H.A., Pereira, C.M.P., Almeida, R.B., Braga, R.C., Guzen, K.P., and Cella, R. (2005) *Tetrahedron Letters*, **46**, 6833.
325 Yin, L., Erdmann, F., and Liebscher, J. (2005) *Journal of Heterocyclic Chemistry*, **42**, 1369.
326 Rodríguez-Franco, M.I., Dorronsoro, I., Hernández-Higueras, A.I., and Antequera, G. (2001) *Tetrahedron Letters*, **42**, 863.
327 Popsavin, M., Torovi, L., Spai, S., Stankov, S., Kapor, A., Tomi, Z., and Popsavin, V. (2002) *Tetrahedron*, **58**, 569.
328 Padmavathi, V., Sarma, M.R., Padmaja, A., and Reddy, D.B. (2003) *Journal of Heterocyclic Chemistry*, **40**, 933.
329 Zolfigol, M.A., Azarifar, D., and Maleki, B. (2004) *Tetrahedron Letters*, **45**, 2181.
330 Nakamichi, N., Kawashita, Y., and Hayashi, M. (2002) *Organic Letters*, **4**, 3955.
331 ung, M.E., Min, S.-J., Houk, K.N., and Ess, D. (2004) *The Journal of Organic Chemistry*, **69**, 9085–9089.
332 Azarifar, D. and Maleki, B. (2005) *Synthetic Communications*, **35**, 2581.
333 Haddach, A.A., Kelleman, A., and Deaton-Rewolinski, M.V. (2002) *Tetrahedron Letters*, **43**, 399.
334 Alkorta, I. and Elguero, J. (2006) *Tetrahedron*, **62**, 8683.
335 Schlosser, M., Volle, J.-N., Leroux, F., and Schenk, K. (2002) *European Journal of Organic Chemistry*, 2913.
336 Elguero, J.C., Jaramillo, C., and Pardo, C. (1997) *Synthesis*, 563.
337 Vasilevsky, S.F., Klyatskaya, S.V., Tretyakov, E.V., and Elguero, J. (2003) *Heterocycles*, **60**, 879.
338 Vasilevsky, S.F., Klyaskaya, S.V., and Elguero, J. (2004) *Tetrahedron*, **60**, 6685.
339 Carra, C., Bally, T., and Albini, A. (2005) *Journal of the American Chemical Society*, **127**, 5552.
340 Konstantinidou, F., Papageorgiou, M., Stephanidou-Stephanatou, J.,

and Tsoleridis, C.A. (2005) *Tetrahedron Letters*, **46**, 4843.

341 Wiley, R.H. and Wiley, P. (1964) Pyrazolones, pyrazolidones, and derivatives, in *The Chemistry of Heterocyclic Compounds. A Series of Monographs, Vol. 20*, Interscience, New York.

342 (a) Varvounis, G., Fiamegos, Y., and Pilidis, G. (2001) *Advances in Heterocyclic Chemistry*, **80**, 74; (b) Varvounis, G., Fiamegos, Y., and Pilidis, G. (2004) *Advances in Heterocyclic Chemistry*, **87**, 142; (c) in preparation.

343 Ballesteros, P., Elguero, J., Claramunt, R.M., Faure, R., Foces-Foces, C., Cano, F.H., and Rousseau, A. (1986) *Journal of the Chemical Society-Perkin Transactions 2*, 1677.

344 Claramunt, R.M. and Elguero, J. (1991) *Organic Preparations and Procedures International*, **23**, 273.

345 Elguero, J., Claramunt, R.M., and Summers, A.J.H. (1978) *Advances in Heterocyclic Chemistry*, **22**, 183.

346 (a) Greenhill, J.V. (1984) *Pyrazoles with Fused Six-membered Heterocyclic Rings* (ed. K.T. Potts), Comprehensive Heterocyclic Chemistry (series eds A.R. Katritzky and C.W. Rees), vol. 5, Pergamon Press, Oxford; (b) Towsend, L.B. and Wise, D.S. (1996) *Fused five- and six-membered rings without ring junction heteroatoms* (ed. C.S. Ramsden), Comprehensive Heterocyclic Chemistry II, (eds A.R. Katritzky, C.W. Rees, and E.F.V. Scriven), vol. 7 Pergamon Press, Oxford.

347 Bogza, S.L., Kobrakov, K.I., Malienko, A.A., Perepichka, I.F., Sujkov, S.Y., Bryce, M.R., Lyubchik, S.B., Batsanov, A.S., and Bogdan, N.M. (2005) *Organic and Biomolecular Chemistry*, **3**, 932.

348 Compton, D.R., Sheng, S.B., Carlson, K.E., Rebacz, N.A., Lee, I.Y., Katzenellenbogen, B.S., and Katzenellenbogen, J.A. (2004) *Journal of Medicinal Chemistry*, **47**, 5872.

349 Quiroga, J., Portilla, J., Insuasty, B., Abonia, R., Nogueras, M., Sortino, M., and Zacchino, S. (2005) *Journal of Heterocyclic Chemistry*, **42**, 61.

350 (a) Diaz, J.A. and Vega, S. (1994) *Journal of Heterocyclic Chemistry*, **31**, 93; (b) Insuasty, B., Rodríguez, R., Quiroga, J., Martínez, R., and Angeles, E. (1997) *Journal of Heterocyclic Chemistry*, **34**, 1131.

351 Klumpp, D.A., Kindelin, P.J., and Li, A. (2005) *Tetrahedron Letters*, **46**, 2931.

352 Vasilevsky, S.F., Tretyakov, E.V., and Elguero, J. (2002) *Advances in Heterocyclic Chemistry*, **82**, 1.

353 Boyer, J.H. (1986) "*Nitroazoles: The C-Nitro Derivatives of Five-Membered N- and N,O- Heterocycles*", *Organic Nitro Chemistry, Vol. 1*, VCH Publishers, Deerfield Beach, FL, USA.

354 Larina, L.I. and Lopyrev, V.A. (1998) *Topics in Heterocyclic Systems*, **1**, 187.

355 Larina, L.I., Lopyrev, V.A., Klyba, L.V., and Bochkarev, V.N. (1998) *Targets in Heterocyclic Systems*, **2**, 443.

356 Ivachtchenko, A.V., Kravchenko, D.V., Zheludeva, V.I., and Pershin, D.G. (2004) *Journal of Heterocyclic Chemistry*, **41**, 931.

357 Furin, G.G. (1998) *Targets in Heterocyclic Systems*, **2**, 355. (see pages 430–435).

358 Furin, G.G. (2004) *Advances in Heterocyclic Chemistry*, **86**, 129.

359 Furin, G.G. (2004) *Advances in Heterocyclic Chemistry*, **87**, 273.

360 Touzot, A., Soufyane, M., Berber, H., Toupet, L., and Mirand, C. (2004) *Journal of Fluorine Chemistry*, **125**, 1299.

361 Attanasi, O.A., Spinelli, D.C., (1996) *Topics in Heterocyclic Systems, Phosphorylated Pyrazoles* (eds A.A. Konovets, A.I. Kostyuk, and A.M. Pinchuk), vol. 1, 239.

362 Antiñolo, A., Carrillo-Hermosilla, F., Díez-Barra, E., Fernández-Baeza, J., Fernández-López, M., Lara-Sánchez, A., Moreno, A., Otero, A., Rodríguez, A.M., and Tejeda, J. (1998) *Journal of The Chemical Society-Dalton Transactions*, 3737.

363 Vega, S., Arranz, M.E., and Arán, V.J. (2005) *Journal of Heterocyclic Chemistry*, **42**, 755.

364 (a) Lopyrev, V.A., Larina, L.I., and Voronkov, M.G. (2001) *Zhurnal Organicheskoi Khimii*, **37**, 165; (b) (2001) *Russian Journal of Organic Chemistry*, **37**, 149.
365 Yang, X.Y. and Knochel, P. (2004) *Synlett*, 2303.
366 Forsyth, S.A., Pringle, J.M., and MacFarlane, D.R. (2004) *Australian Journal of Chemistry*, **57**, 113.
367 Chiappe, C. and Pieraccini, D. (2004) *Journal of Physical Organic Chemistry*, **18**, 275.
368 Schmidt, A. and Habeck, T. (2005) *Letters in Organic Chemistry*, **2**, 37.
369 Enjalbal, C., Sanchez, P., Martinez, J., Aubagnac, J.-L., Sanz, D., Claramunt, R.M., and Elguero, J. (2002) *International Journal of Mass Spectrometry*, **219**, 391.
370 Elguero, J., Foces-Foces, C., Sanz, D., and Claramunt, R.M. (2000) in *Advances in Nitrogen Heterocycles*, Vol. 4, (ed. C.J. Moody), JAI Press, Stamford, USA, p. 295.
371 Pérez-Torralba, M., Claramunt, R.M., Alkorta, I., and Elguero, J. (2007) *Journal of Molecular Structure, Arkivoc*, **xii**, 55.
372 La Monica, G. and Ardizzoia, G.A. (1997) *Progress in Inorganic Chemistry*, **46**, 151.
373 Jalón, F.A., Manzano, B.R., Gómez de la Torre, F., Guerrero, A.M., and Rodríguez, A.M. (1999) *Targets in Heterocyclic Systems*, **3**, 399.
374 Sadimenko, A.P. (2001) *Advances in Heterocyclic Chemistry*, **80**, 158.
375 Sadimenko, A.P. (2001) *Advances in Heterocyclic Chemistry*, **81**, 167.
376 (a) Beck, A., Weibert, B., and Burzlaff, N. (2001) *European Journal of Inorganic Chemistry*, 521; (b) Reger, D.L., Brown, K.J., Gardinier, J.R., and Smith, M.D. (2003) *Organometallics*, **22**, 4973.
377 Chang, W.K., Sheu, S.C., Lee, G.H., Wang, Y., Ho, T.I., and Lin, Y.C. (1993) *Journal of The Chemical Society-Dalton Transactions*, 687.
378 Díaz-Ortiz, A., Elguero, J., Foces-Foces, C., de la Hoz, A., Moreno, A., Moreno, S., Sánchez-Migallón, A., and Valiente, G. (2003) *Organic and Biomolecular Chemistry*, **1**, 4451.
379 Trofimenko, S. (2004) *Polyhedron*, **23**, 197.
380 Togni, A. (1996) *Chimia*, **50**, 86.
381 Herdtweck, E., Peters, F., Scherer, W., and Wagner, M. (1998) *Polyhedron*, **17**, 1149.
382 Ilkhechi, A.H., Bolte, M., Lerner, H.W., and Wagner, M. (2005) *Journal of Organometallic Chemistry*, **690**, 1971.
383 Rossler, K., Kluge, T., Schubert, A., Sun, Y., Herdtweck, E., and Thiel, W.R. (2004) *Zeitschrift für Naturforschung B*, **59**, 1253.
384 Fall, Y., Barreiro, C., Fernández, C., and Mouriño, A. (2002) *Tetrahedron Letters*, **43**, 1433.
385 Hueso-Rodríguez, J.A., Berrocal, J., Gutérrez, B., Farré, A.J., and Frigola, J. (1993) *Bioorganic & Medicinal Chemistry Letters*, **3**, 269.
386 Moreno-Mañas, M., Sebastián, R.M., Vallribera, A., Piniella, J.F., Alvarez-Larena, A., Jimeno, M.L., and Elguero, J. (2001) *New Journal of Chemistry*, **25**, 329.
387 Cudero, J., Pardo, C., Ramos, E., Gutiérrez-Puebla, E., Monge, A., and Elguero, J. (1997) *Tetrahedron*, **53**, 2233.
388 Jacquier, R. and Maury, G. (1967) *Bulletin de la Societe Chimique de France*, 295.
389 Nagai, S.-I., Oda, N., and Ito, I. (1979) *Yakugaku Zasshi*, **99**, 705.
390 Watson, A.A., House, D.A., and Steel, P.J. (1986) *Journal of Organometallic Chemistry*, **311**, 387.
391 Llamas-Saiz, A.L., Foces-Foces, C., Sobrados, I., Elguero, J., and Meutermans, W. (1993) *Acta Crystallographica Section C-Crystal Structure Communications*, **49C**, 724.
392 Watson, A.A., House, D.A., and Steel, P.J. (1995) *Australian Journal of Chemistry*, **48**, 1549.
393 Barz, M., Glas, H., and Thiel, W.R. (1998) *Synthesis*, 1269.
394 (a) LeCloux, D.D., Tokar, C.J., Osawa, M., Houser, R.P., Keyes, M.C., and Tolman, W.B. (1994) *Organometallics*, **13**, 2855; (b) Therrien, B., Konig, A., and Ward, T.R. (1999) *Organometallics*, **18**, 1565.
395 Keyes, M.C., Chamberlain, B.M., Caltagirone, S.A., Halfen, J.A., and

Tolman, W.B. (1998) *Organometallics*, **17**, 1984.

396 (a) Kashima, C., Fukuchi, I., and Hosomi, A. (1994) *The Journal of Organic Chemistry*, **59**, 7821; (b) Kashima, C., Fukuchi, I., Takahashi, K., and Hosomi, A. (1996) *Tetrahedron*, **52**, 10335; (c) Kashima, C., Takahashi, K., Fukusaka, K., and Hosomi, A. (1998) *Journal of Heterocyclic Chemistry*, **35**, 503; (d) Kashima, C., Fukusaka, K., Takahashi, K., and Yokoyama, Y. (1999) *The Journal of Organic Chemistry*, **64**, 1108.

397 LeCloux, D.D., Keyes, M.C., Osawa, M., Reynolds, V., and Tolman, W.B. (1994) *Inorganic Chemistry*, **33**, 6361.

398 Chibiryaev, A.M., Popov, S.A., and Tkachev, A.V. (1996) *Mendeleev Communications*, 18.

399 Faure, R., Frideling, A., Galy, J.-P., Alkorta, I., and Elguero, J. (2002) *Heterocycles*, **57**, 307.

400 Elguero, J., Claramunt, R.M., Shindo, Y., Mukai, M., Roussel, C., Chemlal, A., and Djafri, A. (1987) *Chem Scripta*, **27**, 283.

401 Toda, F., Tanaka, K., Infantes, L., Foces-Foces, C., Claramunt, R.M., and Elguero, J. (1995) *Chemical Communications*, 1453.

402 Wang, P., Onozawa-Komatsuzaki, N., Katoh, R., Himeda, Y., Sugihara, H., Arakawa, H., and Kasuga, K. (2001) *Chemistry Letters*, 940.

403 Barluenga, J., Fernández-Marí, F., Viado, A.L., Aguilar, E., Olano, B., García-Granda, S., and Moya-Rubiera, C. (1999) *Chemistry – A European Journal*, **5**, 883.

404 Barluenga, J., Fernández-Marí, F., González, R., Aguilar, E., Revelli, G.A., Viado, A.L., Fañanás, F.J., and Olano, B. (2000) *European Journal of Organic Chemistry*, 1773.

405 (a) Garanti, L., Molteni, G., and Pilati, T. (2002) *Tetrahedron Asymm*, **13**, 1285; (b) Molteni, G. (2004) *Tetrahedron Letters*, **15**, 1077.

406 Illa, O., Muray, E., Amsallem, D., Moglioni, A.G., Gornitzka, H., Branchadell, V., Bacereido, A., and Ortuño, R.M. (2002) *Tetrahedron Asymm*, **13**, 2593.

407 Shirakawa, S., Lombardi, P.J., and Leighton, J.L. (2005) *Journal of the American Chemical Society*, **127**, 9974.

408 Sibi, M.P., Shay, J.J., and Ji, J. (1997) *Tetrahedron Letters*, **38**, 5955.

409 Sibi, M.P., Zhang, R., and Manyem, S. (2003) *Journal of the American Chemical Society*, **125**, 9306.

410 Sibi, M.P., Ma, Z., and Jasperse, C.P. (2004) *Journal of the American Chemical Society*, **126**, 718.

411 Usachev, S.V., Nikiforov, G.A., Strelenko, Y.A., Chervin, I.I., Lyssenko, K.A., and Kostyanovsky, R.G. (2003) *Mendeleev Communications*, 136.

412 Elguero, J., Marzin, C., and Tizané, D. (1969) *Organic Magnetic Resonance*, **1**, 249.

413 Campayo, L., Bueno, J.M., Navarro, P., Ochoa, C., Jiménez-Barbero, J., Pèpe, G., and Samat, A. (1997) *The Journal of Organic Chemistry*, **62**, 2684.

414 Lamarque, L., Navarro, P., Miranda, C., Arán, V.J., Ochoa, C., Escartí, F., García-España, E., Latorre, J., Luis, S.V., and Miravet, J.F. (2001) *Journal of the American Chemical Society*, **123**, 10560.

415 Miranda, C., Escartí, F., Lamarque, L., Yunta, M.J.R., Navarro, P., García-España, E., and Jimeno, M.L. (2004) *Journal of the American Chemical Society*, **126**, 823.

416 Cafeo, G., Garozzo, D., Kohnke, F.H., Pappalardo, S., Parisi, M.F., Nascone, R.P., and Williams, D.J. (2004) *Tetrahedron*, **60**, 1895.

417 Otting, G., Messerle, B.A., and Soler, L.P. (1997) *Journal of the American Chemical Society*, **119**, 5425.

418 Seltzman, H.H., Carroll, F.I., Burgess, J.P., Wyrick, C.D., and Burch, D.F. (2002) *Journal of Labelled Compounds & Radiopharmaceuticals*, **45**, 59.

419 Meegalla, S.K., Doller, D., Silver, G.M., Wisnewski, N., Soll, R.M., and Dhanoa, D. (2003) *Bioorganic &*

Medicinal Chemistry Letters, **13**, 4035.

420 Kumar, J.S.D., Prabhakaran, J., Arango, V., Parsey, R.V., Underwood, M.D., Simpson, N.R., Kassir, S.A., Majo, V.J., Van Heertum, R.L., and Mann, J.J. (2004) *Bioorganic & Medicinal Chemistry Letters*, **14**, 2393.

421 Vijaykumar, D., Al-Qahtani, M.H., Welch, M.l.J., and Katzenellenbogen, J.A. (2003) *Nuclear Medicine and Biology*, **30**, 397.

422 El-Wetery1, S., El-Azoney1, Kh.M., El-Ghany1, E.A., El-Mohty1, A.A., and Deeb, A. (2001) *Journal of Radioanalytical and Nuclear Chemistry*, **250**, 335.

9
Five-Membered Heterocycles: 1,2-Azoles. Part 2. Isoxazoles and Isothiazoles

Artur M.S. Silva, Augusto C. Tomé, Teresa M.V.D. Pinho e Melo, and José Elguero

9.1
Introduction

Isoxazole **1** (1,2-oxazole) and isothiazole **2** (1,2-thiazole) are five-membered heterocyclic compounds having a pyridine-like N-atom bonded to an O- or an S-atom, respectively. These N–O and N–S σ-bonds are the weakest bonds in each molecule, being their energy much lower than that of N–C, O–C or S–C bonds, and are cleaved in all ring-opening reactions. Isothiazoles react more slowly with nucleophiles than isoxazoles and in both cases the reactions usually originate ring-opening [1].

4 3 5 N 2 O 1 **1**   4 3 5 N 2 S 1 **2**

The first reference to the isoxazole structure **1** was made by Claisen in 1888, for the reaction product of benzoylacetone with hydroxylamine [2]. He proposed the corrected structure (3-methyl-5-phenylisoxazole **3**) for a compound isolated several years before and suggested the name monoazole; however, Hantsch modified it to isoxazole, a name derived from the already know isomeric ring oxazole [3]. Claisen reported the fundamental outline of isoxazole chemistry in 1891 [4] and synthesized the parent compound of the series, isoxazole **1**, in 1903, by oximation of propargylaldehyde acetal [5]. After the fundamental work of Claisen and coworkers and some other contemporary authors [6], the next important contribution to the chemistry of isoxazoles was made by Quilico in 1946, when he began to study the formation of isoxazoles from *N*-oxides and acetylenic compounds [7]. The saturated derivatives had long been know (1892) but it was only in the 1960s that these compounds were studied extensively [8]. The extensive studies on isoxazoles since the 1980s are due to their versatility in the synthesis of various compounds, namely heterocycles and natural products, as well as their applications in several fields, such as agriculture, medicine and industry [1, 6, 8, 9]. In terms of the literature on isoxazole derivatives,

Modern Heterocyclic Chemistry, First Edition.
Edited by Julio Alvarez-Builla, Juan Jose Vaquero, and José Barluenga.

one can find an enormous number of references on their chemistry, physicochemical and biological properties, as well as some book chapters (in *Comprehensive Heterocyclic Chemistry*, 1984 [10] updated in 1996 [8], and *Science of Synthesis* [11–13]) and monographs (*Isoxazoles and Related Compounds*, published in 1962 by Quilico [14] and *Isoxazoles – Part 1 and Part 2*, published in 1991 [6] and 1999 [9] by Grünanger and Vita-Finzi) which collate all this information. Part 1 of the last monograph is restricted to mononuclear isoxazoles and their hydrogenated derivatives (dihydroisoxazoles and tetrahydroisoxazoles), except for the isoxazolones, while Part 2 is devoted to the chemistry of condensed isoxazoles, namely benzisoxazoles and related compounds.

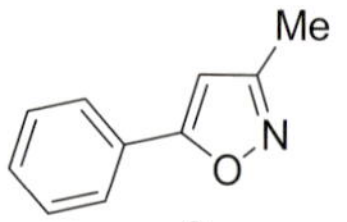

Me
N
O
3

Isothiazole (**2**) was first described in 1956 [15] while the benzisothiazoles have long been known. The most widely known derivative, saccharin (**4**), the first non-carbohydrate sweetening agent discovered in 1879, is 300–500 times as sweet as sucrose [16]. Saccharin (**4**) is manufactured commercially by the cyclization of *ortho*-substituted benzenesulfonamides with strong bases [17]. It still be used in many countries as a non-nutritive sweetener, although it was found that massive doses administered to rats caused bladder cancer, a fact which led to its ban in developing countries [18]. The controversy over its danger when used in small amounts is still unresolved [19]. Although all types of pharmacological activity have been claimed for isothiazoles, some are notable, such as that of thiomuscinol (**5**) on the central nervous system, as a potent agonist of γ-aminobutyrate (GABA) receptors [20], and the significant antifungal activity of isothiazolones (**6**) (marketed under the name Kathon) [21]. Several reviews on isothiazoles and on benzisothiazoles have been published [21–26]; the chapters in *Comprehensive Heterocyclic Chemistry* (1984 [16] and updated in 1996 [27]) and *Science of Synthesis* [28, 29] describe both types of compounds.

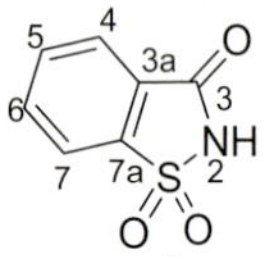

4

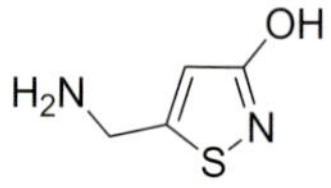

5

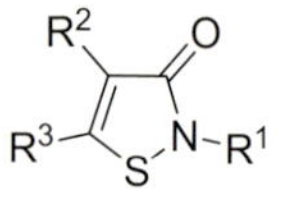

6

6a R^1 = Me, R^2 = R^3 = H
6b R^1 = Me, R^2 = H, R^3 = Cl
6c R^1 = Me, R^2 = R^3 = Cl
6d R^1 = *n*-octyl, R^2 = R^3 =H

9.1.1
Nomenclature

The structure and numbering system of the two mononuclear heterocycles (**1** and **2**) treated in this chapter and some of their best known derivatives is shown above.

However, before discussing their physicochemical properties, synthesis and transformations, it is important to present the structure and nomenclature of all known saturated and benzo-derivatives. For isoxazoles, one can find 4,5-dihydroisoxazoles (**7**) (also known as Δ^2-isoxazoline or 2-isoxazoline), 2,3-dihydroisoxazoles (**8**) (also known as Δ^4-isoxazoline or 4-isoxazoline), isoxazolidines (**9**), 1,2-benzisoxazoles (**10**), 2,1-benzisoxazoles (**11**) and 2,3-dihydro-1,2-benzisoxazoles (**12**). Structure **10** has also been described as indoxazene, 4,5-benzisoxazole, α,β-benzisoxazole and benzo[*d*] isoxazole (IUPAC nomenclature) and benzisoxazole. It is indexed in *Chemical Abstracts* as 1,2-benzisoxazole and numbered as shown below. The first member of the family, 3-phenyl-1,2-benzisoxazole (**13**), was synthesized at the end of the nineteen century (1892) [30] and the 1,2-benzisoxazole itself **10** was obtained in 1908 [31]. Structure **11** has also been described as anthranil, anthroxan, 3,4-benzisoxazole, β,γ-benzisoxazole, benzo[*c*]isoxazole (IUPAC nomenclature) and benzisoxazole. It is indexed in *Chemical Abstracts* as anthranil and numbered as shown below.

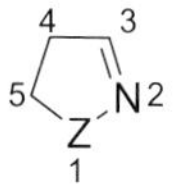

7 Z = O, **14** Z = S; **8** Z = O, **15** Z = S; **9** Z = O, **16** Z = S; **10** Z = O, **17** Z = S; **11** Z = O, **18** Z = S; **12** Z = O, **19** Z = S

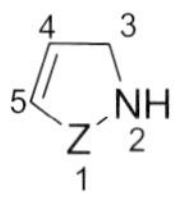

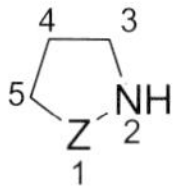

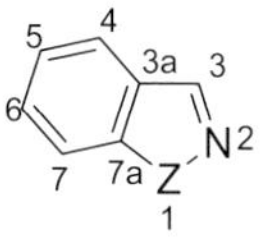

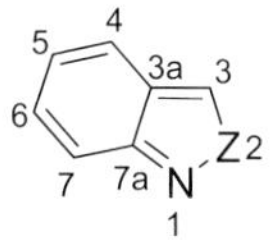

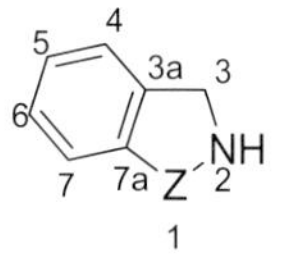

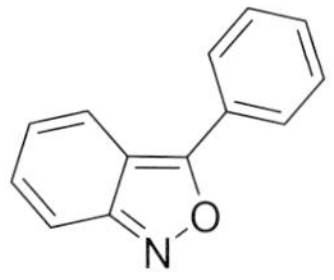

13

For isothiazoles one can find the same type of dihydroisothiazoles (**14** and **15**), tetrahydroisothiazoles (**16**), and benzisothiazoles (**17** and **18**) and their reduced form (**19**). The saturated isothiazole 1,1-dioxides **20** are known as sultams and 1,2-benzisothiazole 1,1-dioxides **21** are called saccharins.

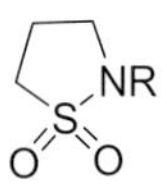

20 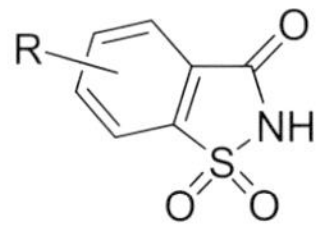 **21**

9.2 General Reactivity

9.2.1 Relevant Physicochemical Data, Computational Chemistry and NMR Data

The calculated π-electron density distributions of isoxazole (**1**) and isothiazole (**2**) are consistent with electrophilic substitution occurring at the 4-position (highest electron density) rather than at other ring positions [32–34]. The positional selectivity in the electrophilic attack on heterocyclic molecules can be explained according to the magnitude of the HOMO electron density of each atomic center [35], which predicts the reactivity order of electrophilic substitution of isoxazole (**1**) as C4 > C5 > C3, in agreement with the experimental data. Comparison of partial rate factors of isothiazole (**2**) with those of benzene and related compounds shows that its 4-position is $\sim 10^4$ times more reactive towards electrophiles than expected on the basis of the calculated π-electron density at carbon atoms and the electronegativity of the heteroatoms [36]. This indicates that when attempting to correlate the theoretical calculations with chemical reactivity one must be careful and consider whether ground or excited states are involved, or whether it is the electron density of intermediates, rather than the original molecule, that determines the product of a chemical transformation.

The π-electron density distribution also suggests that nucleophiles must attack the 3-position of both isoxazole (**1**) and isothiazole (**2**) since it presents the lowest electron density.

The positional reactivity of isothiazole **2** can also be evaluated by ^{1}H NMR. The relative rate of deuterium exchange of H5 and H3, under basic conditions, has been demonstrated to be 400 : 1, with no exchange of H4, whereas those of the hydrogens of the methyl groups of 5-, 4- and 3-methyl-isothiazoles were $100:1:10^{-4}$, respectively [37]. The high reactivity of the 5-position can be due not only to the electron distribution in the ground state but also to the stabilization of the formed anion by the σ-3d (sulfur) bonding. A similar study with isoxazoles provided the same conclusions [36, 38].

The very low π-order of the N–O bond of isoxazole (**1**) (Table 9.1) relative to those of the other ring bonds and the largest localized dipole due to the N–O bond suggests that this bond can be a site of attack of hydrolytic reagents, which is in agreement with the experimental reactivity.

Jug classified isoxazole (**1**) in the range of moderate aromatic compounds (1.548–1.332), based on the index of aromaticity, which corresponds to the value of the lowest bond order (Table 9.1, calculated by MO-SINDO1) [39]. These calcula-

Table 9.1 Calculated π-bond orders of isoxazole (**1**) and isothiazole (**2**).

X(O,S)-N	N-C	C3-C4	C4-C5	C-X(O,S)	Calculation method	Compound, Reference
0.285	0.795	0.546	0.817	0.342	HMO	**1**, [42]
0.412	0.772	0.580	0.776	0.458	MO-LCAO	**1**, [43]
0.404	0.735	0.617	0.738	0.526	PPP-SCF-CI	**1**, [44]
0.296	0.858	0.452	0.833	0.448	CNDO/2	**1**, [45]
1.361	1.957	1.501	1.955	1.498	MO-SINDO1	**1**, [39]
0.502	0.705	0.634	0.707	0.594	HMO	**2**, [21]
0.474	0.707	—	—	—	PPP	**2**, [46]
0.227	0.870	0.410	0.850	0.302	CNDO/2	**2**, [34]

tions indicate that isoxazole (**1**) (1.361) is less aromatic than oxazole (1.392) and imidazole (1.423) but more than pyrazole (1.297). These results have been confirmed by using other quantitative measurements of aromaticity, such as empirical resonance energy values for heteroaromatic systems as well as conjugation energies [40]. The aromaticity of isothiazole (**2**) is greater than that of isoxazole (**1**), just as the aromaticity of thiophene is greater than that of furan.

The tendency of $^1J_{CC}$ coupling constants in ^{13}C NMR to converge towards that of benzene (56 Hz) is another possible criterion of degree of aromaticity, increasing with the convergence [41]. Although an interesting model, these coupling constants also depend on the geometry of the molecule. The fact that isothiazole (**2**) is planar and their $^1J_{CC}$ ($^1J_{C3-C4}$ 52.5 Hz and $^1J_{C4-C5}$ 62.2 Hz) are less divergent than those of the less aromatic isoxazole (**1**) ($^1J_{C3-C4}$ 48.7 Hz and $^1J_{C4-C5}$) 67.7 Hz) seems to support this method.

The proton resonances of isoxazole (**1**) and isothiazole (**2**) are strictly related to electron density distribution on the ring (referred above). The ^{1}H NMR spectrum of isoxazole (**1**), neat or dissolved in various solvents, has been reported in many papers (Table 9.2) [34, 47–49]. The signals of H4 (δ, 6.28–6.41 ppm) appear at lower frequency values than those of H3 (δ, 8.15–8.40 ppm) and H5 (δ, 8.39–8.61 ppm), but all of them are in the aromatic region. For isothiazole (**2**) the same kind of chemical shifts appears, H5 is at a higher frequency than H3, but in some cases it can be reversed. The resonance is deshielded (0.5 ppm) when spectra are acquired in DMSO-d_6 solution, owing to the interaction of this hydrogen with proton acceptors.

Table 9.2 ^{1}H NMR chemical shifts (δ, ppm) of some isoxazole and isothiazole derivatives.

Compound	H3	H4	H5	$^3J_{H3-H4}$ (Hz)	$^3J_{H4-H5}$ (Hz)	$^3J_{H3-H5}$ (Hz)	Solvent
1 [47]	8.34	6.41	8.51	1.5	1.5	—	$CDCl_3$
1 [48]	8.40	6.40	8.61	1.7	1.7	0.5	Neat
1 [49]	8.19	6.32	8.44	1.6	1.6	—	CCl_4
1 [34]	8.15	6.28	8.39	1.78	1.69	0.27	CS_2
22 [49]	(2.28)	6.02	8.13	—	1.6	—	CCl_4
23 [49]	7.90	5.85	(2.42)	1.6	—	—	CCl_4
24 [49]	(2.24)	5.85	(2.41)	—	—	—	CCl_4
25 [49]	—	6.58	8.39	—	1.6	—	CCl_4
26 [49]	8.43	—	8.58	—	—	—	CCl_4
27 [49]	8.15	6.39	—	2.0	—	—	CCl_4
28 [52]	—	—	9.08	—	—	—	$CHCl_3$
29 [49]	—	6.74	—	—	—	—	CCl_4
2 [53]	8.54	7.26	8.72	1.66	4.66	0.15	CCl_4
30 [53]	(2.46)	7.00	8.54	—	4.55	—	CCl_4
31 [53]	8.24	(2.32)	8.21	—	—	0.33	CCl_4
32 [53]	8.24	6.92	(2.56)	1.63	—	—	CCl_4

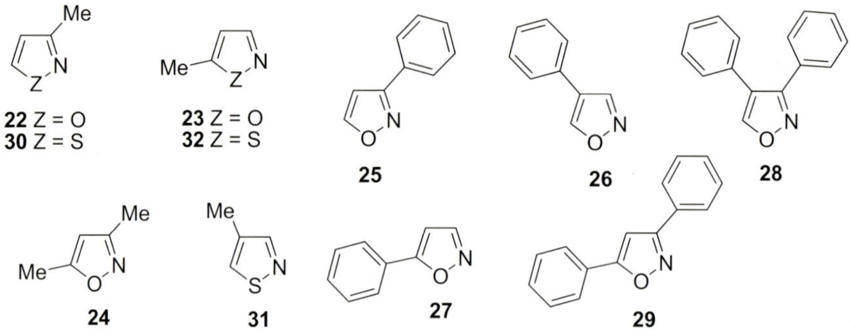

22 Z = O
30 Z = S

23 Z = O
32 Z = S

25 **26** **28** **24** **31** **27** **29**

The interaction between H3 and the ring nitrogen atom, due to the quadrupole relaxation of ^{14}N, is responsible for the broadening of H3 signals of isoxazole and isothiazole derivatives. This broadening is sometimes the diagnostic for the differentiation of H3 and H5 resonances and it is generally reduced in solvents with high viscosity, lower temperatures or nitrogen protonation [$^2J(^{14}$N3-H3) from ~10 to 3 Hz] [47, 50].

As one can see from Table 9.2, the presence of methyl substituents (electron-donating groups) causes a shielding in the resonances of the isoxazole **22–24** and isothiazole **30–32** protons. The ^{1}H NMR spectra of methylisoxazoles can be used as a tool to calculate the ratio of 3- and 5-methylisoxazoles in some reaction mixtures and to determine the isomeric purity of products, since the chemical shift of the corresponding methyl groups are very different.

The ^{1}H NMR chemical shifts of the isoxazole ring of 3-, 4- and 5-phenylisoxazoles **25–27** seem to indicate a different conformation for these compounds, presenting different angles between the planes of the two ring, the most important being that of 5-phenylisoxazole (**27**). This conclusion is based on the deshielding effect of the phenyl ring on the protons lying in the same plane, which decreases with increasing torsional angle.

The resonance of H4 can also be used to distinguish between isomeric 3,5-disubstituted isoxazoles [51]. In the case of unsymmetrical 3(5)-substituted-5(3)-phenyl-isoxazole derivatives bearing one substituent more electron-donating than the phenyl ring, the H4 resonance is shielded ($\Delta\delta$ 0.03–0.80 ppm) for the isomers where the electron-donating substituent group is at the 5-position. Opposite results were obtained for those compounds containing stronger electron-withdrawing groups. In these cases the H4 resonance is deshielded ($\Delta\delta$ 0.11–0.31 ppm) for compounds bearing an electron-withdrawing substituent at the 5-position of the isoxazole ring compared to those of the 3-isomer.

The ^{1}H NMR spectra of 1,2-benzisoxazole and 1,2-benzisothiazole derivatives have the characteristics of both condensed benzene and isoxazole/isothiazole rings. A typical resonance of these compounds is that of H3, which appears around δ 8.7–8.8 ppm and presents long-range coupling with H-7 (^5J 0.9–1.2 Hz) due to the well-known zigzag route [22, 54]. The vicinal coupling constants of the phenyl ring protons are consistent with some degree of *ortho*-quinonoid structure and

Table 9.3 ^{13}C NMR chemical shifts (δ, ppm) of some isoxazole and isothiazole derivatives.

Comp.	C3	C4	C5	Comp.	C3	C4	C5
1 [56]	150.0	140.5	158.9	**2** [58]	157.0	123.4	147.8
22 [57]	159.2	105.7	159.2	**30** [58]	166.7	123.9	148.1
23 [57]	150.9	101.4	169.2	**32** [58]	157.6	123.3	163.0

Table 9.4 Physical properties of isoxazole, isothiazole and their fused benzo-derivatives.

Compound	1	10	11	2	17	18
Mp (°C)	−80	—	—	—	37	
Bp (°C/mmHg)	95/769	84/11	994.5/11	113/760	—	242/760
pK_a	−2.03	—	—	−0.51	—	−0.05
Dipole moment (μ, D)	2.90	—	—	—	2.44	—

correlate with the bond orders of 2,1-benzisoxazoles and 2,1-benzisothiazoles ($^3J_{H4\text{-}H5} \sim 9$ Hz and $^3J_{H5\text{-}H6} \sim 7.5$ Hz) [55].

Table 9.3 presents the ^{13}C resonances of isoxazole (**1**), isothiazole (**2**) and some of their methyl derivatives as examples of the ^{13}C NMR spectra of such compounds. The presence of a methyl group as ring substituent implies a deshielding into the signal of the carbon to which the methyl is bonded.

A comprehensive collection of ^1H and ^{13}C NMR chemical shifts of several isoxazole derivatives is reported in the first monograph of isoxazoles [6].

Unsubstituted isoxazole (**1**) (Table 9.4) and its alkylisoxazole derivatives are usually liquids; the introduction on the ring of more than one substituent with a long chain leads to solid compounds. Phenylisoxazoles are usually solids [6].

Isothiazole (**2**) and their alkylisothiazole and benzisothiazole derivatives are usually liquids or solids with low melting points (Table 9.4). The presence of polar substituents increases the melting points. Isothiazole (**2**) has a low solubility in water (~3.5%) and is miscible with most organic solvents. Benzisothiazoles are insoluble in water, but are soluble in strong acids (salt formation) and in organic solvents [6].

Table 9.4 shows the physical properties of unsubstituted compounds, isoxazole (**1**) and isothiazole (**2**), and their fused benzo-derivatives, 1,2- and 2,1-benzisoxazole (**10**) and (**11**) and 1,2- and 2,1-benzisothiazole (**17**) and (**18**), discussed in this chapter.

9.2.2 Tautomerism

Annular tautomerism does not occur in isoxazoles, benzisoxazoles, isothiazoles and benzisothiazoles, but they present some substituent tautomers. Isoxazolin-3(5)-ones or isothiazolin-3(5)-ones can exist in equilibrium with the corresponding hydroxy-derivatives. Isoxazolin-3-ones or isothiazolin-3-ones have been more studied than the corresponding 5-substituted derivatives. The spectroscopic data indicate that they exist as 3-hydroxy tautomers in solid state or in non-polar solvents, such as cyclohexane or ether, but more polar solvents resulted in a great contribution of the keto tautomers [6, 21, 59, 60]. In the solid state these compounds form hydrogen-bonded dimers **33**. One of the best known 3-hydroxyisoxazoles, due its important neuropharmacological activity, is the natural compound muscinol (**34**), isolated from an *Amanita* species [61, 62].

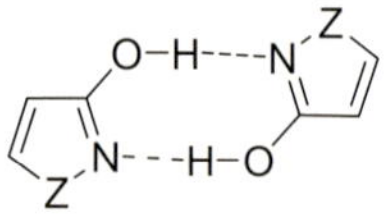

33 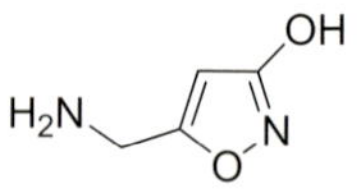 **34**

The infrared (IR) spectra have been useful in establishing the position of the tautomeric equilibrium in 1,2-benzisoxazolin-3-one. The enol form of **35** is preferred in the solid state, as shown by the strong band at 3000–2500 cm^{-1} and the lack of carbonyl band and the C=N band at 1620 cm^{-1}. Both tautomeric forms are present in chloroform solution, since the carbonyl (1670 cm^{-1}) and hydroxyl band (as above) are present [63]. However, X-ray and other spectroscopic data show that 1,2-benzisothiazolin-3-one and its derivatives, 2,1-benzisothiazolin-3-one and saccharin, all exist in the keto form [64].

Studies on tautomerism have shown that in general isoxazolin-4-ones exist preferentially as 4-hydroxy-isoxazoles [65]. The structure of the bioactive natural compound triumferol (**36**) has also been established as a 4-hydroxy-derivative by ^{1}H NMR (δ_{H3} 8.25 ppm, δ_{H5} 8.33 ppm, δ_{OH} 8.18 ppm, acetone-d_6) and *O*-acyl and *O*-methyl derivatives [66].

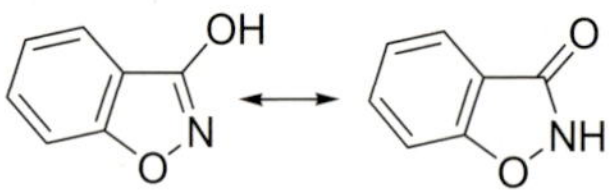

35 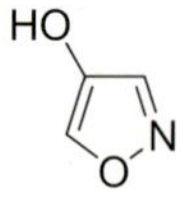 **36**

9.3
Relevant Natural and/or Useful Compounds

Although isoxazole and isothiazole moieties are rarely found in nature, they present important biological applications. Muscinol (**34**), a potent CNS depressant and agonist of the neurotransmitter 4-aminobutyric acid [67], has been isolated from *Amanita muscaria* [61, 62]. The naturally occurring amino acid ibotenic acid (**37**), a widely used neurotoxin and pharmacological tool for studies of glutamic acid receptors [68], has also been isolated from *Amanita muscaria* and from *Amanita pantherina* [62]. Brassilexin (**38a**) and sinalexin (**38b**) are phytoalexins, with fungicidal activity, isolated from the leaves of *Brassica juncea* (Cruciferae) [69, 70]. Aulosirazole (**39**) is the major cytotoxin in the blue-green alga (cyanobacterium) *Aulosira fertilissima* Ghose. It shows selective cytotoxicity against solid tumors [71].

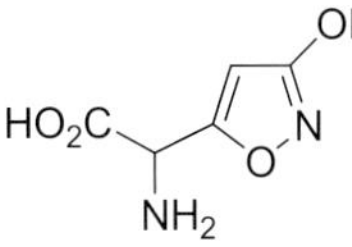

37

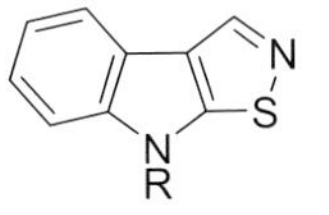

38a, R = H
38b, R = OMe

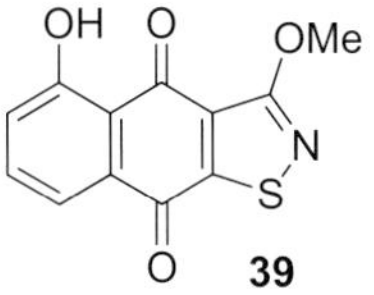

39

Isoxazoles are a large group of heterocyclic compounds that display interesting medicinal, agricultural and some other industrial utilities. Some of the most important are the pharmacologically active isoxazoles, including antibacterial sulfonamides **40–42**, semi-synthetic penicillins **43–46**, semi-synthetic cephalosporin **47**, anabolic steroid **48** and the monoamine oxidase inhibitor **49** (used in psychotherapy) [72].

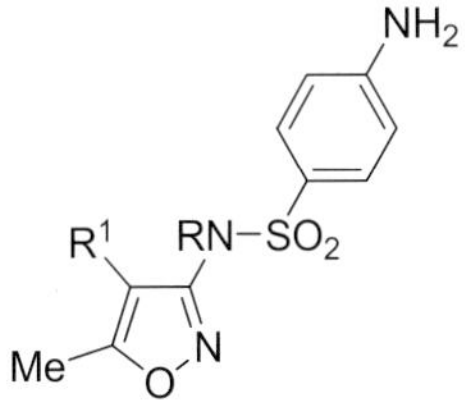

40 R = R^1 = H, Sulfamethoxazole
41 R = H, R^1 = Me, Sulfisoxazole
42 R = Ac, R^1 = Me, Acetylsulfisoxazole

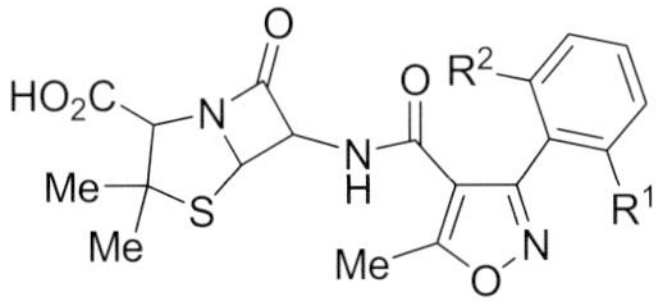

43 R^1 = R^2 = H, Oxacillin
44 R^1 = H, R^2 = Cl, Chloxacillin
45 R^1 = R^2 = Cl, Dichloxacillin
46 R^1 = R^2 = Cl, Floxacillin

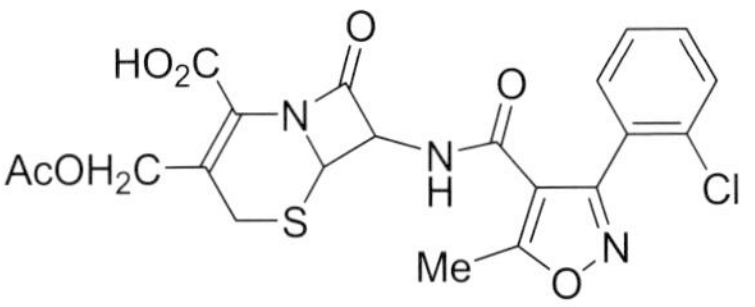

47 Cefoxazole

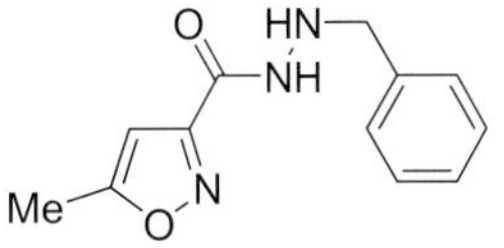

49 Isocarboxazid

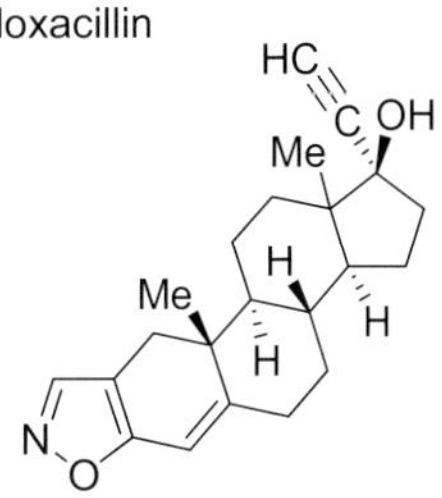

48 Danazol

The isosteric relationship of 1,2-benzisoxazole with that of the indole nucleus has led to its use as a carrier of pharmacophoric moieties in the search for potential drugs. From the many compounds studied, only a few have emerged as candidates for clinical use. 1,2-Benzisoxazole-3-sulfonamide **50** (zonisamide) is a potent antiepileptic drug [73], 6-fluoro-1,2-benzisoxazole **51** (risperidone) is a potent antipsychotic agent with thymosthenic properties [74], its analogue **52** (HRP 913) is a potent dopamine antagonist with antipsychotic properties [75], and 1,2-benzisoxazole-3-acetamidoxime **53** (PF-257) is a psychotropic agent with seemingly new properties [76]. Phosphonate **54** (Bay 52957) is a potent insecticide [77]

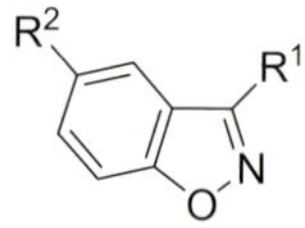

50 $R^1 = CH_2SO_2NH_2$, $R^2 = H$
53 $R^1 = CH_2C(NH_2)NOH$, $R^2 = H$
54 $R^1 = OP(S)(OC_2H_5)_2$, $R^2 = Cl$

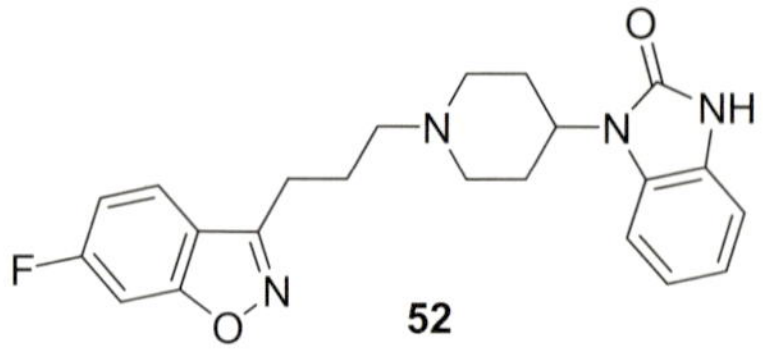

As referred to in the introduction, the most widely known isoxazole derivative is the benzisothiazole saccharin (**4**), the first non-carbohydrate sweetening agent discovered [16, 27]. Although many isothiazole compounds exhibited biological activities (all types of pharmacological activity have been claimed!), the most important are (i) saccharin derivative **55**, which presents potent sedative, hypnotic and anticonvulsant activity [78]; (ii) the adrenergic β-blockers **56** [79]; (iii) the cyclizine analogue **57**, which presents appetite suppressant activity [80]; (iv) the amide **58**, which has potent anti-inflammatory activity [81]; (v) thiomuscinol (**5**), which is active on the CNS as a potent agonist of GABA receptors (Section 9.1) [20]; (vi) the acid **59**, the most interesting compound in the agrochemical sphere, having high herbicidal activity [17, 78]; and (vi) the allyloxy-1,2-benzisothiazole 1,1-dioxide **60**, known as Probenazole or Oryzaemate, which is useful in rice crop protection [82].

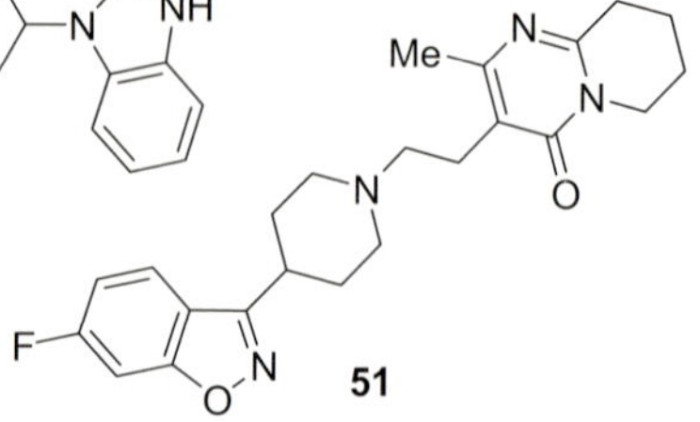

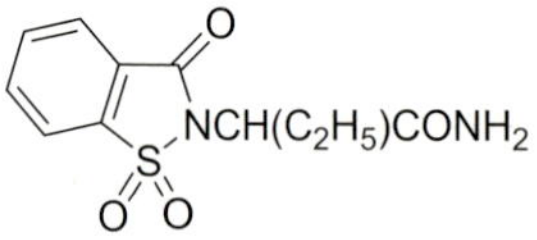

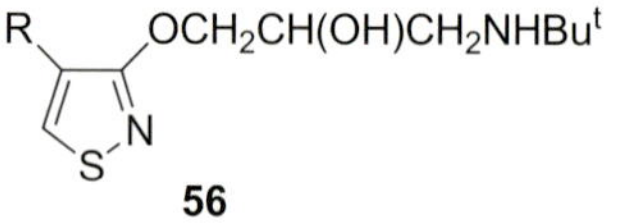

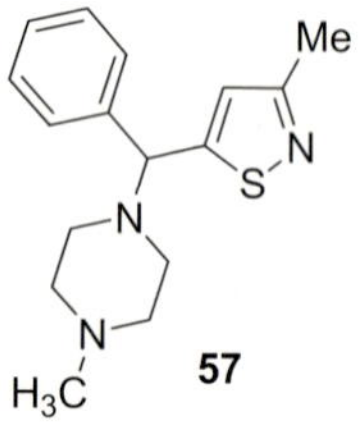

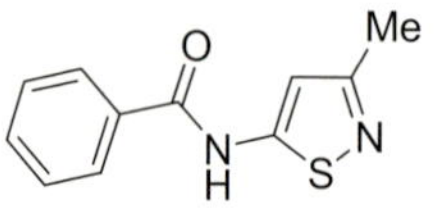

Numerous biological, pharmacological and biocidal activities are fully described in reviews and monographs published about the compounds treated in this chapter [6, 8–10, 16, 26, 27, 83].

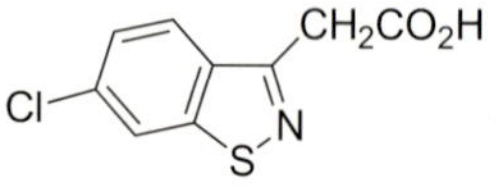

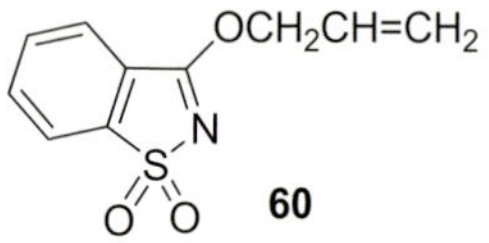

9.4 Synthesis of Isoxazoles and Isothiazoles

9.4.1 Isoxazoles

Synthetic methodologies for the construction of the isoxazole ring can be classified based on the number of atoms of the component synthons, which are subdivided according to the type and arrangement of the atoms in each component. The [3 + 2] approach includes the two major routes to isoxazoles: CCC + NO reactions (reaction of hydroxylamine with a three-carbon atom component) and CNO + CC reactions (1,3-dipolar cycloaddition of nitrile oxides). Isoxazoles can also be obtained via [4 + 1], [5 + 0] and [3 + 1 + 1] routes. Ring transformation reactions also lead to isoxazoles.

9.4.1.1 **[3 + 2] Routes**

9.4.1.1.1 [CCC + NO] Reactions: Reactions of Hydroxylamine with a Three-Carbon Atom Component In 1888 Claisen described the first general synthesis of isoxazoles [2]. The process involved the reaction of β-diketones with hydroxylamine followed by cyclization–dehydration of the intermediate oxime (Scheme 9.1). This became an important route to 4-unsubstituted or 4-substituted isoxazoles bearing the same substituent at 3- and 5-positions. 4-Monosubstituted isoxazoles **62** and unsubstituted isoxazole **61** have also been prepared from the reaction of tetraalkoxypropanes (or β-dialdehydes) with hydroxylamine (Scheme 9.1) [6, 10, 84, 85]. This route was applied to the synthesis of 3,5-disubstituted isoxazole-4-carbaldehydes using also diketones as the three-carbon building-block in the reaction with hydroxylamine [86].

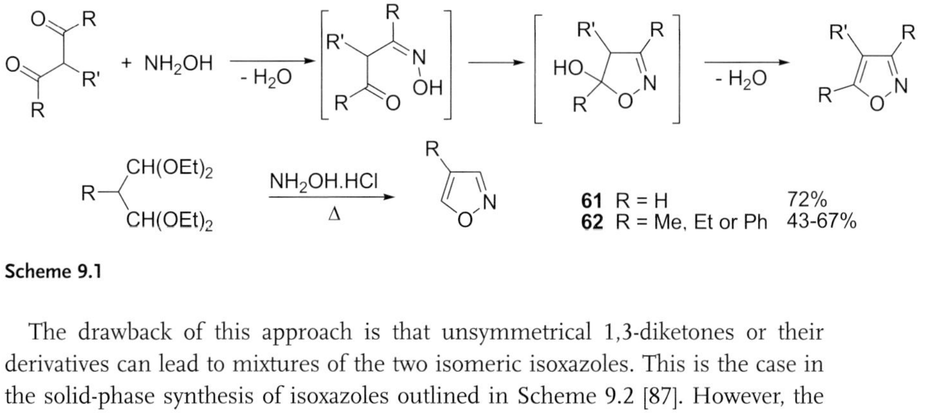

+ NH_2OH ; $- H_2O$; $- H_2O$

$NH_2OH.HCl$, Δ

61 R = H 72%
62 R = Me, Et or Ph 43-67%

Scheme 9.1

The drawback of this approach is that unsymmetrical 1,3-diketones or their derivatives can lead to mixtures of the two isomeric isoxazoles. This is the case in the solid-phase synthesis of isoxazoles outlined in Scheme 9.2 [87]. However, the

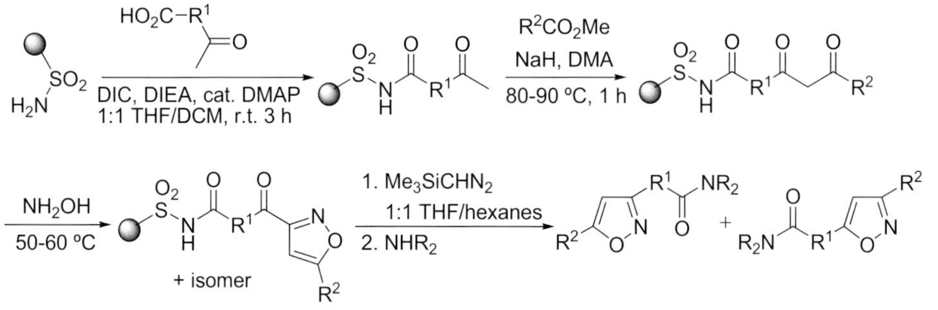

HO_2C-R^1 ; DIC, DIEA, cat. DMAP, 1:1 THF/DCM, r.t. 3 h ; R^2CO_2Me, NaH, DMA, 80-90 °C, 1 h ; NH_2OH, 50-60 °C ; + isomer ; 1. Me_3SiCHN_2 1:1 THF/hexanes 2. NHR_2

Scheme 9.2

selection of a CCC synthon with dissimilar terminal carbon atoms in terms of electronic and/or steric effects, the protection of one terminal carbon or the control of the pH of the medium can lead to selectivity. Nevertheless, many variants of this reaction have been developed and in fact isoxazoles have been prepared from the reaction of hydroxylamine with several three-carbon atom components, namely β-keto aldehydes, β-keto esters, α-acetylenic ketones or aldehydes, α,β-unsaturated ketones, β-imino nitriles and β-keto nitriles [6, 10, 84, 85].

3-Hydroxy-isoxazole is a synthetic unit present in several biologically active compounds. They have been prepared mainly by cyclization of β-keto esters with hydroxylamine. However, this method usually leads to the formation of isoxazol-5-ones as a by-product. By using Meldrum's acids the problem of the lack of regioselectivity in the addition of hydroxylamine can be overcome (Scheme 9.3) [88].

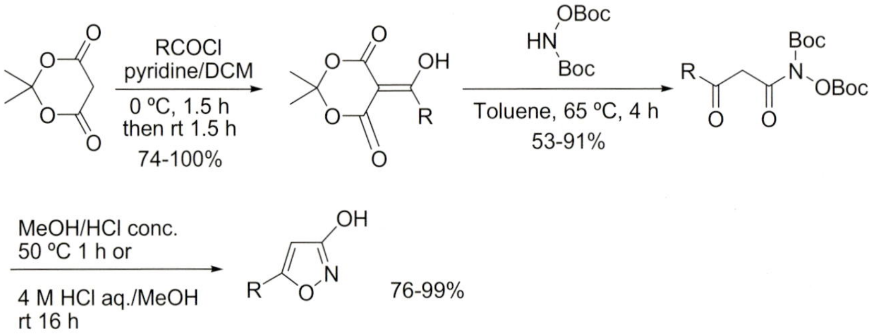

RCOCl
pyridine/DCM
0 °C, 1.5 h
then rt 1.5 h
74-100%

OBoc
HN
Boc
Toluene, 65 °C, 4 h
53-91%

MeOH/HCl conc.
50 °C 1 h or
4 M HCl aq./MeOH
rt 16 h
76-99%

Scheme 9.3

The regioselective synthesis of 5-substituted 3-alkoxyisoxazoles can be achieved using β-oxo thionoesters (Table 9.5, entry 1) [89]. 3-Aryl-5-alkoxyisoxazoles have been obtained in moderate yields from cyclocondensation of acylketene O,S-acetals with hydroxylamine in the presence of sodium alkoxide/alcohol (Table 9.5, entry 2). 5-Alkoxy-3,4-annulated isoxazoles can also be obtained using the same synthetic approach (Table 9.5, entry 3) [90].

The α-keto methylene group in 3,5-diarylcyclohexen-2-ones has been used to obtain fused isoxazoles via Claisen-like condensation with ethyl formate followed by cyclocondensation with hydroxylamine hydrochloride (Table 9.5, entry 4). The same type of approach can be applied to prepare other types of fused isoxazoles [91], including those derived from triterpenoids, namely methyl oleanonate and lanost-8-en-3-one [92].

The reaction of vinyl ketones bearing a potential leaving group at the β-position (such as halogen, alkoxy or dialkylamino) with hydroxylamine has been extensively explored (Scheme 9.4) [6, 10, 84, 85, 93, 94].

The amine exchange reaction of an enamine ketone has also been used for the regioselective synthesis of 4,5-diarylisoxazoles [95–98]. The one-pot reaction of enamino ketones **63** with hydroxylamine hydrochloride leads to the corresponding

Table 9.5 Reaction of hydroxylamine with three-carbon atom components.

Entry	Conditions	Yield	Notes
1 [10]	$NH_2OH.HCl$, rt, 2 h, NEt_3/CH_3CN; then CH_3OH/H_2O, reflux, 3 h, pH 3-5	68-82%; 81-94%	
2 [90]	$NH_2OH.HCl$, $NaOR^1/R^1OH$, reflux, 8-10 h	45-53%	R^1 = Me, Et or *n*-Pr
3 [90]	$NH_2OH.HCl$, $NaOR^1/R^1OH$, reflux, 8-10 h	45-53%	X = CH_2, n = 1, R^1 = Et; X = S, n = 2, R^1 = Me
4 [91]	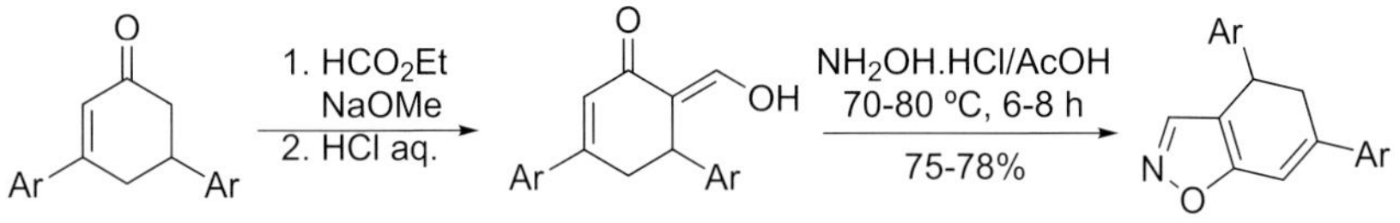 1. HCO_2Et, NaOMe; 2. HCl aq.; then $NH_2OH.HCl/AcOH$, 70-80 °C, 6-8 h	75-78%	
5 [104]	$NH_2OH.HCl/NEt_3$, EtOH, reflux, 5 h	73%	
6 [108]	$NH_2OH.HCl/H_2SO_4$ conc., R^3OH, reflux, 8-12 h	65-95%	R^3 = H, Me, Et or Pr^i
7 [109]	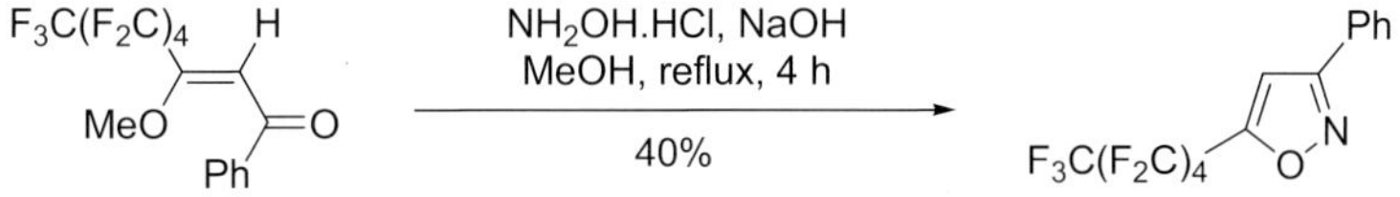 $NH_2OH.HCl$, NaOH, MeOH, reflux, 4 h	40%	
8 [110]	$NH_2OH.HCl$, Na_2CO_3 (→ 3-Ph-5-Ar isoxazole); $NH_2OH.HCl$, pyridine/EtOH, reflux, 3 h (→ 3-Ar-5-Ph isoxazole)		Ar = C_6H_4-OMe-*p*

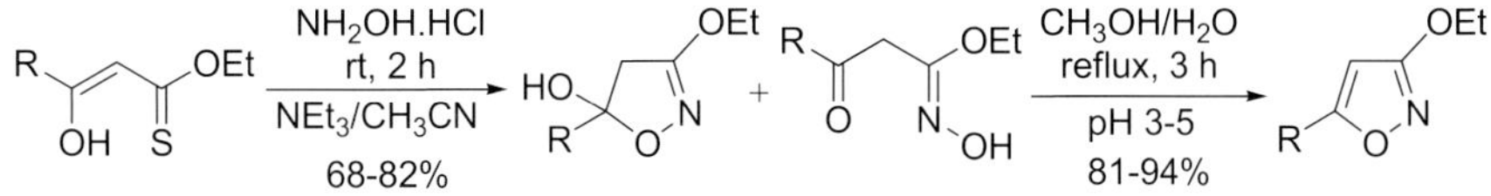

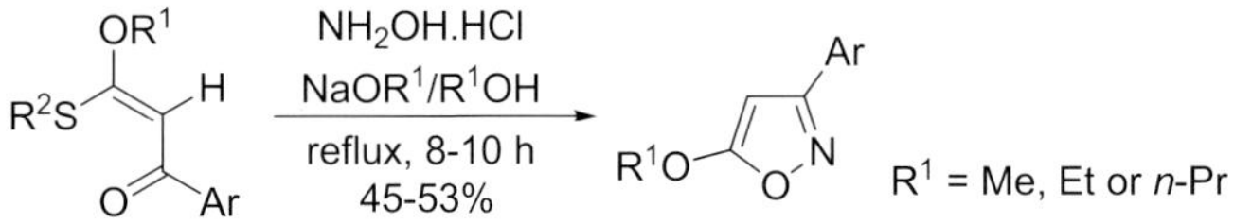

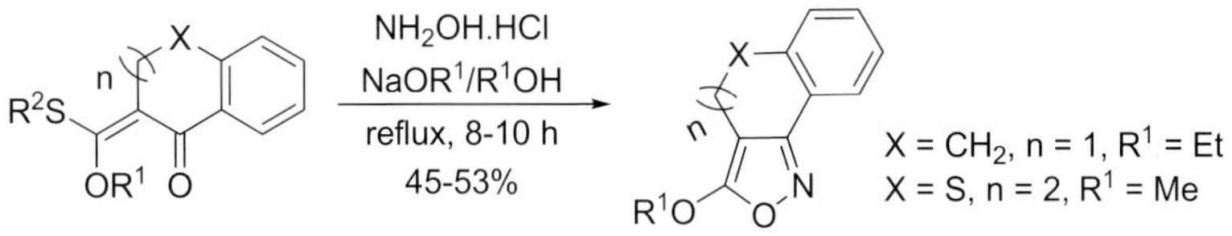

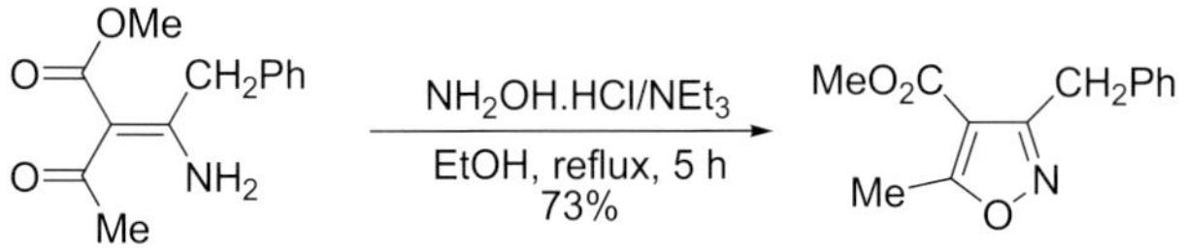

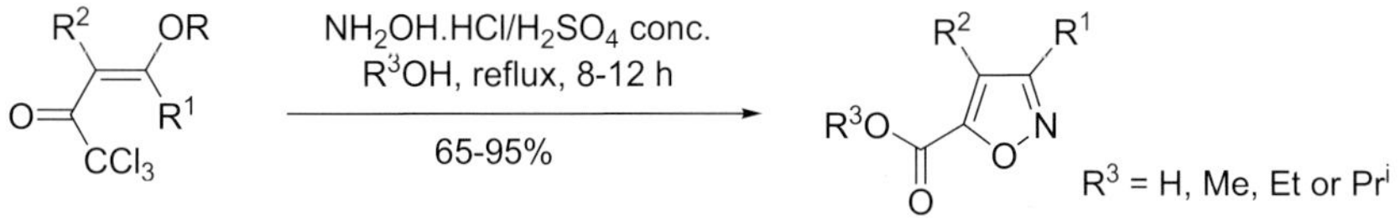

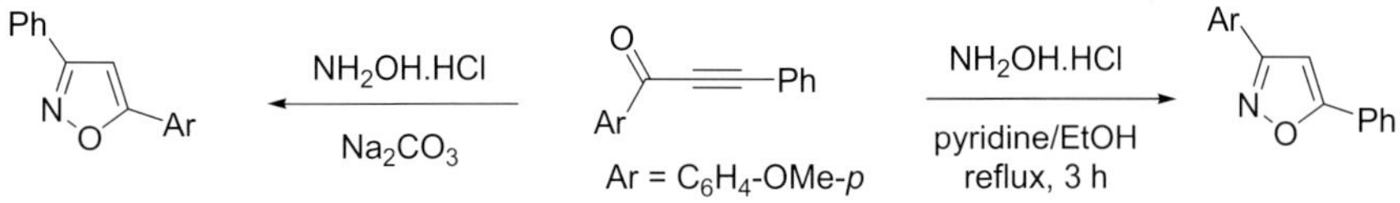

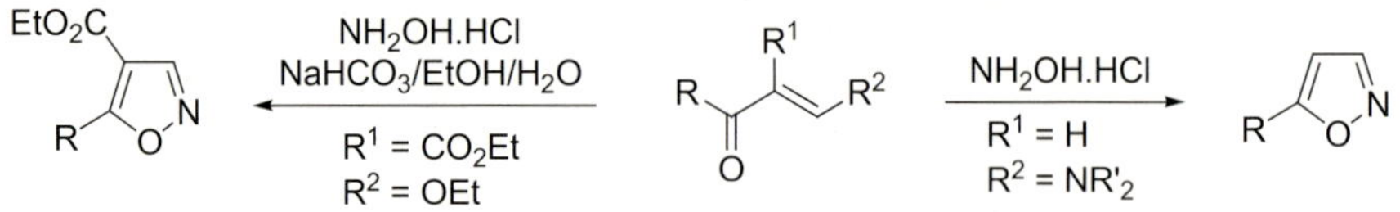

Scheme 9.4

isoxazoles **64** (Scheme 9.5) [95]. This strategy has been applied to the synthesis of the cholinergic channel activator ABT-418 [99].

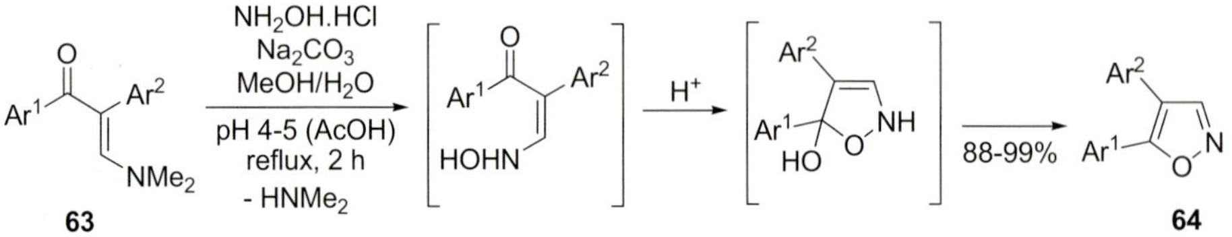

Scheme 9.5

The reaction of 2-acyl-3-(dimethylamino)propenoates with hydroxylamine in refluxing methanol leads to 5-substituted isoxazole-4-carboxylates in high yields (68–90%) [100] and open-chain and cyclohexane *syn*-2-(dimethylamino)ethylene-1,3-diones are converted into 5-substituted 4-acylisoxazoles [101]. A cellulose-based resin has also been used to prepare 5-substituted isoxazole-4-carboxylates via *in situ* generation of a polymer-bound enaminone [101–103]. The same strategy can be used to prepare tri-substituted isoxazoles. In fact, a β-enamino ketoester reacts with hydroxylamine hydrochloride in the presence of triethylamine to give a tri-substituted isoxazole (e.g., Table 9.5, entry 5) [104–107].

The reaction of vinyl ketones bearing a potential leaving group (bromo or benzotriazole) at the α position with hydroxylamine has also been reported (Scheme 9.6) [111, 112]. Depending on the leaving group or the experimental conditions 3(5)-substituted-5(3)-phenylisoxazoles are regioselectively obtained.

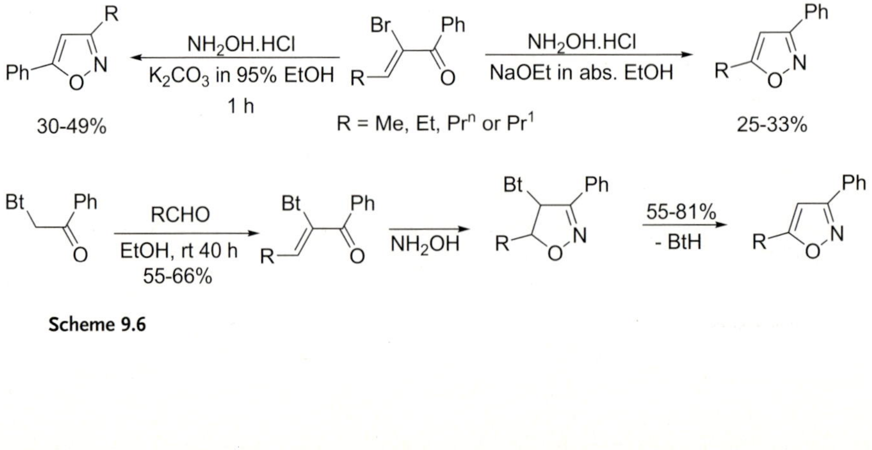

Scheme 9.6

Reaction conditions were found to allow the exclusive formation of isoxazole-5-carboxylic acid derivatives by conjugate addition, in acidic medium, of hydroxylamine to a β-alkoxyvinyl trichloromethyl ketone (Table 9.5, entry 6) [108]. The trichloromethyl group is the carboxyl group precursor: using water as solvent it leads to the formation of carboxylic acids whereas the use of an alcohol leads to ester derivatives. In contrast, the β-perfluoroalkyl-β-alkoxyvinyl phenyl ketone undergoes a selective attack on the carbonyl group upon reacting with hydroxylamine in basic medium, affording an isoxazole bearing a perfluoroalkyl substituent (Table 9.5, entry 7) [109].

Isoxazoles can be obtained via oxidative cyclization of α,β-unsaturated oximes with iodine/potassium iodide [113], with N-bromosuccinimide [114] or with palladium complexes in the presence of sodium phenoxide [115]. 3,5-Diarylisoxazoles can also be obtained using lead(IV) acetate as oxidant, although in moderate yields (24–28%) [116]. The method of preparation of 3,5-disubstituted isoxazoles by oxidative closure of α,β-unsaturated oximes can be carried out using tetrakis(pyridine)cobalt(II) dichromate (TPCD) as oxidant, under mild reaction conditions and very short reaction time (Scheme 9.7) [117]. A route to ABT-418 involving the same type of strategy for the isoxazole ring has been described [118].

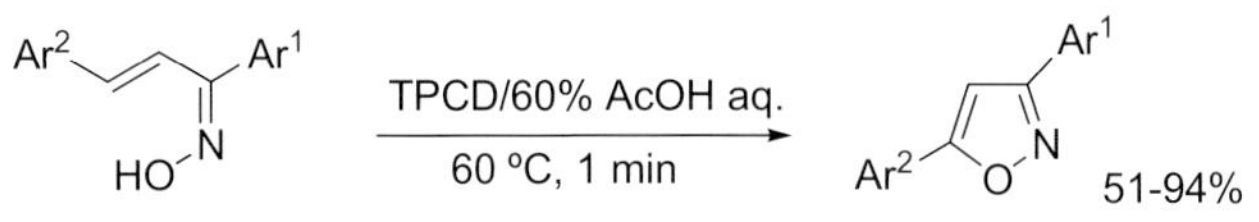

Ar^2, Ar^1, HO, N — TPCD/60% AcOH aq., 60 °C, 1 min → Ar^1, Ar^2, O, N, 51-94%

Scheme 9.7

The synthesis of isoxazoles attached to sugar moieties via oxidative cyclization of α,β-unsaturated oximes has been reported [119]. The isoxazoles were obtained by reacting the oximes with potassium iodide and sodium hydrogen carbonate at 100 °C (64–68% yield).

The reaction of α,β-alkynic ketones with hydroxylamine hydrochloride gives 3- or 5-substituted isoxazole isomers, depending on the conditions used (Table 9.5, entry 8) [110]. This route to isoxazoles has been applied to the synthesis of non-proteinogenic isoxazole substituted α-amino acids [120].

9.4.1.1.2 **[CNO + CC] Reactions: 1,3-Dipolar Cycloaddition of Nitrile Oxides** The study of Quilico *et al.* on the formation of isoxazoles from nitrile oxides and unsaturated compounds is a milestone in the chemistry of isoxazoles [7]. Since then, the 1,3-dipolar cycloaddition of nitrile oxides has became an important approach to isoxazoles [6, 10, 85, 86, 121, 122]. Nitrile oxides can undergo dimerization to give furoxans (1,2,5-oxadiazole-2-oxides), the rate of this process being strongly dependent on the nature of the nitrile oxide substituent. Thus, steric and electronic effects determine the stability of the nitrile oxides, as illustrated by the time required for complete dimerization of some derivatives (Table 9.6) [123]. To avoid dimerization, nitrile oxides are usually generated *in situ*.

Table 9.6 Stability of some nitrile oxides towards dimerization to furoxans [123].

R	Complete dimerization (at 18 °C)	R	Complete dimerization (at 18 °C)
Methyl	<1 min	*p*-Chlorophenyl	10 days
t-Butyl	2–3 days	*p*-Nitrophenyl	Very slow
Phenyl	30–60 min	2,4,6-Trimethylphenyl	Very stable

The 1,3-dipolar cycloaddition of nitrile oxides with mono-substituted alkynes (alkyl/aryl) gives the corresponding 3,5-disubstituted isoxazoles regioselectively and occurs in competition with the 1,3-addition to give acetylenic oximes, which in some cases can be isolated. These oximes can easily be converted into isoxazoles (Scheme 9.8) [124].

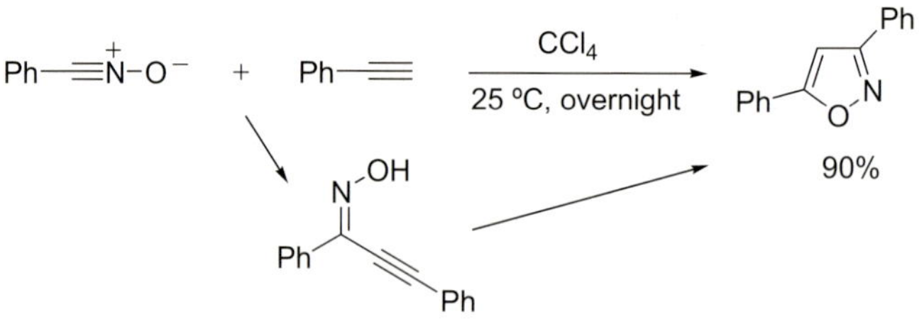

Scheme 9.8

A one-pot synthesis of isoxazoles from monosubstituted acetylenes with nitric acid under biphasic conditions (nitromethane–water, 1 : 1) in the presence of the catalyst $Bu_4N^+AuCl_4^-$ has been described (Table 9.7, entry 1) [125]. Nitrile oxide, generated from α-hydroxyimino carboxylic acids, in the presence of an alkyne furnishes the corresponding isoxazole (Table 9.7, entry 2) [126].

An efficient method for the *in situ* generation and cycloaddition of nitrile oxides from nitroalkanes, using 4-(4,6-dimethoxy[1,3,5]triazin-2-yl)-4-methylmorpholinium (DMTMM) chlorides and DMAP as catalyst through microwave irradiation has been reported. Carrying out the reaction in the presence of the appropriate alkynes, isoxazoles are obtained in high yields (Table 9.7, entries 3 and 4). This approach can also be applied to solid-phase synthesis [127].

Geometry constraints in the intramolecular 1,3-dipolar cycloaddition of nitrile oxides containing internal terminal alkynes leads to the exclusive formation of 4-substituted isoxazoles (Table 9.8).

The reaction of nitrile oxides with disubstituted alkynes leads to isoxazoles exclusively via 1,3-dipolar cycloaddition when the alkyne contains at least one electron-withdrawing substituent [6, 134].

Table 9.7 1,3-Dipolar cycloaddition of nitrile oxides with mono-substituted alkynes.

Entry	Reaction
1 [125]	2 R—≡ + HNO_3 → ($Bu_4N^+AuCl_4^-$ (cat.), CH_3NO_2/H_2O (1:1), 50 °C) → RCO–C≡N^+–O^- → (R—≡, 35-50%) → 5-R-3-COR-isoxazole
2 [126]	R^1C(=N–OH)CO_2H + ≡–R^2 → (CAN(IV), DMF, 0 °C, 20 min) → R^1–C≡N^+–O^- + ≡–R^2 → (41-70%) → 3-R^1-5-R^2-isoxazole R^1 = Ph or R^1 = COPh
3 [127]	3-R^1-5-R^2-isoxazole ← (DMTMM/DMAP, MeCN, rt, 6-12 h, 45-87%) ← R^1–CH_2–NO_2 + R^2–≡ → (DMTMM/DMAP, MeCN, MWI, 3 min., 92-100%) → 3-R^1-5-R^2-isoxazole
4 [127]	N-Boc-2,2-dimethyl-4-ethynyloxazolidine + $EtNO_2$ → (DMTMM/DMAP, MeCN, MWI, 3 min., 95%) → N-Boc-2,2-dimethyl-4-(3-methylisoxazol-5-yl)oxazolidine

Table 9.8 Intramolecular 1,3-dipolar cycloaddition of nitrile oxides.

Entry	Reaction conditions
1 [128]	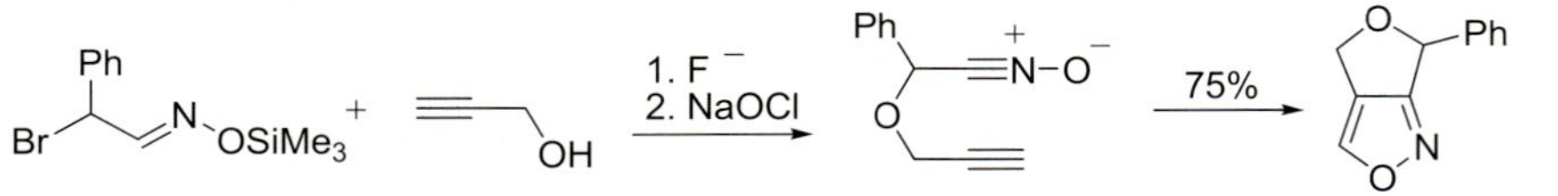 1. F^- 2. NaOCl; 75%
2 [129]	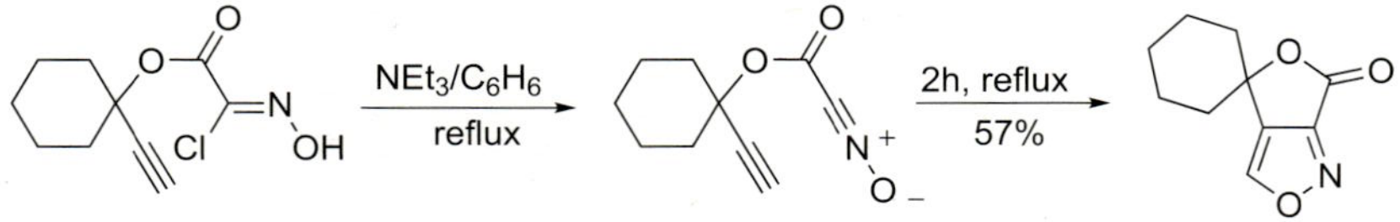 NEt_3/C_6H_6, reflux; 2h, reflux, 57%
3 [130, 131]	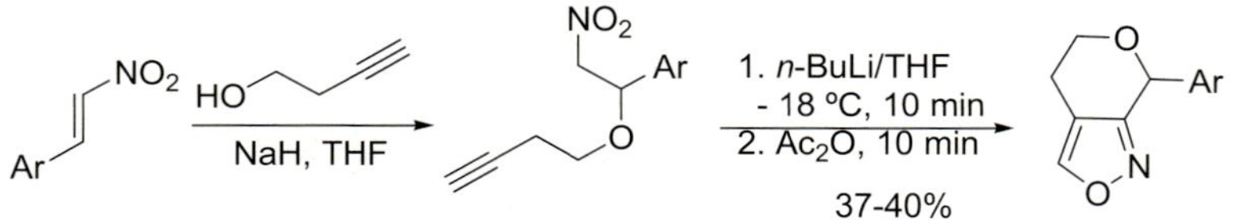 NaH, THF; 1. *n*-BuLi/THF, - 18 °C, 10 min 2. Ac_2O, 10 min, 37-40%
4 [132]	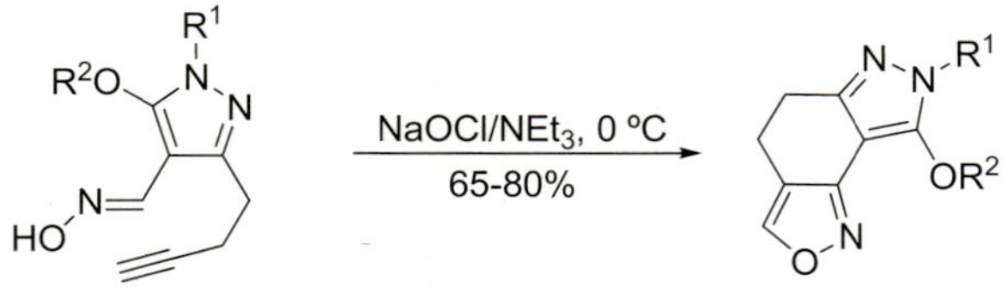 $NaOCl/NEt_3$, 0 °C, 65-80%
5 [133]	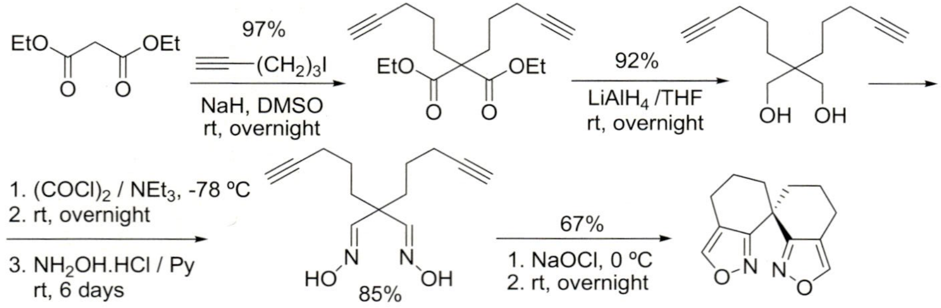 $(CH_2)_3I$ alkyne, NaH, DMSO, rt, overnight, 97%; $LiAlH_4$ /THF, rt, overnight, 92%; 1. $(COCl)_2$ / NEt_3, -78 °C 2. rt, overnight 3. $NH_2OH.HCl$ / Py, rt, 6 days, 85%; 1. NaOCl, 0 °C 2. rt, overnight, 67%

Solid-phase synthesis of 3-hydroxymethyl-4,5-disubstituted isoxazoles **66** has been achieved through a 1,3-dipolar cycloaddition of different alkynes to resin-bound nitrile oxide generated from nitro compound **65** under Mukaiyama conditions [135]. An alternative solid-phase 1,3-dipolar cycloaddition methodology allows the regioselective preparation of 5-hydroxyalkylisoxazoles **67** by anchoring acetylenic compounds on trityl chloride resin and carrying out the cycloaddition with nitrile oxides generated *in situ* from aldoximes (Scheme 9.9) [136].

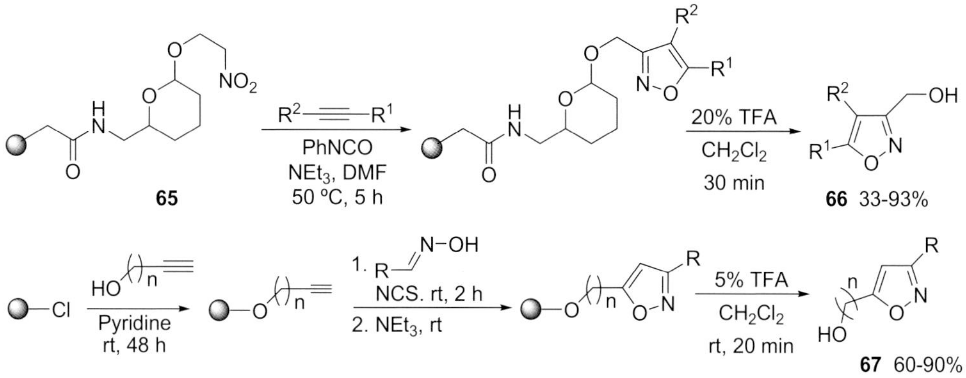

Scheme 9.9

A soluble polymer-supported synthesis of 3-hydroxymethyl-5-arylisoxazole, where construction of the isoxazole ring was based on nitrile oxide cycloaddition reactions, has been reported [137–140]. An alternative route to solid-phase synthesis for the construction of a library of isoxazoles involves a solution phase combinatorial synthesis of isoxazoles via cycloaddition of nitrile oxides with alkynes followed by precipitation of the products as HCl salts [141].

The synthesis of isoxazoles via cycloaddition of nitrile oxides can also be achieved using easily available alkenes instead of alkynes and converting the resulting isoxazolines into the corresponding isoxazoles either by dehydrogenation or by elimination reaction in the case of derivatives bearing a potential leaving group. In many cases the cycloaddition of nitrile oxides to alkenes affords directly the isoxazoles due to the lability of the intermediate isoxazoline under the experimental conditions (Scheme 9.10) [10, 121, 142].

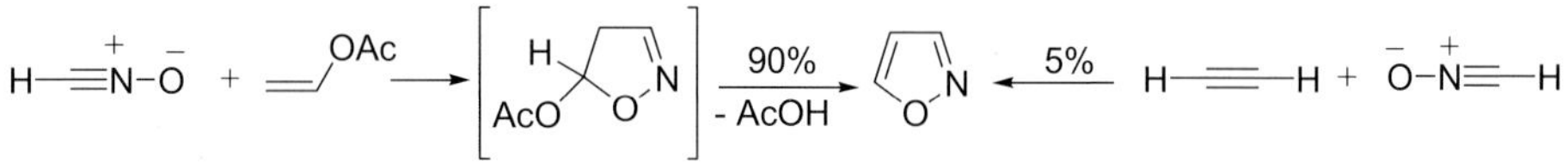

Scheme 9.10

The reaction of alkenes and alkynes with cerium ammonium nitrate (CAN) in acetone or acetophenone under reflux gives 4,5-dihydroisoxazoles or isoxazoles, respectively (Scheme 9.11) [143]. The reaction mechanism involves the nitration of

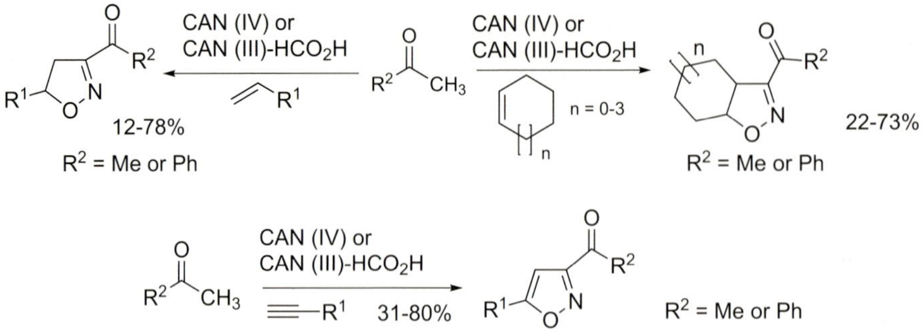

CAN (IV) or CAN (III)-HCO_2H
12-78%
R^2 = Me or Ph
n = 0-3
22-73%
R^2 = Me or Ph
31-80%
R^2 = Me or Ph

Scheme 9.11

acetone or acetophenone mediated by CAN (IV) or CAN (III) followed by the generation of the corresponding nitrile oxide (Scheme 9.12). The 3-acetyl- and 3-benzoylisoxazole derivatives are obtained by 1,3-dipolar cycloaddition of this dipole with the alkenes or alkynes.

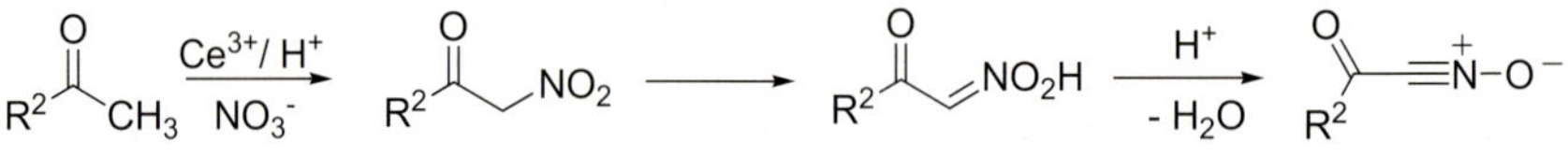

Ce^{3+}/H^+, NO_3^-; H^+, $-H_2O$

Scheme 9.12

Further examples of the use of 1,3-dipolar cycloaddition of nitrile oxides with alkenes and alkynes for the synthesis of isoxazoles have been published [144–155].

An alternative approach to the regioselective construction of the isoxazole ring involves the reaction of nitrile oxides with doubly activated methylene groups containing at least one carbonyl or nitrile substituent. This group ends up in the 5-position of the isoxazole: an acyl group as an alkyl or aryl group, 2-oxoacyl group as an acyl group, an ethoxyoxoacyl as an ethoxycarbonyl group, and a nitrile as an amino group (Scheme 9.13) [10, 156–159].

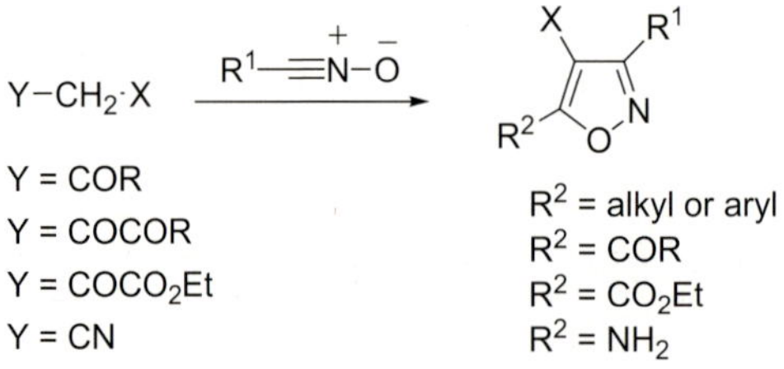

Y−CH_2·X
Y = COR
Y = COCOR
Y = $COCO_2Et$
Y = CN
R^2 = alkyl or aryl
R^2 = COR
R^2 = CO_2Et
R^2 = NH_2

Scheme 9.13

Several 3-aryl-5-alkylisoxazoles were synthesized in good yields by reacting arylnitrile oxides with free enolates, obtained from alkyl methyl ketones, followed by dehydration (Scheme 9.14) [160].

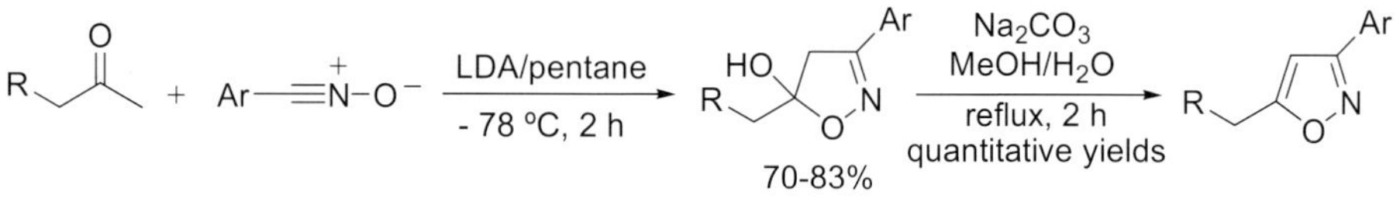

Scheme 9.14

Polyisoxazole systems containing two or more isoxazole rings can be constructed using the 1,3-dipolar cycloaddition of nitrile oxides. Starting from bisnitrile oxides the reaction with alkynes leads to 3,3′-linkage of the isoxazole rings whereas the cycloaddition of nitrile oxides with diynes produces a 5,5′-linkage [6, 161–165].

9.4.1.2 [4 + 1] Routes

9.4.1.2.1 [CCNO + C] Reactions: Reactions of Oxime Dianions with Carbonyl Compounds The reaction of oxime dianions with carbonyl compounds (e.g., esters, amides or aroyl chlorides) is an alternative regioselective method for the construction of the isoxazole ring (Scheme 9.15). The anion is acylated on carbon followed by cyclization–dehydration to give the unsymmetrically substituted isoxazoles. The same type of reaction can be carried out with benzonitriles, benzaldehydes and benzophenones [6, 10]. The condensation of 1,4-dilithiooximes with amides usually leads to higher yields than the reaction with aromatic esters [166–168].

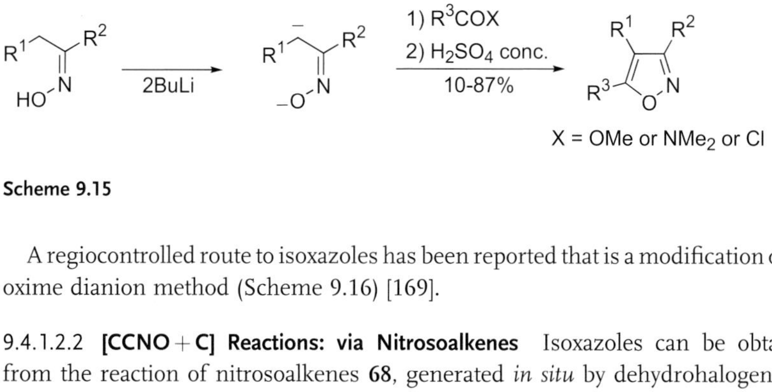

Scheme 9.15

A regiocontrolled route to isoxazoles has been reported that is a modification of the oxime dianion method (Scheme 9.16) [169].

9.4.1.2.2 [CCNO + C] Reactions: via Nitrosoalkenes Isoxazoles can be obtained from the reaction of nitrosoalkenes **68**, generated *in situ* by dehydrohalogenation

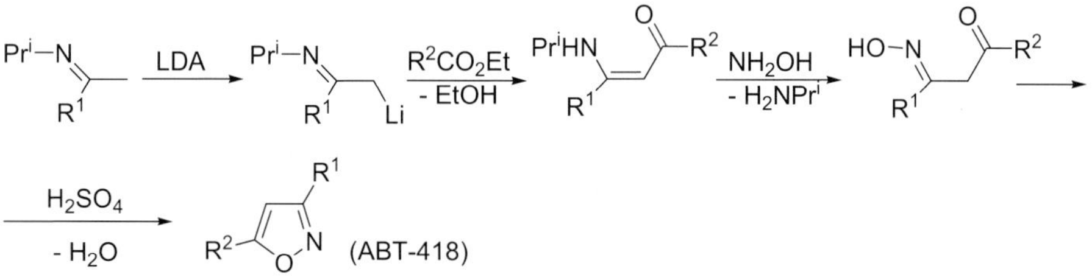

Scheme 9.16

of α-halooximes **69** or of the isomeric nitroso compounds **70**, with a C synthon (Scheme 9.17). The formation of alkene-nitrosyl chloride adduct **71** followed by reaction with cyanide affords 5-aminoisoxazoles **72** [170]. N-Substituted 5-aminoisoxazoles are usually prepared through N-functionalization of 5-aminoisoxazoles. However, such derivatives can also be produced directly from oximes of α-haloketones and isocyanides in the presence of sodium carbonate [171]. The process is thought to involve a [4 + 1] cycloaddition of the transient nitrosoalkene with isocyanides to give **73**. The C synthon can also be a keto-stabilized sulfonium ylide, as illustrated by the reaction of benzoylsulfonium ylide with (2-chloro-2-ethoxy-1-nitrosoethyl)benzene leading to isoxazole **74** [172].

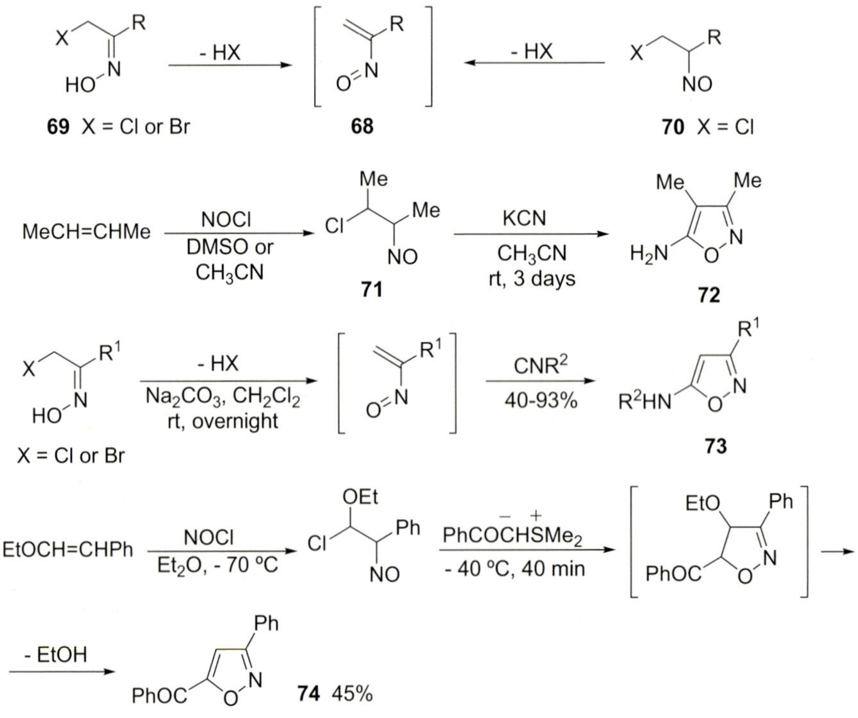

69 X = Cl or Br — $-HX$ → [**68**] ← $-HX$ — **70** X = Cl

MeCH=CHMe — NOCl, DMSO or CH_3CN → **71** — KCN, CH_3CN, rt, 3 days → **72**

X = Cl or Br — $-HX$, Na_2CO_3, CH_2Cl_2, rt, overnight → [ ] — CNR^2, 40-93% → **73**

EtOCH=CHPh — NOCl, Et_2O, - 70 °C → — $PhCO\bar{C}H\overset{+}{S}Me_2$, - 40 °C, 40 min → [ ] → - EtOH → **74** 45%

Scheme 9.17

9.4.1.3 [5 + 0] Routes

9.4.1.3.1 **[CCCNO] Reactions** Reaction of 1,1-dihalo-2-arylcyclopropanes, bearing electron-accepting substituents in the aromatic ring, with a mixture of nitric acid and sulfuric acid leads to halogen-substituted isoxazoles (Scheme 9.18) [173–174].

A simple procedure for the synthesis of 5-aminoisoxazoles takes advantage of the biohydrogenation of nitroacrylonitriles (R=H, Me or Ar) with baker's yeast (Scheme 9.19) [175]. The reduction of nitroacrylonitriles as a route to isoxazoles

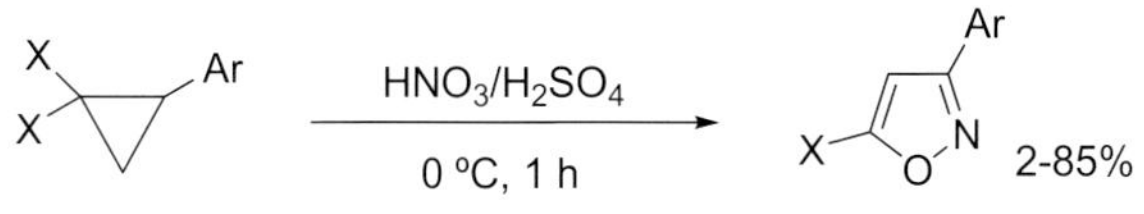

Scheme 9.18

Scheme 9.19

can also be carried out electrochemically or by reduction or by treatment with thiophenol in basic medium [10].

9.4.1.3.2 **[OCCCN] Reactions** Cyclization of α,β-unsaturated ketones or esters bearing an appropriate nitrogen-containing substituent in the β-position can lead to the corresponding isoxazole derivatives (Scheme 9.20). The nitrogen atom can undergo a nucleophilic attack by the carbonyl oxygen atom if a good leaving group is present. However, the formation of an β-acylvinylnitrene as intermediate cannot be excluded. Examples of this type of isoxazole precursors are β-azidovinyl ketones or esters **75**, *N*-(1-pyridinio)acylvinylaminides **76** and acylvinylsulfinimines **77** [6, 10]. The *N*-(1-pyridinio)acylvinylaminides **78** undergo N−N bond cleavage upon heating in benzene to give isoxazoles **79** [176]. Acylvinylsulfinimines **80** [177], generated from the reaction of diphenylsulfilimine with benzoylacetylenes, allows the synthesis of isoxazoles **81** (Scheme 9.21).

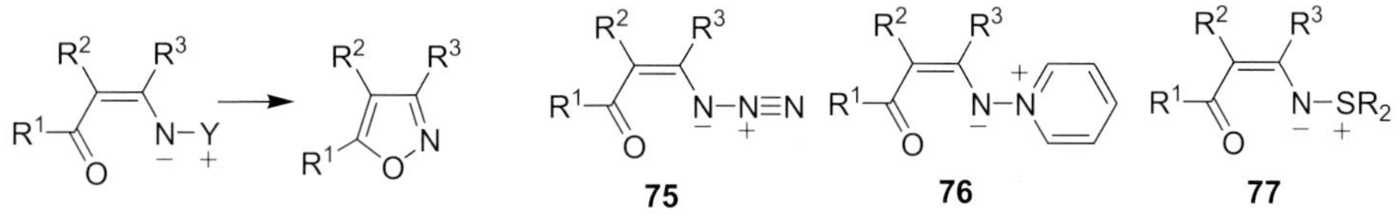

Scheme 9.20

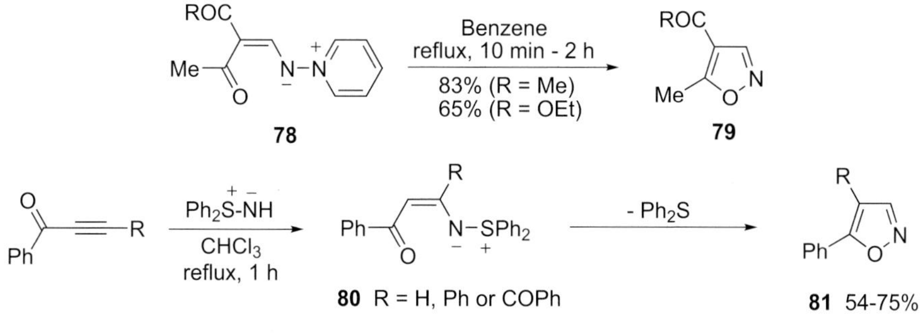

Scheme 9.21

The thermolysis of 3-phenyl-2*H*-azirine-2-carbaldehyde at 200 °C leads to 3-phenylisoxazole in high yield [178]. The same isoxazole can also be obtained in 90% yield by treatment of 3-phenyl-2*H*-azirine-2-carboxaldehyde at 25 °C with Grubbs' catalyst [179]. Furthermore, 2-benzoyl-3-phenyl-2*H*-azirine affords the corresponding isoxazole upon heating in non-hydroxylic solvents [180] (Scheme 9.22).

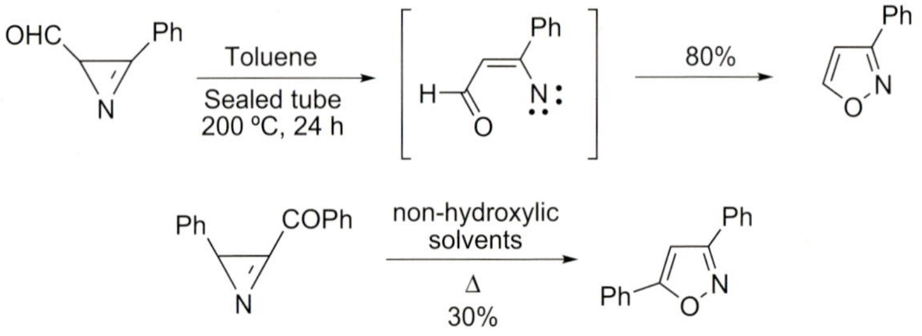

OHC Ph N; Toluene, Sealed tube, 200 °C, 24 h; [H O Ph N:]; 80%; Ph O N

Ph COPh N; non-hydroxylic solvents, Δ, 30%; Ph Ph O N

Scheme 9.22

9.4.1.3.3 **[CONCC] Reactions** Lithium hydroxide-catalyzed cyclization of α-(acylmethoxyimino)nitriles **82** provides a route to 5-acyl-4-aminoisoxazoles **83** [181]. The α-nitro oximes **84** act as an CONCC synthon in the synthesis of benzoyl-protected 3-ribofuranosyl-4-nitroisoxazole-5-carboxylate **85** (Scheme 9.23) [182].

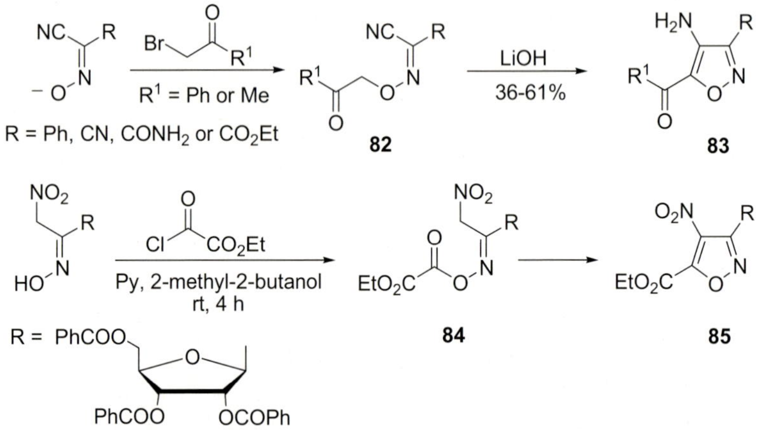

NC R N $^-$O; Br R^1 O; R^1 = Ph or Me; R = Ph, CN, $CONH_2$ or CO_2Et; NC R R^1 O O N **82**; LiOH, 36-61%; H_2N R R^1 O O N **83**

NO_2 R HO N; Cl CO_2Et O; Py, 2-methyl-2-butanol, rt, 4 h; NO_2 R EtO_2C O O N **84**; O_2N R EtO_2C O N **85**

R = PhCOO O PhCOO OCOPh

Scheme 9.23

9.4.1.3.4 **[NOCCC] Reactions** 3-Aminoisoxazole can be synthesized by the hydrolysis and ring closure of vinyl-substituted oximes under acidic conditions

(Scheme 9.24) [183]. Similar results are obtained starting with other oximes (e.g., acetone oxime or ethyl acetate oxime) and the cyanoacetylene can also be replaced by β-chloroacrylonitrile [184].

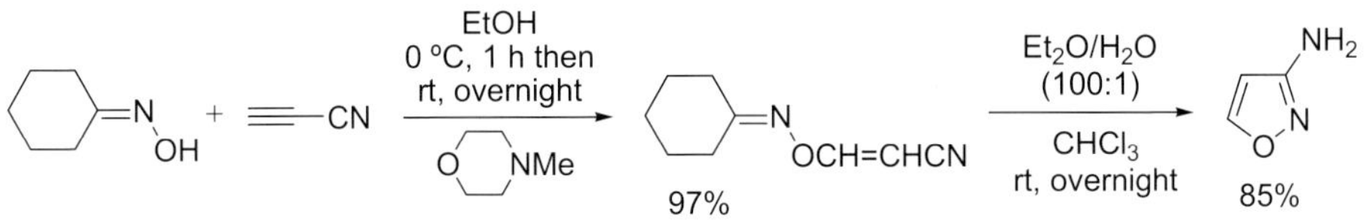

Scheme 9.24

9.4.1.4 [3 + 1 + 1] Routes

9.4.1.4.1 [ONC + C + C] Reactions The reaction of primary nitroalkanes with organic bases affords dioximes **86** which are converted into trialkylisoxazoles **87** in high yields when heated in dilute acids [10]. In contrast, the reaction of phenylnitromethane with *cis*-α-nitrostilbene gives isoxazoline-*N*-oxide **88**. Treatment of **88** with aqueous alcoholic sodium hydroxide allows the synthesis of triphenylisoxazole **89** (Scheme 9.25) [185]. Trisubstituted isoxazoles can also be obtained from the reaction of nitroalkanes with aldehydes in the presence of a base [10, 186].

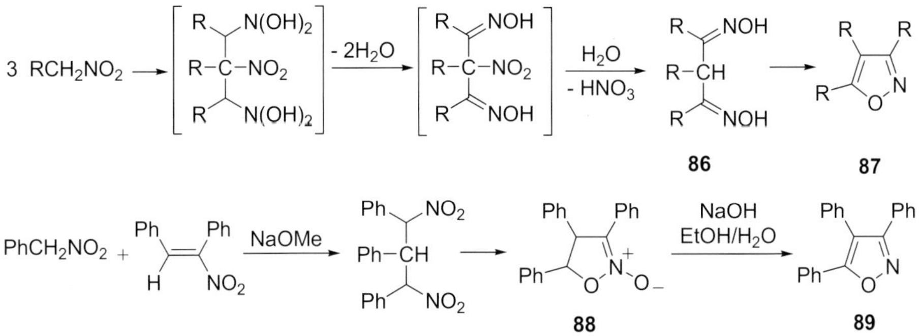

Scheme 9.25

9.4.1.5 Ring Transformations of Heterocycles Leading to Isoxazoles

4-Acylisoxazolin-5-ones **90** rearrange to the isomeric isoxazole-4-carboxylic acids **91** upon treatment with aqueous sodium hydroxide [187]. 4-Benzylidene-3-phenylisoxazolin-5-one (**92**) is converted into isoxazole **93** upon treatment with anhydrous ammonia in ethanol in the presence of benzaldehyde (Scheme 9.26) [188].

The reaction of hydroxylamine with 2-substituted or 3-substituted chromones gives exclusively the corresponding 5-(2-hydroxyphenyl)isoxazoles (Scheme 9.27) [10]. The reaction involves the opening of the chromone ring followed by the formation

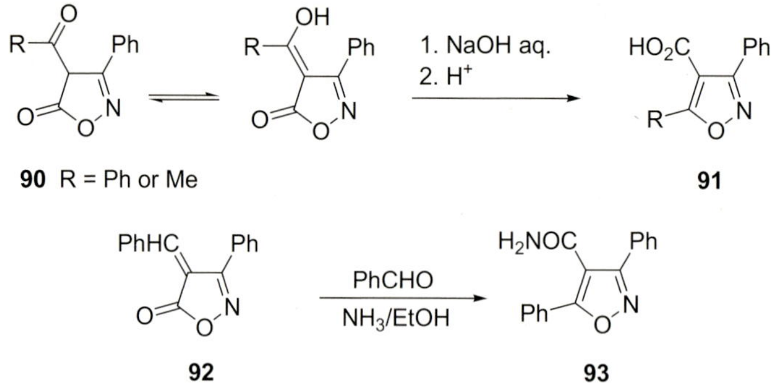

Scheme 9.26

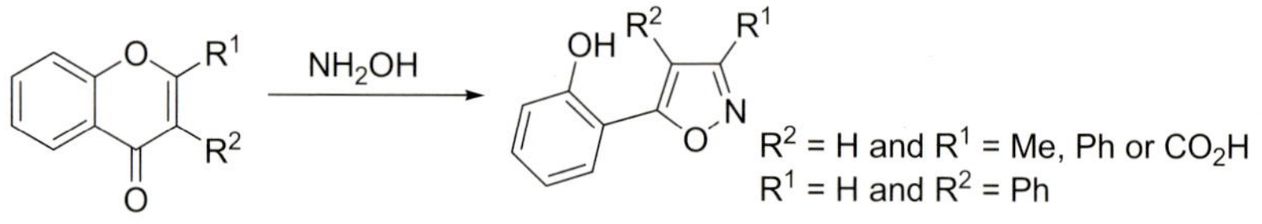

Scheme 9.27

of 5-(2-hydroxyphenyl)isoxazole in 72% overall yield [189, 190]. Some of these isoxazole derivatives show anti-inflammatory related activity.

The 5-amino-4-trifluoroacetyloxazoles **94** can be used in a two-step synthesis of isoxazoles **96** (Scheme 9.28). Nucleophilic attack of hydroxylamine at the five position of **94** leads to ring opening followed by a cyclization to give isoxazoline **95**. Subsequent dehydration in the presence of trifluoroacetic anhydride allows the synthesis of isoxazoles in good yields [191].

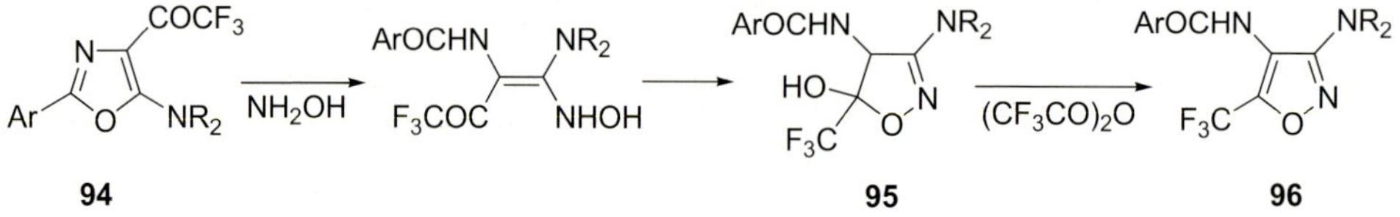

Scheme 9.28

Isoxazolines with the general structure **97** undergo cycloreversion to give isoxazoles **98** (Scheme 9.29) [6]. The parent isoxazole can be prepared in 37% yield by the thermolysis of the cycloadduct obtained from fulminic acid (HCNO) and norbornadiene [192]. This type of approach can also be used to the synthesis of 3-vinylisoxazole (**99**), which is unsubstituted at both the 4- and 5-positions. The two-step procedure involves initial 1,3-dipolar cycloaddition of acrylonitrile oxide to norbornadiene followed by retro-Diels–Alder fragmentation under flash vacuum pyrolysis [193].

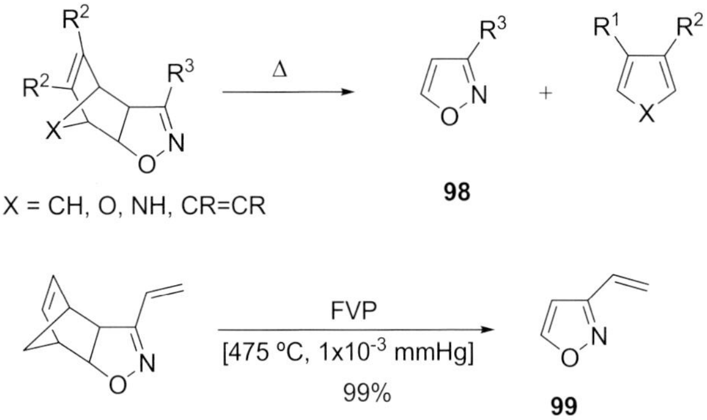

Scheme 9.29

Furazans (1,2,5-oxadiazoles) undergo fragmentation to nitrile and nitrile *N*-oxides by thermolysis or photolysis. Irradiation of benzofurazan in the presence of dimethyl acetylenedicarboxylate (DMAD) to give isoxazole **100** is an illustrative example of this approach (Scheme 9.30) [194, 195].

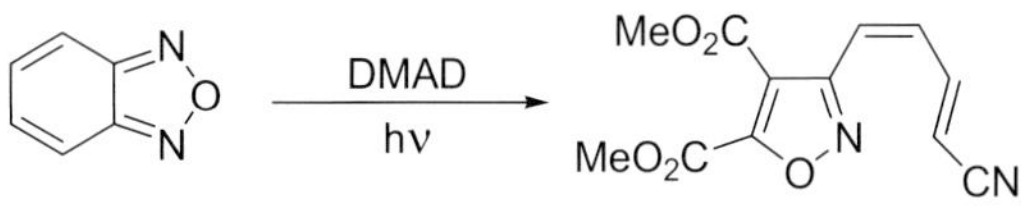

Scheme 9.30

9.4.2 Isothiazoles

The first synthesis of a mononuclear isothiazole ring system was reported in 1956 [15]. Oxidation of 5-amino-1,2-benzisothiazole by alkaline permanganate gives isothiazole-4,5-dicarboxylic acid. Decarboxylation to isothiazole-4-carboxylic acid followed by functional group interconversion leads to the isothiazole itself and a range of monosubstituted isothiazoles [15].

The chemistry of isothiazoles has been reviewed [25–29]. The most convenient methods for the construction of the isothiazole ring involve: (i) oxidative cyclization of a γ-thio amine derivative (formation of the S–N bond), (ii) 1,3-dipolar cycloaddition of nitrile sulfides to alkynes or alkenes and (iii) conversion of other heterocycles into isothiazoles. Examples of such synthetic methodologies are described below.

9.4.2.1 Synthesis from Acyclic Compounds

Isothiazole itself can be prepared from propynal by two distinct routes (Scheme 9.31). It reacts with thiohydroxylamino-*S*-sulfonate (**101**) to give the thiooxime intermediate

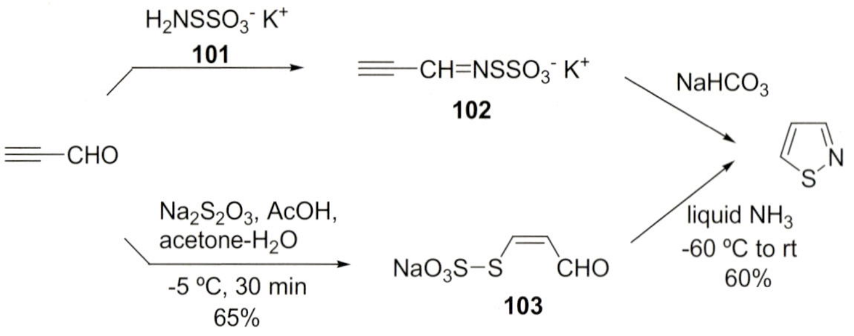

Scheme 9.31

102, which, in the presence of $NaHCO_3$, cyclizes to isothiazole [21]. In the second route, propynal and sodium thiosulfate afford the intermediate **103**, which on treatment with ammonia cyclizes to isothiazole [196]. 3-Methylisothiazole can be obtained by a similar synthetic procedure, changing propynal by but-3-yn-2-one [196].

There are several routes to isothiazoles based on oxidative cyclization reactions in which the N–S bond is formed. A good leaving group attached to the sulfur atom is required, which frequently is introduced by reaction with iodine (Scheme 9.32) [197, 198].

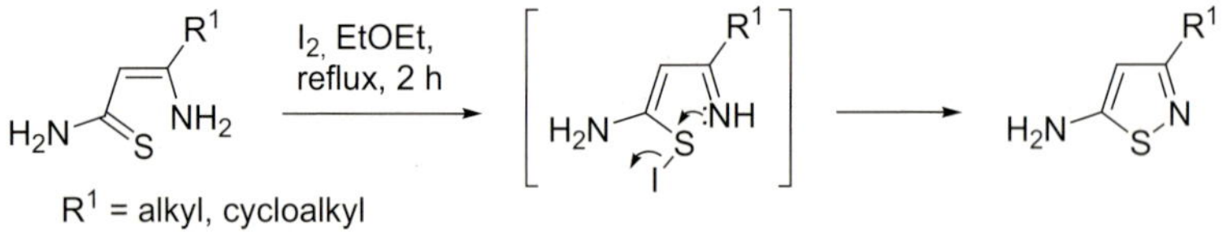

Scheme 9.32

Oxidation of 3-amino-2-cyano-3-phenylpropenedithioates with iodine produces 3-phenyl-5-alkylthioisothiazole-4-carbonitriles in near-quantitative yield (Scheme 9.33) [199].

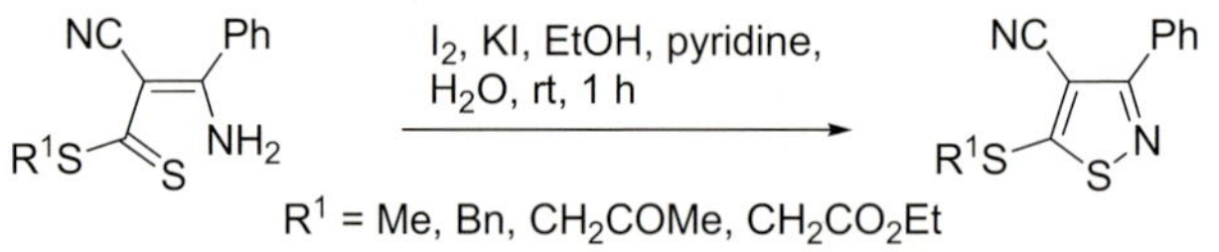

Scheme 9.33

Ketene *S,N*-acetals **104** give the isothiazolium salts **105** in good yields when treated with iodine at room temperature. Dealkylation with KI in DMSO affords the corresponding 5-aryl-3-(arylthio)isoxazoles **106** (Scheme 9.34) [200].

Enamino thioaldehydes **108** can be converted in good yields into 3,4-disubstituted isothiazoles **109** by oxidation with *m*-chloroperbenzoic acid (Scheme 9.35) [201]. Oxidation of enamino thioaldehydes **108** (generated from **107a**) with iodine, at room

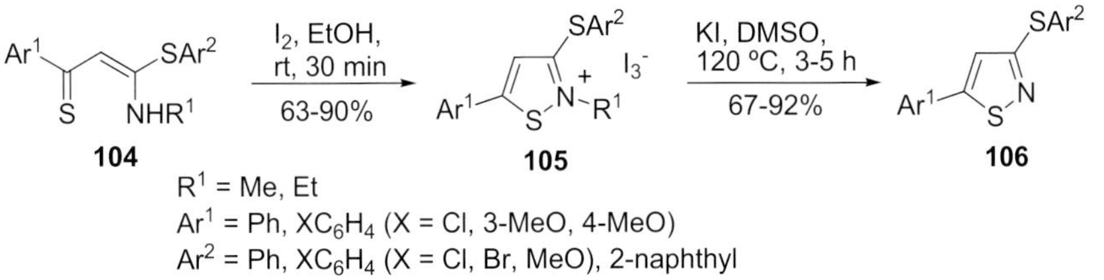

R^1 = Me, Et
Ar^1 = Ph, XC_6H_4 (X = Cl, 3-MeO, 4-MeO)
Ar^2 = Ph, XC_6H_4 (X = Cl, Br, MeO), 2-naphthyl

Scheme 9.34

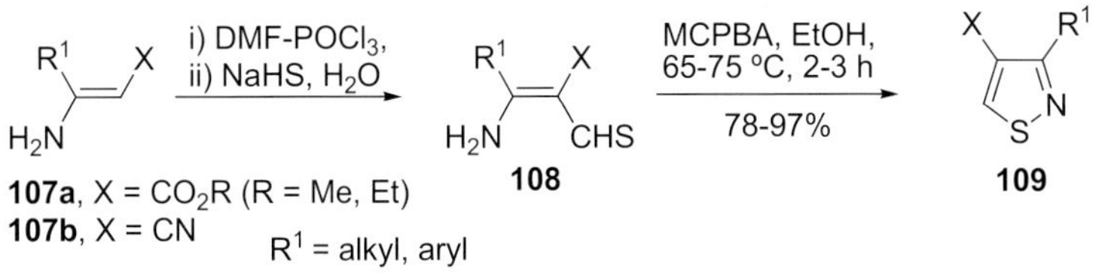

107a, X = CO_2R (R = Me, Et)
107b, X = CN
R^1 = alkyl, aryl

Scheme 9.35

temperature, also affords the corresponding isothiazoles **109** in moderate to good yields [202]. Thioaldehydes **108** may be synthesized from enamines **107** by solvolysis of the corresponding Vilsmeier salts with aqueous or methanolic sodium hydrogen sulfide [201, 202].

3,5-Diaminoisothiazole-4-carboxylate derivatives can be prepared in a one-pot reaction from active methylene nitriles, isothiocyanates and chloramine (Scheme 9.36) [203]. Reactions starting from malononitrile give the corresponding isothiazole in higher yields (41–65%).

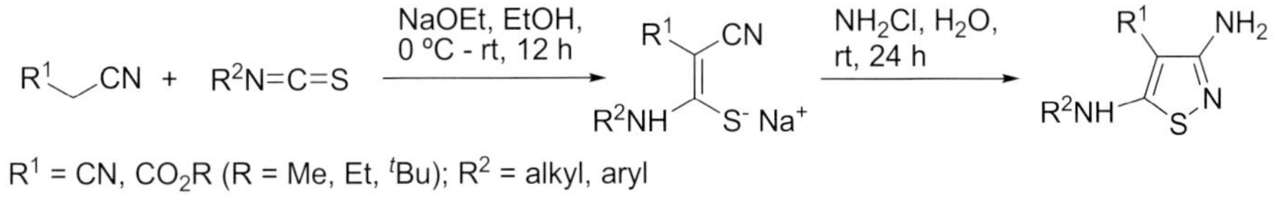

R^1 = CN, CO_2R (R = Me, Et, ^{t}Bu); R^2 = alkyl, aryl

Scheme 9.36

Dithiolate disodium salt **110** reacts with chlorine to give 4-benzoyl-3,5-dichlor-oisothiazole (**111**) in low yield. However, the same compound can be monomethylated and reacted with hydroxylamine-*O*-sulfonic acid to afford isothiazole **112** in a global 75% yield (Scheme 9.37) [204].

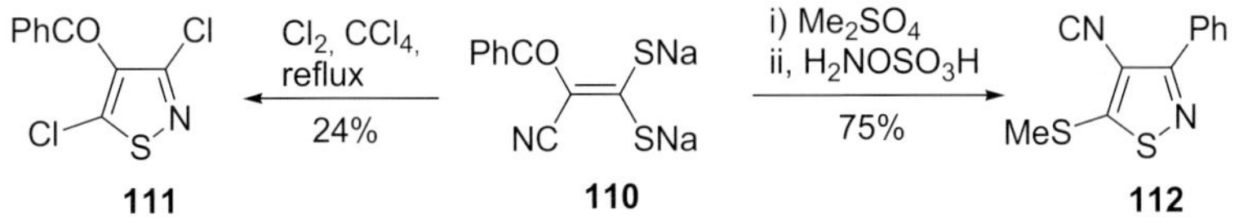

Scheme 9.37

Oximes are also useful intermediates in the synthesis of isothiazoles. For instance, oximes derived from α-oxoketene dithioacetals cyclize to 5-methylthioisothiazoles when treated with thionyl chloride in pyridine (Scheme 9.38) [205].

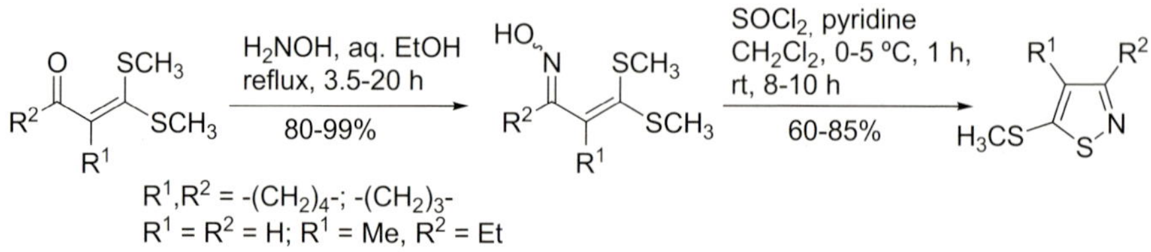

H_2NOH, aq. EtOH, reflux, 3.5-20 h, 80-99%; $SOCl_2$, pyridine, CH_2Cl_2, 0-5 °C, 1 h, rt, 8-10 h, 60-85%

$R^1,R^2 = -(CH_2)_4-; -(CH_2)_3-$
$R^1 = R^2 = H; R^1 = Me, R^2 = Et$

Scheme 9.38

Oxime derivatives **113** are converted into isothiazoles **114** by reaction with a dehydrating agent, namely tosyl isocyanate [206]. Similarly, the 2-(hydroxyimino)alkyl *N,N*-dialkyldithiocarbamates **115** and the analogous trithiocarbonates **117** react with tosyl isocyanate to afford the bis(4-isothiazolyl) disulfides **116** or **119** and the disubstituted isothiazoles **118** (Scheme 9.39) [207].

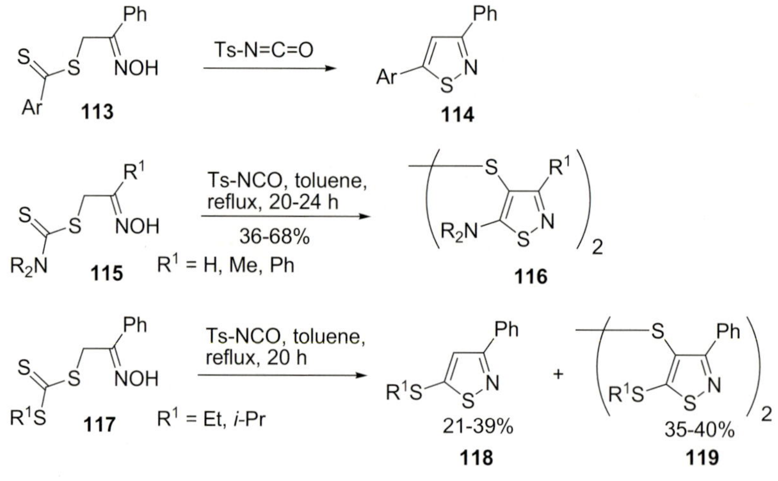

113 → 114: Ts-N=C=O
115 → 116: Ts-NCO, toluene, reflux, 20-24 h, 36-68%; R^1 = H, Me, Ph
117 → 118 (21-39%) + 119 (35-40%): Ts-NCO, toluene, reflux, 20 h; R^1 = Et, *i*-Pr

Scheme 9.39

Oxime tosylates of type **120** react with methyl thioglycolate to give 4-amino or 4-hydroxyisothiazole-5-carboxylate esters (Scheme 9.40) [208, 209]. This synthetic methodology has been used to prepare ethyl 4-aminoisothiazole-5-carboxylate *C*-nucleosides [210].

Photoreaction of arylthioamides with alkenes and alkynes, under aerobic conditions, yields isothiazoles and 1,2,4-thiadiazoles in low to moderate yields (Scheme 9.41). Nitrile sulfides are probable intermediates in these reactions [211].

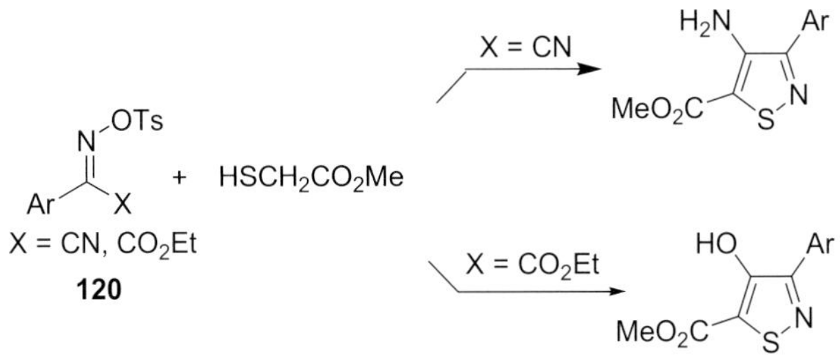

Scheme 9.40

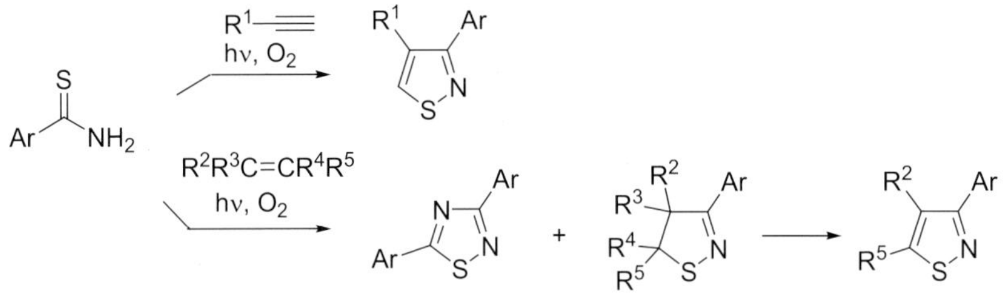

Scheme 9.41

α-Acetylenic aldehydes or ketones are converted into 4-unsubstituted isothiazoles by reaction with hydroxylamine-*O*-sulfonic acid and sodium hydrogen sulfide in buffered aqueous solution in a one-pot procedure (Scheme 9.42) [212].

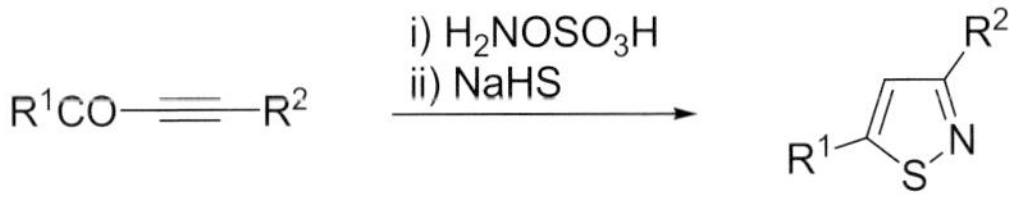

Scheme 9.42

Arylmethylene-malononitriles react with sulfur chloride in the presence of pyridine to give 5-aryl-3-chloroisothiazole-4-carbonitriles in high yields (Scheme 9.43) [213].

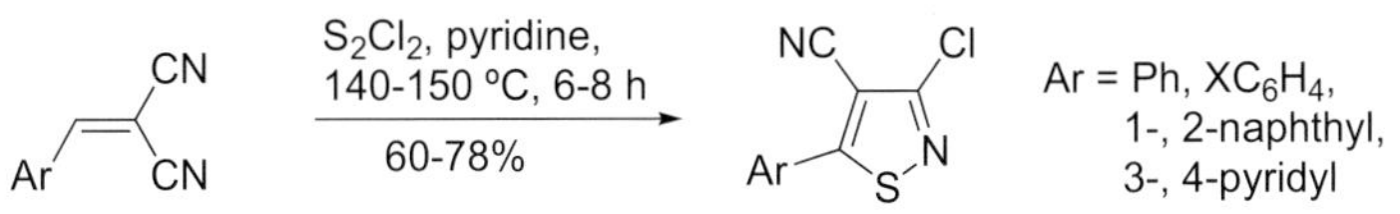

Scheme 9.43

3-Chloroalk-2-enals react with ammonium thiocyanate to afford 4,5-disubstituted isothiazoles **121** (Scheme 9.44) [214, 215]. Cycloalka[*c*]isothiazoles **122** can be prepared by a similar method [216]. 2-Thiocyanatocyclohex-1-ene carbaldehyde (**123**) reacts with anilines to afford isothiazolium salts **124** [217].

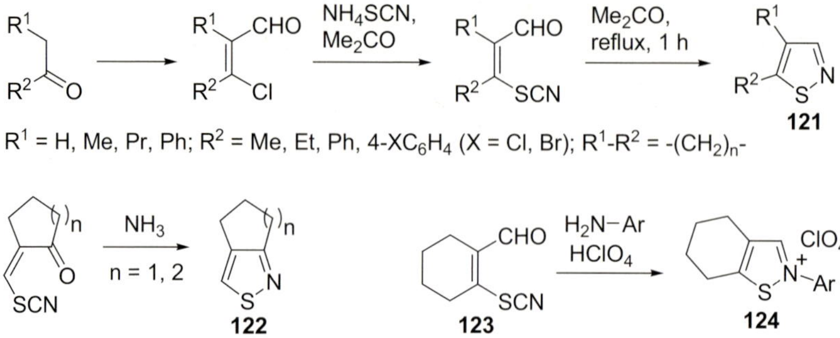

R^1 = H, Me, Pr, Ph; R^2 = Me, Et, Ph, 4-XC_6H_4 (X = Cl, Br); R^1-R^2 = -$(CH_2)_n$-

Scheme 9.44

Methacrylonitrile reacts with trithiazyl trichloride $(NSCl)_3$ in the presence of excess SO_2Cl_2, in refluxing chloroform, to give 4-cyanoisothiazole in 78% yield [218].

9.4.2.2 **Ring Transformations of Heterocycles Leading to Isothiazoles**

Several heterocyclic compounds can, by chemical modification, be converted into isothiazoles. Despite some synthetically interesting exceptions, in most cases the starting heterocycles are not readily available and the routes are unlikely to be general.

A synthetically useful method involves the generation of nitrile sulfides in the presence of alkynes or alkenes, affording isothiazoles or 4,5-dihydroisothiazoles, respectively (Scheme 9.45). The nitrile sulfides are conveniently generated *in situ* by thermal cycloreversion of five-membered heterocycles already containing the C=N-S unit [219]. Decarboxylation of 1,3,4-oxathiazol-2-ones in an inert solvent (e.g., xylene, chlorobenzene) in the presence of an excess of the dipolarophile is one of the most convenient routes [220–223]. Using dimethyl acetylenedicarboxylate as dipolarophile, the isothiazole-4,5-dicarboxylates are obtained in yields as high as 96% [221]. The low regioselectivity observed in the reactions of nitrile sulfides with unsymmetrical alkynes and alkenes is a major disadvantage of this method. Oxidation of 4,5-dihydroisothiazoles with sodium hypochlorite, under phase-transfer conditions, affords the isothiazoles in high yields [223].

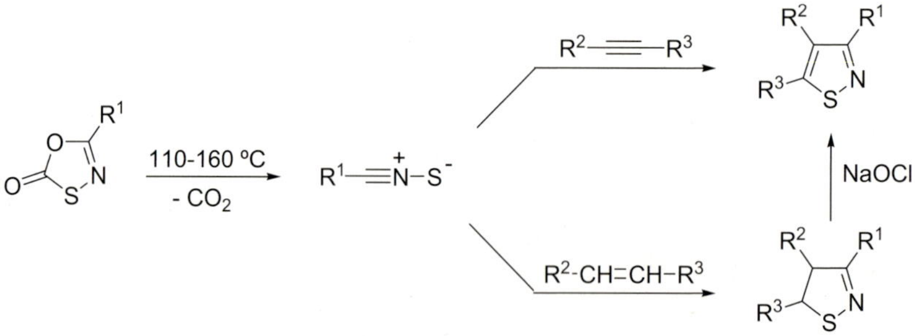

Scheme 9.45

Both 2,5- and 2,3,5-substituted furans react with trithiazyl trichloride $(NSCl)_3$ to afford 5-acylisothiazoles in good yield (Scheme 9.46) [224, 225]. A much simpler procedure makes use of a mixture of ethyl carbamate, thionyl chloride and pyridine in boiling benzene or toluene; this generates the reactive thiazyl chloride, NSCl, *in situ* [226, 227]. Highly polarized 2,5-disubstituted furans (such as **125**) yield only one isothiazole. However, when the electronic properties of the substituents are more balanced, two isomeric isothiazoles are formed [225, 227]. The thiazyl chloride reagent has been used for the direct conversion of calix[*n*]furans into macrocyclic isothiazoles [228].

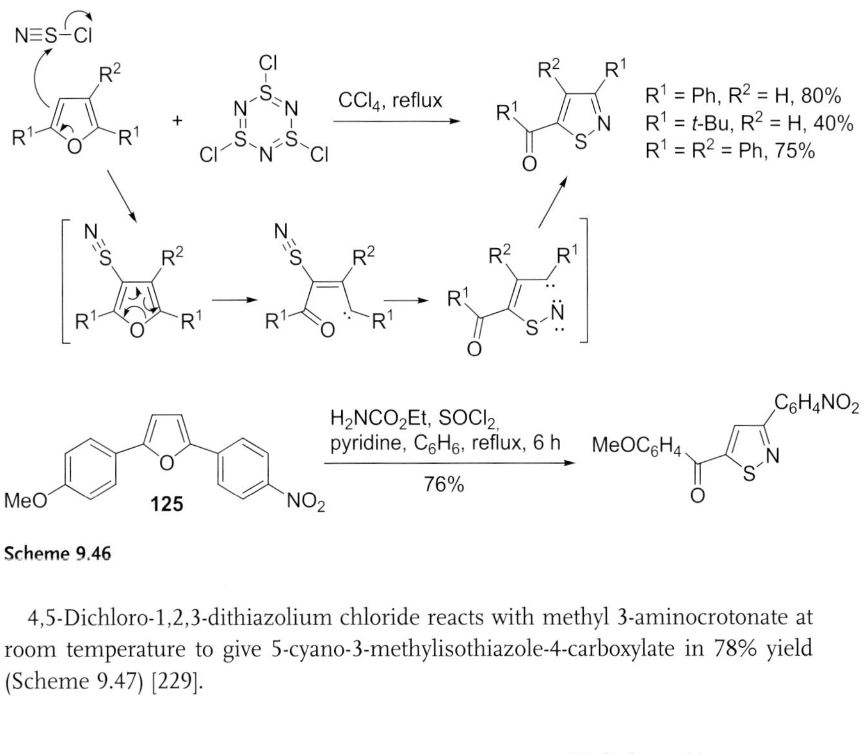

N≡S–Cl
CCl4, reflux
R1 = Ph, R2 = H, 80%
R1 = t-Bu, R2 = H, 40%
R1 = R2 = Ph, 75%
H2NCO2Et, SOCl2,
pyridine, C6H6, reflux, 6 h
76%
MeO
125
NO2
MeOC6H4
C6H4NO2

Scheme 9.46

4,5-Dichloro-1,2,3-dithiazolium chloride reacts with methyl 3-aminocrotonate at room temperature to give 5-cyano-3-methylisothiazole-4-carboxylate in 78% yield (Scheme 9.47) [229].

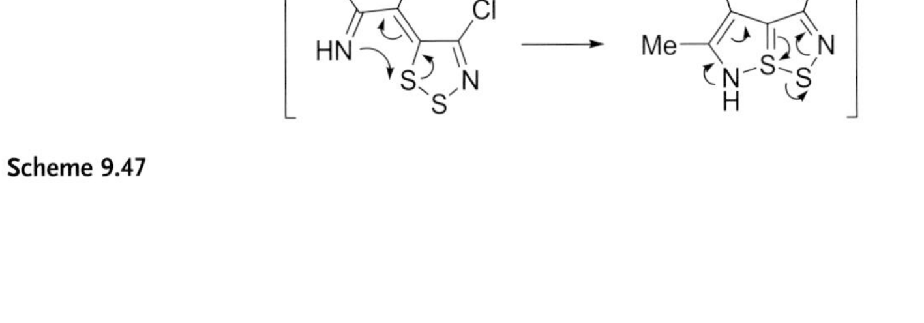

Me
CO2Me
H2N
i) CH2Cl2, rt, 1 h
ii) pyridine, rt, 30 min
78%
MeO2C
Me
NC
- HCl
- 1/8 S8

Scheme 9.47

Thermolysis of triphenyl-1,3-dithiol-2-yl azide (**126**) in refluxing toluene gives 3,4,5-triphenylisothiazole (**127**) in 80% yield. However, under identical conditions, the diphenyl analogue **128** affords a mixture of isothiazoles and 1,4,2-dithiazines (Scheme 9.48). These dithiazines, when refluxed in toluene, extrude the sulfur atom at the 4-position to give, selectively, one isothiazole [230].

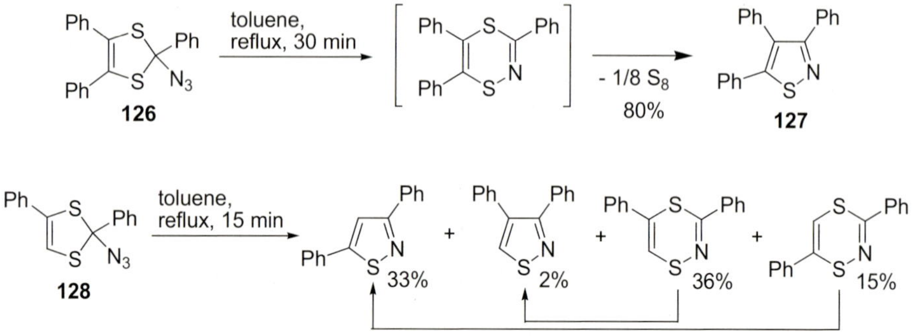

Scheme 9.48

Benzopyran-4-thiones react with diphenylsulfilimine to give the corresponding 5-(2-hydroxyphenyl)isothiazoles in high yields (Scheme 9.49) [231].

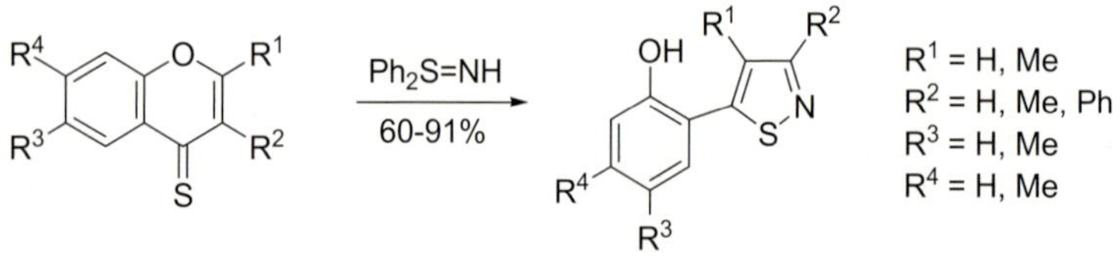

Scheme 9.49

Maleimide derivatives **129** undergo oxidative cyclization to give isothiazole-3,4-dicarboximides **130** (Scheme 9.50) [232]. By ammonolysis, the *N*-unsubstituted derivative affords 5-aminoisothiazole-3,4-dicarboxamide (**131**).

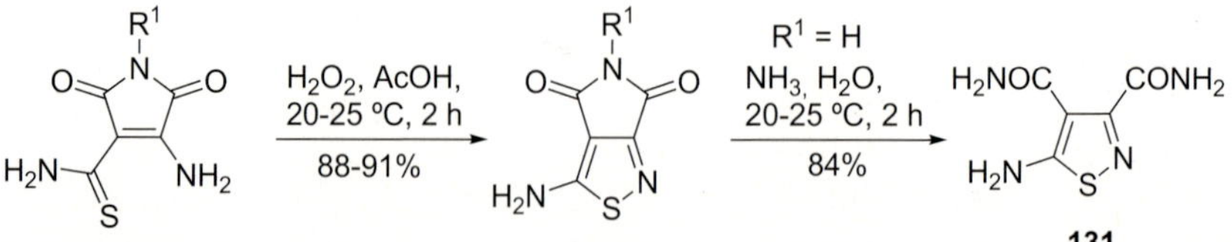

Scheme 9.50

9.5 Synthesis of Benzisoxazoles and Benzisothiazoles

9.5.1 1,2-Benzisoxazoles

Most synthetic approaches to 1,2-benzisoxazoles involve cyclization of a suitable benzene derivative and can be represented as follows:

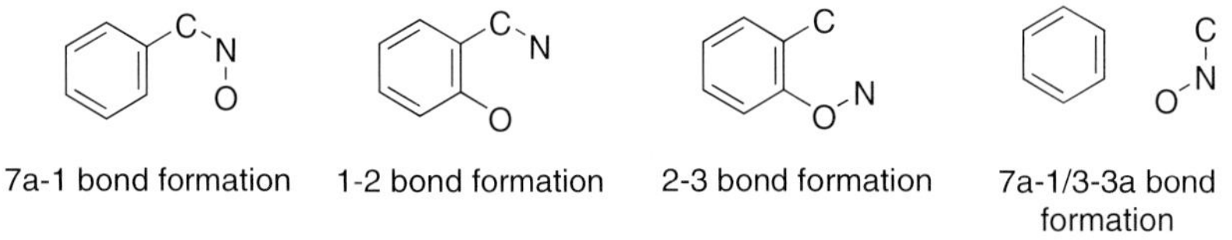

7a-1 bond formation 1-2 bond formation 2-3 bond formation 7a-1/3-3a bond formation

1,2-Benzisoxazoles can also be obtained from 3-3a bond formation in the cyclization processes. The remaining methodology involves heterocyclic rearrangements.

9.5.1.1 Formation of Bond 7a-1

The first synthesis of a 1,2-benzisoxazole was reported in 1892, 3-phenyl-1,2-benzisoxazole **132**, and involved the reaction of hydroxylamine with *ortho*-bromobenzophenone in alkaline medium [30]. However, 1,2-benzisoxazole had been known since 1882, obtained from the reduction of *ortho*-nitrobenzaldehyde with tin and hydrochloric acid [233]. This base-promoted intramolecular displacement reaction for formation of the 7a-1 bond has became an important route to 1,2-benzisoxazoles. Other halogens also undergo this type of displacement, with the reactivity of iodide and fluoride comparable with bromide but chloride less reactive [10, 234]. The displaceable groups also include nitro, amino, methoxy and hydroxyl groups (Scheme 9.51 and Table 9.9).

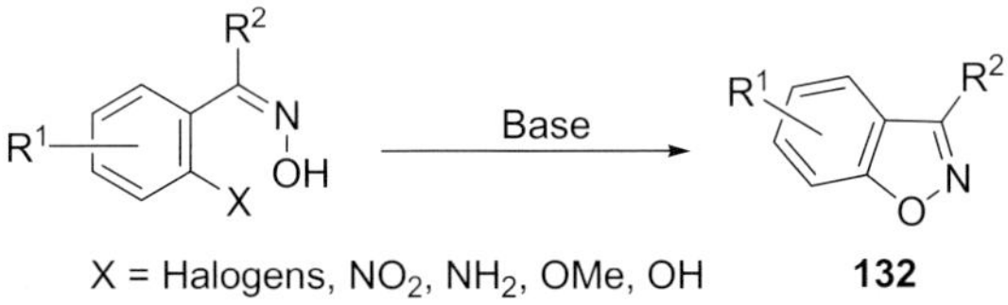

R^2, R^1, N, OH, X, Base, R^1, R^2, N, O

X = Halogens, NO_2, NH_2, OMe, OH **132**

Scheme 9.51

The course of the reaction is determined by the configuration of the oxime. A *syn* relationship of the OH to the aryl substituent bearing the leaving group allows cyclization to 1,2-benzisoxazole whereas the isomeric oximes usually produce Beckman rearrangement products [234]. Amidoximes are configurationally labile, allowing the use of the *anti* oxime as starting material for the synthesis of 1,2-benzisoxazoles. Thus, the amide oxime **133** cyclizes to 3-(4-pyridinylamino)-1,2-benzisoxazole **134** on reacting with potassium *tert*-butoxide via an isomerization/cyclization process (Scheme 9.52) [239].

Table 9.9 Base-promoted intramolecular displacement reactions for 1,2-benzisoxazole 7a-1 bond formation.

1 [235]	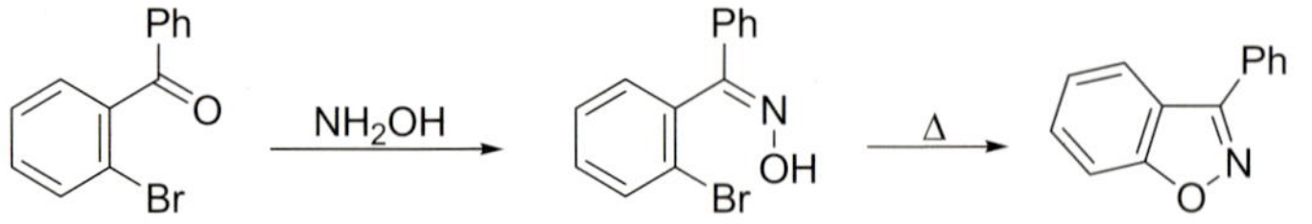 NH_2OH; Δ
2 [236]	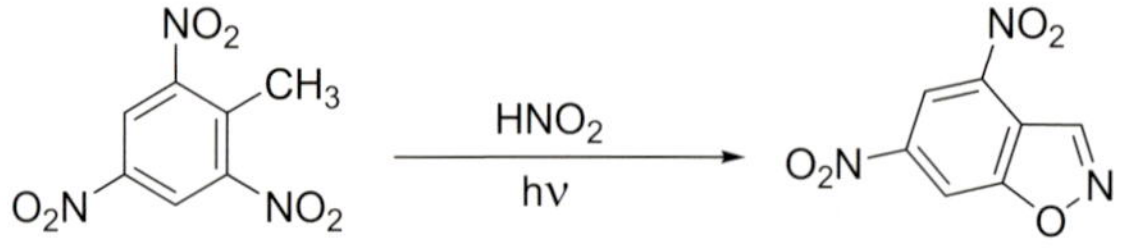 HNO_2, hν Via photonitrosation of trinitrotoluene
3 [237]	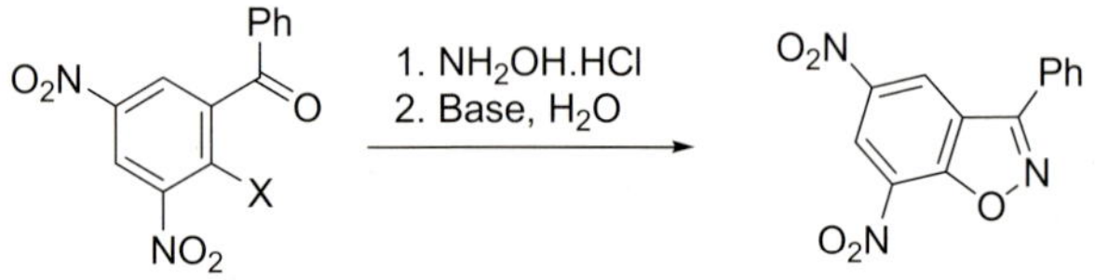 1. $NH_2OH.HCl$ 2. Base, H_2O X = OMe or OH Requires substantial activation such as the presence of nitro groups
4 [238]	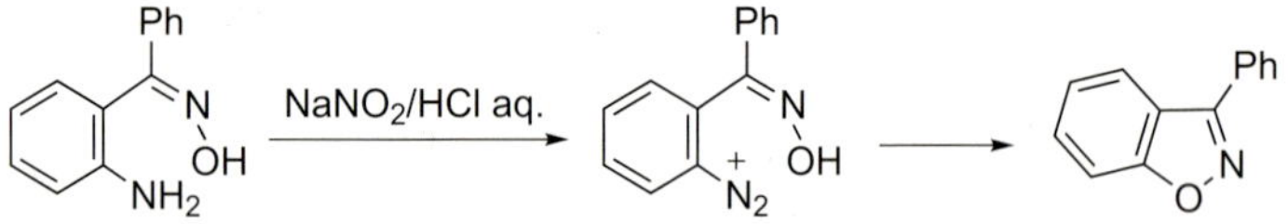 $NaNO_2$/HCl aq. Via diazonium salt

t-BuOK/THF, reflux, 3 h

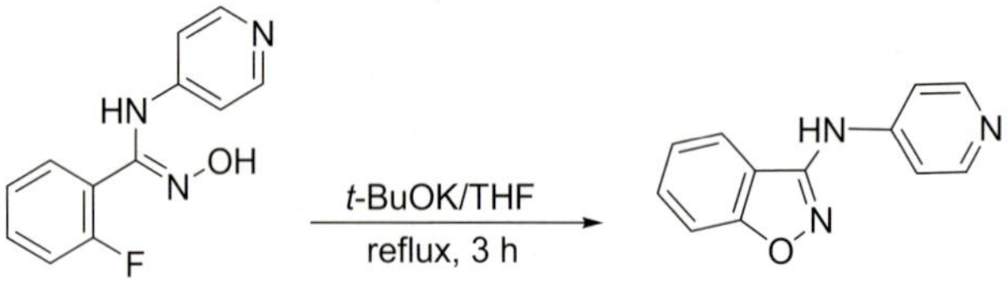

133 → **134** 69%

Scheme 9.52

β-Hydroxyoximes **135**, bearing a phenyl group unsubstituted in the *ortho* positions, are converted into styrylbenzisoxazole **136** upon treatment with phosphoric acid [10]. The base cyclization of dithioacetal **137** followed by desulfurization leads to 3-acyl and 3-aroyl-1,2-benzisoxazoles (**138**) [240] (Scheme 9.53).

9.5.1.2 **Formation of Bond 1-2**

1,2-Benzisoxazoles can be obtained from 2-hydroxybenzophenone oximes by thermolysis, treatment with base or with dehydrating agents (e.g., sulfuric and phosphoric acid). By reacting oxime **139** with thionyl chloride/pyridine the

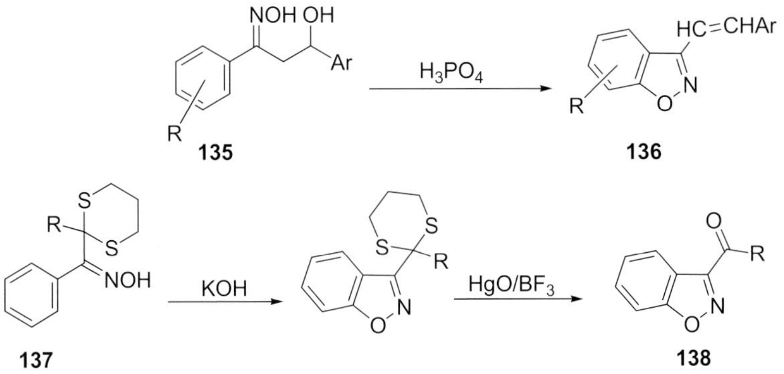

Scheme 9.53

1,2-benzisoxazole **140** can also be prepared [241]. The *O*-sulfonate oxime **141** is converted, in the presence of mild bases, into 1,2-benzoisoxazole in 95% yield [242]. A synthesis of 3-(2-dialkylaminoethyl)-1,2-benzisoxazoles **143** from oxime acetates of 2-hydroxyphenyl ketones **142** has also been reported [243] (Scheme 9.54). A similar synthetic approach has been applied to the synthesis of 3-[2-(1-pyrazolyl)ethyl]-1,2-benzisoxazoles [244].

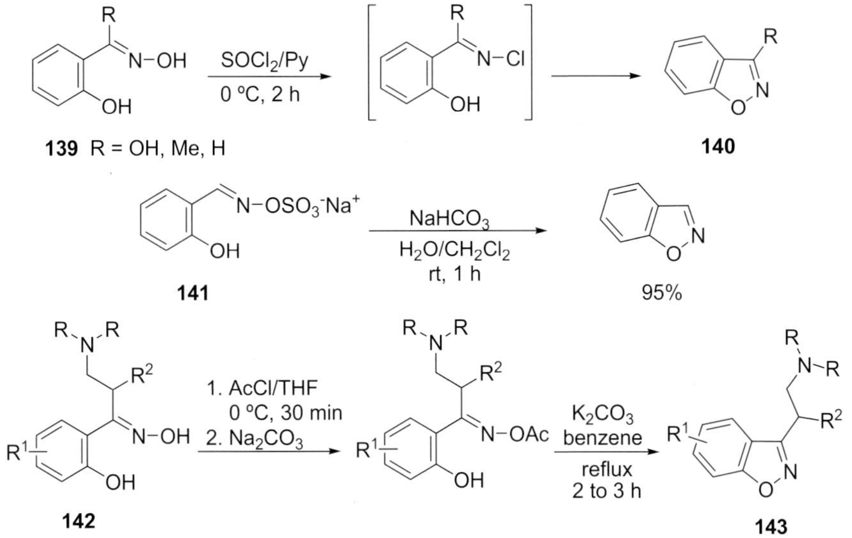

Scheme 9.54

9.5.1.3 **Formation of Bond 2-3**

The synthesis of 3-methyl-1,2-benzisoxazole (**145**) from the reaction of 2-hydroxyacetophenone with hydroxylamine-*O*-sulfonic acid in diluted aqueous base is an example of a 2–3 ring closure. The process occurs via the generation of intermediate **144**, which then undergoes the cyclization [9] (Scheme 9.55).

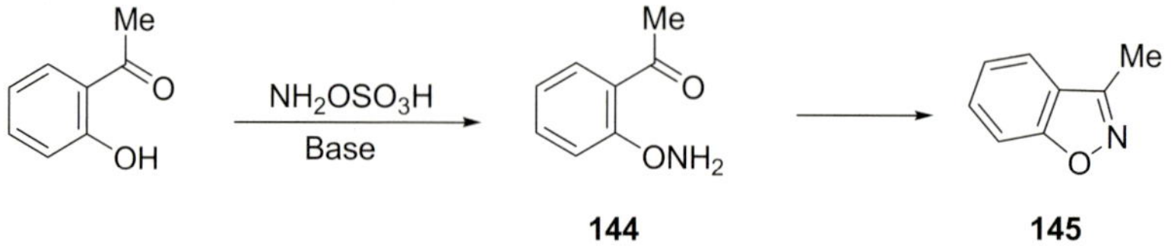

Scheme 9.55

3-Amino-1,2-benzisoxazoles (**147** or **149**) can be obtained from 2-fluorobenzenonitrile [245, 246]. The synthesis involves an S_NAr reaction to give an intermediate (**146** or **148**, respectively) followed by ring-closure to give the 1,2-benzisoxazoles. A solid-phase synthesis of 3-amino-1,2-benzisoxazoles uses a similar synthetic strategy: the displacement of fluoride from 2-fluorobenzonitrile by the Kaiser oxime resin **150** followed by cyclization [247, 248] (Scheme 9.56).

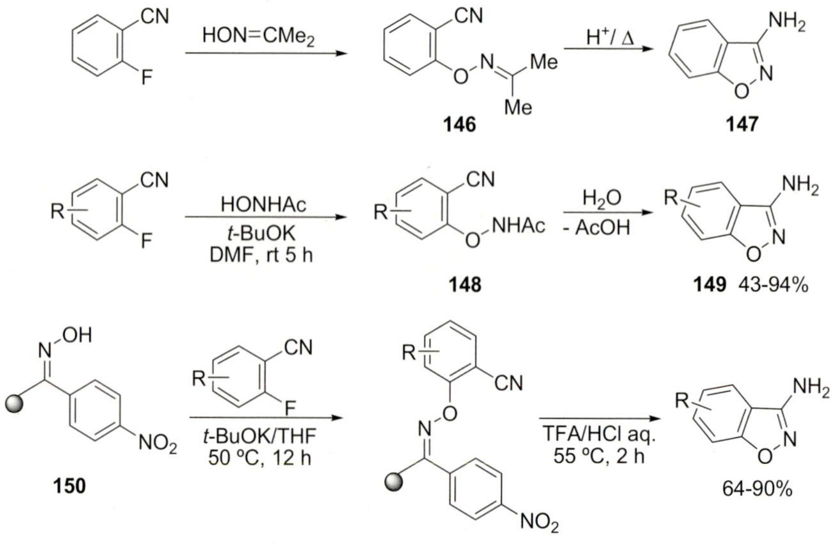

Scheme 9.56

9.5.1.4 **Formation of Bonds 7a-1/3-3a**

1,3-Dipolar cycloaddition of nitrile oxides to benzyne gives 3-substituted 1,2-benzisoxazoles in modest yield (Scheme 9.57) [249]. Other dipolarophiles can also be used for the synthesis of 1,2-benzisoxazole derivatives, namely 1,4-benzoquinones and enamines [9].

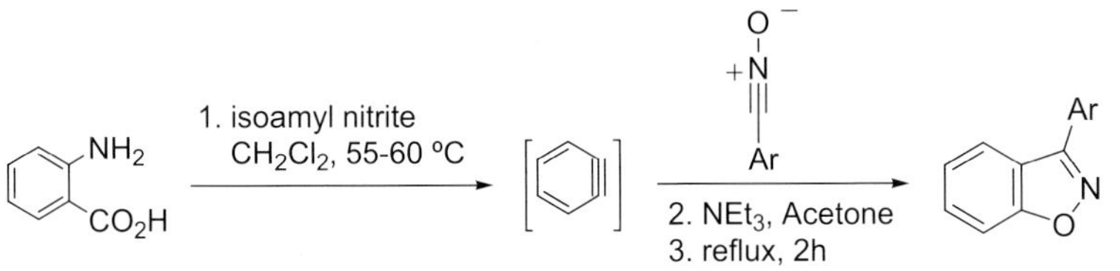

Scheme 9.57

9.5.1.5 From Other Heterocycles

The reaction of 4-hydroxycoumarin **151** (R=H) and 4-hydroxycoumarin substituted derivatives with hydroxylamine leads to 1,2-benzisoxazole-3-acetic acid **152** (Scheme 9.58) [250, 251].

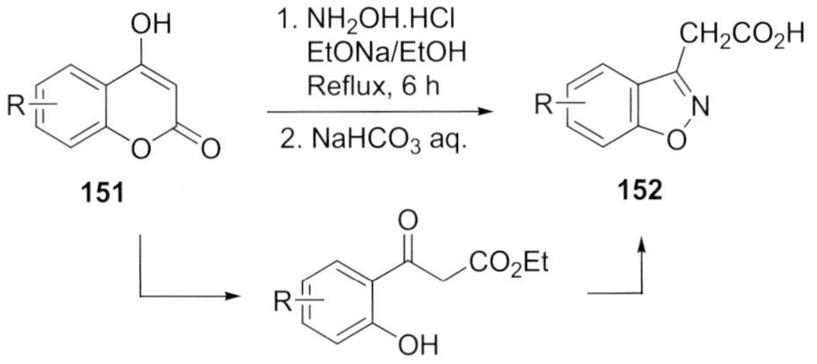

Scheme 9.58

9.5.2
2,1-Benzisoxazoles

2,1-Benzisoxazoles are usually prepared by 1-2 or 2-3 bond formation in the cyclization step or by introduction of atom C3, resulting in the formation of bond 2-3.

9.5.2.1 Formation of Bond 1-2

An important route to 2,1-benzisoxazoles involves reduction of *ortho*-nitrophenones or *ortho*-nitroalkylbenzenes containing an oxygen function on the α-carbon of the alkyl substituent. 3-Phenyl-2,1-benzisoxazole (**154**) can be obtained from **153** in the presence of sulfuric acid. 3-Aryl-2,1-benzisoxazoles are also prepared by the reaction of *ortho*-nitrobenzaldehydes and an aromatic hydrocarbon catalyzed by sulfuric acid (Scheme 9.59) [252].

5-Substituted-2,1-benzisoxazoles **155** have been prepared from 5-substituted-2-nitrobenzaldehydes by the reduction of the nitro group with stannous chloride dihydrate and *in situ* cyclization (Scheme 9.60) [253]. The allyl bromide/Zn mediated reductive cyclization of 2-nitrobenzaldehydes, 2-nitroacetophenone and *N*-(2-nitrobenzylidene)anilines (**156**) leads to 2,1-benzisoxazoles in good to excellent

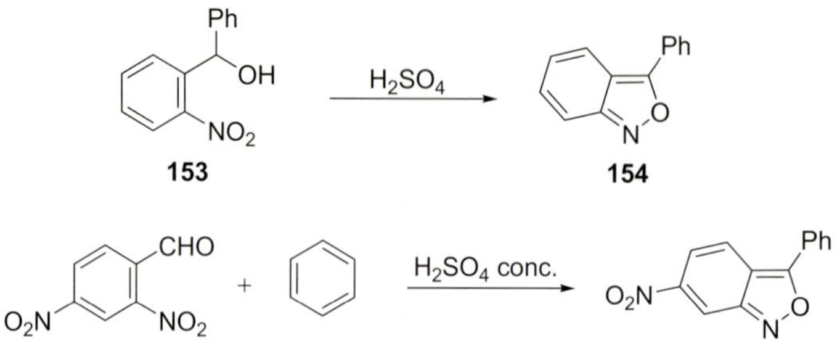

Scheme 9.59

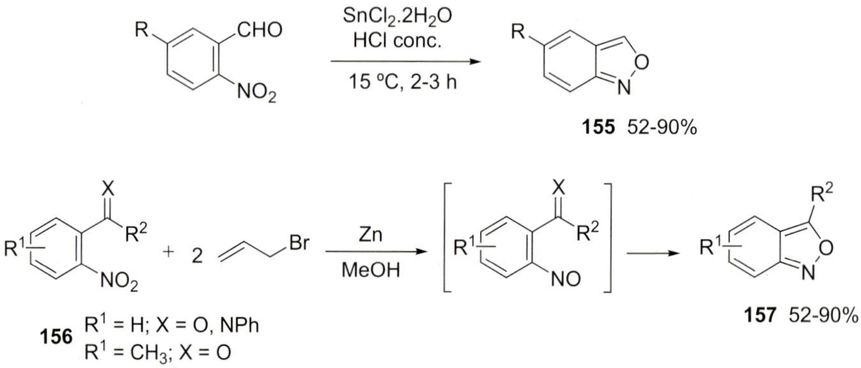

Scheme 9.60

yields [254]. The reductive cyclization of *ortho*-nitrobenzaldehydes and *ortho*-nitroacetophenone can be carried out with 2-bromo-2-nitropropane/Zn in methanol at 50 °C to give 2,1-benzisoxazoles **157** in good yields [255]. Electrochemical synthesis of 2,1-benzisoxazole from nitroarenes by controlled potential cathodic electrolysis has been reported [256].

Many methods of oxidative cyclization of *ortho*-aminoaryl ketones to 2,1-benzisoxazoles are known [9]. An illustrative example is the hypervalent iodine oxidation of *ortho*-aminochalcones **158** to give styryl-2,1-benzisoxazole **159** in good yields (Scheme 9.61) [257]. Under similar reaction conditions, 3-methyl-2,1-benzisoxazole (**160**) is obtained from *ortho*-aminoacetophenone.

The thermolysis of *ortho*-azidoaryl ketones also produces 2,1-benzisoxazoles. For example, the thermal decomposition of azide **161** gives styryl-2,1-benzisoxazoles **162** along with a minor amount of hydroxyquinoline **163** (Scheme 9.62) [258].

9.5.2.2 **Formation of Bond 2-3**

The 2,1-benzisoxazole ring system can be constructed from the acid- or base-catalyzed dehydration of 2-nitrobenzyl compounds. In fact, sulfuric acid cyclization of *ortho*-nitrophenylacetic acid yields a mixture of 2,1-benzisoxazole and

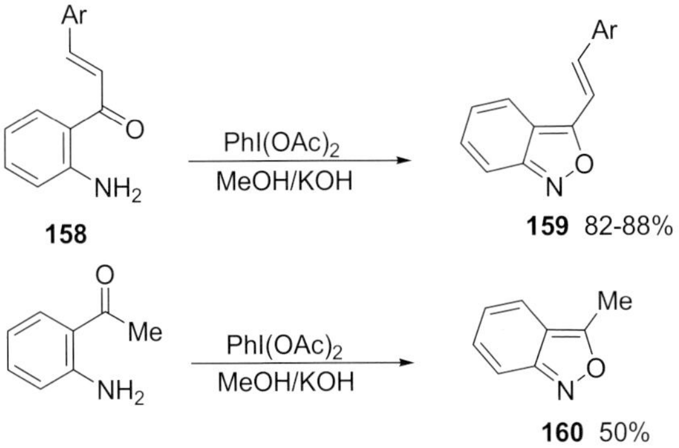

Scheme 9.61

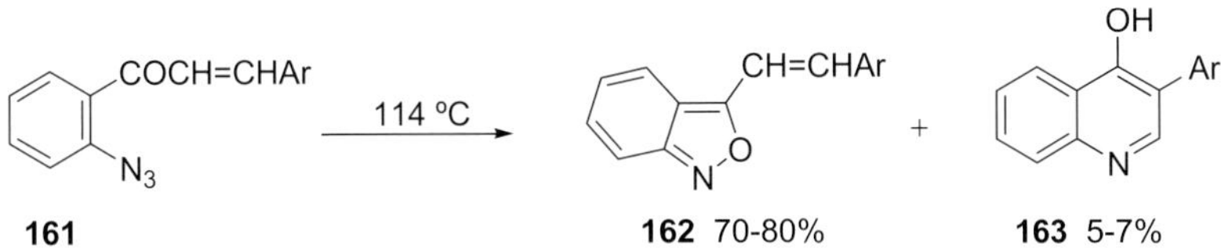

Scheme 9.62

2,1-benzisoxazole-3-carboxylic acid [259]. Me_3SiCl/Et_3N-mediated dehydration of 2-nitrobenzyl derivatives **164** gives sulfones **165** (Scheme 9.63) [260].

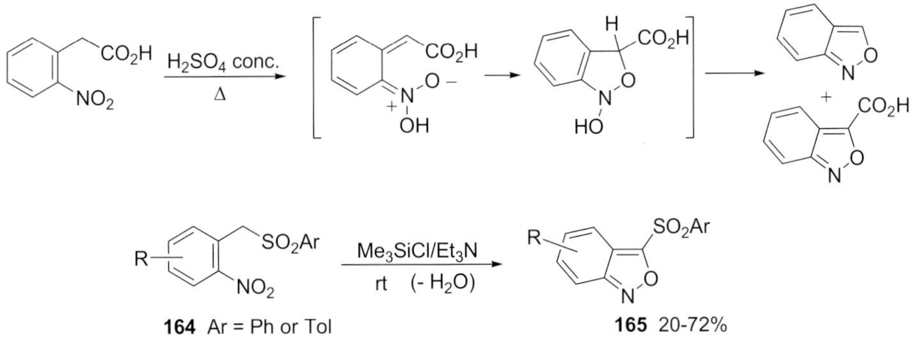

Scheme 9.63

Two research groups claimed the synthesis of 2,1-benzisoxazoles from the reaction of *ortho*-nitrosobenzaldehyde with benzylamine [261] and from treatment of **167a** or **167b** with aqueous sodium hydroxide [262]. However, Kurth *et al.* have demonstrated that the products of these reactions were in fact indazalones [263].

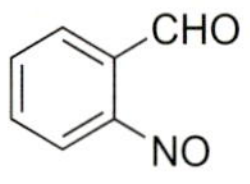

166 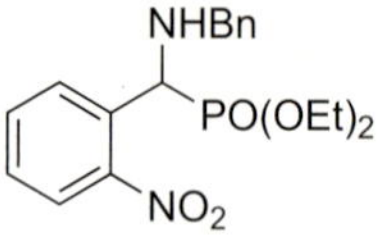 **167a** 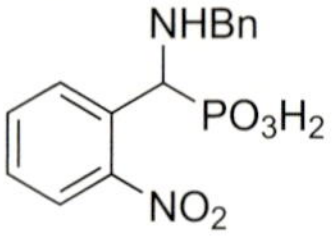 **167b**

9.5.2.3 By introduction of C-3

2,1-Benzisoxazole can be produced from the condensation of nitrobenzenes with benzyl cyanide in the presence of a base. Starting from *para*-chloronitrobenzene the reaction with benzyl cyanide gives 2,1-benzisoxazole **168** in 46% yield via an *ortho*-quinonoid intermediate [264]. The synthesis of 3-substituted 2,1-benzisoxazoles **169** from the reaction of *ortho*-chloronitrobenzene with the sodium salt of malonic ester or ethyl cyanoacetate occurs through an initial nucleophilic displacement (Scheme 9.64).

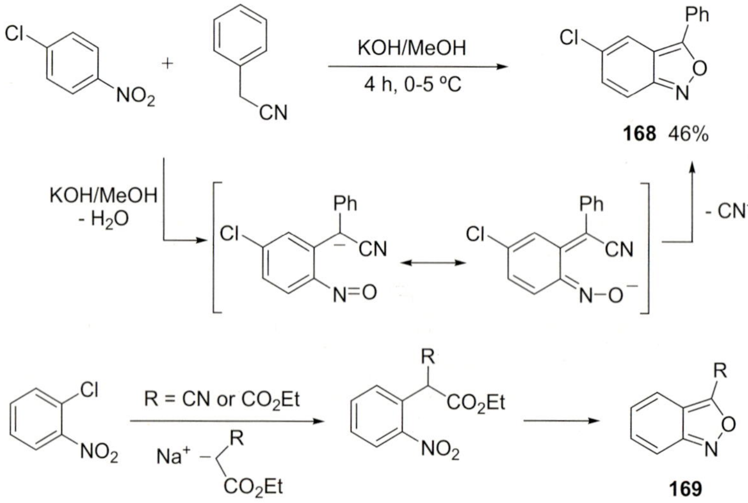

KOH/MeOH, 4 h, 0-5 °C → **168** 46%

KOH/MeOH, $-H_2O$; $-CN^-$

R = CN or CO_2Et → **169**

Scheme 9.64

9.5.3 1,2-Benzisothiazoles

One of the most synthetically appealing methods for 3-substituted 1,2-benzisothiazoles involves the cyclization of the readily accessible oximes of 2-methylthiophenyl ketones **170**. Heating oximes **171** in an acetic anhydride/pyridine mixture converts them into 1,2-benzisothiazoles **172** (Scheme 9.65) [265]. This method has been used to prepare some benzo[*d,d'*]diisothiazoles [266]. The oximes of

R^1 = Me, Et, Ph, $CH_2CH_2CO_2Me$; R^2 = H, Me

Scheme 9.65

2-(*t*-butylthio)benzaldehydes cyclize to 1,2-benzisothiazoles (**172**, R^1=H) when treated with polyphosphoric acid [267].

1,2-Benzisothiazoles **173** and 3-amino-1,2-benzisothiazoles **174** are obtained, respectively, by treatment of 2-sulfanylbenzaldehydes or 2-sulfanylbenzonitriles with chloramine (Scheme 9.66) [268].

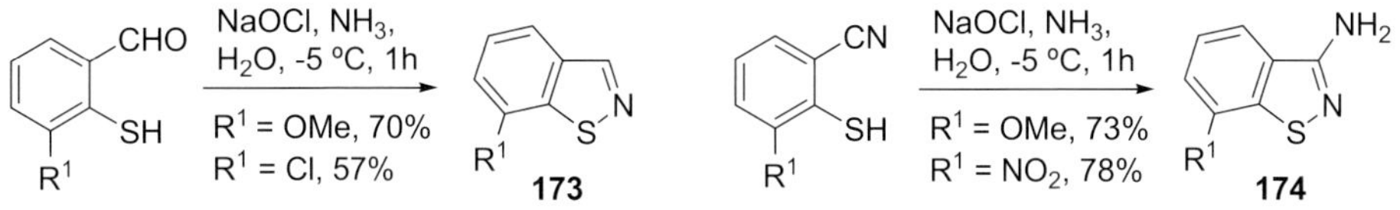

Scheme 9.66

The reaction of 2-benzylthio-4-fluorobenzaldehyde or ketones **175** with sulfuryl chloride gives the corresponding sulfenyl chlorides **176**, which by treatment with ethanol saturated with ammonia afford the 6-fluoro-1,2-benzisothiazoles **177** (Scheme 9.67) [269].

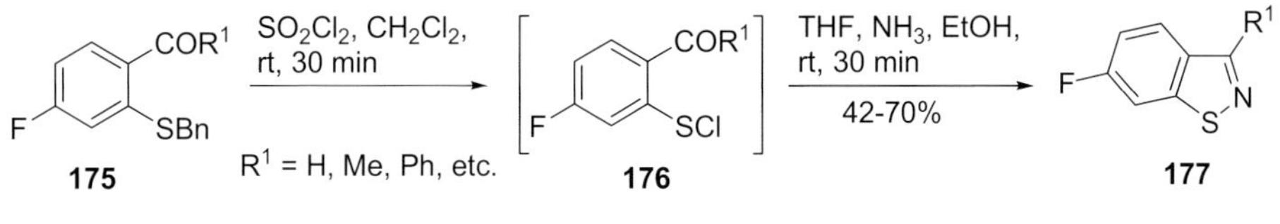

Scheme 9.67

A one-pot procedure for the synthesis of 1,2-benzisothiazoles starting from simple bromobenzenes is shown in Scheme 9.68. It involves the generation of substituted

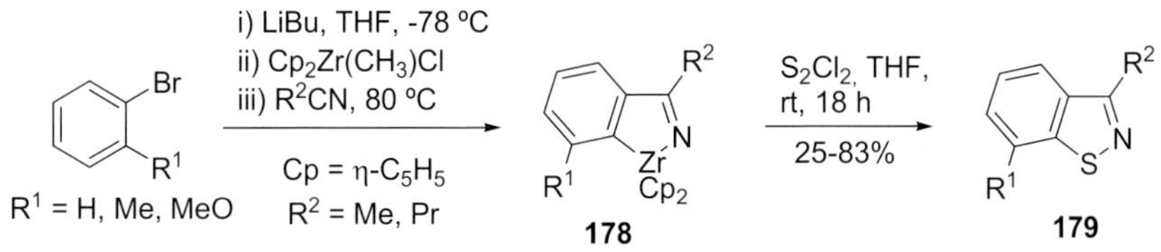

Scheme 9.68

benzynes, formation of zirconium complexes and reaction with nitriles to form metallacyclic compounds **178**. These compounds react with sulfur monochloride to afford regioselectively the benzisothiazoles **179** in moderate to good yields [270].

3-Substituted 1,2-benzisothiazole 1,1-dioxides can be prepared by *ortho*-deprotonation–cyclization of *N*-acylbenzenesulfonamides with 2 equivalents of LDA (Scheme 9.69) [271]. A related method for the synthesis of 3-aryl-1,2-benzisothiazoles involves the ortho-lithiation of *N,N*-diphenylbenzenesulfonamides followed by the addition of aromatic nitriles [272].

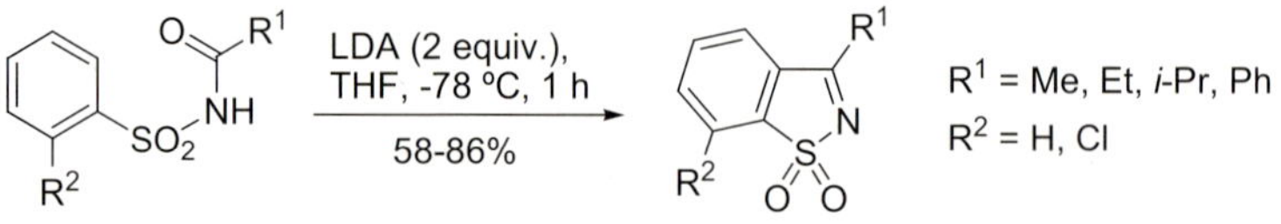

Scheme 9.69

The ortho-lithiation of *N-tert*-butylbenzenesulfonamide followed by reaction with ketones gives the tertiary alcohols **180**, which undergo TMSCl-NaI-MeCN reagent mediated cyclization to afford 3,3-disubstituted 2,3-dihydrobenzisothiazole 1,1-dioxides **181** in high yields (Scheme 9.70) [273].

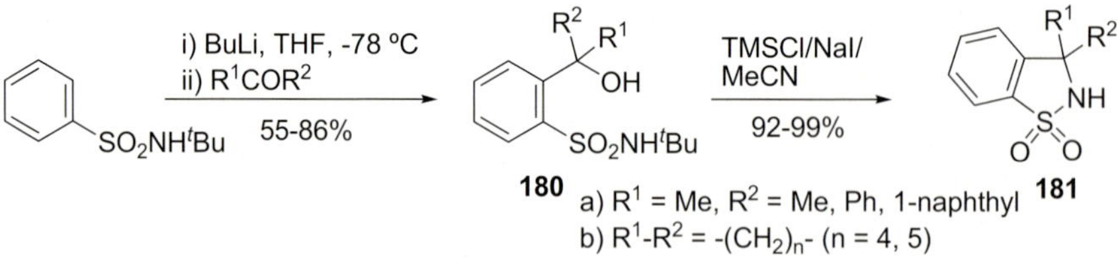

Scheme 9.70

9.5.4
2,1-Benzisothiazoles

ortho-Toluidine and ring substituted *o*-toluidines **182** (R^1=H) react with thionyl chloride in xylene at reflux temperature to yield 2,1-benzisothiazoles **183a** [274]. Similarly, *o*-benzylaniline affords 3-phenyl-2,1-benzisothiazole (**183b**) (Scheme 9.71) [275]. *o*-Toluidines can also be converted into 2,1-benzisothiazoles

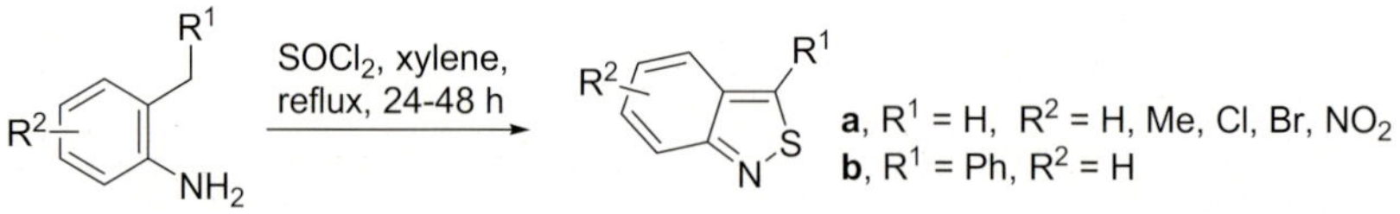

Scheme 9.71

by reaction with *N*-sulfinylmethanesulfonamide (CH_3SO_2NSO) [276]. This method has been used for the synthesis of all the possible angular benzo[*c*]bisisothiazoles and the symmetrical benzo[*c*]trisisothiazole and also benzo[*c:d'*]bisisothiazoles [277, 278].

3-Amino-2,1-benzisothiazoles are easily prepared by the oxidative cyclization of *ortho*-aminothiobenzamides [279, 280]. These compounds can be converted into other 3-substituted derivatives by diazotization and replacement of the diazonium group by halogen atoms or cyanide or nitro groups (Scheme 9.72) [281]. Similarly, oxidative cyclization of *ortho*-aminothiobenzoic acid affords 2,1-benzisothiazol-3-one, which can be converted into 3-chloro-2,1-benzisothiazole (Scheme 9.72) [282]. The chlorine atom is easily and almost quantitatively displaced by nucleophiles, leading to several different 3-substituted 2,1-benzisothiazoles [282].

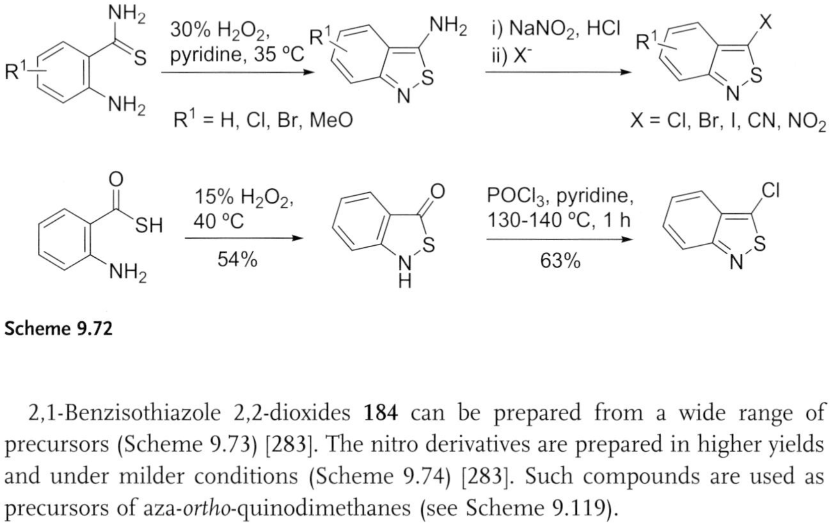

Scheme 9.72

2,1-Benzisothiazole 2,2-dioxides **184** can be prepared from a wide range of precursors (Scheme 9.73) [283]. The nitro derivatives are prepared in higher yields and under milder conditions (Scheme 9.74) [283]. Such compounds are used as precursors of aza-*ortho*-quinodimethanes (see Scheme 9.119).

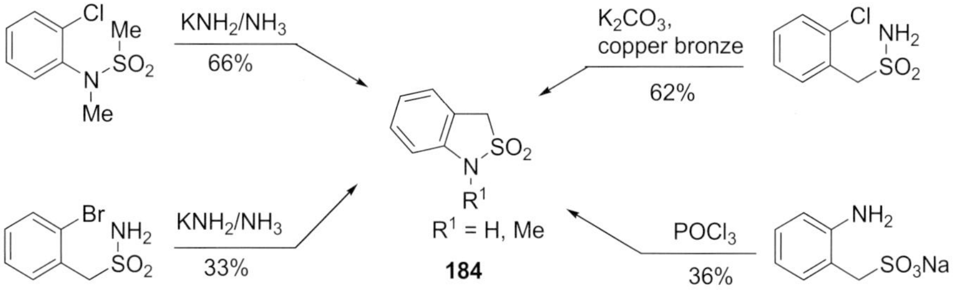

Scheme 9.73

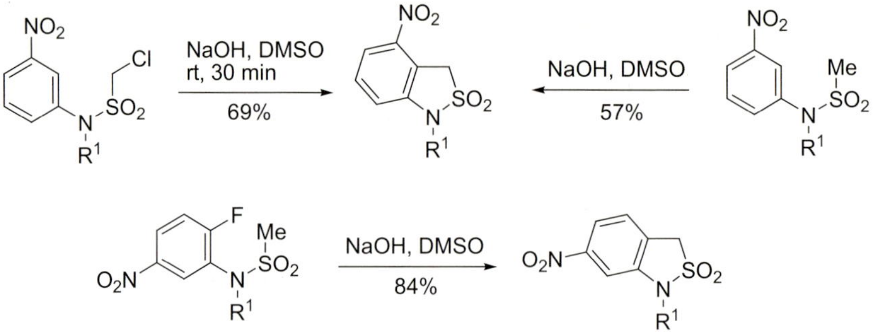

NaOH, DMSO rt, 30 min 69%

NaOH, DMSO 57%

NaOH, DMSO 84%

Scheme 9.74

9.6 Reactivity of 1,2-Azoles

9.6.1 Isoxazoles and Benzisoxazoles

The key feature of these heterocycles is that they possess the typical properties of an aromatic system but contain a weak nitrogen–oxygen bond, which, under certain reaction conditions, particularly in reducing or basic conditions, is a potential site of ring cleavage. Thus, isoxazoles are very useful intermediates since the ring system stability allows the manipulation of substituents to give functionally complex derivatives, yet it is easily cleaved when necessary.

The ring opening provides difunctionalized compounds, namely 1,3-dicarbonyl, enaminoketone, γ-amino alcohol, α,β-unsaturated oxime, β-hydroxy nitrile or β--hydroxy ketone compounds, so that isoxazoles can be considered masked forms of these synthetic units. Consequently, isoxazoles have become an important synthetic tool.

The chemical behavior of benzisoxazoles can, in general, be compared with that of substituted isoxazoles. 1,2-Benzisoxazoles undergo electrophilic substitution in the benzo ring whereas the reaction with nucleophiles involves the isoxazole moiety. Benzisoxazoles readily undergo cleavage of the heterocyclic ring and this feature makes them suitable building blocks for the synthesis of other heterocyclic systems.

9.6.1.1 Thermal and Photochemical Reactions

Scheme 9.75 outlines the reactivity pattern of isoxazoles under thermal reaction conditions. N–O bond cleavage leads to the generation of vinylnitrenes **185**, which rearrange to the corresponding 2*H*-azirines **186**. The 2*H*-azirines can also undergo ring cleavage to give nitrile ylides **187** followed by recyclization to give oxazoles **189** as the final product. However, thermolysis of isoxazoles unsubstituted at C3 usually leads to nitriles **188** (Scheme 9.75) [284, 285]. Thermolysis of 3-unsubstituted 1,2-benzisoxazoles yields the corresponding salicylnitriles **190** [286].

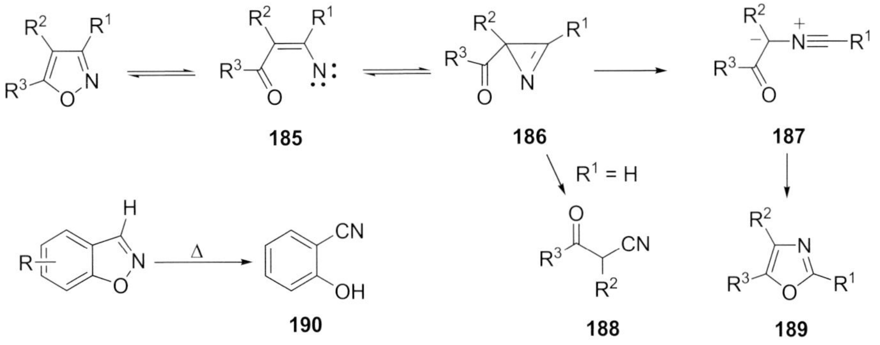

Scheme 9.75

The thermal stability of alkyl or aryl substituted isoxazoles is relatively high. In fact isoxazoles **191** are stable on heating at 280 °C for 10 days [287]. However, isoxazoles having a heteroatom (O, S, N) substituent at C5 undergo ring cleavage at lower temperatures. In fact, 5-alkoxy-3-arylisoxazole **192** is converted into 2*H*-azirine **193** when heated at 200 °C (Scheme 9.76) [288]. High yields of 3-amino-2*H*-azirines **195** are also obtained by both thermolysis and photolysis of 3,5-bis(dimethylamino) isoxazoles **194** [289]. The presence of a carbonyl group in the isoxazole C4 position also favors the cleavage of the N—O bond [287, 290]. Thus, heating 4-acylisoxazoles **196** at 230–240 °C affords the isomeric 4-acyloxazoles **197** in good yields.

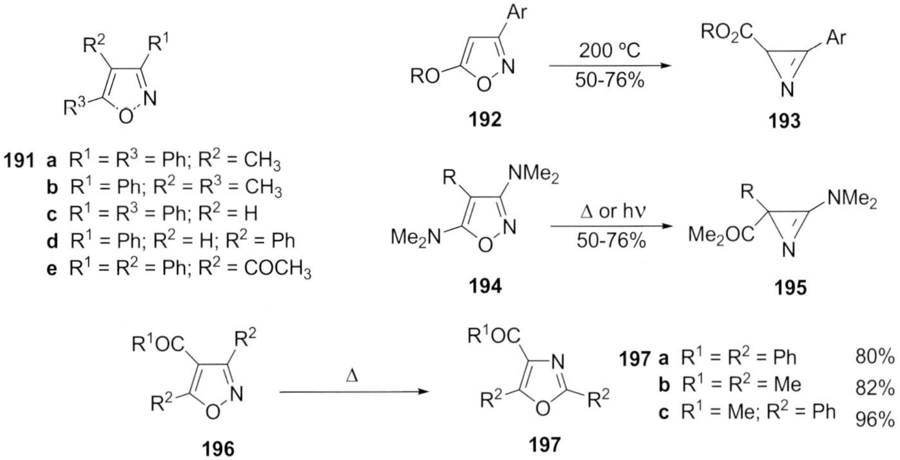

191 **a** $R^1 = R^3 = Ph$; $R^2 = CH_3$
b $R^1 = Ph$; $R^2 = R^3 = CH_3$
c $R^1 = R^3 = Ph$; $R^2 = H$
d $R^1 = Ph$; $R^2 = H$; $R^2 = Ph$
e $R^1 = R^2 = Ph$; $R^2 = COCH_3$

197 **a** $R^1 = R^2 = Ph$ 80%
b $R^1 = R^2 = Me$ 82%
c $R^1 = Me$; $R^2 = Ph$ 96%

Scheme 9.76

Flash vacuum pyrolysis (FVP) of 3-phenyl-1,2-benzisoxazole allows the synthesis of 2-phenylbenzoxazole (**198**) in 80% yield (Scheme 9.77) [291].

Similar chemical behavior is observed when isoxazoles are subjected to photolysis instead of thermolysis (Scheme 9.78). The photochemical rearrangement of **199a**

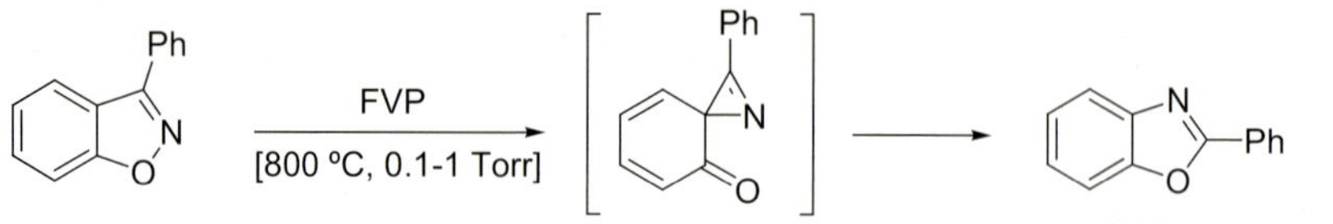

198

Scheme 9.77

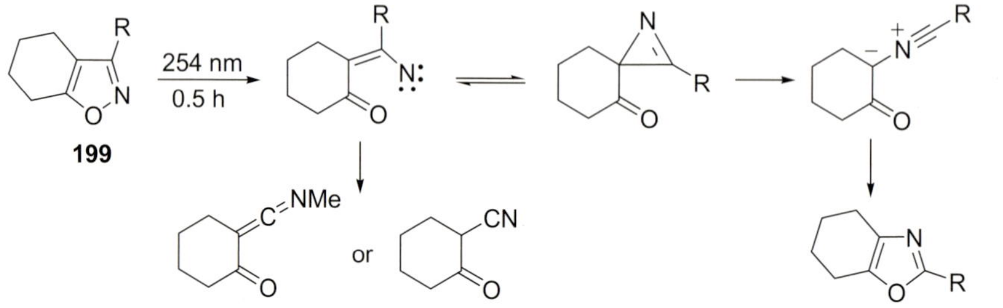

199

201

200 a R = H 99%
b R = Me 63%

Scheme 9.78

(R=H) furnishes **200a** in 99% yield. Irradiation of **199b** (R=Me) gives the corresponding oxazole **200b** in 63% yield. Photolysis of **199b** at $-77\,^{\circ}$C allowed the observation of an IR band at $2050\,\mathrm{cm}^{-1}$, which is assigned to the ketoketenimine **201** [292]. The nature of the solvent used to promote the photochemical reactions can determine the product profile [293].

1,2-Benzisoxazole undergoes photochemical rearrangement to give benzoxazole **202** and salicylnitrile **203**. The synthesis of salicylnitrile can be rationalized by considering the direct cleavage followed by hydrogen transfer, while the formation of benzoxazole occurs via isocyanide intermediate **204**, which can be detected be IR and UV spectroscopy (Scheme 9.79) [294].

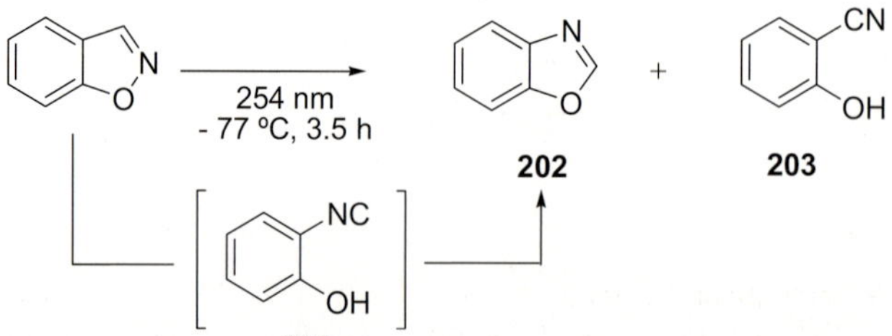

202 203

204

Scheme 9.79

The photolysis of 3-methyl-1,2-benzisoxazole in *n*-hexane/acetonitrile gives a salicylamide, whereas carrying out the irradiation in acetonitrile/methanol (95 : 5) gives an iminoester that is converted into methyl salicylate upon hydrolysis. Photolysis of 3-methyl-1,2-benzisoxazole in 2M H_2SO_4 affords 2-aminophenol via hydrolysis of the benzoxazole intermediate (Scheme 9.80) [10].

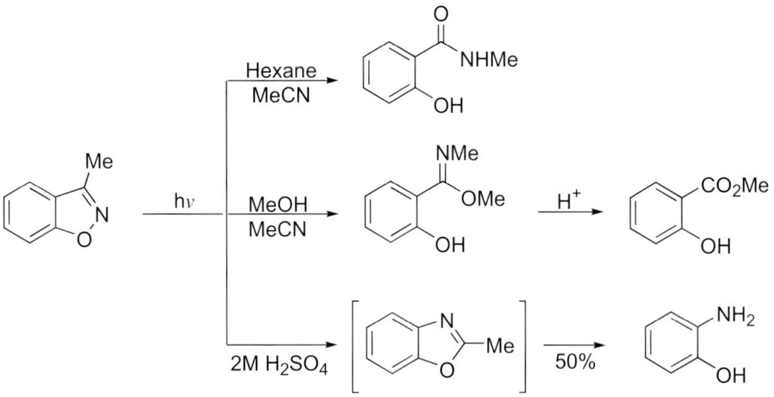

Scheme 9.80

2,1-Benzisoxazoles can undergo ring expansion reactions on photolysis (Scheme 9.81). Carrying out the photochemical reaction in methanol leads to the synthesis of 3-acyl-2-methoxy-3*H*-azepines **205**. The reaction in ether containing water or amines affords the corresponding 2-oxo- or 2-amino-3*H*-azepines (**206** or **207**) [295].

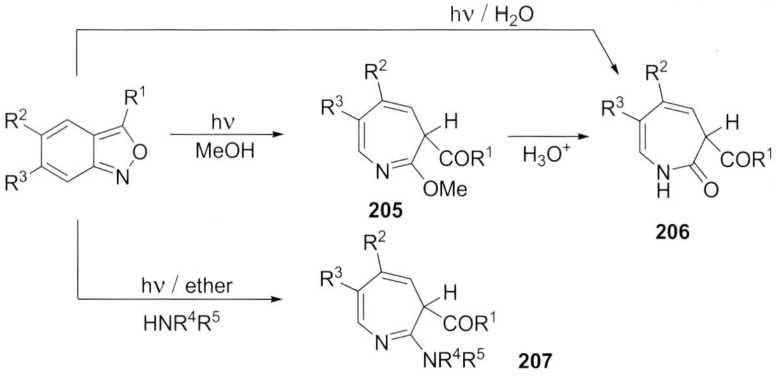

Scheme 9.81

9.6.1.2 **Reactions with Electrophilic Reagents**

Isoxazoles are quaternized by reaction with alkyl iodides or dialkyl sulfates, although special conditions are required due to the low basicity of isoxazoles and their susceptibility to nucleophilic attack. The reactivity of various azoles (1-methylimidazole, thiazole, 1-methylpyrazole, oxazole, isothiazole and isoxazole) towards dimethyl sulfate has been studied and revealed that the parent isoxazole is the least reactive towards quaternization and is also 10^4 times less reactive than pyridine [296]. Thus, direct alkylation with alkyl iodides and sulfates requires relatively vigorous reaction conditions and long reaction times. The rates of quaternization of isoxazole, 1,2-benzisoxazole and 2,1-benzisoxazole with dimethyl sulfate show that 1,2-benzisoxazole is the least reactive towards *N*-methylation whereas 2,1-benzisoxazole reacts faster than isoxazole [297].

The reaction of isoxazoles with secondary and tertiary alcohols and perchloric acid, a efficient source of carbonium ions, is a more convenient route to isoxazolium salts with bulky *N*-substituents. Reaction of 3,5-dimethylisoxazole with a range of alcohols occurs at room temperature, affording isoxazolium salts in 50–90% yield (Scheme 9.82) [298].

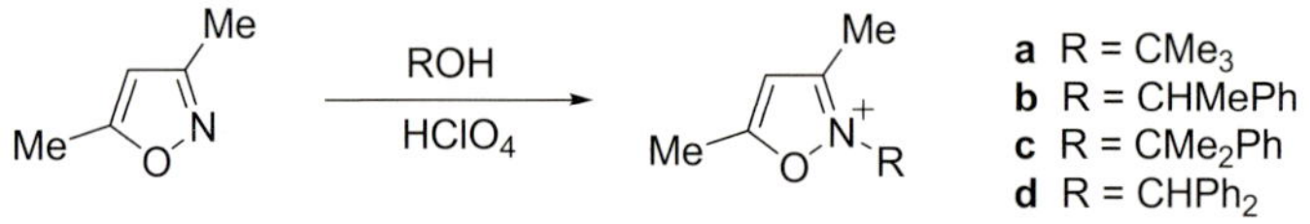

ROH, $HClO_4$

a R = CMe_3
b R = CHMePh
c R = CMe_2Ph
d R = $CHPh_2$

Scheme 9.82

4-Unsubstituted isoxazoles undergo electrophilic substitutions such as nitration, sulfonation, halogenation, chloromethylation and hydroxymethylation, Vilsmeier–Haack formylation and acetoxy mercuration at the 4-position of the ring. Isoxazoles are less reactive towards electrophilic attack than furan but more reactive than pyridine, as expected for a heterocycle having an activating oxygen and a pyridine-like *N*-atom.

The parent isoxazole is nitrated with great difficulty to give 4-nitroisoxazole in 3.5% yield under controlled conditions, with mixed nitric acid and sulfuric acid at 35–40 °C. However, nitration of 3,5-dimethylisoxazole at 100 °C affords the 4-nitro derivative in 86% yield. Both 3-methyl- and 5-methylisoxazole are nitrated regioselectively at the 4-postion. Aryl substituted isoxazoles can be nitrated under mild conditions, although competition between nitration at the C4 of the isoxazole ring and nitration at the aryl group can occur. Under controlled conditions, nitration of 3,5-diphenylisoxazole in Ac_2O/HNO_3 at 20 °C affords only 4-nitro-3,5-diphenylisoxazole. However, the same isoxazole in HNO_3 at 0 °C undergoes nitration at the phenyl groups [9, 10].

1,2-Benzisoxazoles undergo electrophilic substitutions, such as nitration, sulfonation and halogenation, preferentially at the 5-position [9, 10]. Nitration of the parent 1,2-benzisoxazole gives exclusively the 5-nitro-1,2-benzisoxazole [299] and 3-substituted 1,2-benzisoxazoles also lead to the synthesis of the 5-nitro derivatives as the major product.

Isoxazole can act as an activating substituent, as illustrated by the reaction of 5,5′-diisoxazole, which undergoes nitration at the 4-position of both rings (Scheme 9.83) [300].

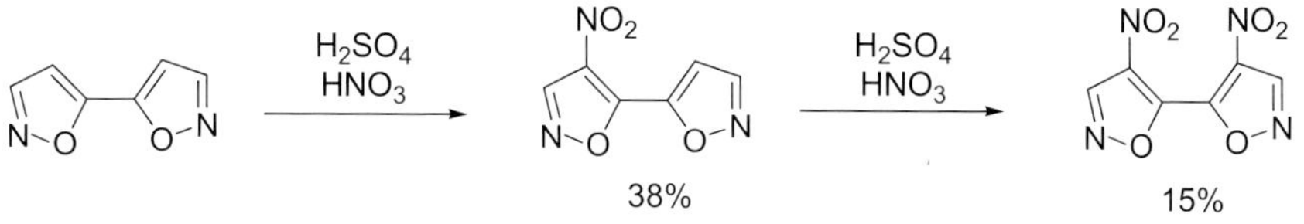

H_2SO_4 / HNO_3 → NO_2 38% ; H_2SO_4 / HNO_3 → NO_2 NO_2 15%

Scheme 9.83

Isoxazoles are rather resistant to sulfonation [6, 10]. This is illustrated by the reaction of 5-phenylisoxazole with chlorosulfonic acid, which undergoes sulfonation only at the phenyl substituent, to give a mixture of *m*- and *p*-phenylsulfonyl chloride isoxazole derivatives. However, on prolonged heating with chlorosulfonic acid, 3-methyl-, 5-methyl, and 3,5-dimethyl-isoxazoles are converted into the corresponding sulfonic and sulfonyl chlorides via electrophilic substitution at C4.

Isoxazoles can be halogenated with various reagents, leading to 4-haloisoxazole derivatives. Treatment of isoxazoles with chlorine or bromine leads to coordination compounds, which afford 4-chloro- or 4-bromoisoxazoles when heated or irradiated. An improved procedure for the C4 halogenation of 3,5-diarylisoxazoles with *N*-halosuccinimide (NBS, NCS and NIS) in acetic acid has been reported. The 4-haloisoxazoles are obtained in yields ranging from 37 to 97% [301].

Other examples of electrophilic substitution of isoxazoles have been reviewed [6, 10].

Ring-opening reactions of isoxazoles can be carried out under acidic conditions. This aspect of the isoxazoles' reactivity has been explored in the conversion of 5-(2,3,5-tri-*O*-benzoyl-β-D-ribofuranosyl)isoxazole-4-carbaldehyde (**208**) into 3-cyano-1,5-benzodiazepine *C*-nucleosides **209** (Scheme 9.84) [302].

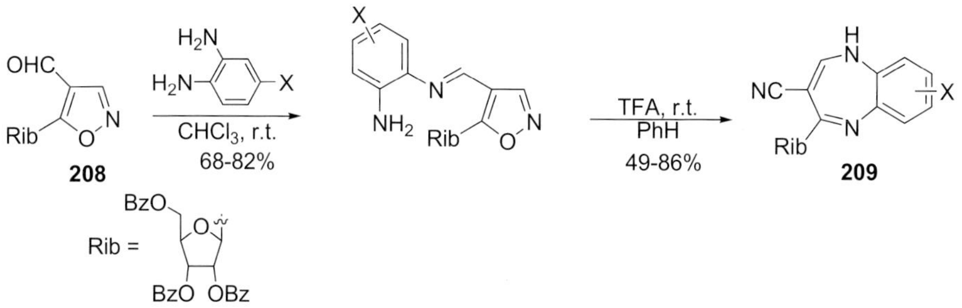

OHC, Rib, **208**; H_2N, H_2N, X; $CHCl_3$, r.t. 68-82%; X, NH_2, Rib; TFA, r.t. PhH 49-86%; NC, H, N, Rib, X, **209**

Rib = BzO, O, BzO OBz

Scheme 9.84

9.6.1.3 Reactions with Nucleophilic Reagents

The lability of isoxazoles towards nucleophiles is a key feature of their reactivity. However, its reactivity depends on the nature and position of the substituents. In general, the stability increases with increasing substitution. In fact, the trisubstituted isoxazoles are usually stable and react with nucleophiles preferentially in the side-chains.

The 3-unsubstituted isoxazoles are cleaved by bases giving cyanoenolates via a one-step concerted E2 type mechanism [303]. Protonation of **210** followed by rearrangement affords β-ketonitriles **211**, which in some cases undergo further reactions (Scheme 9.85). The reaction can be carried out with strong bases such as alkoxides, sodium amide, lithium diisopropylamide, butyllithium and also with hydroxide ion or with weaker bases, such as ammonia and phenyl hydrazine, at higher temperature.

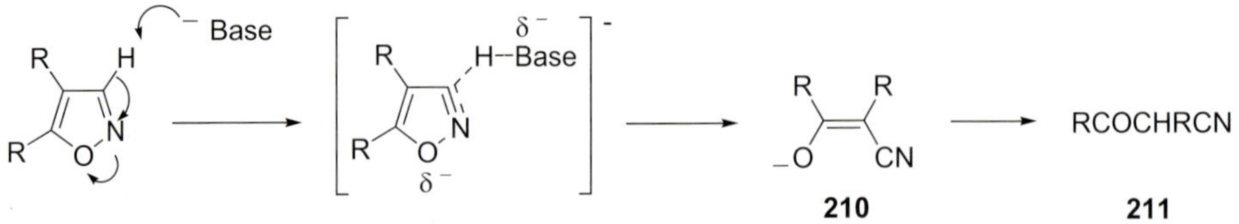

RCOCHRCN

210 **211**

Scheme 9.85

The 3-unsubstituted 1,2-benzisoxazoles present similar behavior, reacting with hydroxide ion and amines to yield 2-cyanophenolates [304]. Reactions of 1,2-benzisoxazoles with sodium borohydride and lithium aluminium hydride usually result also in N–O bond cleavage [9].

The ring opening of 5-unsubstituted 3-alkyl- or 3-arylisoxazoles requires more vigorous reaction conditions, for example heating with alkoxides or use of stronger bases, such as sodium amide or butyllithium. H5 deprotonation of these derivatives leads to N–O and C3/C4 bond cleavage with formation of a nitrile and an ethynolate [305]. The 5-unbstituted isoxazoles bearing a potential leaving group at C3 react with bases without the C3/C4 bond cleavage. This is the case with cyanoisoxazole **212**, which reacts with sodium ethoxide to give **213** as the final product (Scheme 9.86) [306].

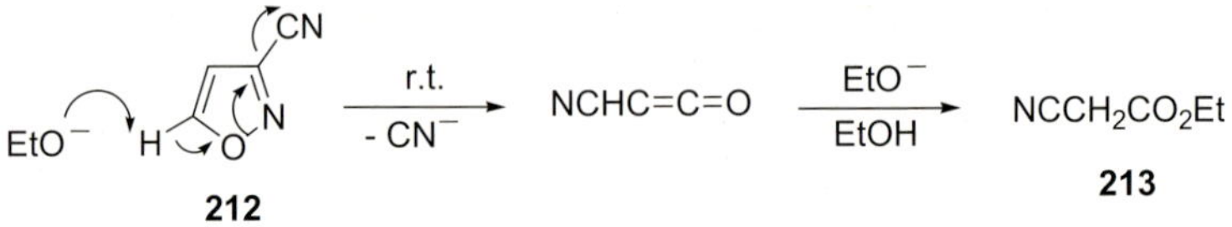

$$\text{EtO}^- \quad \mathbf{212} \xrightarrow[\text{- CN}^-]{\text{r.t.}} \text{NCHC=C=O} \xrightarrow[\text{EtOH}]{\text{EtO}^-} \text{NCCH}_2\text{CO}_2\text{Et}\ \mathbf{213}$$

Scheme 9.86

1,2-Benzisoxazoles can be used as a building block for the synthesis of other heterocycles via a reaction with bases [307, 308]; Scheme 9.87 shows two examples.

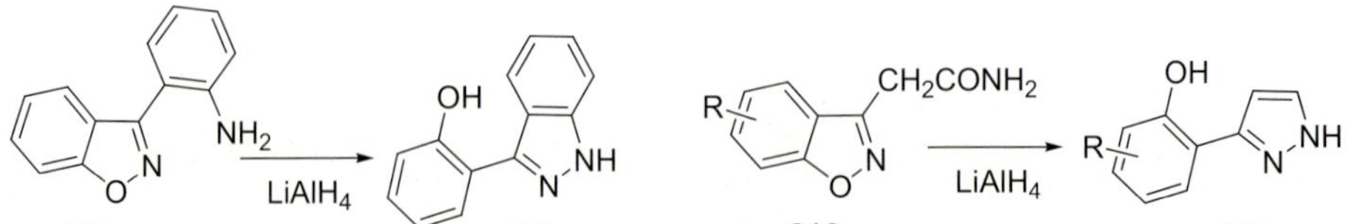

$$\mathbf{214} \xrightarrow{\text{LiAlH}_4} \mathbf{215} \qquad \mathbf{216} \xrightarrow{\text{LiAlH}_4} \mathbf{217}$$

Scheme 9.87

The 1,2-benzisoxazole amine derivative **214** is converted into 3-(2-hydroxyphenyl) indazole (**215**) by treatment with $LiAlH_4$ or NaH [307] and the 1,2-benzisoxazole amide derivatives **216** give 3-(2-hydroxyphenyl)pyrazoles **217** [308].

Trisubstituted isoxazoles with strong electron-withdrawing groups at the 4-position are susceptible to ring cleavage when reacting with nucleophiles (Scheme 9.88). For example, alkaline treatment of 5-methylamino-4-nitroisoxazole **218** leads to oxime **219** along with methylamine [309]. The process proceeds with initial nucleophilic attack at the 5-position of the isoxazole ring. 4-Aroylisoxazole **220** undergoes a rearrangement to give 4-acetylisoxazole **222** via an initial nucleophilic attack at the 5-position followed by cyclization of oxime **221** [310].

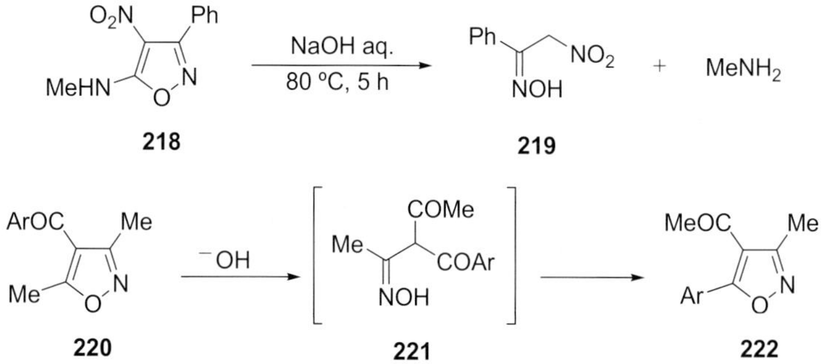

Scheme 9.88

Isoxazolium salts are easily cleaved with nucleophiles. The quaternization of the nitrogen atom increases the lability of the isoxazole ring towards nucleophilic attack. The 3-unsubstituted isoxazolium salts undergo ring cleavage with mild nucleophiles, including carboxylate ions in aqueous solution, which makes these derivatives useful coupling reagents for peptide synthesis. This synthetic strategy is outlined in Scheme 9.89. Deprotonation of isoxazolium salts **223** at the 3-position is followed by ring opening and the ketoketenimines **224** formed react with a carboxylic acid to

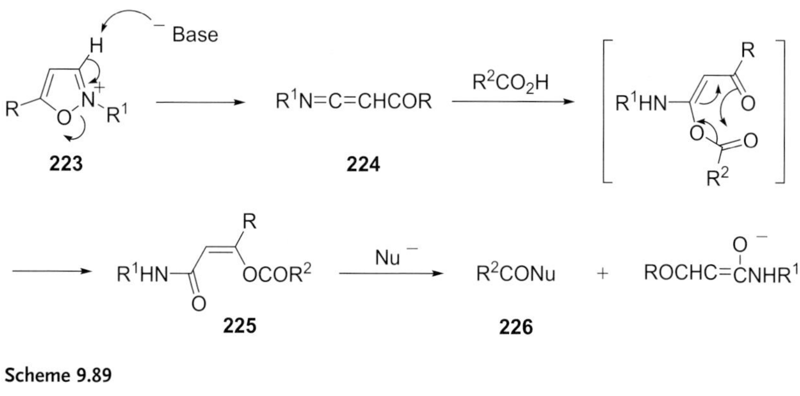

Scheme 9.89

give **225**. Reaction of this enol ester with nucleophiles (e.g., an amino acid) gives the final product **226** [10].

1,2-Benzisoxazolium salts readily undergo ring opening to salicylnitrile derivatives upon treatment with bases [311, 312]. The 3-unsubstituted 2,1-benzisoxazoles show similar instability in the presence of bases, and easily undergo ring opening reactions to give anthranilic acid derivatives. The 2,1-benzisoxazolium salts are particularly reactive towards nucleophilic attack at the 3 position and stable adducts can be obtained from their reaction with a range of nucleophiles. The 3-unsubstituted 2,1-benzisoxazolium salts behave in an analogous manner to their 1,2-isomers. In the reaction of 2,1-benzisoxazolium salt **227** with triethylamine the deprotonation is followed by ring opening to the iminoketene **228**, which undergoes electrocyclization to give *N-tert*-butylbenzoazetinone **229** in 84% yield (Scheme 9.90) [313].

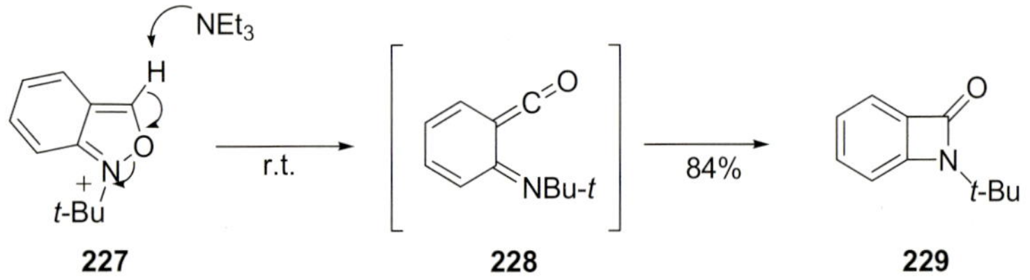

NEt3 / H / O / N / t-Bu / 227 → r.t. → [C=O / NBu-t / 228] → 84% → O / N / t-Bu / 229

Scheme 9.90

The reactivity profile of isoxazoles with nucleophiles also includes nucleophilic addition to the ring and nucleophilic replacement reactions. Halide displacement reactions can be carried out with 3-halo- and 5-haloisoxazoles bearing the appropriate substitution pattern to prevent ring opening. 4-Haloisoxazoles are very stable and their reactivity towards nucleophiles is similar to that of aryl halides.

9.6.1.4 **Cycloaddition Reactions**

4-Nitro-3-phenylisoxazoles **230** act as dienophiles towards 2,3-dimethylbutadiene (Scheme 9.92 below) [314, 315]. The activated C4/C5 double bond undergoes Diels–Alder reaction with 2,3-dimethylbutadiene, acting as a synthetic equivalent of the corresponding didehydro derivative **231** since the activating groups can be easily removed. Thus, isoxazole **230c** undergoes Diels–Alder reaction with 2,3-dimethylbutadiene to give the bicyclic derivatives **232** and **233** in 49 and 37% yields, respectively. The initial adducts **232** can be easily converted into **233** on treatment with DBU and both **232** and **233** are converted into benzisoxazole **234** either by prolonged heating or by oxidation with DDQ.

The reaction of nitroisoxazole **230c** with 1-azadiene **235** affords the isoxazolopyridine **237** in 59% yield, with loss of nitrous acid and dimethylamine from the initial cycloadduct **236** (Scheme 9.91) [315].

The nitroalkene moiety of 4-nitroisoxazoles undergo hetero-Diels–Alder reactions with enol ethers (e.g., ethyl vinyl ether) [316].

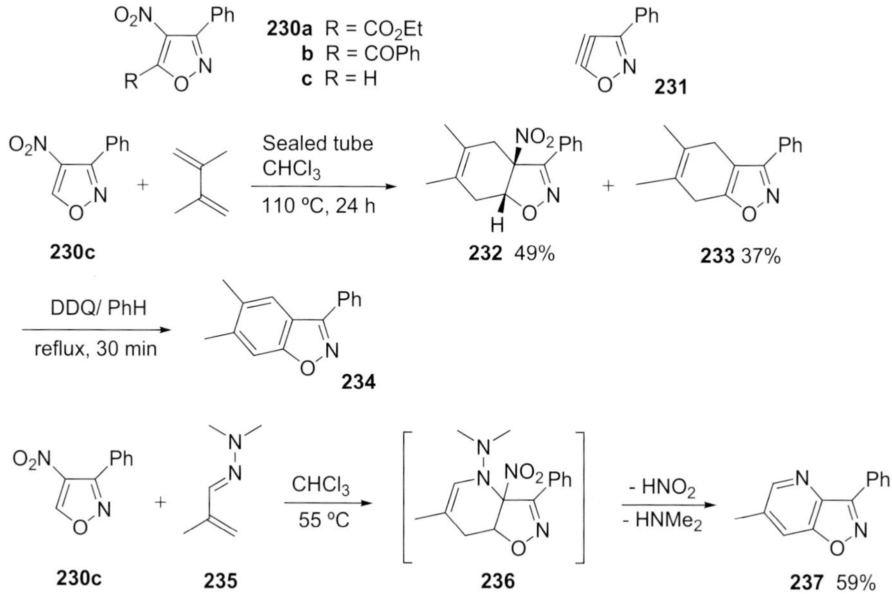

230a R = CO_2Et
b R = COPh
c R = H

231

Sealed tube
$CHCl_3$
110 °C, 24 h

230c

232 49%

233 37%

DDQ/ PhH
reflux, 30 min

234

$CHCl_3$
55 °C

- HNO_2
- $HNMe_2$

230c 235 236 237 59%

Scheme 9.91

2,1-Benzisoxazole participates in Diels–Alder reactions as a diene. Cycloaddition with *N*-phenylmaleimide gives the corresponding *exo* product together with a ring-opened product [317]. From the reaction of 2,1-benzisoxazole with dimethyl acetylenedicarboxylate (DMAD) in refluxing benzene the quinoline 1-oxide is obtained. The mercury sulfate-catalyzed cycloaddition of 2,1-benzisoxazole with cyclic ketones has also been reported (Scheme 9.92) [318].

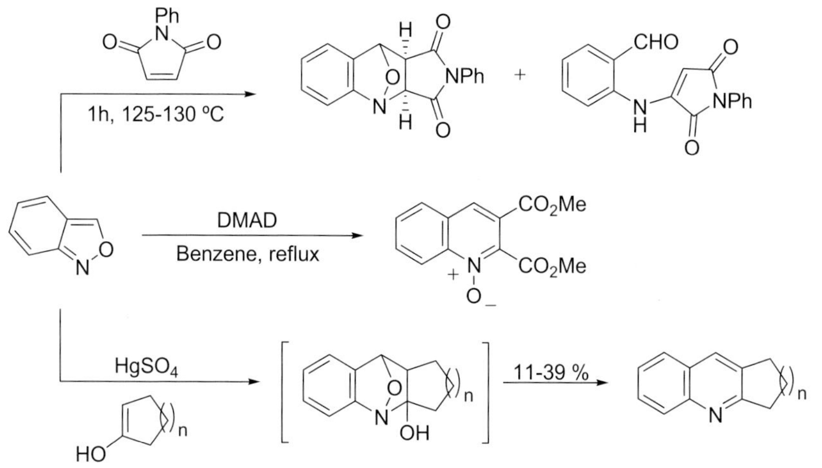

1h, 125-130 °C

DMAD
Benzene, reflux

$HgSO_4$

11-39 %

Scheme 9.92

1,3-Dipolar cycloadditions of 2,1-benzisoxazole are also known, as illustrated by the example presented in Scheme 9.93 [319]. The reaction of 6-nitro-2,1-benzisoxazole-3-carboxylate **238** with excess of diazoacetic esters affords 6*H*-pyrazolo[3,4-g][2,1]benzisoxazoles **240** in good yield. The reaction proceeds via the formation of the 1,3-dipolar cycloadduct **239** followed by elimination of nitrous acid.

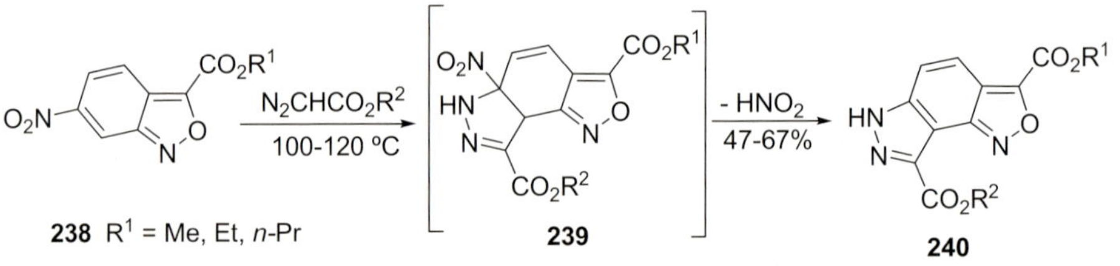

$N_2CHCO_2R^2$, 100-120 °C; $-HNO_2$, 47-67%

238 R^1 = Me, Et, *n*-Pr; **239**; **240**

Scheme 9.93

9.6.1.5 Reactions with Reducing Agents

Isoxazoles are readily cleaved at the N—O bond under reducing conditions and many examples have been reported. Catalytic hydrogenolysis leads to β-enaminoketones or to the corresponding 1,3-diketone obtained by hydrolysis. The reduction with sodium in liquid ammonia in the presence of 3 equivalents of *tert*-butyl alcohol gives β-aminocarbonyl compounds, which are converted into α,β-unsaturated ketones on heating or under acidic conditions. Thus, isoxazoles can be considered masked forms of these important synthetic building blocks. The example shown in Scheme 9.94 illustrates this type of reactivity [113]. The use of isoxazoles as masked β-enaminoketones has been applied in a strategy for the total synthesis of vitamin B_{12} [320].

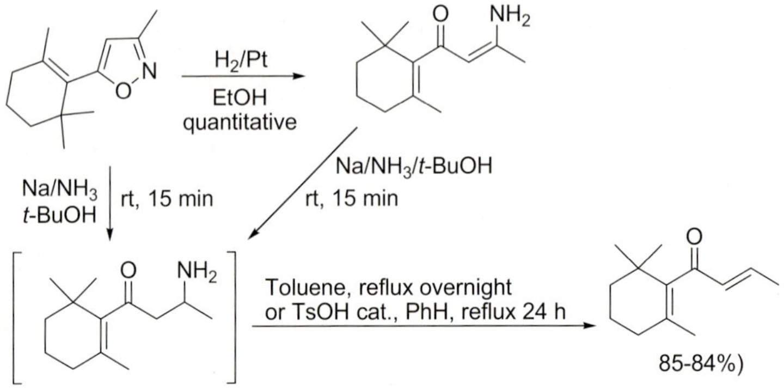

H_2/Pt, EtOH, quantitative; Na/NH_3/*t*-BuOH, rt, 15 min; Na/NH_3/*t*-BuOH, rt, 15 min; Toluene, reflux overnight or TsOH cat., PhH, reflux 24 h; 85-84%)

Scheme 9.94

Samarium diiodide [321] and transition-metal carbonyls [322, 323], such as molybdenum hexacarbonyl in the presence of water, are also efficient reagents for the reductive cleavage of isoxazoles. Nitta *et al.* have reported that 3-methyl-5-(2-oxoalkyl)isoxazoles **241** undergo a $Mo(CO)_6$-induced reductive cleavage to give pyridin-4(1*H*)-ones **244** in a single step [322]. Here, complex formation of isoxazole with $Mo(CO)_6$ is followed by the N–O bond cleavage to give the nitrene complex **242**. Hydrolysis of **242** gives the enamino ketone **243**, which cyclizes to the pyridin-4(1*H*)-ones **244**. Li *et al.* have shown that if the 2-oxoalkyl side-chain is at the 3-position of the isoxazole **245** the corresponding enamino ketone could be isolated after reduction with $Mo(CO)_6$ and could be cyclized to pyran-4-ones **246** under acidic conditions (Scheme 9.95) [323].

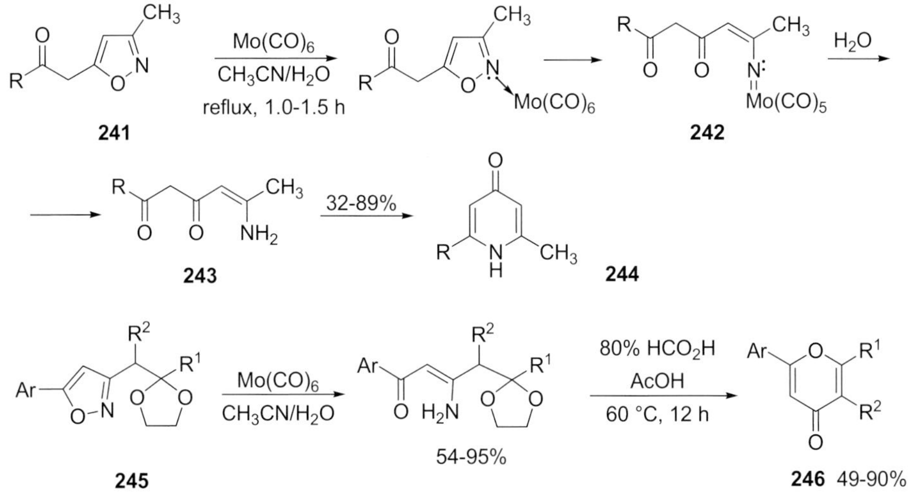

Scheme 9.95

Under reducing conditions the N–O bond of benzisoxazoles is readily cleaved and many examples have been reported. The catalytic hydrogenolysis of 1,2 benzisoxazoles gives 2-iminophenols and/or 2-ketophenols depending on the reaction conditions. The catalytic reduction of 2,1-benzisoxazoles results in the formation of 2-aminophenones.

Hydrogenolysis of 3-aryl-2,1-benzisoxazoles is a useful route to 2-aminobenzophenones (Scheme 9.96) [324]. The efficient reductive cleavage of the N–O bond of

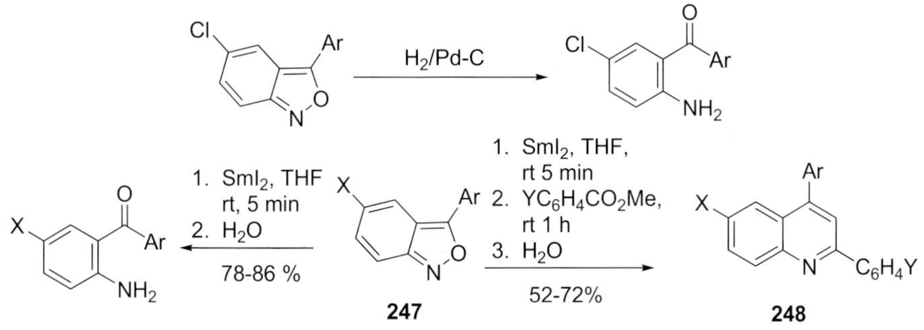

Scheme 9.96

5-substituted-3-aryl-2,1-benzisoxazoles **247** can also be achieved with samarium(II) iodide. If aryl methyl ketones are added to the reactive mixture 2,4-diarylquinolines **248** are obtained [325].

Reductive ring cleavage of isoxazoles to the corresponding β-enaminoketones under treatment with titanium(IV) isopropoxide and ethylmagnesium bromide has been reported (Scheme 9.97). The interaction of equimolar quantities of EtMgBr and $Ti(O\text{-}^iPr)_4$ leads to the formation of titanium(III) isopropoxide. This reagent assists the homolytic cleavage of the N–O bond in isoxazoles **249**, giving the alcoholates **250**, followed by hydrolysis to afford the enaminoketones **251**. From isoxazolines the corresponding β-hydroxyketones are obtained [326]. The isoxazole ring is also cleaved by reactions with LDA [327].

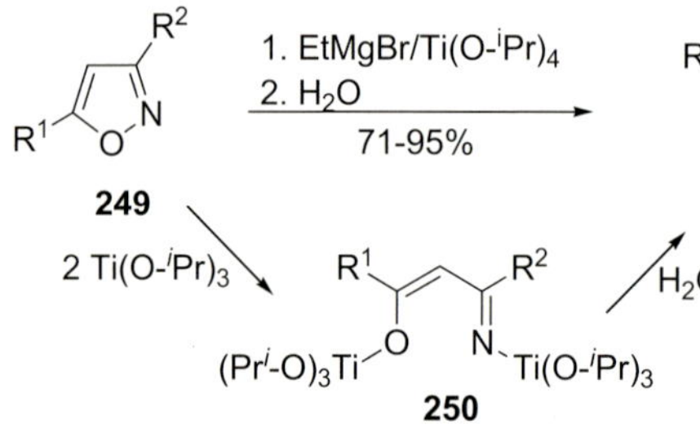

Scheme 9.97

9.6.1.6 Reactions with Oxidizing Agents

Isoxazoles are stable to acidic oxidizing reagents such as peroxyacids, chromic and nitric acids and acidic permanganate. The only general method of oxidation ring cleavage of substituted isoxazoles is ozonolysis. 3-Unsubstituted isoxazoles are also easily converted into cyanoketones with alkaline oxidizing reagents.

Oxidation reactions of isoxazoles bearing heterocyclic substituents allows us to redraw conclusions concerning the relative stability of various heterocyclic compounds under the reaction conditions used (Scheme 9.98) [10]. The isoxazole ring proved to be more stable than furan but less stable than pyrazole and furazan rings.

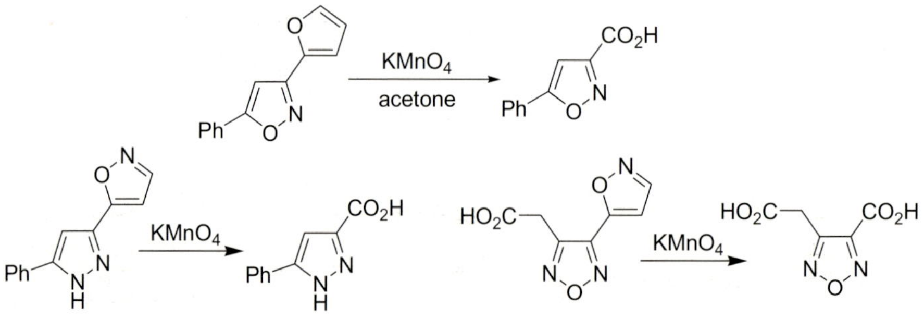

Scheme 9.98

1,2-Benzoisoxazoles are also quite stable towards oxidizing reagents, allowing selective oxidation of substituents of the 1,2-benzisoxazole ring system. Substituted 2,1-benzisoxazole can be oxidized with nitrous acid or with CrO_3/AcOH to generate mixtures of ring-opened products, the rate of which is dependent on the amount of oxidant (Scheme 9.99) [9].

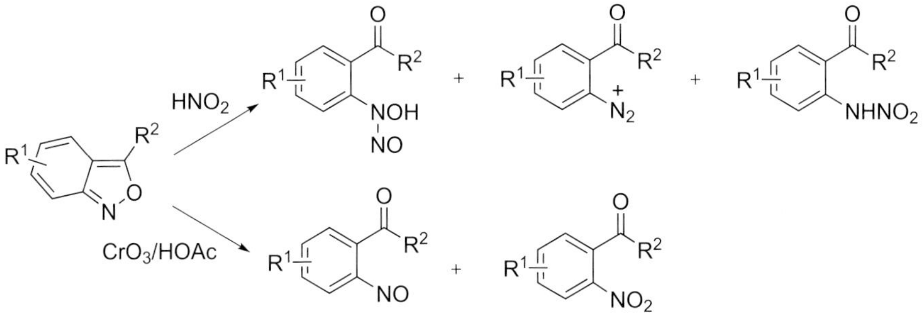

Scheme 9.99

9.6.1.7 **Reactions of Metallated Isoxazoles**

Protons at the α-position of alkylisoxazoles are relatively acidic and can be removed by strong bases (e.g., BuLi, LDA or $NaNH_2/NH_3$), giving carbanionic species. Reaction of these intermediates with electrophiles is an approach to side-chain functionalization and many examples are known [6, 328]. The α-deprotonation of an 5-alkyl substituent is favored over deprotonation of alkyl groups at 3- and 4-positions of the isoxazole ring, allowing regioselective reactions with electrophiles via lateral metallation. Thus, the 3,5-dimethylisoxazole **252a** reacts with BuLi to give specific metallation at C5 methyl group and the subsequent treatment with Me_3SiCl affords **253a** in 70% yield (Scheme 9.100). Regioselective metallation at the same position is observed for trimethylisoxazole **252b** [329, 330]. 4-Metalloisoxazoles are generally prepared by halogen–metal exchange reactions [330].

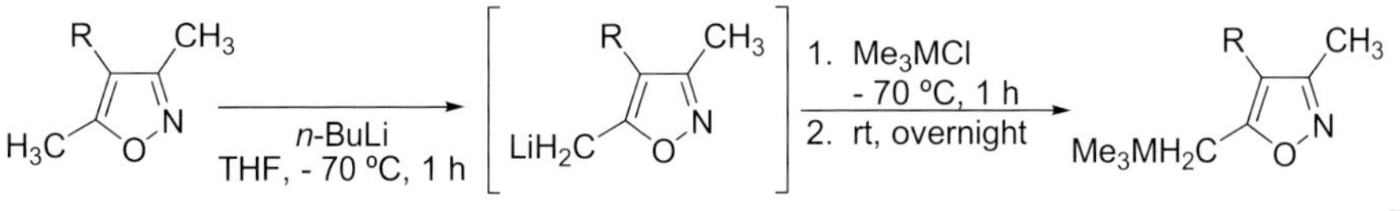

252a R = H
b R = CH_3

253 a R = H; M = Si 70%
b R = CH_3; M = Si 60%
c R = CH_3; M = Sn 75%

Scheme 9.100

Direct lithiation of 3-(Boc-amino)isoxazole **254** and 5-(Boc-amino)isoxazole **256** using BuLi or *s*-BuLi/TMEDA as the lithiating reagents, respectively, has been reported (Scheme 9.101) [331]. The anion intermediates undergo addition to electrophiles to give 4-substituted isoxazoles (**255** and **257**).

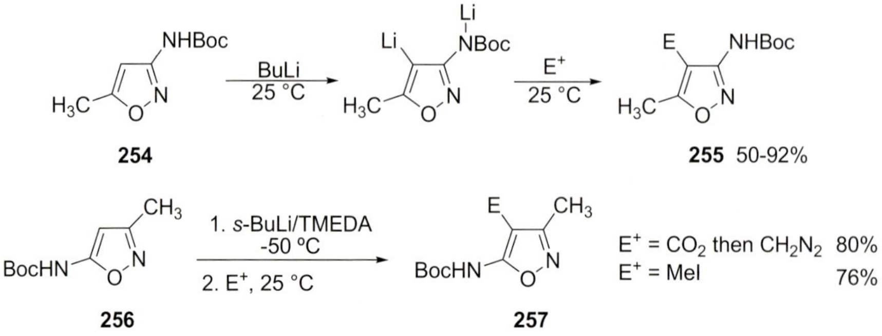

254 → (BuLi, 25 °C) → [Li intermediate] → (E^+, 25 °C) → **255** 50-92%

256 → (1. *s*-BuLi/TMEDA, -50 °C; 2. E^+, 25 °C) → **257**

$E^+ = CO_2$ then CH_2N_2 80%
E^+ = MeI 76%

Scheme 9.101

Palladium-catalyzed coupling reactions (Suzuki–Miyaura, Stille or Heck reactions) of 3,5-disubstituted 4-iodoisoxazole afford in good yields the corresponding 4-substituted derivatives, bearing 4-aryl, 4-heteroaryl, 4-vinyl or 4-acetylenyl groups as substituents [332]. Tin 4-metallated isoxazoles are synthesized by stannylcupration of 4-haloisoxazoles **258** and can also be used as intermediates in the preparation of 4-substituted isoxazoles **259**. The 5-iodoisoxazolylpyridine **261**, obtained from **260** via 1,3-dipolar cycloaddition with iodoacetylene, also undergoes palladium-catalyzed coupling to give 5-substituted isoxazoles **262** (Scheme 9.102) [333, 334].

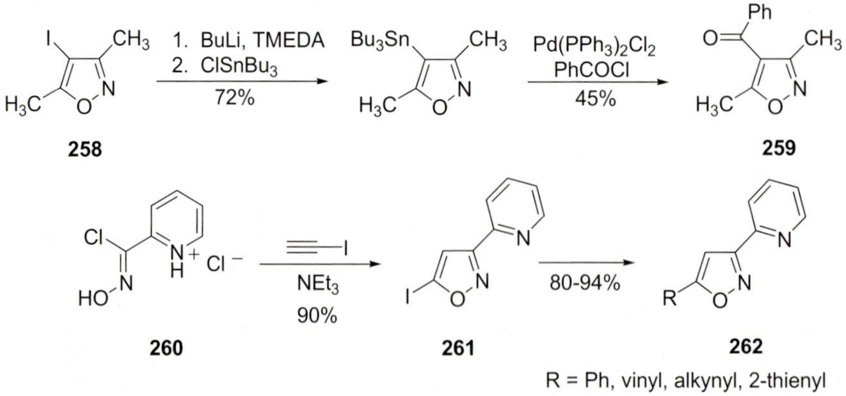

258 → (1. BuLi, TMEDA; 2. $ClSnBu_3$; 72%) → [Bu_3Sn intermediate] → ($Pd(PPh_3)_2Cl_2$, PhCOCl, 45%) → **259**

260 → (iodoacetylene, NEt_3, 90%) → **261** → (80-94%) → **262**

R = Ph, vinyl, alkynyl, 2-thienyl

Scheme 9.102

3,4,5-Trisubstituted isoxazoles can also be prepared via cross-coupling reactions of isoxazolyl-4-sinanols and isoxazole-4-boronic esters with the appropriate halo-compounds [335, 336].

9.6.2
Isothiazoles and Benzisoxazoles

9.6.2.1 Photochemical Reactions

The photochemical reactions of isothiazoles have been reviewed [337]. Isothiazole photoisomerizes to thiazole **264a** (Scheme 9.103) [338]. This photorearrangement is also observed in methylisothiazoles. Each methylisothiazole **263b–d** isomerizes selectively to the corresponding methylthiazole **264**, indicating that the rearrangement occurs via a N2–C3 exchange process [339].

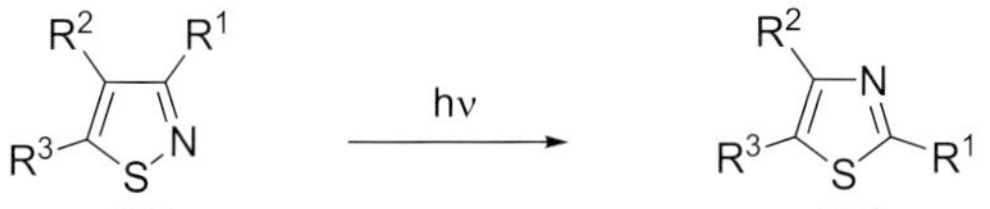

hν

263 **264**

a, $R^1 = R^2 = R^3 = H$
b, $R^1 = Me$, $R^2 = R^3 = H$
c, $R^1 = R^3 = H$, $R^2 = Me$
d, $R^1 = R^2 = H$, $R^3 = Me$

Scheme 9.103

The photochemistry of phenyl substituted isothiazoles has been extensively studied. The resulting products are phenylthiazoles, phenylisothiazoles isomeric of the starting materials and ring-opened compounds. The relative proportions of these products are strongly dependent on the presence of a basc or acid and, in some cases, on the polarity of the solvent used [340–342].

Saccharin derivatives undergo extrusion of sulfur dioxide when irradiated in solution. When irradiated in ethanol or propan-1-ol, the *N*-propyl derivative **265** is converted into *N*-propylbenzamide **266** by hydrogen uptake, while in benzene it gives the *N*-propylbiphenyl-2-carboxamide **267** (Scheme 9.104) [343].

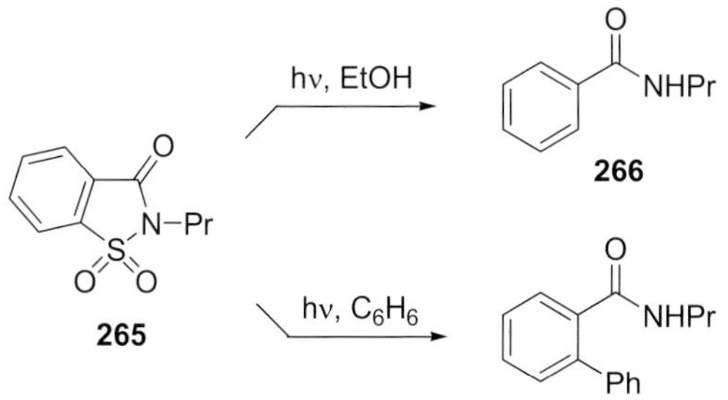

hν, EtOH → **266**
hν, C_6H_6 → **267**
265

Scheme 9.104

In contrast to that observed in the saccharin derivatives, in the 3-mono- and 3,3-disubstituted 2,3-dihydro-1,2-benzisothiazole 1,1-dioxides a net migration of one oxygen atom from sulfur to nitrogen is observed upon irradiation at 254 nm in acetonitrile or methanol (Scheme 9.105) [344].

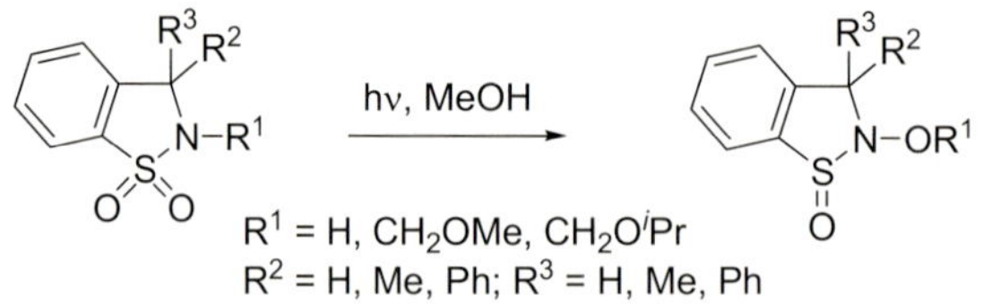

hν, MeOH

R^1 = H, CH_2OMe, CH_2O^iPr
R^2 = H, Me, Ph; R^3 = H, Me, Ph

Scheme 9.105

9.6.2.2 **Reactions with Electrophilic Reagents**

Isothiazoles are quaternized by iodoalkanes, dialkyl sulfate, trialkyloxonium tetrafluoroborate or diazomethane.

Mononuclear isothiazoles give electrophilic substitution at the 4-position, with the 3-position being relatively inert to attack. Nitration also occurs at C4, usually in good yield. 5-Haloisothiazoles give the 4-nitro derivatives in moderate yields by reaction with a mixture of concentrated sulfuric acid and 90% nitric acid (Scheme 9.106) [345].

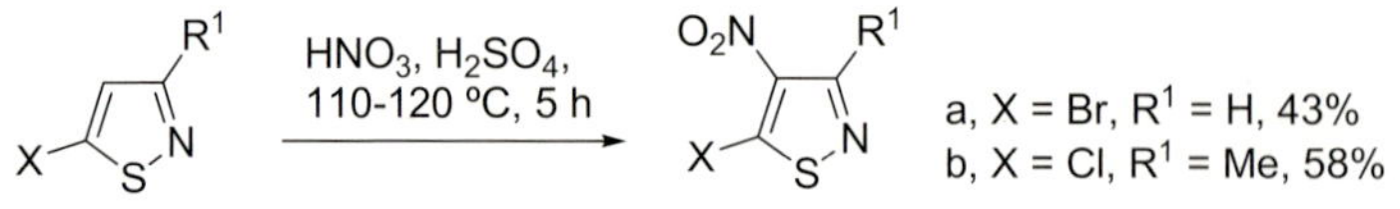

HNO_3, H_2SO_4, 110-120 °C, 5 h

a, X = Br, R^1 = H, 43%
b, X = Cl, R^1 = Me, 58%

Scheme 9.106

Electron-releasing groups in the 3- or 5-position facilitate halogenation. For instance, 3,5-dimethylisothiazole reacts with iodine to give the 4-iodo derivative (Scheme 9.107) [346].

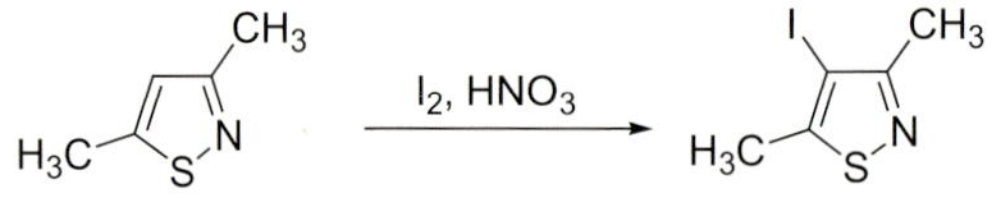

I_2, HNO_3

Scheme 9.107

Bromination of 2-substituted isothiazolin-3-ones **268** affords the 4-bromo derivatives in good yields. The formation of 4,5-dibromo derivatives is much more difficult (Scheme 9.108) [347]. In contrast, even under mild conditions, chlorination of **268** gives primarily 4,5-dichloro derivatives and lesser amounts of the 4-chloro derivatives [347]. 3-Diethylamino-4-(4-methoxyphenyl)isothiazole 1,1-dioxide **269** reacts with bromine to afford the 4,5-dibromo derivative **270**, which, on heating or by treatment with triethylamine, gives the 5-bromoisothiazole **271** [348].

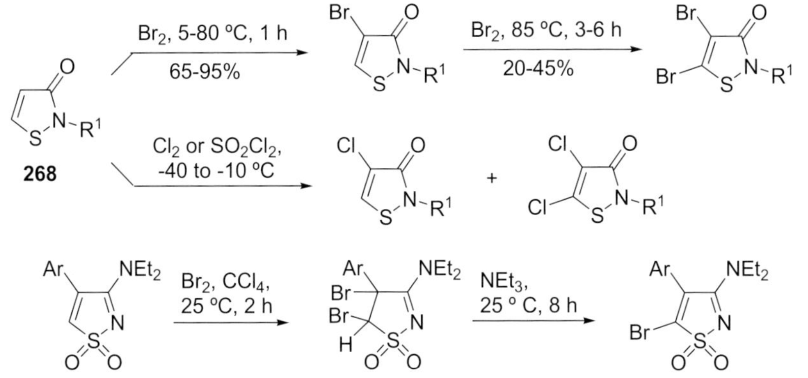

Scheme 9.108

9.6.2.3 Reactions with Nucleophilic Reagents

The displacement of a halogen atom from halo-isothiazoles by nucleophiles is a versatile route to new isothiazole derivatives. As shown in Scheme 9.109, substitution of the halogen by alkylthio groups can be conveniently performed by converting the halo-isothiazole into a thiolate followed by reaction with a suitable alkylating reagent (one-pot procedure) [213].

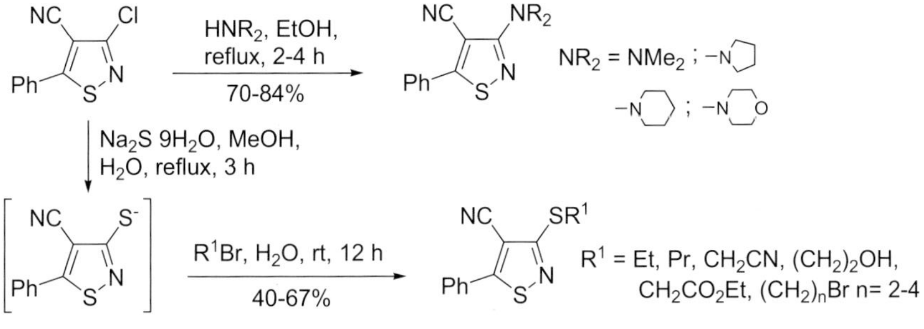

Scheme 9.109

Certain groups at the 5-position of the isothiazole ring can be easily displaced by nucleophiles, affording new functionalized isothiazoles. As an example, starting from 5-methylthioisothiazole **272**, four new isothiazoles have been prepared by successive nucleophilic substitutions (Scheme 9.110) [199]. 5-Unsubstituted isothiazole 1,1-dioxides give addition products with sulfur, oxygen, nitrogen [349] and phosphorus [350] nucleophiles. When 5-bromoisothiazole 1,1-dioxides are reacted with these nucleophiles, the substitution products are obtained [69, 70].

5-Iodoisothiazole **273** can be used as electrophile in Suzuki and Negishi cross-coupling reactions to afford 5-aryl and 5-hetarylisothiazoles **274** in good to excellent yields (48–95%) (Scheme 9.111) [351]. Similarly, 3,5-dichloroisothiazole-4-

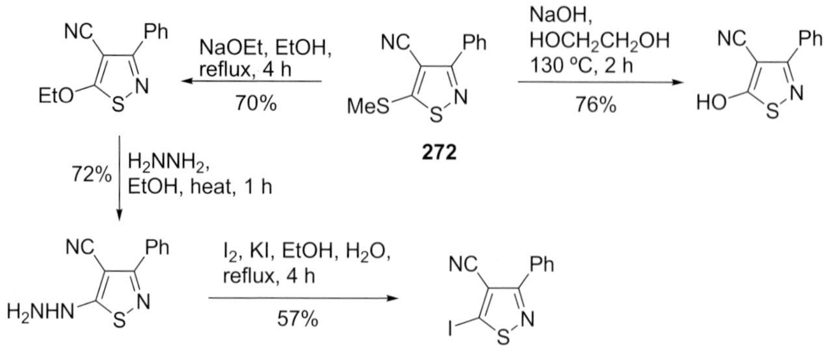

Scheme 9.110

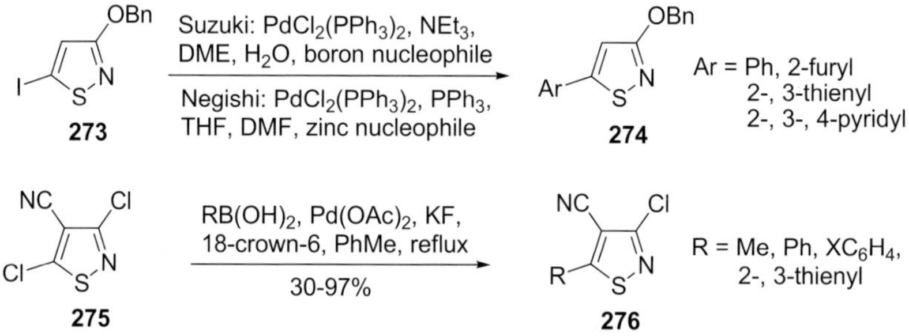

Scheme 9.111

carbonitrile (**275**) reacts with aryl- and methylboronic acids to give in high yields the 3-chloro-5-(aryl or methyl)-isothiazole-carbonitriles **276**. This reaction is totally regioselective [352].

The palladium-catalyzed reaction of 5-bromo-3-diethylamino-4-(4-methoxyphenyl)-isothiazole 1,1-dioxide with vinyl-, aryl-, heteroaryl- and alkynylstannanes provides a general and efficient method for the synthesis of 5-substituted isothiazole 1,1-dioxides [348].

2-Alkylisothiazolium salts are converted into polymeric products by alkali hydroxides or alkoxides. When treated with ethanolic ammonia, isothiazolium salt **277** yields the corresponding demethylated isothiazole **278**. However, under identical conditions, its isomeric compound **279** affords the 3-methylamino derivative **280** (Scheme 9.112) [353]. These transformations result from an initial nucleophilic attack on the ring S atom and recyclization of the initial intermediates.

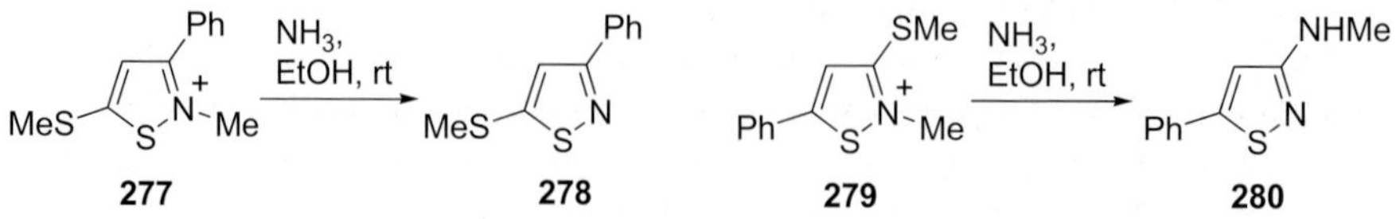

Scheme 9.112

Isothiazolium salts react with other nitrogen nucleophilic reagents such as phenylhydrazine and hydroxylamine to give, respectively, pyrazoles, and isoxazoles; with benzylamine they afford acyclic thiones [21]. They also react with carbanions to give thiophene or benzo[*b*]thiophene derivatives (Scheme 9.113) [354].

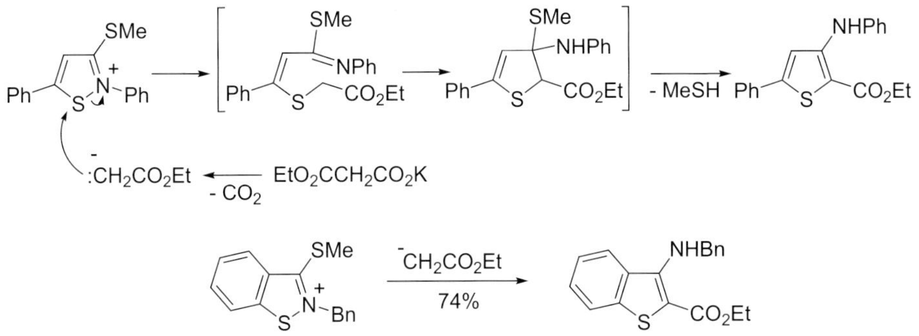

Scheme 9.113

9.6.2.4 **Cycloaddition Reactions**

Isothiazoles participate in Diels–Alder reactions and in 1,3-dipolar cycloadditions as the 2π electrons component [355]. A few examples of their participation in $[2\pi + 2\pi]$ cycloadditions with diphenylketene or inamines have also been reported [356, 357]. Some isothiazole derivatives are used as precursors of reactive dienes (*ortho*-quinodimethanes) that can be trapped in Diels–Alder reactions (see below).

Isothiazol-3(2*H*)-one 1,1-dioxides react with buta-1,3-dienes [358, 359], cyclo-1,3-dienes [356], anthracene [356], 1-azadienes [360], furans [356, 361], 1,3-oxazoles [362] and so on to afford the Diels–Alder adducts in moderate to good yields (Scheme 9.114).

A general method for the synthesis of saccharin derivatives involves the Diels–Alder reaction of 4-bromo-2-*t*-butylisothiazol-3(2*H*)-one 1,1-dioxide with oxi-substituted buta-1,3-dienes (Scheme 9.115) [359]. Dehydrobromination of the cycloadducts and removal of the protecting groups leads to the saccharin derivatives.

Isothiazole derivatives, especially the 1,1-dioxides, undergo 1,3-dipolar cycloadditions with a wide range of dipoles under mild conditions [363]. For instance, isothiazol-3(2*H*)-one 1,1-dioxides react with nitrile imines and nitrile oxides to give the expected cycloadducts (Scheme 9.116) [364]. In the case of reaction with azides and diazo compounds, the presence of a substituent at 4-position makes all the difference in the outcome of the reaction (Scheme 9.117) [360].

The 3-amino-4-arylisothiazole 1,1-dioxides are also very reactive in 1,3-dipolar cycloadditions. They react with a wide range of dipoles, such as oxazolones [365], diazo compounds [366], azides [367], and nitrile oxides [368]. The bicyclic cycloaddition products are versatile intermediates for monocyclic heterocycles by cleavage of one ring.

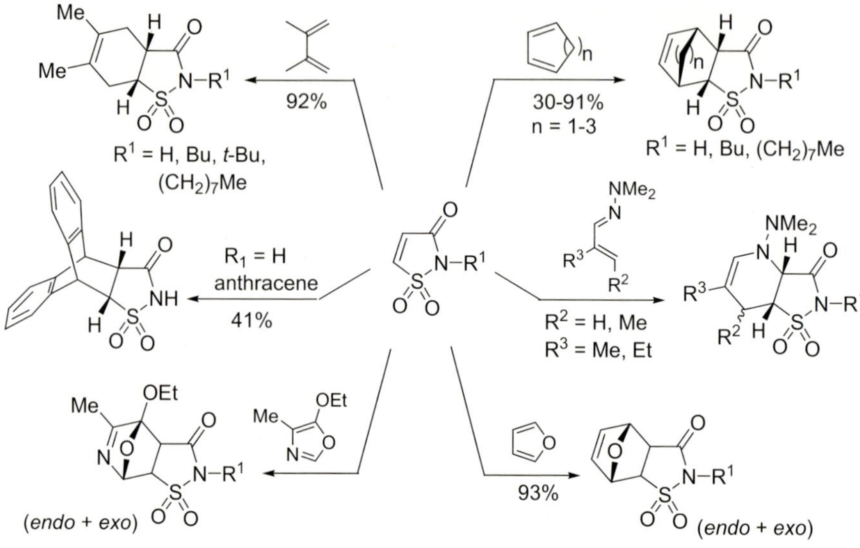

92%

R^1 = H, Bu, *t*-Bu, $(CH_2)_7Me$

30-91%
n = 1-3

R^1 = H, Bu, $(CH_2)_7Me$

R_1 = H
anthracene
41%

R^2 = H, Me
R^3 = Me, Et

(*endo* + *exo*)

93%

(*endo* + *exo*)

Scheme 9.114

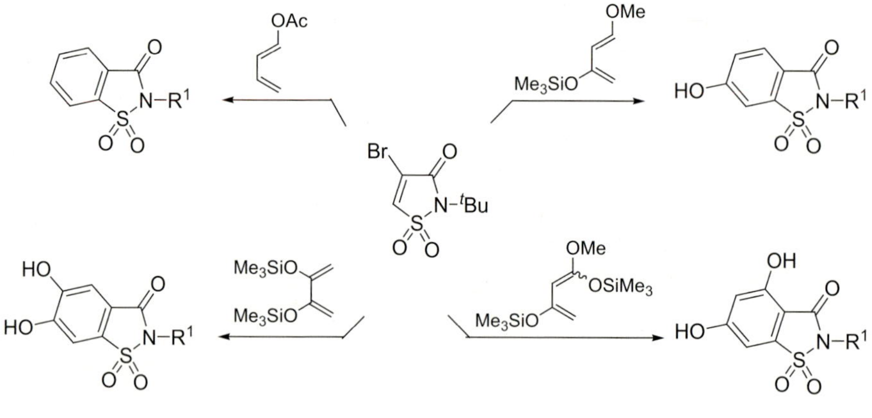

Scheme 9.115

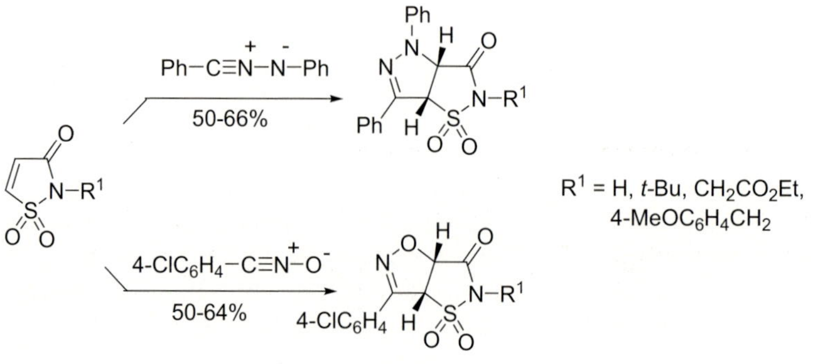

$Ph-C\equiv\overset{+}{N}-\overset{-}{N}-Ph$
50-66%

$4\text{-}ClC_6H_4-C\equiv\overset{+}{N}-O^-$
50-64%

R^1 = H, *t*-Bu, CH_2CO_2Et, 4-$MeOC_6H_4CH_2$

Scheme 9.116

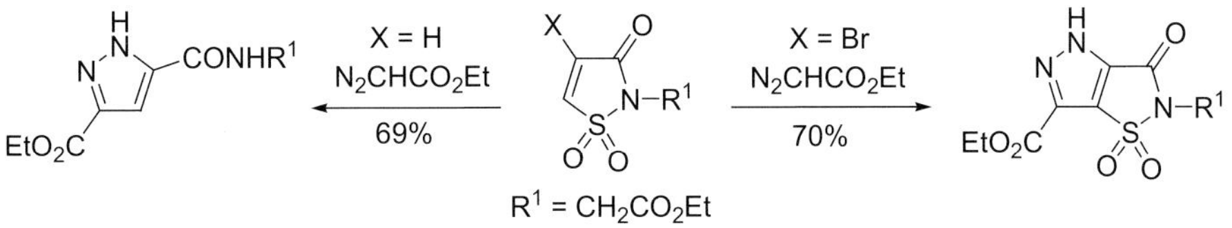

Scheme 9.117

Isothiazole-fused 3-sulfolenes **281** extrude SO_2 when heated at 185 °C in a sealed tube, generating the isothiazole *o*-quinodimethanes **282**. Extrusion of SO_2 in the presence of *N*-phenylmaleimide (NPM) or dimethyl acetylenedicarboxylate affords the corresponding Diels–Alder adducts **283** and **284**, respectively, in good yields (Scheme 9.118) [369]. Benzisothiazole **285** is obtained by the oxidation of the corresponding adduct **284**.

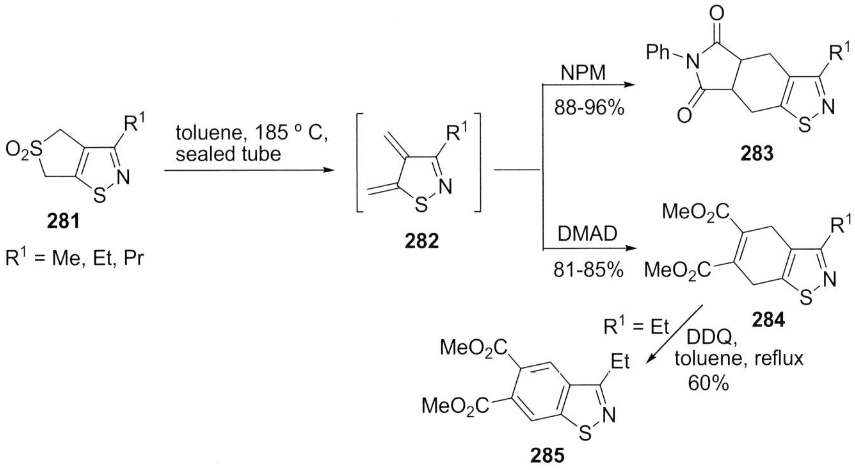

Scheme 9.118

2,1-Benzisothiazole 2,2-dioxides (see Schemes 9.73 and 9.74) extrude SO_2 when heated in refluxing 1,2-dichloro- or 1,2,4-trichlorobenzene to yield aza-*ortho*-quinodimethanes that can be trapped in inter- or intramolecular Diels–Alder cycloadditions (Scheme 9.119) [283, 370].

9.6.2.5 Reactions with Reducing Agents

Catalytic hydrogenation of isothiazol-3-ones **286** at 3.5 atm leads to the *cis*-dihydro derivatives **287** (Scheme 9.120) [214]. The dimethyl derivative **287**, however, isomerizes to the *trans*-isomer **288** via keto-enol tautomerism.

2-Alkyl-3-alkylthio-5-arylisothiazolium halides **289** are reduced to the S,N-acetals **290** by treatment with $NaBH_4$ in ethanol at room temperature (Scheme 9.121) [200, 371].

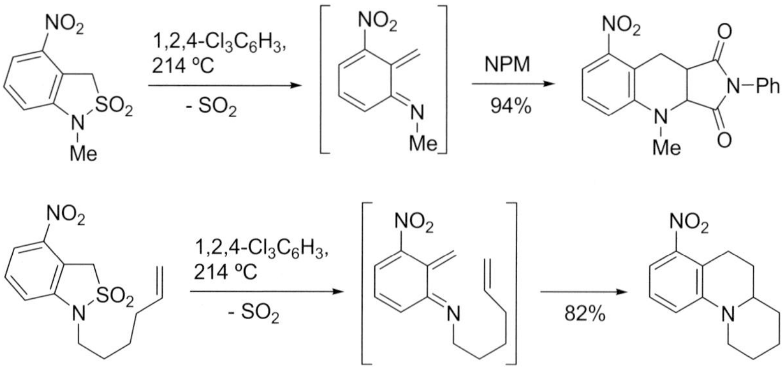

Scheme 9.119

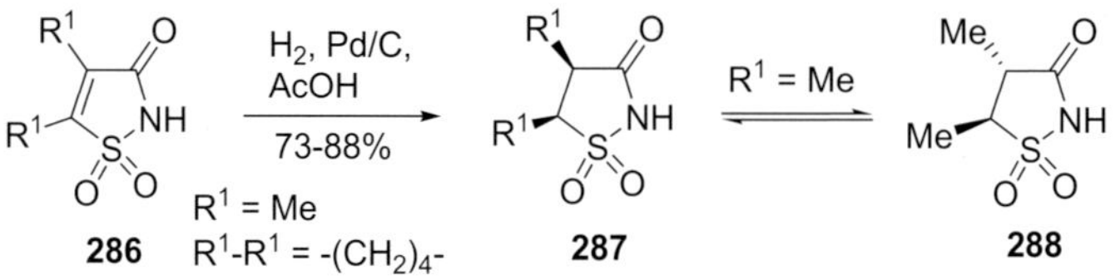

Scheme 9.120

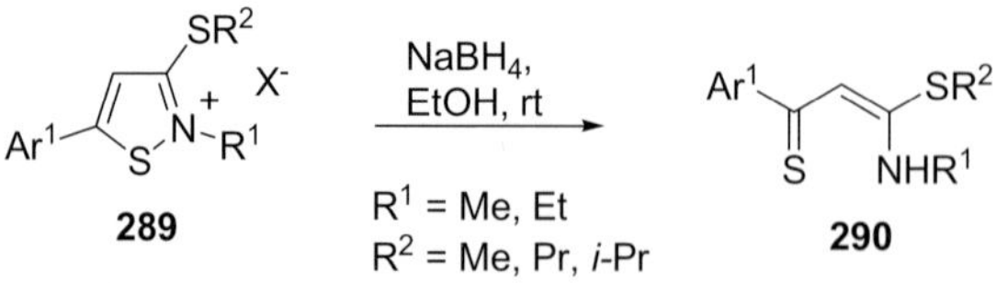

Scheme 9.121

9.6.2.6 **Reactions with Oxidizing Agents**

Isothiazoles unsubstituted at 3-position (**291**) are oxidized with 35% H_2O_2 in acetic acid to 1,2-isothiazol-3(2*H*)-one 1,1-dioxides **292** (Scheme 9.122) [214]. Oxidation of 3-aminoisothiazoles **293** with *m*-chloroperbenzoic acid affords the corresponding 1,1-dioxides **294** [372].

The oxidation of isothiazolium salts **295**, containing electron-withdrawing substituents in the *ortho*-position of the 2-aryl ring, with 30% H_2O_2 in acetic acid gives 3-hydroperoxy derivatives **296** in moderate to good yields (42–70%) and minor amounts of the corresponding 1,1-dioxides **297** (Scheme 9.123). When R^1 is an electron-donating group compounds **297** are the only products (40–63%) [372]. If the reaction is carried out at 80 °C the products are the isothiazol-3(2*H*)-one 1,1-dioxides **298** (up to 81% yield) [372].

Oxidation of 1,2-benzisothiazoles with hydrogen peroxide [373, 374] or perphthalic acid [375] yields the corresponding 1,1-dioxides.

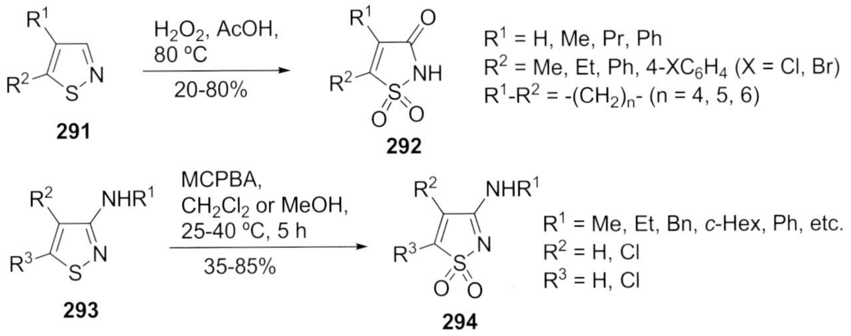

291 → 292: H_2O_2, AcOH, 80 °C, 20-80%

R^1 = H, Me, Pr, Ph
R^2 = Me, Et, Ph, 4-XC_6H_4 (X = Cl, Br)
R^1-R^2 = -$(CH_2)_n$- (n = 4, 5, 6)

293 → 294: MCPBA, CH_2Cl_2 or MeOH, 25-40 °C, 5 h, 35-85%

R^1 = Me, Et, Bn, c-Hex, Ph, etc.
R^2 = H, Cl
R^3 = H, Cl

Scheme 9.122

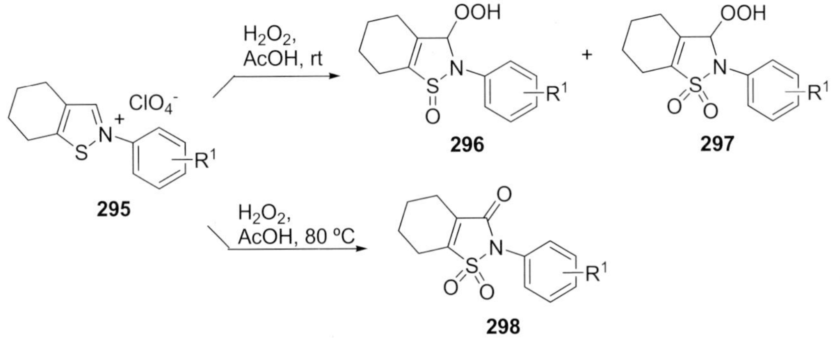

295 (ClO_4^-) → 296 + 297: H_2O_2, AcOH, rt

295 → 298: H_2O_2, AcOH, 80 °C

Scheme 9.123

BuLi, DMF, THF, -75 °C, 1 h, 73% (OHC)

BuLi, THF, -78 °C (Ph, Me, Li)

R^1-X, X = Br, I → 299: R^1 = Me, 83%; R^1 = Bu, 73%; R^1 = Bn, 83%

R^1COR^2, 18-86% → 300 (OH, R^1, R^2)

R^1-CO_2Et → 301 + 302 (OH, R^1, Ph)

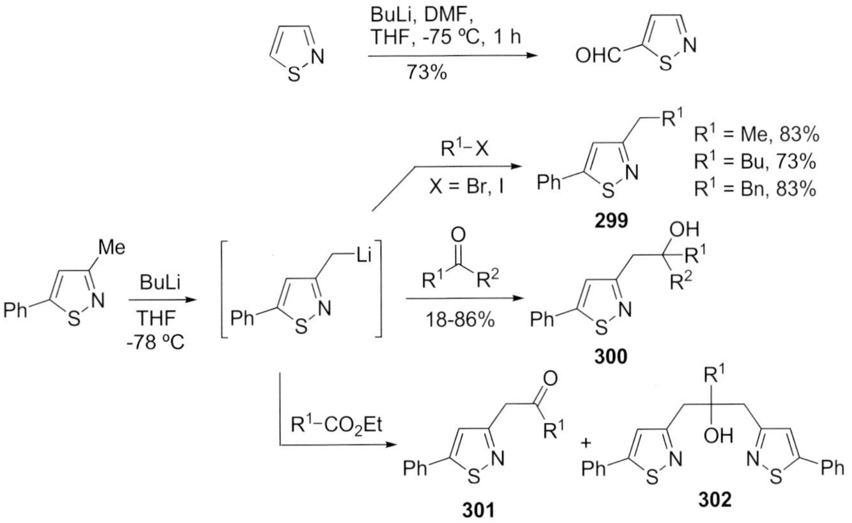

Scheme 9.124

9.6.2.7 Reactions of Metallated Isothiazoles

Isothiazole is selectively lithiated with BuLi at the 5-position. 5-Substituted isothiazoles can be prepared by reacting the 5-lithioisothiazole with electrophiles [376–379]. For instance, it reacts with DMF to afford selectively the isothiazole-5-carbaldehyde in good yield (Scheme 9.124) [196].

3-Benzyloxyisothiazole is also lithiated regioselectively in the 5-position using LDA in diethyl ether. Quenching the lithiated species with electrophiles leads to the corresponding 3,5-disubstituted isothiazoles (Table 9.10).

Table 9.10 Lithiation of 3-benzyloxyisothiazole and reaction with various electrophiles.

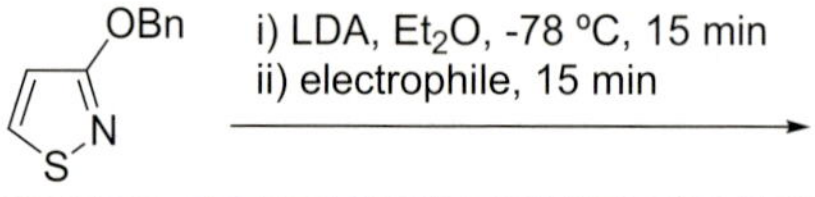

i) LDA, Et_2O, -78 °C, 15 min
ii) electrophile, 15 min

Electrophile	Product	Yield (%)	Reference
MeOD	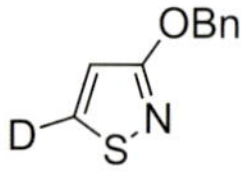 D (5-position), OBn	57	[380]
DMF	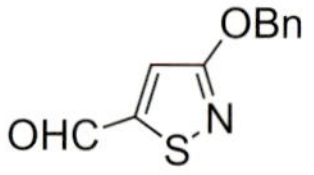 OHC (5-position), OBn	54	[377]
MeOCOCN	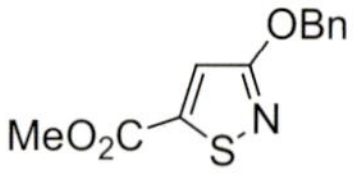 MeO_2C (5-position), OBn	56	[377]
PhCHO	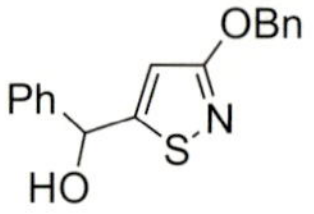 Ph, HO (5-position), OBn	65	[377]
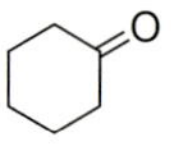 cyclohexanone	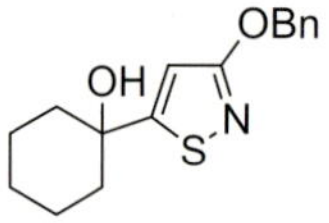 OH, cyclohexyl (5-position), OBn	68	[377]
I_2	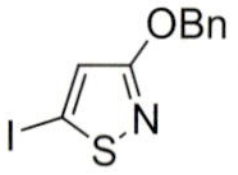 I (5-position), OBn	95	[351]
(i) B(O-*i*-Pr)$_3$ ii) 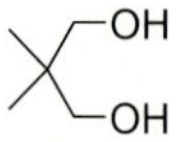 2,2-dimethylpropane-1,3-diol (OH, OH)	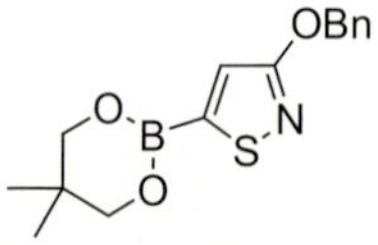 boronate ester (O, B, O) (5-position), OBn	89	[351]

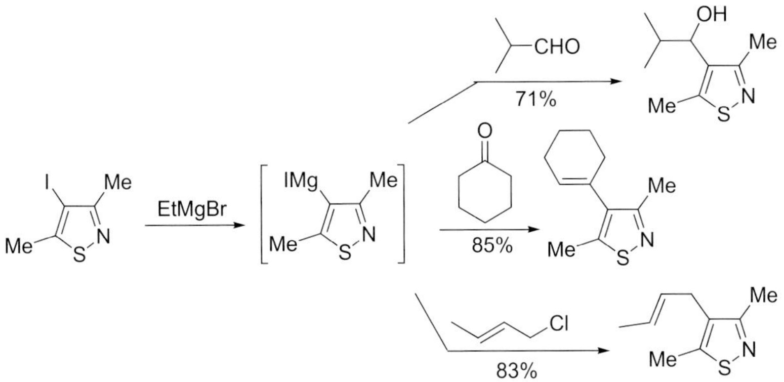

Scheme 9.125

Lithiation of 3-methyl-5-phenylisothiazole with butyllithium occurs selectively at the methyl group (Scheme 9.124). The lithiated species reacts with alkyl halides to give isothiazoles **299** in high yields [363]. It also reacts with a range of aldehydes and ketones to afford the corresponding secondary or tertiary alcohols **300** [381]. When esters are used as electrophiles, mixtures of ketones **301** and tertiary alcohols **302** are obtained [378].

3,5-Dimethylisothiazol-4-ylmagnesium iodide, generated *in situ* from 4-iodo-3,5-dimethylisothiazole and ethylmagnesium bromide, reacts with a range aldehydes, ketones and alkyl halides to afford 4-substituted 3,5-dimethylisothiazoles in good yields (Scheme 9.125) [382].

References

1 Eicher, T. and Hauptmann, S. (1995) *The Chemistry of Heterocycles*, (Translated by H. Suschitzky and J. Suschitzky), George Thieme Verlag, New York, pp. 138 and 160.

2 Claisen, L. and Lowman, O. (1888) *Chemische Berichte*, **21**, 1149.

3 Hantsch, A. (1888) *Justus Liebigs Annalen der Chemie*, **249**, 1.

4 Claisen, L. (1891) *Chemische Berichte*, **24**, 3900.

5 Claisen, L. (1903) *Chemische Berichte*, **36**, 3664.

6 Grünanger, P. and Vita-Finzi, P. (1991), in *Isoxazoles: Part 1* (ed. E.C. Taylor), The Chemistry of Heterocyclic Compounds, Vol. 49, Wiley-Interscience, New York.

7 e.g. Quilico, A. and Speroni, G. (1946) *Gazzetta Chimica Italiana*, **76**, 148.

8 Sutharchanadevi, M. and Murugan, R. (1996) *Isoxazoles* (ed. I. Shinkai) in Comprehensive Heterocyclic Chemistry II, Vol. 3 (eds, A.R. Katritzky, C.W. Rees, and E.F.V. Scriven), Pergamon Press, Oxford, 221 pp.

9 Grünanger, P. and Vita-Finzi, P. (1999) *Isoxazoles: Part 2*, (eds E.C. Taylor and P. Wipf) in The Chemistry of Heterocyclic Compounds, Wiley-Interscience, New York.

10 Lang, S.A., Jr and Lin, Y.-I. (1984) *Isoxazoles and their Benzo Derivatives*, (ed. K.T. Potts) in Comprehensive Heterocyclic Chemistry, Vol. 6 (eds A.R. Katritzky and C.W. Rees), Pergamon Press, Oxford.

11 Wakefield, B.J. (2004) *Isoxazoles*, in Science of Synthesis, Vol. 11 (ed.

E. Schaumann), Georg Thieme Verlag, 229 pp.

12 Smalley, R.K. (2004) *1,2-Benzisoxazoles and Related Compounds*, in Science of Synthesis, Vol. 11 (ed. E. Schaumann), Georg Thieme Verlag, 289 pp.

13 Smalley, R.K. (2004) *2,1-Benzisoxazoles and Related Compounds*, in Science of Synthesis, Vol. 11 (ed. E. Schaumann), Georg Thieme Verlag, 337 pp.

14 Quilico, A. (1962) Isoxazoles and related compounds, in *The Chemistry of Heterocyclic Compounds*, Vol. 17 (ed. A. Weissberger), Interscience, New York.

15 Adams, A. and Slack, R. (1956) *Chemistry and Industry (London)*, 1232.

16 Pain, D.L., Peart, B.J., and Wooldridge, K.R.H. (1984) Isothiazoles and their Benzo Derivatives (ed. K.T. Potts) in Comprehensive Heterocyclic Chemistry, Vol. 6 (eds A.R. Katritzky and C.W. Rees), Pergamon Press, Oxford, 131 pp.

17 Kurzer, F. (1977) *Organic Compounds of Sulphur, Selenium, and Tellurium*, **4**, 339.

18 Hettler, H. (1973) *Advances in Heterocyclic Chemistry, Vol 68*, **15**, 233.

19 (a) Ellwein, L.B. and Cohen, S.M. (1990) *Critical Reviews in Toxicology*, **20**, 311; *Chem. Abstr.*, 1991, **16**, 233942n; (b) Mitchell, M.L. and Pearson, R.L. (1991) *Food Science and Technology*, **48**, 127; (c) Chappel, C.I. (1992) *Regulatory Toxicology and Pharmacology*, **15**, 253;*Chem. Abstr.*, 1992, **117**, 130012a.

20 Matzen, L., Engesgaard, A., Ebert, B., Didriksen, M., Frolund, B., Krogsgaard-Larsen, P., and Jaroszewski, J.W. (1997) *Journal of Medicinal Chemistry*, **40**, 520.

21 (a) Slack, S. and Wooldridge, K.R.H. (1965) *Advances in Heterocyclic Chemistry, Vol 68*, **4**, 107; (b) Wooldridge, K.R.H. (1972) *Advances in Heterocyclic Chemistry, Vol 68*, **14**, 1.

22 Davis, M. (1972) *Advances in Heterocyclic Chemistry, Vol 68*, **14**, 43.

23 Suschitzky, H. and Scriven, E.F.V. (1992) *Progress in Heterocyclic Chemistry*, **4**, 295, 1993, 5, 341.

24 (a) Iddon, B. (1994) *Heterocycles*, **38**, 2487; (b) Iddon, B. (1995) *Heterocycles*, **41**, 533; (c) Iddon, B. (1995) *Heterocycles*, **41**, 1525.

25 (a) Clerici, F. (2002) *Advances in Heterocyclic Chemistry, Vol 68*, **83**, 71; (b) Kaberdin, R.V. and Potkin, V.I. (2002) *Russian Chemical Reviews*, **7**, 673.

26 Elgazwy, A.-S.S.H. (2003) *Tetrahedron*, **59**, 7445.

27 Chapman, R.F. and Peart, B.J. (1996) *Isothiazoles* (ed. I. Shinkai), in Comprehensive Heterocyclic Chemistry II, Vol. 3 (series eds A.R. Katritzky, C.W. Rees, and E.F.V. Scriven), Pergamon Press, Oxford, 319 pp.

28 Brown, D.W. and Sainsbury, M. (2004) *Isothiazoles*, in Science of Synthesis, Vol. 11 (ed. E. Schaumann), Georg Thieme Verlag, 507 pp.

29 Brown, D.W. and Sainsbury, M. (2004) *Benzisothiazoles*, in Science of Synthesis, Vol. 11 (ed. E. Schaumann), Georg Thieme Verlag, 573 pp.

30 Cathcart, W.R. and Meyer, V. (1892) *Chemische Berichte*, **25**, 1498.

31 Conduché, A. (1908) *Annales de Chimie et de Physique*, **13**, 47.

32 (a) Orgel, L.E., Cottrell, T.L., Dick, W., and Sutton, L.E. (1951) *Transactions of the Faraday Society*, **47**, 113; (b) Berthier, G. and Del Re, G. (1965) *Journal of the Chemical Society*, 3109; (c) Wasylishen R.E., Clem T.R., and Becker, E.D. (1975) *Canadian Journal of Chemistry*, **53**, 596.

33 Sokolov, S.D. (1979) *Russian Chemical Reviews (English Translation)*, **48**, 289.

34 Wasylishen, R.E., Rowbothan, J.B., and Schaefer, T. (1974) *Canadian Journal of Chemistry*, **52**, 833.

35 Ha, T.-K. (1979) *Journal of Molecular Structure*, **51**, 87.

36 Clementi, S., Forsythe, P.P., Johnson, C.D., Katritzky, A.R., and Terem, B. (1974) *Journal of the Chemical Society, Perkin Transactions 2*, 399.

37 (a) Olofson, R.A., Landesberg, J.M., Houk, K.N., and Michelman, J.S. (1966) *Journal of the American Chemical Society*, **88**, 4263; (b) White, J.A. and Anderson, R.C. (1969) *Journal of Heterocyclic Chemistry*, **6**, 199.

38 Burton, A.G., Forsythe, P.P., Johnson, C.D., and Katritzky, A.R. (1971) *Journal of the Chemical Society (B)*, 2365.

39 Jug, K. (1983) *The Journal of Organic Chemistry*, **48**, 1394.

40 Katritzky, A.R., Jug, K., and Oniciu, D.C. (2001) *Chemical Reviews*, **101**, 1421.

41 Witanowski, M., and Biedrzycka, Z. (1994) *Magnetic Resonance in Chemistry*, **32**, 62.

42 Zurawski, B. (1966) *Bulletin de l'Academie Polonaise des Sciences, Series des Sciences Chimiques*, **14**, 481.

43 Bochvar, D.A., Bagatur'yants, A.A., and Tutkevich, A.V. (1966) *Izvestiya Akademii Nauk SSSR*, 353.

44 Kamiya, M. (1970) *Bulletin of the Chemical Society of Japan*, **43**, 3344.

45 Matsuura, T. and Ito, Y. (1974) *Bulletin of the Chemical Society of Japan*, **47**, 1724.

46 Witanowski, M., Stefaniak, L., Januszewski, H., Grabowski, Z., and Webb, G.A. (1972) *Tetrahedron*, **28**, 637.

47 Huisgen, R. and Christl, M. (1967) *Angewandte Chemie*, **79**, 471; (1967) *Angewandte Chemie, International Edition*, **6**, 456.

48 Kintzinger, J.P. and Lehn, J.M. (1968) *Molecular Physics*, **14**, 133.

49 Sechi, M., Sannia, L., Orecchioni, M., Carta, F., Paglietti, G., and Neamati, N. (2003) *Journal of Heterocyclic Chemistry*, **40**, 1097.

50 De Munno, A., Ceccarelli, G., and Bertini, V. (1969) *Atti Soc Toscana Sci. Nat. Mem, Series A*, **76**, 408.

51 Sokolov, S.D., Yudintseva, I.M., and Petrovskii, P.V. (1970) *Zhurnal Organicheskoi Khimii*, **6**, 2581.

52 L'Abbé, G. and Mathys, G. (1974) *The Journal of Organic Chemistry*, **39**, 1221.

53 Staab, H.A. and Mannschreck, A. (1965) *Chemische Berichte*, **98**, 1111.

54 (a) Davis, M. and White, A.W. (1969) *Journal of the Chemical Society (C)*, 2189; (b) Rondeau, R.E., Berwick, M.A., and Rosenberg, H.M. (1972) *Journal of Heterocyclic Chemistry*, **7**, 127.

55 Wunsch, K.-H. and Boulton, A.J. (1967) *Advances in Heterocyclic Chemistry, Vol 68*, **8**, 277; (b) Davis, M., Mackay, M.F., and Denne, W.A. (1972) *Journal of the Chemical Society, Perkin Transactions 2* 565.

56 Faure, R., Llinas, J.-R., Vincent, E.-J., and Rajzmann, M. (1975) *Canadian Journal of Chemistry*, **53**, 1677.

57 Gainer, J., Howarth, G.A., Hoyle, W., and Roberts, S.M. (1976) *Organic Magnetic Resonance*, **8**, 226.

58 Plavac, N., Still, I.W.J., Chauhan, M.S., and McKinnon, D.M. (1975) *Canadian Journal of Chemistry*, **53**, 835.

59 Frydenvang, K., Matzen, L., Norrby, P.-O., Sløk, F.A., Liljefors, T., Krogsgaard-Larsen, P., and Jaroszewski, J.W. (1997) *Journal of the Chemical Society, Perkin Transactions 2*, 1783.

60 Boulton, A.J., Katritzky, A.R., Hamid, A.M., and Øksne, S. (1964) *Tetrahedron*, **20**, 2835.

61 (a) Onda, M., Akagawa, M., and Fukushima, H. (1964) *Chemical & Pharmaceutical Bulletin*, **12**, 751; (b) Bowden, K. and Drysdale, A.C. (1965) *Tetrahedron Letters*, **6**, 727.

62 Eugster, C.H., Muller, G.F.R., and Good, R. (1965) *Tetrahedron Letters*, **6**, 1813.

63 (a) Böshagen, H. (1967) *Chemistry in Britain*, **100**, 954; (b) Kinstle, T.H. and Darlage, L.J. (1969) *Journal of Heterocyclic Chemistry*, **6**, 123; (c) Darlage, L.J., Kinstle, T.H., and McIntosh, C.L. (1971) *The Journal of Organic Chemistry*, **36**, 1088.

64 Davis, M., Deady, L.W., Homfeld, E., and Pogany, S. (1975) *Australian Journal of Chemistry*, **28**, 129.

65 Nye, M.J. and Tang, W.P. (1972) *Tetrahedron*, **28**, 455.

66 Kusumi, T., Chang, C.C., Wheeler, M., Kubo, I., Nakanishi, K., and Naoki, H. (1981) *Tetrahedron Letters*, **22**, 3451.

67 (a) Theobald, W., Buch, O., Kunz, H.A., Krupp, P., Stenger, E.G., and Heimann, H. (1968) *Arzneimittel-Forschung*, **18**, 311; (b) Krogsgaard-Larsen, P., Johnston, G.A.R., Curtis, D.R., Game, C.J.A., and McCulloch, R.M. (1975) *Journal of Neurochemistry*, **25**, 803; (c) De Feudis, F.V.1980, *Neurochemical Research*, **5**, 1047.

68 (a) Eugster, C.H. (1969) *In Fortschritte der Chemie Organischer Naturstoffe XXVII* (ed. L. Zechmeister), Springer-Verlag, New York, pp. 261; (b) Krogsgaard-Larsen, P., Honoré, T., Hansen, J.J., Curtis, D.R., and Lodge, D. (1980) *Nature*, **284**, 64.

69 Pedras, M.S.C., Okanga, F.I., Zaharia, I.L., and Khan, A.K. (2000) *Phytochemistry*, **53**, 161.

70 (a) Devys, M., Barbier, M., Loiselet, I., Rouxel, T., Sarniguet, A., Kollmann, A., and Bousquet, J.-F. (1988) *Tetrahedron Letters*, **29**, 6447; (b) Devys, M. and Barbier, M. (1990) *Synthesis*, 214; (c) Pedras, M.S.C. and Zaharia, I.L. (2001) *Organic Letters*, **3**, 1213.

71 Stratmann, K., Belli, J., Jensen, C.M., Moore, R.E., and Patterson, G.M.L. (1994) *The Journal of Organic Chemistry*, **59**, 6279.

72 Griffiths, M.C. (1982) USAN and the USP Dictionary of Drug Names, Pharmacopoeia Convention, Rockville, Md.

73 Masuda, Y., Karasawa, T., Shiraishi, Y., Hori, M., Yoshida, K., and Shimizu, M. (1980) *Arzneimittel-Forschung*, **30**, 477.

74 Janssen, P.A.J., Niemegeers, C.J.E., Awouters, C.J.E., Schellekens, K.H.L., Megens, A.A.H.P., and Meert, T.F. (1988) *The Journal of Pharmacology and Experimental Therapeutics*, **244**, 685.

75 Fielding, S., Novick, W.J., Jr, Geyer, H.M., Petko, W.W., Wilker, J.C., Davis, L., Klein, J.T., and Cornfeldt, M. (1983) *Drug Development Research*, **3**, 233.

76 Karasawa, T., Furukawa, K., Yoshida, K., and Shimizu, M. (1976) *Chemical & Pharmaceutical Bulletin*, **24**, 2673.

77 La Brecque, G.C., Wilson, H.G., Brady, U.E., and Gahan, J.B. (1967) *Journal of Economic Entomology*, **60**, 760.

78 Davis, M. (1979) *Organic Compounds of Sulphur, Selenium, and Tellurium*, **5**, 345.

79 Baldwin, J.J., Engelhardt, E.L., Hirschmann, R., Ponticello, G.S., Atkinson, J.G., Wasson, B.K., Sweet, C.S., and Scriabine, A. (1980) *Journal of Medicinal Chemistry*, **23**, 65.

80 Sykes, A.H. (1983) *Proceedings of the Nutrition Society*, **42**, A93.

81 Gieldanowski, J., Kowalczyk-Bronisz, S.H., Machón, Z., Szary, A., and Blaszczyk, B. (1980) *Archivum Immunologiae Et Therapiae Experimentalis*, **28**, 393[(1980) *Chemical Abstracts*, **93**, 230880].

82 Kurzer, F. (1975) *Organic Compounds of Sulphur, Selenium, and Tellurium*, **3**, 541.

83 (a) e.g. Cutrì, C.C.C., Garozzo, A., Siracusa, M.A., Sarvà, M.C., Castro, A., Geremia, E., Pinizzotto, M.R., and Guerrera, F. (1999) *Bioorganic and Medicinal Chemistry*, **7**, 225; (b) FrØlund, B., Jensen, L.S., Guandalini, L., Canillo, C., Vestergaard, H.T., Kristiansen, U., Nielsen, B., StensbØl, T.B., Madsen, C., Krogsgaard-Larsen, P., and Liljefors, T. (2005) *Journal of Medicinal Chemistry*, **48**, 427.

84 Baraldi, P.G., Barco, A., Benetti, S., Pollini, G.P., and Simon, D. (1987) *Synthesis*, 857.

85 Pinho e Melo, T.M.V.D. (2005) *Current Organic Chemistry*, **9**, 925.

86 Mellor, J.M., Schofield, S.R., and Korn, S.R. (1997) *Tetrahedron*, **53**, 17151.

87 Shen, D.-M., Shu, M., and Chapman, K.T. (2000) *Organic Letters*, **2**, 2789.

88 Sørensen, U.S., Falch, E., and Krogsgaard-Larsen, P. (2000) *The Journal of Organic Chemistry*, **65**, 1003.

89 Ohta, T., Fujisawa, H., Nakai, Y., and Furukawa, I. (2000) *Bulletin of the Chemical Society of Japan*, **73**, 1861.

90 Purkayastha, M.L., Bhat, L., Ila, H., and Junjappa, H. (1995) *Synthesis*, 641.

91 Padmavathi, V., Reddy, B.J.M., Balaiah, A., Reddy, K.V., and Reddy, D.B. (2000) *Molecules*, **5**, 1281.

92 Honda, Y., Honda, T., Roy, S., and Gribble, G.W. (2003) *The Journal of Organic Chemistry*, **68**, 4991.

93 Lin, Y. and Lang, S.A. (1977) *Journal of Heterocyclic Chemistry*, **14**, 345.

94 Jones, R.G. and Whitehead, C.W. (1955) *The Journal of Organic Chemistry*, **20**, 1343.

95 Dominguez, E., Ibeas, E., Marigorta, E.M., Palacios, J.K., and SanMartin, R. (1996) *The Journal of Organic Chemistry*, **61**, 5435.

96 Oliveira, R., SanMartin, R., and Dominguez, E. (2000) *Synlett*, 1028.

97 Oliveira, R., SanMartin, R., Tellito, I., and Dominguez, E. (2002) *Tetrahedron*, **58**, 3021.

98 Oliveira, R., SanMartin, R., Dominguez, E., Solans, X., Urtiaga, M.K., and Arriortua, M.I. (2000) *The Journal of Organic Chemistry*, **65**, 6398.

99 Wittenberger, S.J. (1996) *The Journal of Organic Chemistry*, **61**, 356.

100 Schenone, P., Fossa, P., and Menozzi, G. (1991) *Journal of Heterocyclic Chemistry*, **28**, 453.

101 Menozzi, G., Schenone, P., and Mosti, L. (1983) *Journal of Heterocyclic Chemistry*, **20**, 645.

102 De Luca, L., Giocamelli, G., Porcheddu, A., Salaris, M., and Taddei, M. (2003) *Comptes Rendus Chimie*, **6**, 607.

103 De Luca, L., Giacomelli, G., Porcheddu, A., Salaris, M., and Taddei, M. (2003) *Journal of Combinatorial Chemistry*, **5**, 465.

104 Manferdini, M., Morelli, C.F., and Veronese, A.C. (2000) *Heterocycles*, **53**, 2775.

105 Vicentini, C.B., Mazzanti, M., Morelli, C.F., and Manfrini, M. (2000) *Journal of Heterocyclic Chemistry*, **37**, 175.

106 Makarova, N.V., Zemtsova, M.N., and Moiseev, I.K. (2003) *Chemistry of Heterocyclic Compounds*, **39**, 613.

107 Molteni, V., Hamilton, M.M., Mao, L., Crane, C.M., Termin, A.P., and Wilson, D.M. (2002) *Synthesis*, 1669.

108 Martins, M.A.P., Flores, A.F.C., Bastos, G.P., Sinhorin, A., Bonacorso, H.G., and Zanatta, N. (2000) *Tetrahedron Letters*, **41**, 293.

109 Ohkoshi, M., Yoshida, M., Matsuyama, H., and Iyoda, M. (2001) *Tetrahedron Letters*, **42**, 33.

110 Johnston, K.M. and Shotter, R.G. (1968) *Journal of the Chemical Society (C)*, 1774.

111 Kashima, C., Shirai, S., Yoshiwara, N., and Omote, Y. (1980) *Journal of the Chemical Society. Chemical Communications*, 826.

112 Katritzky, A.R., Wang, M., Zhang, S., and Voronkov, M.V. (2001) *The Journal of Organic Chemistry*, **66**, 6787.

113 Büchi, G. and Vederas, J.C. (1972) *Journal of the American Chemical Society*, **94**, 9128.

114 Hansen, J.F. and Strong, S.A. (1977) *Journal of Heterocyclic Chemistry*, **14**, 1289.

115 Maeda, K., Hosokawa, T., Murahashi, S.-I., and Moritani, I. (1973) *Tetrahedron Letters*, **14**, 5075.

116 Sharma, J.C., Rojinder, S., Berge, D.D., and Kale, A.V. (1986) *Indian Journal of Chemistry*, **25**, 437.

117 Wei, X., Fang, J., Hu, Y., and Hu, H. (1992) *Synthesis*, 1205.

118 Shet, J., Desai, V., and Tilve, S. (2004) *Synthesis*, 1859.

119 Pinheiro, J.M., Ismael, M.I., Figueiredo, J.A., and Silva, A.M.S. (2004) *Journal of Heterocyclic Chemistry*, **41**, 877.

120 Adlington, R.M., Baldwin, J.E., Catterick, D., Pritchard, G.J., and Tang, L.T. (2000) *Journal of the Chemical Society, Perkin Transactions 1*, 2311.

121 Grundmann, C. (1970) *Synthesis*, 344.

122 Huigen, R. (1963) *Angewandte Chemie, International Edition in English*, **2**, 565.

123 Quilico, A. (1970) *Experimenta*, **26**, 1169.

124 Dondoni, A. and Barbaro, G. (1974) *Journal of the Chemical Society, Perkin Transactions 2*, 1691.

125 Gasparrini, F., Giovannoli, M., Misiti, D., Natile, G., Palmieri, G., and Maresca, L. (1993) *Journal of the American Chemical Society*, **115**, 4401.

126 Arai, N., Iwakoshi, M., Tanabe, K., and Narasaka, K. (1999) *Bulletin of the Chemical Society of Japan*, **72**, 2277.

127 Giacomelli, G., De Luca, L., and Porcheddu, A. (2003) *Tetrahedron*, **59**, 5437.

128 Padwa, A., Chiacchio, U., Dean, D.C., and Schoffstall, A.M. (1988) *Tetrahedron Letters*, **29**, 4169.

129 Garanti, L., Sala, A., and Zecchi, G. (1975) *Synthesis*, 666.

130 Yamada, K., Yamada, F., and Somei, M. (2002) *Heterocycles*, **57**, 1231.

131 Yamada, K., Yamada, F., and Somei, M. (2003) *Heterocycles*, **59**, 685.

132 Kizer, D.E., Miller, R.B., and Kurth, M.J. (1999) *Tetrahedron Letters*, **40**, 3535.

133 Wakita, K., Arai, M.A., Kato, T., Shinohara, T., and Sasai, H. (2004) *Heterocycles*, **62**, 831.

134 Quan, C. and Kurth, M. (2004) *The Journal of Organic Chemistry*, **69**, 1470.

135 Cereda, E., Ezhaya, A., Quai, M., and Barbaglia, W. (2001) *Tetrahedron Letters*, **42**, 4951.

136 De Luca, L., Giacomelli, G., and Riu, A. (2001) *The Journal of Organic Chemistry*, **66**, 6823.

137 Shang, Y.-J. and Wang, Y.-G. (2002) *Tetrahedron Letters*, **43**, 2247.

138 Shang, Y.-J. and Wang, Y.-G. (2002) *Synthesis*, 1663.

139 Park, K.-H. and Kurth, M.J. (1999) *The Journal of Organic Chemistry*, **64**, 9297.

140 Haino, T., Tanaka, M., Ideta, K., Kubo, K., Mori, A., and Fukazawa, Y. (2004) *Tetrahedron Letters*, **45**, 2277.

141 Kang, K.H., Pae, A.N., Choi, K., II, Cho, Y.S., Chung, B.Y., Lee, J.E., Jung, S.H., Koh, H.Y., and Lee, H.-Y. (2001) *Tetrahedron Letters*, **42**, 1057.

142 Paul, R. and Tchelitcheff, S. (1962) *Bulletin de la Société chimique de France*, 2215.

143 Itoh, K.-i. and Horiuchi, C.A. (2004) *Tetrahedron*, **60**, 1671.

144 Verbruggen, R. and Viehe, H.G. (1975) *Chimia*, **29**, 350.

145 Mitchell, T.N., El-Farargy, A., Moschref, S.-N., and Gourzoulidou, E. (2000) *Synlett*, 223.

146 Jones, R.C.F., Hollis, S.J., and Iley, J.N. (2000) *Tetrahedron: Asymmetry*, **11**, 3273.

147 Zong, K., Shin, S., II, Jeon, D.J., Lee, J.N., and Ryu, E.K. (2000) *Journal of Heterocyclic Chemistry*, **37**, 75.

148 Han, X. and Natale, N.R. (2001) *Journal of Heterocyclic Chemistry*, **38**, 415.

149 Zamponi, G.W., Stotz, S.C., Staples, R.J., Andro, T.M., Nelson, J.K., Hulubei, V., Blumenfeld, A., and Natale, N.R. (2003) *Journal of Medicinal Chemistry*, **46**, 87.

150 Sammelson, R.E., Ma, T., Galietta, L.J.V., Verkman, A.S., and Kurth, M.J. (2003) *Bioorganic & Medicinal Chemistry Letters*, **13**, 2509.

151 Kaffy, J., Monneret, C., Mailliet, P., Commerçon, A., and Pontikis, R. (2004) *Tetrahedron Letters*, **45**, 3359.

152 Sáez, J.A., Arnó, M., and Domingo, L.R. (2003) *Tetrahedron*, **59**, 9167.

153 Sheng, S.-R., Liu, X.-L., Xu, Q., and Song, C.-S. (2003) *Synthesis*, 2763.

154 Xu, W.M., Tang, E., and Huang, X. (2005) *Tetrahedron*, **61**, 501.

155 Touaux, B., Texier-Boullet, F., and Hamelin, J. (1998) *Heteroatom Chemistry*, **9**, 351.

156 Umesha, K.B., Kumar, K.A., and Rai, K.M.L. (2002) *Synthetic Communications*, **32**, 1841.

157 Bode, J.W., Hachisu, Y., Matsuura, T., and Suzuki, K. (2003) *Organic Letters*, **5**, 391.

158 Bode, J.W., Hachisu, Y., Matsuura, T., and Suzuki, K. (2003) *Tetrahedron Letters*, **44**, 3555.

159 Matsuura, T., Bode, J.W., Hachisu, Y., and Suzuki, K. (2003) *Synlett*, 1746.

160 Di Nunno, L., Scilimati, A., and Vitale, P. (2002) *Tetrahedron*, **58**, 2659.

161 Cramer, R. and McClellas, W.R. (1961) *The Journal of Organic Chemistry*, **26**, 2976.

162 Overberger, C.G. and Fujimoto, S. (1968) *Journal of Polymer Science (C)*, **16**, 4161.

163 Iwakura, Y., Uno, K., Hong, S.-J., and Hongu, T. (1971) *Polymer Journal*, **2**, 36.

164 Casnati, G., Quilico, A., Ricca, A., and Vita Finzi, P. (1966) *Gazzetta Chimica Italiana*, **96**, 1064.

165 Grünanger, P. and Fabbri, E. (1959) *Gazzetta Chimica Italiana*, **89**, 598.

166 Beam, C.F., Dyer, C.D., Schwaez, R.A., and Hauser, C.R. (1970) *The Journal of Organic Chemistry*, **35**, 1806.

167 Barber, G.N. and Olofson, R.A. (1978) *The Journal of Organic Chemistry*, **43**, 3015.

168 He, Y. and Lin, N.-H. (1994) *Synthesis*, 989.

169 Bunnelle, W.H., Singam, P.R., Narayanan, B.A., Bradshaw, C.W., and Liou, J.S. (1997) *Synthesis*, 439.

170 Dines, M. and Scheinbaum, M.L. (1969) *Tetrahedron Letters*, **20**, 4817.

171 Buron, C., Kaim, L.E., and Uslu, A. (1997) *Tetrahedron Letters*, **38**, 8027.

172 Bravo, P., Gaudiano, G., Ponti, P.P., and Ticozzi, C. (1972) *Tetrahedron*, **28**, 3845.

173 Lin, S.-T., Lin, L.-H., and Yao, Y.-F. (1992) *Tetrahedron Letters*, **33**, 3155.

174 Lin, S.-T. and Yang, F.-M. (1996) *Journal of Chemical Research-(S)*, 276.

175 Navarro-Ocaña, A., Jiménez-Estrada, M., González-Paredes, M.B., and Bárzana, E. (1996) *Synlett*, 695.

176 Tamura, Y., Miki, Y., Sumida, Y., and Ikeda, M. (1973) *Journal of the Chemical Society, Perkin Transactions 1*, 2589.

177 Tamura, Y., Sumoto, K., Matsushima, H., Taniguchi, H., and Ikeda, M. (1973) *The Journal of Organic Chemistry*, **38**, 4324.

178 Padwa, A., Smolanoff, J., and Tremper, A. (1975) *Journal of the American Chemical Society*, **97**, 4682.

179 Padwa, A. and Stengel, T. (2004) *Tetrahedron Letters*, **45**, 5991.

180 Singh, B. and Ullman, E.F. (1967) *Journal of the American Chemical Society*, **89**, 6911.

181 Gewald, K., Bellmann, P., and Jänsch, H.-J. (1980) *Annalen Der Chemie-Justus Liebig*, 1623.

182 Deceuninck, D.K., Buffel, D.K., and Hoornaert, G.K. (1980) *Tetrahedron Letters*, **21**, 3613.

183 Morita, K., Hashimoto, N., and Matsumura, K. (1969) Ger, Offen., 1,814,116 [*Chem. Abstr.* (1969) **71**, 124415z].

184 Morita, K., Hashimoto, N., and Matsumura, K. (1970) *Japan*, **70**, 34, 132 [*Chem. Abstr.* (1971) 74, 125679n].

185 Kohler, E.P. and Barrett, G.R. (1924) *Journal of the American Chemical Society*, **46**, 2105.

186 Best, W.M., Ghisalbert, E.L., and Powell, M. (1998) *Journal of Chemical Research-(S)*, 388.

187 Korte, F. and Störiko, K. (1961) *Chemische Berichte*, **94**, 1956.

188 Knowles, A.M. and Lawson, A. (1973) *Journal of the Chemical Society, Perkin Transactions 1*, 537.

189 Mazzei, M., Sottofattori, E., Dondero, R., Ibrahim, M., Melloni, E., and Michetti, M. (1999) *Il Farmaco*, **54**, 452.

190 Mazzei, M., Nieddu, E., Melloni, E., and Minafra, R. (2003) *Il Farmaco*, **58**, 121.

191 Clerin, D., Fleury, J.-P., and Fritz, H. (1976) *Journal of Heterocyclic Chemistry*, **13**, 825.

192 Huisgen, R. and Christl, M. (1973) *Chemische Berichte*, **106**, 3291.

193 Ambler, P.W., Paton, R.M., and Tout, J.M. (1994) *Journal of the Chemical Society, Chemical Communications*, 2661.

194 Boulton, A.J. and Mathur, S.S. (1973) *The Journal of Organic Chemistry*, **38**, 1054.

195 Yavari, I., Esfandiari, S., Mostashari, A.J., and Hunter, P.W.W. (1975) *The Journal of Organic Chemistry*, **40**, 2880.

196 Kang, Y.K., Lee, K.S., Yoo, K.H., Shin, K.J., Kim, D.C., Lee, C.-S., Kong, J.Y., and Kim, D.J. (2003) *Bioorganic & Medicinal Chemistry Letters*, **13**, 463.

197 Goerdeler, J. and Pohland, H.W. (1961) *Chemische Berichte*, **94**, 2950.

198 Hackler, R.E., Burow, K.W., Jr, Kaster, S.V., and Wickiser, D.I. (1989) *Journal of Heterocyclic Chemistry*, **26**, 1575.

199 Krebs, H.-D. (1989) *Australian Journal of Chemistry*, **42**, 1291.

200 Lee, D.J., Kim, B.S., and Kim, K. (2002) *The Journal of Organic Chemistry*, **67**, 5375.

201 Muraoka, M., Yamamoto, T., Enomoto, K., and Takeshima, T. (1989) *Journal of the Chemical Society, Perkin Transactions 1*, 1241.

202 Howe, R.K., Grunew, T.A., Carter, L.G., and Franz, J.E. (1978) *Journal of Heterocyclic Chemistry*, **15**, 1001.

203 Shishoo, C.J., Devani, M.B., Ananthan, S., Bhadti, V.S., and Ullas, G.V. (1988) *Journal of Heterocyclic Chemistry*, **25**, 759.

204 Rudorf, W.-D., Günther, E., and Augustin, M. (1984) *Tetrahedron*, **40**, 381.

205 Dieter, R.K. and Chang, H.J. (1989) *The Journal of Organic Chemistry*, **54**, 1088.

206 Ishida, M., Nakanishi, H., and Kato, S. (1984) *Chemistry Letters*, 1691.

207 Ishida, M., Ichikawa, K., Asano, M., Nakanishi, H., and Kato, S. (1987) *Synthesis*, 349.

208 (a) Beck, J.R., Gajewski, R.P., and Hackler, R.E. (1982). US Pat. 4,346,094[(1982) *Chemical Abstracts*, 97, 55798z]. (b) Beck, J.R. and Gajewski, R.P. (1987) *Journal of Heterocyclic Chemistry*, **24**, 243.

209 Rezessy, B., Toldy, L., and Tóth, G. (1992) *Tetrahedron Letters*, **33**, 6523.

210 Wamhoff, H., Berressem, R., and Nieger, M. (1993) *The Journal of Organic Chemistry*, **58**, 5181.

211 (a) Machida, M., Oda, K., and Kanaoka, Y. (1984) *Tetrahedron Letters*, **25**, 409; (b) Oda, K., Machida, M., and Kanaoka, Y. (1990) *Heterocycles*, **30**, 983.

212 Lucchesini, F., Picci, N., Pocci, M., Munno, A., and Bertini, V. (1989) *Heterocycles*, **29**, 97.

213 Cutrí, C.C.C., Garozzo, A., Siracusa, M.A., Sarvá, M.C., Tempera, G., Geremia, E., Pinizzotto, M.R., and Guerrera, F. (1998) *Bioorganic and Medicinal Chemistry*, **6**, 2271.

214 Schulze, B., Kisrten, G., Kirrbach, S., Rahm, A., and Heimgartner, H. (1991) *Helvetica Chimica Acta*, **74**, 1059.

215 Mühlstädt, M., Brämer, R., and Schulse, B. (1976) *Journal Fur Praktische Chemie*, **318**, 507.

216 Schulse, B., Herre, S., Brämer, R., Laux, C., and Mühlstädt, M. (1977) *Journal Fur Praktische Chemie*, **319**, 305.

217 Hartung, C., Illgen, K., Sieler, J., Schneider, B., and Schulze, B. (1999) *Helvetica Chimica Acta*, **82**, 685.

218 Apblett, A. and Chivers, T. (1990) *Canadian Journal of Chemistry*, **68**, 650.

219 Paton, R.M. (1989) *Chemical Society Reviews*, **18**, 33.

220 Franz, J.E. and Black, L.L. (1970) *Tetrahedron Letters*, **11**, 1381.

221 Howe, R.K., Gruner, T.A., Carter, L.G., Black, L.L., and Franz, J.E. (1978) *The Journal of Organic Chemistry*, **43**, 3736.

222 Howe, R.K. and Franz, J.E. (1978) *The Journal of Organic Chemistry*, **43**, 3742.

223 Crosby, J., McKie, M.C., Paton, R.M., and Ross, J.F. (2000) *Arkivoc*, **1**, 720.

224 Duan, X.-L., Rees, C.W., and Yue, T.-Y. (1997) *Chemical Communications*, 367.

225 Rees, C.W. and Yue, T.-Y. (1997) *Journal of the Chemical Society, Perkin Transactions 1*, 2247.

226 Laaman, S.M., Meth-Cohn, O., and Rees, C.W. (1999) *Synthesis*, 757.

227 Guillard, J., Lamazzi, C., Meth-Cohn, O., Rees, C.W., White, A.J.P., and Williams, D.J. (2001) *Journal of the Chemical Society, Perkin Transactions 2*, 1304.

228 Guillard, J., Meth-Cohn, O., Rees, C.W., White, A.J.P., and Williams, D.J. (2002) *Chemical Communications*, 232.

229 Clarke, D., Emayan, K., and Rees, C.W. (1998) *Journal of the Chemical Society, Perkin Transactions 1*, 77.

230 Nakayama, J., Sakay, A., Tokiyama, A., and Hoshino, M. (1983) *Tetrahedron Letters*, **24**, 3729.

231 Buggle, K. and Fallon, B. (1988) *Journal of Chemical Research-S*, 349; (1988) *Journal of Chemical Research-S*, 2764.

232 (a) Etzbach, K.H. and Eilingsfeld, H. (1988) *Synthesis* 449; (b) Ueda, T., Shibata, Y., and Sakakibara, J. (1986) *Journal of Heterocyclic Chemistry*, **23**, 1773.

233 Friedländer, P. and Henriques, R. (1882) *Chemische Berichte*, **15**, 2105.

234 Bunnett, J.F. and Yih, S.Y. (1961) *Journal of the American Chemical Society*, **83**, 3805.

235 Cathcart, W.R. and Meyer, V. (1892) *Chemische Berichte*, **25**, 3291.

236 Burlinson, N.E., Sitzman, M.E., Kaplan, L.A., and Kayser, E. (1979) *The Journal of Organic Chemistry*, **44**, 3695.

237 Meisenheimer, J., Zimmermann, P., and Kummer, U. (1925) *Annalen Der Chemie-Justus Liebig*, **446**, 205.

238 Meisenheimer, J., Senn, O., and Zimmermann, P. (1927) *Chemische Berichte*, **60**, 1736.

239 Fink, D.M. and Kurys, B.E. (1996) *Tetrahedron Letters*, **37**, 995.

240 Yamamori, T. and Adachi, I. (1981) *Chemical Abstracts*, **94**, 65513.

241 Kalkote, U.R. and Goswani, D.D. (1977) *Australian Journal of Chemistry*, **30**, 1847.

242 Kemp, D.S. and Woodward, R.B. (1965) *Tetrahedron*, **21**, 3019.

243 Comanita, E., Popovici, I., Roman, G., Robertson, G., and Comanita, B. (1999) *Heterocycles*, **51**, 2139.

244 Roman, G., Comanita, E., and Comanita, B. (2002) *Tetrahedron*, **58**, 1617.

245 Shutske, G.M. and Kapples, K.J. (1989) *Journal of Heterocyclic Chemistry*, **26**, 1293.

246 Palermo, M.G. (1996) *Tetrahedron Letters.*, **37**, 2885.

247 Lepore, S.D. and Wiley, M.R. (1999) *The Journal of Organic Chemistry*, **64**, 4547.

248 Lepore, S.D. and Wiley, M.R. (2000) *The Journal of Organic Chemistry*, **65**, 2924.

249 Sasaki, T. and Yoshioka, T. (1969) *Bulletin of the Chemical Society of Japan*, **42**, 826.

250 Casini, G., Gualtieri, F., and Stein, M.L. (1969) *Journal of Heterocyclic Chemistry*, **6**, 279.

251 Melchiorre, C., Giannella, M., and Gualtieri, F. (1972) *Annali di Chimica*, **62**, 216.

252 Lehmstedt, K. (1934) *Chemische Berichte*, **67**, 336.

253 Katritzky, A.R., Wang, Z., Dennis Hall, C., and Akhmedov, N.G. (2003) *Arkivok*, **ii**, 49.

254 Kim, B.H., Kim, T.K., Cheong, J.W., and Lee, S.W. (1999) *Heterocycles*, **51**, 1921.

255 Kim, B.H., Jun, Y.M., Kim, T.K., Lee, S.W., Baik, W., and Lee, B.M. (1997) *Heterocycles*, **45**, 235.

256 Kim, B.H., Jun, Y.M., Choi, Y.R., Lee, D.B., and Baik, W. (1998) *Heterocycles*, **48**, 748.

257 Prakash, O., Saini, R.K., Singh, S.P., and Varma, R.S. (1997) *Tetrahedron Letters*, **38**, 3147.

258 Smalley, R.K., Smith, R.H., and Suschitzky, H. (1978) *Tetrahedron Letters*, **19**, 2309.

259 Eckroth, D.R. and Cochran, T.G. (1970) *Journal of the Chemical Society (C)*, 2660.

260 Wróbel, Z. (1997) *Synthesis*, 753.

261 Cheng, L.-J. and Burka, L.T. (1998) *Tetrahedron Letters*, **39**, 5351.

262 Boduszek, B., Halama, A., and Zon, J. (1997) *Tetrahedron*, **53**, 11399.

263 Kurth, M.J., Olmstead, M.M., and Haddadin, M.J. (2005) *The Journal of Organic Chemistry*, **70**, 1060–1062.

264 Davis, R.B. and Pizzini, L.C. (1960) *The Journal of Organic Chemistry*, **25**, 1884–1888.

265 McKinnon, D.M. and Lee, K.R. (1988) *Canadian Journal of Chemistry*, **66**, 1405.

266 McKinnon, D.M. and Abouzeid, A.A. (1991) *Journal of Heterocyclic Chemistry*, **28**, 445.

267 Meth-Cohn, O. and Tarnowski, B. (1978) *Synthesis*, 58.

268 (a) Rahman, L.K.A. and Scrowston, R.M. (1983) *Journal of the Chemical Society, Perkin Transactions 1*, 2973. (b) Rahman, L.K.A. and Scrowston, R.M. (1984) *Journal of the Chemical Society, Perkin Transactions 1*, 385.

269 Fink, D.M. and Strupczewski, J.T. (1993) *Tetrahedron Letters*, **34**, 6525.

270 Buchwald, S.L., Watson, B.T., Lum, R.T., and Nugent, W.A. (1987) *Journal of the American Chemical Society*, **19**, 7137.

271 Hermann, C.K.F., Campbell, J.A., Greenwood, T.D., Lewis, J.A., and Wolfe, J.F. (1992) *The Journal of Organic Chemistry*, **57**, 5328.

272 Hellwinkel, D. and Karle, R. (1989) *Synthesis*, 394.

273 Liu, Z., Shibata, N., and Takeuchi, Y. (2002) *Journal of the Chemical Society, Perkin Transactions 1*, 302.

274 Davis, M. and White, A.W. (1969) *The Journal of Organic Chemistry*, **34**, 2985.

275 Davis, M., Paproth, T.G., and Stephens, L.J. (1973) *Journal of the Chemical Society, Perkin Transactions 1*, 2057.

276 Singerman, G.M. (1975) *Journal of Heterocyclic Chemistry*, **12**, 877.

277 Danylec, B. and Davis, M. (1980) *Journal of Heterocyclic Chemistry*, **17**, 533.

278 McKinnon, D.M. and Abouzeid, A. (1991) *Journal of Heterocyclic Chemistry*, **28**, 347.

279 Meyer, R.F., Cummings, B.L., Bass, P., and Collier, H.O.J. (1965) *Journal of Medicinal Chemistry*, **8**, 515.

280 Gray, J. and Waring, D.R. (1980) *Journal of Heterocyclic Chemistry*, **17**, 65.

281 Buckley, R.K., Davis, M., and Srivastava, K.S.L. (1971) *Australian Journal of Chemistry*, **24**, 2405.

282 Albert, A.H. and O'Brien, D.E. (1978) *Journal of Heterocyclic Chemistry*, **15**, 529.

283 Review: Wojciechowski, K. (2001) *European Journal of Organic Chemistry*, 3587.

284 Yranzo, G.I., Elguero, J., Flammang, R., and Wentrup, C. (2001) *European Journal of Organic Chemistry*, 2209.

285 Yranzo, G.I. and Moyano, E.L. (2004) *Current Organic Chemistry*, **8**, 1071.

286 Lindemann, H. and Cissée, H. (1929) *Annalen Der Chemie-Justus Liebig*, **469**, 44.

287 Padwa, A., Chen, E., and Ku, A. (1975) *Journal of the American Chemical Society*, **97**, 6484.

288 Szeimies, G., Mannhardt, K., and Mickler, W. (1977) *Chemische Berichte*, **110**, 2922.

289 Voghel, G.J., Eggerichs, T.L., Clamot, B., and Viehe, H.G. (1976) *Chimia*, **30**, 191.

290 Padwa, A. and Chen, E. (1974) *The Journal of Organic Chemistry*, **39**, 1976.

291 Davies, K.L., Storr, R.C., and Whittle, P.J. (1978) *Journal of the Chemical Society. Chemical Communications*, 9.

292 Ferris, J.P. and Trimmer, R.W. (1975) *The Journal of Organic Chemistry*, **41**, 13.

293 Sauers, R.R. and Arnum, S.D. (1987) *Tetrahedron Letters*, **28**, 5797.

294 Ferris, J.P. and Antonucci, F.R. (1974) *Journal of the American Chemical Society*, **96**, 2014.

295 Ogata, M., Matsumoto, H., and Kano, H. (1969) *Tetrahedron*, **25**, 5205.

296 Deady, L.W. (1973) *Australian Journal of Chemistry*, **26**, 1949.

297 Davis, M., Deady, L.W., and Homfeld, E. (1974) *Australian Journal of Chemistry*, **27**, 1221.

298 Woodman, D.J. (1968) *The Journal of Organic Chemistry*, **33**, 2397.

299 Lindemann, H. and Thiele, H. (1926) *Annalen Der Chemie-Justus Liebig*, **449**, 63.

300 Quilico, A. and Simonetta, M. (1946) *Gazzetta Chimica Italiana*, **76**, 255.

301 Day, R.A., Blake, J.A., and Stephens, C.E. (2003) *Synthesis*, 1586.

302 Nishimura, N., Hisamitsu, H., Sugiura, M., and Maeba, I. (2000) *Carbohydrate Research*, **329**, 681.

303 De Munno, A., Bertini, V., and Lucchesini, F. (1977) *Journal of the Chemical Society, Perkin Transactions 1*, 1121.

304 Casey, M.L., Kemp, D.S., Paul, K.G., and Cox, D.D. (1973) *The Journal of Organic Chemistry*, **38**, 2294.

305 Hoppe, I. and Schöllkopf, U. (1979) *Annalen Der Chemie-Justus Liebig*, 219.

306 Musante, C. and Fatutta, S. (1958) *Gazzetta Chimica Italiana*, **88**, 879.

307 Walser, A., Flynn, T., and Fryer, R.I. (1974) *Journal of Heterocyclic Chemistry*, **11**, 885.

308 Pigini, M., Giannella, M., Gualtieri, F., Melchiorre, C., Bolle, P., and Angelucci, L. (1975) *European Journal of Medicinal Chemistry*, **10**, 29.

309 Desimoni, G. and Minoli, G. (1968) *Annali di Chimica (Rome)*, **58**, 562.

310 Eiden, F. and Löwe, W. (1970) *Tetrahedron Letters*, **11**, 1439.

311 Kemp, D.S. (1967) *Tetrahedron*, **23**, 2001.

312 Kemp, D.S., Wang, S.-W., Rebek, J., Jr, Mollan, R.C., Banquer, C., and Subramanyam, G. (1974) *Tetrahedron*, **30**, 3955.

313 Olofson, R.A., Vander Meer, R.K., and Stournas, S. (1971) *Journal of the American Chemical Society*, **93**, 1543.

314 Giomi, D., Nesi, R., Turchi, S., and Fabriani, T. (1994) *The Journal of Organic Chemistry*, **59**, 6840.

315 Turchi, S., Giomi, D., and Nesi, R. (1995) *Tetrahedron*, **51**, 7085.

316 Giomi, D., Turchi, S., Danesi, A., and Faggi, C. (2001) *Tetrahedron*, **57**, 4237.

317 Taylor, E.C., Eckroth, D.R., and Bartulin, J. (1967) *The Journal of Organic Chemistry*, **32**, 1899.

318 Wilk, M., Schwab, H., and Rochlitz, J. (1966) *Annalen Der Chemie-Justus Liebig*, **698**, 44.

319 Boruah, R.C. and Sandhu, J.S. (1982) *Synthesis*, 677.

320 Stevens, R.V., Fitzpatrick, J.M., Germeraad, P.B., Hrrison, B.L., and Lapalme, R. (1976) *Journal of the American Chemical Society*, **98**, 6313.

321 Natale, N.R. (1982) *Tetrahedron Letters*, **23**, 5009.

322 Nitta, M. and Higuchi, T. (1994) *Heterocycles*, **38**, 853.

323 Li, C.-S. and Lacasse, E. (2002) *Tetrahedron Letters*, **43**, 3565.

324 Walker, G.N. (1962) *The Journal of Organic Chemistry*, **27**, 1929.

325 Fan, X. and Zhang, Y. (2002) *Tetrahedron Letters*, **43**, 7001.

326 Churykau, D.H., Zinovich, V.G., and Kulinkovich, O.G. (2004) *Synlett*, 1949–1952.

327 Barbero, A. and Pulido, F.J. (2004) *Synthesis*, 401.

328 Iddon, B. (1994) *Heterocycles*, **37**, 1263.

329 Micetich, R.G. (1970) *Canadian Journal of Chemistry*, **48**, 2006.

330 Nesi, R., Ricci, A., Taddei, M., and Tedeschi, P. (1980) *Journal of Organometallic Chemistry*, **195**, 275.

331 Konoike, T., Kanda, Y., and Araki, Y. (1996) *Tetrahedron Letters*, **37**, 3339.

332 Kromann, H., Sløk, F.A., Johansen, T.N., and Krogsgaard-Larsen, P. (2001) *Tetrahedron*, **57**, 2195.

333 Calle, M., Cuadrado, P., González-Nogal, A.M., and Valero, R. (2001) *Synthesis*, 1949.

334 Ku, Y.-Y., Grieme, T., Sharma, P., Pu, Y.-M., Raje, P., Morton, H., and King, S. (2001) *Organic Letters*, **3**, 4185.

335 Davies, M.W., Wybrow, R.A.J., Johnson, C.N., and Harrity, J.P.A. (2001) *Chemical Communications*, 1558.

336 Denmark, S.E. and Kallemeyn, J.M. (2005) *The Journal of Organic Chemistry*, **70**, 2839.

337 Pavlik, J.W. (2003) *Progress in Heterocyclic Chemistry*, **15**, 37.

338 Catteau, J.P., Lablache-Combier, A., and Pollet, A. (1969) *Journal of the Chemical Society, Chemical Communications*, 1018.

339 Pavlik, J.W., Pandit, C.R., Samuel, C.R., and Day, A.C. (1993) *The Journal of Organic Chemistry*, **58**, 3407.

340 Pavlik, J.W., Tongcharoensirikul, P., Bird, N.P., Day, A.C., and Barltrop, J.A. (1994)

Journal of the American Chemical Society, **116**, 2292.

341 Pavlik, J.W., Tongcharoensirikul, P., and French, K.M. (1998) *The Journal of Organic Chemistry*, **63**, 5592.

342 Pavlik, J.W. and Tongcharoensirikul, P. (2000) *The Journal of Organic Chemistry*, **65**, 3626.

343 (a) Kamigata, N., Saegusa, T., Fujie, S., and Kobayashi, M. (1979) *Chemistry Letters* 9; (b) Ono, I., Sato, S., Fukuda, K., and Inayoshi, T. (1997) *Bulletin of the Chemical Society of Japan*, **70**, 2051.

344 (a) Döpp, D., Krüger, C., Lauterfeld, P., and Raabe, E. (1987) *Angewandte Chemie, International Edition in English*, **26**, 146; (b) Elghamry, I., Döpp, D., and Henkel, G. (2001) *Synthesis*, 1223; (c) Döpp, D., Lauterfeld, P., Schneider, M., Schneider, D., Henkel, G., Issac, Y.A.S., and Elghamry, I. (2001) *Synthesis*, 1228.

345 Albert, A.H., O'Brien, D.E., and Robins, R.K. (1980) *Journal of Heterocyclic Chemistry*, **17**, 385.

346 Alberola, A., Alonso, F., Cuadrado, P., and Sañudo, M.C. (1988) *Journal of Heterocyclic Chemistry*, **25**, 235.

347 Weiler, E.D., Petigara, R.B., Wolfersberger, M.H., and Miller, G.A. (1977) *Journal of Heterocyclic Chemistry*, **14**, 627.

348 Clerici, F., Erba, E., Gelmi, M.L., and Valle, M. (1997) *Tetrahedron*, **53**, 15859.

349 Beccalli, E.M., Clerici, F., and Gelmi, M.L. (1999) *Tetrahedron*, **55**, 2001.

350 Clerici, F., Gelmi, M.L., Pini, E., and Valle, M. (2001) *Tetrahedron*, **57**, 5455.

351 Kaae, B.H., Krogsgaard-Larsen, P., and Johansen, T.N. (2004) *The Journal of Organic Chemistry*, **69**, 1401.

352 Christoforou, I.C., Koutentis, P.A., and Rees, C.W. (2003) *Organic & Biomolecular Chemistry*, **1**, 2900.

353 Hassan, M.E., Magraby, M.A., and Aziz, M.A. (1985) *Tetrahedron*, **41**, 1885.

354 (a) McKinnon, D.M. and Hassan, M.E. (1973) *Canadian Journal of Chemistry*, **51**, 3081; (b) McKinnon, D.M., Hassan, M.E.R., and Chauhan, M.S. (1977) *Canadian Journal of Chemistry*, **55**, 1123; (c) McKinnon, D.M., Duncan, K.A., and Millar, L.M. (1984) *Canadian Journal of Chemistry*, **62**, 1580.

355 Review: Schulze, B., Gidon, D., Siegemund, A., and Rodina, L.L. (2003) *Heterocycles*, **61**, 639.

356 (a) Reinhoudt, D.N. and Kouwenhoven, C.G. (1974) *Tetrahedron Letters*, **15**, 2503; (b) Reinhoudt, D.N. and Kouwenhoven, C.G. (1976) *Recueil des Travaux Chimiques des Pays-Bas*, **95**, 67.

357 Hassan, M.E. (1985) *Bulletin des Sociétés Chimiques Belges*, **94**, 149.

358 Weiler, E.D. and Brennan, J.J. (1978) *Journal of Heterocyclic Chemistry*, **15**, 1299.

359 Abou-Gharbia, M., Moyer, J.A., Patel, U., Webb, M., Schiehser, G., Andree, T., and Haskins, J.T. (1989) *Journal of Medicinal Chemistry*, **32**, 1024.

360 Waldner, A. (1989) *Helvetica Chimica Acta*, **72**, 1435.

361 Yeung, K.S., Meanwell, N.A., Li, Y., and Gao, Q. (1998) *Tetrahedron Letters*, **39**, 1483.

362 Burri, K.F. (1990) *Helvetica Chimica Acta*, **73**, 69.

363 Alberola, A., Calvo, L., Rodríguez, M.T.R., and Sañudo, M.C. (1995) *Journal of Heterocyclic Chemistry*, **32**, 537.

364 Burri, K.F. (1989) *Helvetica Chimica Acta*, **72**, 1416.

365 Baggi, P., Clerici, F., Gelmi, M.L., and Mottadelli, S. (1995) *Tetrahedron*, **51**, 2455.

366 Clerici, F., Ferrario, T., Gelmi, M.L., and Marelli, R. (1994) *Journal of the Chemical Society, Perkin Transactions 1*, 2533.

367 Clerici, F., Gelmi, M.L., Soave, R., and Presti, L.L. (2002) *Tetrahedron*, **58**, 5173.

368 Clerici, F., Ferraris, F., and Gelmi, M.L. (1995) *Tetrahedron*, **51**, 12351.

369 Tso, H.-H. and Chandrasekharam, M. (1996) *Tetrahedron Letters*, **37**, 4189.

370 (a) Wojciechowski, K. (1991) *Synlett*, 571; (b) Wojciechowski, K. (1993) *Tetrahedron*, **49**, 7277.

371 Kim, S.H., Kim, K., Kim, K., Kim, J., and Kim, J.H. (1993) *Journal of Heterocyclic Chemistry*, **30**, 929.

372 Clerici, F., Contini, A., Gelmi, M.L., and Pocar, D. (2003) *Tetrahedron*, **59**, 9399.

373 Shutske, G.M., Allen, R.C., Försch, M.F., Setescak, L.L., and Wilker, J.C. (1983) *Journal of Medicinal Chemistry*, **26**, 1307.

374 Carrington, D.E.L., Clark, K., Hughes, C.G., and Scrowston, R.M. (1972) *Journal* *of the Chemical Society, Perkin Transactions 1*, 3006.

375 Boshagen, H. and Geiger, W. (1979) *Chemische Berichte*, **112**, 3286.

376 Caton, M.P.L., Jones, D.H., Slack, R., and Wooldridge, K.R.H. (1964) *Journal of the Chemical Society*, 446.

377 Jones, D.H., Slack, R., and Wooldridge, K.R.H. (1964) *Journal of the Chemical Society*, 3114.

378 Buttimore, D., Caton, M.P.L., Renwick, J.D., and Slack, R. (1965) *Journal of the Chemical Society*, 7274.

379 Kalish, R., Brogen, E., Field, F.G., Anton, T., Steppe, T.V., and Sternbach, L.H. (1975) *Journal of Heterocyclic Chemistry*, **12**, 49.

380 Bunch, L., Krogsgaard-Larsen, P., and Madsen, U. (2002) *The Journal of Organic Chemistry*, **67**, 2375.

381 Alberola, A., Calvo, L., Rodríguez, M.T.R., and Sañudo, M.C. (1993) *Journal of Heterocyclic Chemistry*, **30**, 393.

382 Alberola, A., Alonso, F., Cuadrado, P., and Sanudo, M.C. (1987) *Synthetic Communications*, **17**, 1207.

10
Five-Membered Heterocycles: 1,3-Azoles

Julia Revuelta, Fabrizio Machetti, and Stefano Cicchi

10.1
Introduction

Imidazole, oxazole, and thiazole, known as 1,3-azoles, are planar five-membered ring systems with three carbons, one nitrogen, and an additional heteroatom (nitrogen, oxygen or sulfur). These compounds and their derivatives have been known since the nineteenth century.

The first reports on imidazoles were published in the 1840s concerning the 2,3,5-triphenylimidazoles [1, 2]. In 1858 Debus [3] reported the reaction between glyoxal and ammonia and pioneered the synthesis of imidazoles. Historically, the molecule was named "glyoxaline" and "iminazole," but imidazole is now used universally.

Oxazole (1,3-oxazole) was first prepared by Cornforth [4] in 1947, about 100 years after the first synthesis of a substituted oxazole, 1,4,5-triphenylisoxazole, reported by Zinin in 1840 [5]. In 1888, Hantzsch [6] gave the name of oxazoles to these class of compounds. Oxazole is not easily synthesized and has been unavailable in large quantities (because of high synthesis costs) and has only recently become commercially available in multigram scale at a reasonable cost. The Cornforth procedure was a lengthy route in which the oxazole was obtained in the final stage by decarboxylation–distillation of the corresponding oxazole-4-carboxylic acid from hot quinoline-CuO. Oxazole may be prepared in the laboratory following the practical method of Brederick and Bangert [7].

Modern oxazole chemistry was stimulated in the 1940s by the synthesis of penicillin, which was presumed to contain the oxazole nucleus. The next important development of oxazole chemistry came from the discovery by Kondrateva in the late 1950s that these compounds acts as dienes in the Diels–Alder reaction and by Huisgen's work on 1,3-dipolar cycloaddition reactions of mesoionic derivatives.

Wallach reported the synthesis of some thiazoles in the 1870s. One of the first thiazoles prepared was the 5-aminothiazole-2-carboxylic acid amide [8].

In terms of literature about imidazoles, oxazoles and thiazoles there are many reviews on specialized topics (chemistry, physicochemical, biological and pharma-

Modern Heterocyclic Chemistry, First Edition.
Edited by Julio Alvarez-Builla, Juan Jose Vaquero, and José Barluenga.

cological properties) but *Comprehensive Heterocyclic Chemistry* (1984 [9] updated in 1996 [10]) and *Science of Synthesis* [11] are the most exhaustive sources of information.

10.1.1 Nomenclature

The general name 1,3-azoles (**1**) indicates, according to the more recent IUPAC nomenclature rules, five-membered (stem -ole) heterocycles with one nitrogen atom (prefix aza-) bearing a second heteroatom in position 3 of the cycle. This name is generally used to indicate thiazole, oxazole and imidazole. The first two names correspond to the IUPAC nomenclature, whereas the name imidazole is a trivial one that is commonly preferred to 1,3-diazole. The state of hydrogenation is indicated either by the prefixes "dihydro-" or in the stem "-olidine". This chapter examines 1,3-oxazole, 1,3-thiazole and imidazole, as well as their saturated derivatives. Figure 10.1 gives an easy, exhaustive description of nomenclature and numbering of the structures.

These heterocyclic rings occur widely in nature, contained in the structure of several secondary metabolites. Some of these compounds are promising candidates as drugs and their synthesis has been performed. The oxazole ring is present in the backbone of diazonamide A (**3**), a complex macrocyclic molecule with anticancer properties, which was recently been synthesized [12]. Phorboxazole A (**4**), which possesses two isolated oxazole rings, is an extremely cytotoxic compound that is active towards numerous human tumor cell lines [13]. Other bis- and trisoxazole oxazole compounds are known, such as muscoride A (**5**) [14] and ulapualide A (**6**) [15]. The thiazole ring, in different degrees of saturation, is present in thiamine pyrophosphate (**7**), 6-aminopenicillanic acid (**8**) and other secondary metabolites such as dendroamide A (**9**) [16] and epothilone A (**10**) [17]. Histidine (**11**), which has an imidazole ring, is a very well known amino acid with a fundamental role in the metabolism and also in the family of alkaloids derived from it. The same ring, in reduced form, is also found in biotin (**12**) and in a large series of alkaloids, such as (−)-agelastatin A (**13**) [18] and fumiquinazoline A (**14**) [19] (Figure 10.2).

10.2 General Reactivity

10.2.1 Relevant Physicochemical Data, Computational Chemistry and NMR Data

These compounds present an additional nitrogen atom in the ring with respect to pyrrole, thiophene and furan. This additional nitrogen atom contributes with one electron to the aromaticity of the ring while a lone pair of electrons remains in the plane of the molecule. This additional nitrogen atom lowers the energy levels of π orbitals, as verified by the ionization potentials, which are higher than those of pyrrole, thiophene and furan, respectively. At the same time the additional nitrogen

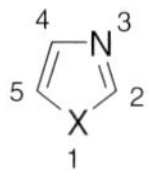

X = O 1,3-oxazole
X = S 1,3-thiazole
X = NH 1*H*-imidazole

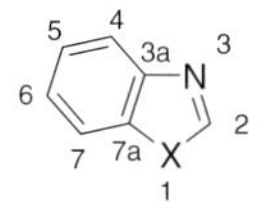

X = O 1,3-benzoxazole
X = S 1,3-benzothiazole
X = NH 1*H*-benzimidazole

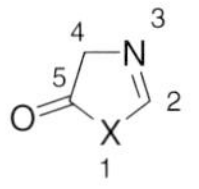

X = O 1,3-oxazol-5(4*H*)-one
X = S 1,3-thiazol-5(4*H*)-one

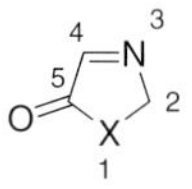

X = O 1,3-oxazol-5(2*H*)-one
X = S 1,3-thiazol-5(2*H*)-one

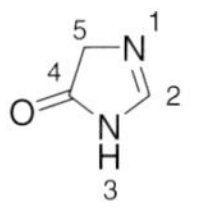

3,5-dihydro-4*H*-imidazol-4-one

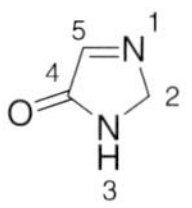

2,3-dihydro-4*H*-imidazol-4-one

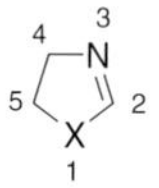

X = O 4,5-dihydro-1,3-oxazole
X = S 4,5-dihydro-1,3-thiazole
X = NH 4,5-dihydro-1*H*-imidazole

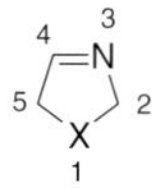

X = O 2,5-dihydro-1,3-oxazole
X = S 2,5-dihydro-1,3-thiazole
X = NH 2,5-dihydro-1*H*-imidazole

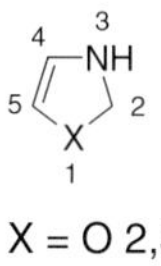

X = O 2,3-dihydro-1,3-oxazole
X = S 2,3-dihydro-1,3-thiazole
X = NH 2,3-dihydro-1*H*-imidazole

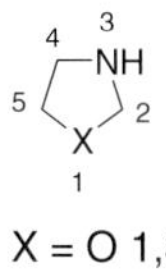

X = O 1,3-oxazolidine
X = S 1,3-thiazolidine
X = NH imidazolidine

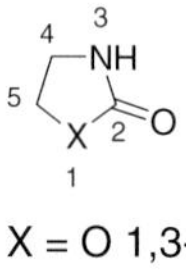

X = O 1,3-oxazolidin-2-one
X = S 1,3-thiazolidin-2-one
X = NH imidazolidin-2-one

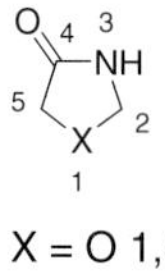

X = O 1,3-oxazolidin-4-one
X = S 1,3-thiazolidin-4-one
X = NH imidazolidin-4-one

Figure 10.1 Nomenclature and numbering of 1,3-oxazole, 1,3-thiazole, imidazole, and their saturated derivatives.

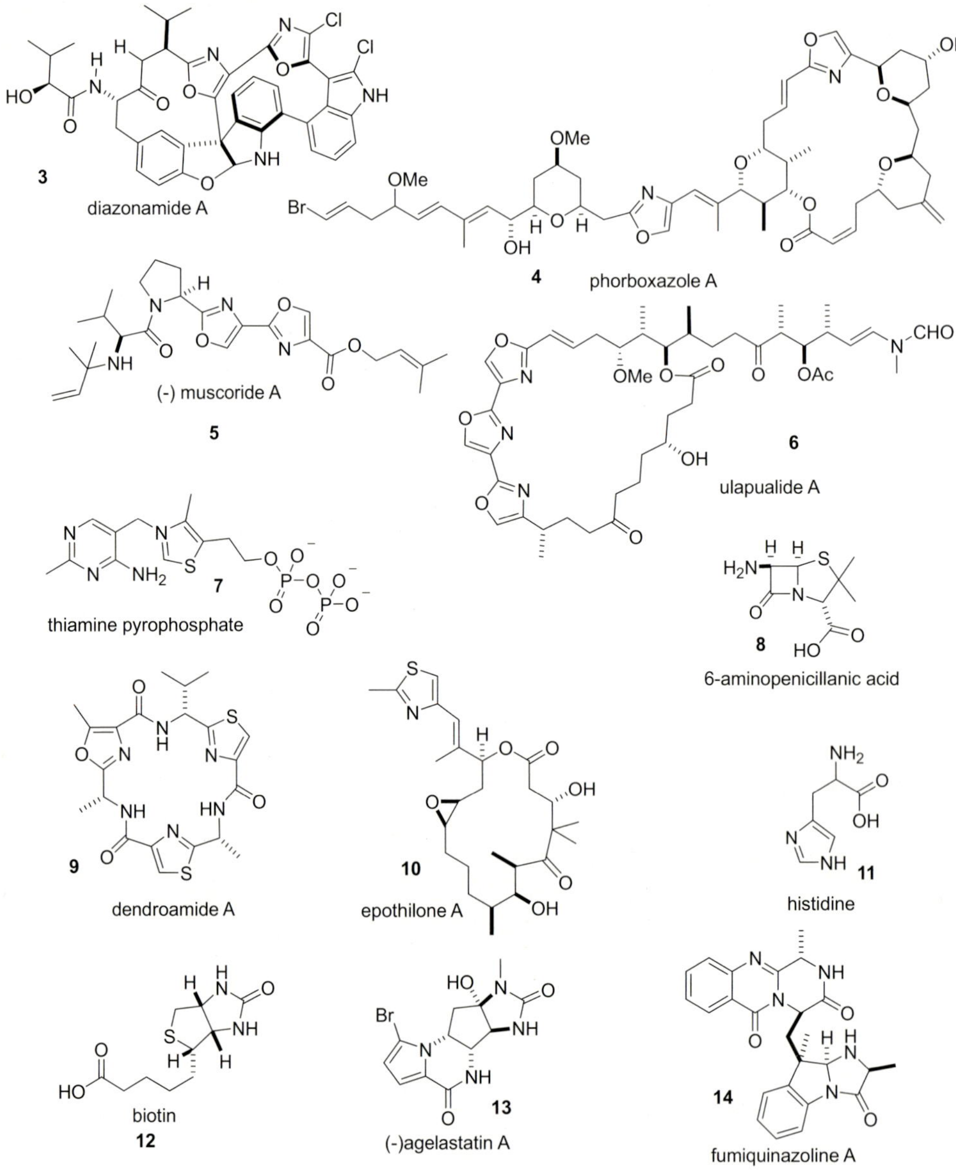

Figure 10.2 Some natural compounds containing the 1,3-azole structure.

atom has an inductive electron-withdrawing effect that provides stabilization to negatively charged reaction intermediates formed in reactions like nucleophilic substitution [Scheme 10.1 (1)] and deprotonation of the methyl group [Scheme 10.1 (2)]. A recent study, based on NMR data of the carbanions of 2-benzyl-1,3-azoles, ranked the π electron-withdrawing power of these heterocycles as thiazole > oxazole

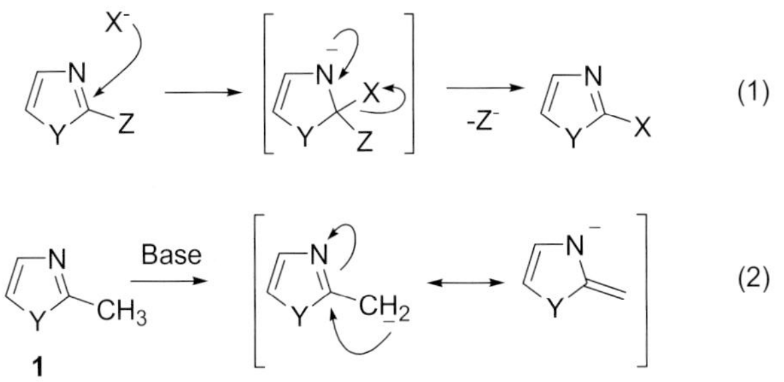

$$\text{(1)}$$

$$\text{(2)}$$

1

Scheme 10.1

> imidazole in terms of charge demand, that is, the fraction of p-negative charge delocalized by the ring [20]

These compounds are aromatic, each carbon atom and one nitrogen atom participating with one electron to the conjugated π system while the other heteroatom participates with a lone pair of electrons. The NMR and UV spectra confirm the aromaticity of the ring (Table 10.1).

The π electrons are delocalized across the ring although the electron density is largely concentrated onto the two heteroatoms. The electron density map also indicates that these compounds are π excessive although this character decreases on passing from N to O to S. From the same map it can be predicted that electrophilic substitution is favored at C4 and C5, while the reduced electron density at C2 should favor nucleophilic attack onto this position (Table 10.2).

The effect of the additional nitrogen atom is evidenced also by the acid/base properties of these compounds. The lone pair on nitrogen atom provide a site for protonation and most azoles are stronger bases than pyrroles. However, the stability of the azolide anions makes imidazole a stronger acid than pyrrole (Figure 10.3).

Outstanding, in this series, is the pK_a of imidazole and of the imidazolium ion. These values are justified by the predominance of a mesomeric effect. The two

Table 10.1 UV and NMR data for 1,3-azoles[a)].

X (Table 10.2)	UV (ethanol) λ (nm) (ε, mol^{-1} dm^3 cm^{-1})	^{1}H (δ, ppm)	^{13}C (δ, ppm)
N	207–208 (3.7)	H2: 7.73	C2: 135.4
		H4: 7.14	C4: 121.9
		H5: 7.14	C5: 121.9
O	205 (3.9)	H2: 7.95	C2: 150.6
		H4: 7.09	C4: 125.4
		H5: 7.69	C5: 138.1
S	207.5 (3.41) 233.0 (3.57)	H2: 8.77	C2: 153.6
		H4: 7.86	C4: 143.3
		H5: 7.27	C5: 119.6

For atom numbering see structure in Table 10.2.

Table 10.2 Physicochemical data for 1,3 azoles.

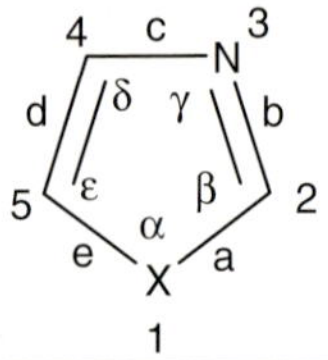

X	Bond angles (°)					Bond distances (Å)				
	α	β	γ	δ	«	a	b	c	d	e
N	107.2	111.3	105.4	106.3	109.8	1.349	1.326	1.378	1.358	1.369
O	103.9	115.0	103.9	109.7	108.1	1.357	1.292	1.395	1.353	1.370
S	89.3	115.2	110.1	115.8	109.6	1.713	1.304	1.372	1.367	1.713

X	Electron densities				
X	1	2	3	4	5
N	1.502	0.884	1.502	1.056	1.052
O	1.730	1.021	1.115	1.058	1.076
S	1.970	0.870	1.190	0.960	1.010

identical resonance structures account for the high stability of these ions (Scheme 10.2). Oxazole, on the other hand, is the least basic of the three heterocycles due to the inductive effect of the oxygen atom.

10.2.2
Tautomerism

A peculiar behavior of *N*-unsubstituted imidazoles is their rapid equilibrium between the two possible tautomers. This rapid equilibrium hampers the separate isolation of 4- and 5-substituted imidazoles. Nevertheless the tautomeric equilibrium can be shifted mainly towards one of the two forms. Imidazoles substituted with electron-

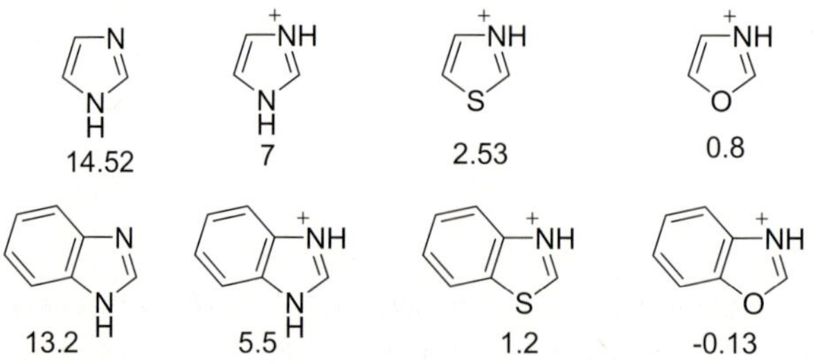

Figure 10.3 Values of pK_a for 1,3-azoles and azolium ions.

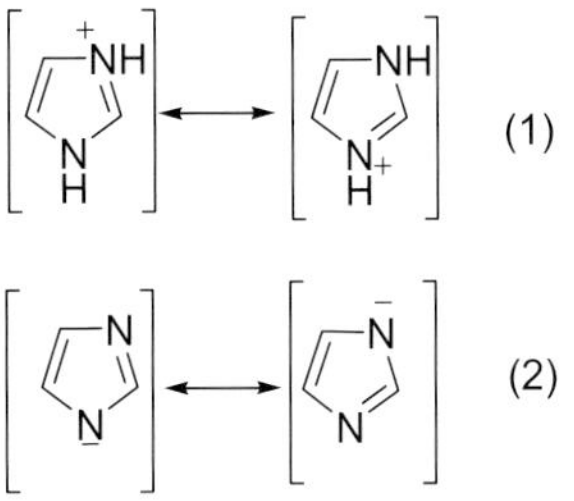

Scheme 10.2 Resonance structures of imidazolium and imidazolide ions.

withdrawing groups are predominantly present as 4-substituted tautomers (e.g., 4-nitroimidazole, **2**) [Scheme 10.3 (1)]. In neutral organic solvents this equilibrium can be an intermolecular process involving two or more imidazole molecules while in protic solvents the solvent itself is involved. The NH proton of imidazole exchanges rapidly in D_2O solution.

The tautomeric equilibrium of 2-amino substituted 1,3-azoles is shifted towards the amino form over the imino form [Scheme 10.3 (2)]. This assumption is confirmed by the basicity of the imino form, which is generally higher than that of the amino. Conversely, 2-hydroxy substituted 1,3-azoles behave as keto derivatives [Scheme 10.3 (3)]. These considerations can generally be extended to other amino and hydroxy derivatives.

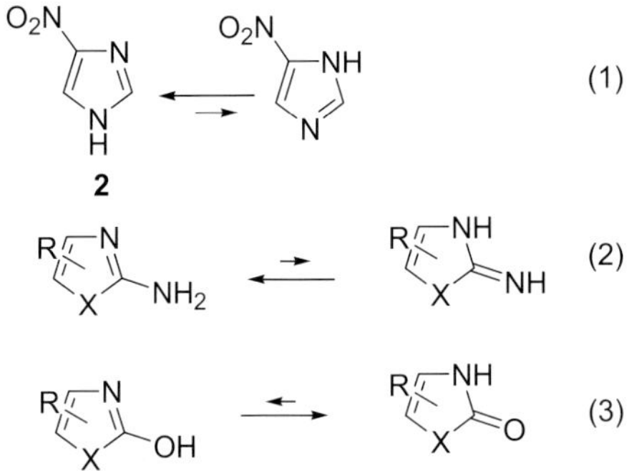

Scheme 10.3 Tautomeric equilibria of substituted 1,3-azoles.

10.3 Synthesis of Aromatic 1,3-Azoles

10.3.1 Synthesis of Imidazoles

A remarkable number of synthetic approaches have been developed for imidazoles, due to is prevalence in natural products and pharmacologically active compounds. Clearly, no a single general synthetic method fulfils all needs in the preparation of functionalized imidazoles, and various cyclization reactions are used to produce specifically substituted imidazoles [21–24]. A detailed compilation of known methods of synthesis of imidazoles has been published [25]. This brief section, while covering classical methods, focuses mainly both on their conceptual extensions and on new synthetic routes. Most classical preparation methods of imidazoles derive from the approach followed by Debus, who pioneered the use of 1,2-dicarbonyl compounds for the synthesis of substituted imidazoles (Scheme 10.4). The reaction was extended using α-ketoaldehydes or α-diketones as substrates. This route in general provides 2-monosubstituted and 2,(4,5- homo and hetero)trisubstituted imidazoles (Scheme 10.4).

NH_3/H_2O or $NH_4Ac/AcOH$

15

$- NH_3$

Debus' reaction: $R^2=R^4=R^5=H$

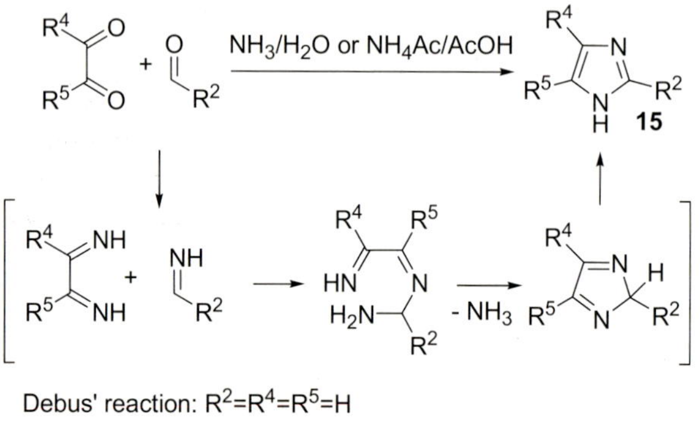

Scheme 10.4

The Radziszewski reaction [26] is a modified version of this method using α-hydroxyketones. A similar synthetic methodology was introduced by Bredereck [27] in which an α-hydroxyketone or an α-haloketone is heated with formamide instead of ammonia or ammonium acetate. Bredereck's reaction provides 4-substituted and 4,5-disubstituted imidazoles (Scheme 10.5).

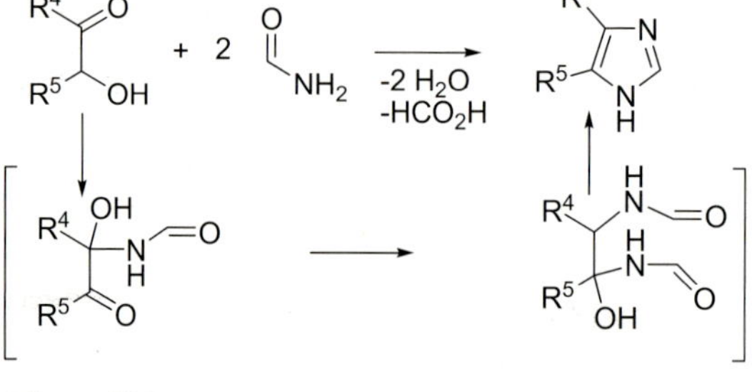

$-2\ H_2O$
$-HCO_2H$

Scheme 10.5

The advent of microwave assisted organic synthesis (MAOS) has allowed the reinvestigation of the classical conditions of Debus synthesis of imidazoles. Thus, synthesis from 1,2-diketones and aldehydes in the presence of various catalysts has been reported. These include silica gel [28], silica gel/Zeolite HY [29], Al_2O_3 [30] and CH_3CO_2H [31]. These recent more reports, with some green chemistry related improvements, utilize solvent-free, silica-gel catalyzed condensation of aromatic aldehydes or benzonitrile derivatives with benzyl (**16**) (as well as other aromatic or heteroaromatic diketones) [31] in the presence of primary amines to obtain the corresponding tetrasubstituted imidazoles (**17**). The reactions proceed with high yields but the tolerated substitution pattern is restricted to symmetrical residues due to a lack of regiocontrol for the 4- and 5-positions (Scheme 10.6) [29].

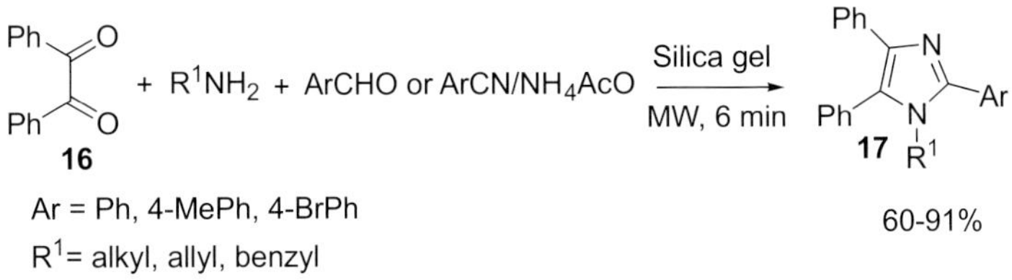

Ph, O, Ph, O, **16** + R^1NH_2 + ArCHO or ArCN/NH_4AcO → Silica gel, MW, 6 min → **17** (Ph, Ph, N, N, Ar, R^1)

Ar = Ph, 4-MePh, 4-BrPh

R^1= alkyl, allyl, benzyl

60-91%

Scheme 10.6

α-Hydroxyketones have also been used in MAOS procedures [32], and both diketones and α-hydroxyketones have found application in ionic-liquid promoted synthesis [33].

The cyclization of N,N′-disubstituted oxamides **18** with PCl_5 to afford 1-substituted 5-chloroimidazoles **19** is another classical method of imidazole ring preparation. This method was discovered by Wallach [34–36] and elaborated by Sarasin [37]. This approach, initially limited to methyl and ethyl symmetrical disubstituted oxamides, was extended to higher symmetrical [34, 38, 39] and unsymmetrical N,N′-disubstituted oxamides (Scheme 10.7) [40]. In the latter case, to obtain appreciable regioselectivity very dissimilar substituent (alkyl versus aryl) should be used. The reaction gives high yields for limited phenyl and chlorinated phenyl substituents but with a 3-pyridyl substituent the yield is low (Scheme 10.7).

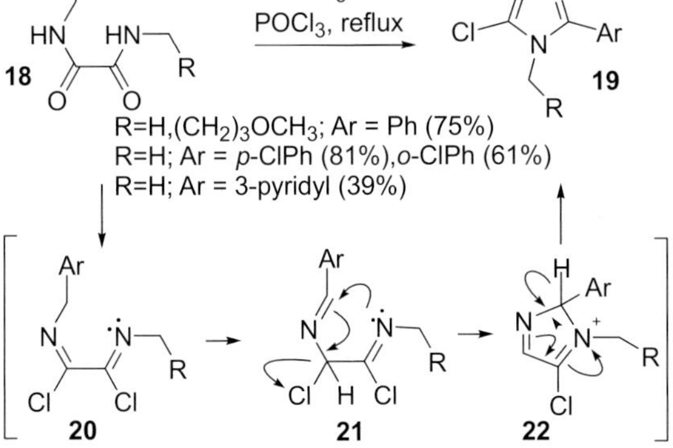

18 (Ar, HN, HN, R, O, O) → PCl_5, $POCl_3$, reflux → **19** (Cl, N, N, Ar, R)

R=H,$(CH_2)_3OCH_3$; Ar = Ph (75%)

R=H; Ar = *p*-ClPh (81%),*o*-ClPh (61%)

R=H; Ar = 3-pyridyl (39%)

[**20** (Ar, N, N, R, Cl, Cl) → **21** (Ar, N, N, R, Cl, H, Cl) → **22** (H, Ar, N, N, +, R, Cl)]

Scheme 10.7

Wallach's reaction proceeds via a dichlorinate adduct (**20**), which after a double bond migration (hydride shift), cyclizes to the final 5-chlorosubsituted imidazole **19** (Scheme 10.7).

Another example of a classical approach to imidazole is represented by the Marckwald synthesis, involving the use of α-aminocarbonyl compounds with cyanates, thiocyanate and isothiocyanates. This approach allows the synthesis of 4,5-disubstituted imidazoles and will be described in more detail in Section 10.4.5 for the synthesis of 2-amino-1,3-azoles. TosMIC (tosylmethyl isocyanide) **25** and other isocyanides, **23**, are key reagents for the preparation of 1,5 substituted imidazole rings, providing the CNC fragment [Scheme 10.8 (1)] [41]. Different species like aldimines, imidoyl chlorides, isothiocyanates, nitriles and imino ethers can undergo cycloadditions with TosMIC giving imidazoles. This methodology provides densely functionalized imidazoles, as **24**, with various substitution patterns in a completely regioselective manner.

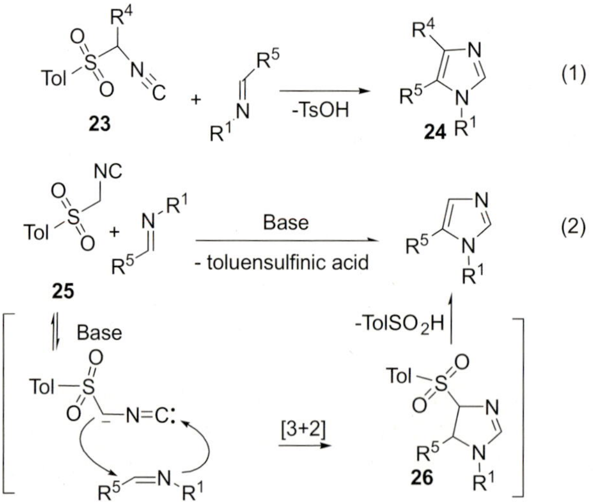

Scheme 10.8

The TosMIC molecule (which accommodates a reactive isocyanide carbon and an activated methylene) can cycloadd its $CH_2N{=}C$ moiety to polarized double bonds under basic conditions [Scheme 10.8 (2)]. When applied to aldimines, this type of reaction affords imidazoles by elimination of *p*-toluenesulfinic acid from the intermediate 4-tosyl-2-imidazolines **26** [42]. This methodology regioselectively provides 1,5 and a limited number of 1,4,5-trisubstituted imidazoles.

This methodology provides regioselectively 1-arylimidazole-5-carboxylates **28** by the addition of anilines to ethyl glyoxylate **27** in methanol followed by reaction with TosMIC (Scheme 10.9) [43].

The related reaction of ethyl glyoxylate or glyoxylic acid with primary amines or ammonia and aryl substituted TosMIC reagents **23** [44] is a versatile method for the synthesis of imidazole-5-carboxylates substituted at C4 (e.g., **29**, Figure 10.4) [45].

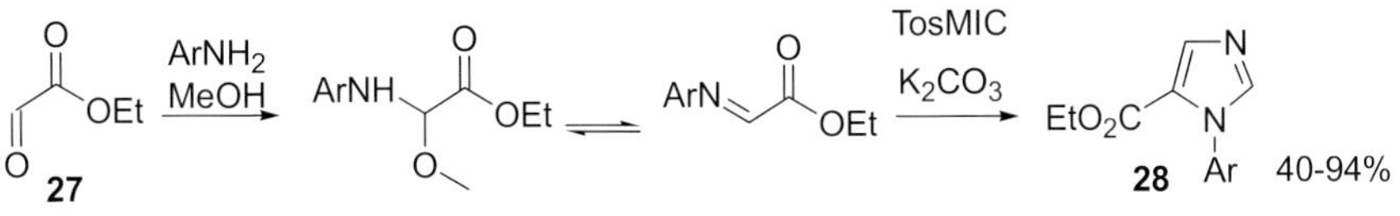

Scheme 10.9

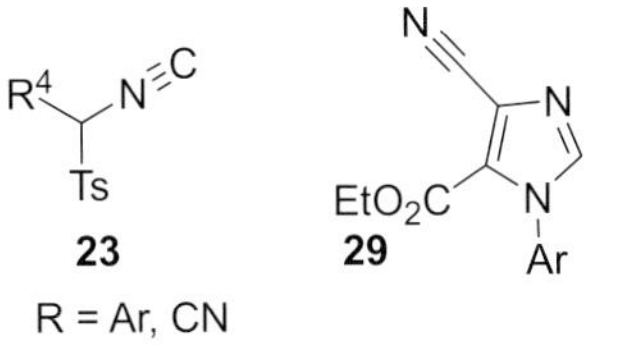

Figure 10.4 Structures of susbstitute TosMIC reagents and of imidazole-5-carboxylates substituted at C4.

The TosMIC chemistry can be extended utilizing an aldehyde as a partner (Scheme 10.10) [46]. Polysubstituted imidazoles are not easily made with this methodology and a mixtures of regioisomeric imidazoles are obtained.

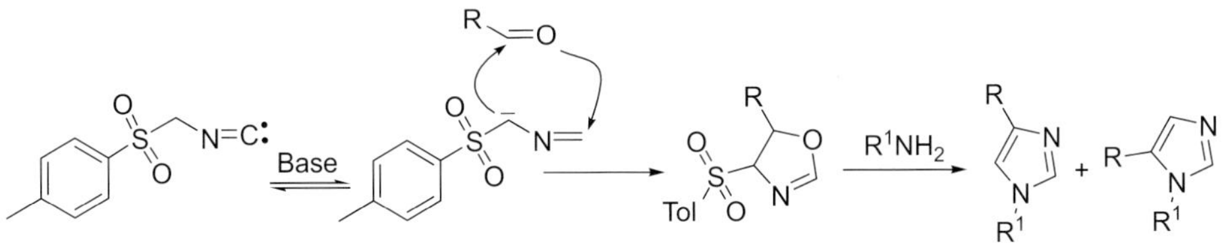

Scheme 10.10

Amidines, guanidines, ureas and thioureas are the most common NCN synthons. With these synthons we can selectively synthesize 1,2,5 substituted imidazoles [47].

Bromo enol ethers **31** react with a range of monosubstituted amidines **30**, giving with high regioselectivity 1,2,5 imidazoles in moderate yields [Scheme 10.11 (1)]. The mechanism that explains the high regioselectivity of this reaction is shown in Scheme 10.11 (2). The unsubstituted nitrogen of amidine **30** adds in Michael fashion to the β-carbon of the ether. Intermediate **32** then cyclizes with extrusion first of HBr and then of ROH to afford imidazole **33**. Initial attack by the monosubstituted nitrogen is disfavored, particularly in cases where R^1 is a bulky group.

1,5 Disubstituted imidazole-4-carboxylates can be efficiently synthesized using the reactivity of BICA [3-bromo-2-isocyanoacrylate (**34**)] [48]. In this strategy, a Michael reaction of a primary amine with α,β-unsaturated ester **34** and subsequent β-elimination of hydrogen bromide occurs as the first step. The resulting enamine **35** undergoes an intramolecular nucleophilic addition to the isonitrile group, affording the final imidazole **36** (Scheme 10.12). This approach has been applied to the synthesis of imidazoles having a range of substituents at the 1- and 5-positions.

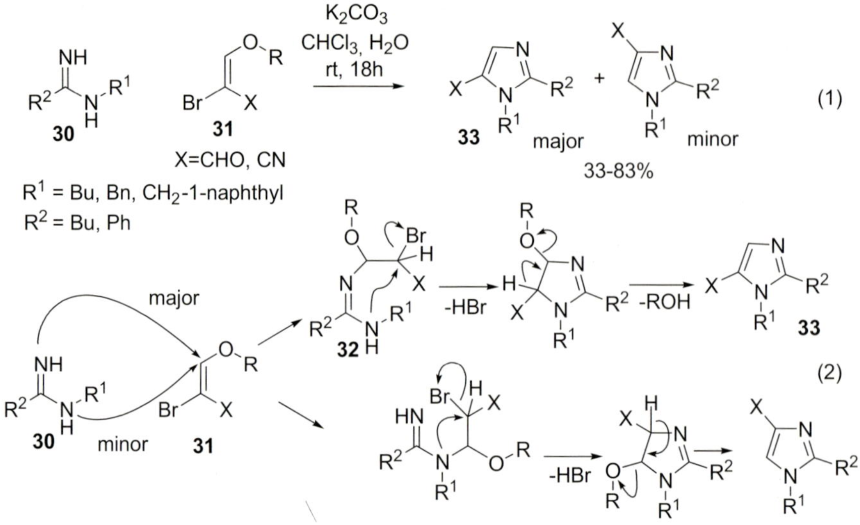

Scheme 10.11

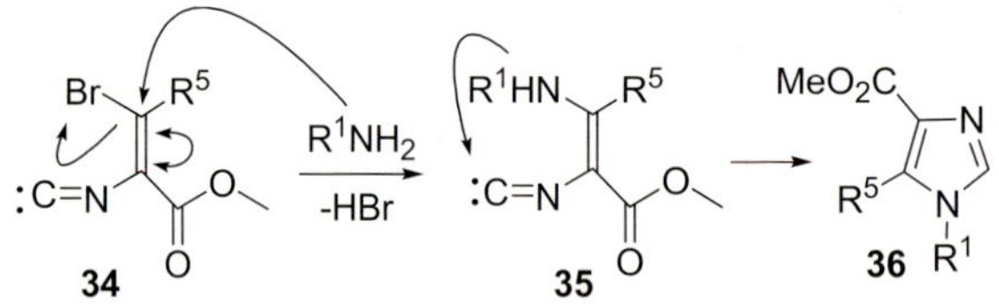

Scheme 10.12

A methodology for the preparation of 2-substituted-4,5-dicyanoimidazoles is the reaction of DAMN (diaminomaleonitrile) with carbonyl compounds under oxidative conditions [49]. A more recent variation is the reaction of DAMN with anhydrides to afford β-aminoenamides that dehydrates to afford the imidazole nucleus [50]. The commercial availability of this reagent gave impulse to its use in the synthesis. Its use and the mechanism of action are discussed in Section 10.5.6 for the synthesis of 2-amino derivatives.

Few methods for the direct construction of tetrasubstituted imidazoles are available and they are often restricted to a fixed pattern of substitution [28, 51, 52]. Popular methodologies for the construction of imidazole rings such as the aforementioned approaches based on van Leusen's TosMIC chemistry are not able to provide direct access to tetrasubstituted imidazoles, because do not allow to insert substituent on C2 of the imidazole ring. Further introduction of substituent requires activation/substitution (Section 10.3). In general the synthesis of tetrasubstituted imidazoles rely on the regiocontrolled synthesis of trisubstituted imidazoles followed by insertion of the fourth substituent [53–56]. This approach have some drawbacks. For example the *N*-alkylation of 2,4,5-trisubstituted imidazoles has as a major drawback the formation of both possible regioisomers that are often difficult to separate.

One of the more versatile intermediates for the synthesis of imidazoles and in particular tetrasubstituted imidazoles are 1,4 dicarbonyl compounds. This approach is limited for two main reasons: (i) the syntheses of these precursors are nontrivial and in many examples involve multistep sequences starting from 1,2-aminoalcohols [57–59]; (ii) the reaction conditions are often drastic. An efficient synthesis of tri- and tetra- substituted imidazoles starting from β-ketoamides under neutral reaction conditions has been reported (Scheme 10.13) [58]. Cyclization was carried out using ammonium trifluoroacetate as a solvent.

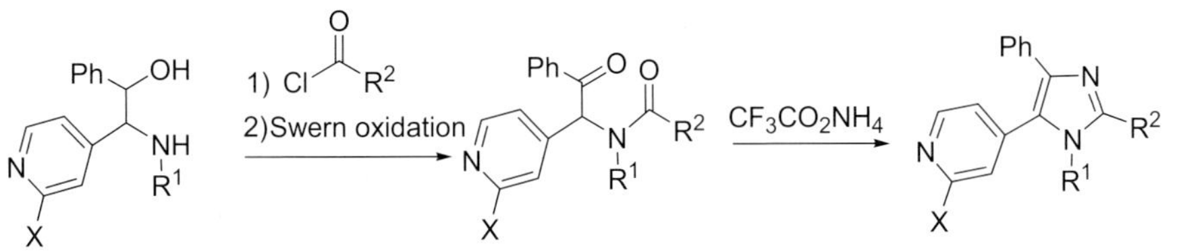

1) $ClC(O)R^2$
2)Swern oxidation
$CF_3CO_2NH_4$

Scheme 10.13

The method tolerates a various array of substituents at N1 and C2. For bulky R^1 the substituent R^2 is limited to small groups due to the difficult synthesis of the corresponding keto amides. With this methodologies it is possible decorate the imidazole scaffold with sufficient flexibility (Figure 10.5) [60].

A very attractive method for the preparation of tetrasubstituted imidazoles is based on an hetero-Cope rearrangement followed by an amidine cyclization [61]. This procedure utilizes as starting materials oximes **37** (bearing R^4 and R^5 imidazole substituents) and imidoyl chlorides **38** (bearing R^1 and R^2 imidazole substituents) (Scheme 10.14). This reaction is highly regioselective, the main limitation being the aromatic nature of the C2 substituents.

Another direct approach towards 1-methyl-tetrasubstituted imidazoles involves the 1,3-dipolar cycloaddition of methylated mesoionic oxazolones (münchnones). The reaction of *N*-methyl-1,3-oxazolium-5-olates **39** with imines **40** involves a 1,3-dipolar cycloaddition to give an unstable primary bicyclic adduct **41** that loses carbon dioxide and benzenesulfinic acid and gains aromaticity (Scheme 10.15) [52]. The phenylsulfonyl leaving group enhances the tendency to aromatize. The yields reported for this reaction are low, at least partly due to self-condensation of münchnones.

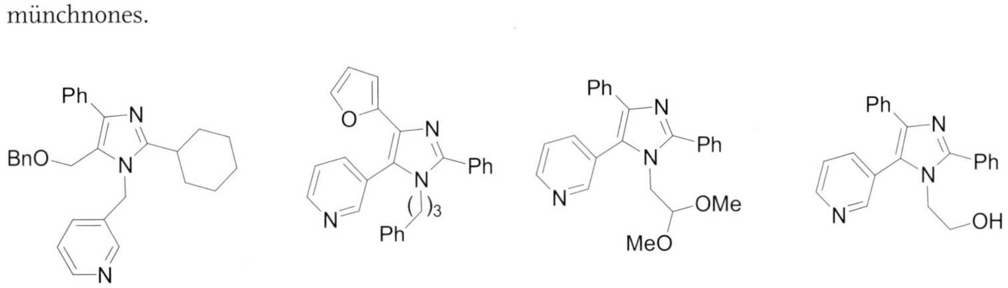

Figure 10.5 Structures of compounds obtained starting from 1,4-dicarbonyl reagents.

R^1, R^4, R^5 = Ar, Alk 32-96%

Scheme 10.14

Scheme 10.15

The problem of self-condensation can be suppressed by using a solid-phase approach towards preparation of imidazoles (Scheme 10.16) The precursor **43** prepared from commercially available resin **42** (ArgoGel-MB-CHO) and treated with ethyl-diisopropylcarbodiimide (EDC) led to the münchnone **44**. Subsequent cycloaddition with tosyl imine **45** followed by elimination of toluenesulfinic acid and CO provided the resin linked imidazoles **46**. The imidazoles were liberated from the resin by treatment with glacial acetic acid at 100 °C and obtained in good yield and purities [62].

There are a few examples that utilize the concept of a multicomponent reaction (MCR), especially the Ugi type reaction, for the synthesis of imidazoles. However, most syntheses have been performed in a stepwise fashion and were not set-up as MCR. A three component reaction utilizing aldehydes, *o*-picolilamines and isocyanides is shown in Scheme 10.17 [63].

The reaction proceeds through *in situ* formation of imine **49** from aldehyde **48** and amine **47** followed by the attack of the α-acidic isocyanide **51** and subsequent ring

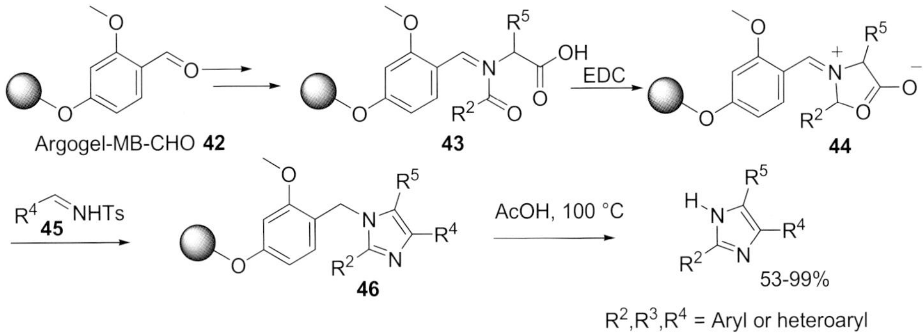

Argogel-MB-CHO **42**

43

EDC

44

R^4 **45** NHTs

46

AcOH, 100 °C

53-99%

R^2,R^3,R^4 = Aryl or heteroaryl

Scheme 10.16

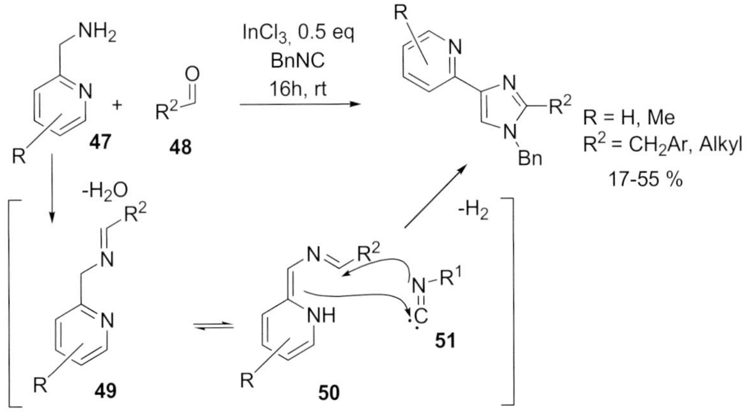

47 + **48**

$InCl_3$, 0.5 eq
BnNC
16h, rt

R = H, Me
R^2 = CH_2Ar, Alkyl
17-55 %

$-H_2O$

$-H_2$

49 **50** **51**

Scheme 10.17

closure. Subsequent aromatization affords the final imidazole. The nucleophilicity of the (2-pyridyl)methyl carbon derives from the enamine tautomer **50**.

Scheme 10.18 depicts the synthesis of 1-substituted-4-carboxylic acid imidazoles **53** through a resin-bound isonitrile [64]. The solid supported isonitrile

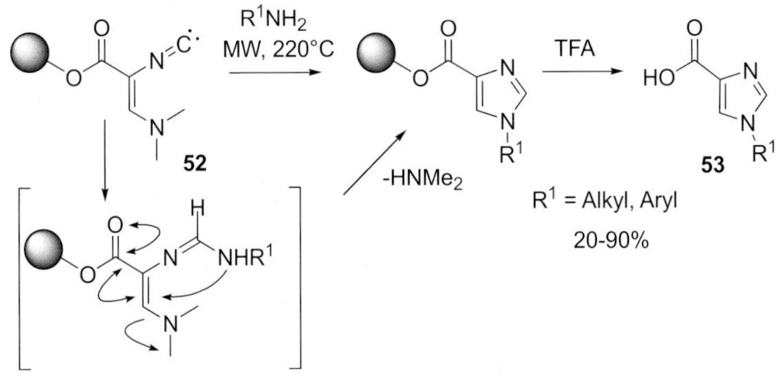

52

R^1NH_2
MW, 220°C

TFA

53

$-HNMe_2$

R^1 = Alkyl, Aryl
20-90%

Scheme 10.18

(Wang resin) **52** was treated with alkyl and aryl amines to give, in variable yields, the imidazoles. The reactions can be speeded up using microwaves. The reaction proceeds with a mechanism involving α-addition of the amine to the isocyanide.

10.3.2
Synthesis of Oxazoles

Oxazole are common substructures in several biologically active compounds, synthetic intermediates and pharmaceuticals, and consequently there is a continuing stimulus for the development of more general and versatile synthetic methodologies for this class of compounds [65].

The cyclodehydration of 2-acylamino ketones **54**, known as the Robinson–Gabriel reaction, is one of the oldest and most widely used synthesis of 2,5-disubstituted and 2,4,5-trisubstituted oxazoles (Scheme 10.19) [66, 67].

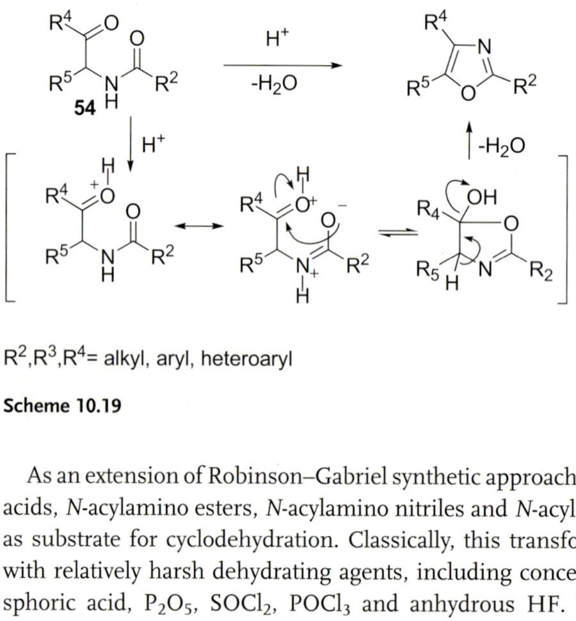

R^2,R^3,R^4= alkyl, aryl, heteroaryl

Scheme 10.19

As an extension of Robinson–Gabriel synthetic approach to oxazoles, *N*-acylamino acids, *N*-acylamino esters, *N*-acylamino nitriles and *N*-acyl peptides have been used as substrate for cyclodehydration. Classically, this transformation was carried out with relatively harsh dehydrating agents, including concentrated H_2SO_4, polyphosphoric acid, P_2O_5, $SOCl_2$, $POCl_3$ and anhydrous HF. These classical reagents continue to be used to prepare a wide variety of oxazole derivatives and some examples are shown in Table 10.3.

2-Monosubstituted and 2,4-disubstituted oxazoles are generally inaccessible by the Robinson–Gabriel method owing to the sensitivity of 2-acylamino aldehydes to oxidative and dehydrating conditions. In addition, the same problems can be suffered by α-acylamino ketones containing stereochemically sensitive side chains. However, milder dehydrating agent can be used, offering broader functional group compatibility [76]. A protocol based on triphenylphosphine/iodine in the presence of triethylamine has been introduced by Wipf and Miller (Scheme 10.20) [77].

Table 10.3 Robinson–Gabriel synthesis of substituted oxazoles.

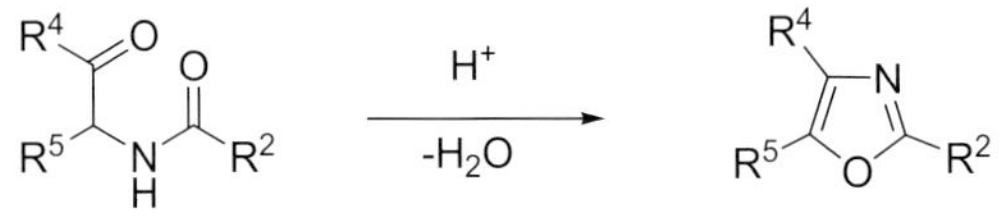

R4, O, O, R5, N, H, R2; H^+ / $-H_2O$ → R4, N, R5, O, R2

R^2	R^4	R^5	Yield (%)	Reference
CH_2Cl	Et	H	31[a)]	[68]
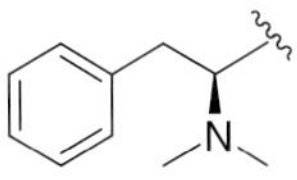 N	H	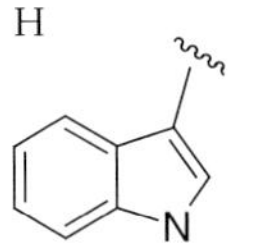 N	77[a)]	[69]
Me	CO_2Me	CH_2CO_2Me	19[b)]	[70]
Pr	Ph	Ph	87[c)]	[71]
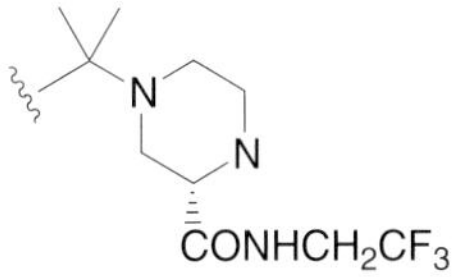 N, N, $CONHCH_2CF_3$	H	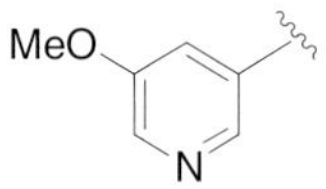 MeO, N	82[c)]	[72]
$(CH_2)_6CO_2Me$	H	Ph	84[d)]	[73]
i-Pr	H	3,4-di-MeOPh	79[e)]	[74]
Me	*i*-Pr	$N(Bn)_2$	86[f)]	[75]

Dehydrating agents:
a) $POCl_3$.
b) $SOCl_2$.
c) H_2SO_4.
d) P_2O_5.
e) TFA/TFAA.
f) PPA.

The authors have proposed a mechanism in which an enol phosphonium salt **56** loses Ph_3PO to generate acylimino carbene **57**, which cyclizes to the oxazole ring (Scheme 10.20). The same result can be obtained using the Burgess reagent **55** (Scheme 10.20), in combination with microwave irradiation [78].

The Robinson–Gabriel synthesis of oxazoles can be extended to solid-phase synthesis utilizing solid supported α-acylaminoketones with TFAA as dehydrating agent [79], (Scheme 10.21) or the Burgess reagent [80].

Following the Robinson–Gabriel [66, 67, 81] approach new methodologies have been developed mainly with the aim of improving the generation of the α-aminoacyl ketone precursors. A synthesis described by Moody and coworkers is based on Rh-catalyzed insertion of carbenoids. The use, as counterparts, of nitriles [82] affords directly the oxazole, while the use of primary amides goes through a preliminary insertion of the carbenoid in the N–H bond and subsequent cyclization following the methodologies introduced by Wipf [83]. The source of carbenoids are diazoesters **58** (Scheme 10.22).

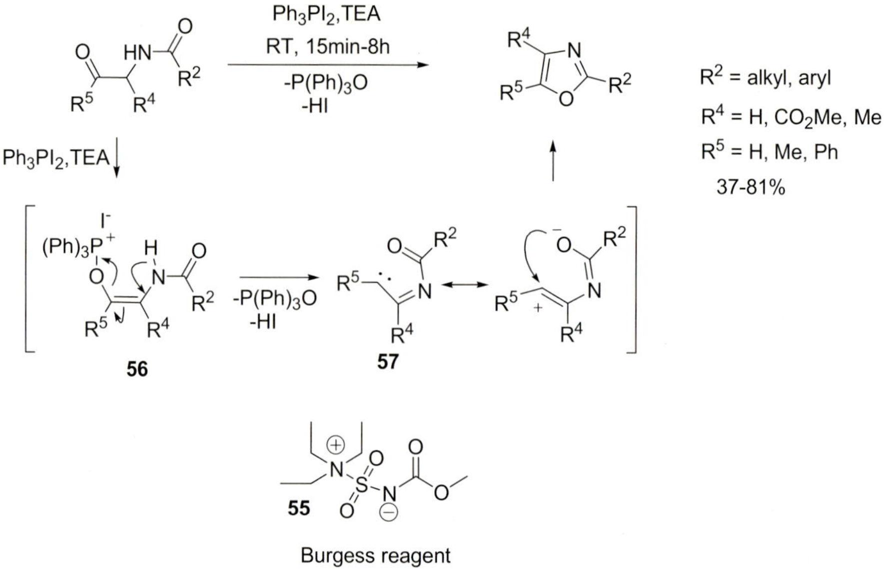

Ph_3PI_2, TEA; RT, 15min-8h; $-P(Ph)_3O$; -HI

R^2 = alkyl, aryl
R^4 = H, CO_2Me, Me
R^5 = H, Me, Ph
37-81%

56 **57**

55 Burgess reagent

Scheme 10.20 Wipf-Miller cyclodehydration synthesis.

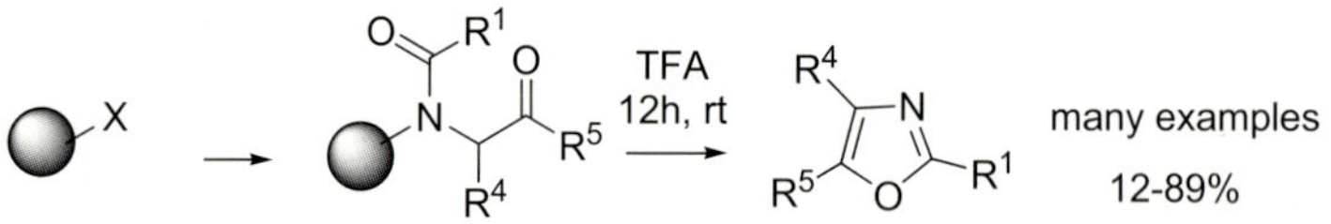

TFA 12h, rt — many examples 12-89%

Scheme 10.21

$Rh_2(OAc)_4$; Ph_3P, I_2, TEA; **58**; 31-88%

R^1 = aminoacid residues
R^5 = alkyl, Ph, OMe

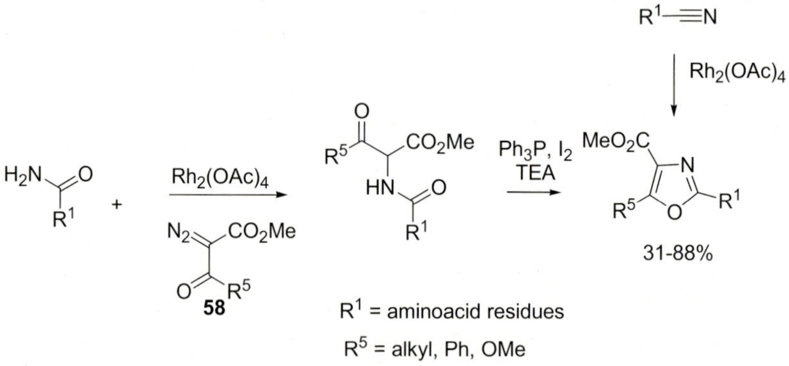

Scheme 10.22

Another well established and widely used method to prepare oxazoles is the cyclocondensation of α-haloketones **59** with primary amides **60** (Hantzsch type reaction). The reaction is typically driven by heating the mixture of **59** and **60** in the presence of a buffer to remove the generated acid. Both 2,4-disubstituted oxazoles and

2-amino-4-substituted oxazoles can be accessed with this methodology (Scheme 10.23). This approach can be extended to synthesize 2-aminooxazoles using urea and its derivatives as the partner of α-haloketones (Section 10.4.4). Furthermore, this approach, using thioamides, is the most general method for the synthesis of thiazoles and it is discussed in details in the next section.

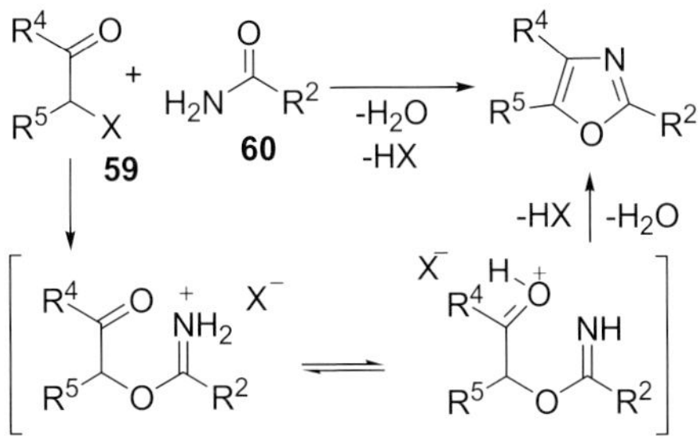

Scheme 10.23

As we have previously seen for the synthesis of imidazoles, TosMIC (**23**) is also a useful reagent for the synthesis of oxazoles [45]. Aldehydes condense, in presence of a base, with TosMIC to afford 4,5-disubstituted oxazolines that eliminate toluenesulfinic acid to yield 5-monosubstituted oxazoles, as already outlined for the imidazole series. The reaction can be extended to substituted TosMIC, giving 4,5-disubstituted oxazole (Table 10.4). The reaction proceeds with the nucleophilic addition of TosMIC to aldehyde followed by cyclization–aromatization.

The TosMIC route has been extended to solid-phase synthesis. The use of a polymer-supported version of TosMIC offers the advantage, compared with the homogeneous counterpart, of easy recovery of pure products. Resins that are unstable in basic reaction conditions are avoided. Polystyrene-SO_2-CH_2-NC resin **61**, prepared from Merrifield resin, can be used as a solid-phase equivalent of TosMIC and has found application in the synthesis of 5-aryloxazoles using tetrabutylammonium hydroxide as base (Figure 10.6) [88].

Resin **62** has also been used for the preparation of 5-aryloxazoles in conjunction with t-Bu-tetramethylguanidine [89].

An alternative method to the cyclodehydration of keto amide is the base-promoted [90] or palladium-catalyzed cycloisomerization of propargyl amides. The first reaction in Scheme 10.24 is an example of the latter, in which 2,5 disubstituted oxazoles **63** are prepared. The reaction proceeds through a palladium-catalyzed coupling step followed by cyclization [91]. In the same manner [Scheme 10.24 (2)], 2,4,5 trisubstituted oxazol-5-yl carbonyl derivatives **64** can be prepared using cheap and easily removable silica gel as the mediator of cycloisomerization. These oxazol-5-yl carbonyl compounds were inaccessible utilizing the base- or palladium catalyst-mediated procedure [92].

Oxazoles can also be prepared from β-(acyloxy)vinyl azides by reaction with phosphorous(III) reagents (Table 10.5). The intermediate iminophosphorane gives substituted oxazoles through an intramolecular version of the aza-Wittig reaction.

Table 10.4 Substituted oxazoles prepared using the van Leusen–TosMIC route.

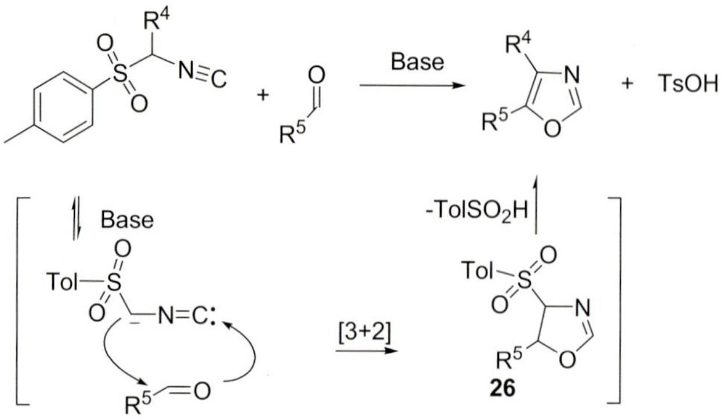

R^4	R^5	Yield (%)	Reference	R^4	R^5	Yield (%)	Reference
H	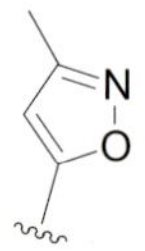 3-methylisoxazol-5-yl	65	[84]	H	p-CF_3Ph	79	[74]
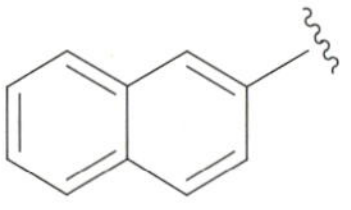 2-naphthyl	CO_2Et	80	[45]	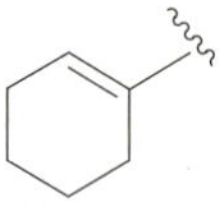 cyclohex-1-enyl	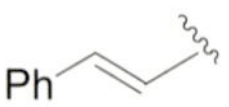 Ph-CH=CH-	90	[85]
3,4,5-$(MeO)_3$Ph	3-BnO-4-MeOPh	89	[86]	Me	p-NO_2Ph	96	[87]

This route is particularly useful for the synthesis of acid-labile oxazole derivatives (Table 10.5) [93].

An intermolecular version of the aza-Wittig reaction, using aroyl chlorides, affords imidoyl chloride derivatives **65** which easily cyclize to afford derivatives **66** (Scheme 10.25) [94, 95].

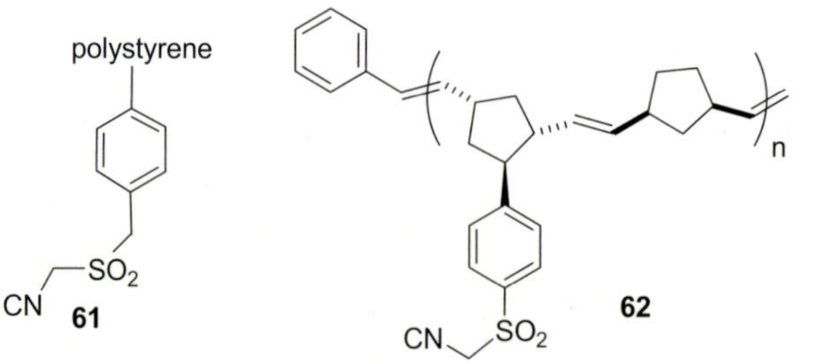

Figure 10.6 Resins used for the preparation of 5-aryloxazoles.

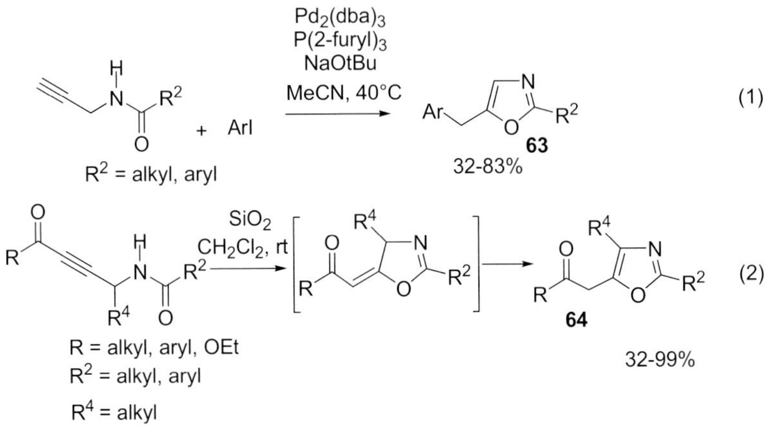

Scheme 10.24

Iminophosphorane joined with isothiocyanate are also useful intermediates for the synthesis of 2-amino substituted oxazoles. This tandem iminophosphorane/heterocumulene mediated annulation is described in Section 10.4.4 [96]

Table 10.5 Oxazoles via intramolecular aza-Wittig rearrangement.

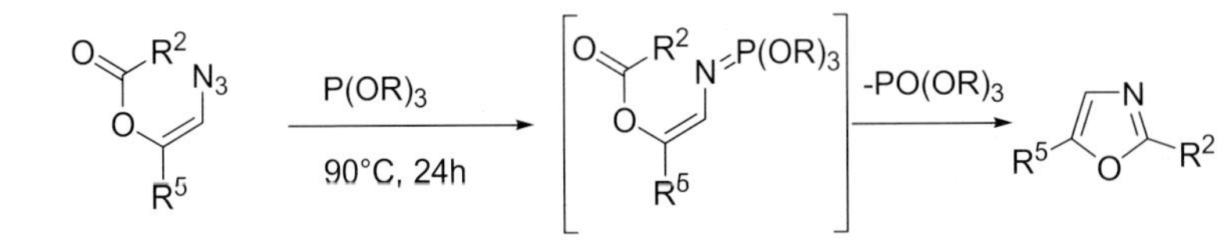

R^2	R^5	Yield (%)
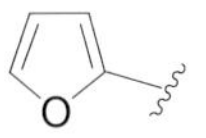	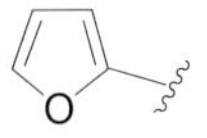	55
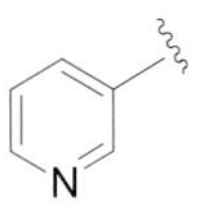	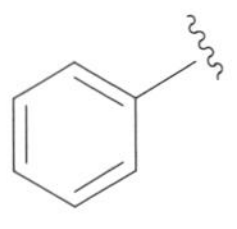	74
	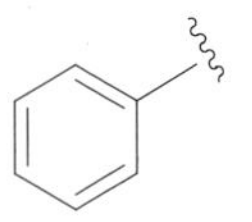	61
Me	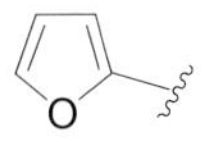	61

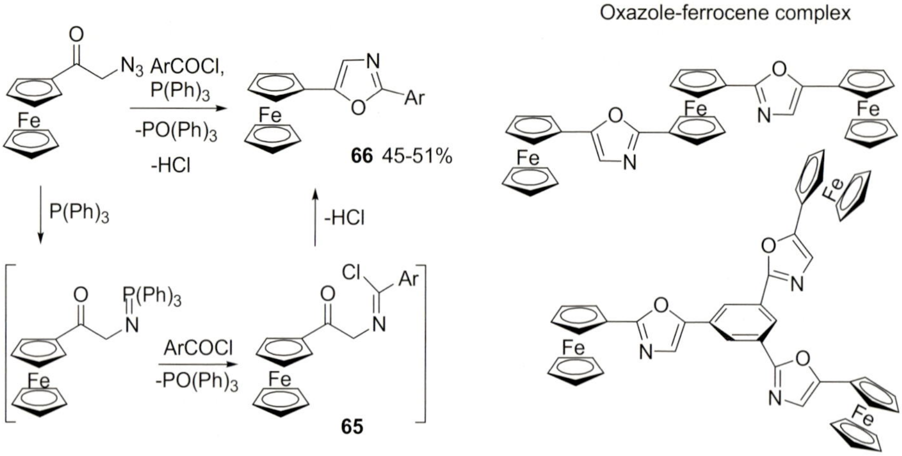

Scheme 10.25

10.3.3
Synthesis of Thiazoles

The classical method for the synthesis of thiazole is the Hantzsch process in which an α-halocarbonyl compound **59** (or the corresponding α-haloacetal) is condensed with a primary thioamide **67** (or a thiourea for the 2-amino derivatives) [97, 98]. The reaction proceeds with the nucleophilic attack of sulfur on the carbon atom bearing the halogen. The acyclic intermediate (isolated in few cases) α-S-alkyliminium salt **68**, after a proton transfer, undergoes cyclization and subsequent acid-catalyzed elimination of water (Scheme 10.26).

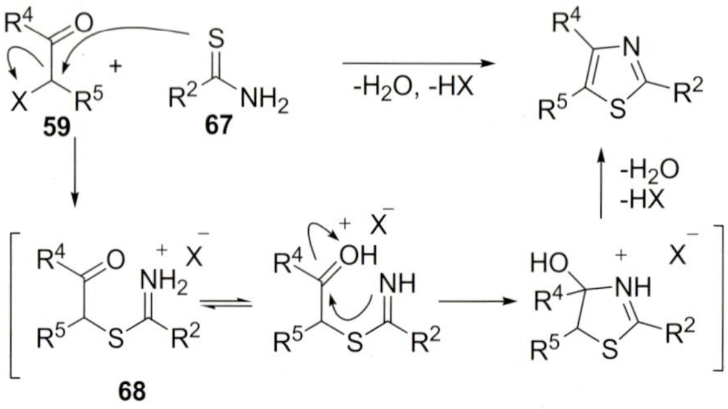

Scheme 10.26

This reaction usually proceeds smoothly to yield the desired thiazole and excellent yields have been obtained for simple thiazoles. However, for some types of substituent the range of pattern is limited, and low yields occur, as a result of dehalogenation of the α-haloketone during the reaction. Using thioamides and unsubstituted α-halocarbonyl it is possible to access 2-substituted thiazoles. In this case the range of substitution pattern is in almost every case abundant. Table 10.6 lists some examples.

Dimethyl and diethyl acetals as well as halohydrine derivatives frequently replace the aldehydes.

Table 10.6 2-Substituted thiazoles prepared following the Hantzsch procedure.

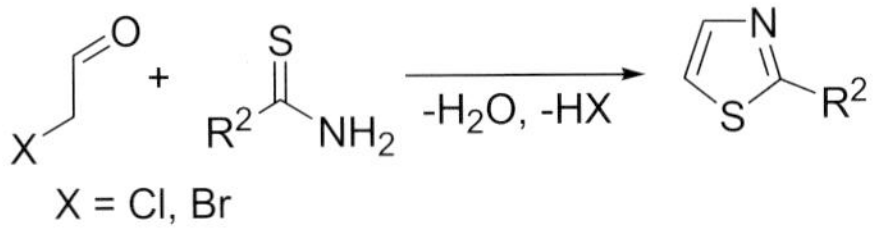

X = Cl, Br; $-H_2O$, $-HX$; R^2

R^2	Yield (%)	Reference	R^2	Yield (%)	Reference
Me	49	[99]	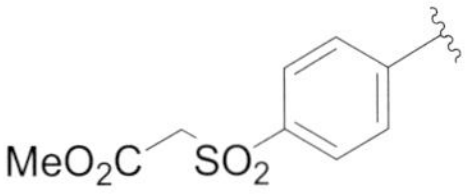 MeO_2C SO_2	33	[100]
Ph	79	[101]	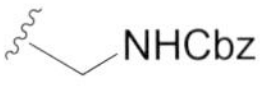 NHCbz	71	[102]
NH_2	91	[103]	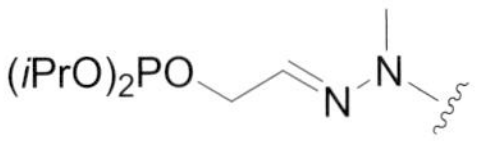 $(iPrO)_2PO$ N N	41	[104]
SH	50	[105]	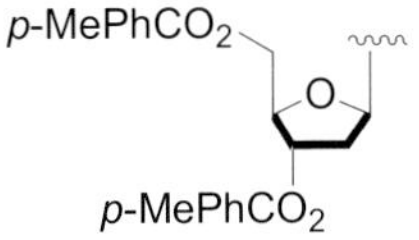 p-$MePhCO_2$ O p-$MePhCO_2$	70	[106]
SMe	88	[107]	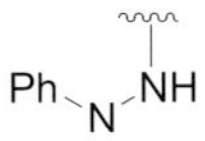 Ph N NH	78	[108]
Me_2N	85	[109]	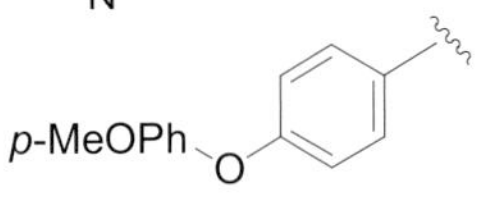 p-MeOPh O	50	[110]
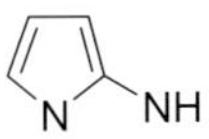 N NH	80	[111]	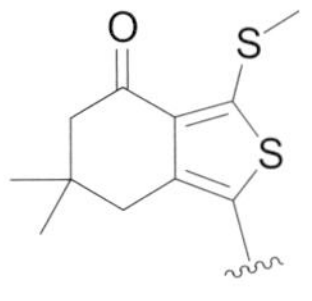 O S S	65	[112]
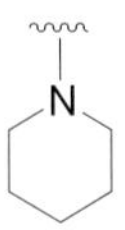 N	50	[113]	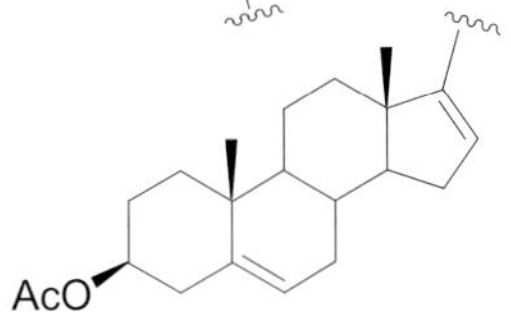 AcO	15	[114]
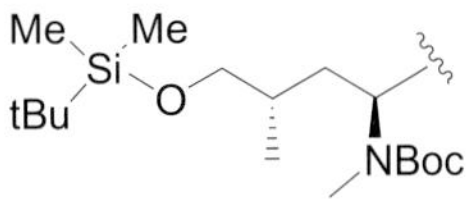 Me Me Si tBu O NBoc	73	[115]	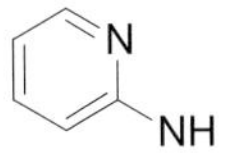 N NH	79	[116]

For the synthesis of 4-substituted thiazole the thioformamide is the reagent of choice, while using substituted thioamides and substituted α-haloketones affords 2,4,5-trisubstituted azoles (Table 10.7).

A modification of the Hantzsch synthesis utilizes α-tosylketones instead of α-halocarbonyl compounds [128]. One of the advantages of this modification is to

Table 10.7 Synthesis of 2,4,5-trisubstituted azoles.

X = Cl, Br; -H_2O, -HX

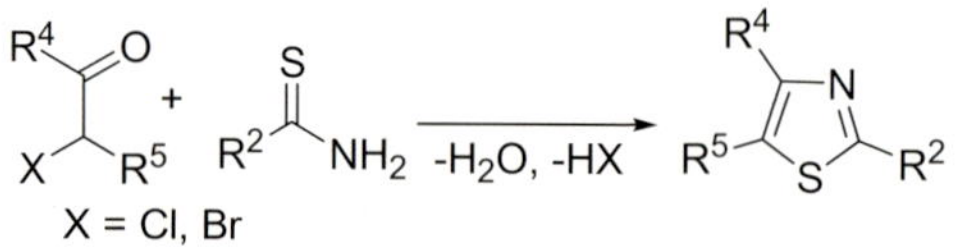

R^2	R^4	R^5	Yield (%)	Reference
H	CH_2CH_2Phth	H	78	[117]
H	$CH_2CH_2CO_2Et$	H	40	[118]
H	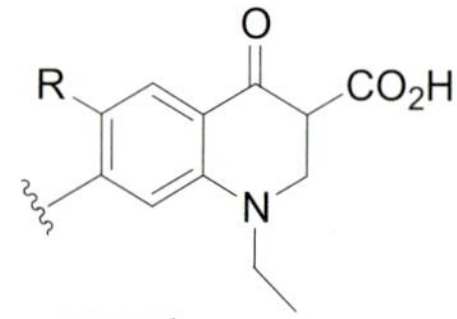 [structure: R, CO_2H, N]	H	83	[119]
H	*m*-NH_2Ph	H	84	[119]
H	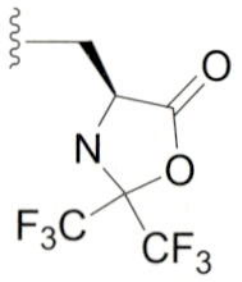 [structure: N, O, O, F_3C, CF_3]	H	52	[120]
Me	3,4-diOMe-Ph	CO_2Me	80	[121]
Bt	Me	Me	66	[122]
	Ph	Ph	50	
	Ph	Me	59	
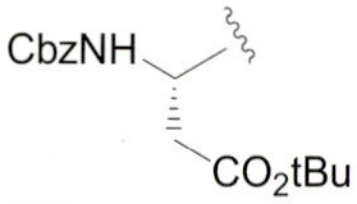 CbzNH CO_2tBu	CO_2Et	Me	81	[123]
NH_2	Me	CO_2Me	53	[124]
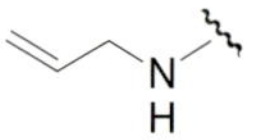 N H	CO_2Me	CO_2Me	65	[125]
Ph	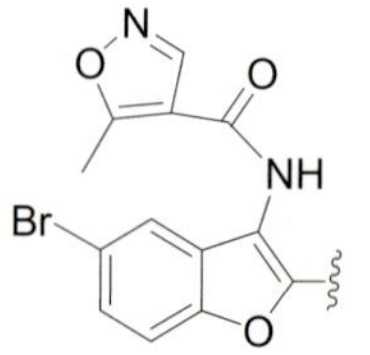 [structure: N, O, O, NH, Br, O]	H	86	[126]
NH_2	CO_2Et	H	76	[127]

avoid the use of lachrymatory and toxic α-halocarbonyl compounds. This method involves a reaction of ketones with the hypervalent iodine reagent HTIB [hydroxy-(tosyloxy)iodobenzene] (**69**) to produce the α-tosylketone **71** through the intermediate formation of α-λ^3-iodanil ketone **70** (Scheme 10.27).

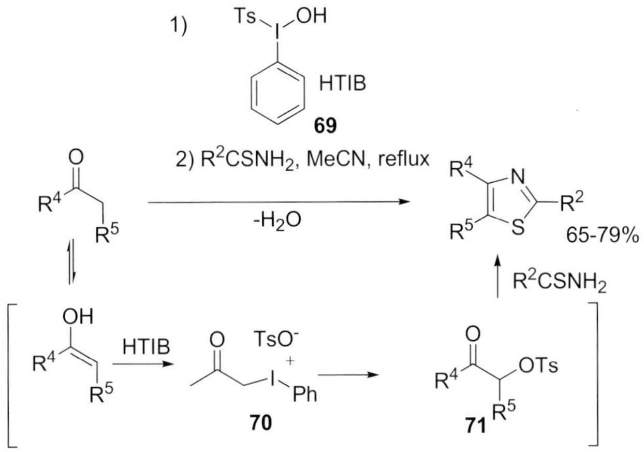

$R^2 = NH_2$, NHAr
R^4 = Ar
R^5 = H, COMe, CO_2Me

Scheme 10.27

It was subsequently noted that α-λ^3-iodanil ketones, obtained *in situ*, could also undergo direct cyclization by the reaction with thioamides, avoiding the isolation of the α-tosylketone **71** [129].

Another important synthetic method for the synthesis of thiazoles involves treating α-acylamino ketones **54** with phosphorous pentasulfide or Lawesson reagent [130] (Gabriel synthesis) (Scheme 10.28).

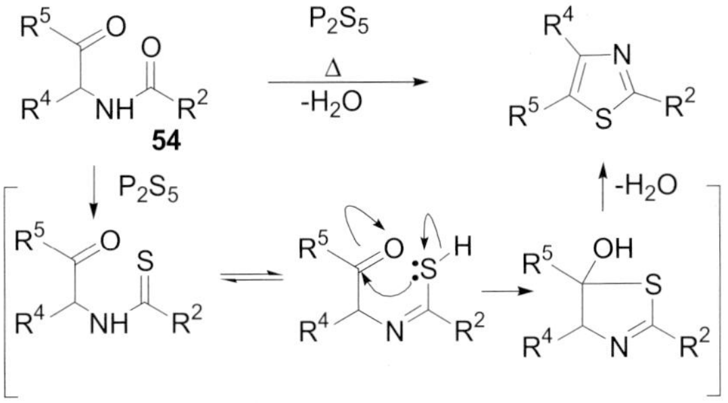

Scheme 10.28

An alternative thiazole synthesis, which is applied to combinatorial chemistry, is shown in Scheme 10.29. These isocyanide based four-component reactions provides

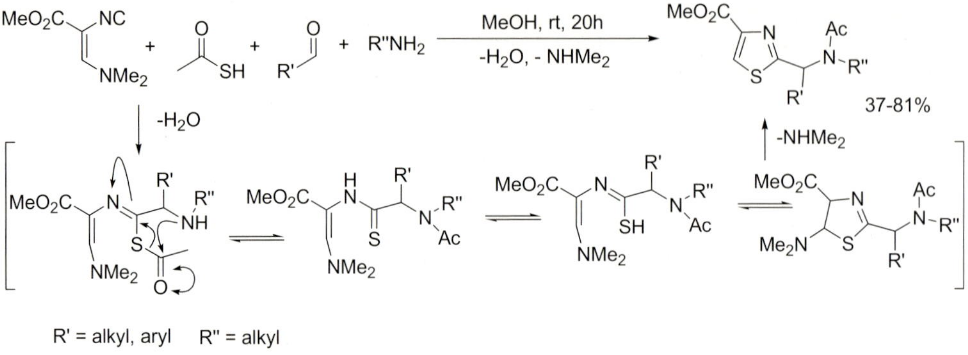

R' = alkyl, aryl R'' = alkyl

Scheme 10.29

2,4-disubstituted thiazoles. The substituent on C4 is limited to carbomethoxy (Scheme 10.29) [131].

Methods involving a regioselective metal-catalyzed coupling reaction have also been utilized to construct highly substituted thiazoles in view of the fact that coupling reactions would be selective for the more electron-deficient C2 (see below).

10.4 Reactivity

This section considers the reactivities of 1,3-azoles in detail and, when possible, the reactions of these heteroaromatic systems are compared among themselves. These reactions can be rationalized with reference to the tautomeric and acid–base equilibria shown by these compounds that are discussed in Section 10.1.

10.4.1 Reactions with Electrophilic Reagents

10.4.1.1 Electrophilic Attack at N3

Electrophilic attack at the azomethine nitrogen (pyridine-type) depends upon (i) the electron density on the pyridine-type nitrogen and (ii) the substituents present on the azole ring. The nature and position of the heteroatom other than azomethine nitrogen determine the electron density on the former. The interaction of pyridine-type nitrogen with pyrrole-type nitrogen, oxygen and sulfur atoms has a considerable influence due to the presence of two opposite electronic effects: (i) the mesomeric effect and (ii) the inductive effect. The balancing of both these effects determines the electron density at the pyridine-type nitrogen atom. The inductive effect is stronger when the second atom is oxygen or sulfur and thus lowers the electron density on nitrogen. In contrast, with two nitrogen atoms the mesomeric effect is stronger and increases the electron density at position 3. Imidazoles substituted at N1 [132], oxazoles [133], and thiazoles [134] **72** are alkylated/acylated with the formation of quaternary salts **73** (Scheme 10.30a) and alkylation/acylation of imidazole **74** (with free NH group) produces protonated N-alkyl-/acyl-imidazolium salt **75** that can be deprotonated by a base to afford N-alkyl imidazole **76** [135]. In this case, if the electrophilic reagent is a proton, this reaction sequence is a simple tautomer interconversion (Scheme 10.30b).

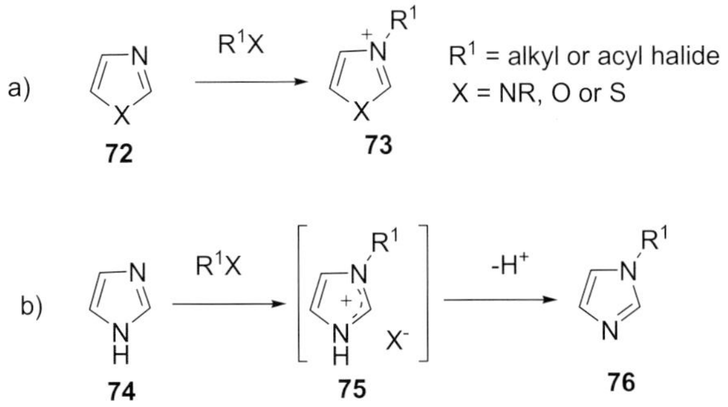

R^1 = alkyl or acyl halide
X = NR, O or S

Scheme 10.30

There are several examples of quaternizing alkylations of imidazoles using diverse reagents like alkyl, alkenyl or arylalkyl halides, ethyl chloroacetate, phenacyl bromide or dimethyl sulfate [136]. Although quaternization at the already substituted nitrogen atom has been reported and one of the products of the reaction of imidazole with 2,2-dichlorodiethyl sulfide was identified as **77**, it is more likely to be **78** or **79** (Figure 10.7). The observations of Pinner and Schwarz in 1902 that the quaternary salt obtained from 1-methylimidazole and 1-bromopentane was decomposed by alkali to give both aminomethane and 1 aminopentane was the first piece of evidence (more recently confirmed by NMR studies) to support the accepted view that quaternization takes places at the unsubstituted ring nitrogen [137].

The oxazole and thiazole nitrogen atom also reacts with various alkylating and acylating agents [138, 139]. Oxazoles react with bromine in methanol to give a mixture of products via initial attack of bromine on nitrogen atom to form the charged complex **80**. Subsequent rapid reactions with methanol lead to intermediates **81** or **82** depending on the C2 and C4 substituents ability to stabilize the C=N bond. Thus, a phenyl substituent on C2 favors the 2,5-dihydrooxazole structure **81**, whereas a phenyl group on C4 favors the 4,5-dihydrooxazole structure **82**. The intermediates give cyclic or ring-opened products **83–86** (Scheme 10.31) [140].

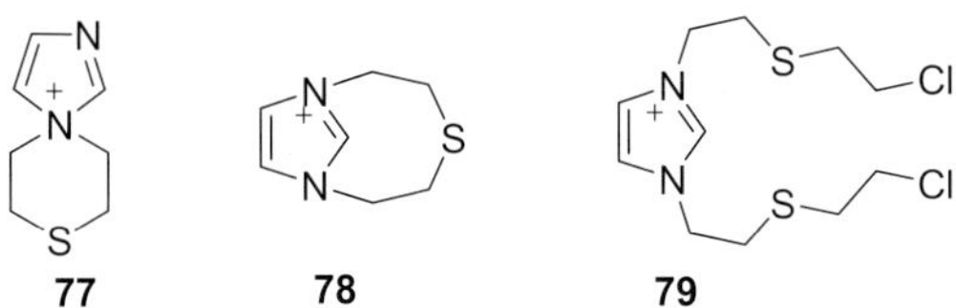

Figure 10.7 Possible products of the reaction of imidazole with 2,2-dichlorodiethyl sulfide.

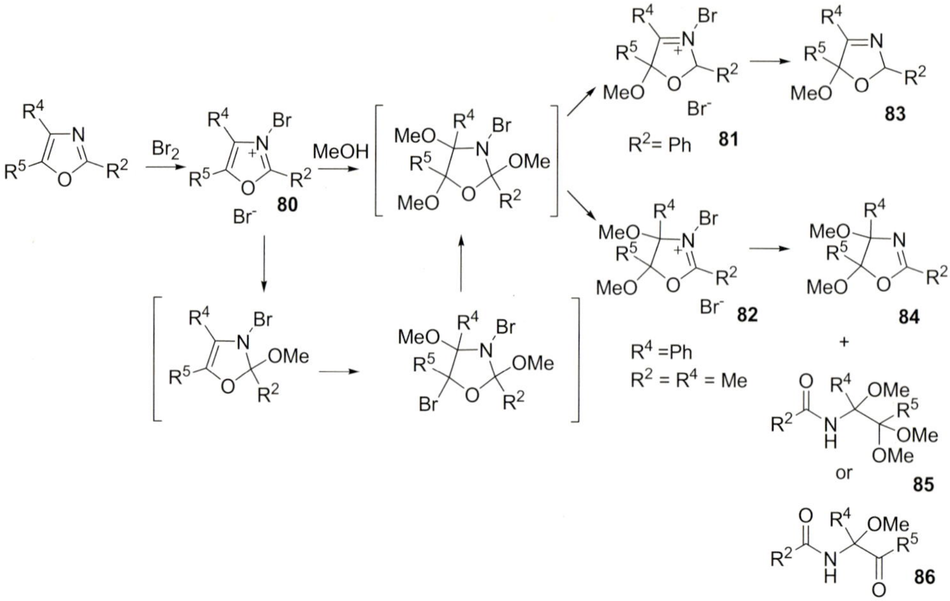

Scheme 10.31

The electronic density on pyridine-type nitrogen is also affected by the substituents on the azole ring and may be rationalized as follows:

1) Strongly electron-withdrawing substituents (e.g., NO_2, COR, CHO) make these reactions less favorable by decreasing the electron density on the nitrogen atom(s). The effect is largely inductive and, therefore, is particularly strong for the α-position.
2) Strongly electron-donating substituents (e.g., NH_2, OR) facilitate electrophilic attack by increasing the electron density on the nitrogen atom. This is due to the mesomeric effect and is, therefore, strongest for the α- and γ-positions.
3) Groups with relatively weak electronic effects have a relatively small electronic influence.

More recently, an efficient and straightforward copper-catalyzed method allowing vinylation of imidazoles in high yields and selectivities with di- or trisubstituted vinyl bromides has been described. The reaction can be performed with catalytic amounts of copper iodide and inexpensive nitrogen ligands under very mild temperature conditions (35–110 °C) [141]. For example, the viynilation of imidazole **74** by β-bromostyrene afforded compound **87** with a yield of 93% (Scheme 10.32).

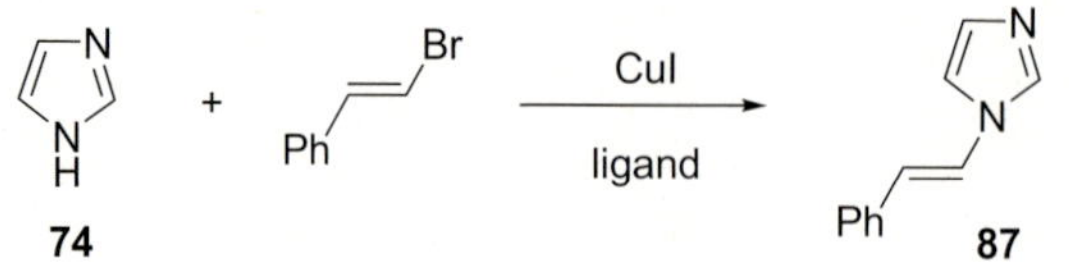

Scheme 10.32

In this context, triphenylphosphine has been used as nucleophilic catalyst for umpolung addition of azoles to electron-deficient allenes. This strategy offers a simple and efficient method for functional allylation of azoles under neutral conditions and affords heterocyclic substituted Michael olefins [142]. The catalytic cycle might be initiated by a nucleophilic addition of triphenylphosphine to the electron-deficient allene **88**. The enolate **89** then deprotonates the azole, generating the vinylphosphonium **90**. Subsequently, nucleophilic addition to vinylphosphonium **90** leads to the ylide. Finally, the enolate is obtained by prototropy and then undergoes a β-elimination, affording the final allylazole **91** and regenerating the nucleophilic catalyst. (Scheme 10.33).

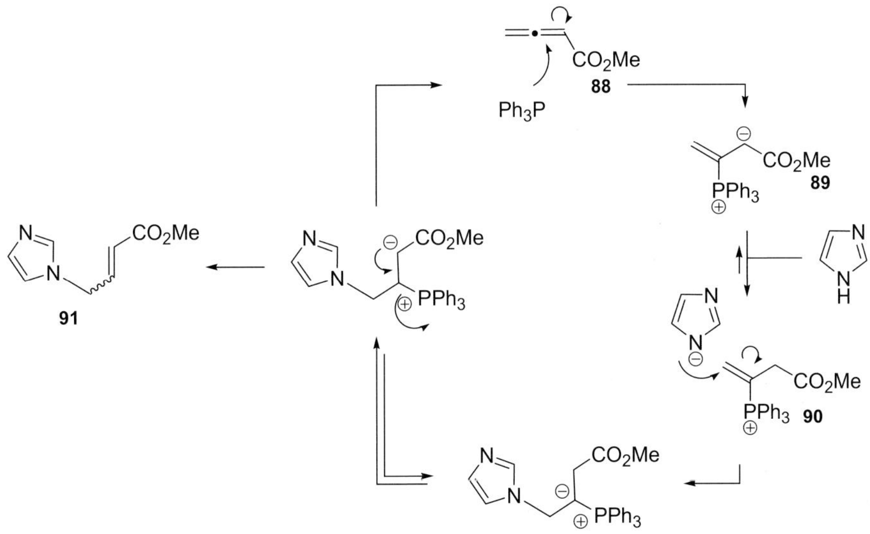

Scheme 10.33

Finally, many metal complexes with azoles or alkyl derivatives are known. The sulfur, oxygen and pyrrolidine-type nitrogen atoms are less nucleophilic than a pyridine-type nitrogen atom and the latter is expected to be the dominant donor. For example, a palladium(II) bisimidazole complex was obtained from the reaction of Pd (cod)ClMe (cod=cyclooctadiene) with equimolar amounts of **92** and proved to be an effective catalyst for the Heck reaction under phosphine-free conditions using ionic liquids as solvent (Scheme 10.34) [143].

10.4.1.2 **Electrophilic Attack at Carbon**

Figure 10.8 depicts the reactivity order in 1,3-azoles.

Imidazoles containing an unsubstituted NH group are easily chlorinated (Cl_2/H_2O or *N*-chlorosuccinimide/$CHCl_3$), brominated ($Br_2/CHCl_3$ or $KOBr/H_2O$) and iodinated (I_2/HIO_3). Substitution generally occurs first at the 4-position but further reaction at the other available positions takes place readily, providing

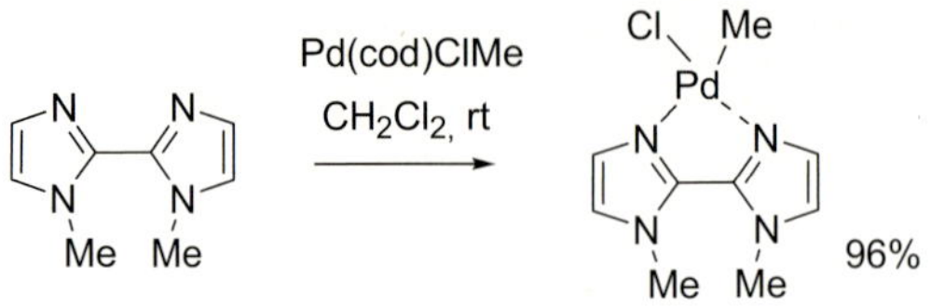

Scheme 10.34

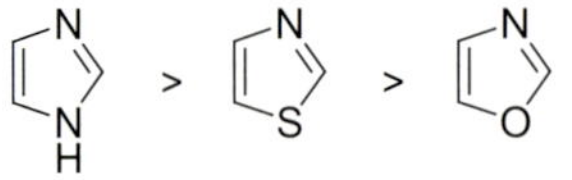

Figure 10.8 Order of reactivity in 1,3-azoles.

2,4,5-tribromoimidazole [Scheme 10.35 (1)] [144]. Ring bromination of oxazole **90** with bromine or NBS (*N*-bromosuccinimide) occurs preferentially at the 5-position to afford 5-bromooxazole and, if this is occupied, at the 4-position [Scheme 10.35 (2)] [145].

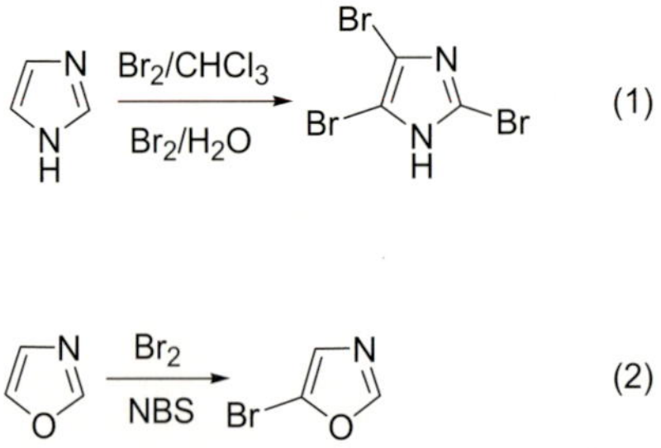

Scheme 10.35

Thiazole does not react with bromine or chlorine in an inert solvent, but thiazoles with an electron-releasing substituent in the 2 or 4-position are brominated at C5 [146]. For example, 2-aminothiazole afforded compound **93** in a bromination reaction (Scheme 10.36).

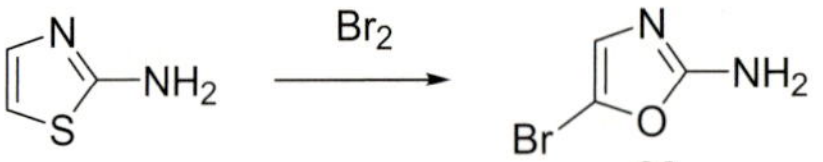

Scheme 10.36

Nitration of imidazoles [137] and thiazoles [139] has usually been carried out using either a mixture of concentrated (or fuming) nitric acid and concentrated sulfuric acid, or in some cases with concentrated nitric acid and acetic anhydride. Nitration

(HNO_3/H_2SO_4, 160 °C) and sulfonation (H_2SO_4/SO_3, 160 °C) of imidazoles proceeds at the 4-position very slowly, because the reaction takes places in acidic medium, with formation of imidazolium cations (e.g., **74** yields **94**) (Scheme 10.37) [137]. The deactivating effect of a protonated nitrogen atom is considerably greater than, for example, the two nitro groups in *m*-dinitrobenzene.

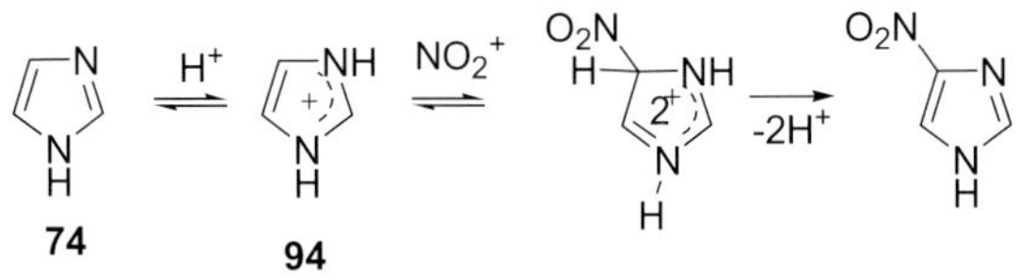

Scheme 10.37

With the object of finding milder nitration conditions, the use of cerium (IV) ammonium nitrate [147], montmorillonite impregnated with bismuth nitrate [148], and nitrations with dinitrogen pentoxide [149, 150] have been studied. Direct nitration of various imidazoles and thiazoles with nitric acid/trifluoroacetic anhydride, which involves N_2O_5, affords mononitro derivatives with an average yield of 60% [151]. Finally, dinitrothiazoles have been obtained by direct nitration of the corresponding mononitro derivative with acetyl nitrate [152].

In the case of thiazole, the reaction only occurs at the 5-position under forcing conditions (H_2SO_4/SO_3 in the presence of $HgSO_4$, 250 °C). As expected, these reactions are facilitated by activating groups such as an amino group; for example, 2-aminothiazole **92** is sulfonated at low temperature with the formation of sulfamic acid **95**, which on heating rearranges to 2-aminothiazole-5-sulfonic acid **96** (Scheme 10.38) [153].

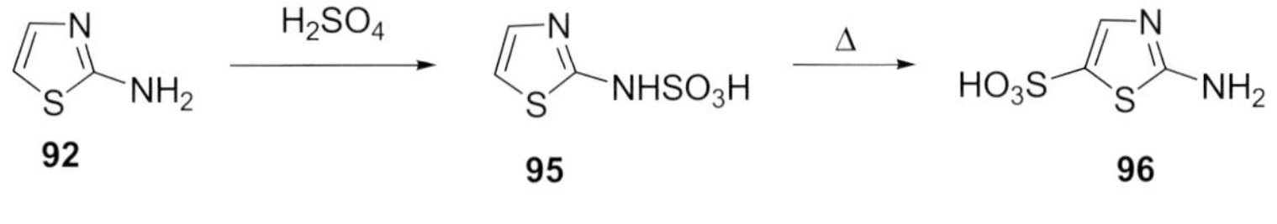

Scheme 10.38

Mercuration of oxazole with mercury(II) acetate occurs at C4 or C5 depending upon the available unsubstituted position. If both positions are substituted, mercuration occurs at the 2-position. Thiazole **97** is mercurated at positions 2, 4 and 5 in the order: C5 > C4 > C2, providing 2,4,5-tris(acetoxymercury)thiazole **98** on treatment with mercury acetate in the presence of aqueous acetic acid (Scheme 10.39) [153].

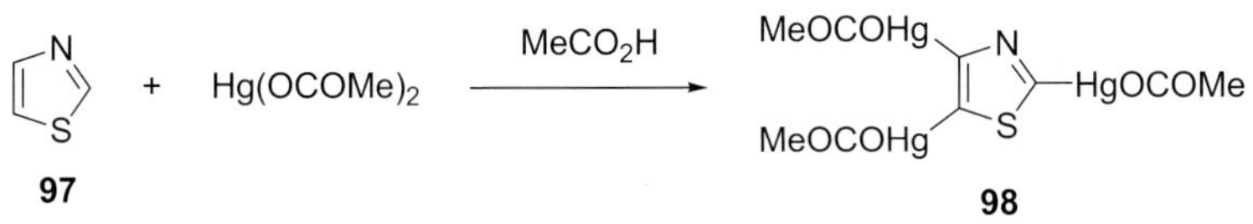

Scheme 10.39

Diazonium ions couple with the anions of N-unsubstituted imidazoles at the 2-position (e.g., **74** affords **99**) [Scheme 10.40 (1)]. In general, other azoles react only when they contain an amino or an hydroxy group. For example, 2-aminothiazole **92** undergoes diazo coupling with diazonium salts at C5 to afford **100** [Scheme 10.40 (2)] [154].

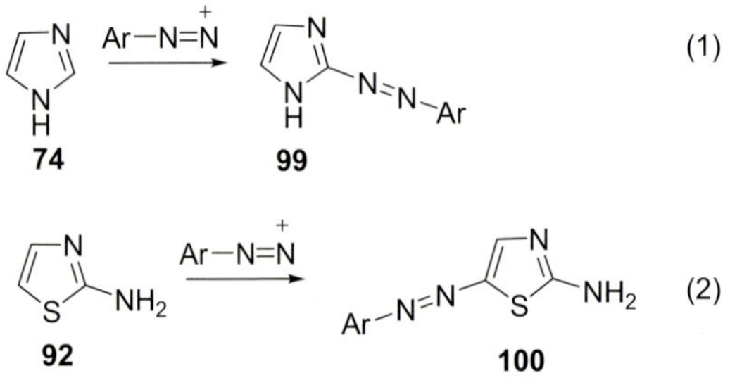

Scheme 10.40

Phosphorylation of 1-substituted imidazoles has been achieved under basic conditions by treatment with phosphorous(V) acid chlorides and provided good yields of the corresponding phosphinic acid salts after treatment with aqueous base [155].

Direct electrophilic silylation of 1,3-azoles with silane (II) under basic conditions afforded C-trimethylsilylazoles in good yields. The silylation occur at C2 [156].

On the other hand, electrophilic substitution is by far the most common method for substitution at the 2-position of substituted 1,3-azoles [157]. With this aim the addition of electrophiles as acid chlorides [158] or aldehydes [159] has been studied. In general, the reactions are run with an amine base for acid chlorides and are proposed to proceed via an intermediate carbene/ylide species [160]. In this context, the addition of 2,2,2-trichloroacetyl chloride to **76** affords **101** (Scheme 10.41).

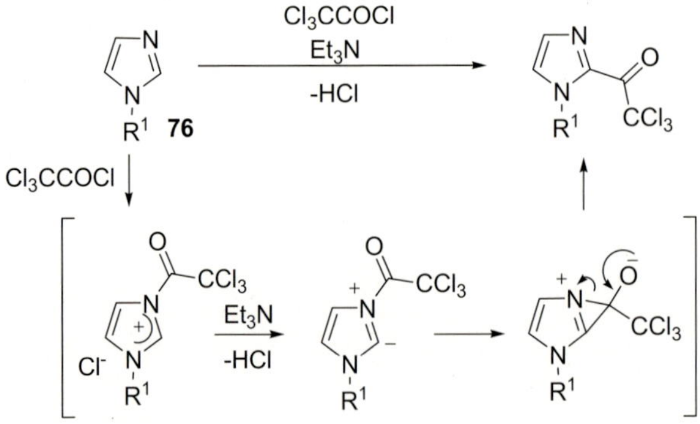

Scheme 10.41

Imidazoles are sufficiently nucleophilic to condense with aldehydes under thermal conditions in the presence of acid to give 2-hydroxymethylimidazoles. For example, the addition of formaldehyde to *N*-methylimidazole yields **102** (Scheme 10.42). The reaction, however, is highly variable and substrate dependent.

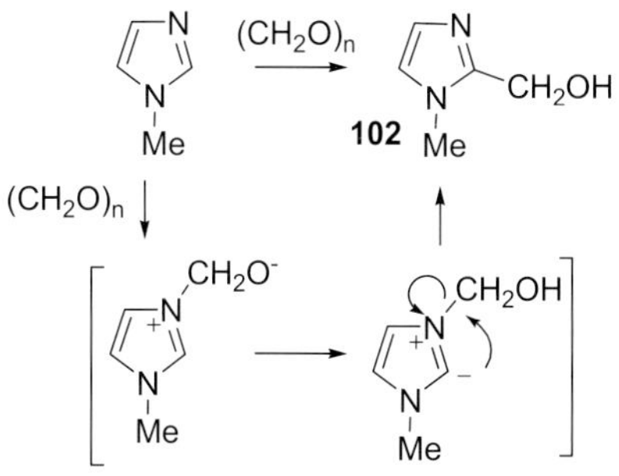

Scheme 10.42

The addition of imidazole-1-carboxamides **103** to phenyl isocyanate gives the corresponding amides **104** in modest yield (Scheme 10.43) [161].

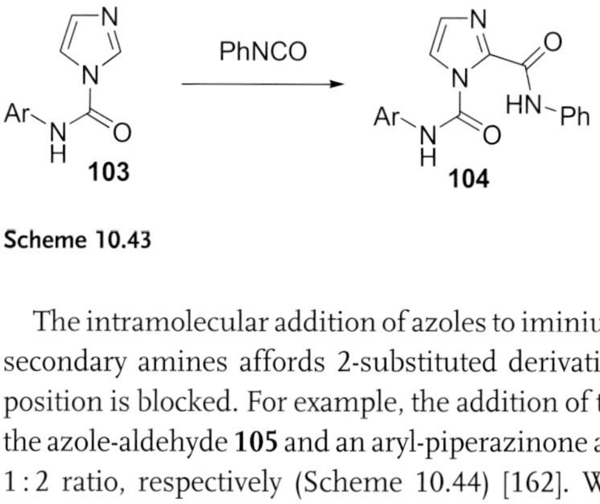

Scheme 10.43

The intramolecular addition of azoles to iminium ions formed from aldehydes and secondary amines affords 2-substituted derivatives cleanly only when the 4- or 5-position is blocked. For example, the addition of the iminium ion **106** obtained from the azole-aldehyde **105** and an aryl-piperazinone affords a mixture of **107** and **108** in a 1 : 2 ratio, respectively (Scheme 10.44) [162]. When position 2 is substituted the reaction yields the 4-substituted product [137, 163].

The preformation of 1-cyano-4-(*N,N*-dimethylamino)-pyridinium bromide (**109**), from DMAP (4-*N,N*-dimethylaminopyridine) and BrCN, allows for the selective 2-cyanation of *N*-methylimidazole to yield **110** (Scheme 10.45). In the absence of DMAP, the 2-position is brominated [164].

Finally, although the Lewis acid-promoted Friedel–Crafts acylation is the most commonly used method for the acylation of an aromatic ring, it is impracticable for imidazoles, due to deactivation of the Lewis acids. Other azoles, such as oxazoles and thiazoles, are generally also not amenable, because of their electron-deficient aromatic character. This type of acylation, however, proceeds in the presence of strong activation from electron-donating groups, such as alkoxy, amino or arylthio groups, in the substrates [165].

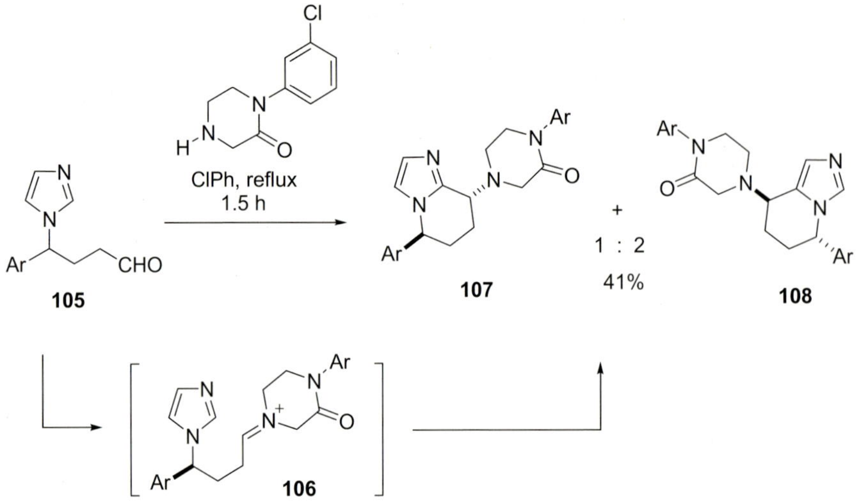

Scheme 10.44

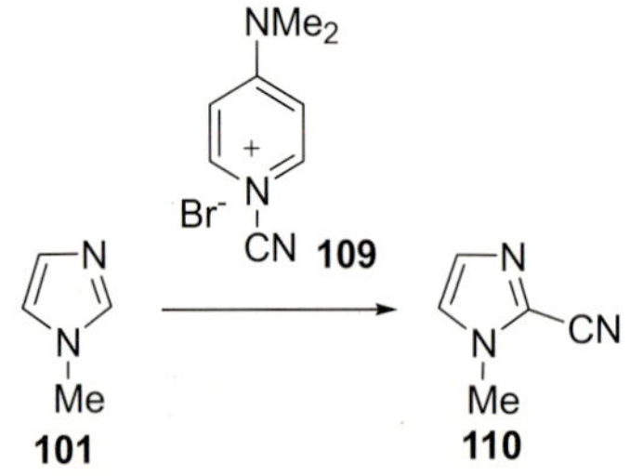

Scheme 10.45

10.4.2
Reaction with Oxidizing Agents

Imidazoles and thiazoles are resistant to oxidation, but they can be oxidized by hydrogen peroxide and peracids as, for example, perbenzoic or peracetic acid. In the case of imidazoles the ring is degraded [166] to afford **111a** and **111b** [Scheme 10.46 (1)] and with thiazoles the 3-oxide **112** is obtained [Scheme 10.46 (2)] [167].

Photosensitized oxidation of imidazole (**74**) with singlet oxygen produces imidazolidin-2-one **114** via the cyclic peroxide **113** (Scheme 10.47) [168].

Trisubstituted imidazoles and thiazoles undergo photosensitized oxidation with the formation of ring cleaved products depending on the solvent and the sensitizer used [169].

The oxazole ring is cleaved by oxidizing agents such us permanganate, chromic acid or hydrogen peroxide to give acids or amides [170].

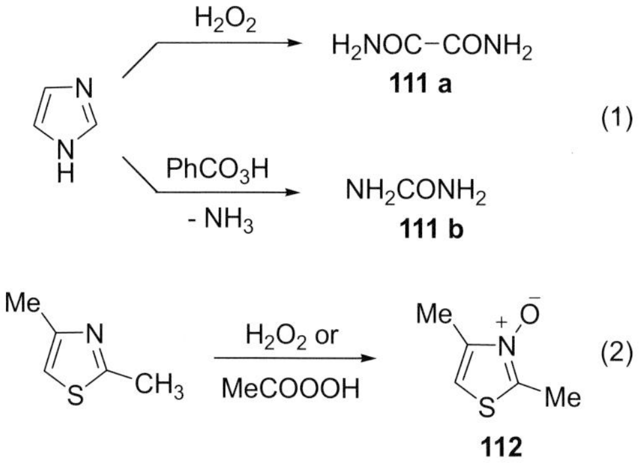

Scheme 10.46

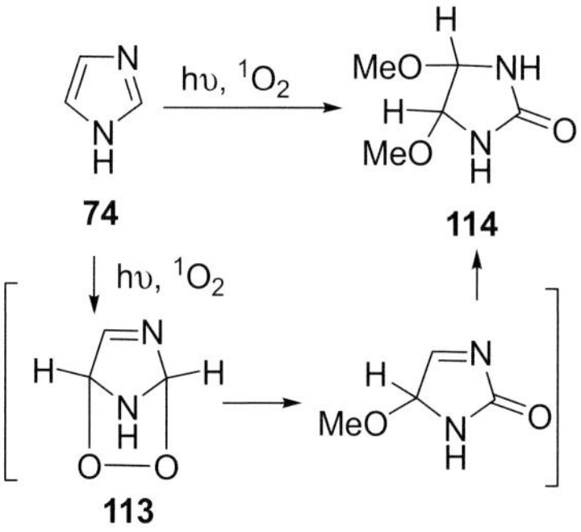

Scheme 10.47

10.4.3
Reactions with Nucleophilic Reagents

Substitution at C2 of 1,3-azoles with a leaving group generates a heterocycle that can be functionalized via displacement of the C2 substituent. Heteroatom nucleophiles add either as the deprotonated species (e.g., alkoxides, thiolates) or under milder basic conditions. Diazo salts (e.g., **115**) can be displaced by bromide ion (Scheme 10.48) [171] or by alcohols [172].

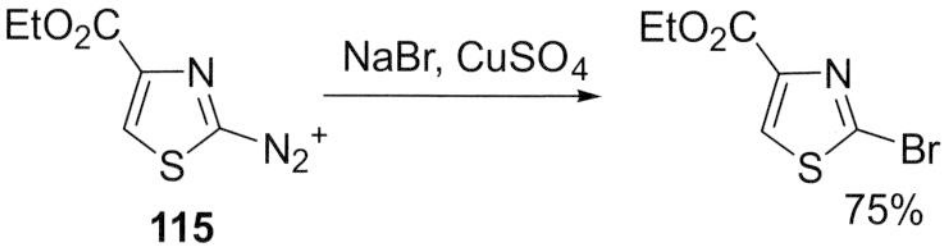

Scheme 10.48

2-Haloazoles **116** undergo nucleophilic substitution reactions with replacement of the halogen atom by nucleophiles [Scheme 10.49 (1)]. Halogen atoms at the 4- and

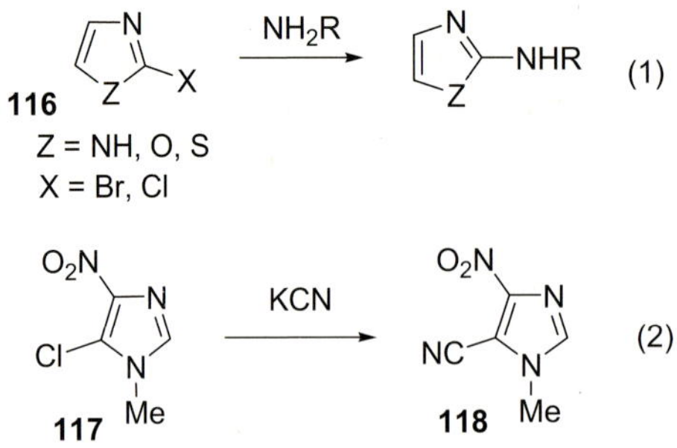

Scheme 10.49

5- positions of imidazoles are normally unreactive but can be activated by an α- or γ- electron-withdrawing substituent (e.g., **117** yields **118**) [Scheme 10.49 (2)] [173].

Addition of carbon nucleophiles to halogenated thiazoles includes the use of sodium cyanide [174], indolyl Grignard reagent [Scheme 10.50 (1)] [175], 2-lithio-2-nitropropane [Scheme 10.50 (2)] [176] and ester enolates [177].

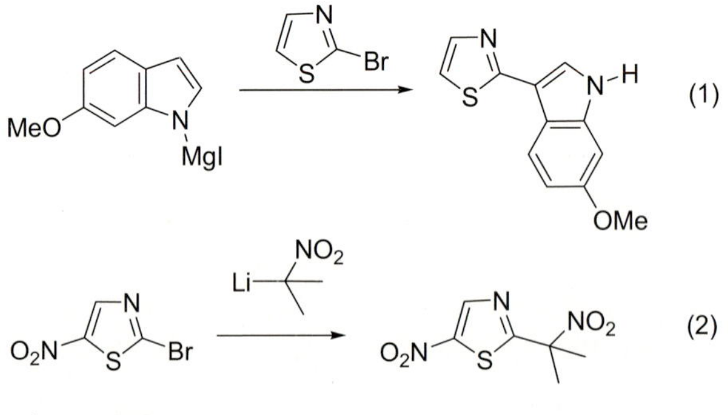

Scheme 10.50

Imidazole (**74**) has been used as electrophile with silyl enol ethers in the presence of alkyl chloroformates to provide the 2-substituted 2,3-dihydroimidazole **119** (Scheme 10.51) [178]. The same reaction can be performed with thiazole.

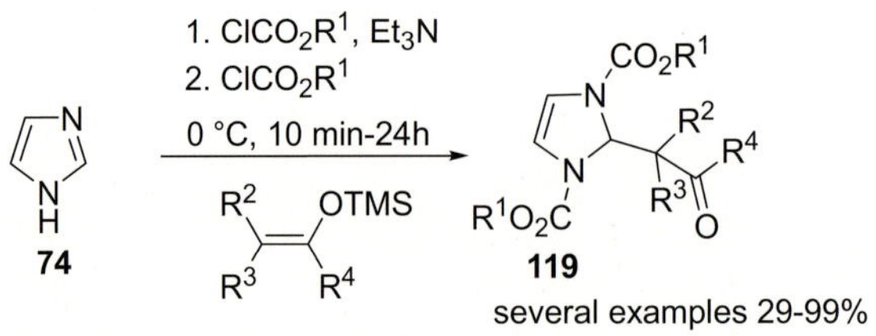

Scheme 10.51

Imidazo[1,2-*b*]thiazolines **120** undergo nucleophilic displacement with allylic, benzylic and alkyl Grignard reagents to yield **121** (Scheme 10.52) while alkyl or aryl lithium reagents result in nucleophilic attack on sulfur and loss of ethylene to afford 2-thioalkyl-1*H*-imidazoles **122** (Scheme 10.52) [179].

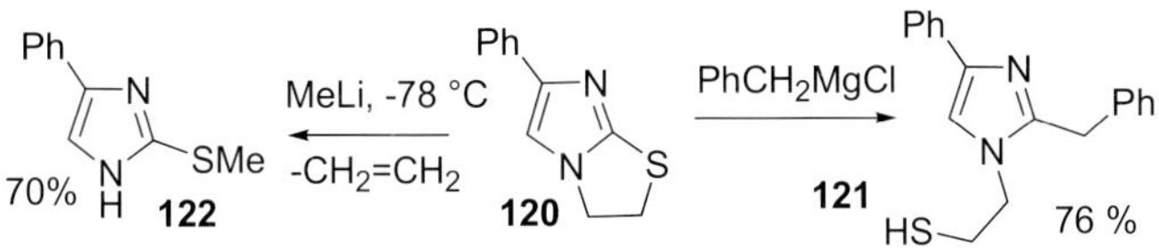

Scheme 10.52

In some cases nucleophilic attack results in the cleavage of the ring. For example, oxazoles **123** when treated with ammonia at 200 °C undergo nucleophilic attack at the 2-position and are transformed into the corresponding imidazoles **124** (Scheme 10.53) [180].

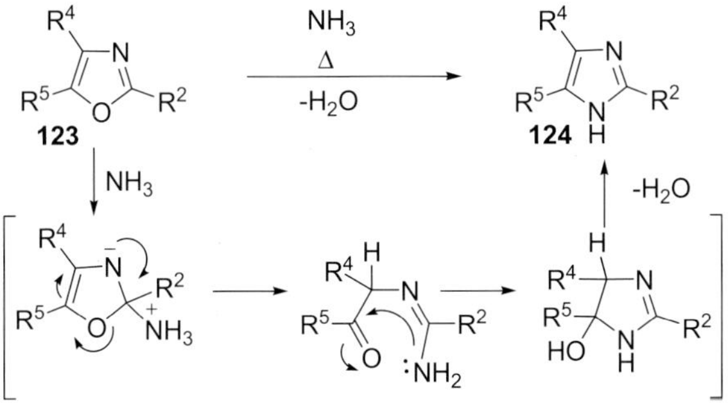

Scheme 10.53

Imidazole (**74**) reacts with acid chlorides to give salts that in the presence of alkali react to afford compound **125**, derived from the ring cleavage (Scheme 10.54) [181].

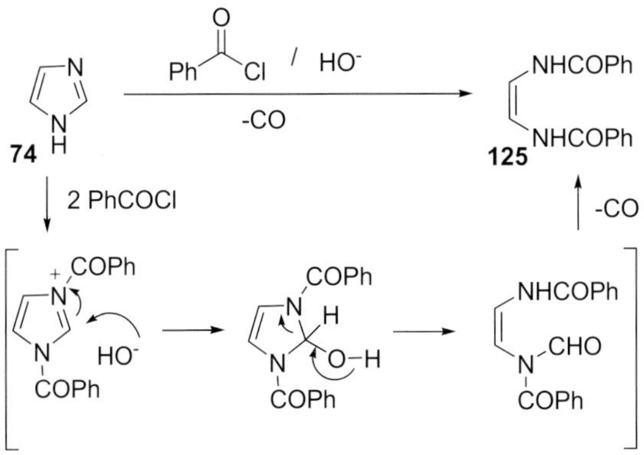

Scheme 10.54

Finally, thiazoles are quite inert in the presence of hydroxide and alkaline ions. However, two proteases were found to catalyze the enantioselective hydrolysis of several 5-hydroxythiazoles (**126**) to give α-amino acids (**127**) (Scheme 10.55) [182].

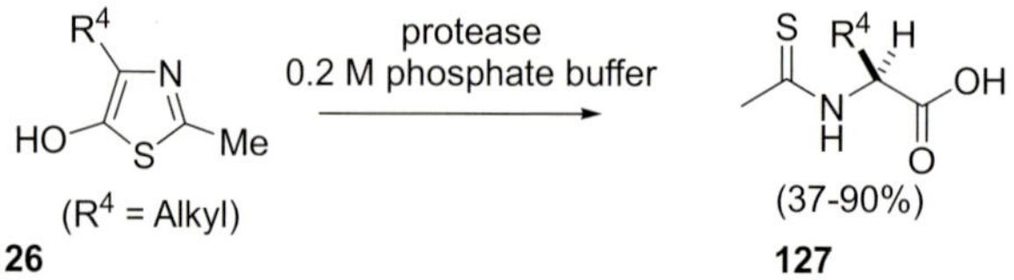

R^4 N HO S Me (R^4 = Alkyl) **126** — protease, 0.2 M phosphate buffer → S R^4 H N H OH O (37-90%) **127**

Scheme 10.55

10.4.4
Reactions of N-Metallated Imidazoles

Although the direct alkylation of imidazoles at N3 affords the 1-substituted imidazoles through a tautomer interconversion (Scheme 10.56, route a), in some cases the synthesis of these compounds has been reported using N-metallated imidazoles (Scheme 10.56, route b).

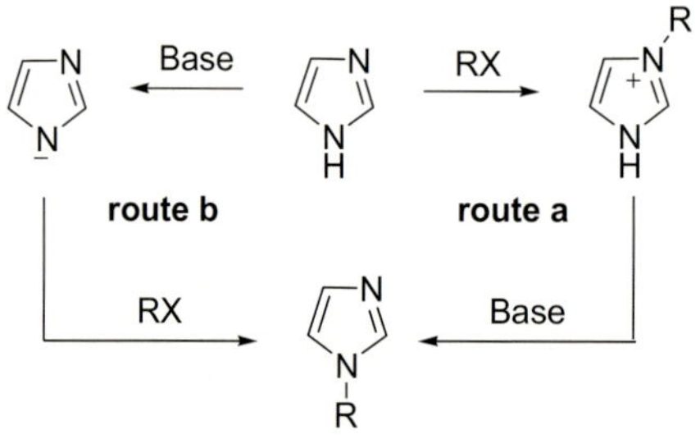

Base, RX, route b, route a, RX, Base

Scheme 10.56

Route b can be considered as a complementary way for the preparation of 1-substituted imidazoles and has permitted the introduction of the imidazolyl unit in a large number of structures for the synthesis of natural products or analogues [183].

In general, N-metallated imidazoles **128** are obtained using a base such as NaH or BuLi (Scheme 10.57).

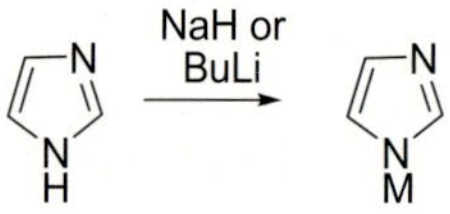

NaH or BuLi → **128**

M = Na^+ or Li^+

Scheme 10.57

10.4.5
Reactions of C-Metallated Azoles

10.4.5.1 Lithium Azoles

These derivatives are prepared by direct deprotonation [184, 185] using strong bases or, particularly useful in the case of the less acidic sites in aromatic rings, by halogen exchange [186] between an halogenated heterocycle and an organolithium compound or lithium metal.

1,3-Azoles are prone to be lithiated at C2, but if this position is already occupied, lithiation occurs at C5. If a C4 metallation is required, usually the halogen–lithium exchange methodology is employed. The combination of all these techniques allows the selective lithiation at any position in the azole nucleus.

N-Substituted imidazoles tend to lithiate with alkyllithiums at C2, affording a carbenoid species that can be used as a bulky base, as in the case of 2-lithio-1-methylimidazole, which has been used in the stereoselective deprotonation of cyclohexene oxides when combined with a chiral lithium amide [187]. However, 2-lithioimidazoles are employed normally as nucleophiles, for instance in addition reactions to aldehydes [188], ketones [189], esters and isocyanates [190], as well as in silylation [191], sulfenylation [192], and cyclic sulfate ring-opening [193] reactions.

2-Lithiated N-substituted imidazoles such as 2-lithio-*N*-methylimidazole (**129**), prepared by direct deprotonation using *n*-butyllithium, has recently been used in the reaction with the diester **130** for the preparation of compound **131** as a zinc ligand (Scheme 10.58) [194].

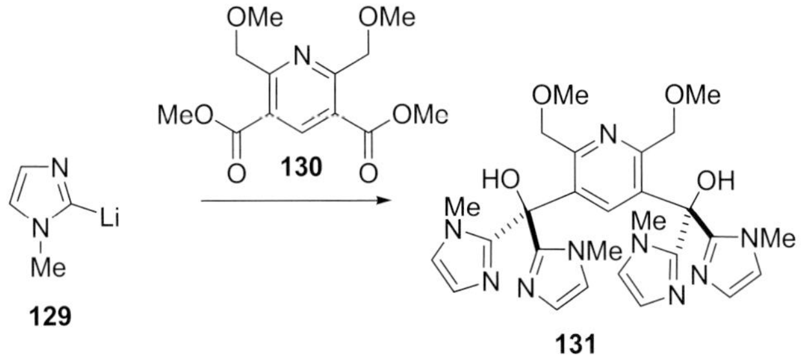

Scheme 10.58

2-Lithioimidazoles can also be generated by treatment of 1-substituted imidazoles with an excess of lithium powder in the presence of a catalytic amount of an electron-carrier [195] such as isoprene [196]. The isoprene-catalysed lithiation of different 1-substituted imidazoles **132** (PG = trityl, allyl, benzyl, vinyl, *N*,*N*-dimethylsulfamoyl, *para*-toluenesulfonyl, *tert*-butoxycarbonyl, acetyl, trimethylsilyl, *tert*-butyldimethylsilyl) leads to the cleavage of the protecting group producing 1*H*-imidazole (Scheme 10.59). However, the use of 1-(diethoxymethyl)imidazole (**133**) in the same lithiation reaction allows the preparation of the corresponding 2-lithio intermediate, which by reacting with different electrophiles such as, for example, benzaldehyde or *N*-phenyl-benzyl imine leads to 2-functionalized imidazoles (Scheme 10.60) [197].

PG = Tr, Allyl, Bn, vinyl, $-SO_2NMe_2$, Ts, Boc, Acetyl, $-SiMe_3$, $-SiBu^tMe_2$

Scheme 10.59

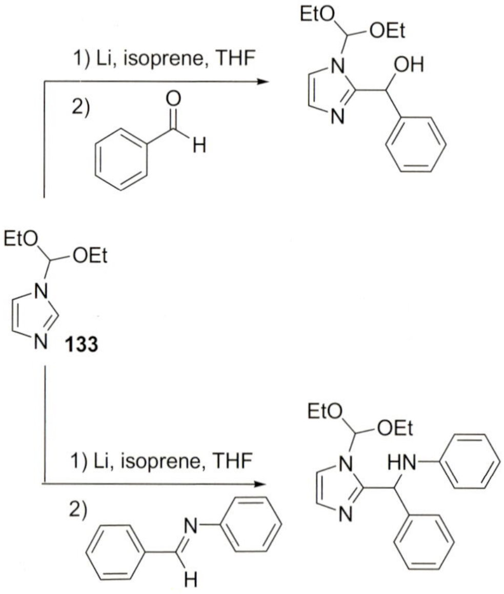

Scheme 10.60

As mentioned above, 5-lithiumimidazoles can be generated by direct deprotonation with an alkyllithium if the C2 position of the ring is blocked. When the substituent at C2 is a trialkylsilyl group, introduced by deprotonation and reaction with a trialkylsilyl halide, lithiation at C5 takes place. Examples of the use of these 2-sylilated imidazol-5-yllithiums (**132**) are in the synthesis of imidazolsugars **133** (Scheme 10.61) [198].

As in the case of any 1,3-azole, oxazoles are readily lithiated at C2. However, attempts to trap 2-lithioxazoles with electrophiles must contend with complications due to the ring opening of the anion that equilibriates by an elimination/ring-opening to produce the β-cyano enolate **135**, which, according to NMR data, is the dominant form. In many cases enolate **135** cyclizes back after the C-electrophilic attack, affording mixtures of C2 and C4 substituted oxazoles, **136** and **137** (Scheme 10.62) [200].

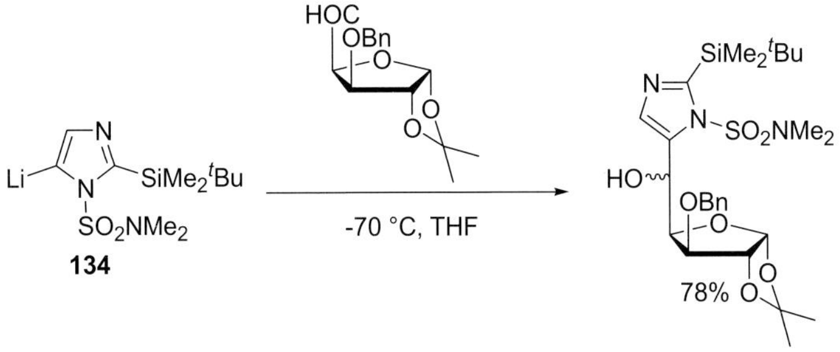

Scheme 10.61

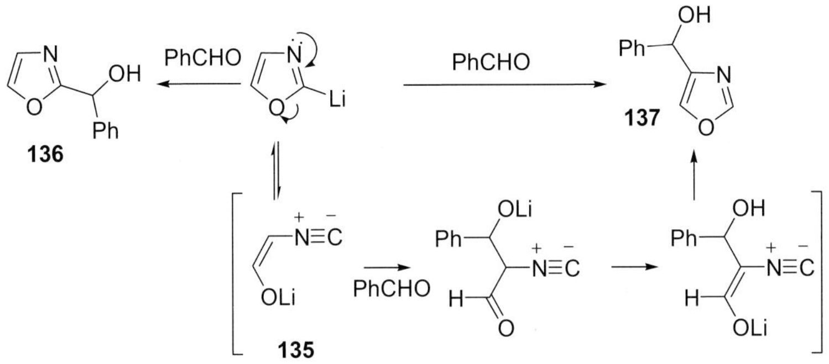

Scheme 10.62

In this electrophilic ring opening, it is possible to lock the electron pair at the oxazole nitrogen by complexation with a Lewis acid, such as borane, thus allowing C2-lithiation (Scheme 10.63) [201].

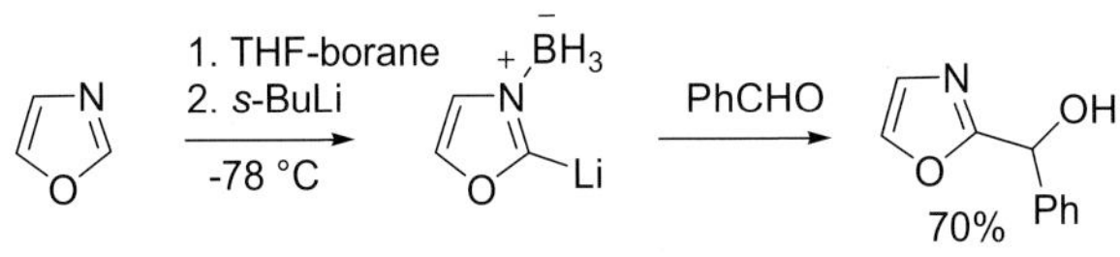

Scheme 10.63

In 2-substitued oxazoles, direct C5-lithiation can be carried out, allowing further reaction with electrophiles [202], although the bromine–lithium exchange methodology has also been used [203]. Remarkably, in 2 methyl-4-substituted oxazoles, a selectivity for lithiation at C5 to give compound **138**, versus lithiation at the methyl group to give compound **139**, has been observed, depending on the lithium base (Scheme 10.64) [204]. 5-Lithiation of 2-substituted oxazoles has also been achieved by ortho-lithiation to a triflate group [205]. Concerning 5-bromo-2-phenyloxazole, 5-lithiation and further reaction with electrophiles has been achieved through an initial LDA-promoted 4-lithiation followed by halogen migration, leading to a 4-bromo-5-lithio-2-phenyloxazole intermediate [206].

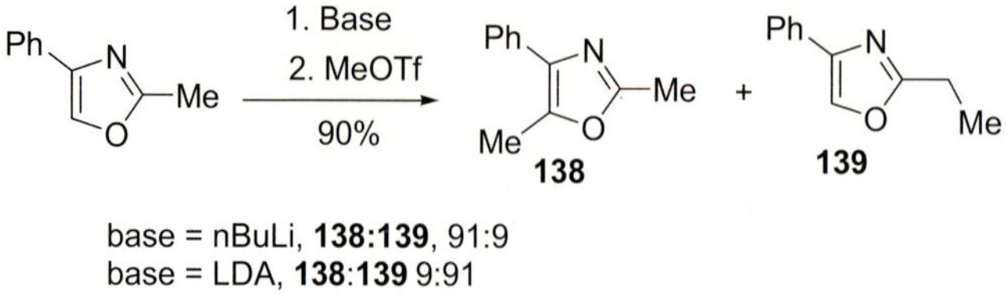

base = nBuLi, **138:139**, 91:9
base = LDA, **138:139** 9:91

Scheme 10.64

2-Lithiothiazoles [207] have been used as nucleophiles, for instance addition to aldehydes [208], the thiazole moiety being considered as a formyl equivalent [209], for example in addition reactions to lactones [210], to benzyloxyacetaldoximes [211] or in reactions with nitrones **140** for the synthesis of amino sugars (e.g., **141**) (Scheme 10.65) [212].

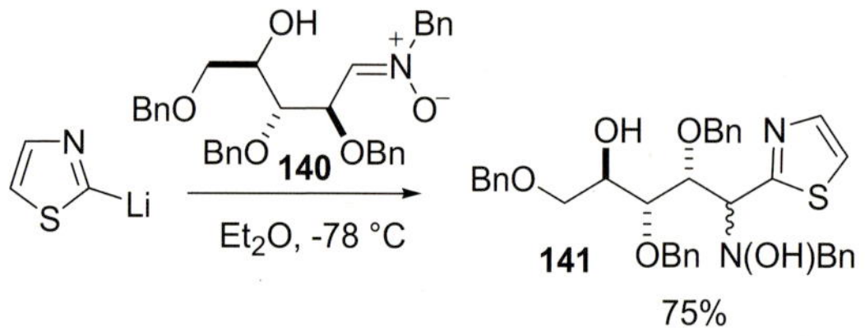

Scheme 10.65

Lithiation at C5 in thiazoles takes place directly if the C2 position is blocked, an example being the lithiation of 2-(methylthio)thiazole (**142**) to give intermediate **143**, which can react further with a nitrile such as *p*-chlorobenzonitrile, affording 5-(arylcarbonyl)thiazole **144** after hydrolysis (Scheme 10.66) [213].

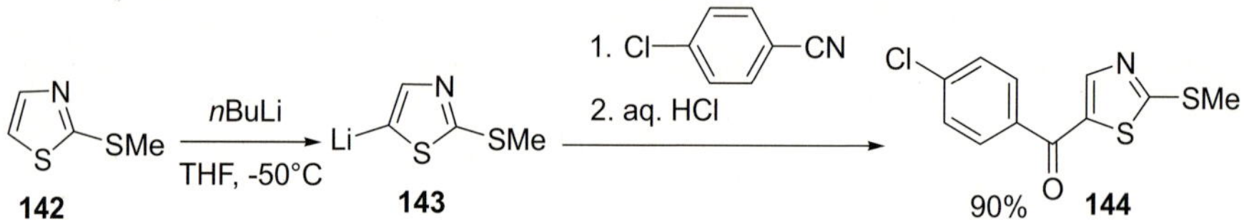

Scheme 10.66

5-Lithiation has also been achieved in 2-thiazolamines bearing a bromine atom at C5 through a halogen migration process starting from a LDA-promoted 4-lithiation [214].

4-Lithiated thiazoles have usually been generated by bromine–lithium exchange, an example of their use being the synthesis of some photochromic dithiazolylethenes [215].

10.4.5.2 Magnesium Azoles

The direct preparation of azolic organomagnesium reagents using the standard reaction between a halogenated derivative and magnesium is rather difficult due to the presence of a basic nitrogen. In these cases, the usual preparative procedure is to treat the azole with an alkyl Grignard reagent (generally EtMgBr, *i*PrMgBr, or $i\text{Pr}_2\text{Mg}$) or to perform a halogen–magnesium exchange by treating bromo and iodo azoles with the mentioned alkyl Grignards [216]. This procedure tolerates the presence of other functionalities [217]. Furthermore, the preparation of the organolithium derivative followed by interchange using magnesium dibromide can also be used.

The generated imidazolylmagnesium halide has been employed in addition reactions to carbonyl compounds for the preparation, for example, of ligands for the α_{2D} adrenergic receptor [218], sugar-mimic glycosidase inhibitors [219] or C-nucleosides [220]. It has also been used in acylation reactions with esters in the synthesis of pilocarpine analogues [221] or with Weinreb amides such as **145** (Scheme 10.67) [222].

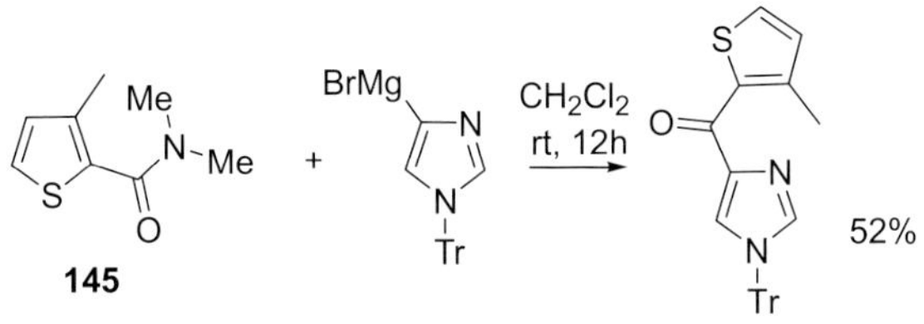

Scheme 10.67

In addition, examples of the use of oxazolylmagnesiums can be found in the addition of 2-(methylthio)-5-oxazolylmagnesium bromide (**146**) to the aldehyde **147** to give compound **148**, which has been employed for the synthesis of conformationally locked C-nucleosides (Scheme 10.68) [223].

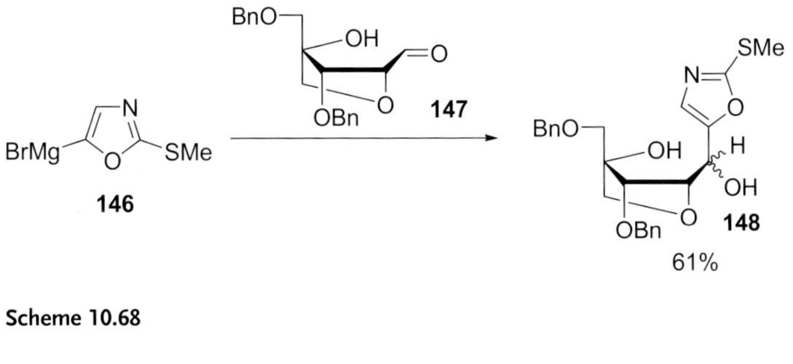

Scheme 10.68

Finally, thiazolylmagnesiums metalated at C2 have been obtained by the usual bromine-magnesium exchange using alkyl Grignards, even regioselectively. For example, 2-thiazolylmagnesium bromide **149** has been obtained from 2,4-dibromothiazole. This reagent has been used in an addition reaction to the chiral nitrile **150**, affording, after reduction, the corresponding amine, which is a building block for the synthesis of thiazolyl peptides (Scheme 10.69) [224]. The lithium–magnesium transmetalation can also be used for the generation of thiazolylmagnesiums, an example being the preparation of 2-methylthiazol-4-ylmagnesium bromide, which is

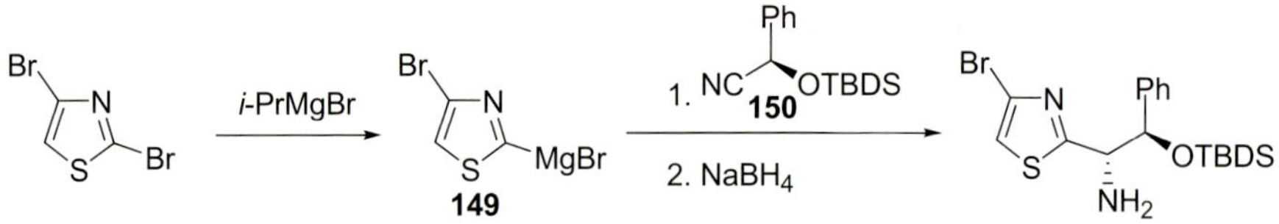

Scheme 10.69

useful in the preparation of a fragment of epothilone via a copper(I)-catalyzed coupling to an allylic bromide [225].

10.4.5.3 **Silicon Azoles**

Azole silanes are usually prepared by reaction of the corresponding heterocyclic organolithiums with alkylhalosilanes [226].

Imidazoles have been silylated at C2 using the conventional lithium–silicon exchange, although it has been observed that 2-*tert*-butyldimethylsilylimidazole can be obtained by reaction of *O-tert*-butyldimethylsilylimidazolyl aminals with organolithium reagents through a retro-[1,4]-Brook rearrangement [227]. Usually, 2-silylimidazoles are employed in a subsequent lithiation at the 3-position and further reaction with electrophiles such as aldehydes [228] or tosyl azide [229].

The introduction of a silyl group at the 2-position in N-protected imidazoles was used as a logical way of changing the acidic proton by an easily removable group, thus allowing deprotonation at C5 and further transformations (Section 10.4.5.1). Examples are 2-silylated imidazoles, which are lithiated at C5 and act as nucleophiles [230].

The preparation of 2-silylated oxazoles is not obvious, since the usual 2-lithiation–silylation sequence drives the above mentioned ring opening to give an isocyano enolate (Section 10.4.5.1) after the lithiation step. This problem was overcome by O-silylation of the isocyano enolate followed by a base-promoted insertion to give the corresponding 2-silyloxazole **151** (Scheme 10.70) [231]. The procedure can be simplified by a heat-induced cyclization in the final distillation step [232]. These derivatives have also been prepared by reaction with trimethylsilylbromide in the presence of triethylamine [233].

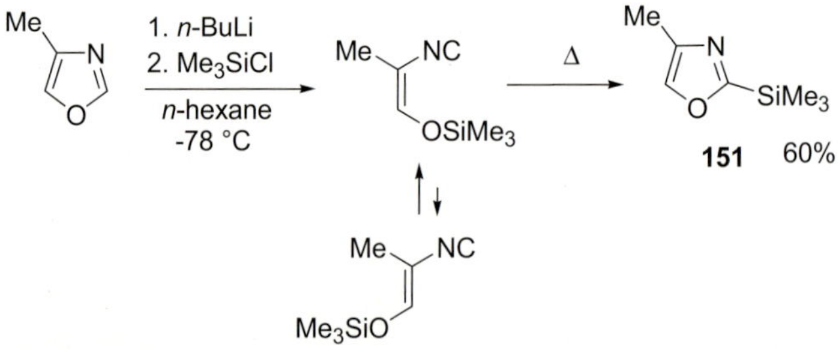

Scheme 10.70

These 2-silylated oxazoles can be used as nucleophiles in additions to aldehydes [234], such as to the tripeptide-derived aldehyde **152** (Scheme 10.71) [232].

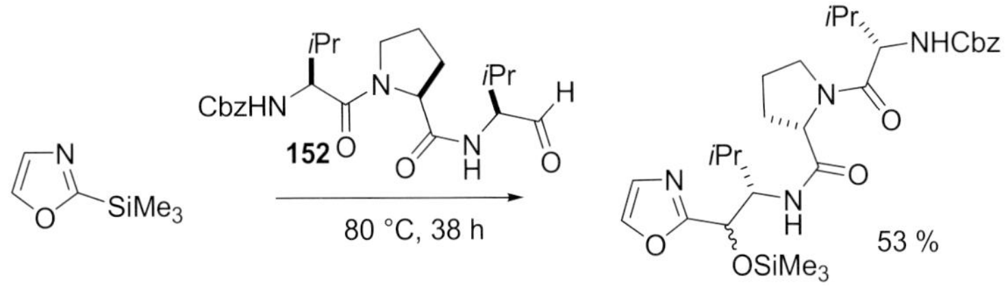

Scheme 10.71

Recently, 4-(triethylsilyl)oxazoles have been prepared, by treatment of (triethylsilyl) diazoacetates with rhodium(II) octanoate and nitriles, as precursors of 4-halogenated oxazoles after treatment with *N*-bromosuccinimides [235]. On the other hand, these derivatives have been used to perform a 4-litiation followed by reaction with electrophiles [236].

2-(Trimethylsilyl)thiazole (**153**), which is prepared by the conventional lithiation–-silylation sequence, has been frequently used for addition reactions to aldehydes [237], mainly for chain elongation due to consideration of the thiazole moiety as an equivalent of the formyl synthon. An example of the use of **153** is its diastereoselective addition to the chiral aldehyde **153**, yielding the corresponding protected alcohol **155**, an intermediate in the synthesis of the pseudopeptide microbial agent AI-77-B (Scheme 10.72) [238].

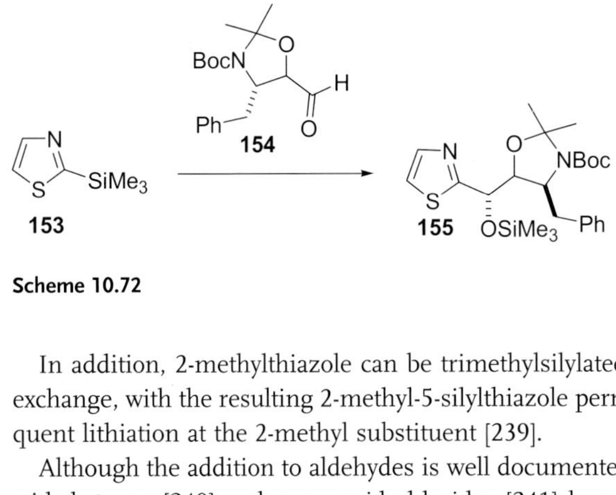

Scheme 10.72

In addition, 2-methylthiazole can be trimethylsilylated at C5 by lithium–silicon exchange, with the resulting 2-methyl-5-silylthiazole permitting, therefore, a subsequent lithiation at the 2-methyl substituent [239].

Although the addition to aldehydes is well documented, the less known reaction with ketones [240] and some acid chlorides [241] has also been reported. Other examples of the use of 2-(trimethylsilyl)thiazole are the ring expansion of a cyclopropanated carbohydrate [242], the copper(I) salt-mediated coupling to iodobenzene [243] and the *ipso*-substitution with iodine [244].

10.4.5.4 **Tin Azoles**

In general, azolic stannanes have been obtained by reaction of their corresponding azolic organolithiums with a chlorostannane [245, 246]. These metallated azoles have found application mainly in palladium-catalyzed cross-coupling reactions (the so-called Stille coupling) (see Section 10.4.6.31).

2-Azolylstannanes and 2-substituted-5-stannylazoles have been prepared following the usual stanylation sequence. 5-Stannylimidazoles have also been prepared by a 2,5-dilithiation, followed by a double stannylation and a 2-hydrodestannylation sequence [247].

The preparation of 4-stannylated azoles is not obvious. 4-Stannylated thiazoles are usually obtained by a sequential halogen–lithium–tin interchange [248], although after lithiation of 4-bromo-2-stannylthiazoles, to give the 4-stannylated heterocycles, rearrangements were observed [249]. In some cases, 4-stannylthiazoles are prepared by palladium-catalyzed cross-coupling of the corresponding bromide using bis(trimethyltin) [250].

The same methodology has been employed when 4-stannylated oxazoles bearing labile groups are required [246].

10.4.5.5 Zinc Azoles

Organozincs are a useful class of organometallic reagents due to their tolerance of several functional groups [251]. Azolic zinc derivatives are generally prepared by exchange reactions of the corresponding organolithium or magnesium with zinc halides, being stable at higher temperatures than their precursors [251]. Other methods for their preparation employ zinc dust [251], active Rieke zinc [251] or electrochemical methods [252].

2-Zincated 1,3-azoles are used mainly in palladium-catalyzed Negishi cross-couplings, as in the reaction of the triflate **157** with the *N*-silylated imidazolylzinc chloride **156** to afford compound **158** (Scheme 10.73), in studies toward the synthesis of anxiolytics.

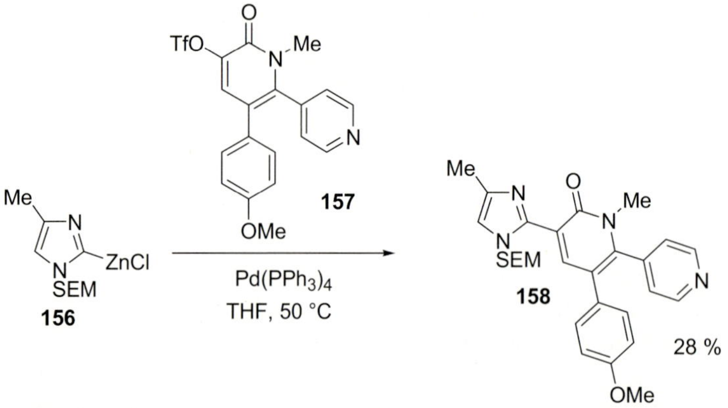

Me N ZnCl N SEM **156**

TfO O N Me N OMe **157**

$Pd(PPh_3)_4$

THF, 50 °C

Me N N SEM O N Me N OMe **158** 28 %

Scheme 10.73

More recently, the coupling of the organozinc reagent **159** with the bromopyridine **160** afforded compound **161**, which is an intermediate in the synthesis of the heterocycle core of the GE 2270 antibiotic (Scheme 10.74) [253].

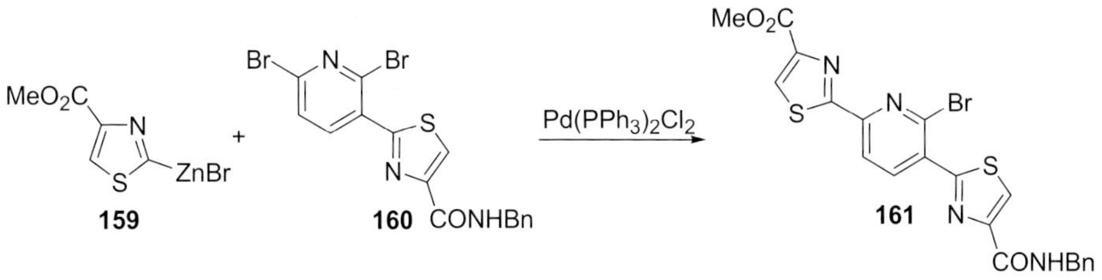

Scheme 10.74

Imidazol-4-yl-zinc chloride has been used in several synthesis [218], whereas oxazol-2-yl-zinc [254] and thiazol-2-yl-zinc [255] derivatives are also employed in Negishi cross-couplings. Furthermore, copper-catalyzed cross-coupling reactions have been performed using *N*-methylimidazol-2-yl-zinc iodide [256]. Finally, thiazol-4-yl-zinc bromide is used in additions to nitrones [209].

10.4.5.6 Copper Azoles

Azolyl copper reagents are usually obtained from the corresponding organolithium reagents (2 equivalents) by reaction with a copper(I) salt [239, 245].

For example, *N*-substituted 4,5-diiodoimidazole **162** has been regioselectively transformed into the 5-cuprated imidazole **163** after reaction with $(PhMe_2CCH_2)_2$-CuLi. These organocopper reagents reacted with electrophiles such as allyl bromide to give the corresponding 5-functionalized imidazole **164**(Scheme 10.75) [257].

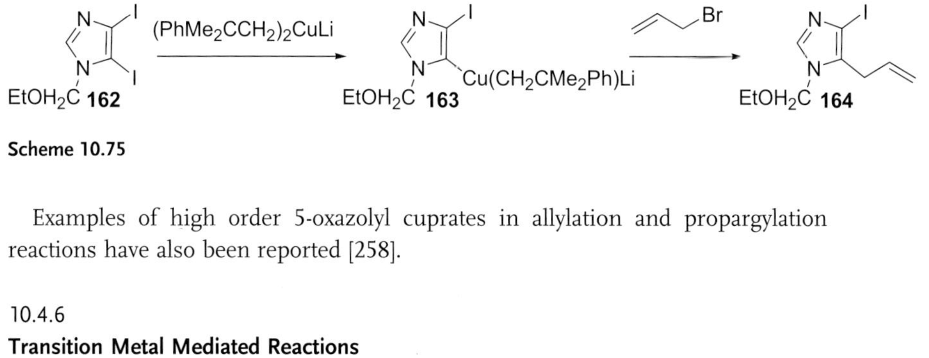

Scheme 10.75

Examples of high order 5-oxazolyl cuprates in allylation and propargylation reactions have also been reported [258].

10.4.6 Transition Metal Mediated Reactions

10.4.6.1 Metal-Mediated Functionalization

Recently, several new methods for the functionalization of 1,3-azoles under selective conditions of C–H activation have been developed, particularly through the use of transition metal catalysts.

The arylation of 1,3-azole derivatives (oxazoles, thiazoles and protected imidazoles) proceeds using catalytic palladium, rhodium, and/or copper in the presence of an inorganic base [259]. However, these reactions often suffer from low yields and poor selectivity, and a major disadvantage is the need to apply a relatively high reaction temperature. In general, the addition of copper(I) iodide facilitates the reactions and in specific cases can promote the reaction without a source of palladium (Scheme 10.76).

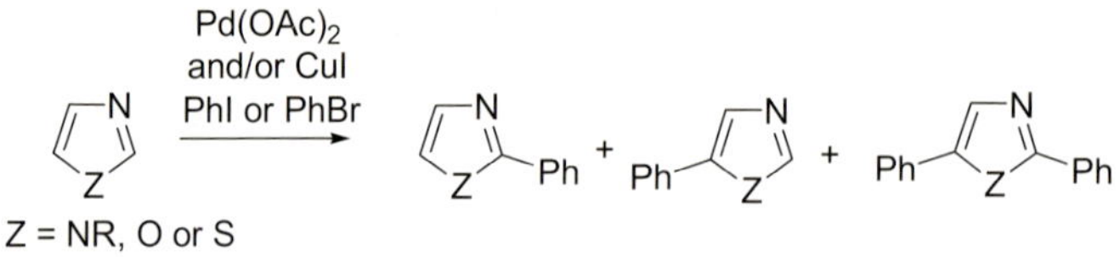

Pd(OAc)$_2$ and/or CuI PhI or PhBr

Z = NR, O or S

Scheme 10.76

Employing solid-supported aryl iodides, regioselective arylation at C2 is obtained (Scheme 10.77) [260]. The observed selective monofunctionalization can be attributed to the solid-phase pseudodilution effect which prohibits a second equivalent of the iodide interacting with the already coupled product.

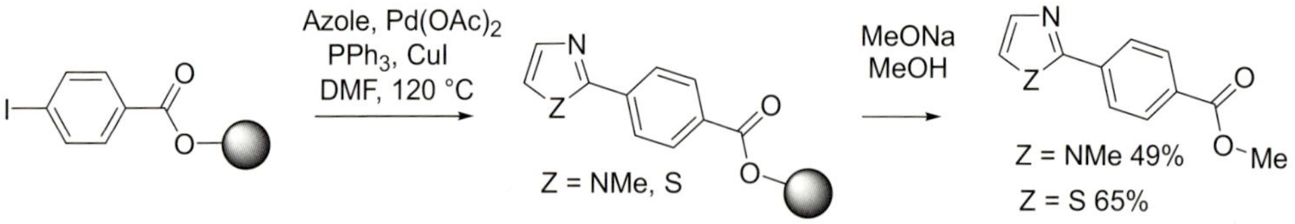

Azole, Pd(OAc)$_2$, PPh$_3$, CuI, DMF, 120 °C; Z = NMe, S; MeONa, MeOH; Z = NMe 49%, Z = S 65%

Scheme 10.77

In contrast, in the absence of CuI the 5-arylated products are obtained (Scheme 10.78) [260].

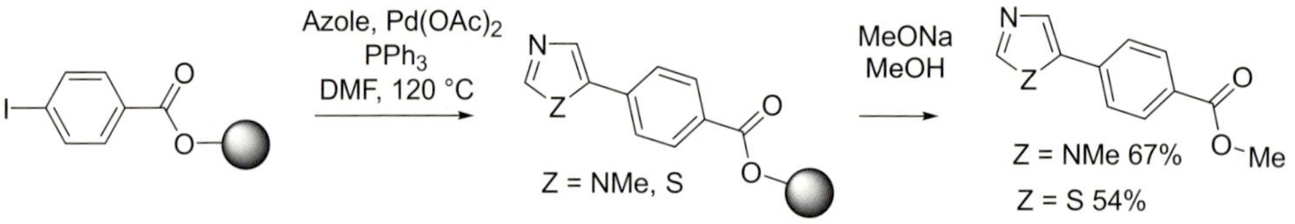

Azole, Pd(OAc)$_2$, PPh$_3$, DMF, 120 °C; Z = NMe, S; MeONa, MeOH; Z = NMe 67%, Z = S 54%

Scheme 10.78

However, although in particular cases it is possible to obtain the arylation with palladium in a regioselective form, in general, non-fused azoles gave mixtures of substitution at the 2- and/or 5-position and the use of bulky phosphines as ligands gives improved yields of diarylation [261].

As an alternative to problematic organometallic azole functionalization, thiazole N-oxides have been investigated as alternatives [262].The N-oxide group not only imparts a dramatic increase in reactivity in direct arylation at all positions of the azole ring but also changes the weak azole bias for C5 > C2 arylation to a reliable C2 > C5 > C4 reactivity profile (Scheme 10.79). This permits high yielding, regioselective, and room temperature arylation at C2, high yielding arylation at C5, and the first

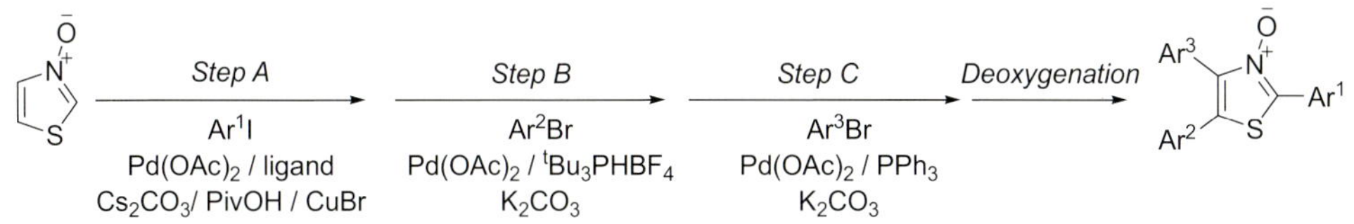

Scheme 10.79

examples of arylation at C4, providing a unique opportunity for exhaustive functionalization of the azole core (Scheme 10.80).

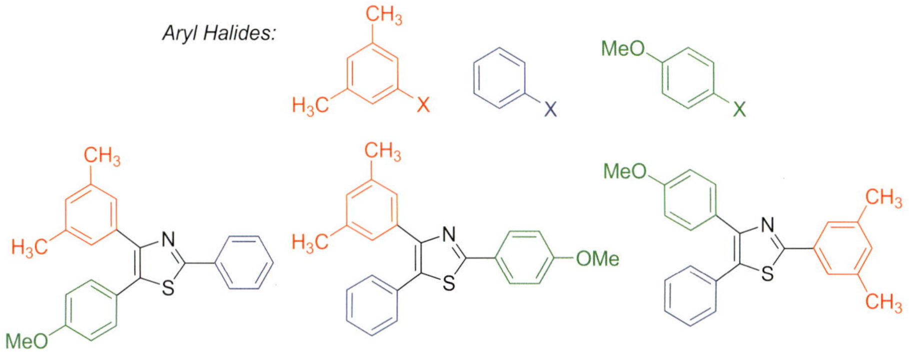

Scheme 10.80

In recent years, phosphine-free and even base-free conditions, palladium-catalyzed C–H bond arylation of azoles have been reported [263]. However, these methods are typically restricted to only one type of heterocycle, even under harsh conditions, such as elevated temperature and/or microwave assistance. Moreover, stoichiometric amounts of copper salts (1–3 equiv) or silver additives are generally required to improve these phosphine-free, Pd-catalyzed coupling reactions.

The direct C-arylation of azoles with a broad spectrum of aryl bromides without the presence of phosphines, the aid of CuI, or other metal additives has been described by using pivalic acid as a cocatalyst. Particularly noteworthy is than this protocol can tolerate an array of functional groups such as ester, nitrile, nitro, aldehyde, methoxy, trifluoromethyl, fluoro, and chloro substituents [264].

In addition to such carbon–carbon bond-forming reactions, carbon–heteroatom bond formation is also an important issue. Limited examples of intra- [265] and intermolecular [266] C–H funcionalization with amines have been described. For example, C–H, N–H coupling of azoles takes place with several amines in the presence of a copper catalyst to undergo amination at the 2-position (Scheme 10.81) [267].

Finally, several metal-promoted N–C cross-coupling reactions have been developed [268]. For example, copper-catalyzed N-phenylation of imidazoles **165** with diphenyliodonium tetrafluoroborate affords *N*-phenylimidazoles **166** (Scheme 10.82) [269].

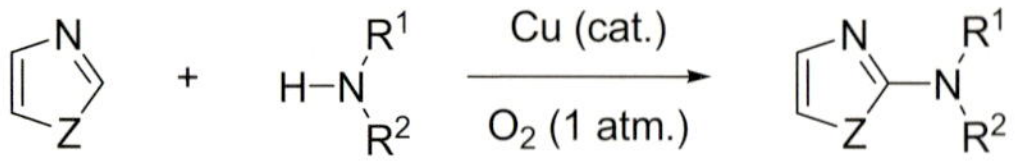

Scheme 10.81

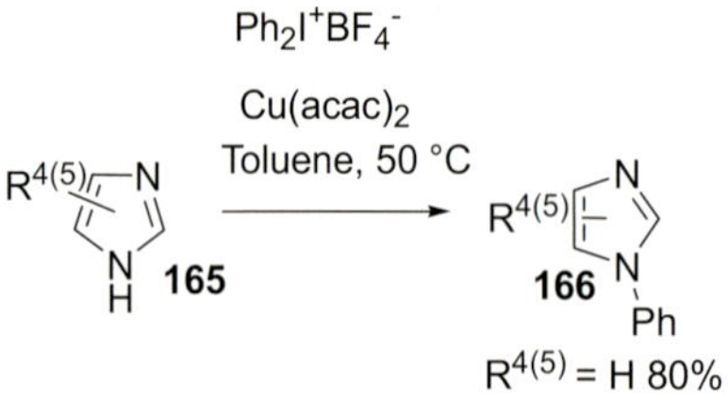

Scheme 10.82

The N-arylation of azoles, in lower nitrile solvents, with aryl halides has been achieved efficiently in the presence of copper powder without any additional ligands. Thus, the first nitrile type of monodentate ligand-mediated, "ligand-free-like" copper catalyzed N-arylation procedure was established [270].

Arylboronic acids **167** react efficiently with imidazoles **165** in the presence of a novel diamine-copper complex to give various *N*-arylimidazoles **168** (Scheme 10.83) [271].

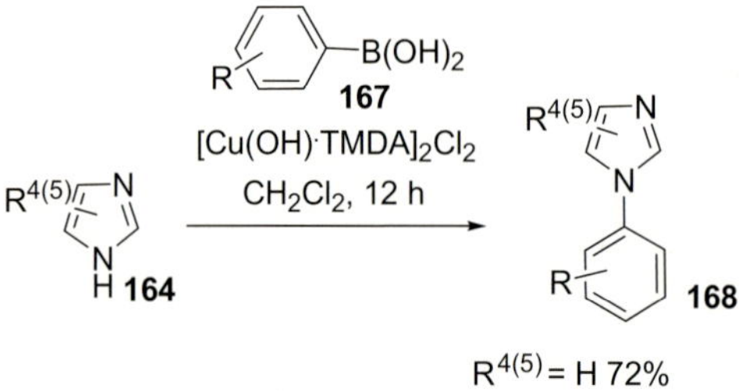

Scheme 10.83

Alkoxydienyl and alkoxystynyl boronates have also been used in various copper acetate mediated cross-coupling reactions with imidazole, affording various *N*-alkoxydienyl- and *N*-styrylimidazoles in good yields under mild conditions [272].

10.4.6.2 **Catalytic Transition-Metal Mediated Reactions of Halogenated Azoles**

2-Halogenated azoles are among the most valuable synthons for further functionalization of 1,3-azoles using catalytic transition-metal mediated reactions. They are readily prepared by direct halogenation (Br_2, I_2, or *N*-halosuccinimides) or trapping of C2-metalated (Li, Mg, Zn) azoles. Numerous methods have therefore been made for their further functionalization. Coupling reactions mediated by transition-metal catalysts allow for bond formation between halogenated azoles and unsubstituted olefins and acetylenes, as well as dimerization reactions.

10.4.6.2.1 **Heck and Ullmann Couplings** Classical Heck-type couplings (using Pd-catalyst and a base) are one of the most common bond-forming reactions of aromatic halides. Its use in the chemistry of 1,3-azoles is also very common. A σ-azolylpalladium complex, generated *in situ* from 2-halogenated azole, undergoes an insertion with the alkene/alkyne cosubstrate, followed by β elimination to give the product.

A useful example of this reaction is the coupling of methyl 3-oxo-6-heptynoate **170** with 2-bromothiazole (**169**) in the presence of K_2CO_3 and catalytic amounts of Pd $(PPh_3)_4$ that provide the derivative **171** (Scheme 10.84) [273].

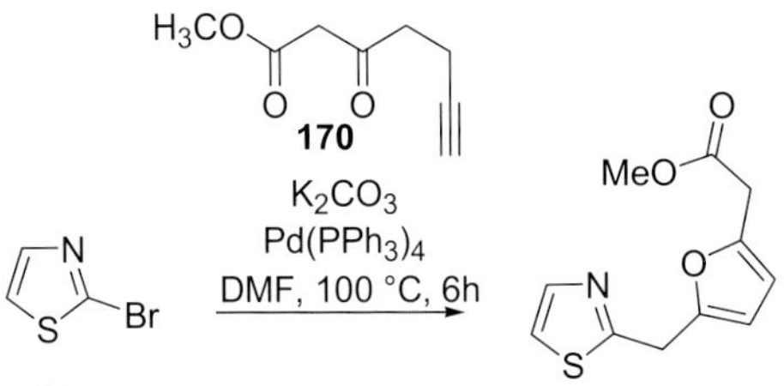

Scheme 10.84

This reaction has also been utilized to catalyze the dimerization of bromothiazole **169** [274], isolating the product **172** in an improved yield (Scheme 10.85).

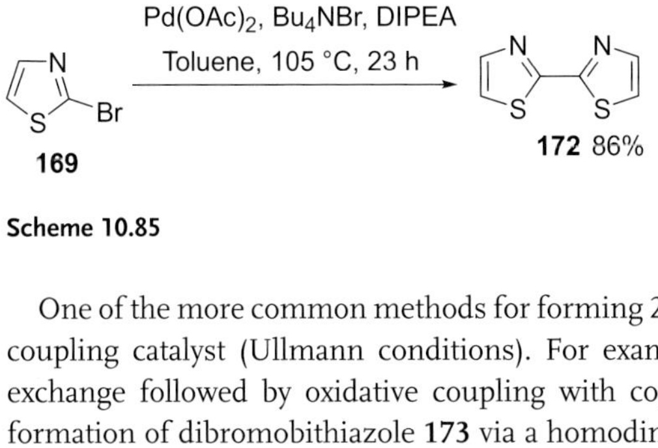

Scheme 10.85

One of the more common methods for forming 2,2′-azoles is to utilize copper as a coupling catalyst (Ullmann conditions). For example, selective lithium–bromine exchange followed by oxidative coupling with copper(II) chloride allows for the formation of dibromobithiazole **173** via a homodimerization (Scheme 10.86) [275].

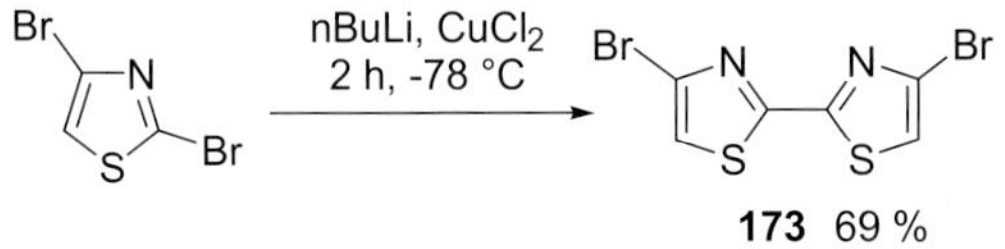

Scheme 10.86

Another example of this reaction is the cyclization of the symmetric derivative **174** in modest yield utilizing copper to furnish a tetracyanobisimidazole **175** (Scheme 10.87) [276].

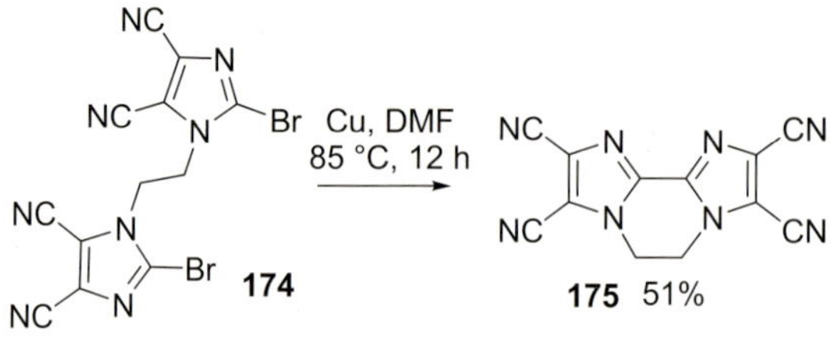

Scheme 10.87

However, despite extensive research in this area, the classical Ullmann reactions were invariably plagued by the need for large amounts of copper (in the form of salts, oxides, or finely divided metal) and for very harsh reaction conditions, most notably for a high reaction temperature. In this context, very recently it was shown that simple copper(I) complexes formed *in situ* with chelating nitrogen- and/or oxygen-containing ligands could effect N- and O-arylations of many different organic compounds with azol halides in good yields at temperatures in the range 80–150 °C [277].

The range of employed ligands is continuously increasing [278], and other kinds of copper-containing catalysts, such as Cu_2O-coated soluble copper nanoparticles [279], copper-exchanged apatites [280], and copper-containing perovskites [281] have been successfully tested as well.

10.4.6.2.2 **Sonogashira Reaction** The Sonogashira reaction [using Pd(0)/Cu(I)-catalyst] is a useful method for the cross-coupling of a terminal alkyne to an 2-halogenated azole. In this case the σ-azolylpalladium complex reacts with a copper acetylide, generated *in situ*, and after a β-elimination the final product is obtained.

This reaction has been used, for example, to prepare histidine derivatives wherein diiodide **176** undergoes selective coupling as well as dehalogenation in the presence of excess phenylacetylene (Scheme 10.88) [282].

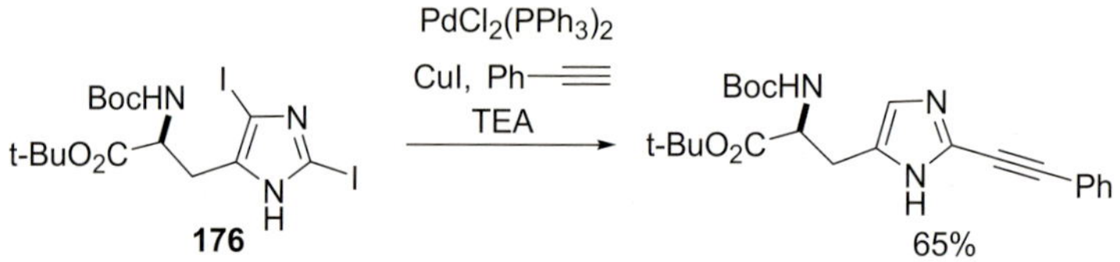

Scheme 10.88

Imidazole derivatives [283] and thiazole analogs (Scheme 10.89) [284] have been investigated, and regioselective alkynylation is often feasible.

Finally, although in general 2-halogenated azoles are employed as starting materials for this type of reaction, oxazolyl and thiazolyl triflates **177** are also effective substrates [285] for substitution (Scheme 10.90).

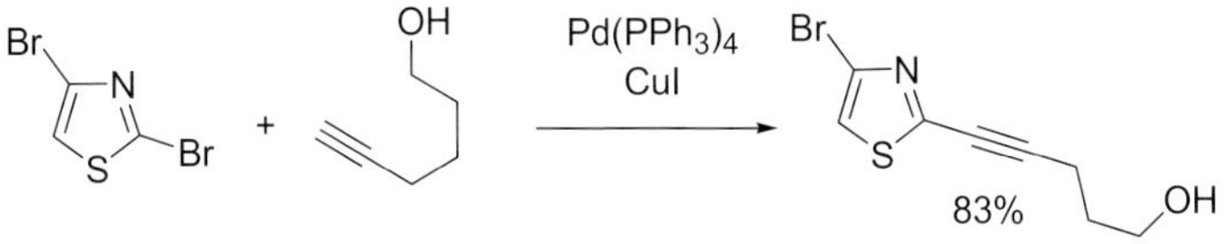

Scheme 10.89

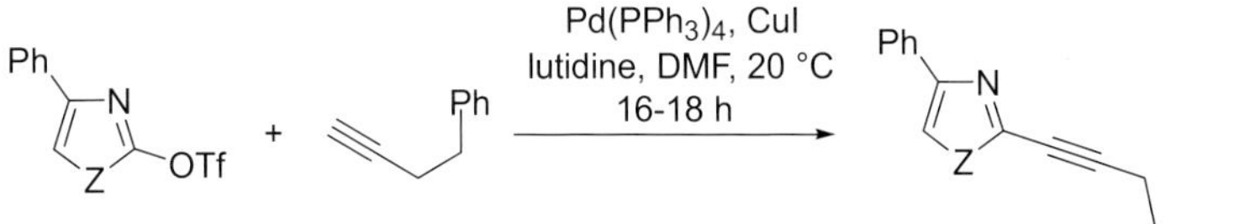

Scheme 10.90

10.4.6.3 **Stoichiometric Organometallic/Transition Metal Mediated Reactions**

Whereas the use of 1,3-azoles in coupling reactions with unsubstituted partners (Section 10.4.6.1) is somewhat limited, the use of stoichiometric organometallic reagents is very broad. The use of organostannanes, boronates and other metal-substituted reagents in transition-metal mediated coupling reactions has been widely examined.

10.4.6.3.1 **Stille Cross-Coupling Reactions** The availability of organostannanes and their well-understood cross-coupling reactions has been applied to the coupling with 2-halogenated azoles.

The cross coupling of the organostannane **178** with 2-bromothiazole **169** has been used to construct alkylidene-cephalosporine derivatives **179** as β-lactamase inhibitors (Scheme 10.91) [241].

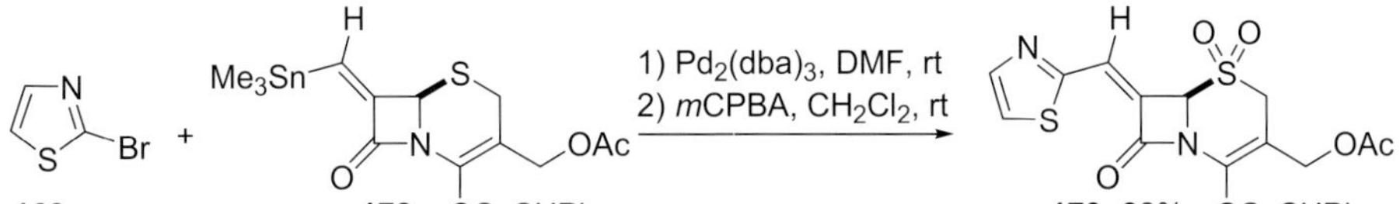

Scheme 10.91

Trialkylstannyl-1,3-azoles have also been utilized as partners in the Stille reaction with aromatic or heteroaromatic halides and triflates.

This reaction constitutes a useful solution for the selective arylation of azoles. For example, a good yield of the arylated imidazole derivative **181** is obtained from dibromide **180**, with good selectively for C2 substitution (Scheme 10.92) [287].

Examples using N-protected imidazolyl-stannanes or thiazolyl-stannanes can be found in the coupling with iodouracil derivatives [288]. Standard conditions were

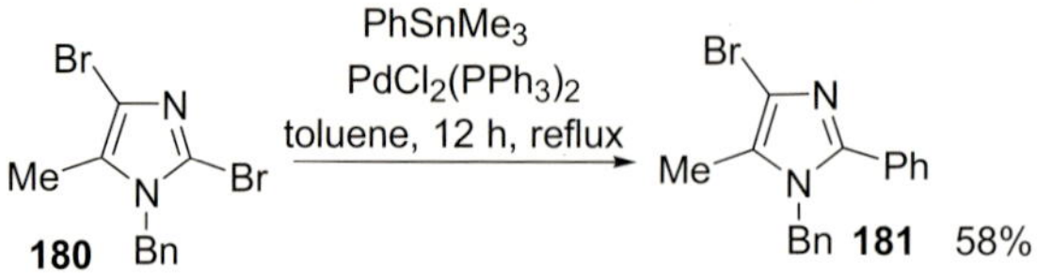

Scheme 10.92

effective for incorporation of the thiazole moiety; however, the reaction of imidazole **182** required the use of stoichiometric silver(I) oxide (Scheme 10.93).

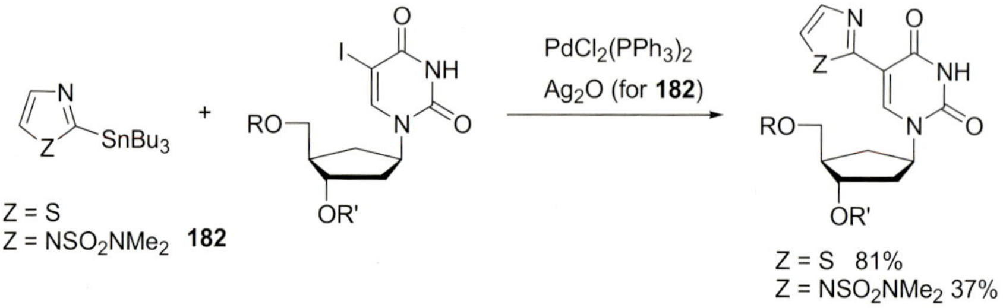

Scheme 10.93

More recently, these conditions have been employed in the synthesis of RNA containing imidazole attached directly to 5-position of uracyl heterocycles of tandem G–U wobble base pairs. The modified uridine was prepared using a palladium-catalyzed coupling of 5-iodouridine and 4-tributylstannyl imidazole [289].

Other examples of Stille couplings using 2-substituted 5-stannylimidazoles [290] are applicable to the synthesis of cytotoxic agents [291] and an imidazolyl isomer of the alkaloid didemnimide C [292]. 5-Stannylimidazoles have also been prepared by a 2,5-dilithiation, followed by a double stannylation and a 2-hydrodestannylation sequence [291]. In addition, the stannyl group on imidazoles has been employed for *ipso*-iodination reactions, as in the synthesis of inhibitors of phophodiesterase PDE4 [293].

Finally, sometimes attempts to generate either coupling partner as a discrete stannane lead to dimeric or decomposition products. In some cases it is necessary to form the stannane *in situ*, such as in the synthesis of dimethyl sulfomycinamate, which was prepared via *in situ* stannane formation from triflate **184** followed by addition of bromide **183** to the reaction (Scheme 10.94) [294].

10.4.6.3.2 **Suzuki–Miyaura Cross-Coupling Reactions** Reactions of azole derivatives that take advantage of a boron-containing partner in the coupling reaction are also very common, but generally use an halogenated azole and an aryl or heteroaryl boronic acid, many of which are now commercially available. The Suzuki reaction is exceptionally tolerant of functional groups that often need protection under other coupling conditions.

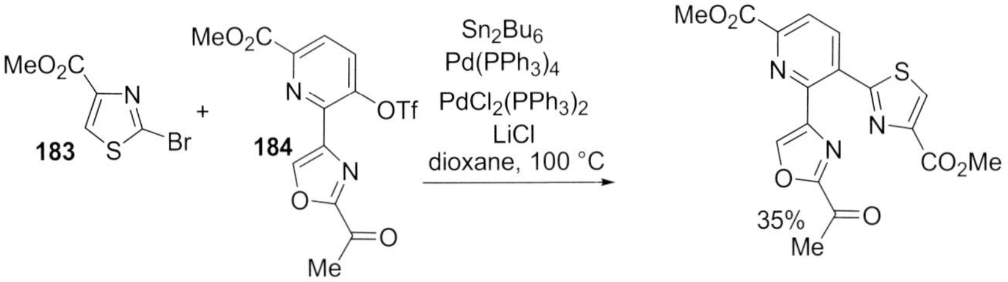

Scheme 10.94

Also in this case, 2-bromothiazole (**169**) [295a–c] is a common substrate for this type of reactions (Scheme 10.95) [295a].

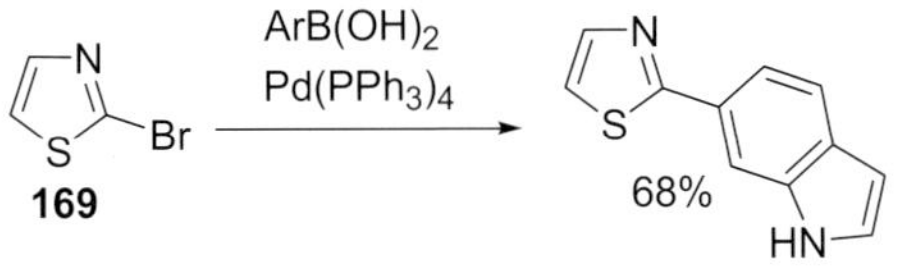

Scheme 10.95

These cross-coupling reactions are a useful solution for the selective arylation of azoles. For example, chemoselective reaction of 2,4-dibromothiazole **185** proceeds selectively at C2 [Scheme 10.96 (1)] [296], as does the 2-bromo-5-chloro derivative **186** [Scheme 10.96 (2)] [297].

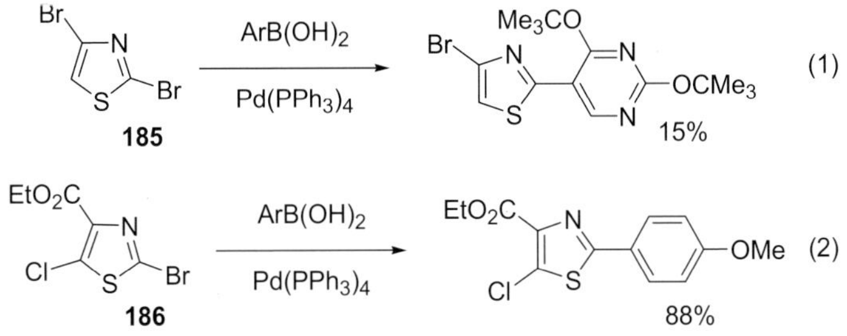

Scheme 10.96

10.4.6.4 **Zinc, Magnesium and Other Metal Mediated Couplings**

Metal-substituted coupling partners for transition-metal mediated reaction, other than tin and boron, are of interest for environmental (tin toxicity) or synthetic (availability of boronates) reasons.

Zinc-mediated couplings have been developed. For example, highly active zinc [298] generates **188**, which efficiently couples with 2-iodothiazole (**187**) to afford indolyl-thiazole adduct **189** (Scheme 10.97).

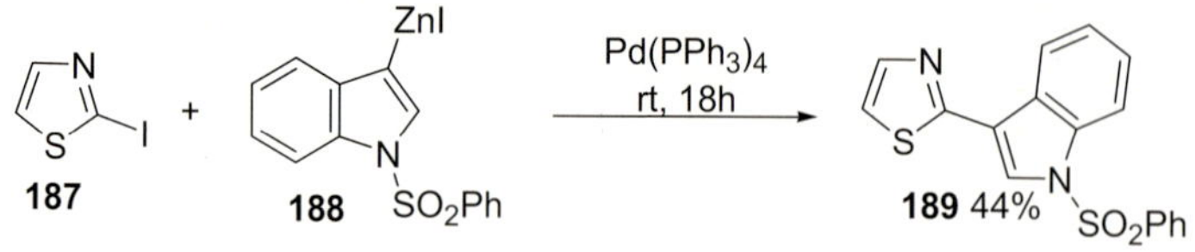

Scheme 10.97

Another class of cross-coupling reactions recently described are the Grignard couplings. For example, the combination of zinc bromide and a Grignard reagent allows for installation of the isobutylene fragment in thiazole **190** in the first step (Scheme 10.98) towards the fungicidal natural product hectochlorin [299].

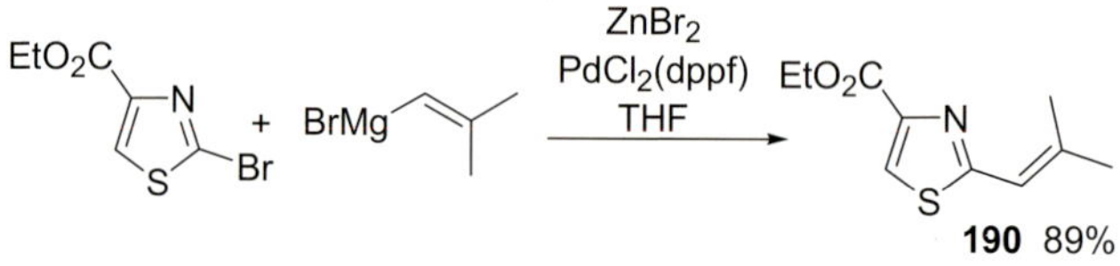

Scheme 10.98

New methods for the coupling of thioether derivatives have been reported. Nickel catalysis in combination with both aryl and alkyl Grignard reagents affords derivative **191** (Scheme 10.99) [300].

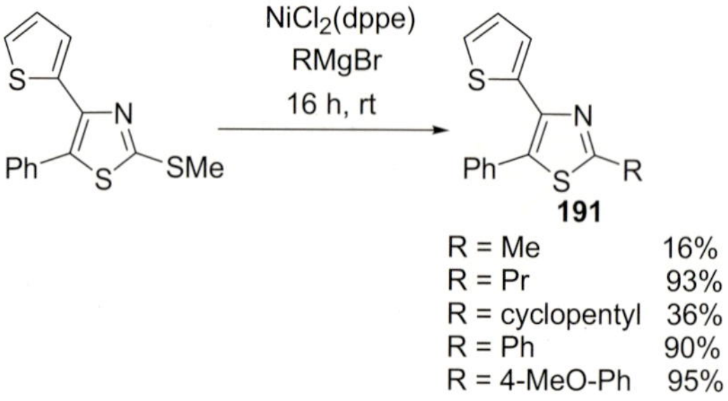

Scheme 10.99

Finally, the use of other organometallic reagents has proven useful with halogenated azoles, and include organoaluminates [301] with palladium catalysis as well as Grignard reagents with the catalysis of nickel [302] and iron [303].

10.4.7
Reactions with Radicals

Free radical reactions are still very much less common in azole chemistry than those involving, for example, electrophilic or nucleophilic reagents. In reactions involving free radicals, substituents have little orienting effect; however, rather selective radical reactions are now known.

Phenyl radicals attack azoles unselectively to form a mixture of phenylated products. The phenyl radicals may be prepared from the usual precursors: PhN(NO)COMe, $Pb(OCOPh)_4$, $(PhCO_2)_2$ or $PhI(OCOPh)_2$. For example, the three monophenylthiazoles are obtained in the practically constant ratio of 6 : 3 : 1 (2 : 5 : 4, respectively) using the photolysis of benzyl peroxide as source of radicals in the presence of thiazole. In the case of 1-methylimidazole, phenylation, using the decomposition of benzoyl peroxide at 118 °C, no change in the overall yield is reported whether the solvent is 1-methylimidazole itself or acetic acid. In acetic acid, however, only 1-methyl-2-phenylimidazole was formed, while with the excess of heterocycle 1-methyl-2-phenyl- and -5-phenyl-imidazoles were isolated in the ratio 79 : 21 [304].

In contrast, alkyl radicals produced by oxidative decarboxylation of carboxylic acids are nucleophilic and attack azoles at the most electro-deficient sites. Thus, imidazole and 1-alkylimidazoles are alkylated exclusively at the C2. Similarly, thiazoles are attacked in acidic media by methyl and propyl radicals to give 2-substituted derivatives in moderate yields, with smaller amounts of 5-substitution. Similar reactions occur with acyl radicals, for example with the $CONH_2$ radical from formamide [305].

Recently developed are alkyl [Scheme 10.100 (1)] and aryl [Scheme 10.100 (2)] radical cyclization onto azoles for the synthesis of bi-or tricyclic heterocycles [306].

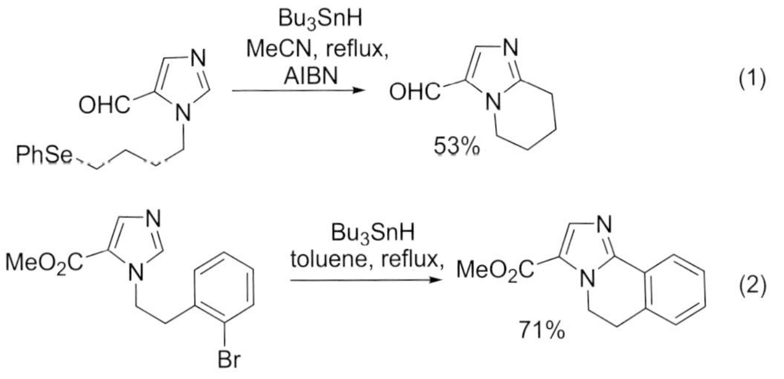

Bu_3SnH, MeCN, reflux, AIBN; OHC, PhSe, N; 53% (1)

Bu_3SnH, toluene, reflux; MeO_2C, N, Br; 71% (2)

Scheme 10.100

10.4.8 Reactions with Reducing Agents

Oxazoles are readily reduced, usually with ring scission (Scheme 10.101). Only acyclic products have been reported from the reductions with complex metal hydrides of oxazoles. Similar results have been obtained using catalytic hydrogenation or reduction by dissolving metals [307].

Imidazoles are generally resistant to reduction. Unless the NH of imidazole is substituted, the preferential reaction with a complex hydride will be salt formation,

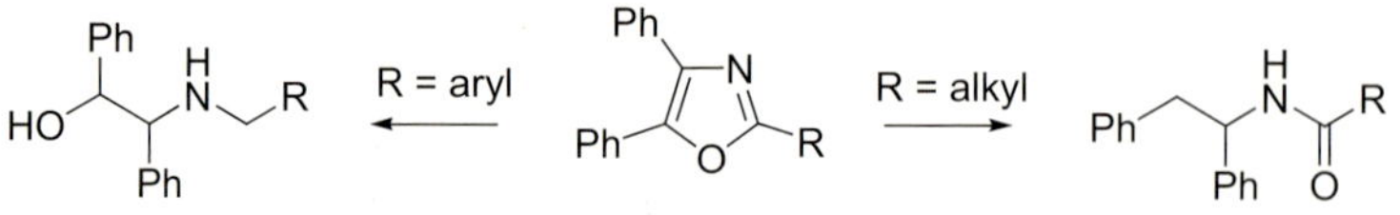

Scheme 10.101

which leaves a negative charge on the ring nitrogen. Consequently, the species becomes resistant to reduction.

Finally, thiazole reduction is a very useful method for the synthesis of aldehydes [207]. The aldehyde is prepared via a three-step reaction sequence that consists of N-methylation of the thiazole ring **192**, reduction of the resulting *N*-thiazolium salt (not shown) to the thiazolidine **193** and, finally, $HgCl_2$-promoted hydrolysis of the heterocyclic nucleus of **193** (Scheme 10.102).

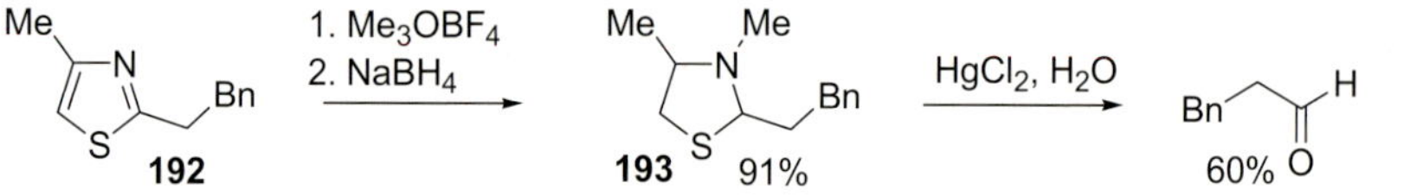

Scheme 10.102

One of the first examples of this reaction was the thiazole-based one-carbon homologation of 2,3-*O*-isopropylidine-D-glyceraldehyde **194** to protected D-erythrose **195** (Scheme 10.103) [308].

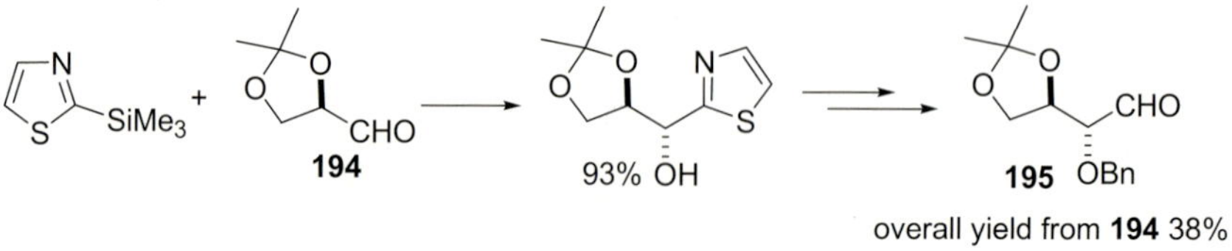

Scheme 10.103

10.4.9
Electrocyclic and Photochemical Reactions

10.4.9.1 Diels–Alder Reactions and 1,3-Dipolar Cycloadditions

The distinction between these two classes of reactions is simply semantic for the five-membered rings: Diels–Alder reaction at the F/B positions in **196** (four-atom fragment) is equivalent to 1,3-dipolar cycloaddition in **197** across the three-atom fragment, both providing the four π-electron component of the cycloaddition (Figure 10.9) [309].

Oxazoles exhibit diene-type characteristics and undergo Diels–Alder reactions with alkenic and alkynic dienophiles (HOMO-oxazole, LUMO-dienophile) (Scheme 10.104) [310]. The presence of electron-releasing substituents on the oxazole ring facilitates the reaction with dienophiles.

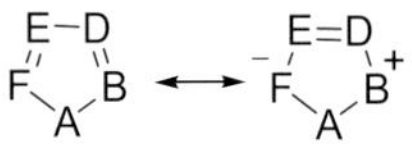

Figure 10.9 A Diels–Alder reaction at the F/B positions in **196** (four-atom fragment) is equivalent to 1,3-dipolar cycloaddition in **197** across the three-atom fragment.

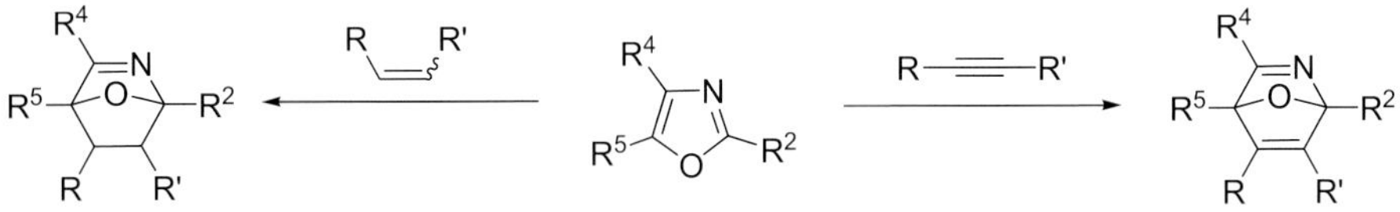

Scheme 10.104

The primary adducts of oxazoles with alkenes and alkynes are usually too unstable to be isolated. An exception, for example, is compound **200**, obtained from 5-ethoxy-4-methyloxazole **198** and 4,7-dihydro-1,3-dioxepine **199**, which has been separated into its *endo* and *exo* components (Scheme 10.105) [310].

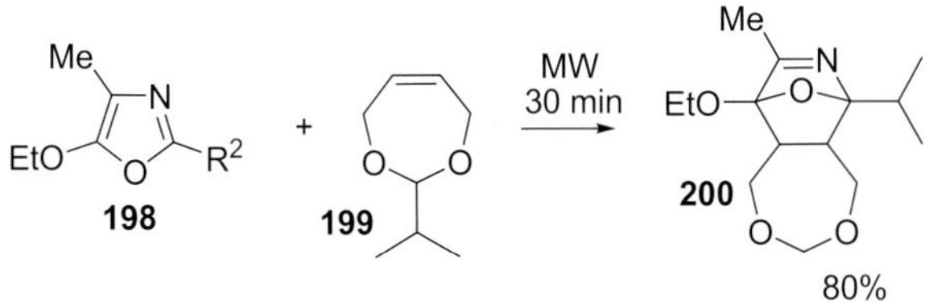

Scheme 10.105

If the dienophile is unsymmetrical the cycloaddition can afford the two regioisomers. This is usually the case in the reactions of oxazoles with monosubstituted alkynes; while with alkenes regioselectivity is observed. A general rule for the reactions of alkyl- and alkoxy-substituted oxazoles is that in the adducts the more electronegative substituent of the dienophile R^4 occupies the position shown in Scheme 10.104.

Acid- or base-catalyzed cleavage of the ether bridge in primary cycloadducts leads to pyridine derivatives (Scheme 10.106) [312]. The intermediates **201** cleave to unstable dihydropyridinols **202**, which aromatize in four ways:

- path A: pyridines are formed by dehydration;
- path B: 3-hydroxypyridines results from elimination of R^3H;
- path C: elimination of R^4H if R^3 is hydrogen;
- path D: dehydrogenation if R^3 is hydrogen.

Generally, more than one path is followed and a mixture of products results.

However, the reaction of oxazoles with alkyne dienophiles affords furans **203** with the elimination of cyanide in a retro-Diels–Alder reaction (Scheme 10.107) [313].

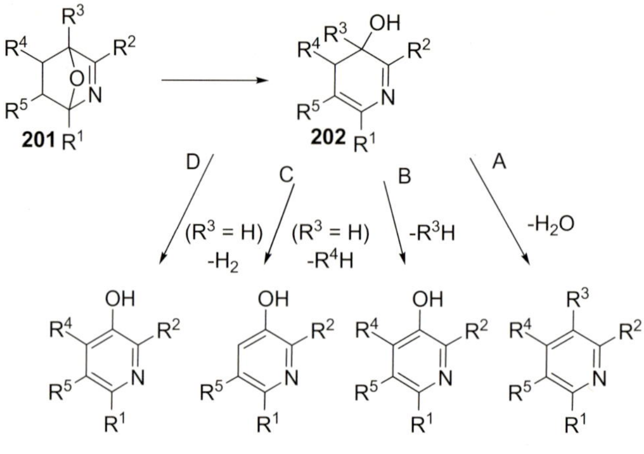

Scheme 10.106

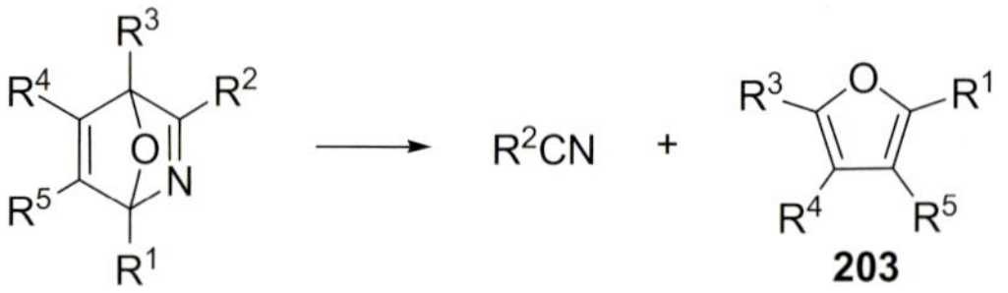

Scheme 10.107

In contrast to oxazole, thiazole does not undergo Diels–Alder cycloaddition reaction. This behavior can be correlated with the more dienic character of oxazole relative to thiazole.

Dipolarophiles like DMAD (dimethyl acetylenedicarboxylate), dibenzoylacetylene and ethyl propiolate condense with ylides **204** resulting from quaternization of 4-methylthiazole with an α-bromo ketone or ester and subsequent deprotonation. The 1 : 1 molar adduct **205** rearranges to a pyrrolothiazine **206** (Scheme 10.108) [341].

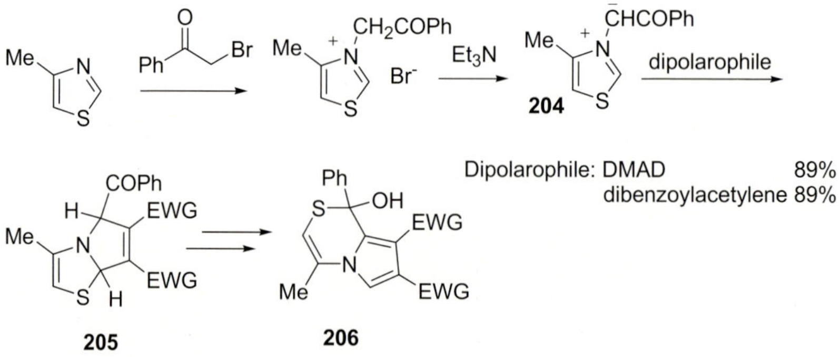

Dipolarophile: DMAD 89%
dibenzoylacetylene 89%

Scheme 10.108

Thiazoles itself reacts with DMAD at room temperature in DMF; the initially formed adduct **207** rearranges either via a concerted suprafacial 1,5-sigmatropic shift or by a non-concerted pathway proceeding via zwitterions **208** to **209** ($R{=}R^1{=}H$) (Scheme 10.109) [315].

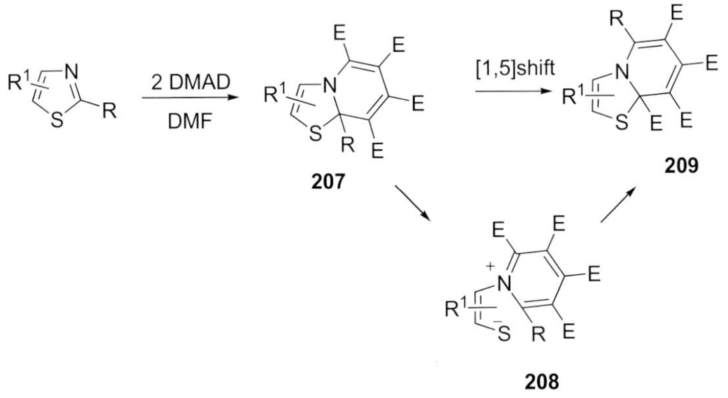

Scheme 10.109

2-Isopropylthiazole (**209**) reacts with dichloroketene in a [2 + 2 + 2] manner to give **210** as the major product (Scheme 10.110) [316].

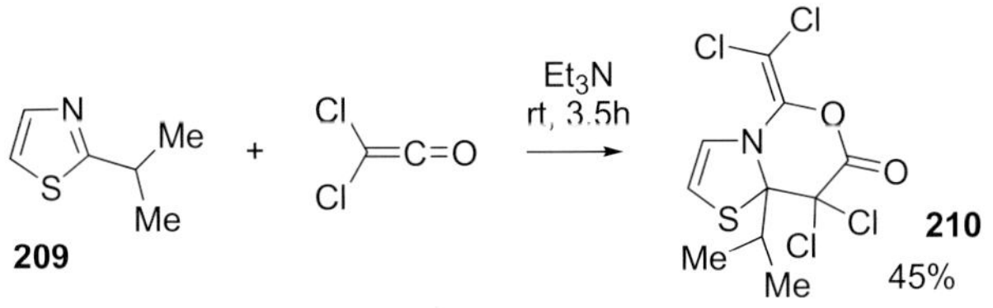

Scheme 10.110

Finally, reactions of imidazoles **211** with DMAD usually do not lead to normal Diels–Alder adducts but to products of N-alkylation (**212**) (Scheme 10.111) [317].

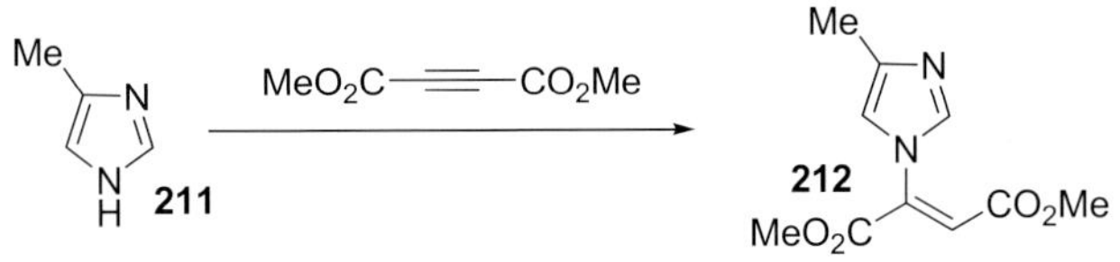

Scheme 10.111

There are instances, nevertheless, in which some form of addition takes place. For example, 1,2-dimethylimidazole gives the adduct **213** (Figure 10.10) [318].

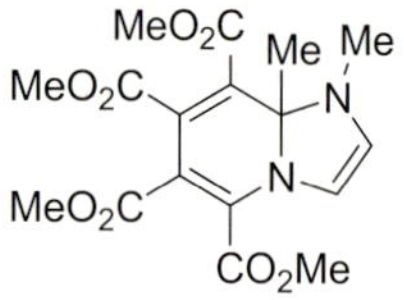

213

Figure 10.10 Diels–Alder adduct between 1,2-dimethylimidazole and DMAD.

10.4.9.2 **Photochemical Reactions**

Oxazoles are generally photostable and are, indeed, produced by light-induced rearrangements of isoxazoles [319]. However, irradiation of 2,5-diphenyloxazole in ethanol gives a mixture of 3,5-diphenylisoxazole (**214**), 4,5-diphenyloxazole (**215**), the phenanthrooxazole (**216**) and traces of benzoic acid. This reaction proceed by two distinct paths, which are rationalized as shown in Scheme 10.112.

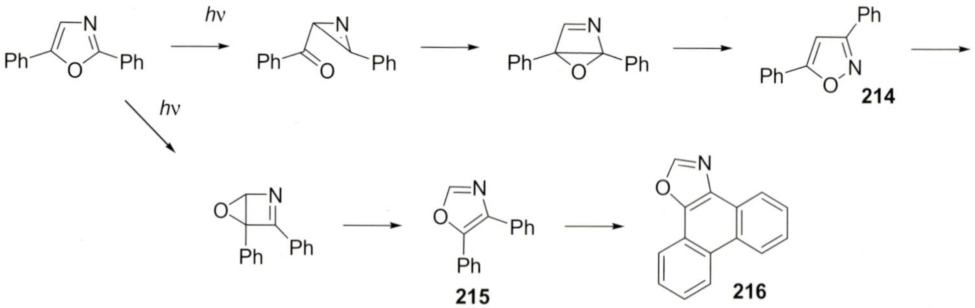

Scheme 10.112

In the case of photoaddition of acetone and other ketones to 1-, 2- and 1,2-dimethylimidazoles the products are α-hydroxyalkylimidazoles **217**, which are derived from the selective attack of excited carbonyl oxygen at C5 (Scheme 10.113) [320]. Imidazole itself does not react.

Benzophenone reacts differently with 1,2-dimethylimidazole. The diarylketone adds at the 2-methyl group [Scheme 10.114 (1)]; 1-benzylimidazole reacts at the exocyclic methylene group [Scheme 10.114 (2)] [321].

More recently, a photoinduced procedure for the intermolecular hydroamination of alkenes using azoles has been described. This reaction occurs in modest to good yield for six- and seven-membered cyclic alkenes. Upon irradiation at 254 nm in the presence of methyl benzoate and a small amount of triflic acid as an additive (20 mol. %), imidazoles can react with the alkene to afford complex Markovnikov adducts [322].

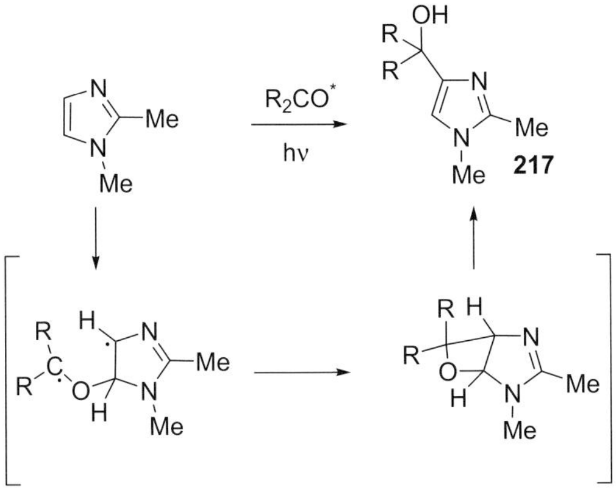

Scheme 10.113

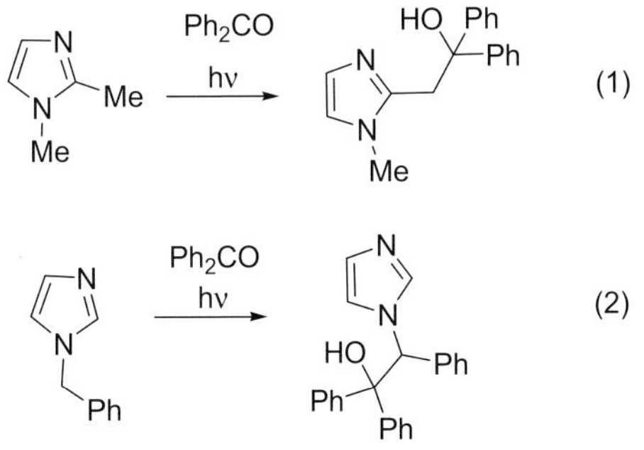

Scheme 10.114

10.5 Derivatives

10.5.1 Dihydro-1,3-Azoles

The most important partially saturated derivatives of 1,3-azoles are 4,5-dihydroazoles, and several methods have been developed to obtain substituted derivatives both achiral and chiral. The impetus to this development was surely given by the wide use oxazolines (another common name for 4,5-dihydroxazoles) have found in homogeneous catalysis [323]. For this reason, synthetic approaches to 4,5-dihydrooxazoles are the blueprint of this section. The synthesis of 4,5-dihydroimidazoles and -thiazoles, which also have important application in catalysis [324] and natural compounds chemistry, are discussed in comparison with 4,5-dihydrooxazoles.

The most important and widely used approach is the cyclic dehydration of a β-hydroxyamide derivative. Several reagents are commonly used to obtain this dehydration and new ones are developed every year. Simple heating of the amide can in some cases afford the final oxazoline derivative but a high temperature is generally required and efforts have been devoted to using milder reaction conditions. All known reactions can be divided into two main classes [Scheme 10.115 (1) and (2)]: (i) reactions that give retention of configuration on the stereocenter that bears the hydroxyl group; (ii) reactions that cause inversion of configuration on the same stereocenter [325]. The first class of reactions resembles the biosynthetic process.

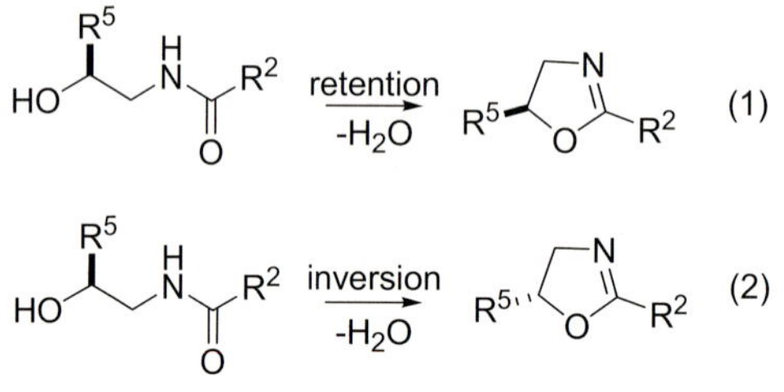

Scheme 10.115

In the first class the common mechanism is the activation of the carbonyl group of the amide moiety to nucleophilic attack of the hydroxyl group, followed by elimination of water. The activation is performed through the use of Lewis or Brønsted acids (Scheme 10.116).

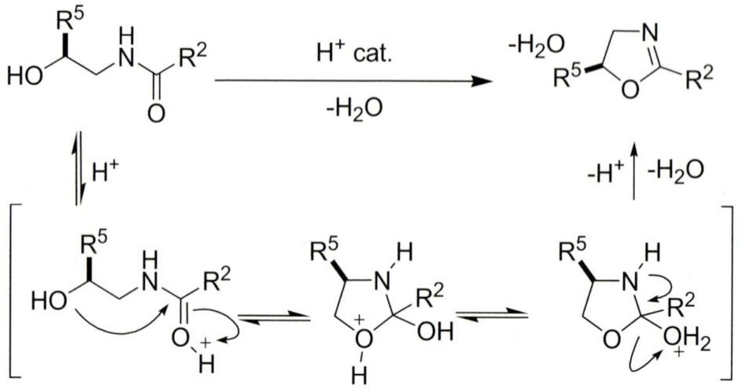

Scheme 10.116

This class of reagents includes $TiCl_4$ [326], TsOH [327], Ph_3PO-Tf_2O [328] and recently also Mo(IV) and Mo(VI) oxides [325].

The second class of reagents is instead based on the transformation of the hydroxyl group into a good leaving group to perform a nucleophilic substitution by the amide oxygen through an S_N2 mechanism. This mechanism is described, using TsCl as reagent, in Scheme 10.117.

This class of reagents includes Martin's sulfrane [329], the Burgess reagent [330], Mitsunobu reagents [331], DAST (diethylaminosulfur trifluoride) [332], polymer-supported TsCl [333] and $SOCl_2$ [334].

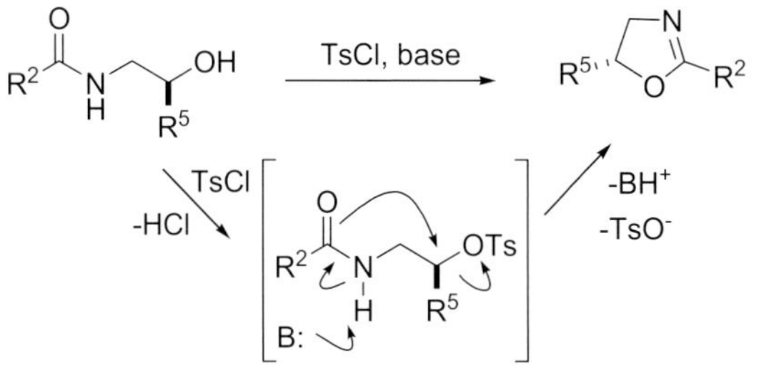

Scheme 10.117

In a similar process, a solid-phase supported synthesis of 4,5-dihydrooxazoles has been performed, transforming the hydroxyl group into a iodide through the use of PPh_3, I_2, imidazole [335].

Several of these methods can be extended to the synthesis of 4,5-dihydrothiazoles. Thiazoles derivatives can be obtained by cyclodehydration of β-hydroxythioamides with $SOCl_2$-Py [336], and with $MsCl/NEt_3$ [337], as well as by $TiCl_4$-induced dehydration of amides derived from vicinal aminothiols [338], cyclization of a serine-derived thiolamide with [339] or without [340] the use of Burgess' reagent, dehydrocyclization of thioamides with deoxofluor or DAST [332, 341] or with Mitsunobu reagents [342], and reaction of aminothiols with carboxylic acids [346].

Concerning 4,5-dihydroimidazoles, *N*-aminoethyl amides dehydrocyclize in the presence of $POCl_3$ [343], and this procedure has been applied also for the solid-phase synthesis of imidazoline **218** (Scheme 10.118) [344].

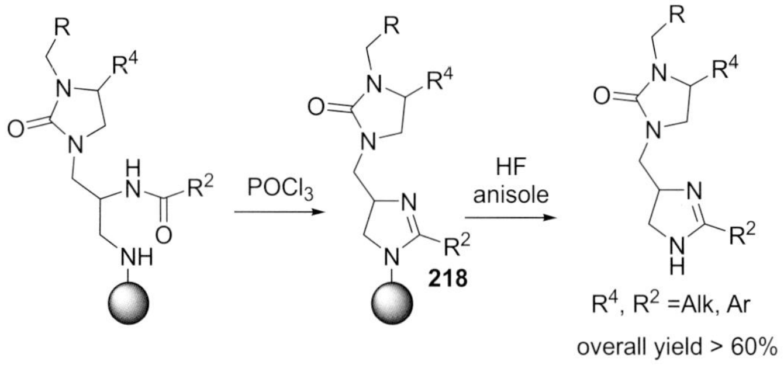

Scheme 10.118

A somewhat related procedure is used to produce imidazoline derivatives. N-hydroxyethylamides are treated with excess thionyl chlorides, or thionyl chlorides followed by PCl_5, to afford N-chloroethylimidoyl chlorides **219**. These intermediates are treated with amines and anilines to produce *N*-chloroethylamidines **220**, which are converted into imidazolines upon workup with aqueous sodium hydroxide (Scheme 10.119) [345].

Direct condensation of carboxylic acids with β-aminoalcohols is quite a drastic procedure but works nicely with substituted aminoalcohols in the presence of acid

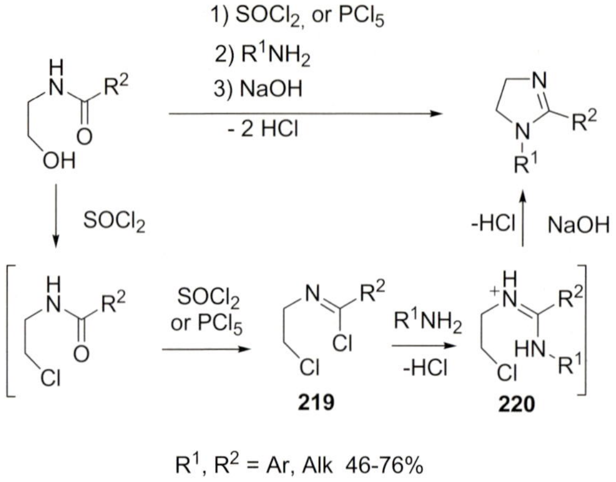

Scheme 10.119

catalysts, such as boric acids, to afford polysubstituted oxazolines such as **221** (Scheme 10.120) [346].

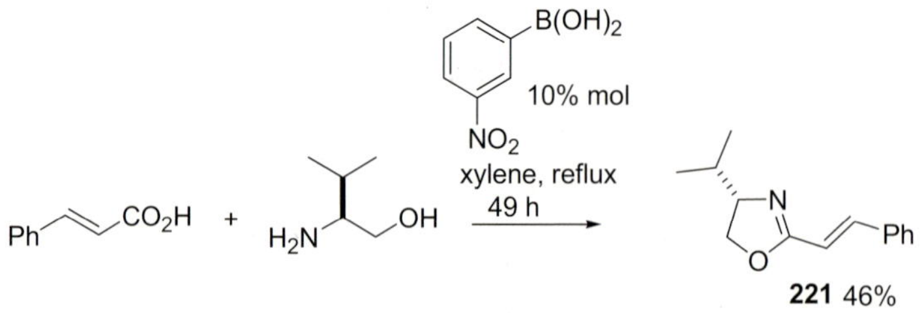

Scheme 10.120

The use of a combination of PPh_3, a base and CCl_4 [347] allows the one-pot condensation and cyclic dehydration of amino acids and β-aminoalcohols, in a process that affords 2-aminomethyl-4,5-dihydrooxazoles **222** (Scheme 10.121) [348].

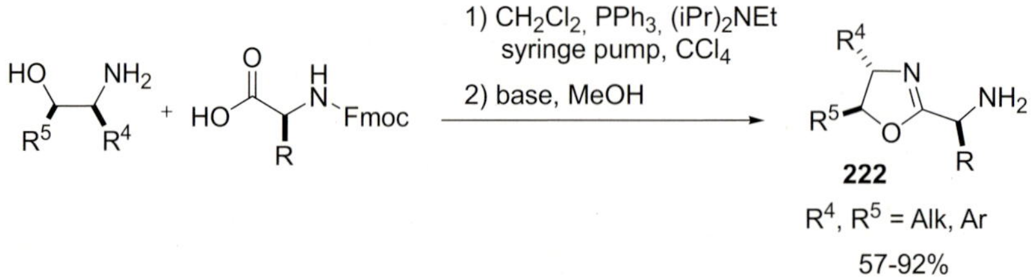

Scheme 10.121

A microwave accelerated two-step, one-pot procedure has been proposed, using 1-acyl benzotriazole, $SOCl_2$ and the aminoalcohol. The $SOCl_2$ was needed to complete the cyclization of the intermediate amide formed. This procedure is also convenient for the synthesis of thiazoline derivatives [349].

Several other derivatives of carboxylic acids can be used for the synthesis of oxazoline. The reaction of esters with vicinal diamines affords the expected imidazolines and, in the same way, also thiazolines and oxazolines. The reaction conditions are refluxing toluene and $Al(Me)_3$ [350].

Vicinal diamines, as well as aminoalcohols and aminothiols, also react with orthoesters in the presence of HCl to afford the corresponding heterocycles. Often the orthoester is the solvent [351].

The cyano group reacts with aminoalcohols in the presence of metal catalysts, such as Cd salts [352] or $ZnCl_2$ [Scheme 10.122 (1)] [353]. The same reaction can be performed using aminothiols to afford thiazoline derivatives [354].

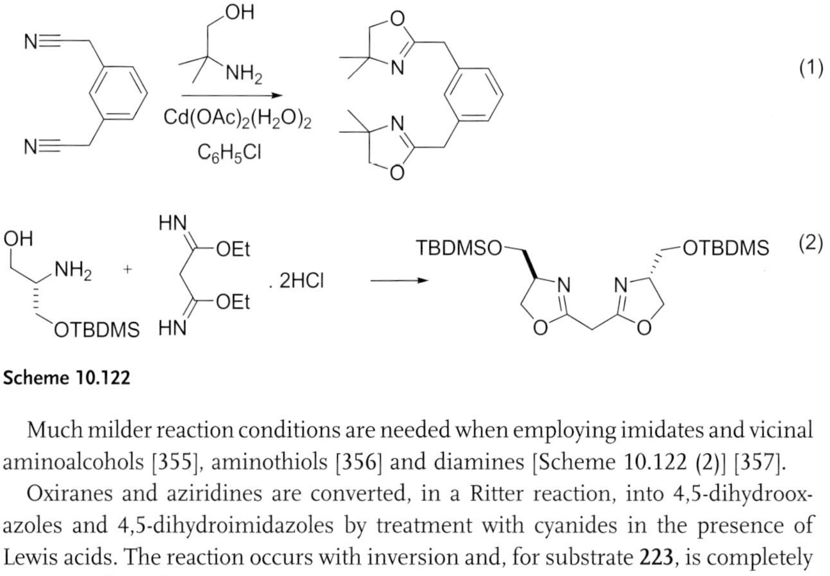

$Cd(OAc)_2(H_2O)_2$, C_6H_5Cl (1)

. 2HCl (2)

Scheme 10.122

Much milder reaction conditions are needed when employing imidates and vicinal aminoalcohols [355], aminothiols [356] and diamines [Scheme 10.122 (2)] [357].

Oxiranes and aziridines are converted, in a Ritter reaction, into 4,5-dihydrooxazoles and 4,5-dihydroimidazoles by treatment with cyanides in the presence of Lewis acids. The reaction occurs with inversion and, for substrate **223**, is completely regiospecific (Scheme 10.123) [358].

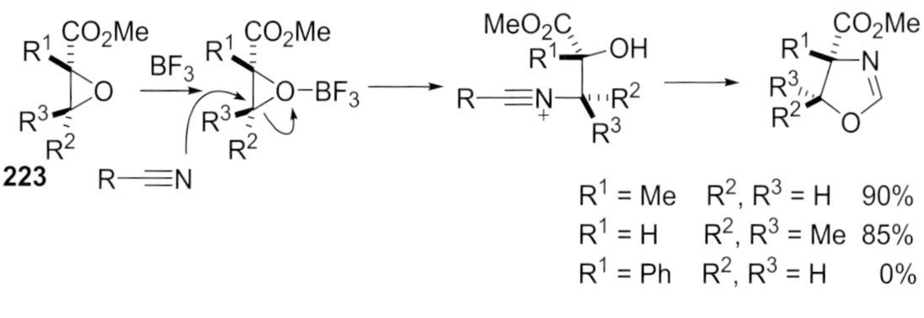

R^1 = Me	R^2, R^3 = H	90%
R^1 = H	R^2, R^3 = Me	85%
R^1 = Ph	R^2, R^3 = H	0%

Scheme 10.123

Analogously the Ritter reaction of enantiopure 2-(1-aminoalkyl)aziridines **224** with different nitriles affords enantiopure tetrasubstituted imidazolines. Again the opening of the aziridine ring takes place with total regio- and stereoselectivity, by the mechanism proposed in Scheme 10.124 [359].

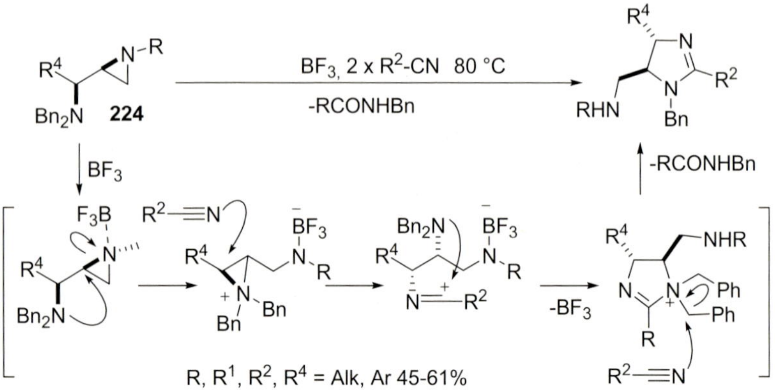

Scheme 10.124

A very general route to 1,3-azolines is represented by the cyclization obtained by the aza-Wittig reaction of substituted azides. The reaction occurs under very mild conditions and the synthesis appears particularly versatile. Triphenylphosphine reacts with the azido group of **225** to afford the corresponding iminophosphorane **226**, which then react with the vicinal ester group to give the ring closure (Scheme 10.125) [360].

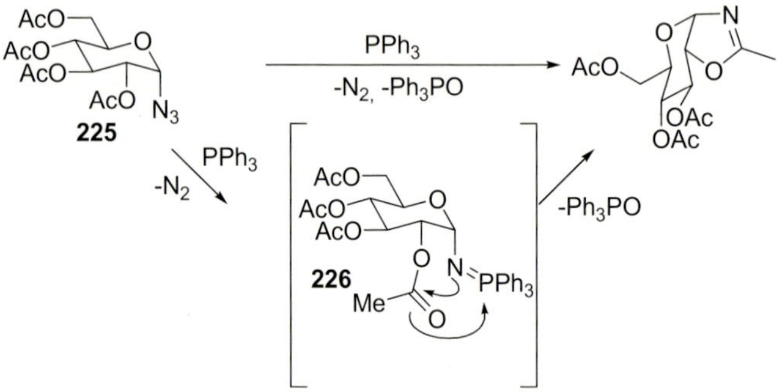

Scheme 10.125

The use of polymer-supported triphenylphosphine makes the purification easier.

The same reaction can be performed using thioester to obtain the corresponding thiazolidine [361].

The equivalent reaction for the synthesis of imidazolidine is more limited since it is necessary to use activated amides to obtain the ring closure. For example, azido imide **227** reacts to give the corresponding imidazolinone **228** (Scheme 10.126) [362].

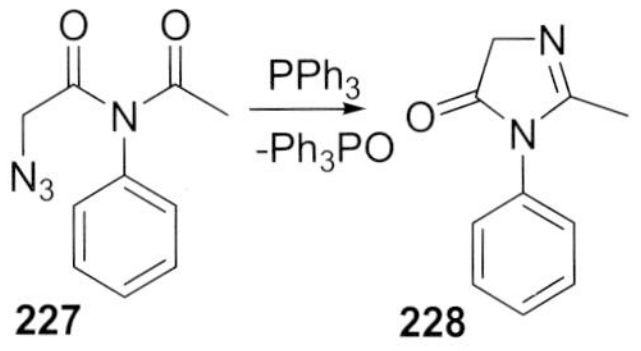

Scheme 10.126

A more general route to imidazoline is obtained by treating the intermediate iminophosphorane **229** with an acyl chloride and subsequently with NH_3. Imidoyl chloride **230** is the proposed reactive intermediate (Scheme 10.127) [363].

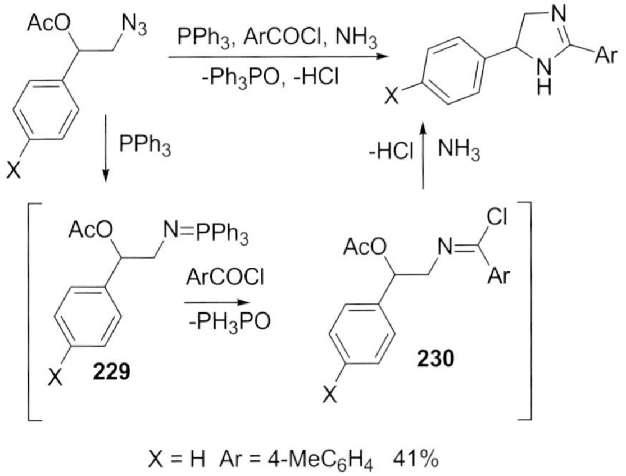

Scheme 10.127

Propargylamides **231** are transformed into the correspondent 5-carboxymethylene-4,5-dihydrooxazoles **232** in a Pd-catalyzed process in the presence of CO, an alcohol and molecular oxygen (Scheme 10.128) [364].

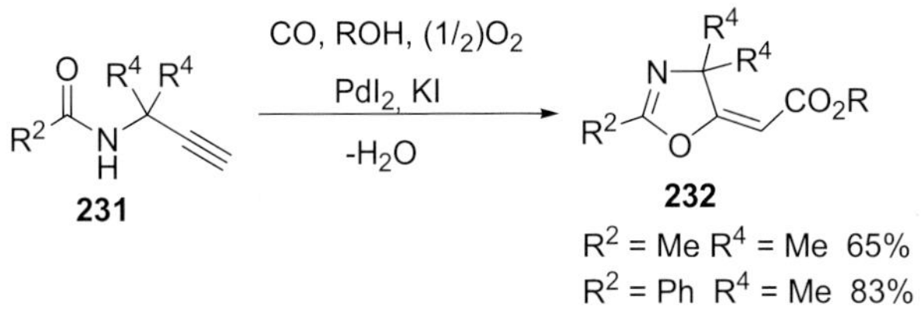

Scheme 10.128

A multicomponent reaction of oxazolones with aldehydes, primary amines and TMSCl affords diastereoselectively highly substituted imidazolines. The proposed mechanism proceeds through a 1,3-dipolar cycloaddition reaction of the mesoionic intermediate **233** (Scheme 10.129) [365].

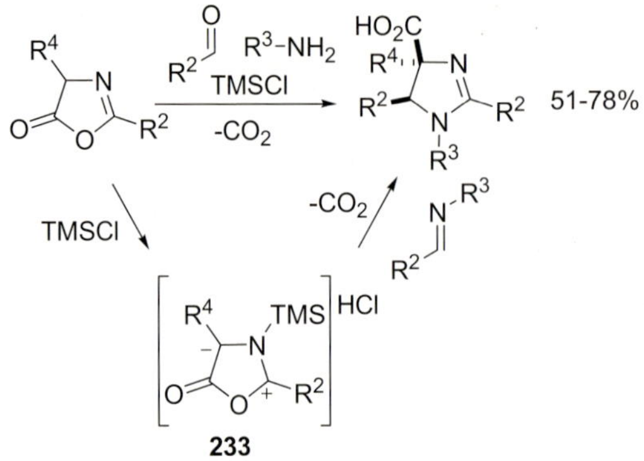

Scheme 10.129

Another multicomponent reaction (MCR) has been developed using aldehydes, primary amines and isonitriles with AgOAc as catalyst. This MCR probably involves an aldol-type addition of the isocyanide **234** to the *in situ* generated imine followed by ring closure of the intermediate **235** (Scheme 10.130). The role of AgOAc is to increase the acidity of the proton α to the isonitrile group through complexation [366].

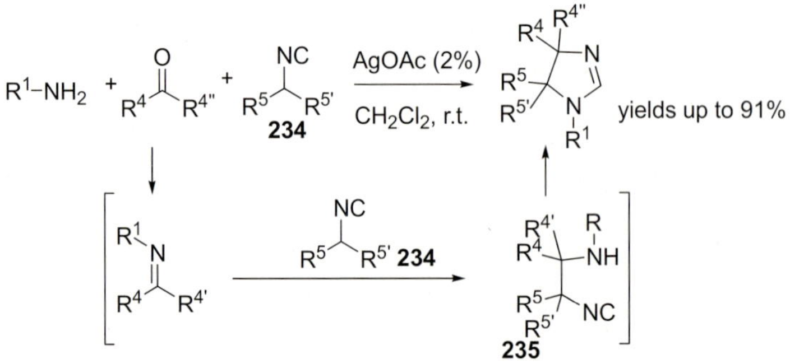

Scheme 10.130

Isonitrile derivatives have also found application in reactions with N-sulfonylimines, as **236**, catalyzed by Ru or Gd complexes to afford stereoselectively the corresponding *N*-sulfonyl-2-imidazolines **237** (Scheme 10.131).

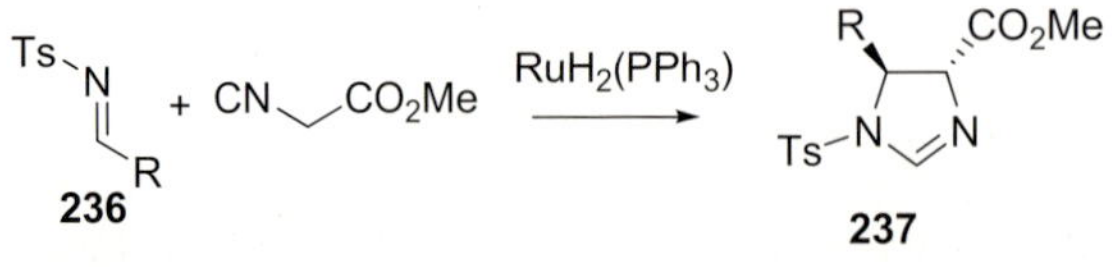

Scheme 10.131

The salt derived from Ph_3PO/Tf_2O is effective in the dehydrocyclization of N-aminoethyl amides obtained by condensation of amino acids. This procedure affords imidazolines that still contain the amino acid functionality and preserves the stereochemical integrity [367]. The same procedure has been used by the same authors for the biomimetic synthesis of thiazoline starting from cystein-containing dipeptides [368].

N,N-Dichloro-*o*-nitrobenzenesulfonamide (2-$NsNCl_2$) is an effective electrophilic nitrogen source for the direct diamination of α,β-unsaturated ketones in the presence of acetonitrile, without the use of any metal catalysts (Scheme 10.132) [369].

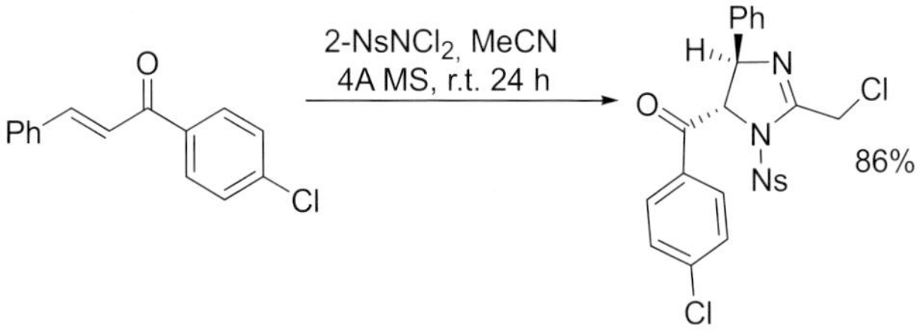

2-$NsNCl_2$, MeCN
4A MS, r.t. 24 h
86%

Scheme 10.132

Peculiar to the synthesis of thiazolines is the phosphine-induced annulation of thiolamides and 2-alkynoates. The proposed mechanism, depicted in Scheme 10.133, is based on the bielectrophilic character imparted by the phosphine to the triple bond [370]. The addition of tributylphosphine to the polarized triple bond of **238** induces, after migration of the double bond, the formation of the ylide **239**, which undergoes cyclization to afford thiazoline **240**.

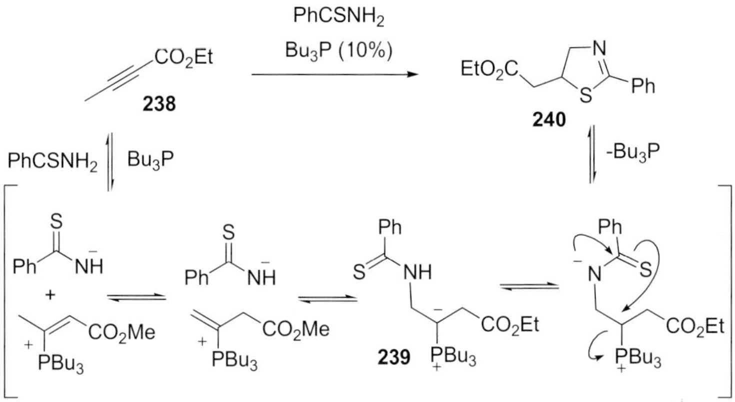

$PhCSNH_2$
Bu_3P (10%)
238 **239** **240**
$PhCSNH_2$ Bu_3P $-Bu_3P$

Scheme 10.133

Direct oxazoline–thiazoline conversion can be realized by thiolysis of oxazolines with H_2S in methanol/triethylamine, followed by cyclodehydration with Burgess reagent **55** (for structure see Scheme 10.20). This protocol is high yielding, chemoselective and essentially free of racemization for C(5)-unsubstituted and trans-4,5-disubstituted peptide oxazolines (Scheme 10.134) [371].

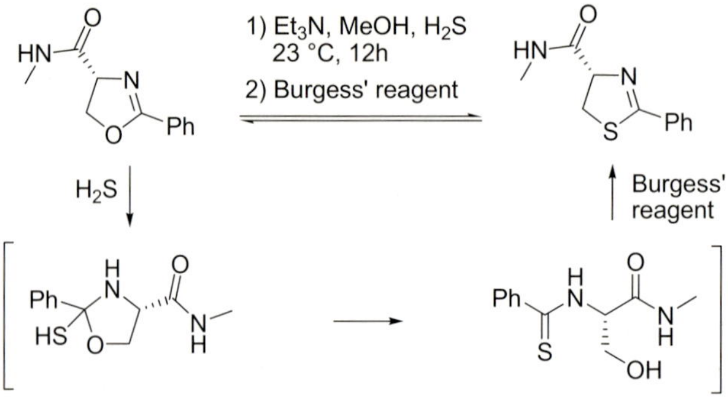

Scheme 10.134

A very mild transformation of secondary and tertiary amides into thiazolines has been described using iminium triflates. The reaction proceeds at very low temperature and is tolerant towards several functional groups [372]. The iminium triflate **241** undergoes nucleophilic addition–elimination by 2-aminoethanthiol to afford **242**, which subsequently cyclizes with loss of the secondary amine to give **243** (Scheme 10.135).

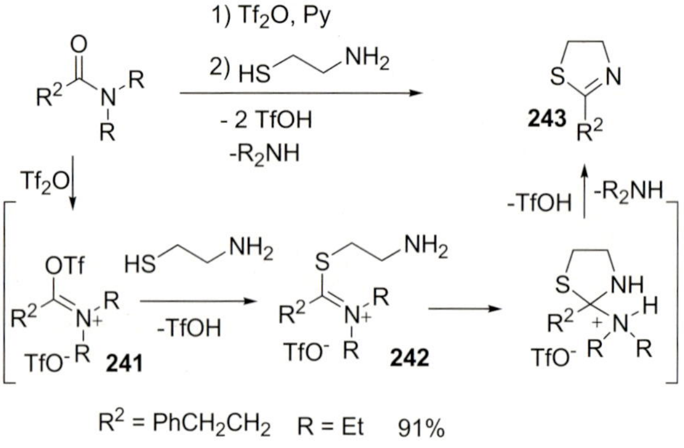

Scheme 10.135

10.5.2
Benzo-1,3-Azoles

The synthesis of this class of compounds presents peculiar approaches, since the structure forces the syntheses towards cyclization reactions on a preformed 1,2 disubstituted benzene ring. Extremely rare are synthesis in which the benzene ring is formed during the synthesis. However, this implies that many synthetic approaches are common for benzoimidazoles, benzoxazoles and benzothiazoles. For this reason all the procedures presented here are applied to the three different kinds of

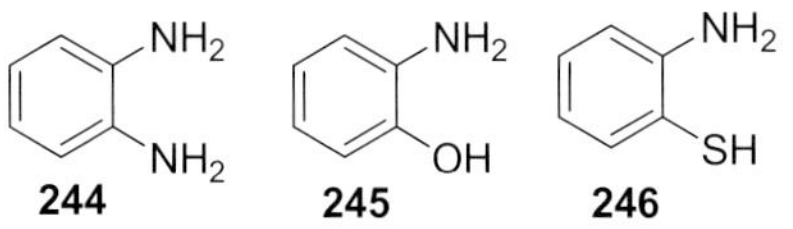

Figure 10.11 Starting materials used for the formation of benzo-1,3-azoles.

derivatives. Special syntheses that concern just one heterocycle are presented together with similar generic procedures.

The starting material are usually the ortho-substituted anilines **244–246** (Figure 10.11).

These compounds, as well as their derivatives bearing other substituents on the benzene ring, are generally available and stable. Only 2-aminobenzenethiol and its derivatives are not very stable since they are oxygen sensitive. To circumvent this problem they are often used in the form of derivatives such as acid salts, alkaline salts, zinc salts or disulfides [373].

A general synthetic approach is the intramolecular nucleophilic addition of the heteroatom substituent onto an imine moiety followed by oxidation to afford the aromatic derivative. The imine can be used as starting material but often the reaction is performed on a mixture of the 2-substituted aniline **247** and the aldehyde. The saturated intermediate **248** is subsequently oxidized to afford the final benzo-derivative **249** (Scheme 10.136).

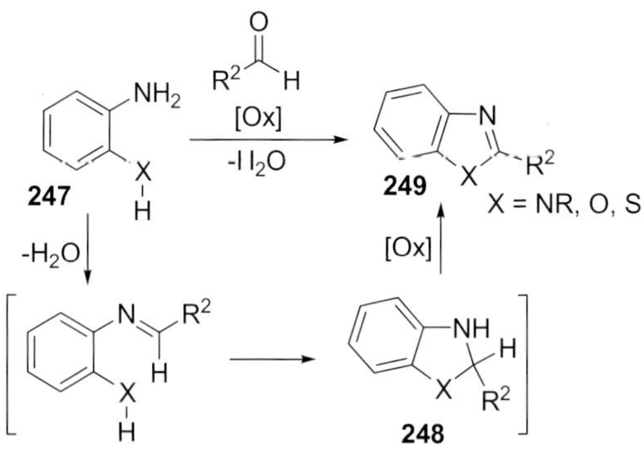

Scheme 10.136

Several oxidants have been used, such as DDQ or the simple 1,4 benzoquinone [374, 375], MnO_2 [376], $Mn(OAc)_3$ [377], NBS [378], Ag_2O [379], and oxone [380]; however, reactions in which the mixture of reagents is heated at high temperature in DMSO [381], or even under solvent-free conditions with Yb catalyst have also been described [382]. Remarkable is the reported synthesis of substituted benzoxazoles by heating the aniline and the aldehyde in xylenes in the presence of activated carbon is; the oxidant is presumed to be atmospheric oxygen [383]. Benzimidazoles can be obtained at room temperature by condensing *ortho*-phenylenediamine and aldehydes with silica-supported thionyl chloride [384]. Examples of reactions performed under microwave irradiation, optimizing the time and the yields [385], have been reported, as have several examples of reactions applied on the solid phase for the production of libraries [374, 376, 386, 387] as well as for the synthesis of libraries in solution phase [388].

The most important synthetic methods for the preparation of a wide range of benzoazoles is the condensation of 2-substituted anilines with carboxylic acids or derivatives (Scheme 10.137). Benzimidazole can be made in 80% yield by merely standing a mixture of *o*-phenylenediamine and formic acid at room temperature for 5 days; however, at 100 °C the process takes only 2 h and it is applicable to a wide range of 2-substituted benzimidazole. Careful choice of reaction conditions is, however, essential to obtain good yield for each substrate [389]. The most widely used conditions (Phillip's method [390]) involve heating the reagents in the presence of hydrochloric acid, usually around 4 M concentration. However, the range of reaction conditions that has been used is wide: from merely heating the diamines with a carboxylic acid [391], to heating in the presence of acids such as HCl [392], PPA [393], and $POCl_3$ [394].

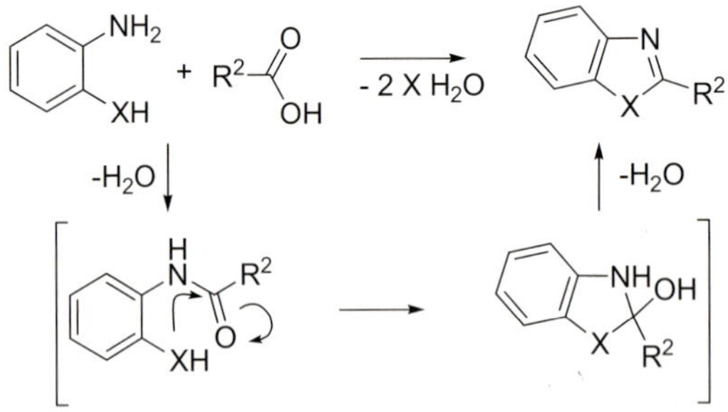

Scheme 10.137

If the *o*-diaminoarene has one of the amino groups substituted by an alkyl or aryl group, 1-substituted benzimidazoles are formed [395].

As described in Section 10.5.1 the complex $Tf_2O/POPh_3$ acts as a dehydrating agent favoring the formation of the benzoimidazole **250** in common solvents, giving high yields in 30–60 min at room temperatures (Scheme 10.138) [344].

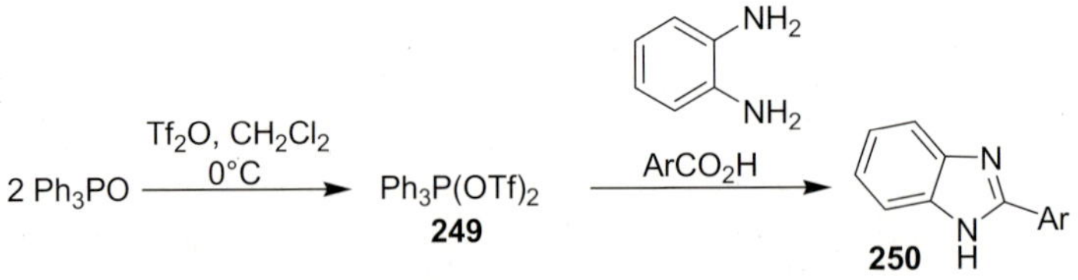

Scheme 10.138

A range of acid derivatives can substitute for carboxylic acids (Scheme 10.139): esters [397], orthoesters [398], nitriles [399], imidates [400], acid chlorides (including

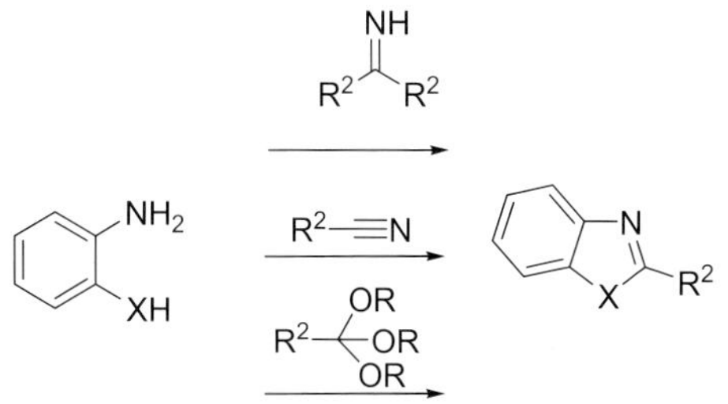

Scheme 10.139

phosgene) [401] and anhydrides [402]. Concerning the use of orthoester the reaction conditions may vary considerably depending on the reactivity of the 2-substituted aniline. In many cases an excess of orthoester is used – it is often the solvent – and the reaction is conducted in the presence of catalytic TsOH [402], of a base [404] or of KSF clay [405].

A few examples of the synthesis of benzimidazoles using amides have been reported in the literature. Usually, these reactions occur at high temperature [406].

A procedure strictly related to the previous method is the reaction of N-acyl derivatives of **244–246**, which undergo thermal dehydration to afford the corresponding benzoazoles (Scheme 10.140) [407–410].

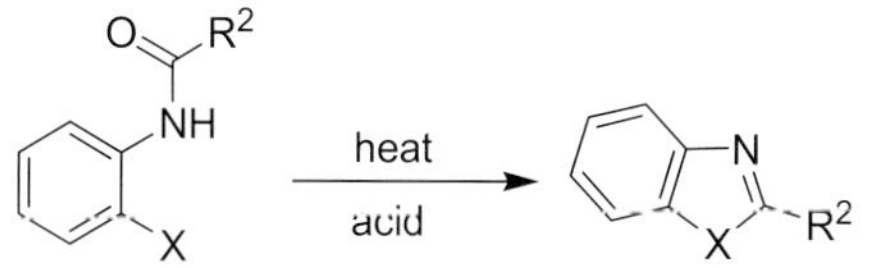

$X = NHR^1$, OH, SH $\qquad R^2 = CF_2Bn$ X = OH 73%

Scheme 10.140

The cyclization may occur by simple uncatalyzed thermolysis or with aqueous or ethanolic acid as well as phosphoryl chloride. Recently, Mitsunobu reaction conditions were also used [411].

Since the amide intermediate **252** is formed, the Beckmann rearrangement of *o*-hydroxybenzophenone oxime **251** leads directly to benzoxazole **253** (Scheme 10.141) [412].

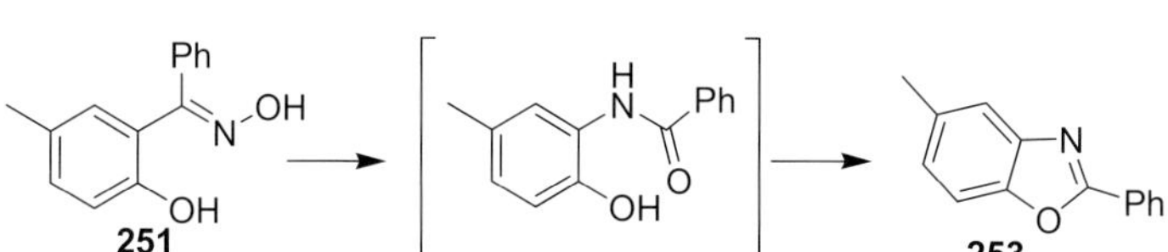

251 **252** **253**

Scheme 10.141

In situ reduction (tin and acetic or hydrochloric acids, hydrogen/palladium carbon, hydrogen/Raney nickel/hydrochloric acid) of *o*-nitroacylaminoarenes is followed by cyclization to afford 1,3-benzoimidazoles usually in good yields (Scheme 10.142) [413].

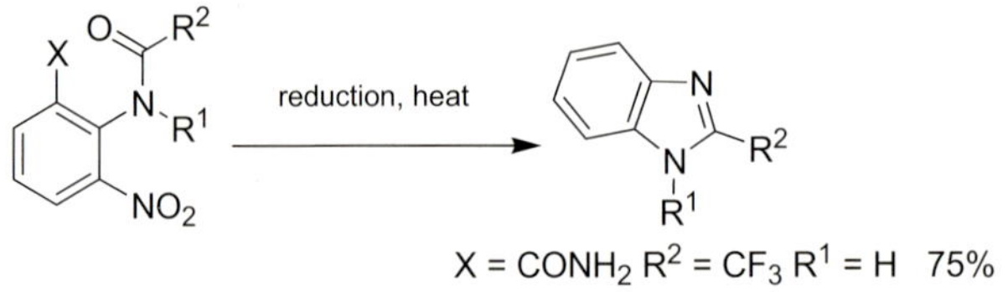

reduction, heat

$X = CONH_2$ $R^2 = CF_3$ $R^1 = H$ 75%

Scheme 10.142

This general method can be applied to the synthesis of 2-unsubstituted benzimidazoles by cyclization of an *o*-formamidoarylamine and to 1-aminobenzidimazoles when *o*-acylaminophenylhydrazines are the substrates.

An example of a MCR can be included in this methodology since the reaction of *o*-phenylenediamine derivative **254** affords, in the Ugi reaction, compound **255**, which upon deprotection cyclizes to afford benzimidazole derivatives **256** (Scheme 10.143) [414].

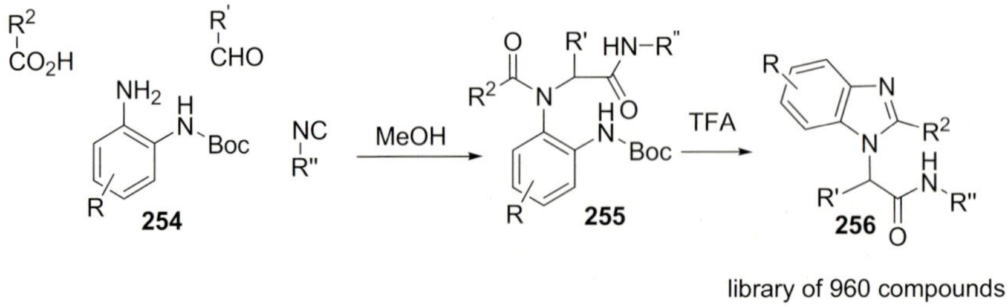

library of 960 compounds

Scheme 10.143

o-Nitrochlorobenzene easily undergoes nucleophilic substitution reactions with sulfurated reagents. The adducts can successively cyclize, after reduction of the nitro group, to afford the corresponding benzothiazole derivative (Scheme 10.144).

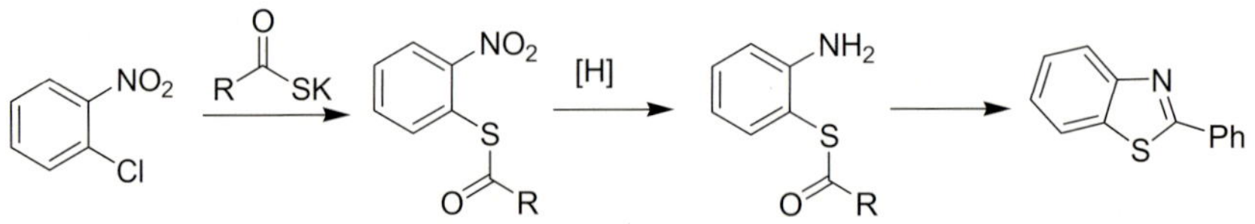

Scheme 10.144

O,N- [415], N, N′-, S,N- [416] diacylated compounds can also cyclize under different conditions, ranging from simple heating at 200 °C [417] to microwaves irradiation of mixtures with montmorillonite K10 [418].

A modification has been introduced to prepare 1-acetyl-2-methylbenzimidazole (**257**) in quantitative yield (Scheme 10.145) [419].

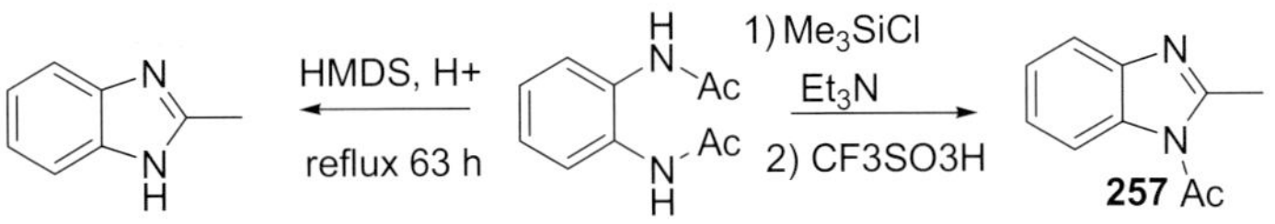

HMDS, H+
reflux 63 h
1) Me_3SiCl
Et_3N
2) CF3SO3H
257

Scheme 10.145

Simple aniline derivatives, such as thioanilides (Jacobson method) [420] or arylmonothiocarbamates (Jacobson–Hunter method) [421], cyclize to afford the corresponding 2-substituted benzothiazole derivative in the presence of potassium ferricyanide (Scheme 10.46).

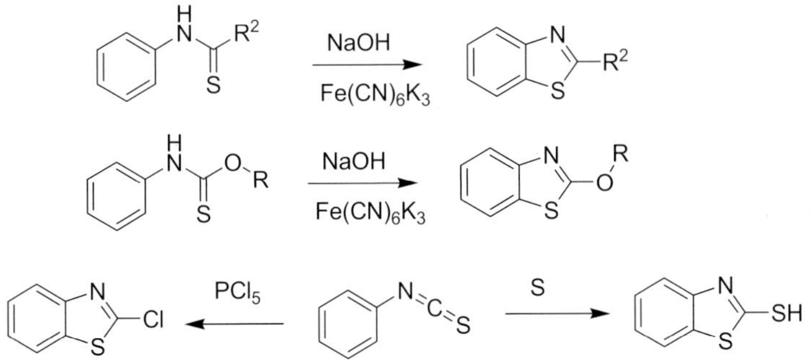

NaOH
$Fe(CN)_6K_3$
PCl_5
S

Scheme 10.146

Aryl isothiocyanate can be cyclized by heating with PCl_5 to 2-chlorobenzothiazoles (Hunter's method) or with sulfur to 2-mercaptobenzothiazoles.

o-Halothioanilides undergo ring closure, presumably through an intermolecular aromatic nucleophilic substitution, under basic conditions [422, 423] in the presence of catalysts. The thioamide can also be formed *in situ* (Scheme 10.147) [424].

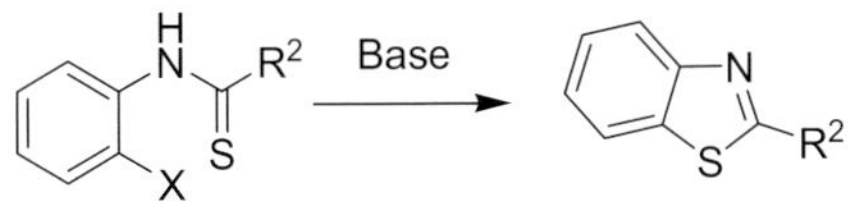

Base

Scheme 10.147

A novel palladium-catalyzed carbonylation of iodobenzene has recently been linked to base-induced coupling and cyclization with *o*-phenylenediamine, to give 2-arylbenzimidazoles without having to use an aryl carboxylic acid (Scheme 10.148).

Provided that bases with pK_a values around 6.6 are used, the yields of 2-arylbenzimidazole lie in the range 70–98%. This route is tolerant of various functional

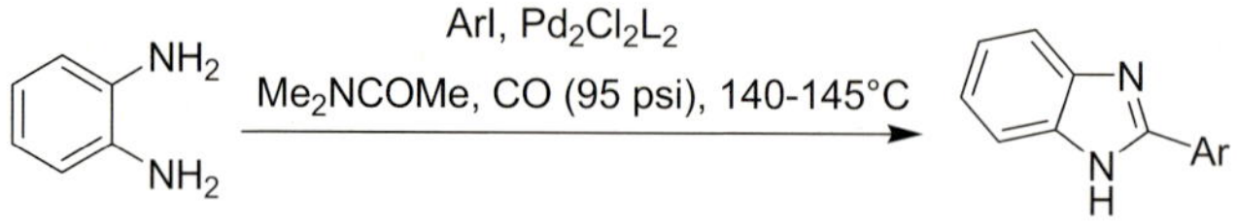

ArI, $Pd_2Cl_2L_2$

NH_2 NH_2 Me_2NCOMe, CO (95 psi), 140-145°C N N H Ar

Scheme 10.148

groups and complements the classical approaches where the required benzoic acid is not readily available [425].

The presence of a fused aromatic ring on the five-membered ring induces some modification of the reactivity.

As expected from the nature of the heterocyclic portion, nucleophilic reaction concerns essentially the five-membered ring in the lone reactive position, that is, 2. Conversely, electrophiles attack the benzenoid ring. The fused aryl ring appears to exhibit less aromatic stability than the hetero-ring, as evidenced by the easy oxidation of benzimidazole to imidazole-4,5-dicarboxylic acid, and by its catalytic reduction over platinum oxide to give 4,5,6,7-tetrahydrobenzimidazole.

Benzimidazole is subject to N-alkylation and N-acylation as imidazoles, although the benzene ring reduces the reactivity as well as the basicity (Section 10.5.1). Furthermore, the electron-withdrawing nature of the benzene ring increases the facility with which nucleophilic substitution occurs at C2.

Since most electrophilic substitution reactions involve Lewis acids it is the benzoazolium species that is involved and in this substrate it is the benzene ring that is the more reactive.

Only a few electrophilic substitutions take place at the C2 of benzoazoles, for example acylation in the presence of triethylamine (compare reactivity of 1,3-azoles, Section 10.4).

Sulfonation with oleum at 100 °C affords 4-, 6- and 7-benzothiazole sulfonic acids in the ratio 70 : 25 : 5%, respectively, while bromination at 100 °C in acetic acid gives 4,6-dibromobenzothiazole. Nitration of 2-methylbenzoxazole gives a 4 : 1 mixture of 6- and 5-nitro derivatives.

10.5.3
Tetrahydro-1,3-Azoles

Imidazolines, oxazolidines, and thiazolidines are easily obtained by reactions of vicinal diamines, aminoalcohols and aminothiols, respectively, with carbonyl compounds (Scheme 10.149).

The typical condensation is conveniently conducted in boiling benzene with continuous removal of water [426] as described for the synthesis of oxazolidine **258**, which was then transformed into the Garner aldehyde **259** (a useful synthetic intermediate) (Scheme 10.150) [427].

However, the most recent syntheses can be achieved at rt. For example, imidazolidine **262** is obtained by simply reacting at rt the aldehyde **260** with the vicinal

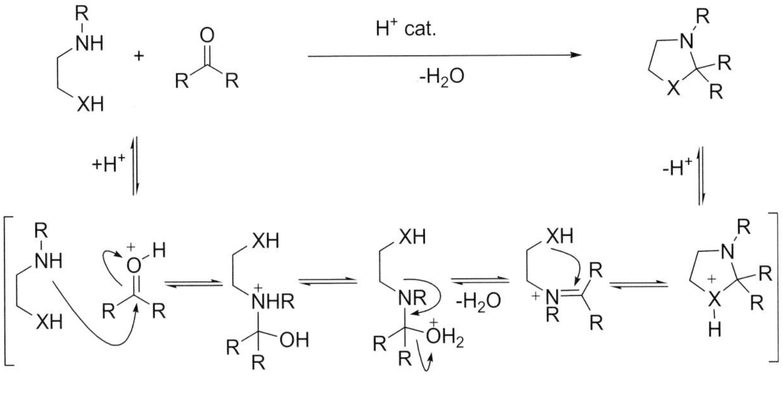

Scheme 10.149

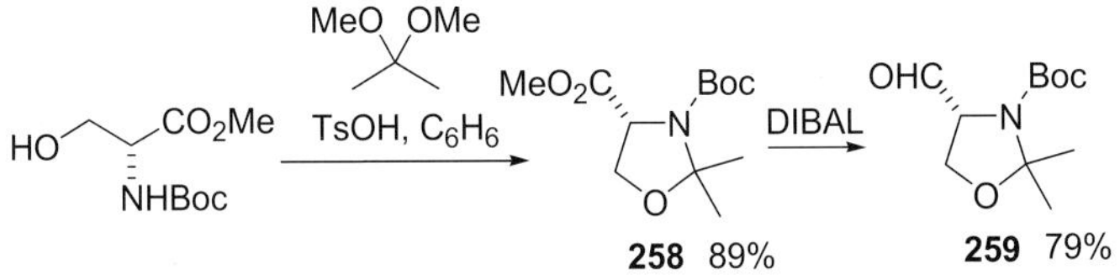

Scheme 10.150

diamine **261** in the presence of molecular sieves [Scheme 10.151 (1)] [428]. In some cases less reactive substrates, such as **263**, may require activation, for example, the use of montmorillonite KSF [Scheme 10.151 (2)] [429].

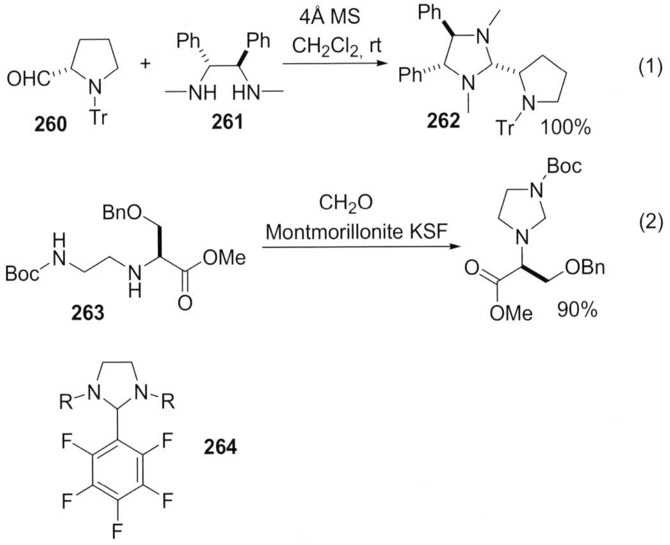

Scheme 10.151

Oxazolidines have also been obtained by these procedures [430], although an application to the production of a library of 96 oxazolidines through a parallel solid-phase synthesis required $MgSO_4$ as dehydrating agent and a temperature of 60 °C [431].

Particularly interesting is the synthesis under acid catalysis of substituted imidazolidines bearing a perfluorinated phenyl ring on C2, such as **264** (Scheme 10.151). On heating at temperatures ranging from 65 to 144 °C these compounds afford the corresponding carbene even the in absence of any transition metal [432].

The easy nucleophilic substitution of a benzotriazolyl group by C-nucleophiles allows the ready synthesis of unsymmetrical imidazolines. The process proceeds through Mannich reactions between 1,2-ethanediamines, benzotriazole (BtH) and formaldehyde at rt to produce imidazolidine **265**. Finally, reaction of **265** with different nucleophiles affords the products of general structure **266** (Scheme 10.152) [433].

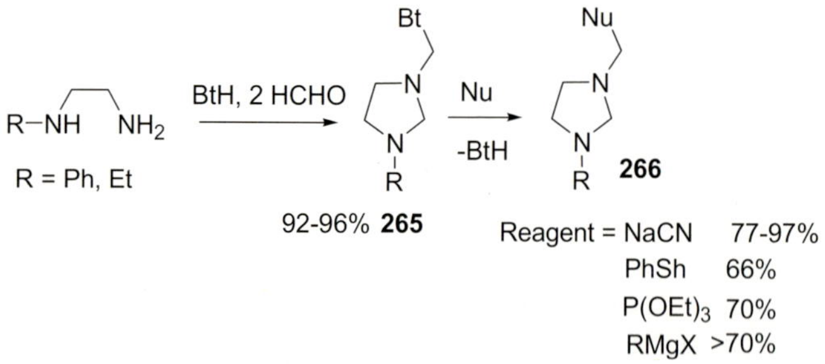

Scheme 10.152

Simple thiazolidine has been obtained by reacting at rt cysteamine hydrochloride with formaldehyde. Under these reaction conditions cysteamine can be replaced by aziridine and hydrogen sulfide [434]. Diastereomeric mixtures have obtained by reacting N-protected amino glyoxals **267** with L-cysteine methyl ester (**268**) (Scheme 10.153) [435].

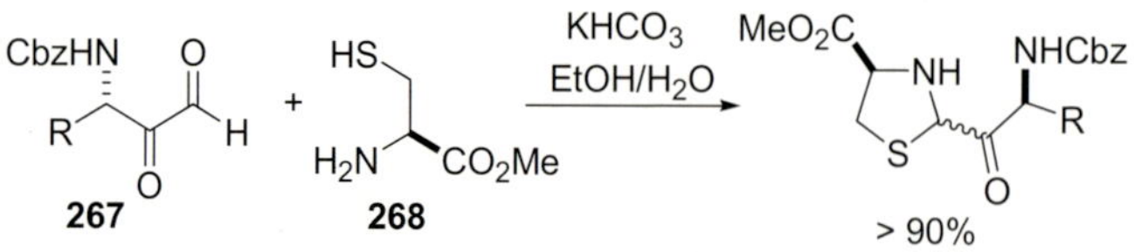

Scheme 10.153

Several derivatives of carbonyl compounds can be used, such as acetals [436], hemiacetals [437], Schiff bases and orthoesters. The reaction of N-tosylaminoalcohols with orthoformates affords 2-methoxy-oxazolidines **269** that react with allyltrimethylsilane or trimethylsililenolether at 0 °C in a zinc chloride or trimethylsilyl triflate-catalyzed reaction to afford new oxazolidine derivatives [438]. The kinetic derivative

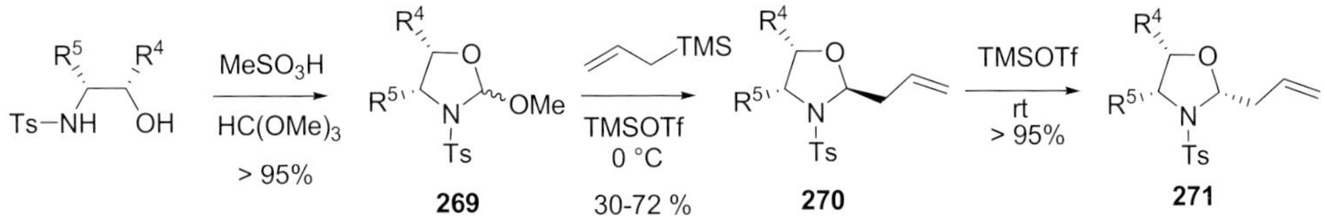

Scheme 10.154

270 isomerizes to the all-cis oxazolidine **271** when treated at room temperature with TMSOTf (Scheme 10.154).

Similar reactivity is found with titanium enolates [439].

Similar substrates, in the enantiopure form, have been obtained in the solid phase by reaction of enantiopure aminoalcohols with solid-supported aldehydes and subsequent reaction with a sulfonyl chloride [440].

A new procedure for the formation of oxazolidines derived from ketones has been reported. It is based on the use of isopropoxytrimethylsilane and a catalytic amount of trimethylsilyl trifluoromethanesulfonate and has found application in the synthesis of a polymer-supported oxazolidine aldehyde [441].

A vicinal aminoalcohol moiety is a perfect starting point for the construction of an oxazolidine ring; however, if the oxazolidine ring is built up in a different manner, hydrolysis of the ring is an efficient way to obtain aminoalcohols in stereoselective fashion.

Intramolecular conjugate addition of the N-hydroxymethyl moiety onto an α,β-unsaturated ester affords the corresponding oxazolidine **272** with high stereoselectivity. Finally, the oxazolidine ring was cleaved (Scheme 10.155) [442].

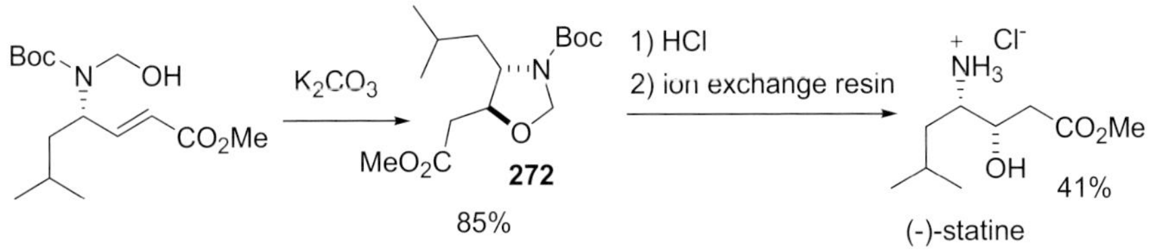

Scheme 10.155

Recently, a novel MCR has been developed using four components for the synthesis of highly substituted 1,3-oxazolidines. The reaction may be catalyzed by transition metals; however, the use of microwaves can efficiently substitute the metal catalysis (Scheme 10.156) [443].

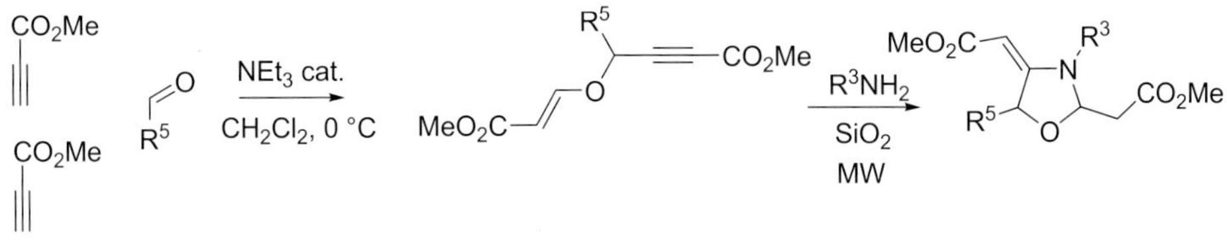

Scheme 10.156

10.5.4
Alkyl-1,3-Azoles

The synthesis of alkyl-1,3-azoles is described throughout Section 10.3, for this reason this short section will deal only with the most characteristic reactivity of the alkyl residues linked to the heteroaromatic nuclei.

Alkyl groups attached to heterocyclic systems undergo many of the same reactions as those on benzenoid rings. For example, free radical bromination with NBS is often performed on these substrates. Bromination with NBS of 2-alkylthiazoles affords α-dibromothiazoles in good yields. The hydrolysis of these compounds leads to 2-acylthiazoles [444]. The side chains can, under certain circumstances, be oxidized. For example, methyl thiazoles with SeO_2 give thiazolecarbaldehydes but oxazole derivatives cannot be easily oxidized since the oxazole itself is reactive towards strong oxidants. However, it is stable to the Sharpless oxidation conditions with certain precautions [445].

In addition to these reaction, alkyl groups in the 2-positions of imidazole, oxazole and thiazole rings show reactivity that results from the easy loss of a proton from the carbon atoms of the alkyl group adjacent to the ring [compare Scheme 10.1 (2)]. Very strong bases, such as sodamide, LDA or butyllithium convert 2-methyl-oxazole and -thiazole and 1,2-dimethylimidazole essentially completely into the corresponding anions, although this transformation is not always straightforward since it is very sensitive to reaction conditions and the nature of the substituents [446].

Butyllithium reacts with 1,2-dimethylimidazole at −80 °C to lithiate the 2-methyl group, but at higher temperatures some 5-metalation also occurs [447], while treatment with LDA at −78 °C gives 84 % of 2-methyllithiation and 18% of 5-lithiation [448]. Much better control was obtained with 1-dimethylaminomethyl-2-methyl-imidazole, which was converted by butyllithium into the stabilized anion **273**, which reacted with benzyl chloride to form **274** (Scheme 10.157) [449].

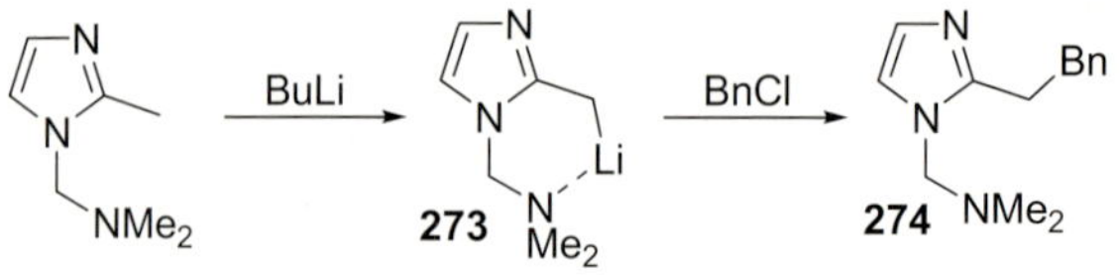

Scheme 10.157

BuLi reacts at −78 °C with 2-methylthiazole to give a mixture of 2-lithiomethyl and 5-lithio derivatives in a ratio of 1 : 4 [450], while when the 2-position is blocked 4-or 5- methyl groups can be lithiated [451]. These anions all react readily even with mild electrophilic reagents; thus the original alkyl groups can be modified through alkylation, acylation, carboxylation and reaction with aldehydes [452].

Extremely useful are the reactions of alkyl-1,3-azoles in which traces of the reactive anion is involved.

In aqueous or alcoholic solutions, many 2-alkyl-1,3-azoles react with bases to give traces of anions. With suitable electrophilic reagents these anions undergo reasonably rapid and essentially non-reversible reaction. 2-Methyl and 2-ethylbenzothiazoles condense with aromatic aldehydes at room temperatures in 50% aqueous sodium hydroxide under phase-transfer catalysis conditions to afford secondary carbinols [453], while 2-methyl thiazole heated at 150°C with $ZnCl_2$ and benzaldehyde gives the styryl derivatives. To confirm the peculiar reactivity of the 2-alkyl group, neither 4- nor 5-methyl thiazole undergo such condensation.

1-Benzyl-2-methylimidazole **275** reacts with benzoyl chloride in the presence of a tertiary amine to give the phenacyl derivative **276**, but with 1-benzyl-5-methylimidazole the product is the 2-benzoyl derivative due to the substitution at C2 (Section 10.4.1.2) [454]. The reaction has also been extended to 2-methylimidazole [Scheme 10.158 (1)] [455].

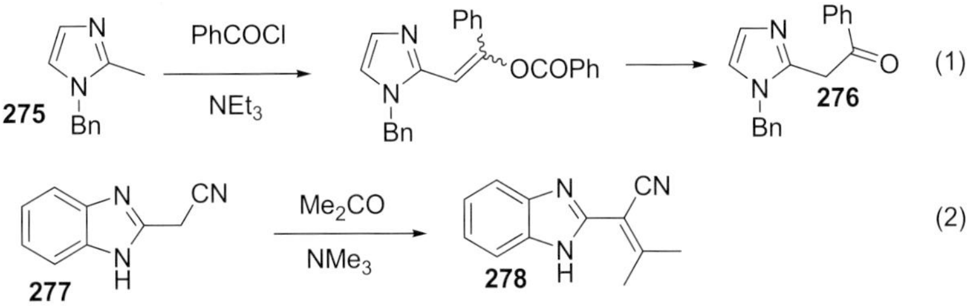

Scheme 10.158

The presence of other activating group makes the condensation reaction even easier. In the presence of trimethylamine, 2-cyanomethylbenzimidazole (**277**) condenses with acetone to give the unsaturated derivative **278** [Scheme 10.158 (2)] [456].

10.5.5 Quaternary 1,3-Azolium Salts

In the past literature methods for the synthesis of azolium and azolinium salts have generally focused on N-alkylation reactions (Section 10.4.1.1). However, these reactions are limited to reactive halides and are inappropriate for introduction of chiral substituents. Since this class of compounds has very important applications, new general synthesis of them are of substantial interest. In this context, imidazolium salts are precursors of carbene ligands used for the synthesis of metathesis catalysts [457] and have been used also as ionic liquids (Scheme 10.159) [458].

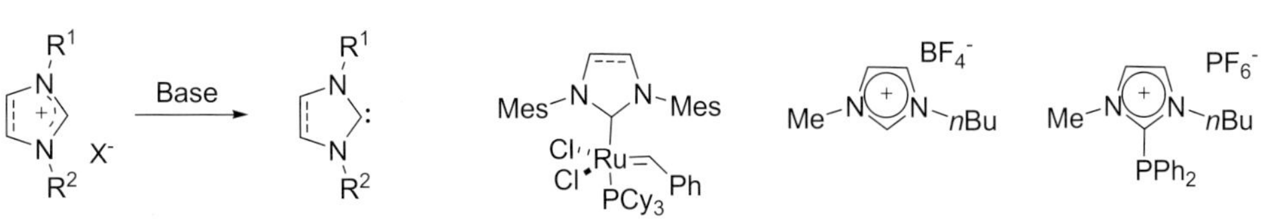

Scheme 10.159

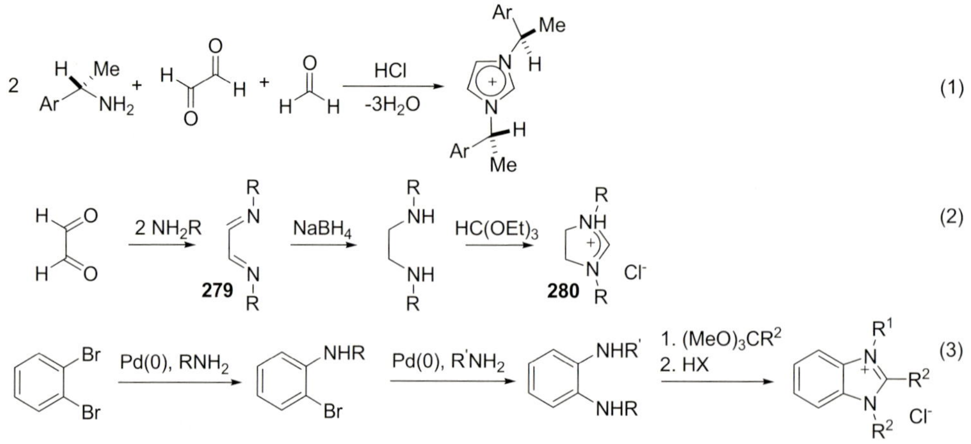

Scheme 10.160

Chiral imidazolium salts have been synthesized in one step (Scheme 10.160). Starting from racemic amines the *meso*-forms were produced. In contrast, only the C_2-symmetric imidazolium salts are formed when enantiomerically pure amines are employed [Scheme 10.160 (1)] [459].

In a similar reaction, starting from glyoxal, 1,3-diarylimidazolium chlorides are obtained in a three-step sequence via the diimine **279**, which is then reduced. Finally, cyclization of the diamines afforded the imidazolinium salt **280** [Scheme 10.160 (2)] [460].

Benzimidazolium salts have been prepared through a subsequent Buchwald–Hartwig amination and ring closure. This method is suitable for the preparation of 2-substituted salts and benzimidazolium salts that bear chiral substituents on one or both the nitrogen atoms [Scheme 10.160 (3)] [409].

Solid-supported azolium and benzoalium salts have been prepared. Azole derivatives react with bromoacetic acid to give azolium acetic acids that have been anchored to a Wang resin [Scheme 10.161 (1)].

Compounds **281–283** have been employed in the Westphal reaction, obtaining cycloiminium salts [Scheme 10.161 (2)].

The synthesis of 1-amino-3-alkylimidazolium **284** [461] and *N*-imidazolium-*N*-methyl-amides **286** [462] has been reported. In the first case, the salts are obtained by direct amination of the corresponding alkoxycarbonylazoles using mesitylenesulfonylhydroxylamine (MSH) as the aminating agents (Scheme 10.162).

In the second case, the amino derivatives **284** are acylated with acyl chlorides, and the resulting betaines **285** are alkylated with methyl iodide to give the salts **286**. Azolium *N*-aminides **287**, generated from the corresponding salts in the presence of *N*-ethyldiisopropylamine, and cycloimmonium N-ylides are 1,3-dipoles, usually involved in 1,3-dipolar cycloaddition reactions (Figure 10.12) [463]. 2-Alkyl- and 2-amino-substituted structures **288** have the potential to function as 1,4-dinucleophiles through deprotonation and can react

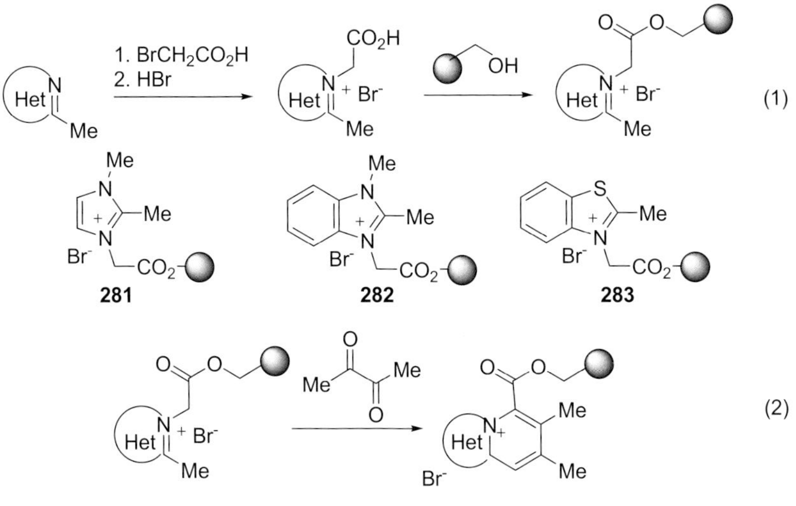

Scheme 10.161

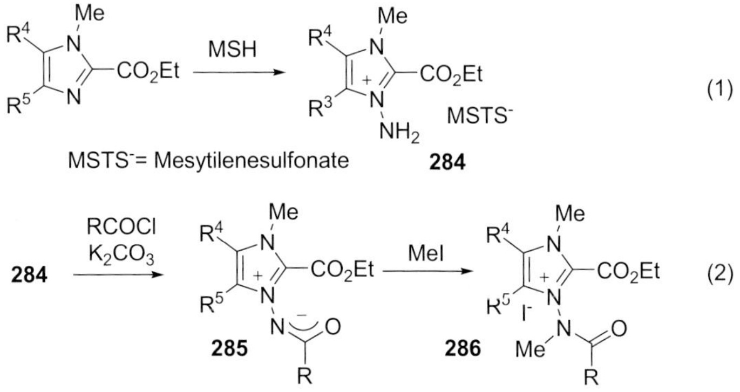

Scheme 10.162

with 1,2-dicarbonyl compounds to afford a great variety of derivatives [464]. 2-Alkoxycarbonylazolium *N*-ylides-*N*-aminides **289** are species that have the potential to act as efficient 1,4-dipole equivalents when they react with heterocumulenes such as iso(thio)cyanates and carbodiimides [465].

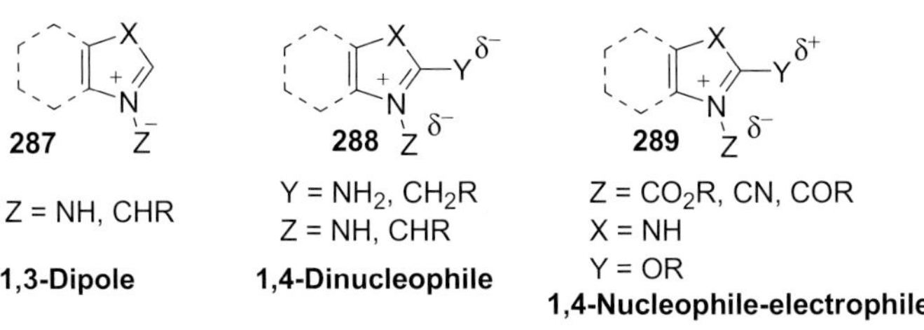

Figure 10.12

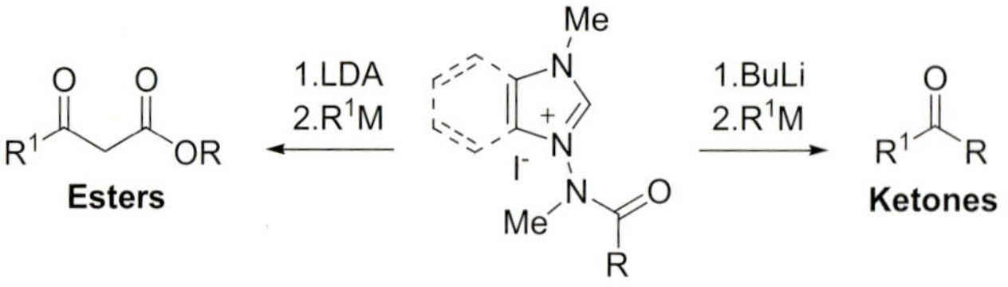

Scheme 10.163

N-Imidazolium-*N*-methylamides and bis-amides behave as highly selective acylating reagents towards organometallics, leading to ketones [462] and diketones. The metallation of alkoxycarbonyl-*N*-imidazolium-*N*-methyl amides with LDA followed by the addition of a Grignard reagent affords 4-oxo and homologous esters (Scheme 10.163).

10.5.6
Oxy- and Amino-1,3-Azoles

Cyanogen chloride (or bromide) as well as cyanamide are the reagents of choice for the synthesis of 2-aminoderivatives. These reagents have found application in the synthesis of the simple heterocycles as well as their benzofused derivatives by reaction with α-hydroxy [466] or α-amino ketones (an extension of the Marckwald synthesis in which cyanamide replaces cyanate, thiocyanate and isothiocyanate as a counterpart of α-aminoaldehydes or ketones) (Scheme 10.164) [467, 468].

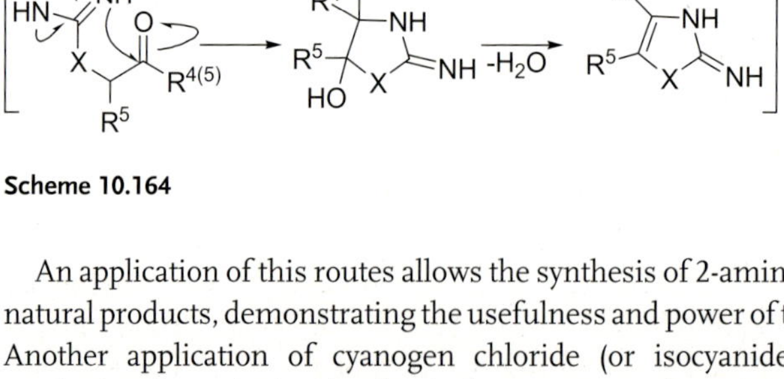

Scheme 10.164

An application of this routes allows the synthesis of 2-aminoimidazole-containing natural products, demonstrating the usefulness and power of this methodology [469]. Another application of cyanogen chloride (or isocyanide dichlorides) for the synthesis of 2-aminoimidazoles is the reaction with DAMN, in which the nucleophilic addition of DAMN proceeds through a mechanism analogous to that reported above [470].

The reaction of *o*-aminophenols with cyanogen bromide or cyanamide affords 2-aminobenzoxazoles **290** or benzoxazoleimines **291**; yields are high and the general method can be adopted to give 2-substituted amino derivatives (Scheme 10.165).

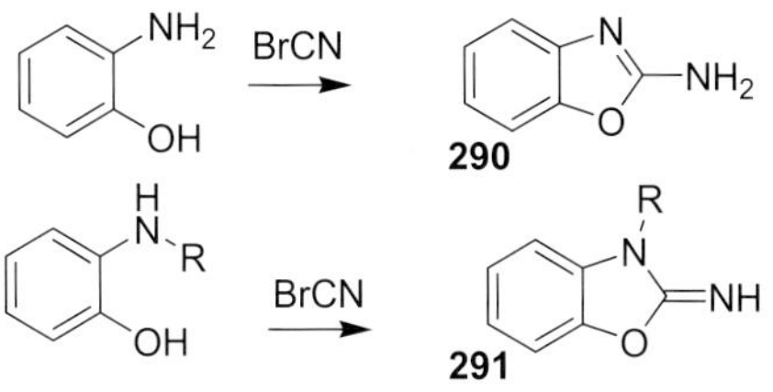

Scheme 10.165

Alternatively, the reactivity of 2-bromo derivatives towards amines in the presence of CuBr can be used as well as the direct nucleophilic amination of benzimidazole by sodium amide.

The reaction of a β-aminoalcohol with BrCN affords the correspondent 2-amino-oxazoline, whose reactivity is nicely exploited in the synthesis of the bis-oxazoline **292** (Scheme 10.166) [471].

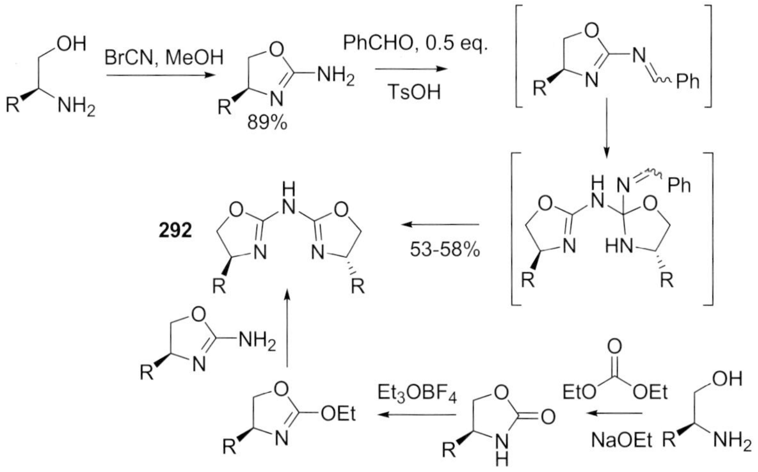

Scheme 10.166

The cycloaddition of azides across a double bond provides another method of imidazole preparation. In this reaction an iminium species, **293**, generated *in situ* under Vilsmeier conditions, attacks an azide functionality [472]. The method is intrinsically limited to 2-dimethylamino substituted imidazoles **294** (Scheme 10.167).

A variation of the Hantzsch procedure uses thiourea with α-halocarbonyl compounds to produce 2-aminothiazoles, while α-halocarboxylic acids afford the correspondent 2-amino-4-hydroxythiazoles [Scheme 10.168 (1)].

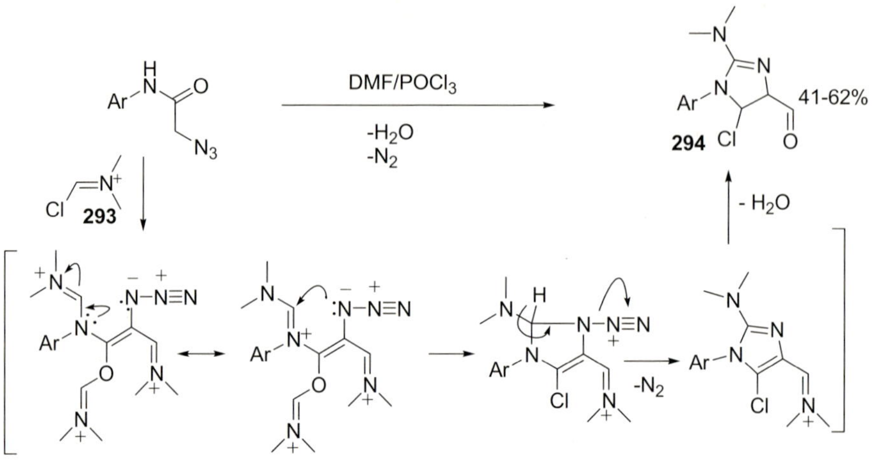

Scheme 10.167

This reaction has found application for the development of a solid-supported synthesis of 2-aminothiazole using the amino group as a convenient, traceless point of attachment to acid-sensitive resins [Scheme 10.168 (2)] [473].

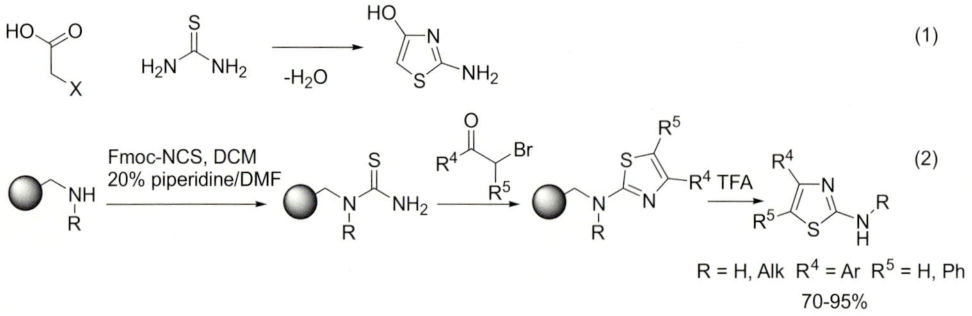

Scheme 10.168

2,4-Diaminothiazole (**295**) can be prepared by the reaction of thiourea with chloroacetonitrile (Scheme 10.169).

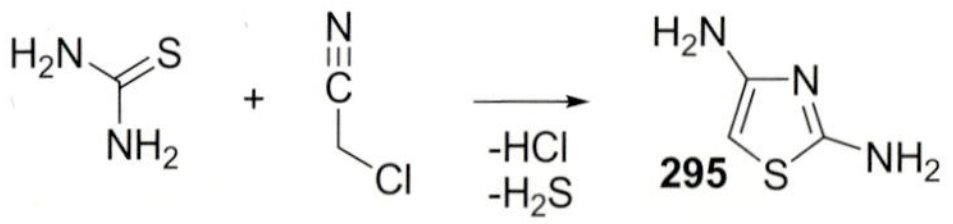

Scheme 10.169

Mono-, di-, and tri-substituted arylthioureas **296** are very easily cyclized to 2-aminobenzothiazoles by the action of bromine in a solvent such as $CHCl_3$, CCl_4

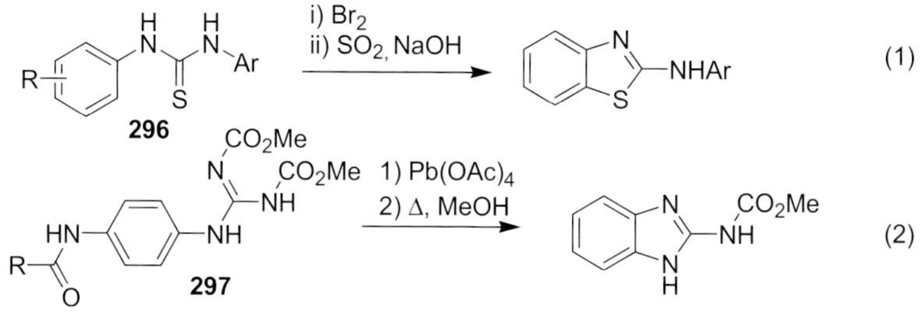

Scheme 10.170

(Hugershoff's method) followed by a treatment with SO_2 and with a base [Scheme 10.170 (1)].

Under oxidative or acidic conditions, or merely by heating the reagents, appropriately functionalized guanidines **297** cyclize to 2-aminobenzimidazoles [Scheme 10.170 (2)] [474].

A variation of the Hantzsch synthesis has been introduced for the preparation of 2-aminoimidazoles using α-haloketones and *N*-acetylguanidines to afford the corresponding 4(5)-substituted *N*-(1*H*-imidazol-2-yl)acetamides **298**. These compounds are then hydrolyzed to their corresponding 2-aminoimidazoles **299** (Scheme 10.171) [475].

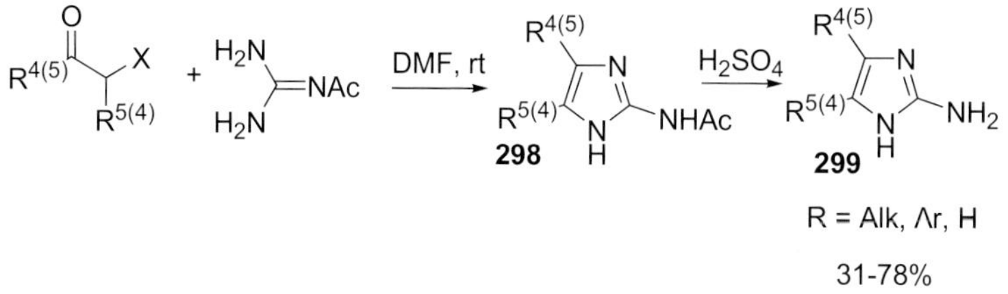

Scheme 10.171

Iminophosphoranes have found several applications in the synthesis of 2-aminoimidazole derivatives and particularly for the synthesis of natural product [476]. Aza-Wittig-type reactions of properly substituted iminophosphorane **300** with CO_2, isocyanates or isothiocyanates afford heterocumulenes that undergo

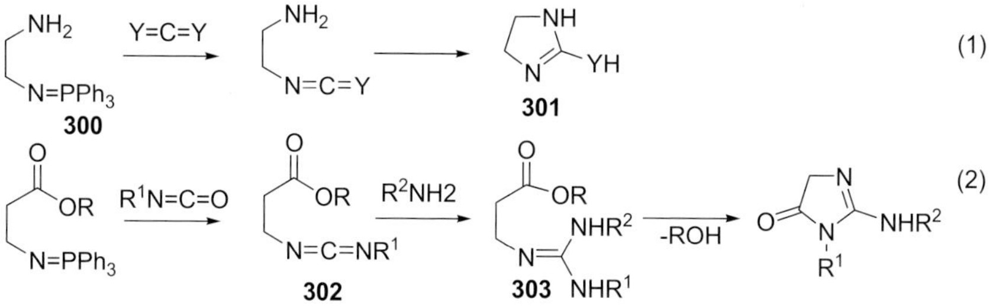

Scheme 10.172

nucleophilic attack of the amino group to give the five-membered heterocycles **301** [Scheme 10.172 (1)].

A variation to this method is the use of azido esters in combination with isocyanates. The intermediate carbodiimide **302** reacts with a primary amine to afford the guanidine derivative **303** that finally cyclizes on the ester group [Scheme 10.172 (2)] [476].

Concerning the reactivity of 2-aminoazoles, some general points can be made. In aminoazoles with the amino group α to C=N, the imino resonance structure justifies the increased reactivity of the pyridine-like nitrogen atom towards electrophilic reagents, but decreases that of the amino group. Consequently, protons, alkylating agents and metal ions usually react with amino azoles at the annular nitrogen. This is exemplified by the behavior of 2-aminothiazole. If the thiazole reacts in its neutral form, the ring nitrogen atom is the more reactive center, except when bulky substituents are present at the C4 position. If the thiazole reacts in the form of its conjugate base, the ambident anion leads to a mixture of products resulting from N-ring and N-exocyclic reactivity [Scheme 10.173 (1)].

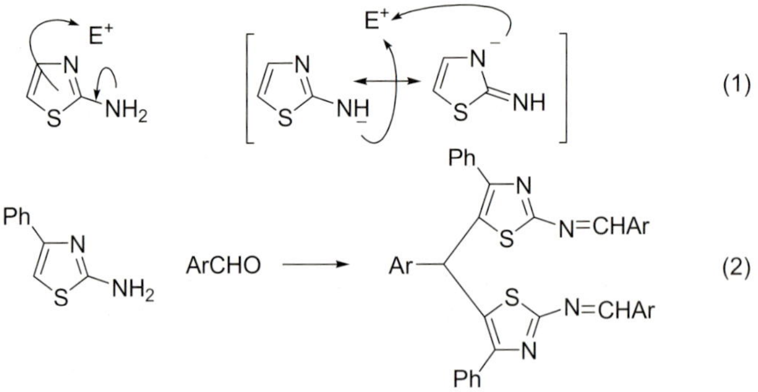

Scheme 10.173

Under mild conditions, 2-aminothiazoles react at their exocyclic nitrogen atom with aromatic aldehyde, yielding Schiff bases. Under more forcing conditions, however, the 5 position can also react [Scheme 10.173 (2)].

Acylation of 2-, 4- and 5-aminothiazoles takes place on the exocyclic nitrogen atom. Acetic anhydrides acetylate aminothiazoles and benzothiazoles on the exocyclic nitrogen atom. The reaction of 2-aminothiazoles with alkyl or aryl isocyanates or isothiocyanates gives the corresponding thiazolylureas or thioureas.

Alkylation of Δ^4-thiazolin-2-ones may yield O−R or N−R derivatives according to experimental conditions. With diazomethane in ethanol, *O*-methylation takes place whereas *N*-methylation occurs when a basic solution of the thiazolinone reacts with methyl iodide. Alkylation of 2-amino imidazole occurs at the exocyclic N atom.

Azolidin-2-ones are popular tools in asymmetric synthesis and as synthetic intermediates, and new methods for their synthesis are described in the literature. Azolidin-2-ones are commonly prepared from aminoalcohols, vicinal diamines and aminothiols by incorporation of a carbonyl unit. Additions to the list of reagents that effect the transformation are bis(trichloromethylcarbonate) [477, 478] and trichloromethyl chloroformate [479], which offer the advantages of easier handling and reduced risk of exposure compared to phosgene [480]. Another reagent of choice is carbonyldiimidazole, which by reaction with cysteine affords derivative **304** [Scheme 10.174 (1)] [481].

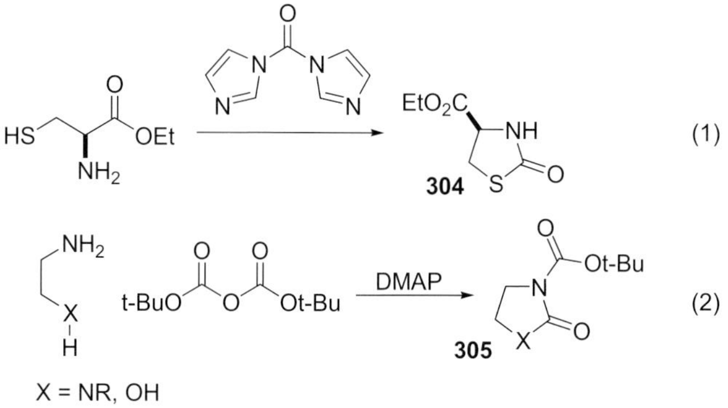

Scheme 10.174

Vicinal aminoalcohols and diamines react with Boc_2O/DMAP to afford the corresponding azolidinones **305** substituted on the N atom with Boc groups, while aminothiols are less efficient in this transformation [Scheme 10.174 (2)] [482].

A recent method for the effective synthesis of imidazolidinones and oxazolidinones is MW irradiation. The proposed mechanism is shown in Scheme 10.175 [483].

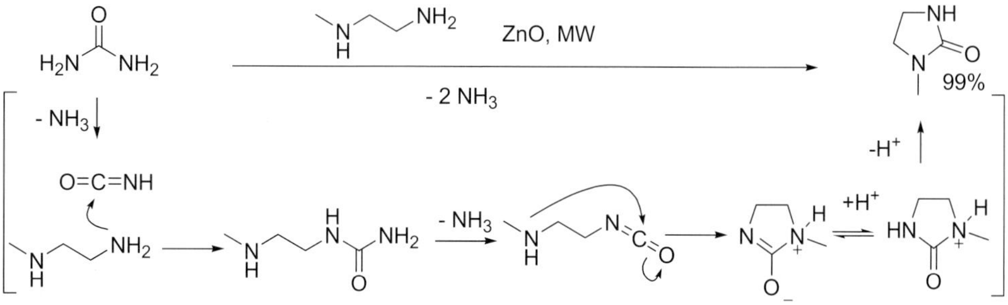

Scheme 10.175

The ability of the benzotriazole nucleus to act as a good leaving group has found another application in the synthesis of polysubstituted imidazolidinones. Treating amine **306** with *s*-BuLi and subsequently with an aldimine or an aldehyde, affords imidazolidinones or oxazolidinones bearing a benzotriazolyl group on C4.

With imidazolidinones **307** it is possible to substitute the benzotriazolyl group with various *C*-nucleophiles to obtain differently substituted imidazolinones (Scheme 10.176) [484].

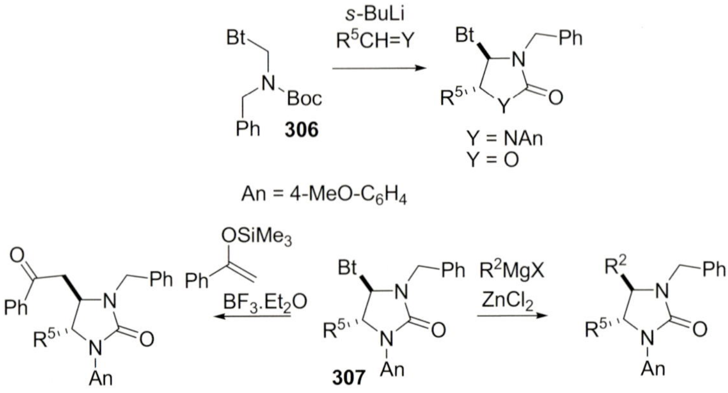

Scheme 10.176

Aziridines in the presence of Lewis acids rearrange to afford substituted oxazolidinones. This approach has found interesting application for the production of enantiopure 4- and 5-carboxymethyl oxazolidinones, which are important precursors for the synthesis of amino acid analogs or as starting material for the synthesis of natural products. In the first case, aziridine **308** was treated with methyl chloroformate to afford the corresponding oxazolidinone **309** with retention of configuration on the reacting stereocenter due to a double inversion of configuration as described in Scheme 10.177 [485].

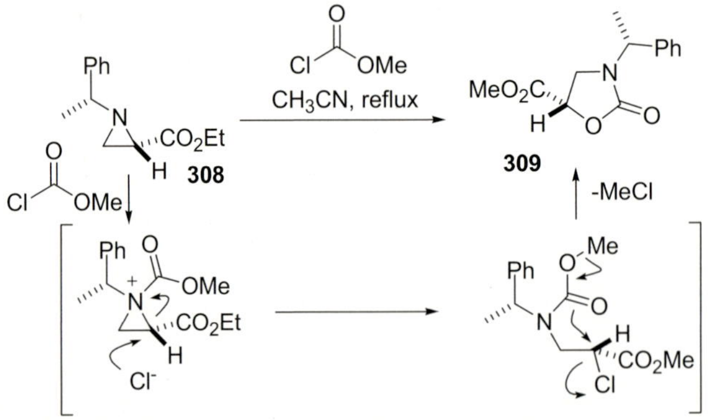

Scheme 10.177

In contrast, treating aziridine **310** with Lewis acids induces a rearrangement in which it is the carbonyl group of the carbamoyl moiety, activated by the exit of isobutene, that induces the enlargement of the three-membered ring (Scheme 10.178) [486].

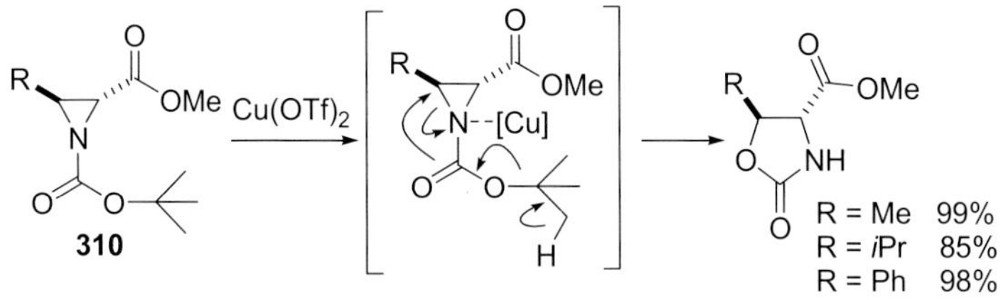

Scheme 10.178

The reaction of vinyl epoxides with aryl isocyanates is facilitated by Pd catalysis [487]. The intermediate π-allyl complex equilibrates to afford stereoselectively the cis derivative from either isomer of the epoxide [Scheme 10.179 (1)].

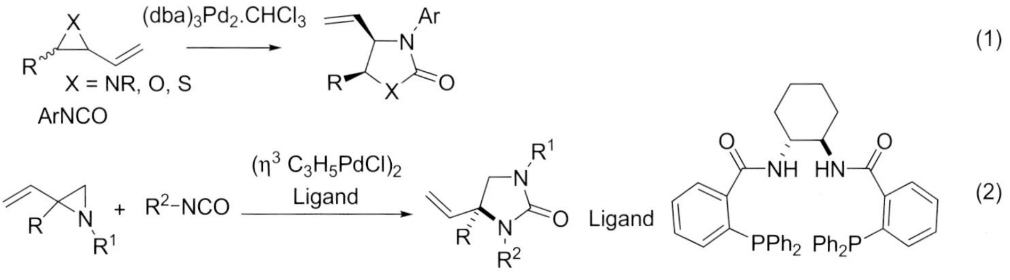

Scheme 10.179

The same reactivity has been extended to aziridines [488] and to thiiranes [489]; for every substrates the enantioselective version of the transformation has also been described [Scheme 10.179 (2)].

Oxazolidinones can be obtained also through a palladium-catalyzed oxidative carbonylation of β-aminoalcohols [490]. In an analogous procedure vicinal diamines are converted into imidazolidinones through an oxidative carbonylation catalyzed by $W(CO)_6$ [491].

Particularly important, especially for the breakthrough that it allowed in the synthesis of tetrodotoxin [492], is the Rh-catalyzed C—H insertion reaction for the oxidative conversion of carbamates into oxazolidinones [493] and the more recent expansion to the synthesis of imidazolidines (Scheme 10.180) [494].

Some specific reported syntheses for the obtainment of benzo-1,3-azol-2-ones have found recent application. The most direct approach to these derivatives is the reaction of a 2-substituted aniline with the general reagent **311** (Scheme 10.181).

The reagent of general formula **311** stands for a series of different reagents in which the Z groups represents good leaving groups. For this reason, good reagents

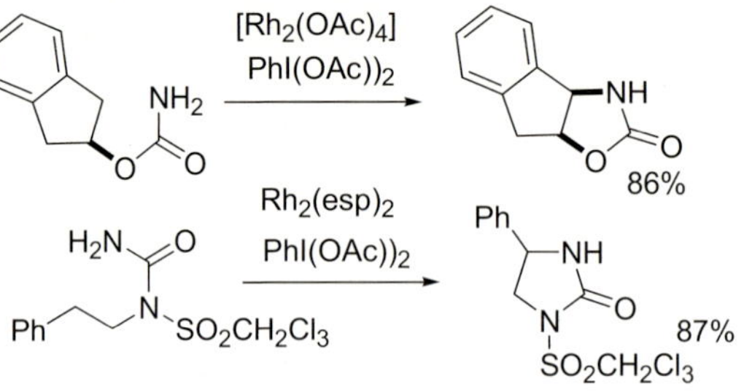

Scheme 10.180

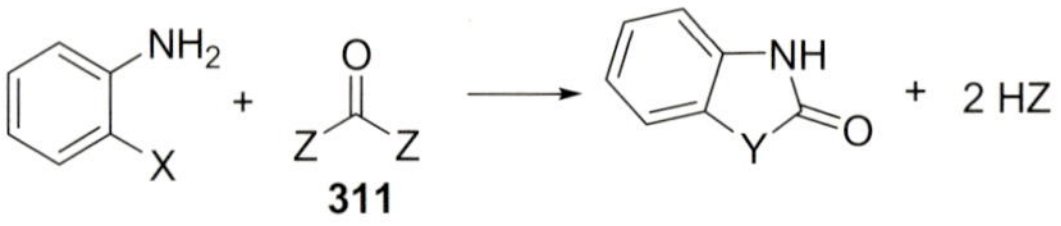

Scheme 10.181

are phosgene [495], urea [496],carbonyldiimidazole [497], dimethyl carbonate [498], *N,N*-diethylcarbamyl chloride and carbon dioxide [499].

10.5.7 Azole N-Oxides and Azoline N-Oxides

Oxazole and imidazole N-oxides cannot be synthesized by oxygenation of oxazoles or imidazoles, respectively. The only method described for the synthesis of oxazole *N*-oxides is the condensation of monooximes of 1,2-dicarbonyl compounds with aldehydes in acidic medium [Scheme 10.182 (1)]. The aldehyde may be aromatic or aliphatic (including formaldehyde) and the oxime may be derived from an aromatic

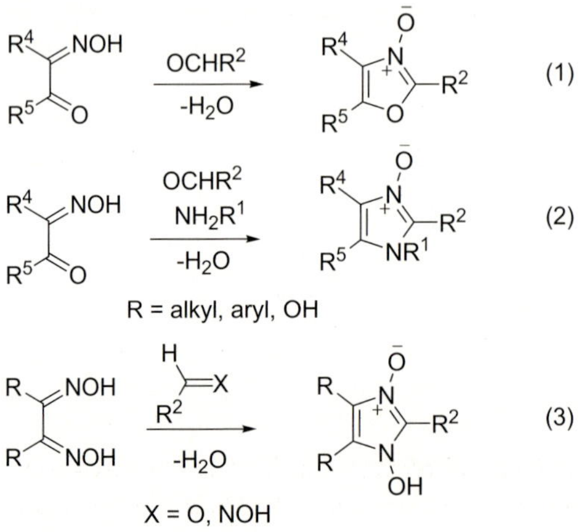

Scheme 10.182

diketone or it may be an α-keto aldoxime, leading to a 2,5-disubstituted oxazole *N*-oxide. For imidazole N-oxides the most common approach is the condensation of an α-oximinoketone with an aldehyde and a primary amine. The use of an hydroxylamine in this condensation [Scheme 10.182 (2)] or the reaction of the aldehyde oxime or aldehyde with a 1,2-dioxime [Scheme 10.183 (3)] afford 1-hydroxyimidazole-3-oxides.

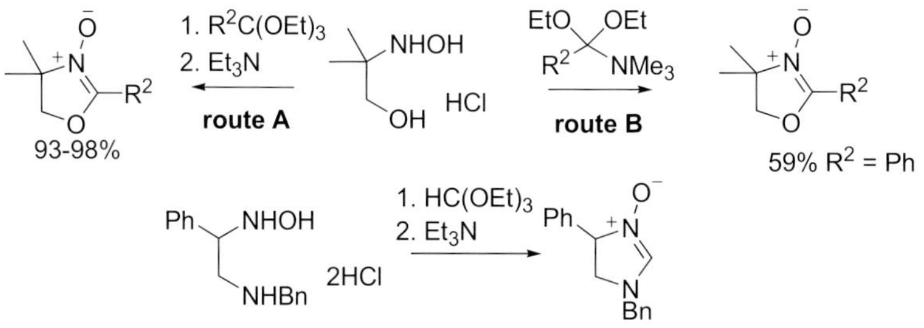

1. $R^2C(OEt)_3$ 2. Et_3N route A; 93-98%; HCl; EtO OEt R^2 NMe_3 route B; 59% R^2 = Ph; Ph NHOH NHBn 2HCl; 1. $HC(OEt)_3$ 2. Et_3N; Ph N Bn

Scheme 10.183

A similar method has been used for the synthesis of 4,5-dihydrooxazole-3-oxides (cyclic *C*-alkoxynitrones). In this case the condensation of an β-hydroxyamino alcohol hydrochloride with an ortho ester (Scheme 10.183, route A) [500] or an amide acetal (Scheme 10.183, route B) [501] affords the desired compound. 4,5-Dihydro-1*H*-imidazole-3-oxides (cyclic *C*-aminonitrones) have been prepared using route A with N-(2-aminoethyl)hydroxylamine dihydrochloride as starting material.

Treatment of 2-(hydroxyamino)alkan-1-one oximes with phenyl or methylglyoxal affords 4,5-dihydro-1*H*-imidazole 3-oxides. The reaction is occurs through the formation of intermediates **312–314** (Scheme 10.184).

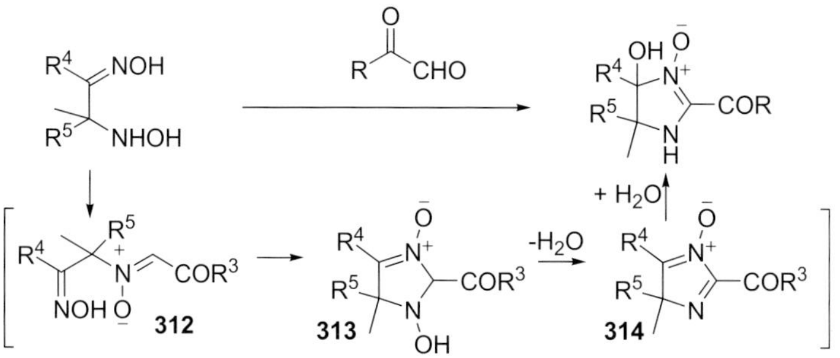

R^4 NOH R^5 NHOH; R CHO; OH R^4 R^5 COR; + H_2O; R^4 R^5 COR^3 NOH **312**; **313** OH; $-H_2O$; **314**

Scheme 10.184

On the other hand, oxidation of 4,5-dihydrooxazoles with 3-chloroperoxybenzoic acid (MCPBA) produces oxaziridines that undergo isomerization to the corresponding nitrones upon treatment with trifluoromethanesulfonic acid (TfOH) [Scheme 10.185 (1)] [502].

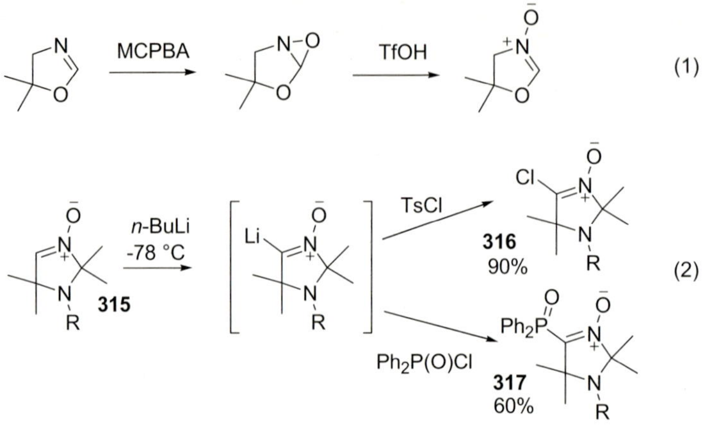

Scheme 10.185

Lithiation of **315** followed by treatment with 4-toluensulfonyl chloride or diphenylphosphoryl chloride [503] affords the *C*-chloronitrone or the *C*-phosphorylnitrone, **316** and **317**, respectively [Scheme 10.185 (2)].

Oxidation of **315** in methanol with lead(IV) acetate or manganese(IV) or lead(IV) oxides affords predominantly the corresponding *C*-methoxynitrones **318** [504], which react with potassium hydrosulfide to afford the thiohydroxamic acids **319** [505]. Compound **319** has also been obtained by treatment of the nitrone **320** with sodium sulfide (Scheme 10.186) [506].

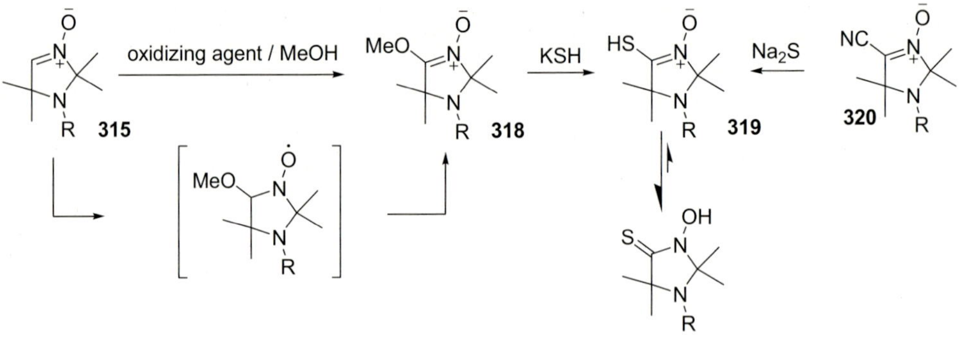

Scheme 10.186

Finally, the reactivity of azole-*N*-oxides is somewhat similar to that of the azolonium ions, particularly when the cationic species is involved. In the case of imidazoline-*N*-oxides (cyclic *C*-amino nitrones) and oxazoline-*N*-oxides (cyclic *C*-hydroxy nitrones) the reactivity is well known. These compounds are versatile synthetic intermediates that readily undergo 1,3-dipolar cycloaddition [507] and addition of nucleophiles [508] and are also useful as radical spin traps [509].

References

1 Laurent, A. (1845) *Journal Fur Praktische Chemie*, **35**, 455.
2 Rochleder, F. (1842) *Justus Liebigs Annalen der Chemie*, **41**, 89.
3 Debus, H. (1858) *Justus Liebigs Annalen der Chemie*, **107**, 199.
4 Cornforth, J.W. and Cornforth, R.H. (1947) *Journal of the Chemical Society*, 96.
5 Zinin, N. (1840) *Annalen*, **34**, 186.
6 Hantzsch, A.R. (1888) *Chemische Berichte*, **21**, 924.
7 Bredereck, H. and Bangert, R. (1962) *Angewandte Chemie International Edition*, **1**, 662.
8 Wallach, O. (1874) *Chemische Berichte*, **7**, 903.
9 Imidazoles: Grimmett, M.R. (1984) *Comprehensive Heterocyclic Chemistry I*, Vol. 5, Pergamon, Oxford, Ch 4.06, p. 345; Oxazoles: Boyd, G.V. (1984) *Comprehensive Heterocyclic Chemistry I*, Vol. 6, Pergamon, Oxford, Ch. 4.18, p. 177; Thiazoles: Metzger, J. (1984) *Comprehensive Heterocyclic Chemistry I*, Vol. 6, Pergamon, Oxford, Ch. 4.19, p. 235.
10 Imidazoles: Grimmett, M.R. (1996), in *Comprehensive Heterocyclic II Chemistry*, Vol. 3, Pergamon, Oxford, Ch. 2, p. 77; Oxazoles: Hartner, F.V., Jr (1996) in *Comprehensive Heterocyclic Chemistry II*, Vol. 3, Pergamon, Oxford, Ch. 4, p. 261; Thiazoles: Dondoni, A. and Merino, P. (1996) *Comprehensive Heterocyclic Chemistry II*, Vol. 3, Pergamon, Oxford, Ch. 6, p. 373.
11 Imidazoles: Grimmett, M.R. (2002) *Science of Synthesis*, Vol. 12, Georg Thieme Verlag, Stuttgart, Ch.3, p. 325; Oxazoles: Boyd, G.V. (2002) *Science of Synthesis*, Vol. 11, Georg Thieme Verlag, Stuttgart, Ch. 12, p. 383; Thiazoles: Kikelj, D. and Urleb, U. (2002) *Science of Synthesis*, Vol. 11, Georg Thieme Verlag, Stuttgart, Ch, 17, p. 627.
12 Nicolau, K.C., Rao, P.S., Hao, J., Reddy, M.V., Rassias, G., Huang, X., Chen, D.Y.K., and Snyder, S.A. (2003) *Angewandte Chemie International Edition*, **42**, 1753; Burgets, A.W.G., Li, Q., Wei, Q., and Harran, P.G. (2003) *Angewandte Chemie International Edition*, **42**, 4961.
13 Pattenden, G., González, M.A., Little, P.B., Millan, D.S., Plowright, A.T., Tornos, J.A., and Ye, T. (2003) *Organic and Biomolecular Chemistry*, **1**, 4173.
14 Coqueron, P.Y., Didier, C., and Ciufolini, M.A. (2003) *Angewandte Chemie International Edition*, **42**, 1411.
15 Chattopaday, S.K., Kempson, J., McNeil, A., Pattenden, G., Reader, M., Rippon, D.E., and Waite, D. (2000) *Journal of the Chemical Society-Perkin Transactions 1*, 2415.
16 You, S.-L. and Kelly, J.W. (2003) *The Journal of Organic Chemistry*, **68**, 9506.
17 Storer, R.I., Takemoto, T., Jackson, P.S., and Ley, S.V. (2003) *Angewandte Chemie International Edition*, **42**, 2521.
18 Hale, K.J., Domostoj, M.M., Tocher, D.E., Irving, E., and Scheinmann, F. (2003) *Organic Letters*, **5**, 2927.
19 Spider, B.B. and Zeng, H. (2003) *The Journal of Organic Chemistry*, **68**, 545.
20 Abbotto, A., Bradamante, S., and Pagani, G.A. (1996) *The Journal of Organic Chemistry*, **61**, 1761.
21 Grimmett, M.R. (1970) in *Advances in Heterocyclic Chemistry*, Vol. 12 (eds A.R. Katritzky, and A.J. Boulton), Academic, New York, pp. 103.
22 Grimmett, M.R. (1980) in *Advances in Heterocyclic Chemistry*, Vol. 27 (eds A.R. Katritzky and A.J. Boulton), Academic, New York, p. 241.
23 Ebel, K., Koehler, H., Garner, A.O., and JSckh, R. (1989) *In Ullmann's Encyclopedia of Industrial Chemistry*, Vol. A13, VCH, Weinheim, p. 661.
24 Ebel, K. (1994) Methoden der Organischen Chemie (Houben-Weyl), in *Band E8c Hetarene III/Tell 3* (ed. E. Schaumann), Georg-Thieme Verlag, Stuttgart, p. 1.
25 Grimmett, M.R. (1997) *Imidazole and Benzimidazole Synthesis*, Academic Press, London.
26 Radziszwski, B. (1882) *Berichte C der deutschen chemischen Gesellschaft* , **15**, 1493.

27 Bredereck, H. and Theilig, G. (1953) *Chemische Berichte*, **86**, 88.
28 Balalaie, S., Hashemi, M.M., and Akhbari, M. (2003) *Tetrahedron Letters*, **44**, 1709.
29 Balalaie, S. and Arabanian, A. (2000) *Green Chemistry*, **2**, 274.
30 Usyatinsky, A.Y. and Khemelnitsky, Y.L. (2000) *Tetrahedron Letters*, **41**, 5031.
31 Wolkenberg, S.E., Wisnoski, D.D., Leister, W.H., Wang, Y., Zhao, Z., and Lindsley, C.W. (2004) *Organic Letters*, **6**, 1453.
32 Xu, L., Wan, L.-F., Salehi, H., Deng, W., and Guo, Q.-X. (2004) *Heterocycles*, **63**, 1613.
33 Siddiqui, S.A., Narkede, U.C., Palimkar, T.D., Lahoti, R.J., and Srinivasan, Q.Q. (2005) *Tetrahedron*, **61**, 3539.
34 Wallach, O. and Boehringer, A. (1877) *Annalen*, **184**, 50.
35 Wallach, O. (1882) *Annalen*, **214**, 257.
36 Kochergin, P.M. (1964) *Journal of General Chemistry of the USSR (English Translation)*, **84**, 2758.
37 Sarasin, J. and Wegmann, E. (1924) *Helvetica Chimica Acta*, **7**, 713.
38 Trout, G.E. and Levy, P.R. (1965) *Recueil des Travaux Chimiques des Pays*, **84**, 1257.
39 Trout, G.E. and Levy, P.R. (1966) *Recueil des Travaux Chimiques des Pays*, **85**, 765.
40 Godefroi, E.F., van der Eychen, C.A.M., and Janssen, P.A.J. (1967) *The Journal of Organic Chemistry*, **32**, 1259.
41 van Leusen, A.M., Schaart, F.J., and van Leusen, D. (1979) *Recueil des Travaux Chimiques des Pays*, **98**, 258.
42 van Leusen, A.M., Wildeman, J., and Oldenziel, O.H. (1977) *The Journal of Organic Chemistry*, **42**, 1153.
43 Chen, B.-C., Bednarz, M.S., Zhao, R., Sundeen, J.E., Chen, P., Shen, Z., Skoumbourdis, A.P., and Barrish, J.C. (2000) *Tetrahedron Letters*, **41**, 5453.
44 Sisko, J., Mellinger, M., Sheldrake, P.W., and Baine, N.H. (2000) *Organic Syntheses*, **77**, 198.
45 Sisko, J., Kassick, A.J., Mellinger, R., Filan, J.J., Allen, A., and Olsen, M.A. (2000) *The Journal of Organic Chemistry*, **65**, 1516.
46 Horne, D.A., Yakushijin, K., and Buechi, G. (1994) *Heterocycles*, **39**, 139.
47 Susan, C., Shilcrat, S.C., Mokhallalati, M.K., Fortunak, J.M.D., and Pridgen, L.N. (1997) *The Journal of Organic Chemistry*, **62**, 8449.
48 Nunami, K.-I., Yamada, M., Fukui, T., and Matsumoto, K. (1994) *The Journal of Organic Chemistry*, **59**, 7635.
49 Begland, W., Hartter, D.R., Jones, F.N., Sam, D.J., Shepperd, W.A., Webster, O.W., and Weigert, F.J. (1974) *The Journal of Organic Chemistry*, **39**, 2341.
50 Al-azmi, A., Booth, B.L., Pritchard, R.G., and Proença, F.G.R.P. (2001) *Journal of the Chemical Society-Perkin Transactions 1*, 485.
51 Rolfs, A. and Liebscher, J. (1997) *The Journal of Organic Chemistry*, **62**, 3480.
52 Consonni, R., Croce, P.D., Ferraccioli, R., and La Rosa, C. (1991) *Journal of Chemical Research*, 188.
53 For *N*-alkylation: Liverton, N.J., Butcher, J.W., Claiborne, C.F., Claremon, D.A., Libby, B.E., Nguyen, K.T., Pitzenberger, S.M., Selnick, H.G., Smith, G.R., Tebben, A., Vacca, J.P., Varga, S.L., Agarwal, L., Dancheck, K., Forsyth, A.J., Fletcher, D.S., Frantz, B., Hanlon, W.A., Harper, C.F., Hofsess, S.J., Kostura, M., Lin, J., Luell, S., O'Neil, E.A., Orvillo, C.J., Pang, M., Parsons, J., Rolando, A., Sahly, Y., Visco, D.M., and O'Keefe, S.J. (1999) *Journal of Medicinal Chemistry*, **42**, 2180; Wagner, G.K., Kotschenreuther, D., Zimmermann, W., and Laufer, S.A. (2003) *The Journal of Organic Chemistry*, **68**, 4527.
54 For metal-activated coupling: Fukumoto, Y., Sawada, K., Hagihara, M., Chatani, N., and Murai, S. (2002) *Angewandte Chemie International Edition*, **41**, 2779; Lipshutz, B.H. and Hagen, W. (1992) *Tetrahedron Letters*, **33**, 5865.
55 From imidazole *N*-oxides: Laufer, S., Wagner, G., and Kotschenreuther, D. (2002) *Angewandte Chemie International Edition*, **41**, 2290.
56 For the use of imidazolium ylides to introduce substituents in the 2-position: Hlasta, D.J. (2001) *Organic Letters*, **3**, 157.

57 Bleicher, K.H., Gerber, F., Wührich, Y., Alanine, A., and Capretta, A. (2002) *Tetrahedron Letters*, **43**, 7687.
58 Claiborne, C.F., Liverton, N.J., and Nguyen, K.T. (1998) *Tetrahedron Letters*, **39**, 8939.
59 Lee, H.B. and Balasubramanian, S. (2000) *Organic Letters*, **2**, 323.
60 Davies, J.R., Kane, P.D., and Moody, C.J. (2004) *Tetrahedron*, **60**, 3967.
61 Lantos, I., Zhangt, W.-Y., Shui, X., and Eggleston, D.S. (1993) *The Journal of Organic Chemistry*, **58**, 7092.
62 Bilodeau, M.T. and Cunningham, A.M. (1998) *The Journal of Organic Chemistry*, **63**, 2800.
63 Illgen, K., Nerdinger, S., Behnke, D., and Friedrich, C. (2005) *Organic Letters*, **7**, 39.
64 Henkel, B. (2004) *Tetrahedron Letters*, **45**, 2219.
65 D.C. Palmer (ed.) (2003) *Oxazoles: Synthesis, reactions, and spectroscopy*, Part A, Vol. 60, John Wiley & Sons, Inc., Hoboken, NJ.
66 Gabriel, S. (1907) *Chemische Berichte*, **40**, 2647.
67 Robinson, R. (1909) *Journal of the Chemical Society*, **95**, 2167.
68 Kim, K.S., Kimball, S.D., Misra, R.N., Rawlins, D.B., Hunt, J.T., Xiao, H.-Y., Lu, S., Qian, L., Han, W.-C., Shan, W., Mitt, T., Cai, Z.-W., Poss, M.A., Zhu, H., Sack, J.S., Tokarski, J.S., Chang, C.Y., Pavletich, N., Kamath, A., Humphreys, W.G., Marathe, P., Bursuker, I., Kellar, K.A., Roongta, U., Batorsky, R., Mulheron, J.G., Bol, D., Fairchild, C.R., Lee, F.Y., and Webster, K.R. (2002) *Journal of Medicinal Chemistry*, **45**, 3905.
69 Guella, G., Mancini, I., N'Diaye, I., and Pietra, F. (1994) *Helvetica Chimica Acta*, **77**, 1999.
70 Reck, S. and Friedrichsen, W. (1998) *The Journal of Organic Chemistry*, **63**, 7680.
71 Bal'on, Y.G. and Smirnov, V.A. (1990) *Journal of Organic Chemistry USSR (English Translation)*, **26**, 1712; (1990) *Zhurnal Organicheskoi Khimii*, **26**, 1983.
72 Ikemoto, N., Miller, R.A., Fleitz, F.J., Liu, J., Petrillo, D.E., Leone, J.F., Laquidara, J., Marcune, B., Karady, S., Armstrong, J.D.JIII, and Volante, R.P. (2005) *Tetrahedron Letters*, **46**, 1867.
73 Dai, Y., Guo, Y., Curtin, M.L., Li, J., Pease, L.J., Guo, J., Marcotte, P.A., Glaser, K.B., Davidsen, S.K., and Michaelides, M.R. (2003) *Bioorganic and Medicinal Chemistry Letters*, **13**, 3817.
74 Huth, A., Rosenberg, D., Schumann, I., and Thielert, K. (1984) *Annalen Der Chemie-Justus Liebig*, 641.
75 Lipshutz, B.H., Hungate, R.W., and NcCarthy, K.E. (1983) *Journal of the American Chemical Society*, **105**, 7703.
76 Bagley, M.C., Buck, R.T., Hind, S.L., and Moody, C.J. (1998) *Journal of the Chemical Society-Perkin Transactions 1*, 591.
77 Wipf, P. and Miller, C.P. (1993) *The Journal of Organic Chemistry*, **58**, 3604.
78 Brain, C.T. and Paul, J.M. (1999) *Synlett*, 1642.
79 Pulici, M., Quartieri, F., and Felder, E.R. (2005) *Journal of Combinatorial Chemistry*, **3**, 463.
80 Spanka, C., Clapham, B., and Janda, K.D. (2002) *The Journal of Organic Chemistry*, **67**, 3045.
81 Lister, J. and Robinson, R. (1912) *Journal of the Chemical Society*, 1297.
82 Doyle, K.J. and Moody, C.J. (1994) *Tetrahedron*, **50**, 3761.
83 Bagley, M.C., Buck, R.T., Hind, S.L., Moody, C.J., and Slawin, A.M.Z. (1996) *Synlett*, 825.
84 Crowe, E., Hossner, F., and Hughes, M.J. (1995) *Tetrahedron*, **32**, 8889.
85 Moskal, J., van Stralen, R., Postma, D., and van Leusen, A.M. (1986) *Tetrahedron Letters*, **27**, 2173.
86 Wang, L., Woods, K.W., Li, Q., Barr, K.J., McCroskey, R.W., Hannick, S.M., Gherke, L., Credo, R.B., Hui, Y.-H., Marsh, K., Warner, R., Lee, J.Y., Zielinski-Mozng, N., Frost, D., Rosenberg, S.H., and Sham, H.L. (2002) *Journal of Medicinal Chemistry*, **45**, 1697.
87 Iwanowicz, E.J., Watterson, S.H., Guo, J., Pitts, W.J., Dhar, T.G.M., Shen, Z., Chen, P., Gu, H.H., Fleener, C.A., Rouleau, K.A., Cheney, D.L., Townsend, R.M., and Hollenbaugh, D.L. (2003) *Bioorganic and Medicinal Chemistry Letters*, **13**, 2059.

88 Kulkarni, B.A. and Ganesan, A. (1999) *Tetrahedron Letters*, **40**, 5633.

89 Barrett, A.G.M., Cramp, S.M., Hennessy, A.J., Procopiou, P.A., and Roberts, R.S. (2001) *Organic Letters*, **3**, 271.

90 Wipf, P., Rahman, L.T., and Rector, S.R. (1998) *The Journal of Organic Chemistry*, **63**, 7132.

91 Arcadi, A., Cacchi, S., Cascia, L., Fabrizi, G., and Marinelli, F. (2001) *Organic Letters*, **3**, 2501.

92 Wipf, P., Aoyama, Y., and Benedum, T.E. (2004) *Organic Letters*, **6**, 3593.

93 Takeuchi, H., Yanagida, S.-I., Ozaki, T., Hagiwara, S., and Eguchi, S. (1989) *The Journal of Organic Chemistry*, **54**, 431.

94 Tárraga, A., Molina, P., Curiel, D., and Velasco, D.M. (2001) *Organometallics*, **20**, 2145.

95 Tárraga, A., Molina, P., Curiel, D., Velasco, D.M., and López, J.L. (1999) *Tetrahedron*, **55**, 14701.

96 Dhar, T.G.M., Guo, J., Shen, Z., Pitts, W.J., Gu, H.H., Chen, B.-C., Zhao, R., Bednarz, M.S., and Iwanowicz, E.J. (2002) *Organic Letters*, **4**, 2091.

97 Hantzsch, A. and Weber, J.H. (1887) *Berichte der Deutschen Chemischen Gesellschaft*, **20**, 3118.

98 Hantzsch, A. (1888) *Berichte der Deutschen Chemischen Gesellschaft*, **21**, 942.

99 Kurkjy, Y.R. and Brown, E.V. (1952) *Journal of the American Chemical Society*, **74**, 5778.

100 Faucher, A.-M., White, P.W., Brochu, C., Grand-Maître, C., Rancourt, J., and Fazal, G. (2004) *Journal of Medicinal Chemistry*, **47**, 18.

101 Begtrup, M. and Hansen, L.B.L. (1992) *Acta Chemica Scandinavica (Copenhagen, Denmark: 1989)*, **46**, 372.

102 Irako, N., Hamada, Y., and Shioiri, T. (1995) *Tetrahedron*, **51** (46), 2731.

103 Astle, E.J. and Pierce, J.B. (1955) *The Journal of Organic Chemistry*, **20**, 178.

104 Pavlov, V.A. and Smith, J.A.S. (1996) *Chemistry of Heterocyclic Compounds (English Translation)*, **32**, 721; (1996) *Khim Geterotsikl Soedin*, **32**, 837.

105 Mathes, R.A. and Beber, A.J. (1948) *Journal of the American Chemical Society*, **70**, 1451.

106 Miller, T.J., Farquar, H.D., Sheybani, A., Tallini, C.E., Saurage, A.S., Fronczek, F.R., and Hammer, R.P. (2002) *Organic Letters*, **6**, 877.

107 Brandsma, L., Jong, R.L.P., and VerKruijsse, H.D. (1985) *Synthesis*, 948.

108 Lee, B.W. and Lee, S.D. (2000) *Tetrahedron Letters*, **41**, 3883.

109 Gillon, D.W., Forrest, I.J., Meakins, G.D., Tirel, M.D., and Wallis, J.D. (1983) *Journal of the Chemical Society-Perkin Transactions 1*, 341.

110 Penning, T.D., Russell, M.A., Chen, B.B., Chen, H.Y., Liang, Ch.-D., Mahoney, M.W., Malecha, J.W., Miyashiro, J.M., Yu, S.S., Askonas, L.J., Gierse, J.K., Harding, E.I., Highkin, M.K., Kachur, J.F., Kim, S.H., Villani-Price, D., Pyla, E.Y., Ghoreishi-Haack, N.S., and Smith, W.G. (2002) *Journal of Medicinal Chemistry*, **45**, 3482.

111 Nußbaumer, T., Krieger, C., and Neidlein, R. (2000) *European Journal of Organic Chemistry*, 2449.

112 Chambers, M.S., Atack, J.R., Broughton, H.B., Collinson, N., Cook, S., Dawson, G.R., Hobbs, S.C., Marshall, G., Maubach, K.A., Pillai, G.V., Reeve, A.J., and MacLeod, A.M. (2003) *Journal of Medicinal Chemistry*, **46**, 2227.

113 Keil, D. and Hartmann, H. (1995) *Liebigs Annalen der Chemie*, **6**, 979.

114 Zhu, N., Ling, Y., Lei, X., Handratta, V., and Brodie, A.M.H. (2003) *Steroids*, **68** (7–8), 603.

115 Ardá, A., Soengas, R.G., Nieto, M.I., Jiménez, C., and Rodríguez, J. (2008) *Organic Letters*, **10**, 2175.

116 Kim, S.-H., Tokarski, J.S., Leavitt, K.J., Fink, B.E., Salvati, M.E., Moquin, R., Obermeier, M.T., Trainor, G.L., Vite, G.G., Stadnick, L.K., Lippy, J.S., You, D., Lorenzi, M.V., and Chen, P. (2008) *Bioorganic and Medicinal Chemistry Letters*, **18**, 634.

117 Walczynski, K., Timmerman, H., Zuiderveld, O.P., Zhang, M.Q., and Glinka, R. (1999) *Il Farmaco*, **54**, 533.

118 Nishide, K.u., Yamamura, M., Yamazaki, M., Kobori, T., Tunemoto, D.,

and Kondo, K. (1988) *Chemical and Pharmaceutical Bulletin*, **36**, 2346.

119 Zhang, M.Q., Haemers, A., Berghe, D.V., Pattyn, S.R., Bollaert, W., and Levshin, I. (1991) *Journal of Heterocyclic Chemistry*, **28**, 673.

120 Burger, K., Gold, M., Neuhauser, H., Rudolph, M., and Hoess, E. (1992) *Synthesis*, 1145.

121 Liu, C.-L., Li, L., and Li, Z.-M. (2004) *Bioorganic and Medicinal Chemistry*, **12**, 2825.

122 Katritzky, A.R., Chen, J., and Yang, Z. (1995) *The Journal of Organic Chemistry*, **60**, 5638.

123 Tavecchia, P., Gentili, P., Kurz, M., Sottani, C., Bonfichi, R., Selva, E., Lociuro, S., Restelli, E., and Ciabatti, R. (1995) *Tetrahedron*, **51**, 4867.

124 Atkins, E.F., Dabbs, S., Guy, R.G., Mahomed, A.A., and Mountford, P. (1994) *Tetrahedron*, **50**, 7253.

125 Hartenstein, H., Blitzke, T., Sicker, D., and Wilde, H. (1993) *Journal Fur Praktische Chemie-Chemiker-Zeitung*, **335** (2), 176.

126 Kuramoto, M., Sakata, Y., Terai, K., Kawasaki, I., Kunitomo, J., Ohishi, T., Yokomizo, T., Takeda, S., Tanaka, S., and Ohishi, Y. (2008) *Organic and Biomolecular Chemistry*, **6**, 2772.

127 Delgado, O., Müller, H.M., and Bach, T. (2008) *Chemistry–A European Journal*, **14**, 2322.

128 Moriarty, R.M., Vaid, B.K., Duncan, M.P., Levy, S.G., Prasash, O., and Goval, S. (1992) *Synthesis*, 845.

129 Ochiai, M., Nishi, Y., Hashimoto, S., Tsuchimoto, Y., and Chen, D.-W. (2003) *The Journal of Organic Chemistry*, **68**, 7887.

130 Belyuga, A.G., Brovarets, V.S., and B.S. Drach. (2004) *Russian Journal of General Chemistry*, **74**, 1418; *Zhurnal Obshchei Khimii*, **74**, 1529.

131 Heck, S. and Dömling, A. (2000) *Synlett*, 424.

132 Zoltewicz, J.A. and Deaby, L.W. (1978) *Advances in Heterocyclic Chemistry*, **22**, 71.

133 Kujundzic, N. and Gluncic, B. (1990) *Croatica Chemica Acta*, **63**, 215.

134 Dondoni, A., Dall'Occo, T., Fantin, G., Fogagnolo, M., and Medici, A. (1984) *Tetrahedron Letters*, **25**, 3637.

135 Benjes, P.A. and Grimmet, M.R. (1994) *Advances in Detailed Reaction Mechanisms*, vol. 3 (ed. J.M. Coxon), JAI Press, Greenwith, pp. 199.

136 Arduengo, A.J. (1999) *Accounts of Chemical Research*, **32**, 913 and references citated therein.

137 Grimmett, M.R. (1996) *Comprenhesive Heterocyclic Chemistry II*, Vol. 3 (eds A.R. Katritzky, C. Rees, and E.F.V. Scriven), Pergamon, Oxford, pp. 7.

138 Hartner, F.W. (1996) *Comprehensive Heterocyclic Chemistry II*, Vol. 3 (eds A.R. Katritzky, C. Rees, and E.F.V. Scriven), Oxford, Pergamon, pp. 261.

139 Dondoni, A. and Merino, P. (1996) *Comprehensive Heterocyclic Chemistry II*, Vol. 3 (eds A.R. Katritzky, C. Rees, and E.F.V. Scriven), Pergamon, Oxford, pp. 373.

140 Hassner, A. and Fischer, B. (1989) *Tetrahedron*, **45**, 6249.

141 Taillefer, A., Ouali, A., Renard, B., and Spindler, J.-F. (2006) *Chemistry - A European Journal*, **12**, 5301.

142 Virieux, D., Guillouzic, A-F., and Cristau, H-J. (2006) *Tetrahedron*, **62**, 3710.

143 Park, S.B. and Alper, H. (2003) *Organic Letters*, **5**, 3209.

144 Grimmett, M.R. (1993) *Advances in Heterocyclic Chemistry*, **57**, 291.

145 Lakhan, R. and Ternai, B. (1974) *Advances in Heterocyclic Chemistry*, **17**, 99.

146 Hetzger, J.V. (1979) *Chemistry of Heterocyclic Compounds*, **34–1**, 1.

147 Sathunuru, R., Rao, U.N., and Biehl, E. (2003) *Arkivoc*, **124**.

148 Samajdar, S., Becker, F.F., and Banik, B.K. (2001) *Arkivoc*, **27**.

149 Bakke, J.M. (2003) *Pure and Applied Chemistry*, **75**, 1403.

150 Bakke, J.M., Gautun, H.S.H., Rømming, C., and Sletvold, I. (2001) *Arkivoc*, **26**.

151 Katritzky, A.R.., Scriven, E.F.V., Majumder, S., Akhmedova, R.G., Akhmedov, N.G., and Vakulenko, A.V. (2005) *Arkivoc*, **179**.

152 Katritzky, A.R., Vakulenko, A.V., Sivapackiam, J., Draghici, B., and Damavarapub, R. (2008) *Synthesis*, **5**, 699.

153 Metzger, J.V. (1984) *Comprehensive Heterocyclic Chemistry*, Vol. 6 (eds A.R. Katritzky, C. Rees, and E.F.V. Scriven), Pergamon, Oxford, pp. 256.

154 Schofield, K., Grimmett, M.R., and Keene, B.R.T. (1976) *Heteroaromatic nitrogen compounds: The azoles*, Cambridge University Press, London, pp. 60.

155 Komarov, I.V., Strizhak, A.V., Kornilov, M.Y., Kostyuk, A.N., and Tolmachev, A.A. (1998) *Synthetic Communications*, **28**, 2355.

156 Zarudnitskii, E.V., Pervak, I.I., Merkulov, A.S., Yurchenko, A.A., Tolmachev, A.A., and Pinchuk, A.M. (2006) *Synthesis*, **8**, 1279.

157 Zificsak, C.A. and Hlasta, D.J. (2004) *Tetrahedron*, **60**, 8991.

158 Stefancich, G., Silvestri, R., and Artico, M. (1993) *Heterocyclic Chemistry*, **30**, 529; Castellano, S., Zorzin, L., Florio, C., Frausin, F., and Stefancich, G. (2001) *Il Farmaco (Societa Chimica Italiana: 1989)*, **56**, 771; Caliendo, G., Cirino, G., Greco, G., Novellino, E., Perissutti, E., Pinto, A., Santagada, V., Silipo, C., and Sorrentino, L. (1994) *European Journal of Medicinal Chemistry*, **29**, 381; Caliendo, G., Cirino, G., Greco, G., Novellino, E., Perissutti, E., Pinto, A., Santagada, V., and Sorrentino, L. (1995) *European Journal of Medicinal Chemistry*, **30**, 315; Baird, E.E. and Dervan, P.B. (1996) *Journal of the American Chemical Society*, **118**, 6141; Shafiee, A., Zarghi, A., and Dehpour, A.R. (1997) *Journal of Pharmaceutical Sciences*, **3**, 461; Sharma, S.K., Tandon, M., and Lown, J.W. (2000) *The Journal of Organic Chemistry*, **65**, 1102; Sawada, K., Okada, S., Kuroda, A., Watanabe, S., Sawada, Y., and Tanaka, H. (2001) *Chemical and Pharmaceutical Bulletin*, **49**, 799; Collins, I., Moyes, C., Davey, W.B., Rowley, M., Bromidge, F.A., Quirk, K., Atack, J.R., McKernan, R.M., Thompson, S.A., Wafford, K., Dawson, G.R., Pike, A., Sohal, B., Tsou, N.N., Ball, R.G., and Castro, J.L. (2002) *Journal of Medicinal Chemistry*, **45**, 1887.

159 Roe, A.M. (1963) *Journal of the Chemical Society*, 2195; Vanelle, P., Maldonado, J., Crozet, M.P., Savornin, B., Delmas, F., and Timon-David, P. (1992) *European Journal of Medicinal Chemistry*, **27**, 551; Martin, A., Diaz, J.A., and Vega, S. (1995) *Anales de Quimica*, **91**, 290; Urban, F.J. and Breitenbach, R. (1999) *Synthetic Communications*, **29**, 645; Basso, D., Broggini, G., Passerella, D., Pilati, T., Terranno, A., and Zecchi, G. (2002) *Tetrahedron*, **58**, 4445.

160 Regel, E. (1977) *Annalen Der Chemie-Justus Liebig*, 159.

161 Burak, K. (1991) *Die Pharmazie*, **46**, 668.

162 Dinsmore, C.J., Zartman, C.B., Baginsky, W.F., O'Neill, T.J., Koblan, K.S., Chen, I.W., McLoughlin, D.A., Olah, T.V., and Huff, J.R. (2000) *Organic Letters*, **2**, 3473.

163 Kundu, B., Sawant, D., Partani, P., and Kesarwani, A.P. (2005) *The Journal of Organic Chemistry*, **70**, 4889.

164 Reese, C.B. and Pei-Zhuo, Z. (1993) *Journal of the Chemical Society-Perkin Transactions 1*, 2291; Ulhaq, S., Chinje, E.C., Naylor, M.A., Jaffar, M., Stratford, I.J., and Threadgill, M.D. (1998) *Bioorganic and Medicinal Chemistry*, **6**, 2139.

165 Bossio, R., Marcaccini, S., Pepino, R., Polo, C., and Torroba, T. (1991) *Organic Preparations and Procedures International*, **23**, 670; Mekonnen, B. and Cranck, G. (1997) *Journal of Heterocyclic Chemistry*, **34**, 567; Dondoni, A., Medici, A., Venturoli, C., Forlani, L., and Bertolasi, V. (1980) *The Journal of Organic Chemistry*, **45**, 621; Medici, A., Pedrini, P., Venturoli, C., and Dondoni, A. (1981) *The Journal of Organic Chemistry*, **46**, 2790.

166 Balasubramain, P.N., Sinha, A., and Bruice, T.C. (1987) *Journal of the American Chemical Society*, **109**, 1456.

167 Katritzky, A.R. and Lagowski, J.M. (1971) *Chemistry of the Heterocyclic N-oxides*, Academic, New York, pp. 51.

168 Wasserman, H.H., Wolff, M.S., Stiller, K., Saito, I., and Pickett, J.E. (1981) *Tetrahedron*, **37**, 191.

169 Beak, P. and Messer, W. (1969) *Tetrahedron*, **25**, 3287.

170 Katritzky, A.R. (1985) *Handbook of Heterocyclic Chemistry*, Pergamon, Oxford, pp. 313.

171 Kelly, T.R. and Lang, F. (1995) *Tetrahedron Letters*, **36**, 5319.

172 Lai, Y.H. and Jiang, J. (1997) *The Journal of Organic Chemistry*, **62**, 4412.

173 Sarasin, W. (1924) *Helvetica Chimica Acta*, **7**, 714.

174 Chauvière, G., Viodé, C., and Périé, J. (2000) *Journal of Heterocyclic Chemistry*, **37**, 119.

175 Ayer, W.A., Craw, P.A., Ma, Y.T., and Miao, S. (1992) *Tetrahedron*, **48**, 2919.

176 Gellis, A., Vanelle, P., Maldonado, J., and Crozet, M.P. (1997) *Tetrahedron Letters*, **38**, 2085.

177 Swain, C.J., Baker, R., Kneen, C., Moseley, J., Saunders, J., Seward, E.M., Stevenson, G., Beer, M., Stanton, J., and Watling, K. (1991) *Journal of Medicinal Chemistry*, **34**, 140.

178 Itoh, T., Miyazaki, M., Nagata, K., and Ohsawa, A. (2000) *Tetrahedron*, **56**, 4383.

179 Sisko, J., Kassick, A.J., and Shetzline, S.B. (2000) *Organic Letters*, **2**, 2877.

180 Boyd, G.V. (1984) *Comprehensive Heterocyclic Chemistry*, Vol. 6 (eds A.R. Katritzky, C. Rees, and E.F.V. Scriven), Pergamon, Oxford, pp. 191.

181 Grimmett, M.R. (1996) *Comprenhesive Heterocyclic Chemistry II*, Vol. 5 (eds A.R. Katritzky, C. Rees, and E.F.V. Scriven), Pergamon, Oxford, pp. 408.

182 Crich, J.Z., Brieva, R., Marquat, P., Gu, R.L., Flemming, S., and Sik, C.J. (1993) *The Journal of Organic Chemistry*, **58**, 3252.

183 Sorbera, L.A., Fernandez, R., and Castanar, J. (2001) *Drugs of the Future*, **26**, 453.

184 Wakefield, B.J. (1988) *Organolithium Methods*, Academic Press, London; Brandsma, L. (1990) *Preparative Polar Organometallic Chemistry*, Springer-Verlag, Berlin; Rewcastle, G.W. and Katritzky, A.R. (1993) *Advances in Heterocyclic Chemistry*, **56**, 155; Clayden, J. (2002) Organolithiums: Selectivity for Synthesis, in *Tetrahedron Organic Chemistry Series*, Vol. 23 (eds J.E. Baldwin and R.M. Williams), Pergamon, Oxford.

185 Undheim, K. and Benneche, T. (1990) *Heterocycles*, **30**, 1155; Snieckus, V. (1990) *Chemical Reviews*, **90**, 879; Quéguiner, G., Marsais, F., Snieckus, V., and Epsztajn, J. (1991) *Advances in Heterocyclic Chemistry*, **52**, 187; Mongin, F. and Quéguiner, G. (2001) *Tetrahedron*, **57**, 4059; Turck, A., Plé, N., Monguin, F., and Quéguiner, G. (2001) *Tetrahedron*, **57**, 4489; Godard, A., Rocca, P., Guillier, F., Durvey, G., Nivoliers, F., Marsais, F., and Quéguiner, G. (2001) *Canadian Journal of Chemistry*, **79**, 1754.

186 Sotomayor, N. and Lete, E. (2003) *Current Organic Chemistry*, **7**, 1; Ardeo, A., Collado, M.I., Osante, I., Ruiz, J., Sotomayor, N., and Lete, E. (2001) *Targets in Heterocyclic Chemistry*, **5**, 393.

187 Pettersen, D., Dinér, P., Amedjkouh, M., and Ahlberg, P. (2004) *Tetrahedron: Asymmetry*, **15**, 1607; Dinér, P., Pettersen, D., Lill, S.O.N., and Ahlberg, P. (2005) *Tetrahedron: Asymmetry*, **16**, 2665.

188 Montagne, C., Fournet, G., and Joseph, B. (2003) *Synlett*, 1533; Richardson, T.I., Ornstein, P.L., Briner, K., Fisher, M.J., Backer, R.T., Biggers, C.K., Clay, M.P., Emmerson, P.J., Hertel, L.W., Hsiung, H.M., Husain, S., Kahl, S.D., Lee, J.A., Lindstrom, T.D., Martinelli, M.J., Mayer, J.P., Mullaney, J.T., O'Brien, T.P., Pawlak, J.M., Revell, K.D., Shah, J., Zgombick, J.M., Herr, R.J., Melekhov, A., Sampson, P.B., and King, C.-H.R. (2004) *Journal of Medicinal Chemistry*, **47**, 744.

189 Janssens, F., Leenaerts, J., Diels, G., De Boeck, B., Megens, A., Langlois, X., van Rossem, K., Beetens, J., and Borgers, M. (2005) *Journal of Medicinal Chemistry*, **48**, 2154.

190 Chittiboyina, A.G., Reddy, C.R., Watkins, E.B., and Avery, M.A. (2004) *Tetrahedron Letters*, **45**, 1869.

191 Wang, L., Wang, G.T., Wang, X., Tong, Y., Sullivan, G., Park, D., Leonard, N.M., Li, Q., Cohen, J., Gu, W.-Z., Zhang, H., Bauch, J.L., Jakob, C.G., Hutchins, C.W., Stoll, V.S., Marsh, K., Rosenberg, S.H., Sham, H.L., and Lin, N.-H. (2004) *Journal of Medicinal Chemistry*, **47**, 612.

192 Feldman, K.S. and Skoumbourdis, A.P. (2005) *Organic Letters*, **7**, 929.

193 Jalil, M.A. and Hui, E.B. (2006) *Tetrahedron Letters*, **47**, 1473.

194 Worm, K., Chu, F., Matsumoto, K., Best, M.D., Lynch, V., and Anslyn, E.V. (2003) *Chemistry – A European Journal*, **9**, 741.

195 Yus, M. (2004 & 2006) in *The Chemistry of Organolithium Compounds, Vols* 1 & 2 (eds Z. Rappoport and I. Marek), John Wiley & Sons, Ltd., Chichester.

196 Yus, M. (1996) *Chemical Society Reviews*, **25**, 155; Ramón, D.J., and Yus, M. (2000) *European Journal of Organic Chemistry*, **2**, 225; Yus, M. (2001) *Synlett*, 1197; Torregrosa, R., Pastor, I.M., and Yus, M. (2005) *Tetrahedron*, **61**, 11148.

197 Torregrosa, R., Pastor, I.M., and Yus, M. (2007) *Tetrahedron*, **63**, 947.

198 Tschamber, T., Siendt, H., Boiron, A., Gessier, F., Deredas, D., Frankowski, A., Picasso, S., Steiner, H., Aubertin, A.M., and Streith, J. (2001) *European Journal of Organic Chemistry*, 1331; Weinberg, K., Jankowski, S., Le Nouen, D., and Frankowski, A. (2002) *Tetrahedron Letters*, **43**, 1089.

199 Bayh, O., Awad, H., and Mongin, F. (2005) *The Journal of Organic Chemistry*, **70**, 5190.

200 Iddon, B. (1994) *Heterocycles*, **37**, 1321; Hilf, C., Bosold, F., Harms, K., Marsch, M., and Boche, G. (1997) *Chemische Berichte-Recueil*, **130**, 1230.

201 Vedejs, E. and Monahan, S.D. (1996) *The Journal of Organic Chemistry*, **61**, 5192.

202 Williams, D.R., Brooks, D.A., and Meyer, K.G. (1998) *Tetrahedron Letters*, **39**, 8023.

203 Swaleh, S. and Liebscher, J. (2002) *The Journal of Organic Chemistry*, **67**, 3184.

204 Evans, D.A., Cee, V.J., Smith, T.E., and Santiago, K.J. (1999) *Organic Letters*, **1**, 87.

205 Smith, A.B., Minbiole, K.P., and Freeze, S. (2001) *Synlett*, 1739; Smith, A.B., Minibiole, K.P., Verhoest, P.R., and Schelhass, M. (2001) *Journal of the American Chemical Society*, **123**, 10942.

206 Stanetty, P., Spina, M., and Mihovilovic, M.D. (2005) *Synlett*, 1433.

207 Dondoni, A. and Marra, A. (2004) *Chemical Reviews*, **104**, 2557; Altman, L.J. and Richheimer, S.L. (1971) *Tetrahedron Letters*, **12**, 4709.

208 Busscher, G.F., Rutjes, F.P.J.T., and van Delft, F.L. (2004) *Tetrahedron Letters*, **45**, 3629; Dondoni, A., Catozzi, N., and Marra, A. (2004) *The Journal of Organic Chemistry*, **69**, 5023.

209 Dondoni, A. (1998) *Synthesis*, 1681; Dondoni, A., Franco, S., Junquera, F., Merchán, F.L., Merino, P., Tejero, T., and Bertolasi, V. (1995) *Chemistry – A European Journal*, **1**, 505.

210 Dondoni, A., Catozzi, N., and Marra, A. (2005) *The Journal of Organic Chemistry*, **70**, 9257.

211 Swanson, D.M., Dubin, A.E., Shah, C., Nasser, N., Chang, L., Dax, S.L., Jetter, M., Breitenbucher, J.G., Liu, Ch., Mazur, C., Lord, B., Gonzales, L., Hoey, K., Rizzolio, M., Bogenstaetter, M., Codd, E.E., Lee, D.H., Zhang, S.-P. Chaplau, S.R., and Carruthers, N.I. (2005) *Journal of Medicinal Chemistry*, **48**, 1857.

212 Sasaki, S., Hamada, Y., and Shioiri, T. (1997) *Tetrahedron Letters*, **38**, 3013; Dondoni, A., Junquera, F., Merchán, F.L., Merino, P., Schermann, M.C., and Tejero, T. (1997) *The Journal of Organic Chemistry*, **62**, 5484; Dondoni, A. and Perrone, D. (1999) *Tetrahedron Letters*, **40**, 9375; Dondoni, A., Giovannini, G., and Perrone, D. (2002) *The Journal of Organic Chemistry*, **67**, 7203.

213 Marcantonio, K.M., Frey, L.F., Murry, J.A., and Chen, C. (2002) *Tetrahedron Letters*, **43**, 8845.

214 Stanetty, P., Schnürch, M., Mereiter, K., and Mihovilovic M.D. (2005) *The Journal of Organic Chemistry*, **70**, 567.

215 Takami, S., Kawai, T., and Irle, M. (2002) *European Journal of Organic Chemistry*, 3796.

216 Silvermann, G.S. and Rakita, P.E. (eds) (1996) *Handbook of Grignard Reagents*, Marcel Dekker, New York.

217 Boymond, L., Rottländer, M., Cahiez, G., and Knochel, P. (1998) *Angewandte Chemie International Edition*, **37**, 1701; Abarbri, M., Dehmel, F., and Knochel, P. (1999) *Tetrahedron Letters*, **40**, 7499; Abarbri, M., Thibonnet, J., Bèrillon, L., Dehmel, F., Rottländer, M., and Knochel, P. (2000) *The Journal of Organic Chemistry*, **65**, 4618; Kneisel, F.F. and Knochel, P. (2002) *Synlett*, 1799.

218 Ross, T.M., Setter, M.C., McDonnell, M.E., Boyd, R.E., Connelly, C.D., Martínez, R.P., Lewis, M.A., Codd, E.E., Raffa, R.B., and Reitz, A.B. (2000) *Journal of Medicinal Chemistry*, **43**, 765.

219 Tschamber, T., Siendt, H., Tarnus, C., Deredas, D., Frankowski, A., Kohler, S., and Streith, J. (2002) *European Journal of Organic Chemistry*, 702.

220 Hari, Y., Obika, S., Sasaki, M., Morio, K., Yamagata, Y., and Imanishi, T. (2002) *Tetrahedron*, **58**, 3051; Obika, S., Hari, Y., Morio, K., and Imanishi, T. (2000) *Tetrahedron Letters*, **41**, 215.

221 Golden, K.G., Mattson, M.N., Cha, K.H., and Rapoport, H. (2002) *The Journal of Organic Chemistry*, **67**, 5913.

222 Boyd, R.E., Rasmussen, R., Press, J.B., Raffa, R.B., Codd, E., Connelly, C.D., Li, Q.S., Martinez, R.P., Lewis, M.A., Almond, H.R., and Reitz, A.B. (2001) *Journal of Medicinal Chemistry*, **44**, 863.

223 Obika, S., Hari, Y., Morio, K., and Imanishi, T. (2000) *Tetrahedron Letters*, **41**, 221.

224 Spieß, A., Heckmann, G., and Bach, T. (2004) *Synlett*, **131**

225 Bekish, A.V., Isakov, V.E., and Kulinkovich, O.G. (2005) *Tetrahedron Letters*, **46**, 6979.

226 Häbich, D. and Effenberger, F. (1979) *Synthesis*, 841; Rappoport, Z. and Apeloig, Y. (eds) (1998) *The Chemistry of Organosilicon Compounds*, Vol. 2, John Wiley & Sons, Ltd., Chichester, U.K, Part 2. Chapter 29.

227 Gimisis, T., Arsenyan, P., Georganakis, D., and Leondiadis, L. (2003) *Synlett*, 1451.

228 Tong, Y., Lin, N.-H., Wang, L, Hasvold, L., Wang, W., Leonard, N., Li, T., Qun-Li, J.C., Cohen, J., Gu, W.-Z., Zhang, H., Stoll, V., Bauch, J., Marsch, K., Rosenberg, S.H., and Sharm, H.L. (2003) *Bioorganic and Medicinal Chemistry Letters*, **13**, 1571; Frankowski, A., Deredas, D., Dubost, E., Gessier, F., Jankowski, S., Neuburger, M., Seliga, C., Tschamber, T., and Weinberg, K. (2003) *Tetrahedron*, **59**, 6503.

229 Zanirato, P. and Cerini, S. (2005) *Organic and Biomolecular Chemistry*, **3**, 1508.

230 Harusawa, S., Murai, Y., Moriyama, H., Ohishi, H., Yoneda, R., and Kurihaba, T. (1995) *Tetrahedron Letters*, **36**, 3165; Harusawa, S., Murai, Y., Moriyama, H., Imazu, T., Ohishi, H., Yoneda, R., and Kurihara, T. (1996) *The Journal of Organic Chemistry*, **61**, 4405; Ganellin, C.R., Fkyeray, A., Bang-Andersen, B., Athmani, S., Tertiuk, W., Garbarg, M., Ligneau, X., and Schwartz, J.C. (1996) *Journal of Medicinal Chemistry*, **39**, 3806; Erikse, B.L., Vedse, P., and Begtrup, M. (2001) *The Journal of Organic Chemistry*, **66**, 8344.

231 Dondoni, A., Fantin, G., Fogagnolo, M., Medici, M., and Pedrini, P. (1987) *The Journal of Organic Chemistry*, **52**, 3413.

232 Edwards, P.D., Wolanin, D.J., Andisik, D.W., and Davis, M.W. (1995) *Journal of Medicinal Chemistry*, **38**, 76.

233 Zarudnitskii, E., Perkav, I.I., Merkulov, A.S., Yurchenko, A.A., Tolmachev, A.A., and Pinchuk, A.M. (2006) *Synthesis*, 1279.

234 Wu, Y.-D., Lee, J.H., Houk, K.N., and Dondoni, A. (1996) *The Journal of Organic Chemistry*, **61**, 1922.

235 Ducept, P.C. and Marsden, S.P. (2000) *Synlett*, 692.

236 Miller, R.A., Smith, R.M., and Marcune, B. (2005) *The Journal of Organic Chemistry*, **70**, 9074.

237 Dondoni, A., Orduna, J., and Merino, P. (1992) *Synthesis*, 201; Wagner, A. and Mollath, M. (1993) *Tetrahedron Letters*, **34**, 619; Tejero, T., Dondoni, A., Rojo, I., Merchán, F.L., and Merino, P. (1997) *Tetrahedron*, **53**, 3301; Touwé, D., Piron, J., Defreyn, P., and Van Binst, G. (1993) *Tetrahedron Letters*, **34**, 5499; Ghosh, A.K., Bischoff, A., and Cappiello, J. (2001) *Organic Letters*, **3**, 2677; Khare, N.K., Sood, R.K., and Aspinall, G.O. (1994) *Canadian Journal of Chemistry*, **72**, 237; Nicolau, K.C., Rodríguez, R.M., Fylaktakidou, K.C., Suzuki, H., and Mitchell, H.J. (1999) *Angewandte Chemie International Edition*, **38**, 3340.

238 Dondoni, A., Perrone, D., and Merino, P. (1995) *The Journal of Organic Chemistry*, **60**, 8074.

239 Bagley, M.C., Dale, J.W., Xiong and X., Bower, J. (2003) *Organic Letters*, **5**, 4421.

240 Carcano, M. and Vasella, A. (1998) *Helvetica Chimica Acta*, **81**, 889.

241 Fürstner, A. and Ernst, A. (1995) *Tetrahedron*, **51**, 773.

242 Hoebert, J. (1997) *The Journal of Organic Chemistry*, **62**, 6615.

243 Ito, H., Sensul, H., Arimoto, K., Miura, K., and Hosomi, A. (1997) *Chemistry Letters*, 639.

244 Beard, R.L., Colon, D.F., Song, T.K., Davies, P.J.A., Kochlar, D.M., and Chandratna, R.A.S. (1996) *Journal of Medicinal Chemistry*, **39**, 3556.

245 Chinchilla, R., Nájera, C., and Yus, M. (2004) *Chemical Reviews*, **104**, 2667

246 Knochel, Paul (eds.) (2005) *Handbook of Functionalized Organometallics*, Wiley-VCH Verlag GmbH, Weinheim; Chinchilla, R., Nájera and C., Yus, M. (2007) *Arkivoc*, **152**.

247 Achab, S., Guyot, M., and Potier, P. (1995) *Tetrahedron Letters*, **36**, 2615; Gaare, K., Repstad, T., Beneche, T., and Undhim, K. (1993) *Acta Chemica Scandinavica (Copenhagen, Denmark: 1989)*, **47**, 57.

248 Bach, T. and Heuser, S. (2002) *The Journal of Organic Chemistry*, **67**, 5789.

249 Kelly, T.R. and Lang, F. (1995) *Tetrahedron Letters*, **36**, 9293.

250 Heckmann, G. and Bach, T. (2005) *Angewandte Chemie International Edition*, **44**, 1199.

251 Erdik, E. (1996) *Organozinc Reagents in Organic Synthesis*, CRC Press, New York; Knochel, P. and Jones, P. (1999) *Organozinc Reagents*, Oxford University Press, Oxford; Fürstner, A. (ed.) (1996) *Active Metals*, VCH, Weinheim, Germany.

252 Gosmini, C., Nédeléc, J.Y., and Périchon, J. (1997) *Tetrahedron Letters*, **38**, 1941.

253 Heckmann, G. and Bach, T. (2005) *Angewandte Chemie International Edition*, **44**, 1199.

254 Anderson, B.A. and Harn, N.K. (1996) *Synlett*, 583.

255 Rottländer, M. and Knochel, P. (1998) *The Journal of Organic Chemistry*, **63**, 203.

256 Prasad, A.S., Stevenson, T.M., Citineni, J.R., Nyzam, V., and Knochel, P. (1997) *Tetrahedron*, **53**, 7237.

257 Yang, X. and Knochel, P. (2006) *Chemical Communications*, 2170.

258 Marino, J.P. and Nguyen, H.N. (2003) *Tetrahedron Letters*, **44**, 7395; Alberico, D., Scott, M.E., and Lautens, M. (2007) *Chemical Reviews*, **107**, 174; Lewis, J.C., Bergman, R.G., and Ellman, J.A. (2008) *Accounts Chemical Research*, **41**, 1013; Campeau, L.-C. and Fagnou, K. (2006) *Chemical Communications*, 1253; Campeau, L.-C., Stuart, D.R., and Fagnou, K. (2007) *Aldrichimica Acta*, **40**, 35; Seregin, I.V. and Gevorgyan, V. (2007) *Chemical Society Reviews*, **36**, 1173.

259 Pivsa-Art, S., Satoh, T., Kawamura, Y., Miura, M., and Nomura, M. (1998) *Bulletin of the Chemical Society of Japan*, **71**, 467; Akita, Y., Itagaki, Y., Takizawa, S., and Ohta, A. (1989) *Chemical and Pharmaceutical Bulletin*, **37**, 1477; A review on arylation of arenes: Miura, M. and Nomura, M. (2002) *Topics in Current Chemistry*, **219**, 211.

260 Kondo, Y., Nomine, T., and Sakamoto, T. (2000) *Organic Letters*, **2**, 3111.

261 Yokooji, A., Okazawa, T., Satoh, T., Miura, M., and Nomura, M. (2003) *Tetrahedron*, **59**, 5685.

262 Campeau, L-Ch., Bertrand-Laperle, M., Leclerc, J-P., Villemure, E., Gorelsky, S., and Fagnou, K. (2008) *Journal of the American Chemical Society*, **130**, 3276.

263 Wang, X., Gribkov, D.V., and Sames, D. (2007) *The Journal of Organic Chemistry*, **72**, 1476; Cerna, I., Pohl, R., Klepetarova B., and Hocek, M. (2006) *Organic Letters*, **8**, 5389; Besselievre, F., Mahuteau-Betzer, F., Grierson, D.S., and Piguel, S. (2008) *The Journal of Organic Chemistry*, **73**, 3278; Bellina, F., Cauteruccio, S., Mannina, L., Rossi, R., and Viel, S. (2006) *European Journal of Organic Chemistry*, **693**; Bellina, F., Cauteruccio, S., and Rossi, R. (2007) *The Journal of Organic Chemistry*, **72**, 8543; Lebrasseur, N. and Larrosa, I. (2008) *Journal of the American Chemical Society*, **130**, 2926.

264 Zhao, D., Wang, W., Lian, S., Yang, F., Lan, J., and You, J. (2009) *Chemistry - A European Journal*, **15**, 1337.

265 Brasche, G. and Buchwald, S.L. (2008) *Angewandte Chemie International Edition*, **47**, 1932; Inamoto, K., Hasegawa, C., Hiroya, K., and Doi, T. (2008) *Organic Letters*, **10**, 5147; Thu, H.-Y., Yu, W.-Y., and Che, C.-M. (2006) *Journal of the American Chemical Society*, **128**, 9048; Wasa, M. and Yu, J.-Q. (2008) *Journal of the American Chemical Society*, **130**, 14058; Lebel, H., Huard, K., and Lectard, S. (2005) *Journal of the American Chemical Society*, **127**, 14198.

266 Hamada, T., Ye, X., and Stahl, S.S. (2008) *Journal of the American Chemical Society*, **130**, 833; Chen, X., Hao, X.-S., Goodhue, C.E., and Yu, J.-Q. (2006) *Journal of the American Chemical Society*,

128, 6791; Thu, H.-Y., Yu, W.-Y., and Che, C.-M. (2006) *Journal of the American Chemical Society*, **128**, 9048.

267 Monguchi, D., Fujiwara, T., Furukawa, H., and Mori, A. (2009) *Organic Letters*, **11**, 1607.

268 Klapars, A., Antilla, J.C., Huang, X., and Buchwald, S.L. (2001) *Journal of the American Chemical Society*, **123**, 7727; Lam, P.Y.S., Clark, C.G., Saubern, S., Adams, J., Winters, M.P., Chan, D.M.T., and Combs, A. (1998) *Tetrahedron Letters*, **39**, 2941; Lam, P.Y.S., Vicent, G., Clark, C.G., Deudon, S., and Jadhav, P.K. (2001) *Tetrahedron Letters*, **42**, 3415.

269 Kang, S.K., Lee, S.H., and Lee, D. (2000) *Synlett*, 1022.

270 Zhu, R., Xing, L., Wang, X., Cheng, C., Su, D., and Hua, Y. (2008) *Advanced Synthesis & Catalysis*, **350**, 1253.

271 Collman, J.P. and Zhong, M. (2000) *Organic Letters*, **2**, 1233.

272 Deagostino, A., Prandi, C., Zavattaro, C., and Venturello, P. (2007) *European Journal of Organic Chemistry*, **8**, 1318.

273 Cacchi, S., Fabrizi, G., and Moro, L. (1997) *The Journal of Organic Chemistry*, **62**, 5327.

274 Hassan, J., Lavenot, L., Gozzi, C., and Lemaire, M. (1999) *Tetrahedron Letters*, **40**, 857.

275 Nuβbaumer, T. and Neidlein, R. (2000) *Heterocycles*, **52**, 349.

276 Apen, P.G. and Rasmussen, P.G. (1991) *Journal of the American Chemical Society*, **113**, 6178.

277 Tubaro, C., Biffis, A., Scattolin, E., and Basato, M. (2008) *Tetrahedron*, **64**, 4187.

278 Beletskaya, I.P. and Cheprakov, A.V. (2004) *Coordination Chemistry Reviews*, **248**, 2337; Shafir, A. and Buchwald, S.L. (2006) *Journal of the American Chemical Society*, **128**, 8742; Shafir, A., Lichtor, P.A., and Buchwald, S.L. (2007) *Journal of the American Chemical Society*, **129**, 3490; Zhang, H., Cai, Q., and Ma, D. (2005) *The Journal of Organic Chemistry*, **70**, 5164; Cai, Q., Zou, B., and Ma, D. (2006) *Angewandte Chemie International Edition*, **45**, 1276; Ouali, A., Spindler, J.-F., Cristau, H.-J., and Taillefer, M. (2006), *Advanced Synthesis & Catalysis*, **348**, 499; Wang, Z., Bao, W., and Jiang, Y. (2005) *Chemical Communications, 2849;* Manbeck, G.F., Lipman, A.J., Stockland, R. A., Jr., Freidl, A.L., Hasler, A.F., Stone, J.J., and Guzei, I. A, (2005) *The Journal of Organic Chemistry*, **70**, 244; Liu, L., Frohn, M., Xi, N., Dominguez, C., Hungate, R., and Reider, P.J. (2005) *The Journal of Organic Chemistry*, **70**, 10135; Guo, X., Rao, H.H., Fu, H., Jiang, Y.Y., and Zhao, Y.F. (2006) *Advanced Synthesis & Catalysis*, **348**, 2197; de Lange, B., Lambers-Verstappen, M.H., Schmieder-van de Vondervoort, L., Sereinig, N., De Rijk, R., De Vries, A. H. M., and De Vries, J.G. (2006) *Synlett*, 3105; Jiang, D., Fu, H., Jiang, Y., and Zhao, Y. (2007) *The Journal of Organic Chemistry*, **72**, 672; Zhu, L.; S Cheng, L., Zhang, Y., Xie, R., and You, J. (2007) *The Journal of Organic Chemistry*, **72**, 2737; Lv, X., and Bao, W. (2007) *The Journal of Organic Chemistry*, **72**, 3863.

279 Son, S.U., Park, I.K., Park, J., and Hyeon, T. (2004) *Chemical Communications*, 778.

280 Choudary, B.C., Sridhar, C., Kantam, M.L., Venkanna, G.T., and Sreedhar, B. (2005) *Journal of the American Chemical Society*, **127**, 9948; Kantam, M.L., Yadav, J., Laha, S., Sreedhar, B., and Jhab, S. (2007) *Advanced Synthesis & Catalysis*, **349**, 1938.

281 Lohmann, S., Andrews, S.P., Burke, B.J., Smith, M.D., Attfield, J.P., Tanaka, H., Kancko, K., and Ley, S.V. (2005) *Synlett*, 1291.

282 Evans, D.A. and Bach, T. (1993) *Angewandte Chemie International Edition*, **32**, 1326.

283 David, W.M., Kumar, D., and Kerwin, S.M. (2000) *Bioorganic and Medicinal Chemistry Letters*, **10**, 2509.

284 Nicolau, K.C., King, N.P., Finlay, M.R.V., He, Y., Roschangar, F., Vourlumis, D., Valber, H., Sarabia, F., Ninkovic, S., and Hepworh, D. (1999) *Bioorganic and Medicinal Chemistry*, **7**, 665; Nicolau, K.C., He, Y., Roschangar, F., King, N.P., Vourloumis, D., and Li, T. (1998) *Angewandte Chemie International Edition*, **37**, 84; Evans, O.R. and Lin, W. (2000) *Journal of The Chemical Society-Dalton Transactions*, 3949; Frackenpohl, J., Braje, W.M., and Hoffmann, H.M.R. (2001) *Journal of the Chemical Society-Perkin Transactions* 1, 47; Siebeneicher, H. and

Doye, S. (2002) *European Journal of Organic Chemistry*, 1213; Kumar, D., David, W.M., and Kerwin, S.M. (2001) *Bioorganic and Medicinal Chemistry Letters*, **11**, 2971.

285 Langille, N.F., Dakin, L.A., and Panek, J.S. (2002) *Organic Letters*, **4**, 2485.

286 Buynak, J.D., Doppalapudi, V.R., Frotan, M., Kumar, R., and Chambers, A. (2000) *Tetrahedron*, **56**, 5709.

287 Wang, D. and Haseltine, J. (1994) *Journal of Heterocyclic Chemistry*, **31**, 1637.

288 Brauer, D.J., Kottsieper, K.W., Lick, C., Stelzer, O., Waffenschmidt, H., and Wasserscheid, P. (2001) *Journal of Organometallic Chemistry*, **630**, 177.

289 Kosugi, M., Shimizu, Y., and Migina, T. (1977) *Chemistry Letters*, 1423; Milstein, D. and Stille, J.K. (1979) *Journal of the American Chemical Society*, **101**, 4992; Stille, J.K. (1986) *Angewandte Chemie International Edition*, **25**, 508; Mitchell, T.N. (1992) *Synthesis*, 803; Farina, V., Krishnamurthy, V., and Scott, W.J. (1997) *Organic Reactions*, **50**, 1; Davies, A.G. (1997) *Organothin Chemistry*, Wiley-VCH, New York; Farina, V., Krishnamurthy, V., and Scott, W.J. (1998) *The Stille Reaction*, John Wiley & Sons, Inc., New York; Kosugi, M. and Fugami, K. (2002) *Journal of Organometallic Chemistry*, **653**, 50.

290 Revesz, L., Bonne, F., and Makavou, P. (1998) *Tetrahedron Letters*, **39**, 5171.

291 Achab, S., Guyot, M., and Potier, P. (1995) *Tetrahedron Letters*, **36**, 2615; Gaare, K., Repstad, T., Beneche, T., and Undhim, K. (1993) *Acta Chemica Scandinavica (Copenhagen, Denmark: 1989)*, **47**, 57.

292 Terpin, A., Winklhofer, C., Schumann, S., and Steglich, W. (1998) *Tetrahedron*, **54**, 1745.

293 McDonald, D., Perrier, H., Liu, S., Laliberté, F., Rasori, R., Robichaud, A., Masson, P., and Huang, Z. (2000) *Journal of Medicinal Chemistry*, **43**, 3820.

294 Kelly, T.R. and Lang, F. (1996) *The Journal of Organic Chemistry*, **61**, 4623.

295 Yang, Y. and Martin, A.R. (1992) *Heterocycles*, **34**, 1395; Ward, P., Armour, D.R., Bays, D.E., Evans, B., Giblin, G.M.P., Heron, N., Hubbard, T., Liang, K., Middlemiss, D., Mordaunt, J., Naylor, A., Pegg, N.A., Vinader, M.V., Watson, S.P., Bountra, C., and Evans, D. (1995) *Journal of Medicinal Chemistry*, **38**, 4985; Kranich, R., Eis, K., Geis, O., Mühle, S., Bats, J.W., and Schmalz, H.-G. (2000) *Chemistry – A European Journal*, **6**, 2874.

296 Wellmar, U., Hörnfeldt, A.-B., and Gronowitz, S. (1995) *Journal of Heterocyclic Chemistry*, **32**, 1159.

297 Hodgetts, K.J. and Kershaw, M.T. (2002) *Organic Letters*, **4**, 1363.

298 Sakamoto, T., Kondo, Y., Takazawa, N., and Yamanaka, H. (1993) *Tetrahedron Letters*, **34**, 5955; Sakamoto, T., Kondo, Y., Takazawa, N., and Yamanaka, H. (1996) *Journal of the Chemical Society-Perkin Transactions 1*, 1927.

299 Cetusic, J.R.P., Green, F.R., Graupner, P.R., and Oliver, M.P. (2002) *Organic Letters*, **4**, 1307.

300 Alvarez-Ibarra, C., Asperilla, R., de Dios-Corredor, C., Martinez-Santos, E., and Quiroga, M.L. (1991) *Heterocycles*, **32**, 2127.

301 Kitade, Y., Kozaki, A., Miwa, T., and Nakanishi, M. (2002) *Tetrahedron*, **58**, 1271; Hirota, K., Kitade, Y., Kanbe, Y., and Maki, Y. (1992) *The Journal of Organic Chemistry*, **57**, 5268.

302 Iwamura, T., Okamoto, Y., Yokomoto, M., Shimizu, H., Hori, M., and Kataoka, T. (1994) *Synthesis*, 203; Bold, G., Faessler, A., Capraro, H.-G., Cozens, R., Klimkait, T., Lazdins, J., Mestan, J., Poncioni, B., Rösel, J., Stover, D., Tintelnot-Blomley, M., Acemoglu, F., Beck, W., Boss, E., Eschbach, M., Hürlimann, T., Masso, E.L., Roussel, S., Ucci-Stoll, K., Wyss, D., and Lang, M. (1998) *Journal of Medicinal Chemistry*, **41**, 3387.

303 Fürstner, A. and Leitner, A. (2002) *Angewandte Chemie International Edition*, **41**, 609.

304 Minisci, F. and Porta, O. (1974) *Advances in Heterocyclic Chemistry*, **16**, 123.

305 Grimmett, M.R. (1980) *Advances in Heterocyclic Chemistry*, **27**, 241.

306 Allin, S.M., Bowman, W.R., Elsegood, M.R.J., McKee, V., Karim, R., and Rahman, S.S. (2005) *Tetrahedron*, **61**, 2689; Aldabbagh, F., Bowman, W.R.,

Mann, E., and Slawin, A.M.Z. (1999) *Tetrahedron*, **55**, 8111; Aldabbagh, F., Bowman, W.R., and Mann, E. (1997) *Tetrahedron Letters*, **38**, 7937.

307 Lakhan, R. and Ternai, B. (1974) *Advances in Heterocyclic Chemistry*, **17**, 99.

308 Dondoni, A., Fogagnolo, M., Medici, A., and Pedrini, P. (1985) *Tetrahedron Letters*, **26**, 5477.

309 Katritzky, A.R. (1985) *Handbook of Comprehensive Heterocyclic Chemistry*, Pergamon, Oxford, pp. 328.

310 Kondrat'eva, G.Y. (1957) *Khimicheskaya Nauka i Promyshlennost*, **2**, 666; Karspeiskii, M.Y. and Florentev, V.L. (1969) *Russian Chemical Reviews (English Translation)*, **38**, 540.

311 Alla, V.R., Gurjar, M.K., Devi, T.R., and Ramanaiam, K. Indian Patent IN175617 A 19950715; (2003) *Chemical Abstracts*, **139**, 52798. 261.

312 For recent examples see: Ohba, M., Kubo, H., and Ishibashi, H. (2000) *Tetrahedron*, **56**, 7751; Whitney, S.E. and Rickborn, B. (1988) *The Journal of Organic Chemistry*, **53**, 5595.

313 Jaworski, T., Mizerski, T., and Krokilowska, A. (1979) *Polish Journal of Chemistry*, **53**, 1799; Reddy, G.S. and Bhatt, M.V. (1980) *Tetrahedron Letters*, **21**, 3627.

314 Potts, K.T., Choudhury, D.R., and Westby, T.R. (1976) *The Journal of Organic Chemistry*, **41**, 187.

315 Abbot, P.J., Acheson, R.M., Eisner, U., Watkin, D.J., and Carruthers, J.R.J. (1976) *Journal of the Chemical Society-Perkin Transactions 1*, 1269; Maeda, M., Ito, S., and Kajima, M. (1976) *Tetrahedron Letters*, **38**, 3463.

316 Elliott, M.C., Kruiswijk, E., and Long, M.S. (2001) *Tetrahedron*, **57**, 6651; Medici, A., Fantin, G., Fogagnolo, M., Pedrin, P., Dondoni, A., and Andreetti, G.D. (1984) *The Journal of Organic Chemistry*, **49**, 590.

317 Kawahara, T., Nakajima, T., Ito, T., and Ogura, H. (1982) *Heterocycles*, **9**, 1623.

318 Acheson, R.M. and Taylor, G.A. (1960) *Journal of the Chemical Society*, 4600.

319 Sauers, R.R. and Van Arnum, S.D. (1987) *Tetrahedron Letters*, **28**, 5797.

320 Matsuura, T., Banba, A., and Ogura, K. (1971) *Tetrahedron*, **27**, 1211.

321 Nakano, T., Rodríguez, W., de Roche, S.Z., Larrauri, J.M., Rivas, C., and Pérez, C. (1980) *Journal of Heterocyclic Chemistry*, **17**, 1777.

322 Allin, S.M., Barton, W.R.S., Bowmana, W.R., Bridge, E., Elsegood, M.R.J., McInally, T., and McKee, V. (2008) *Tetrahedron*, **64**, 7745.

323 Desimoni, G., Faita, G., and Quadrelli, P. (2003) *Chemical Reviews*, **123**, 3119; Rechavi, D. and Lemaire, M. (2002) *Chemical Reviews*, **102**, 3467.

324 Helmchen, G., Krotz, A., Ganz, K.-T., and Hansen, D. (1991) *Synlett*, 257.

325 Sakakura, A., Kondo, R., and Ishihara, K. (2005) *Organic Letters*, **7**, 1971.

326 Raman, P., Razavi, H., and Kelly, J.K. (2000) *Organic Letters*, **2**, 3289.

327 Reddy, L.R., Saravanan, P., and Corey, E.J. (2004) *Journal of the American Chemical Society*, **126**, 6230.

328 You, S.-L., Razavi, H., and Kelly, J.K. (2003) *Angewandte Chemie International Edition*, **42**, 83.

329 Yokokawa, F. and Shioiri, T. (2002) *Tetrahedron Letters*, **43**, 8679.

330 Wipf, P. and Miller, C.P. (1992) *Tetrahedron Letters*, **33**, 907.

331 Wipf, P. and Miller, C.P. (1992) *Tetrahedron Letters*, **33**, 6267.

332 Phillips, A.J., Uto, Y., Wipf, P., Reno, M.J., and Williams, D.R. (2000) *Organic Letters*, **2**, 1165.

333 Pirrung, M.C., Turney, L.N., Mc Clerren, A.L., and Raetz, C.R.H. (2003) *Journal of the American Chemical Society*, **125**, 1575.

334 Vastila, P., Pastor, I.M., and Adolfsson, H. (2005) *The Journal of Organic Chemistry*, **70**, 2921.

335 Benito, J.M., Christensen, C.A., and Meldal, M. (2005) *Organic Letters*, **7**, 581.

336 Pfund, E., Lequeux, T., Masson, S., and Vazeux, M. (2002) *Organic Letters*, **4**, 843; Crawhall, J.C. and Elliott, D.F. (1951) *Journal of the Chemical Society*, 2071.

337 Abrunhosa, I., Gulea, M., Levillain, J., and Masson, S. (2001) *Tetrahedron : Asymmetry*, **12**, 2851.

338 Walzer, M.A. and Heathcock, C.H. (1992) *The Journal of Organic Chemistry*, **57**, 5566; Kedrowski, B.L. and Heathcock, C.H. (2002) *Heterocycles*, **58**, 601.

339 Wipf, P. and Fritch, P.C. (1994) *Tetrahedron Letters*, **35**, 5397.

340 Zarantonello, P., Lesile, C.P., Ferretto, R., and Kazmierski, W.M. (2002) *Bioorganic and Medicinal Chemistry Letters*, **12**, 561.

341 Wipf, P. and Wang, X. (2002) *Organic Letters*, **4**, 1197.

342 Le Flemme, N., Marchand, P., Gulea, M., and Masson, S. (2000) *Synthesis*, **8**, 1143.

343 Fodor, G. and Nagubandi, S. (1980) *Tetrahedron*, **36**, 1279.

344 Acharaya, A.N., Ostresh, J.M., and Houghten, R.A. (2001) *The Journal of Organic Chemistry*, **66**, 8673; Acharaya, A.N., Ostresh, J.M., and Houghten, R.A. (2001) *Journal of Combinatorial Chemistry*, **3**, 612.

345 Boland, N.A., Casey, M., Hynes, S.J., Matthews, J.W., and Smyth, M.P. (2002) *The Journal of Organic Chemistry*, **67**, 3919.

346 Wipf, P. and Wang, X. (2002) *Journal of Combinatorial Chemistry*, **4**, 656.

347 Vorbruggen, H. and Krolikiewicz, K. (1993) *Tetrahedron*, **49**, 9353.

348 Rajaram, S. and Sigman, M.S. (2002) *Organic Letters*, **4**, 3399.

349 Katritzky, A.R., Cai, C., Suzuki, K., and Sing, S.K. (2004) *The Journal of Organic Chemistry*, **69**, 811.

350 Ferretti, G., Dukat, M., Giannella, M., Piergentili, A., Pigini, M., Quaglia, W., Damaj, M.I., Martin, B.R., and Glennon, R.A. (2002) *Journal of Medicinal Chemistry*, **45**, 4724.

351 Martin, P.K., Matthews, H.R., Rapoport, H., and Thyagarajan, G. (1968) *The Journal of Organic Chemistry*, **33**, 3758.

352 Gerische, M., Krumper, J.R., Bergman, R.G., and Don Tilley, T. (2003) *Organometallics*, **22**, 47.

353 Bolm, C., Weickhardt, K., Zender, M., and Rauff, X. (1991) *Chemische Berichte*, **124**, 1173.

354 Ehrler, J. and Farooq, S. (1994) *Synlett*, 702.

355 Bayardon, J. and Sinou, D. (2004) *The Journal of Organic Chemistry*, **69**, 3121.

356 Pattenden, G. and Thom, S.M. (1993) *Journal of the Chemical Society-Perkin Transactions 1*, 1629.

357 Baati, R., Gouverneur, V., and Mioskowski, C. (1999) *Synthesis*, **6**, 927.

358 Garcìa Ruano, J.L. and Garcìa Paredes, C. (2000) *Tetrahedron Letters*, **41**, 5357.

359 Concellòn, J.M., Riego, E., Suàrez, J.R., Garcìa-Granda, S., and Diaz, M.R. (2004) *Organic Letters*, **6**, 4499.

360 Damkaci, F. and DeShong, P. (2003) *Journal of the American Chemical Society*, **125**, 4408.

361 Chen, J. and Forsyth, C.J. (2003) *Organic Letters*, **5**, 1281.

362 Takeuchi, H., Hagiwara, S., and Educhi, S. (1989) *Tetrahedron*, **45**, 6375.

363 Molina, P., Diaz, I., and Tàrraga, A. (1995) *Synlett*, 1031.

364 Bacchi, A., Costa, M., Gabriele, B., Pelizzi, G., and Salerno, G. (2002) *The Journal of Organic Chemistry*, **67**, 4450.

365 Peddibhotla, S., Jayakumar, S., and Tepe, J.J. (2002) *Organic Letters*, **4**, 3533.

366 Bon, R.S., van Vliet, B., Sprenkels, N.E., Scmitz, R.F., de Kanter, F.J.J., Stevens, C.V., Swart, M., Bickelhaupt, M., Groen, M.B., and Orru, R.V. (2005) *The Journal of Organic Chemistry*, **70**, 3542.

367 You, S.-L. and Kelly, J.W. (2004) *Organic Letters*, **6**, 1681.

368 You, S.-L., Razavi, H., and Kelly, J.W. (2003) *Angewandte Chemie International Edition*, **42**, 83.

369 Pei, W., Wei, H.-X., Chen, D., Headley, A.D., and Li, G. (2003) *The Journal of Organic Chemistry*, **68**, 8404.

370 Liu, B., Davis, R., Joshi, B., and Reynolds, D.W. (2002) *The Journal of Organic Chemistry*, **67**, 4595 and references cited therein.

371 Wipf, P., Miller, C.P., Venkatraman, S., and Fritch, P.C. (1995) *Tetrahedron Letters*, **36**, 3695.

372 Charette, A.B. and Chua, P. (1998) *The Journal of Organic Chemistry*, **63**, 908.

373 Mylari, B., Larson, E.R., Beyer, T., Zembrowski, W.J., Aldinger, C.E., Dee, M.F., Siegel, T.W., and Singleton, D.H.

(1991) *Journal of Medicinal Chemistry*, **34**, 108.

374 Yokum, T.S., Alsina, J., and Barany, G. (2000) *Journal of Combinatorial Chemistry*, **2**, 282.

375 Ismail, M.A., Brun, R., Wenzler, T., Tanious, F.A., Wilson, W.D., and Boykin, D.W. (2004) *Bioorganic and Medicinal Chemistry*, **12**, 5405.

376 Bougrin, K., Loupy, A., and Souflaoui, M. (1998) *Tetrahedron*, **54**, 8055.

377 Varam, S.R. and Kumar, D. (1998) *Journal of Heterocyclic Chemistry*, **35**, 1539; Ouyang, J., Ouyang, C., Fujii, Y., Nakano, Y., Shoda, T., and Nagano, T. (2004) *Journal of Heterocyclic Chemistry*, **41**, 359.

378 Guzow, K., Szabelski, M., Malicka, J., and Wicz, W. (2001) *Helvetica Chimica Acta*, **84**, 1086.

379 Yoshifujii, M., Nagase, R., Kawashima, T., and Inamoto, N. (1978) *Heterocycles*, **10**, 57.

380 Beaulieu, P.L., Hache, B., and Von Moos, E. (2003) *Synthesis*, 1683.

381 Mathis, A.C., Wang, Y.M., Holt, D.P., Huang, G.-F., Debnath, M.L., and Klunk, W.E. (2003) *Journal of Medicinal Chemistry*, **46**, 2740.

382 Curini, M., Epifano, F., Montanari, F., Rosati, O., and Taccone, S. (2004) *Synlett*, 1832.

383 Kawashita, Y., Nakamichi, N., Kawabata, H., and Hayashi, M. (2003) *Organic Letters*, **5**, 3713.

384 Alloum, A.B., Bougrin, K., and Soufiaoui, M. (2003) *Tetrahedron Letters*, **44**, 5935.

385 Kodomari, M., Tamaru, Y., and Aoyama, T. (2004) *Synthetic Communications*, **34**, 3029.

386 Hisashi, A., Fukase, K., and Kusumoto, S. (2002) *Journal of Combinatorial Chemistry*, **4**, 475.

387 Krchňák, V. and Holladay, M.W. (2002) *Chemical Reviews*, **102**, 61.

388 Chang, J., Zhao, K., and Pan, S. (2002) *Tetrahedron Letters*, **43**, 951.

389 Wright, J.B. (1951) *Chemical Reviews*, **48**, 397.

390 Phillip, M.A. (1931) *Journal of the Chemical Society*, 1143.

391 Alvarez, F., Gherardi, A., Nebois, P., Sarciron, M.-E., Peatvy, A.-F., and Walschofer, N. (2002) *Bioorganic and Medicinal Chemistry Letters*, **12**, 977.

392 Ding, Y., Hofstadler, S.A., Swayze, E.E., Risen, L., and Griffey, R.H. (2003) *Angewandte Chemie International Edition*, **42**, 3409.

393 Yu, H., Kawanishi, H., and Koshima, H. (2003) *Heterocycles*, **60**, 1457.

394 Cao, D.-X., Fang, Q., Wang, D., Liu, Z.-Q., Xue, G., Xu, G.-B., and Yu, W.-T. (2003) *European Journal of Organic Chemistry*, 3628.

395 Freyer, A.J., Lowe-Ma, C.K., Nissan, R.A., and Wilson, W.S. (1992) *Australian Journal of Chemistry*, **45**, 525.

396 Hendrickson, J.B. and Hussoin, M.S. (1987) *The Journal of Organic Chemistry*, **52**, 4137.

397 Padilla-Martinez, I., Martinez-Martinez, F., Garcia-Baez, E.V., Torres-Valencia, J.M., Rojas-Lima, S., and Hoepfl, H. (2001) *Journal of the Chemical Society-Perkin Transactions 2*, 1817; Suzuki, N., Yamabayashi, T., and Izawa, Y. (1976) *Bulletin of the Chemical Society of Japan*, **49**, 353.

398 Katrizky, A.R., Rachwal, B., Rachwal, S., and Zaklika, K.A. (1994) *Heterocycles*, **38**, 2415; Katrizky, A.R., Musgrave, R.P., Rachwal, B., and Zaklika, K.A. (1995) *Heterocycles*, **41**, 345.

399 Abbotto, A., Bradamante, S., Facchetti, A., and Pagani, G.A. (2002) *The Journal of Organic Chemistry*, **67**, 5753; Hunger, A., Kebrle, J., Rossi, A., and Hoffmann, K. (1960) *Helvetica Chimica Acta*, **43**, 800.

400 Nabulsi, N.A.R. and Gandour, R.D. (1991) *The Journal of Organic Chemistry*, **56**, 2260.

401 Grimshaw, J. and Trocha-Grimshaw, J. (1975) *Tetrahedron Letters*, **16**, 2601.

402 Lu, L., Lachicotte, R., Penner, T.L., Perlstein, J., and Whitten, D.G. (1999) *Journal of the American Chemical Society*, **121**, 8146; Gillespie, H.B., Spano, F., and Graaf, S. (1960) *The Journal of Organic Chemistry*, **25**, 942.

403 Rivas, F.M., Giessert, A.J., and Diver, S.T. (2002) *The Journal of Organic Chemistry*, **67**, 1708.

404 Victor, F., Loncharich, R., Tang, J., and Spitzer, W.A. (1997) *Journal of Medicinal Chemistry*, **40**, 3478.

405 Wang, B.B. and Smith, P.J. (2003) *Tetrahedron Letters*, **44**, 8967.

406 Von Galhn, B., Kramer, W., Neidlein, R., and Suschtzky, H. (1999) *Journal of Heterocyclic Chemistry*, **36**, 1001.

407 Subhashini, N.J.P. and Hanumanthu, P. (1991) *Indian Journal of Chemistry Section B-Organic Chemistry Including Medicinal Chemistry*, **30**, 427.

408 Burkholder, C.R., Dolbier, W.R., and Medebielle, M. (2000) *Journal of Fluorine Chemistry*, **102**, 369.

409 Rivas, F.M., Riaz, U., Giessert, A., Smulik, J.A., and Diver, S.T. (2001) *Organic Letters*, **3**, 2673.

410 Abbotto, A., Bradamante, S., Facchetti, A., and Pagani, G. (2002) *The Journal of Organic Chemistry*, **67**, 5753.

411 Schreiner, E.P., Wolff, B., Winiski, A.P., and Billich, A. (2003) *Bioorganic and Medicinal Chemistry Letters*, **13**, 4313.

412 Auwers, J. (1925) *Chemische Berichte*, **58**, 34; Blatt, H.A. (1938) *Journal of the American Chemical Society*, **60**, 205; Bhawal, B.M., Mayabhate, S.P., and Likhite, A.P. (1995) *Synthetic Communications*, **25**, 3315.

413 White, A.W., Almassy, R., Calvert, A.H., Curtin, N.J., Griffin, R.J., Hostomsky, Z., Maegley, K., Newell, D.R., Srinivasan, S., and Golding, B.T. (2000) *Journal of Medicinal Chemistry*, **43**, 4084.

414 Tempest, P., Ma, V., Thomas, S., Hua, Z., Kelly, M.G., and Hulme, C. (2001) *Tetrahedron Letters*, **42**, 4959.

415 Takeuchi, H. and Koyama, K. (1982) *Journal of the Chemical Society-Perkin Transactions 1*, 1269.

416 Barany, X. and Pianka, X. (1953) *Journal of the Chemical Society*, 2217.

417 Raman, D.V. and Kantharaj, E. (1994) *Tetrahedron*, **50**, 2485.

418 Bougrin, K., Loupy, A., Petit, A., Daou, B., and Soufiaoui, M. (2001) *Tetrahedron*, **57**, 163.

419 Rigo, B., Valligny, D., and Tisne, S. (1988) *Synthetic Communications*, **18**, 167.

420 Paizs, C., Tosa, M., Majdik, C., Taehtinen, P., Irimie, F.D., and Kanerva, L.T. (2003) *Tetrahedron : Asymmetry*, **14**, 1943.

421 Desai, R.D., Hunter, R.F., and Kureishy, M.J. (1936) *Journal of the Chemical Society*, 1668.

422 Yoshino, K., Hori, N., Hori, M., Morita, T., and Tosukamoto, G. (1989) *Journal of Heterocyclic Chemistry*, **26**, 1039; Bowman, W.R., Heaney, H., and Smith, P.H.G. (1982) *Tetrahedron Letters*, **23**, 5093.

423 Benedi, C., Bravo, F., Uriz, P., Fernandez, E., Claver, C., and Castillon, S. (2003) *Tetrahedron Letters*, **44**, 6073.

424 Boggust, W., Cocker, W., Schwarz, J., and Stuart, E. (1950) *Journal of the Chemical Society*, 680.

425 Perry, R.J. and Wilson, B.D. (1993) *The Journal of Organic Chemistry*, **58**, 7016.

426 Garner, P. and Park, J.M. (1991) *Organic Syntheses*, **70**, 18.

427 Liang, X., Andersch, J., and Bols, M.J. (2001) *Journal of the Chemical Society-Perkin Transactions 1*, 2136.

428 Bejjani, J., Chemla, F., and Audouin, M. (2003) *The Journal of Organic Chemistry*, **68**, 9747.

429 Zhao, J., Pattaropong, V., Jiang, Y., and Hu, L. (2003) *Tetrahedron Letters*, **44**, 229.

430 Nishiyama, T., Nishikawa, T., and Yamada, F. (1989) *Journal of Heterocyclic Chemistry*, **26**, 1687.

431 Tremblay, M.R., Wentworth, P., Jr, Lee, G.E., Jr, and Janda, K.D. (2000) *Journal of Combinatorial Chemistry*, **2**, 698.

432 Nyce, G.W., Csihony, S., Waymouth, R.M., and Hedrick, J.L. (2004) *Chemistry – A European Journal*, **10**, 4073.

433 Katritzky, A.R., Suzuki, K., and He, H.-Y. (2002) *The Journal of Organic Chemistry*, **67**, 3109.

434 Barbry, O. and Couturier, D. (1987) *Chemische Berichte*, **120**, 1073.

435 Groarke, M., McKervey, M.A., Moncrieff, H., and Niewehhuyzen, M. (2000) *Tetrahedron Letters*, **41**, 1279.

436 Hayashi, Y., Skwarczynski, M., Hamada, Y., Sohma, Y., Rimura, T., and Kiso, Y. (2003) *Journal of Medicinal Chemistry*, **46**, 3782.

437 Gosselin, F., Roy, A., O'Shea, P.D., Chen, C., and Volante, R.P. (2004) *Organic Letters*, **6**, 641.

438 Conde-Frieboes, K. and Hoppe, D. (1990) *Synlett*, 99.

439 Bruggemann, M., Frohlic, R., Wibbeling, B., Holst, C., and Hoppe, D. (2002) *Tetrahedron*, **58**, 321.

440 Conde-Fireboes, K., Schjeltved, R.K., and Breinholt, J. (2002) *The Journal of Organic Chemistry*, **67**, 8952.

441 Wills, A.J., Krishnan-Ghosh, Y., and Balasubramanian, S. (2002) *The Journal of Organic Chemistry*, **67**, 6646.

442 Yoo, D., Oh, J.S., and Kim, Y.G. (2002) *Organic Letters*, **4**, 1213.

443 Tejedor, D., Santos-Exposito, A., Gonzalez-Cruz, D., Marrero-Tellado, J.J., and Garcia-Tellado, F. (2005) *The Journal of Organic Chemistry*, **70**, 1042.

444 Jones, G., Ollivierre, H., Fuller, L.S., and Young, J.H. (1991) *Tetrahedron*, **47**, 2263.

445 Pridgen, L.N., Shjilcrat, S.C., and Lantos, I. (1984) *Tetrahedron Letters*, **25**, 2835.

446 Wasserman, H.H. and Gambale, R.J. (1985) *Journal of the American Chemical Society*, **107**, 1423.

447 Koyce, D.S., Stowe, G.T., and Wong, W. (1974) *The Journal of Organic Chemistry*, **39**, 2301.

448 Ngochindo, R.I. (1990) *Journal of the Chemical Society, Perkin Transactions 1*, 1645.

449 Tarnchompoo, B., Thebtaranorth, C., and Thebtaranorth, Y. (1990) *Tetrahedron Letters*, **31**, 5779.

450 Metzger, J.V. (1984) *Comprehensive Heterocyclic Chemistry*, Vol. 6 (eds A.R. Katritzky, C. Rees, and E.F.V. Scriven), Pergamon, Oxford, pp. 275.

451 Cornwall, P., Dell, C.P., and Knight, D.W. (1987) *Tetrahedron Letters*, **28**, 3585.

452 Katritzky, A.R. (1985) *Handbook of Heterocyclic Chemistry*, Pergamon, Oxford, pp. 341.

453 Dryanska, V. and Ivanov, C. (1975) *Tetrahedron Letters*, **16**, 3519.

454 Macco, A.A., Godefroi, E.F., and Drouen, J.J.M. (1975) *The Journal of Organic Chemistry*, **40**, 252.

455 Kim, J.-W., Davis, F., Huang, L.F., Abdelaal, S.M., and Upadhyaya, S.P. (1996) *Journal of Heterocyclic Chemistry*, **33**, 65.

456 Okamoto, Y. and Ueda, T. (1977) *Chemical and Pharmaceutical Bulletin*, **25**, 3087.

457 Stragies, R., Voightmann, U., and Blechert, S. (2000) *Tetrahedron Letters*, **41**, 5465; Smulik, J.A. and Diver, S.T. (2000) *Organic Letters*, **2**, 2271; Morgan, J.P. and Grubbs, R.H. (2000) *Organic Letters*, **2**, 3153; Lee, C.W. and Grubbs, R.H. (2000) *Organic Letters*, **2**, 2145; Briot, A., Bujard, M., Gouverneur, V., Nolan, S.P., and Mioskowski, C. (2000) *Organic Letters*, **2**, 1517; Fürstner, A., Thiel, O.R., Ackermann, L., Schanz, H.-J., and Nolan, S.P. (2000) *The Journal of Organic Chemistry*, **65**, 2204.

458 Mathews, C.J., Smith, P.J., and Welton, T. (2000) *Journal of the Chemical Society. Chemical Communications*, 1246; McCluskey, A., Garner, J., Young, D.J., and Caballero, S. (2000) *Tetrahedron Letters*, **41**, 8147; Lau, R.M., van Rantwijk, F., Seddon, K.R., and Sheldon, R.A. (2000) *Organic Letters*, **2**, 4189; Sirieix, J., Oβberger, M., Betzemeier, B., and Knochel, P. (2000) *Synlett*, 1613; Howarth, J., James, P., and Dai, J. (2000) *Tetrahedron Letters*, **41**, 10319.

459 Herrmann, W.A., Goossen, L.J., Artus, G.R.J., and Köcher, C. (1997) *Organometallics*, **16**, 2472.

460 Arduengo, A.J., Krafczyk, R., and Schmutzler, R. (1999) *Tetrahedron*, **55**, 14523.

461 Tamura, Y., Minamikawa, J., and Ykeda, M. (1977) *Synthesis*, 1.

462 de las Heras, M.A., Molina, A., Vaquero, J.J., García-Navio, J.L., and Alvarez-Builla, J. (1993) *The Journal of Organic Chemistry*, **58**, 5862.

463 Rodina, L.L., Kolberg, A., and Schilze, B. (1998) *Heterocycles*, **49**, 587.

464 Molina, A., Vaquero, J.J., García-Navio, J.L., Alvarez-Builla, J., de Pascual-Teresa, B., Gago, F., and Rodrigo, M.M. (1999) *The Journal of Organic Chemistry*, **64**, 3907 and references citated therein.

465 Valenciano, J., Sánchez-Pavón, E., Cuadro, A.M., Vaquero, J.J., and Alvarez-Builla, J. (2001) *The Journal of Organic Chemistry*, **66**, 8528.

466 Cockerill, A.F., Deacon, A., Harrison, R.G., Osborte, D.J.,

Prime, D.M., Ross, W.J., Todd, A., and Verge, J.P. (1976) *Synthesis*, 591.
467 Lawson, A. (1956) *Journal of the Chemical Society*, 307.
468 Baran, P.S., Zografos, A.L., and O'Malley, D.P. (2004) *Journal of the American Chemical Society*, **126**, 3726.
469 Garg, N.K., Sarpong, R., and Stoltz, B.M. (2002) *Journal of the American Chemical Society*, **124**, 13179.
470 Merchan, F.L., Garin, J., and Tejero, T. (1982) *Synthesis*, 984.
471 Werner, H., Vicha, R., Gissibl, A., and Reiser, O. (2003) *The Journal of Organic Chemistry*, **68**, 10166 and reference cited therein.
472 Majo, V.J., and Perumal, P.T. (1998) *The Journal of Organic Chemistry*, **63**, 7136.
473 Kearney, P.C., Fernandez, M., and Flygare, J.A. (1998) *The Journal of Organic Chemistry*, **63**, 196.
474 Rajappa, S., Sreenivasan, R., and Khawadekar, A. (1986) *Journal of Chemical Research*, **5**, 158.
475 Little, T.L. and Webber, S.E. (1994) *The Journal of Organic Chemistry*, **59**, 7299.
476 Fresneda, P.M. and Molina, P. (2004) *Synlett*, 1 and references cited therein.
477 Sicker, D. (1989) *Synthesis*, 875.
478 Schmitz, W.D. and Romo, D. (1996) *Tetrahedron Letters*, **37**, 4857.
479 Pridgen, L.N., Prol, J., Alexander, B., and Gillyard, L. (1989) *The Journal of Organic Chemistry*, **54**, 3231.
480 Gaupp, S. and Effenberg, F. (1999) *Tetrahedron : Asymmetry*, **10**, 1777.
481 D'Ischia, M., Prota, G., and Rottevele, R.C. (1987) *Synthetic Communications*, **17**, 1577.
482 Hassner, A. and Basel, Y. (2000) *The Journal of Organic Chemistry*, **65**, 6368.
483 Kim, Y.J. and Varma, R.S. (2004) *Tetrahedron Letters*, **45**, 7205.
484 Katritzky, A.R., Luo, Z., Fang, Y., and Steel, P.J. (2001) *The Journal of Organic Chemistry*, **66**, 2858.
485 Sim, T.B., Kang, S.H., Lee, K.S., and Lee, W.K. (2003) *The Journal of Organic Chemistry*, **68**, 104.
486 Tomasini, C. and Secchione, A. (1999) *Organic Letters*, **1**, 2153; Lucarini, S. and Tomasini, C. (2001) *The Journal of Organic Chemistry*, **66**, 727.
487 Trost, B.M. and Hurnaus, R. (1989) *Tetrahedron Letters*, **30**, 3893; Larksarp, C. and Alper, H. (1997) *Journal of the American Chemical Society*, **119**, 3709.
488 Trost, B.M. and Fandrick, D.R. (2003) *Journal of the American Chemical Society*, **125**, 11836; Dong, C. and Alper, H. (2004) *Tetrahedron : Asymmetry*, **15**, 1537.
489 Larksarp, C., Sellier, O., and Alper, H. (2001) *The Journal of Organic Chemistry*, **66**, 3502.
490 Gabriele, B., Mancuso, R., Salerno, G., and Costa, M. (2003) *The Journal of Organic Chemistry*, **68**, 601.
491 Qian, F., McCusker, J.E., Zhang, Y., Main, A.D., Chlebowski, M., Kokka, M., and McElwee-White, L. (2002) *The Journal of Organic Chemistry*, **67**, 4086.
492 Hinman, A. and Du Bois, J. (2003) *Journal of the American Chemical Society*, **125**, 11510.
493 Espino, C.G. and Du Bois, J. (2001) *Angewandte Chemie International Edition*, **40**, 598.
494 Espino, C.G., Fiori, K.W., Kim, M., and Du Bois, J. (2004) *Journal of the American Chemical Society*, **126**, 15378.
495 Weinstock, J., Gaitanopoulos, D.E., Stringer, O.D., Franz, R.G., Hieble, J.P., Kinter, L.B., Mann, W.A., Flaim, K.E., and Gessners, G. (1987) *Journal of Medicinal Chemistry*, **30**, 1166.
496 Khajavi, M.S., Hajihadi, M., and Naderi, R. (1996) *Journal of Chemical Research (S)*, 92.
497 Mewshaw, R.E., Zhao, R., Shi, X., Marquis, K., Brennan, J.A., Mazandarani, H., Coupet, J., and Andree, T.H. (2002) *Bioorganic and Medicinal Chemistry Letters*, **12**, 271.
498 Fu, Y., Baba, T., and Ono, Y. (2001) *Journal of Catalysis*, **197**, 91.
499 Preston, P.N. (1981) *Benzimidazoles and Congeneric Tricyclic Compounds* (ed. P.N. Preston), Interscience-Wiley, New York.
500 Ashburn, S.P. and Coates, R.M. (1984) *The Journal of Organic Chemistry*, **49**, 3127.
501 Ashburn, S.P. and Coates, R.M. (1985) *The Journal of Organic Chemistry*, **50**, 3076.

502 Hendrickson, J.B. and Pearson, D.A. (1983) *Tetrahedron Letters*, **24**, 4657.

503 Voinov, M.A. and Grigor'ev, I.A. (2002) *Tetrahedron Letters*, **43**, 2445.

504 Shchukin, G.I., Starichennko, V.F., Grigor'ev, I.A., Dikanov, S.A., Gulin, V.I., and Volodarskii, L.B. (1987) *Izvestiya Akademii Nauk SSSR Seriya Khimicheskaya*, 125 [*Bulletin Academy of Sciences USSR, Division of Chemical Sciences (English Translation)*, (1987) **36**, 125.]

505 Grigor'ev, I.A., Bakunova, S.M., and Kirilyuk, I.A. (2000) *Russian Chemical Bulletin*, **49**, 2031.

506 Shchukin, G.I., Grigor'ev, I.A., and Volodarskii, L.B. (1990) *Chemistry of Heterocyclic Compounds (English Translation)*, **26**, 409.

507 Tufariello, J.J. (1984) *1,3-Dipolar Cycloaddition Chemistry*, Vol. 5 (ed. A. Padwa), John Wiley & Sons, Inc., New York, p. 83; Breuer, E., Aurich, H.G., and Nielsen, A. (1989) *Nitrones, Nitronates and Nitroxides*, John Wiley & Sons, Ltd., Chichester, UK; Dirat, O., Kouklovsky, C., Mauduit, M., and Langlois, Y. (2000) *Pure and Applied Chemistry*, **72**, 1721.

508 Frederickson, M. (1997) *Tetrahedron*, **53**, 403; Gothelf, K.V. and Jorgensen, K.A. (1998) *Chemical Reviews*, **98**, 863; Merino, P., Franco, S., Merchan, F.L., and Tejero, T. (2000) *Synlett*, 442.

509 Janzen, E.G. and Haire, D.L. (1990) *Advances in Free Radical. Chemistry*, **1**, 253; Free Radical Research Janzen, E.G. (1971) *Accounts of Chemical Research*, **4**, 31.

11
Five-Membered Heterocycles with Two Heteroatoms: O and S Derivatives

David J. Wilkins

11.1
Introduction

A general review of chemistry for all the derivatives described in this chapter covering the period up to 1995 can be found in *Comprehensive Heterocyclic Chemistry* first edition (Volume 6) and second edition (Volume 3).

11.2
1,2-Dioxoles and 1,2-Dioxolanes

11.2.1
Introduction

Most work on 1, 2-dioxole systems has been on derivatives of the fully saturated 1, 2-dioxolane (**1**) and as cyclic peroxides these have been the subject of three major reviews [1–3]. 1,2-Dioxoles are unstable and they have only been detected spectroscopically at temperatures below −60 °C [4].

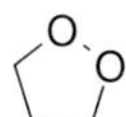

1

11.2.2
Relevant Physicochemical Data

11.2.2.1 NMR Spectroscopy

^1H and ^{13}C NMR data for *trans*- and *cis*-3, 5-dimethyl-1,2-dioxolane **2**, **3** have been published [5]; proton resonances appear (in ppm) at $\delta_H = 4.25$ cis and 4.3 trans for H3, at $\delta_H = 2.77$ cis and 2.19 trans for H4, and at $\delta_H = 4.25$ cis and 4.30 trans for H5.

Modern Heterocyclic Chemistry, First Edition.
Edited by Julio Alvarez-Builla, Juan Jose Vaquero, and José Barluenga.

The cis and trans isomers are readily identified by ^{13}C NMR, which shows cis δ 19.25 Me, 49.34 C4 and 77.30 C3,5; trans δ 18.40 Me, 48.61 C4 and 77.04 C3,5.

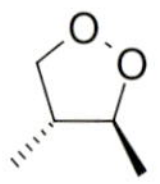

2 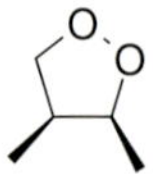 3

11.2.2.2 **Electron Diffraction Studies**

Electron diffraction has been used to obtain the following dimensions for the heterocyclic ring of perfluorinated 1,2-dioxolane (**4**) [6]: O–O, 1.443 Å; C–O, 1.377 Å and C–C, 1.531 Å; C–C–C, 98.1°; C–C–O, 107.3° and C–O–O, 102.9°.

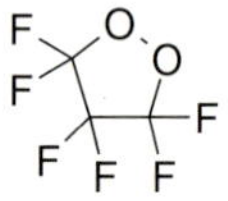

4

11.2.3
Synthesis

11.2.3.1 **Synthesis by Ring Construction**

Oxidative addition of elemental fluorine to appropriate 1, 3-dicarbonyl compounds provides a convenient synthesis of perfluorinated 1,2-dioxolanes. Compound **4** can be synthesized by treatment of difluoromalonyl difluoride with fluorine [6] and **5** is similarly prepared from either hexafluoroacetylacetone or the copper(II) or nickel(II) chelate of trifluoroacetylacetone [7].

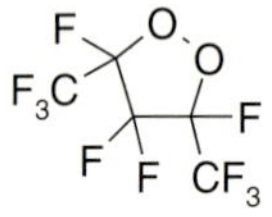

5

Several studies have appeared on the formation of 1, 2-dioxolanes either by *endo* cyclization of allylic hydroperoxides [8] or by *exo* cyclization of homoallylic hydroperoxides upon treatment with various electrophilic reagents [9]. For example, cyclization of the homoallylic peroxide **6** with Bu^tOCl affords the 1, 2-dioxolane **7** (Scheme 11.1) [10].

Photooxygenation of arylcyclopropanes **8** (R = Ph) provides direct access to the corresponding 1,2-dioxolanes **9** (R = Ph), although the reaction is non-stereospecific [11]. The addition can also be radical mediated, as in the conversion of **8** (R = H) to **9** (R = H) by treatment with O_2 in the presence of PhSeSePh and AIBN (Scheme 11.2) [12].

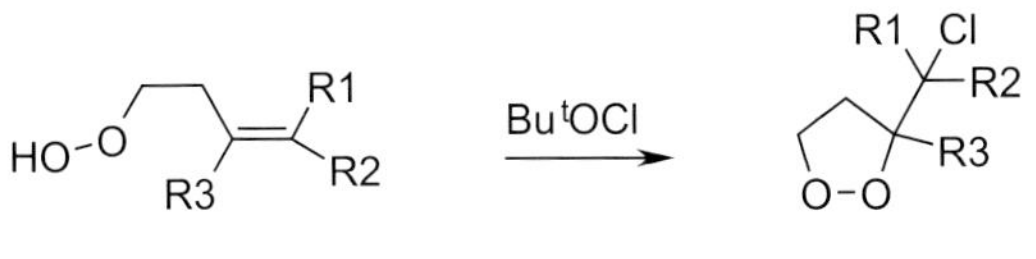

Bu^{t}OCl

6 **7**

Scheme 11.1

8, R = H or Ph **9, R = H or Ph**

Scheme 11.2

α,β-Unsaturated imines **10** react directly with singlet oxygen to give 3-amino-1,2-dioxolanes **11** (Scheme 11.3) [13].

10 **11**

Scheme 11.3

11.2.3.2 Ring Transformations of Heterocycles Leading to 1,2-Dioxoles and 1,2-Dioxolanes

3-Hydroperoxypyrazolines **12** react readily with oxygen with the loss of N_2 to give 3-hydroxy-1,2-dioxolanes **13** (Scheme 11.4) [14].

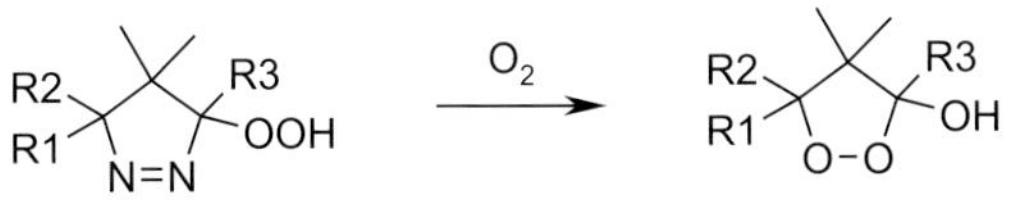

O_2

12 **13**

Scheme 11.4

Lewis acid treatment of 1,2,4-trioxolanes gives metallated carbonyl oxides that may be trapped by cycloaddition to allylsilanes to give 1,2-dixolanes **14** [15].

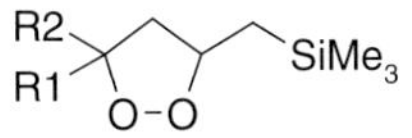

14

11.2.4
Reactivity of 1,2-Dioxoles and 1,2-Dioxolanes

Solution thermolysis of the bicyclic dioxolane **15** gives epoxy aldehyde **16** in non-polar solvents and the diketone **17** in polar solvents [16]. Flash vacuum pyrolysis of **15** at 450 °C and 10^{-3} Torr gave only **16** [17]. Monocyclic dioxolanes such as **18** [17] and **19** [18] give under similar conditions a mixture of epoxide and carbonyl compounds (Scheme 11.5).

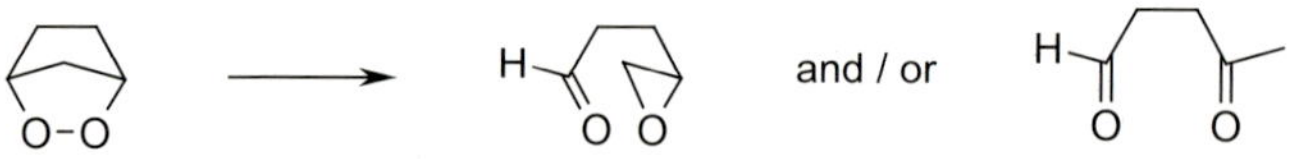

O–O H O O and / or H O O

15 **16** **17**

O O R2 O O R1 Ph

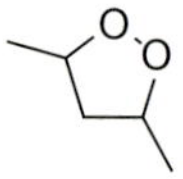

18 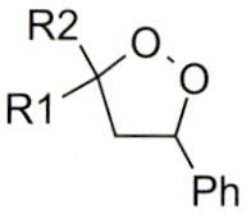 **19**

Scheme 11.5

11.3
1,3-Dioxoles and 1,3-Dioxolanes

11.3.1
Introduction

There has been relatively little work published on the fully conjugated system, 1,3-dioxolium salts **20** and their benzo analogues **21**, but there are reviews on these compounds [19, 20]. More recently a comprehensive review chapter on 1,3-dioxolium salts has appeared [21]. A much larger volume of work has been published on 1,3-dioxoles **22**, although most work published in this area is on the fully saturated compounds, 1,3-dioxolanes **23**. A major review of these compounds has been published [22].

O O + O O + O O O O

20 **21** **22** **23**

Table 11.1 X-ray structural data for **24**.

Bond length (Å)				
O1–C2	C2–O3	O3–C4	C4–C5	C5–O1
1.281	1.282	1.472	1.505	1.480
Internal angle (°)				
O1	C2	O3	C4	C5
108.1	103.4	103.1	108.6	116.8

11.3.2 Relevant Physicochemical Data

11.3.2.1 X-Ray Diffraction Studies

X-Ray diffraction studies on the 1,2-dioxolan-2-yl cation **24** has given molecular dimensions (Table 11.1). The data show the structure to be essentially planar at C2 [23].

24

11.3.2.2 NMR Spectroscopy

Table 11.2 presents the ^1H NMR spectra of 1,3-dioxolane derivatives.

^{13}C NMR data for 1,3-dioxolium salts **24–26** and 1,3-dioxolane **27** are shown in Table 11.3. The high frequency for C2 signals in 1,3-dioxolium salts is notable when compared to the dioxolane **27**.

Table 11.2 ^1H NMR data for ring protons of 1,3-dioxolane derivatives.

Compound	δ_H (ppm)			
	2H	4H	5H	Reference
4,4-Dimethyl-1,3-dioxolane	4.9		3.51	[24]
2-Imino-1,3-dioxolane		4.40	4.40	[25]
cis-2,2,4,5-Tetramethyl-1,3-dioxolane		4.15	4.15	[26]
trans-2,2,4,5-Tetramethyl-1,3-dioxolane		3.38	3.38	[26]

Table 11.3 ^{13}C NMR data for ring carbons of 1,3-dioxole derivatives.

Compound	δ_C (ppm)			
	C2	C4	C5	Reference
24	174.7	146.6	146.6	[27]
25	175.4	147.9	147.9	[27]
26	181.9	90.6	90.6	[27]
27	101.5	153.2	69.1	[28]

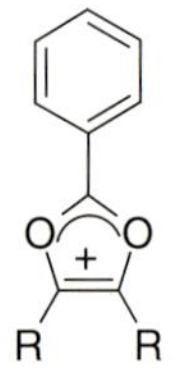

24, R = Ph
25, R = Me

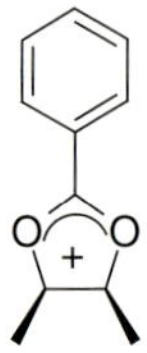

26

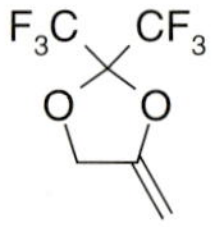

27

^{17}O NMR studies have been carried out on 1,3-dioxolanes **28** and a few simple derivatives. The chemical shift for the parent compound has been given variously as δ_O (with respect to $H_2{}^{17}O$): +34 [29], +35.8 [30] and +34.8 [31].

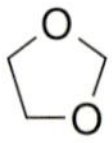

28

11.3.3 Synthesis

11.3.3.1 Synthesis by Ring Construction

1,3-Dioxolium salts have been prepared from α-diazoanhydrides **29** by palladium-catalyzed decomposition to give a carbene intermediate that undergoes electrocyclization to give **30** (Scheme 11.6) [32].

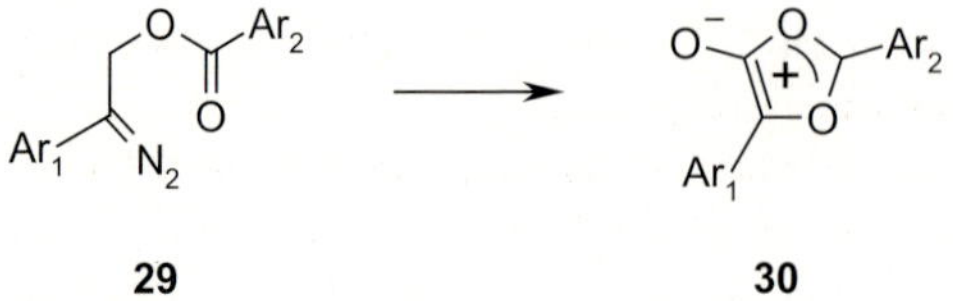

29 → **30**

Scheme 11.6

Other methods that involve the decomposition of diazo compounds include the reaction of the diazoketone **31** with $PhCO_2Tf$ to afford the diazolium salt **32** [27] and the copper-catalyzed decomposition of the 2-diazo-1,3-dicarbonyl compound **33**, which in the presence of aldehydes or ketones gives 4-acyl-1,3-dioxoles **34** (Scheme 11.7) [33].

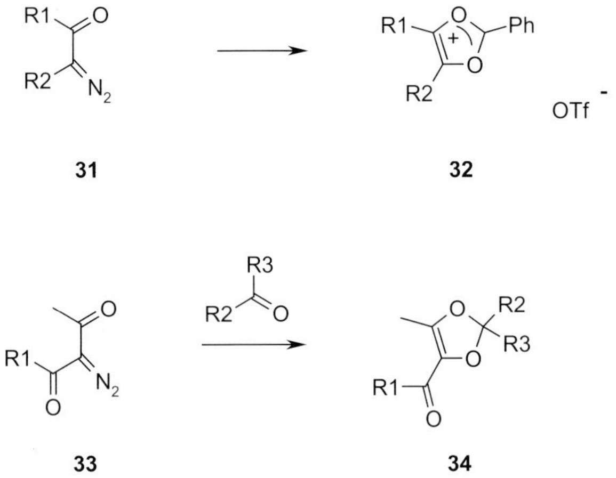

Scheme 11.7

Rhodium-catalyzed reaction of the diazo esters **35** with carbonyl compounds, R^1COR^2, provides a new route to silyl dioxolanones **36** in a process involving an intermediate carbonyl ylide (Scheme 11.8) [34].

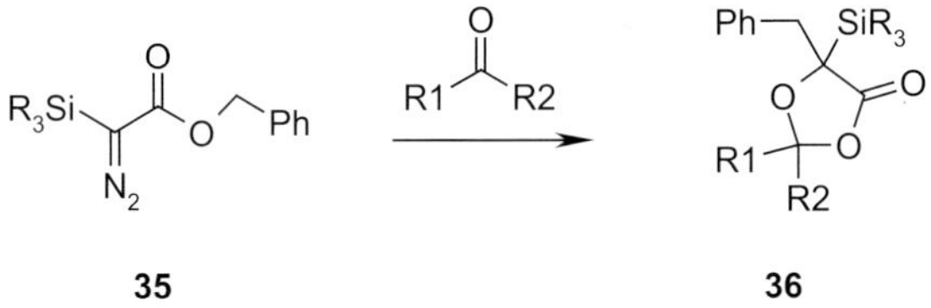

Scheme 11.8

The treatment of chloroesters **37** and **38** with $SbCl_5$ results in cyclization with rearrangement of the carbon skeleton to give, respectively, the salts **39** [35] and **40** [36] (Scheme 11.9).

Vinyl esters such as **41** react with benzoyl or *t*-butyl hexachloroantiminoate to give the 1,3-dioxolium salt **42** (Scheme 11.10) [37, 38].

A simple preparation of 2-methyl-1,3-dioxolane (**44**) involves heating the vinyl ether **43** with KOH to give **44** in 66% yield [39]. A similar preparation involving enol ethers starts with **45**, which reacts with either EtMgBr or $Bu_2{}^iAlH$ followed by benzaldehyde to give hydroxyalkyldioxolanes **46** (Scheme 11.11) [40].

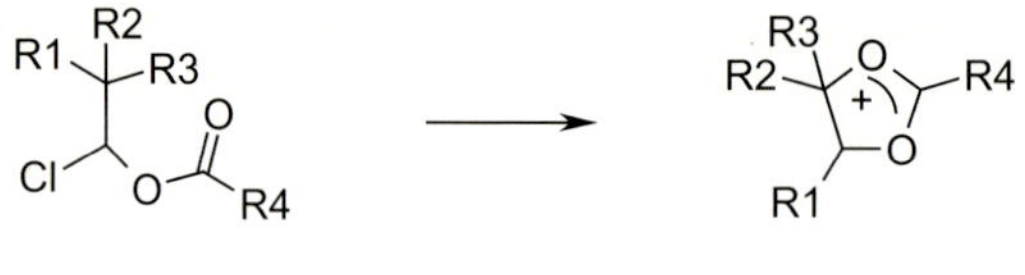

37 **39**

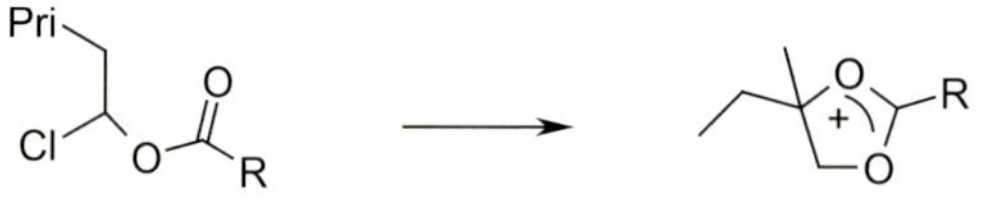

38 **40**

Scheme 11.9

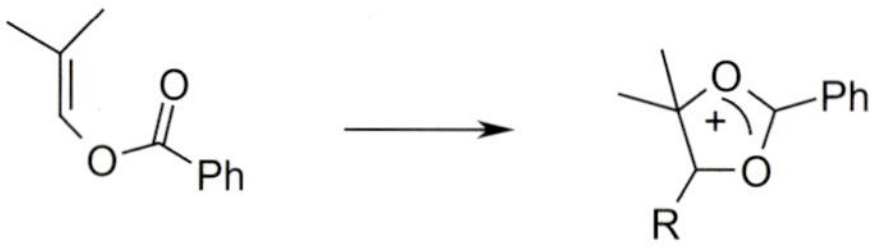

41 **42**

Scheme 11.10

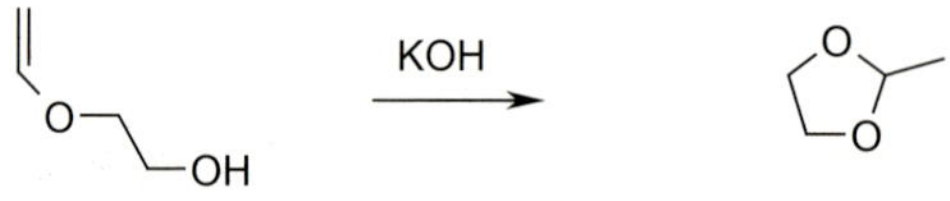

43 **44**

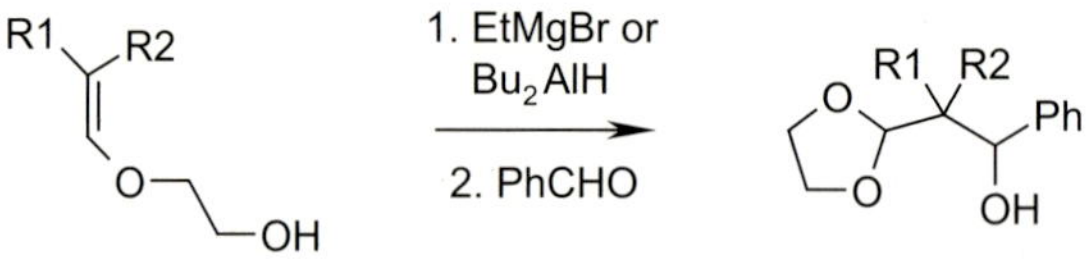

45 **46**

Scheme 11.11

Treatment of glycidic esters like **47** with cationic zirconium species and $AgClO_4$ affords the dioxolane **48** [41]. Epoxides like **49** react with CO_2 to give dioxolanones **50** (Scheme 11.12) The reaction may be catalyzed either by mixed alkali metal or manganese halides [42] or alkali metal/lead/indium halides [43]. The use of ionic liquids to catalyze this reaction has also been described [44].

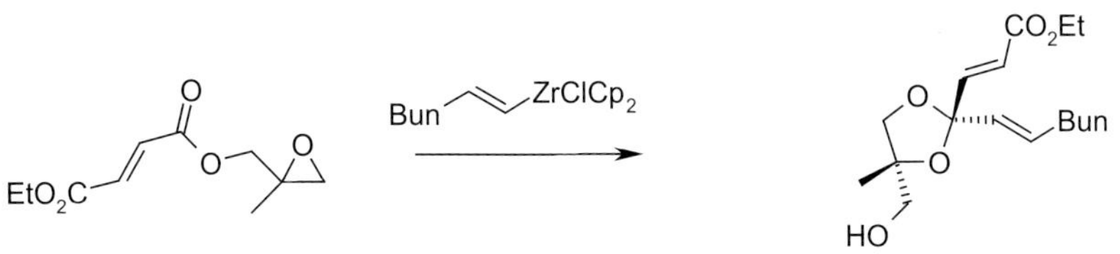

Scheme 11.12

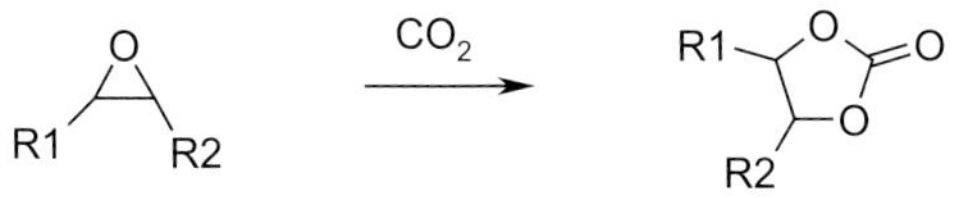

Treatment of the acetylenic alcohol **51** with aq LiOH affords the dioxolane **52**; the yield of this reaction is increased by the addition of acetone (Scheme 11.13) [45].

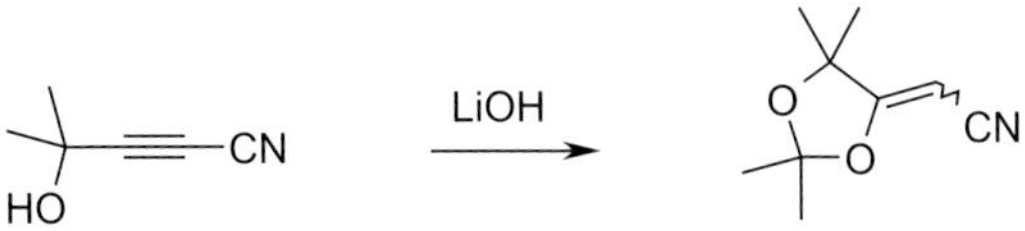

Scheme 11.13

Reaction of the stabilized iodonium ylide **53** with ketones affords 1,3-dioxoles **54** (Scheme 11.14) [46].

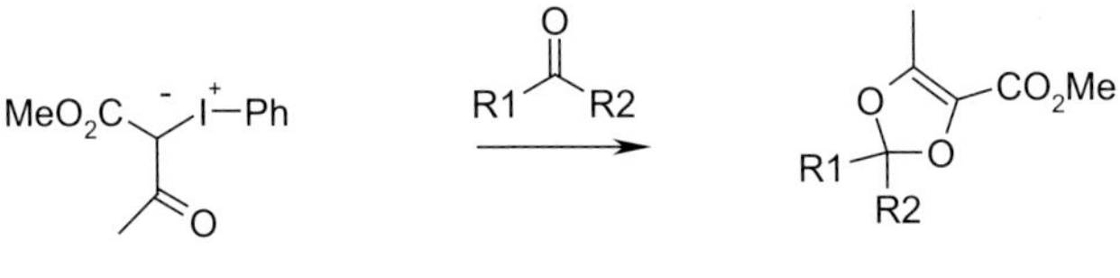

Scheme 11.14

The palladium-catalyzed reaction of propargyl acetates with CO in methanol results in cyclization to give dioxolanes **55** (Scheme 11.15) [47].

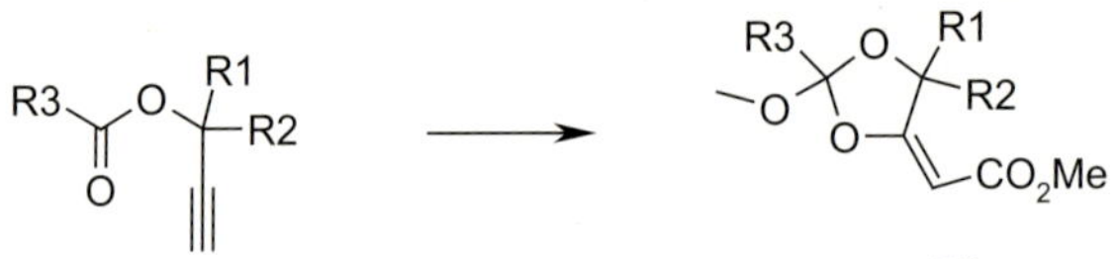

55

Scheme 11.15

Treatment of α-hydroxyketones, $R^1CH\ OH\ COR^2$, with triphosgene, Cl_3CO_2CO, gives the 1,3-dioxol-2-ones **56** in moderate yield [48].

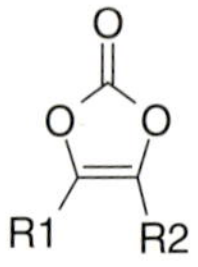

56

2-Imino-1,3-dioxoles **57** can be prepared either by electrochemical reduction of benzils **58** in the presence of $Ar_2N{=}CCl_2$ [49] or by the treatment of bis tin derivatives such as **59** with Ar_2NCS (Scheme 11.16); **59** also reacts with CS_2 to give 1,3-dioxole-2-thiones [50].

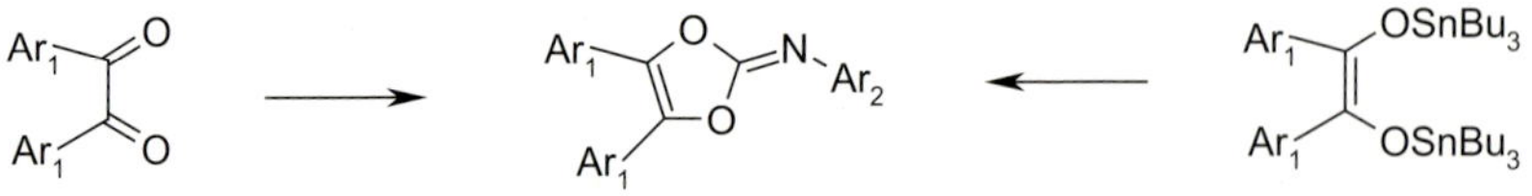

58 **57** **59**

Scheme 11.16

2,2-Disubstituted-1,3-dioxolanes **61** may be formed by nucleophilic attack with stabilized carbanions on 2-haloalkyl esters **60** (Scheme 11.17) [51].

60 **61**

Scheme 11.17

The transfer hydrogenation of the tosylacetophenone **62** using a chiral ruthenium catalyst and formic acid as the hydrogen source unexpectedly gives the chiral dioxolanone **63** in 94% ee (Scheme 11.18) [52].

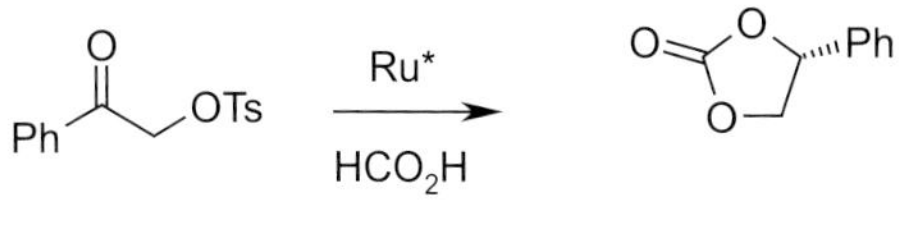

Ru*

HCO_2H

62 63

Scheme 11.18

1,2-Diols like **64** react with oxalyl chloride and triethylamine to give the 1,3-dioxolanone **65** rather than the expected six-membered ring products (Scheme 11.19) [53].

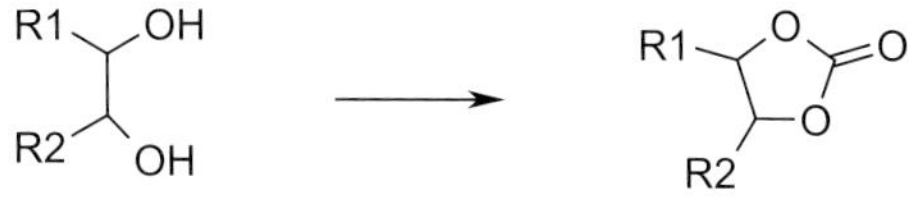

64 65

Scheme 11.19

The treatment of styrene oxide with ruthenium trichloride in acetone gives the dioxolane **66** [54]. Titanium-based catalysts for the reaction of epoxides with acetone to give 2,2-dimethyl-1,3-dioxolanes are TiO CF_3CO_2 and TiCl CF_3SO_3 [55].

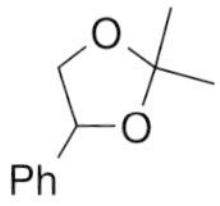

66

The most widely used method for the preparation of 1,3-dioxolanes involves the reaction of carbonyl compounds with 1,2-diols. A wide range of catalysts has been used in this reaction, trimethylsilyl triflate in the presence of a trimethylsilyl ether [56], K_{10} montmorillonite under solvent-free conditions [57, 58], scandium triflate [59] and *N*-benzoylhydrazinium salts [60].

11.3.3.2 **Ring Transformations of Heterocycles Leading to 1,3-Dioxoles and 1,3-Dioxolanes**

The treatment of 1,3-dioxane **67** with mCPBA provides an interesting route into 1,3-dioxolane aldehydes **68**, presumably via an epoxide rearrangement (Scheme 11.20) [61].

11.3.4 Reactivity of 1,3-Dioxoles and 1,3-Dioxolanes

The phenyliminodioxolane **69** (R=Ph) undergoes isomerization to the oxazolidinone **70** upon heating at 200 °C with LiCl [62]. The parent compound **69** (R=H) undergoes

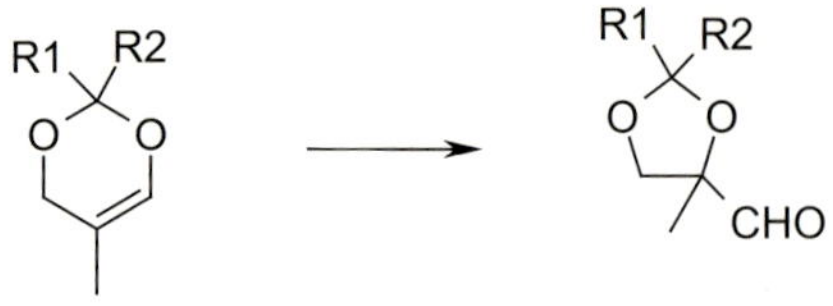

67 68

Scheme 11.20

spontaneous isomerization and trimerization to give the 1,3,5-triazine **71** (Scheme 11.21) [63].

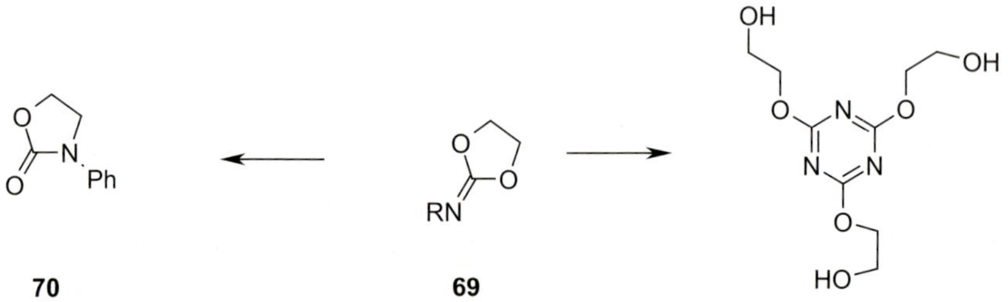

70 69 71

Scheme 11.21

11.3.4.1 Reactions with Nucleophiles

There has been several studies on stable cations such as **72** and their reactions with nucleophiles; **72** reacts with NaOMe at C2, but with $NaHCO_3$ or LiCl at C4 to give ring open 2-hydroxyethyl or 2-chloroethyl methyl carbonates, respectively [64]. The reaction of salts like **73** with acetylenic Grignard reagents takes place at C2 [65].

72 73

Efficient hydrolysis of 2,2-disubstituted-1,3-dioxolanes to carbonyl compounds has been an area of much interest. Methods for the hydrolysis of 2,2-dimethyl-1,3-dioxolanes include cerium ammonium nitrate and oxalic acid [66] and a polymer-supported dicyanoketene acetal [67]. Cleavage may also be achieved using Ph_3P/CBr_4 [68] or $CpTiCl_3$ [69], other reagents used include DDQ in aq. CH_3CN [70] and $TeCl_4$ [71].

11.3.4.2 Uses in Asymmetric Synthesis

There has been a large volume of work published on the use of 1,3-dioxolanes in asymmetric synthesis. TADDOL complexes **74** have been used widely in asymmetric synthesis and have been the subject of a major review [72]. Further developments in this area involve the synthesis of polymer-supported TADDOLs [73] and TADDOL crown ethers [74]. Benzylcyclohexanone has been deracemized by the formation of an inclusion complex with TADDOL compound **74**, R,R=cyclohexyl. The X-ray structure of this complex has also been published [75]. The TADDOL complex **74**, R,R=cyclopentyl, has been used to direct enantioselective Diels–Alder reactions in aqueous solution [76].

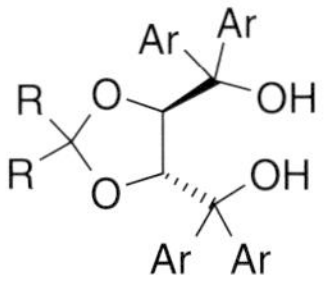

74

There has also been considerable work on C–C bond formation by attack on suitable dioxolane systems by carbanions, and of particular interest has been the use of chiral dioxolanes in asymmetric synthesis. Conjugate addition of organozinc compounds to chiral dioxolanes such as **75** and **76** affords adducts with a high degree of enantioselectivity [77, 78].

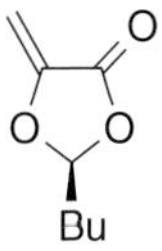

75 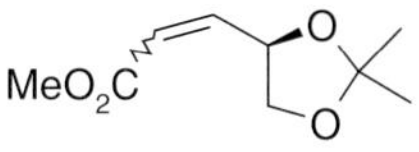 **76**

11.3.4.3 Reactions with Carbenes and Radicals

Treatment of 1,3-dioxolane (**23**) with ferrous sulfate and a peroxide in the presence of pyridine gives predominantly **77** together with small amounts of products resulting from reaction at the 4-position of the dioxolane and the 2-position of the pyridine [79].

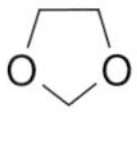

23 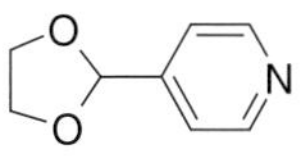 **77**

The reaction of 1,3-dioxolanes with carbenes generally proceeds by insertion into the C2–H bond and this has been examined for phase-transfer generated Cl_2C: and Br_2C: [80, 81] and for various arylchlorocarbenes (ArClC:) [82, 83]. Ethoxycarbonylcarbene behaves differently and reacts with **78** by insertion into the C2–O bond to give **79** [84].

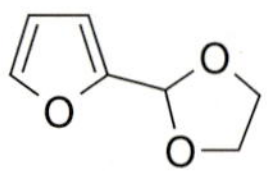

78

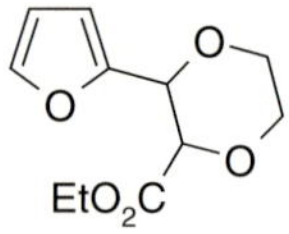

79

11.3.5
Compounds of Interest

1,3-Dioxolane derivatives such as **80** have been used as a fungicide [85] and related derivatives have anti-fungal activity [86–88]. The structurally related compounds **81** and **82** have been prepared as intermediates for ketoconazole synthesis by lipase-catalyzed kinetic resolution [89].

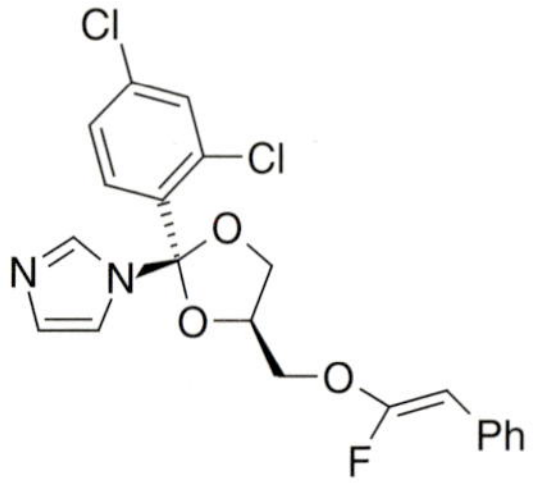

80

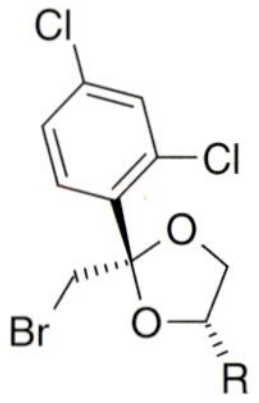

81, R = CH_2OH
82, R = CO_2H

Tertiary ammonium salts containing a 1,3-dioxolane ring (**83**) have been described as muscarinic acetylcholine antagonists [90].

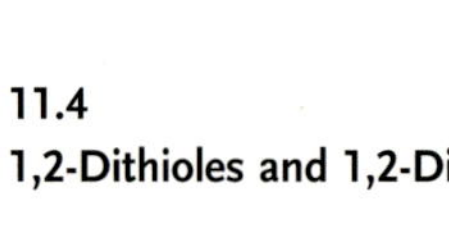

83

11.4
1,2-Dithioles and 1,2-Dithiolanes

11.4.1
Introduction

The 1,2-dithiolane system **84** is known but the partially unsaturated 1,2-dithiole system **85** is very common, along with its derivatives **85** (X=O, S, NR). There are only a few references to systems where X=H,H or alkyl. Cationic species such as **86** and **87** are also known. 1,2-Dithiolanes have a greater reactivity than acyclic disulfides, which is attributed to some repulsion between the lone pairs on the adjacent sulfur atoms [91]. 1,2-Dithiolium cations are aromatic 6π systems and are very stable except to nucleophilic reagents. A comprehensive review of 1,2-dithiolium salts has been published [92].

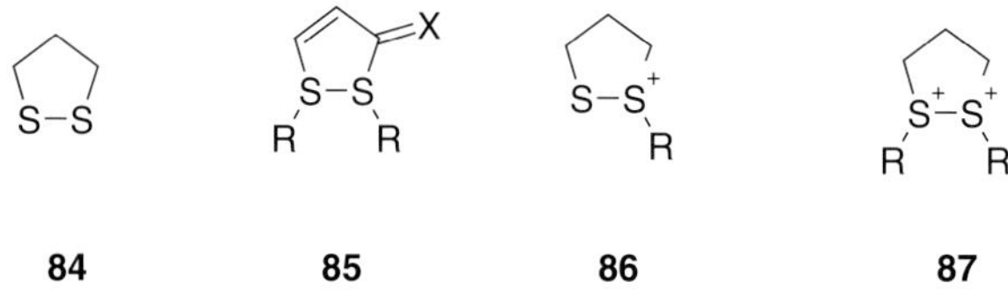

11.4.2
Relevant Physicochemical Data

X-Ray methods for 1,3-dithiolanes have shown the S–S bond distances to be in the range of acyclic disulfides [91]. Figure 11.1 shows the bond lengths and angles of 1,2-dithiol-3-thione [93].

The ^{1}H NMR spectrum of 1,2-dithiolane **84** has signals at δ2.09 ppm for H4 and δ3.36 ppm for H3 and H5 [91]. Protons in 1,2-dithiole-3-ones absorb at higher field than the corresponding 1,2-dithiole-3-thiones. In the unsubstituted 1,2-dithiolium ion **88**, the H3 and H5 protons absorb at δ10.7 ppm and the H4 proton at δ8.88 ppm [91].

S–S$^+$

88

The ^{13}C NMR chemical shifts for 1,2-dithiolanes are ~41 ppm for C3 and C5 and 56 ppm for C4; the chemical shifts for 1,2-dithiole-3-one are 216 ppm for C3, 140.2 ppm for C4 and 155.1 ppm for C5 [94].

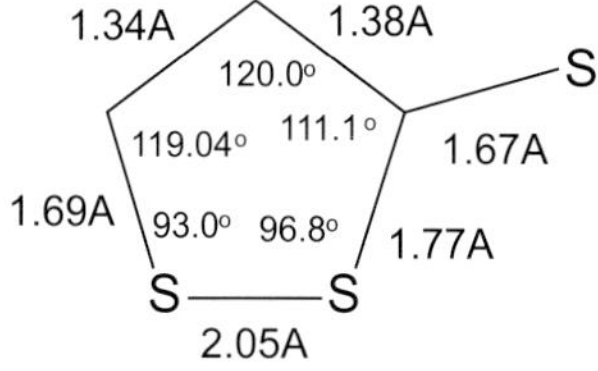

Figure 11.1 Geometry of 1,2-dithiol-3-thione.

11.4.3 Synthesis

11.4.3.1 Synthesis by Ring Construction

The synthesis of 1,2-dithioles and 1,2-dithiolanes has been the subject of a review [95]. The most practical method for the synthesis of 1,2-dithiols involves direct S−S bond formation, usually via oxidation or displacement of a leaving group on one of the sulfur atoms. For example, treatment of dithiols with bromine on hydrated silica has affords 1,2-dithiolanes **84** and **89** [96].

S–S OH OH

84 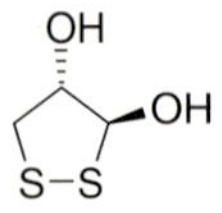 **89**

A mild method for the preparation of 1,2-dithiolanes involves the treatment of 1,3-dithiocyanates with $Bu^n{}_4N^+F^-$ [97].

Treatment of β-ketoamides such as **90** with Lawesson's reagent gives the 1,2-dithiole-3-thione **91** [98]. In a similar preparation, treatment of malonates **92** with Lawesson's reagent and sulfur in xylene in the presence of a catalytic amount of 2-mercaptobenzothiazole and ZnO affords the 1,2-dithiol-3-thione **93** [99].

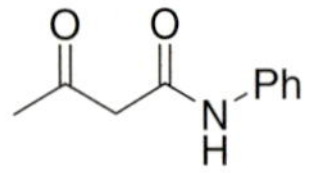

O O N H Ph S S–S

90 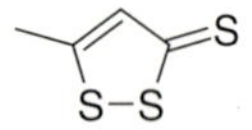 **91**

O O RO OR R1 R1 RS S S–S

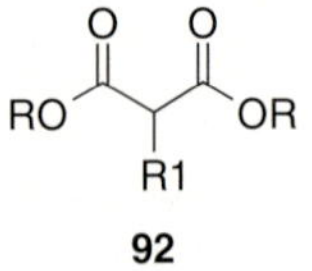

92 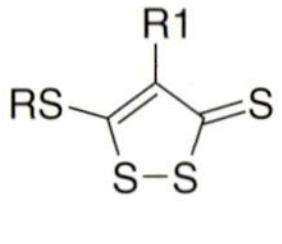 **93**

Dicyclic 1,2-dithiolane **94** has been prepared by treating either the ylide $Ph_3{=}Ar_2$ or the thione $Ar_2C{=}S$ with sulfur in boiling xylene in the presence of maleic anhydride [100]. The cycloaddition of thiocarbonyl sulfides with reactive alkynes gives access to 1,2-dithioles **95** [101].

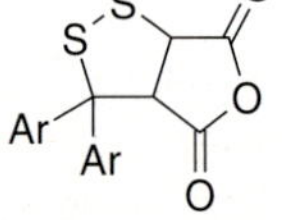

S S O O O Ar Ar Ph Ph S S R R

94 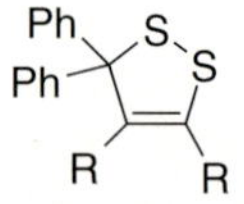 **95**

1,2-Dithiolanes are formed by the reaction of 1,3-halopropanes or haloalkenes with disulfide ions. Hydrogen disulfide provides the source of disulfide ion and it readily condenses with phenylpropynyl chloride to form 5-phenyl-1,2-dithiol-3-one (**96**) [102]. Similarly 1,3-diketones condense with hydrogen disulfide to give 1,2-dithiolium salts **97** (Scheme 11.22) [102].

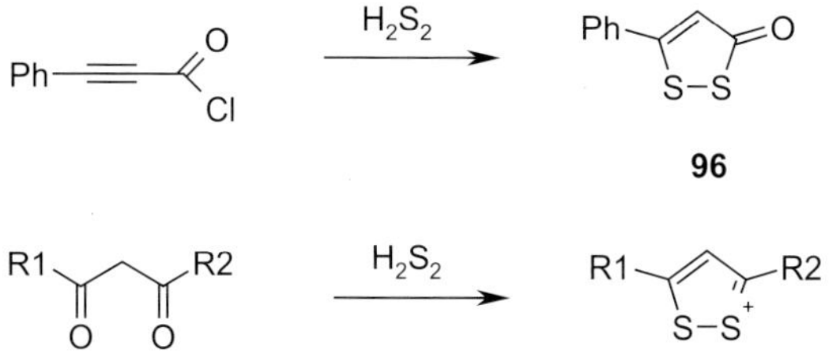

H_2S_2 **96**

H_2S_2 **97**

Scheme 11.22

α-Thioderivatives of enamines or enol ethers react with carbon disulfide to form 5-amino or 5-alkoxy-1,2-dithiole-3-thiones [103]. An interesting reaction of the oxime **98** with S_2Cl_2 gives the bicyclic 1,2-dithiole isomers **99** and **100** (Scheme 11.23) [104].

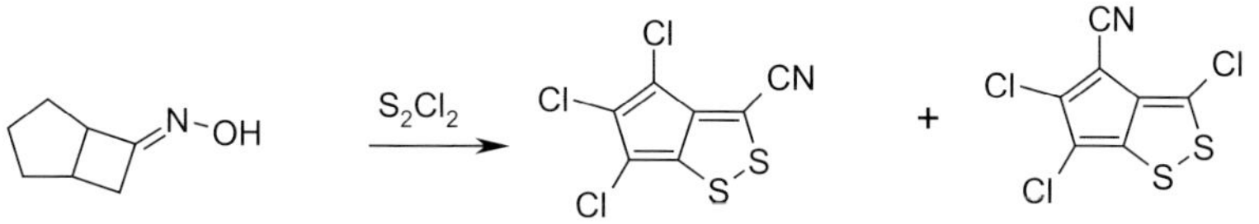

98 S_2Cl_2 **99** + **100**

Scheme 11.23

A common method for the synthesis of 1,2-dithioles is the sulfurization of various 3-carbon units by elemental sulfur. The conversion of **101** into **102** and **103** into **104** is usually carried out under thermal conditions (180–250 °C) (Scheme 11.24) [105].

11.4.3.2 Ring Transformations of Heterocycles Leading to 1,2-Dithioles and 1,2-Dithiolanes

The synthesis of 1,2-dithioles by the transformation of other heterocyclic rings is a common method of preparation. Thiophene derivative **105**, when reacted with Na_2S in air affords the 1,2-dithiole **106** (Scheme 11.25) [106].

Ring contraction of 1,3-dithianes or 1,3-dithienes is also a well reported method for the preparation of 1,3-dithioles and 1,3-dithiolanes. The reaction is usually carried out under oxidizing conditions in the presence of acid or from 1, 3-dithiane-S-oxides with acids. Bromine has been used to transform the 1, 3-dithiane **107** into the 1,2-dithiolane **108** (Scheme 11.26) [107].

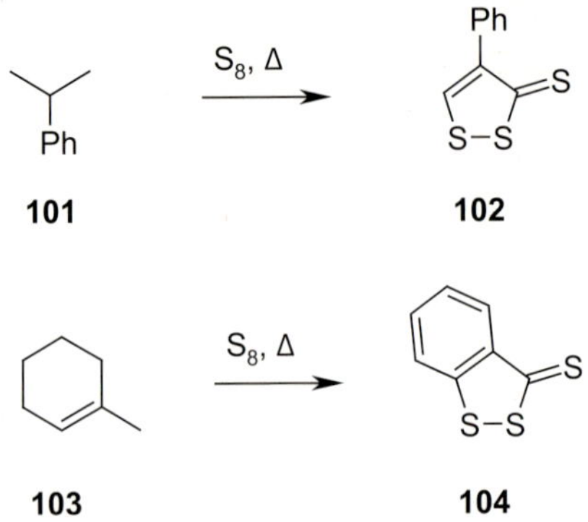

Scheme 11.24

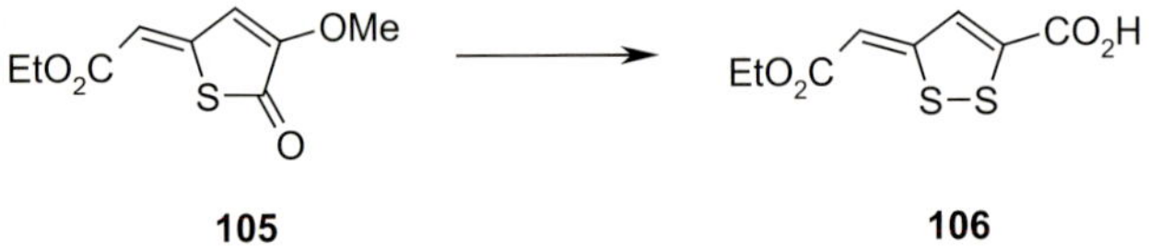

Scheme 11.25

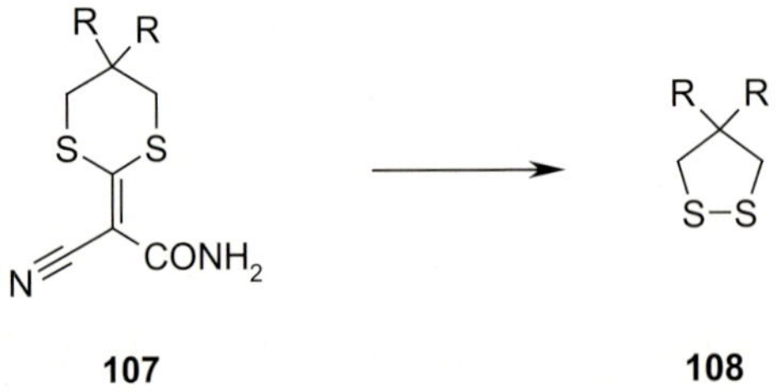

Scheme 11.26

Isoxazoles possess a three-carbon unit suitable for conversion into the 1,2-dithiole skeleton. Thus, 5-phenylisoxazole (**109**) is thionated to afford 5-phenyl-1,2-dithiole-3-thione (**110**) [108] and isoxazoline-3-thiones **111** affords 1,2-dithiole-3-imines **112** on treatment with H_2S (Scheme 11.27) [109]. Thiazolidinones **113** react with Lawesson's reagent to give 3-imino-1,2-dithioles **114** [110].

11.4.4 Reactivity of 1,2-Dithioles and 1,2-Dithiolanes

11.4.4.1 1,2-Dithiolium Salts

11.4.4.1.1 Reactions with Nucleophiles 1,2-Dithiolium salts react readily only with nucleophiles and often at the least hindered 3 or 5 position to form 1,2-dithioles.

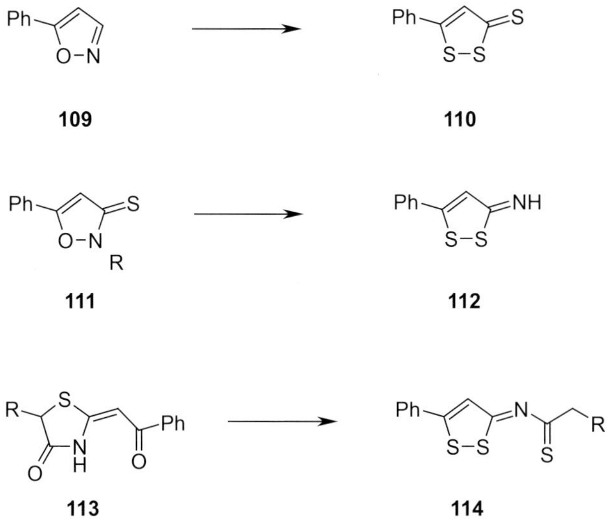

Scheme 11.27

These 1,2-dithioles may reform 1,2-dithiolium salts if there is a suitable leaving group at the 3-position Scheme 11.28.

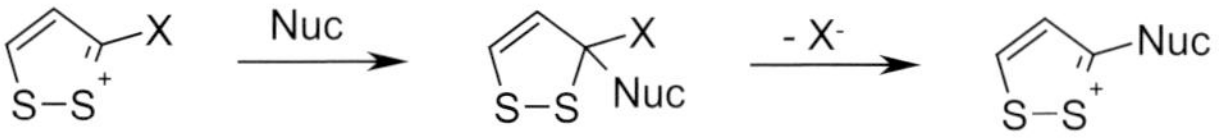

Scheme 11.28

Oxygen and sulfur nucleophiles react in a similar fashion to give 3-alkoxy or 3-alkylthio substituted 1,2-dithioles, again if there is a suitable leaving group at the 3-position the products obtained are 1,2-dithiolium salts [111]. 1,2-Dithiolium salts react with nitrogen nucleophiles at the 3-position but the final product depends on the substitution pattern on the dithiolium ring. Dithiolium salts **115**, without a leaving group at C3, react with ammonia to form isothiazoles **116** [111] and with primary and secondary amines to form aminothiones **117** (Scheme 11.29) [111].

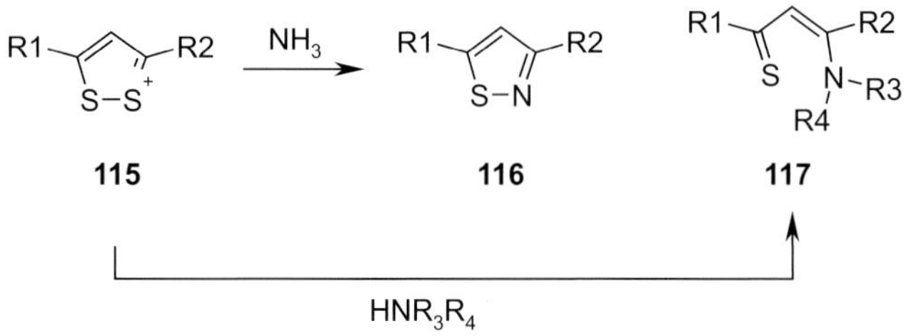

Scheme 11.29

The reaction of 1,2-dithiolium salts with carbon nucleophiles usually leads to ring opening and recyclization to give thiopyrans or thiophenes – these reactions are similar to those involving nitrogen nucleophiles by attack at C3 or it may involve an initial attack of the carbanion on a ring sulfur [112, 113].

11.4.4.1.2 **Reduction** Electrochemical reduction of 1,2-dithiolium salts **118** gives bis-1,2-dithiole dimers **120** via radical intermediates **119** whose stability depends on the substitution pattern (Scheme 11.30) [114].

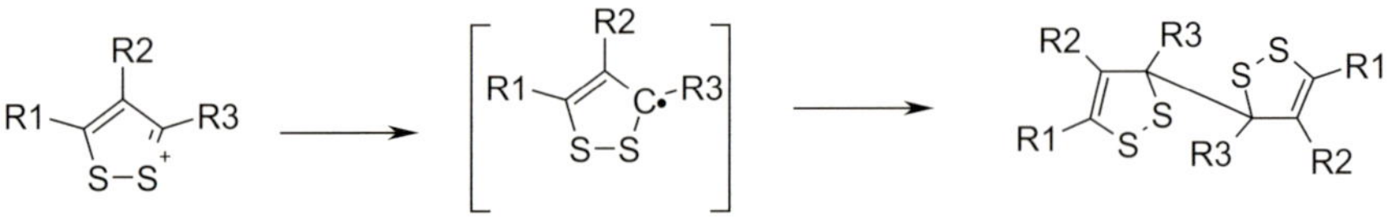

118 **119** **120**

Scheme 11.30

11.4.4.2 **1,2-Dithioles**

To a certain extent 1,2-dithioles behave like acyclic disulfides in their reactions. 1,2-Dithioles with exocyclic double bonds are potentially aromatic and readily undergo reactions that result in the formation of 1,2-dithiolium salts.

11.4.4.2.1 **Reactions with Electrophiles** 1,2-Dithioles react with electrophiles at the ring sulfur atoms; thus chlorine, sulfuryl chloride and sulfenyl chlorides react at the ring sulfur to afford acyclic products [115].

11.4.4.2.2 **Reactions with Nucleophiles** 1,2-Dithiole-3-thiones react with nucleophiles at S2, C3, C4 or C5, the exact position depending on the substitution pattern on the ring and on the hard/soft character of the nucleophile [116–118]. 1,2-Dithioles are attacked at C3 by hydroxide ion. Ethoxide ion attacks Oltipraz (**121**) at C5; subsequent ring opening and recyclization forms pyrrolodiazine **122** [116].

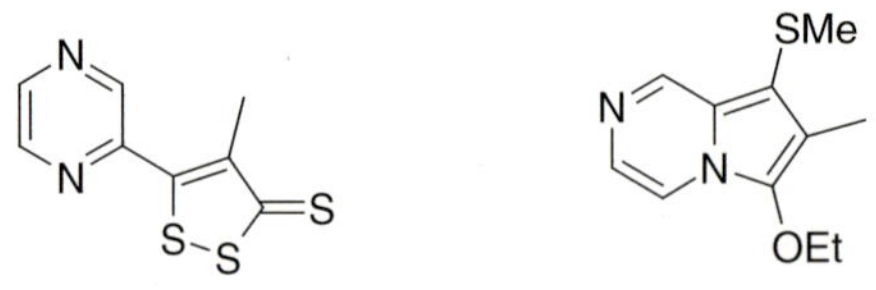

121 **122**

1,2-Dithiole-3-ones and 3-thiones behave like acyclic ketones and thiones, forming imines such as **125**, oximes and hydrazones with nitrogen nucleophiles [119]. 1,2-Dithioles **123** can also undergo ring opening and recyclization to afford isothiazoles **124** (Scheme 11.31). This pathway is dependant on the substituents at R1 and R2 and is particularly common with fused dithioles [120].

1,2-Dithioles react with carbon nucleophiles at a range of sites. Grignard reagents react as thiophiles, attacking at S2 of 1,2-dithiol-3-ones. Phosphonium ylides and

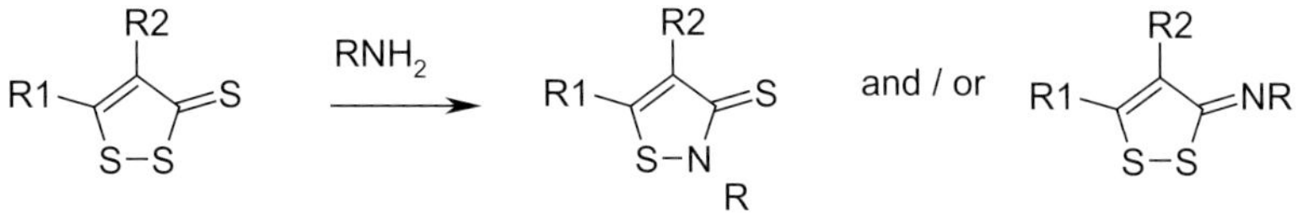

Scheme 11.31

other carbanions attack at C3 of 1,2-dithiole-3-ones and 3-thiones to form 3-alkylidene-1,2-dithioles [121].

Reaction of the 1,2-dithiole-3-thione **126** with trimethyl phosphate affords some of the desired coupling product but also the phosphonate derivative **127** (Scheme 11.32) [122].

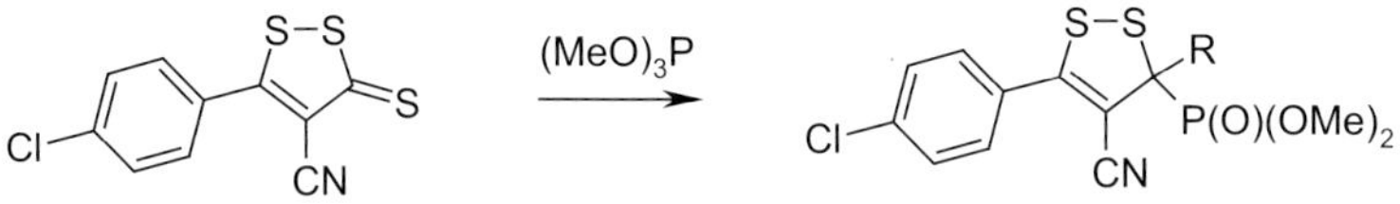

Scheme 11.32

11.4.4.2.3 **Reactions with Carbenes and Nitrenes** 1,2-Dithioles react with carbene and nitrenes at the S–S bond, leading to insertion and sometimes with loss of one sulfur atom to form thiophenes (**128**) or isothiazoles (**129**), respectively, Scheme 11.33 [123, 124].

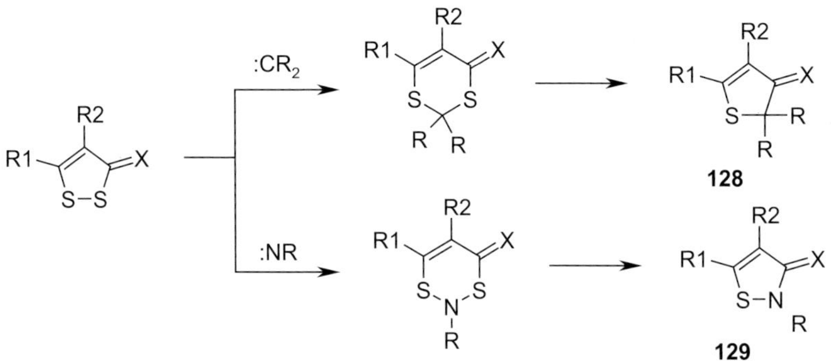

Scheme 11.33

11.4.4.3 **1,2-Dithiolanes**

11.4.4.3.1 **Reaction with Electrophiles** Reactions of 1,2-dithiolanes all occur at the ring sulfur atoms. The sulfur is nucleophilic and is readily alkylated to form 1,2-dithiolium salts **130** (Scheme 11.34) [125].

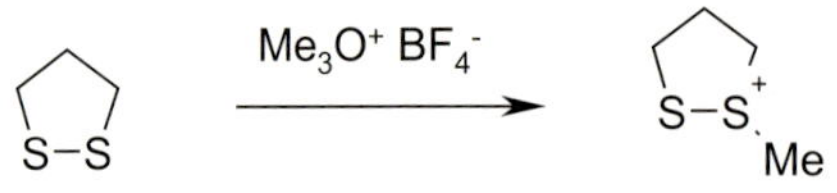

Scheme 11.34

11.4.4.3.2 **Reaction with Nucleophiles** Alkyl lithium reagents, Grignard reagents and cyanide ion attack 1,2-dithiolanes at a sulfur atom to form ring open products [126, 127].

1,2-Dithiolanes **131** react with lithium acetylides to give the ring expanded product **132** [128]. If the 1,2-dithiolane **131** is reacted with dimethylsulfoxonium methylide the 1,3-dithiane **133** is formed (Scheme 11.35) [129].

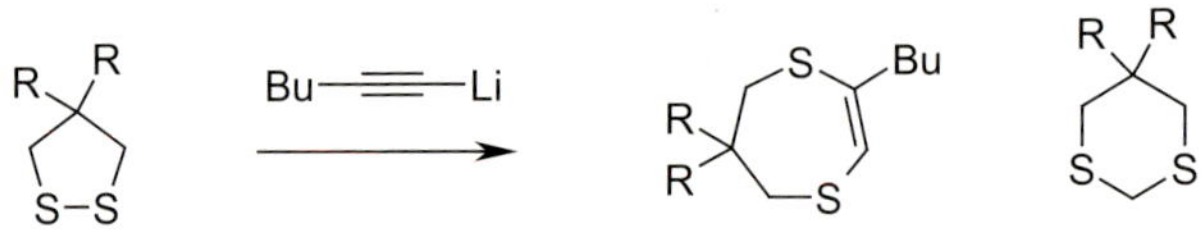

Scheme 11.35

11.4.4.3.3 **Reactions with Carbenes** Carbenes react with 1,2-dithiolanes to afford insertion products at the S–S bond. The 1,2-dithiolane **134** reacts with diphenylcarbene to afford a mixture of the 1,3-dithiane insertion product **135** and the desulfurization product **136** (Scheme 11.36) [130].

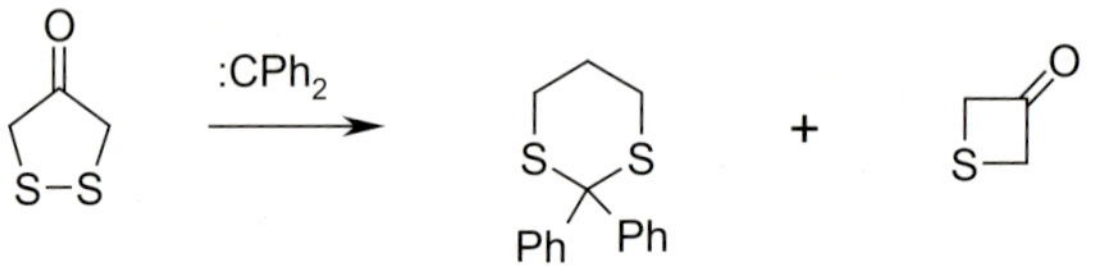

Scheme 11.36

11.4.4.3.4 **Compounds of Interest** 1,2-Dithiolane-4-carboxylic acid (asparagusic, **137**) is thought to act in biological systems as a substrate of a dehydrogenating enzyme and to stimulate pyruvate oxidation. Lipoic acid **138** is involved in the oxidative decarboxylation of 3-ketoacids, oxidative phosphorylation and in photosynthesis [91]. The sulfonamide derivative of lipoic acid (**139**) is a glutathione reductase enhancer [131]. Oltipraz (**121**) has schistosomicidal activity [116].

137

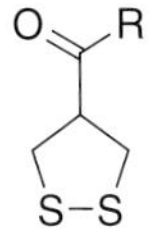

138, R = OH
139, R = $NHSO_2Me$

11.5 1,3-Dithioles and 1,3-Dithiolanes

11.5.1 Introduction

This section deals with 1,3-dithioles derivatives such as 1,3-dithiolylium ions **140**, mesoionic 1,3-dithiol-4-ones **141**, 1,3-dithioles **142**, 1,3-dithiolanes **143** and the tetrathiafulvalene (TTF) system **144**. The latter system has been the subject of a large number of publications because of its organic conducting properties.

140 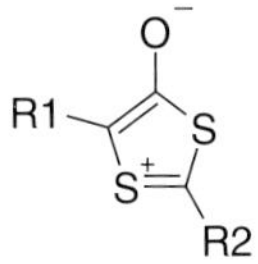 **141** 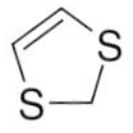 **142**  **143** 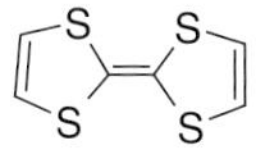 **144**

A comprehensive review chapter on 1,3-dithiolium salts has appeared [132]. The synthesis and reactions of 1,3-dithioles and dithiolanes have also been reviewed [95]. There have also been several reviews on the properties of TTFs [133–135].

11.5.2 Relevant Physicochemical Data

11.5.2.1 X-Ray Diffraction Studies

Several X-ray crystal structure determinations on 1,3-dithioles and 1,3-dithiolanes have been published; bond lengths and angles are presented in Tables 12.4 and 12.5, respectively.

Interestingly, the TTF system **144** is not planar but is slightly distorted into a chair conformation [136].

11.5.2.2 NMR Spectroscopy

^1H NMR data of variously substituted 1,3-dithiolylium ions have been published [113, 138].

Table 11.6 contains ^1H NMR data for several derivatives.

Table 11.4 Bond lengths (Å) for 1,3-dithiolane derivatives.

Bond	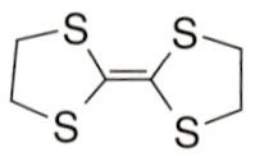	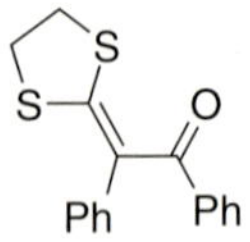
S1–C2	1.756	1.72
S3–C2	1.758	1.76
S3–C4	1.729	1.80
C4–C5	1.314	1.54
S1–C5	1.732	1.83
Reference	[136]	[137]

Table 11.5 Bond angles (°) for 1,3-dithiolane derivatives.

Bond	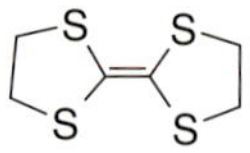	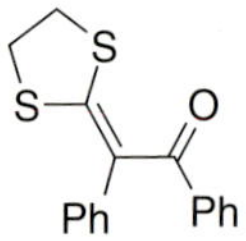
S1–C2–S3	114.5	115.5
C2–S3–C4	94.3	96.7
S3–C4–C5	118.6	115.4
C4–C5–S1	118.0	108.3
C5–S1–C2	94.5	94.6
Reference	[136]	[137]

Table 11.6 ^{1}HNMR data for 1,3-dithiole derivatives.

Compound	H2 (δ, ppm)	H4 & 5 (δ, ppm)	Reference
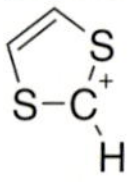	11.65	9.67	[139]
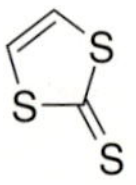		7.2	[138]
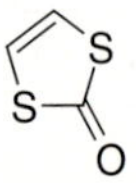		6.74	[138]

Table 11.7 ^{13}C NMR data for 1,3-dithiole derivatives.

Compound	C2 (δ, ppm)	C4 (δ, ppm)	Reference
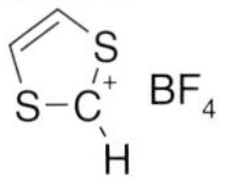 S, S–C⁺ H BF_4	179.5	146.2	[140]
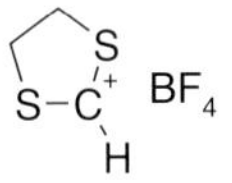 S, S–C⁺ H BF_4	221.2	46.4	[140]
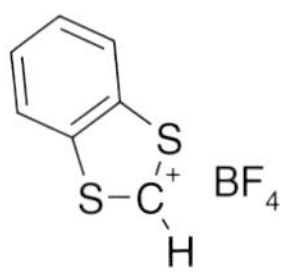 S, S–C⁺ H BF_4	182.4	146.0	[140]
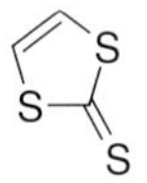 S, S, S	140.7	133.7	[141]

^{13}C NMR data for 1,3-dithiole derivatives are presented in Table 11.7. Calculated electron densities have been correlated with observed ^{13}C chemical shifts for the benzo-1,3-dithiolylium ion [140].

11.5.2.3 **Theoretical Methods**

Simple LCAO-MO calculations for 1,3-dithiolc-2-thione, benzo-1,3-dithiolylium ion,1,3-dithiole-2-one and 1,3-dithiolylium ions indicate that the lowest electron density is found at the 2-position of the 1,3-dithiolylium cation and showed that the C4–C5 bond order corresponded approximately to that of an isolated double bond [138].

11.5.3 Synthesis

11.5.3.1 **1,3-Dithiolium Salts, 1,3-Dithiolones and 1,3-Dithioles**

Several methods are known for the preparation of 1,3-dithiolylium salts from acyclic precursors by the formation of one bond. Acid-catalyzed cyclization of thioesters **145**, dithioesters **146** or thio analogues **147** using either mixtures of perchloric acid and glacial acetic acid or sulfuric acid in the presence of H_2S [138] or H_2S/BF_3 [113] results in the formation of **148** (Scheme 11.37).

Dithiocarbamates like **149** afford the perchlorate salt **150** upon reaction with perchloric acid (Scheme 11.38). Reduction of **150** with sodium borohydride gives the 1,3-dithiole **151**, which can then be treated again with perchloric acid to afford the 1,2-dithiolylium salt **152** [142].

145, X = O, Y = O
146, X = O, Y = S
147, X = S, Y = O

148

Scheme 11.37

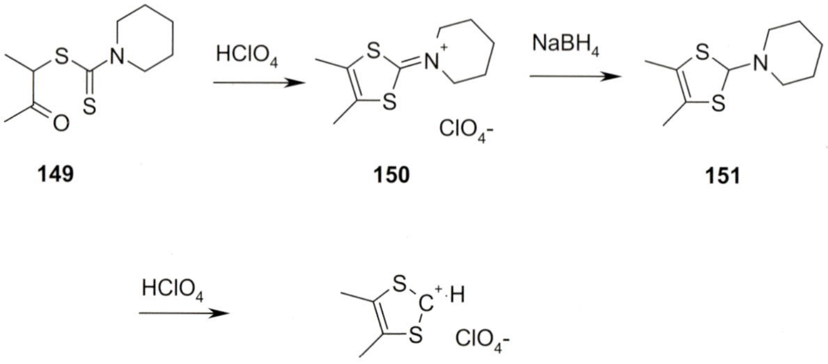

$HClO_4$ $NaBH_4$ $HClO_4$

149 **150** **151** **152**

Scheme 11.38

1,3-Dithiol-2-ones **154** are also obtained by acid-catalyzed cyclization of thioxodithiocarbamates **153** (Scheme 11.39) [143].

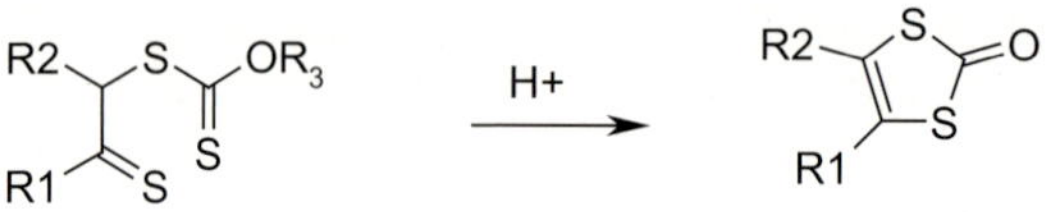

H+

153 **154**

Scheme 11.39

In a similar approach *S*-vinyl-*N*,*N*-dialkyldithiocarbamates **155** afford, upon reaction with bromine, the 1,3-dithiolylium bromides **156** (Scheme 11.40) [144, 145].

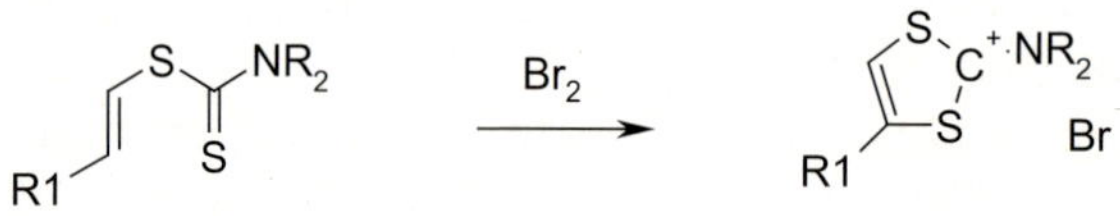

Br_2

155 **156**

Scheme 11.40

Thiobenzoylthioglycolic acid **157** reacts with acetic anhydride in the presence of boron trifluoride results to form dithiolylium salt **158** (Scheme 11.41) [146].

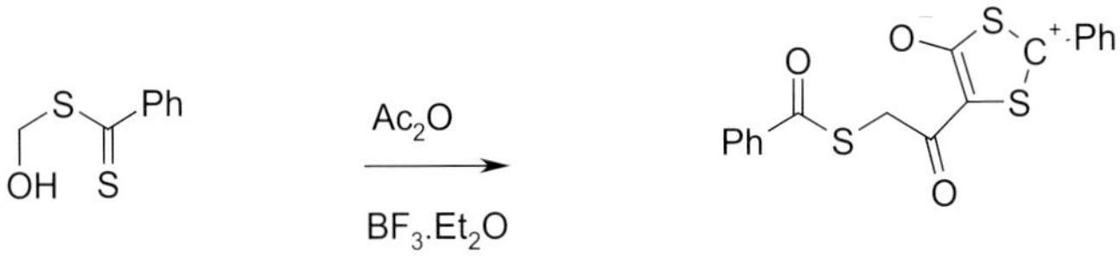

157 —(Ac_2O, $BF_3.Et_2O$)→ 158

Scheme 11.41

Various methods have been described for the synthesis of 1,3-dithiole-2-thiones, which are widely used as precursors for the preparation of 1,3-dithiole derivatives.

Dithiocarbamates **159** are easily cyclized with conc. sulfuric acid or 70% perchloric acid to yield the 1,3-dithiolylium salts **160**, which upon treatment with H_2S afford the 1,3-dithiole-2-thiones **161** (Scheme 11.42) [147].

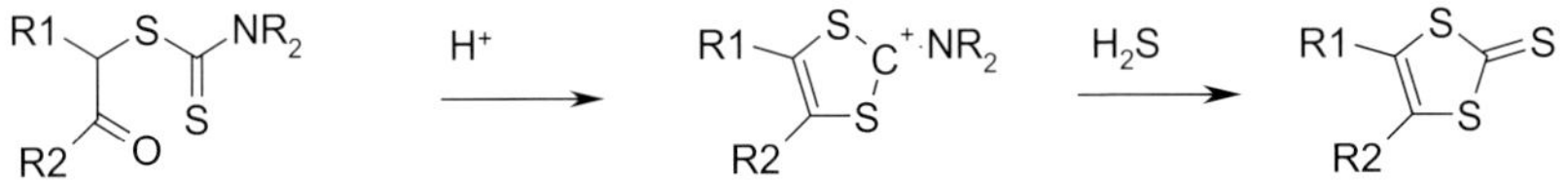

159 —(H^+)→ 160 —(H_2S)→ 161

Scheme 11.42

An alternative approach uses propargyl-*t*-butyltrithiocarbamates **162**, which afford mono substituted 1,3-dithiole-2-thiones **163** on treatment with trifluoroacetic acid/ glacial acetic acid (Scheme 11.43) [148].

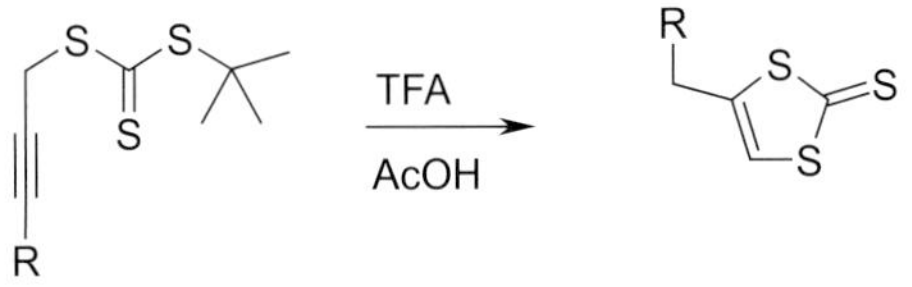

162 —(TFA, AcOH)→ 163

Scheme 11.43

A two-step conversion of alkynes into dithiolones **164** and dithiolethiones **165** has been reported. The initial step involves palladium-catalyzed addition of Pr^i_3Si-S-S-SiPr^i_3 followed by treatment with fluoride ion and either PhSOCl or $CSCl_2$ [149].

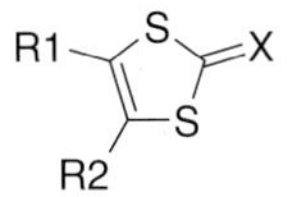

164, X = O
165, X = S

A convenient synthesis of 1,3-dithiole-2-thiones **166** involves the treatment of a terminal alkyne with butyllithium, sulfur and then CS_2 (Scheme 11.44) [150]. In similar fashion, the reaction of phenylacetylene with sulfur and KOH in DMSO leads to direct formation of the dithiole **167**, albeit in low yield [151].

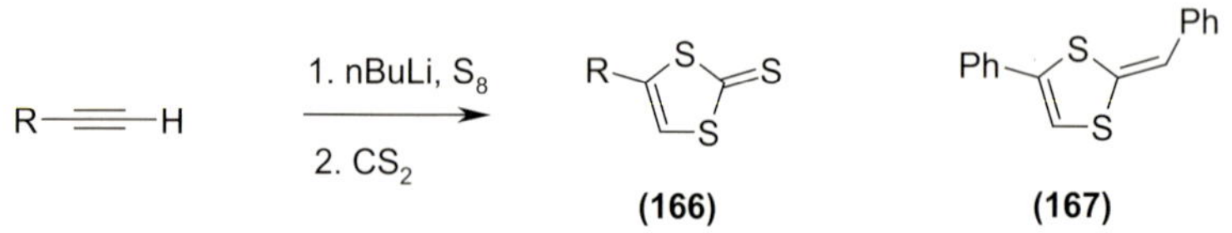

Scheme 11.44

The bis-dithiole salt **168** reacts with long-chain alkyl iodides to give dithioles **169** (Scheme 11.45) [152].

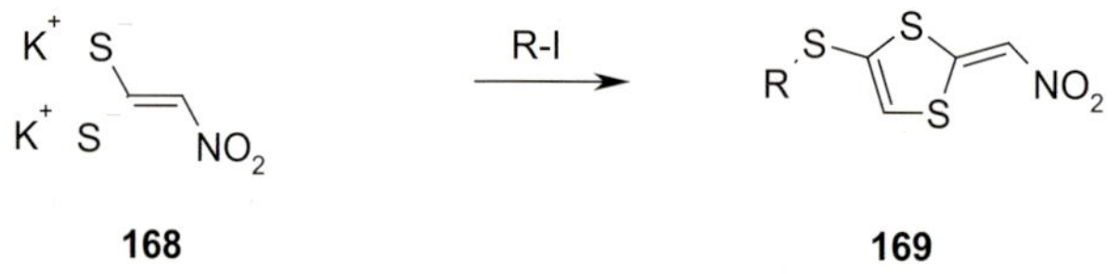

Scheme 11.45

Unsubstituted 1,3-dithiole-2-thione (**172**) can be prepared from 1,2-dichloroethyl ethyl ether (**170**) and potassium trithiocarbonate (Scheme 11.46). The intermediate 4-ethoxy-1,3-dithiolane-2-thione **171** is treated with *p*-toluene sulfonic acid to give **172** [153].

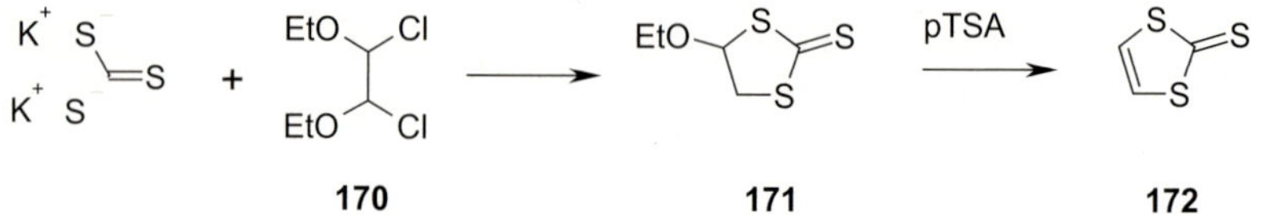

Scheme 11.46

1,3-Dithiolylium salts **175** have also been prepared by the reaction of α-haloketones **173** with an excess of dithio-carboxylic acids **174** in the presence of strong mineral acids at 60–80 °C (Scheme 11.47) [154].

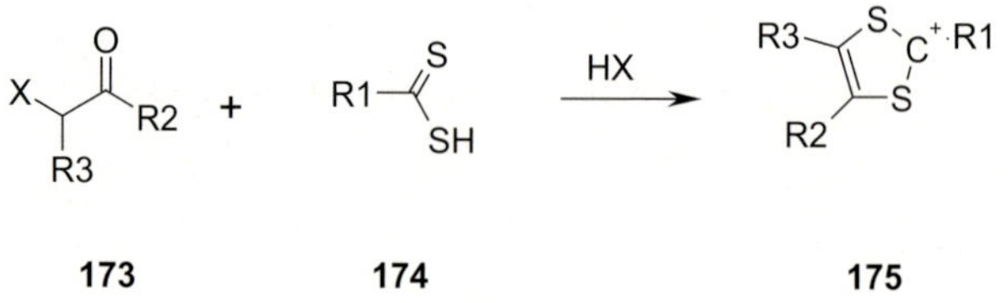

Scheme 11.47

α-Haloketones **176** also react with N,N-dialkyldithiocarbamidates **177** in the presence of strong acid to afford 2-alkylamino-1,3-dithiolylium salts **178** (Scheme 11.48) [155, 156].

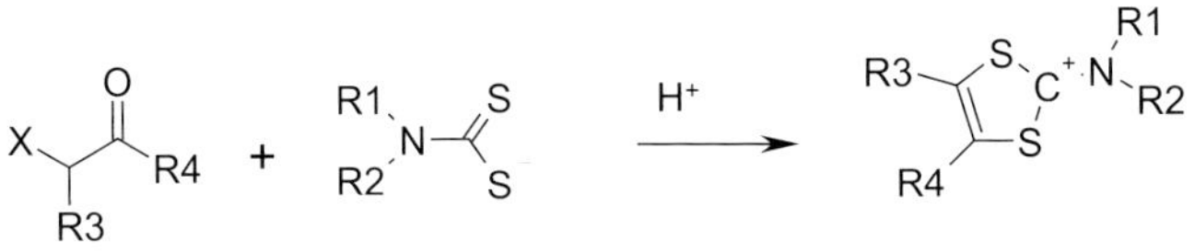

176 **177** **178**

Scheme 11.48

Benzo-annellated 1,3-dithiole-2-thiones (**180**) can be prepared by the reaction of benzene-1,2-dithiol (**179**) with thiocarbonyldiimidazole in glacial acetic acid (Scheme 11.49) [157].

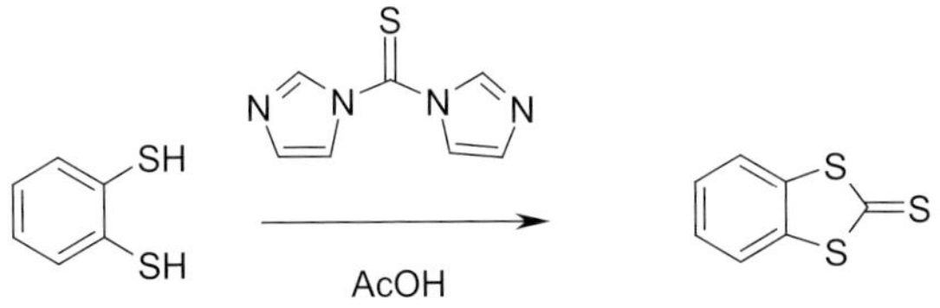

179 **180**

Scheme 11.49

11.5.3.2 **Ring Transformations of Heterocycles Leading to 1,3-Dithiole Derivatives**

Thermolysis of 1,2,3-thiadiazoles **181** in carbon disulfide provides a useful route to 1,3-dithiole-2-thiones **182** (Scheme 11.50) [158].

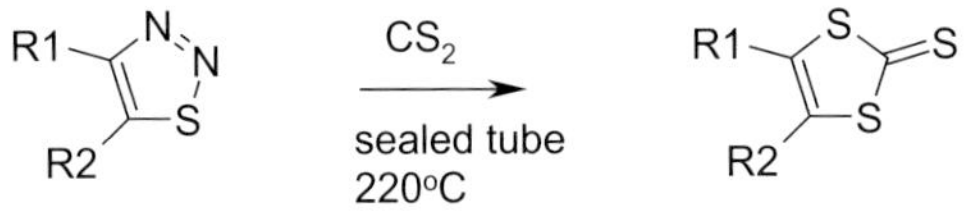

181 **182**

Scheme 11.50

1,2-Dithiole-3-thione derivatives can also be used as starting compounds for 1,3-dithiole synthesis. 1,2-Dithiole-3-thiones **183** react with alkynes or even benzyne to afford 1,3-dithioles **184** (Scheme 11.51) [159].

Alkynes bearing electron-withdrawing groups also react with O,S-ethylene dithiocarbonate **185** (X=O) or ethylene trithiocarbonate **185** (X=S) to afford 1,3-dithiole-2-ones **186** (X=O) or 1,3-dithiole-2-thiones **186** (X=S) (Scheme 11.52) [160, 161].

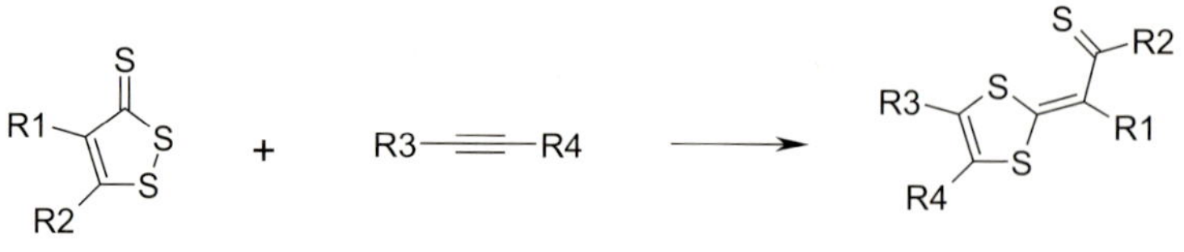

183 184

Scheme 11.51

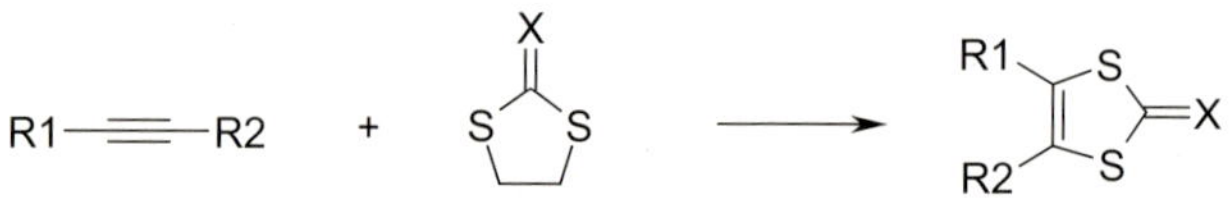

185 186

Scheme 11.52

11.5.3.3 **Ring Synthesis of 1,3-Dithiolanes**

The most widely used method for the synthesis of 1,3-dithiolanes involves the condensation reaction of an aldehyde or ketone with suitably substituted 1,2-dithiols in the presence of a catalyst. Catalysts for the conversion of aldehydes and ketones into 1,3-dithiolanes with ethanediol include HCl [162], BF_3Et_2O [162], iodine [163], $MoCl_5$ [164], indium triflate [165] and $Cu(CF_3SO_3)_2$ [166].

Iodine and indium triflate and indium chloride are also good catalysts for effecting the transthioacetalization of 1,3-dioxolanes [167]. $ZrCl_4$ has also been used in this transformation [168].

The use of $BF_3{\cdot}Et_2O$ as a catalyst allows the synthesis of some very sensitive systems; the cyclopropanone **187** reacts with 1,2-dithiols **188** in 24–58% yield to give the spiro derivatives **189** (Scheme 11.53) [169].

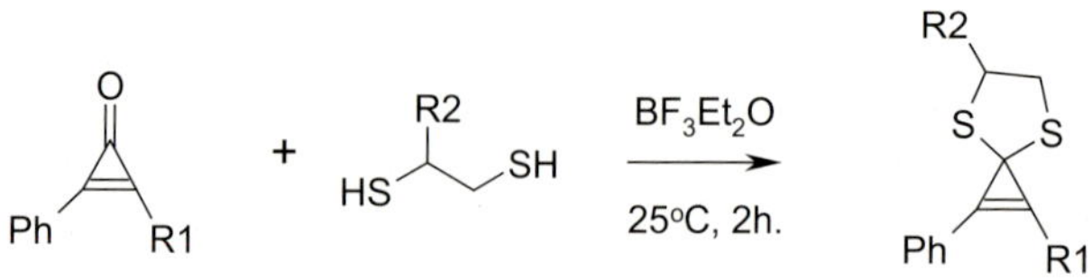

187 188 189

Scheme 11.53

Carbonyl compounds possessing an α-methylene group (**190**) react with 1,2-bis(chlorosulfenyl)alkanes **191** to give 2- formyl-1,3-dithiolanes **192** (Scheme 11.54) [170].

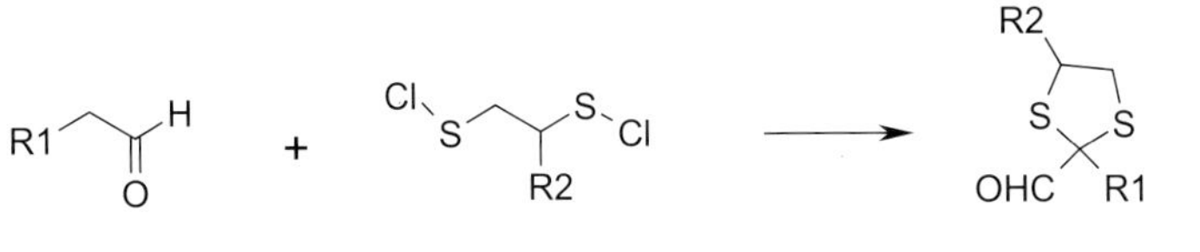

190 **191** **192**

Scheme 11.54

Compound **191** also reacts with ethyl acetoacetate to give 2-acetyl-2-ethoxycarbonyl-1,3-dioxolanes **193**, which are readily hydrolyzed and decarboxylated to give 2-acetyl-1,3-dioxolanes **194** (Scheme 11.55) [170].

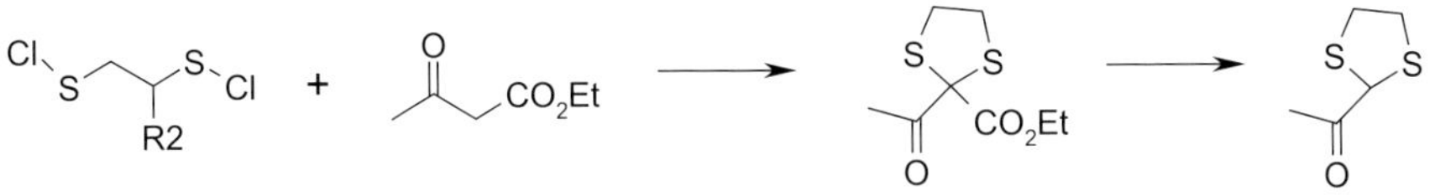

191 **193** **194**

Scheme 11.55

1,2-Bis(triphenylphosphonium)ethane dibromides **195** react with 1,2-ethanedithiol in the presence of a base to give an almost quantitative yield of 2-triphenylphosphoniomethyl-1,3-dithiolane (**196**), which can then undergo Wittig reactions (Scheme 11.56) [171].

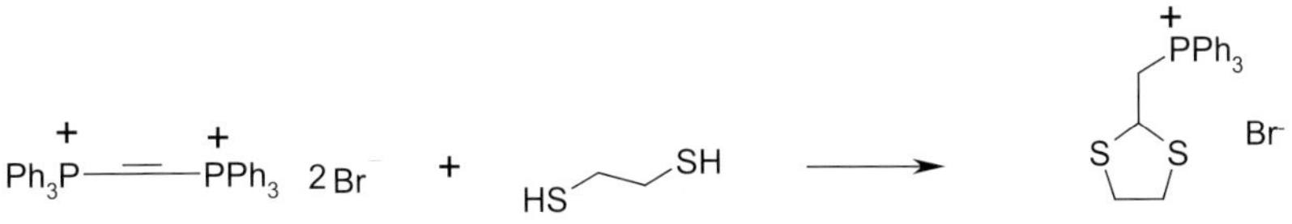

195 **196**

Scheme 11.56

2-Amino substituted 1,3-dithiolanes **198** can be obtained from amide acetals such as **197** and 1,2-ethanedithiol (Scheme 11.57) [172].

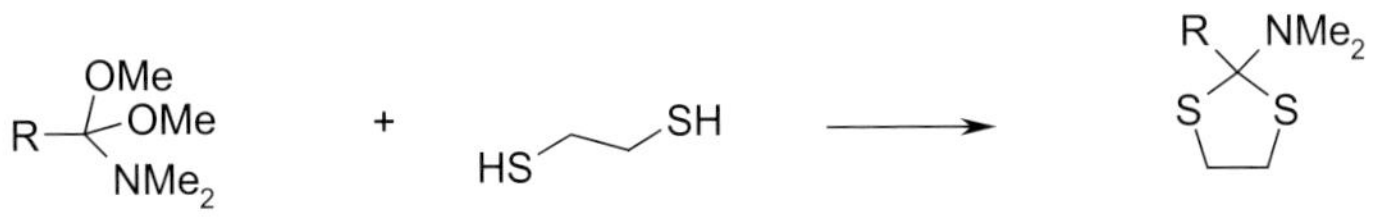

197 **198**

Scheme 11.57

Ethane-1,3-dithiol also reacts with dichloromethyl methyl ether in the presence of sodium to give 2-methoxy-1,3-dithiolane **199** (Scheme 11.58) [173]. It also reacts with phosgene to give 1,3-dithiolan-2-one **200** [174].

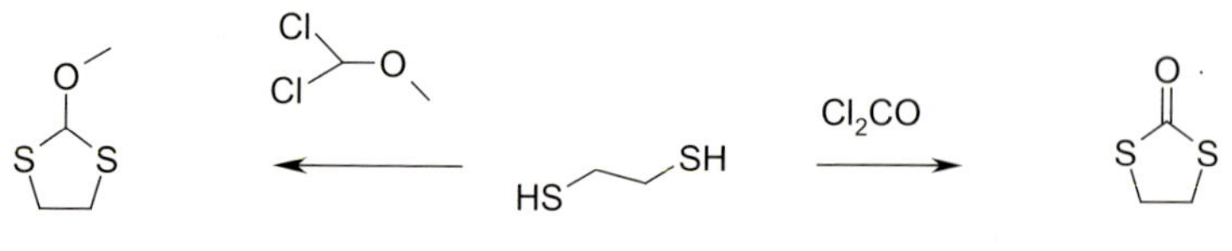

Cl_2CO

199 **200**

Scheme 11.58

The reaction of ethylene dibromide with sodium trithiocarbonate affords 1,3-dithiolan-2-thione (**201**) (Scheme 11.59) [175].

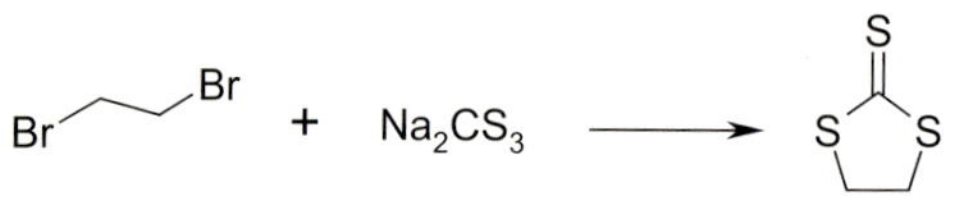

Na_2CS_3

201

Scheme 11.59

11.5.3.4 **Ring Transformations of Heterocycles Leading to 1,3-Dithiolane Derivatives**

2-Substituted 1,3-dioxolanes react with ethanediol in an ionic liquid – transthioacetalization occurs to afford the corresponding 2-substituted 1,3-dithiolanes [176]. The asymmetric synthesis of various nucleoside analogues in which the ribose base is replaced by a 1,3-dithiolane ring has been described [177].

The transformation of complex and sensitive carbonyl compounds under neutral conditions into 1,3-dithiolanes **203** can be achieved using 2-phenyl or 2-halogeno 1,3,2-dithiaborolanes **202** (Scheme 11.60) [178].

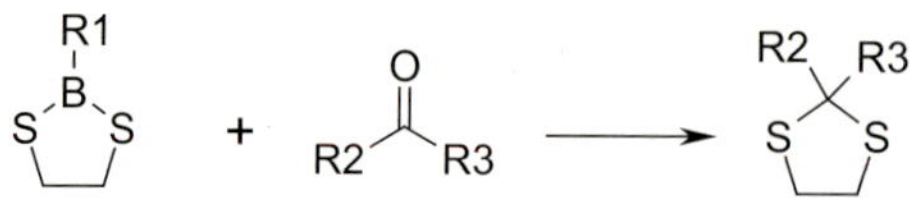

202, R1 = Ph or Cl **203**

Scheme 11.60

An alternative method for the synthesis of 1,3-dithiolanes **205** from acid-sensitive carbonyl compounds involves the reaction of 2-ethoxy-1,3-dithiolane (**204**) in the presence of mercury(II) chloride or other Lewis acids (Scheme 11.61) [179].

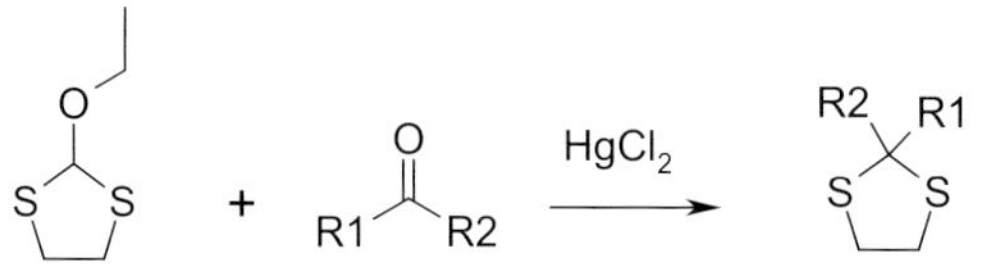

Scheme 11.61

11.5.3.5 **Synthesis of Tetrathiafulvalenes**

Tetrathiafulvalenes (TTFs) are electron donors that easily form charge-transfer complexes with electron acceptors such as radical salts. The discovery of the exceptional electron conductivity of TTFs has encouraged a great deal of research into their synthesis.

The vast majority of syntheses of TTFs involve the preparation of the π-bond in between the two 1,3-dithiole rings as the final step. This bond can be formed via an elimination reaction involving either two-proton electrochemical oxidation or the σ- and π- bonds can be formed simultaneously by a coupling reaction between two carbenes.

The electrochemical oxidation of the 1,3-dithioles **206** and **208** in the presence of pyridine has been described. The respective TTFs **207** and **209** were obtained in low yield, 30 and 40%, respectively (Scheme 11.62) [180].

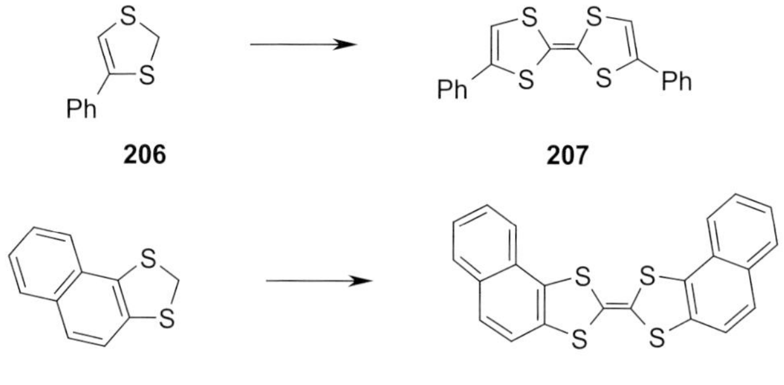

Scheme 11.62

The synthesis of TTF via elimination of a proton from 1,3-dithiolylium salts in the final synthetic step involves the reaction of a carbene or phosphonium ylide on a 1,3-dithiolylium salt bearing a hydrogen at C2. These precursors can be prepared either by an alkylation of 2-alkylthio-1,3-dithioles **210** [181] or by oxidation of 1,3-dithiole-2-thiones **211** either with peracid or hydrogen peroxide (Scheme 11.63) [182]. Treatment of these precursors with a tertiary amine base affords TTF (**144**). The presence of alkyl groups in the 4 or 5 position does not interfere with this reaction; however, electron-withdrawing groups in these positions can lead to no reaction.

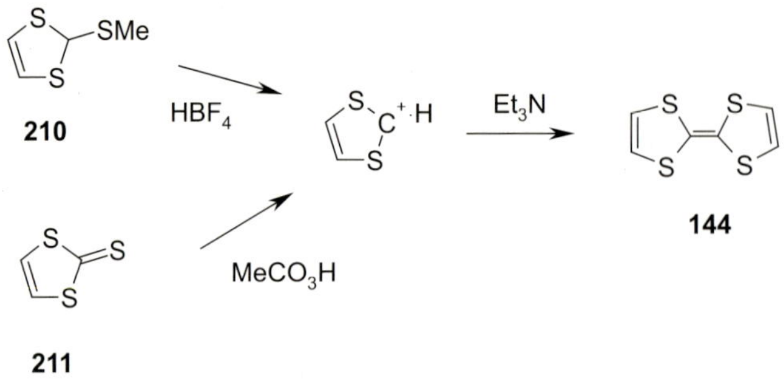

Scheme 11.63

The reaction of carbon disulfide with either a strained or electron-rich acetylene derivatives in the presence of an acid or under high pressure is another common method for the preparation of TTFs. This method can be used to afford TTFs with electron-withdrawing substituents in the ring. Acetylenes **212** and **213** react with CS_2 to give the TTFs **214** [183] and **215** (Scheme 11.64) [184].

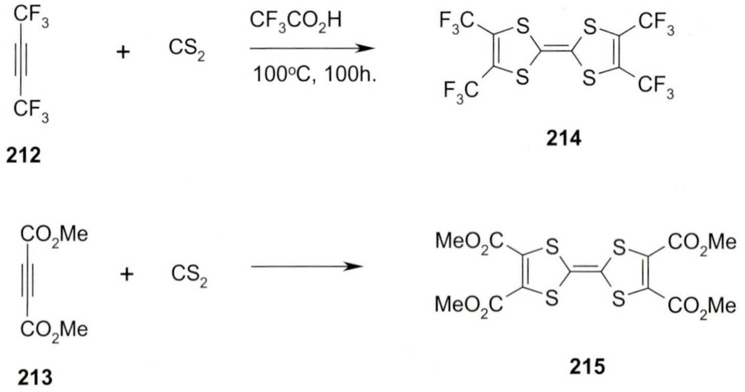

Scheme 11.64

1,3-Dithiolylium salts **217** react with 2-triphenylphosphino-1,3-dithioles **216** to afford an intermediate that eliminates a proton and triphenylphosphine on treatment with base to give the TTF **218** (Scheme 11.65) [185].

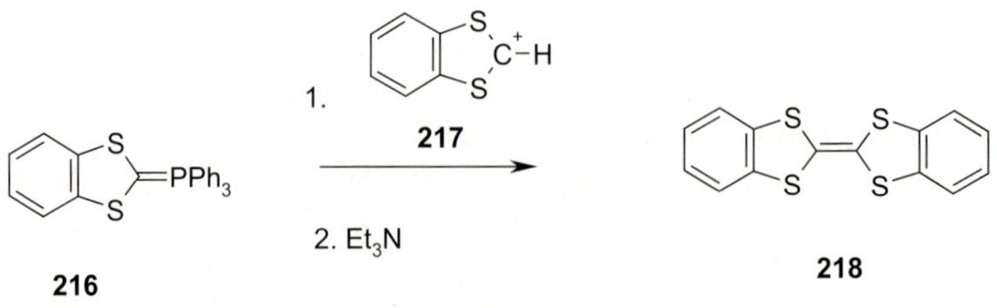

Scheme 11.65

Similarly, TTF (**144**) is obtained from a mixture of trialkylphosphanes and alkynes in the presence of carbon disulfide (Scheme 11.66) [186].

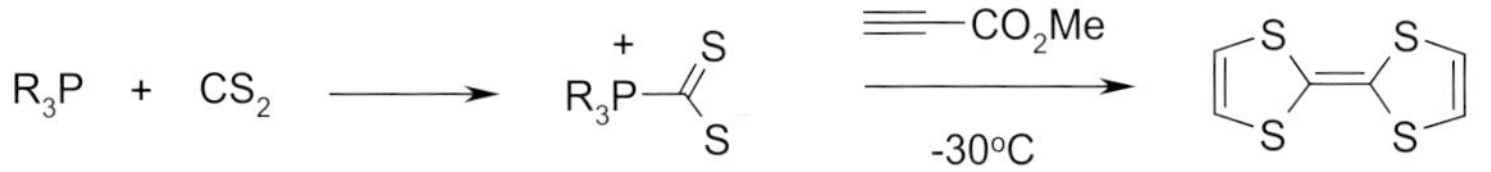

Scheme 11.66

TTF may be obtained by dethioxygenation of 2-thioxo-1,3-dithioles by transition metal complexes. The reaction of 2-thioxo-1,3-dithioles **219** with $Fe_3(CO)_{12}$ or $Co_2(CO)_8$ furnishes TTFs **220** with aryl, alkyl or electron-withdrawing groups in the 4- and 5-positions of the dithiole ring (Scheme 11.67) [187, 188].

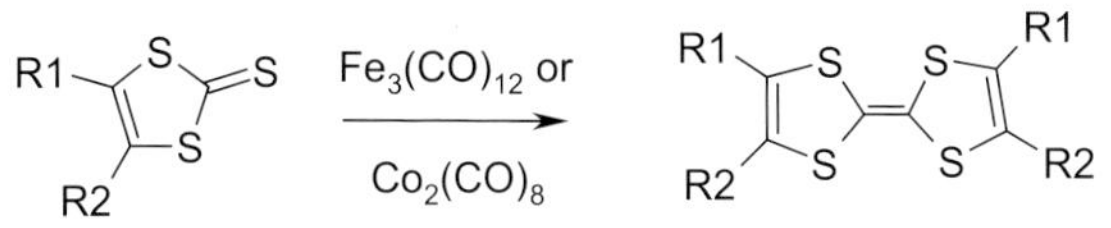

Scheme 11.67

2-Thioxo-1,3-dithioles bearing electron-withdrawing groups such as in **221** also react with trivalent phosphorus compounds to form the corresponding TTFs (**222**) as a mixture of regioisomers (Scheme 11.68) [189].

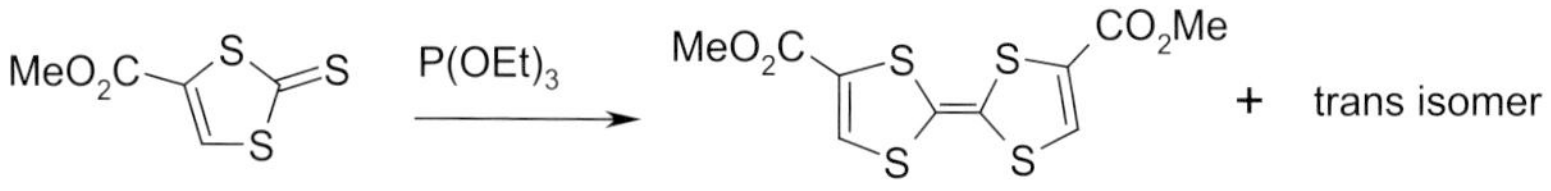

Scheme 11.68

Tetrachloroethylene has also been used as a central building block for TTF synthesis. This method is suitable for the synthesis of symmetrical and unsymmetrical TTFs. The reaction of tetrachloroethylene with 1,2-benzenethiols **223** and **224** leads to a mixture of benzotetrathiafulvalenes **225** (Scheme 11.69) [190].

TTF (**144**) has been synthesized in two steps (85% overall yield), starting from 4,5-bis(benzoylthio)-1,3-dithiole-2-thione (**226**) (Scheme 11.70). The intermediate tetrathianaphthalene **227** is treated with base to afford TTF. This method is suitable for the large-scale synthesis of TTF as no chromatography is required [191].

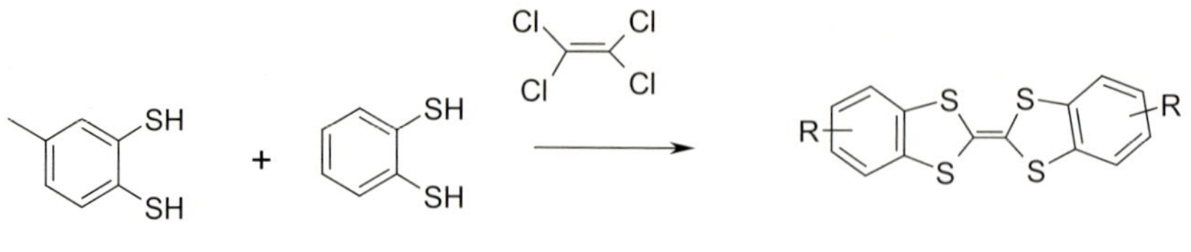

223 224 225; R = H or Me

Scheme 11.69

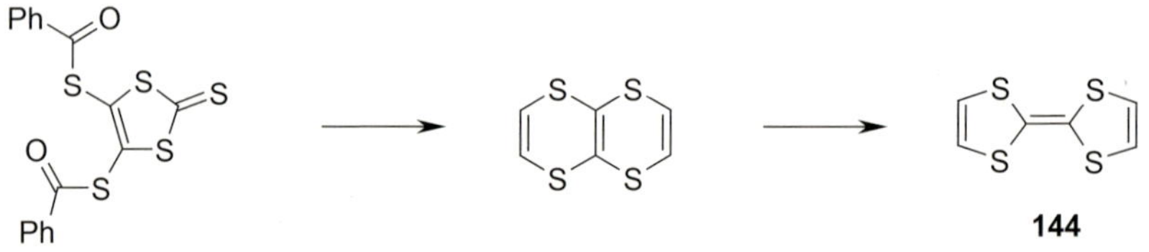

226 227 144

Scheme 11.70

11.5.4 Reactivity of 1,3-Dithiolylium Ions, Mesoionic 1,3-Dithiol-4-ones and 1,3-Dithioles

11.5.4.1 Thermal and Photochemical Reactions

1,3-Dithiole-2-one **228** undergoes a photochemically induced decarbonylation to afford the dithione derivative **229**, which may react further to give the dithiete **230** (Scheme 11.71) [192].

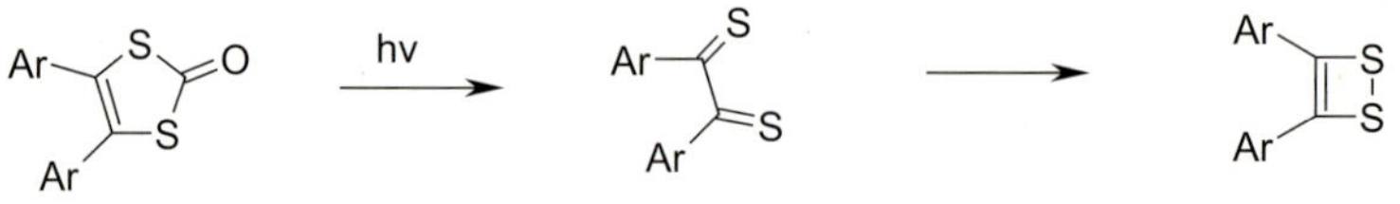

228 229 230

Scheme 11.71

Photolysis of the 1,3-dithiolylium-4-oate **231** gives a mixture of the 1,2-dithiol-3-one **232** along with diazine **233** and thiophene **234** (Scheme 11.72) [193].

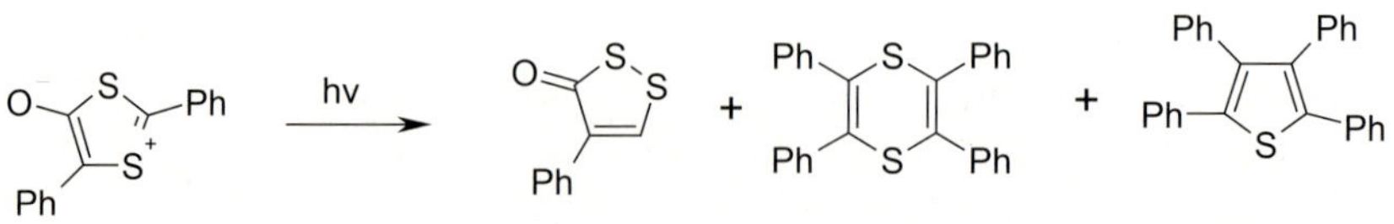

231 232 233 234

Scheme 11.72

11.5.4.2 Reactions with Electrophiles

The attack of electrophiles at the ring carbons of 1,3-dithiolylium ions is seldom observed. 1,3-Dithiole-2-thiones can be alkylated on the ring sulfur atom to give a 1,3-dithiolylium system – strong alkylating agents such as triethyloxonium tetrafluoroborate must be used [194].

11.5.4.3 Reactions with Nucleophiles

1,3-Dithiolylium salts react with nucleophiles at the 2-position while 1,3-dithiolylium-4-oates undergo nucleophilic attack at the 4-position. 1,3-Dithiolylium cations are hydrolyzed to the corresponding 2-hydroxy-1,3-dithioles, which upon treatment with acid reform the 1,3-dithiolylium cation [195]. The reaction of 1,3-dithiolylium salts **235** with alcohols leads to stable 2-alkoxy-1,3-dithioles **236** (Scheme 11.73) [196]. These alkoxy derivatives are useful precursors of 2-aryl-1,3-dithioles [197]. Sulfur nucleophiles react in a similar fashion to alcohols with 1,3-dithiolylium salts to afford 2-alkylthio or 2-arylsulfanyl-1,3-dithioles [196].

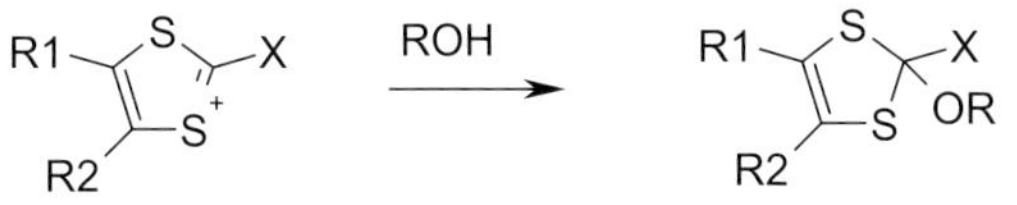

R1, R2, S, S+, X — ROH → R1, R2, S, S, X, OR

235 **236**

Scheme 11.73

Secondary amines react with 1,3-dithiolylium salts that are unsubstituted at the 2-position (**237**) to afford 2-amino-1,3-dithioles **238** (Scheme 11.74) [196]. 2-Methylsulfanyl-1,3-dithiolylium salts **239** give the corresponding 2-amino-1,3-dithiolylium salts **240** with secondary amines [198]. Primary amines react with **239** to afford 2-imino-1,3-dithioles **241** [199].

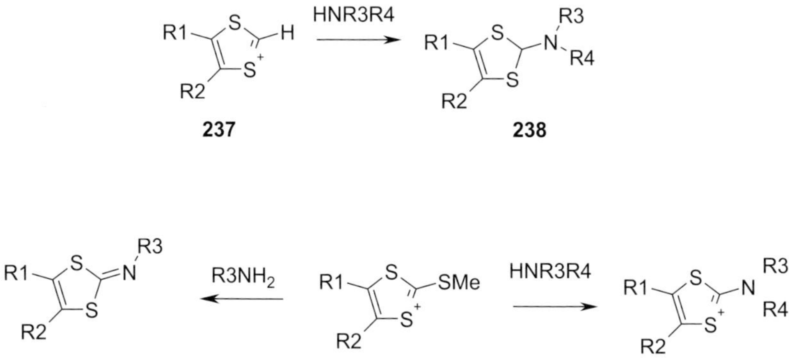

R1, R2, S, S+, H — HNR3R4 → R1, R2, S, S, N, R3, R4

237 **238**

R1, R2, S, S, N, R3 ← $R3NH_2$ — R1, R2, S, S+, SMe — HNR3R4 → R1, R2, S, S+, N, R3, R4

241 **239** **240**

Scheme 11.74

2-Methylsulfanyl-1,3-dithiolylium salts **239** also react with Grignard reagents to form 2-alkyl-2-methylsulfanyl-1,3-dithiolylium salts **242**, which may react with excess Grignard reagent to give 2,2-dialkyl-1,3-dithioles **243** (Scheme 11.75) [200].

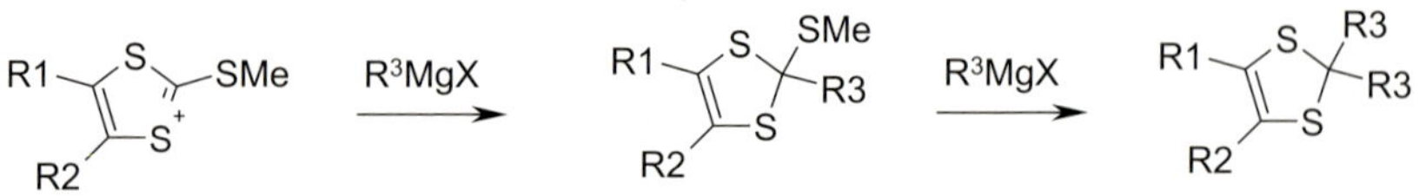

Scheme 11.75

11.5.4.4 **Reductions**

1,3-Dithiolylium salts are easily reduced by $NaBH_4$, $LiAlH_4$ or NaSH to give the corresponding 1,3-dithiole [196]. If 1,3-dithiolylium salts **244** are reduced with zinc [201] or $V(CO)_6$ [202] the dimer **245** is formed (Scheme 11.76). 2-Methylsulfanyl-1,3-dithiolylium iodide (**246**) forms TTF (**144**) when reduced with zinc in the presence of bromine [203].

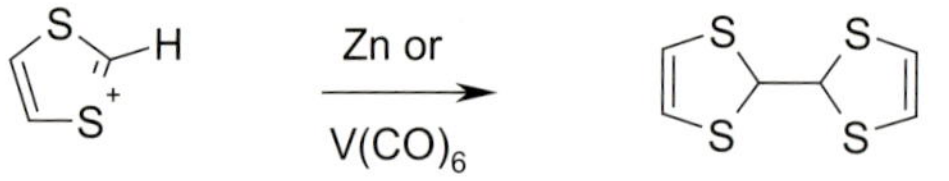

Scheme 11.76

1,3-Dithiolylium-4-olates **248** can be regarded as masked 1,3-dipole thiocarbonyl ylides; they react with dipolarophiles to give cycloadducts. Thiophenes **249** are obtained when **248** are reacted with alkynes (Scheme 11.77) [204].

Stable adducts such as **250** are formed when **248** is reacted with electron-deficient alkenes [205].

11.5.4.5 **Coupling Reactions**

The palladium-catalyzed coupling of the tributyl tin 1,3-dithiole derivative **251** with 2-iodoquinoline to give the coupled product **252** has been described (Scheme 11.78) [206].

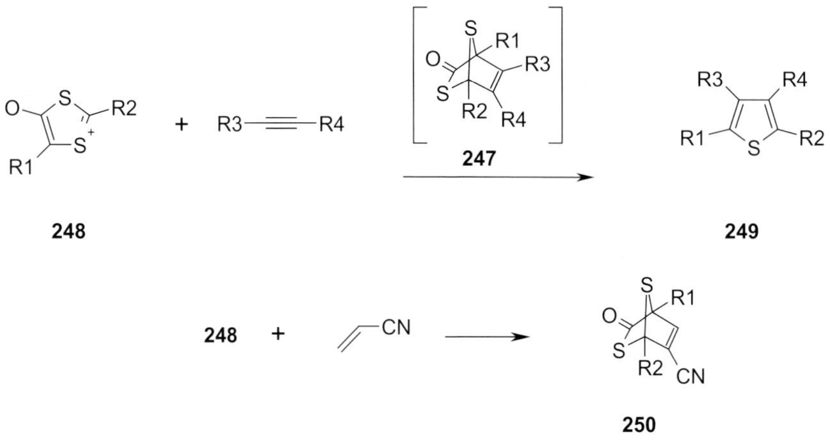

Scheme 11.77

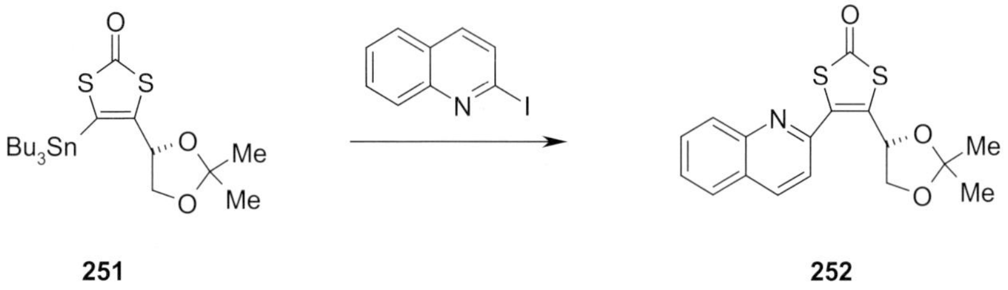

Scheme 11.78

11.5.5 Reactivity of 1,3-Dithiolanes

1,3-Dithiolanes are resistant to both acid and alkaline hydrolysis, they are also resistant to nucleophilic attack by nucleophiles such as hydride ions.

11.5.5.1 Cleavage Reactions

A review of methods for cleaving 1,3-dithiolanes to aldehydes or ketones has been published [207]. The ring can be cleaved by a mixture of $HgCl_2/CdCO_3$ quite effectively [208]; alternatively, sodium in ethanol in liquid ammonia [209], NBS in acetone [210], or thallium(III) nitrate in methanol can be used. The latter method has been used for the selective cleavage of a mixture of thioketals **253** to afford the unsaturated ketone **254** (Scheme 11.79) [211].

$SOCl_2$-treated silica and DMSO is a good combination with which to cleave 2-substituted-1,3-dithiolanes; however, under similar conditions 2,2-disubstituted-1,3-dithiolanes undergo ring expansion to give dihydro-1,4-dithiins [212].

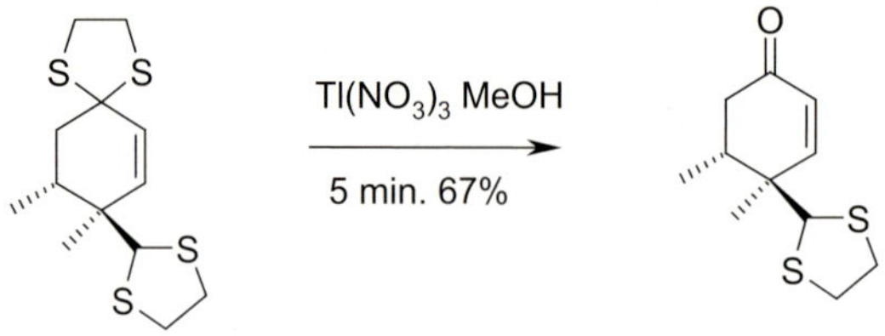

Scheme 11.79

Recent methods for the cleavage of 1,3-dithiolanes include the use of 2,4,6-trichlorotriazine and DMSO [213], oxone on wet alumina [214] and ferric nitrate on K-10 montmorillonite clay [215].

Treatment of 2-substituted-1,3-dithiolanes with NBS followed by 1,2-ethanediol affords 1,3-dioxolanes [216].

11.5.5.2 **Electrophilic Attack at Carbon**

1,3-Dithiolanes can be deprotonated at C2 with equimolar amounts of butyllithium. The resultant carbanions can then be trapped by various electrophiles [217]. 2-Alkynyl-1,3-dithiolanes **255** affords allenes **256**when treated with an organocuprate compound followed by an electrophile and then a Grignard reagent with nickel catalysis (Scheme 11.80) [218]. Similarly, the reaction of 2-alkenyl-1,3-dithiolanes **257** with Bu_2CuLi or BuLi and an electrophile followed by treatment with a Grignard reagent under nickel catalysis affords alkenes **258** [219].

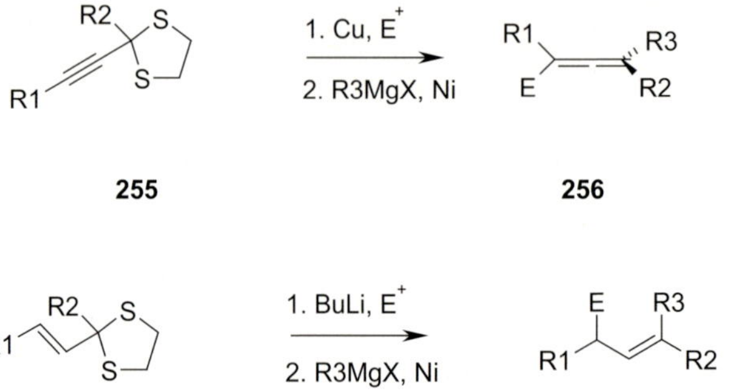

Scheme 11.80

11.5.5.3 **Oxidations**

1,3-Dithiolanes can be oxidized to afford 1,3-dithiolane-1-oxides by photooxidation [220], while oxidation under more vigorous conditions leads to mono and bis sulfones [221]. Enantioselective oxidations of 1,3-dithiolane (**143**) by microbial oxidation gives the corresponding *S*-oxide **259** in high enantiomeric excess (Scheme 11.81) [222]. A review of the preparation and use of the C2 symmetric bis-sulfoxide **260** has been published [223].

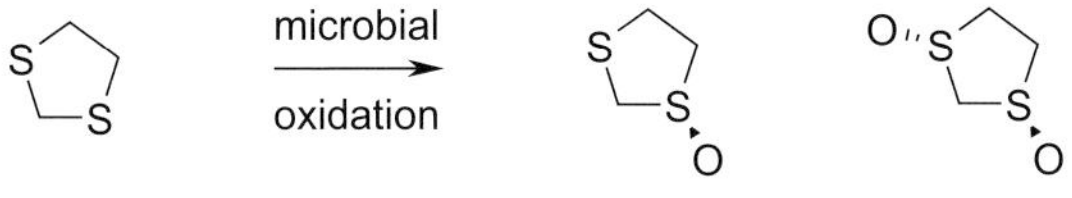

microbial oxidation

143 **259** **260**

Scheme 11.81

On a similar theme the 2-phenyl monosulfoxide derivative **262** has been prepared in a diastereoselective manner by the reaction of 2-phenyl-1,3-dithiolane (**261**) with Bu^{t}OOH and Cp_2TiCl_2 (Scheme 11.82) [224].

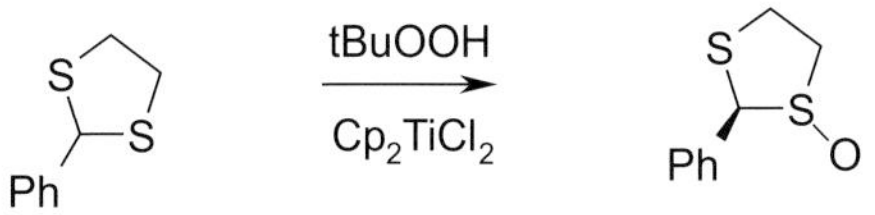

tBuOOH Cp_2TiCl_2

261 **262**

Scheme 11.82

11.5.5.4 **Radical Reactions**

Intramolecular addition of the 1,3-dithiolan-2-yl radical generated from **263** by photolysis affords the spirocyclic derivative **264** (Scheme 11.83) [225].

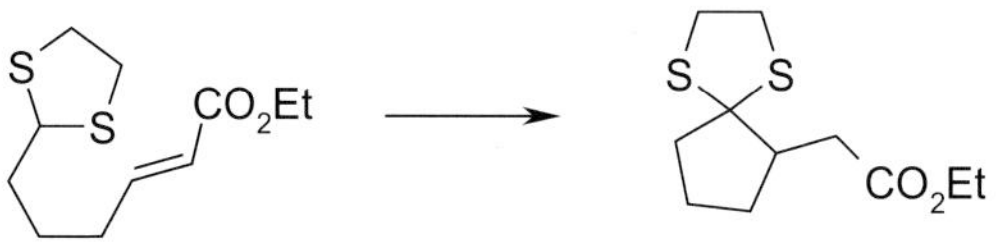

263 **264**

Scheme 11.83

11.5.5.5 **Ring Transformation Reactions**

Treatment of dithiolane **265** with WCl_6 in DMSO gives the ring-expanded dithiin **266** (Scheme 11.84) [226].

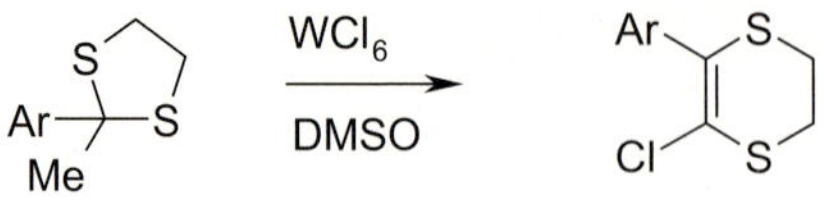

WCl_6, DMSO

265 → 266

Scheme 11.84

Ethylenediamine reacts with 1,3-dithiolane derivatives **267** to give imidazolines **268** (Scheme 11.85) [227].

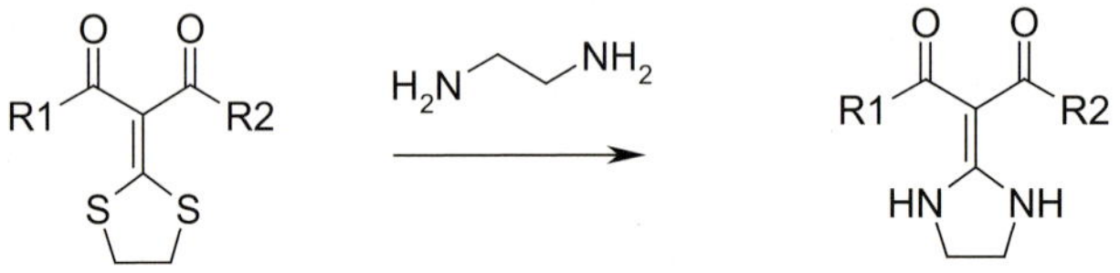

$H_2N\!\frown\!NH_2$

267 → 268

Scheme 11.85

11.5.6
Compounds of Interest

TTF derivatives have been the focus of scientific interest since the mid-1970s. Indeed, their organic metal properties and superconductivity has prompted the publication of several reviews discussed earlier in this section [133–135].

11.6
1,2-Oxathioles and 1,2-Oxathiolanes

11.6.1
Introduction

The parent oxathiolone **269** is known but most work published has been on the S-oxidized derivatives like **270** and **271**. The corresponding saturated derivatives **272** and **273** have also been described. A comprehensive review of 1,2-oxathiolium salts has been published [228].

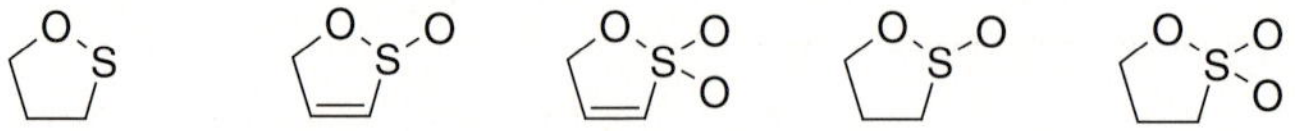

269 270 271 272 273

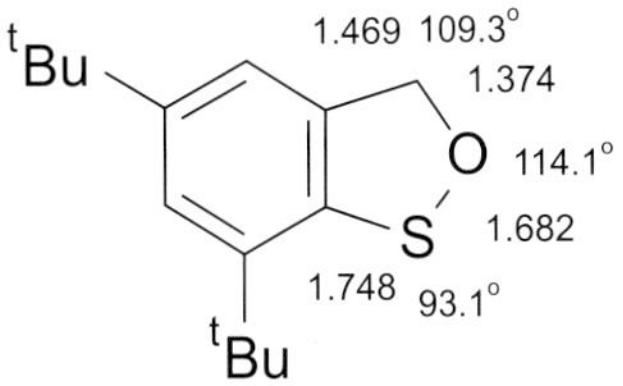

Figure 11.2 Geometry of a 1,2-oxathiolone derivative.

Table 11.8 ^{1}H NMR data for 1,2-oxathiolanes derivatives.

Compound	3-H (δ, ppm)	4-H (δ, ppm)	5-H (δ, ppm)	Reference
1,2-Oxathiolan-2-oxide (**272**)	1.35	1.35	3.45	[230]
1,2-Oxathiolan-2,2-dioxide (**273**)	3.1	2.5	4.82	[231]

11.6.2 Relevant Physicochemical Data

11.6.2.1 X-Ray Diffraction

The X-ray structure of **274** has been published to give the dimensions [229] shown in Figure 11.2.

11.6.2.2 NMR Spectroscopy

Table 11.8 gives ^{1}H NMR data for S-oxidized 1,2-oxathiolane systems **272** and **273**.

11.6.3 Synthesis

11.6.3.1 Ring Synthesis of 1,2-Oxathioles

Chlorination of 1,3-thioalcohols using either Cl_2 or SO_2Cl_2 affords 1,2-oxathiolan-2-oxide **272** by way of an intermediate 3-hydroxysulfinyl chloride (**275**) (Scheme 11.86) [232].

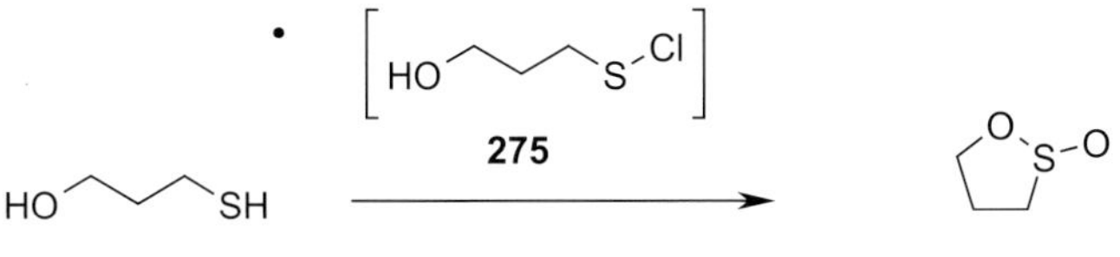

Scheme 11.86

Oxathiolane **277** is prepared by treatment of *N*-alkylcystinol **276** with NCS or NBS (Scheme 11.87) [233].

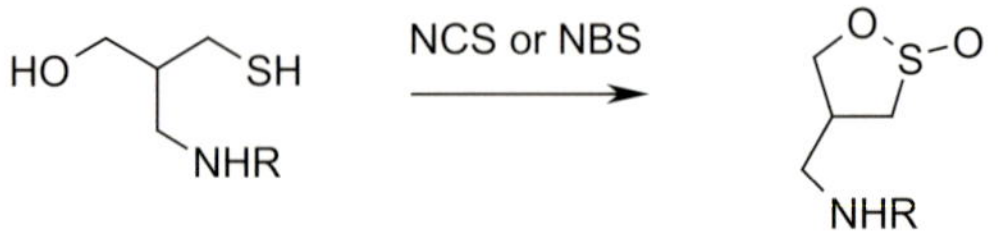

NCS or NBS

276 **277**

Scheme 11.87

Substituted 1,3-thioalcohols **278** can also be oxidized with $NaIO_4$ to give the corresponding 1,2-oxathiolane-2-oxides **279** (Scheme 11.88) [234].

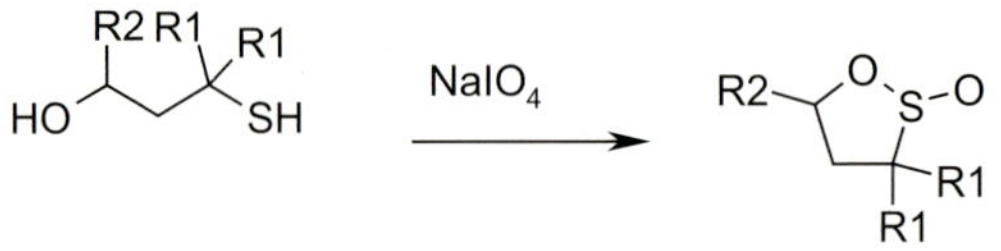

$NaIO_4$

278 **279**

Scheme 11.88

A ring-closing metathesis reaction of the allylvinyl sulfonate **280** using a second-generation Grubb's catalyst gives the oxathiole **270** (Scheme 11.89) [235].

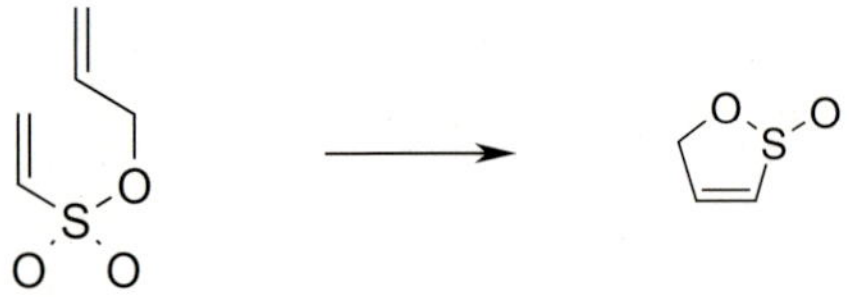

280 **270**

Scheme 11.89

Full details of the asymmetric synthesis of chiral γ-sultones **281** have been described [236].

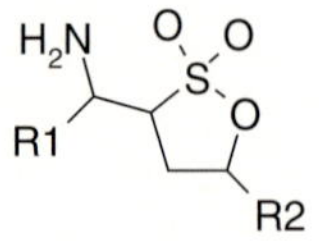

281

The direct reaction of cyclopropanes **282** with SO_2 in TFA to give 1,2-oxathiolane-2-oxides **283** and **284** has been studied extensively to determine the regioselectivity of this reaction (Scheme 11.90) [237].

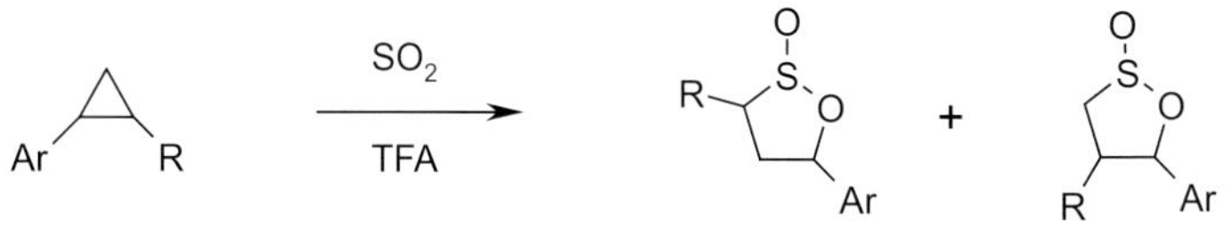

282 **283** **284**

Scheme 11.90

Similarly, treatment of the methylene cyclopropane **285** with SO_3 affords **286** directly (Scheme 11.91) [238].

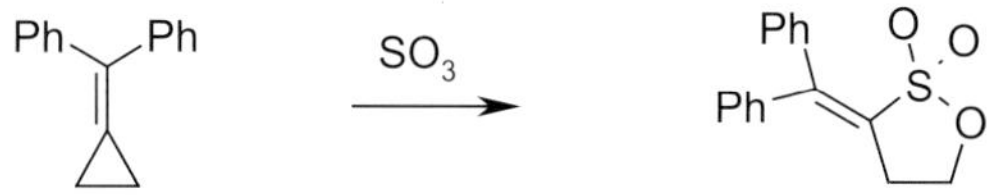

285 **286**

Scheme 11.91

The adamantyl alcohols **287** can also be treated with SO_3 to afford 1,2-oxathiolane-2,2-dioxides **288** (Scheme 11.92) [239].

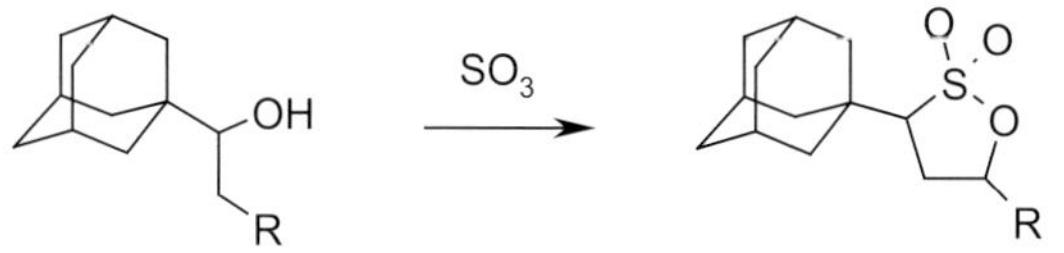

287 **288**

Scheme 11.92

The epoxide **289** can be cyclized with triethylamine to give the 1,2-oxathiolane-2,2-dioxides **290** (Scheme 11.93) [240].

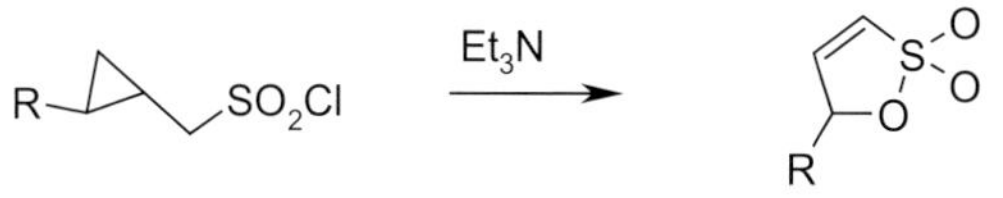

289 **290**

Scheme 11.93

11.6.3.2 Ring Transformations of Heterocycles Leading to 1,2-Oxathiole Derivatives

Oxidative ring expansion of thietan-2-ones **291** by treatment with chlorine or SO_2Cl_2 and acetic anhydride gives the 1,2-oxathiolan-5-one **292** (Scheme 11.94) [241].

Scheme 11.94

Cyclic sulfides **293** undergo ring contraction upon treatment with $BF_3{\cdot}Et_2O$ to give the 1,2-oxathiolane-2,2-dioxides **294** (Scheme 11.95) [242].

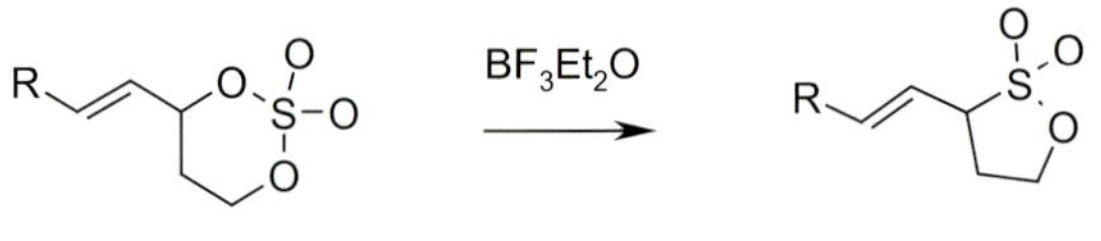

Scheme 11.95

11.6.4
Reactivity of 1,2-Oxathioles and 1,2-Oxathiolanes

Most reactions in this area have involved either 1,2-oxathiolane 2-oxides or 2,2-dioxides. The type of reactions involve either thermal extrusion of SO or SO_2 or nucleophilic attack at the cyclic sulfinate or sulfonate functions.

11.6.4.1 Thermal or Photochemical Reactions

1,2-Oxathiolane 2-oxides **295** readily undergo thermal extrusion of SO to afford α,β-unsaturated carbonyl compounds. The reaction is most likely to proceed via an initial tautomerization from the 5H to the 3H form **296**, which then undergoes an electrocyclic process to give cinnamaldehyde (Scheme 11.96) [243].

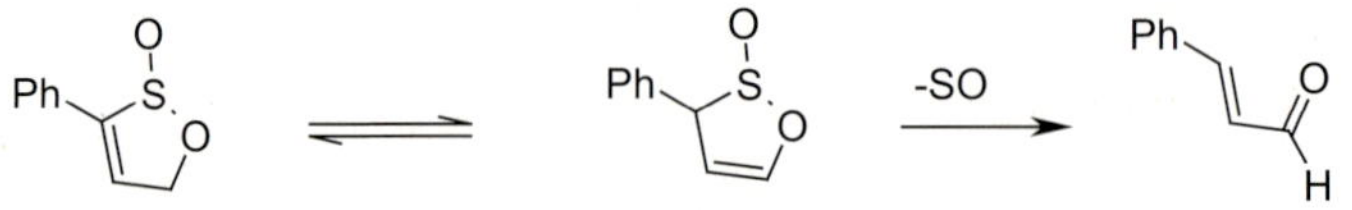

Scheme 11.96

1,2-Oxathiolan-5-one 2,2-dioxide derivative **297** undergoes thermally induced elimination of SO_2 at 180 °C to give methacrylic acid (Scheme 11.97) [244].

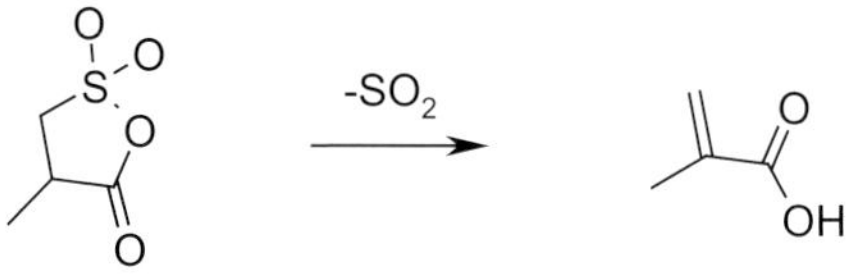

297

Scheme 11.97

11.6.4.2 Nucleophilic Attack

Nucleophilic attack on the cysteine-derived chiral 4-amino-1,2-oxathiolane 2-oxide **298** by alkyllithiums takes place on the sulfur atom with inversion of configuration to give **299** (Scheme 11.98) [245].

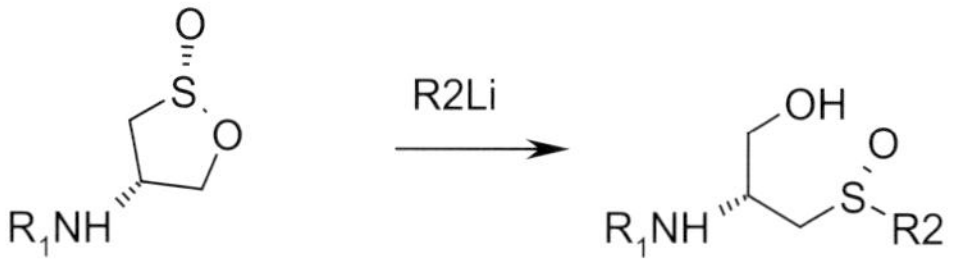

298 299

Scheme 11.98

Reaction of the 1,2-oxathiolane 2-oxide **300** with bromine results in the loss of SO_2 to give the 1,3-dibromo derivative **301** (Scheme 11.99) [246]. Compound **300** also reacts with PCl_3 to give the ring open product **302** [247].

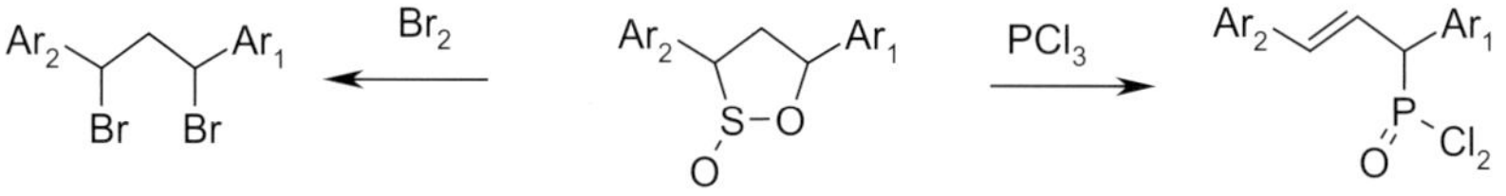

301 300 302

Scheme 11.99

11.7 1,3-Oxathioles and 1,3-Oxathiolanes

11.7.1 Introduction

Little work has been published on 1,3-oxathiolium salts **303** and their mesoionic system **304** [20]. A review on 1,3-oxathiolium salts has appeared [248]. A larger amount of material has been published on 1,3-oxathioles **305**. Most studies have been

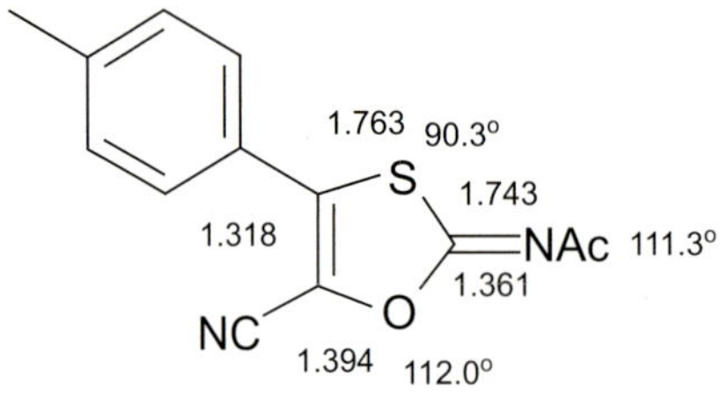

307

Figure 11.3 Geometry of a 1,3-oxathiolone derivative.

carried out on the fully saturated systems, 1,3-oxathiolanes (**306**), and reviews covering this system have appeared [249, 250].

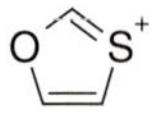

303 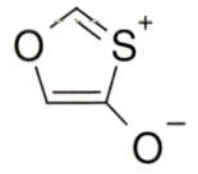 **304** 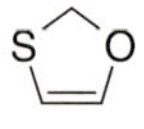 **305** 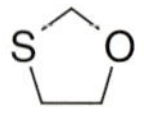 **306**

11.7.2
Relevant Physicochemical Data

11.7.2.1 X-Ray Diffraction

The X-ray structure of **307** has been published to give the dimensions [251] shown in Figure 11.3.

11.7.2.2 NMR Spectroscopy

Table 11.9 gives ^{1}H NMR data for 1,3-oxathiolane systems **308** and **309**.

308 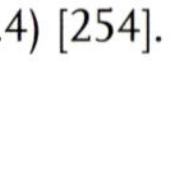 **309**

^{13}C NMR data for the mesoionic 1,3-oxathiolium 4-oxide compound **310** has been published (Figure 11.4) [254].

Table 11.9 ^{1}H NMR data for 1,3-oxathiolane derivatives.

Compound	2-H	4-H	5-H	Reference
5-Methyl-1,3-oxathiolane (**308**)	4.72/4.89	2.5/3.0	3.96	[252]
1,3-Oxathiolan-2-one (**309**)		3.59	4.53	[253]

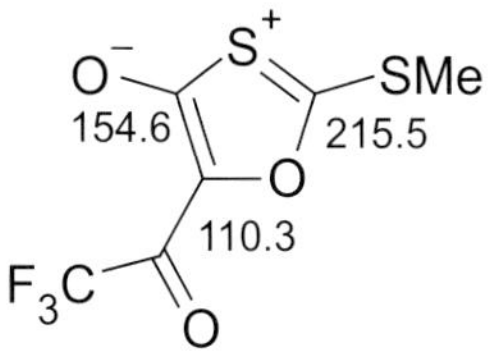

Figure 11.4 ^{13}C NMR data for a mesoionic 1,3-oxathiolium-4-oxide compound.

11.7.3 Synthesis

11.7.3.1 Ring Synthesis of 1,3-Oxathioles and 1,3-Oxathiolanes

An important method for the preparation of 1,3-oxathiolium salts has been published. Phenacyldithiocarbamates **311** can be desulfurized with a silver, mercury or copper salt to afford 2-amino-1,3-oxathiolium salt **312** (Scheme 11.100) [255].

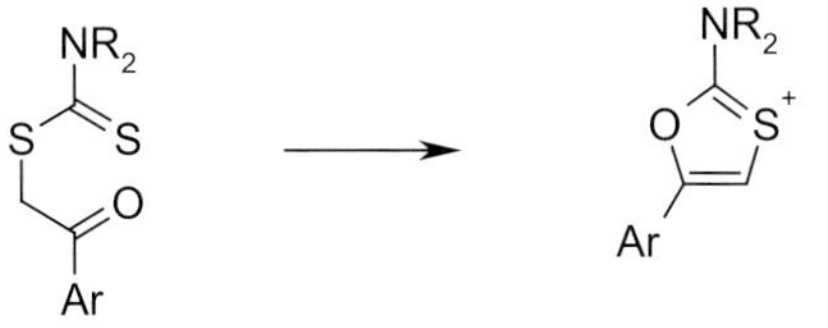

Scheme 11.100

Thermolysis of the bis-dithiocarbonates **313** results in loss of COS to afford the 1,3-oxathiolane **314** in up to 90% yield (Scheme 11.101) [256].

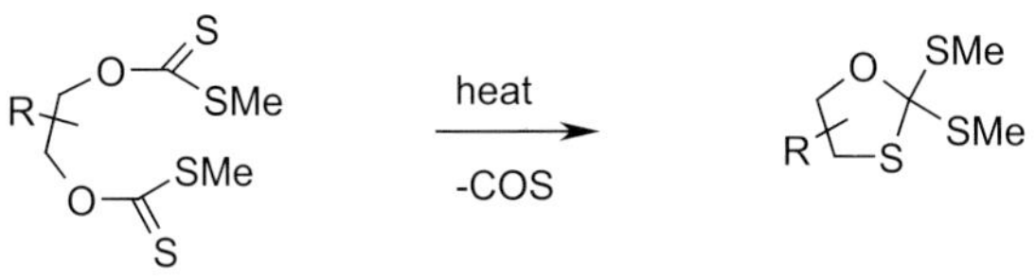

Scheme 11.101

The 1,3-oxathiolane 2,2-dioxide **316** has been prepared by treatment of the diazosulfone **315** with BF_3 (Scheme 11.102) [257]. Similarly, rhodium-catalyzed decomposition of α-diazoketones **318** in the presence of CS_2 affords the 1,3-

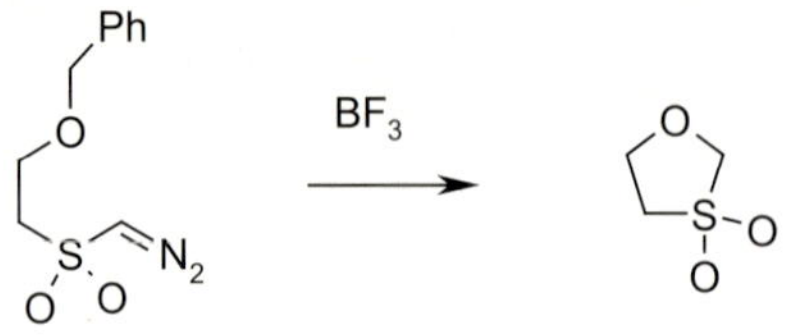

Ph BF$_3$ O S N$_2$ O O

315 316

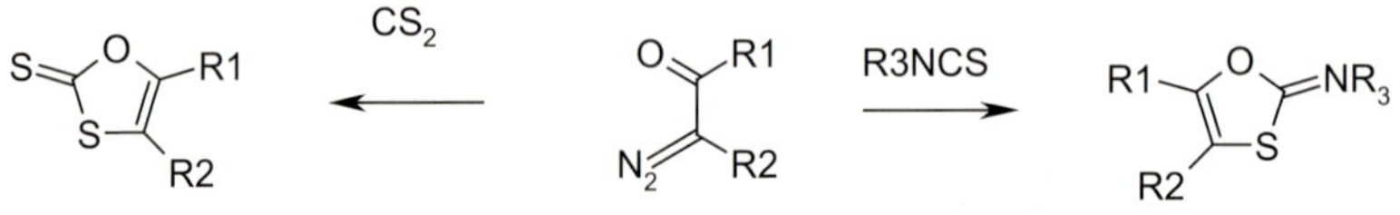

CS$_2$ R3NCS R1 R2 N$_2$ NR$_3$

317 318 319

Scheme 11.102

oxathiolane-2-thione **317** [258]. If the decomposition is carried out in the presence of an isothiocyanate then 2-imino derivatives **319** are formed [259].

The reaction of hydroxyethylthiocyanates **320** with a tertiary alcohol in sulfuric acid gives the 2-iminooxathiolane **321** (Scheme 11.103) [260].

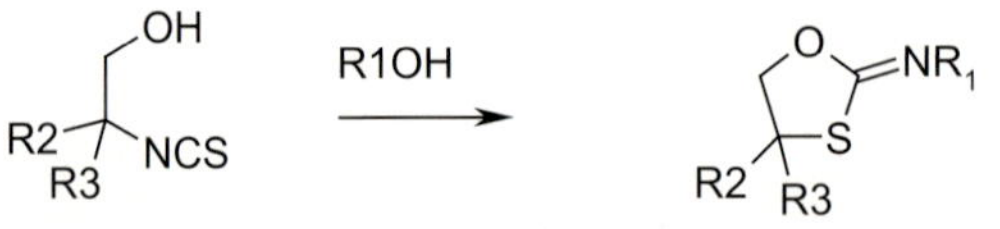

OH R1OH NR$_1$ R2 R3 NCS

320 321

Scheme 11.103

Routes to fully conjugated systems include the reaction of the phenacyl benzoate **322** with H_2S and iodine to give the oxathiolium salt **323** in 88% yield (Scheme 11.104) [261].

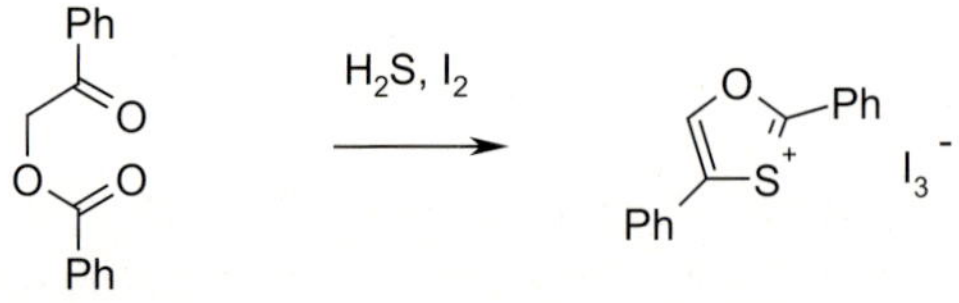

Ph H_2S, I_2 Ph I_3^-

322 323

Scheme 11.104

The addition of diazomethane to an α-oxothioester **324** affords 4-alkylthio substituted 1,3-oxathioles **325** (Scheme 11.105) [262].

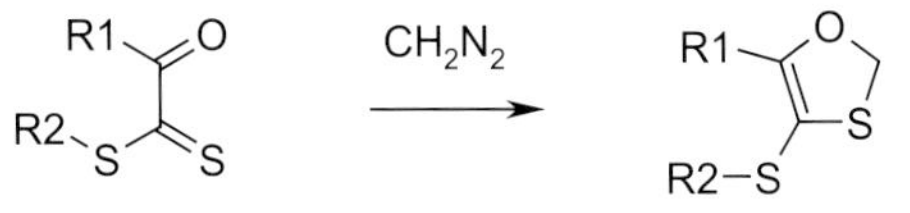

R1 O R2 S S CH_2N_2 R1 O S R2−S

324 **325**

Scheme 11.105

The standard method of preparation of 1,3-oxathiolanes is to condense a carbonyl compound with 2-thioethanol. Several catalysts have been used for this reaction – the most common is $BF_3{\cdot}Et_2O$ [263]. More recently, other catalysts used for this reaction include triisopropyl triflate [264], indium triflate [265], NBS [266], tetrabutylammonium tribromide [267] and scandium triflate [268]. A catalyst that is selective for aldehydes in the presence of acyclic ketones is $LiBF_4$ [269]. A catalyst that is suitable for α,β-unsaturated ketones is an aminopropyl functionalized silica [270]. Dimethyl acetals can also be used in place of the carbonyl component for this reaction [271].

The iodonium salt **326** reacts with either carbon disulfide or a thioketone to form 2-thione 1,3-oxathiolane derivative **327** (Scheme 11.106) [46].

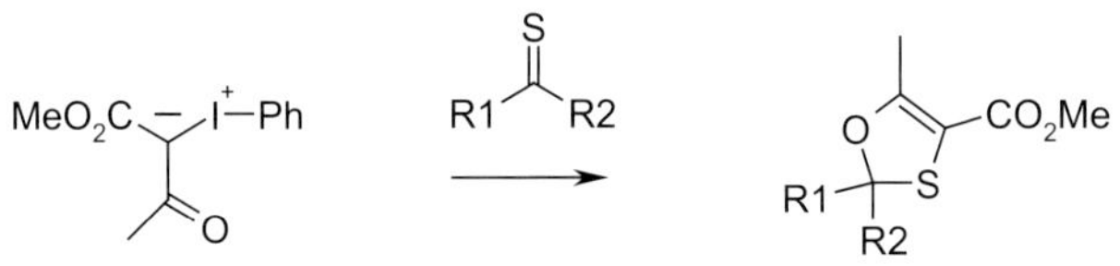

MeO_2C I^+−Ph O S R1 R2 O CO_2Me R1 S R2

326 **327**

Scheme 11.106

The phosphonate derivative **329** is prepared by the condensation of the α-haloaldehyde **328** with KSCN (Scheme 11.107) [272]. Acetylenic alcohols **330** also react with KSCN by a more complex course to give the 1,3-oxathiolane derivative **331** [273].

The standard method for the preparation of 2-imino-1,3-oxathioles **333** is to condense a phenacyl bromide with either a dithiocarbamate **332** (X=SR) [274] or a thiourea **332** ($X{=}NMe_2$) (Scheme 11.108) [275].

Silyl enol ethers **334** and α-halosulfonyl bromides when treated with DBN afford 1,3-oxathiole 3,3-dioxide **335** (Scheme 11.109) [276].

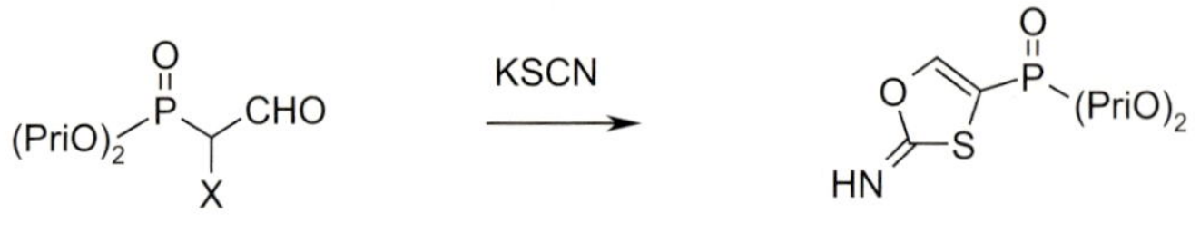

328 **329**

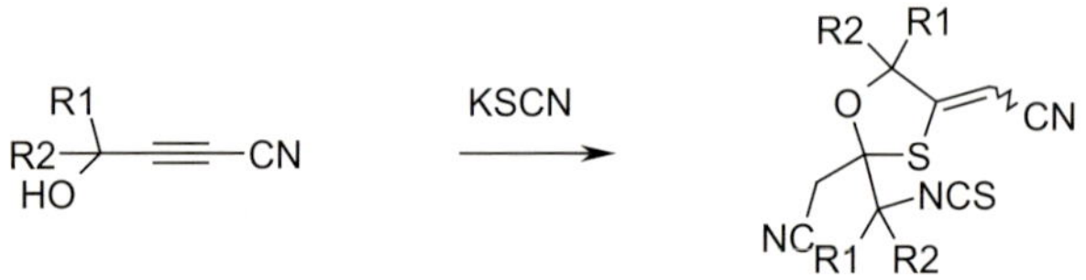

330 **331**

Scheme 11.107

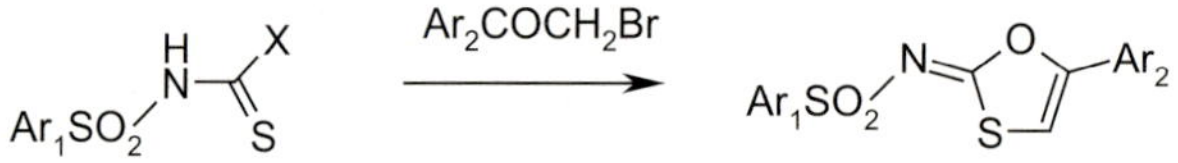

332 **333**

Scheme 11.108

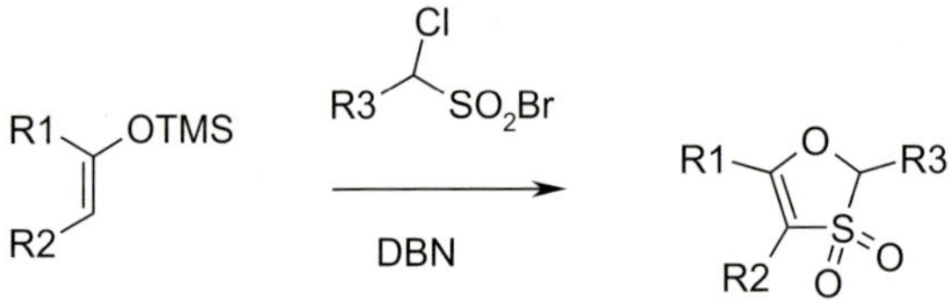

334 **335**

Scheme 11.109

Treatment of the silyl derivative **336** with CsF generates a thiocarbonyl ylide that can then be trapped by addition of aldehydes to give 5-substituted 1,3-oxathiolanes **337** (Scheme 11.110) [277].

11.7.3.2 **Ring Transformations of Heterocycles Leading to 1,3-Oxathiole Derivatives**
The 1,2,4-dithiazol-3-one **338** undergoes cycloaddition to ynamines **339** to give 2-imino-1,3-oxathioles **340** (Scheme 11.111) [278]. Similarly, the 1,2,4-thiadiazol-3-one **341** reacts with enolates **342** to also give 2-imino-1,3-oxathioles **343** [279].

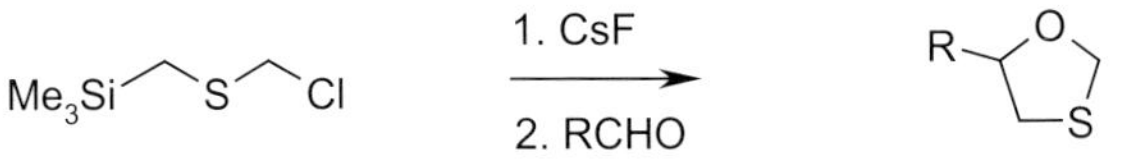

336 **337**

Scheme 11.110

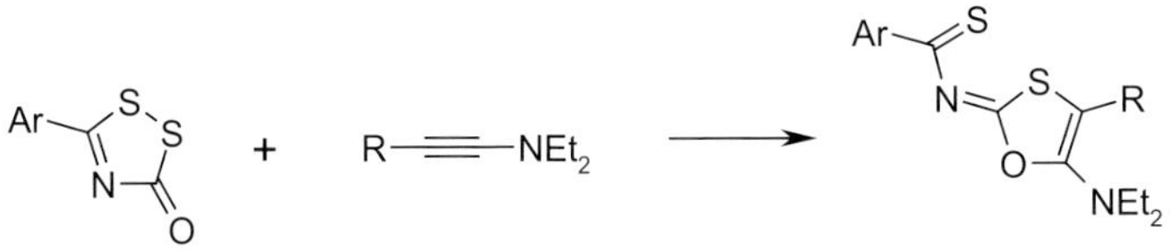

338 **339** **340**

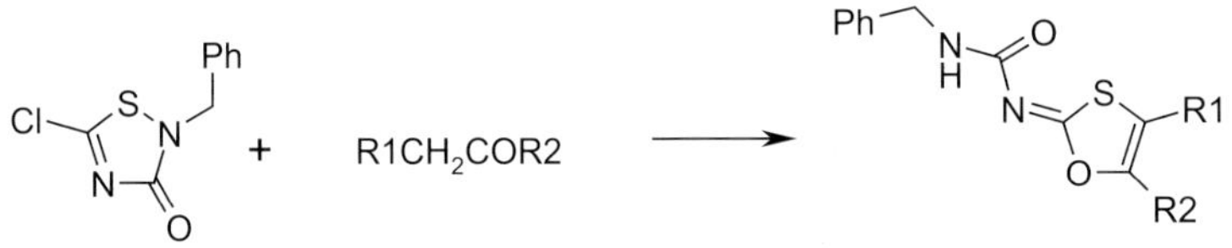

341 **342** **343**

Scheme 11.111

The 1,4-oxathiane **344** undergoes rearrangement to the 2-acyl-1,3-oxathiolane 3-oxide derivatives **345** and **346** on treatment with singlet oxygen (Scheme 11.112) [280].

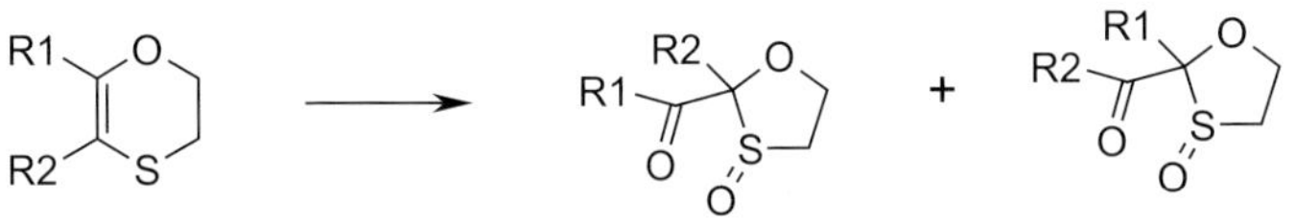

344 **345** **346**

Scheme 11.112

11.7.4 Reactivity of 1,3-Oxathioles

11.7.4.1 Reactions with Electrophiles

Electrophilic substitution on the heterocyclic ring of 1,3-oxathioles is essentially unknown.

11.7.4.2 **Reactions with Nucleophiles**

1,3-Oxathiolium salts **347** react with NaN_3 by attack at C2, followed by loss of nitrogen and rearrangement to give the 1,4,2-oxathiazine **348** (Scheme 11.113) [281].

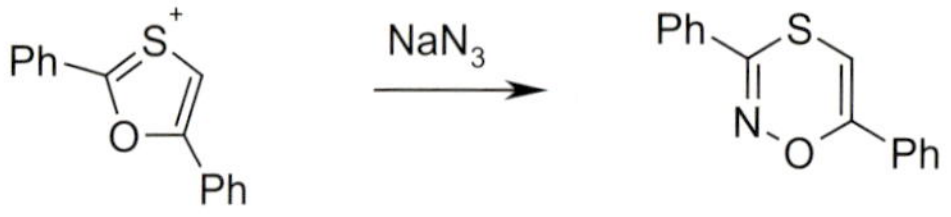

347 **348**

Scheme 11.113

11.7.4.3 **Cycloaddition Reactions**

Mesoionic 1,3-oxathiolium-4-olates **349** undergo ready cycloadditions with alkene and alkyne dipolarophiles. Addition of **349** to CS_2 or PhNCS proceeds by formation of a bicyclic adduct that then loses COS to give a new mesoionic system (**350**) (Scheme 11.114) [282].

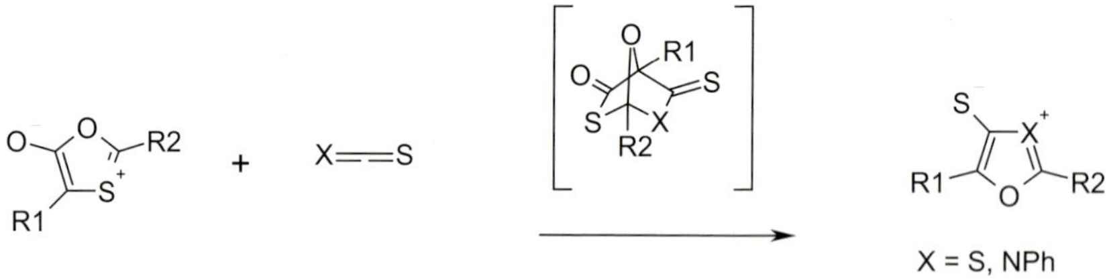

X = S, NPh

349 **350**

Scheme 11.114

11.7.5
Reactivity of 1,3-Oxathiolanes

11.7.5.1 **Thermal Reactions**

Pyrolytic extrusion of CO_2 from both 1,3-oxathiolan-2-ones and 5-ones affords thiiranes. 2-Imino-1,3-oxathiolanes **351** rearrange to the isomeric thiazolidinones **352** at 80 °C where R=alkyl; where R=Ph an alternative pathway leads to the formation of RNCO and a thiirane (Scheme 11.115) [283].

11.7.5.2 **Reactions with Electrophiles**

Oxidation of 1,3-oxathiolanes to the corresponding 3-oxides is readily achieved in high yield using peroxyacetic acid [284]; *m*CPBA can also be used [285]. Asymmetric 3-oxidation using $Bu^tOOH/TiOPr^i_4$ and diethyl tartrate has been examined. The diastereoselectivity observed was moderate but the enantioselectivity observed was very poor [286].

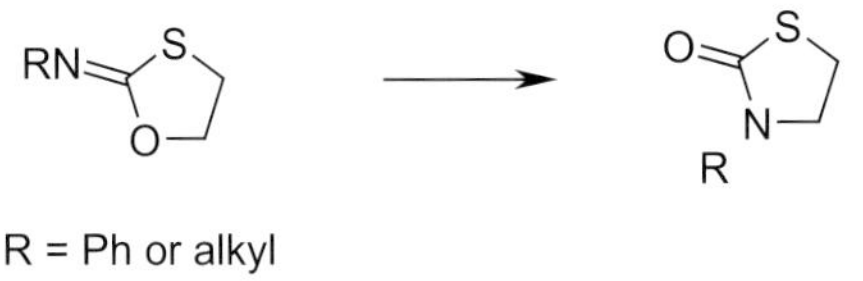

RN=, S, O → O=, S, N, R

R = Ph or alkyl

351 **352**

Scheme 11.115

11.7.5.3 Reactions with Nucleophiles

New catalysts for the efficient hydrolysis of 1,3-oxathiolanes to the corresponding aldehydes or ketones include H_2O_2/CH_3CN [287], $NaNO_2/AcCl$ [288] and Amberlyst-15/glyoxylic acid under solvent-free conditions [289]. A method that is selective for 1,3-oxathiolanes in the presence of 1,3-dioxolanes is NBS in aqueous acetone [290].

The 1,3-oxathiolan-2-one (**353**) reacts with secondary amines in toluene to afford, by loss of CO_2, thioethylamines **354** [291], when the same reaction is performed in dioxan the thiopropyl carbamate is formed (Scheme 11.116) [292].

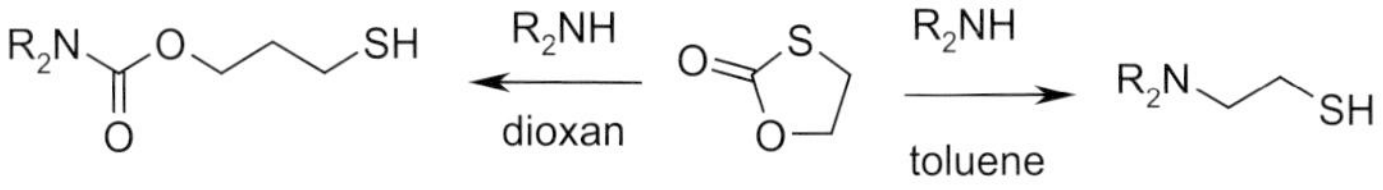

R_2N, O, SH, O ← R_2NH / dioxan — O=, S, O — R_2NH / toluene → R_2N, SH

355 **353** **354**

Scheme 11.116

1,3-Oxathiolane-5-ones such as **356** can be deprotonated and alkylated with reactive electrophiles at C4 when R=Me but not when R=H [293].

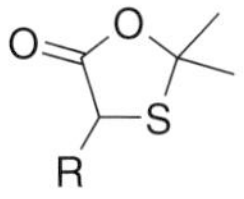

O=, O, S, R

356

11.7.5.4 Radical, Electrochemical Reactions

2-Substituted 1,3-oxathiolanes **357** can undergo reductive cleavage when treated with trimethylsilyl hydride. Reduction occurs at the C2−S bond to give **358** (Scheme 11.117) [294]. 2-Substituted 1,3-oxathiolan-5-ones react in a similar way. They also undergo selective anodic fluorination to give the 4-fluoro derivatives [295].

11.7.5.5 Ring Expansion

2-Substituted 1,3-oxathiolanes **357** react with carbenes to give ring-expanded 1,4-oxathianes **359** via insertion into the C2−S bond (Scheme 11.118) [296].

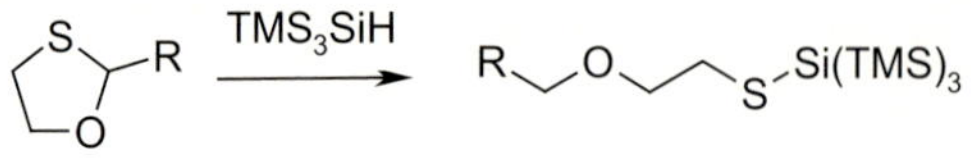

357 358

Scheme 11.117

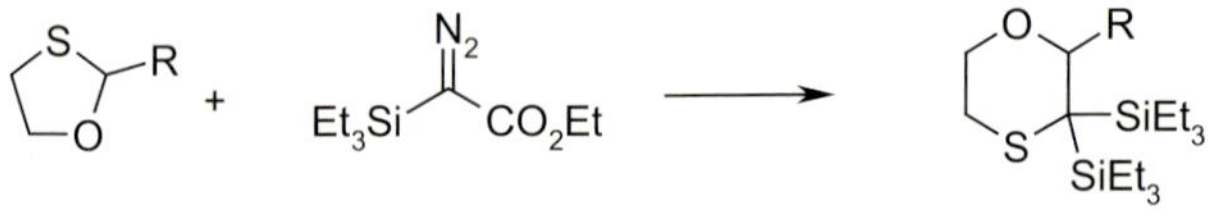

357 359

Scheme 11.118

The titanium tetrachloride mediated reaction of α,β-unsaturated oxathiolanes **360** with styrene or α-methylstyrene gives the ring-expanded dihydrothiapyrans **361** (Scheme 11.119) [297].

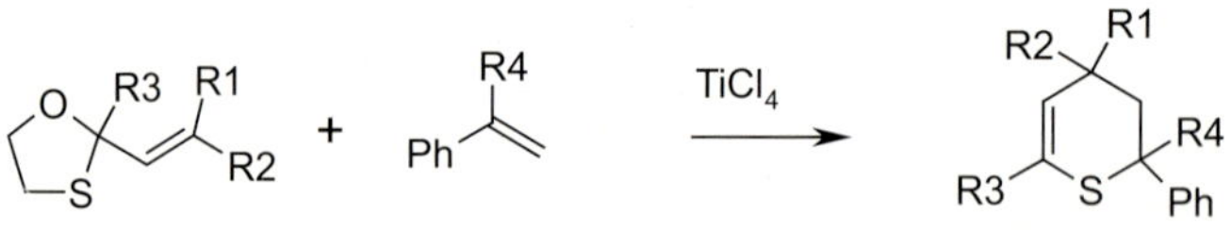

360 361

Scheme 11.119

11.7.6
Other Compounds of Interest

1,3-Oxathiolanes bearing a 2-propenyl, 2-furyl- or 2-phenyl group have been used as flavorings [298]. Anti-viral activity has been claimed for the nucleoside analogue **362** [299].

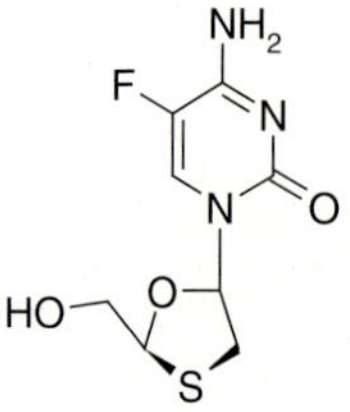

362

References

1 Kropf, H. (1988) *Methoden der Organischen Chemie (Houben-Weyl)*, **E13**, 424.
2 McCullough, K.J. (1995) *Contemporary Organic Synthesis*, **2**, 225.
3 McCullough, K.J. and Nojima, M. (2001) *Current Organic Chemistry*, **6**, 601.
4 Graziano, M.L., Iesce, M.L., Cermola, F., Cimminiello, G., and Scarpati, R.J. (1991) *Journal of the Chemical Society, Perkin Transactions 1*, 1479.
5 Bloodworth, A.J. and Loveitt, M.E. (1978) *Journal of the Chemical Society, Perkin Transactions 1*, 522.
6 Jin, A., Mack, H.-G., Waterfeld, A., Dakkouri, M., and Oberhammer, H. (1992) *Journal of Molecular Structure*, **274**, 163.
7 Talbott, R.L. (1965) *The Journal of Organic Chemistry*, **30**, 1429.
8 Courtneidge, J.L., Bush, M., and Loh, L.S. (1992) *Tetrahedron*, **48**, 3835.
9 Bascetta, E. and Gunstone, F.D. (1984) *Journal of the Chemical Society, Perkin Transactions 1*, 2207.
10 Bloodworth, A.J. and Tallant, N.A. (1990) *Tetrahedron Letters*, **31**, 7077.
11 Shim, S.C. and Song, J.S. (1986) *The Journal of Organic Chemistry*, **51**, 2871.
12 Feldman, K.S., Simpson, R.E., and Parvez, M. (1986) *Journal of the American Chemical Society*, **108**, 1328.
13 Akasaka, T., Takeuchi, K., Misawa, Y., and Ando, W. (1989) *Heterocycles*, **28**, 445.
14 Baumstark, A.L. and Vasquez, P.C. (1992) *The Journal of Organic Chemistry*, **57**, 393.
15 Dissault, P.H. and Liu, X. (1999) *Tetrahedron Letters*, **40**, 6553.
16 Salomon, R.G., Salomon, M.F., and Coughlin, D.J. (1978) *Journal of the American Chemical Society*, **100**, 660.
17 Bloodworth, A.J. and Baker, D.S. (1981) *Journal of the Chemical Society, Chemical Communications*, 547.
18 Yoshida, M., Miura, M., Nojima, M., and Kusabayashi, S. (1983) *Journal of the American Chemical Society*, **105**, 1753.
19 Hartmann, H. (1993) *Methoden der Organischen Chemie (Houben-Weyl)*, **E8a**, 1.
20 Hartmann, H. (1993) *Methoden der Organischen Chemie (Houben-Weyl)*, **E8a**, 10.
21 Perst, H. (2002) Science of Synthesis, Vol. 11, Georg Thieme Verlag, Ch 1, 13.
22 Klausener, A., Frauenrath, H., Lange, W., Mikhail, G.K., Schneider, S., and Shroder, D. (1991) *Methoden der Organischen Chemie (Houben-Weyl)*, **E14a/1**, 1.
23 Childs, R.F., Orgias, R.M., Lock, C.J.L., and Mahendran, M. (1993) *Canadian Journal of Chemistry*, **71**, 836.
24 Jones, R.A.Y., Katritzky, A.R., Lehman, P.G., Record, K.A.F., and Shapiro, B.B. (1971) *Journal of the Chemical Society-B*, 1302.
25 Flynn, C.R. and Michl, J. (1974) *The Journal of Organic Chemistry*, **39**, 3442.
26 Anet, F.A.L. (1962) *Journal of the American Chemical Society*, **84**, 747.
27 Lorenz, W. and Maas, G. (1987) *The Journal of Organic Chemistry*, **52**, 375.
28 Hung, M.H., Rozen, S., Feiring, A.F., and Resnick, P.R. (1993) *The Journal of Organic Chemistry*, **58**, 972.
29 Sugarowa, T., Kawada, Y., Katoh, M., and Iwamura, H. (1979) *Bulletin of the Chemical Society of Japan*, **52**, 3391.
30 Gerothanassis, I.P. and Lauterwein, J. (1986) *Magnetic Resonance in Chemistry*, **24**, 1034.
31 Eliel, E.L., Petrusiewicz, K.M., and Jewell, L.M. (1979) *Tetrahedron Letters*, **19**, 3649.
32 Hamaguchi, M. and Nagai, T. (1985) *Journal of the Chemical Society, Chemical Communications*, 190.
33 Alonso, M.E. and Chitty, A.W. (1981) *Tetrahedron Letters*, **22**, 4181.
34 Bolm, C., Saladin, S., and Kaysan, A. (2002) *Organic Letters*, **4**, 4631.
35 Luk'yanov, S.M., Borodaev, S.V., and Borodaeva, S.V. (1983) *Zhurnal Organicheskoi Khimii*, **19**, 2154.

36 Luk'yanov, S.M., Borodaev, S.V., and Dorofeenko, G.N. (1981) *Zhurnal Organicheskoi Khimii*, **17**, 2233.

37 Luk'yanov, S.M., Borodaev, S.V., and Dorofeenko, G.N. (1981) *Zhurnal Organicheskoi Khimii*, **17**, 2234.

38 Luk'yanov, S.M., Borodaev, S.V., and Zhdanov, Y.A. (1985) *Zhurnal Organicheskoi Khimii*, **21**, 2067.

39 Trofimov, B.A., Oparina, L.A., Parshina, L.N., Lavrov, V.I., Grigorenko, V.I., and Zhumabekov, M.K. (1986) *Zhurnal Organicheskoi Khimii*, **22**, 1583.

40 Maier, P. and Redlich, H. (2000) *Synlett*, 257.

41 Wipf, P. and Xu, W. (1993) *The Journal of Organic Chemistry*, **58**, 5880.

42 Kim, H.S., Kim, J.J., Lee, B.G., and Kwon, Y.S. (2000) US Pat. 6,160,130.

43 Kim, H.S., Kim, J.J., Lee, S.D., Park, K.Y., and Kim, H.G. (2000) US Pat. 6,156,909.

44 Peng, J. and Deng, Y. (2001) *New Journal of Chemistry*, **25**, 639.

45 Andriankova, L.V., Abramova, N.D., Mal'kina, A.G., and Skvortsov, Yu.M. (1989) *Izvestiya Akademii Nauk SSSR, Seriya Khimia*, 1421.

46 Batsila, C., Kostakis, G., and Hadjiarapoglou, L.P. (2002) *Tetrahedron Letters*, **43**, 5997.

47 Kato, K., Yamamoto, Y., and Akita, H. (2002) *Tetrahedron Letters*, **43**, 6587.

48 Sahu, D.P. (2002) *Indian Journal of Chemistry Section B-Organic Chemistry Including Medicinal Chemistry*, **41**, 1722.

49 Guirado, A., Zapata, A., and Galvez, J. (1994) *Tetrahedron Letters*, **35**, 2365.

50 Sakai, S., Marata, M., Wada, N., and Fujinami, T. (1983) *Bulletin of the Chemical Society of Japan*, **56**, 1873.

51 Kwiatkowski, S. and Danikiewicz, W. (1986) *Polish Journal of Chemistry*, **59**, 1285.

52 Cross, D.J., Kenny, J.A., Houston, I., Campbell, L., Walsgrove, T., and Wills, M. (2001) *Tetrahedron Asymmetry*, **12**, 1801.

53 Itaya, T., Iida, T., and Eguchi, H. (1993) *Chemical & Pharmaceutical Bulletin*, **41**, 408.

54 Iranpoor, N. and Kazemi, F. (1998) *Synthetic Communications*, **28**, 3189.

55 Iranpoor, N. and Zeynizadeh, B. (1998) *Journal of Chemical Research-S*, 466.

56 Kurihara, M. and Hakamata, W. (2003) *The Journal of Organic Chemistry*, **68**, 3413.

57 Dewan, S.K., Singh, R., and Kumar, A. (2003) *Oriental Journal of Chemistry*, **19**, 119.

58 Cramarossa, M.R., Forti, L., and Ghelfi, F. (1997) *Tetrahedron*, **53**, 15889.

59 Ishihara, K., Karumi, Y., Kubota, M., and Yamamoto, H. (1996) *Synlett*, 839.

60 Lee, S.B., Jung, H., and Lee, K.W. (1996) *Bulletin of the Korean Chemical Society*, **17**, 362.

61 Wattenbach, C., Maurer, M., and Frauenrath, H. (1999) *Synlett*, 303.

62 Gulbins, K. and Hamann, K. (1961) *Chemische Berichte*, **94**, 3287.

63 Flynn, C.R. and Michl, J. (1974) *The Journal of Organic Chemistry*, **39**, 3442.

64 Akgun, E. and Tunali, M. (1985) *Doga Bilim Derg, Ser A1*, **9**, 258.

65 Safiev, O.G., Kruglov, D.E., Zlotskii, S.S., and Rakhmankulov, D.L. (1985) *Izvestiya Vyssh Uchebn Zaved Khim Khim Tekhnol*, **28**, 33.

66 Xiao, X. and Bai, D. (2001) *Synlett*, 535.

67 Masaki, Y., Yamada, T., and Tanaka, N. (2001) *Synlett*, 1311.

68 Johnstone, C., Kerr, W.J., and Scott, J.S. (1996) *Chemical Communications*, 341.

69 Yan, S., Chen, N., Li, J., and Zhang, Y. (1996) *Hecheng Huaxue*, **4**, 184.

70 Tanemura, K., Suzuki, T., and Horaguchi, T. (1992) *Chemical Communications*, 979.

71 Tani, H., Inamasu, T., Masumoto, K., Tamura, R., Shimazu, H., and Suzuki, H. (1992) *Phosphorus Sulfur*, **67**, 261.

72 Seebach, D., Beck, A.K., and Heckel, H. (2001) *Angewandte Chemie, International Edition*, **40**, 92.

73 Degni, S., Wilen, C.-E., and Leino, R. (2001) *Organic Letters*, **3**, 2551.

74 Irurre, J., Riera, M., and Cintora, M.A. (2001) *Synthesis*, 647.

75 Kaku, H., Tokaoka, S., and Tsunoda, T. (2002) *Tetrahedron*, **58**, 3401.

76 Miyamoto, H., Kimura, T., Daikaura, N., and Tanaka, K. (2003) *Greem Chemistry*, **5**, 57.

77 Suarez, R.M., Sostelo, J.P., and Saraideses, L.A. (2002) *Synlett*, 1435.

78 Suarez, R.M., Sostelo, J.P., and Saraideses, L.A. (2003) *Chemistry - A European Journal*, **9**, 4179.

79 Zorin, V.V., Zelechonok, Yu.B., Zlotskii, S.S., and Rakhmankulov, D.L. (1985) *Zhurnal Organicheskoi Khimii*, **21**, 193.

80 Safiev, O.G., Grazov, O.G., Zorin, V.V., Rakhmankulov, D.L., and Paushkin, Ya.M. (1989) *Doklady Akademii Nauk SSSR*, **308**, 135.

81 Tkachenko, T.K., Klyavlin, M.S., Zlotskii, S.S., and Rakhmankulov, D.L. (1992) *Zhurnal Organicheskoi Khimii*, **28**, 1301.

82 Safiev, O.G., Nazarov, D.D., Zorin, V.V., Rakhmankulov, D.L., and Paushkin, Ya.M. (1990) *Doklady Akademii Nauk SSSR*, **310**, 889.

83 Dement'eva, L.P. and Kostikov, R.R. (1990) *Zhurnal Organicheskoi Khimii*, **26**, 138.

84 Molchanov, A.P., Serkinov, T.G., and Badovskaya, L.A. (1992) *Zhurnal Organicheskoi Khimii*, **28**, 2320.

85 Kim, B.T., Han, S.Y., and Pak, C.S. (2000) PCT Int. Appl. WO 43390.

86 Koch, P., McCullough, J.R., Senanayake, C.H., Tanoury, G.J., and Hong, Y. (1998) PCT Int. Appl. WO 21204.

87 Koch, P., McCullough, J.R., Senanayake, C.H., Tanoury, G.J., and Hong, Y. (1998) PCT Int. Appl. WO 21205.

88 Koch, P., McCullough, J.R., Senanayake, C.H., Tanoury, G.J., and Hong, Y. (1998) PCT Int. Appl. WO 21213.

89 Kim, Y.K., Cheong, C.S., Lee, S.L., Jun, S.J., Kim, K.S., and Cho, H.-S. (2002) *Tetrahedron Asymmetry*, **13**, 2501.

90 Angeli, P., Brasili, L., Franchini, S., Giardina, D., Gulin, U., and Marucci, G. (1999) *Medicinal Chemistry Research*, **9**, 89.

91 Teuber, L. (1990) *Sulfur Reports*, **9**, 257.

92 Pedersen, C.Th. (2002) *Science of Synthesis*, **11**, Chapter 7 107.

93 Yang, Y., Liu, H.C., and Wei, C.S. (1985) *Acta Crystallographica Section C: Crystal Structure Communications*, **C41**, 1242.

94 Plavac, N., Still, I.W.J., Chauhan, M.S., and McKinnon, D.M. (1975) *Canadian Journal of Chemistry*, **53**, 836.

95 Elgemeie, G.H. and Sayed, S.H. (2001) *Synthesis*, 1747.

96 Ali, M.H. and McDermott, M. (2002) *Tetrahedron Letters*, **43**, 6271.

97 Burns, C.J., Field, L.D., Morgan, J., Ridley, D.D., and Vignevich, V. (1999) *Tetrahedron Letters*, **40**, 6489.

98 Nishio, T. (1998) *Helvetica Chimica Acta*, **81**, 1207.

99 Aimar, M.L. and De Rossi, R.H. (2000) *Synthesis*, 1749.

100 Okuma, K., Kojima, K., and Shibata, S. (2000) *Heterocycles*, **53**, 2753.

101 Huisgen, R. and Rapp, J. (1997) *Tetrahedron*, **53**, 939.

102 Leaver, D., Robertson, W.A.H., and McKinnon, D.M. (1962) *Journal of the Chemical Society*, 5104.

103 Bobylev, V.A., Petrov, M.L., and Petrov, A.A. (1981) *Zhurnal Organicheskoi Khimii*, **17**, 139.

104 Rakitin, O.A., Rees, C.W., Williams, D.J., and Torroba, T. (1996) *The Journal of Organic Chemistry*, **61**, 9178.

105 Adelaere, B. and Guemas, J.P. (1989) *Sulfur Letters*, **10**, 31.

106 Stachel, H.-D. and Zeitler, K. (1995) *Liebigs Annalen der Chemie*, 2011.

107 Glass, R.S., Petson, A., Wilson, G.S., Martinez, R., and Juaristi, E. (1987) *The Journal of Organic Chemistry*, **51**, 4337.

108 Tornetta, B. (1958) *Annali di Chimica-Rome*, **48**, 577.

109 Sugai, S. and Tomita, K. (1980) *Chemical & Pharmaceutical Bulletin*, **28**, 487.

110 Markovic, R., Baranac, M. and Jovetic, S. (2003) *Tetrahedron Letters*, **44**, 7087.

111 Leaver, D., Mc Kinnon, D.M., and Robertson, W.A.H. (1965) *Journal of the Chemical Society*, 32.

112 Bartho, B., Faust, J., Pohl, R., and Mayer, R. (1976) *Journal fur Praktische Chemie*, **318**, 221.

113 Losac'h, N. and Stavaux, M. (1980) *Advances in Heterocyclic Chemistry*, **27**, 151.

114 Pedersen, C.T. (1980) *Sulfur Reports*, **1**, 1.

115 Vasil'eva, T.P., Lin'kova, M.G., Kil'disheva, O.V., and Knunyants, I.L. (1974) *Izvestiya Akademii Nauk SSSR, Ser Khim*, 643.

116 Fleury, M.B., Largeron, M., Barreau, M., and Vuilhorgne, M. (1985) *Tetrahedron*, **41**, 3705.

117 Fleury, M.B., Largeron, M., and Martens, T. (1987) *Tetrahedron*, **43**, 3421.

118 Fleury, M.B., Largeron, M., and Martens, T. (1988) *Journal of Heterocyclic Chemistry*, **25**, 1223.

119 Landis, P.S. (1965) *Chemical Reviews*, **65**, 237.

120 Hoffmann, R.W. and Goldmann, A.S. (1978) *Chemische Berichte*, **111**, 2716.

121 Losac'h, N. (1971) *Advances in Heterocyclic Chemistry*, **13**, 151.

122 Abdou, W.M., Hennawy, I.T., and Khoshnich, O.E. (1996) *Phosphorus Sulfur and Silicon and the Related Elements*, **109–110**, 557.

123 Elkaschef, M.A., Abdel-Megeid, F.M.E., and El-Barbary, A.A. (1974) *Tetrahedron*, **30**, 4113.

124 Chauhan, M.S. and McKinnon, D.M. (1976) *Canadian Journal of Chemistry*, **54**, 3879.

125 Caserio, M. and Kim, J.J. (1985) *Phosphorus Sulfur*, **23**, 169.

126 Smith, E.H. (1984) *Journal of the Chemical Society, Perkin Transactions 1*, 523.

127 Tazaki, M., Nagahama, S., and Tagaki, M. (1988) *Chemistry Letters*, 1339.

128 Demchuk, D.V. and Nikishin, G.I. (1997) *Russian Chemical Bulletin*, **46**, 199.

129 Tazaki, M. and Yamada, M. (1996) *Phosphorus Sulfur and Silicon and the Related Elements*, **116**, 253.

130 Ando, W., Kumamoto, Y., and Takata, T. (1985) *Tetrahedron Letters*, **26**, 5817.

131 Fujita, T. and Yokoyama, T. (1998) Eur. Pat. 869,126

132 Schukat, G. and Fanghänel, E. (2002) *Science of Synthesis*, **11**, Chapter 8 191.

133 Becher, J., Jeppesen, J.O., and Nielsen, K. (2003) *Synthetic Metals*, **133–134**, 309.

134 Garin, J., Orduna, J., and Andreu, R. (2001) *Recent Research Developments in Organic Chemistry*, **5**, Part 1, 77.

135 Gorgues, A., Kreher, D., Gautier, N., Dumur, F., Allard, E., Liu, S.-G., Cariou, M., Hudhomme, P., Cousseau, J., Levillain, E.J., Delaunay, J., and Gallego-Planas, N. 2002 *NATO Science Series, II: Mathematics, Physics and Chemistry*, **59**, 169.

136 Cooper, W.F., Kenney, N.C., Edmonds, J.W., Nagel, A., Wudl, F., and Coppens, P. (1971) *Journal of the Chemical Society. Chemical Communications*, 889.

137 Schmitt, W.H. and Tulinsky, A. (1967) *Tetrahedron Letters*, **8**, 5311.

138 Prinzbach, H. and Futterer, E. (1966) *Advances in Heterocyclic Chemistry*, **7**, 39.

139 Nakayama, J., Fujiwara, K., and Hoshino, M. (1976) *Bulletin of the Chemical Society of Japan*, **49**, 3567.

140 Sakamoto, K., Nakamura, N., Oki, M., Nakayama, J., and Hishino, M. (1977) *Chemistry Letters*, 1133.

141 Olah, G. and Grant, J.L. (1977) *The Journal of Organic Chemistry*, **42**, 2237.

142 Buza, D. and Gradwska, W. (1980) *Polish Journal of Chemistry*, **54**, 2379.

143 Narita, M. and Pittman, C.U., Jr. (1976) *Synthesis*, 489.

144 Hiratani, K., Shiono, H., and Okawara, M. (1973) *Chemistry Letters*, 867.

145 Ueno, Y., Nakayama, A., and Okawara, M. (1975) *Synthesis*, 277.

146 Potts, K.T., Choudhury, D.R., Elliot, A.J., and Singh, U.P. (1976) *The Journal of Organic Chemistry*, **41**, 1724.

147 Mas, A., Fabre, J.M., Torreilles, E., Giral, L., and Brun, G. (1977) *Tetrahedron Letters*, **18**, 2579.

148 Naley, N.F. (1978) *Tetrahedron Letters*, **19**, 5161.

149 Gareau, Y., Tremblay, M., Gauvreau, D., and Juteau, H. (2001) *Tetrahedron*, **57**, 5739.

150 Takimiya, K., Morikami, A., and Otsubo, T. (1997) *Synlett*, 319.

151 Gusarova, N.K., Chernysheva, N.A., Sukhov, B.G., Afonin, A.V., Fedorov, G.A., Yakimova, S.V., and Trofimov, B.A. (2003) *Chemistry of Heterocyclic Compounds (English Translation)*, **39**, 128.

152 Rao, H.S.P., Sakthikumar, L., Vanitha, S., and Kumar, S.S. (2003) *Tetrahedron Letters*, **44**, 4701.

153 Chen, C.H. (1976) *Journal of the Chemical Society, Chemical Communications*, 920.

154 Fabian, K. and Hartmann, H. (1971) *Journal fur Praktische Chemie*, **313**, 722.

155 Takamizawa, A. and Hirai, K. (1969) *Chemical & Pharmaceutical Bulletin*, **17**, 1924.

156 Ueno, Y., Masuyama, Y., and Okawara, M. (1975) *Chemistry Letters*, 603.

157 Bajwa, G.S., Berlin, K.D., and Pohl, H.A. (1976) *The Journal of Organic Chemistry*, **41**, 145.

158 Spencer, H.K., Cava, M.P., Yamagishi, F.G., and Garito, A.F. (1976) *The Journal of Organic Chemistry*, **41**, 730.

159 Drozd, V.N., Udachin, Yu.M., Bogomolova, G.S., and Sergeichuk, V.V. (1980) *Zhurnal Organicheskoi Khimii*, **16**, 883.

160 O'Connor, B.R. and Jones, F.N. (1970) *The Journal of Organic Chemistry*, **35**, 2002.

161 Melby, L.R., Hartzler, H.D., and Shepard, W.A. (1976) *The Journal of Organic Chemistry*, **39**, 2456.

162 Lindsey, J.S., Shreimany, I.C., Hsu, H.C., Kearney, P.C., and Marguerettoz, A.M. (1987) *The Journal of Organic Chemistry*, **52**, 827.

163 Firouzabadi, H., Iranpoor, N., and Hazarkhani, H. (2001) *The Journal of Organic Chemistry*, **66**, 7527.

164 Firouzabadi, H. and Karimi, B. (2001) *Phosphorus Sulfur and Silicon and the Related Elements*, **175**, 207.

165 Muthusamy, S., Babu, S.A., and Gunanathan, C. (2002) *Tetrahedron*, **58**, 7897.

166 Anand, R.V., Saravanan, P., and Singh, V.K. (1999) *Synlett*, 415.

167 Ranu, B.C. and Chouhan, G. (2002) *Synlett*, 727.

168 Firouzabadi, H., Iranpoor, N., and Karimi, B. (1999) *Synlett*, 319.

169 Yoshida, H., Kinoshita, H., Kato, J., Kanehira, N., Ogata, T., and Matsumoto, K. (1987) *Synthesis*, 393.

170 Leir, C.M. (1972) *The Journal of Organic Chemistry*, **37**, 887.

171 Christau, H.-J., Christol, H., and Bottaro, D. (1978) *Synthesis*, 826.

172 Feugeas, C. and Olschwang, D. (1969) *Bulletin de la Societe Chimique de France*, 332.

173 Stutz, P. and Stadler, P.A. (1972) *Helvetica Chimica Acta*, **55**, 75.

174 Backer, H.J. and Wiggerink, G.L. (1941) *Recueil des Travaux Chimiques des Pays-Bas*, **60**, 453.

175 Husemann, A. (1852) *Justus Liebigs Annalen der Chemie*, **81**, 96.

176 Ranu, B.C., Das, A., and Samanta, S. (2002) *Journal of the Chemical Society, Perkin Transactions 1*, 1520.

177 Caputo, R., Guaragna, A., Palumbo, G., and Pedatella, S. (2003) *European Journal of Organic Chemistry*, 346.

178 Morton, D.R. and Hobbs, S.J. (1979) *The Journal of Organic Chemistry*, **44**, 656.

179 Jo, S., Tanimoto, S., Oida, T., and Okano, M. (1981) *Bulletin of the Chemical Society of Japan*, **54**, 1434.

180 Bryce, M.R. (1984) *Tetrahedron Letters*, **25**, 2403.

181 Chiang, L.-Y., Shu, P., Holt, D., and Cowan, D. (1983) *The Journal of Organic Chemistry*, **48**, 4713.

182 Demetriadis, N.G., Huang, S.J., and Samulski, E.T. (1977) *Tetrahedron Letters*, **18**, 2223.

183 Hartzler, H.D. (1973) *Journal of the American Chemical Society*, **95**, 3422.

184 Rice, J.E. and Okamoto, Y. (1981) *The Journal of Organic Chemistry*, **46**, 446.

185 Gonella, N.C. and Cava, M.P. (1978) *The Journal of Organic Chemistry*, **43**, 369.

186 Melby, L.R., Hatzler, H.D., and Shepard, W.A. (1974) *The Journal of Organic Chemistry*, **39**, 2456.

187 Miles, M.G., Wagner, J.S., Wilson, J.D., and Siedle, A.R. (1975) *The Journal of Organic Chemistry*, **40**, 3577.

188 Hansen, T.K., Hawkins, I., Varma, K.S., Edge, S., Larsen, S., Becher, J., and Underhill, A.E. (1991) *Journal of the Chemical Society-Perkin Transactions 2*, 963.

189 Lambert, C. and Christiaens, L. (1984) *Tetrahedron Letters*, **25**, 833.

190 Bajwa, G.S., Berlin, K.D., and Pohl, H.A. (1976) *The Journal of Organic Chemistry*, **41**, 145.

191 Meline, R.L. and Elsenbaumer, R.L. (1998) *Journal of the Chemical Society, Perkin Transactions 1*, 2467–2469.

192 Kusters, W. and de Mayo, P. (1974) *Journal of the American Chemical Society*, **96**, 3502.

193 Kato, H., Shiba, T., Aoki, N., Iijima, H., and Tezuka, H. (1982) *Journal of the Chemical Society, Perkin Transactions 1*, 1885.

194 Moses, P.R. and Chambers, J.Q. (1974) *Journal of the American Chemical Society*, **96**, 945.

195 Buza, D. and Gradowska, W. (1982) *Polish Journal of Chemistry*, **56**, 1313.

196 Nakayama, J., Fujiwara, K., and Hoshino, M. (1976) *Bulletin of the Chemical Society of Japan*, **49**, 3567.

197 Mocerino, M. and Stick, R.V. (1990) *Tetrahedron Letters*, **31**, 3051.

198 Campaigne, E. and Hamilton, R.D. (1964) *The Journal of Organic Chemistry*, **29**, 2877.

199 Hunig, S., Kiesslich, G., Oette, K.-H., and Quast, H. (1971) *Justus Liebigs Annalen der Chemie*, **754**, 46.

200 Fanghanel, E. and Mayer, R. (1964) *Zeitschrift fur Chemie*, **4**, 384.

201 Kruger, A. and Wudl, F. (1977) *The Journal of Organic Chemistry*, **42**, 2778.

202 Seidle, A.R. and Johannesen, R.B. (1975) *The Journal of Organic Chemistry*, **40**, 2002.

203 Fanghanel, E., van Hinh, L., and Schukat, G. (1976) *Zeitschrift fur Chemie*, **16**, 317.

204 Gotthardt, H. and Weisshuhn, C.M. (1978) *Chemische Berichte*, **111**, 2028.

205 Lukac, J. and Heimgartner, H. (1979) *Helvetica Chimica Acta*, **62**, 1236.

206 Dinsmore, A., Garner, C.D., and Joule, J.A. (1998) *Tetrahedron*, **54**, 3291.

207 Banerjee, A.K. and Laya, M.S. (2000) *Russian Chemical Reviews*, **69**, 947.

208 Hach, V. (1953) *Chemicke Listy*, **47**, 227.

209 Stocken, L.A. (1947) *Journal of the Chemical Society*, 592.

210 Cain, E.N. and Welling, L.L. (1975) *Tetrahedron Letters*, **16**, 1353.

211 Lipshutz, B.H., Moretti, R., and Crow, R. (1989) *Tetrahedron Letters*, **30**, 15.

212 Firouzabadi, H., Iranpoor, N., and Hazarkhani, H. (2002) *The Journal of Organic Chemistry*, **67**, 2572.

213 Karimi, B. and Hazarkhani, H. (2003) *Synthesis*, 2547.

214 Ceccherelli, P., Curini, M., Marcotullio, M.C., Epifano, F., and Rosati, O. (1996) *Synlett*, 767.

215 Hirano, M., Ukawa, K., Yakabe, S., Lark, J.H., and Morimoto, T. (1997) *Synthesis*, 858.

216 Karimi, B., Seradj, H., and Maleki, J. (2002) *Tetrahedron*, **58**, 4513.

217 Corey, E.J. and Seebach, D. (1965) *Angewandte Chemie*, **77**, 1134.

218 Tseng, H.-R. and Luh, T.-Y. (1997) *The Journal of Organic Chemistry*, **62**, 4568.

219 Chiangand, C.-C. and Lah, T.-Y. (2001) *Synlett*, 977.

220 Pandey, B., Bal, S.Y., and Khire, U.R. (1989) *Tetrahedron Letters*, **30**, 4007.

221 Baliah, V., Prema, S., Jawahringh, C.B., and Jeyaraman, R. (1981) *Synthesis*, 995.

222 Alphand, V., Gaggero, N., Colonna, S., Pasta, P., and Furstoss, R. (1997) *Tetrahedron*, **53**, 9695.

223 Delouvrie, B., Fensterbank, L., Najera, F., and Malacria, M. (2002) *European Journal of Organic Chemistry*, 3507.

224 Della Sala, G., Labano, S., Lattanzi, A., Tedesco, C., and Scettri, A. (2002) *Synthesis*, 505.

225 Nishida, A., Nishida, M., and Ynemitsu, O. (1990) *Tetrahedron Letters*, **31**, 4007.

226 Firouzabadi, H., Iranpoor, N., and Karimi, B. (1999) *Synlett*, 413.

227 Zhu, Z.M., Wang, Y., Zu, Y.T., Mei, Z.M., Liu, Q., and Hu, J.H. (1997) *Chinese Chemical Letters*, **8**, 367.

228 Perst, H. (2002) *Science of Synthesis*, **11**, Chapter 6, 97.

229 Krische, B., Walter, W., and Adiwidjaja, G. (1982) *Chemische Berichte*, **115**, 3842.

230 Harpp, D.N., Gleason, J.G., and Ash, D.K. (1971) *The Journal of Organic Chemistry*, **36**, 322.

231 Ohline, R.W., Allred, A.L., and Bordwell, F.G. (1964) *Journal of the American Chemical Society*, **86**, 4641.

232 King, J.F. and Rathore, R. (1989) *Tetrahedron Letters*, **30**, 2763.

233 Liskamp, R.M.J., Zeegers, H.J., and Ottenheijm, H.C.J. (1981) *The Journal of Organic Chemistry*, **46**, 5408.

234 Yolka, S., Fellous, R., Lizzani-Cuvelier, L., and Loiseau, M. (1998) *Tetrahedron Letters*, **39**, 991.

235 Karsch, S., Schwab, P., and Metz, P. (2002) *Synlett*, 2019.

236 Enders, D., Wallert, S., and Runsink, J. (2003) *Synthesis*, 1856.

237 Grigoriev, E.V., Yatsenko, A.V., Novozhilov, N.V., Saginova, L.G., and Petrosyan, V.S. (1993) *Vestnik Moskovskogo Universiteta, Seria 2, Khimia*, **34**, 87.

238 Bakker, B.H., Cerfontain, H., and Tomassen, H.P.M. (1989) *The Journal of Organic Chemistry*, **54**, 1680.

239 Kovalev, V.V. and Shokova, E.A. (1988) *Zhurnal Organicheskoi Khimii*, **24**, 738.

240 Bonini, B.F., Kemperman, C., Willems, S.T.H., Fochi, M., Mazzanti, G., and Zwanenburg, B. (1998) *Synlett*, 1411.

241 Vasil'eva, T.P. (1992) *Izvestiya Akademii Nauk SSSR, Seria Khimia*, 2153.

242 Duffy, D.E., Condit, F.H., Teleha, C.A., Wang, C.-L.J., and Calabrese, J.C. (1993) *Tetrahedron Letters*, **34**, 3667.

243 Langendries, R.F.J. and De Schryver, F.C. (1972) *Tetrahedron Letters*, **13**, 4781.

244 Nishitomi, K., Nagai, T., and Tokura, N. (1968) *Bulletin of the Chemical Society of Japan*, **41**, 1388.

245 Liskamp, R.M.J., Zeegers, H.J., and Ottenheijm, H.C.J. (1981) *The Journal of Organic Chemistry*, **46**, 5408.

246 Tian, L., Xu, G.-Y., Ye, Y., and Liu, L.-Z. (2003) *Synthesis*, 1329.

247 Grigor'ev, E.V. and Saginova, L.G. (2001) *Chemistry of Heterocyclic Compounds (English Translation)*, **37**, 649.

248 Perst, H. and Klenke, C. (2002) *Science of Synthesis*, **11**, Chapter 3 35.

249 Rakhmankulov, D.L., Zorin, V.V., Latypova, F.N., Zlotskii, S.S., and Karakhanov, R.A. (1983) *Russian Chemical Reviews (English Translation)*, **52**, 350.

250 Wimmer, P. (1991) *Methoden der Organischen Chemie (Houben-Weyl)*, **E14a/1**, 794.

251 Le Marechal, A.M., Robert, A., and Leban, I. (1993) *Journal of the Chemical Society-Perkin Transactions*, **1**, 351.

252 Keskinen, R., Nikkila, A., and Pihlaja, K. (1972) *Tetrahedron*, **28**, 3943.

253 Jones, F.N. and Andreades, S. (1969) *The Journal of Organic Chemistry*, **34**, 3011.

254 Gotthardt, H., Fiest, U., and Schoy-Tribbensee, G. (1985) *Chemische Berichte*, **118**, 774.

255 Shibuya, I., Yonemoto, K., Tsuchiya, T., and Yasumoto, M. (1992) Jpn. Pat. 0,436,417,8.

256 Faure, A. and Descotes, G. (1978) *Synthesis*, 286.

257 Van Leusen, A.M., Richters, P., and Strating, J. (1966) *Recueil des Travaux Chimiques des Pays-Bas*, **85**, 323.

258 Ibata, T. and Nakano, H. (1990) *Bulletin of the Chemical Society of Japan*, **63**, 2450.

259 Nakano, H. (1992) *Bulletin of the Chemical Society of Japan*, **65**, 3088.

260 Shiryaev, A.K., Moiseev, I.K., and Popov, V.A. (1992) *Zhurnal Organicheskoi Khimii*, **28**, 418.

261 Gerstenberger, M.R.C., Haas, A., Wille, R., and Yazdanbakhsch, M. (1986) *Revue de Chimie Minerale*, **23**, 485.

262 Moran, J.R., Tapia, I., and Alcazar, V. (1990) *Tetrahedron*, **46**, 1783.

263 Wilson, G.E., Huang, M.G., and Schlomann, W.W. (1968) *The Journal of Organic Chemistry*, **33**, 2133.

264 Streinz, L., Koutek, B., and Saman, D. (1997) *Collection of Czechoslovak Chemical Communication*, 2293.

265 Kazahaya, K., Hamada, N., Ito, S., and Sato, T. (2002) *Synlett*, 1535.

266 Kamal, A., Chouhan, G., and Ahmed, K. (2002) *Tetrahedron Letters*, **43**, 6947.

267 Mondal, E., Sahu, P.R., Bose, G., and Khan, A.T. (2002) *Tetrahedron Letters*, **43**, 2843.

268 Karimi, B. and Ma'mani, L. (2003) *Synthesis*, 2503.

269 Yadav, J.S., Reddy, B.V.S., and Pandey, S.K. (2001) *Synlett*, 238.

270 Kerverdo, S., Lizzani-Cuvelier, L., and Dunach, E. (2002) *Tetrahedron*, **58**, 10455.

271 Castro, P.P., Tikomirov, S., and Gutierrez, C.G. (1988) *The Journal of Organic Chemistry*, **53**, 5179.

272 Guseinov, F.I., Asadov, Kh.A., and Burangulova, R.N. (2002) *Russian Journal of Organic Chemistry (English Translation)*, **38**, 1216.

273 Trofimov, B.A., Skvortsov, Yu.M., Moshchevitina, E.I., Mal'kina, A.G., and Bel'skii, V.K. (1991) *Zhurnal Organicheskoi Khimii*, **27**, 188.

274 Hans, M. and Dehne, H. (1983) *Die Pharmazie*, **38**, 441.

275 Gutsu, Ya., Boy, L.V., Maiga, S.B., Inthapanya, P., and Barba, N.A. (1992) *Bul Acad Stiinte Repub Mold, Stiinte Biol Chim.*, 56.

276 Block, E., Aslam, M., Iyerand, R., and Hutchinson, J. (1984) *The Journal of Organic Chemistry*, **49**, 3664.

277 Hosomi, A., Hayashi, S., Hoashi, K., Kohra, S., and Tominaga, Y. (1987) *Journal of the Chemical Society, Chemical Communications*, 1442.

278 Dibo, A., Stavaux, M., Lozac'h, N., and Hordvik, A. (1985) *Acta Chemica Scandinavica. Series B: Organic Chemistry and Biochemistry*, **39**, 103.

279 L'abbe, G., Buelens, J., Dehaen, W., Toppet, S., and Van Meervelt, L. (1994) *Journal of the Chemical Society, Perkin Transactions 1*, 1263.

280 Cermola, F., De Lorenzo, F., Giordano, F., Graziano, M.L., Iesce, M.R., and Palumbo, G. (2000) *Organic Letters*, **2**, 1205.

281 Yonemoto, K., Honda, K., Shibuya, I., Tsuchiya, T., and Yasumoto, M. (1992) *Bulletin of the Chemical Society of Japan*, **65**, 668.

282 Gotthardt, H. and Opperman, M. (1985) *Journal of the Chemical Society, Chemical Communications*, 1145.

283 Sakai, S., Niimi, H., Kobayashi, Y., and Ishii, Y. (1977) *Bulletin of the Chemical Society of Japan*, **50**, 3271.

284 Lee, W.S., Hahn, H.E., and Nam, K.D. (1986) *The Journal of Organic Chemistry*, **51**, 2789.

285 Ogawa, K., Yamada, S., Terada, T., Yamazaki, T., and Honna, T. (1985) *Chemical & Pharmaceutical Bulletin*, **33**, 2256.

286 Bortolini, O., Di Furia, F., Licini, G., Modena, G., and Rossi, M. (1986) *Tetrahedron Letters*, **27**, 6257.

287 Chavan, S.P., Dantale, S.W., Pasupathy, K., Tejwani, R.B., Kamat, S.K., and Ravindranathan, T. (2002) *Greem Chemistry*, **4**, 337.

288 Khan, A.T., Mondal, E., and Sahu, P.R. (2003) *Synlett*, 377.

289 Chavan, S.P., Soni, P., and Kamat, S.K. (2001) *Synlett*, 1251.

290 Karimi, B., Seradj, H., and Tabaei, M.H. (2000) *Synlett*, 1798.

291 Reynolds, D.D., Massad, M.K., Fields, D.L., and Johnson, D.L. (1961) *The Journal of Organic Chemistry*, **26**, 5109.

292 Reynolds, D.D., Fields, D.L., and Johnson, D.L. (1961) *The Journal of Organic Chemistry*, **26**, 5111.

293 McIntosh, J.M., Mishra, P., and Siddiqui, M.A. (1984) *The Journal of Organic Chemistry*, **49**, 1036.

294 Arya, P., Lesage, M., and Wagner, D.D.M. (1991) *Tetrahedron Letters*, **32**, 2853.

295 Fuchigami, T. (1997) *Phosphorus Sulfur and Silicon and the Related Elements*, **120–121**, 343.

296 Ionnou, M., Porter, M.J., and Saez, F. (2002) *Journal of the Chemical Society, Chemical Communications*, 346.

297 Kerverdo, S., Lizzani-Cuvelier, L., and Duñach, E. (2003) *Tetrahedron Letters*, **44**, 853.

298 Yang, Y., Zheng, F., Sun, B., Ding, F., Liu, Y., and Ren, Y. (2001) *Chemical Journal on Internet*, **3**, 57.

299 Mansour, T.S. and Jin, H. (1995) PCT Int. Appl. WO 29 176.

12
Five-Membered Heterocycles with Three Heteroatoms: Triazoles

Larry Yet

12.1
1,2,3-Triazoles

12.1.1
Introduction

The chemistry of 1,2,3-triazoles is reviewed extensively elsewhere [1]. This chapter will focus on the general synthesis and reactivity of the monocyclic 1,2,3-triazole system with recent methods, which include solid-phase and microwave-assisted reactions.

12.1.2
General Reactivity

12.1.2.1 Relevant Physicochemical Data and NMR Data

N-Unsubstituted 1,2,3-triazole can be shown as either 1*H*- or as 2*H*-triazoles since these two tautomeric forms are in equilibrium in the solution phase (Figure 12.1). In this chapter, for simplication, this type of compound will be represented as 1*H*-triazoles, independent of the predominant tautomer. In the gas phase, the 2*H*-tautomer of the 1,2,3-triazole represents more than 99.9% of the equilibrium mixture [2]. 1*H*-1,2,3-Triazole is both a weak base ($pK_a = 1.17$) and a weak acid ($pK_a = 9.40$). The basicity of *N*-unsubstituted and *N*-methyl-1,2,3-triazoles in the gas phase, in solution, and in the solid state has been determined [3].

The ^1H and ^{13}C NMR spectra of the parent 1,2,3-triazole for the protons and carbons at the 4- and 5-positions are identical because the compound exists as both 1*H*- and 2*H*-triazoles in solution at room temperature (Table 12.1). Other ^1H and ^{13}C NMR data of the methyl group attached to the 1,2,3-triazole in the 1- and 2-positions are listed for comparison.

Modern Heterocyclic Chemistry, First Edition.
Edited by Julio Alvarez-Builla, Juan Jose Vaquero, and José Barluenga.

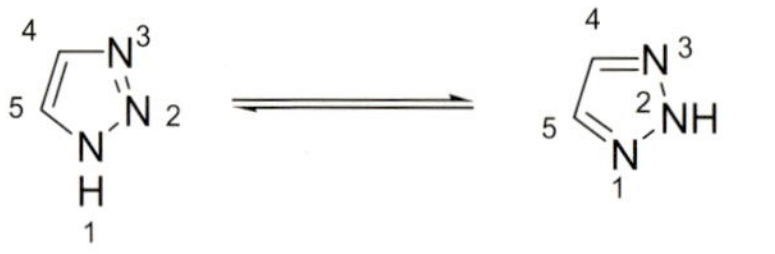

1H-1,2,3-triazole ⇌ **2H-1,2,3-triazole**

stable, water-soluble, colorless crystals
mp 120-121 °C
pK_a = 9.40
pK_a = 1.17 of the protonated species

Figure 12.1 Tautomeric structures of 1,2,3-triazoles.

12.1.3 Relevant Natural and/or Useful Compounds

1,2,3-Triazoles are not present in natural products and are remarkably stable to metabolic transformations such as oxidation, reduction, and both basic and acidic hydrolysis. 1,2,3-Triazoles have found broad use in industrial applications such as dyes and brighteners for fibers, corrosion inhibitors for many metals and alloys, light stabilizers for organic materials and polymers, and agrochemicals such as herbicides, fungicides, and antibacterial agents [1d]. They have been considered as an interesting component from the viewpoint of biological activity and are seen in many drugs such as potent HIV-inhibitors [6], antimicrobial agents [7], and selective β_3-adrenergic receptor agonists [8]. Figure 12.2 shows the structures of β-lactam antibiotics tazobactam [9] and cefatrizine [10].

12.1.4 Synthesis of 1,2,3-Triazoles

The most important general approach to the synthesis of 1,2,3-triazoles involves the use of azide reagents. Azides that have been employed in these syntheses can be alkyl,

Table 12.1 ^{1}H and ^{13}C NMR data (ppm) of 1,2,3-triazoles.

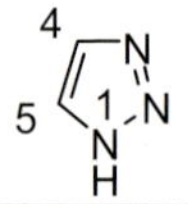

	^{1}H NMR (DMSO-d_6)		Reference	^{13}C NMR (DMSO-d_6)		Reference
Substituents	H4	H5		C4	C5	
—	7.91	7.91	[4]	130.3	130.3	[5]
1-Me	7.72	8.08	[4]	134.3	125.5	[5]
2-Me	7.77	7.77	[4]	133.2	133.2	[5]

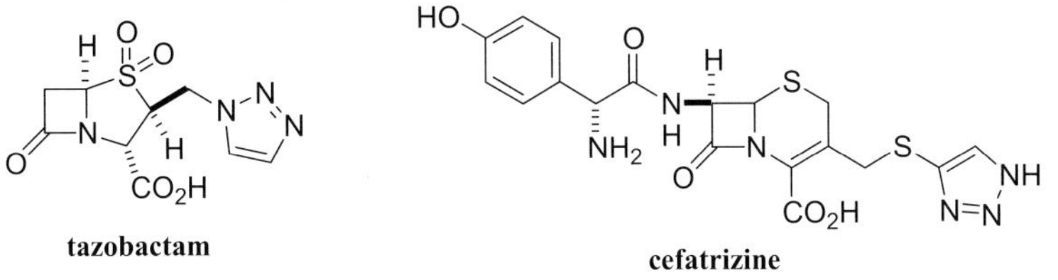

tazobactam

cefatrizine

Figure 12.2 Structure of β-lactam antibiotics containing 1,2,3-triazole rings.

aryl, heteroaryl, acyl, alkoxycarbonyl and sulfonyl azides, trimethylsilyl azide, hydrazoic acid, and sodium azide. Azides can react with substituted alkynes and alkenes and with activated methylene compounds to yield various 1,2,3-triazoles.

12.1.4.1 **1,3-Dipolar Cycloadditions of Alkynes with Azide Reagents**

1,3-Dipolar cycloaddition of azides to alkynes is the most popular method for the syntheses of various 1,2,3-triazoles since it provides the desired product directly. A review on 1,2,3-triazole formation via 1,3-dipolar cycloaddition of acetylenes with azides under mild conditions has been published [11]. When unsymmetrical alkynes are used, two possible regioisomers are usually obtained. Isomers with the electron-withdrawing groups at the C4 position and the electron-donating groups at C5 are usually the major products.

N-Unsubstituted-1,2,3-triazoles are prepared by direct addition of hydrazoic acid [12] or with azide ions [13] to alkynes, such as the reaction of α,β-acetylenic aldehydes **1** with sodium azide in dimethyl sulfoxide (DMSO) followed by hydrolysis to give 5-substituted-4-carbaldehyde-1,2,3-triazole derivatives **2** (Scheme 12.1) [14]. The major disadvantage of this method is that often thermal conditions are required for these reactions and that sodium azide can be explosive.

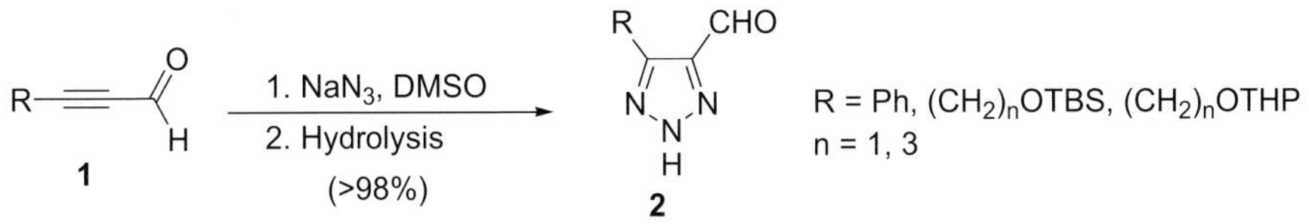

1. NaN_3, DMSO
2. Hydrolysis
(>98%)

R = Ph, $(CH_2)_nOTBS$, $(CH_2)_nOTHP$
n = 1, 3

Scheme 12.1

The most prominent method employed for the synthesis of 1,2,3-triazoles is the addition of alkyl, aryl, and heteroaryl azides to alkynes. Azides can add to acetylene [15] and symmetrically substituted alkynes [16] to give only 4,5-unsubstituted- and 4,5-disubstituted-1,2,3-triazoles, respectively. Addition of azides to monosubstituted alkynes afford mixtures of 1,4- and 1,5-disubstituted 1*H*-1,2,3-triazoles **3** and **4**, respectively (Table 12.2). The ratio of the two products depends on the structure of the monosubstituted alkyne; alkynes with the electron-withdrawing groups preferentially give products substituted at C4 like **3**, while alkynes with electron-donating groups provide major products substituted at C5 like **4**.

Table 12.2 Addition of azides to monosubstituted alkynes.

$R^1{-}N_3$ + $\equiv{-}R^2$ → **3** + **4**

R^1	R^2	Yield (%)		Reference
		3	4	
Ph	Ph	43	52	[17]
Ph	CO_2Me	88	12	[18]
Bn	CONHBn	65	22	[19]
CF_2CFHCF_3	Bu	37	58	[20]
$CH_2PO(OEt)_2$	CH_2OH	30	69	[21]
4-Tol	Bz	60	11	[22]
Benzotriazolylmethyl	Ph	40	60	[23]

Recent advances in this area are the acceleration and regioselectivity of these azide additions to alkynes, which can sometimes be slow. "Click chemistry" is a recently coined term to denote a growing family of powerful chemical reactions that are based on "spring-loaded" energy-intensive substrates that can, under the right conditions, unload their energy to form stable products in high selectivity [24]. A microreview on copper(I)-catalyzed alkyne–azide "click" cycloadditions from a mechanistic and synthetic perspective has been written [25]. A mini-review has been published on the 1,3-dipolar cycloadditions of azides and alkynes as a universal ligation tool in polymer and materials science [26]. A highlight has been published on the click reaction in the luminescent probing of metal ions and its implications on biolabeling techniques [27]. A perspective on the Cu(I)-catalyzed 1,3-dipolar cycloaddition of azides and alkynes in carbohydrate chemistry, highlighting developments in the preparation of simple glycoside and oligosaccharide mimetics, glyco-macrocycles, glycopeptides, glyco-clusters, and carbohydrate arrays has been reported [28]. A review titled "Click chemistry – What's in a name? Triazole synthesis and beyond" has been published [29]. A highlight on the copper-free azide–alkyne cycloadditions with new insights and perspectives has been written [30]. Substituent effects in the 1,3-dipolar cycloadditions of azides with alkenes and alkynes have been investigated with the high accuracy CBS-QB3 method [31].

For example, a high-yielding copper-catalyzed reaction, which involves the reaction of azides to terminal alkynes in the presence of copper(II) sulfate with ascorbic acid or sodium ascorbate, gave 1,4-disubstituted-1,2,3-triazoles **5** regardless of the group on the alkynes or azides (Scheme 12.2) [32]. The mechanism of the ligand-free Cu(I)-catalyzed azide–alkyne cycloaddition reaction has been proposed in a literature report [33]. Triazole-linked glycopeptides have been obtained by Cu(I)-catalyzed cycloadditions of either azide-functionalized glycosides and acetylenic amino acids

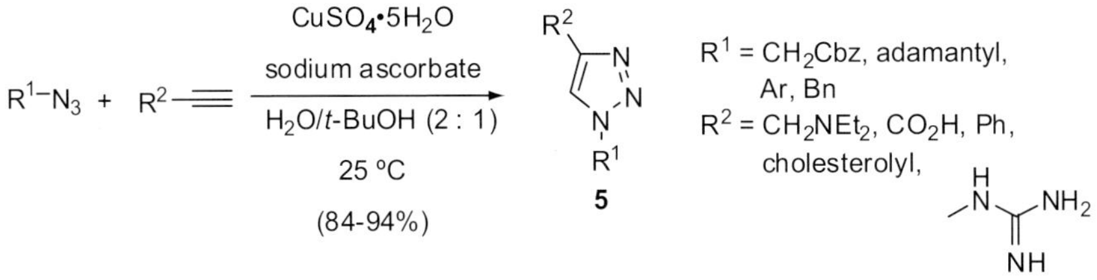

Scheme 12.2

or acetylenic glycosides and azide-containing amino acids in the presence of sodium ascorbate [34]. A highly efficient one-pot synthesis of 1,2,3-triazole-linked glycoconjugates involving a Cu(I)-catalyzed 1,3-dipolar cycloaddition as a key step has been reported [35].

Microwave irradiation has also become a recent method for the synthesis of 1,2,3-triazoles under solvent-free conditions. For example, 1,3-dipolar cycloaddition of organic azides **6** with acetylenic amides **7** under solvent-free microwave irradiation produced *N*1-substituted-4*C*-carbamoyl-1,2,3-triazoles **8** (Scheme 12.3) [19]. Microwave irradiation allowed a substantial decrease in reaction times and offered greater simplicity in the purification step.

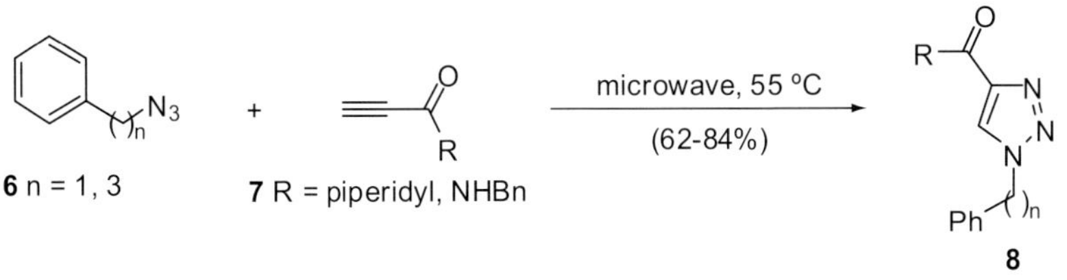

Scheme 12.3

Addition of azides to unsymmetrical disubstituted alkynes often yields mixtures of isomeric 1,2,3-triazoles. The relative ratios of the two products depend strongly on the nature of the substituents on the alkyne. A more recent example showed the 1,3-dipolar cycloadditions of azido ester **10** with electron-deficient alkynes **9** to give 1,4,5-trisubstituted 1,2,3-triazoles **11** under mild conditions in water (Scheme 12.4) [36]. 1,3-Dipolar cycloaddition of tributyl(3,3,3-trifluoro-1-propynyl)stannane (**12**) with

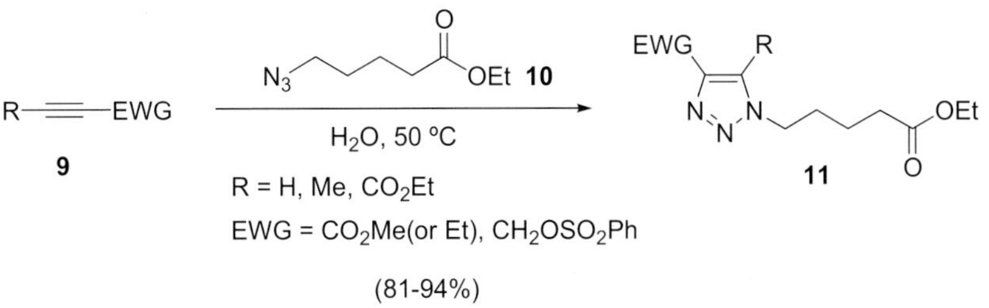

Scheme 12.4

phenyl azide gave the corresponding 1,2,3-triazole **13**, which was a useful building block for further functionalization (Scheme 12.5) [37]. An interesting approach to prepare regiospecifically 4,5-disubstituted-1,2,3-triazoles is the addition of bromomagnesium acetylides **15** to aryl azides **14** to yield 1,5-disubstituted-1,2,3-triazoles **16**, which could be trapped with various electrophiles to form 1,4,5-trisubstituted 1,2,3-triazoles **17** (Scheme 12.6) [38]. In addition, copper(I) iodide promoted reaction of alkyl azides and terminal alkynes in the presence of iodine monochloride led to a regiospecific synthesis of 5-iodo-1,4-disubstituted-1,2,3-triazole **18**, which could be further elaborated to a range of 1,4,5-trisubstituted-1,2,3-triazole derivatives (Scheme 12.7) [39]. A series of 4,5-disubstituted-1,2,3-triazoles has been regiospecifically prepared directly from propargyl halides and sodium azide via the Banert cascade [40].

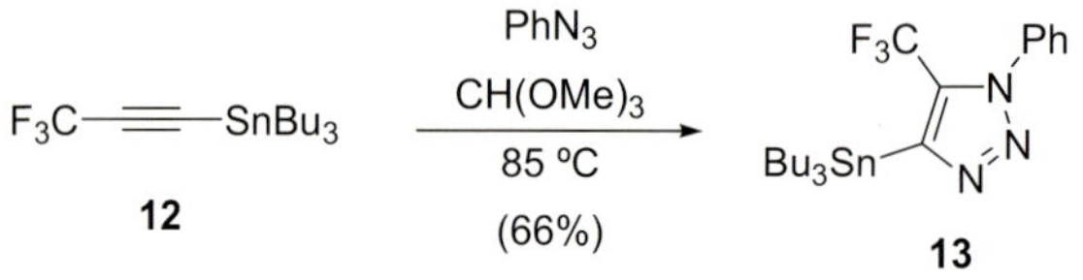

F_3C—≡—$SnBu_3$ **12** → (PhN_3, $CH(OMe)_3$, 85 °C, (66%)) → **13** (F_3C, Ph, Bu_3Sn)

Scheme 12.5

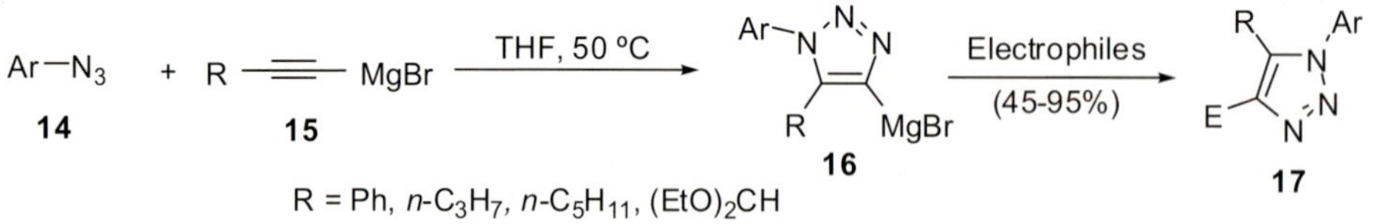

Ar—N_3 **14** + R—≡—MgBr **15** → (THF, 50 °C) → **16** (Ar, R, MgBr) → (Electrophiles, (45-95%)) → **17** (R, Ar, E)

R = Ph, n-C_3H_7, n-C_5H_{11}, $(EtO)_2CH$

Scheme 12.6

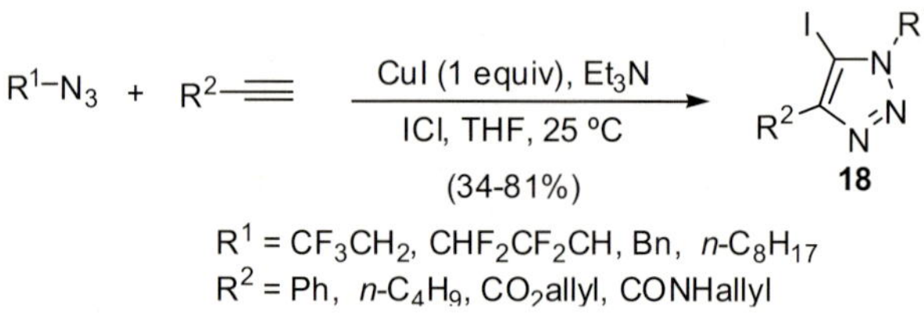

R^1–N_3 + R^2—≡ → (CuI (1 equiv), Et_3N, ICl, THF, 25 °C, (34-81%)) → **18** (I, R^1, R^2)

R^1 = CF_3CH_2, CHF_2CF_2CH, Bn, n-C_8H_{17}
R^2 = Ph, n-C_4H_9, CO_2allyl, CONHallyl

Scheme 12.7

Recently, metal-mediated methodologies have been published where 1,2,3-triazoles can be prepared regiospecifically, with non-activated terminal alkynes and trimethylsilyl azide as the safe synthetic equivalent of the highly explosive hydrazoic acid or sodium azide. Various triazoles (**19**) were synthesized from non-activated terminal alkynes, allyl methyl carbonate, and trimethylsilyl azide ($TMSN_3$) in a [3 + 2] cycloaddition with the use of a Pd(0)-Cu(I) bimetallic catalyst (Scheme 12.8) [41]. The

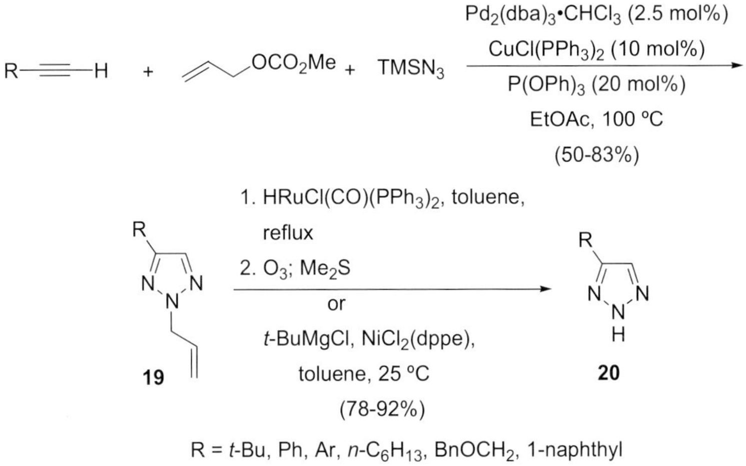

R—≡—H + $\diagup\!\!\diagdown$OCO$_2$Me + TMSN$_3$ → ($Pd_2(dba)_3$•$CHCl_3$ (2.5 mol%), $CuCl(PPh_3)_2$ (10 mol%), $P(OPh)_3$ (20 mol%), EtOAc, 100 °C (50-83%))

19 → (1. $HRuCl(CO)(PPh_3)_2$, toluene, reflux; 2. O_3; Me_2S or t-BuMgCl, $NiCl_2$(dppe), toluene, 25 °C (78-92%)) → **20**

R = t-Bu, Ph, Ar, n-C_6H_{13}, $BnOCH_2$, 1-naphthyl

Scheme 12.8

allyl group of **19** is efficiently deprotected by ruthenium-catalyzed isomerization followed by ozonolysis or by nickel-catalyzed Grignard addition reaction to give 4-substituted triazoles **20** [42]. 2-Allyl-1,2,3-triazoles **21** have been prepared regiospecifically by the palladium-catalyzed three-component coupling reaction of alkynes, allyl methyl carbonate, and trimethylsilyl azide (Scheme 12.9) [43]. Similarly, the four-component coupling reactions of silylacetylenes, allyl carbonates, and trimethylsilyl azide catalyzed by a Pd(0)-Cu(I) bimetallic catalyst led to trisubstituted 1,2,3-triazoles [44]. The [3 + 2] cycloaddition of non-activated terminal alkynes and trimethylsilyl azide proceeded smoothly in the presence of copper catalyst and *N,N*-dimethylformamide and methanol to give the corresponding *N*-unsubstituted-1,2,3-triazoles in good to high yields [45]. [3 + 2]-Cycloadditions of alkyl azides with various unsymmetrical internal alkynes in the presence of Cp*$RuCl(PPh_3)_2$ as catalyst in refluxing benzene led to 1,4,5-trisubstituted-1,2,3-triazoles, whereas alkyl phenyl and dialkyl acetylenes underwent cycloadditions to afford mixtures of regioisomeric 1,2,3-triazoles and acyl-substituted internal alkynes reacted with complete regioselectivity [46]. In the presence of catalytic Cp*$RuCl(PPh_3)_2$ or Cp*RuCl(COD), primary and secondary azides reacted with a broad range of terminal alkynes containing a range of functionalities to produce selectively 1,5-disubstituted-1,2,3-triazoles [47]. 1,3-Dipolar

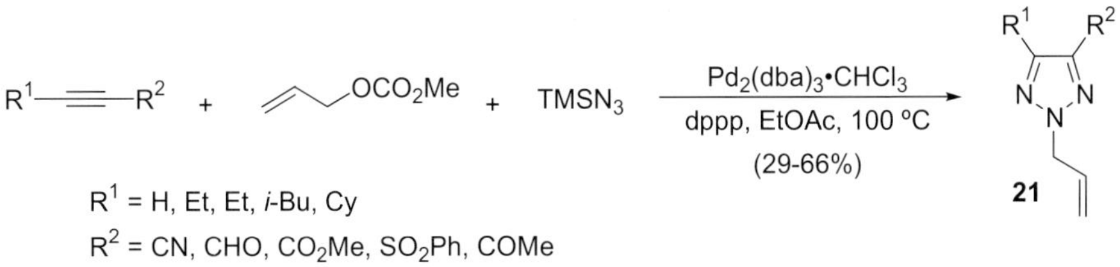

R^1—≡—R^2 + $\diagup\!\!\diagdown$OCO$_2$Me + TMSN$_3$ → ($Pd_2(dba)_3$•$CHCl_3$, dppp, EtOAc, 100 °C (29-66%)) → **21**

R^1 = H, Et, Et, i-Bu, Cy

R^2 = CN, CHO, CO_2Me, SO_2Ph, COMe

Scheme 12.9

cycloaddition of trifluoromethylated propargylic alcohols with azides in the presence of catalytic $[Cp^*RuCl_2]_n$ afforded exclusively 4-trifluoromethyl-1,4,5-trisubstituted-1,2,3-triazoles in high yields [48]. Copper-catalyzed [3 + 2] cycloaddition of azides to mono- and disubstituted alkynes with N-heterocyclic carbene ligands has been found to be a versatile and highly efficient reaction in which an internal alkyne was successfully shown to work for the first time [49].

Azides can be prepared *in situ* from their respective halides and then reacted with their alkyne partners. 1,4-Disubstituted-1,2,3-triazoles **22** have been obtained in excellent yields by a convenient one-pot procedure from various aryl and alkyl iodides and terminal alkynes without isolation of potentially unstable organic azide intermediates (Scheme 12.10) [50]. The microwave-assisted synthesis of this version has been published by the same group [51]. A similar one-pot procedure uses benzyl and alkyl halides for generation of the azides with a series of terminal and disubstituted alkynes [52]. A one-pot sequential and cascade sequence involving the formation of allylic azides, from aryl/heteroaryl/vinyl halides, allene, and sodium azide, by palladium-catalyzed anion capture, and cyclization-anion capture, followed by 1,3-dipolar cycloaddition provided various 1,2,3-triazoles in good yields [53]. A copper(I)-catalyzed three-component reaction with amines, propargyl halides and azides in water affords 1-substituted-1*H*-1,2,3-triazol-4-ylmethyl)dialkylamines [54]. Terminal alkynes reacted with benzyl- or alkyl halides and sodium azide in the presence of a copper(I) catalyst immobilized on 3-aminopropyl- or 3-[(2-aminoethyl)amino]propyl-functionalized silica gel in ethanol to generate exclusively the corresponding regiospecific 1,4-disubstituted-1,2,3-triazoles in good to excellent yields [55]. An efficient and improved procedure for the preparation of aromatic azides from the corresponding aromatic amines **23** is accomplished under mild conditions with *tert*-butyl nitrite and trimethylsilyl azide and their application in the Cu(I)-catalyzed azide–alkyne 1,3-dipolar cycloaddition gives 1,4-disubstituted-1,2,3-triazoles **24** without the need for isolation of the azide intermediates (Scheme 12.11) [56].

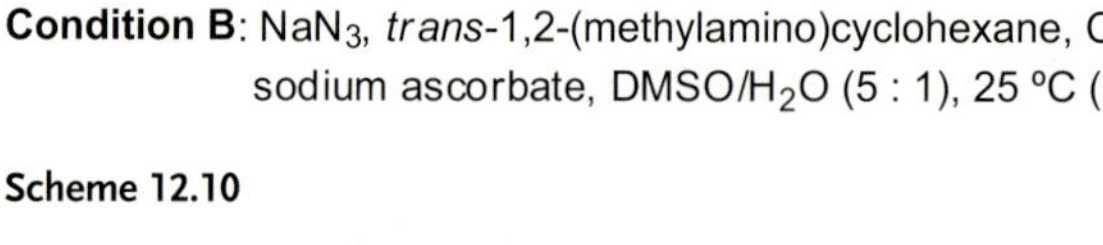

Ar-I + ≡—R —Conditions→ **22** (1-Ar-4-R-1,2,3-triazole)

R = alkyl, TMS, aryl, CH_2OAr, CH_2NEt_2

Condition A: NaN_3, L-Proline, $CuSO_4{\cdot}5H_2O$, sodium ascorbate, Na_2CO_3, DMSO/H_2O (9 : 1), 60 °C (66-98%)

Condition B: NaN_3, *trans*-1,2-(methylamino)cyclohexane, CuI, sodium ascorbate, DMSO/H_2O (5 : 1), 25 °C (38-99%)

Scheme 12.10

Many recent reports have shown that polymer-supported azides and alkynes can be employed in the synthesis of 1,2,3-triazole derivatives. Functionalized 1,2,3-triazoles **26** and **27** were prepared by [3 + 2] cycloaddition of resin-bound α-azido esters **25** with terminal alkynes (Scheme 12.12) [57]. Polystyrene resin-bound azide **28** reacted

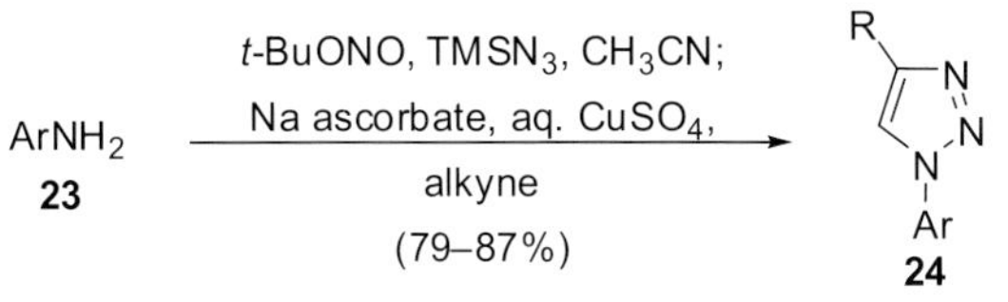

Scheme 12.11

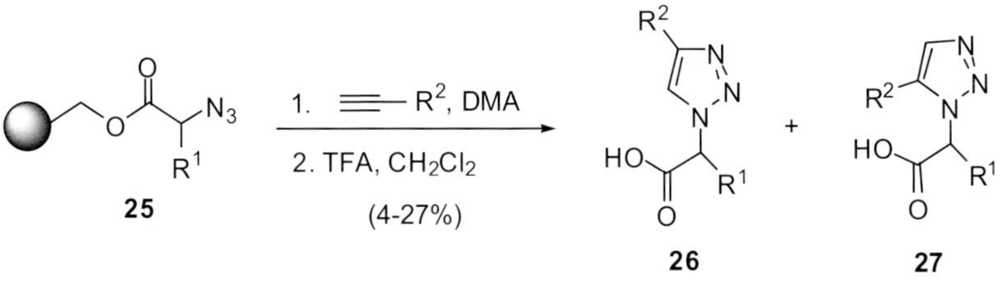

Scheme 12.12

with disubstituted alkynes followed by acidic cleavage to give 1,2,3-triazoles **29** (Scheme 12.13) [58]. Regiospecific copper(I)-catalyzed 1,3-dipolar cycloadditions of resin-bound alkynes **30** to azides afforded solid-supported 1,2,3-triazoles **31**, which were ligated further to give 1,4-substituted-1,2,3-triazole-peptide compounds (Scheme 12.14) [59]. Immobilized REM resin azides have provided a regioselective method for the preparation of 1,5-trisubstituted-1*H*-1,2,3-triazoles via a 1,3-dipolar cycloaddition of trimethylsilyl-propynoic acid [60]. A library of peptidotriazoles have been prepared by solid-phase peptide synthesis combined with a regiospecific copper (I)-catalyzed 1,3-dipolar cycloaddition between resin-bound alkynes and protected amino azides [61].

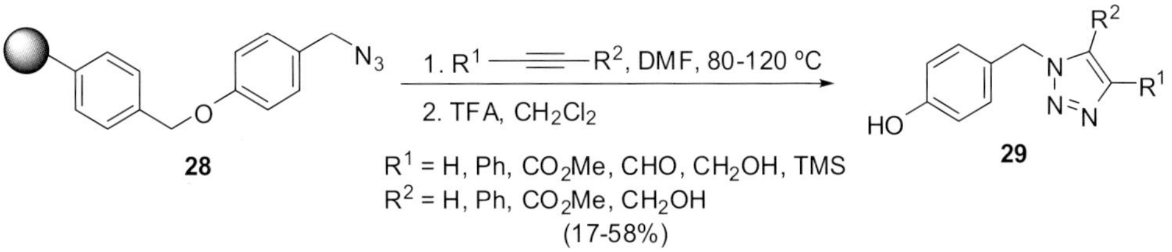

Scheme 12.13

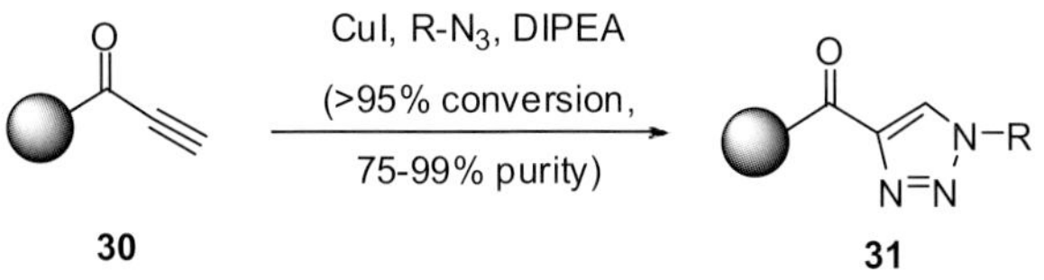

Scheme 12.14

There are published reports on the use of polymer-supported azide reagents in the 1,3-dipolar cycloadditions of alkynes. Different alkyl bromides reacted with Merrifield resin supported ammonium azide (**32**) to give various alkyl azides, which were reacted with methyl propiolate to give 1,2,3-triazoles **33** in excellent yields (Scheme 12.15) [62]. The monomethyl ether of poly(ethylene glycol) (PEG)- or MeOPEG-bound azide **34** has been utilized in the 1,3-dipolar cycloadditions with various alkynes to afford regioisomeric mixtures of **35** and **36** (Scheme 12.16) [63]. The 1,2,3-triazoles could be cleaved with formic acid in dioxane in one example ($R = CO_2Me$). 1,3-Dipolar cycloaddition of poly(ethylene glycol)-supported azide with various dipolarophiles followed by acidic cleavage afforded 4- and 5-substituted-1,2,3-triazoles [64].

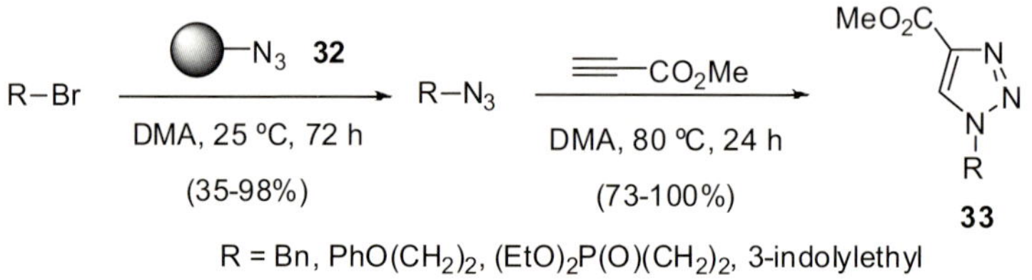

R–Br → (N_3 **32**; DMA, 25 °C, 72 h; (35-98%)) → $R–N_3$ → (≡–CO_2Me; DMA, 80 °C, 24 h; (73-100%)) → **33**

R = Bn, $PhO(CH_2)_2$, $(EtO)_2P(O)(CH_2)_2$, 3-indolylethyl

Scheme 12.15

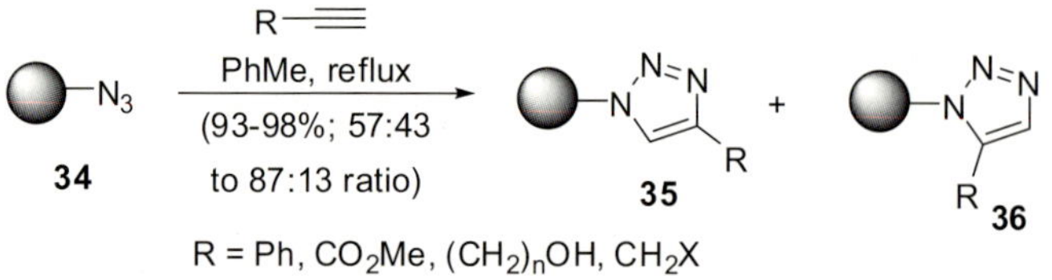

N_3 **34** → (R–≡; PhMe, reflux; (93-98%; 57:43 to 87:13 ratio)) → **35** + **36**

R = Ph, CO_2Me, $(CH_2)_nOH$, CH_2X

Scheme 12.16

12.1.4.2 Reactions of α,β-Unsaturated Systems with Azide Reagents

α,β-Unsaturated systems are good substrates for azide additions to prepare 1,2,3-triazole derivatives. Frequently, the azides undergo addition to unactivated or activated alkenes with electron-withdrawing or electron-rich substituents, such as enamines, enamides, enol ethers, and ketene acetals, to give 4,5-dihydro-1*H*-1,2,3-triazoles, which are unstable and by elimination of a stable fragment functional group aromatizes to 1,2,3-triazoles. Generally, the addition of azides to alkenes is regioselective and only one isomer is obtained.

Several reports have been published on sodium azide additions to alkenes with strongly electron-withdrawing substituents to give *N*-unsubstituted-1,2,3-triazoles [65]. For example, tetrabutylammonium fluoride-catalyzed [3 + 2] cycloaddition reactions of 2-aryl-1-cyano(or carbethoxy)-1-nitroethenes **37** with trimethysilyl azide under solvent-free conditions provided 4-aryl-5-cyano(or carbethoxy)-1*H*-1,2,3-triazoles **38** in good to excellent yields under mild conditions (Scheme 12.17) [66a].

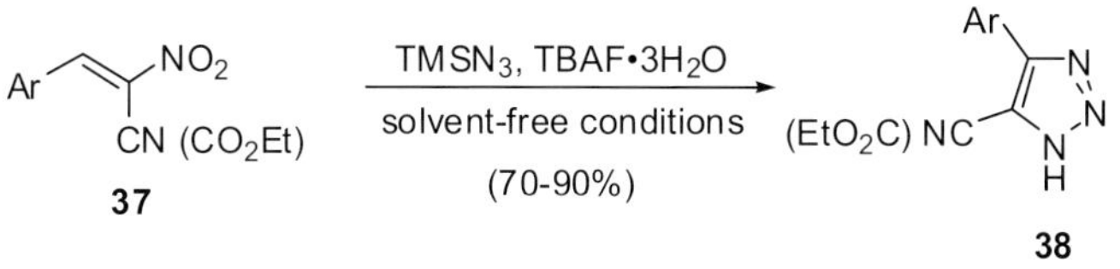

Ar
NO2
CN (CO2Et)
37
TMSN3, TBAF•3H2O
solvent-free conditions
(70-90%)
Ar
(EtO2C) NC
N
N
N
H
38

Scheme 12.17

Similarly, nitroalkenes or vicinal acetoxy nitro derivatives underwent a clean reaction with sodium azide in hot dimethyl sulfoxide to give the corresponding 1,2,3-triazoles in good yield [66b]. There are reports of additions of azide reagents to enol ethers [67], vinyl acetates [68], α-acylphosphorus ylides [69], allenes [70], and alkenes with very strong electron-withdrawing groups like nitro and sulfonyl [71] to produce the corresponding 1,2,3-triazoles.

There are several interesting reports on the additions of various azide additions to enamine-type alkenes to give 1,2,3-triazoles where the amine portion either becomes part of the product or is eliminated as a fragment via the unstable 4,5-dihydro-1*H*-1,2,3-triazole intermediate. For example, aroyl-substituted ketene aminals **39** react with aryl azides to provide polysubstituted 1,2,3-triazoles **40** (Scheme 12.18) [72]. A series of 5-fluoroalkylated 1*H*-1,2,3-triazoles **42** have been prepared in good yield by the regiospecific 1,3-dipolar cycloaddition reaction of (*Z*)-ethyl-3-fluoroalkyl-3-pyrrolidinoacrylates **42** with aryl azides (Scheme 12.19) [73]. Benzyl azides also participated in these reactions but sodium carbonate was required to provide good yields of the triazoles. Condensation of enaminones **43** with mesyl azide (MsN_3) gave 1,4,5-trisubstituted-1,2,3-triazoles **44** (Scheme 12.20) [74]. A solid-phase version of this reaction has also been reported [75].

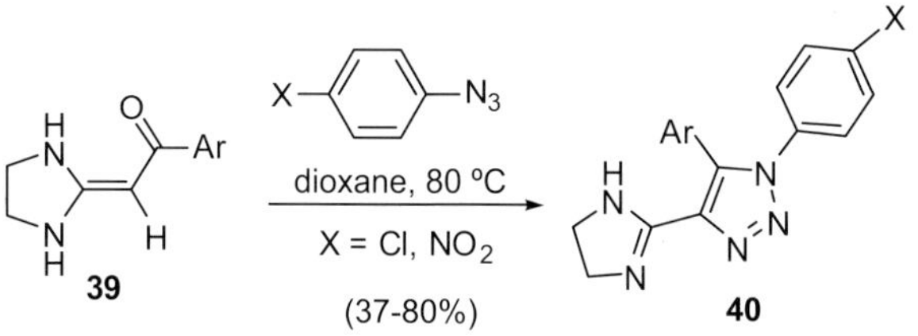

X
N3
dioxane, 80 °C
X = Cl, NO2
(37-80%)
Ar
H
39
40

Scheme 12.18

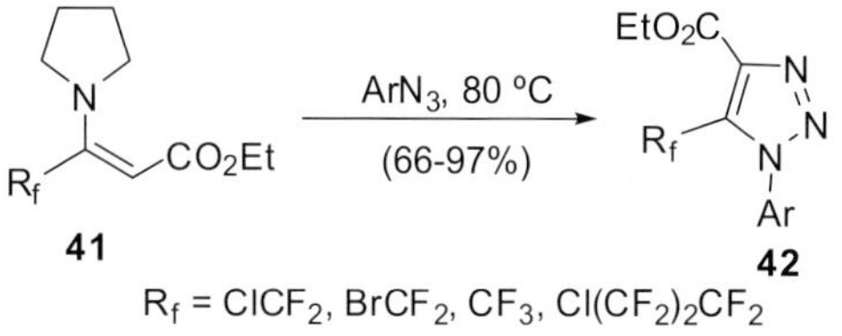

N
Rf
CO2Et
41
ArN3, 80 °C
(66-97%)
EtO2C
Rf
Ar
42
Rf = ClCF2, BrCF2, CF3, Cl(CF2)2CF2

Scheme 12.19

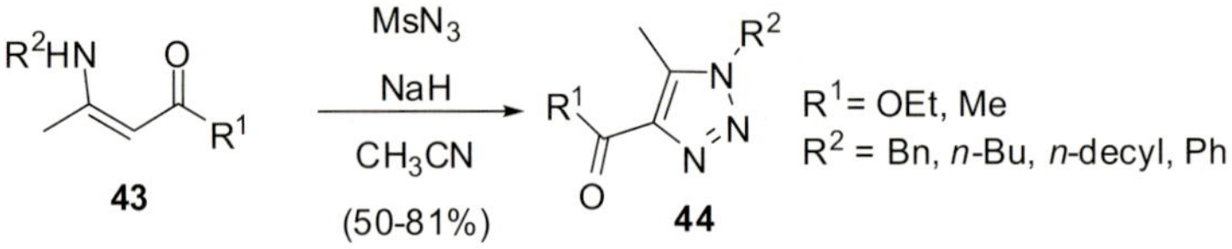

Scheme 12.20

4-Acyl-1*H*-1,2,3-triazoles **46** have been formed from diethylaluminium azide and α,β-unsaturated ketones **45** by [3 + 2] cycloaddition of azide, followed by 1,5-hydride transfer to the β carbon of the triazoline side chain and fragmentation of the tertiary amino group (Scheme 12.21) [76].

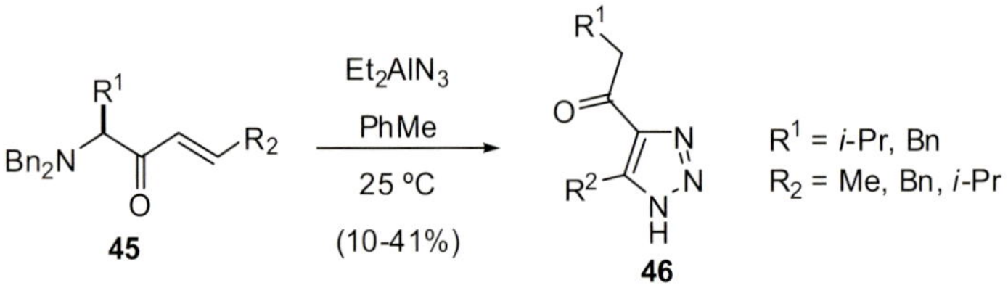

Scheme 12.21

12.1.4.3 Reactions of Azides and Hydrazines with Active Methylene Compounds

Base-catalyzed condensation of azides with active methylene compounds, known as the Dimroth reaction, is a versatile method for the preparation of 1,2,3-triazoles regioselectively. The 5-position of the 1,2,3-triazole can be an alkyl, aryl, hydroxyl, alkoxycarbonyl, or an amino group depending on the functional group of the active methylene compound used. Various azides can react with 1,3-diketones, 3-oxoesters, and 3-oxoamides to give 4-carboxy-1,2,3-triazoles **47** in good yields (Table 12.3) [77–79].

Table 12.3 Reactions of azides with active methylene compounds to give 4-carboxy-1,2,3-triazoles.

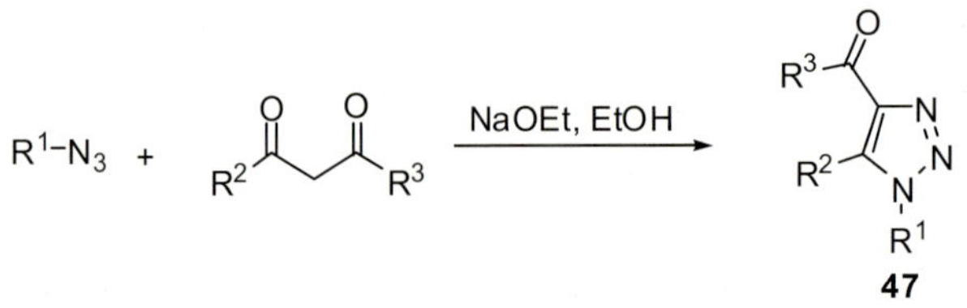

R^1	R^2	R^3	Yield (%)	Reference
$4\text{-}O_2NC_6H_4$	Me	Me	83	[77]
$3,5\text{-}Cl_2C_6H_3CH_2$	Me	OEt	89	[78]
$2\text{-}O_2N\text{-}4\text{-}ClC_6H_3$	Me	NHPh	80	[79]

Table 12.4 Reactions of organic azides with malonic esters and amides to give 1*H*-1,2,3-triazol-5-ols.

$R^1{-}N_3$ + $R^2C(O)CH_2C(O)R^2$ —NaOMe / MeOH→ **48**

R^1	R^2	Yield (%)	Reference
Bn	OEt	88	[80]
4-pyridyl	OEt	75	[81]
Ph	NHPh	72	[82]

Furthermore, the base-catalyzed reaction of organic azides with malonic esters and malonamides gave the best synthesis of 1*H*-1,2,3-triazol-5-ols **48** with an alkyloxy(or aryloxy)carbonyl or carbamoyl functions at C4 (Table 12.4) [80–82]. Similarly, reactions of azides under base-catalyzed conditions with acetonitrile derivatives yielded 1-substituted-1*H*-1,2,3-triazol-5-amines **49** (Table 12.5) [83–85].

There are also reports where activated acyl compounds can react with hydrazines instead of azides to give substituted 1,2,3-triazoles. For example, α-aminoacetophenones **50** react with hydrazines in acetic acid to give an efficient preparation of 2,4-disubstituted-1,2,3-triazoles **51** (Scheme 12.22) [86]. Various phenacyl halides **52** react with excess tosyl hydrazide in refluxing methanol to provide 4-aryl-1-(*p*-toluenesulfonylamido)-1,2-3-triazoles **53** (Scheme 12.23) [87]. A general and efficient

Table 12.5 Reactions of organic azides with acetonitrile derivatives to give 1*H*-1,2,3-triazol-5-amines.

$R^1{-}N_3$ + R^2CH_2CN —Base/Solvent→ **49**

R^1	R^2	Base/solvent	Yield (%)	Reference
$PhSCH_2$	$2\text{-}FC_6H_4$	K_2CO_3/DMSO	29	[83]
Bn	Ph	K_2CO_3/DMSO	84	[84]
Bn	CN	K_2CO_3/DMSO	48	[84]
Bn	$CONH_2$	K_2CO_3/DMSO	84	[84]
$2\text{-}O_2NC_6H_4$	$CONH_2$	NaOEt/EtOH	55	[85]

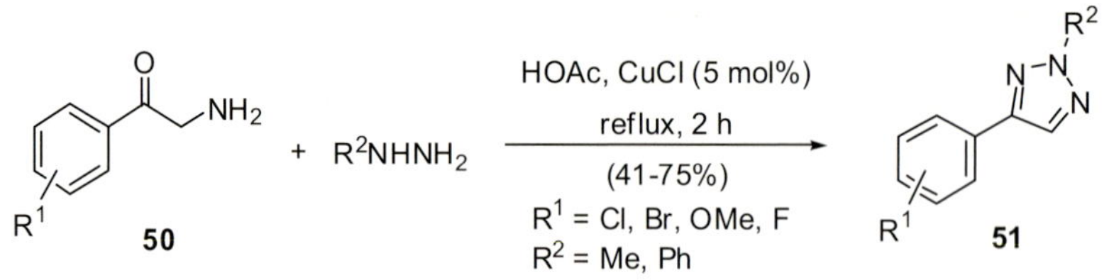
HOAc, CuCl (5 mol%)
reflux, 2 h
(41-75%)
R1 = Cl, Br, OMe, F
R2 = Me, Ph
+ R2NHNH2
50
51

Scheme 12.22

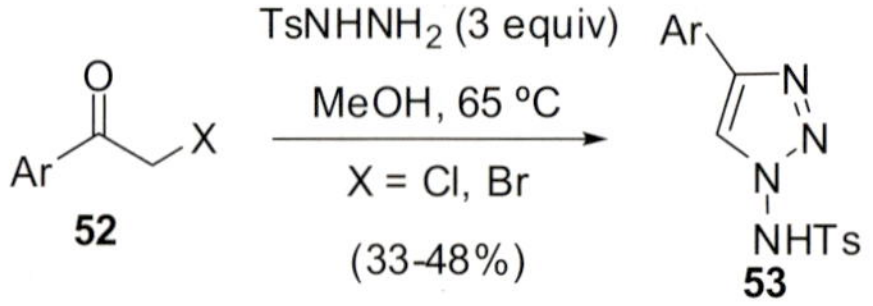
TsNHNH2 (3 equiv)
MeOH, 65 °C
X = Cl, Br
(33-48%)
52
NHTs
53

Scheme 12.23

method for the preparation of 2,4-diaryl-1,2,3-triazoles **55** from α-hydroxyacetophenones **54** and arylhydrazines has been reported (Scheme 12.24) [88].

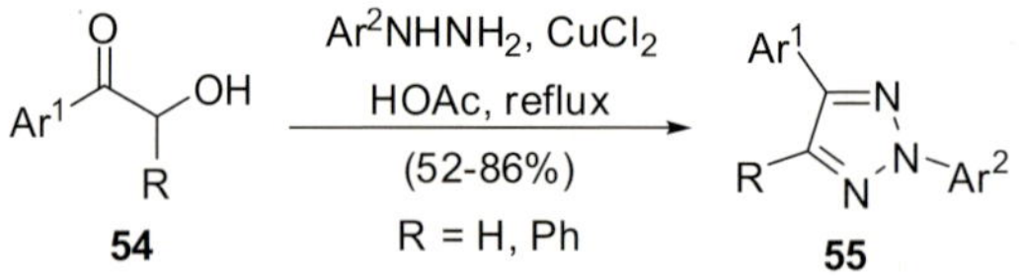
Ar2NHNH2, CuCl2
HOAc, reflux
(52-86%)
R = H, Ph
54
55

Scheme 12.24

12.1.4.4 **Oxidation/Cyclization of Hydrazones**

Oxidation of bis(hydrazones) **56** with manganese dioxide or mercury(II) oxide affords 4-aryl-1*H*-1,2,3-triazol-1-amines **57** as the only regioisomer (Scheme 12.25) [89]. However, other vicinal bis(hydrazones) have generated regioisomeric 1*H*-1,2,3-triazoles [90, 91]. Unsymmetrical vicinal bis(arylsulfonylhydrazones) have been cyclized either with acid or base to give 1-(arylsulfonylamino)-1*H*-1,2,3-triazoles as a mixture of regioisomers [92].

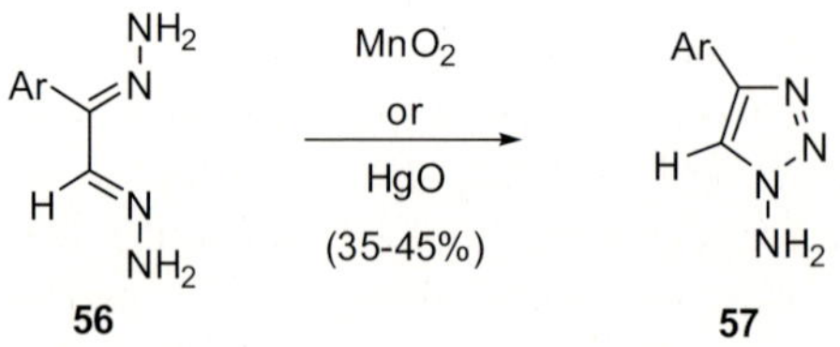
MnO2
or
HgO
(35-45%)
56
57

Scheme 12.25

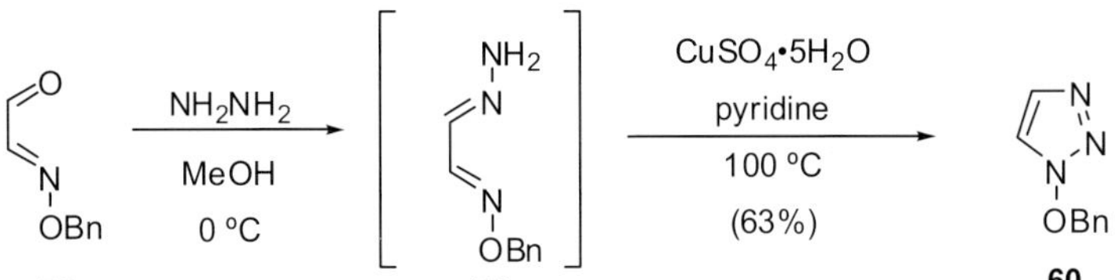

Scheme 12.26

Imine hydrazones can be cyclized in the presence of oxidants to give 1,2,3-triazole derivatives. Glyoxal *O*-benzyloxime hydrazone **59**, which is generated *in situ* from the reaction of glyoxal *O*-benzyloxime **58** with excess hydrazine, afforded 1-(benzyloxy)-1*H*-1,2,3-triazole (**60**) by oxidative cyclization (Scheme 12.26) [93]. Similarly, iodobenzene diacetate-mediated oxidation of hydrazones **61** furnished fused 1,2,3-triazoloheterocycles **62** (Scheme 12.27) [94]. α-Hydroxyimino hydrazones **63** undergo intramolecular cyclization with elimination of water to generate 2*H*-1,2,3-triazole derivatives **64** in the presence of acetic anhydride or phosphorus pentachloride (Scheme 12.28) [95].

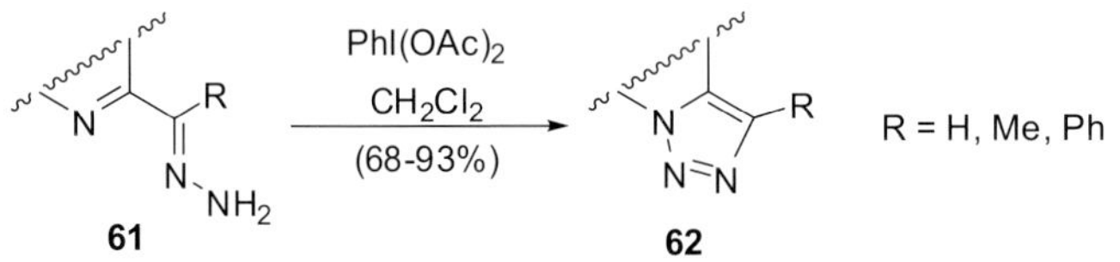

Scheme 12.27

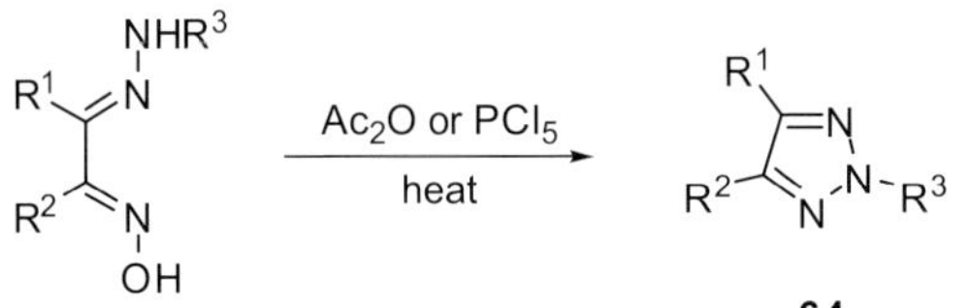

Scheme 12.28

12.1.4.5 **Other Methods for Preparations of 1,2,3-Triazoles**

Several other recent methods for preparation of 1,2,3-triazoles do not fall into the above categories. Treatment of oxazolone **65** with *iso*-pentyl nitrite in the presence of acetic acid gives 1,2,3-triazole **66**, a precursor to β-(*N*-1,2,3-triazolyl)-substituted-α,β-unsaturated-α-amino acid derivatives (Scheme 12.29) [96]. 2-Aryl-2*H*,4*H*-imidazo[4,5-*d*][1,2,3]triazoles **68** have been prepared from the reaction of triethyl *N*-1-ethyl-2-methyl-4-nitro-1*H*-imidazol-5-yl phosphoramidate (**67**) with aryl isocyanates (Scheme 12.30) [97]. *N*-(Uracil-6-yl)-*S*,*S*-diphenylsulfilimine (**69**) reacted with aryldiazonium salts to give arylsulfilimines **70**, which were thermolyzed to the 1,2,3-

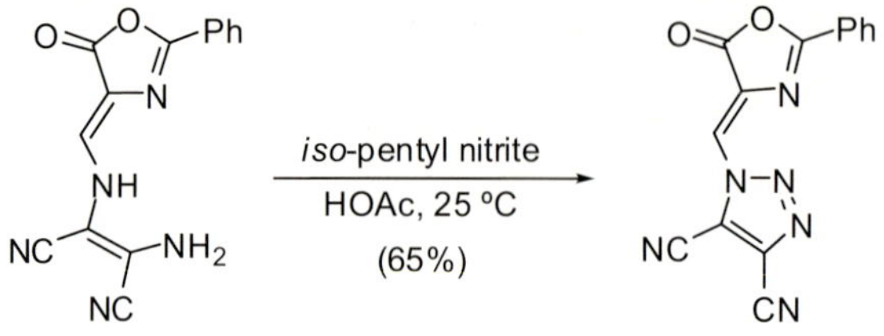

65 66

Scheme 12.29

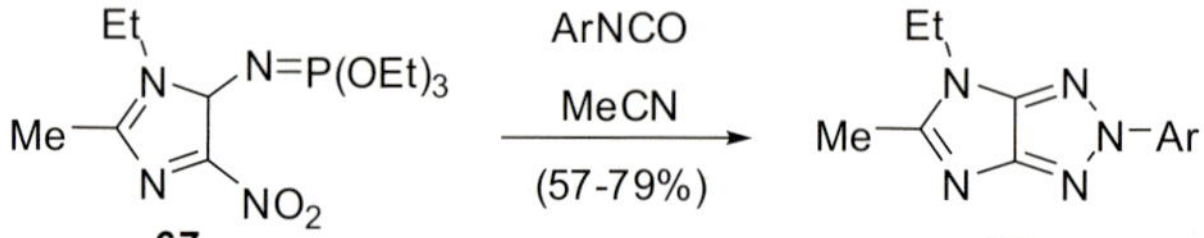

67 68

Scheme 12.30

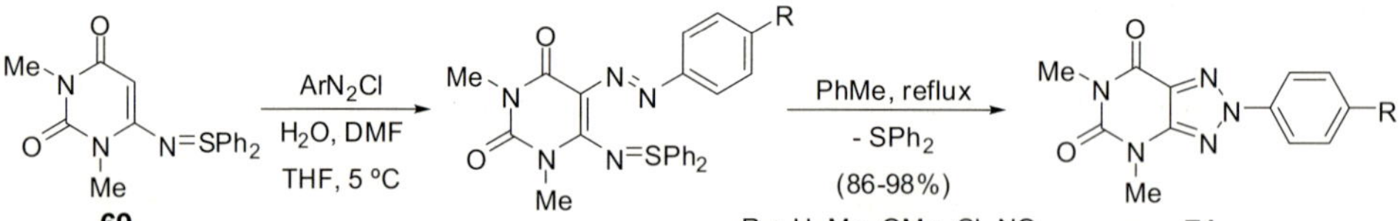

69 70 R = H, Me, OMe, Cl, NO_2 71

Scheme 12.31

triazolopyrimidine diones **71** in good yields (Scheme 12.31) [98]. Polystyrene-sulfonyl hydrazide resins **72** reacted with various amines to give regiospecifically 1,4-disubstituted-1,2,3-triazoles **73** via traceless cleavage reactions (Scheme 12.32) [99]. Palladium-catalyzed synthesis of 1*H*-triazoles **75** from alkenyl bromides **74** and sodium azide in the presence of xantphos ligand has been reported (Scheme 12.33) [100]. A new one-pot procedure has been developed to synthesize 1-aryl- and 1-vinyl-1,2,3-triazoles directly from boronic acids and alkynes, which avoids the need to isolate unstable azide intermediates [101].

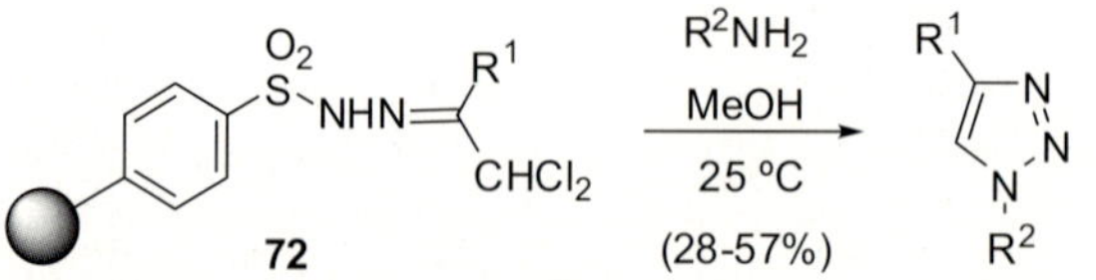

72 73

Scheme 12.32

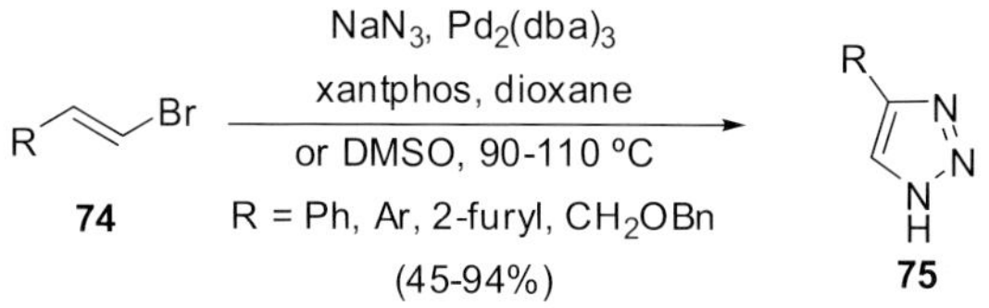

NaN_3, $Pd_2(dba)_3$, xantphos, dioxane or DMSO, 90-110 °C; R = Ph, Ar, 2-furyl, CH_2OBn (45-94%); **74** → **75**

Scheme 12.33

12.1.5 Reactions of 1,2,3-Triazoles

12.1.5.1 Reactions of Carbon of 1,2,3-Triazoles

In general, the 1,2,3-triazole ring system is relatively resistant to both oxidation and reduction conditions but many synthetically useful reactions can still be achieved with this system. For example, lithiation of *N*-protected 1,2,3-triazoles is one of the most general methods for introduction of carbon or hetero substituents onto the C5 position of the ring. 1-Substituted-1,2,3-triazoles **76** react easily with *n*-butyllithium, lithium diisopropylamide (LDA), or with lithium tetramethylpiperidine (LTMP) to generate lithio species **77**, which are quenched with various electrophiles to give 1,5-disubstituted-1,2,3-triazoles **78** (Table 12.6) [93, 102, 103]. Low temperatures must be maintained, otherwise cycloreversion to the alkyne species can result. Alkyl, alcohol,

Table 12.6 Lithiation of 1*H*-1,2,3-triazoles and quenching with electrophiles.

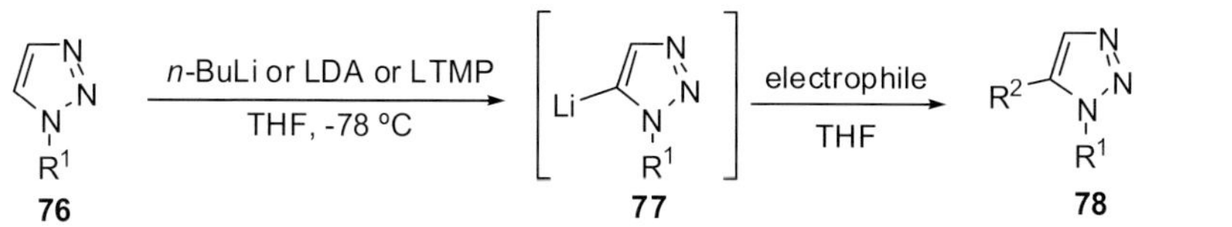

76 → (*n*-BuLi or LDA or LTMP, THF, -78 °C) → [**77**] → (electrophile, THF) → **78**

R^1	R^2	Electrophile	Yield (%)	Reference
SEM	CH(OH)Ph	PhCHO	45	[102]
SEM	Me	MeI	30	[102]
SEM	Cl	Cl_3CCCl_3	50	[102]
SEM	SPh	PhSSPh	80	[102]
OBn	Me	MeI	93	[93]
OBn	CHO	DMF	87	[93]
OBn	Cl	Cl_3CCCl_3	88	[93]
OBn	Br	Br_2	86	[93]
OBn	$SnBu_3$	Bu_3SnCl	91	[93]
OBn	CO_2Me	$ClCO_2Me$	76	[93]
OBn	TMS	TMSCl	93	[93]
Me	Me	MeI	56	[103]
Me	CH(OH)Ph	PhCHO	61	[103]
Me	PhC(O)	$PhCONMe_2$	78	[103]
Me	$PhCOCH_2$	$PhCOCH_2Br$	66	[103]

halogen, sulfur, silicon, tin, aldehyde, and ester products can be formed from these reactions. The 2-(trimethylsilyl)ethoxymethyl (SEM) and benzyloxyl (OBn) N-protecting groups both stabilize the intermediate triazol-5-yllithium species by intramolecular coordination. The SEM group can be deprotected easily with dilute hydrochloric acid or with tetrabutylammonium fluoride to give the 5-substituted-1*H*-1,2,3-triazoles, while the Bn group can be deprotected with catalytic hydrogenation to give the corresponding 5-substituted-1*H*-1,2,3-triazol-1-ols. From Table 12.6, higher yields are generally obtained for the same type of reaction (Cl, Me) when the nitrogen was protected with the benzyloxy group [93]. Lithiation of 1-methyl-1*H*-1,2,3-triazole with *n*-butyllithium or with LTMP followed by addition of electrophiles gives moderate yields of 5-alkyl- or 5-acyl-1-methyl-1,2,3-triazoles [103]. Bromine–lithium exchange can be carried out on 4,5-dibromo-1-(methoxymethyl)-1*H*-1,2,3-triazole with *n*-butyllithium and subsequent quenching with various electrophiles gives high yields of the corresponding 5-substituted-1,2,3-triazoles [104].

1-(Benzyloxy)-1*H*-1,2,3-triazoles **79** have been lithiated followed by transmetalation to the zinc species **80**, which undergoes Negishi cross-coupling with aryl iodides to generate 5-aryl-1,2,3-triazoles **81** (Scheme 12.34) [105].

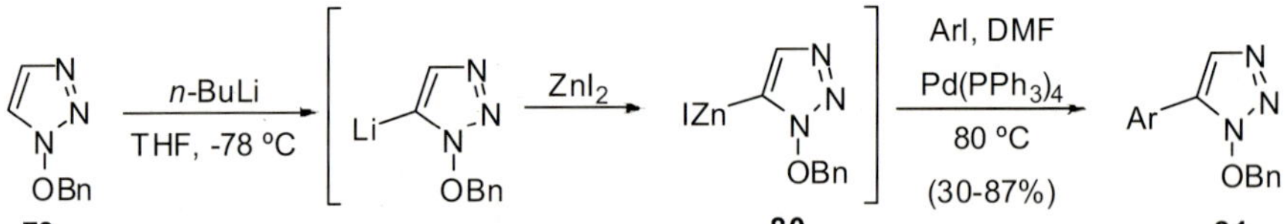

Scheme 12.34

Substitution of halogens by nucleophiles is possible in *N*-substituted-1,2,3-triazoles. Displacement of the chloro group in *N*-substituted-1,2,3-triazoles **82** by nucleophiles, such as cyanide, phenolates, arene, and furylthiolates, gave moderate to good yields of derivatives **83** (Table 12.7) [106–109]. Heating 1-aryl-5-chloro-1*H*-1,2,3-triazoles with hydrazine gave 5-substituted-1*H*-1,2,3-triazol-1,5-diamines, via the 5-hydrazinotriazole intermediates which spontaneously rearrange under the reaction conditions [110]. 5-Bromo-1-methyl-1*H*-1,2,3-triazole reacts with aniline to give the corresponding 5-anilino derivative [111].

12.1.5.2 **Reactions of Nitrogen of 1,2,3-Triazoles**

1*H*-1,2,3-Triazoles can be N-alkylated with alkyl halides in the presence of bases such as sodium alkoxide, sodium hydride, or sodium hydroxide to give mixtures of 1- and 2-alkylated-1*H*-1,2,3-triazoles [112]. Selectivity for alkylation at the 1-position has been achieved in the presence of silver or thallium salts of 1,2,3-triazoles on reaction with alkyl halides [113]. Reaction of 2-(trimethylsilyl)-2*H*-1,2,3-triazoles with primary alkyl halides afforded products of selective alkylation at N1 [103]. 1*H*-1,2,3-Triazoles can be N-acylated with acyl halides and anhydrides to give exclusively 1-acyl-1*H*-1,2,3-triazoles; however, the acyl group can migrate to the 2-position on heating or on treatment with base [114].

Table 12.7 Nucleophilic displacement of 5-chloro-*N*-substituted-1,2,3-triazoles with nucleophiles.

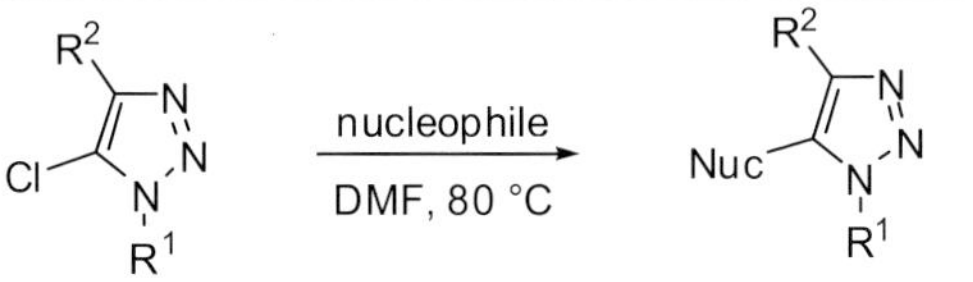

R^1	R^2	Nucleophile	Nuc	Yield (%)	Reference
PMB	CO_2Et	NaCN	CN	66	[106]
PMB	CO_2Et	4-OMeC_6H_4SNa	4-OMeC_6H_4S	74	[106]
Ph	Ph	NaCN	CN	75	[107]
Bn	CO_2Et	NaOPh	OPh	82	[108]
Bn	CO_2Me	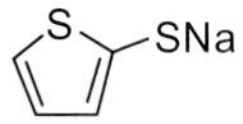 2-thienyl-SNa	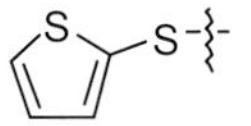 2-thienyl-S–	48	[109]
PMB	CO_2Me	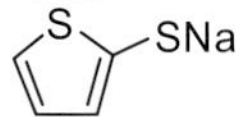 2-thienyl-SNa	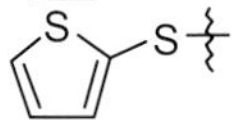 2-thienyl-S–	34	[109]

N-Arylation of 1,2,3-triazoles is possible with activated aryl halides. Activated aryl halides such as 1-fluoro-2-nitrobenzene and 1-fluoro-4-nitrobenzene with 1*H*-1,2,3-triazoles afforded mixtures of the corresponding 1- and 2-nitrophenyl-1,2,3-triazoles [115, 116]. However, reactions with even more activated halides such as 1-fluoro-2,4-dinitrobenzene and 2-chloro(or fluoro)-1,3,5-trinitrobenzene provided only 1-substituted-1*H*-1,2,3-triazoles **84** (Scheme 12.35) [116]. The copper-catalyzed arylation of 1,2,3-triazole with iodobenzene proved to be problematic as competitive *N*-1/*N*-2 arylation products were observed [117]. Efficient post-triazole regioselective *N*-2 arylation to give **86** has been developed from C4, C5 disubstituted-1,2,3-NH-triazoles **85** (Scheme 12.36) [118].

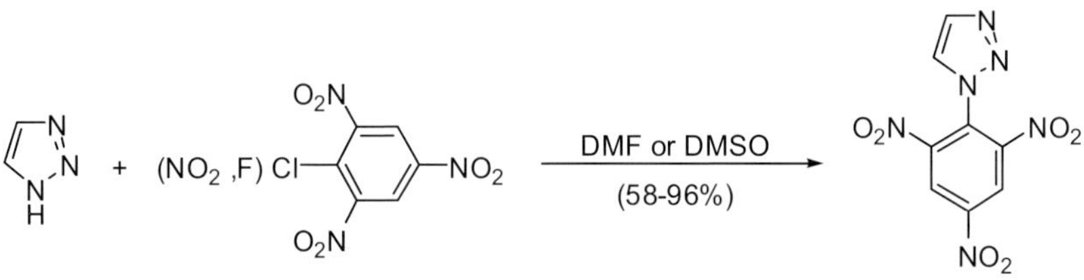

Scheme 12.35

1*H*-1,2,3-Triazoles can also N-substituted with heteroatoms. For example, 4,5-diphenyl-1*H*-1,2,3-triazole has been aminated with hydroxylamine-*O*-sulfonic acid to yield mixtures of *N*-aminotriazoles substituted in the 1- and 2-positions [119]. 1*H*-1,2,3-Triazoles can be oxidized by peracids such as 3-chloroperoxybenzoic acid and hydrogen peroxide to give 1*H*-1,2,3-triazol-1-ols [92, 120], which are also obtained by catalytic hydrogenation of 5-substituted-1-(benzyloxy)-1*H*-1,2,3-triazoles [92, 121].

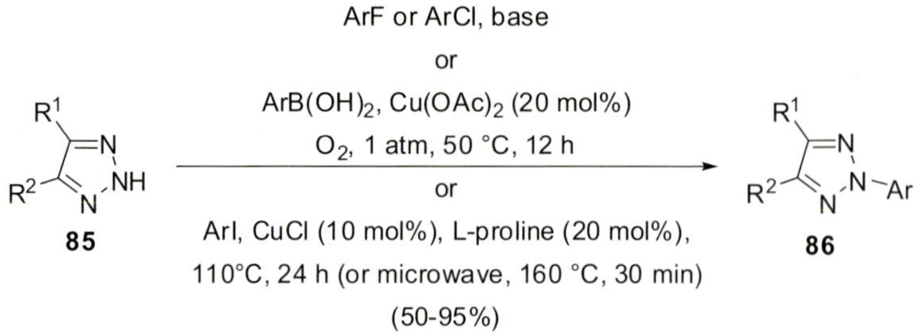

ArF or ArCl, base
or
$ArB(OH)_2$, $Cu(OAc)_2$ (20 mol%)
O_2, 1 atm, 50 °C, 12 h
or
ArI, CuCl (10 mol%), L-proline (20 mol%),
110°C, 24 h (or microwave, 160 °C, 30 min)
(50-95%)

85 → 86

Scheme 12.36

12.1.5.3 **Electrophilic Reactions of 1,2,3-Triazoles**

Halogenations are the most common type of electrophilic reactions of 1,2,3-triazoles. For example, 1*H*-1,2,3-triazole reacts with bromine to afford 4,5-dibromo-1*H*-1,2,3-triazoles (**87**) in almost quantitative yield (Scheme 12.37) [122]. 4-Bromo- and 5-bromo-1*H*-1,2,3-triazoles are obtained indirectly by bromination of a triazole with a protecting group at N1 [123]. In general, most halogens are introduced into the carbons of the ring system by a lithiation/electrophilic sequence (Section 12.1.5.1).

Direct nitration of 1*H*-1,2,3-triazole is not possible. Nitration of 1-phenyl and 4-phenyl-1*H*-1,2,3-triazole was also unsuccessful as nitration occured only on the phenyl ring [124]. However, nitration of 2-methyl-2*H*-1,2,3-triazole (**88**) with a mixture of fuming nitric acid and concentrated sulfuric acid afforded 2-methyl-4-nitro-2*H*-1,2,3-triazole (**89**), which can be nitrated further to **90** under more vigorous conditions (Scheme 12.38) [125].

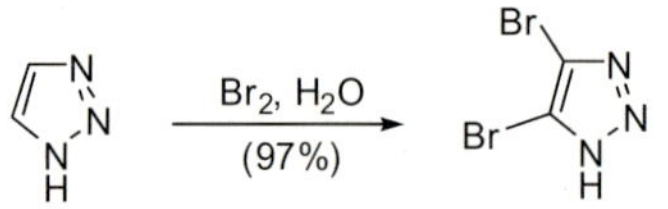

Br_2, H_2O (97%) → 87

Scheme 12.37

88 → HNO_3, H_2SO_4, 25 °C, 3 h (98%) → 89 → HNO_3, H_2SO_4, 100 °C, 10 h (97%) → 90

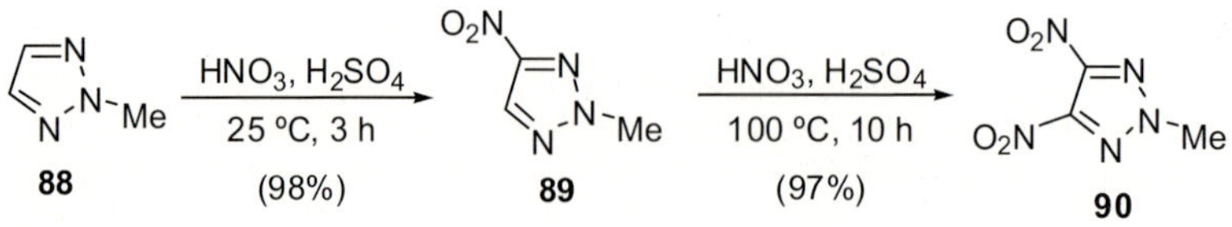

Scheme 12.38

12.2 Benzotriazole

12.2.1 Introduction

12.2.2 General Reactivity

12.2.2.1 Relevant Physicochemical Data and NMR Data

The parent benzotriazole is a weak base with a pK_a of 8.2, which is a stronger NH acid than indazole, benzimidazole, or 1,2,3-triazole (Figure 12.3). The 1-substituted 1*H*-benzotriazole is remarkably stable to strong acid and bases and to oxidative and reductive conditions. The two tautomeric forms of benzotriazole (Figure 12.4) are in equilibrium, but 1*H*-benzotriazole is the predominant species (99.9%) in both the gas and solution phases [2].

The ^1H and ^{13}C NMR spectra of the parent benzotriazole for the protons and carbons at the 4-/7- and 5-/6-positions are identical because benzotriazole undergoes rapid proton exchange between the tautomeric forms at room temperature (Table 12.8). Other ^1H and ^{13}C NMR data of the methyl group attached to the benzotriazole in the 1- and 2- positions are listed for comparison.

12.2.3 Synthesis of Benzotriazoles

12.2.3.1 Synthesis by Ring-Closure Reactions

Diazotization of benzene-1,2-diamine derivatives is the most common synthetic route to 1-substituted benzotriazoles. Diazotization is mostly commonly performed with nitrous acid, generated *in situ* from sodium nitrite and a mineral acid source such as nitric, sulfuric, or acetic acids. A range of various 1-substituted benzotriazoles **92** can be prepared from the corresponding benzene-1,2-diamine derivatives **91** (Table 12.9).

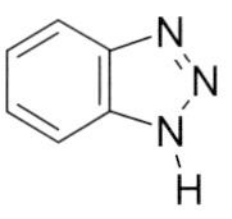

N N N H

acid pKa 8.2 for proton loss
very weak Bronsted base (pKa for proton addition)
Lewis base of appreciable strength
non-volatile, crystalline, odorless, nontoxic
almost insoluble in water, soluble in sodium carbonate solution

Chemical Stability of Benzotriazole Ring System

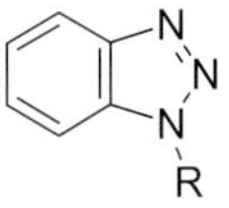

N N N R

Stable: thermally to 400 °C
to hot strong sulfuric acid
to fused potassium hydroxide
to oxidation
to reduction

Figure 12.3 Physical and chemical properties of the parent and 1-substituted 1*H*-benzotriazole.

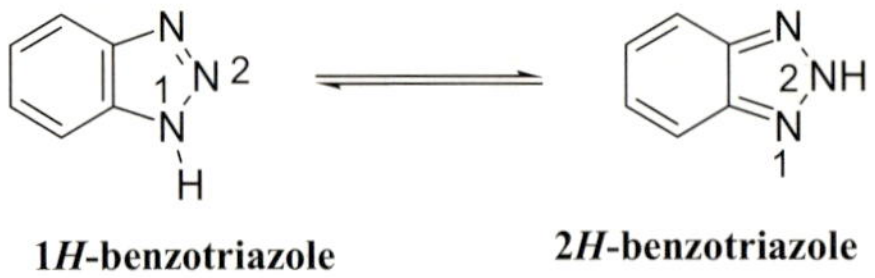

1*H*-benzotriazole 2*H*-benzotriazole

Figure 12.4 Tautomerisim in benzotriazoles.

Table 12.8 ^{1}H and ^{13}C NMR data (ppm) of benzotriazoles.

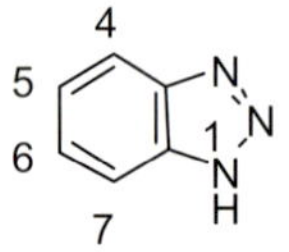

Substituents	^{1}H NMR (DMSO-d_6)					^{13}C NMR (DMSO-d_6)				Reference
	H4	H5	H6	H7	Reference	C4	C5	C6	C7	
—	8.00	7.44	7.44	8.00	[126]	130.3	130.3	130.3	130.3	[128]
1-Me	8.02	7.47	7.47	7.36	[127]	119.3	123.4	126.8	108.8	[129]
2-Me	7.85	7.34	7.34	7.85	[127]	117.5	125.9	125.9	117.5	[129]

1-Substituted benzotriazoles can be prepared from various azides with a benzyne intermediate generated *in situ* from 2-aminobenzoic acid (Table 12.10). 1-Chloro-2-nitrobenzenes **93** react with hydrazine to yield benzotriazol-1-ols **95** via the 2-nitrophenylhydrazines **94** (Table 12.11). The polymer-supported synthesis versions of benzotriazol-1-ols has also been published [141, 142]. These benzotriazol-1-ols can be readily deoxygenated to their NH derivatives by reductive cleavage with either phosphorus trichloride or samarium(II) iodide [142]. Various substituted benzotria-

Table 12.9 Diazotization of benzene-1,2-diamine derivatives to give 1-substituted benzotriazoles.

91 → $NaNO_2$, HCl or H_2SO_4 or HOAc, 0 °C → **92**

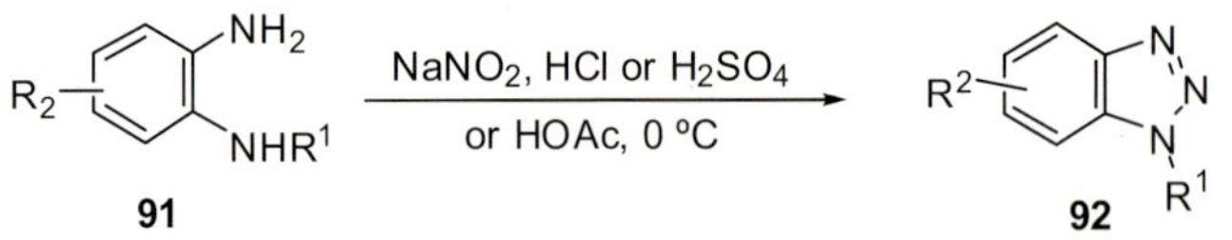

R^1	R^2	Yield (%)	Reference
H	H	81	[130]
Me	5,6-$(NO_2)_2$	83	[131]
Ac	5-6-Me_2	63	[132]
CO_2Et	7-Cl-4-OEt-5-CO_2Me	68	[133]
SO_2Ph	5-Me	100	[134]
Benzotriazol-2-yl	H	63	[135]

Table 12.10 Synthesis of 1-substituted benzotriazoles from azides and a benzyne intermediate.

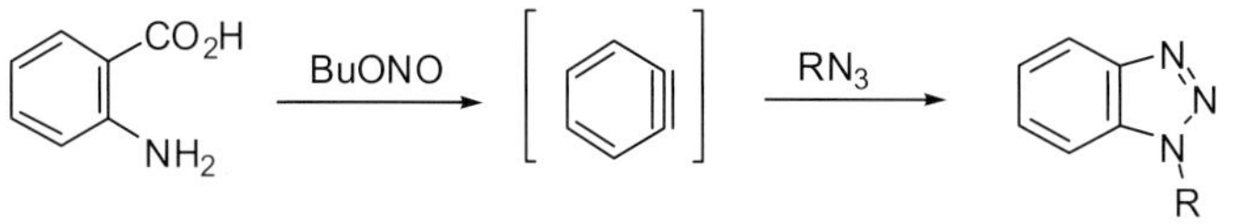

R	Yield (%)	Reference
Ph	52	[136]
4-$O_2NC_6H_4$	62	[137]
Bz	63	[137]
SO_2Ph	52	[137]
4-$OMeC_6H_4CO$	60	[137]
1-Naphthalenyl	75	[138]

Table 12.11 Benzotriazol-1-ols from 1-chloro-2-nitrobenzenes and hydrazines.

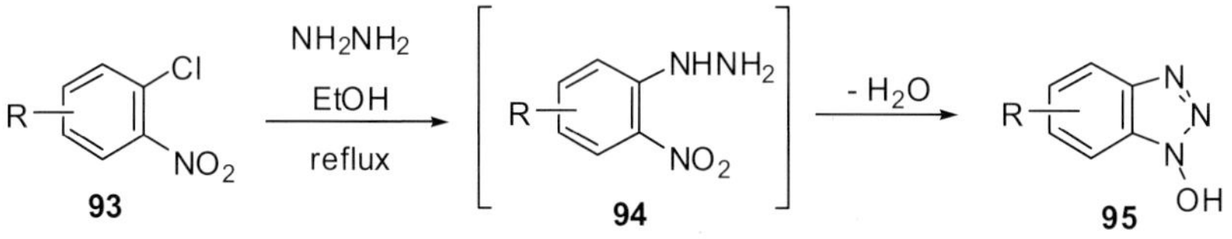

R	Yield (%)	Reference
H	90	[139]
4,5,6-Cl_3	67	[139]
6-CF_3	90	[140]
6-OMe	6	[140]
6-$CONH_2$	39	[140]
6-SO_2NHBn	96	[135]

zoles **97** have been prepared by the [3 + 2] cycloaddition of azides to benzynes generated from aryl triflates **96** and cesium fluoride (Scheme 12.39) [143].

2-Substituted benzotriazoles can be prepared by several methods. For example, 2-aminoazobenzenes **98** can be converted into their 2-aryl-2*H*-benzotriazoles **99**

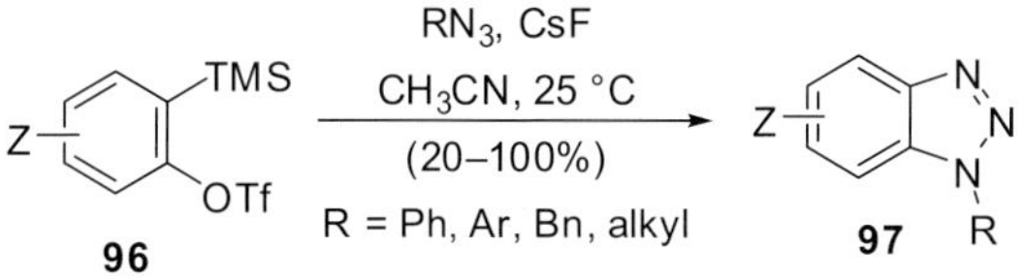

Scheme 12.39

by oxidation with copper(II) sulfate in refluxing pyridine [144], copper(II) acetate in air [145], or by refluxing the azo compound in thionyl chloride (Scheme 12.40) [145a]. 2-Aryl-2*H*-benzotriazoles **101** can be prepared directly by reaction of 1-halo(or nitro)-2-nitrobenzenes **100** with an excess of an arylhydrazine (Scheme 12.41) [146].

$CuSO_4$, pyridine, reflux or $Cu(OAc)_2$, air, DMF, reflux or $SOCl_2$, PhH, reflux; **98** → **99**

Scheme 12.40

$ArNHNH_2$, heat; **100** → **101**

X = Cl, Br, NO_2

Scheme 12.41

12.2.4 Reactions of Benzotriazoles

The growing applications of benzotriazole methodology as a versatile synthetic tool have been reviewed extensively [147]. Practically all the chemistry occurs on the N1-position of the benzotriazole. The benzotriazole group conveys multiple activating influences such as a leaving group, proton activator, ambident anion directing group, cation stabilizer, radical precursor, and anion precursor. The benzotriazole group can also easily be eliminated by radical-type reactions, by hydrolysis, by palladium-catalyzed S_N2' substitution, and by reductive metal reductive reactions. This section will not describe the chemistry needed to give the benzotriazole derivatives; however, the intermediates of these substitution reactions will be used to explain the myriad of useful synthetic reactions.

12.2.4.1 Acylation of 1-Benzotriazoles and Benzotriazole Methodology

The classical preparation of *N*-acylbenzotriazoles uses the corresponding acid chlorides (Table 12.12) [148]. More recently, two methods have been developed for the preparation of *N*-acylbenzotriazoles directly from carboxylic acid without the necessity of isolating the acid chlorides. Carboxylic acids are converted into the mixed carboxylic sulfonic anhydride, which is then attack by the benzotriazole anion with methanesulfonylbenzotriazole as the reagent [149]. Treatment of carboxylic acids

Table 12.12 Synthesis of *N*-acylbenzotriazoles.

Acyl Reagent + Benzotriazole Reagent $\xrightarrow{\text{Base}}$ (*N*-acylbenzotriazole, Bt–C(O)R)

Acyl reagent	Benzotriazole reagent	Reference
RCOCl	BtH	[148]
RCO_2H	$BtSO_2Me$	[149]
$RCO_2H/SOCl_2$	BtH (4 equiv)	[150]
CO/iodonium salts	BtH	[151]

with thionyl chloride in the presence of excess benzotriazole provided *N*-acylbenzotriazoles in high yields [150]. *N*-Acylbenzotriazoles can be synthesized by palladium-catalyzed carbonylation of benzotriazole and hypervalent iodonium salts [151].

N-Acylbenzotriazoles are useful intermediates in several synthetically valuable reactions (Table 12.13) [152]. *N*-Acylbenzotriazoles can react with ammonia and primary and secondary amines to give high yields of their respective amides [149]. *O*-Alkyl, *N*-alkyl, and *O*,*N*-dialkylhydroxamic acids have been synthesized from *N*-acylbenzotriazoles [153]. C-Acylation of *N*-acylbenzotriazoles with furan, thiophene, pyrrole, and indole under Friedel–Crafts conditions gave products in high yields [154]. β-Diketones have been prepared from monoketones and *N*-acylbenzotriazoles in the

Table 12.13 Benzotriazole-mediated methodology of *N*-acylbenzotriazoles.

(*N*-acylbenzotriazole, Bt–C(O)R)

Reactants	Products	Reference
Amines (ammonia, primary, secondary)	Amides (primary, secondary, tertiary)	[149]
R^1NHOR^2HCl	*O*-Alkyl, *N*-alkyl, *O*,*N*-dialkylhydroxamic acids	[153]
Five-membered heterocycles	C2-acylated heterocycles	[154]
Cyclic and acyclic ketones	β-Diketones	[155]
Nitriles (primary, secondary)	α-Substituted β-ketonitriles	[156]
Sulfones	β-Ketosulfones	[157]
Acetoacetic esters	β-Ketoesters/β-diketones	[158]
Grignards/heteroaryllithiums	Ketones	[159]
Sodium azide	Acyl azides	[160]
Indoles	Aroylindoles	[161]

presence of base [155]. α-Substituted β-ketonitriles have been synthesized from *N*-acylbenzotriazoles with primary and secondary alkyl nitriles [156]. C-Acylation of sulfones with *N*-benzotriazoles affords β-keto sulfones [157]. β-Ketoesters and β-diketones have been prepared by an acylative-deacylative sequence [158]. Stable and easily accessible *N*-acylbenzotriazoles, derived from various aliphatic, unsaturated, (hetero)aromatic, and *N*-protected-*R*-amino carboxylic acids, have been reacted with Grignard and heteroaryllithium reagents to afford the corresponding ketones [159]. A general synthesis of acyl azides from the corresponding *N*-acyl benzotriazoles have been described [160]. Stable and easily accessible *N*-aroylbenzotriazoles react with indoles in the presence of a base to afford the corresponding *N*-aroylindoles [161].

12.2.4.2 **Benzotriazole-Mediated Imidoylation**

N-(Imidoyl)benzotriazoles have found synthetic applications in the syntheses of various substituted guanidines. For example, benzotriazole-1-carboxamidinium tosylate (**102**) was found to be an efficient reagent for the synthesis of mono- and disubstituted guanidines **103** in moderate to good yields and offers advantages over previous procedures (Table 12.14) [162]. Introduction of Boc groups on both nitrogens of the amidine moiety and nitro or chloro group on the benzotriazole enhances the ability of the benzotriazole moiety as a leaving group [163]. Bis(benzotriazolyl) carboximidamide (**104**) has been developed as a new guanylating agent for the synthesis of tri- and tetrasubstituted guandines **105** [163, 164]. Benzotriazolyl carboximidoyl chlorides (**106**) are stable, colorless, and conveniently handled reagents for the synthesis of unsymmetrical guanidines **107** [165]. Polysubstituted acylguanidines and guanylureas **109** [166] have been prepared from *N*-acyl-*N,N*-disubstituted benzotriazolyl carboximidates **108**.

Table 12.14 Synthesis of substituted guanidines with various benzotriazole imidates.

Benzotriazole imidate	Reagents	Products	Reference
$BtC(=\overset{+}{N}H_2)NH_2\ ^{-}OTs$ **102**	R^1R^2NH	$H_2NC(=NH)NR^1R^2$ **103**	[162]
$BtC(=NH)Bt$ **104**	R^1R^2NH, R^3R^4NH	$R^2R^1NC(=NH)NR^3R^4$ **105**	[163, 164]
$BtC(=NR^1)Cl$ **106**	R^2R^3NH, R^4R^5NH	$R^2R^3NC(=NR^1)NR^4R^5$ **107**	[165]
$R^1C(=O)N{=}C(Bt)NR^2R^3$ **108** R^1 = aryl, alkyl, NHAr	R^4R^5NH	$R^1C(=O)N{=}C(NR^4R^5)NR^2R^3$ **109**	[166]

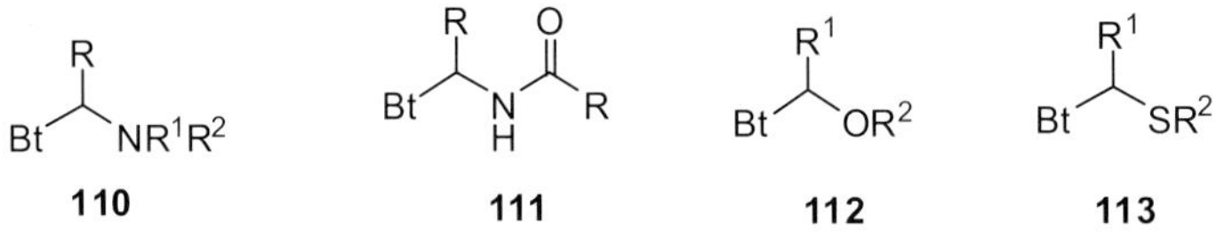

Figure 12.5 Structures of amino- (**110**), amido- (**111**), alkoxy- (**112**), and alkylthio- (**113**) methylbenzotriazoles [167].

12.2.4.3 Benzotriazole-Mediated Amino-, Amido-, Alkoxy-, and Alkylthio-Alkylations

A very detailed recent review describes extensively the synthesis and the broad utility of aminomethylbenzotriazoles **110**, amidomethylbenzotriazoles **111**, alkoxymethylbenzotriazoles **112**, and alkylthiomethylbenzotriazoles **113** (Figure 12.5) and so this chemistry will not be presented here [167].

12.2.5 Benzotriazole-Containing Reagents

Substituents located in the N1 position of 1,2,3-benzotriazoles have become useful reagents in various reactions (Table 12.15). 1-Hydroxybenzotriazole (**114**) is a useful co-reagent in peptide coupling reactions in the activation of carboxylic acids [168]. Aminium-based **115** [169] and phosphonium-based **116** [170] benzotriazoles are currently utilized as peptide coupling reagents. 1-Aminobenzotriazole (**117**) is a useful reagent in the generation of benzyne intermediate, which can be trapped with various dienes [171]. 1-Cyanobenzotriazole (**118**) has been found to participate in electrophilic cyanations of sp^2 and sp carbanions [172]. 1*H*-Benzotriazole-1-yl methanesulfonate (**119**) has been explored as a regioselective *N*-mesylating reagent [173]. Reagent **119** mesylated molecules containing both primary and secondary amines on the primary amino position and mesylation occurred on the amino group in molecules containing both amino and hydroxy groups. 1*H*-Benzotriazole-1-yl alkyl carbonates (**120**) are convenient and inexpensive coupling agents in the preparation of active esters for the synthesis of amides [174]. A general and efficient route to thionoesters via thionoacyl nitrobenzotriazoles **121** has been reported [175]. The Vilsmeier-type reagent **122** with β-enaminonitriles provides a regioselective route to the preparation of nicotinonitriles [176] and was employed in the direct and efficient synthesis of dimethylformamidrazones from hydrazines [177]. 1-(Chloromethyl)benzotriazole reacted with sodium dialkyl phosphites to give dialkyl-(1-benzotriazolmethyl)phosphonates **123**, which are potential Horner–Emmons reagents [178] for the stereoselective preparation of (*E*)-1-(1-alkenyl)benzotriazoles [179]. 2-Benzotriazolyl-1,3-dioxolane (**124**) has been utilized as a novel formyl cation equivalent [180]. The novel three-carbon synthon 1-(1*H*-1,2,3-benzotriazol-1-yl)-3-chloroacetone (**125**) has been used for the synthesis of benzothiazoles, pyrido[1,2-*a*]indoles, styryl-substituted indolizines, and imidazo[1,2-*a*]pyridines [181]. Various functionalized *N*-allylamines and *N*-allylsulfonamides have been synthesized by Pd(II)-catalyzed intermolecular amination of the corresponding *N*-allylbenzotriazoles **126** [182]. *S*-(1*H*-1,2,3-Benzotriazol-1-ylmethyl)-*O*-ethylcarbo-

Table 12.15 Synthetic utility of 1-substituted benzotriazole reagents.

R	Number	Synthetic utility	Reference
OH	**114**	Co-reagent in peptide coupling	[168]
$^{\oplus}CHNR_2\,X^{\ominus}$	**115**	Peptide coupling agent	[169]
$\overset{\oplus}{O}P(NHR_2)_3\,\overset{\ominus}{X}$	**116**	Peptide coupling agent	[170]
NH_2	**117**	Benzyne intermediate	[171]
CN	**118**	Electrophilic cyanations	[172]
OMs	**119**	N-Mesylating reagent	[173]
OCO_2R	**120**	Synthesis of active esters	[174]
C(S)R*	**121**	Synthesis of thioesters	[175]
$CH{=}\overset{\oplus}{N}Me_2\overset{\ominus}{Cl}$	**122**	Vilsmeier-type reagent	[176, 177]
$CH_2P(O)(OR)_2$	**123**	Horner–Emmons reagent	[178, 179]
1,3-dioxolan-2-yl	**124**	Formyl cation equivalent	[180]
$CH_2C(O)CH_2Cl$	**125**	Three-carbon synthon	[181]
$CHRCH{=}CH_2$	**126**	*N*-Allylating reagent	[182]
$CH_2SC(S)OEt$	**127**	Benzotriazolylmethyl radical	[183]
CH_2TMS	**128**	One-carbon synthon	[184]
SO_2R	**129**	Sulfonylating reagents	[185]
C(S)X X = R, OR, SR, HetNH	**130**	Thioacylating reagents	[186]
$PhC(OCH_3)(CF_3)C(O))$	**131**	Mosher-Bt reagent	[187]
CH_2OH	**132**	Formaldehyde generation	[188]
$CONH_2$	**133**	Carbamoyl chloride reagent	[189]

nodithioate (**127**) has been used to generate the benzotriazolylmethyl radical, which was trapped by various olefins [183]. 1-(Trimethylsilylmethyl)benzotriazole (**128**) has been utilized as a one-carbon synthon in the conversion of alkyl and aryl carboxylic acids into their corresponding homologated acids or esters [184]. *N*-Alkane-, *N*-arene-, and *N*-heteroenesulfonylbenzenetriazoles **129** have been exploited as efficient sulfonylating agents [185]. Several benzotriazole sulfur reagents **130** have been prepared and used for thioacylations, thiocarbamoylations, alkyl/alkoxythioacylations, and aryl/alkylthioacylations [186]. Benzotriazole of 3,3,3-trifluoro-2-methoxy-2-phenylpropionic acid (**131**) reacted with water-soluble amino acids and peptides in an acetonitrile/water (2 : 1) mixture to give the corresponding Mosher derivative in quantitative yield [187]. Anionic *in situ* generation of formaldehyde from benzotriazolylmethanol **132** has proved to be a very useful and versatile tool in synthesis [188]. Carbamoyl-1*H*-benzotriazole **133**, an effective carbamoyl chloride substitute, and a range of its analogs have been synthesized in good yields in two very simple steps from 1,2-diaminobenzene [189].

12.3 1,2,4-Triazoles

12.3.1 Introduction

The chemistry of 1,2,4-triazoles is extensively reviewed elsewhere [190]. A comprehensive review on the chemistry of mercapto- and thione- substituted 1,2,4-triazoles and their utility in heterocyclic synthesis has been published [191]. This section will focus on methods for the synthesis of the monocyclic 1,2,4-triazole system, including solid-phase and microwave-assisted reactions.

12.3.2 General Reactivity

12.3.2.1 Relevant Physicochemical Data and NMR Data

The parent 1,2,4-triazole consists of a five-membered aromatic ring containing three nitrogen atoms, two of which are adjacent; it is a stable, water-soluble solid. Two tautomeric forms, 1*H*-tautomer and 4*H*-tautomer, can be envisaged (Figure 12.6). Theoretical and analytical methods show that the 1*H*-tautomer is the preferred structure. Every carbon atom in 1,2,4-triazole is linked to two nitrogen atoms and, thus, this ring system is electron deficient. The ring is deactivated towards electrophilic attack so nitration and other reactions at carbon typical of aromatic chemistry do not apply to the parent compound. However, electrophilic attack at nitrogen is found in abundance in the literature and this will be discussed later. The parent compound has a pK_a of 10.26 and so alkali metal salts form readily at the N1 position. The pK_a of the protonated species is 2.19 and the weakly basic nature allows electrophilic attack at the N4 position in 1-substituted-1,2,4-triazoles.

The ^{1}H NMR spectrum of the parent 1,2,4-triazole in HMPT is temperature dependent; the H3 and H5 protons show one broad singlet at slightly above or below room temperature due to the rapid proton exchange between the tautomeric forms

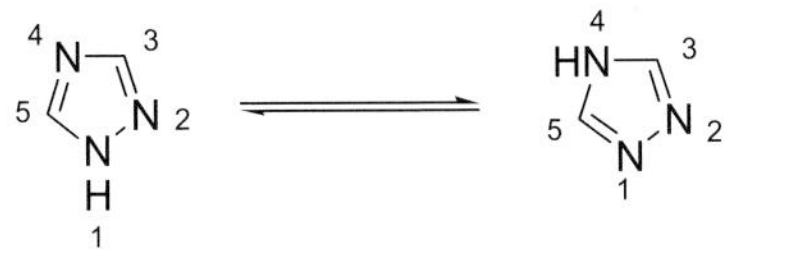

1*H*-tautomer
stable, water-soluble, colorless crystals
mp 120-121 °C
pK_a = 10.26
pK_a = 2.19 of the protonated species

4*H*-tautomer

Figure 12.6 Tautomerism of 1,2,4-triazoles.

Table 12.16 ^{1}H and ^{13}C NMR data (ppm) of 1,2,4-triazole.

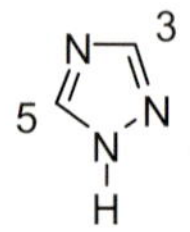

	^{1}H NMR (HMPT)		^{13}C NMR (CD_3OD-d_4)	
Temperature (°C)	H3	H5	C3	C5
37	8.03	8.03	147.4	147.4
10	8.17	8.17		
−34	7.92	8.85		

(Table 12.16) [192]. However, at −34 °C, the H3 protons and H5 protons are observed separately. The ^{13}C NMR spectrum shows a single peak for the parent triazole.

12.3.3
Relevant Natural and/or Useful Compounds

1,2,4-Triazoles have several applications in analytical chemistry, in industrial, and in molecular recognition processes [190d,190f]. The 1,2,4-triazole ring is also a component of a wide range of biologically active pharmaceutical products. For example, rizatriptan benzoate, marketed as Maxalt™, was launched in 1998 by Merck & Co. as an antimigraine medication (Figure 12.7) [193]. Voriconazole is sold as Vfend™ by Pfizer for treatment of fungal infections [194]. Aprepitant is sold as Emend™ for the treatment of chemotherapy-induced nausea and vomiting [195].

12.3.4
Synthesis of 1,2,4-Triazoles

12.3.4.1 Reactions of Acylhydrazines with Various Nitrogen-Containing Reagents

One of the most common methods of preparing 1,2,4-triazoles is the reaction of acylhydrazines with various nitrogen-containing reagents. For example, reactions of

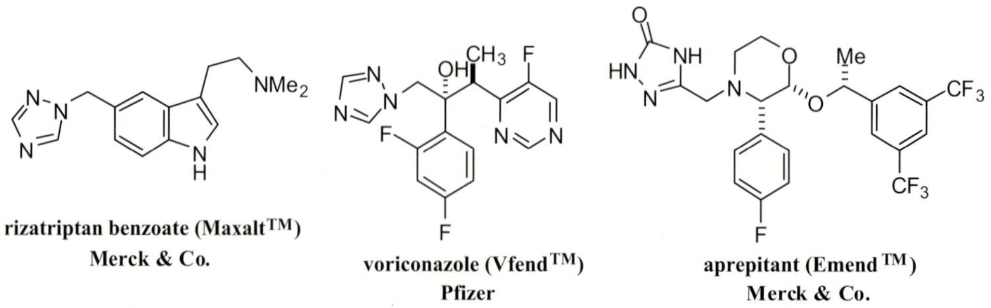

rizatriptan benzoate (Maxalt™)
Merck & Co.

voriconazole (Vfend™)
Pfizer

aprepitant (Emend™)
Merck & Co.

Figure 12.7 Some biologically active pharmaceutical products that contain a 1,2,4-triazole ring.

acylhydrazines with isothiocyanates or isocyanates to give 1,2,4-triazoles can be seen in the following examples. Acylhydrazines **134** and isothiocyanates **135** afforded 1,2,4-triazole-3-thiones **136**, which were intercepted by alkyl halides to give substituted 3-thio-1,2,4-triazoles **137** (Scheme 12.42) [196]. Solid-supported acylhydrazines **138** react with isocyanates or isothiocyanates followed by base-induced cyclization/cleavage to provide 1,2,4-trisubstituted urazoles and thiourazoles **139** (Scheme 12.43) [197]. A traceless liquid-phase synthesis of 3-alkylamino-4,5-disubstituted-1,2,4-triazoles on poly(ethylene glycol)-supported thioureas and acylhydrazines has been reported [198].

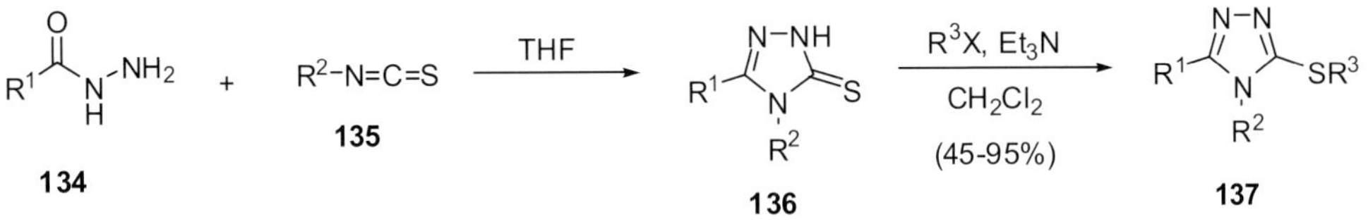

134 + R^2–N=C=S **135** —THF→ **136** —R^3X, Et_3N, CH_2Cl_2 (45-95%)→ **137**

Scheme 12.42

138 —1. R^2NCX, $ClCH_2CH_2Cl$ (X = O,S); 2. Et_3N, PhMe or KO*t*-Bu, THF (18-44%)→ **139** X = O,S

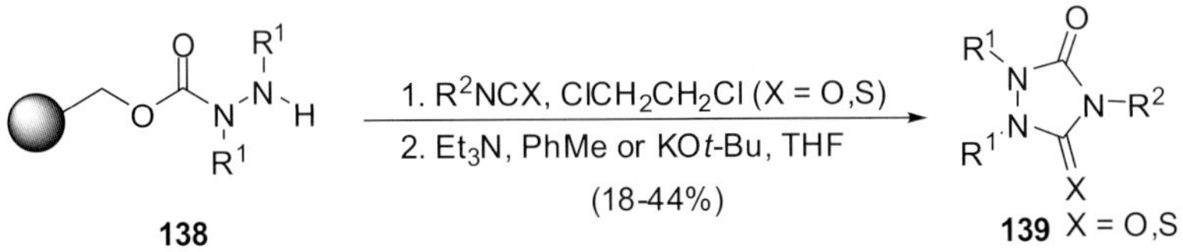

Scheme 12.43

Acylhydrazines can also be used in conjunction with substituted imidates to give 1,2,4-triazole derivatives. The three-component condensation of acylhydrazines in the presence of *S*-methyl isothioamide hydroiodide **140**, silica gel, and ammonium acetate under microwave irradiation afforded 1,2,4-triazoles **141** in good yields (Scheme 12.44) [199]. Acylhydrazines **142** react with imidates **143** to yield 1,2,4-triazoles **144** (Scheme 12.45) [200].

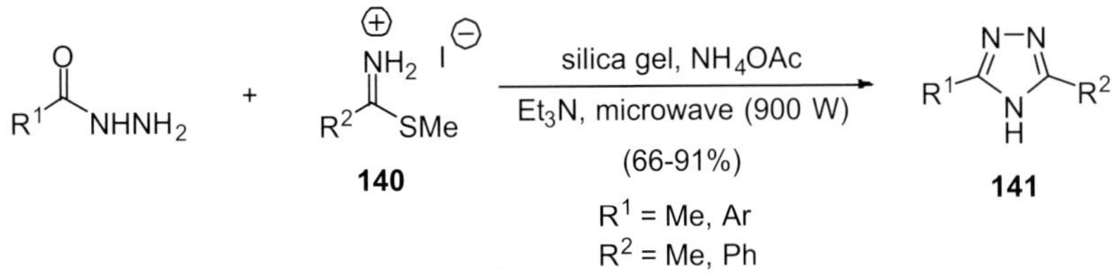

$R^1C(O)NHNH_2$ + **140** —silica gel, NH_4OAc, Et_3N, microwave (900 W) (66-91%); R^1 = Me, Ar; R^2 = Me, Ph→ **141**

Scheme 12.44

Other non-traditional reactions of acylhydrazines with nitrogen reagents for the synthesis of 1,2,4-triazoles are also available. An efficient one-pot, three-component synthesis of substituted 1,2,4-triazoles **146** has been prepared from primary acylhy-

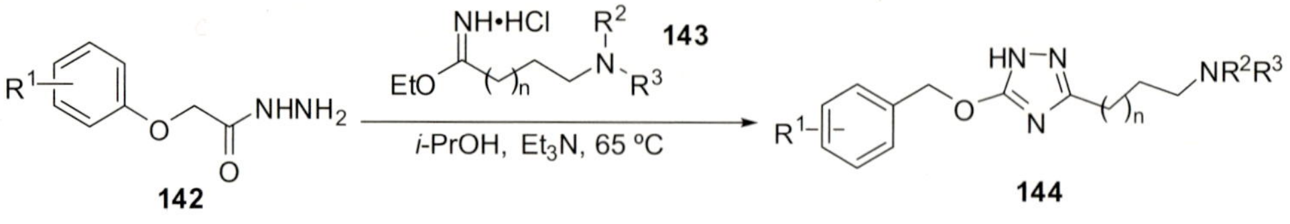

Scheme 12.45

drazines, dimethylamino acetals **145**, and amines (Scheme 12.46) [201]. 1,3-Benzoxazine **147** reacts with acylhydrazines in refluxing methanol to give 1,2,4-triazoles **148** (Scheme 12.47) [202]. Diethoxyphosphinyl acetic acylhydrazine **150** was found to be a unique reagent that provided a convenient and efficient process to prepare fused [5,5]-, [5,6]-, and [5,7]-3-[(*E*)-2-(arylvinyl)]-1,2,4-triazoles **151** from aldehydes and alkoxyimines **149** (Scheme 12.48) [203]. A convenient and efficient one-step, base-catalyzed microwave-assisted synthesis of 3,5-disubstituted-1,2,4-triazoles by condensation of a nitrile and acyl hydrazide has been reported [204].

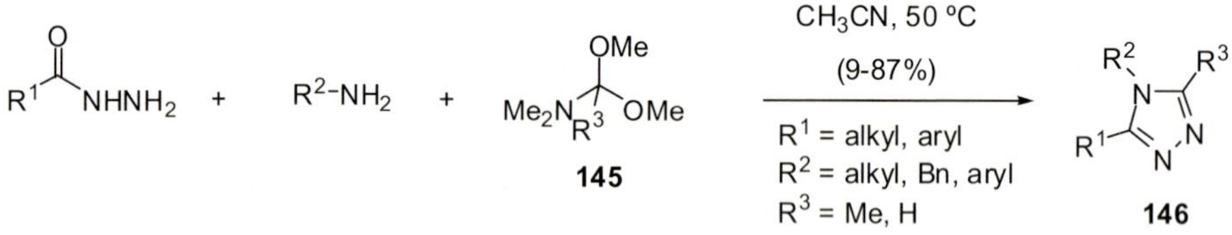

Scheme 12.46

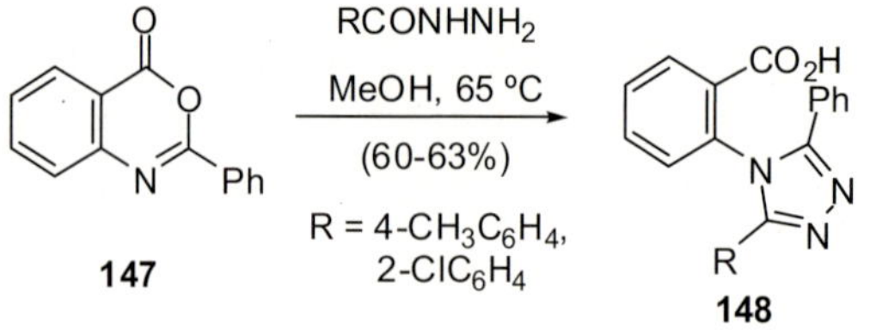

Scheme 12.47

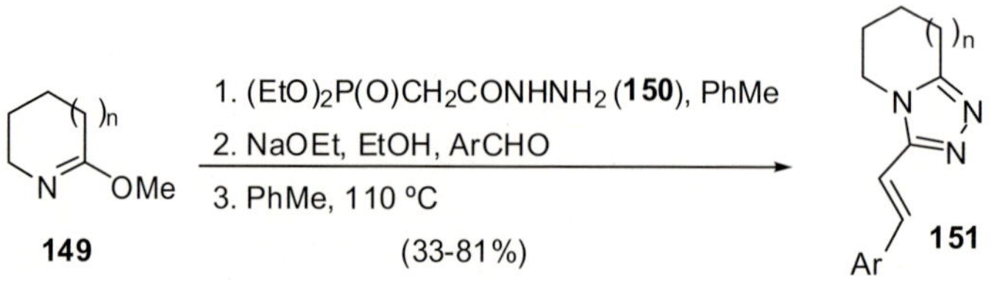

Scheme 12.48

12.3.4.2 **Reactions of Hydrazones**

Substituted hydrazones are a rich source of precursors for the syntheses of 1,2,4-triazoles. For example, hydrazonyl chlorides can be used as partners in reactions with compounds containing a C–N multiple bond that lead to 1,2,4-triazoles. Aryl-substituted hydrazonyl chlorides **152** reacted with cycloalkanone oximes **153** to give 1,2,4-triazolospiro compounds **154** (Scheme 12.49) [205]. Intermolecular cyclization of hydrazonyl chlorides **155** with nitriles catalyzed by ytterbium(III) triflate afforded a series of 1,3,5-trisubstituted-1,2,4-triazoles **156** in good yields (Scheme 12.50) [206]. Dipolar cycloadditions between hydrazonyl chlorides **157** and nitriles in aqueous sodium bicarbonate in the presence of a surfactant provided mild conditions for the synthesis of 1-aryl-5-substituted-1,2,4-triazoles **159** via intermediate **158** (Scheme 12.51) [207]. A series of 1,2,4-triazoles have been prepared by oxidative intramolecular cyclization of heterocyclic hydrazones with copper dichloride [208]. Other C–N multiple partners include aryl cyanides [209], amidines [210], and cyanamides [211].

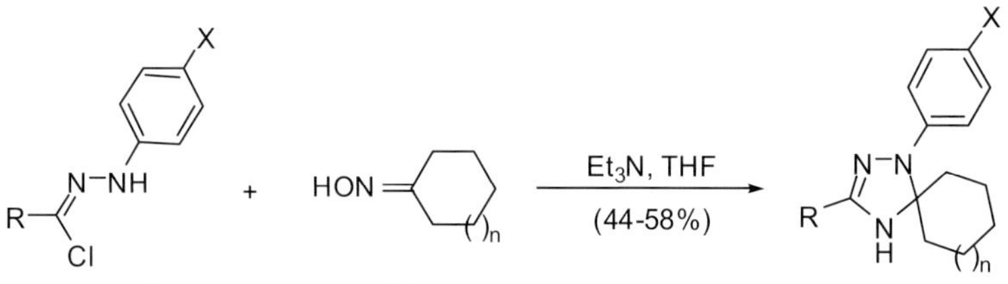

X
N–NH
R
Cl
152
+
HON
153
Et_3N, THF
(44-58%)
N–N
R
N
H
154

Scheme 12.49

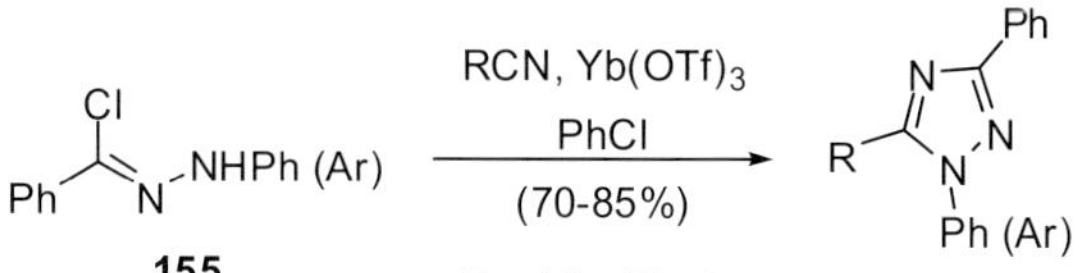

Cl
Ph
N
NHPh (Ar)
155
RCN, $Yb(OTf)_3$
PhCl
(70-85%)
R = Me, Ph, Ar
Ph
N
N
R
N
Ph (Ar)
156

Scheme 12.50

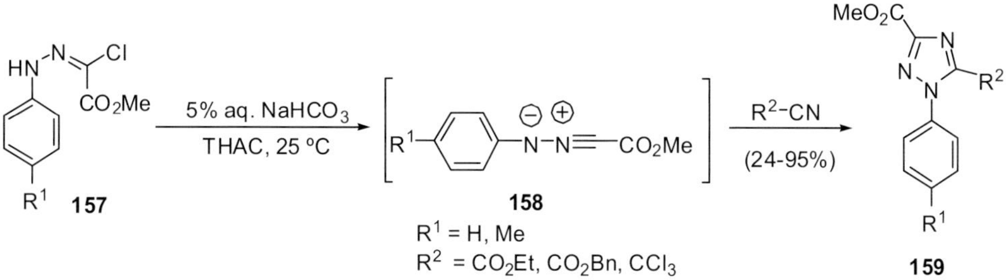

HN
N
Cl
CO_2Me
R^1
157
5% aq. $NaHCO_3$
THAC, 25 °C
R^1–N–N≡–CO_2Me
158
R^1 = H, Me
R^2 = CO_2Et, CO_2Bn, CCl_3
R^2-CN
(24-95%)
MeO_2C
N
N
N
R^2
R^1
159

Scheme 12.51

Aminohydrazones or amidrazones are versatile reagents that can react with various electrophilic carbon compounds to give 1,2,4-triazoles. As an example, *N*-tosylamidrazones **160** can react either with acid chlorides or with ethyl chloroformate to give tosylated 1,2,4-triazoles **161** or 1,2,4-triazole-3-ones **162**, respectively (Scheme 12.52) [212]. Other amidrazone reactions can occur with carboxylic acids [213], cyanogen bromide [214], aldehydes [215], and orthoesters [215, 216]. Amidrazones **163** have been oxidized to 1,3,5-trisubstituted 1,2,4-triazoles **164** in good yields by silver carbonate, Dess–Martin periodinane, sodium and calcium hypochlorites, and tetrapropylammonium perruthenate (TPAP)/*N*-methylmorpholine *N*-oxide (NMO) combination (Scheme 12.53) [217–219].

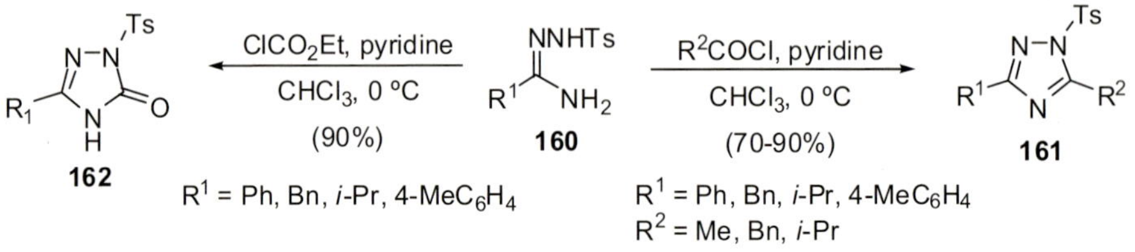
ClCO2Et, pyridine
CHCl3, 0 °C
(90%)
162
160
R2COCl, pyridine
CHCl3, 0 °C
(70-90%)
161
R1 = Ph, Bn, i-Pr, 4-MeC6H4
R1 = Ph, Bn, i-Pr, 4-MeC6H4
R2 = Me, Bn, i-Pr

Scheme 12.52

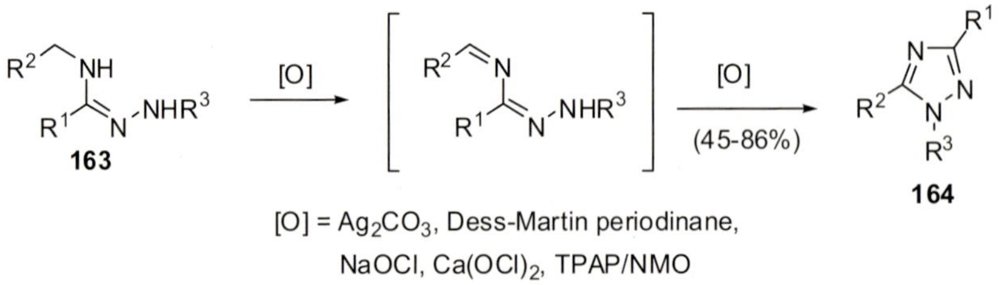
[O]
163
[O]
(45-86%)
164
[O] = Ag2CO3, Dess-Martin periodinane,
NaOCl, Ca(OCl)2, TPAP/NMO

Scheme 12.53

Other hydrazone intermediates have been utilized in the synthesis of 1,2,4-triazoles. Addition of primary amines to α-nitrohydrazones **165** followed by addition of sodium nitrite affords 1,3,5-trisubstituted-1,2,4-triazole **166** (Scheme 12.54) [220]. 1,3,-Dipolar cyclocondensation of *C*-acetyl-*N*-arylnitrilimines **167** with benzoylhydrazones **168** furnished 1,2,4-triazoles **169** (Scheme 12.55) [221]. Iodobenzene

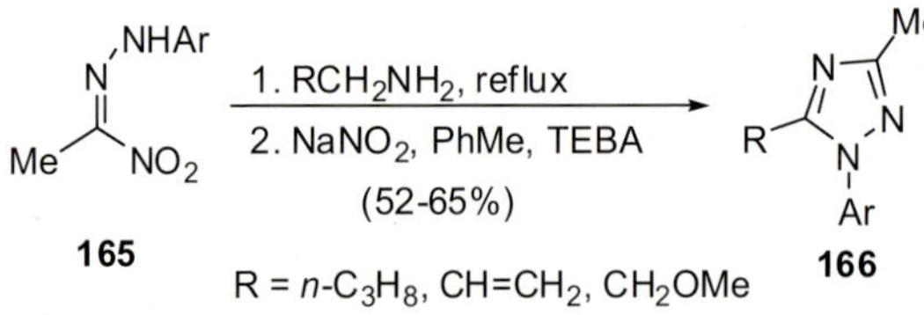
1. RCH2NH2, reflux
2. NaNO2, PhMe, TEBA
(52-65%)
165
166
R = n-C3H8, CH=CH2, CH2OMe

Scheme 12.54

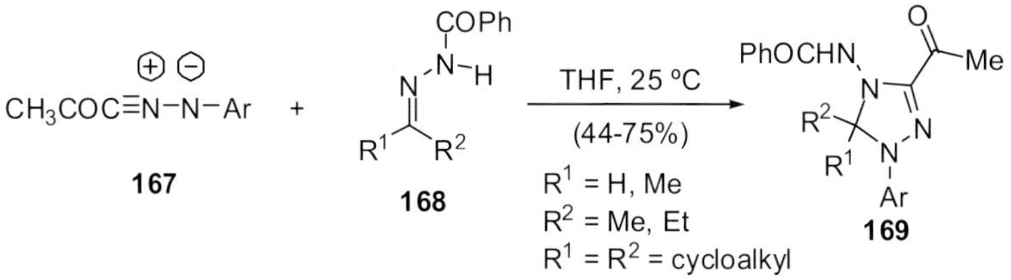

Scheme 12.55

diacetate or lead tetraacetate cyclization of hydrazones **170** [222] or **172** [223] afforded fused 1,2,4-triazoles **171** and **173**, respectively (Scheme 12.56).

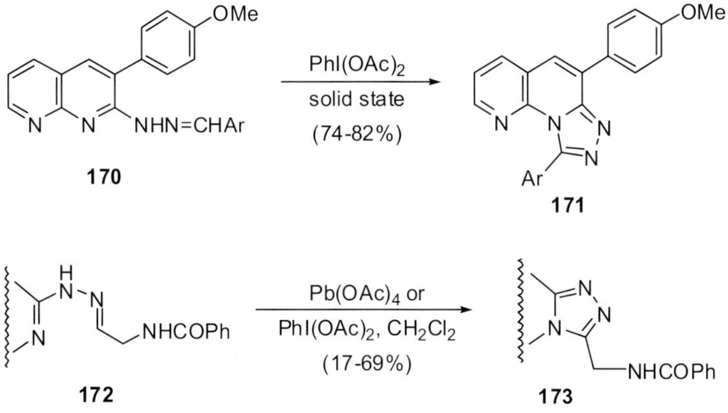

Scheme 12.56

12.3.4.3 **Reactions of Oxadiazoles or Thiadiazoles**

1,2,4-Triazoles can be synthesized from oxadiazoles. Photolysis of 1,2,4-oxadiazoles in the presence of nucleophiles led to 1,2,4-triazole products [224] and 1,3,4-oxadiazoles can undergo ring-cleavage with nitrogen nucleophiles followed by recyclization of the intermediates to give 1,2,4-triazoles. For example, the unusual hydrazinolysis of 5-perfluoroalkyl-1,2,4-oxadiazoles **174** provided an expedient route to 5-perfluoroalkyl-1,2,4-triazoles **175** (Scheme 12.57) [225]. Similarly, 3,5-bis(trifluoromethyl)-1,3,4-oxadiazole is particularly activated towards nucleophilic

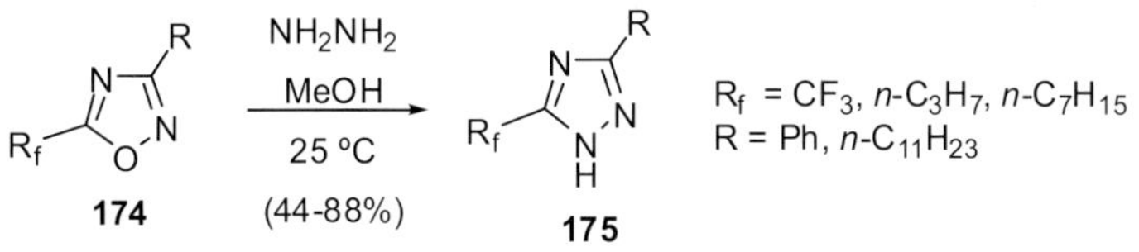

Scheme 12.57

attack by primary amines to yield 4-substituted-1,2,4-triazoles [226]. Microwave-assisted rate acceleration of reactions between 2-aminothiadiazoles **177** with oxadiazoles **176** on alumina support affords thiadiazolyl-substituted-1,2,4-triazoles **178** (Scheme 12.58) [227]. Photochemistry of some fluorinated oxadiazoles gives rise to mixtures of fluorinated 1,3,4-oxadiazoles and 1,2,4-triazoles [228]. 2-Amino-1,3,4-oxadiazoles **179** reacted with alcohols followed by subsequent ring-cleavage and ring-cyclization to give 3-alkoxy-1,2,4-triazoles **180** (Scheme 12.59) [229], while amines and hydrazines react with 2-amino-1,3,4-oxadiazoles to afford 3-amino- and 3,5-diamino-1,2,4-triazoles, respectively [230]. Condensation of highly reactive chloromethyloxadiazoles with ethylenediamines provides a concise synthesis of [1,2,4]triazolo[4,3-*a*]piperazines [231]. Reaction of some fluorinated 1,2,4-oxadiazoles in the presence of methylamine or propylamine under photochemical irradiation in methanol or acetonitrile led to the corresponding fluorinated 1-methyl- or 1-propyl-1,2,4-triazoles [232].

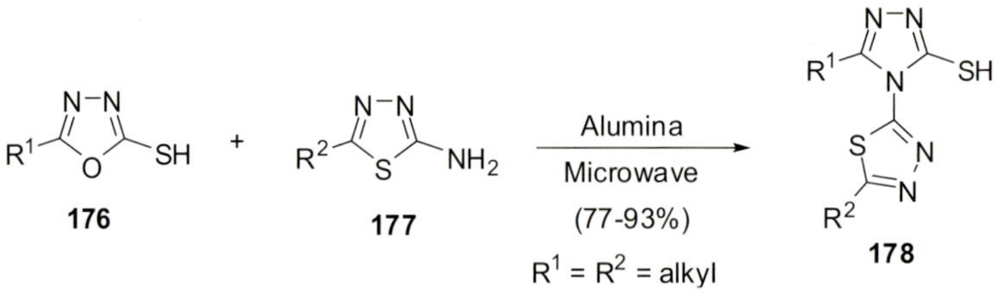

176 + **177** → Alumina, Microwave (77-93%), $R^1 = R^2$ = alkyl → **178**

Scheme 12.58

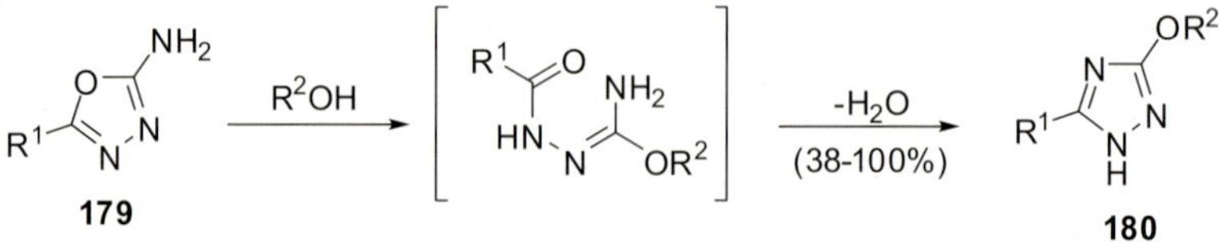

179 → R^2OH → [intermediate] → $-H_2O$ (38-100%) → **180**

Scheme 12.59

12.3.4.4 Synthesis of 1,2,4-Triazoles from Thioureas, Thiocyanates, and Thioamides

Unsaturated and saturated thio compounds have been employed in the syntheses of 1,2,4-triazoles. Δ^2-1,2,4-Triazolin-5-ones **182** have been prepared from 1-aryl/alkyl-6-phenyl-2-thiobioureas **181** in the presence of benzyl chloride and aqueous ethanol (Scheme 12.60) [233]. A novel one-pot synthesis of 1,2,4-triazole-3,5-diamine derivatives **184** and **185** from isothiocyanates **183** and monosubstituted hydrazines has

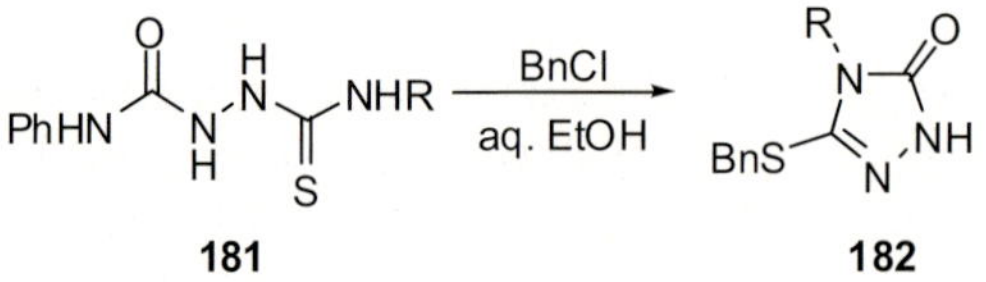

181 → BnCl, aq. EtOH → **182**

Scheme 12.60

been reported; derivatives **185** were obtained with higher regioselectivity when aromatic and sterically bulky hydrazines were used (Scheme 12.61) [234]. *S*-Ethyl thioamides **186** react with acyl hydrazides **187** in refluxing *n*-butanol to give 3,4,5-trisubstituted 4*H*-1,2,4-triazoles **188** (Scheme 12.62) [235]. An efficient synthesis of substituted 1,2,4-triazoles involved condensation of benzoyl hydrazides with thioamides under microwave irradiation [236]. 3-*N*,*N*-Dialkylamino-1,2,4-triazoles **191** have been synthesized from *S*-methylisothioureas **189** and acyl hydrazides **190** in moderate to good yields (Scheme 12.63) [237].

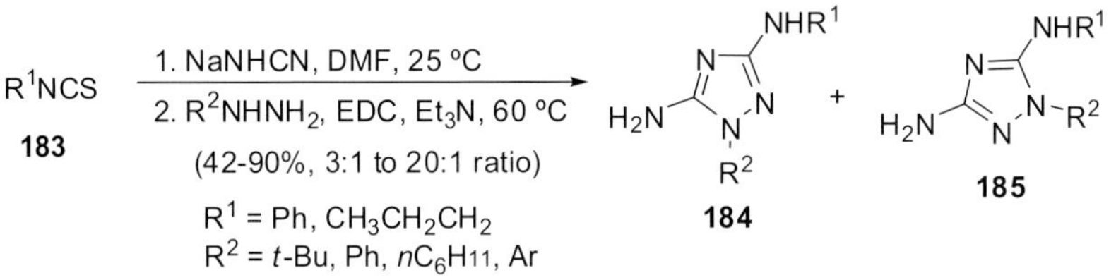

R^1NCS **183** → 1. NaNHCN, DMF, 25 °C; 2. R^2NHNH_2, EDC, Et_3N, 60 °C (42-90%, 3:1 to 20:1 ratio) → **184** + **185**

R^1 = Ph, $CH_3CH_2CH_2$
R^2 = *t*-Bu, Ph, nC_6H_{11}, Ar

Scheme 12.61

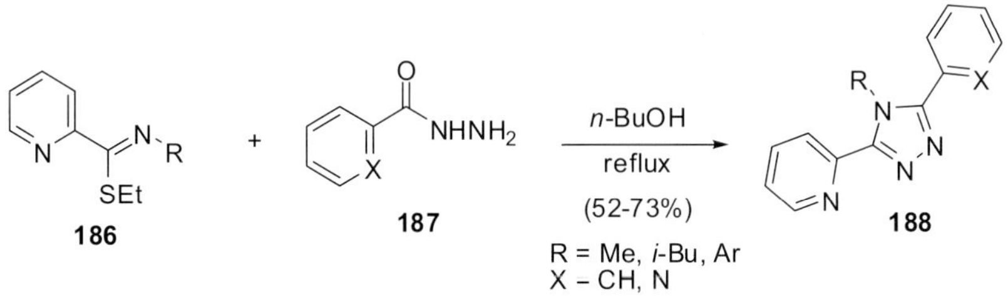

186 + **187** → *n*-BuOH, reflux (52-73%) → **188**

R = Me, *i*-Bu, Ar
X - CH, N

Scheme 12.62

189 + **190** → THF/HOAc (10:1), reflux (33-72%) → **191**

R^1 = R^2 = piperidinyl derivatives
R^3 = Ar, CH_2Ar, pyridyl
R^4 = Me, CH_2NHBoc, CH_2Ar, CH_2OMe

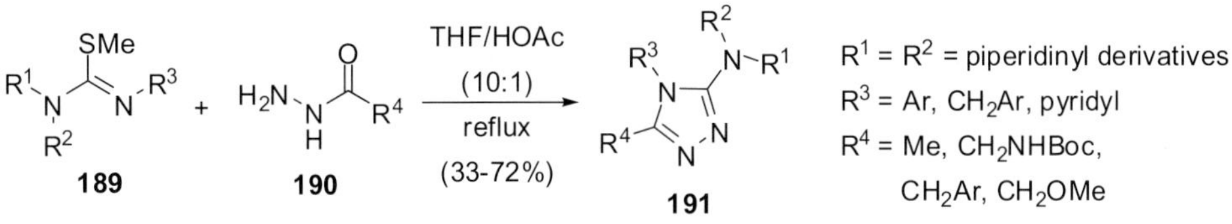

Scheme 12.63

Combinatorial solid-phase reactions have been used in the synthesis of libraries of 1,2,4-triazole compounds. Reaction of resin bound *S*-methyl-*N*-acylisothioureas **192** with hydrazines followed by acidic cleavage yielded 3-amino-1,2,4-triazoles **193** under mild conditions (Scheme 12.64) [238]. 3,4,5-Trisubstituted 1,2,4-triazoles **195** have been synthesized on solid-phase from various thioamides **194** and hydrazides, leading to peptidomimetic scaffolds (Scheme 12.65) [239]. A robust "catch,

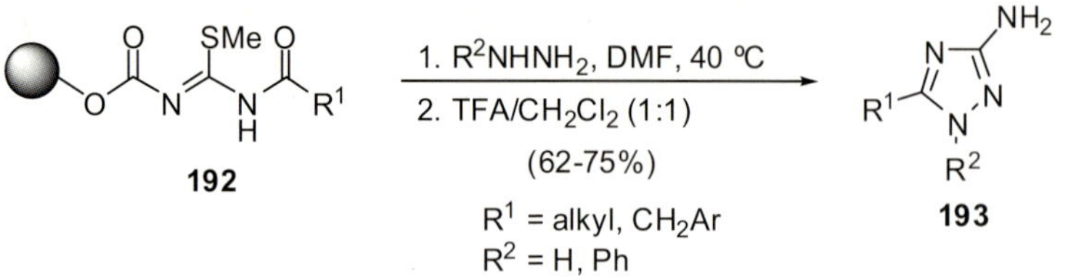

1. R^2NHNH_2, DMF, 40 °C
2. TFA/CH_2Cl_2 (1:1)
(62-75%)
R^1 = alkyl, CH_2Ar
R^2 = H, Ph

192 → 193

Scheme 12.64

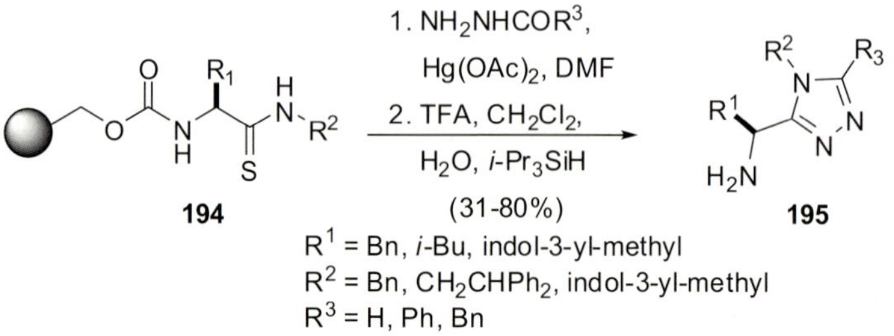

1. NH_2NHCOR^3, $Hg(OAc)_2$, DMF
2. TFA, CH_2Cl_2, H_2O, *i*-Pr_3SiH
(31-80%)
R^1 = Bn, *i*-Bu, indol-3-yl-methyl
R^2 = Bn, CH_2CHPh_2, indol-3-yl-methyl
R^3 = H, Ph, Bn

194 → 195

Scheme 12.65

cyclize, and release" preparation of 3-thioalkyl-1,2,4-triazoles mediated by the polymer-bound base P-BEMP has been described [240].

12.3.4.5 Reactions of Semicarbazides

A couple of reports present the use of semicarbazides in the synthesis of 1,2,4-triazolones. Condensation of semicarbazide hydrochloride **196** with orthoester **197** resulted in a simple synthesis of chlorotriazolinone **198** (Scheme 12.66), and the method was applied to the convergent synthesis of an NK_1 antagonist [241]. Amines have been converted into 1-formyl semicarbazides **199**, which were cyclized smoothly to 2,4-dihydro-3*H*-1,2,4-triazolin-3-ones **200** with hexamethyldisilazane (HMDS), bromotrimethylsilane, and a catalytic amount of ammonium sulfate (Scheme 12.67) [242].

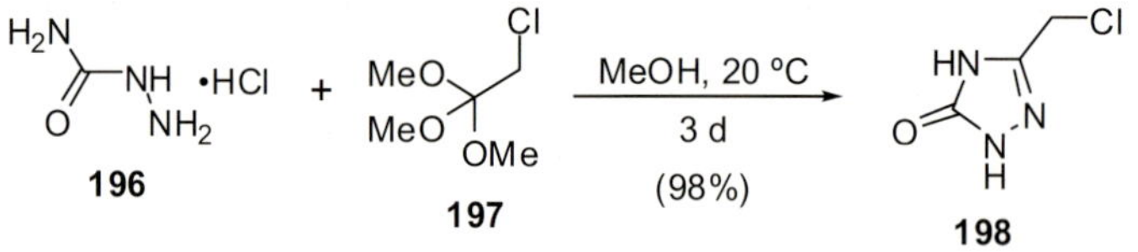

196 + 197 → 198: MeOH, 20 °C, 3 d (98%)

Scheme 12.66

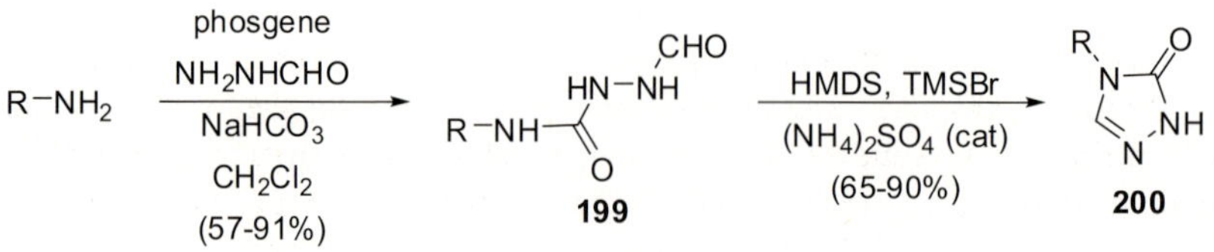

$R-NH_2$ → 199: phosgene, NH_2NHCHO, $NaHCO_3$, CH_2Cl_2 (57-91%)
199 → 200: HMDS, TMSBr, $(NH_4)_2SO_4$ (cat) (65-90%)

Scheme 12.67

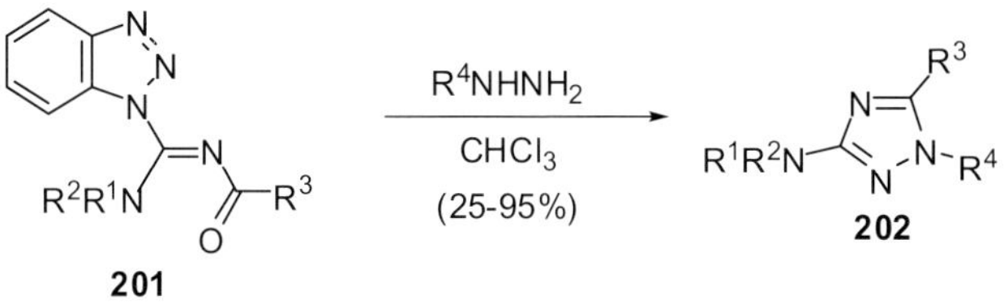

Scheme 12.68

12.3.4.6 **Synthesis of 1,2,4-Triazoles via Benzotriazole Methodology**

Solution- and solid-phase benzotriazole-mediated methodologies are employed in the synthesis of 1,2,4-triazoles. Acyl 1*H*-benzotriazol-1-carboximidamides **201** and hydrazines have been employed in a general synthesis of *N,N*-disubstituted 3-amino-1,2,4-triazoles **202** (Scheme 12.68) [243]. Polymer-supported *N*-acyl-1*H*-benzotriazole-1-carboximidamides **203** reacted with hydrazines followed by acidic cyclizative release to give 3-alkylamino-1,2,4-triazoles **204** (Scheme 12.69) [244]. Reaction of acyl hydrazides **206** with imidoylbenzotriazoles **205** in the presence of catalytic amounts of acetic acid under microwave irradiation afforded 3,4,5-trisubstituted triazoles **207** (Scheme 12.70) [245].

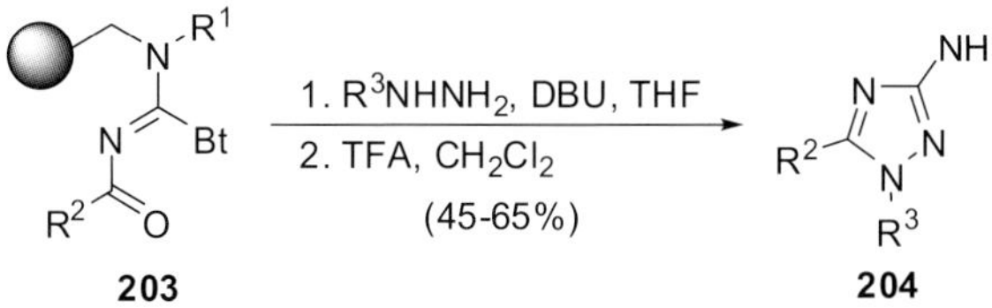

Scheme 12.69

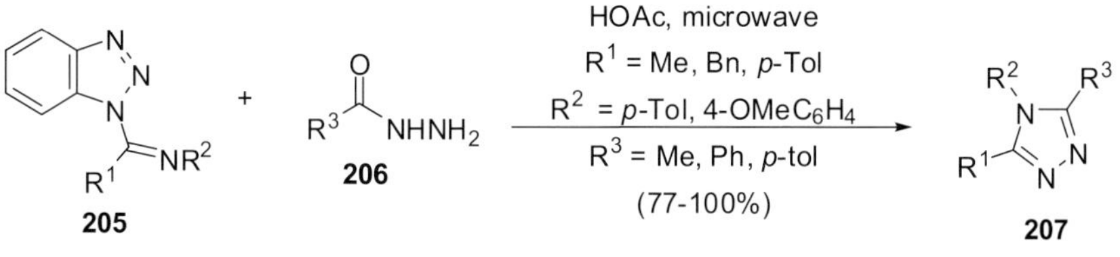

Scheme 12.70

12.3.4.7 **Other Synthesis of 1,2,4-Triazoles**

Three-component condensation of ethyl trifluoroacetate (**208**), hydrazine, and amidines in the presence of sodium hydroxide gave 3-trifluoromethyl-5-substituted-1,2,4-triazoles **209** (Scheme 12.71) [246, 247]. The amidine supplies a C–N bond to the new ring system while the other two nitrogen atoms are derived from hydrazine. Reaction of aromatic nitriles with hydrazine dihydrochloride in the presence of hydrazine hydrate in ethylene glycol under microwave irradiation gave 3,5-disubstituted-4-amino-1,2,4-triazoles **210** (Scheme 12.72) [248]. 1,3-Dipolar cycloadditions between poly(ethylene glycol) supported münchnones and diethyl azodicarboxylate led to synthesis of 3,5-disubstituted-1,2,4-triazoles [249].

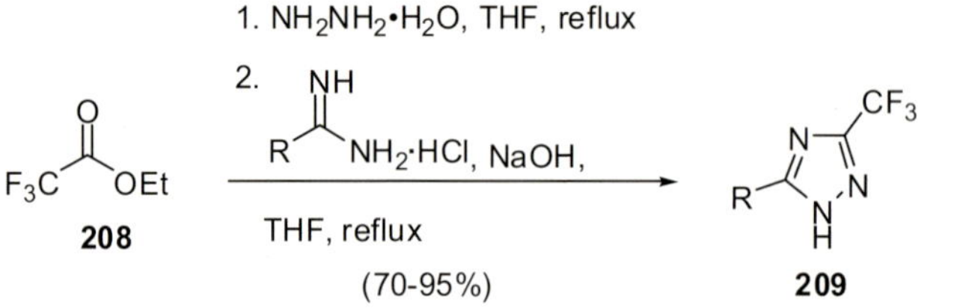

Scheme 12.71

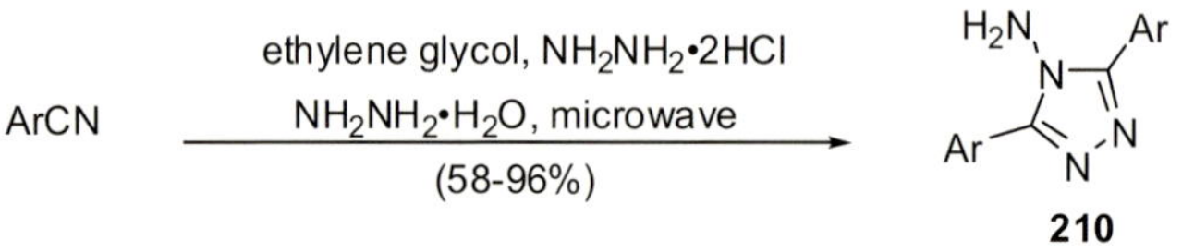

Scheme 12.72

12.3.5
Reactions of 1,2,4-Triazoles

12.3.5.1 Reactions on the Nitrogen of 1,2,4-Triazoles

12.3.5.1.1 N-Alkylation of 1,2,4-Triazoles 1,2,4-Triazoles that are unsubstituted on nitrogen can be readily alkylated. 1-Substituted-1,2,4-triazoles are the predominant products of these base-catalyzed alkylation reactions; 4-substituted-1,2,4-triazoles are rarely isolated after purification. DBU is found to be a mild and convenient base for the alkylation of 1,2,4-triazole with alkyl halides in the high-yielding syntheses of 1-substituted-1,2,4-triazoles **211** (Scheme 12.73) [250]. Sodium hydroxide [251], sodium methoxide [252], and sodium hydride [253] are other bases that have been successfully employed in these reactions.

The synthesis of 1,2,4-triazole-functionalized solid-support **212** and its use in the solid-phase synthesis of various N1 and N2 trisubstituted-1,2,4-triazoles **213** has been reported (Scheme 12.74) [254].

1,2,4-Triazoles can be alkylated at N4 by using a removable protecting group at N1 and then forming a quaternary salt with the 1-substituted triazole. 1-Acetyl and 1-cyanoethyl groups have been used as removable protecting groups [255].

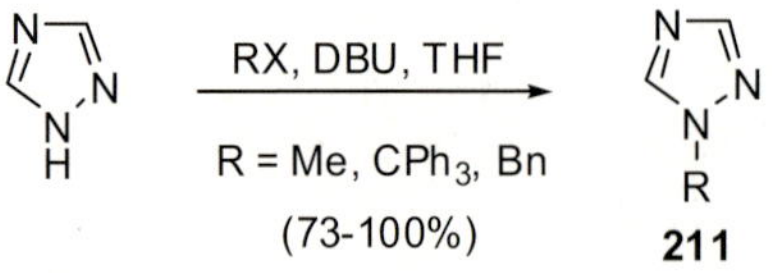

Scheme 12.73

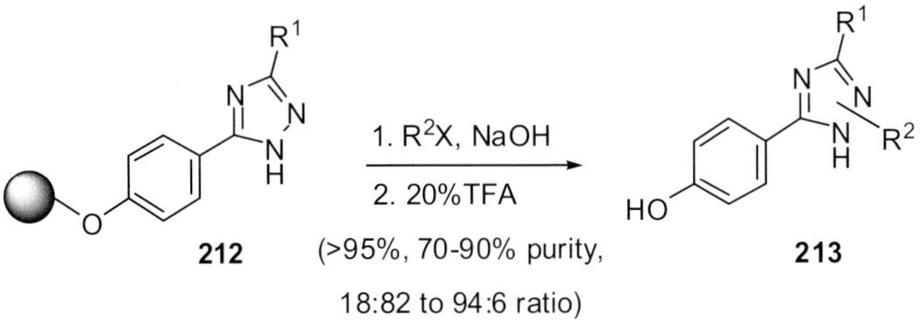

Scheme 12.74

The sequence involving quaternization/dealkylation with 1,1′-methylenebis(triazolium) salts **214** linked at the N1 position is shown in Scheme 12.75 [256]. Other similar sequences have also been reported [257]. Alternatively, 3-phenylthio-1,2,4-triazoles **215** were alkylated to their triazolium salts **216**, which under aqueous basic conditions provided 2,4-disubstituted-1,2,4-triazol-3-ones **217** (Scheme 12.76) [258].

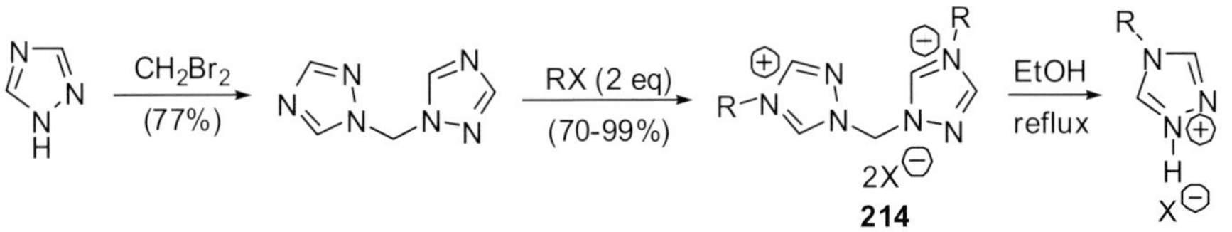

Scheme 12.75

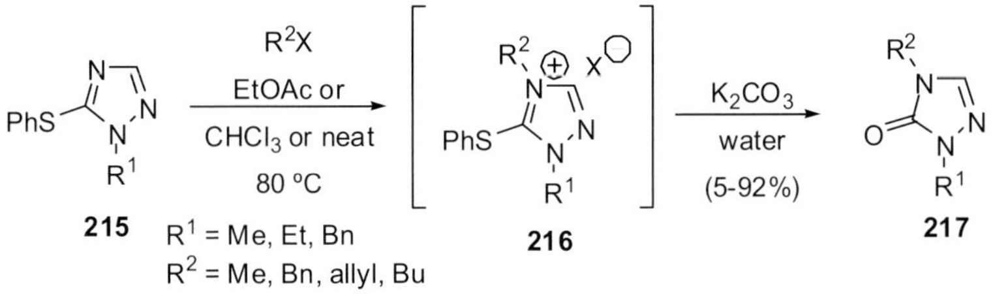

Scheme 12.76

12.3.5.1.2 **N-Acylation of 1,2,4-Triazoles** N-Unsubstituted-1,2,4-triazoles can be readily acylated at N1 by common acylating reagents such as acetyl chloride or acetic anhydride under standard conditions to give 1-acyl-1,2,4-triazoles [259]. 1*H*-1,2,4-Triazol-3-amines are acetylated first on N1 and then on the 3-amino group [260].

12.3.5.1.3 **N-Arylation of 1,2,4-Triazoles** 1,2,4-Triazole can be N-arylated at the 1-position by activated aryl halides such as 1-fluoro-2-nitrobenzene or 1-chloro-2-nitro-4-(trifluoromethyl)benzene [261]. N1-Phenylation can be achieved with triphenylbismuth diacetate and copper(II) acetate [262].

Recent protocols of copper-catalyzed N-arylations of aryl iodides with 1,2,4-triazoles represent a new landmark in the field of Ullmann-type arylation couplings of nitrogen-containing heterocycles. The aryl iodides **218** reacted efficiently and regioselectively at the N1 position with 1,2,4-triazole in the presence of copper(I) iodide and *trans*-diamine ligand **219**, with potassium phosphate as a base, to give **220** (Scheme 12.77) [117]. Another similar procedure employs copper(I) oxide, Chxn-Py-Al as the ligand, and cesium carbonate as the base [263]. However, these procedures are limited to aryl iodides; aryl bromides and aryl chlorides are not efficiently cross-coupled under these conditions.

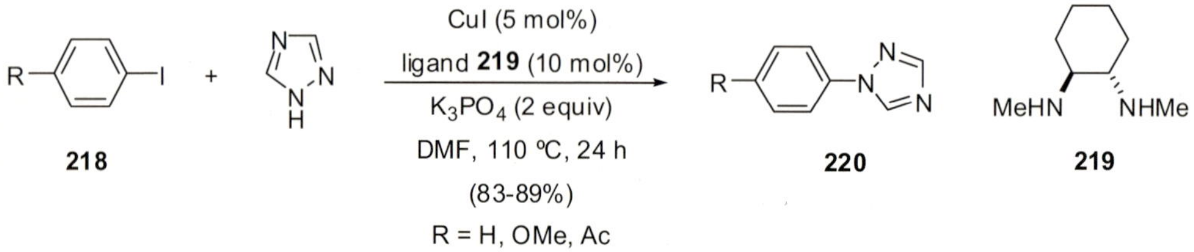

CuI (5 mol%), ligand **219** (10 mol%), K_3PO_4 (2 equiv), DMF, 110 °C, 24 h (83-89%); R = H, OMe, Ac

218 → **220**; **219**: MeHN, NHMe

Scheme 12.77

12.3.5.2 Reactions on the Carbons of 1,2,4-Triazoles

1,2,4-Triazoles are electron-deficient aromatic systems and so conventional electrophilic substitution reactions are not a practical method for the introduction of carbon substituents at C3 or C5. The most common methods for introduction of groups to C3 or C5 are by triazolyllithium intermediates or more recently by radical or carbene species.

12.3.5.2.1 C-Substitution by Triazolyllithium

Hydrogen–metal exchange by *n*-butyllithium can occur at C5 if N1 is substituted to give organolithium species, which then react quickly with alkylating agents or with other electrophiles (Scheme 12.78) [264]. If the N1 is a removable protecting group, then monosubstituted 1,2,4-triazoles can be prepared by this method [264b,264c]. 1-(Methoxymethyl)-1*H*-1,2,4-triazole is converted directly into 5-acyl derivatives by reaction with acyl chlorides and triethylamine and the methoxymethyl protecting group could be removed in a subsequent step [265].

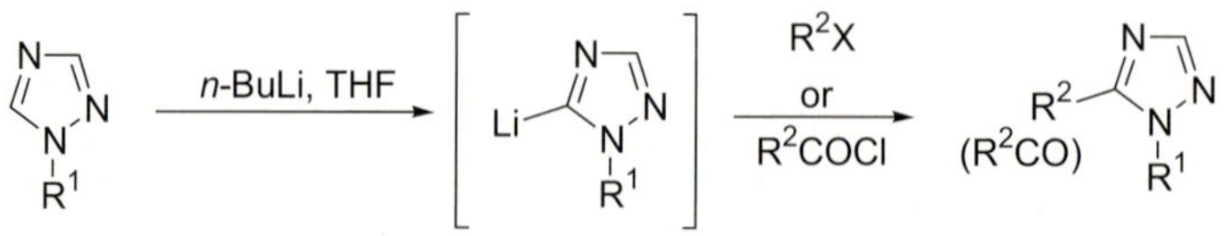

n-BuLi, THF; R^2X or R^2COCl → R^2 (R^2CO)

Scheme 12.78

Methods are available for preparation of 3-substituted-1,2,4-triazoles. A C5 removable group such as a phenylthio is employed in a protection/deprotection sequence to give 1-substituted 3-acyl-1,2,4-triazoles **221** (Scheme 12.79) [266]. 1*H*-1,2,4- and

n-BuLi, PhSSPh, THF (97%); 1. LiTMP, $RCONMe_2$ (35-67%); 2. RaNi (31-52%)

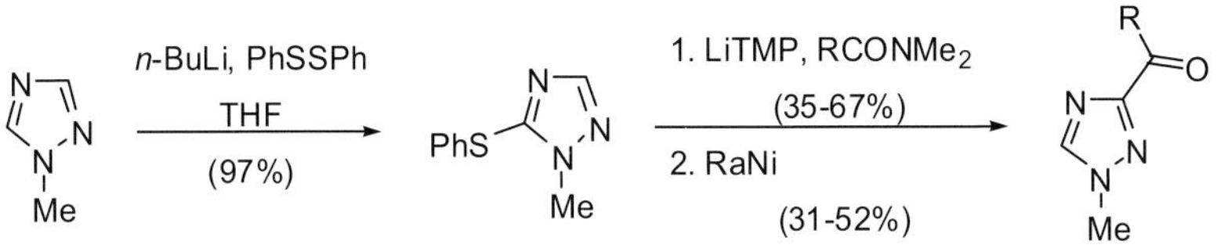

221

Scheme 12.79

1-methyl-1*H*-1,2,4-triazoles have been transformed directly into their 3(5)-arylcarbamyoyl derivatives by heating with aryl isocyanates [267]. These C-acylations are suggested to proceed by formation of N-acylated triazoles, followed by thermal rearrangements.

12.3.5.2.2 **Radical Reactions of 1,2,4-Triazoles** Reaction of 1-*N*-alkyl triazoles **222** with an alkyl radical generated from the corresponding secondary carboxylic acid in the presence of silver nitrate affords the triazole ring **223** alkylated selectively in the 5-position (Scheme 12.80) [268].

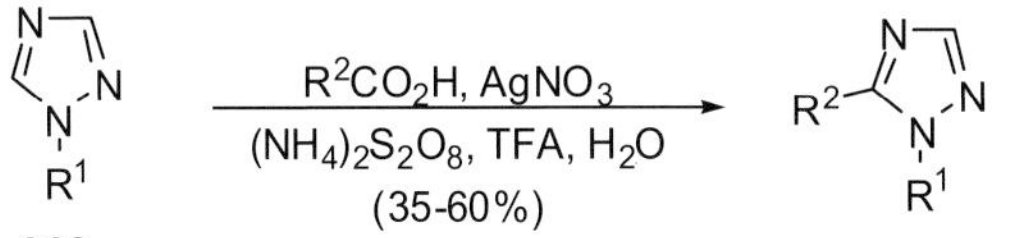

222 → R^2CO_2H, $AgNO_3$, $(NH_4)_2S_2O_8$, TFA, H_2O (35-60%) → **223**

Scheme 12.80

12.3.5.2.3 **Carbene Reactions of 1,2,4-Triazoles** There are two published reports on the syntheses of stable 1,2,4-triazolyl carbenes. Thermal decomposition *in vacuo* of 5-methoxytriazoline **224** provided in quantitative yield 1,2,4-triazol-5-ylidene **225**, a stable carbene in the absence of oxygen and moisture (Scheme 12.81) [269]. Nucleophilic carbene **225** could react with various alcohols, thiols, amines, oxygen, sulfur, selenium, isocyananates, and metal carbonyls to form a myriad of addition products. Reactions of 1,2,4-triazolyl perchlorate salts **226** with base afforded stable nucleophilic 1,2,4-triazol-5-ylidenes **227**, which could react with acetonitrile and elemental sulfur and selenium to yield addition products (Scheme 12.82) [270].

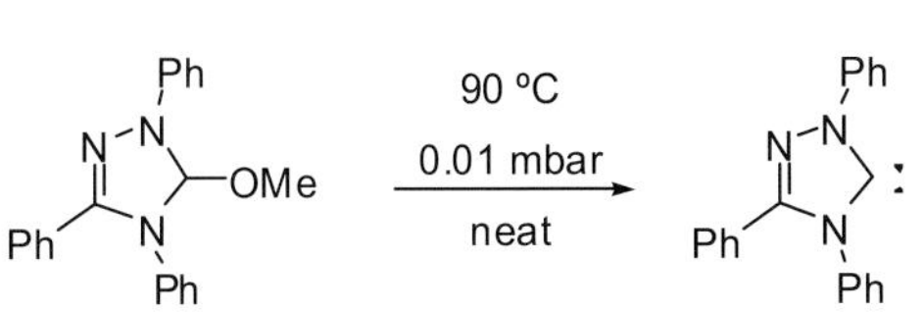

224 → 90 °C, 0.01 mbar, neat → **225**

Scheme 12.81

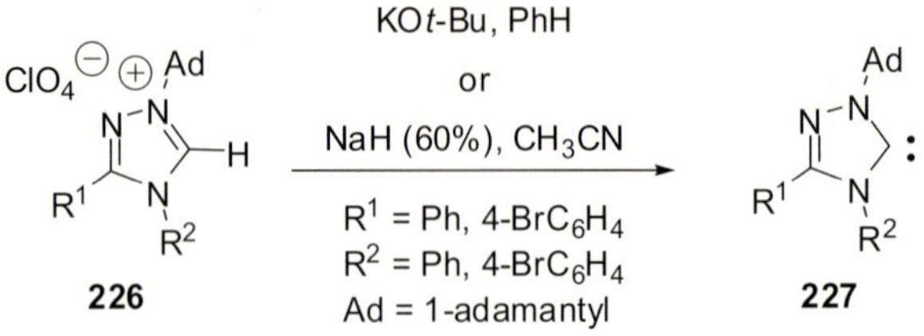

KO*t*-Bu, PhH or NaH (60%), CH_3CN

R^1 = Ph, 4-BrC_6H_4
R^2 = Ph, 4-BrC_6H_4
Ad = 1-adamantyl

226 → 227

Scheme 12.82

12.3.5.2.4 **Halogenations of 1,2,4-Triazoles** Most halogenation reactions of 1,2, 4-triazoles are *N*-chlorotriazoles, kinetic products that rearranges to 3-halo-1,2,4-triazoles slowly upon storage or heating in water [271]. 5-Chloro derivatives of 1-substituted-1,2,4-triazoles have been obtained by C5-lithio derivatives [272]. *C*-Halo-1,2,4-triazoles can be prepared more directly without isolation of the *N*-halo-1,2,4-triazoles. For example, bromination of 1*H*-1,2,4-triazole in excess aqueous sodium hydroxide afforded 3,5-dibromo-1*H*-1,2,4-triazole (**228**), which selectively exchanges a bromine with a fluorine to give **229** (Scheme 12.83) [273].

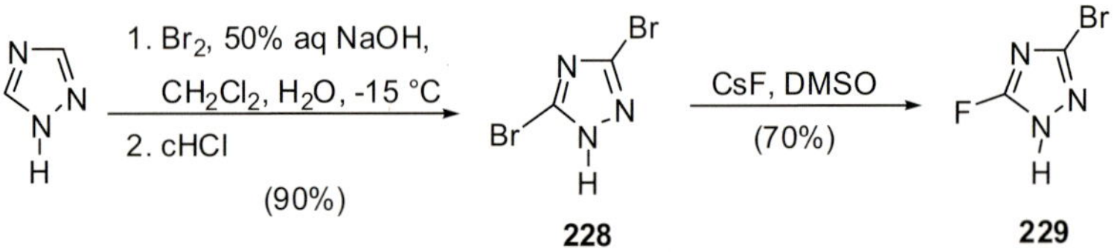

1. Br_2, 50% aq NaOH, CH_2Cl_2, H_2O, -15 °C
2. cHCl
(90%)

228

CsF, DMSO
(70%)

229

Scheme 12.83

12.3.5.2.5 **Other Reactions of 1,2,4-Triazoles** Urazoles **230** can be converted into the corresponding 1,2,4-triazol-3,5-diones **231** by various oxidizing reagents (Table 12.17). Trichloroisocyanuric acid [274], a silica gel/sodium nitrite combination [275], an ionic complex, obtained from N_2O_4 and 18-crown-6 [276], Oxone/

Table 12.17 Dehydrogenation of urazoles to 1,2,4-triazol-3,5-diones with various oxidants.

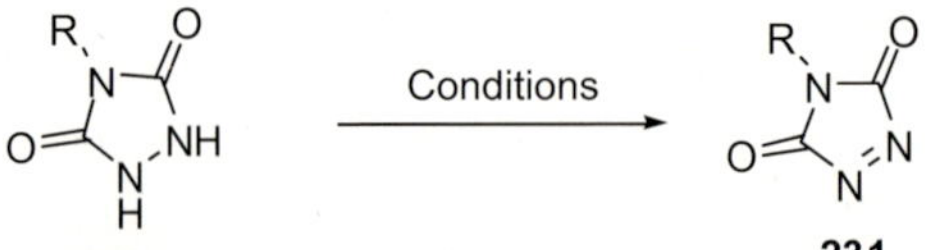

230 → Conditions → 231

Conditions	Reference
Trichloroisocyanuric acid	[274]
Silica gel/$NaNO_2$	[275]
N_2O_4/18-crown-6	[276]
Oxone/$NaNO_2$/wet silica gel	[277]
Silica sulfuric acid/$NaNO_2$	[278]

sodium nitrite in the presence of wet silica [277], and silica sulfuric acid/sodium nitrite [278] have all been reported as oxidants in this reaction.

1,2,4-Triazoline-3,5-dione **233** underwent an ene reaction with olefins **232** to yield trialkylated allylic urazoles **234**, which were further elaborated into allylic amines **235** (Scheme 12.84) [279].

1-Methyl-1,2,4-triazole **236** participated in a palladium-catalyzed C–H arylation reaction with 3,5-dimethoxychlorobenzene (**237**) to give coupled product **238** (Scheme 12.85) [280].

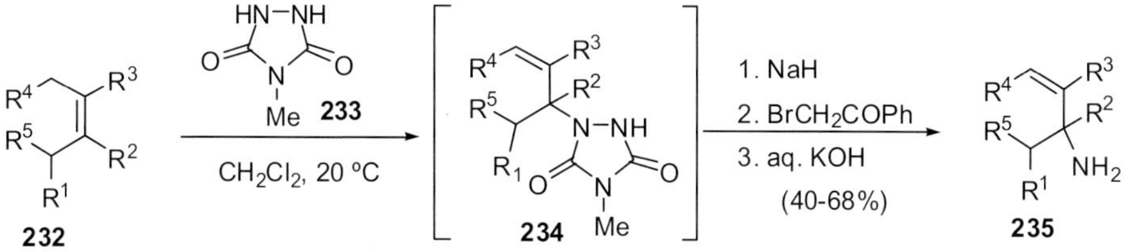

Scheme 12.84

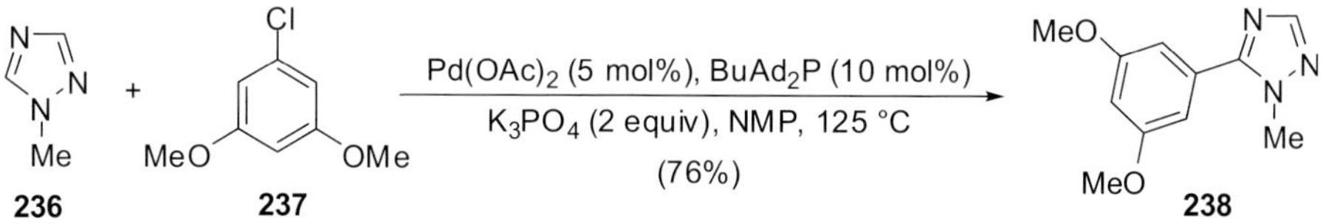

Scheme 12.85

12.3.6
1,2,4-Triazole-Containing Reagents

Monocyclic 1,2,4-triazole-containing structures have found synthetic utility. For example, 4-phenyl-1,2,4-triazole-3,5-dione (**239**) was found to be a novel and reusable reagent for the aromatization of 1,4-dihydropyridines under mild conditions [281] and to be an efficient and chemoselective reagent for the oxidation of thiols to their corresponding symmetrical disulfides [282]. *N*-4-(*p*-Chloro)phenyl-1,2,4-triazole-3,5-dione **240** has been used as an effective oxidizing agent for the oxidation of 1,3,5-trisubstituted pyrazolines to their corresponding pyrazoles under mild conditions at room temperature [283]. 1-Benzylsulfanyl-1,2,4-triazole (**241**) is a useful electrophilic sulfur source in the organocatalyzed α-sulfenylation of aldehydes [284]. Catalyst **242** catalyzed the oxidation of allylic alcohols to allylic esters with manganese(IV) oxide in excellent yields [285] and the oxidation of unactivated aldehydes to esters with manganese(IV) oxide in excellent yields [286]. The asymmetric synthesis of hydrobenzofuranones via desymmetrization of cyclohexadienones using the intramolecular Stetter reaction has been accomplished with 1,2,4-triazolium salt catalyst **243** [287].

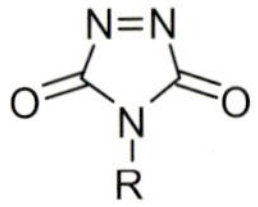

239 R = Ph

240 R = 4-ClC_6H_4

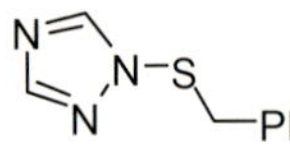

241

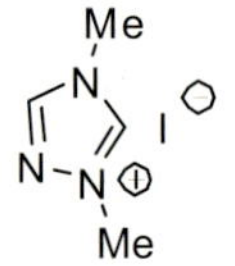

242

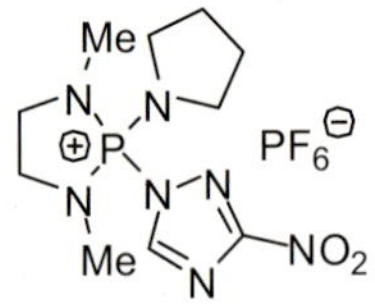

243

1,2,4-Triazolium salt catalysts **244** and **246** have been employed in the highly enantioselective azadiene Diels–Alder reactions [288]. Chiral catalyst **245** promoted the intramolecular Stetter cyclization of an aldehyde onto a vinylphosphine oxide or vinylphosphonate Michael acceptor [289]. Chiral triazolium salt **246** has been employed successfully in the hetero Diels–Alder reactions of α-chloroaldehyde bisulfite adducts with various oxodienes under biphasic reaction conditions with high levels of enantioselectivity [290] and in the highly enantioselective *cis*-cyclopentene-forming annulation reactions [291].

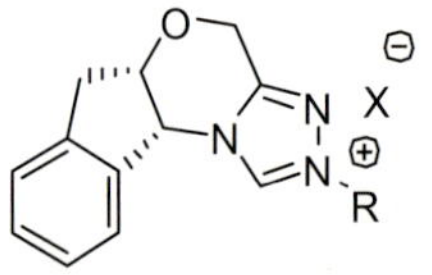

244 R = 4-$OMeC_6H_4$, X = BF_4

245 R = C_6F_5, X = BF_4

246 R = Mes, X = Cl

Bicyclic 1,2,4-triazolium salts have varied synthetic utility in a host of reactions. 1,2,4-Triazolium salts **247** have been identified as a new family of stable annulated N-heterocyclic carebenes that found applications in catalytic benzoin condensations and transesterifications at ambient temperature [292]. *N*-Pentafluorophenyl triazolium tetrafluoroborate salts **248** were found to be useful catalysts in the macrocyclization of α,ω-dialdehydes to α-hydroxyketones [293] and in the synthesis of 1,2-amino alcohols via azidation of epoxy aldehydes (where modest asymmetric induction was achieved) [294]. *N*-Pentafluorophenyl triazolium tetrafluoroborate salt **249** was found to be useful catalyst in the asymmetric intermolecular Stetter reaction of glyoxamides with alkylidenemalonates [295]. Chiral catalyst **250** has been utilized in the N-heterocyclic carbene-catalyzed redox amidations of α-functionalized aldehydes with amines [296]. The N-heterocyclic catalyst **251** promoted *O* to *C* carboxyl transfer on a range of indolyl and benzofuranyl carbonates [297] and also promoted the formal [2 + 2] cycloaddition of ketenes with *N*-tosyl imines to give the corresponding β-lactams [298]. Chiral triazolium catalyst **252** has been found to be efficient in the formal [2 + 2] cycloaddition reactions of alkyl(aryl)ketenes with 2-oxoaldehydes to afford β-lactones with α-quaternary-β-tertiary stereocenters in high yields with good diastereoselectivities and excellent enantioselectivities [299]. The asymmetric Michael addition of aromatic heterocyclic aldehydes to arylidenemalonates catalyzed

247 X = BF_4, PF_6

248 R = H

249 R = Bn

250

251 R^1 = H, R^2 = Ph

252 R^1 = CPh_2OTBS, R^2 = Ph

253 R^1 = $CH_2OTBDPS$, R^2 = Bn

by N-heterocyclic carbene **253** has been disclosed [300]. Catalyst **253** was also effective in an intermolecular Stetter reaction to give 1,4-diketones [301].

References

1 (a) Wamhoff, H. (1984) in *Comprehensive Heterocyclic Chemistry*, vol. 5 (eds A.R. Katritzky and C.W. Rees), Pergamon, Oxford, pp. 669–732; (b) Dehne, H. (1994) in *Methoden der Organischen Chemie (Houben-Weyl)*, vol. E8d (ed. E. Schumann), Georg Thieme Verlag, Stuttgart, pp. 305–405; (c) Fan, W-.Q. and Katritzky, A.R. (1996) in *Comprehensive Heterocyclic Chemistry II*, vol. 4 (eds A.R. Katritzky, C.W. Rees, and E.F.V. Scriven), Pergamon, Oxford, pp. 1–126; (d) Tomé, A.C. (2004) in *Science of Synthesis, Vol. 13, Five-Membered Hetarenes with Three or More Heteroatoms* (eds R.C. Storr and T.L. Gilchrist), Georg Thieme Verlag, Stuttgart, New York, pp. 415–602; (e) Rachwal, S. and Katritzky, A.R. (2008) in *Comprehensive Heterocyclic Chemistry III*, vol. 5 (eds A.R. Katritzky, C.A. Ramseden, E.F.V. Scriven, and R.J.K. Taylor), Elsevier, Oxford, pp. 1–158.

2 Tomas F., Abboud, J.-L.M., Laynez, J., Notario, R., Santos, L., Nilsson, S.O., Catalan, J., Claramunt, R.M., and Elguero, J. (1989) *Journal of the American Chemical Society*, **111**, 7348–7353.

3 Abboud, J.-L.M., Foces-Foces, C., Notario, R., Trifonov, R.E., Volovodenko, A.P., Ostrovskii, V.A., Alkorta, I., and Elguero, J. (2001) *European Journal of Organic Chemistry*, 3013–3024.

4 Elguero, J., Gonzales, E., and Jacquier, R. (1967) *Bulletin de la Société Chimique de France*, 2998.

5 Elguero, J., Marzin, C., and Roberts, J.D. (1976) *The Journal of Organic Chemistry*, **39**, 357–363.

6 (a) Alvarez, R., Velazquez, S., -Felix, A.S., Aquaro, S., De Clercq, E., Perno, C.-F., Karlsson, A., Balzarini, J., and Camarasa, M.J. (1994) *Journal of Medicinal Chemistry*, **37**, 4185–4194; (b) Velaquez, S., Alvarez, R., Perez, C., Gago, F., De Clercq, E., Balzarini, J., and Camarasa, M.-J. (1998) *Antiviral Chemistry & Chemotherapy*, **9**, 481–489; (c) Brik, A., Alexandratos, J., Lin, Y.-C., Elder, J.H., Olson, A.J., Wlodawer, A., Goodsell, D.S., and Wong, C.-H. (2005) *ChemBioChem*, **6**, 1167–1169.

7 Genin, M.J., Allwine, D.A., Anderson, D.J., Barbachyn, M.R., Emmert, D.E., Garmon, S.A., Graber, D.R., Grega, K.C., Hester, J.B., Hutchinson, D.K., Morris, J., Reischer, R.J., Ford, C.W., Zurenko, G.E., Hamel, J.C., Schaadt, R.D., Stapert, D., and Yagi, B.H. (2000) *Journal of Medicinal Chemistry*, **43** 953–970.

8 Brockunier, L.L., Parmee, E.R., Ok, H.O., Candelore, M.R., Cascieri, M.A., Colwell, Jr., L.F., Deng, L., Feeney, W.P., Forrest, M.J., Hom, G.J., MacIntyre, D.E., Tota, L., Wyvratt, M.J., Fisher, M.H., and Weber, A.E. (2000) *Bioorganic & Medicinal Chemistry Letters*, **10**, 2111–2114.

9 Micetich, R.G., Maiti, S.N., Spevak, P., Hall, T.W., Yamabe, S., Ishida, N., Tanaka, M., Yamazaki, T., Nakai, A., and Ogawa, K. (1987) *Journal of Medicinal Chemistry*, **30**, 1469–1474.

10 Weinstein, A.J. (1980) *Drugs*, **20**, 137.

11 Katritzky, A.R., Zhang, Y., and Singh, S.K. (2003) *Heterocycles*, **60**, 1225–1239.

12 (a) Birkofer, L. and Richtzenhain, K. (1979) *Chemische Berichte*, **112**, 2829–2836; (b) Hartzel, L.W. and Benson, F.R. (1954) *Journal of the American Chemical Society*, **76**, 667–670.

13 (a) Woerner, F.P. and Reimlinger, H. (1970) *Chemische Berichte*, **103**, 1908–1917; (b) Marei, M.G., El-Ghanam, M., and Salem, M.M. (1994) *Bulletin of the Chemical Society of Japan*, **67**, 144–148.

14 Journet, M., Cai, D., Kowal, J.J., and Larsen, R.D. (2001) *Tetrahedron Letters*, **42**, 9117–9118.

15 (a) Dimroth, O. and Fester, G. (1910) *Chemische Berichte*, **43**, 2219–2223; (b) Gold, H. (1965) *Justus Liebigs Annalen der Chemie*, **688**, 205–216; (c) Hubert, A. (1970) *Bulletin des Sociétés Chimiques Belges*, **79**, 195–202; (d) Fournier, J.O. and Miller, J.B. (1965) *Journal of Heterocyclic Chemistry*, **2**, 488–490.

16 (a) Huisgen, R., Knorr, R., Mobius, L., and Szeimies, G. (1965) *Chemische Berichte*, **98**, 4014–4021; (b) Sasaki, T., Eguchi, S., Yamaguchi, M., and Esaki, T. (1981) *The Journal of Organic Chemistry*, **46**, 1800–1804; (c) Mitchell, G., and Rees, C.W. (1987) *Journal of the Chemical Society-Perkin Transactions 1*, 413–422; (d) Abu-Orabi, S., Atfah, M.A., Jibril, I., Mari'i, F., and Ali, A.A. (1989) *Journal of Heterocyclic Chemistry*, **26**, 1461–1468; (e) Malet, R., Serra, N., Abramovitch, R.A., Moreno-Manas, M., and Pleixats, R. (1993) *Journal of Heterocyclic Chemistry*, **30**, 317–321.

17 Kirmse, W. and Horner, L. (1958) *Justus Liebigs Annalen der Chemie*, **614**, 1–4.

18 Crandall, J.K., Conover, W.W., and Komin, J.B. (1975) *The Journal of Organic Chemistry*, **40**, 2042–2044.

19 Katritzky, A.R. and Singh, S.K. (2002) *The Journal of Organic Chemistry*, **67**, 9077–9079.

20 Lermontov, S.A., Shkavrov, S.V., and Pushin, A.N. (2000) *Journal of Fluorine Chemistry*, **105**, 141–147.

21 Louerat, F., Bougrin, K., Loupy, A., Retana, A.M.O., Pagalday, J., and Palacios, F. (1998) *Heterocycles*, **48**, 161–170.

22 Biagai, G., Giorgi, I., Livi, O., Lucacchini, A., Martin, C., and Scartoni, V. (1993) *Journal of Pharmaceutical Sciences*, **82**, 893.

23 Katritzky, A.R., Falli, C.N., Shcherbakova, I.V., and Verin, S.V. (1996) *Journal of Heterocyclic Chemistry*, **33**, 335–339.

24 Kolb, H.C., Finn, M.G., and Sharpless, K.B. (2001) *Angewandte Chemie-International Edition*, **40**, 2004–2021.

25 Bock, V.D., Hiemstra, H., and van Maarseveen, J.H. (2006) *European Journal of Organic Chemistry*, 51–68.

26 Lutz, J.F. (2007) *Angewandte Chemie-International Edition*, **46**, 1018–1025.

27 Wolfbeis, O.S. (2007) *Angewandte Chemie-International Edition*, **46**, 2980–2982.

28 Dedola, S., Nepogodiev, S.A., and Field, R.A. (2007) *Organic and Biomolecular Chemistry*, **5**, 1006–1017.

29 Gil, M.V., Arevalo, M.J., and López, Ó. (2007) *Synthesis*, 1589–1620.

30 Lutz, J.-F. (2008) *Angewandte Chemie-International Edition*, **47**, 2182–2184.

31 Jones, G.O. and Houk, K.N. (2008) *The Journal of Organic Chemistry*, **73**, 1333–1342.

32 Rostovtsev, V.V., Green, L.G., Fokin, V.V., and Sharpless, K.B. (2002) *Angewandte Chemie-International Edition*, **41**, 2596–2599.

33 Rodionov, V.O., Fokin, V.V., and Finn, M.G. (2005) *Angewandte Chemie-International Edition*, **44**, 2211–2215.

34 Kuijpers, B.H.M., Groothuys, S., Keereweer, A.R., Quaedflieg, P.J.L.M., Blaauw, R.H., van Delft, F.L., and Rutjes, F.P.J.T. (2004) *Organic Letters*, **6**, 3123–3126.

35 Chittaboina, S., Xie, F., and Wang, Q. (2005) *Tetrahedron Letters*, **46**, 2331–2336.

36 Li, Z., Seo, T.S., and Ju, J. (2004) *Tetrahedron Letters*, **45**, 3143–3146.

37 Hanamoto, T., Hakoshima, Y., and Egashira, M. (2004) *Tetrahedron Letters*, **45**, 7573–7576.

38 Krasinski, A., Fokin, V.V., and Sharpless, K.B. (2004) *Organic Letters*, **6**, 1237–1240.

39 Wu, Y.-M., Deng, J., Li, Y., and Chen, Q.-Y. (2005) *Synthesis*, 1314–1318.

40 Loren, J.C. and Sharpless, K.B. (2005) *Synthesis*, 1514–1520.

41 (a) Kamijo, S., Jin, T., Huo, Z., and Yamamoto, Y. (2003) *Journal of the American Chemical Society*, **125**, 7786–7787; (b) Kamijo, S., Jin, T., Huo, Z., and Yamamoto, Y. (2004) *The Journal of Organic Chemistry*, **69**, 2386–2393.

42 Kamijo, S., Huo, Z., Jin, T., Kanazawa, C., and Yamamoto, Y. (2005) *The Journal of Organic Chemistry*, **70**, 6389–6397.

43 Kamijo, S., Jin, T., Huo, Z., and Yamamoto, Y. (2002) *Tetrahedron Letters*, **43**, 9707–9710.

44 Kamijo, S., Jin, T., and Yamamoto, Y. (2004) *Tetrahedron Letters*, **45**, 689–691.

45 Jin, T., Kamijo, S., and Yamamoto, Y. (2004) *European Journal of Organic Chemistry*, 3789–3791.

46 Majireck, M.M. and Weinreb, S.M. (2006) *The Journal of Organic Chemistry*, **71**, 8680–8683.

47 Boren, B.C., Narayan, S., Rasmussen, L.K., Zhang, L., Zhao, H., Lin, Z., Jia, G., and Fokin, V.V. (2008) *Journal of the American Chemical Society*, **130**, 8923–8930.

48 Zhang, C.-T., Zhang, X., and Qing, F.-L. (2008) *Tetrahedron Letters*, **49**, 3927–3930.

49 Diez-Gonzalez, S., Correa, A., Cavallo, L., and Nolan, S.P. (2006) *Chemistry - A European Journal*, **12**, 7558–7564.

50 (a) Feldman, A.K., Colasson, B., and Fokin, V.V. (2004) *Organic Letters*, **6**, 3897–3899; (b) Anderson, J., Bolvig, S., and Liang, X. (2005) *Synlett*, 2941–2947.

51 Appukkuttan, P., Dehaen, W., Fokin, V.V., and der Eycken, E.V. (2004) *Organic Letters*, **6**, 4223–4225.

52 Kacprzak, K. (2005) *Synlett*, 943–946.

53 Gardiner, M., Grigg, R., Kordes, M., Sridharan, V., and Vicker, N. (2001) *Tetrahedron*, **57**, 7729–7735.

54 Yan, Z.-Y., Zhao, Y.-B., Fan, M.-J., Liu, W.-M., and Liang, Y.-M. (2005) *Tetrahedron*, **61**, 9331–9337.

55 Miao, T. and Wang, L. (2008) *Synthesis*, 363–368.

56 (a) Barral, K., Moorhouse, A.D., and Moses, J.E. (2007) *Organic Letters*, **9**, 1809–1811; (b) Moorhouse, A.D. and Moses, J.E. (2008) *Synlett*, 2089–2092.

57 Blass, B.E., Coburn, K.R., Faulkner, A.L., Hunn, C.L., Natchus, M.G., Parker, M.S., Portlock, D.E., Tullis, J.S., and Wood, R. (2002) *Tetrahedron Letters*, **43**, 4059–4061.

58 Harju, K., Vahermo, M., Mutikainen, I., and Kauhaluoma, J.Y. (2003) *Journal of Combinatorial Chemistry*, **5**, 826–833.

59 Tornoe, C.W., Christensen, C., and Meldal, M. (2002) *The Journal of Organic Chemistry*, **67**, 3057–3064.

60 Coats, S.J., Link, J.S., Gauthier, D., and Hlasta, D.J. (2005) *Organic Letters*, **7**, 1469–1472.

61 Tomøe, G.W., Sanderson, S.J., Mottram, J.C., Coombs, G.H., and Meldal, M. (2004) *Journal of Combinatorial Chemistry*, **6**, 312–324.

62 Blass, B.E., Coburn, K.R., Faulkner, A.L., Seibel, W.L., and Srivastava, A. (2003) *Tetrahedron Letters*, **44**, 2153–2155.

63 Garanti, L. and Molteni, G. (2003) *Tetrahedron Letters*, **44**, 1133–1135.

64 Molteni, G. and Buttero, P.D. (2005) *Tetrahedron*, **61**, 4983–4987.

65 (a) Meek, J.S. and Fowler, J.S. (1967) *Journal of the American Chemical Society*, **89**, 1967–1967; (b) Meek, J.S. and Fowler, J.S. (1968) *The Journal of Organic Chemistry*, **33**, 985–991; (c) Tanaka, Y. and Miller, S.I. (1972) *The Journal of Organic Chemistry*, **37**, 3370–3372; (d) Velezheva, V.S., Erofeev, Y.V., and Suvorov, N.N. (1980) *The Journal of Organic Chemistry USSR*, **16**, 1839–1844; (e) Dong, Z., Hellmund, K.A., and Pyne, S.G. (1993) *Australian Journal of Chemistry*, **46**, 1431–1436; (f) Prager, R.H. and Razzino, P. (1994) *Australian Journal of Chemistry*, **47**, 1375–1385; (g) Bajpai, I.K. and Bhaduri, A.P. (1996) *Synthetic Communications*, **26**, 1849–1859.

66 (a) Amantini, D., Fringuelli, F., Piermatti, O., Pizzo, F., Zunino, E., and Vaccaro, L. (2005) *The Journal of Organic*

Chemistry, **70**, 6526–6529; (b) Quiclet-Sire, B. and Zard, S.Z. (2005) *Synthesis*, 3319.

67 (a) Munk, M.E. and Kim, Y.K. (1964) *Journal of the American Chemical Society*, **86**, 2213–2217; (b) Hüisgen, R., Mobius, L., and Szeimies, G. (1965) *Chemische Berichte*, **98**, 1138–1152; (c) Hüisgen, R. and Szeimies, G. (1965) *Chemische Berichte*, **98**, 1153–1158; (d) Roque, D.R., Neill, J.L., Antoon, J.W., and Stevens, E.P. (2005) *Synthesis*, 2497–2502.

68 (a) Biagi, G., Livi, O., Ramacciotti, G.L., Scartoni, V., Bazzichi, I., Mazzoni, M.R., and Lucacchini, A. (1990) *Farmaco (Societa Chimica Italiana: 1989)*, **45**, 49; (b) Biagi, G., Dell'Omodarme, G., Giorgi, I., Livi, O., and Scartoni, V. (1992) *Farmaco (Societa Chimica Italiana: 1989)*, **47**, 91.

69 (a) Harvey, G.R. (1966) *The Journal of Organic Chemistry*, **31**, 1587–1590; (b) Ykman, P., L'abbe, G., and Smets, G. (1971) *Tetrahedron*, **27**, 845–849; (c) Ykman, P., L'abbe, G., and Smets, G. (1971) *Tetrahedron*, **27**, 5623–5629.

70 (a) Bleiholder, R.F. and Shechter, H. (1968) *Journal of the American Chemical Society*, **90**, 2131–2137; (b) Wedegaertner, D.K., Kattak, R.K., Harrison, I., and Cristie, S.K. (1991) *The Journal of Organic Chemistry*, **56**, 4463–4467.

71 (a) Cailleux, P., Piet, J.C., Benhaoua, H., and Carrie, R. (1996) *Bulletin des Sociétés Chimiques Belges*, **105**, 45–51; (b) Hager, C., Miethchen, R., and Reinke, H. (2000) *Journal of Fluorine Chemistry*, **104**, 135–142; (c) Cafici, L., Pirali, T., Condorelli, F., Del Grosso, E., Massarotti, A., Sorba, G., Canonico, P.L., Tron, G.C., and Genazzani, A.A. (2008) *Journal of Combinatorial Chemistry*, **10**, 732–740.

72 Liu, B., Wang, M.-X., Wang, L.-B., and Huang, Z.-T. (2000) *Heteroatom Chemistry*, **11**, 387–391.

73 (a) Peng, W. and Zhu, S. (2003) *Synlett*, 187–190; (b) Peng, W. and Zhu, S. (2003) *Tetrahedron*, **59**, 4395–4404.

74 Melo, J.O.F., Ratton, P.M., Augusti, R., and Donnici, C.L. (2004) *Synthetic Communications*, **34**, 369–376.

75 Zaragoza, F. and Petersen, S.V. (1996) *Tetrahedron*, **52**, 10823–10826.

76 Adamo, G., Benedetti, F., Berti, F., Nardin, G., and Norbedo, S. (2003) *Tetrahedron Letters*, **44**, 9095–9097.

77 Biagi, G., Livi, O., Ramacciotti, G.L., Scartoni, V., Bazzichi, L., Mazzoni, M.R., and Lucacchini, A. (1990) *Farmaco (Societa Chimica Italiana: 1989)*, **45**, 49.

78 Cottrell, I.F., Hands, D., Houghton, P.G., Humphrey, G.R., and Wright, S.H.B. (1991) *Journal of Heterocyclic Chemistry*, **28**, 301–304.

79 Biagi, G., Giorgi, I., Livi, O., Manera, C., and Scartoni, V. (1997) *Journal of Heterocyclic Chemistry*, **34**, 845–851.

80 Olesen, P.H., Nielsen, F.E., Pedersen, E.B., and Becher, J. (1984) *Journal of Heterocyclic Chemistry*, **21**, 1603–1608.

81 L'abbe, G. and Beenaerts, L. (1989) *Tetrahedron*, **45**, 749–756.

82 Begtrup, M. and Pedersen, C. (1964) *Acta Chemica Scandinavica (Copenhagen, Denmark: 1989)*, **18**, 1333–1336.

83 Gibson, K.R., Thomas, S.R., and Rowley, M. (2001) *Synlett*, 712–714.

84 Cottrell, I.F., Hands, D., Houghton, P.G., Humphrey, G.R., and Wright, S.H.B. (1991) *Journal of Heterocyclic Chemistry*, **28**, 301–304.

85 Biagi, G., Giorgi, I., Livi, O., Scartoni, V., Velo, S., and Baril, P.L. (1996) *Journal of Heterocyclic Chemistry*, **33**, 1847–1853.

86 Luo, Y. and Hu, Y. (2003) *Synthetic Communications*, **33**, 3513–3517.

87 Batanero, D.B. and Barba, F. (2004) *Heterocycles*, **63**, 1175–1180.

88 Tang, W.-J. and Hu, Y.-Z. (2006) *Synthetic Communications*, **36**, 2461–2468.

89 Hauptmann, S., Wilde, H., and Moser, K. (1967) *Tetrahedron Letters*, **8**, 3295–3297.

90 Wittig, G. and Krebs, A. (1961) *Chemische Berichte*, **94**, 3260–3268.

91 Hauptmann, S., Wilde, H., and Moser, K. (1971) *Journal Fur Praktische Chemie*, **313**, 882–888.

92 (a) Wittig, G. and Dorsch, H.-L. (1968) *Justus Liebigs Annalen der Chemie*, **711**, 46–54; (b) Wittig, G. and Meske-Schuller, J. (1968) *Justus Liebigs Annalen der Chemie*, **711**, 65–75.

93 Uhlmann, P., Felding, J., Vedso, P., and Begtrup, M. (1997) *The Journal of Organic Chemistry*, **62**, 9177–9181.

94 Prakash, O., Gujral, H.K., Rani, N., and Singh, S.P. (2000) *Synthetic Communications*, **30**, 417–425.

95 Biagi, G., Livi, O., Lucacchini, A., Martini, C., and Scartoni, V. (1992) *Journal of Pharmaceutical Sciences*, **81**, 543.

96 Polak, M. and Vercek, B. (2000) *Synthetic Communications*, **30**, 2863–2871.

97 Taher, A., Eichenseher, S., and Weaver, G.W. (2000) *Tetrahedron Letters*, **41**, 9889–9891.

98 Matsumoto, N. and Takahashi, M. (2003) *Heterocycles*, **60**, 2677–2684.

99 Raghavendra, M.S. and Lam, Y. (2004) *Tetrahedron Letters*, **45**, 6129–6132.

100 Barluenga, J., Valdes, C., Beltran, G., Escribano, M., and Aznar, F. (2006) *Angewandte Chemie-International Edition*, **45**, 6893–6896.

101 Tao, C.-Z., Cui, X., Li, J., Liu, A.-X., Liu, L., and Guo, Q.-X. (2007) *Tetrahedron Letters*, **48**, 3525–3529.

102 Holzer, W. and Ruso, K. (1992) *Journal of Heterocyclic Chemistry*, **29**, 1203–1207.

103 Ohta, S., Kawasaki, I., Uemura, T., Yamashita, M., Yoshioka, T., and Yamaguchi, S. (1997) *Chemical & Pharmaceutical Bulletin*, **45**, 1140–1145.

104 Iddon, B. and Nicholas, M. (1996) *Journal of the Chemical Society-Perkin Transactions 1*, 1341–1348.

105 Felding, J., Uhlmann, P., Kristensen, J., Vedso, P., and Begtrup, M. (1998) *Synthesis*, 1181–1184.

106 Buckle, D.R. and Rockell, C.J.M. (1982) *Journal of the Chemical Society-Perkin Transactions 1*, 627–630.

107 Smith, P.A.S. and Wirth, J.G. (1968) *The Journal of Organic Chemistry*, **33**, 1145–1155.

108 Buckle, D.R., Rockell, C.J.M., and Oliver, R.S. (1982) *Journal of Heterocyclic Chemistry*, **19**, 1147–1152.

109 Iddon, B. and Nicholas, M. (1996) *Journal of Chemical Research-S*, 512–513.

110 L'abbe, G., Bruynseels, M., Beenaerts, L., Vandendriessche, A., Delbeke, P., and Toppet, S. (1989) *Bulletin des Sociétés Chimiques Belges*, **98**, 343–347.

111 Pedersen, C. (1959) *Acta Chemica Scandinavica (Copenhagen, Denmark: 1989)*, **13**, 888–892.

112 Tanaka, Y. and Miller, S.I. (1973) *Tetrahedron*, **29**, 3285–3296.

113 Gilchrist, T.L., Gymer, G.E., and Rees, C.W. (1975) *Journal of the Chemical Society-Perkin Transactions 1*, 1–8.

114 (a) Huttel, R. and Kratzer, J. (1959) *Chemische Berichte*, **92**, 2014–2021; (b) Birkofer, L. and Wegner, P. (1966) *Chemische Berichte*, **99**, 2512–2517; (c) Birkofer, L. and Wegner, P. (1967) *Chemische Berichte*, **100**, 3485–3494.

115 Carboni, R.A., Kauer, J.C., Hatchard, W.R., and Harder, R.J. (1967) *Journal of the American Chemical Society*, **89**, 2626–2633.

116 Elguero, J., Gonzalez, E., and Jacquier, R. (1967) *Bulletin de la Societe Chimique de France*, 2998–3003.

117 Antilla, J.C., Baskin, J.M., Barder, T.E., and Buchwald, S.L. (2004) *The Journal of Organic Chemistry*, **69**, 5578–5587.

118 Liu, Y., Yan, W., Chen, Y., Petersen, J.L., and Shi, X. (2008) *Organic Letters*, **10**, 5389–5392.

119 Gilchrist, T.L. and Gymer, G.E. (1974) *Advances in Heterocyclic Chemistry, Vol 68*, **16**, 33–85.

120 Begtrup, M. and Vedso, P. (1995) *Journal of the Chemical Society-Perkin Transactions 1*, 243–247.

121 Spetzler, J.C., Meldal, M., Feldig, J., Vedso, P., and Begtrup, M. (1998) *Journal of the Chemical Society-Perkin Transactions 1*, 1727–1732.

122 Iddon, B. and Nicholas, M. (1996) *Journal of the Chemical Society-Perkin Transactions 1*, 1341–1347.

123 Begtrup, M. (1988) *Bulletin des Sociétés Chimiques Belges*, **97**, 573–597.

124 Lynch, B.M. and Chan, T.-L. (1963) *Canadian Journal of Chemistry*, **41**, 274–277.

125 Begtrup, M. and Nytoft, H.P. (1986) *Acta Chemica Scandinavica. Series B: Organic Chemistry and Biochemistry*, **40**, 262–269.

126 Jagerovic, N., Jimeno, M.L., Alkorta, I., Elguero, J., and Claramunt, R.M. (2002) *Tetrahedron*, **58**, 9089–9094.

127 Katritzky, A.R., Kuzmierkiewicz, W., and Greenhill, J.V. (1991) *Recueil des Travaux Chimiques des Pays-Bas*, **110**, 369.

128 Elguero, J., Marzin, C., and Roberts, J.D. (1974) *The Journal of Organic Chemistry*, **39**, 357–363.

129 Begtrup, M., Elguero, J., Favre, R., Camps, P., Estopa, C., Ilarsky, D., Fruchier, A., Marzin, C., and De Mendoza, J. (1988) *Magnetic Resonance in Chemistry*, **26**, 134.

130 Damschroder, R.E. and Peterson, W.D. (1955) *Organic Syntheses, Coll Vol III*, 106–108.

131 Coburn, M.D. (1973) *Journal of Heterocyclic Chemistry*, **10**, 743–746.

132 Benson, F.R., Hartzel, L.W., and Saell, W.L. (1952) *Journal of the American Chemical Society*, **74**, 4917–4920.

133 Kato, S. and Morie, T. (1996) *Journal of Heterocyclic Chemistry*, **33**, 1171–1178.

134 Morgan, G.T. and Scharff, G.E. (1914) *Journal of the Chemical Society*, **105**, 117–123.

135 Harder, R.J., Carboni, R.A., and Castle, J.E. (1967) *Journal of the American Chemical Society*, **89**, 2643–2647.

136 Reynolds, G.A. (1964) *The Journal of Organic Chemistry*, **29**, 3733–3734.

137 Reid, W. and Schon, M. (1965) *Chemische Berichte*, **98**, 3142–3144.

138 Mitchell, G. and Rees, C.W. (1987) *Journal of the Chemical Society-Perkin Transactions 1*, 403–412.

139 Leonard, N.J. and Golankiewicz, K. (1969) *The Journal of Organic Chemistry*, **34**, 359–365.

140 Konig, W. and Geiger, R. (1970) *Chemische Berichte*, **103**, 788–798.

141 Pop, I.E., Deprez, B.P., and Tartar, A.L. (1997) *The Journal of Organic Chemistry*, **62**, 2594–2603.

142 Schiemann, K. and Showalter, H.D.H. (1999) *The Journal of Organic Chemistry*, **64**, 4972–4975.

143 (a) Shi, F., Waldo, J.P., Chen, Y., and Larock, R.C. (2008) *Organic Letters*, **10**, 2409–2412; (b) Chandrasekhar, S., Seenaiah, M., Rao, C.L., and Reddy, C.R. (2008) *Tetrahedron*, **64**, 11325–11327.

144 Carboni, R.A., Kauer, J.C., Castle, J.E., and Simmons, H.E. (1967) *Journal of the American Chemical Society*, **89**, 2618–2625.

145 (a) Rangnekar, D.W. and Dhamnaskar, S.V. (1988) *Journal of Heterocyclic Chemistry*, **25**, 1663–1664; (b) Sabnis, R.W. and Rangnekar, D.W. (1990) *Journal of Heterocyclic Chemistry*, **27**, 417–420.

146 (a) Mattaar, J.F. (1922) *Recueil des Travaux Chimiques des Pays-Bas*, **41**, 24–37; (b) Kamel, M., Ali, M.I., and Kamel, M.M. (1967) *Tetrahedron*, **23**, 2863–2868.

147 (a) Katritzky, A.R., Rachwal, S., and Hitchings, G.J. (1991) *Tetrahedron*, **47**, 2683–2732; (b) Katritzky, A.R., Lan, X., and Fan, W.-Q. (1994) *Synthesis*, 445–456; (c) Katritzky, A.R., Yang, Z., and Cundy, D.J. (1994) *Aldrichim Acta*, **27**, 31–38; (d) Katritzky, A.R., Lan, X., Yang, J.Z., and Denisko, O.V. (1998) *Chemical Reviews*, **98**, 409–548; (e) Katritzky, A.R. and Belyakov, S.A. (1998) *Aldrichim Acta*, **31**, 35–45;(f)Katritzky, A.R. and Rogovoy, B.V. (2003) *Chemistry - A European Journal*, **9**, 4586–4593.

148 Stuab, H.A., Bauer, H., and Schneider, K.M. (1988) in *Azolides in Organic Synthesis and Biochemistry*, Wiley-VCH Verlag GmbH, Weinheim, pp. 129–205.

149 Katritzky, A.R., He, H.-Y., and Suzuki, K. (2000) *The Journal of Organic Chemistry*, **65**, 8210–8213.

150 Katritzky, A.R., Zhang, Y., and Singh, S.K. (2003) *Synthesis*, 2795–2798.

151 Wang, L. and Chen, Z.-C. (2001) *Synthetic Communications*, **31**, 1633–1638.

152 Katritzky, A.R., Suzuki, K., and Wang, Z. (2005) *Synlett*, 1656–1665.

153 Katritzky, A.R., Kirichenko, N., and Rogovoy, B.V. (2003) *Synthesis*, 2777–2780.

154 (a) Katritzky, A.R., Suzuki, K., and Singh, S.K. (2003) *The Journal of Organic Chemistry*, **68**, 5720–5723; (b) Katritzky, A.R., Suzuki, K., and Singh, S.K. (2004) *Croatica Chemica Acta*, **77**, 175–178.

155 Katritzky, A.R. and Pastor, A. (2000) *The Journal of Organic Chemistry*, **65**, 3679–3682.

156 Katritzky, A.R., Abdel-Fattah, A.A.A., and Wang, M. (2003) *The Journal of Organic Chemistry*, **68**, 4932–4934.

157 Katritzky, A.R., Abdel-Fattah, A.A.A., and Wang, M. (2003) *The Journal of Organic Chemistry*, **68**, 1443–1446.

158 Katritzky, A.R., Wang, Z., Wang, M., Wilkerson, C.R., Hall, C.D., and Akhmedov, N.G. (2004) *The Journal of Organic Chemistry*, **69**, 6617–6622.

159 Katritzky, A.R., Le, K.N.B., Khelashvili, L., and Mohapatra, P.P. (2006) *The Journal of Organic Chemistry*, **71**, 9861–9864.

160 Katritzky, A.R., Widyan, K., and Kirichenko, K. (2007) *The Journal of Organic Chemistry*, **72**, 5802–5804.

161 Katritzky, A.R., Khelashvili, L., Mohapatra, P.P., and Steel, P.J. (2007) *Synthesis*, 3673–3677.

162 Katritzky, A.R., Parris, R.L., and Allin, S.M. (1995) *Synthetic Communications*, **25**, 1173–1186.

163 Musiol, J.-J. and Moroder, L. (2001) *Organic Letters*, **3**, 3859–3861.

164 Katritzky, A.R., Rogovoy, B.V., Chassaing, C., and Vvedensky, V. (2000) *The Journal of Organic Chemistry*, **65**, 8080–8082.

165 Katritzky, A.R., Rogovoy, B., Klein, C., Insuasty, H., Vvedensky, V., and Insuasty, B. (2001) *The Journal of Organic Chemistry*, **66**, 2854–2857.

166 Katritzky, A.R., Rogovoy, B.V., Cai, X., Kirichenko, N., and Kovalenko, K.V. (2004) *The Journal of Organic Chemistry*, **69**, 309–313.

167 Katritzky, A.R., Manju, K., Singh, S.K., and Meher, N.K. (2005) *Tetrahedron*, **61**, 2555–2581.

168 (a) König, W. and Geiger, R. (1970) *Chemische Berichte*, **103**, 788–798; (b) Windridge, G.C. and Jorgensen, E.C. (1971) *Journal of the American Chemical Society*, **93**, 6318–6319; (c) Bosshard, H.R., Schechter, I., and Berger, A. (1973) *Helvetica Chimica Acta*, **56**, 717–723.

169 (a) Ehrlich, A., Rothemund, S., Brudel, M., Beyermann, M., Carpino, L.A., and Bienert, M. (1993) *Tetrahedron Letters*, **34**, 4781–4784; (b) Carpino, L.A. (1993) *Journal of the American Chemical Society*, **115**, 4397–4398.

170 (a) Kim, S., Chang, H., and Ko, Y.K. (1985) *Tetrahedron Letters*, **26**, 1341–1342; (b) Castro, B., Dormoy, J.R., Evin, G., and Selve, C. (1975) *Tetrahedron Letters*, **14**, 1219–1222; (c) Coste, J., Le-Nguyen, D., and Castro, B. (1990) *Tetrahedron Letters*, **31**, 205–208.

171 Rigby, J.H., Holsworth, D.D., and James, K. (1989) *The Journal of Organic Chemistry*, **54**, 4019–4021.

172 Hughes, T.V. and Cava, M.P. (1999) *The Journal of Organic Chemistry*, **64**, 313–315.

173 Kim, S.Y., Sung, N.-D., Choi, J.-K., and Kim, S.S. (1999) *Tetrahedron Letters*, **40**, 117–120.

174 Lee, J.S., Oh, Y.S., Lim, J.K., Yang, W.Y., Kim, I.H., Lee, C.W., Chung, Y.H., and Yoon, S.J. (1999) *Synthetic Communications*, **29**, 2547–2557.

175 Shalaby, A. and Rapoport, H. (1999) *The Journal of Organic Chemistry*, **64**, 1065–1070.

176 Katritzky, A.R., Denisenko, A., and Arend, M. (1999) *The Journal of Organic Chemistry*, **64**, 6076–6079.

177 Katritzky, A.R., Huang, T.-B., and Voronkov, M.V. (2000) *The Journal of Organic Chemistry*, **65**, 2246–2248.

178 Huang, X. and Qian, H. (1999) *Synthetic Communications*, **29**, 803–808.

179 Qian, H. and Huang, X. (2000) *Synthetic Communications*, **30**, 1413–1417.

180 Katritzky, A.R., Odens, H.H., and Voronkov, M.V. (2000) *The Journal of Organic Chemistry*, **65**, 1886–1888.

181 Katritzky, A.R., Ymoshenko, D.O., Monteux, D., Vvedensky, V., Nikonov, G., Cooer, C.B., and Deshpande, M. (2000) *The Journal of Organic Chemistry*, **65**, 8059–8062.

182 Katritzky, A.R., Yao, J., and Denisko, O.V. (2000) *The Journal of Organic Chemistry*, **65**, 8063–8065.

183 Katritzky, A.R., Button, M.A.C., and Denisenko, S.N. (2001) *Heterocycles*, **54**, 301–308.

184 Katritzky, A.R., Zhang, S., Hussein, A.H.M., and Fang, Y. (2001) *The Journal of Organic Chemistry*, **66**, 5606–5612.

185 (a) Katritzky, A.R., Rodriguez-Garcia, V., and Nair, S.K. (2004) *The Journal of Organic Chemistry*, **69**, 1849–1852; (b) Katritzky, A.R., Abdel-Fattah, A.A.A.,

Vakulenko, A.V., and Tao, H. (2005) *The Journal of Organic Chemistry*, **70**, 9191–9197.

186 Katritzky, A.R., Witek, R.M., Rodriguez-Garcia, V., Mohapatra, P.P., Rogers, J.W., Cusdio, J., Abdel-Fattah, A.A.A., and Steel, P.J. (2005) *The Journal of Organic Chemistry*, **70**, 7866–7881.

187 Katritzky, A.R., Mohapatra, P.P., Fedoseyenko, D., Duncton, M., and Steel, P.J. (2007) *The Journal of Organic Chemistry*, **72**, 4268–4271.

188 Deguest, G., Bischoff, L., Fruit, C., and Marsais, F. (2007) *Organic Letters*, **9**, 1165–1167.

189 Perry, C.J., Holding, K., and Tyrrell, E. (2008) *Synthetic Communications*, **38**, 3354–3365.

190 (a) Potts, K.T. (1961) *Chemical Reviews*, **61**, 87–127; (b) Temple, C. (1981) in *1,2,4-Triazole: The Chemistry of Heterocyclic Compounds*, vol. 37 (ed. A. Weissberger), John Wiley & Sons, Inc., New York, (c) Polya, J.B. (1984) in *Comprehensive Heterocyclic Chemistry*, vol. 5 (eds A.R. Katritzky and C.W. Rees), Pergamon, Oxford, pp. 733–790; (d) Garratt, P.J. (1996) in *Comprehensive Heterocyclic Chemistry II*, vol. 4 (eds A.R. Katritzky, C.W. Rees, and E.F.V. Scriven), Elsevier, Oxford, pp. 127–163; (e) Curtis, A.D.M. (2004) in *Science of Synthesis, Vol. 13, Five-Membered Hetarenes with Three or More Heteroatoms* (eds R.C. Storr and T.L. Gilchrist), Georg Thieme Verlag, Stuttgart, New York, pp. 603–640; (f) Curtis, A.D.M. and Jennings, N. (2008) in *Comprehensive Heterocyclic Chemistry III*, vol. 5 (eds A.R. Katritzky, C.A. Ramseden, E.F.V. Scriven, and R.J.K. Taylor), Elsevier, Oxford, pp. 159–208.

191 Shaker, R.M. (2006) *Arkivoc (Arkive for Organic Chemistry)*, **9**, 59–112.

192 Creagh, L.T. and Truitt, P. (1969) *The Journal of Organic Chemistry*, **33**, 2956–2957.

193 Street, L., Baker, R., Davey, W., Guiblin, A., Jelley, R., Reeve, A., Routledge, H., Sternfeld, F., Watt, A., Beer, M., Middlemiss, D., Noble, A., Stanton, J., Scholey, K., Hargreaves, R., Sohal, B., Graham, M., and Matassa, V. (1995) *Journal of Medicinal Chemistry*, **38** 1799–1810.

194 Hossain, M.A. and Ghannoum, M.A. (2000) *Expert Opinion on Investigational Drugs*, **9**, 1797–1813.

195 Patel, L. and Lindley, C. (2003) *Expert Opinion on Pharmacotherapy*, **4**, 2279–2296.

196 Theoclitou, M.-E., Delaet, N.G.J., and Robinson, L.A. (2002) *Journal of Combinatorial Chemistry*, **4**, 315–319.

197 (a) Phoon, C.W. and Sim, M.M. (2002) *Journal of Combinatorial Chemistry*, **4**, 491–495; (b) Park, K.-H. and Cox, L.J. (2002) *Tetrahedron Letters*, **43**, 3899–3901.

198 Zong, Y.-X., Wang, J.-Ke, Yue, G.-Ren, Feng, L., Song, Z.-En, Song, H., and Han, Y.-Qi (2005) *Tetrahedron Letters*, **46**, 5139–5141.

199 Rostamizadeh, S., Tajik, H., and Yazdanfarahi, S. (2003) *Synthetic Communications*, **33**, 113–117.

200 Martin, S.W., Romine, J.L., Chen, L., Mattson, G., Antal-Zimanyi, I.A., and Poindexter, G.S. (2004) *Journal of Combinatorial Chemistry*, **6**, 35–37.

201 Stocks, M.J., Cheshire, D.R., and Reynolds, R. (2004) *Organic Letters*, **6**, 2969–2971.

202 Deshmukh, M.B., Suryawanshi, A.W., Mali, A.R., and Desai, S.R.D. (2004) *Synthetic Communications*, **34**, 2655–2658.

203 Liu, F., Palmer, D.C., and Sorgi, K.L. (2004) *Tetrahedron Letters*, **45**, 1877–1880.

204 Yeung, K.-S., Farkas, M.E., Kadow, J.F., and Meanwell, N.A. (2005) *Tetrahedron Letters*, **46**, 3429–3432.

205 Ferwanah, A.-R.S., Kandile, N.G., Awadallah, A.M., and Miqdad, O.A. (2002) *Synthetic Communications*, **32**, 2017–2025.

206 Su, W., Yang, D., and Li, J. (2005) *Synthetic Communications*, **35**, 1435–1440.

207 Molteni, G. and Del Buttero, P. (2005) *Heterocycles*, **65**, 1183–1188.

208 Ciesielski, M., Pufky, D., and Döring, M. (2005) *Tetrahedron*, **61**, 5942–5947.

209 Hüisgen, R., Grashey, R., Seidel, M., Wallbillich, G., Knupfer, H., and Schmidt, R. (1962) *Justus Liebigs Annalen der Chemie*, **653**, 105–113.

210 Anzani, F., Croce, P.D., and Stradi, R., *Journal of Heterocyclic Chemistry*, **17**, 311–313.
211 Peronnet, J., Girault, P. (1980) *Bulletin de la Societe Chimique de France* (1973) 2843–2847.
212 Chouaieb, H., Mosbah, M.B., Kossentini, M., and Salem, M. (2003) *Synthetic Communications*, **33**, 3861–3868.
213 Atkinson, M.R. and Polya, J.B. (1954) *Journal of the Chemical Society*, 3319–3324.
214 Davidson, J.S. (1979) *Synthesis*, 359–360.
215 Fraser, J.K., Neilson, D.G., Newlands, L.R., and Watson, K.M. (1975) *Journal of the Chemical Society-Perkin Transactions 1*, 2280–2284.
216 Paul, H., Hilgetag, G., and Jahnchen, G. (1968) *Chemische Berichte*, **101**, 2033–2036.
217 El Kaim, L., Grimaud, L., Jana, N.K., Mettetal, F., and Tirla, C. (2002) *Tetrahedron Letters*, **43**, 8925–8933.
218 Paulvannan, K., Chen, T., and Hale, R. (2000) *Tetrahedron*, **56**, 8071–8076.
219 Paulvannan, K., Hale, R., Sedehi, D., and Chen, T. (2001) *Tetrahedron*, **57**, 9677–9682.
220 El Kaim, L., Grimaud, L., Jana, N.K., Mettetal, F., and Tirla, C. (2002) *Tetrahedron Letters*, **43**, 8925–8927.
221 Ferwanah, A.-R.S. (2003) *Synthetic Communications*, **33**, 243–251.
222 Mogilaiah, K., Babu, H.R., and Reddy, N.V. (2002) *Synthetic Communications*, **32**, 2377–2384.
223 Music, I. and Vercek, B. (2001) *Synthetic Communications*, **31**, 1511–1519.
224 Buscemi, S., Vivona, N., and Caronna, T. (1996) *The Journal of Organic Chemistry*, **61**, 8397–8401.
225 Buscemi, S., Pace, A., Pibiri, I., and Vivona, N. (2003) *The Journal of Organic Chemistry*, **68**, 605–608.
226 Reitz, D.B. and Finkes, M.J. (1989) *Journal of Heterocyclic Chemistry*, **26**, 225–230.
227 Kidwai, M., Misra, P., Bhushan, K.R., and Dave, B. (2000) *Synthetic Communications*, **30**, 3031–3040.
228 Pace, A., Pibiri, I., Buscemi, S., and Vivona, N. (2004) *The Journal of Organic Chemistry*, **69**, 4108–4115.
229 Gehlen, H. and Blankenstein, G. (1962) *Justus Liebigs Annalen der Chemie*, **651**, 137–141.
230 (a) Gehlen, H. and Blankenstein, G. (1962) *Justus Liebigs Annalen der Chemie*, **651**, 128–132; (b) Gehlen, H. and Robisch, G. (1963) *Justus Liebigs Annalen der Chemie*, **663**, 119–123.
231 Balsells, J., DiMichele, L., Liu, J., Kubryk, M., Hansen, K., and Armstrong III, J.D. (2005) *Organic Letters*, **7**, 1039–1042.
232 Buscemi, S., Pace, A., Piccionello, A.P., Pibiri, I., and Vivona, N. (2005) *Heterocycles*, **65**, 387–394.
233 (a) Suni, M.M., Nair, V.A., and Joshua, C.P. (2001) *Synthetic Communications*, **31**, 1599–1605; (b) Suni, M.M., Nair, V.A., and Joshua, C.P. (2001) *Tetrahedron*, **57**, 2003–2009.
234 Liu, C. and Iwanowicz, E.J. (2003) *Tetrahedron Letters*, **44**, 1409–1411.
235 Klingele, M.H. and Brooker, S. (2004) *European Journal of Organic Chemistry*, 3422–3434.
236 Wu, D.-Q., He, J.-L., Wang, J.-K., Wang, X.-C., and Zong, Y.-X. (2006) *Journal of Chemical Research*, 293–294.
237 Batchelor, D.V., Beal, D.M., Brown, T.B., Ellis, D., Gordon, D.W., Johnson, P.S., Mason, H.J., Ralph, M.J., Underwood, T.J., and Wheeler, S. (2008) *Synlett*, 2421–2424.
238 Yu, Y., Ostresh, J.M., and Houghten, R.A. (2003) *Tetrahedron Letters*, **44**, 7841–7843.
239 Boeglin, D., Cantel, S., Heitz, A., Martinez, J., and Fehrentz, J.-A. (2003) *Organic Letters*, **5**, 4465–4468.
240 Graybill, T.L., Thomas, S., and Wang, M.A. (2002) *Tetrahedron Letters*, **43**, 5305–5309.
241 Cowden, C.J., Wilson, R.D., Bishop, B.C., Cottrell, I.F., Davies, A.J., and Dolling, U.-H. (2000) *Tetrahedron Letters*, **41**, 8661–8664.
242 Huang, X., Palani, A., Xiao, D., Asianian, R., and Shih, N.-Y. (2004) *Organic Letters*, **6**, 4795–4798.

243 Katritzky, A.R., Rogovoy, B.V., Vvedensky, V.Y., Kovalenko, K., Steel, P.J., Markov, V.I., and Forood, B. (2001) *Synthesis*, 897–903.

244 Makara, G.M., Ma, Y., and Margarida, L. (2002) *Organic Letters*, **4**, 1751–1754.

245 Katritzky, A.R., Khashab, N.M., Kirichenko, N., and Singh, A. (2006) *The Journal of Organic Chemistry*, **71**, 9051–9056.

246 Funabiki, K., Noma, N., Kuzuya, G., Matsui, M., and Shibata, K. (1999) *Journal of Chemical Research-S*, 300–301.

247 Xue, H., Twamley, B., and Shreeve, J.M. (2004) *The Journal of Organic Chemistry*, **69**, 1397–1400.

248 (a) Bentiss, F., Lagrenée, M., and Barby, D. (2000) *Tetrahedron Letters*, **41**, 1539–1541; (b) Bentiss, F., Lagrenee, M., Traisnel, M., Mernari, B., and Elattari, H. (1999) *Journal of Heterocyclic Chemistry*, **36**, 149–152.

249 Wang, J.-K., Zong, Y.-X., and Yue, G.-R. (2005) *Synlett*, 1135–1136.

250 Bulger, P.G., Cottrell, I.F., Cowden, C.J., Davies, A.J., and Dolling, U.-H. (2000) *Tetrahedron Letters*, **41**, 1297–1301.

251 Katritzky, A.R., Kuzmierkiewicz, W., and Greenhill, J.V. (1991) *Recueil des Travaux Chimiques des Pays-Bas*, **110**, 369–373.

252 Mirzaei, Y.R., Twamley, B., and Shreeve, J.M. (2002) *The Journal of Organic Chemistry*, **67**, 9340–9345.

253 Takahashi, K., Shimizu, S., and Ogata, M. (1987) *Synthetic Communications*, **17**, 809–815.

254 Katritzky, A.R., Qi, M., Feng, D., Zhang, G., Griffith, M.C., and Watson, K. (1999) *Organic Letters*, **1**, 1189–1191.

255 Olofson, R.A. and Kendall, R.V. (1970) *The Journal of Organic Chemistry*, **35**, 2246–2248.

256 Diez-Barra, E., de la Hoz, A., Rodriguez-Curiel, R.I., and Tejeda, J. (1997) *Tetrahedron*, **53**, 2253–2260.

257 (a) Astleford, B.A., Goe, G.L., Keay, J.G., and Scriven, E.F.V. (1989) *The Journal of Organic Chemistry*, **54**, 731–732; (b) Smith, K., Small, A., and Hutchings, M.G. (1990) *Chemistry Letters*, **19** 347–350.

258 Kawasaki, I., Domen, A., Kataoika, S.-Y., Yamauchi, K., Yamashita, M., and Ohta, S. (2003) *Heterocycles*, **60**, 351–363.

259 (a) Staab, H.A. (1956) *Chemische Berichte*, **89**, 1927–1940; (b) Woodruff, M., and Polya, J.B. (1975) *Australian Journal of Chemistry*, **28**, 133–141.

260 (a) Van den Bos, B.G. (1960) *Recueil des Travaux Chimiques des Pays-Bas*, **79**, 836–842; (b) Hirata, T., Wood, H.B., and Driscoll, J.G. (1973) *Journal of the Chemical Society-Perkin Transactions 1*, 1209–1212.

261 (a) Yuxiong, O., Boren, C., Jiarong, L., Shuan, D., Jianjun, L., and Huiping, J. (1994) *Heterocycles*, **38**, 1651–1664; (b) Chen, M.J., Chi, C.S., and Chen, Q.Y. (1990) *Phosphorus, Sulfur Silicon and Related Elements*, **54**, 87–93; (c) Mackay, M.F., Trantino, G.J., and Wilshire, J.F.K. (1993) *Australian Journal of Chemistry*, **46**, 417–425.

262 Fedorov, A.Y. and Finet, J.P. (1999) *Tetrahedron Letters*, **40**, 2747–2748.

263 Cristau, H.-J., Cellier, P.P., Spindler, J.-F., and Taillefer, M. (2004) *Chemistry - A European Journal*, **10**, 5607–5622.

264 (a) Hamburg, G. and Mildenberger, H. (1982) *Annalen Der Chemie-Justus Liebig*, 1387–1393; (b) Katritzky, A.R., Lue, P., and Yannakopoulou, K. (1990) *Tetrahedron*, **46**, 641–648; (c) Gugina, N., Holzer, W., and Wasicky, M. (1992) *Heterocycles*, **34**, 303–314; (d) Katritzky, A.R., Darabantu, M., Aslan, D.C., and Oniciu, D.C. (1998) *The Journal of Organic Chemistry*, **63**, 4323–4331.

265 Regel, E. (1977) *Justus Liebigs Annalen der Chemie*, 159–168.

266 Ohta, S., Kawasaki, I., Fukuno, A., Yamashita, M., Tada, T., and Kawabata, T. (1993) *Chemical & Pharmaceutical Bulletin*, **41**, 1226–1231.

267 Papadopoulos, E.P. and Schupbach, C.M. (1979) *The Journal of Organic Chemistry*, **44**, 99–104.

268 Hansen, K.B., Springfield, S.A., Desmond, R., Devine, P.N., Grabowski, E.J.J., and Reider, P.J. (2001) *Tetrahedron Letters*, **42**, 7353–7355.

269 Enders, D., Breuer, K., Kallfass, U., and Balensiefer, T. (2003) *Synthesis*, 1292–1295.

270 Korotkikh, N.I., Rayenko, G.F., Shvaika, O.P., Pekhtereva, T.M., Cowley, A.H., Jones, J.N., and Macdonald, C.L.B. (2003) *The Journal of Organic Chemistry*, **68**, 5762–5765.

271 Becker, H.G.O. and Eibsch, R. (1972) *Journal Fur Praktische Chemie*, **314**, 923–935.

272 Fugina, N., Holzer, W., and Wasicky, M. (1992) *Heterocycles*, **34**, 303–314.

273 Zumbrunn, A. (1998) *Synthesis*, 1357–1361.

274 Zolfigol, M.A., Madrakian, E., Ghaemi, E., and Mallakpour, S.E. (2002) *Synlett*, 1633–1636.

275 Zolfigol, M.A., Torabi, M., and Mallakpour, S.E. (2001) *Tetrahedron*, **57**, 8381–8384.

276 Zolfigol, M.A., Zebarjadian, M.H., Chehardoli, G., Mallakpour, S.E., and Shamsipur, M. (2001) *Tetrahedron*, **57**, 1627–1629.

277 Zolfigol, M.A., Bagherzadeh, M., Chehardoli, G., and Mallakpour, S.E. (2001) *Synthetic Communications*, **31**, 1149–1154.

278 Zolfigol, M.A., Chehardoli, G., and Mallakpour, S.E. (2003) *Synthetic Communications*, **33**, 833–841.

279 Adam, W., Pastor, A., and Wirth, T. (2000) *Organic Letters*, **2**, 1295–1297.

280 Chiong, H.A. and Daugulis, O. (2007) *Organic Letters*, **9**, 1449–1451.

281 Zolfigol, M.A., Choghamarani, A.G., Shahamirian, M., Safaiee, M., Mohammadpoor-Baltork, I., Mallakpour, S., and Abdollahi-Alibeik, M. (2005) *Tetrahedron Letters*, **46**, 5581–5584.

282 Christoforou, A., Nicolaou, G., and Elemes, Y. (2006) *Tetrahedron Letters*, **47**, 9211–0213.

283 Zolfigol, M.A., Azarifar, D., Mallakpour, S., Mohammadpoor-Baltork, I., Forghaniha, A., Malekia, B., and Abdollahi-Alibeik, M. (2006) *Tetrahedron Letters*, **47**, 833–836.

284 Marigo, M., Wabnitz, T.C., Fielenbach, D., and Jorgensen, K.A. (2005) *Angewandte Chemie-International Edition*, **44**, 794–797.

285 Maki, B.E., Chan, A., Phillips, E.M., and Scheidt, K.A. (2007) *Organic Letters*, **9**, 371–374.

286 Maki, B.E. and Scheidt, K.A. (2008) *Organic Letters*, **10**, 4331–4334.

287 Liu, Q. and Rovis, T. (2006) *Journal of the American Chemical Society*, **128**, 2552–2553.

288 He, M., Struble, J.R., and Bode, J.W. (2006) *Journal of the American Chemical Society*, **128**, 8418–8420.

289 Cullen, S.C. and Rovis, T. (2008) *Organic Letters*, **10**, 3141–3144.

290 He, M., Beahm, B.J., and Bode, J.W. (2008) *Organic Letters*, **10**, 3817–3820.

291 Chiang, P.-C., Kaeobamrung, J., and Bode, J.W. (2007) *Journal of the American Chemical Society*, **129**, 3520–3521.

292 Ma, Y., Wei, S., Lan, J., Wang, J., Xie, R., and You, J. (2008) *The Journal of Organic Chemistry*, **73**, 8256–8264.

293 Mennen, S.M. and Miller, S.J. (2007) *The Journal of Organic Chemistry*, **72**, 5260–5269.

294 Vora, H.U., Moncecchi, J.R., Epstein, O., and Rovis, T. (2008) *The Journal of Organic Chemistry*, **73**, 9727–9731.

295 Liu, Q., Perreault, S., and Rovis, T. (2008) *Journal of the American Chemical Society*, **130**, 14066–14067.

296 Bode, J.W. and Sohn, S.S. (2007) *Journal of the American Chemical Society*, **129**, 13798 13799.

297 Thomson, J.E., Kyle, A.F., Gallagher, K.A., Lenden, P., Concellon, C., Morrill, L.C., Miller, A.J., Joannesse, C., Slawin, A.M.Z., and Smith, A.D. (2008) *Synthesis*, 2805–2818.

298 Duguet, N., Campbell, C.D., Slawin, A.M.Z., and Smith, A.D. (2008) *Organic and Biomolecular Chemistry*, **6**, 1108–1113.

299 He, L., Lv, H., Zhang, Y.-R., and Ye, S. (2008) *The Journal of Organic Chemistry*, **73**, 8101–8103.

300 Enders, D. and Han, J. (2008) *Synthesis*, 3864–3868.

301 Enders, D., Han, J., and Henseler, A. (2008) *Chemical Communications*, **38** 3989–3991.

13
Oxadiazoles

Giovanni Romeo and Ugo Chiacchio

13.1
Introduction

There are four isomeric types of oxadiazoles (**1–4**).

1 **2** **3** **4**

Examples of all these ring derivatives are reported; the 1,2,3-oxadiazole system is well represented by the mesoionic sydnones (**5**, X = O) and sydnonimines (**5**, X = NR). In fact, potential 1,2,3-oxadiazoles **6** are not known: when formed in some reactions, they isomerize, instantaneously, into the open α-diazoketone tautomeric forms **7** (Figure 13.1).

Arguments concerning the existence of **6** and **7** have been summarized [1], but the firmly established 1,2,3-oxadiazole ring system is of the sydnone type.

Ring systems of type **2** are commonly termed azoximes and the 1,2,5-oxadiazoles **3** are often referred to by the trivial name furazan, while 1,2,5-oxadiazole-2-oxide, a well-known derivative, has the trivial name furoxan.

Notably, 1,2,4-oxadiazoles have received great attention in the pharmaceutical industry. In contrast, 1,3,4-oxadiazoles have recently found extensive application in the field of new materials for the development of electric as well as optical devices.

The chemistry of 1,2,3- [2], 1,2,4- [3], 1,2,5- [4], and 1,3,4-oxadiazoles [5] has been widely reported in a series of books and reviews. We refer here to the cited references for general aspects of reactivity of these heterocycles. Particular attention is paid to the literature published after 1995.

Modern Heterocyclic Chemistry, First Edition.
Edited by Julio Alvarez-Builla, Juan Jose Vaquero, and José Barluenga.

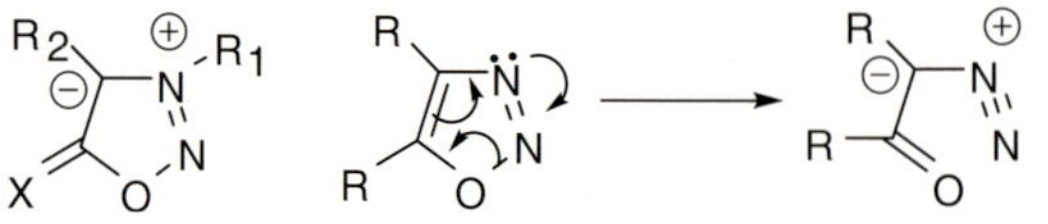

Figure 13.1 Mesoionic sydnones (**5**, X = O) and sydnone imines (**5**, X = NR) are well represented, while potential 1,2,3-oxadiazoles **6** are not known since when formed in some reactions they isomerize instantaneously into the open α-diazoketone tautomeric forms **7**.

13.2
1,2,3-Oxadiazoles

These are the least common of the oxadiazole group of heterocycles and the literature relating to them is rather sparse. With a very limited and still not perfectly defined number of exceptions, simple 1,2,3-oxadiazoles **6** are not isolable because they isomerize immediately to the their more stable open-chain tautomers, the α-diazoketones **7**. The sterically protected 1,2,3-oxadiazole **8** is the only known oxadiazole, bearing alkyl substituents, which exists in the cyclic form in the crystalline state but as diazoketone in chloroform solution [6].

1,2,3-Oxadiazoles have been proposed [7] as not-isolated intermediates in the oxidation of alkenes with nitrous oxide: a recent DFT analysis predicts that the reaction consists of two steps, with the formation of 1,2,3-oxadiazole in the first and its decomposition in the second, leading to carbonyl compounds [8] (Scheme 13.1).

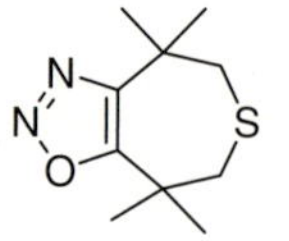

8

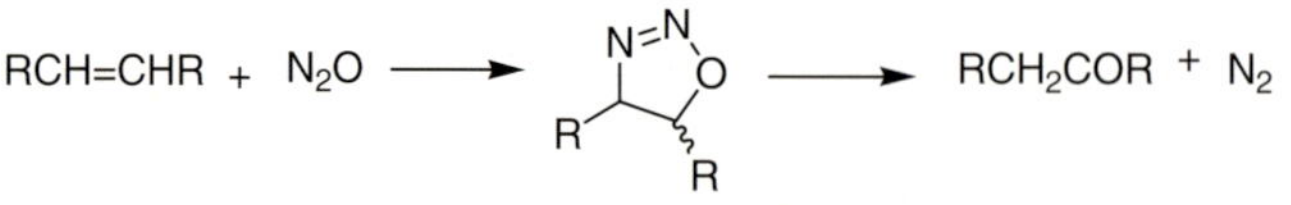

RCH=CHR + N_2O ⟶ [1,2,3-oxadiazoline] ⟶ $RCH_2COR + N_2$

Scheme 13.1

Fusion with an aromatic ring does not stabilize the system. 1,2,3-Benzoxadiazole (**9**) is more stable as an *o*-quinone diazide (**10**↔**11**): the ionization potentials, measured by MS for **10** and **11** and estimated for **9**, indicate that the stabilizing influence of the zwitterionic structures is more important than the gain in aromaticity [9, 10].

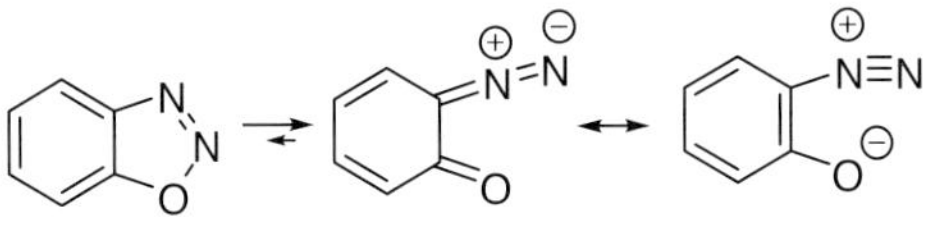

9 10 11

Some substituted 1,2,3-benzoxadiazoles have been shown to exist in equilibrium with their open chain tautomers. The relative concentration of the species at the equilibrium is strongly dependent upon solvent and substitution effects: the diazoketone structure is stabilized by hydrogen bonding and polar interactions. The most stable of these compounds is 5,7-di-*t*-butyl-1,2,3-benzoxadiazole (**12**) which is 6.3 kJ mol^{-1} more stable than its diazocyclohexadienone valence isomer **13** in the vapor phase.

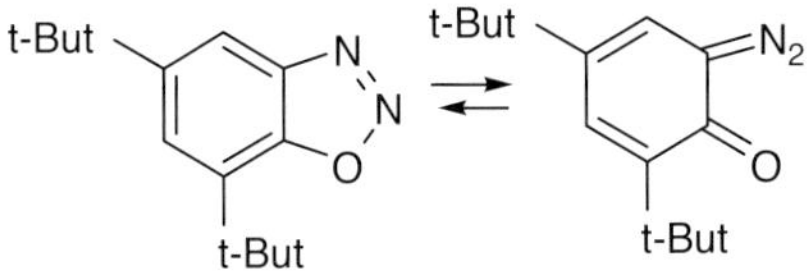

12 13

Several examples in the literature report on compounds that have been incorrectly formulated as 1,2,3-oxadiazoles [11]: the previously formulated diazoesters **14** were successively established as non-cyclic **15**.

14 15

13.2.1
Sydnones and Sydnonimines

The 1,2,3-oxadiazole ring system is present in the stable mesoionic sydnones (**5a**, X = O and sydnone imines (**5b**, X = NR). The trivial term "sydnone" comes from the University of Sidney, where the first example of these compounds, the 3-phenylsydnone **16**, was synthesized (by Earl and Mackney) by cyclodehydration of *N*-nitroso-*N*-phenylglycine with acetic anhydride [12] (Scheme 13.2).

Baker and Ollis coined the term mesoionic [13] to describe the structure of such compounds, for which a totally covalent structure cannot be written, and which cannot be represented satisfactorily by any one polar structure. The term was then extended to several compounds that can be depicted only as resonance hybrids of dipolar structures. Structure **17**, in which the positive charge is delocalized, can be represented as the summary of three canonical forms **18–20** [14]. The formal positive charge is associated with the ring atoms, and the formal negative charge is associated

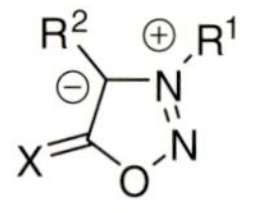

5a: X = O

5b: X = NR_1

R^1 = Me, Ph, Bn
R^2 = Me, Et, Ph, Bn

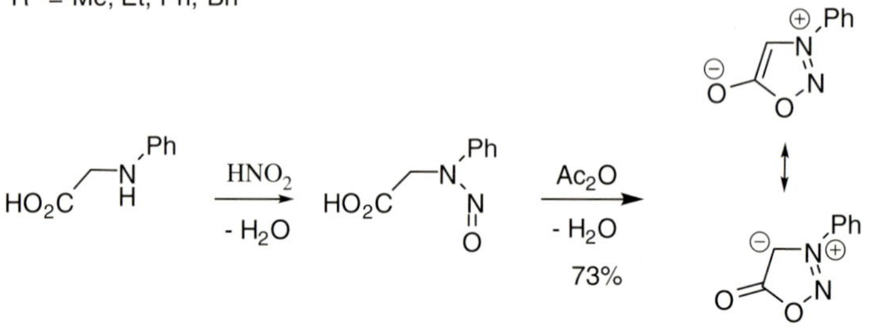

HNO_2, - H_2O; Ac_2O, - H_2O, 73%

16

Scheme 13.2

with ring atoms or an exocyclic nitrogen or chalcogen atom. X-Ray evidence shows that the valence tautomer **21** should also be considered in discussions of the structure of the sydnones [15].

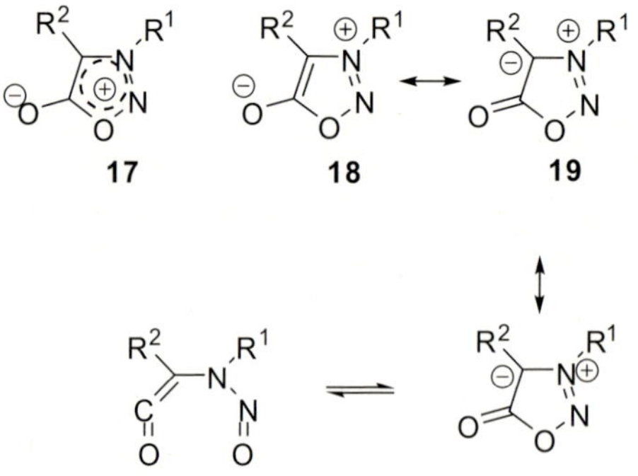

17 18 19

21 20

13.2.2
1,2-3-Oxadiazolines

The partially reduced 4,5-dihydro-1,2,3-oxadiazole system, theoretically assemblable by dipolar cycloaddition of diazoalkanes to carbonyl compounds, has never been detected directly in such reactions, although dihydro-1,2,3-oxadiazoline structures have been proposed in the literature [16]. *Ab initio* and DFT calculations indicate that, in the reaction of diazomethane with formaldehyde, the kinetically most favorable cycloadduct is less stable than the reactants and has a lower barrier for nitrogen elimination [17]. Derivative **22** has been postulated as intermediate in the metabolism of some (2-hydroxyethyl)- or (2-haloethyl)nitrosoureas, a class of highly active antitumor agents (Scheme 13.3) [18].

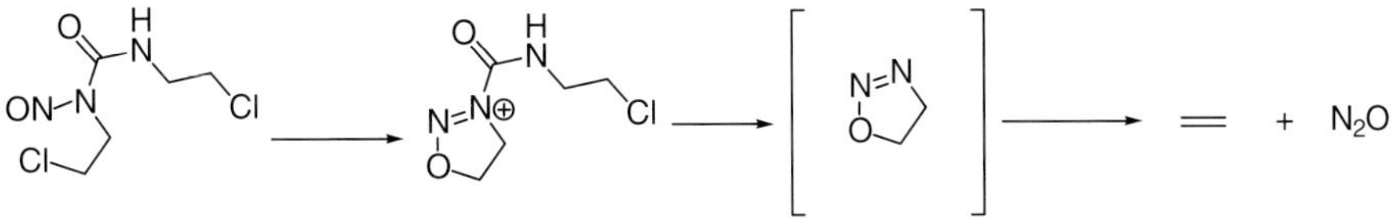

Scheme 13.3

One derivative, the salt **23**, has been isolated as a crystalline solid [19].

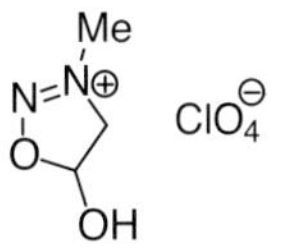

23

An unambiguous characterization of the so-called "Traube's oxazomalonic acid," obtained from the condensation of dimethyl malonate with nitric oxide, has demonstrated that the compound is really an unusual five-membered heterocycle and corresponds to 3-hydroxy-2-carboxysydnone dianion **24** (Scheme 13.4): the synthesis, structure and spectroscopic analysis of the potassium salt and methyl ester have also been reported [20].

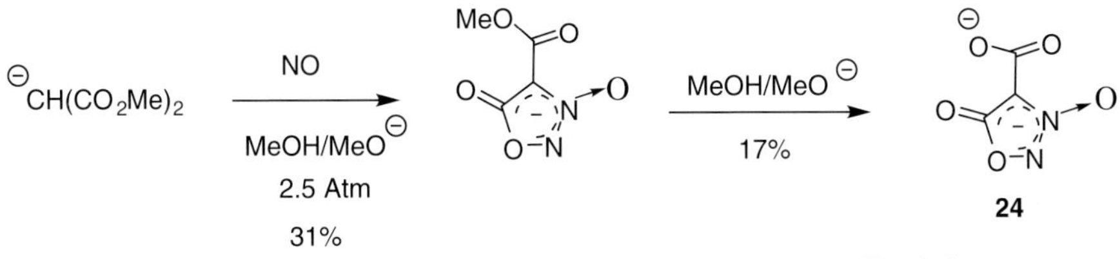

Scheme 13.4

The oxidation state represented by an N-oxide lends stability to the 1,2,3-oxadiazole system [21]. Thus, the reaction of diene **25** with nitrosyl chloride afforded 4,4-dimethyl-5-(2-methylpropenyl)-Δ^2-1,2,3-oxadiazoline 3-oxide **26**, whose structure was confirmed by chemical and spectroscopic data (Scheme 13.5).

A wide variety of 1,2,3-oxadiazole 3-oxides, valuable candidates for drug frameworks, have been synthesized by the reaction of nitric oxide with functionalized

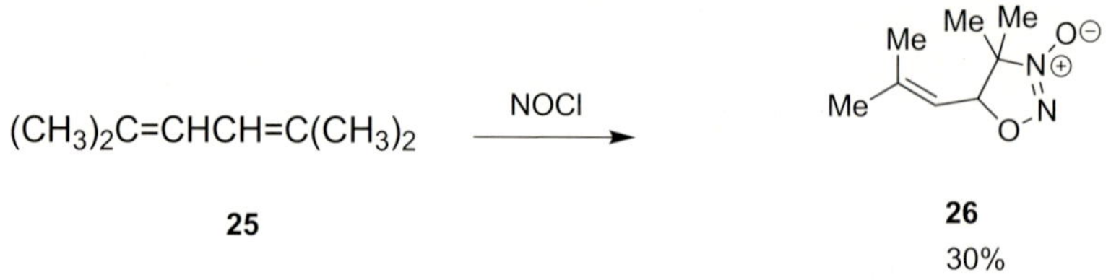

$(CH_3)_2C{=}CHCH{=}C(CH_3)_2$ —NOCl→ 26 (30%)

25

Scheme 13.5

alkynyllithium derivatives [22]. A theoretical study [23] indicates that the overall reaction is stepwise and is considered to include two processes. In the first, the nitrogen atom in nitric oxide at first attacks the C1 atom in alkynyllithium to afford the intermediate **27**. In the second, another nitric oxide reacts with **27** to produce **28**. Then, attack of the oxygen atom at C2 to form a five-membered-ring geometry (**29**) is followed by addition of water, leading to the final 5-alkyl-1,2,3-oxadiazole 3-oxides **30** (Scheme 13.6).

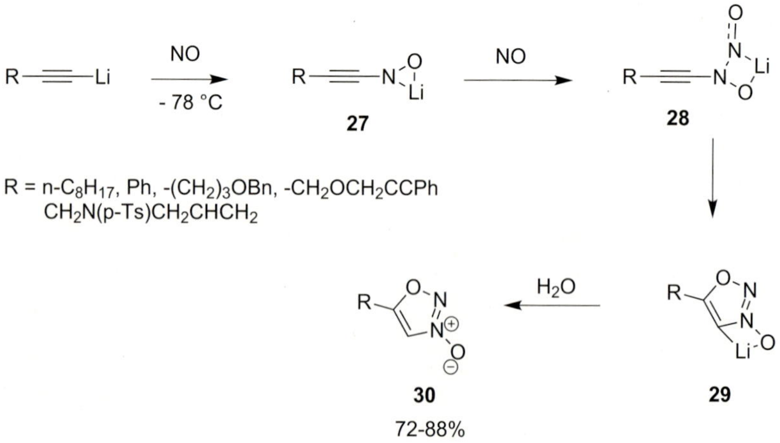

R = n-C_8H_{17}, Ph, -$(CH_2)_3$OBn, -CH_2OCH_2CCPh
CH_2N(p-Ts)CH_2CHCH_2

NO, - 78 °C; NO; H_2O

27 **28** **29** **30** 72-88%

Scheme 13.6

The structure of some of the obtained compounds was confirmed by X ray crystallography and spectroscopic data.

The isomeric 2-oxide system is present in **32**, isolated as a crystalline solid from the reaction of the nitroazo-compounds **31** with bases [24]. As a general process, the intramolecular alkylation of 2-halo- or 2-cyano-substituted nitramines proceeds through an O-alkylation and leads to 4,5-dihydro-1,2,3-oxadiazole 2-oxides **32** (Scheme 13.7) [25].

Recently, a new synthetic approach to functionally substituted 4,5-dihydro-1,2,3-oxadiazolo 2-oxides **34** has been described, starting from sulfamic acid derivatives **33** (Scheme 13.8) [26].

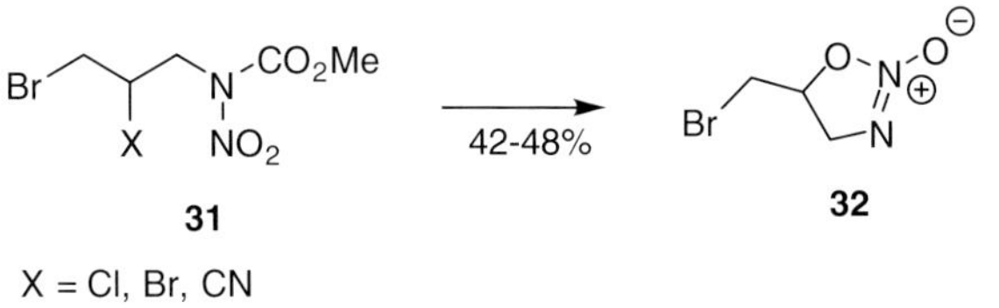

31

X = Cl, Br, CN

42-48%

32

Scheme 13.7

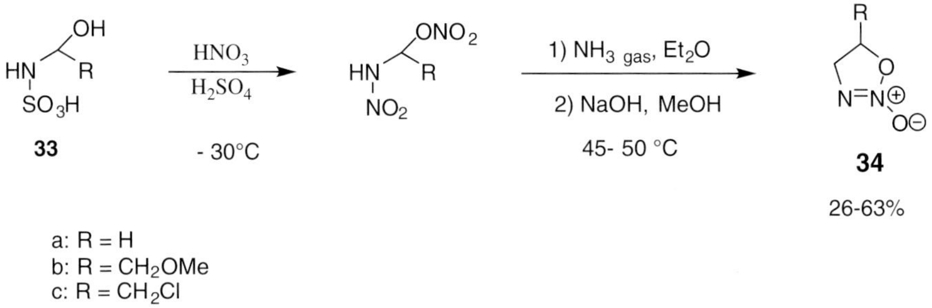

33

HNO_3 / H_2SO_4 / - 30°C

1) $NH_{3\ gas}$, Et_2O / 2) NaOH, MeOH / 45- 50 °C

34

26-63%

a: R = H
b: R = CH_2OMe
c: R = CH_2Cl

Scheme 13.8

Stable derivatives of the 1,2,3-oxadiazolidine ring system **35** are unknown.

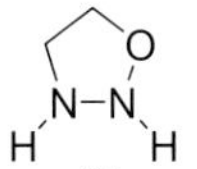

35

13.2.3 Theoretical Aspects

Ab initio theoretical studies and semiempirical MNDO calculations [27], performed on 1,2,3-oxadiazoles, indicate that the heterocycle is too unstable to be isolated and predict a major stability for its tautomer the diazoacetaldehyde. MNDO calculations on substituted 1,2,3-benzoxadiazoles and the isomeric diazocyclohexadienones reach the same conclusion [10].

Ab initio methods have also been used to calculate the geometry and the energy of 4,5-dihydro-1,2,3-oxadiazole (**22**): the molecule is predicted to be unstable, its most favorable mode of decomposition being a retro 1,3-dipolar addition to diazomethane and formaldehyde [28].

Several theoretical studies have addressed the structure and aromaticity of sydnones [20–34]. Sydnones could be regarded as aromatic because their structures could be represented as cyclic arrays of p-orbitals containing six p-electrons, with four from the C=N−N system and two from the lone pair on oxygen. However, sydnones are not a delocalized "aromatic" ring system, as confirmed by their chemical reactivity: the reactions of sydnones include both substitutions and additions

Semiempirical and *ab initio* calculations provide the same overall description of the bond lengths. The bond from C5 to the exocyclic oxygen atom is essentially a double bond, while the bonds O1–N2, O1–C5, and C4–C5 are approximately single bonds. N2–N3 and N3–C4 are partial double bonds (**36**). On this basis, these compounds could be regarded as 1,3-dipolar azomethyne imines bearing a conjugative carbonyl group at C5 [29].

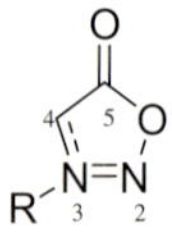

36 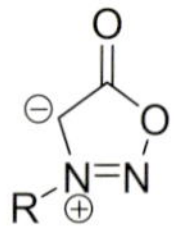 **37**

X-Ray structural measurements confirm that the exocyclic C–O bond is close in length to that of a normal carbonyl group [35–38]. Therefore, according to the values of the net atomic charges on the ring, determined by semiempirical methods, the resonance structure **37** appears to be the best single representation. The large dipole moments of sydnones (>6 D) are consistent with their strongly polar character and the charge separation shown in structure **37** [29, 36].

Frontier orbital energies and coefficients for sydnones and for some substituted sydnones, performed by the MINDO/3 method, show that the HOMO is a pure p-orbital with a large coefficient on N2 and C4 (Figure 13.2) [29, 33]. On this basis, 1,3-dipolar cycloadditions of sydnones to electron-deficient alkenes should be controlled by the HOMO of the sydnone and the LUMO of the dipolarophile.

13.2.4
Structural Aspects

Structural parameters, IR spectra, ionization potentials, relative energies, isomerization barriers, and solvation energies have been calculated for sydnones and for the aromatic benzo-1,2,3-oxadiazole (prevalent tautomer in the gas phase) and zwitterionic 6-diazocyclohexa-2,4-dienone (prevalent tautomer in a polar solvent) molecules. The calculations indicate that unsubstituted 1,2,3-oxadiazole is unstable in all solvents [39].

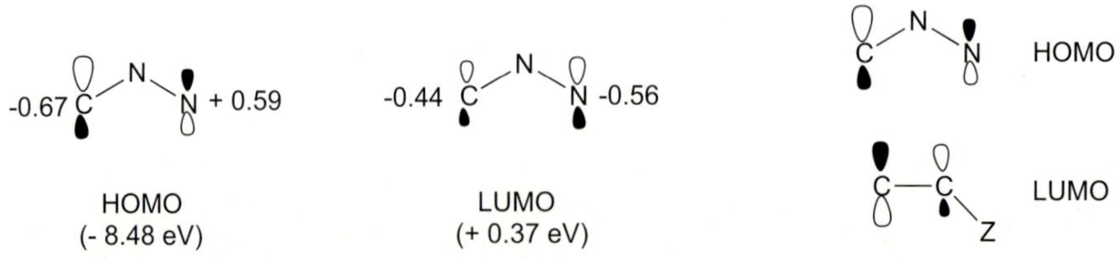

-0.67 C N N + 0.59

HOMO
(- 8.48 eV)

-0.44 C N N -0.56

LUMO
(+ 0.37 eV)

HOMO

LUMO

Figure 13.2 Frontier orbital energies and coefficients for sydnones and for some substituted sydnones show that the HOMO is a pure p-orbital with a large coefficient on N2 and C4.

13.2.4.1 X-Ray Diffraction

Crystal structure data of several sydnones have been reported [34, 40]. The ring is nearly planar and the values for bond lengths are in the range illustrated in structure **38**. The exocyclic carbonyl bond length is 1.21 Å; the N2–N3 and N3–C4 bonds are somewhat shorter than a single bond, while the only C–C bond present is longer than a double bond.

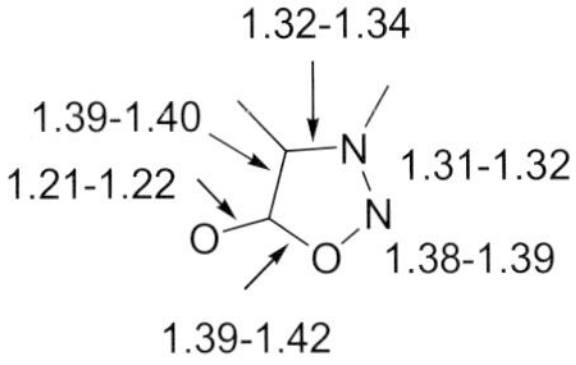

1.32-1.34
1.39-1.40
1.21-1.22
1.31-1.32
1.38-1.39
1.39-1.42

38

13.2.4.2 UV and IR Spectra

UV and IR spectroscopy have afforded useful information in studies of the equilibrium between 1,2,3-benzoxadiazole and *o*-quinone diazide structures. The IR spectra of 1,2,3-benzoxadiazoles show absorptions at 1626, 1611, 1464, and 1457 cm^{-1}, whereas the isomeric quinone diazides show strong absorptions at 2090 and 1718 cm^{-1} [41]. The UV spectrum of 1,2,3-benzoxadiazole in an argon matrix shows maxima at 201, 243, and 289 nm [10].

Carbonyl stretching frequencies of sydnones are in the range 1720–1790 cm^{-1}. Alkylsydnones show a single maximum at 290 nm in the UV spectrum.

13.2.4.3 NMR Spectra

The ^{1}H and ^{13}C spectra of several sydnones and sydnonimines have been reported [42]. In accord with the dipolar structure **5**, the H4 proton resonates upfield in the range 6.2–6.8 ppm. Analogously, for the strong deshielding effect of positively charged N3 atom, the 3-alkyl protons are shifted downfield (4.10–4.40 ppm) with respect to 4-alkyl protons (2.20–2.50 ppm).

For 3-methylsydnone, the ^{14}N and ^{17}O spectra have also been determined and a complete set of chemical shift values for all the atoms have been reported [45]. The ^{13}C chemical shifts are reported on structures **39** and **40**.

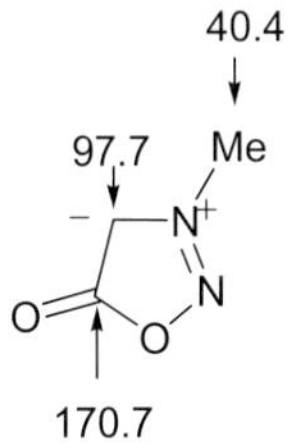

40.4
97.7
170.7

39

109.8
33.9
174.2

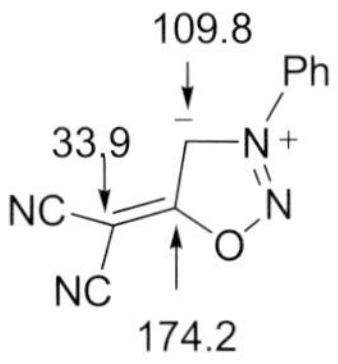

40

According to a synthetic approach that allowed for the independent labeling of the nitrogen atoms (Section 13.2.5.1), the NMR chemical shift for each ^{15}N has been determined unambiguously.

13.2.4.4 Mass Spectra

Electron impact mass spectra of the sydnone ring are characterized by the loss of NO (M–30) and CO (M–28) fragments, which can occur consecutively or simultaneously. The fragment M-58 represents generally the base peak and the molecular ion is often distinguishable [44].

Reported CI spectra indicate the same pattern of fragmentation. In the fused sydnone **41**, the initial loss of NO is followed by CO, HCN, acetylene, and finally Ph, as the principal fragment ion [45].

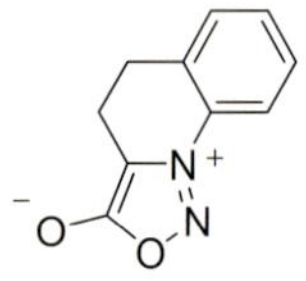

41

13.2.4.5 Other Properties

The highly polarized yet neutral electrical character and the high dipolar moments of sydnones have been exploited for the design of technologically interesting thermotropic liquid crystals (LCs) with properties between those of covalent and ionic LC. The molecular design, synthesis, and characterization of the first examples of both classical and non-conventional chiral mesoionic (mesomeric + ionic) liquid crystals derived from sydnones have been reported (Scheme 13.9) [46, 47].

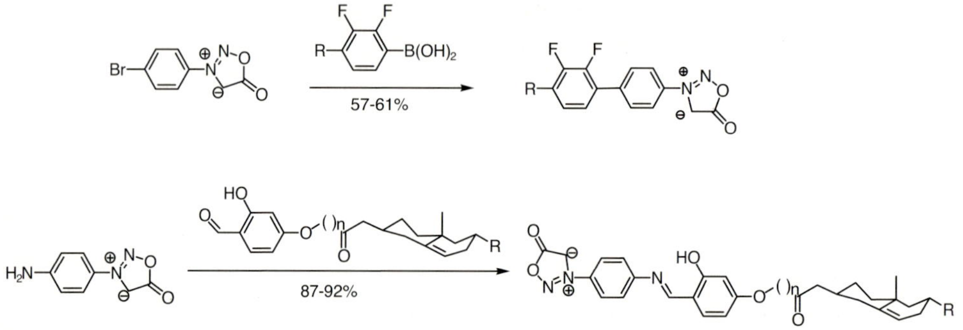

57-61%

87-92%

Scheme 13.9

The occurrence of chiral smectic phases in these novel compounds was evidenced by optical microscopy, calorimetry, and X-ray studies.

A side-chain polysiloxane containing 3-(4-aminophenyl)sydnone moieties at terminal and aliphatic spacer has been prepared and its structure was confirmed by IR and NMR measurements. By introducing sydnone into polysiloxane, the polymer displays a high electrorheological effect due to the increased interaction between sydnone moieties [48].

13.2.5 Synthesis of 1,2,3-Oxadiazoles

The synthetic approach towards substituted 1,2.3-benzoxadiazoles is based on the synthesis of the tautomeric open-chain 6-diazo-1,2-cyclohexadienones [49, 50]. Thus, the diazotization of 2-aminophenols by treatment with sodium nitrite or isoamyl nitrite, followed by careful neutralization with potassium carbonate, afforded the diazoketones (Scheme 13.10), which is in equilibrium with the cyclic tautomer benzoxadiazole (**42**) (Section 13.2). An alternative route exploited the reaction of a substituted *o*-benzoquinone with tosyl hydrazine [43]. Naphthoxadiazole (**43**), stable in the solid state at $-19\,^{\circ}$C, has been prepared according to this synthetic route.

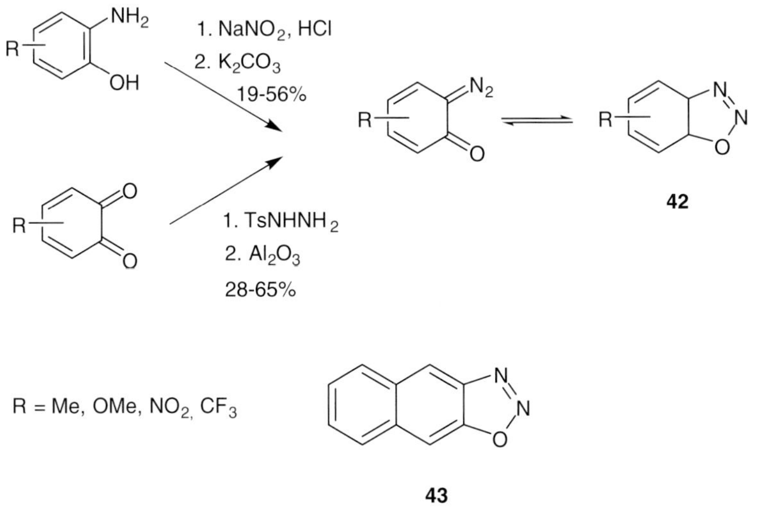

1. $NaNO_2$, HCl
2. K_2CO_3
19-56%

1. $TsNHNH_2$
2. Al_2O_3
28-65%

42

R = Me, OMe, NO_2, CF_3

43

Scheme 13.10

13.2.5.1 Sydnones and Sydnonimines

Despite extensive studies of the sydnone ring, practically only one general synthetic entry is available [51]. The method involves (a) nitrosation of amino acids to give **44**; (b) formation of a mixed anhydride **45** and (c) cyclization to the sydnone ring **46** (Scheme 13.11).

The nitrosation step has been carried out under neutral conditions, using isoamyl nitrite. Among the dehydrating agents, trifluoroacetic anhydride gives the most rapid

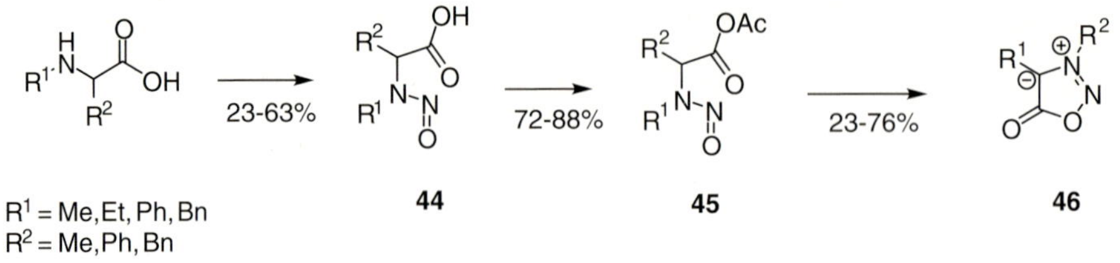

Scheme 13.11

results; thionyl chloride, phosphorus oxychloride, phosphoric anhydride, and carbodiimides have also been used successfully.

The cyclization step can be aided by ultrasonic irradiation [52, 53]: in this way, functionalized 3-aryl sydnones have been prepared in good yields.

A similar general method towards sydnonimines involves the nitrosation of the corresponding α-aminonitriles **47** and cyclization of the intermediate **48** (Scheme 13.12) [54]. Substituents can be introduced at the exocyclic nitrogen atom by normal methods in acidic or buffered solutions: sydnone imines are more stable in acid and less stable in base than sydnones.

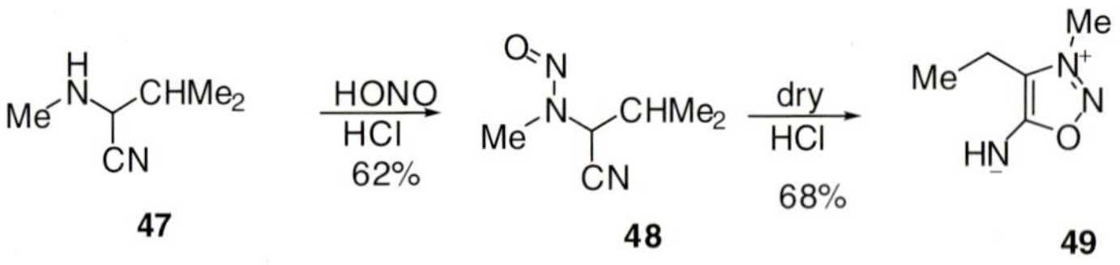

Scheme 13.12

A three-component reaction of the Mannich type has been exploited to prepare 3-*N*-hydroxy- (**50**) and 3-*N*-amino- (**51**) substituted sydnone imines (Scheme 13.13) [55–57].

13.2.5.2 4,5-Dihydro-1,2,3-Oxadiazolines

Methods for the synthesis of the few reported compounds of this type have been described in Section 13.2.2. Scheme 13.14 describes the synthesis of 4,5-dihyro-3-methyl-1,2,3-oxadiazolium tosylate (**52**) [58]. Accordingly, 5-alkoxy-substituted derivatives can be prepared by cyclization of 2,2-dialkoxy-*N*-methyl-*N*-nitrosoethylamines [59].

13.2.6 Reactivity of 1,2,3-Oxadiazoles

With the exception of some benzo-fused derivatives and 4,5-dihydro-1,2,3-oxazolidinium salts, the chemistry of the 1,2,3-oxadiazole system is nearly confined to the mesoionic sydnones or sydnonimines.

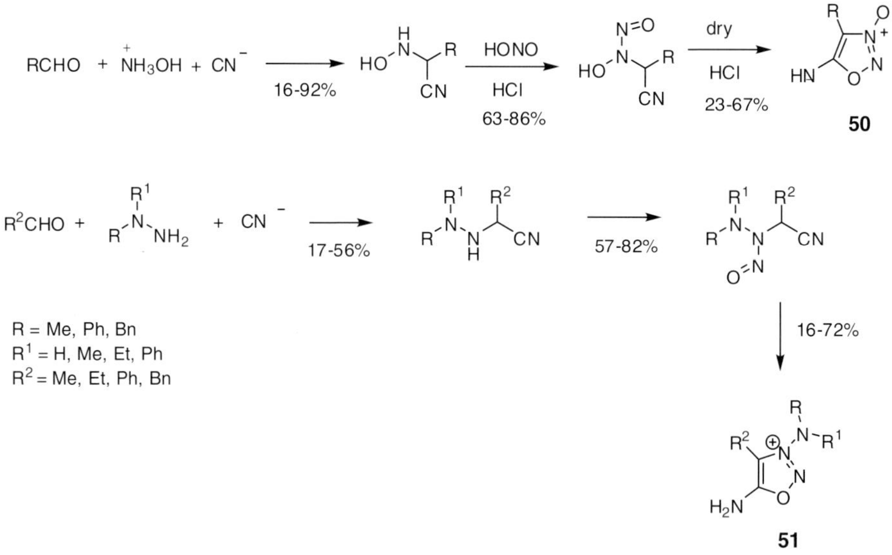

Scheme 13.13

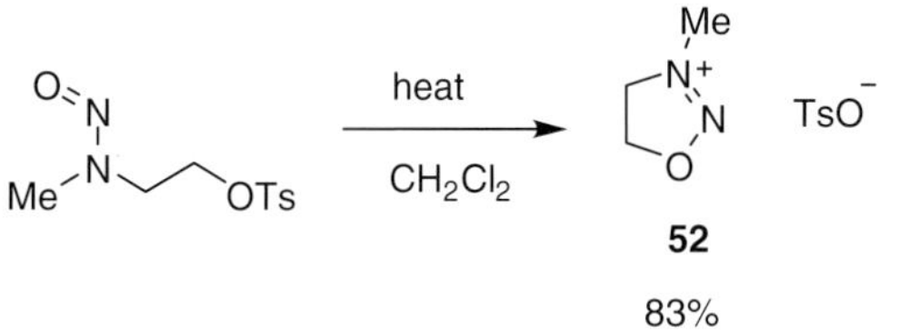

Scheme 13.14

13.2.6.1 **Benzo-1,2,3-Oxadiazoles**

UV irradiation cleaves the benzoxadiazole ring to 2-diazocyclohexadienones: subsequent loss of nitrogen and Wolff rearrangement leads to ketene **53** (Scheme 13.15) [10]. The formation of 2-naphthol **54** and methyl indene-2-carboxylate **55** by irradiation of naphthoxadiazole **43** is amenable to the loss of nitrogen from the diazocarbonyl tautomer [60].

13.2.6.2 **4,5-Dihydro-1,2,3-Oxadiazoles**

The known chemistry is limited to 4,5-dihydro-3-methyl-1,2,3-oxadiazolinium salts. The cation reacts with nucleophiles at the methyl group (methylation of the nucleophile) or at C5, with opening of the ring.

13.2.6.3 **Sydnones**

13.2.6.3.1 **General Aspects** Sydnones are crystalline compounds that are sensitive to hydrolysis, especially in basic media where they are rapidly cleaved. The ring is also

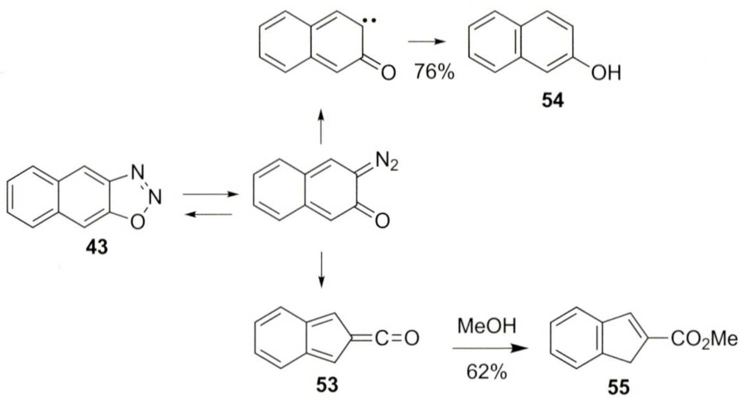

Scheme 13.15

cleaved by catalytic reduction and, oxidatively, by reaction with nitric acid, potassium permanganate, and other oxidants.

The chemical reactivity of the sydnone system is displayed in ring cleavage reactions and in processes in which the ring system is retained such as substitution or addition reactions (at the 4-position). Substituents can be introduced into the 4-position by conventional electrophilic substitution or after metallation at C4. Standard transformation of functional groups at C4 of sydnone have also been investigated extensively and targeted to the synthesis of various 4-substituted sydnones.

Sydnones can act as 1,3-dipoles in dipolar cycloaddition reactions.

13.2.6.3.2 Ring Cleavage

Hydrolysis (Acid and Basic Ring Cleavage) The alkaline hydrolysis of sydnonimines (Scheme 13.16) proceeds through an experimentally ascertained third-order kinetics, and leads to ring cleavage that affords the nitrosonitrile **56** [61].

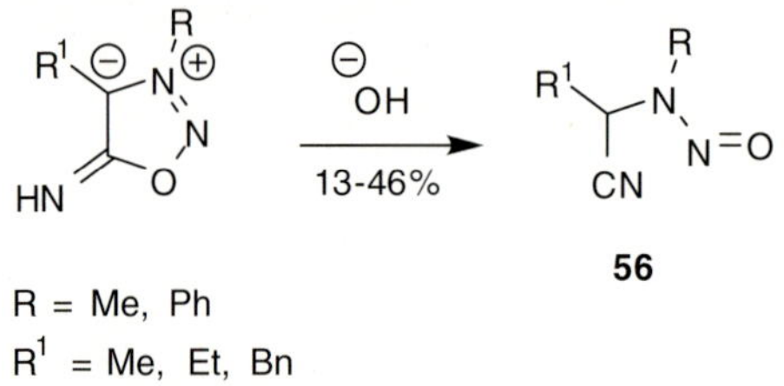

R = Me, Ph
R^1 = Me, Et, Bn

Scheme 13.16

Acid hydrolysis of sydnones, which occurs at elevated temperatures, has been exploited as a synthetic path to alkyl- and arylhydrazines (Scheme 13.17) [62–64].

Oxidative Ring Cleavage Ring cleavage of sydnones with oxygen in the dark affords a mixture of products; thus, oxidation of 3-phenylsydnone gives benzaldehyde, benzyl

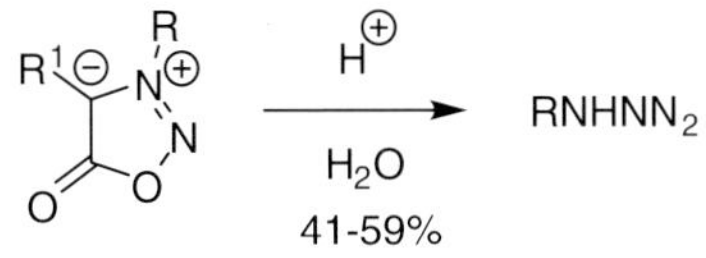

$\xrightarrow[H_2O\ 41\text{-}59\%]{H^{\oplus}}$ $RNHNN_2$

R = Ph, Pyridyl
R^1 = Me, Ph, Bn

Scheme 13.17

alcohol, and benzyl formate, while 3-benzyl-4-phenylsydnone affords benzyl phenylglyoxylate, benzyl benzoate, diphenylmethane, benzyl alcohol benzaldehyde, and benzoic acid (Scheme 13.18) [65].

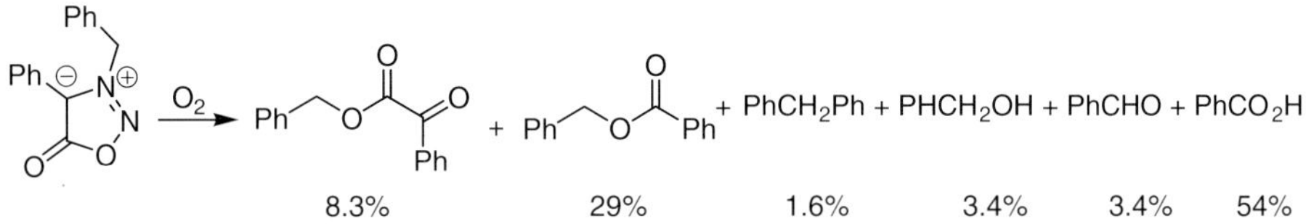

$\xrightarrow{O_2}$ + $PhCH_2Ph$ + $PHCH_2OH$ + PhCHO + $PhCO_2H$

8.3% 29% 1.6% 3.4% 3.4% 54%

Scheme 13.18

The proposed mechanism, which implies radical intermediates, arises from an initial electron-transfer reaction of the sydnone with oxygen. Recombination of the radical ion with $^{\bullet}O_2^{-}$ would lead to the hydroxyperoxy zwitterion **57**, which could then cyclize at the 3- or 2-position to give **58** and **59**, respectively. Further collapse of **58** and **59** afforded the obtained mixtures of compounds (Scheme 13.19).

Oxidation with ozone of 4-methylsydnones leads to pyruvate esters (Scheme 13.20) [66].

Thermal and Photochemical Ring Cleavage Thermochemical and photoinduced decomposition of sydnones give different products as a function of the nature of substituents present in the ring.

3,4-Diarylsydnones lose carbon dioxide by UV irradiation or by flash photolysis and give transient nitrile imines, which can be intercepted by external or internal dipolarophiles [67, 68]. For example, the photochemically induced reaction of 3,4-diphenylsydnone affords, in the presence of DMAD, the dimethyl 2,5-diphenylpyrazole-3,4,dicarboxylate (**61**) (Scheme 13.21), which originates from the loss of CO_2 and the addition of the dipolarophile to the dipolar intermediate **60** (Scheme 13.21) [69].

In the thermochemically induced process, the addition of DMAD occurred first to give **62**, followed by elimination of carbon dioxide to form the dimethyl 1,5-diphenylpyrazole-3,4-dicarboxylate **63**.

Alkenes and other trapping agents have been used to capture the dipolar nitrilimine [70–72]. When 1,3-butadiene was used as a dipolarophile, 1,3-diphenyl-5-vinylpyrazole (**64**) was obtained, so confirming that **60** is the trapped fragment (Scheme 13.22) [73].

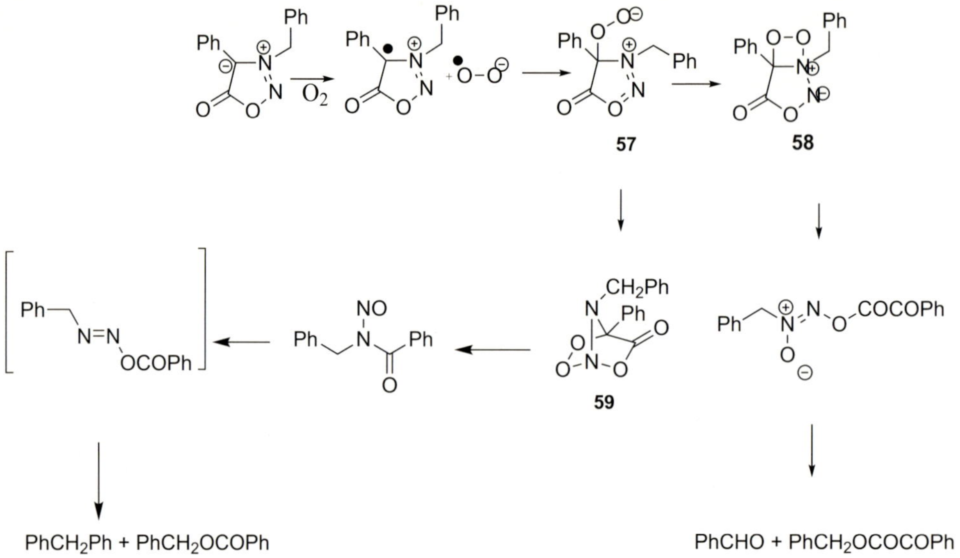

PhCH$_2$Ph + PhCH$_2$OCOPh

PhCHO + PhCH$_2$OCOCOPh

Scheme 13.19

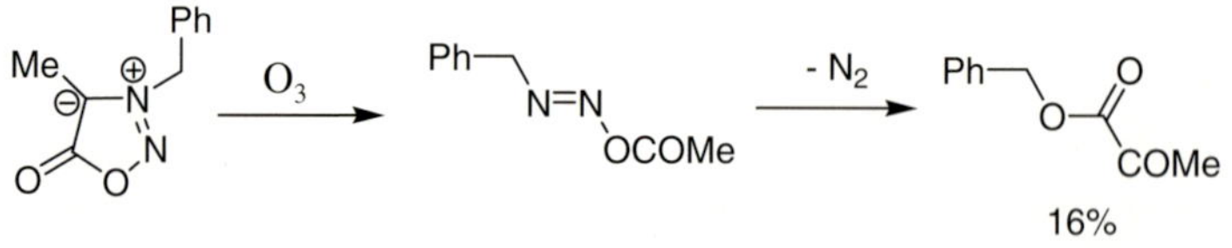

16%

Scheme 13.20

47%

98%

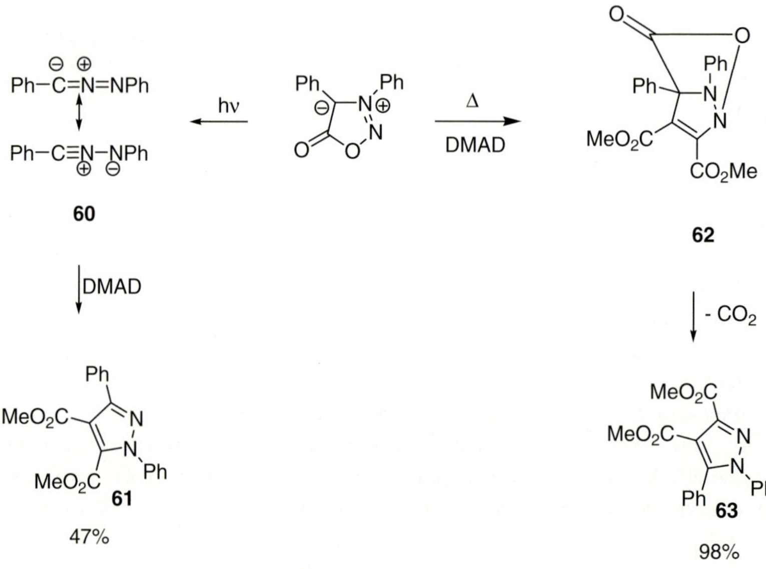

Scheme 13.21

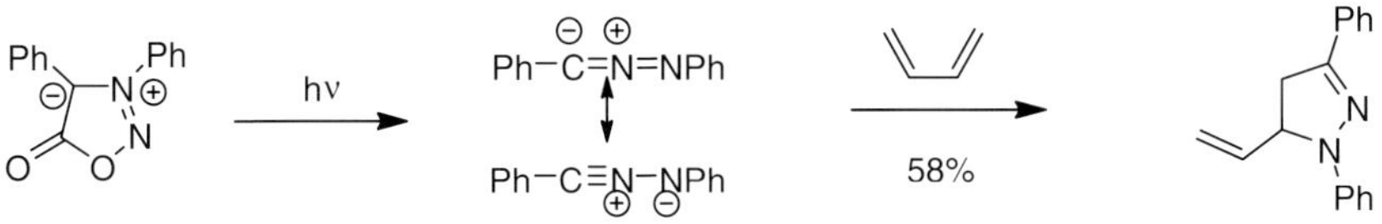

Scheme 13.22

3-Aryl-4-[2-(2-vinylphenyl)ethenyl]sydnones undergo fast isomerization to the trans isomer and competitive photolysis of the sydnone moiety, giving the corresponding nitrile imine, which cannot react intramolecularly. In the presence of acrolein, a [3 + 2] cycloaddition takes place to give the *trans*-styrylpyrazoline derivative **65**, which during isolation aromatizes to the pyrazoles **66** and **67** (Scheme 13.23) [74].

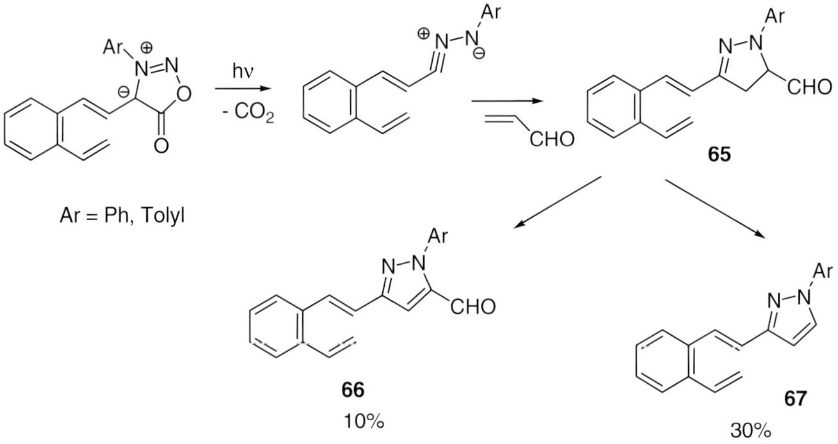

Scheme 13.23

Nitrile imines have not been detected from 3-arylsydnones unsubstituted at C4; ESR techniques have, however, revealed the presence of the radical species **68**, which originates from the photolysis of 3-phenylsydnone [75].

Nitrile imines have also been claimed as intermediates in the formation of triazole derivatives by photochemical decomposition of a sydnone in dioxane. By labeling the nitrogen atoms in **69**, the mechanism for the formation of the triazole **71**, based on **70** as key intermediate, has been supported (Scheme 13.24) [76].

The photosensitized oxidation of sydnones with singlet oxygen has also been reported to give a mixture of products. In the presence of Rose Bengal as a sensitizer, singlet oxygen adds to a sydnone as a dipolarophile. The identification of benzoic acid and dibenzoylphenylhydrazine among the reaction products has been rationalized on the basis of two simultaneous reaction pathways (Scheme 13.25) [77].

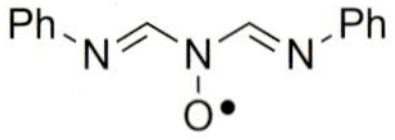

Ph–N=CH–N(O•)–CH=N–Ph

68

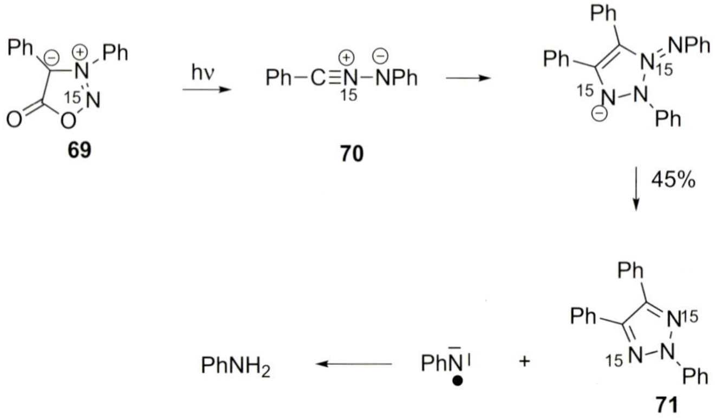

Scheme 13.24

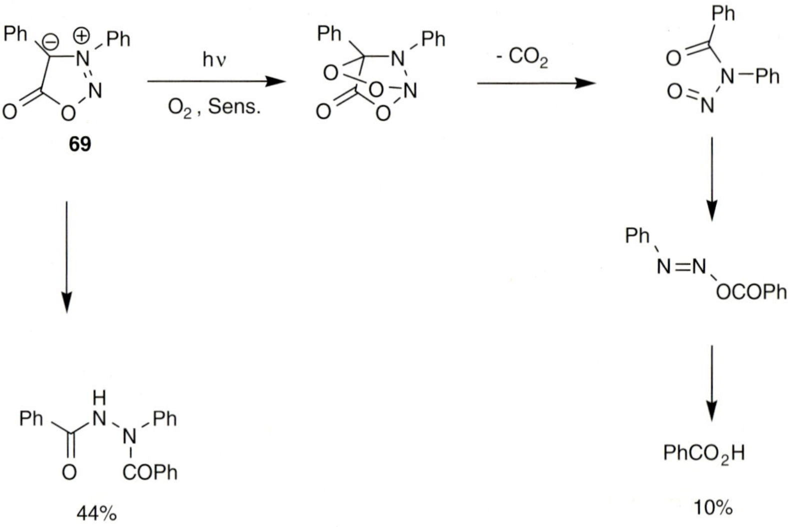

Scheme 13.25

Several sydnones develop a color when irradiated by UV light; for instance, a blue color has been observed by irradiation of the 3-(3-pyridyl)sydnone. The phenomenon has been explained through the formation of diketene **72**, which has been identified as the blue species [78].

72

13.2.6.3.3 **Nucleophilic Substitution at C4** Replacement reactions at C4 in sydnones have been reviewed. Butyllithium has been exploited to displace the bromine atom from a 3-phenylsydnone [46b]: the resulting organometallic compound has been carbonylated, added to ketones and converted into a silyl derivative [79]. Grignard compounds have also been prepared from 3-bromosydnones, and subsequently reacted with ketones to give the corresponding alcohols (Table 13.1) [80].

Metallation reactions have been exploited as a synthetic tool for effecting electrophilic substitution at C4 (Section 13.2.6.3.4) [81].

13.2.6.3.4 **Electrophilic Substitution at C4** Electrophiles can be directly introduced at C4 in the sydnone ring. Table 13.2 summarizes a series of such reactions reported in literature [82, 83].

With 4-unsubstituted 3-alkyl- or 3-phenylsydnones, substitution occurs only at the electron-rich C4, while when aryl substituents are present at C3 the position of the electrophilic attack depends upon the nature of the aryl group, Exclusive aryl ring nitration occurs with electron donors on the aryl group [84]. Thus, 3-(2-aminophenyl) sydnone is brominated in the benzene ring *para* to the amino group [85] while the nitration of 3,4-diphenylsydnone affords the 4-nitrophenyl derivative. As further

Table 13.1 Nucleophilic replacement reactions at C4 in sydnones.

Reagent A → Reagent B → 60-84%

X	Reagent A	Y	Reagent B	Z
Br, H	BuLi	Li	CO_2 $COCl_2$ $MeCOCHMe_2$	COOH $(CO)_{1/2}$ $MeC(OH)CHMe_2$
Br	Mg, ether, MeI	MgBr	I_2 Ac_2O RCHO	I COMe CH(OH)R
SMe	H_2O_2	SO_2Me	$NaBH_4$	H

Table 13.2 Electrophilic replacement reactions at C4 in sydnones.

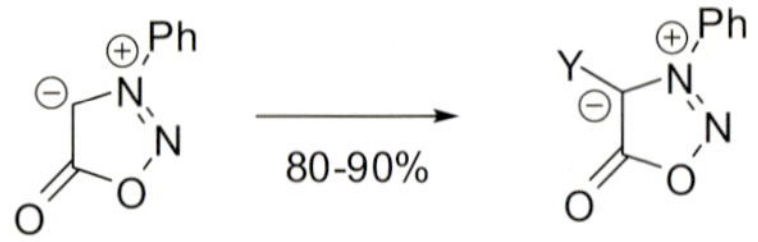

80-90%

Reagent	Y
(a) Br_2, ether, $NaHCO_3$	Br
(b) $HONO_2$ + $HOSO_3H$, 0 °C	NO_2
(c) SO_3 (dioxane)	SO_3H
(d) $ClSO_3H$ + H_3PO_4	SO_2Cl
(e) Ac_2O + BF_3 (ether)	COMe
(f) $HCONMe_2$ + $POCl_3$	CHO
(g) $Hg(OAc)_2$	HgOAc
(h) DMSO + AcCl	SMe

examples, the sydnone ring is brominated in preference to a pyrazolyl system at C3, while nitration of 3-methyl-4-phenyl sydnone affords the 4-nitrophenyl substituted derivative.

Intramolecular electrophilic substitutions at C4 provide a route to fused sydnones such as **73** [86] and **74** [87].

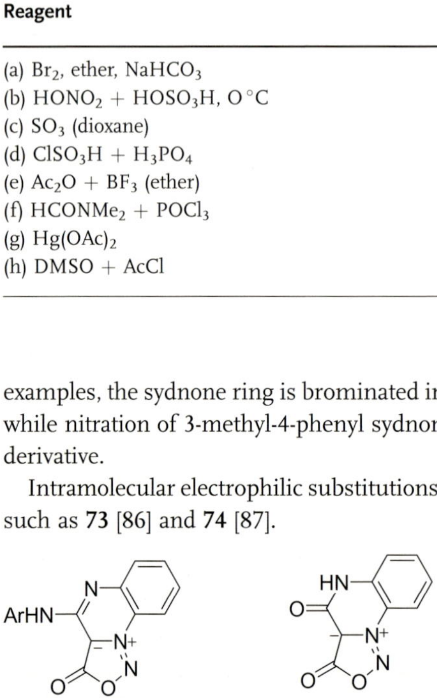

73 **74**

Electrophiles have also been introduced at the 4-position through organometallic derivatives. 4-Lithio intermediates [46b] (Section 13.2.6.3.3) have been used to introduce several S, Se, and Te electrophiles [88, 89], and also formyl or acetyl substituents (Scheme 13.26) [90]. This strategy has been exploited for the synthesis of 3-amino-4-benzoylsydnone (**75**), the first example of a sydnone containing an amino group at C3 [91].

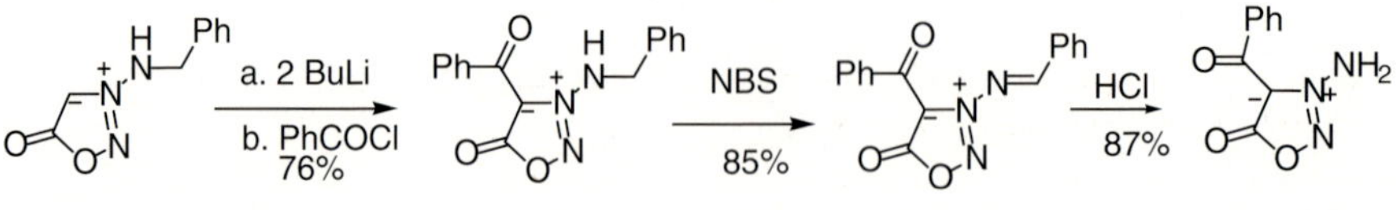

a. 2 BuLi, b. PhCOCl, 76%; NBS, 85%; HCl, 87%

75

Scheme 13.26

Vinyl and aryl substituents at C4 have been introduced by means of other organometallic species. Thus, the reaction of 4-lithio-3-phenylsydnone with copper(I) bromide affords the stable copper derivative **76**, which gives palladium (0)-catalyzed coupling to iodobenzenes and vinyl bromides [92]. The reaction of the lithium intermediate with copper(II) bromide leads to the dimer **77** (Scheme 13.27).

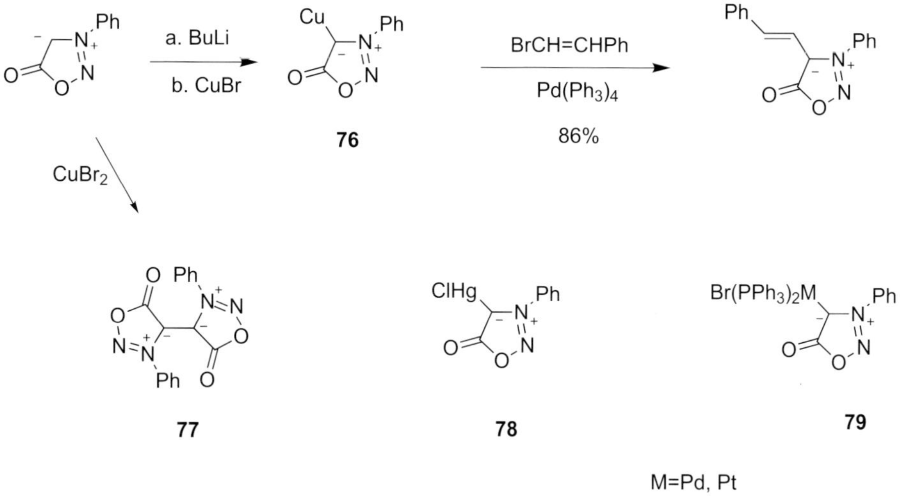

a. BuLi
b. CuBr
76
BrCH=CHPh
$Pd(Ph_3)_4$
86%
$CuBr_2$
77
78
79
M=Pd, Pt

Scheme 13.27

Various 4-arylethynyl sydnones have been prepared in good yields by the reaction of 4-bromo-3-phenylsydnone with aryl acetylenes under palladium catalysis [93].

Chloromercuro derivatives **78** have also been used in Heck coupling reactions with vinyl halides, and sydnones with platinum or palladium substituents **79** have been prepared from 4-bromo-3-phenylsydnone and $M(PPh)_3$ [94].

13.2.6.3.5 **Reactions of Substituents** Standard transformations occur in various functional groups present on the sydnone ring. Thus, 3-phenylsydnone-4-carboxylic acids can be easily converted into the corresponding esters, amides, and hydrazides; tertiary alcohols can be dehydrated to alkenes and ketones can be condensed with benzaldehyde. Aldehyde **80** can be converted into the corresponding (*E*)- and (*Z*)-alkenes by a Wittig reaction (Scheme 13.28) [95].

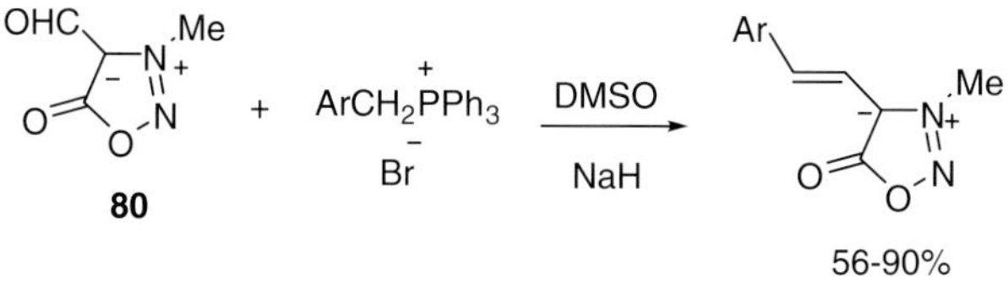

80
$ArCH_2\overset{+}{P}Ph_3\ \overset{-}{Br}$
DMSO
NaH
56-90%

Scheme 13.28

Sydnonyl-substituted α,β-unsaturated ketones have been synthesized by Claisen–Schmidt condensation of 4-acetyl-3-arylsydnones with aryl aldehydes. An easy, eco-friendly synthetic version has also been reported that involves grinding 4-acetyl-3-aylsydnones with aryl aldehydes in a mortar [96].

Further reaction of sydnonyl-substituted α,β-unsaturated ketones with hydrazine hydrate afforded sydnonyl-substituted pyrazolines **81**, which possess useful applications in medicine (Scheme 13.29) [97].

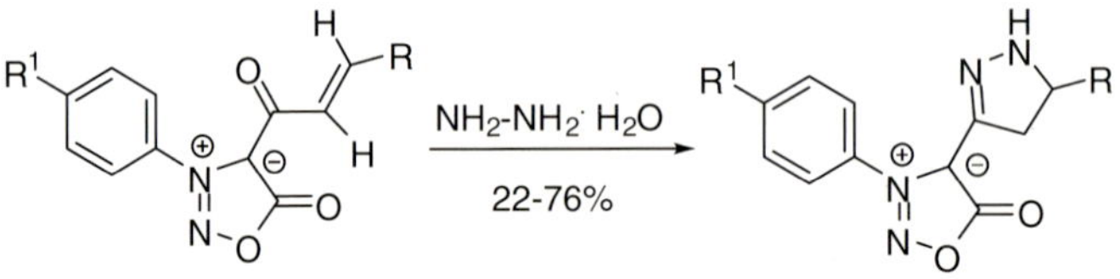

NH_2-$NH_2 \cdot H_2O$

22-76%

81

R = Me, Et, Ph, Bn
R^1 = Me, OMe, NO_2

Scheme 13.29

Moreover, the reaction of 3-aryl-2-bromo-1-sydnonylpropenones with 3-arylaminomethyl-4-amino-5-mercapto-1,2,4-triazoles gives 3-arylaminomethyl-6-(3-arylsydnon-4-yl)-8-aryl-1,2,4-triazolo[3,4-*b*][1,3,4]thiadiazepines **82** with antibacterial activity (Scheme 13.30) [98].

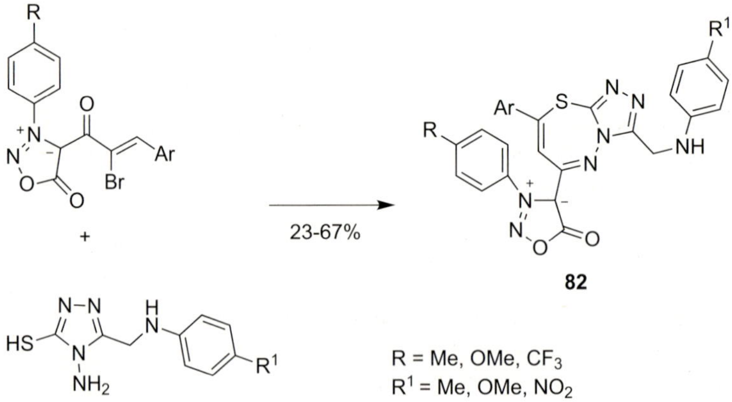

23-67%

82

R = Me, OMe, CF_3
R^1 = Me, OMe, NO_2

Scheme 13.30

4-Aryl-3-formylsydnones can be easily converted into their oximes, from which other functionalized sydnones such as the nitrile **83** [99] and the nitrile oxide **84** [100] can be obtained.

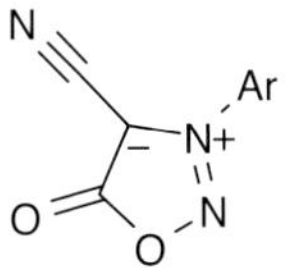

83

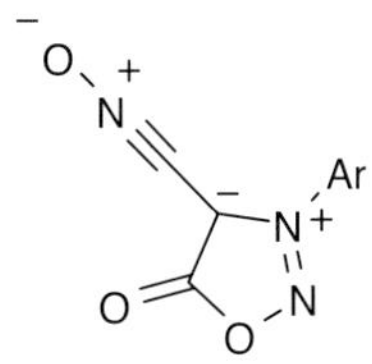

84

The reaction of 3-aryl-4-carbohydroximic acid chlorides with hydrazine hydrate gives hydrazino(3-arylsydnon-4-yl)methanone oximes, which by reaction with aldehydes are good precursors of 4-triazolyl-sydnones (**85**, Scheme 13.31) [101].

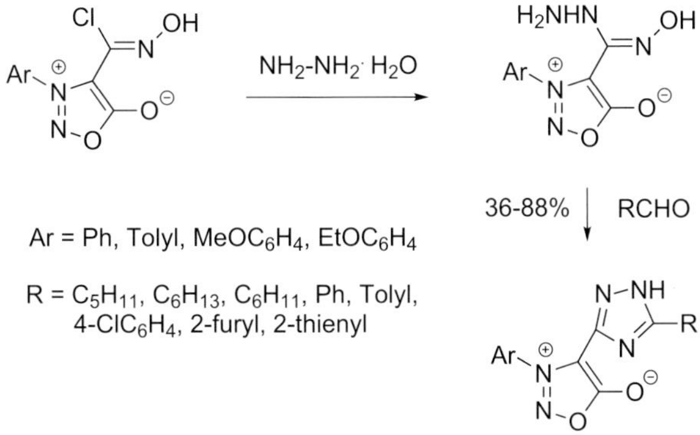

NH_2-$NH_2 \cdot H_2O$

36-88% RCHO

Ar = Ph, Tolyl, $MeOC_6H_4$, $EtOC_6H_4$

R = C_5H_{11}, C_6H_{13}, C_6H_{11}, Ph, Tolyl, 4-ClC_6H_4, 2-furyl, 2-thienyl

85

Scheme 13.31

4-Formylsydnones undergo reduction, Claisen condensation with acetophenone, and condensation with nitroalkanes and active methylene compounds. Thus, the Knoevenagel reaction of 3-aryl-4-formylsydnones affords multifunctional derivatives [102].

3-Aryl-4-formylsydnone-4′-phenyl-thiosemicarbazones and 3′ aryl-4-formylthiosemicarbazones **86** react with ethyl chloroacetates, ethyl 2-chloroacetoacetate, and 2-bromoacetophenone to produce heterocyclic substituted sydnone derivatives **87a–c** that possess 4-oxo-thiazolidine and thiazoline groups (Scheme 13.32) [103]. The antioxidant activity of the synthesized compounds was evaluated. Among these compounds, 4-methyl-2-[(3-arylsydnon-4-yl-methylene)hydrazono]-2,3-dihydro-thiazole-5-carboxylic acid ethyl ester and 4-phenyl-2-[(3-arylsydnon-4-yl-methylene) hydrazono]-2,3-dihydro-thiazoles exhibit potent DPPH (1,1-diphenyl-2-picrylhydrazyl) radical scavenging activity, comparable to that of vitamin E.

A suitable substituent at C4 can be used as a temporary blocking group to allow reaction to take place at another side of the sydnone. For example, 4-acetyl-

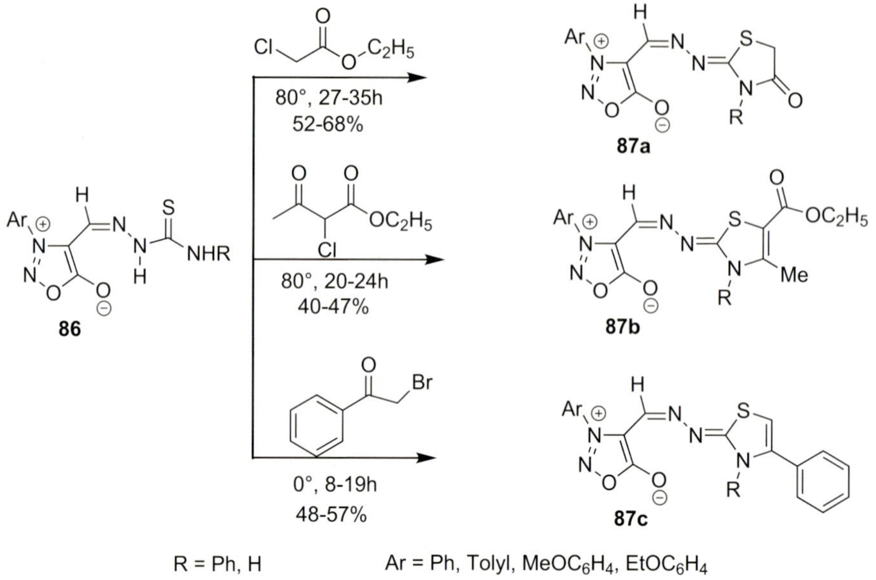

R = Ph, H Ar = Ph, Tolyl, $MeOC_6H_4$, $EtOC_6H_4$

Scheme 13.32

or 4-formyl-3-phenylsydnones can be nitrated in the aromatic ring (in the *meta* position) and subsequently the acyl group can be removed under basic conditions.

Similarly, a thioether group at C4 can be removed by oxidation to sydnone sulfone and subsequent reduction with sodium borohydride.

13.2.6.3.6 **1,3-Dipolar Cycloaddition Reactions** Sydnones can be regarded as cyclic azomethine imines and as such they undergo thermal cycloaddition reactions with a range of dipolarophiles. As previously discussed (Section 13.2.6.3.2), on photolysis 3,4-diaryl-sydnones lose carbon dioxide and afford transient nitrile imines, which can be trapped by alkynes to give pyrazole derivatives.

Thermal reactions with acetylenic dipolarophiles also lead to pyrazoles by spontaneous loss of carbon dioxide from the cycloadducts. According to this reaction route, a series of 5-halopyrazoles (**89**) with potential pharmacological activity has been synthesized in good yields by 1,3-dipolar cycloaddition of 4-halogenated sydnones **88** with dimethyl acetylenedicarboxylate (DMAD) (Scheme 13.33) [104].

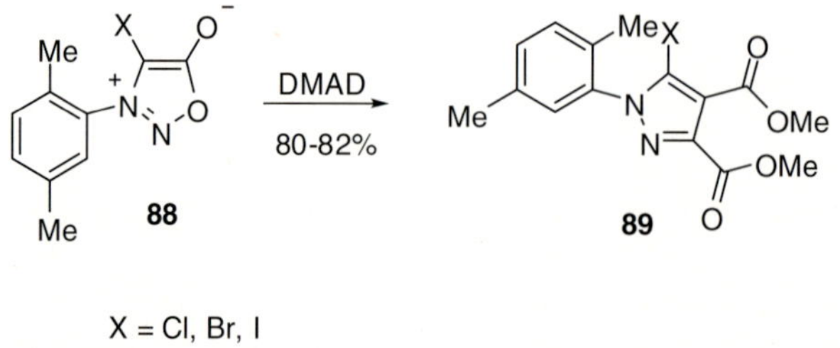

X = Cl, Br, I

Scheme 13.33

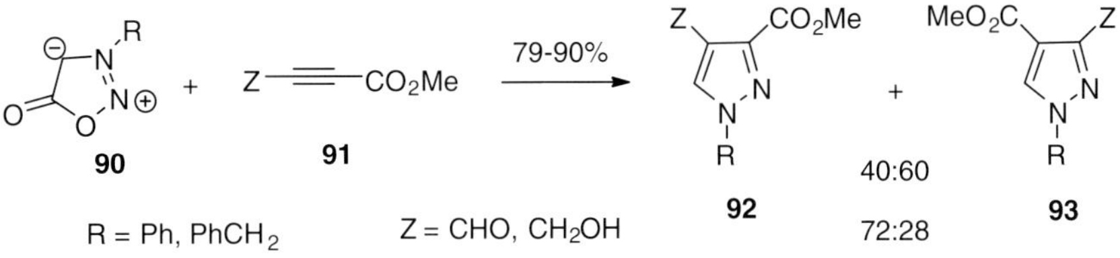

90 + 91 → (79-90%) 92 + 93

R = Ph, $PhCH_2$ Z = CHO, CH_2OH 40:60 72:28

Scheme 13.34

With unsymmetrical alkynes **91**, the cycloaddition reactions of sydnones **90** rarely show a good regioselectivity (Scheme 13.34) [105].

With monosubstituted alkenes bearing conjugative electron-withdrawing groups, the regioselectivity of the reaction is that predicted by frontier orbital analysis (Figure 13.2), that is, with the carbon bearing the electron-withdrawing group next to nitrogen. The obtained products are usually dihydropyrazoles or pyrazoles formed by oxidation of the intermediate dihydropyrazoles.

The unstable species formed by loss of carbon dioxide are also azomethine ylides: thus, in the reaction of 3-phenylsydnone with *N*-phenylmaleimide, a second dipolar cycloaddition reaction can take place (Scheme 13.35) [106].

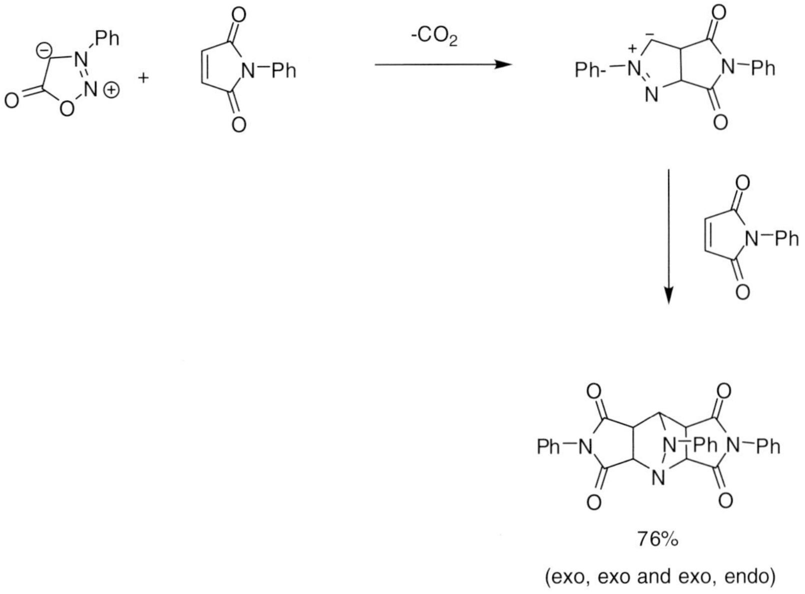

$-CO_2$

76%

(exo, exo and exo, endo)

Scheme 13.35

The tandem 1,3-dipolar cycloaddition between sydnones and 1,5-cyclooctadiene afforded 9,10-diazatetracyclo[6.3.0.0.^{4,11}0.^{5,9}]undecanes (Weintraub reaction) [107].

13.2.7
Important Compounds and Applications

1,2,3-Oxadiazole derivatives show a wide range of biological activities. In particular, two most important and studied compounds are the sydnonimines molsidomine (**94**) and sydnocarb (**95**).

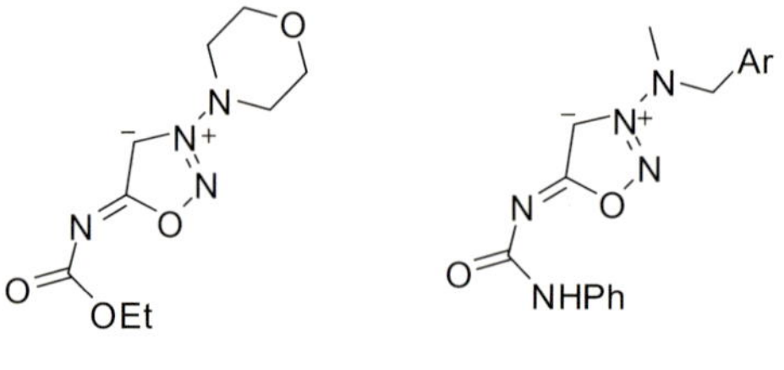

94 **95**

Molsidomine (**94**), endowed with very low toxicity, has a long-term effect in vasodilation, thus exerting a positive effect in cases of ischemic heart diseases. In combination with the β-blocker propanolol, molsidamine has shown a high efficacy for the treatment of portal hypotension [108].

Molsidomine is also used in treating angina pectoris [109].

The pharmacological activity is correlated to the formation of a metabolite, the *N*-morpholino-*N*-nitrosoaminoacetonitrile, which acts as a nitric oxide donor (Scheme 13.36) [110].

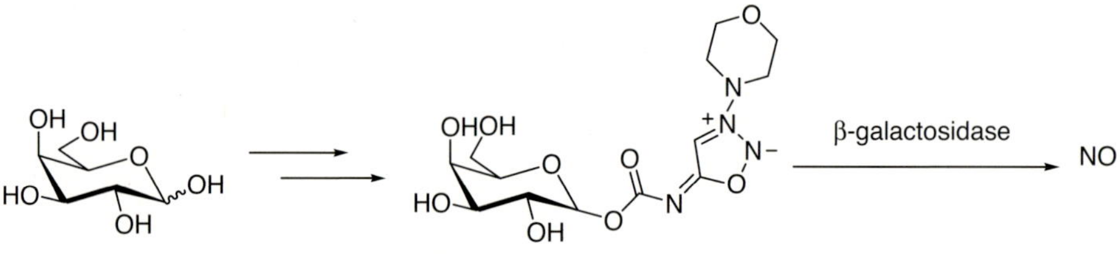

β-galactosidase → NO

Scheme 13.36

Accordingly, to achieve site-specific delivery of nitric oxide (NO), a new class of glycosidase activated NO donors has been developed, in which glucose, galactose, and *N*-acetylneuraminic acid were covalently coupled to 3-morphorlinosydnonimine, via a carbamate linkage at the anomeric position [111]. The β-glycosides were successfully prepared for these conjugates, while the α-glycosidic compounds were very unstable. The new stable sugar–NO conjugates could release NO in the presence of glycosidases (Scheme 13.37). Such NO prodrugs may be used as enzyme-activated NO donors in biomedical research.

With analogous aim, conjugates of cephalosporin with 3-morpholinosydnonimine have been designed and evaluated [112]. The obtained compounds demonstrated promising β-lactamase dependent NO releasing ability.

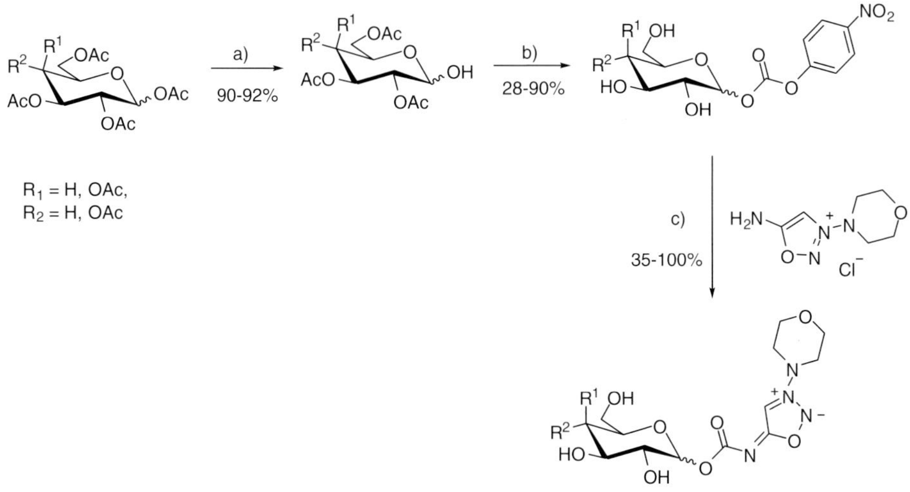

R_1 = H, OAc,
R_2 = H, OAc

a) $BnNH_2$, THF, rt, 30h; b) p-$NO_2C_6H_4OCOCl$, Et_3N, CH_2Cl_2, rt, 4.5h; c) pyridine, rt, 12h.

Scheme 13.37

Sydnocarb acts on the central nervous system and has been used as a psychostimulant and an antidepressive. Nitrososydnonimines **96** and **97** showed potent antithrombotic activity [113, 114].

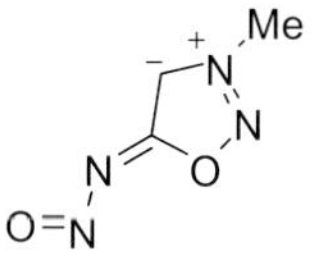

96

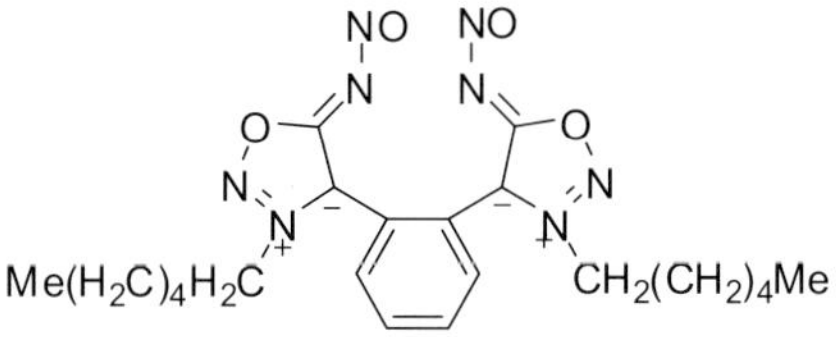

97

A series of derivatives – **98** and **99** – prepared by manipulation of the carboxylic group of 3-(3-carboxyphenyl)- and 3-(3-carboxyphenyl)sydnones, or by Claisen–Schmidt condensation of 3-(4-acetylphenyl)sydnone with aldehydes or malononitrile, showed high antibacterial activity against both Gram-positive and Gram-negative organisms [115].

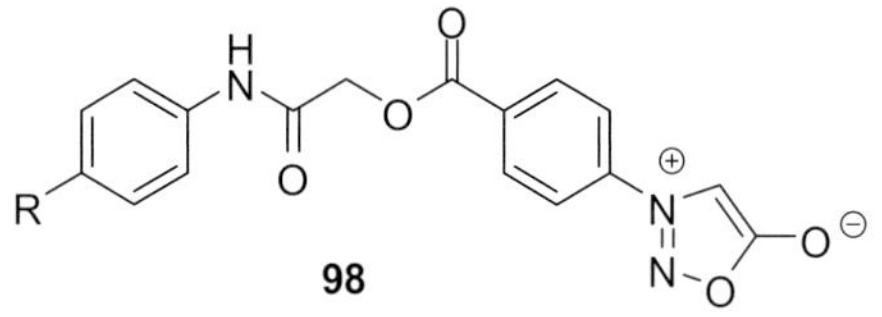

98

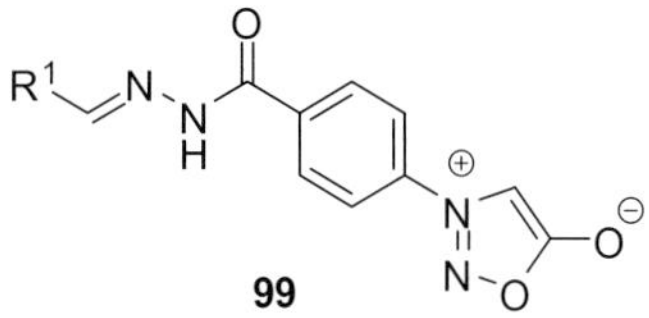

99

R = H, EtO, Ac, CO_2H, CO_2Et

R^1 = Ph, 2-Furyl, 4-Cl-C_6H_4, 4-NO_2-C_6H_4, 2-Cl-4-NO_2-C_6H_3

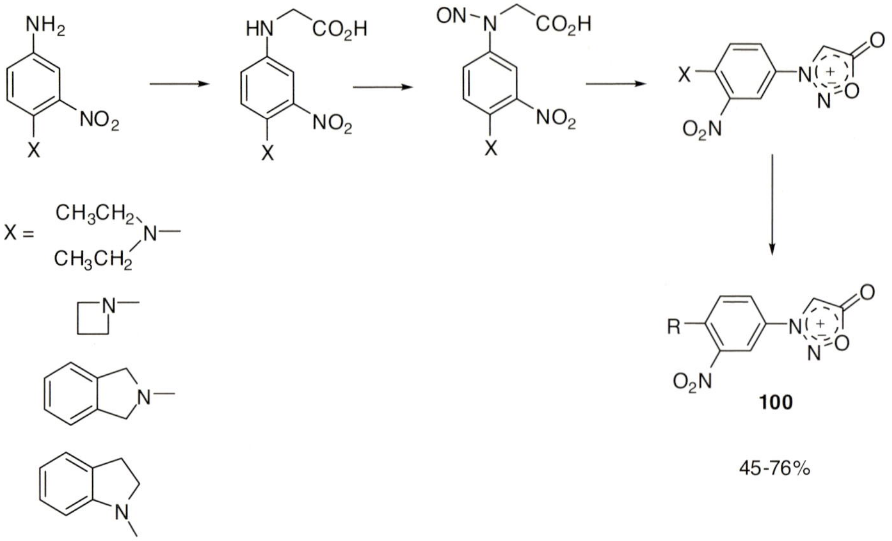

Scheme 13.38

A series of 4′-substituted-3′-nitrophenylsydnones **100** have been synthesized (Scheme 13.38) and evaluated [116, 117] for anticancer activity and it was found that the 4′-chloro, 4′-fluoro and 4′-pyrrolidino compounds significantly enhanced the survival of Sarcoma 180 (S180), Ehrlich carcinoma (Ehrlich), and Fibrous histiocytoma (B10MCII) tumor bearing mice.

Many other sydnones have been tested for antioxidant, antimicrobial, antifungal, analgesic, anti-inflammatory, and antipyretic activities [118].

4-Styrylcarbonyl-3-phenylsydnone derivatives **101** and **102** showed activity similar to that of aspirin, at the same dosage [119].

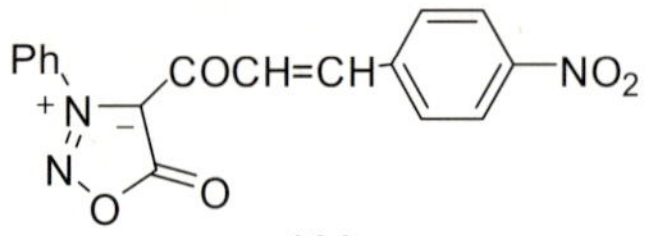

13.3 1,2,4-Oxadiazoles

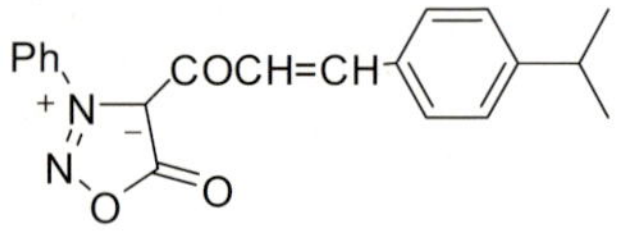

The chemistry of 1,2,4-oxadiazoles **103** has been extensively reported [120]. Research on this class of heterocycles has registered great interest in medicinal chemistry. Many derivatives possess diverse biological activities [121–123]. Some 1,2,4-oxadiazoles can reduce pain and inflammation in rats and mice [124, 125]; for example, *N*-[3-aryl-1,2,4-oxadiazol-5-yl-methyl]phthalimides have been found to be analgesic, and one of them, namely, *N*-[3-phenyl-1,2,4-oxadiazol-5-yl-methyl]phthalimide, possesses highly enhanced analgesic activity compared to aspirin [124].

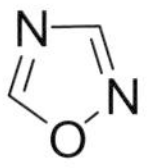

103

Furthermore, the 1,2,4-oxadiazole ring has been exploited as a peptidomimetic, as a stable ester and amide isostere; specific 1,2,4-oxadiazoles have been used as inhibitors in several biological systems [126, 127]. In particular, numerous papers deal with applications of the soluble guanylyl cyclase inhibitor 1*H*-[1,2,4]-oxadiazole[4,3-*a*] quinoxalin-1-one (ODQ) (**104**) and with applications of the neuroexcitatory quisqualic acid (**105**), the only naturally occurring 1,2,4-oxadiazole known hitherto [128].

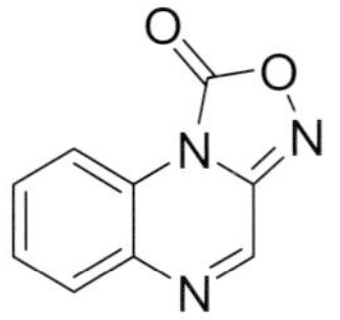

104 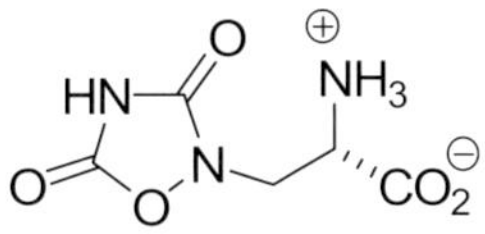 **105**

Partially or fully saturated 1,2,4-oxadiazoles (**106–109**) have also been reported. In particular, 4,5-dihydro-1,2,4-oxadiazoles **106** have been evaluated very little for biological activities compared to 1,2,4-oxadiazoles, but some of them have shown interesting pharmacological results. For example, some are fungicides [129, 130], and other 3,4,5-triaryl-4,5-dihydro-1,2,4-oxadiazoles demonstrated bronchodilator, anticholinergic, hypertensive, analgesic, anti-inflammatory, diuretic, antiulcer, vasodilatatory, and sedative properties [131]. Some 4-adamantyl-5-aryl-3-phenyl-1,2,4-oxadiazolines have been evaluated *in vitro* for antiviral activity against human immunodeficiency virus (HIV), where the 5-phenyl substituent produced a reduction of more than 50% of viral cytopathic effects [132]. The present chapter updates the previous work and reviews the literature published since, with reference to new advances, preparations, reactions, and uses.

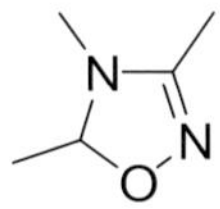

106

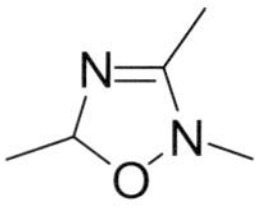

107 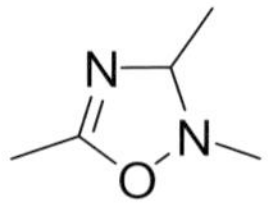 **108** 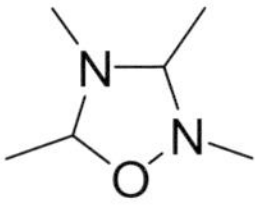 **109**

13.3.1
Structure

The 1,2,4-oxadiazole ring is planar and described as having little aromatic character [133] – lower than furan on the Bird index [134]. Dipole moments and Kerr constants of certain oxadiazoles seem to indicate some ability of the ring oxygen atom to donate π electrons into the ring. This heterocyclic system has an appreciable heterodiene character, as suggested by X-ray analysis, which indicates, for both C–N distances, conjugated double bond character. The low aromaticity manifests itself by allowing rearrangement to more thermodynamically stable ring systems, thus making 1,2,4-oxadiazoles good substrates for ring-to-ring transformations [135].

The ring of 4,5-dihydro-1,2,4-oxadiazoles, according to CNDO/2 calculations, is nonplanar, adopting an envelope conformation with one atom sitting above the plane described by the four others, and, in contrast to the 1,2,4-oxadiazole ring, it is quite polar [136].

The parent compound **103** is an extremely volatile liquid, very soluble in water and organic solvents, but it is unstable at room temperature. 3,5-Dialkyl and 3,5-diaryl 1,2,4-oxadiazoles are thermally stable and do not hydrolyze by treatment with aqueous sodium hydroxide or hydrochloric acid. In contrast, **103** and monosubstituted oxadiazoles **110** and **111** are thermally and hydrolytically markedly less stable (Scheme 13.39) [137].

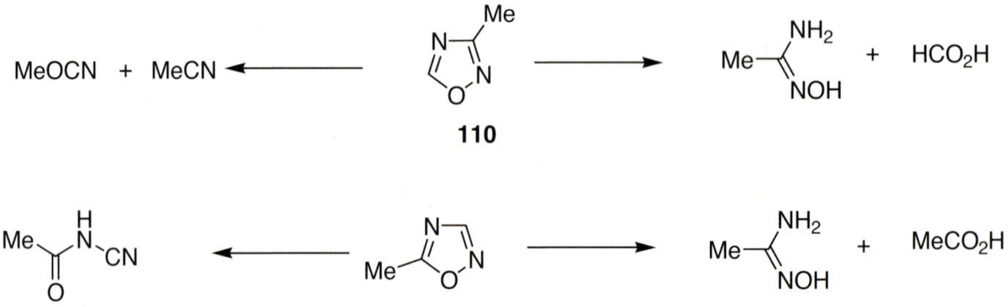

MeOCN + MeCN ← **110** → Me–C(=NOH)–NH_2 + HCO_2H

Me–C(=O)–NH–CN ← **111** → Me–C(=NOH)–NH_2 + $MeCO_2H$

Scheme 13.39

Tautomerism of 3- and 5-hydroxy, 3- and 5-amino, and 3- and 5-sulfur analogues has been recently reviewed [138]. In 5-hydroxy-3-phenyl-1,2,4-oxadiazole (**112a**), the keto forms **112b** and **112c** predominate according to NMR data [139]. The tautomer **113b** is more important in the 5-phenyl isomer in solution, but in acetone and oxygenated solvents **113a** allows for an effective hydrogen bonded dimer **114** (Figure 13.3).

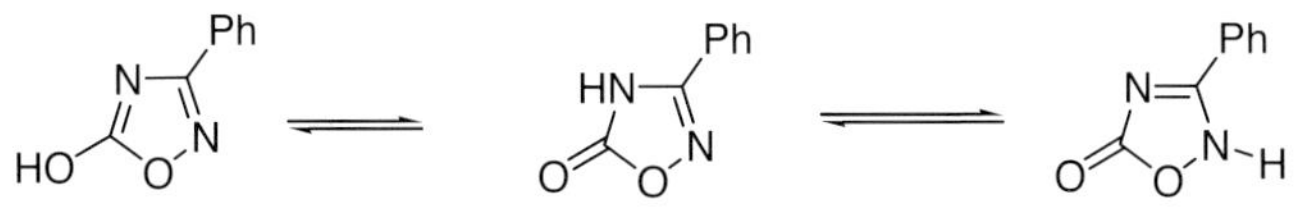

112a 112b 112c

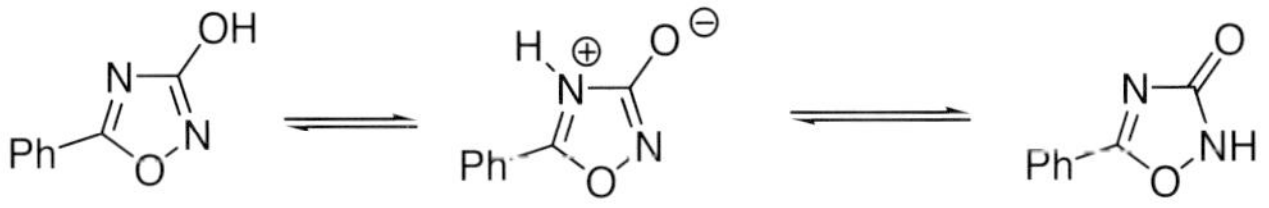

113a 113b 113c

114a 114b

Figure 13.3 Tautomerism exhibited by hydroxy derivatives.

In aminooxadiazole derivatives the tautomeric imino form **115b** is less significant, since **115a** is more basic; in the corresponding sulfur analog there is only evidence for the thione form **116** with the hydrogen at N2.

115a 115b

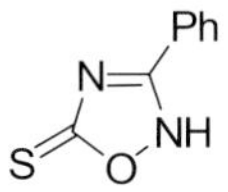

116

Interestingly, the 5-aryl-4,5-dihydro-1,2,4-oxadiazole **117a** undergoes formal tautomerism with the 4-aryl-1,3-diaza-1,3-butadiene **117b**, which in turn can undergo ring closure with loss of water to form the quinoxaline **118** (Scheme 13.40) [140].

13.3.2
Theoretical Aspects

Theoretical studies on the structure and properties of 1,2,4-oxadiazoles have been reported. Semiempirical (PM3 and AM1) and *ab initio* molecular orbital

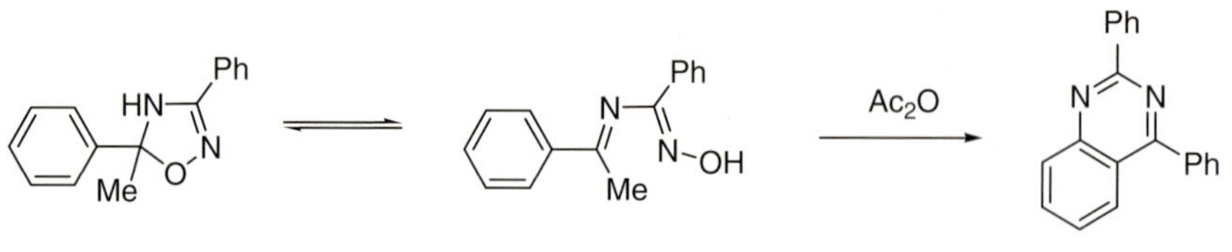

Ac_2O

117a 117b 118

Scheme 13.40

calculations have been performed for diaryl-1,2,4-oxadiazoles and 4,5-dihydro-1,2,4-oxadiazoles to determine bond orders, total energies, ionization potentials, and dipole moments [141]. In particular, *ab initio* molecular calculations give values close to those obtained by crystallographic techniques and NMR spectroscopy. Proton affinities and pK_a values of amino-substituted oxadiazoles have been calculated [142]. INDO studies on 3-phenyl-1,2,4-oxadiazole and its 5-methyl analog suggest that nucleophilic attack should occur on C3 and C5 [143]. In connection with pharmacological structure–activity relationships, semiempirical and *ab initio* molecular orbital calculations have been reported for a series of analgesic compounds, leading to new suggestions for their mechanism of activity [120, 144]. A new model of interaction between the drug and the enzyme has been proposed that involves an electron transfer from the amino acid residue of the enzyme to the drug.

A theoretical study of photoinduced ring-isomerization of 3-amino-5-methyl- and 3-amino-5-phenyl-1,2,4-oxadiazoles has been reported. The results agree well with experimental data and explain the ring-photoisomerization into the corresponding 2-amino-1,3,4-oxadiazoles through a ring contraction–ring expansion route [145] (see Scheme 13.19 in 1,3,4-oxadiazoles). On the same basis a theoretical study of degenerate Boulton–Katritzky rearrangements concerning the anion of the 3-formylamino-1,2,4-oxadiazole has been carried out by using semiempirical MNDO and *ab initio* Hartree–Fock procedures [146].

A combined kinetic and theoretical study of the monocyclic rearrangements of the *(Z)*-hydrazone of 3-benzoyl-5-phenyl-1,2,4-oxadiazole into the corresponding triazole (Scheme 13.41) has been investigated at the DFT level [147].

The synthetic approach towards 1,2,4-oxadiazoles, based on the BH_3- or BF_3-mediated cycloaddition of benzonitrile oxide to nitriles, has been investigated theoretically according to quantum chemical methods (MP2 and B3LYP) together with a topological analysis of the charge density (Section 13.3.4.2) [148]. Activation by the Lewis acid occurs via two different mechanisms: if the Lewis acid is coordinated to the nitrile oxide, the reactant is activated, so that the reaction is expected to be catalytic. If the Lewis acid is coordinated to the nitrile and strong enough, the process requires a stoichiometric amount of Lewis acid and forms a stable Lewis acid–product complex.

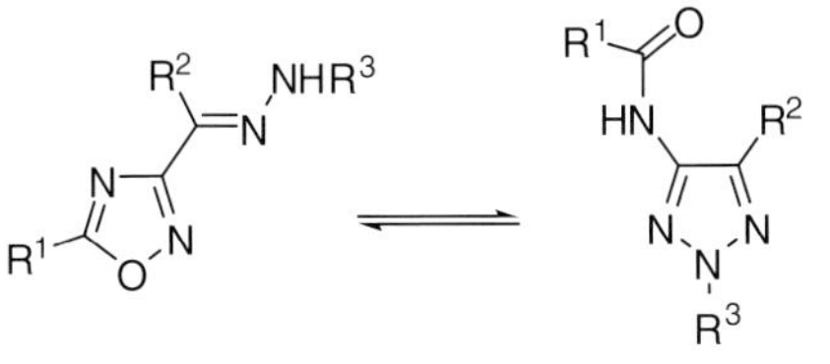

R^1 = Ph, H R^2 = Ph R^3 = Ph, 2,4-NO_2-C_6H_3

119 **120**

Scheme 13.41

13.3.3
Structural Aspects

13.3.3.1 X-Ray Diffraction

X-Ray data of many 1,2.4-oxadiazoles confirms that the ring is planar [149–154]. Values of C–N bond lengths are consistent with a heterodiene character and account for the low aromaticity of the system. Table 13.3 shows the reported bond lengths and bond angles for methyl 2-[3-(4-methylphenyl)-1,2,4-oxadiazol-5-yl]benzoate (**121**), a compound used as spacer in the synthesis of a potential non-peptide angiotensin receptor antagonist [155].

Table 13.3 Molecular dimensions for methyl 2-[3-(4-methylphenyl)-1,2,4-oxadiazol-5-yl]benzoate (**121**).

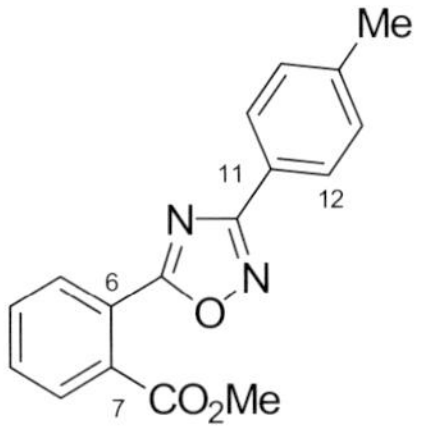

121

Bonds	Distances (Å)	Bond angles	(°)
O1–N2	1.415	O1–N2–C3	103.51
N2–C3	1.310	N2–C3–N4	114.10
C3–N4	1.325	C3–N4–C5	102.83
N4–C5	1.298	N4–C5–O1	113.30
C5–O1	1,347	C5–O1–N2	106.25

The 3,5-diphenyl-oxadiazole fragment is almost coplanar. The angles between the planes of the rings C5–N4/C6–C7 and C5–N4/C11–C12 are 11.13 and 2.28°, respectively. The phenyl rings are tilted to the same side with respect to the oxadiazole ring and the angle between them is 8.86°. An interesting aspect of the crystal structure is the presence of two weak C–H–O bonds between two neighboring molecules in the same layer. Each molecule behaves as both donor and acceptor, leading to a dimer formation [155].

Crystal structures for a series of 2,3-dihydro-1,2,4-oxadiazoles have been reported [156–158]. The 4,5-dihydro-1,2,4-oxadiazole ring in compounds **122** [156] and **123** [151] are in an envelope conformation, with the oxygen atom above the plane occupied by other atoms.

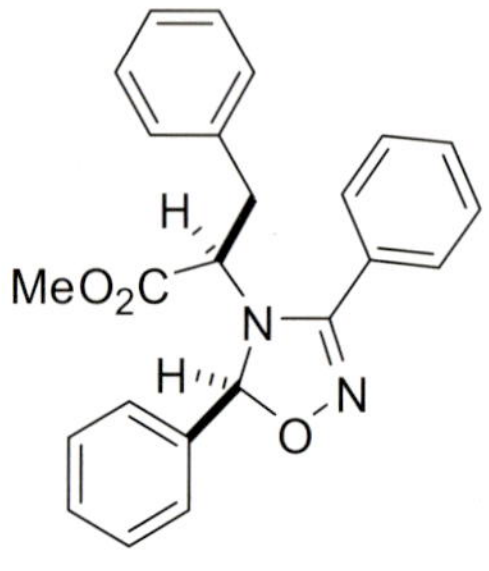

122 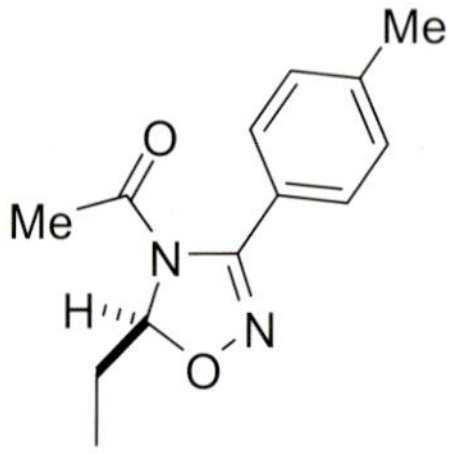 **123**

Yu and coworkers have reported the preparation and spectroscopic and X-ray diffraction studies of two diastereoisomeric Δ^2-1,2,4-oxadiazolines (**124**, showing the assigned configuration of the diastereoisomer) having a spiral junction at C3 of fructopyranose. These compounds show extensive applications as drugs [159, 160].

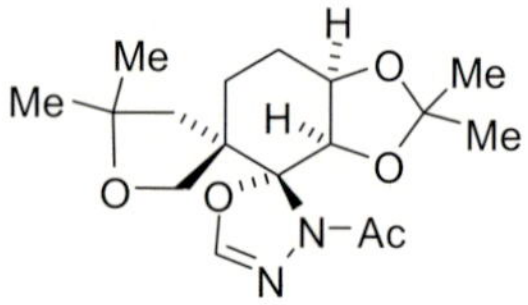

124

According to the biological activities of 1,2,4-oxadiazole derivatives, the combination of an oxadiazole moiety with a sugar framework has been performed. Unsaturated glycosides having an 1,2,4-oxadiazole part as an aglycone, **125** and **126**, have been reported [153]: crystallographic data, providing precise information regarding the configuration at C8 and also about the molecular conformation, have shown that compound **125** has a torsion angle H(15)–C(15)–C(10)–H(10) of −43.2°, which clearly shows that the anomeric proton is disposed equatorially. The ring oxygen atom is a little above the C(10)-C(15)-C(14)-C(13) plane; the C(12) atom is slightly below this plane. The *p*-tolyl ring and the 1,2,4-oxadiazole rings are coplanar [torsion angle N(2)–C(3)–C(16)–C(17)Z10.618°]. The bond distances C(13)–C(12) and C(12)–O(11) are 1.54 and 1.43 Å, respectively.

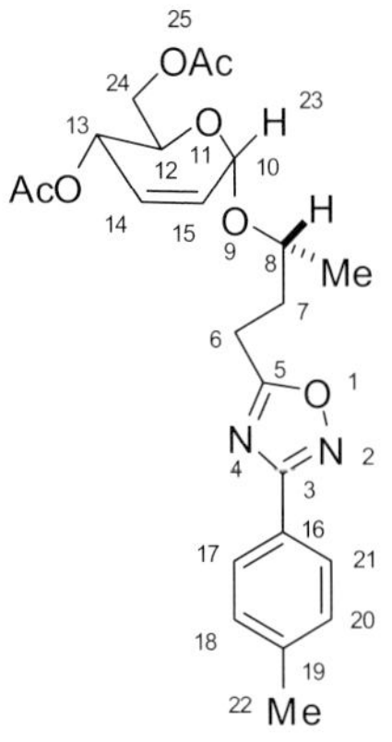

125

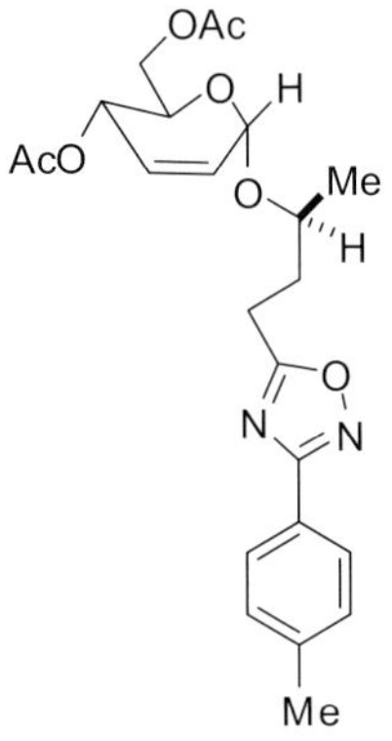

126

Based on the recent observation that Pt(II) mono- and bis-1,2,4-oxadiazoline complexes exhibit *in vitro* cytotoxicity against a series of platinum-sensitive and resistant human cancer cell lines, with a potency comparable to that of cisplatin and superior to carboplatin [161], a series of PtX2(nitrile)(oxadiazoline) (**127**) and PtX2 bis-1,2,4-oxadiazoline (**128**) complexes have been prepared.

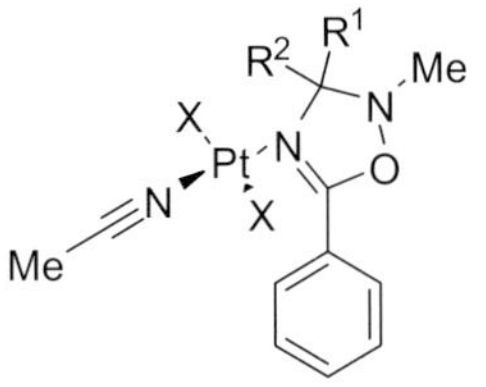

127

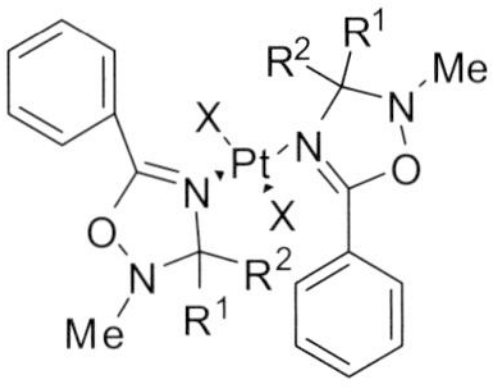

128

13.3.3.2 NMR Spectroscopy

Protons on the 1,2,4-oxadiazole ring are shifted downfield with respect to protons in benzene, according to the electron deficiency of the heterocyclic ring. In the parent compound **103**, the signal for the C3 proton is at 8.99 ppm, while the C5 proton resonates at 9.49 ppm. The presence of an alkyl or aryl substituent shifts the resonance upfield; for 3-phenyl-1,2,4-oxadiazole, the H5 in CCl_4 is at 8.70 ppm [120]. Resonances for H5 of 4-unsubstituted 5-alkyl-4,5-dihydro-1,2,4-oxadiazoles appear at 5.4–5.7 ppm [125, 162]. Chemical shifts for H5 of 5-alkyl-2,5-dihydro-1,2,4-oxadiazoles have been found at 6.1–6.3 ppm [163].

For 3-substituted 2,3-dihydro-1,2,4-oxadiazoles, the H3 shift is in the range 5.7–6.2 and 7.2–7.6 ppm, according to the presence of N-alkyl or N-aryl substituents, respectively [164].

Many fully assigned ^{13}C data for C3/C5 disubstituted 1,2,4-oxadiazoles have been reported [165–168]. C3 resonances are in the range 148–169, while chemical shifts for

Table 13.4 ^{13}C NMR shifts (ppm) for C3/C5-disubstituted 1,2,4-oxadiazoles **129–132**.

1,2,4-Oxadiazole	129	130	131	132
C3	148.6	168.8	176.9	169.3
C5	164.1	174.7	182.7	173.4
Solvent	CD_3CN	$CDCl_3$	$CDCl_3$	DMSO

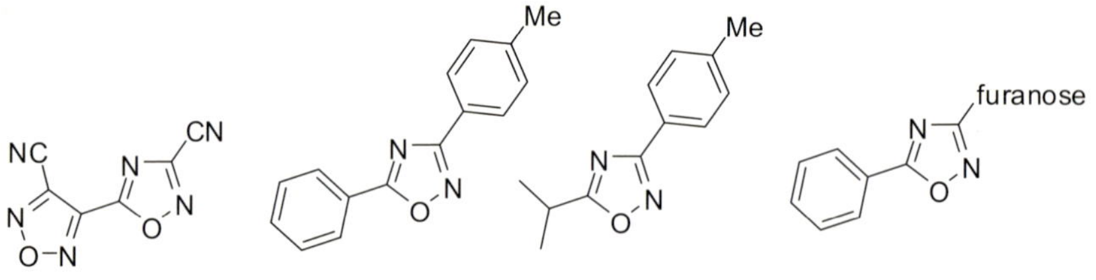

NC CN N N O N O N Me Me furanose

129 **130** **131** **132**

C5 are downfield, in the range 165–185 ppm. Table 13.4 summarizes the data for oxadiazoles **129–132**.

13.3.3.3 **UV and IR Spectroscopy**

UV spectra of aryl-substituted 1,2,4-oxadiazoles have been reported [162, 169]; non-aryl 1,2,4-oxadiazoles have no UV absorption. UV and fluorescence spectra of Cu(II) complexes of 5-(2-hydroxyphenyl)-3-phenyl-1,2,4-oxadiazole (**133**) have been reported [170]: Cu (II) binds to monodentate oxadiazole via N4 and the OH group in a 2 : 1 complex.

A detailed IR analysis exists for the parent compound and a series of fully conjugated 1,2,4-oxadiazoles [162]. Diagnostic absorptions are at 1590–1560 (C=N), 1119–1218 (C–O) and 895–910 (N–O) cm^{-1} [171, 172]. For 4,5-dihydro-1,2,4-oxadiazole **134**, the C=N absorption is shifted to around 1600 cm^{-1} [162, 173]. 2,3-Dihydro-1,2,4-oxadiazoles exhibit a $\nu_{C=N}$ between 1670 and 1676 cm^{-1} [156, 183], while in 2,5-dihydro the same absorption is at 1622–1640 cm^{-1} [162, 173].

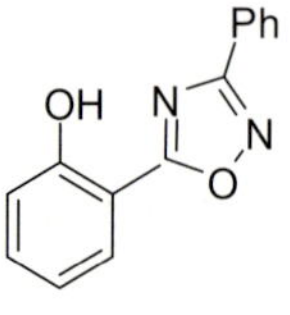

Ph OH N N O Ar HN N H O R^1

133 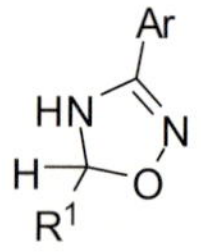 **134**

13.3.3.4 **Mass Spectrometry**

The diagnostic fragmentation pattern of 1,2,4-oxadiazoles is a 1,3-dipolar cycloreversion process, which proceeds via initial cleavage of the 1,5 (C–O) and 3,4 (C–N) bonds: the positive charge is retained in the predominant nitrile oxide fragment (Scheme 13.42) [175, 176].

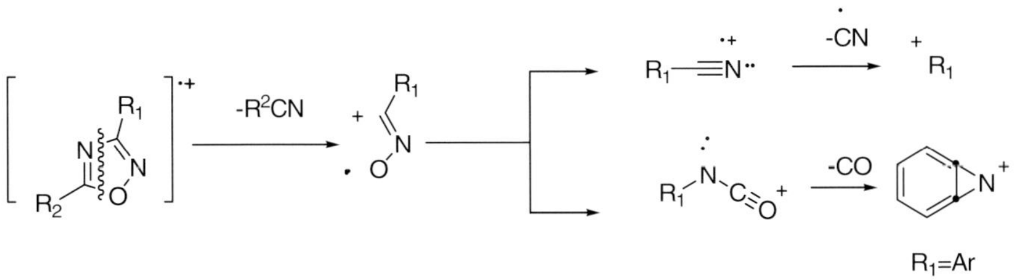

Scheme 13.42

The nitrile oxide fragment itself fragments further, either by expulsion of oxygen to give a nitrile, which may then lose a CN fragment, or via rearrangement and expulsion of CO.

A recent review reports a detailed mass spectrometric analysis of a series of 1,2,4-oxadiazoles and 4,5-dihydro-1,2,4-oxadiazoles [177]. The fragmentation mode of the latter compounds differs from that of 1,2,4-oxadiazoles. For example, the electron impact dissociation of compounds **135** is reported in Scheme 13.43.

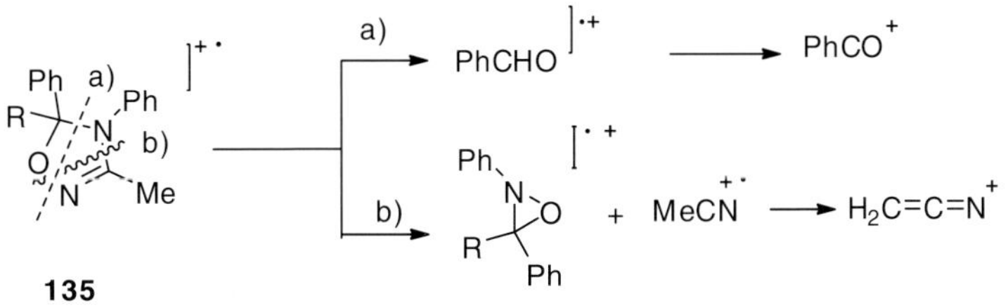

Scheme 13.43

13.3.4 Synthesis

Many synthetic routes for the 1,2,4-oxadiazole system have been reported [120]. 1,2,4-Oxadiazoles can be achieved from open-chain precursors through conventional heterocyclization reactions: the best represented approach exploits the cyclodehydration of *O*-acyl-amidoximes, a method first used by Tiemann and Kruger [178], or N-acylamidoximes, a method developed by Beckmann and Sandel (amidoxime route) [179]. Another different general synthetic route is based on the 1,3-dipolar cycloaddition of nitrile oxides to nitriles, developed by Leandri (cycloaddition route) [180].

13.3.4.1 **Amidoxime Route**

13.3.4.1.1 **Cyclization of O-Acylamidoximes** According to the most represented route, 1,2,4-oxadiazoles **138** can be prepared by cyclization of *O*-acylamidoximes **137**, which are obtained from the appropriate amidoxime **136** (easily prepared by reaction of the corresponding nitrile with hydroxylamine) and an acylating reagent [120] (generally acyl halides [181, 182], esters [183], or anhydrides [184]) (Scheme 13.44).

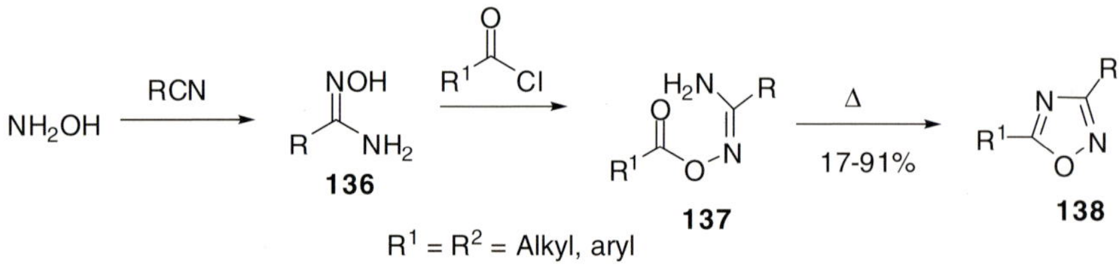

$R^1 = R^2$ = Alkyl, aryl

Scheme 13.44

The cyclization of *O*-acylamidoximes is performed by heating them at their melting point [185], or at reflux in a high-boiling solvent (DMF [186], toluene [187], pyridine [188], ethanol [189], acetonitrile [190], glacial acetic acid at reflux) [191, 192], eventually in the presence of a dehydrating agent (phosphorous pentoxide, phosphorus oxychloride, or acetic anhydride).

The experimental conditions required to realize the ring closure of the corresponding *O*-acylamidoximes vary as a function of their structures. In some cases, depending on the substrates, the cyclodehydration reaction occurs under the same conditions as the acylation reaction and the open-chain intermediate is not isolated.

An efficient one-pot method based on the reaction of nitriles with hydroxylamine hydrochloride in the presence of magnesia-supported sodium carbonate, followed by reaction with acyl halides under solvent-free conditions and microwave irradiation, has been reported [193a]. The use of microreactors (microfluidic chips) as an alternative to "in flask" chemistry has been exploited for a rapid synthesis of bis-substituted 1,2,4-oxadiazoles from aryl nitriles and acyl chlorides or succinic anhydride in a single continuous microreactor sequence. In this way, a multiday, multistep sequence has been amended to a highly efficient procedure lasting less than 30 min (Scheme 13.45) [193b].

Cyclization can be performed under mild conditions if the weak nucleophilic amide group is converted into the more nucleophilic amide ion. Thus, 1,2,4-oxadiazoles **140** have been obtained by treatment of **139** with DBU at 70 °C [205]. A significant advance is the use of TBAF at room temperature as a cyclization media, a process that occurs in high yields in the presence of 0.1–1 eq of TBAF, with the fluoride ion acting as both a homogeneous and strongly basic reagent (Scheme 13.46) [194, 195].

A wide variety of carboxylic acid derivatives can be used for the formation of *O*-acylated amidoximes, such as esters [196], acid chlorides [152, 154, 171, 175, 187, 197],

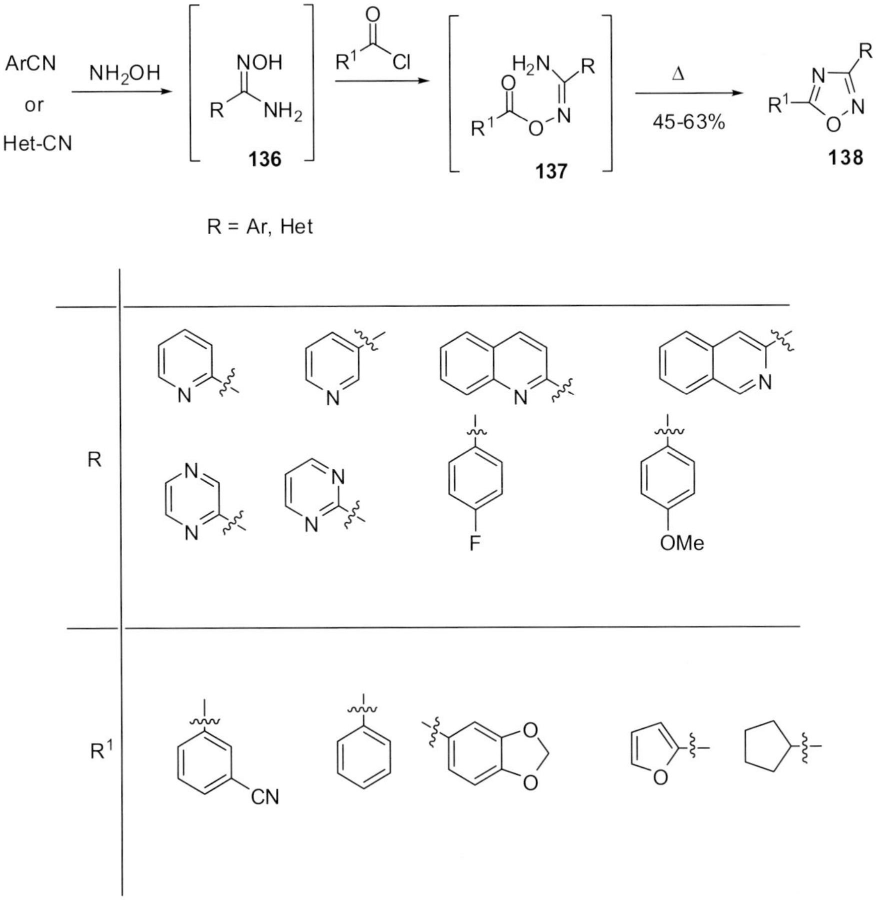

ArCN or Het-CN → NH_2OH → [136] → R^1COCl → [137] → Δ, 45-63% → 138

R = Ar, Het

Scheme 13.45

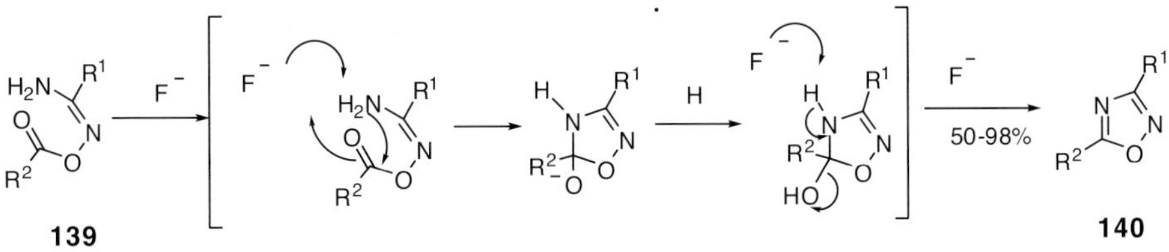

139 → F^- → [ ] → F^-, 50-98% → 140

R^1 = Ph,2-,3- or 4-Tol,2-,3- or 4 $MeOC_6H_4$,2-,3-, or 4-$NO_2C_6H_4$, Me, 4-$BocHNC_6H_4$, 2-Cl, 5-1,4-$BocHNC_6H_2$
R^2 = Me, Ph, OMe, Bu^t, CH_2Cl, $CH_2OCH_2CH_3$, CF_3, Pr^i, 2-, 3-, or 4-$NO_2C_6H_4$, CH_2Ph

Scheme 13.46

acid anhydrides [197b, 152, 171, 198], including symmetrical acid anhydrides derived from amino acids [184c], and amino-acid activated as succinimides [183].

Succinic [199] and glutaric anhydrides [200] are excellent substrates for reaction with amidoximes, giving 1,2,4-oxadiazol-5-yl carboxylic acids **141** and **142**, respectively: the obtained compounds are excellent substrates for coupling to amino acid derivatives (Scheme 13.47).

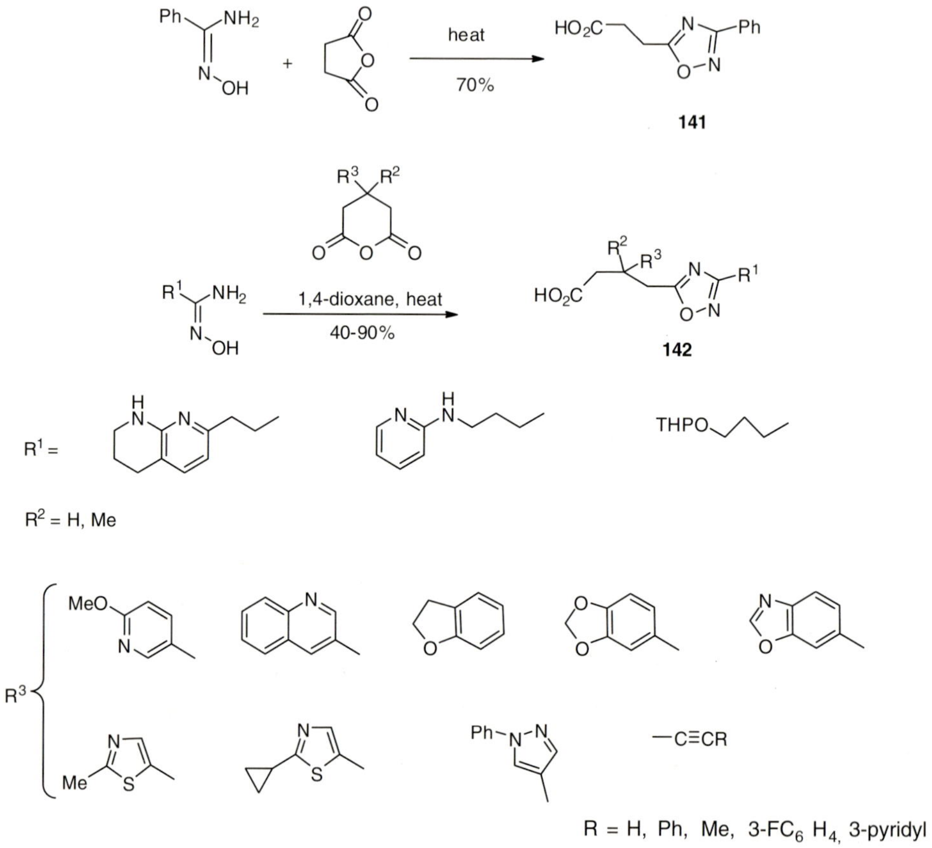

heat

70%

141

1,4-dioxane, heat

40-90%

142

R^1 =

THPO

R^2 = H, Me

R^3

—C≡CR

R = H, Ph, Me, 3-FC_6H_4, 3-pyridyl

Scheme 13.47

Examples of the amidoxime route, by which two 1,2,4-oxadiazole moieties can be linearly joined by alkyl chains through the annular 5,5′-positions, have been reported. Thus, the 5,5′-bis-1,2,4-oxadiazolyl system **144** has been obtained by reaction of malonates **143** with two equivalents of an amidoxime in the presence of potassium carbonate (Scheme 13.48) [196f].

The use of carboxylic acids, activated *in situ* and reacted with an amidoxime, has also been exploited, using various coupling reagents, including dicyclohexylcarbodiimide (DCC) [184a–f], 1-[3-(dimethylamino)propyl]-3-ethylcarbodiimide(EDC) [184a,d, 201c], (EDC)/HOBt [201b,c, 202], bis(2-oxo-3-oxazolidinyl)phosphinic

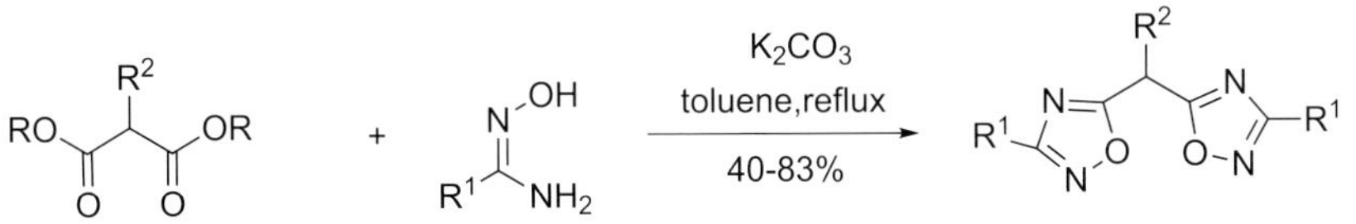

143

144

R^1 = Me, Et, Bn, 4-FC_6H_4, 4-$MeOC_6H_4$, cyclopropylmethyl, CH_2CH_2OMe
R^2 = H, Bn
R = Me, Et

Scheme 13.48

chloride (BOP-Cl) [184a], 2-(1*H*-benzotriazole-1-yl)-1,1,3,3-tetramethyl-uronium tetrafluoroborate (TBTU) [186, 201c], 1,1′-carbonyldiimidazole (CDI) [184a, 203], and/or high-speed microwave irradiation (Scheme 13.49) [186,197a, 201].

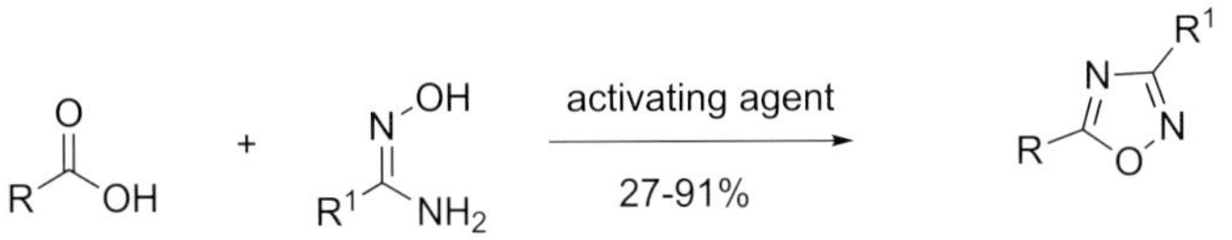

145

R^1 = 4-MeC_6H_4, Me, 4-FC_6H_4, 3-pyridyl
R = Ph, 4-MeC_6H_4, 4-$MeOC_6H_4$, 4-$NO_2C_6H_4$, 4-$EtOC_6H_4$, 3-$MeOC_6H_4$

Scheme 13.49

Chiral 1,2,4-oxadiazoles **147** have been synthesized from amino acids by reaction of the readily available *N*-protected (α-aminoacyl)benzotriazoles **146** with amidoximes in ethanol (Scheme 13.50) [189].

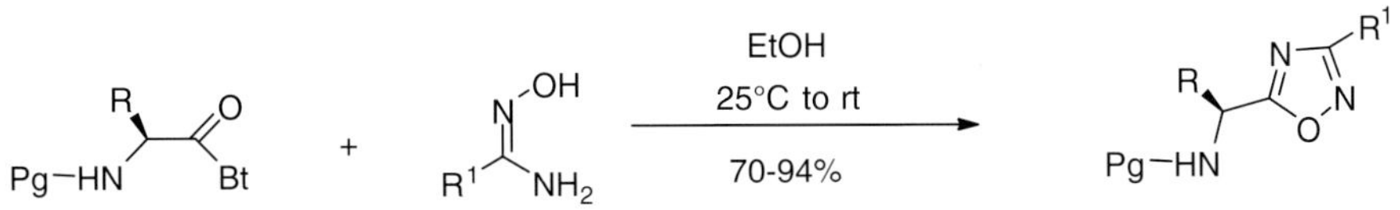

146

147

R^1 = 4-Tol, 4-pyridyl, Bn
R = Me, Me_2CH, $PhCH_2$, $MeSCH_2CH_2$, $NH_2COCH_2CH_2$, $C_8H_6NCH_2$
Pg = Boc, Fmoc

Scheme 13.50

In some cases, nitriles can be used as acylating reagent for amidoximes, and the subsequent heterocyclization involves loss of ammonia in the final step, with formation of 1,2,4-oxadiazoles **148** (Scheme 13.51) [204].

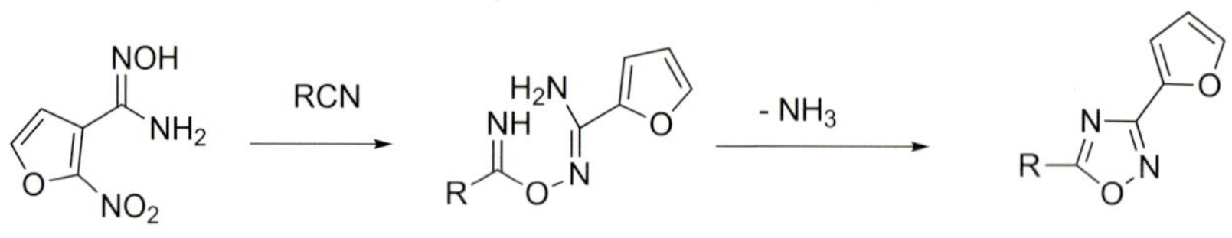
NOH
NH2
NO2
O
RCN
H2N
NH
R
O
N
- NH3
N
R
O
N
148

R = Alky, Aryl

Scheme 13.51

For this purpose the reaction is carried out in the presence of an ammonia acceptor reagent (a carboxylic acid or an excess of the nitrile). For example, from the reaction of benzamidoximes with perfluoroalkyl nitriles, a series of fluorinated 5-alkyl-1,2,4-oxadiazoles can be obtained [205].

Disubstituted 1,2,4-oxadiazoles have been synthesized in good yields and good purity in a one-pot procedure by reaction of aromatic nitriles, hydroxylamine hydrochloride, and sodium carbonate in ethylene glycol under heating at 195 °C [168].

Microwave irradiation of nitriles in the presence of hydroxylamine and different aromatic aldehydes, under solvent-free conditions, affords fully conjugated 1,2,4-oxadiazoles **149** in high yields (Scheme 13.52) [206].

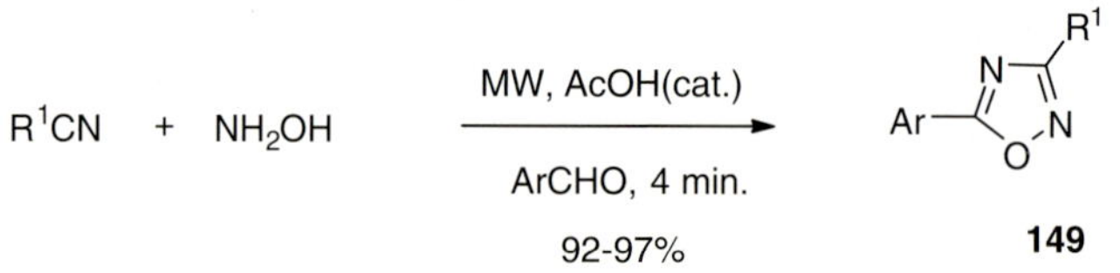
R1CN + NH2OH
MW, AcOH(cat.)
ArCHO, 4 min.
92-97%
R1
N
Ar
O
N
149

R^1 = Ph, 4-MeC$_6$H4, 3-ClC$_6$H$_4$

Ar = Ph, 4-MeC$_6$H4, 4-ClC$_6$H$_4$, 4-MeOC$_6$H4

Scheme 13.52

Other methods to obtain 1,2,4-oxadiazoles **150** include the palladium-mediated coupling of an aryl iodide with an amidoxime in the presence of carbon monoxide (Scheme 13.53) [207].

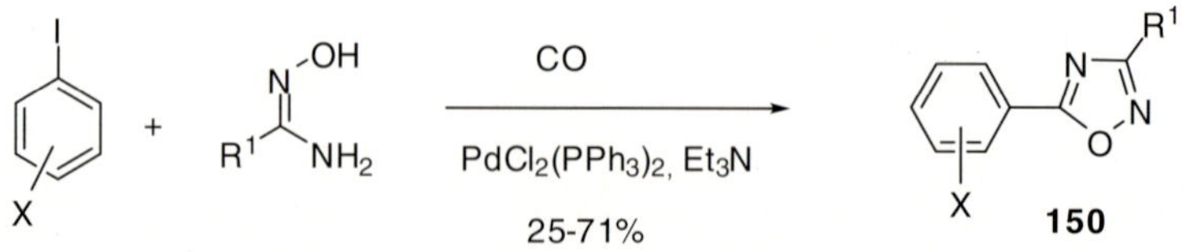
I
X
+
N
OH
R1
NH2
CO
PdCl2(PPh3)2, Et3N
25-71%
R1
N
O
N
X
150

R^1 = Me, CO$_2$Et X = 2-OMe, 4-OMe, 4-Br, 4-NO$_2$, 4-CO$_2$Me

Scheme 13.53

Similarly, 1,2,4-oxadiazoles **152** have been prepared by palladium-catalyzed reactions of diaryliodonium salts (**151**) with amidoximes in the presence of carbon monoxide (Scheme 13.54) [208].

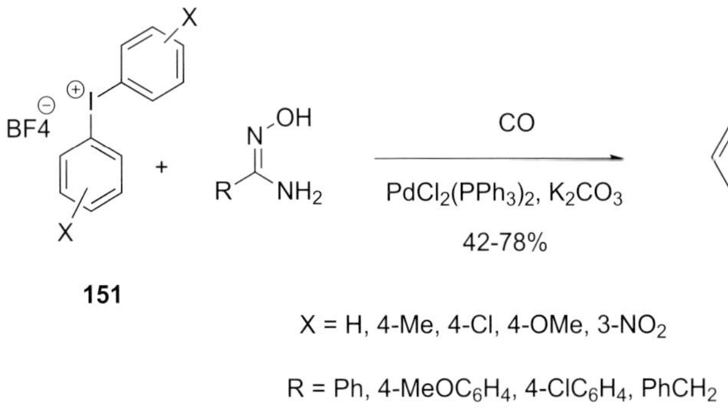

CO

$PdCl_2(PPh_3)_2$, K_2CO_3

42-78%

X = H, 4-Me, 4-Cl, 4-OMe, 3-NO_2

R = Ph, 4-$MeOC_6H_4$, 4-ClC_6H_4, $PhCH_2$

Scheme 13.54

13.3.4.1.2 **Cyclization of N-Acylamidoximes** N-Acylamidoximes **158** cannot be prepared by acylation of amidoximes because O-acylation is faster. Suitable starting materials for N-acylamidoximes can be found in *N*-acylimidic chlorides **153** (X = Cl), cyanides **154** (X = CN), acylamidines **155** (X = NHR, NR_2), *N*-acyl(alkylthio)imides **156** (X = SR), or *N*-acyl(alkoxy)imides **157** (X = OR), which, by reaction with hydroxylamine, lead to 1,2,4-oxadiazoles **159** (Scheme 13.55) [209].

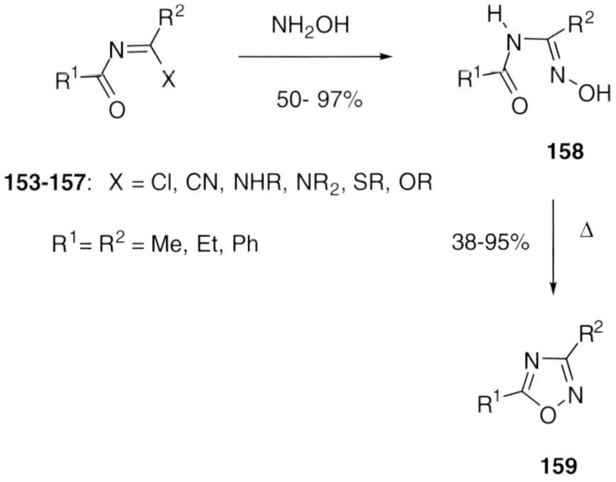

NH_2OH

50- 97%

153-157: X = Cl, CN, NHR, NR_2, SR, OR

$R^1 = R^2$ = Me, Et, Ph

38-95% Δ

Scheme 13.55

Imidates such as **160** react with cyanamide to give *N*-cyanoamidines **161**, while the hydrochloride **162**, is transformed into **163** [210]. Both **161** and **163** give 3-amino-1,2,4-oxadiazoles **164** on treatment with hydroxylamine (Scheme 13.56).

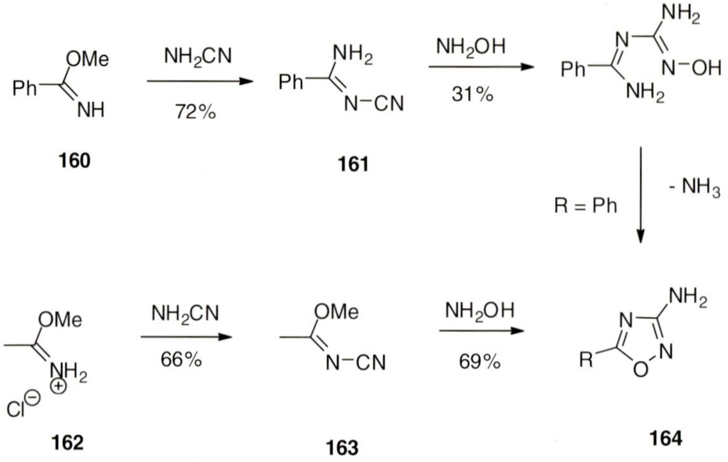

Scheme 13.56

A different approach involves the nitrosation of dimethylaminopropenoates **165**, with formation of the corresponding oximes **166**, which undergo cyclization to give the 5-substituted 1,2,4-oxadiazoles 3-carboxylates **167** (Scheme 13.57) [211].

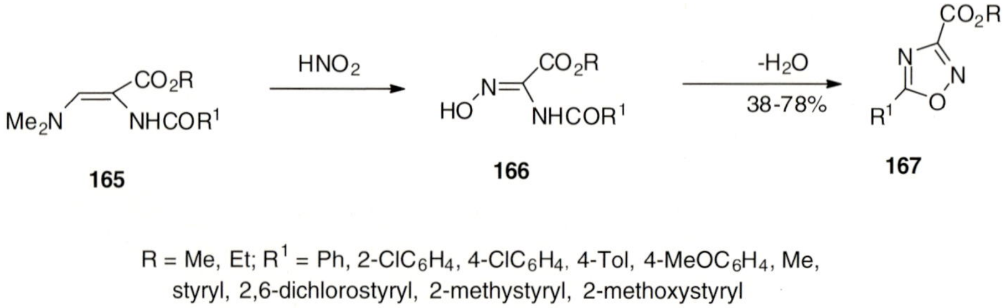

R = Me, Et; R^1 = Ph, 2-ClC_6H_4, 4-ClC_6H_4, 4-Tol, 4-$MeOC_6H_4$, Me, styryl, 2,6-dichlorostyryl, 2-methystyryl, 2-methoxystyryl

Scheme 13.57

The reaction of cyanohydrines **168** with hydroxylamine leads to the non-isolable amidoximes **169**, which, through intramolecular acylation to **170**, cyclize to epimeric 1,2,4-oxadiazoles **171** (Scheme 13.58) [212].

A series of substituted 1,2,4-oxadiazoles **176** have been synthesized through a new and versatile solid-phase synthesis protocol using resin-bound nitriles (**172**). This resin was treated with hydroxylamine and converted into resin-bound amidoximes **173**, which were transformed into the polymer supported *O*-acylamidoximes **174** upon treatment with acyl chloride. These compounds were subsequently converted into immobilized 1,2,4-oxadiazoles **175**, and then to **176** by treatment with 95% aqueous TFA in 15–52% overall yield (Scheme 13.59) [213].

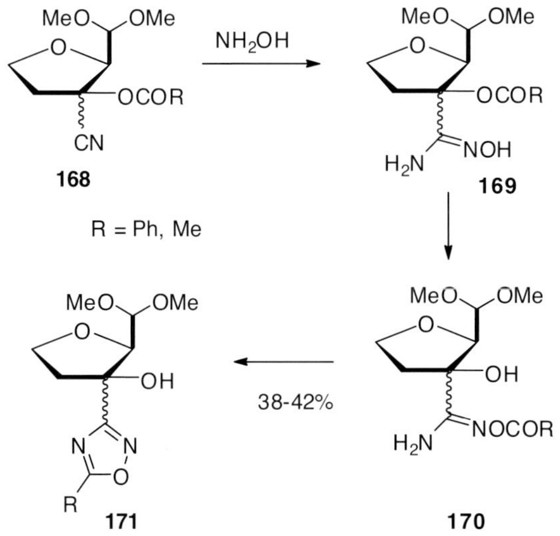

MeO OMe, O, OCOR, CN, 168 — NH_2OH → MeO OMe, O, OCOR, H_2N, NOH, 169

R = Ph, Me

169 → MeO OMe, O, OH, H_2N, NOCOR, 170 — 38-42% → MeO OMe, O, OH, N, N, O, R, 171

Scheme 13.58

172 (CN) — NH_2OH, Δ → 173 (H_2N, N, OH) — RCOCl, Py → 174 (H_2N, N, O, R, O) — TBAF → 175 (N, N, O, R) — 95 % aqeous TFA (15-52% overall yield) → 176 (Ph, H, N, O, N, N, O, R)

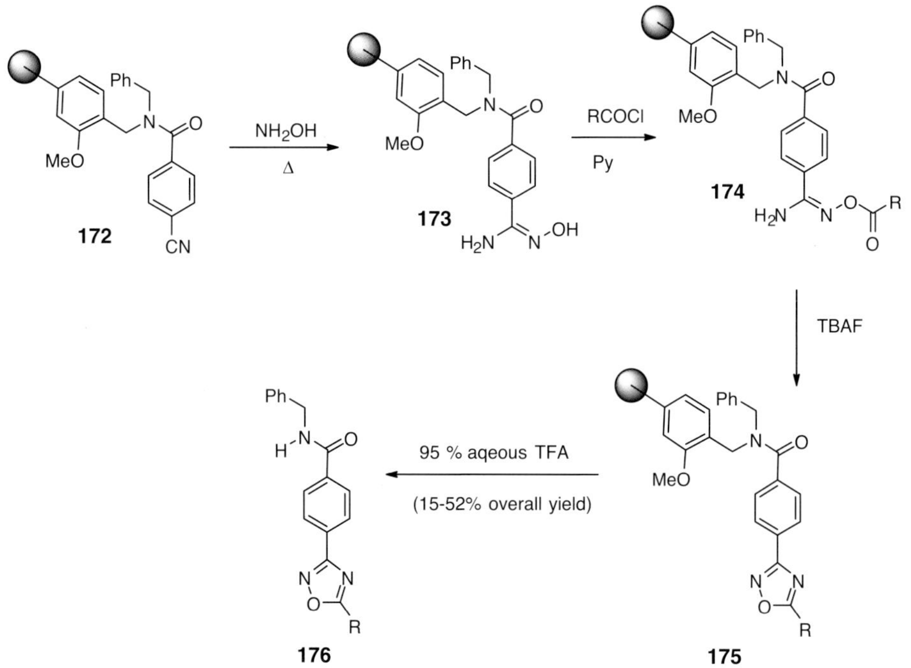

R = Me, *t*-butyl, Ph, 1-Naphthyl, 2-Naphthyl, 2-furyl, 2-thienyl, 4-pyridyl, 3-pyridyl, 2-MeO-C_6H_4, 3-MeO-C_6H_4, 4-MeO-C_6H_4, 2-F-C_6H_4, 3-F-C_6H_4, 4-F-C_6H_4, Cyclopropyl, Cyclobutyl, Cyclopentyl, Cyclohexyl, Phenylacetyl, hydrocinnamyl

Scheme 13.59

13.3.4.2 Cycloaddition Route

Another general and well-established route to 1,2,4-oxadiazoles **159** relies on the 1,3-dipolar cycloaddition between a nitrile **176** and a nitrile oxide **177** (Scheme 13.60) [120]. Aromatic and electron-deficient nitriles showed good reactivities, while aliphatic nitriles do not undergo cycloaddition to the oxadiazole derivative. However, under Lewis acid catalysis even aliphatic nitriles form cycloadducts [148].

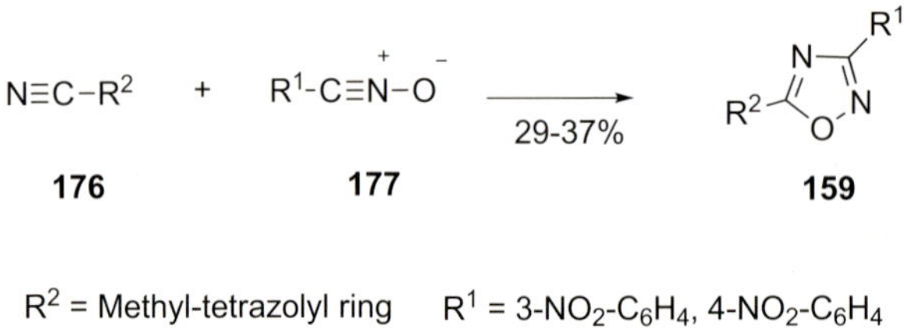

$N{\equiv}C{-}R^2$ (**176**) + $R^1{-}C{\equiv}\overset{+}{N}{-}\overset{-}{O}$ (**177**) → (29-37%) **159**

R^2 = Methyl-tetrazolyl ring R^1 = 3-NO_2-C_6H_4, 4-NO_2-C_6H_4

Scheme 13.60

Non-activated nitriles undergo cycloaddition with especially reactive nitrile oxides such as bromo- and chlorocyanogen oxide (Scheme 13.61) [214].

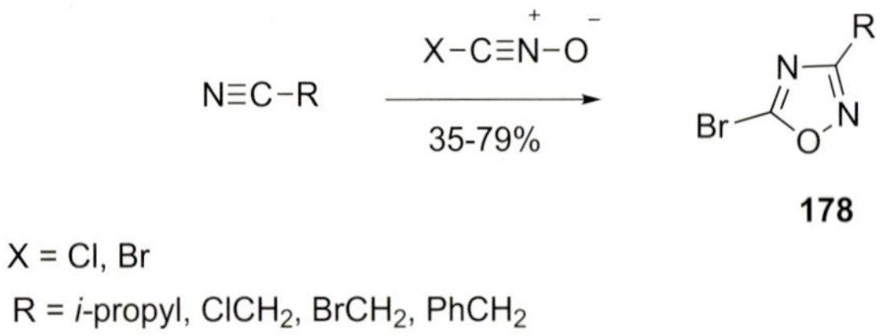

$N{\equiv}C{-}R$ → ($X{-}C{\equiv}\overset{+}{N}{-}\overset{-}{O}$, 35-79%) **178**

X = Cl, Br
R = *i*-propyl, $ClCH_2$, $BrCH_2$, $PhCH_2$

Scheme 13.61

Several methods are reported in the literature for the *in situ* generation of nitrile oxides. Huisgen's base-induced dehydrohalogenation of hydroximoyl chlorides [215] and Mukaiyama's dehydration of primary nitro compounds, using phenyl isocyanate with a catalytic amount of triethylamine [216], are the most frequently used routes to generate nitrile oxides. Thus, the loss of HCl from imidoyl chloride **179** leads to the nitrile oxide **180**, which undergoes cycloaddition to the dicyanoketene acetal **181**, producing the 1,2,4-oxadiazole **182** (Scheme 13.62) [217].

The ultrasound cycloaddition of nitrile oxide, formed by Mukaiyama's dehydration of nitroethane, with trichloroacetonitrile **183** affords the 1,2,4-oxadiazole **184** whose remarkable reactivity towards nucleophilic substitution by amines has been widely exploited (Scheme 13.63) [218].

Treatment of nitriles with acetone or acetophenone in the presence of iron(III) nitrate affords 3-acetyl- or 3-benzoyl-oxadiazoles **186**; the reaction proceeds through enolization and nitration to give an α-nitroketone, which undergoes an acid-catalyzed dehydration to the intermediate nitrile oxide **185** (Scheme 13.64) [219].

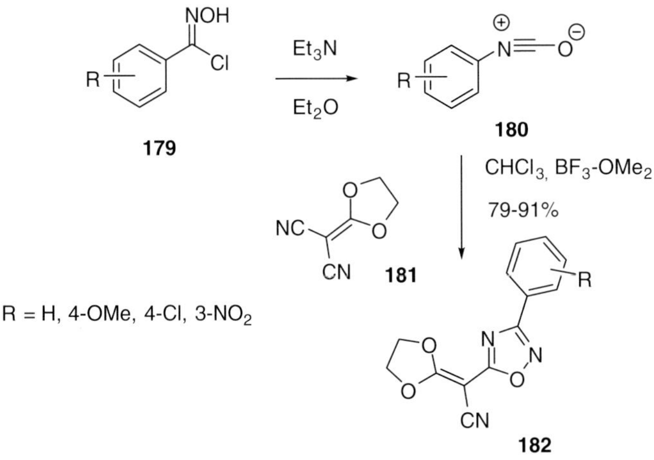

179 → 180: Et_3N, Et_2O

180 + 181 → 182: $CHCl_3$, BF_3-OMe_2, 79-91%

R = H, 4-OMe, 4-Cl, 3-NO_2

Scheme 13.62

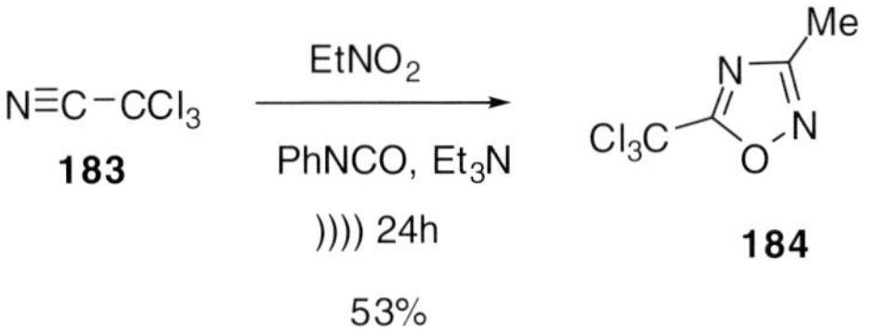

183 ($N\equiv C-CCl_3$) → 184: $EtNO_2$, PhNCO, Et_3N,)))) 24h, 53%

Scheme 13.63

Me(C=O)R → [$H_2C(NO_2)C(=O)R$ → 185] → 186: $Fe(NO_3)_3$, 80 °C; R^1–C≡N

R= Me, Ph
R^1 = Me, Et, Pr, i-Pr

186 25-95%

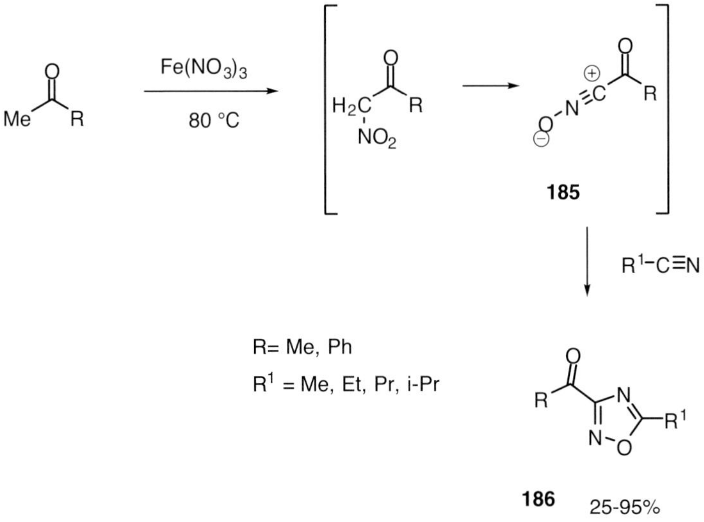

Scheme 13.64

A similar reaction leading to 1,2,4-oxadiazoles from ketones, nitriles, and nitric acid has been described using yttrium triflate as catalyst (Scheme 13.65) [220].

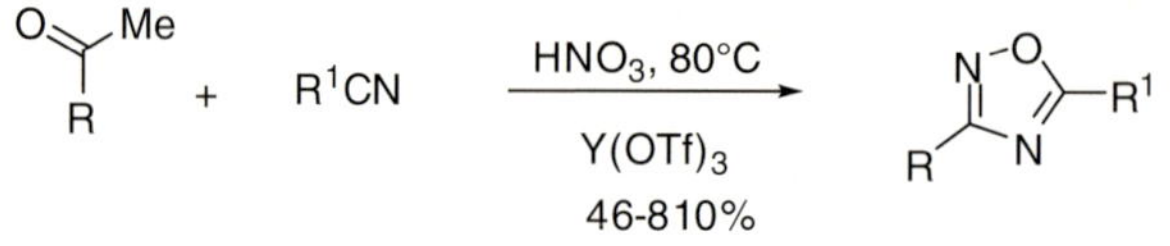

R = Me, 4-FC_6H_4, 4-ClC_6H_4, 3-$NO_2C_6H_4$, 2,4-Cl_2-5FC_6H_2, 3-NO_2-4MeC_6H_3

R^1 = Me, Ph, 2-FC_6H_4

Scheme 13.65

The reaction mechanism involves the 1,3 dipolar cycloaddition of nitriles with nitrile oxide **187**, which is obtained by enolization of the ketones promoted by yttrium triflate, followed by nitration and subsequent dehydration (Scheme 13.66).

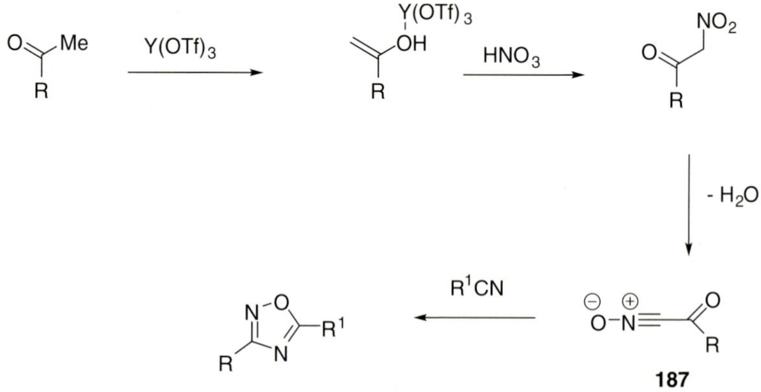

Scheme 13.66

A less common method for the formation of nitrile oxides is the oxidation of aromatic aldoximes with ceric ammonium nitrate (Scheme 13.67) [221]; the subsequent cycloaddition to nitriles leads to 1,2,4-oxadiazoles **188**.

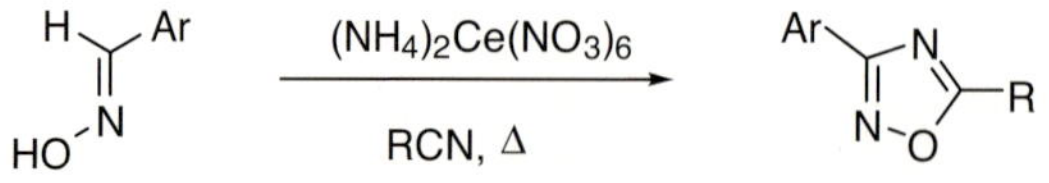

R = Me, Et

Ar = Ph, 4-MeC_6H_4, 4-$MeOC_6H_4$, 4-Cl, 4-$NO_2C_6H_4$

Scheme 13.67

The cycloaddition methodology has been employed for the synthesis of complex systems. The reaction between nitrile oxides **189** and *trans*-[$PdCl_2(RCN)_2$], or RCN (R = Me, Et, CH_2CN, NMe_2, Ph) in the presence of $PdCl_2$, proceeded smoothly under mild conditions and allowed isolation of the *trans*-[-$PdCl_2$]-1,2,4-oxadiazole complexes (**190–197**) in 40–85% yields. (Scheme 13.68) [222].

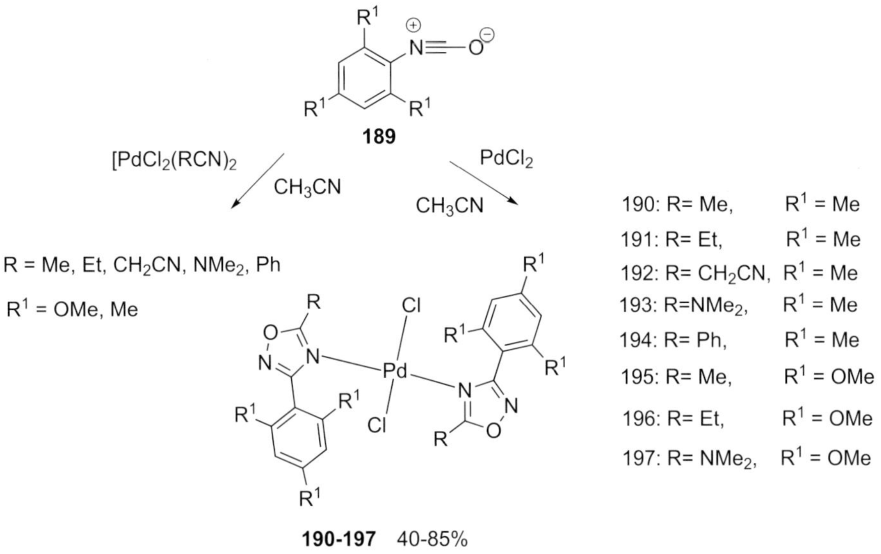

Scheme 13.68

3-Aryl-5-*C*-glucosyl-1,2,4-oxadiazoles **199** and **200**, assayed as glycogen phosphorylase inhibitors, have been prepared in high yield by 1,3-dipolar cycloaddition of aryl nitrile oxides to benzoylated glucosyl cyanide **198** and subsequent cleavage of the protecting group (Scheme 13.69) [223].

1,2,4-Oxadiazoles have been prepared by cycloaddition of nitrile oxides to different dipolarophiles. Thus, the cycloaddition of nitrile oxides to amidoximes **201** proceeds with loss of diethylamine to give the 1,2,4-oxadiazole-4-oxide **202**, which can be deoxygenated with trimethyl phosphite to give 1,2,4-oxadiazole **203** (Scheme 13.70) [224].

The reaction of nitrile oxide **204** with imine **205** affords the 1,2,4-oxadiazole **207** via the non-isolable intermediate **206** (Scheme 13.71) [225].

13.3.4.3 Miscellaneous Synthesis of 1,2,4-Oxadiazoles

Fully conjugated 1,2,4-oxadiazoles have been prepared by oxidation of 4,5-dihydro-1,2,4-oxadiazoles **208** (Section 13.3.4.4.1), containing hydrogen atoms in the 4- and 5-positions: the oxidation can be performed by MnO_2 [125], nitric acid [125], NaOCl [162], or *N*-chlorosuccinimide (NCS) [169] (Scheme 13.72).

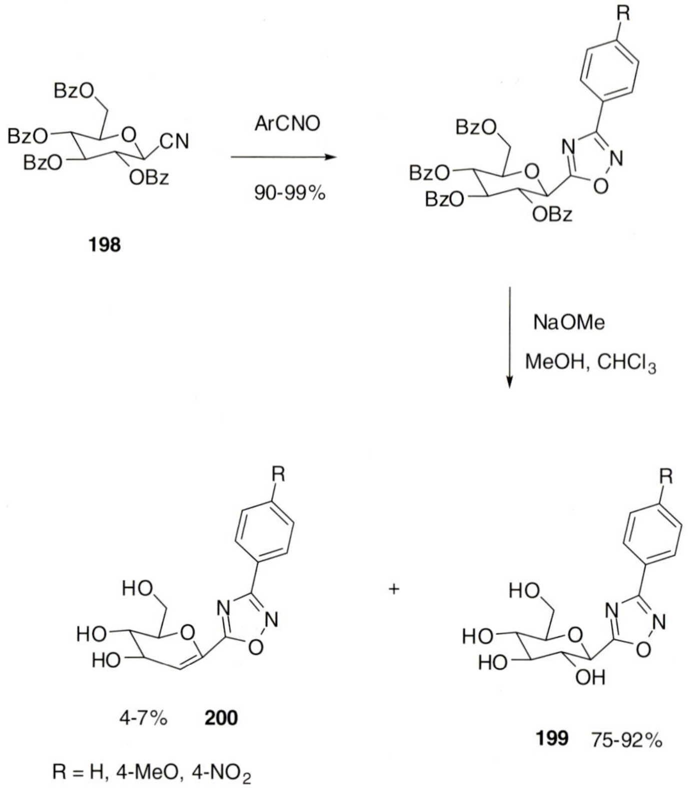

BzO, BzO, BzO, CN, OBz
198
ArCNO
90-99%
R, BzO, BzO, BzO, N, N, O, OBz
NaOMe
MeOH, $CHCl_3$
R, HO, HO, HO, N, N, O
4-7% **200**
+
R, HO, HO, HO, N, N, O, OH
199 75-92%
R = H, 4-MeO, 4-NO_2

Scheme 13.69

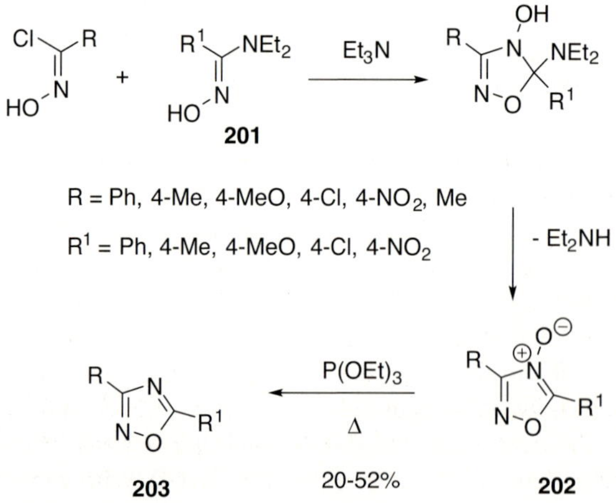

Cl, R, N, HO
+
R^1, NEt_2, N, HO
201
Et_3N
OH, R, N, NEt_2, N, O, R^1
R = Ph, 4-Me, 4-MeO, 4-Cl, 4-NO_2, Me
R^1 = Ph, 4-Me, 4-MeO, 4-Cl, 4-NO_2
- Et_2NH
R, N, N, O, R^1
203
$P(OEt)_3$
Δ
20-52%
R, $N^{\oplus}$, $O^{\ominus}$, N, O, R^1
202

Scheme 13.70

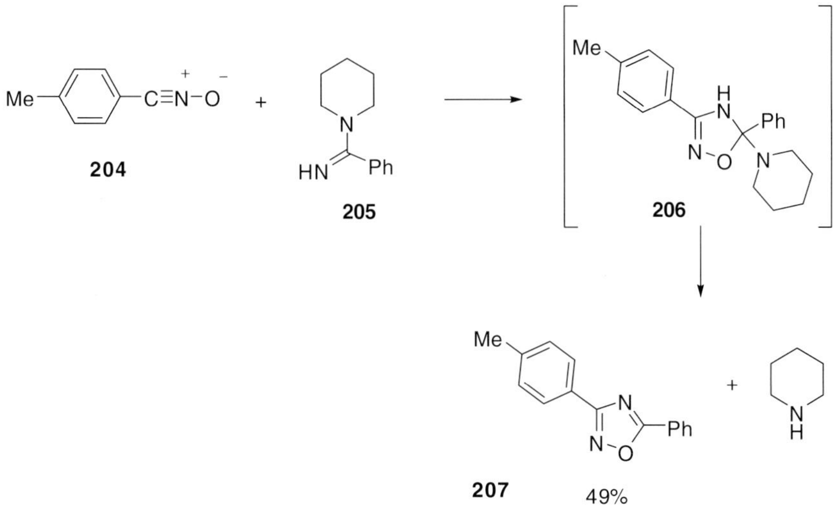

Scheme 13.71

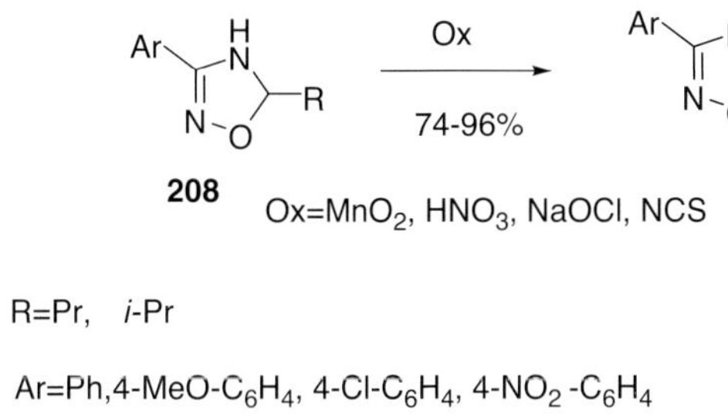

Ox=MnO_2, HNO_3, NaOCl, NCS

R=Pr, *i*-Pr

Ar=Ph,4-MeO-C_6H_4, 4-Cl-C_6H_4, 4-NO_2-C_6H_4

Scheme 13.72

Oxidation of *N*-benzylamidoxime **209** with $KMnO_4$ affords 1,2,4-oxadiazole **211**, through the intermediate 4,5-dihydro-1,2,4-oxadiazole **210** (Scheme 13.73) [226].

Bicyclic 4,5-dihydro-1,2,4-oxadiazole **212** leads to **213** through a retro-[2 + 2] cycloaddition via loss of styrene in toluene at reflux (Scheme 13.74) [227].

Some recent synthetic procedures concern ANRORC-like reactions that consist of the Addition of a Nucleophile to a π-deficient heterocycle, followed by Ring-Opening and Ring-Closure steps. By this approach, a heterocycle can be transformed into a different one containing the heteroatoms originally belonging to the nucleophilic reagent. Thus, the reaction of 5-fluoroalkyl-1,2,4-oxadiazoles **214** with hydroxylamine furnishes high yields of 3-fluoroalkyl-1,2,4-oxadiazole **217** in a virtual C5–C3 annular shift (Scheme 13.75) [228]. The reaction is promoted by nucleophilic attack of the hydroxylamine to the electron-deficient C5 to produce **215**. Heterocyclization of the dioxime intermediate **216** and removal of hydroxylamine leads to the more stable oxadiazole **217**, in an irreversible ring-degenerate process.

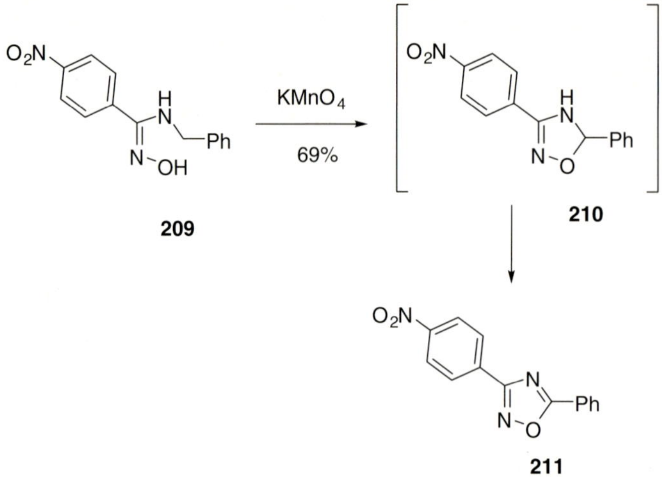

$KMnO_4$, 69%: **209** → [**210**] → **211**

Scheme 13.73

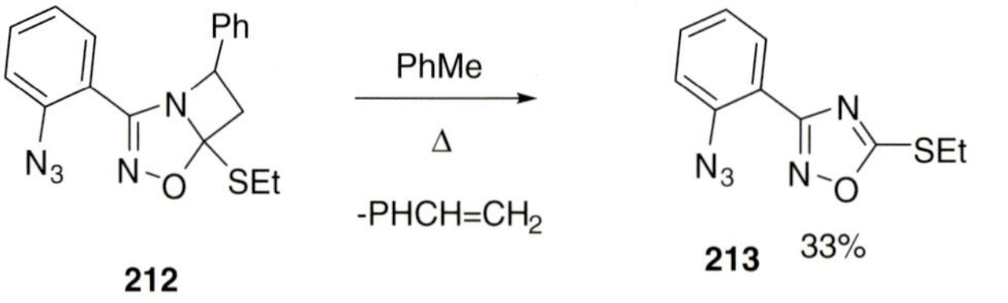

PhMe, Δ, $-PHCH{=}CH_2$: **212** → **213** (33%)

Scheme 13.74

NH_2OH, 70-94%: **214** → **217**; **214** → [**215** ⇌ intermediate ⇌ **216**] → **217**

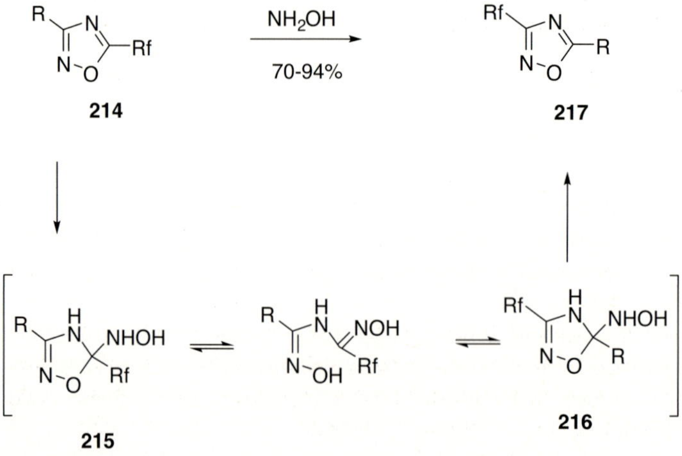

Scheme 13.75

Accordingly, photoinduced rearrangements of O–N bond containing azoles can be useful for the synthesis of 3-amino-5-alkyl-1,2,4-oxadiazoles [229]. This procedure exploits the photofragmentation pattern of the furazan heterocycle into a nitrile and a nitrile oxide. Thus, irradiation of 3-alkanoylamino **217** at $\lambda = 313$ nm in methanol and in the presence of ammonia or primary aliphatic amines gives the corresponding 3-amino- or 3-*N*-alkylamino-5-alkyl-1,2,4-oxadiazoles **219** as a result of the heterocyclization of the intermediate **218** (Scheme 13.76) [230, 231].

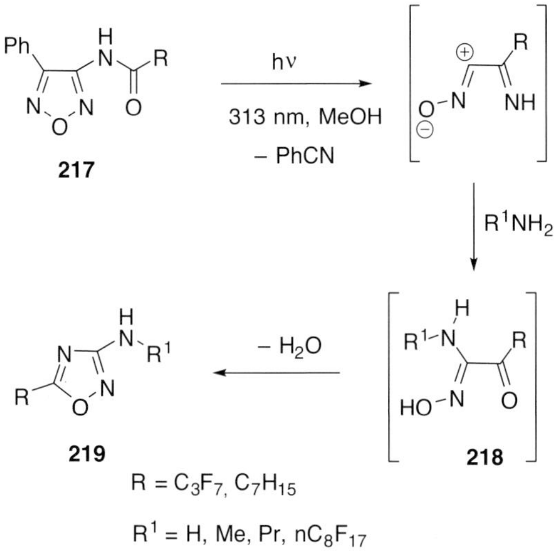

hν, 313 nm, MeOH, – PhCN; R^1NH_2; – H_2O

217, **218**, **219**

R = C_3F_7, C_7H_{15}

R^1 = H, Me, Pr, nC_8F_{17}

Scheme 13.76

Unfortunately, yields of isolated products (about 30–40%) were not very good because of the photoreactivity of oxadiazoles under irradiation conditions; this, however, appears to be the only method that allows these derivatives to be obtained.

13.3.4.4 Synthesis of Dihydro-1,2,4-Oxadiazoles

13.3.4.4.1 **4,5-Dihydro-1,2,4-Oxadiazoles** The main methodology towards the synthesis of 4,5-dihydro-1,2,4-oxadiazoles **220** relies on the reaction of carbonyl compounds with amidoximes **136** under acidic conditions (Scheme 13.77) [120, 162, 232].

The use of chloroformate or diethyl carbonate leads to 4,5-dihydro-1,2,4-oxadiazolones **222** via an intermediate acetamidoxime, which cyclizes under base treatment (Scheme 13.78) [233].

The reaction with phosgene or thiophosgene constitutes an alternative route towards the 4,5-dihydro-1,2,4-oxadiazol-5-ones or -5-thiones **223** (Scheme 13.79) [234].

A widely exploited route to 4,5-dihydro-1,2,4-oxadiazoles is the 1,3-dipolar cycloaddition of nitrile oxides to azomethines [120, 235].

Thus, the reaction of imines **224** with hydroxyimoyl chlorides **225** in the presence of triethylamine gives 4,5-dihydro-1,2,4-oxadiazoles **226** (Scheme 13.80) [236].

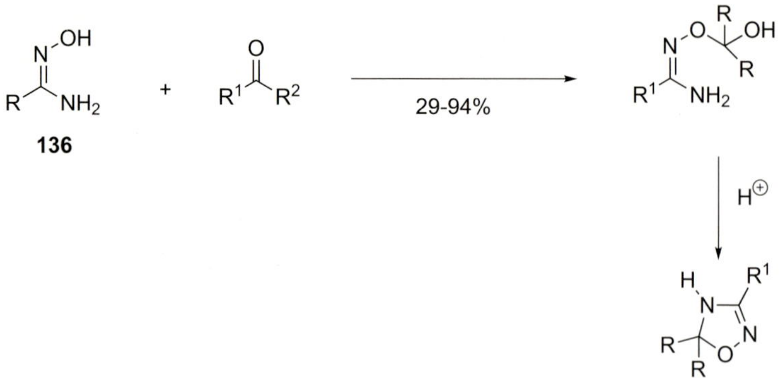

R = Ph, Bn, 2-furyl, 2-thienyl, 2-Tol, 3-Tol, 4-Tol, 4-pyridyl, 2-MeO-$C_6H_4CH_2$,
R^1 = H, Me, Me_2CH, Et, 4-MeO-C_6H_5,
R^2 = H, Me;
R^1, R^2 = -$(CH_2)_5$-

Scheme 13.77

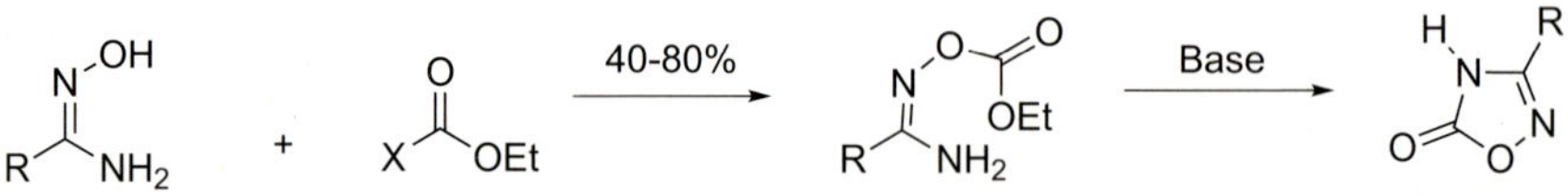

R = Me, 4-Tol
X = Cl, OEt

Scheme 13.78

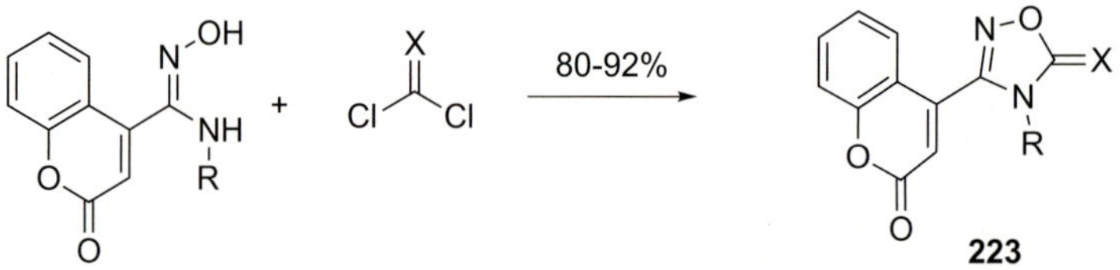

R = H, Ph, 4-Tol,
X = O, S

Scheme 13.79

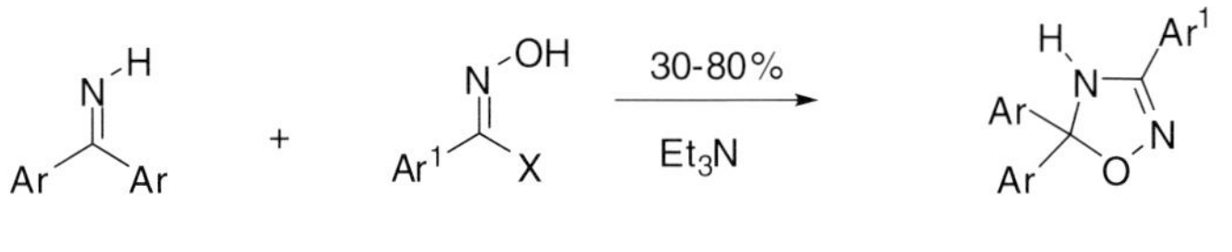

224 + 225 $\xrightarrow[\text{Et}_3\text{N}]{30\text{-}80\%}$ 226

Ar = Ph, 4-MeC$_6$H$_4$, 4-ClC$_6$H$_4$, 4-BrC$_6$H$_4$, 3NO$_2$-C$_6$H$_4$, 4-C$_6$H$_4$, 4-C$_6$H$_4$, 4-C$_6$H$_4$,

Ar^1 = Ph, 4-MeC$_6$H$_4$

Scheme 13.80

Scheme 13.81 reports a parallel synthesis of 4,5-dihydro-1,2,4-oxadiazoles, through a cycloaddition reaction of imines with poly(ethylene glycol) (PEG) supported nitrile oxide. 4-Formylbenzoic acid (**227**) was attached to the dihydroxylated PEG by esterification in the presence of DCC. The PEG-bound derivative **228** was converted into oxime **229** by treatment with hydroxylamine hydrochloride, in the presence of trioctylamine, which with *N*-chlorosuccinimide afforded the PEG-bound chlorooxime **230**. This derivative was then treated with several imines to give the corresponding cycloadducts **231**, which were released from the PEG by treatment with sodium methoxide in methanol, to afford 1,2,4-oxadiazolines **232** in 71–91% overall yield [237].

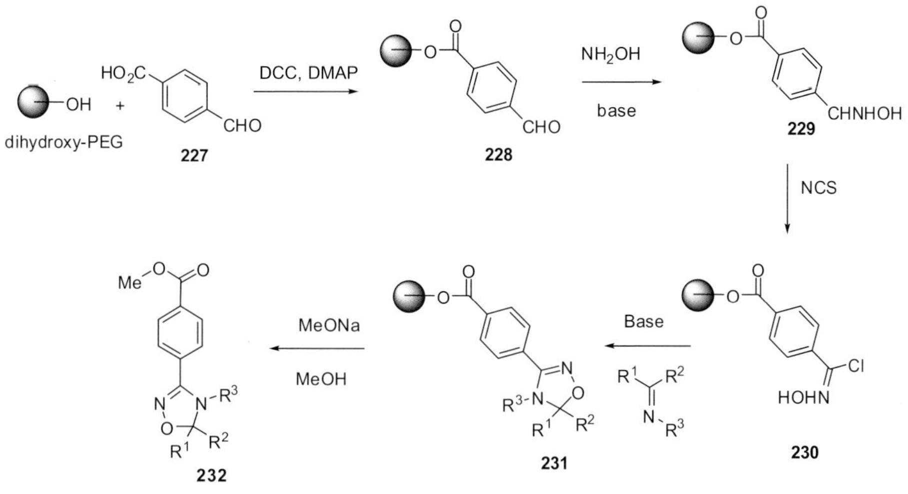

dihydroxy-PEG + 227 (HO$_2$C, CHO) $\xrightarrow{\text{DCC, DMAP}}$ 228 $\xrightarrow[\text{base}]{\text{NH}_2\text{OH}}$ 229 (CHNHOH) $\xrightarrow{\text{NCS}}$ 230 (Cl, HOHN) $\xrightarrow{\text{Base}}$ (R^1, R^2, N-R^3 imine) 231 $\xrightarrow[\text{MeOH}]{\text{MeONa}}$ 232

R^1 = H, Me

R^2 = Ph, 4-MeO-C$_6$H$_4$, 4-F-C$_6$H$_4$, 4-Me-C$_6$H$_4$, 3-NO$_2$-C$_6$H$_4$, 4-C$_7$H$_5$O$_2$,

R^3 = Ph, 4-F-C$_6$H$_4$, 4-Me-C$_6$H$_4$, PhCH$_2$, butyl

Scheme 13.81

The use of cyclic imines in the cycloaddition reaction is a useful route to various fused 4,5-dihydro-1,2,4-oxadiazoles. In this way, oxadiazolo-1,4-diazepines, **233** [238], oxadiazole-1,5-benzodiazepines **234** [239], and oxadiazolotriazole-1,5-benzodiazepines **235** have been prepared [240] (Scheme 13.82).

27-30%

R = Ph, Me

233

Et_3N

91%

R = 3-Cl,4-Me-C_6H_3,

R^1 = SMe

234

Et_3N

47-58%

R = NO_2, Cl,

R^1 = 4- NO_2-C_6H_4, 4- Me-C_6H_4,4- Cl-C_6H_4

235

Scheme 13.82

13.3.4.4.2 **2,5-Dihydro-1,2,4-Oxadiazoles** Aminonitrones **236**, prepared by reaction of hydroxylamines with ethyl cyanoformate, cyclize by treatment with triphosgene to 2,5-dihydro-1,2,4-oxadiazin-5-ones **237** (Scheme 13.83).

A related reaction involves the acylations of 2-aminopyridine *N*-oxides **238** and **239** with ethyl chloropyruvate or phosgene, respectively (Scheme 13.84) [241].

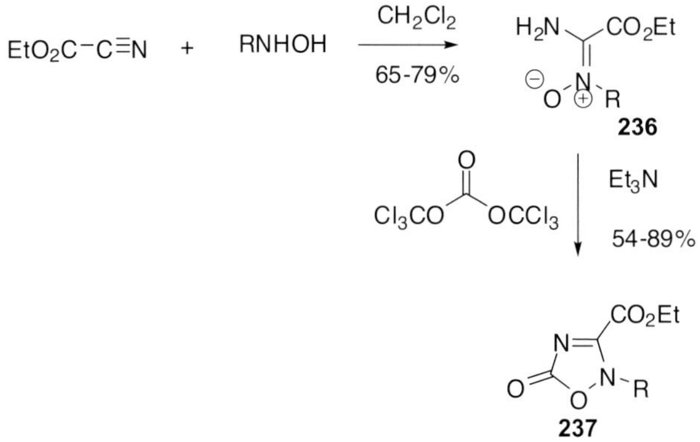

$EtO_2C{-}C{\equiv}N$ + RNHOH → (CH_2Cl_2, 65-79%) 236 → ($Cl_3CO(CO)OCCl_3$, Et_3N, 54-89%) 237

R = Me, i-propyl, Ph, 4-BrC_6H_4, 4-MeC_6H_4, 3-MeC_6H_4

Scheme 13.83

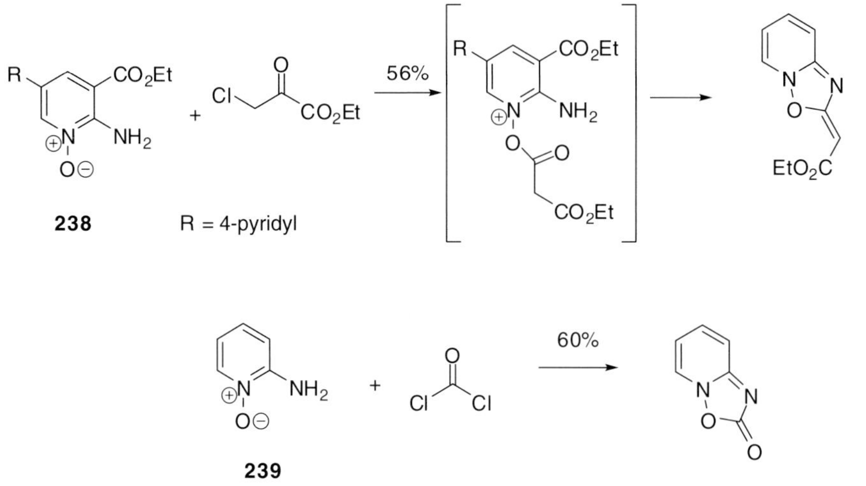

238 + $ClCH_2COCH_2CO_2Et$ → (56%) [intermediate] →

R = 4-pyridyl

239 + $ClCOCl$ → (60%)

Scheme 13.84

The reaction of substituted oxazoles **241** with arylnitroso derivatives **240** affords 2-aryl-2,5-dihydro-1,2,4-oxadiazoles **242** regioselectively, through a formal [3 + 2] cycloaddition, proceeding via a ring opening of oxazoles promoted by a nucleophilic attack of the nitroso compound at the 2-position of the penta-atomic ring (Scheme 13.85) [242].

A facile one-pot synthesis of 2,3,5-substituted 1,2,4-oxadiazolines from nitriles in aqueous solution has been reported [243]. Thus, alkyl/aryl amidoximes, prepared from the corresponding nitriles and *N*-alkylhydroxylamines, readily undergo consecutive double Michael additions to electron-deficient alkynes and provide highly

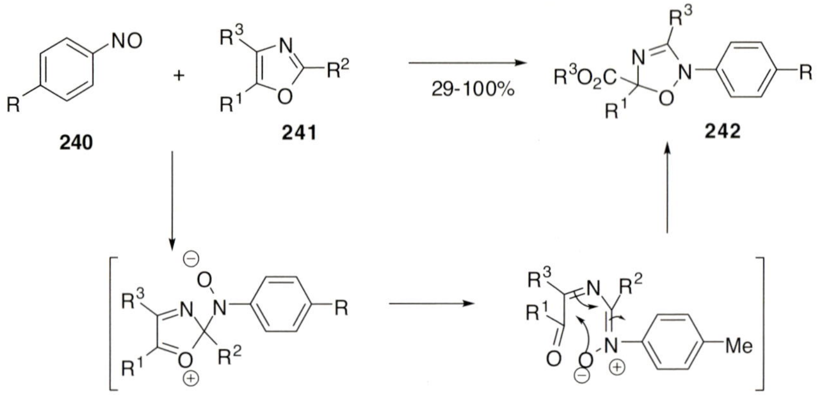

R = H, Cl, Me $\quad$ R^1 = Me, MeO, EtO, $\quad$ R^2 = Me, nonyl, 4-MeC_6H_4, 4-$MeOC_6H_4$

R^3 = H, Me, CO_2Et, Ph, 4-$NO_2C_6H_4$

Scheme 13.85

substituted 1,2,4-oxadiazolines **243** in good yields in homogeneous aqueous solution (Scheme 13.86).

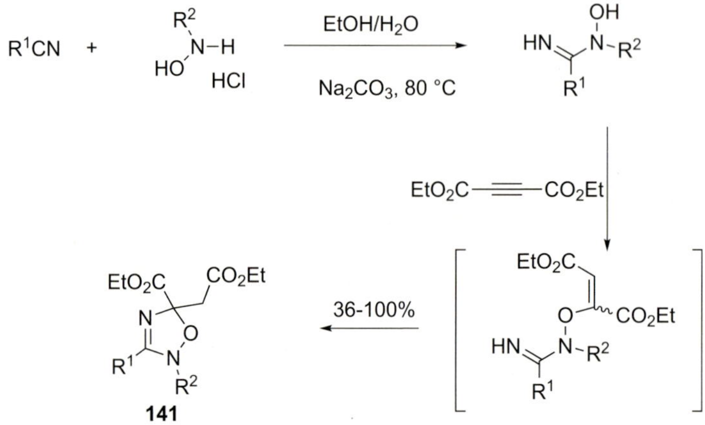

R^1 = Me, *i*-pr- Ph, $PhCH_2$, 2-furyl, 2-pyridyl, 3-pyridyl, 4-pyridyl

R^2 = Me, Cy, $PhCH_2$

Scheme 13.86

13.3.4.4.3 **2,3-Dihydro-1,2,4-Oxadiazoles** A general route to 2,3-dihydro-1,2,4-oxadiazoles is based on the 1,3-dipolar cycloaddition of nitrones to nitriles. Thus, 3-*t*-butyl-2,3-dihydro-1,2,4-oxadiazoles **246** have been prepared through cycloaddition between butylnitrone **244** and different activated nitriles **245** (Scheme 13.87) [244].

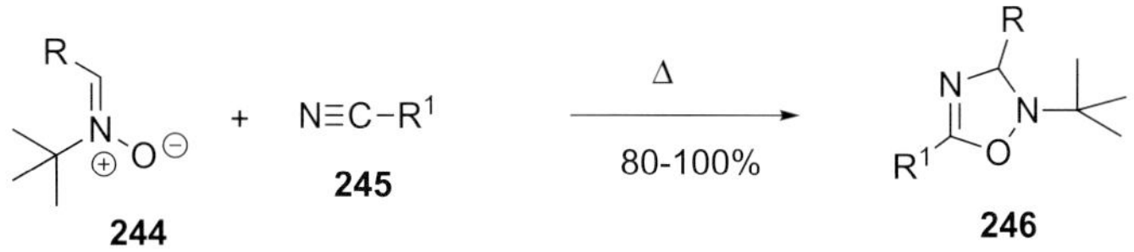

R = Ph, 4-NO_2-C_6H_4

R^1 = ClC(CN)$_2$, Br_2(CN), C(CN)$_3$, CCl_3

Scheme 13.87

The cycloadditions have been performed in the absence of solvent, under microwave irradiation within 2–10 min [245].

Recently, a series of 5-trichloro- and 5-(2-methylpropanenitrile)-Δ^4-1,2,4-oxadiazolines **247** have been synthesized by 1,3-dipolar cycloaddition of nitrones to trichloroacetonitrile and 2,2-dimethylmalononitrile, respectively. These oxadiazolines rearrange into formamidine derivatives **248** by prolonged heating, via ring opening and a 1,2-aryl shift from carbon to the adjacent amino nitrogen (Scheme 13.88) [246].

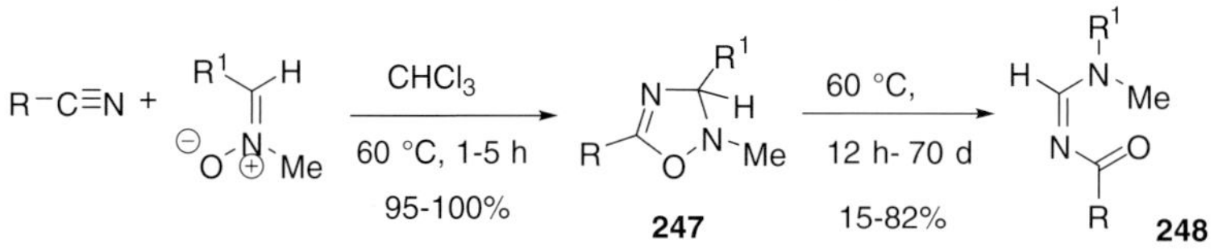

R = CCl_3, Me_2CCN

R^1 = 2-$MeOC_6H_4$, 2,3-(MeO)$_2C_6H_3$, 2,4-(MeO)$_2C_6H_3$, 2,5-(MeO)$_2C_6H_3$, 2,6-(MeO)$_2C_6H_3$, 3,4-(MeO)$_2C_6H_3$, 2,3,4-(MeO)$_3C_6H_2$, 3,4,5-(MeO)$_3C_6H_2$, 2,4,5-(MeO)$_3C_6H_2$, 2,4,6-(MeO)$_3C_6H_2$

Scheme 13.88

Based on a recent observation that Pt(II) mono- and bis-oxadiazoline complexes exhibit *in vitro* cytotoxicity against a series of platinum-sensitive and resistant human cancer cell lines with a potency comparable to that of cisplatin and superior to carboplatin [247] a series of 2,3-dihydro-1,2,4-oxadiazoles have been synthesized by 1,3-dipolar cycloaddition of coordinated dinitriles **249a,d** to nitrones **250a–c** (Scheme 13.89). Moreover, Pt(II)oxadiazoline complexes **251a** and **251d**, having only one of the coordinated nitriles, have been used for various new mixed ligand complexes. Thus, **252–254** have been obtained by reaction of the corresponding oxadiazolines with pyridine, 4-*N*-dimethylpyridine, and 1-benzyl-2-methylimidazole, respectively (Scheme 13.90) [248].

A novel type of heterocycle, 2,3a-disubstituted 5,6-dihydro-3aH-[1,3]oxazolo[3,2-*b*][1,2,4]oxadiazoles **258a–g**, has been generated by an intermolecular Pt(II)-mediated 1,3-dipolar cycloaddition between the oxazoline *N*-oxide **256** and coordinated nitriles

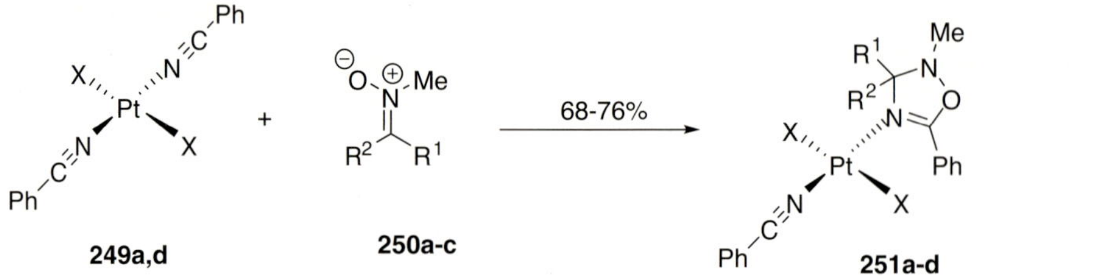

68-76%

249a,d

250a-c

251a-d

46-71%

250a-c

261a-d

	X	R^1	R^2
a	Cl	H	Ph
b	Br	H	Ph
c	I	H	Ph
d	Cl	CO_2E	CH_2CO_2Et

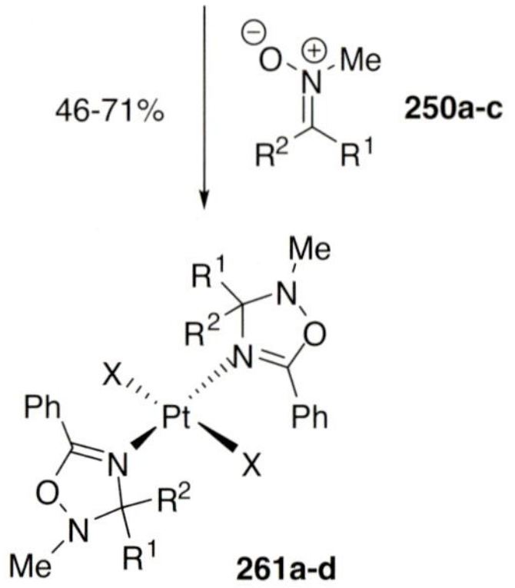

Scheme 13.89

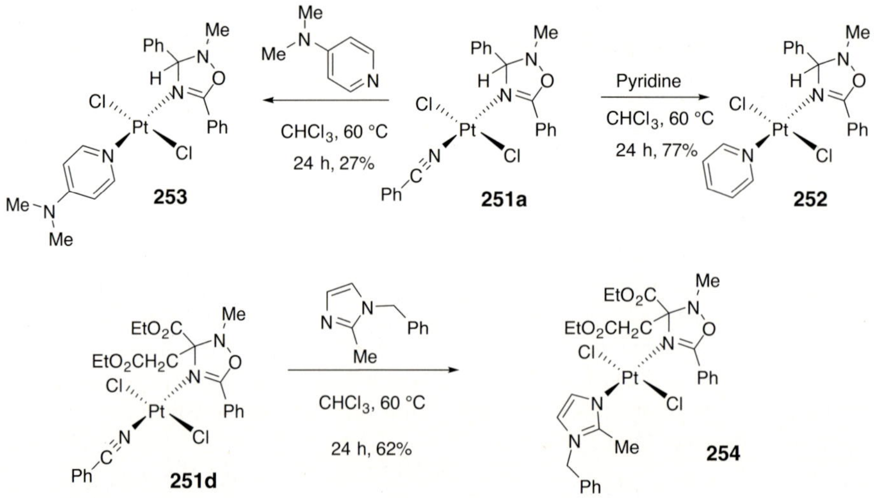

$CHCl_3$, 60 °C
24 h, 27%

Pyridine
$CHCl_3$, 60 °C
24 h, 77%

253

251a

252

$CHCl_3$, 60 °C
24 h, 62%

251d

254

Scheme 13.90

in the complexes *trans/cis*-[$PtCl_2$(R-CN)$_2$] **255**. The reaction is unknown for free RCN and oxazoline *N*-oxides, but under PtII-mediated conditions the reaction proceeds smoothly and gives pure complexes **257a–g** in 42–79% yields (Scheme 13.91) [249].

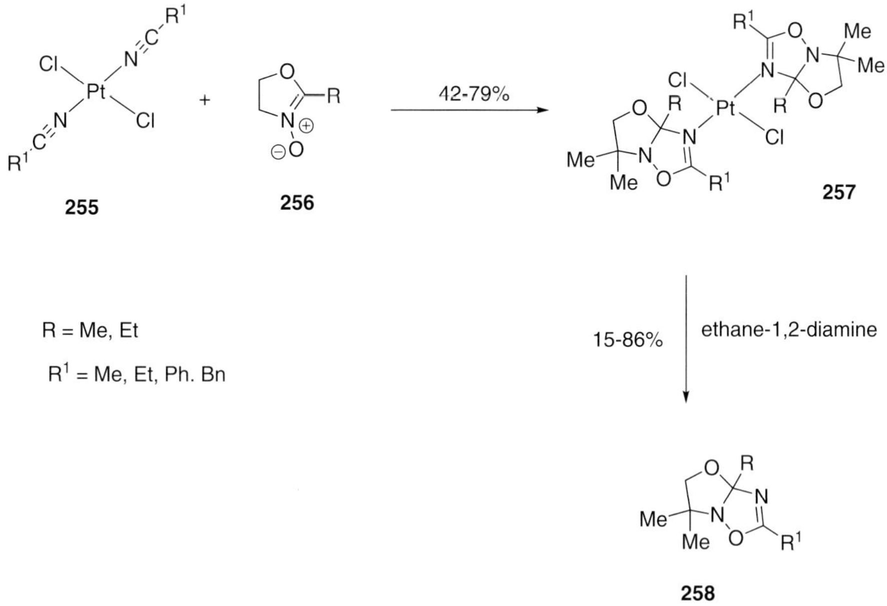

Scheme 13.91

13.3.4.5 **Synthesis of 1,2,4-Oxadiazolidines**

A general synthetic approach to 1,2,4-oxadiazolidines **259** exploits the 1,3-dipolar cycloaddition of nitrones to a C=N double bond, a method first used by Beckmann (Scheme 13.92) [250].

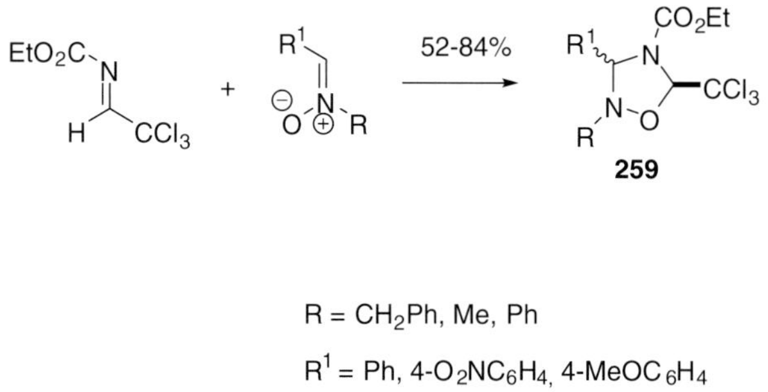

Scheme 13.92

The use of isocyanates and isothiocyanates as dipolarophiles affords an easy entry to 1,2,4-oxadiazolidinones and thiones **260** (Scheme 13.93) [251].

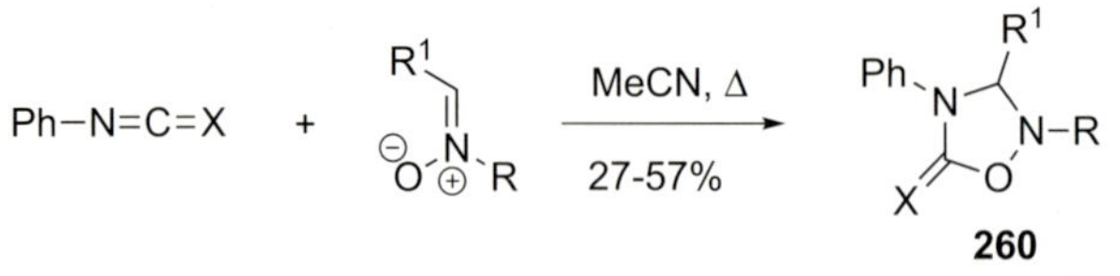

Ph−N=C=X + nitrone (MeCN, Δ; 27-57%) → **260**

R= Me, $PhCH_2$, 4-$O_2NC_6H_4$, 2,3-$(MeO)_2C_6H_3CH_2$ X = O, S

R^1 = Ph, 2-$O_2NC_6H_4$, 3-$O_2NC_6H_4$, 2,3-$(MeO)_2C_6H_4$, 3,4-$(MeO)_2C_6H_4$

Scheme 13.93

On this basis, 1,2,4-oxadiazolidinones as stable chiral building blocks have been prepared by 1,3-dipolar cycloaddition of isocyanates with mannosyl- or erythrosyl derived nitrones **261**. The reaction proceeds with a good diastereoselectivity, giving enantiopure 1,2,4-oxadiazolidin-5-ones **262** after removal of the auxiliary (Scheme 13.94) [252].

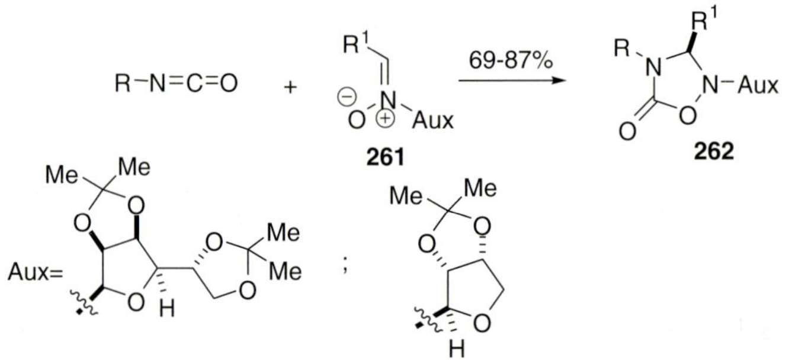

R−N=C=O + **261** (69-87%) → **262**

R= Ph, $PhCH_2$, PhCO, 4-$O_2NC_6H_4$, 4-$CF_3C_6H_4$, 4-FC_6H_4, 4-$MeOC_6H_4$, 2,6-$Cl_2C_6H_3$,

R^1 =Me, Cy, *t*-Bu, Ph, 3-Py, 2-furyl, 1- Naph, 4-$O_2NC_6H_4$, 4-$CF_3C_6H_4$, 4-BrC_6H_4, 4-$MeOC_6H_4$

Scheme 13.94

The reaction of oxaziridines **263** with isothiocyanates affords 1,2,4-oxadiazolidin-5-thiones **264**; interestingly, the reaction of **263** with chlorosulfonyl isocyanate **265** leads to 1,2,4-oxadiazolidin-3-ones **266**, as established by X-ray crystallography (Scheme 13.95) [253].

13.3.4.6 Synthesis of 1,2,4-Oxadiazole-N-Oxides

The 1,3-dipolar cycloaddition of amidoximes **267** with nitrile oxides affords an easy entry to 1,2,4-oxadiazole-*N*-oxides **268**, through elimination of an amine (Scheme 13.96). A version of this approach exploited the use of Wang-supported nitrile oxide [224, 254].

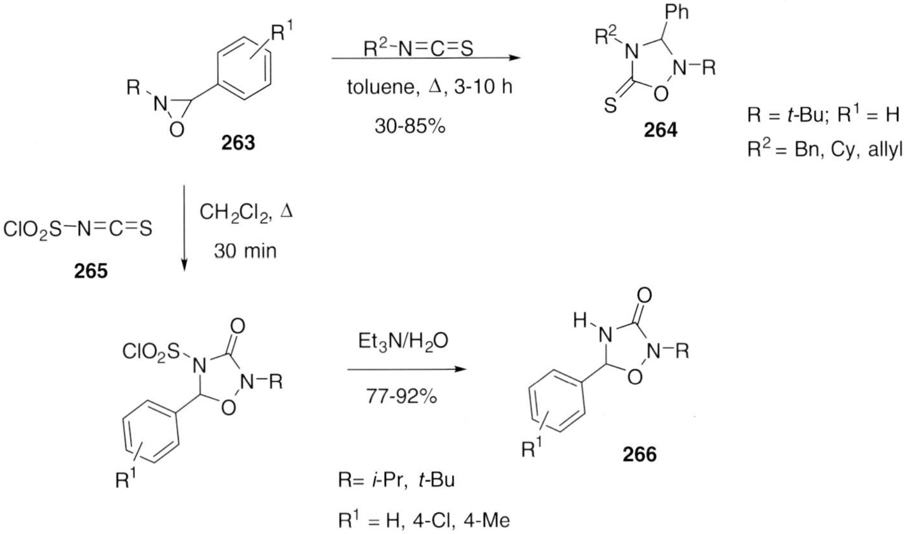

R = *t*-Bu; R^1 = H
R^2 = Bn, Cy, allyl

R= *i*-Pr, *t*-Bu
R^1 = H, 4-Cl, 4-Me

Scheme 13.95

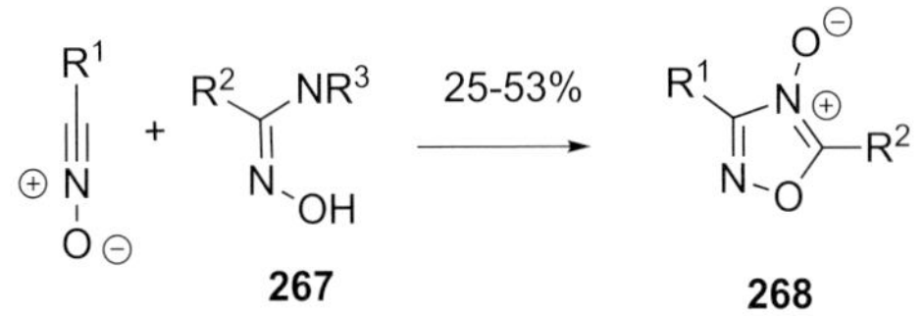

R^1 = Ph, 4-$McOC_6H_4$, 4-ClC_6H_4, 4-$O_2NC_6H_4$, 4-MeC_6H_4

R^2= Me, Ph, 4-$O_2NC_6H_4$, 4-$MeOC_6H_4$, 4-ClC_6H_4

R^3 = H, Et

Scheme 13.96

13.3.5 Reactivity of 1,2,4-Oxadiazoles

13.3.5.1 Reactions with Electrophiles

1,2,4-Oxadiazoles are rather inert against electrophilic attack. Halogenation, nitration, Friedel–Crafts alkylation, and acylation do not occur in this ring system. However, electrophilic mercuration of 5-unsubstituted oxadiazoles **269** is possible (Scheme 13.97) [120].

3,5-Diaryl-substituted 1,2,4-oxadiazoles serve as monodentate ligands for some transition metal complexes. The reaction of 1,2,4-oxadiazole **270** with Cu(II)acetate, to give **271**, occurs selectively on N4 (Scheme 13.98) [170].

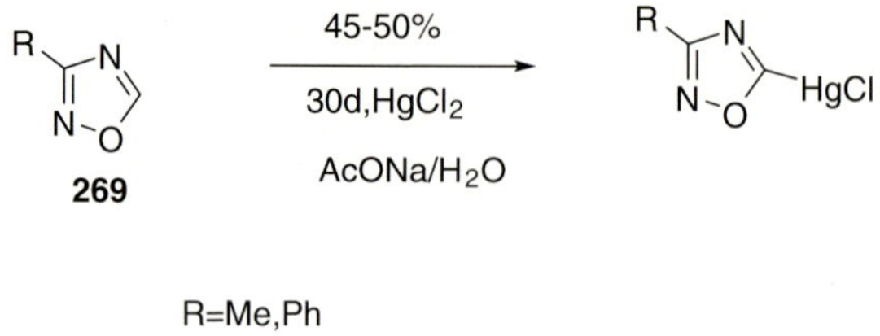

45-50%

30d,$HgCl_2$

$AcONa/H_2O$

269

R=Me,Ph

Scheme 13.97

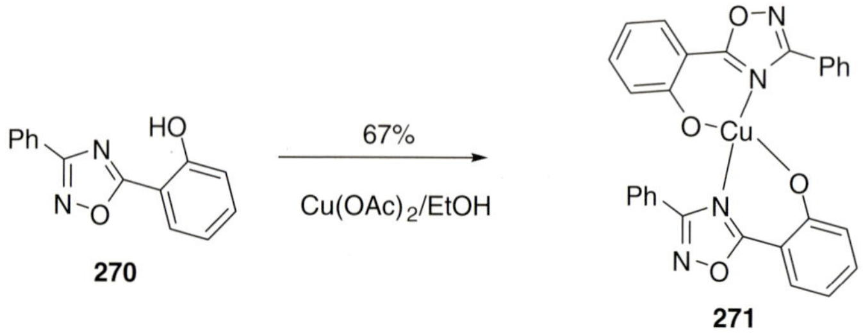

67%

$Cu(OAc)_2/EtOH$

270

271

Scheme 13.98

The closely correlated oxadiazole **272** is, by treatment with dimethyl sulfate and perchloric acid, methylated at N2 to give the 1,2,4-oxadiazolium salt **273** (Scheme 13.99).

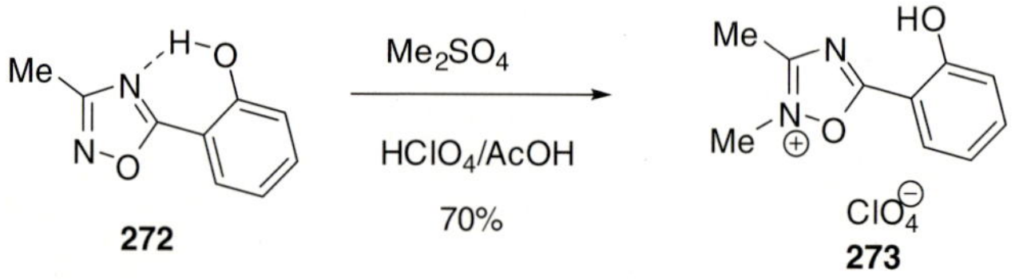

Me_2SO_4

$HClO_4/AcOH$

70%

272

273

Scheme 13.99

An interesting example of an intramolecular electrophilic attack at the N2, reported in Scheme 13.100, yields oxadiazolopyrimidinium salts **274** [255].

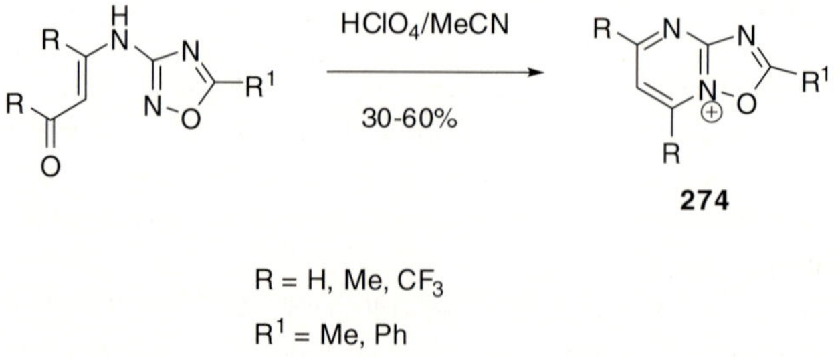

$HClO_4/MeCN$

30-60%

274

R = H, Me, CF_3

R^1 = Me, Ph

Scheme 13.100

Acetylation of 1,2,4-oxadiazoline **275** with acetic anhydride in pyridine furnishes the N2-acetylated compound **276**, while the treatment with potassium hydride in 1,2-dimethoxyethane (DME) gives mixtures of the N2 and the N4-acetylated heterocycles **276** and **277**, respectively (Scheme 13.101) [120b].

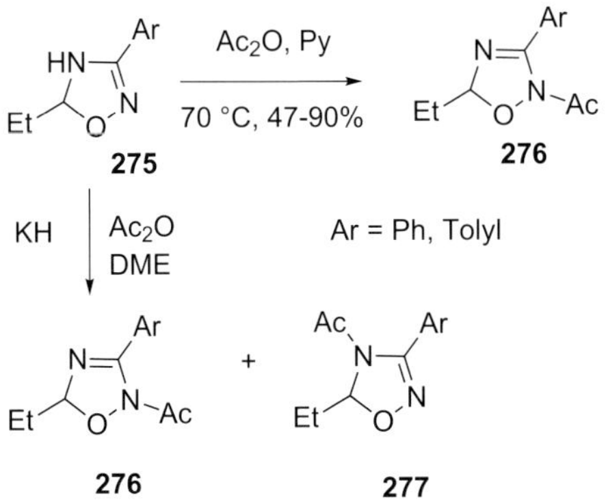

Ar
HN
Et O N
275
Ac_2O, Py
70 °C, 47-90%
Ar
N
Et O N Ac
276
KH
Ac_2O
DME
Ar = Ph, Tolyl
Ar
N
Et O N Ac
276
+
Ac Ar
N
Et O N
277

Scheme 13.101

4,5-Dihydro-1,2,4-oxadiazol-5-one **278** can be N-alkylated with alkyl halides or with epoxides in the presence of bases to give **279** and **280**, respectively (Scheme 13.102) [256].

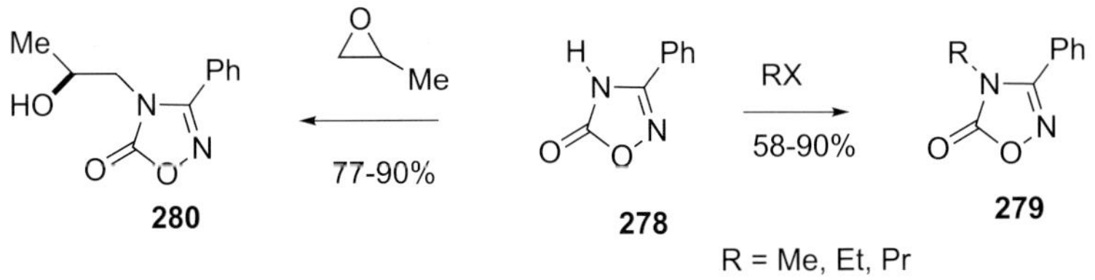

Me
HO N Ph
O O N
280
O
Me
77-90%
H Ph
N
O O N
278
RX
58-90%
R Ph
N
O O N
279
R = Me, Et, Pr

Scheme 13.102

Similarly, 4,5-dihydro-1,2,4-oxadiazol-5-one **278** reacts with alkyl halides **281** and **283** to give the N4 substituted derivatives **282** and **284**, respectively. Compound **278** also reacts with acrolein, via Michael addition, to give **285** (Scheme 13.103) [257].

Unsubstituted 1,2,4-oxadiazolidine 3,5-dione (**286**) undergoes alkylation preferentially at N2. Thus, the reaction with benzyl bromide leads to 2-benzyl derivative **289**, which can be further methylated at *N*-4 to give **290** [258]. Similarly, the reaction with (*S*)-aziridine **287** produced the protected (*S*)-quisqualic acid **288** (Scheme 13.104) [259].

13.3.5.2 **Reactions with Nucleophiles**

Nucleophilic attack on 1,2,4-oxadiazole systems occurs mainly at C5 with nucleophilic displacements of good leaving groups. Table 13.5 summarizes some of these reactions [120, 121, 198, 218, 260].

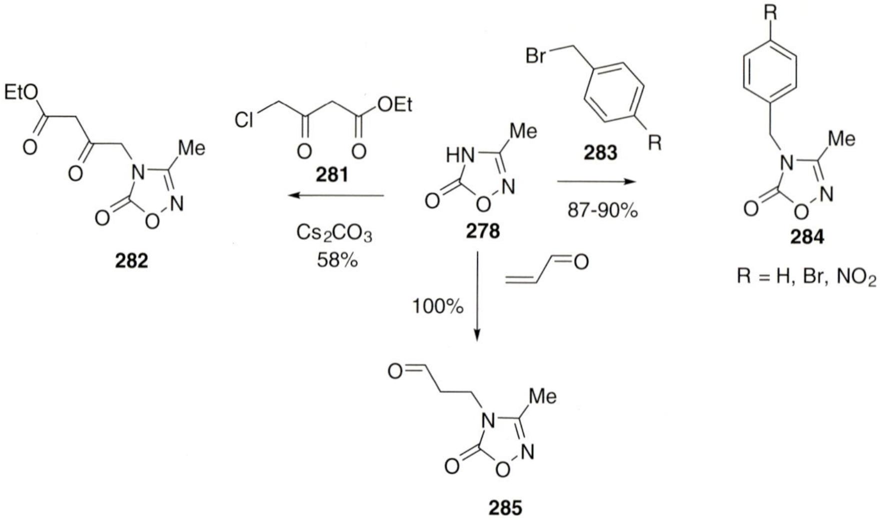

Scheme 13.103

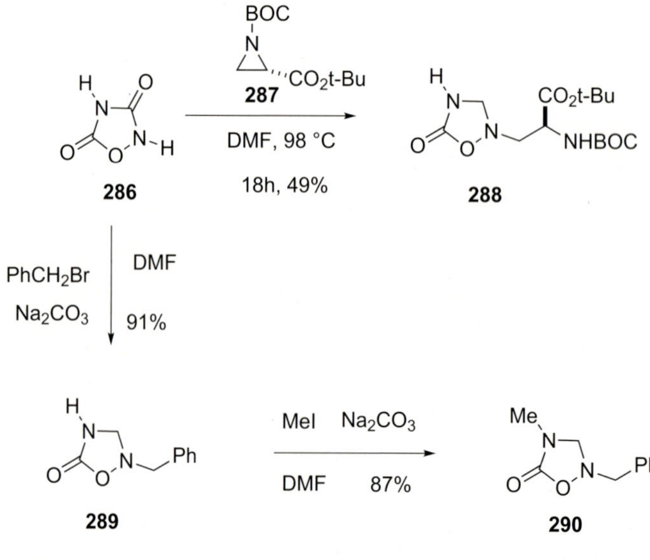

Scheme 13.104

The 3-position is remarkably stable to nucleophilic attack. While the 5-trichloromethyl group of **291** leads, by treatment with KOH, to the 5-oxo compound **292**, the 3-trichloromethyl isomer **293** gives carboxylic acid **294**, which rearranges with loss of CO_2 to acetyl cyanamide **295** (Scheme 13.105) [261].

Table 13.5 Nucleophilic displacements on 1,2,4-oxadiazoles.

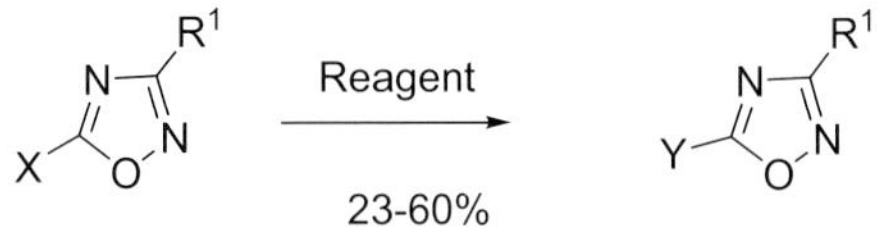

Reagent
23-60%

R^1	X	Reagent	Y
Me	Cl	OH	OH
Me	Cl	$MeNH_2$	MeNH
Me	OEt	Me_2NH	Me_2N
C_6H_4-p-NO_2	S-C_6H_2-[2-Cl-4,6-$(NO_2)_2$]	$(CH_2)_5NH$	$(CH_2)_5N$
C_6H_4-p-Me	Cl	$PhCH_2ONa$	$PhCH_2O$
C_6H_4-p-NO_2	CCl_3	NH_3	NH_2

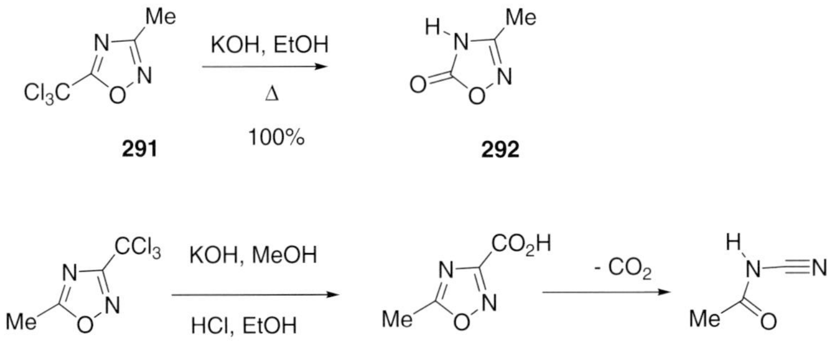

KOH, EtOH, Δ, 100%

291 → 292

KOH, MeOH; HCl, EtOH; $-CO_2$

293 → 294 → 295

Scheme 13.105

Nucleophilic addition of hydrazine or hydroxylamine to the 5-position of 5-fluoroalkyl-1,2,4-oxadiazoles leads to triazole or oxadiazole derivatives **296** and **297** (Scheme 13.106) [135, 228, 262].

The reaction of fully conjugated 3,5-diaryl-1,2,4-oxadiazoles **298** with butyllithium allows facile access to 5-butyl-3,5-diaryl-4,5-dihydro-1,2,4-oxadiazoles **299** (Scheme 13.107) [263].

A similar reaction occurs with 3-methyl derivative **300** to produce **301** [120a], while the 5-methyl of **302** is deprotonated by butyllithium to afford the anion **303**, which produces, after CO_2 treatment, the corresponding acid **304** (Scheme 13.108) [264].

4,5-Dihydro-1,2,4-oxadiazol-5-one **305** hydrolyzes by treatment with NaOH to give amidoximes **306** (Scheme 13.109) [153].

13.3.5.3 Reductions and Oxidations of 1,2,4-Oxadiazoles

Catalytic hydrogenation of 1,2,4-oxadiazoles **307** have been reported, and begins with N-O fission, leading to the corresponding iminoamides intermediates **308** that under

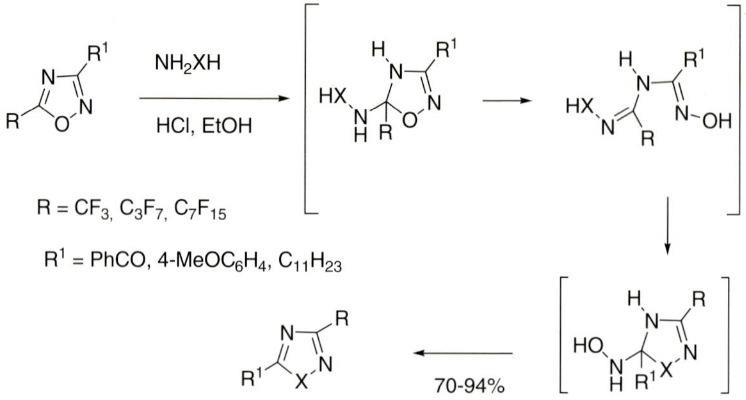

NH_2XH

HCl, EtOH

70-94%

R = CF_3, C_3F_7, C_7F_{15}

R^1 = PhCO, 4-$MeOC_6H_4$, $C_{11}H_{23}$

296 : X = NH, NMe

297 : X = O

Scheme 13.106

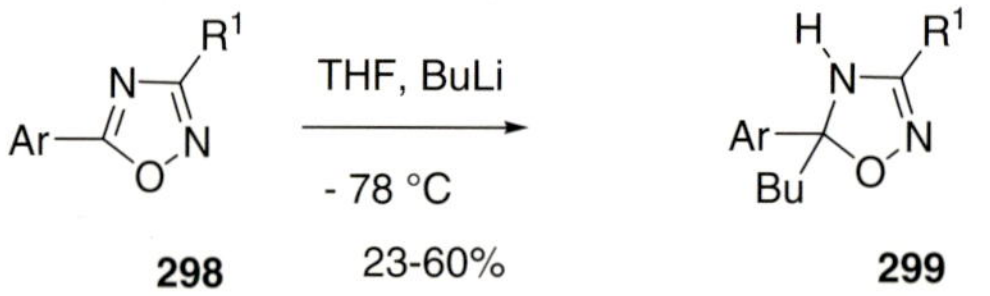

THF, BuLi

- 78 °C

23-60%

298 **299**

Ar = Ph. 2-ClC_6H_4, 2-OHC_6H_4,

R^1 = Ph, 2-ClC_6H_4

Scheme 13.107

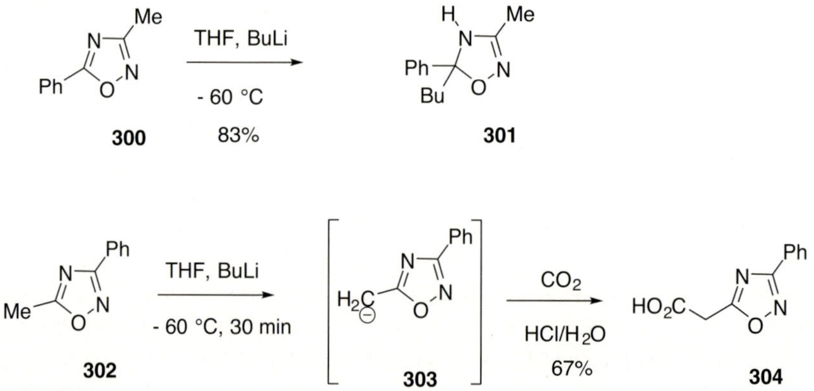

THF, BuLi

- 60 °C

83%

300 **301**

THF, BuLi

- 60 °C, 30 min

CO_2

HCl/H_2O

67%

302 **303** **304**

Scheme 13.108

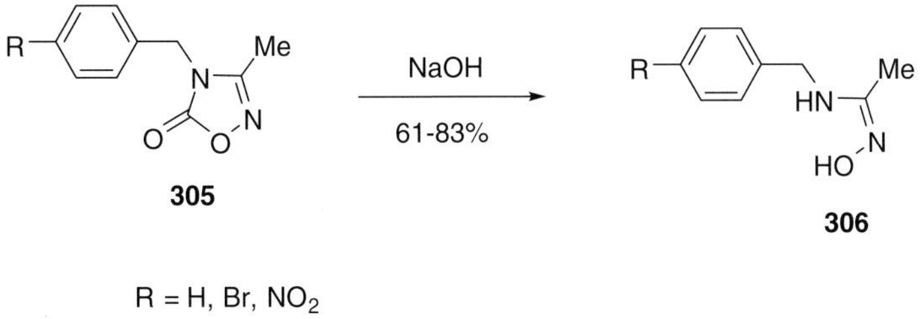

R = H, Br, NO_2

Scheme 13.109

the reaction conditions adopted are converted into amidines **309** (Scheme 13.110) [265].

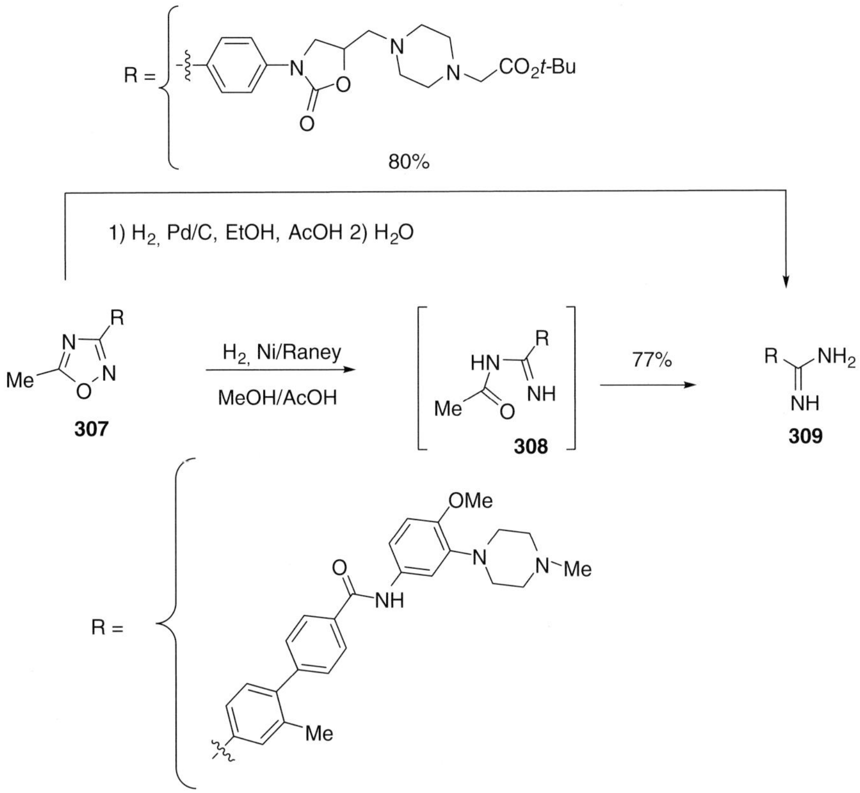

Scheme 13.110

$LiAlH_4$ reduction furnishes N-substituted amidoximes [121, 266]. An application of this reaction concerns the reduction of the Wang resin-bound 1,2,4-oxadiazole **310** to furnish directly the amidooxime **311** via a reductive cleavage from the resin followed by a reductive ring opening of the 1,2,4-oxadiazole ring (Scheme 13.111) [267].

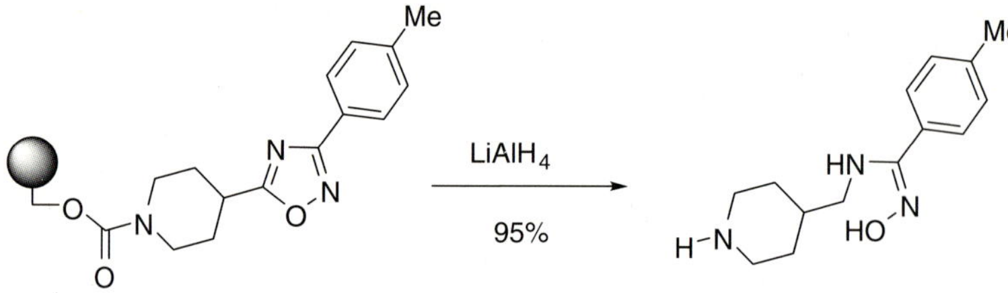

Me

N O N

LiAlH4

95%

Me

HN N HO

H N

310 311

Scheme 13.111

Amidines are also obtained by catalytic hydrogenation of 4,5-dihydro-1,2,4-oxadiazoles. Thus, 1,2,4-oxadiazolo[4,5-*a*]indolines **312** are catalytic hydrogenated over Raney nickel to furnish the corresponding amidines **313** that under tautomerization reaction lead to the opened structures **314** (Scheme 13.112) [268].

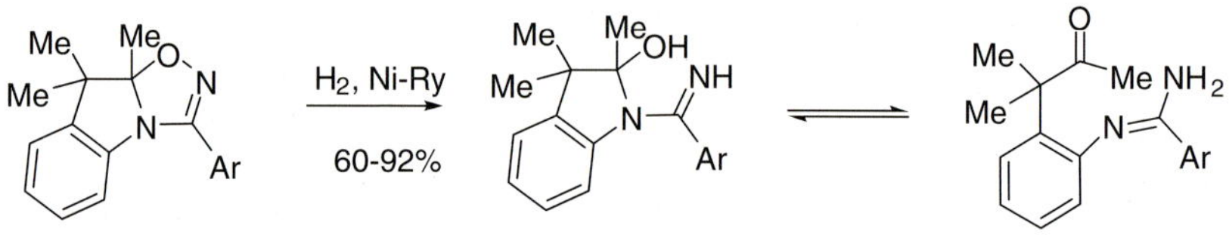

H_2, Ni-Ry

60-92%

312 313 314

Ar= Ph, 4-Cl-C_6H_4,2-Cl-C_6H_4, 2,6-Cl_2-C_6H_3, 2,4,6-MeI_3-C_6H_2

Scheme 13.112

Similar reactions have been reported for 4,5-dihydro-1,2,4-oxadiazole 5-ones **315** (Scheme 13.113) [269].

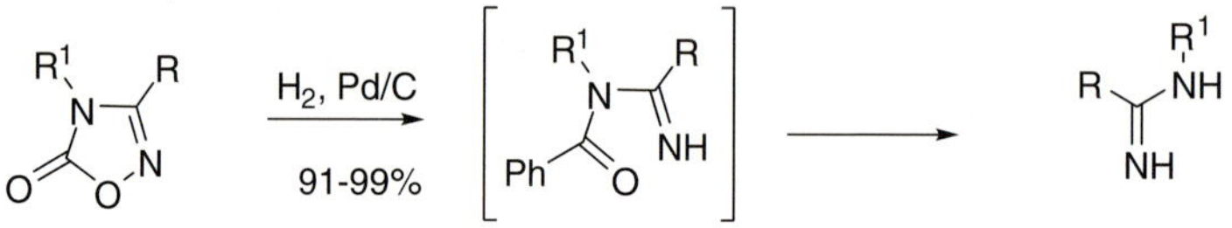

H_2, Pd/C

91-99%

315

R = Me, 4-Me-C_6H4

R^1 = H, Me

Scheme 13.113

Oxidation of 4,5-dihydro-1,2,4-oxadiazoles **316** leads to fully conjugated 1,2,4-oxadiazoles **317**; the oxidation has been performed with different oxidants such as *N*-chlorosuccinimide, manganese dioxide, and concentrated HNO_3 (Scheme 13.114) [125, 162, 169].

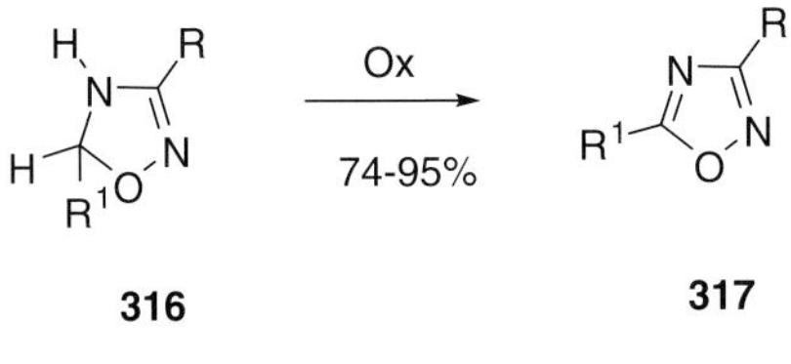

Scheme 13.114

13.3.5.4 Thermal and Photochemical Ring Cleavage

Owing their low aromaticity, 1,2,4-oxadiazoles undergo, by thermal or base-treatment, an easy ring rearrangement known as the Cusmano–Ruccia or Boulton–Katritzky rearrangement. This rearrangement involves a nucleophilic attack on N2 by the oxygen, sulfur, selenium, nitrogen atoms, or carbon anion of a side chain (W) linked at the 3-position of the heterocycle. The generalized rearrangement is reversible only when atom W is oxygen. Table 13.6 summarizes the more common rearrangements [270].

The ring degenerate version of the process has also been reported and investigated as a function of substituent effects and experimental conditions [271]. Thus, for the interconversion **318–319**, mixtures enriched in compound **318** are obtained, while in neutral conditions compound **319** predominates. The effect of substituent X is significant in basic media (Scheme 13.115).

The rearrangement has been used for a synthesis of a series of 3-amino-5-aryl-, 3-amino-5-alkyl-, and 3-amino-5-polyfluorophenyl-1,2,4-oxadiazoles **321** starting from 3-amino-5-methyl-1,2,4-oxadiazoles **320** (Scheme 13.116).

Detailed studies of experimental and theoretical aspects of the rearrangement of phenylhydrazones of 3-benzoyl-1,2,4-oxadiazoles **322** and **323** into the corresponding triazoles **324** and **325** have been performed (Scheme 13.117) [272].

Table 13.6 Rearrangement reactions of 1,2,4-oxadiazole rings.

Sequence atoms XYW	Rearranged products	Sequence atoms XYW	Rearranged products
NCC	Imidazole	CNN	1,2,3-Triazole
CNO	1,2,5-Oxadiazole	NCN	1,2,4-Triazole
NCS	1,2,4-Thiadiazole	CCO	Isoxazole
NCSe	1,2,4-Selenadiazole	CNC	Imidazole

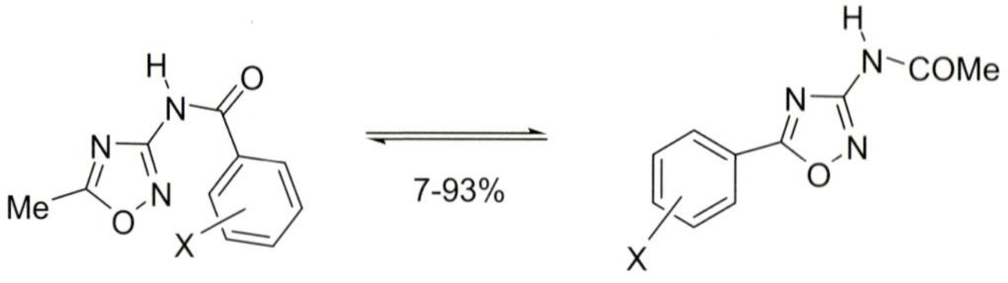

X=H, *p*-Me, *p*-OMe, *p*-Cl, *p*-CF_3, *p*-CN, *p*-NO_2, *m*-NO_2, *m*-Me, *m*-Cl, *m*-CF_3, *p*-CN, *m*-OMe

Scheme 13.115

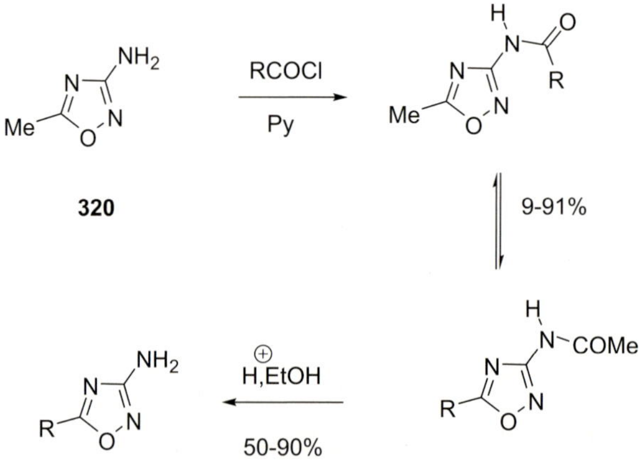

R = *n*-Pr, *t*-Bu, *n*-$C_{11}H_{23}$, Ph, 4-CF_3-C_6H_4, 2-NO_2-C_6H_4, 2-CF_3-C_6H_4, 3-CF_3-C_6H_4, 2-Furyl, 2-Thienyl, 2,3,4,5-Tetrafluorophenyl, 2,3,4-Trifluorophenyl

Scheme 13.116

1,2,4-Oxadiazoles undergo photochemically induced azole to azole interconversions, similar to the previously mentioned thermal rearrangements (Scheme 13.118) [273].

Thus, photolysis of the 3-acetamino-1,2,4-oxadiazole **326** involves cleavage of the N—O bond and the formation of the new oxadiazole **327** (Scheme 13.119).

Similarly, the irradiation of 3-(*o*-aminophenyl)-1,2,4-oxadiazole **328** affords the indazole **331** from the photolytic species **329** and the benzimidazole **332** as

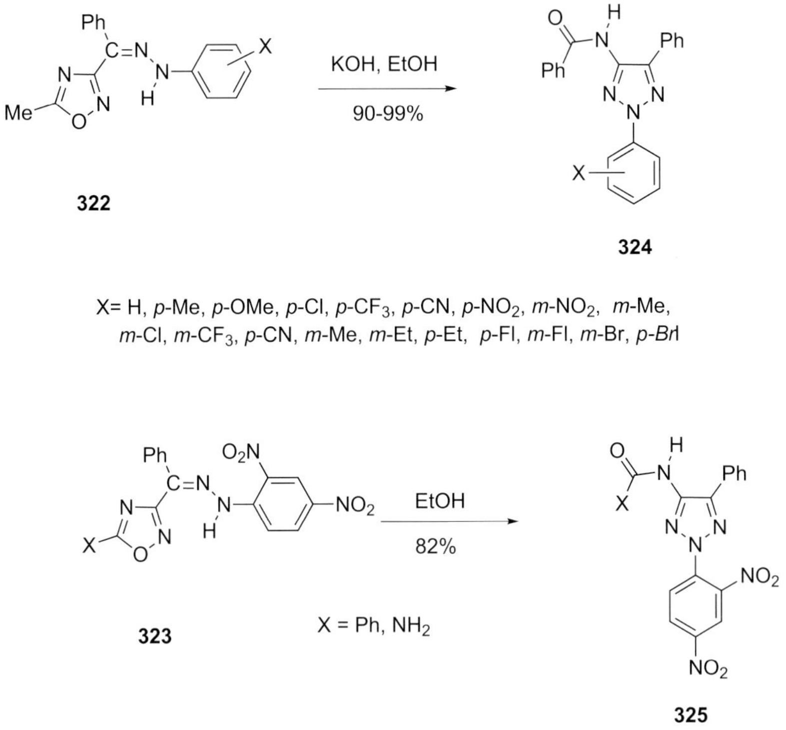

322 → (KOH, EtOH; 90-99%) → 324

X= H, *p*-Me, *p*-OMe, *p*-Cl, *p*-CF_3, *p*-CN, *p*-NO_2, *m*-NO_2, *m*-Me, *m*-Cl, *m*-CF_3, *p*-CN, *m*-Me, *m*-Et, *p*-Et, *p*-Fl, *m*-Fl, *m*-Br, *p*-Br

323 → (EtOH; 82%) → 325

X = Ph, NH_2

Scheme 13.117

hν

X=O,NR;Y=W=COR

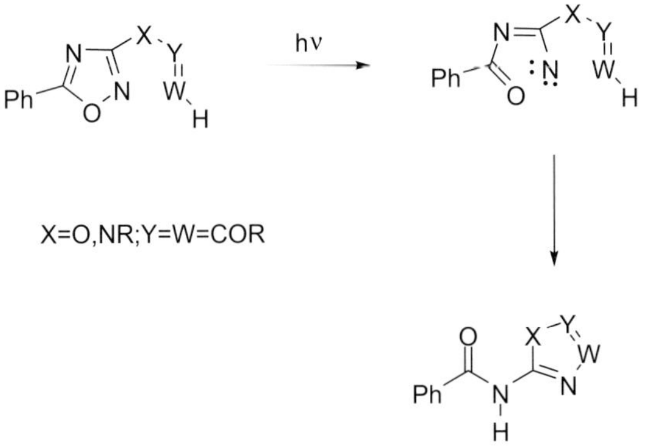

Scheme 13.118

a byproduct originating from the carbodiimide **330**, the rearrangement product of **329** (Scheme 13.120) [274].

UV irradiation of 1,2,4-oxadiazoles **333** in the presence of nucleophilic nitrogen sources (primary amines, ammonia, hydrazine) affords triazoles **335**, via cleavage of the N—O bond and addition of the nucleophile to the intermediate **334** (Scheme 13.121) [275].

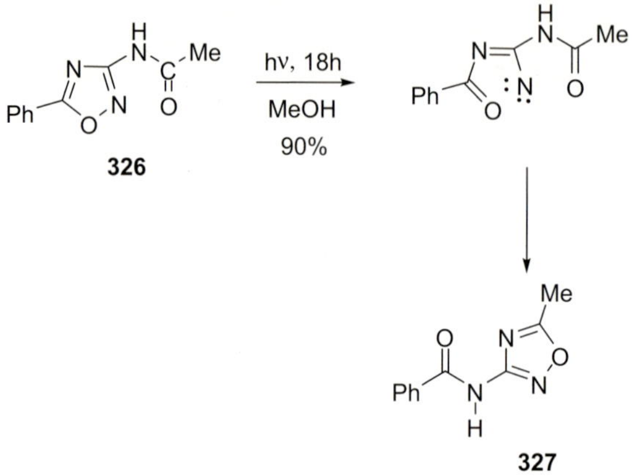

Scheme 13.119

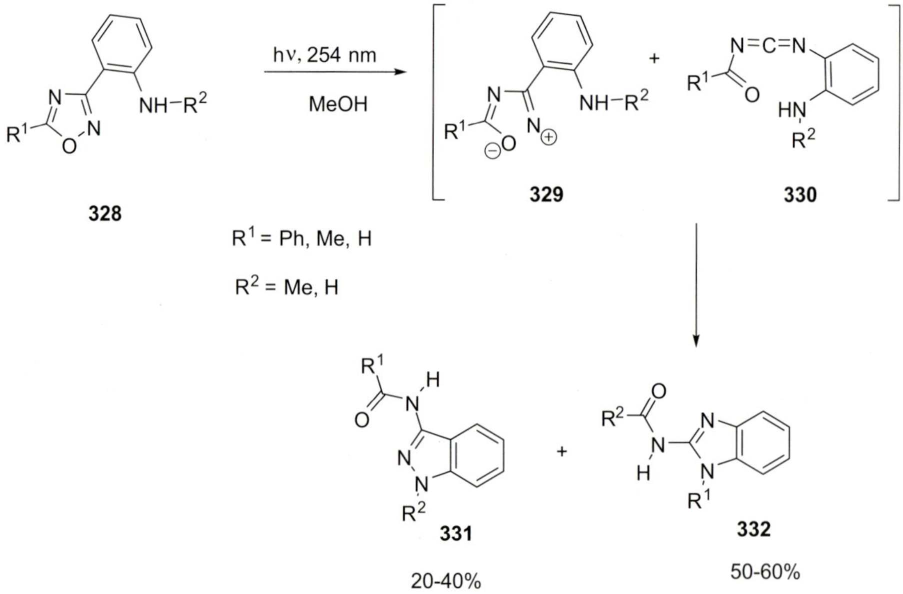

Scheme 13.120

The use of methanol as nucleophile leads to a 1 : 1 mixture of 1,3,4-oxadiazole **337** and triazole **336** (Scheme 13.122).

When an amino group is present at C5, the reaction with a sulfur nucleophile leads to 1,2,4-thiadiazoles **338** (Scheme 13.123) [276].

Irradiation of 1,2,4-oxadiazoles-4-oxides **339** in methanol causes the initial formation of an isolable nitrile together with the nitrosocarbonyl derivative **340**, which

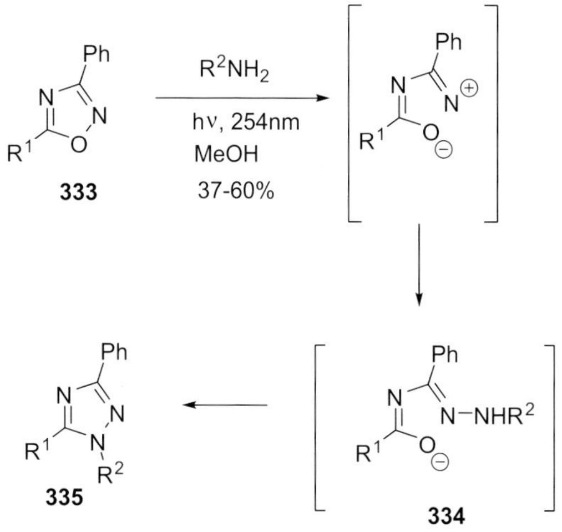

Scheme 13.121

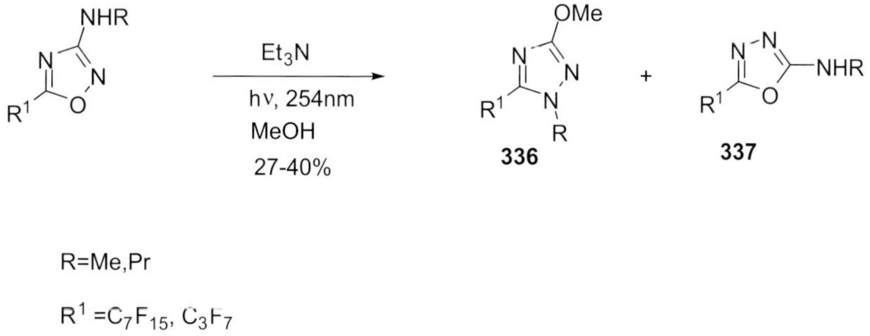

R=Me,Pr

$R^1 = C_7F_{15}, C_3F_7$

Scheme 13.122

can be trapped by cyclohexadiene to give the hetero-Diels–Alder adducts **341** in good yields (Scheme 13.124) [277].

The intermediate nitrosocarbonyls **340** can also be trapped in ene reactions to give adducts **342** and **343** [254, 278].

The 1,2,4-oxadiazol-5-ones **344** undergo, when heated in vacuum, a retro-1,3-dipolar cycloaddition to give nitrones **345** (Scheme 13.125) [251].

13.3.5.5 **Reactivity of Substituents**

Reactions involving substituents attached to ring carbons reveal the particular stability of the heterocyclic system. A series of examples are related to the formation and reactivity of α-anions. Thus, 5-methyl-3-phenyl-1,2,4-oxadiazole (**302**) is deprotonated by bases to the corresponding anion, which adds to carbonyl group of ketones or CO_2 [264]. Conversely, the methyl group of 3-methyl-5-phe-

hν, 254 nm

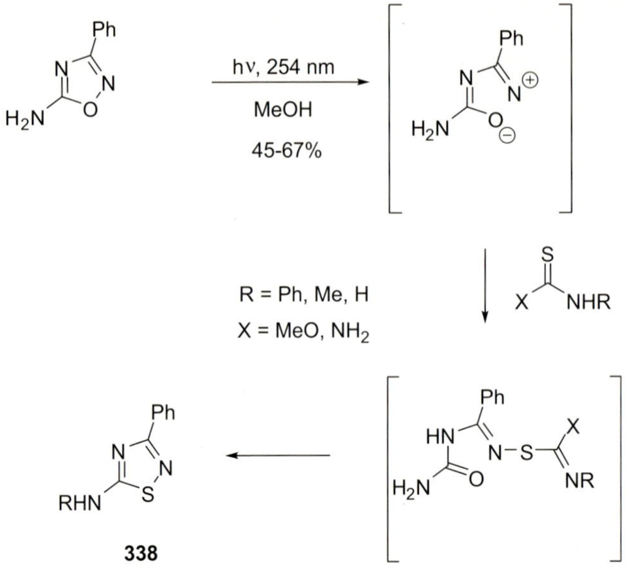

MeOH

45-67%

R = Ph, Me, H

X = MeO, NH_2

338

Scheme 13.123

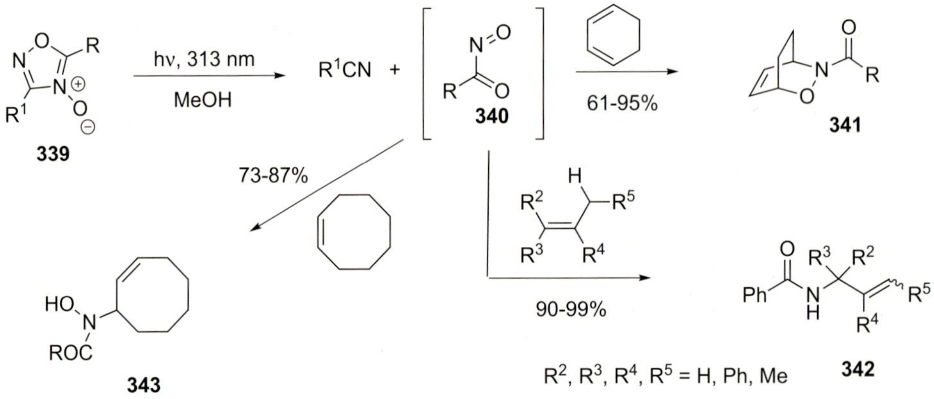

hν, 313 nm

MeOH

R^1CN +

340

61-95%

341

339

73-87%

90-99%

342

343

R^2, R^3, R^4, R^5 = H, Ph, Me

R = Ph, 2,4,6-MePh, p-Cl-Ph, p-MeOPh

R^1 = Ph, 2,4,6-MePh, p-Cl-Ph

Scheme 13.124

nyl-1,2,4-oxadiazole (**300**) by treatment with butyllithium does not form the anion: the reagent adds to the 4,5-bond (Scheme 13.108) [120a].

1,2,4-Oxadiazole **302** also undergoes the aldol condensation with benzaldehyde (Scheme 13.126) [120a].

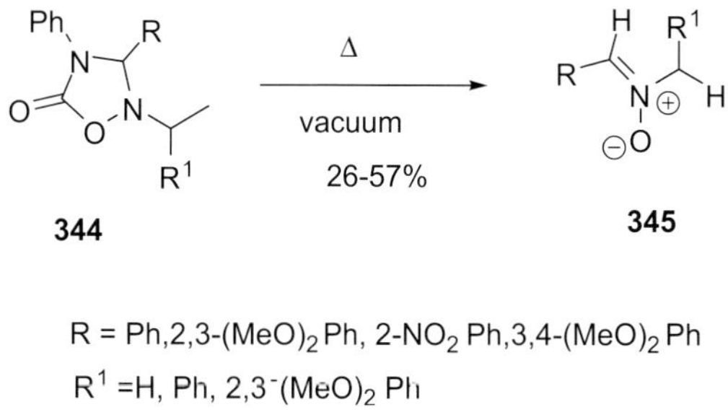

R = Ph,2,3-$(MeO)_2$Ph, 2-NO_2 Ph,3,4-$(MeO)_2$ Ph
R^1 =H, Ph, 2,3-$(MeO)_2$ Ph

Scheme 13.125

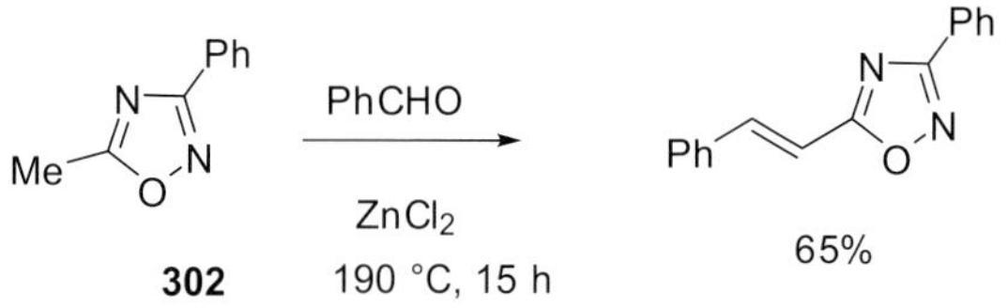

Scheme 13.126

Phosphonate **347**, obtained by Arbuzov reaction of 3-chloromethyl-1,2,4-oxadiazole **346**, has been used in Wadsworth–Emmons reactions: the methodology provides useful access to 3-alkenyl-1,2,4-oxadiazoles **348** (Scheme 13.127) [264].

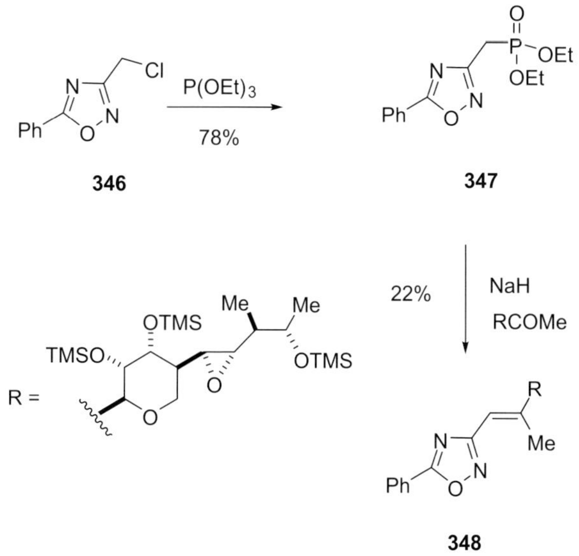

Scheme 13.127

The formation of α-anions has been exploited for the synthesis of a library of 5-alkenyl-substituted 1,2,4-oxadiazoles **353**. Compounds **353** have been prepared starting from a polystyrene-supported oxadiazolyl-substituted selenium resin **350**, prepared by reaction of polystyrene-supported selenyl acetic acid **349**, with amidoxime and DDC through Porco's two-step, one-pot condensation. Alkylation of **350** by base treatment and addition of allyl bromides produced the α-alkylated selenium resins **351**, which were used as dipolarophiles in a 1,3-dipolar cycloaddition to furnish polystyrene-supported oxadiazolyl- and isoxazolinyl-substituted selenium resins **352**. Oxadiazolyl and isoxazolinyl substituted olefins **353** were then obtained stereoselectively through selenoxide *syn*-elimination from resins **352** by H_2O_2 treatment (Scheme 13.128) [279].

R = Ph, *p*-Me-C_6H_4, *p*-F-C_6H_4, *p*-Cl-C_6H_4,
R^1 = Ph, H
R^2 = CO_2Et, *p*-MeO-C_6H_4, *p*-Br-C_6H_4, *p*-Cl-C_6H_4, *p*-NO_2-C_6H_4

Scheme 13.128

The stability of the oxadiazole ring is pointed out in many reactions that substituents at C3 or C5 undergo. Among the more recent reactions, Scheme 13.129 shows the synthesis of aryl ethers **354** [280], the nucleophilic attack of methanol to a pentafluorophenyl or tetrafluorophenyl 1,2,4-oxadiazole (**355**) [275b], a Sonogashira coupling to afford **356** [195a], and the synthesis of amino compounds **357** by tetrapropylammonium perruthenate (TPAP) oxidation of the hydroxyl group followed by reductive amination of the resulting aldehyde [200].

The reaction of 5- and 3-chloromethyl-1,2,4-oxadiazoles **358** and **359**, respectively, with pyrazolyl-purine affords the corresponding derivatives **360** and **361** (Scheme 13.130) [187].

Analogously, the polymer supported-1,2,4-oxadiazole **362** reacts with primary amines to give 5-aminomethyl oxadiazoles **363** (Scheme 13.131) [281].

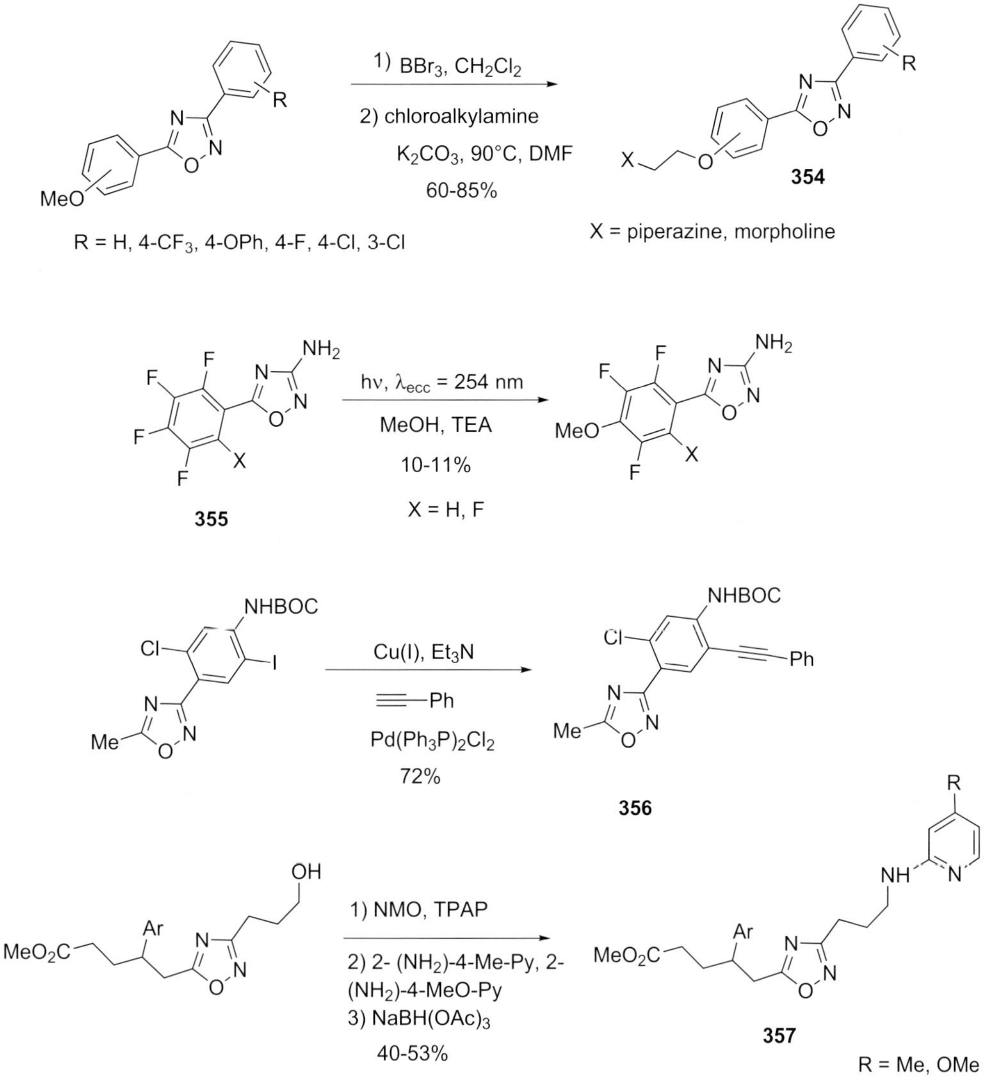

1) BBr_3, CH_2Cl_2
2) chloroalkylamine
K_2CO_3, 90°C, DMF
60-85%

R = H, 4-CF_3, 4-OPh, 4-F, 4-Cl, 3-Cl

354
X = piperazine, morpholine

355
hν, λ_{ecc} = 254 nm
MeOH, TEA
10-11%
X = H, F

Cu(I), Et_3N
≡—Ph
$Pd(Ph_3P)_2Cl_2$
72%

356

1) NMO, TPAP
2) 2- (NH_2)-4-Me-Py, 2-(NH_2)-4-MeO-Py
3) $NaBH(OAc)_3$
40-53%

357
R = Me, OMe

Scheme 13.129

A series of highly π-conjugated nonsymmetrical liquid crystals, based on the core 3,5-(disubstituted)-1,2,4-oxadiazole with a shape similar to a hockey stick, have been synthesized by Sonogashira coupling reaction. Thus, compounds **366** were prepared in 74–82% yield, by reaction of an aryl iodide containing the 1,2,4-oxadiazole ring (**364**) with the terminal arylacetylenes **365**, using 10 mol.% of dichlorobis(triphenylphosphine)palladium, 5 mol.% of the co-catalyst copper(I) iodide, in a triethylamine–tetrahydrofuran mixture (7 : 3). The obtained compounds showed liquid crystal phases, in particular smectic and nematic typical of calamitic structures, and moreover exhibit strong blue fluorescence in solution (Scheme 13.132) [282].

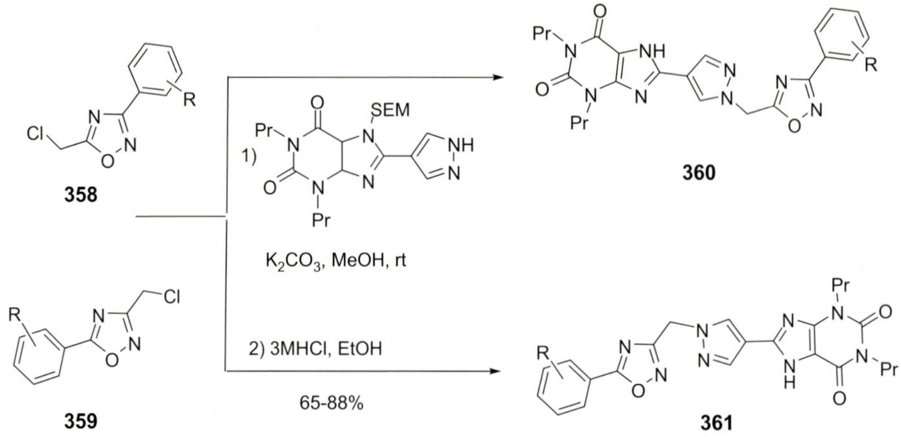

358

359

360

361

R = H,4-Cl,4-CF$_3$,4-CN,4-OMe,4-Me,
3-Cl,3-CF$_3$,3-OMe,3-Me,
2-Cl,2-CF$_3$,2-OMe,

Scheme 13.130

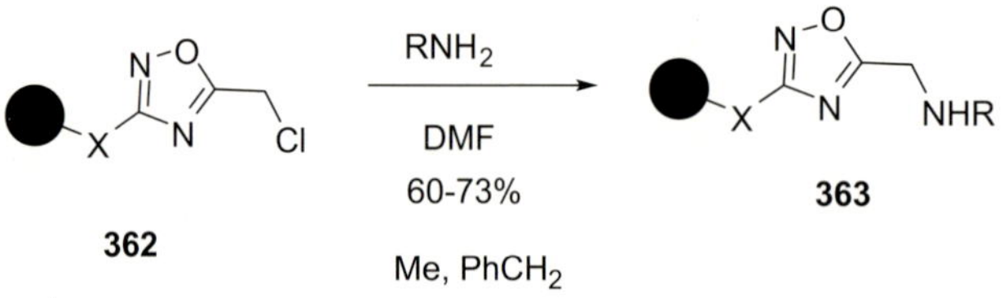

362

363

Scheme 13.131

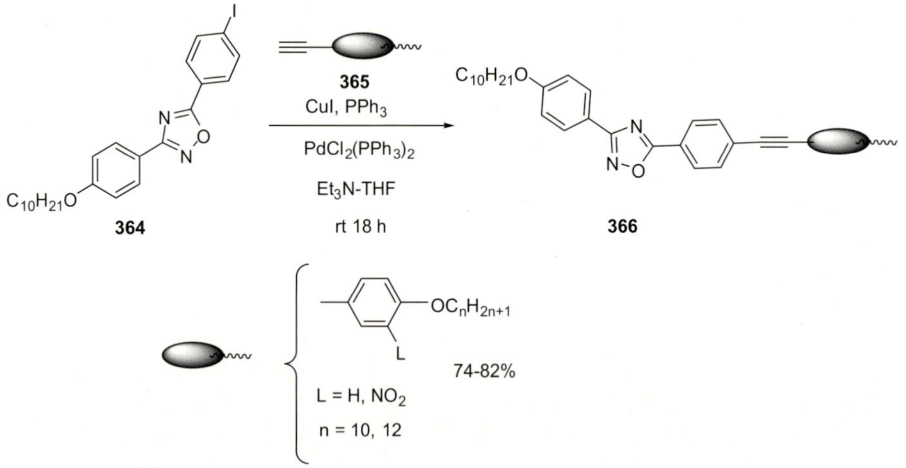

364

365

366

Scheme 13.132

13.3.6
1,2,4-Oxadiazoles in Medicine

1,2,4-Oxadiazole ring occurs widely in biologically active synthetic compounds, and is often used in drug discovery as a hydrolysis-resisting bioisosteric replacement for amide or ester functionalities [283] because of its electronic properties. Its derivatives can be found in a vast number of compounds exerting biological activity, such as ligands of benzodiazepine receptors [284, 285], anti-inflammatory agents [131, 199, 234], antiviral agents [283], inhibitors of protein tyrosine phosphatases [286], agonists of muscarinic receptors [287], inhibitors of Src SH2 [183], antagonists of histamine H_3-receptors [288], integrin receptor antagonists [200], angiotensin II receptor antagonists [289], and HIV-1 reverse transcriptase inhibitors [290]. 1,2,4-Oxadiazole moieties have been used in the design of dipeptidomimetics as peptide building blocks [184a,b]. Compounds **367** contain, at C5 of the 1,2,4-oxadiazole nucleus, a residue of an amino group linked to peptide moieties, and a carboxyl or ester functionality attached at C3 directly or through a methylene chain [291]

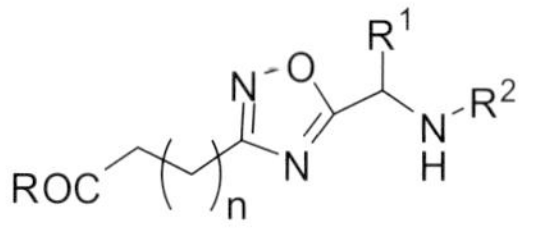

367

n =0, 1,2

R = OH, OEt, NHMe R^1 = $PhCH_2$, Me, *i*-Prop

R^2 = Ac, Boc, H-Tyr-D-Ala, H-Arg-Pro--Lys-Pro-Gln-Gln-Phe

Numerous 1,2,4-oxadiazoles have been suggested as potential agonists for cortical muscarinic [292], benzodiazepine [293], and 5-HT1D (5-hydroxytryptamine) receptors [294], and as antagonists for 5-HT [295] or histamine H3 receptors [296]. They show activity as antirhinoviral agents [297], growth hormone secretagogues [298], anti-inflammatory agents [234], and antitumor agents [183, 188, 299]. They also inhibit the SH2 domain of tyrosine kinase [300], monoamine oxidase [301], human neutrophil elastase [302], and human DNA topoisomerases. Finally, tropane derivatives of 1,2,4-oxadiazoles display high affinity for the cocaine binding site of the dopamine transporter [303].

More recently, it has been reported that the ring opening of N-oxides of adenosines **368**, followed by exocyclic ring closure, in the presence of carboxylic anhydrides and thiophenol, followed by ammonia treatment, generates 1,2,4-oxadiazolyl imidazoles **369** (Scheme 13.133) [304]. The so-obtained separation of the fused imidazole and pyrimidine rings of purine nucleosides increases the conformation flexibility: these shape-modified analogues have been used to investigate triple helix formation and as probes for the study of enzyme interactions [305a].

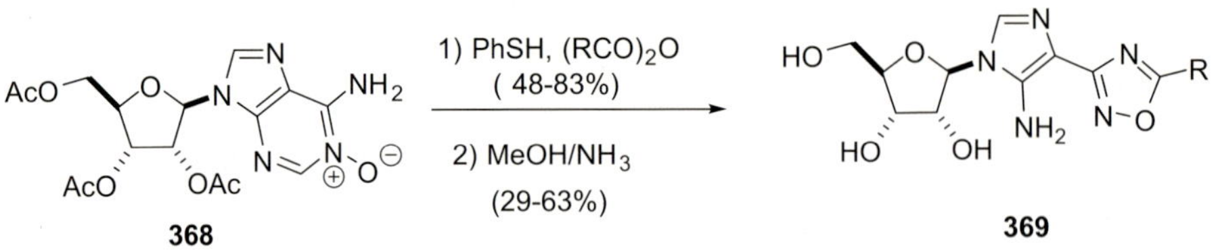

1) PhSH, $(RCO)_2O$ (48-83%)
2) $MeOH/NH_3$ (29-63%)

368 → 369

R = Me, Et, pr, *i*-pr, *t*-bu, C_5H_{11}, C_7H_{15}, C_9H_{19}, Ph

Scheme 13.133

Furthermore, a series of keto-1,2,4-oxadiazoles (**370**) have been prepared that have been shown to be potent inhibitors of human mast cell tryptase and useful in the treatment of asthma and allergic diseases.

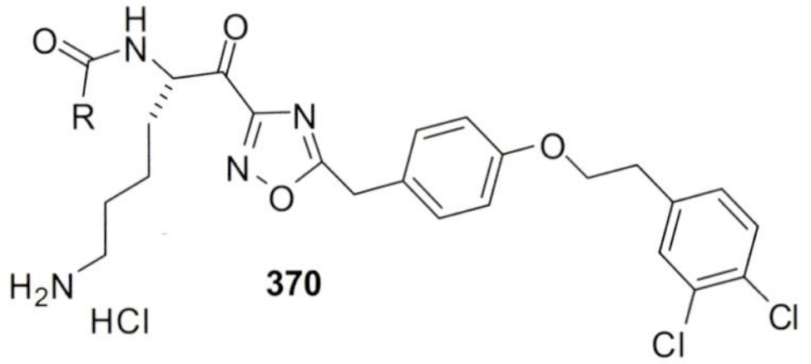

370

R = CF_3, C_3H_5, OEt, $N(CH_2)_4O$, *t*-bu, *i*-pr, *n*-pentyl, 2,4-$F_2C_6H_3$, 3,4-$F_2C_6H_3$, 4-FC_6H_4, 4-ClC_6H_4

The 1,2,4.oxadiazol-5-one moiety can act as a bioisostere of the carboxylic acid function in retinoid structures. Recently, the solid-phase or solution-phase syntheses of a new series of non-carboxylic acid retinoic acid receptor ligands (RARs) bioisosteres of Am580 or tazarotene-like retinoids has been reported [305b].

In particular, the retinoidal activity of compound **371** (RAR-β,γ selective) is significant. These non-carboxylic acid type RAR ligands may exhibit different pharmacological behaviors from classical carboxylic acid compounds, as well as unique biological activity, and they may provide further scope for clinical applications.

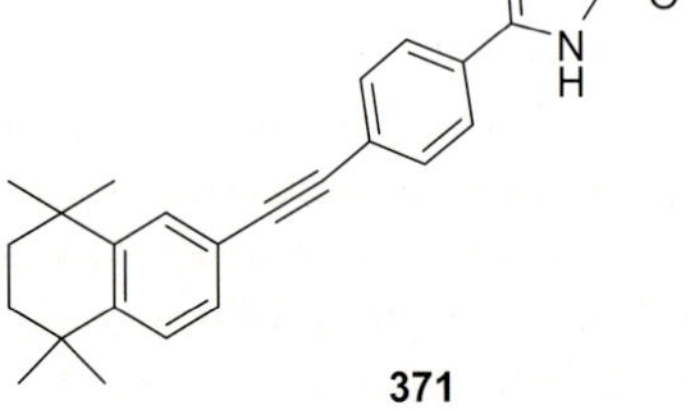

371

13.4
1,2,5-Oxadiazoles

1,2,5-Oxadiazole (**372**) is often referred to by the trivial name furazan; for 1,2,5-oxadiazole-2-oxide (**373**), a common derivative, the trivial name furoxan is still in wide usage. The first report on a 1,2,5-oxadiazole ring system appeared in the 1850s: the parent compound **1** was prepared in 1964 by treatment of glyoxime with succinic anhydride [306]. For 2,1,3-benzoxadiazoles (**374**) and the corresponding N-oxide **375**, the terms benzofurazan and benzofuroxan are commonly used. The partially reduced dihydro- (Δ^2, Δ^3) and tetrahydro-derivatives **376–378** are very rare.

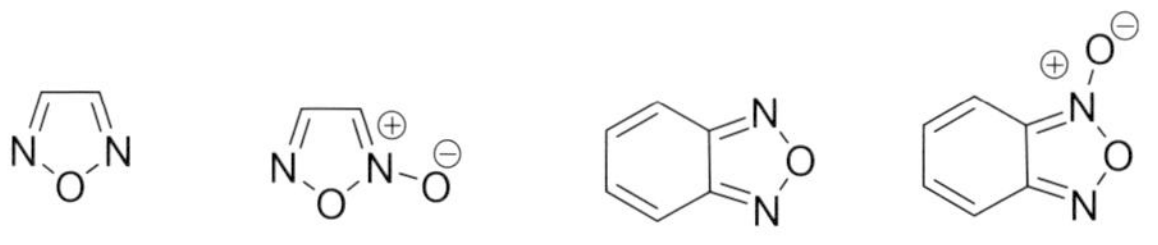

372 **373** **374** **375**

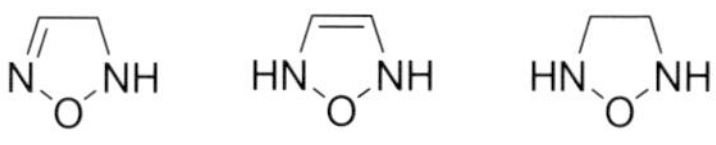

376 **377** **378**

1,2,5-Oxadiazoles, their N-oxides, as well as their benzo-fused systems are biologically active compounds. Some of their derivatives are important because of their anthelmintic, fungicidal, bactericidal, and herbicidal action. They have also been found to possess antitumoral activity.

Several reviews have been published on 1,2,5-oxadiazoles [307–311]: the most comprehensive account of the chemistry of furoxans and benzofuroxans is that by Gasco and Boulton [312], and the more recent one by Paton [313].

13.4.1
Structure

1,2,5-Oxadiazole is a heteroaromatic compound; strictly, it should be regarded as a π-excessive heterocycle with six electrons distributed over five atoms. However, the π-electron density on the heteroatoms is so great that the values for the C-atoms are smaller than one, where π-deficiency prevails, thereby influencing the reactivity [314]. Despite low π-electron density on the C-atoms, 1,2,5-oxadiazoles do not react at all or only slowly with nucleophiles: treatment with strong bases, such as NaOH in methanol, causes ring opening to form sodium salts of α-oximinonitriles **380** (Scheme 13.134).

The 1,2,5-oxadiazole ring is a stable system and annular-group tautomerism is not favored [308]. Thus, the two theoretically possible tautomeric forms **381a** and **381b** for 3-hydroxyfurazans **381** can be discarded on the basis of IR and NMR data, which show the exclusive presence of the hydroxy compound in chloroform solution (Scheme 13.135).

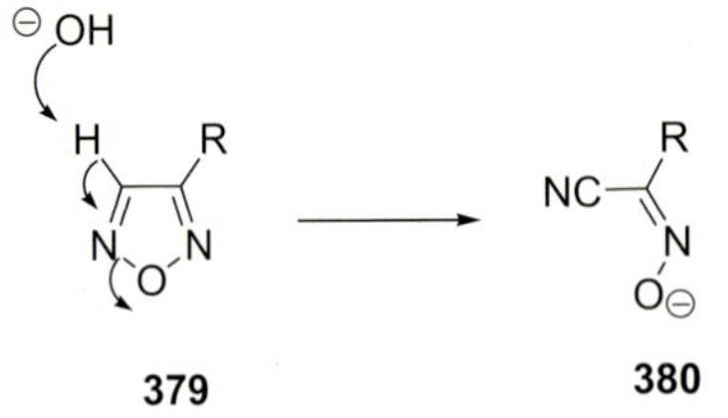

379 380

Scheme 13.134

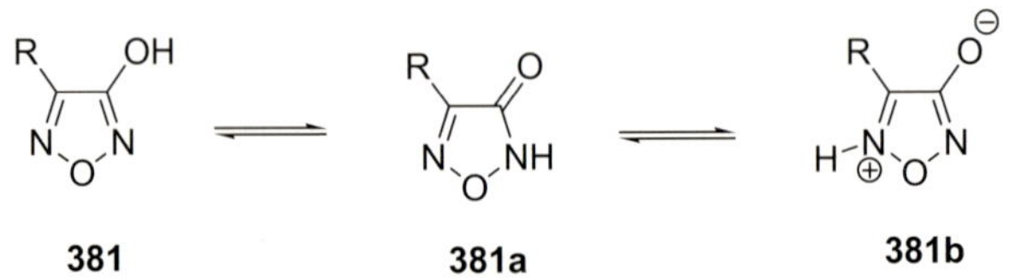

381 381a 381b

Scheme 13.135

However, the formation of 2-alkyl-1,2,5-oxadiazol-3(2*H*)-ones **382** by alkylation of trimethylsilyl derivatives of 3-hydroxyfurazans using triethyl orthoformate has been reported [315]. The compounds were characterized by NMR and MS measurements.

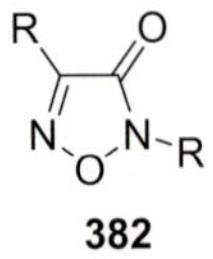

382

The N-oxide structure for furoxans and their benzo-derivatives was ascertained by Wieland [316] and Werner [317]. Ring-chain tautomerism is a distinctive feature of furoxan chemistry, as evidenced in the interconversion of 2-oxide **383** and 5-oxide isomers **384** (Figure 13.4: see Section 13.4.2).

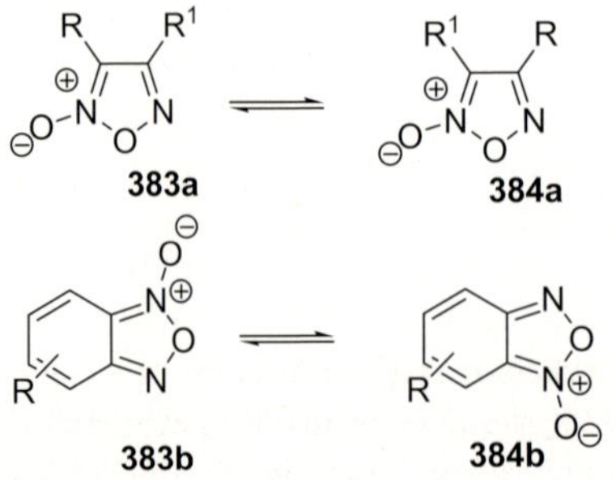

383a 384a

383b 384b

Figure 13.4 Example of ring-chain tautomerism shown by the interconversion of 2-oxide **383** and 5-oxide isomers **384**.

A furazan fused to a five-membered heterocycle was first described in 1908 [318]: annelation in 5/5-byciclic systems suggests the presence of strain energy in the molecules, which is manifested in the difficulty to form these compounds. Destabilization of the distorted aromatic oxadiazoles results not only from bond stretching, angular distortion, and torsional effects, but also from the decreased resonance stabilization [319].

5,7-Dimethyl-3-phenyl-furazano- and –furoxano[5,4-*a*]pyridinium perchlorates (**385** and **386**), a new type of condensed system, have been obtained by cyclocondensation of aminophenylfuroxan and aminophenylfurazan with acetylacetone in the presence of $HClO_4$. The structure of these compounds is supported by crystallographic analysis and CNDO/2 calculations [320].

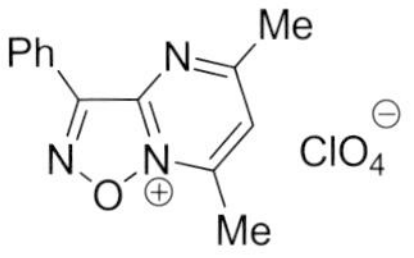

385 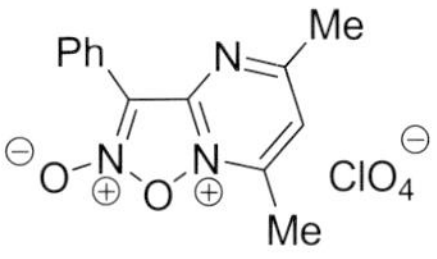 **386**

Very few partially or totally reduced 1,2,5-oxadiazole ring systems has been reported. For instance, a Japanese patent [321] describes the synthesis of 5-(4-oxo-2,5-diphenyl-1,2,5-oxadiazolidine-3-yl)-2,4(1*H*,3*H*)-pyrimidinedione (**387**), the first representative of 1,2,5-oxadiazolidines.

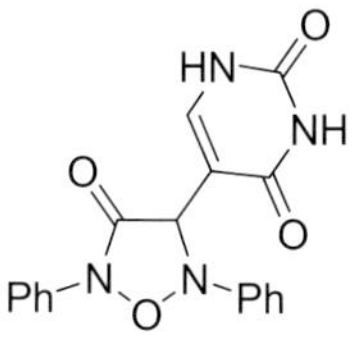

387

As a result of the low aromatic character of the benzofuroxan system, recent studies have revealed that nitrobenzofuroxan acts as very versatile Diels–Alder reagent, with the carbocyclic ring being capable of acting as a dienophile [322], a heterodiene [322], or a carbodiene [323], depending upon the experimental conditions. Thus, treatment of 4-nitro-6-trifluoromethylfurazan (**388**) and -furoxan (**389**) with 1,3-cycloexadiene afforded a mixture of fused derivatives **390** and **391**, respectively (Scheme 13.136) [324].

13.4.2 Theoretical Aspects

Theoretical studies on the structure and properties of 1,2,5-oxadiazoles have been reported. Molecular orbital calculations [325, 326] and *ab initio* quantum mechanical methodologies have been used to determine bond orders, total energies, ionization potentials, and dipole moments [327]: a good match with experimental data has been obtained.

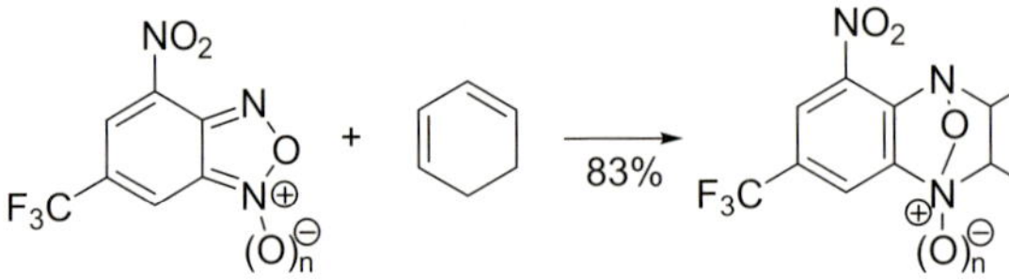

388 n = 0 **390**

389 n = 1 **391**

Scheme 13.136

Dipole moments for a set of substituted 1,2,5-oxadiazoles and 1,2,5-oxadiazole 2-oxides have been measured in benzene solution [328]. The dipole moment of furazans is oriented with the negative end towards the oxygen atom, while in furoxans the data revealed a strong electron shift from the exocyclic oxygen back into the heterocyclic system, corresponding to a mesomeric moment of approximately 3 D. The molecule of furoxan is well characterized as electron-overcrowded, particularly near the nitrogen atom N2.

Dipole moments of 3-amino-4R-furazans **392** have been determined experimentally and also calculated by HF *ab initio* (STO-3G, 3-21G, 4-31G, 6-31G, 6-31G**/4-31G, 6-31G** levels) and semiempirical (MNDO, AM1, PM3) quantum chemical methods; good agreement with the experimental values has been found.

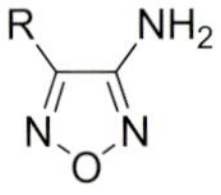

392

R = H, NH_2, OMe, N_3, COOH, COOMe, NO_2

For these compounds, the amino–imino tautomeric equilibrium is strongly shifted towards the amino-form [329].

A great deal of interest has been taken in the 2-oxide and 5-oxide tautomers and the pathway of their interconversion (**393**↔**395**) has been studied in some detail [330–335]. Structures and relative stabilities of furoxan and its open-chain tautomers have been calculated by semiempirical and *ab initio* procedures. The obtained results support a mechanism that involves the *cis*-1,2-dinitroethene **394** as intermediate/transition state, with an energy about 120 kJ mol^{-1} above that of furoxan.

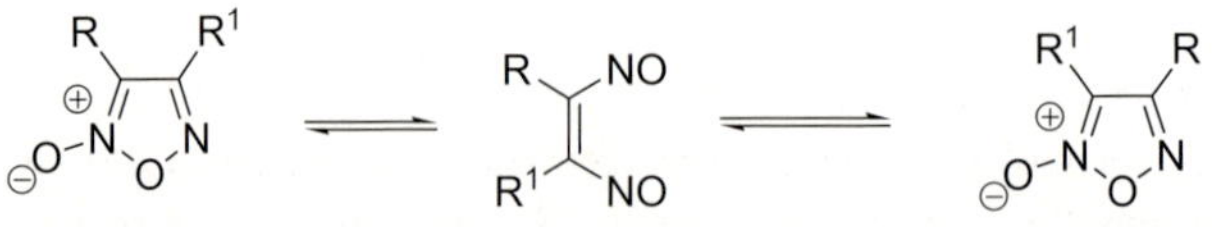

393 **394** **395**

Analogously, in the benzofuroxan series, MP2 calculations afford a correct prediction of the structure [336]: the most likely intermediate in the interconversion

396↔**398** is *anti*-1,2-dinitrosobenzene (**397**) with an energy of 50 kJ mol^{-1} above that of benzofuroxan [337].

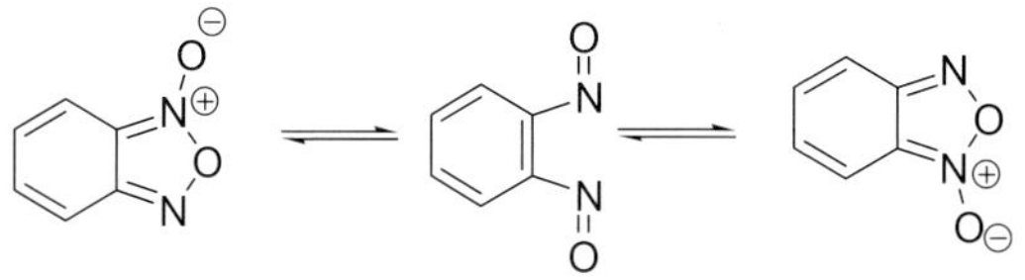

396 **397** **398**

The involvement of 1,2-nitrosoarenes as intermediates in the equilibration process has been established by matrix isolation experiments [338–340]: photolysis (360 nm) of benzofuran in an argon matrix at 14 K generated 1,2-nitrosobenzene, which was characterized by UV and IR spectroscopy, and subsequent thermolysis or photolysis (320 nm) afforded benzofuroxan. The gain of resonance energy of the benzene ring does not compensate for the energy needed to open the furoxan ring, and therefore dinitrosobenzene is less stable than benzofuroxan.

More recently, the first experimental evidence for the formation of 2,3-dinitroso-2-butene as a reactive intermediate, during the photolytically induced decomposition of dimethylfuroxan, has been reported by matrix isolation experiments [341a]. DFT calculations gave the geometry of the dinitrosoalkene intermediate [341b]. However, this species is photolabile and decomposes upon prolonged photolysis time to give acetonitrile *N*-oxide as the final, photostable product. The two photoproducts were characterized using a combination of experimental and quantum chemical results.

Factors influencing the equilibrium constants and rates have been reviewed [312, 342–344].

The equilibrium between annelated furoxans and the isomeric dinitroso derivatives, for example, **399** and **400**, has been investigated theoretically by semiempirical (AM1, PM3) and *ab initio* methods (MP2/6-31G*//6-31G* and RHF/6-31G*) [345]. For both 1,2- and 2,3-dinitrosonaphthalene, several conformers exist as minimum on the potential energy surface (PES). Calculations of the energy difference between [1,2,5]thiadiazolo[3,4-*e*]benzoxadiazole-1-oxide and -3-oxide are in agreement with experimental data.

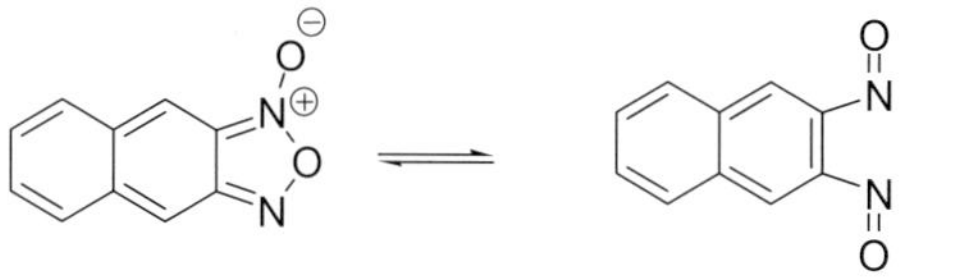

399 **400**

Ab initio and density functional theoretical studies on non-classical furoxans **401** (Y = O, NH, S for each of the following combinations of X,Z: CH,CH; N, CH; CH,N; and N,N) and their open-chain *anti*-1,2-dinitroso isomers **402** have been reported [346, 347]. Calculations indicate that, in all cases considered, the non-classical furoxans are less stable than the corresponding open-chain isomers.

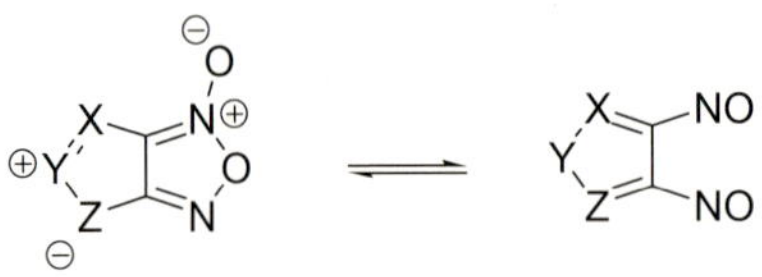

401 ⇌ **402**

Density functional theory (DFT) has been used to calculate the heats of formation and IR active vibrational frequencies of 12 furazan compounds [348]. The assignments of the vibrational motions to IR frequencies based on a force field analysis are given to clarify the complex coupling in these molecules.

Dissociation enthalpies of terminal (N–O) bonds, ΔH°(N–O), in furoxans have been calculated from enthalpy of formation, enthalpy of sublimation, and enthalpy of vaporization data [349].

13.4.3 Structural Aspects

13.4.3.1 X-Ray Diffraction

X-Ray crystal structures have been reported for various furazans and furoxans [350–357]. Bond lengths and bond angles have been determined also by double resonance modulation microwave spectroscopy [358]. For furazans, crystallographic data show that the heterocyclic ring is essentially planar and possess C_{2v} symmetry. π-Bond orders are 0.72–0.82 for N2–C3 and C4–N5 and 0.45–0.52 for C3–C4. These data suggest a significant π-delocalization; in contrast, O1–N2 and N5–O1 are essentially single bonds (0.32–0.36). Benzofurazans show similar parameters and a significant double bond fixation in the fused ring. The molecular geometry for the parent furazan (**403**) has been determined by microwave spectroscopy.

0.5051 0.8116 0.3563

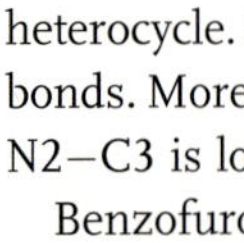

403

As for furazans, the oxadiazole ring of furoxans in nearly planar, but the exocyclic oxygen at N2 causes substantial distortion, lying 0.05 Å out of the plane of the heterocycle. Structures are characterized by the long O1–N2 and the short N2–O_{exo} bonds. Moreover, C3–C4 is shortened, with about 30% double bond-character, while N2–C3 is longer than C4–N5.

Benzofuroxans show a similar pattern of bond lengths and angles. In the homocyclic ring, significant bond localization is supported by the consideration that the C4–C5 and C6–C7 bonds are notably shorter than C5–C6, in accord with their chemical reactivity (Section 13.4.5.2).

Crystal structure simulations for three azoxyfurazans, 4,2′-dichloroazoxyfurazan (**404**), 4,4′-di(morpholin-1-yl)azoxyfurazan (**405**), and 4,4′-dimethylazoxyfurazan (**406**), have been carried out to test the reliability of standard force fields for furazan derivatives [359]. The predicted crystal structures were compared with experimental ones, obtained by X-ray diffraction analysis.

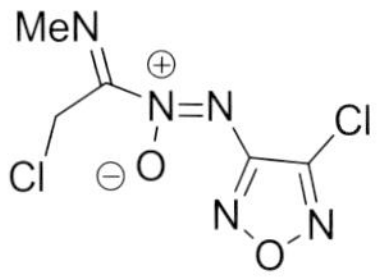

404 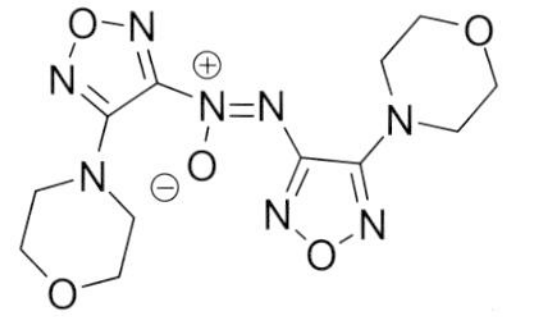 405 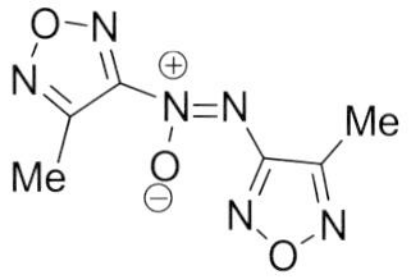 406

X-Ray data for 1,2,5-oxadiazoles fused with a pyrazine ring have been reported recently [360].

Azofurazan annulated macrocycles **407** [361], **408** [362], and **409** [363] have been synthesized and then characterized by X-ray analysis. Lactam **407** show two furazan rings linked to a piperazine system, while compound **408** contains four furazan rings bonded by three azo bonds. The ion binding ability of compounds **409** was tested.

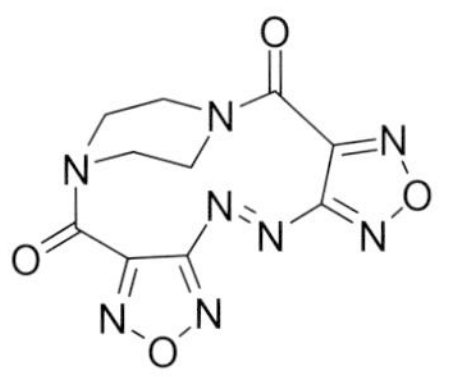

407 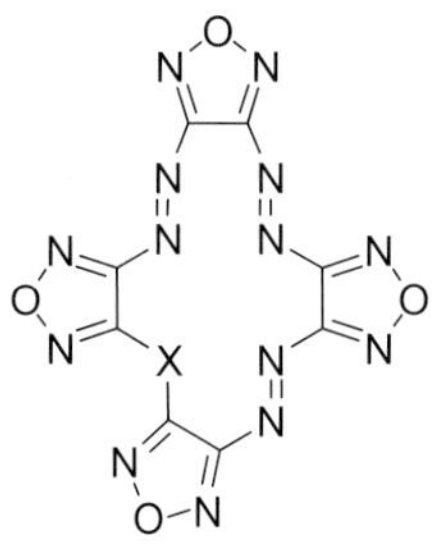 **408** X = O, S, $O(CH_2)_2O$ 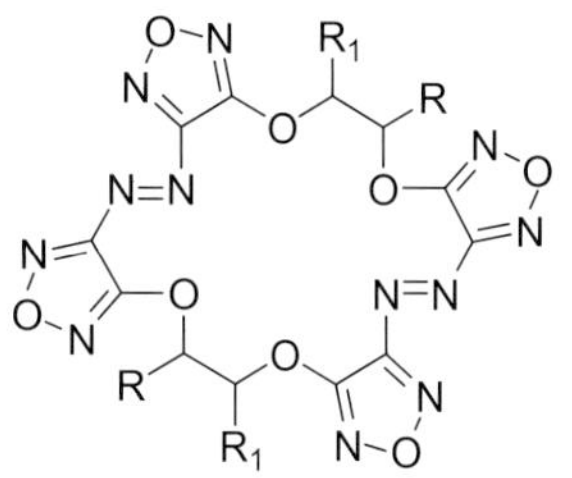 **409**

13.4.3.2 NMR Spectroscopy

Monosubstituted furazans [364], phenylfurazans [365], azoxyfurazans [366], and hydroxy-, alkoxy- and phenoxyfurazans [367] have been studied by NMR spectroscopy. Additive schemes have been developed and spectrum–structure correlations have been elucidated.

Table 13.7 reports the NMR chemical shifts (^1H, ^{13}C, ^{15}N, and ^{17}O) of furazan, furoxan, and their benzo derivatives [308].

Table 13.7 NMR chemical shifts (ppm) for furazan, furoxan, and their benzo-fused derivatives.

Compound	H3	H4	C3	C4	N2	N5	O1	O-exo
Furazan	7.92	7.92	142.0	142.0	−33.5	−33.5	450	
Benzofurazan			148.6	148.6	−35.6	−35.6		
Furoxan	7.44	8.50	105.2	146.8	−15.4	3.9	507.5	364.0
Benzofuroxan			113.7	152.2	−25.3	−13.2		

In the furoxan series the lower chemical shifts of 3H with respect to 4H is due to the contribution of the resonance structure **411**.

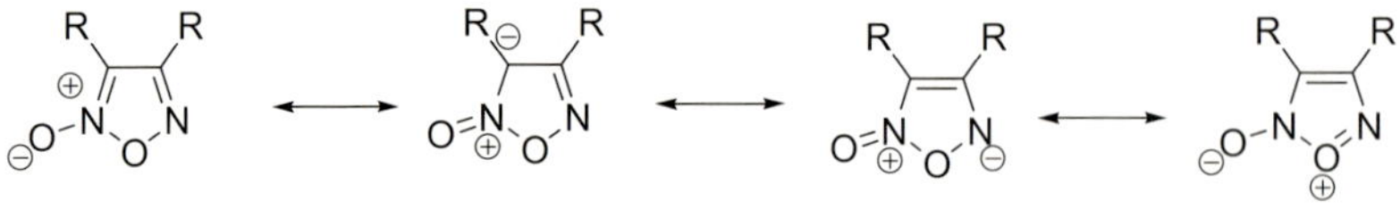

410 **411** **412** **413**

The strong shielding effect exerted by the exocyclic oxygen atom on the proton attached to the substituents at C3 allows one to identify the individual isomer **383a** and **384a.** Thus, for dimethylfuroxan the resonance at 2.16 ppm is assigned to the 3-methyl group, while that at 2.38 ppm is attributable to the 4-methyl group [368].

In benzofuroxans, the shielding effect exerted by N-oxide shifts all the signals to higher frequency with respect to the resonances of benzofurazans, with the larger shifts for H7 and for H4 [369]. The ring-chain tautomerism of these compounds has been investigated by ^{1}H spectroscopy. The unsymmetrical ABCD pattern for the homocyclic protons, observed at −40 °C, changes into a symmetrical A_2B_2 pattern at 100 °C, as a consequence of the rapid equilibration of two isomers. In this way, exchange rates over a range of temperatures have been determined and thermodynamic activation parameters calculated [344, 370].

The most noteworthy feature of the ^{13}C NMR of furoxans is the large difference in chemical shifts of C3 and C4 resonances. As indicated in Table 13.7, C3 resonates at higher field in the range 100–123 ppm, while C4 appears in the range 140–160 ppm [368].

The chemical shifts and multiplicities of the two bridgehead carbons in the ^{13}C NMR spectra of various fused furoxans have been shown to provide a general method for assigning structure in these tautomeric systems [371].

3-Methylfurazans with nitrogen-containing substituents at C4 have been studied by ^{1}H, ^{13}C and ^{14}N NMR spectroscopy [372]. A correlation between the chemical shifts in ^{13}C NMR spectra of these furazans and monosubstituted benzenes with the same substituents was found. The influence of substituents on the NMR data of the iodofurazans was also investigated [356].

The ^{15}N NMR spectra of furazan and benzofurazan show a single absorption at −33 and −36 ppm, respectively [373, 374]. The nitrogen atoms of furoxans show distinct signals in the range −26 to −15 for N2 and −14 to +4 ppm for N5: these resonances coalesce on heating.

The parent furoxan **373** has ^{17}O signals at 508 and 364 ppm, while furazan **372** shows a single resonance at 450 ppm [368]. The corresponding figures for dimethylfuroxan are 460 and 350 ppm, and 475 ppm for dimethylfurazan.

13.4.3.3 **UV and IR Spectroscopy**

Characteristic peaks in the IR spectra of furazans are in the ranges 1525–1560 (C=N−O), 1430–1385 (N−O), and 1040–1030 and 890–880 cm^{-1} (heterocyclic ring). Furoxans show diagnostic peaks at 1625–1600 (CN−O), 1490–1400 ($C{=}NO_2$),

1360–1280 (N–O), and at 1190–1150, 1030–1000, and 890–875 cm^{-1} (heterocyclic ring) [308].

MINDO/3 and DFT methods have been used to calculate the IR active vibrational frequencies of a series of furazans and furoxans [348]. The 1605 cm^{-1} band of furoxans was assigned to vibrations of the C:N(O) group [326].

Monocyclic furazans and dimethylfurazan have UV absorption bands at 228–241 nm, while typical monocyclic furoxans have a peak at 255–295 nm [375].

The extended conjugation in the chromophores of benzofurazans and benzofuroxans results in a shifts of λ_{max} to longer wavelengths (350–410 nm); the energy band can extend into the visible region when conjugating groups are present.

7-Halo-4-nitro and 7-halodinitrobenzofurazans are used as analytical reagents because of the strong visible fluorescence in the region 525–545 nm when reacted with ethers, thioethers, and amines. Thus, tyrosil, cysteinyl, and amino residues of proteins can be labeled by this method, providing access to a fluorescence probe incorporating various biological interesting molecules [376].

13.4.3.4 **Mass Spectrometry**

Two general patterns of fragmentation under ionizing radiation characterize monocyclic furazans (Scheme 13.137) [377–379]. Initial ring opening by cleavage of the weak O1–N2 bond is followed by the C3–C4 bond breaking to yield nitrile and nitrile oxide (path a) or by the extrusion of NO (path b) [380]. Peaks attributable to RC^+ are usually observed.

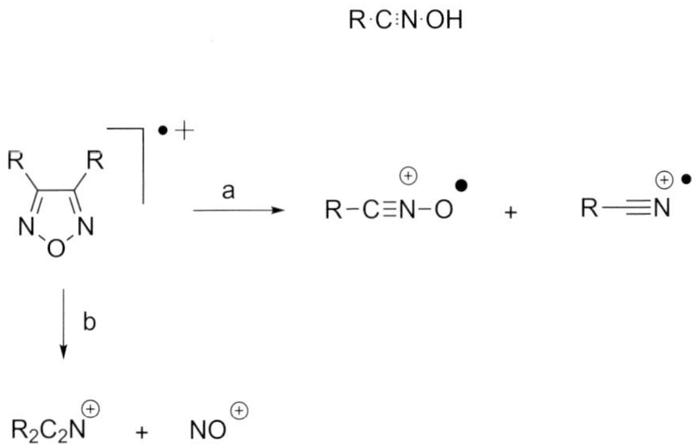

R·C:N·OH

R R N O N •+ a → R–C≡N–O⊕• + R≡N⊕•

b ↓

$R_2C_2N^{\oplus}$ + $NO^{\oplus}$

Scheme 13.137

Unlike many aromatic N-oxides, the $(M–16)^+$ peak for furoxans is weak. The electron impact mass spectra of furazans are characterized by diagnostic peaks at $(M–30)^+$ and $(M–60)^+$, due to the loss of NO and two NO molecules, respectively [381, 382]. The fragmentation pattern (Scheme 13.138) is consistent with the O1–N2 bond cleavage, with formation of the 1,2-dinitrosoethene tautomers, followed by sequential expulsion of NO. In parallel with this route, cleavage of the C3–C4 bond yields two nitrile oxides.

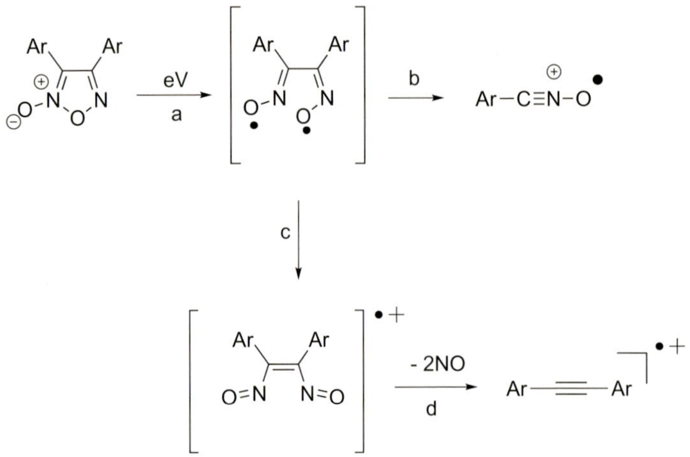

Scheme 13.138

The acetylenic structure of the (M–60)$^+$ fragment has been ascertained by mass-analyzed ion kinetic energy (MIKE) spectroscopy performed under high energy collision activation conditions [312].

The direct generation of NO by a chemical decomposition suggests that the furoxan derivatives can be utilized as a potential NO-related biological probe [381].

13.4.4 Synthesis

Many synthetic routes for the 1,2,5-oxadiazole system have been reported [308, 313, 384]. Different approaches are generally required for furazans, furoxans, and their benzo-fused analogues; thus separate subsections are devoted to the synthesis of furazans, benzofurazans, furoxans and benzofuroxans.

Moreover, according to the stability of the heterocyclic ring, numerous different 1,2,5-oxadiazole derivatives can be generally prepared by exploiting appropriate interconversion reactions of the substituents present on the five-membered ring.

13.4.4.1 Furazans

Three main routes have been designed for the synthesis of furazans: (i) dehydrative cyclization of 1,2-dioximes, (ii) deoxygenation of furoxans, and (iii) Boulton–Katritzky rearrangement of other five-membered heterocycles.

13.4.4.1.1 **Dehydration of 1,2-Dioximes** Furazan **372** was first prepared in 1964, by melting glyoxime with succinic anhydride, in 57% yield [306]. Cyclization of substituted glyoximes **413** is the most exploited methodology for the preparation of mono and disubstituted furazans **414** (Scheme 13.139). The starting material is prepared by reaction of 1,2-diketones **411** with hydroxylamine or by α-nitrosation of an alkylketone **412**, followed by oximation of the resulting 1,2-dione monooxime. A "one-pot" method for the synthesis of 3-alkyl-, 3-aryl-, and 3-hetaryl-4-aminofurazans

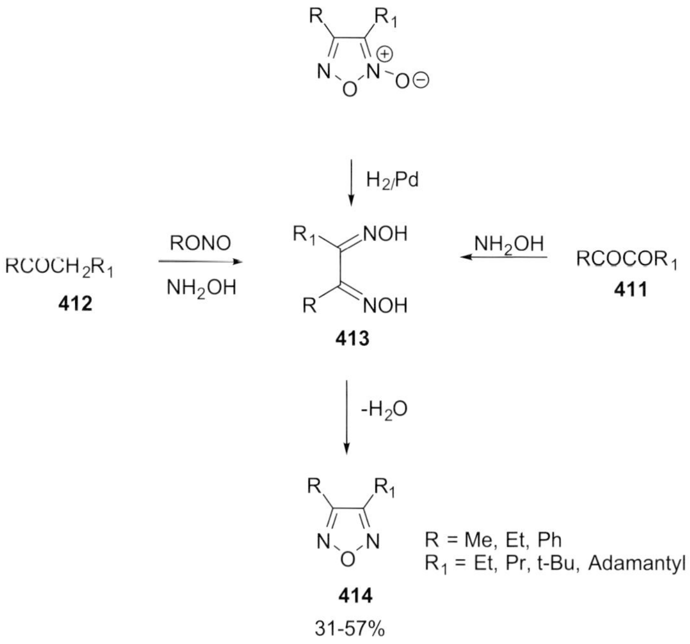

Scheme 13.139

from β-alkyl- or β-aryl and β-hetaryl-β-oxoesters has been reported recently. The multistep process involves hydrolysis of the ester, nitrosation at the activated methylene group, and treatment of the resulting intermediate with an alkaline solution of hydroxylamine in the presence of urea [385].

Dioximes can be also obtained by reduction of furoxans with H_2, Pd/C (Section 13.4.4.1.2); thus, when furoxan is easily available, as for example by dimerization of nitrile oxides, the sequence furoxan–glyoxime–furazan constitutes a valuable synthetic route for symmetrically substituted furoxans. Various dehydrating agents have been utilized, such as acetic, succinic and phthalic anhydrides, sulfuric acid, dicyclohexylcarbodiimide, phosphorus oxychloride, thionyl chloride, and alcoholic sodium hydroxide. According to this procedure, diaminofurazan **418** has been prepared in good yield by reaction of glyoxime **416** and hydroxylamine hydrochloride in aqueous NaOH, to give diaminoglyoxime **417**, followed by KOH mediated dehydration; the reaction has also been performed under microwave irradiation in 2/3 min (Scheme 13.140) [386].

For monosubstituted furazans such as **419**, basic dehydrating agents must be avoided because most of these compounds are isomerized to oximes of α-ketonitriles (**421**), according to a sequence that originates from the initial deprotonation at C4 (Scheme 13.141) [387]. Thus, in these cases, dehydration of the corresponding dioximes is conveniently carried out with anhydrides or sulfuric acid.

In contrast, disubstituted furazans are stable to both heat and chemical conditions and a wide range of dehydrating agents may be used. An unusual synthesis of

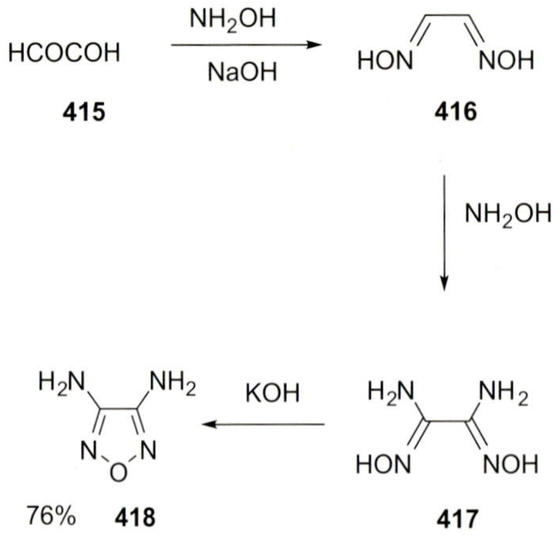

Scheme 13.140

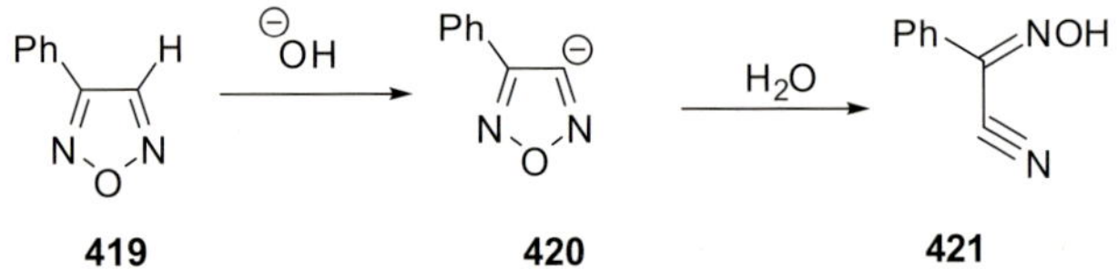

Scheme 13.141

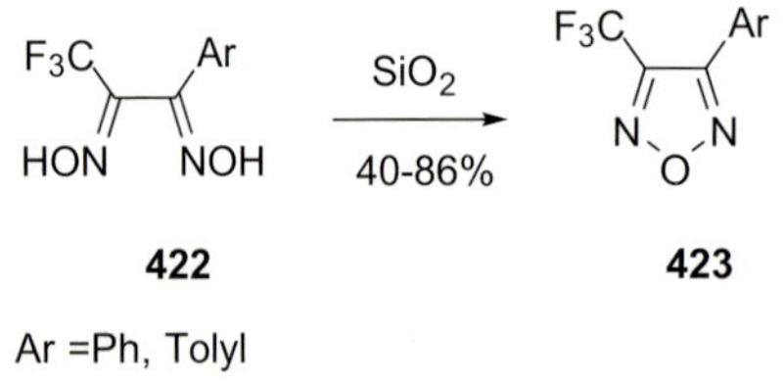

Ar =Ph, Tolyl

Scheme 13.142

3-(trifluoromethyl)-4-aryl-furazans **423** has been reported (Scheme 13.142) [388]: dehydration of 1,1,1-trifluoromethyl-2,3-dione dioximes **422**, which failed with traditional methods, was performed on heating with silica gel.

A modification of the dehydration route involves the conversion of dioximes into diesters, followed by cyclization via distillation or reaction with alkali [389].

The 1,2-dioxime dehydration route is compatible with various substituents, such as alkyl, aryl, acyl, carboxyl, and amino groups. For example, 3-amino-4-phenylfurazans (**425**) are obtained by treatment of aroyl cyanides **424** with hydroxylamine and sodium acetate in ethanol or on heating N-hydroxy-2-(hydroxyimino)-2-arylacetimidamide (**426**) with sodium acetate in ethanol (Scheme 13.143) [390].

13.4.4.1.2 **Deoxygenation of Furoxans** Furazans have been prepared by deoxygenation of furoxans: this method is suitable for furazans bearing different substituents, including alkyl, aryl, acyl, cyano, and amino groups (Scheme 13.144) [313].

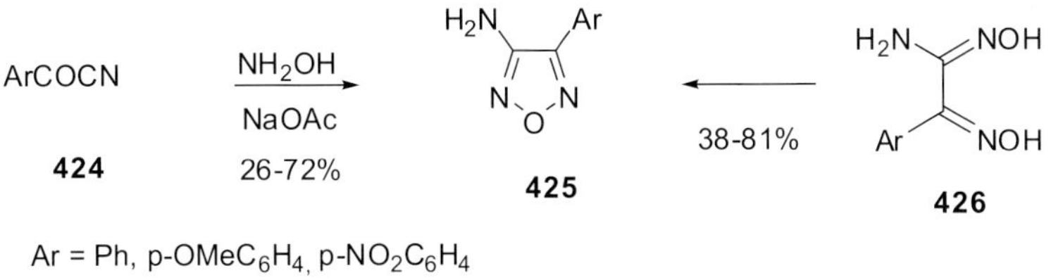

ArCOCN 424 → (NH_2OH, NaOAc, 26-72%) 425 ← (38-81%) 426

Ar = Ph, p-$OMeC_6H_4$, p-$NO_2C_6H_4$

Scheme 13.143

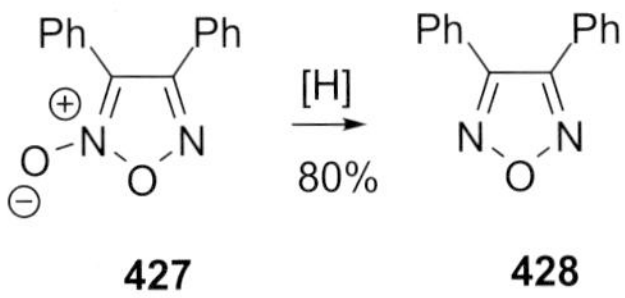

427 → ([H], 80%) 428

Scheme 13.144

However, the reduction process must be carried out so as to avoid over-reduction and, when the furazan is thermally labile, the formation of by-products by ring opening [319]. The most employed reducing agents include trialkyl and triarylphosphites and phosphines, phosphorous pentachloride, stannous chloride in acetic acid, and Zn/acetic acid [308].

The strategy is particularly efficient for the preparation of symmetrical substituted furazans because the corresponding furoxans can be easily prepared via dimerization of the appropriate nitrile oxides (Section 13.4.4.3.3) (Scheme 13.145) [391].

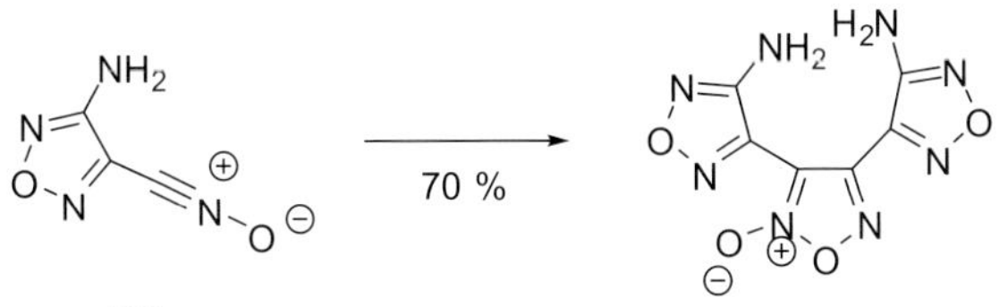

429 → (70 %) 430

Scheme 13.145

Vinyl azides **431** [392] and vicinal vinyl nitro compounds **433** [393] can be precursors for furoxans and furazans (Scheme 13.146).

13.4.4.1.3 **Boulton–Katritzky Rearrangement** Oximes of several classes of 3-acyl-1-oxa-2-azoles, such as isoxazoles, 1,2,4-oxadiazoles, and furazans, undergo a thermal or base-catalyzed rearrangement, known as the Boulton–Katritzky rearrangement, which leads to furazans (Scheme 13.147) [342, 394].

The reaction can proceed by a concerted electrocyclic mechanism or, in the presence of a base catalyst, in two steps by an intramolecular nucleophilic attack at the nitrogen atom of an anionic intermediate.

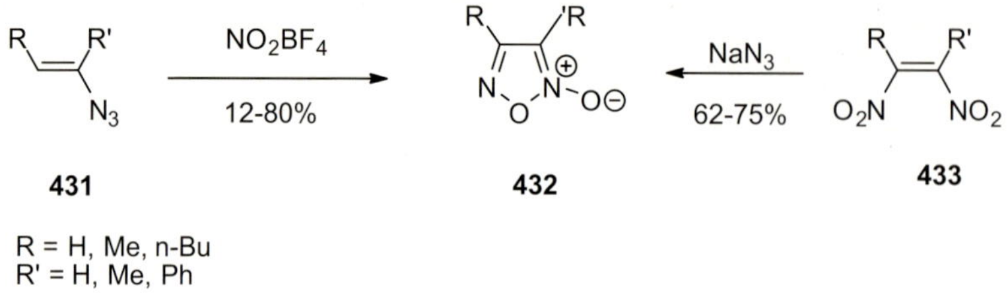

R = H, Me, n-Bu
R' = H, Me, Ph

Scheme 13.146

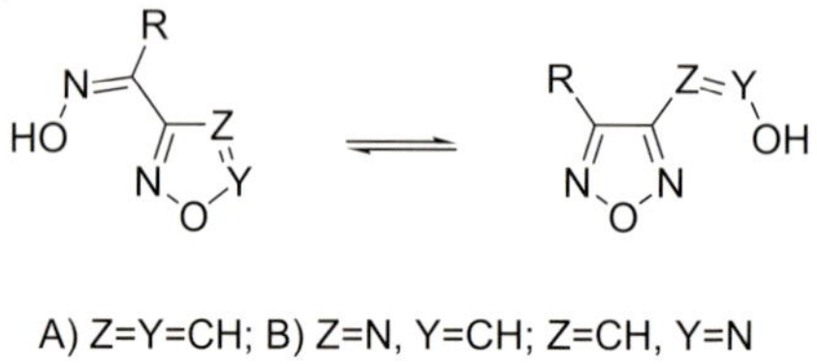

A) Z=Y=CH; B) Z=N, Y=CH; Z=CH, Y=N

Scheme 13.147

The process is geometry dependent, with the *(Z)*-isomer rapidly transformed, while the *(E)*-isomer is generally stable. Thus, the reaction of 3-benzoyl-5-phenyl-1,2,4-oxadiazole (**434**) with hydroxylamine gives rise to a mixture of *(E)*-oxime **436** and the amido furazan **437** resulting from the rearrangement of *(Z)*-oxime **435** [395]. The addition of an acid improves the process because of the *(Z/E)*-oxime isomerization. Under this condition the amidofurazan **437** is hydrolyzed to its amino derivative **438** (Scheme 13.148).

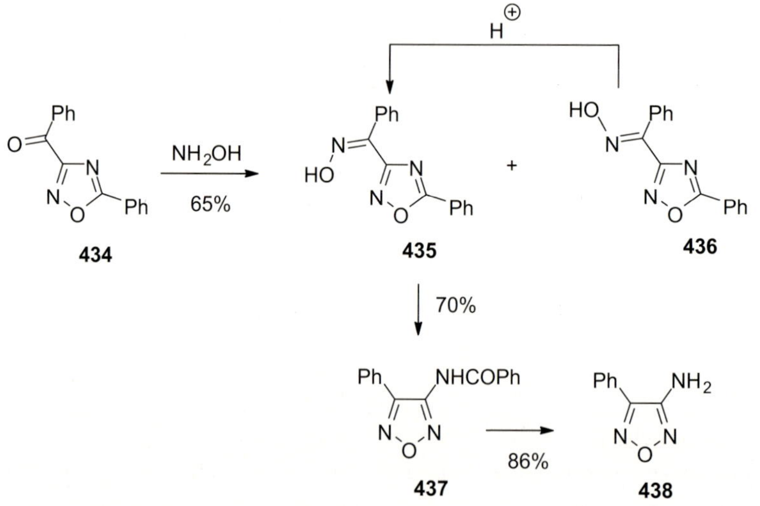

Scheme 13.148

The transformation is not limited to oximes, but also amidoximes and hydrazidoximes can give the corresponding furazans [396].

The *(Z)*-oximes of 3-acylisoxazoles **439**, which are more stable than the corresponding 1,2,4-oxadiazole derivatives, do not react in the absence of a catalyst; the rearrangement occurs easily by treatment with bases, leading to the formation of β-ketoalkylfurazans **440** (Scheme 13.149).

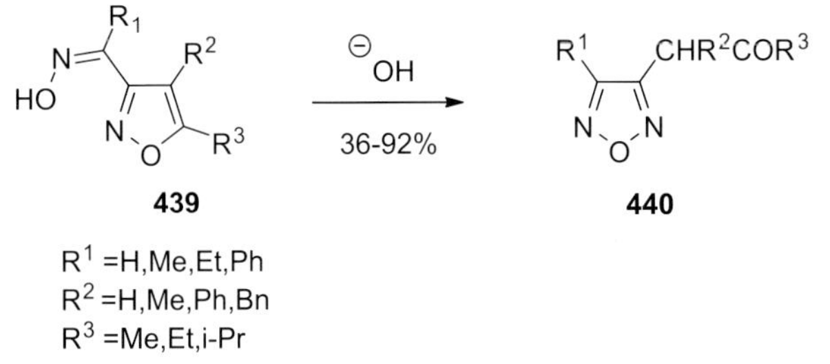

R^1 =H,Me,Et,Ph
R^2 =H,Me,Ph,Bn
R^3 =Me,Et,i-Pr

Scheme 13.149

Various oximes of 3-acyl-substituted furazans **441** undergo the Boulton–Katritzky rearrangement in which the oxadiazole ring is converted into a new furazan system bearing a hydroxyiminoalkyl group (**442**). Several examples of *N*-mono and *N,N*-dialkylfurazanamidoximes (**443**↔**444**) have been reported (Scheme 13.150) [397].

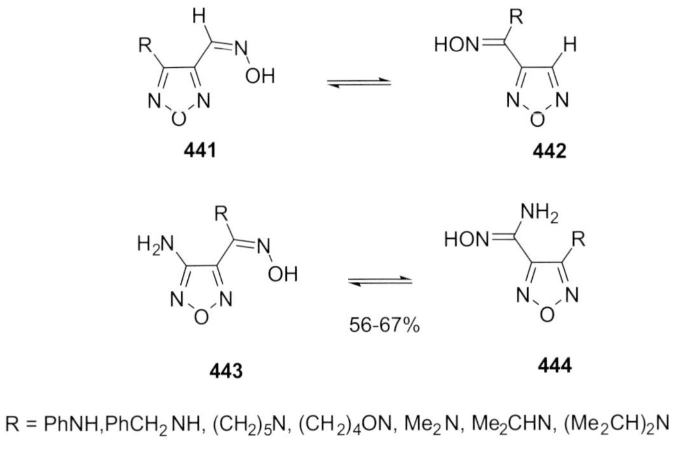

R = PhNH, $PhCH_2NH$, $(CH_2)_5N$, $(CH_2)_4ON$, Me_2N, Me_2CHN, $(Me_2CH)_2N$

Scheme 13.150

13.4.4.2 **Benzofurazans**

The main synthetic routes towards benzofurazans are (i) dehydration of *o*-quinone dioximes, (ii) cyclization of *o*-substituted nitrosoarenes, and (iii) deoxygenation of benzofuroxans (Scheme 13.151) [307, 309].

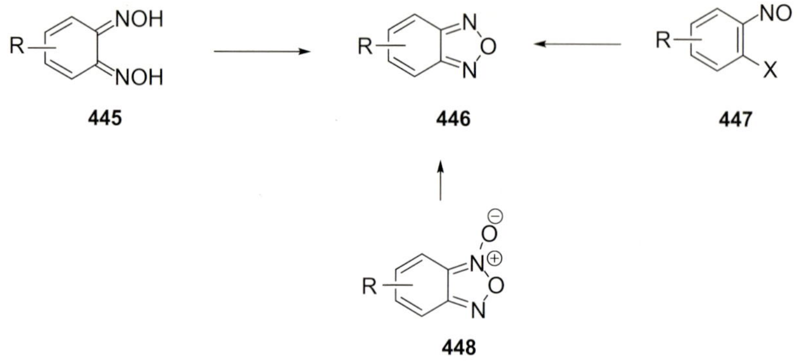

Scheme 13.151

Various dehydration conditions have been used for the conversion of *o*-quinone dioximes into benzofurazans, such as the use of acetic anhydride, thionyl chloride, sulfuric acid, phenyl isocyanate, and alcoholic sodium hydroxide. Alternatively, cyclization may be performed by thermolysis of the corresponding dioxime diacetates or dibenzoates [307, 309].

The utility of this synthetic approach is linked to the availability of dioxime precursors, which can be prepared by direct oximation of *o*-quinones or by reduction of the corresponding benzofuroxans, although, in many cases, direct deoxygenation to benzofurazans can occur.

A different widely exploited methodology starts from *o*-nitrosoarenes: thus, *o*-azido nitroso derivatives, generated from the *o*-chloro analogues, can be converted into benzofurazans by thermolysis [398], while 1-amino-2-nitrosoarene affords benzofurazan by oxidation with ferricyanide or hypochlorite, probably through an *o*-quinone dioxime intermediate. In addition, *o*-nitrosophenol heated in the presence of hydroxylamine leads to furazans, presumably by oximation of the tautomeric *o*-quinone monooxime followed by dehydration. Other approaches involve the thermolysis of *o*-nitroanilines or the reaction of *o*-dinitroarenes with sodium azide [399] and the reduction of *o*-dinitroarenes with sodium borohydride [400] (Scheme 13.152)

Moreover, benzofurazans can be synthesized by a Boulton–Katritzky rearrangement. Thus, 7-nitrosobenzofuroxan or 3-methyl-7-nitroso-2,1-benzisoxazole afford the corresponding 4-nitrofurazan [398] and 4-acetylbenzofurazan [401]. Analogously, benzofurazan is formed by photolysis of 2,1,3-benzoselenadiazole *N*-oxide. This reaction involves cleavage of the heterocyclic ring, and the extrusion of selenium followed by ring closure [402].

13.4.4.3 **Furoxans**

The most exploited routes towards furoxans are (i) oxidative cyclization of 1,2-dioximes, (ii) dehydration of α-nitroketoximes, and (iii) dimerization of nitrile oxides for symmetrically substituted furoxans. For unsymmetrical furoxans, the possibility

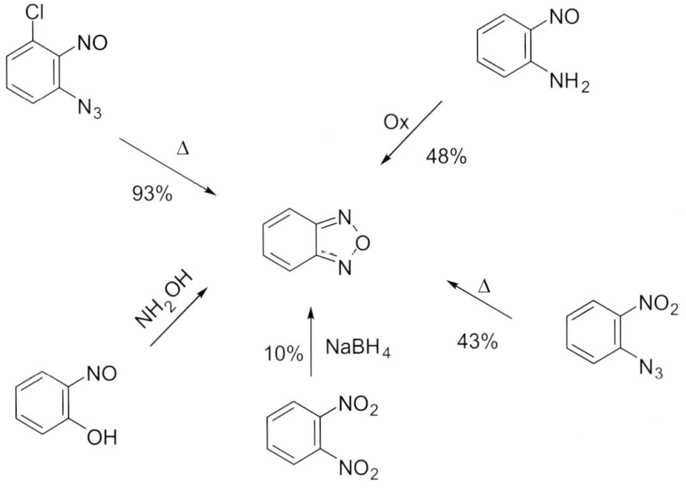

Scheme 13.152

of formation of mixtures of 2- and 5-isomers must be taken in account in choosing a suitable synthetic strategy. No data have been reported on the direct oxidation of furazans to furoxans.

13.4.4.3.1 **Oxidation of 1,2-Dioximes** The oxidation of 1,2 dioximes offers a valuable route towards furoxans. The oxidation can be carried out with *t*-butyl hypochlorite [403], lead tetraacetate, dinitrogen tetroxide [404], as well as electrochemically [405]; as an example, Scheme 13.153 shows the oxidation reaction of compound **449** to give **450**.

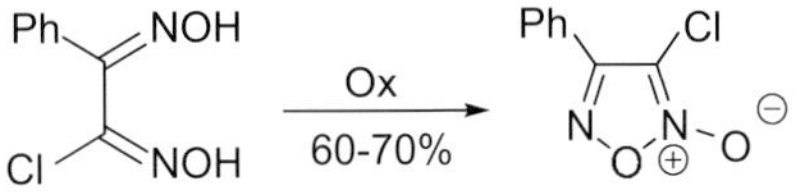

Scheme 13.153

Ring closure can be achieved stereospecifically, thus allowing the formation of individual isomers for asymmetrically substituted furoxans **452** and **453** (Scheme 13.154) [406].

This approach is suitable for the synthesis of 1,2,5-oxadiazoles fused to other carbocyclic and heterocyclic systems [406, 407].

13.4.4.3.2 **Dehydration of α-Nitro Ketoximes** Mono- and polycyclic furoxans have been easily prepared by a synthetic strategy that starts from readily available alkenes [312, 408]. The process involves the initial reaction of alkenes **454** with

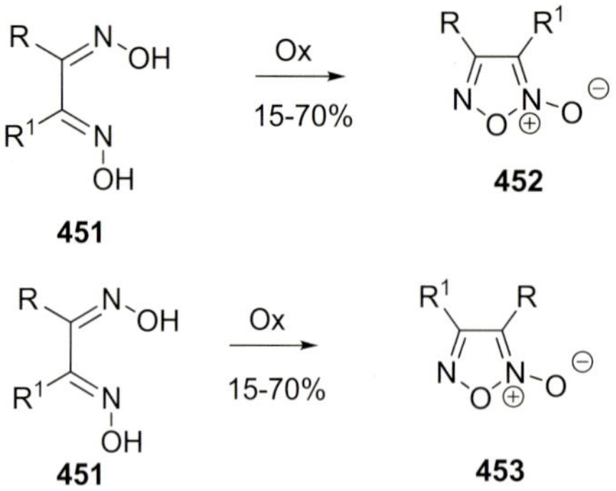

$R = NH_2$, CN, SPh, t-Bu

R^1 = Me, Ph, Me-C_6H_4, SO_2-C_6H_4-Cl

Scheme 13.154

nitrogen trioxide to afford the 1-nitro-2-nitroso adduct **455** – isolable as its nitroso dimer **456** – followed by thermal isomerization to the α-nitro ketoxime tautomer **457** and dehydration with cyclization to the target furoxans **458** (Scheme 13.155).

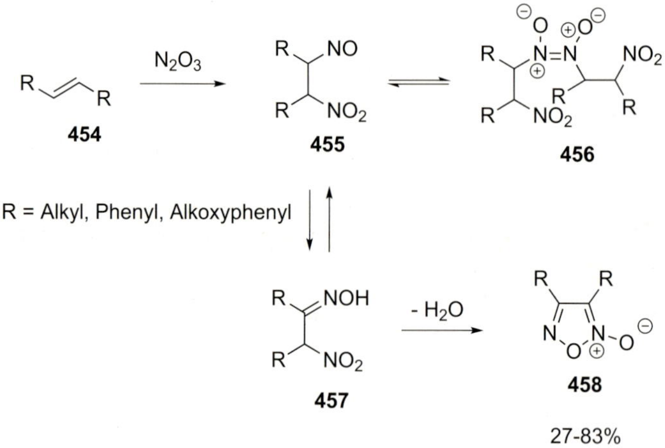

R = Alkyl, Phenyl, Alkoxyphenyl

27-83%

Scheme 13.155

Mixtures of isomers are obtained from non-symmetrical alkenes. The reaction route is compatible with a wide range of functional groups; thus, nitrosation of crotonoaldehyde leads to 4-formyl-3-methylfuroxan [409] and similarly the reaction of β-nitrostyrene with N_2O_3 yields the corresponding aryl nitrofuroxan isomers [410].

Recently, it has been reported that $AgNO_2$/TSMCl reacts with olefins to afford nitrosonitrates that are then converted into furoxans in high yields (Scheme 13.156): the reaction of $AgNO_2$ with TSMCl furnishes first hexamethylsiloxane and N_2O_3

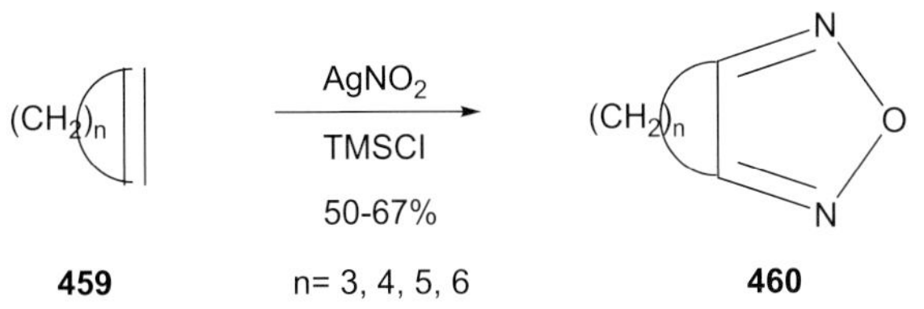

(CH$_2$)$_n$ 459 —AgNO$_2$, TMSCl, 50-67%, n= 3, 4, 5, 6→ (CH$_2$)$_n$ 460

Scheme 13.156

which *in situ* add to alkenes **459** [411]. The approach has been applied to the synthesis of furoxans fused to penta-, six-, seven-, and eight-membered saturated rings **460**.

A general method for the synthesis of furoxans **463** starts from α-nitroketones **461** [412], by conversion into the corresponding α-nitro-ketoximes **462**, followed by treatment with acidic alumina (Scheme 13.157) [413].

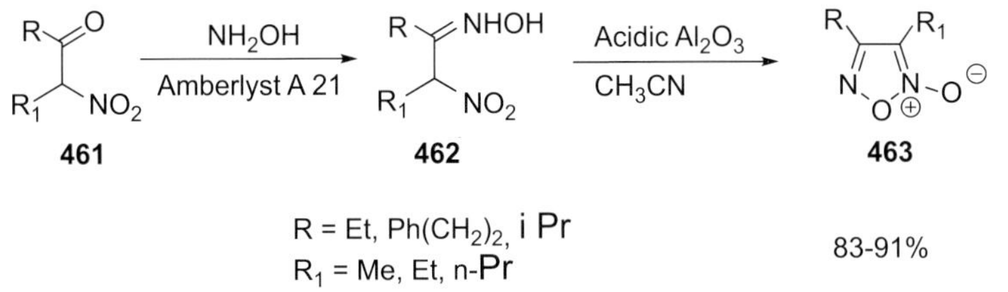

461 —NH_2OH, Amberlyst A 21→ **462** —Acidic Al_2O_3, CH_3CN→ **463**

R = Et, Ph(CH$_2$)$_2$, i Pr
R$_1$ = Me, Et, n-Pr

83-91%

Scheme 13.157

A number of symmetrically substituted dibenzoylfuroxans have been synthesized by treating substituted acetophenones with nitric acid distilled from sulfuric acid [414].

13.4.4.3.3 **Dimerization of Nitrile Oxides** Besides their characteristic and synthetically important 1,3-dipolar cycloaddition reactions, nitrile oxides undergo spontaneous [3 + 2] dimerization, usually regarded as an unwanted side reaction, which can be exploited for the preparation of furoxans (Scheme 13.158).

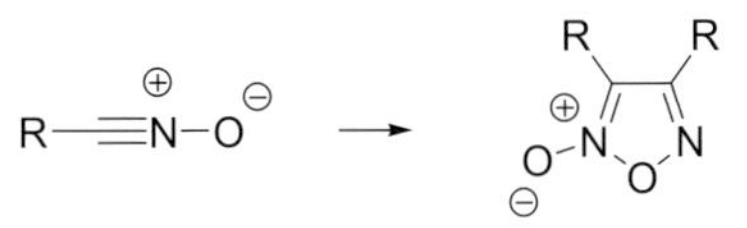

$R-C\equiv N^{+}-O^{-}$ →

Scheme 13.158

Two paths have been proposed for the dimerization of nitrile oxide to furoxans, but the detailed mechanism is unknown. The most widely accepted mechanism is a concerted 1,3-dipolar cycloaddition process, where one nitrile oxide acts as a dipole, while the C–N multiple bond in the other nitrile oxide acts as a dipolarophile (Scheme 13.159) [415, 416]. A (closed-shell) stepwise mechanism, often called the carbene mechanism, has also been proposed [417–419]. In the carbene mechanism, the first step corresponds to bond formation between the carbenoid carbons of

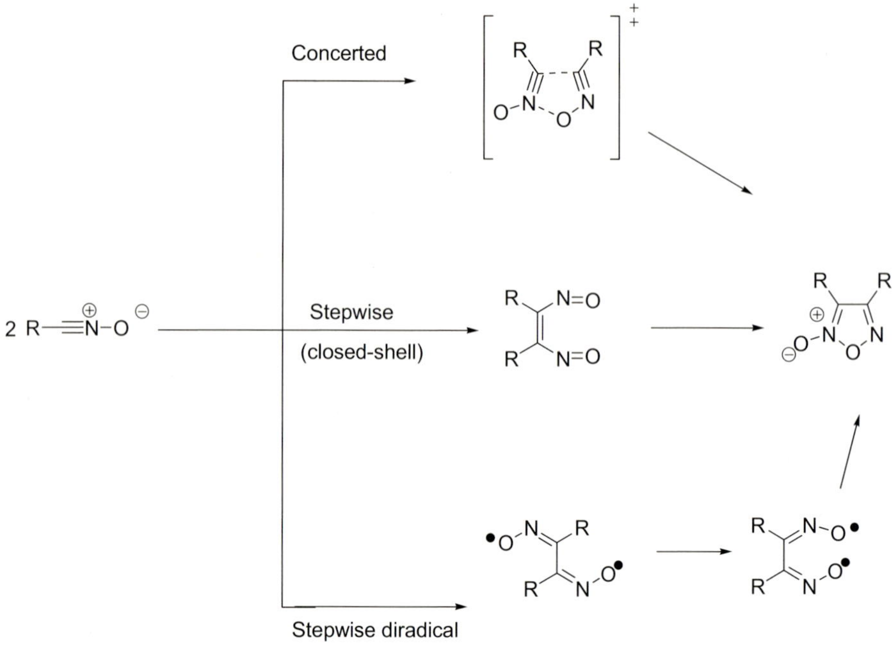

Scheme 13.159

two nitrile oxides to form a dinitroso alkene intermediate, which then cyclizes to the furoxan.

DFT calculations performed at the B3LYP/6-31G* level on the dimerization reactions of acetonitrile oxide and *para*-chlorobenzonitrile oxide to form furoxans indicate that these processes are stepwise, involving dinitrosoalkene intermediates that have considerable diradical character (stepwise diradical mechanism in Scheme 13.159). The rate-determining steps for these two reactions correspond to C−C bond formation [420].

The retardation of dimerization in aromatic nitrile oxides arises from the interruption of conjugation between the nitrile oxide and aryl groups in the C−C bond formation step. The reluctance of aromatic nitrile oxides to dimerize with respect to aliphatic nitrile oxides is attributed to conjugative stabilization of the former. The dimerization processes in solution are slower than in the gas phase, and polar solvents retard the reaction rates.

The method is not appropriate for bicyclic furoxans; a few examples of intramolecular dimerization have been reported. The bimolecular dimerization competes with the unimolecular rearrangement to the isomeric isocyanate, with the latter dominant at higher temperature. Therefore, the preparation of furoxans from nitrile oxides is carried out in concentrated solution at room temperature.

Nitrile oxides are conveniently accessible from many compounds such as oximes, hydroximoyl halides, nitromethyl compounds, alkyl esters of α-nitroalkanoic acids, and others.

The most useful methods have been widely reported in reviews and specialized books [421, 422].

13.4.4.4 Benzofuroxans

The benzofuroxan system can be constructed by suitable modification of the synthetic methods used to synthesize benzofurazans. Thus, the main routes involve the thermolysis of *o*-nitroaryl azides, the oxidation of *o*-quinone dioximes, the oxidation of *o*-nitroanilines (Scheme 13.160), and the Boulton–Katritzky rearrangement.

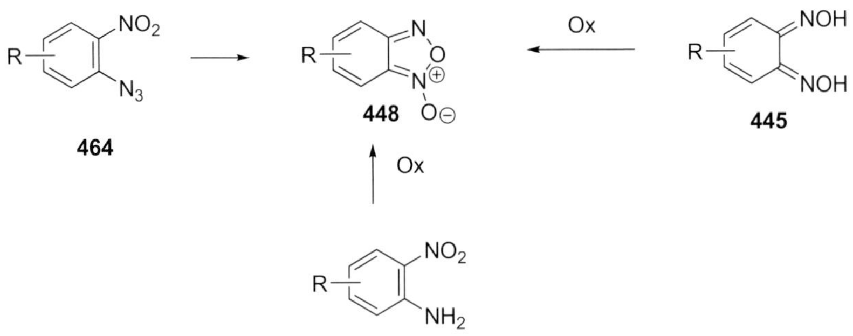

Scheme 13.160

The most exploited methodology for the synthesis of benzofuroxans and hetero-substituted analogues is the thermolysis or photolysis of *o*-nitroarylazides, which can be easily generated from the *o*-nitrohaloarene and sodium azide [423]. The reaction mechanism involves the intramolecular displacement of the nitrogen by the oxygen of the adjacent nitro group.

The method can be used to prepare various hetero-substituted analogues, such as thieno-, imidazo-, oxadiazolo-, thiadiazolo-, pyrido-, quinilino-, pyridazino-, and pyrimidino-derivatives. Thus, photolysis of 4-azido-5-nitrothiophene-2-carboxylic acid ester **465** gives a mixture of thieno[2,3-*c*]furoxan isomers **466** and **467** (Scheme 13.161) [424].

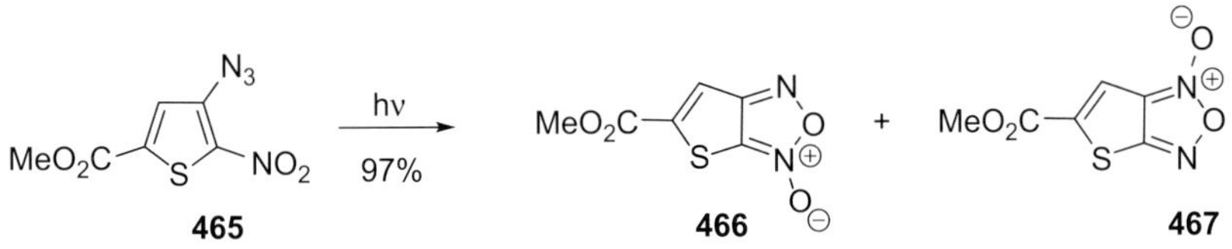

Scheme 13.161

In analogy with the formation of furoxans by oxidation of 1,2-dioximes, benzofurazans can be prepared by oxidation of *o*-quinone dioximes. The method is, however, limited by the availability of the starting materials.

Oxidative ring closure of *o*-nitroanilines constitutes a preferable and commonly used alternative route towards benzofuroxans [425, 426]. Alkaline hypochlorite is the most used reagent: the mechanism involves an initial N-chlorination, followed by deprotonation and loss of chloride ion (Scheme 13.162).

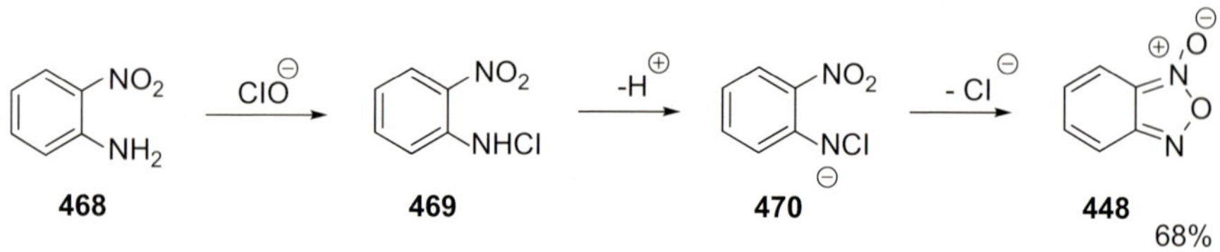

Scheme 13.162

13.4.5
Reactivity of the Heterocyclic Ring

The parent 1,2,5-oxadiazole, with pK_a about −5, is less basic than isoxazole (pK_a = −2.97). 1,2,5-Oxadiazoles are aromatic in nature and are to be considered as π-excessive heterocycles with relatively π-deficient C-atoms. Despite the electron deficiency of carbon atoms, nucleophilic substitution reactions are not common; however, when good leaving groups are present, then reactions can take place. Generally, electrophilic substitutions at the C-atoms cannot be achieved.

13.4.5.1 Furazans and Benzofurazans

13.4.5.1.1 Reactions with Electrophiles and Oxidizing Agents The heterocyclic ring of 1,2,5-oxadiazoles is particularly resistant to attack by electrophilic reagents; thus, halogenation, nitration, and oxidation take place at substituent groups. Electrophilic substitutions in benzofurazan and phenylfurazan occur on the aromatic ring, predominantly at the 4-position and in ortho–para positions, respectively. For example, bromination of phenylfurazan with bromine in the presence of Ag_2SO_4/H_2SO_4 gave the *p*-bromophenyl derivative in 82% yield [427].

Similar results have been obtained on nitration with fuming nitric acid, which gives 4-nitro- or 2,4-dinitro products [428, 429].

There is a single example of an electrophilic reaction at the ring carbon of furazans: insertion of methoxymethylcarbene in the C−H bond of a furazan occurred on thermolysis of furazans **471** with methyl diazoacetate in the presence of copper stearate to give the corresponding methoxycarbonylmethylfurazans **472** in 9–12% yield (Scheme 13.163) [430, 431].

For benzofurazans and benzofuroxans the most facile electrophilic substitution is nitration, which occurs preferentially at the 4-position; a second nitro group can sometimes be inserted at C6. Other electrophiles react less readily, with nitrosation and diazo-coupling occurring only in the presence of activating groups. 5-Methylbenzofurazan reacts with bromine to give substitution at the 4-position; however,

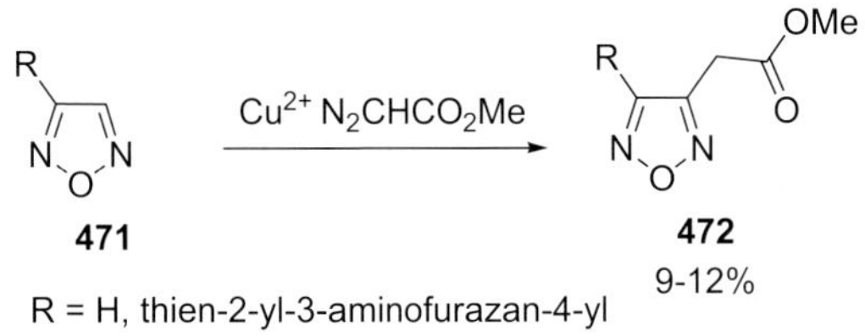

R
Cu^{2+} N_2CHCO_2Me
471
472
9-12%
R = H, thien-2-yl-3-aminofurazan-4-yl

Scheme 13.163

bromine in the presence of sun light undergoes electrophilic addition to the 4,5,6,7-tetra-adduct rather than substitution.

Direct oxidation of furazans to furoxans has not been achieved, as can be expected from the high ionization energy. For instance, oxidation of 3,4-dimethyl-1,2,5-oxadiazole (**473**) [432] and 3-methyl-4-phenylfurazan (**474**) [433] with potassium permanganate occurs at the alkyl substituents, giving rise to 1,2,5-oxadiazole-3,4-dicarboxylic acid (**475**) and 4-phenylfurazan-3-carboxylic acid (**476**), respectively. Under mild conditions, oxidation of fused furazan **477** afforded the dicarboxylic acid **478** (Scheme 13.164) [434].

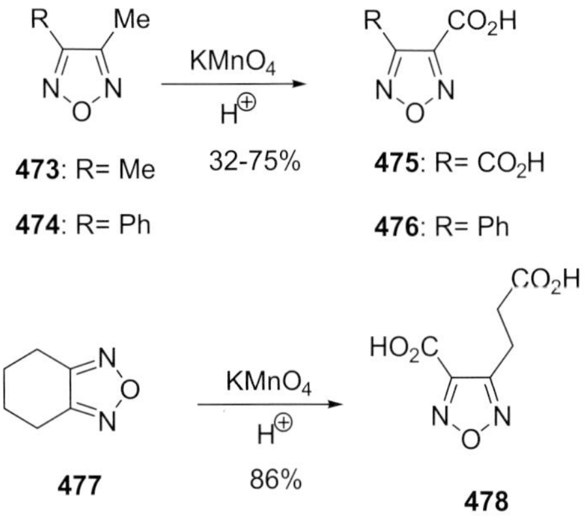

$KMnO_4$, $H^{\oplus}$, 32-75%
473: R= Me
474: R= Ph
475: R= CO_2H
476: R= Ph

$KMnO_4$, $H^{\oplus}$, 86%
477
478

Scheme 13.164

The heterocyclic ring is also resistant to acid attack; pK_a values for protonation of methylphenylfurazan and benzofurazan are −4.9 and −8.4, respectively.

Quaternization with dimethyl sulfate in sulfolane proceeds under forcing conditions, more slowly than that of isoxazoles with iodomethane, to give the corresponding N-methylfurazinium salt [435]. N-Ethyl salts of furazan and 3-phenylfurazan have been obtained by reaction with triethyloxonium tetrafluoroborate.

The reaction of the silyloxy-furazan derivative **479** with triethyl orthoformate led to a mixture of 2-ethyl-1,2,5-oxadiazole-3(2*H*)-one (**480**) and O-ethyl compound **481** (Scheme 13.165). Compound **480** is a rare example of a tricoordinate N-substituted 1,2,5-oxadiazole [436].

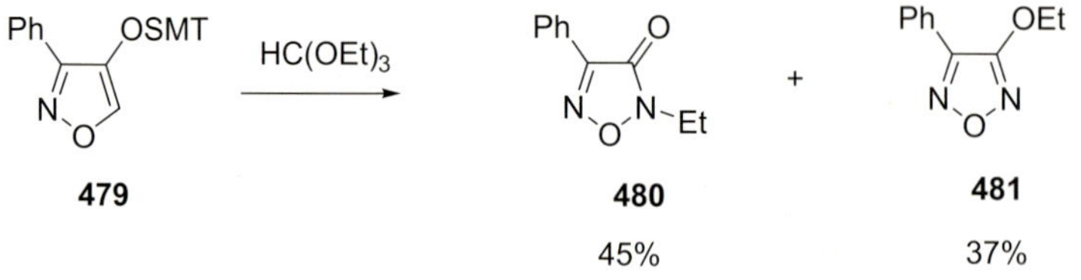

Scheme 13.165

13.4.5.1.2 **Reactions with Nucleophiles and Reducing Agents** Furazans and benzofurazans are generally resistant to attack by nucleophiles. As reported in Scheme 13.134, treatment of the parent compound and monosubstituted furazans with strong bases, as NaOH in methanol, causes the ring opening to form sodium salts of α-oximinonitriles. Disubstituted furazans are comparatively inert.

Furazan ring cleavage occurs also when **482** is treated with Ac_2O at elevated temperatures to produce acylated derivatives **483** [387]. Ring opening with subsequent recyclization has been observed by nucleophilic attack of hydroxylamine on monosubstituted furazans **482**, leading to aminofurazans **485** (Scheme 13.166) [437].

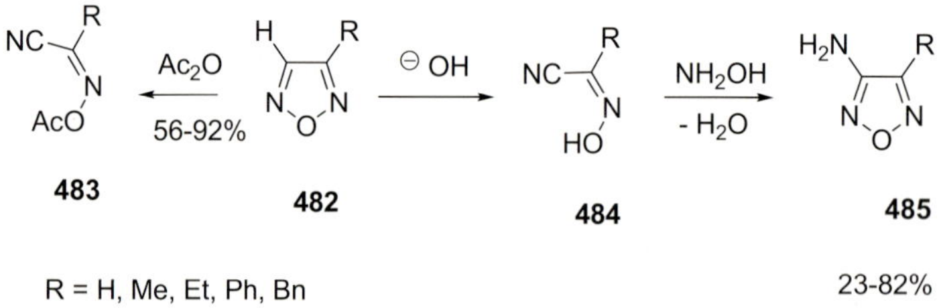

Scheme 13.166

However, when a good leaving group is present, then substitution can occur. Thus, displacement of the nitro group by a hydroxy group has been observed on heating 3-nitro-4-phenylfurazan **486** with sodium hydroxide (Scheme 13.167) [436]; displacement of nitrite or phenylsulfonyl group by alkoxy nucleophiles and by ammonia [438] has also been reported.

Similarly, the homocyclic ring of benzofurazans is susceptible to analogous nucleophilic substitutions. Thus, halides are displaced by various nucleophiles such as alkoxides, fenoxides, cyanide, amines, and thiolates. 4-Halogenobenzofurazans give 4- or 5-substitued products generated from normal *ipso* or *cine* reactions; the *cine* products are amenable to an addition–elimination mechanism (AE), while *ipso* substitution can result from both *AE* and S_NAr (Scheme 13.168) [439].

However, substitution reaction on the homocyclic ring can take place even in the absence of a leaving group: benzofurazan has been converted into its 4-formyl derivative by treatment with LDA in DMF.

The furazan ring is susceptible to reduction. Thus, benzofurazans give 1,2-diaminoarenes by treatment with tin and hydrochloric acid, while catalytic reduction

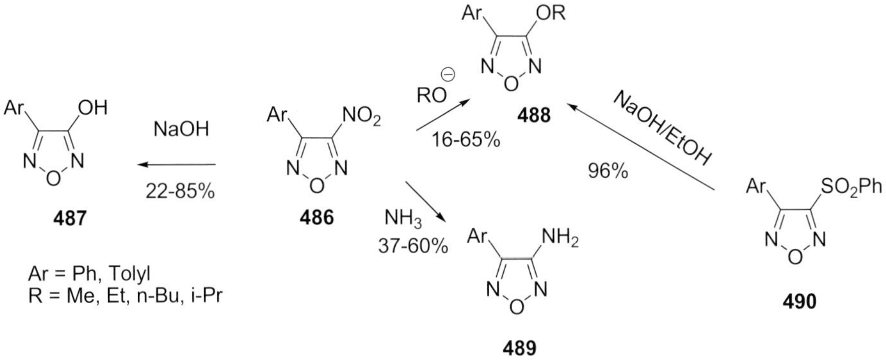

Scheme 13.167

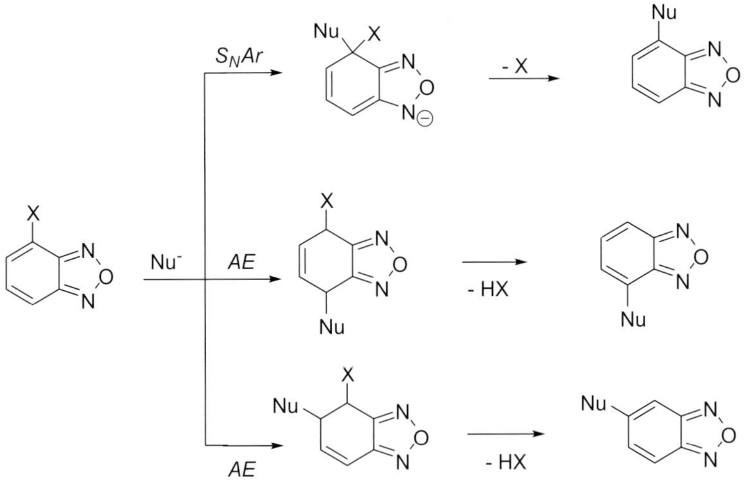

Scheme 13.168

takes place at the homocyclic ring to afford tetramethylenefurazans. 1,2-Diamines are also formed by reduction of furazans with sodium borohydride, whereas $LiAlH_4$ causes fragmentation of the C3–C4 bond, yielding primary amines as final products. Treatment with phosphites results in both fragmentation and deoxygenation to nitriles. Zinc and acetic acid can lead to a selective reduction of the oxadiazolo moiety in the presence of other heterocyclic systems: furazano[3,4-*d*]pyrimidines **491** and furazano[3,4-*e*]pyrazines **492** have been converted into the corresponding *o*-amino compounds (Scheme 13.169) [440].

4,6-Dinitrobenzofurazan, which is strongly electrophilic, undergoes facile σ-complexation with weak nucleophiles to form stable Meisenheimer complexes (Section 13.4.5.3).

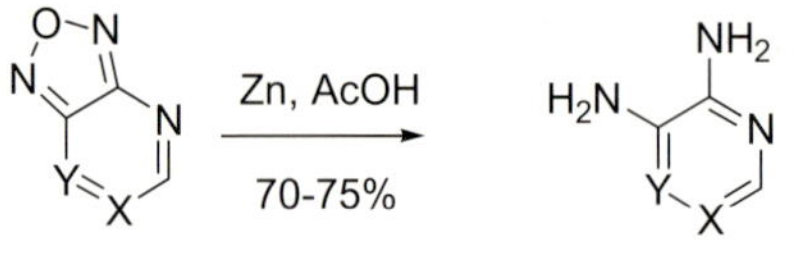

491: X= N, Y=CH

492: X= CH, Y=N

Scheme 13.169

13.4.5.1.3 **Thermal and Photochemical Ring Cleavage** Thermolysis and photolysis of 1,2,5-oxadiazoles proceeds by cleavage of the O1–N2 and C3–C4 bonds to give nitrile and nitrile oxides, together with products derived therefrom.

The thermal process requires temperatures above 200 °C, except for ring-strained derivatives, where less drastic conditions are needed. Thus, diphenylfurazan decomposes at 250 °C to give benzonitrile, phenyl isocyanate, and 3,5-diphenyl-1,2.4-oxadiazole, with the latter two products arising from the rearrangement and 1,3-dipolar cycloaddition with benzonitrile of the initially formed benzonitrile oxide. In contrast, the ring-strained acenaphthofurazan **493** fragments at 120–150 °C to produce the transient **494**, which in the presence of phenylacetylene gives **495** in 53% yield (Scheme 13.170) [441].

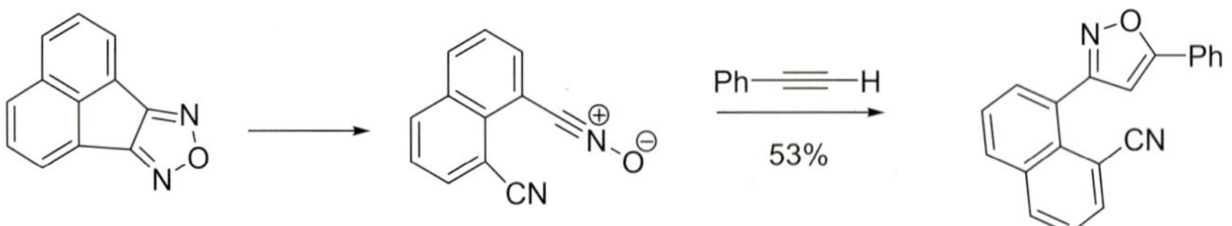

Scheme 13.170

The kinetics of the gas-phase thermolysis of several furazans to phenyl isocyanate and 3,5-diphenyl-1,2.4-oxadiazole have also been examined [442], and a biradical mechanism has been proposed.

Benzofurazans, which are thermally more stable, may be cleaved photochemically. For example, benzofurazan **446** in benzene affords cyanoisocyanate **498** and azepine **499** (Scheme 13.171). Compound **499** probably originates from the reaction of the solvent with the acylnitrene intermediate **497** [443].

13.4.5.2 **Furoxans and Benzofuroxans**

13.4.5.2.1 **Reactions with Electrophiles and Oxidizing Agents** As reported for furazans, the furoxan nucleus shows low reactivity towards electrophiles; reactions occur at the substituents or at the homocyclic ring of benzofuroxans [444]. Reaction with acids is also slow: benzofuroxans have pK_a values of about −8, similar to those of

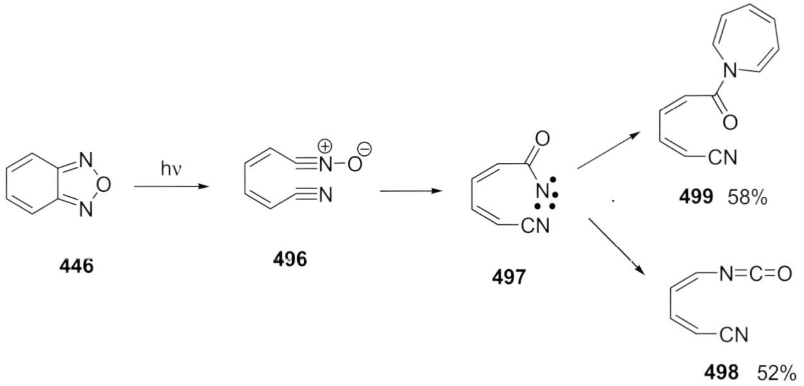

Scheme 13.171

benzofurazans. Treatment of the parent furoxan **372** with concentrated H_2SO_4 proceeds with ring-cleavage to (hydroxyimino)acetonitrile oxide **500**, followed by dimerization to bis(hydroxyiminomethyl)furoxan **501** in nearly quantitative yield (Scheme 13.172) [375].

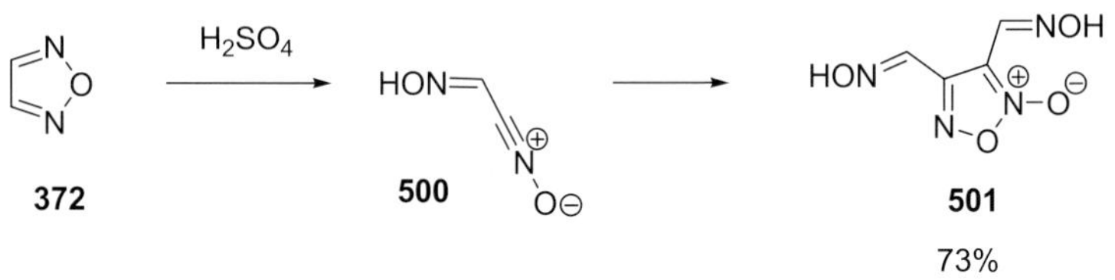

Scheme 13.172

Quaternization is difficult for all furoxans: benzofuroxan does not react with triethyloxonium tetrafluoroborate.

The heterocyclic ring of furazans is also resistant to attack by oxidizing agents, with reactions occurring preferentially at the substituents groups. However, benzofuroxan is oxidized by persulfuric or trifluoroperacetic acid to 1,2-dinitrobenzene, while the 4,6-dinitro compound affords the 1,2,3,4-tetranitrobenzene [445].

13.4.5.2.2 **Reaction with Nucleophiles and Reducing Agents** The reactivity of furoxans with nucleophiles and reducing agents is good. In fact, Grignard reagents react with disubstituted furoxans, primarily at C3, leading to nitrile and nitronate fragments, which, in the presence of an excess of Grignard reagent, yield the corresponding ketones. Monosubstituted furoxans give glyoximes.

Nucleophilic substitution of substituents at C3 and C4 is an easy and valuable pathway towards the synthesis of a wide series of derivatives. The nitro group in particular is readily displaced by numerous nucleophiles such as amines, alkoxides, thiols, azide, halides, and sulfonyl groups [375, 446]. In the benzofuroxans series, nucleophilic reactions take place preferentially at the homocyclic ring; the reactivity is enhanced by the presence of nitro groups [447]. Numerous biologically interesting applications of this kind of reaction have recently appeared in the literature (Section 13.4.6).

In the absence of a good leaving group, nucleophilic attack occurs at N5 of the oxadiazole ring: in this case, the reaction with secondary amines proceeds via ring opening to furnish *o*-nitroarylhydrazines. However, a substitution reaction on the homocyclic ring can take place even in the absence of a leaving group: with 4-nitrobenzofuroxan, carbanions of the form RSO_2ClCH^- cause the displacement of the hydrogen at C5 and at C7 [448].

All monosubstituted furoxans are quite sensitive to bases, which causes ring-opening reactions, with the formation of nitrile oxides **503** from 4-substituted furoxans **502** and *aci*-nitro compounds **505** from 3-substituted furoxans **504** (Scheme 13.173) [313].

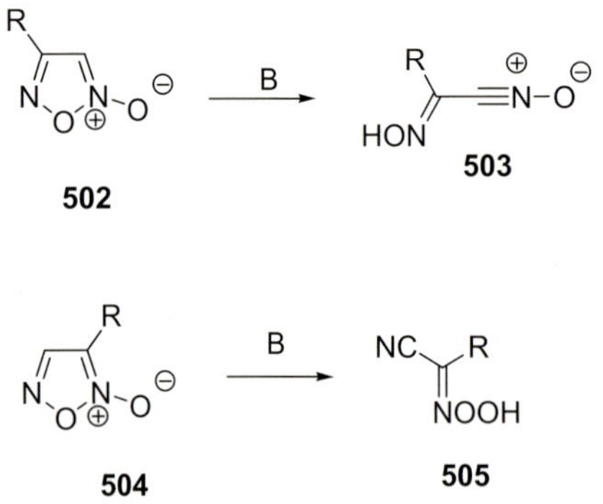

Scheme 13.173

Base attack is favored at the position adjacent to the more highly electron-withdrawing substituent.

Ring opening with subsequent recyclization has been observed by nucleophilic attack of hydroxylamine in aqueous KOH on 3-thien-2-yl and 4-thien-2-yl furoxans (**506**, **507**), leading to aminofurazans **508** (Scheme 13.174) [449].

Furoxans and benzofuroxans can be reduced by various reagents to yield furazans α-dioximes, 1,2-diamines, and nitriles, according to the experimental conditions. Catalytic hydrogenation usually leads to dioximes, but, under forcing conditions, ring cleavage at C3–C4 and N1–O2 can occur. For example, tetramethylenefuroxane **509** affords cyclohexane-1,2-dione dioxime (**510**) by treatment with H_2 and Pd/C at room temperature, while the use of Raney nickel at 100 °C leads to 1,6-diaminohexane (**511**) (Scheme 13.175).

$NaBH_4$ behaves in a similar way, while $LiAlH_4$ reduction is accompanied by ring cleavage to give primary amine fragments. Benzofuroxans are reduced to *o*-nitroaniline derivatives by ferrous salts [450] and to *o*-phenylenediamines by ammonium sulfate–sodium borohydride [451].

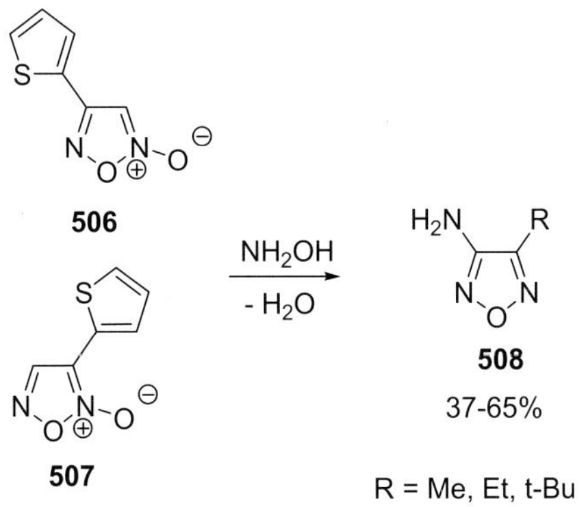

506

507

NH_2OH
$- H_2O$

508
37-65%

R = Me, Et, t-Bu

Scheme 13.174

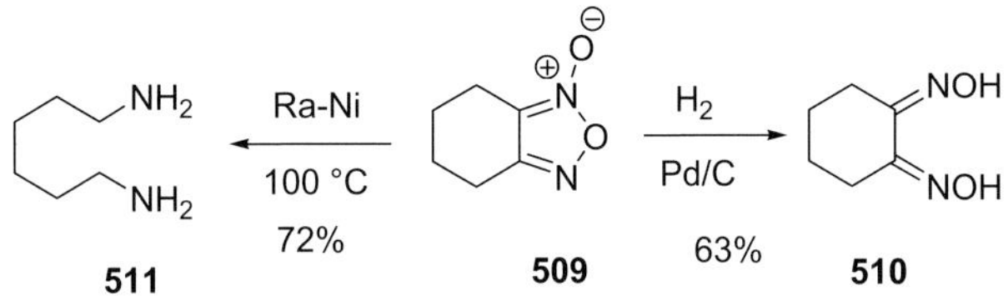

511

Ra-Ni
100 °C
72%

509

H_2
Pd/C
63%

510

Scheme 13.175

Reduction with tervalent phosphorus compounds, such as trialkyl and triaryl phosphites and phosphines, causes deoxygenation of furoxans and benzofuroxans to give furazans and benzofurazans, respectively, leaving the heterocycle intact.

13.4.5.2.3 **Thermal and Photochemical Ring Cleavage** Monocyclic furoxans undergo by thermolysis ring cleavage at the O1–N2 and C3–C4 bonds to give two nitrile oxides fragments, in a formal retro 1,3-dipolar cycloaddition reaction. In the presence of a dipolarophile, the nitrile oxide can be trapped as its 1,3-dipolar cycloadduct; otherwise, the nitrile oxide rearranges to the isomeric isocyanate (Scheme 13.176).

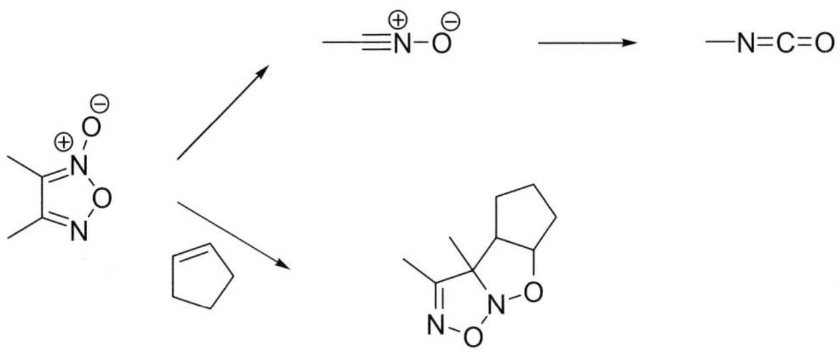

—N=C=O

Scheme 13.176

Under flash vacuum pyrolysis conditions, the nitrile oxides can be isolated [452a].

Bicyclic furoxans afford bisnitrile oxides and diisocyanates. The process is sensitive to the ring strain: for decamethylenefuroxan **513** a temperature above 200 °C is required, while the trimethylene analog **512** reacts at 80–100 °C (Scheme 13.177) [452b,c].

In general, benzofuroxans are much less susceptible to decomposition.

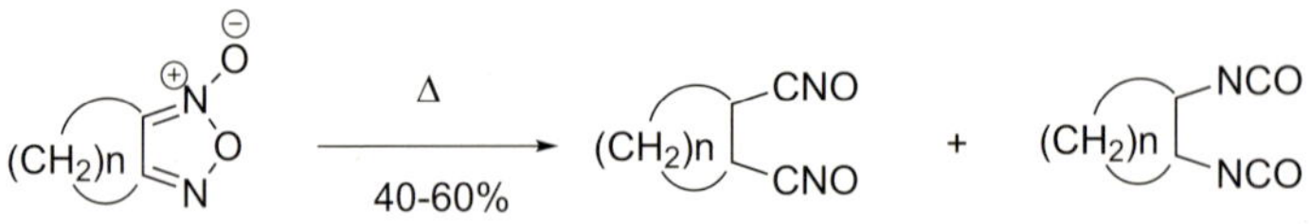

Δ

40-60%

512 n= 3

513 n= 10

Scheme 13.177

13.4.5.3 **Meisenheimer Complex Formation**

4,6-Dinitro compounds **513** and **514** are strongly electrophilic and form stable Meisenheimer complexes when treated even with weak nucleophiles under mild conditions. In particular, **514** is regarded as a super-electrophile – more powerful than 1,3,5-trinitrobenzene. Thus, **514** reacts with methanol, enols, phenols, anilines, thiophenes, pyrroles and indoles, and nitroalkanes to form, in the absence of base, stable C-bonded σ-adducts **516**. Moreover, the electrophilic character of **515** is also confirmed by its reaction with 1,8-bis(dimethylamino)naphthalene, the so-called "proton sponge," yielding carbon-linked compound **517** (Scheme 13.178) [453].

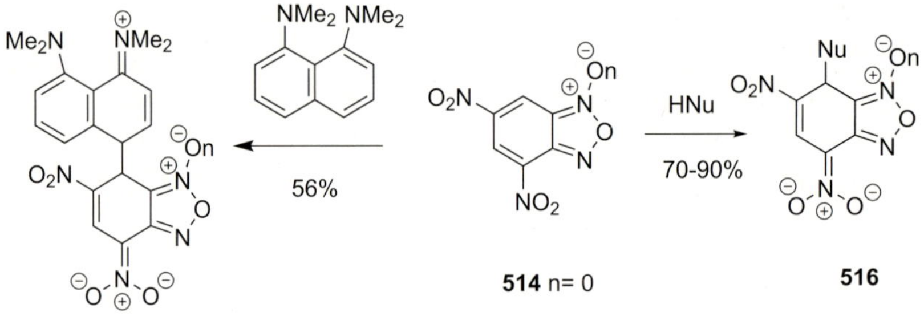

56%

HNu

70-90%

514 n= 0

515 n= 1

516

517

Scheme 13.178

13.4.5.4 **Heterocyclic Ring Rearrangements of Furoxans and Benzofuroxans**

Furoxans have been extensively used as starting material for synthetic conversions into various other heterocyclic systems, some of which show interesting biological activity. The Boulton–Katritzky rearrangement is the most exploited reaction route, affording an easy entry towards isoxazoles, pyrazoles, and furazans. Other conversion reactions give access to isoxazolines, quinaxoline, and benzimidazole *N*-oxides.

13.4.5.5 **Rearrangements of Furoxans**

The Boulton–Katritzky rearrangement [342, 393] of non-condensed furoxan derivatives has been reported for oximes of 4-furoxanylcarbonyl compounds [454]; in particular, the base-catalyzed rearrangement of the *(Z)*-isomer of 4-benzoyl-3-methylfuroxan oxime (**518**) leads to 3-(1-nitroethyl)-4-phenyl-1,2,5-oxadiazole **519** (Scheme 13.179).

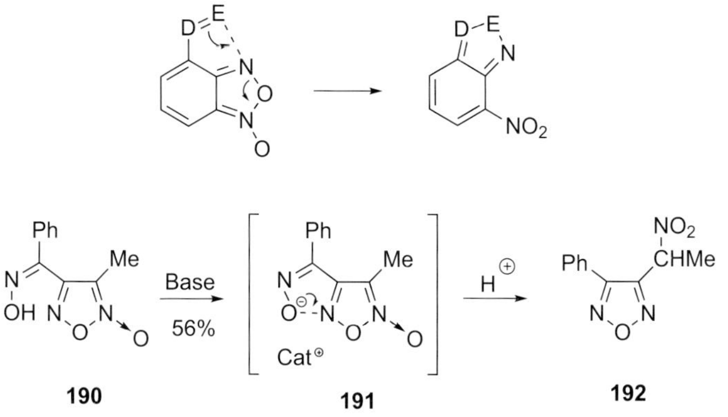

Scheme 13.179

Other variants of rearrangement of monocyclic furoxans have been performed for derivatives involving different side chains: C-N-N (phenylhydrazones), N-C-N (amidines), and N-C-S (thioureides) [455]. Thus, treatment of *(Z)*-isomers of phenylhydrazones **520** with Bu^{t}OK and heating yielded 1,2,3-triazoles **523**, formed with a Nef-type [456] reaction via 5-(1-nitroethyl)-1,2,3-triazole **522** intermediate (Scheme 13.180).

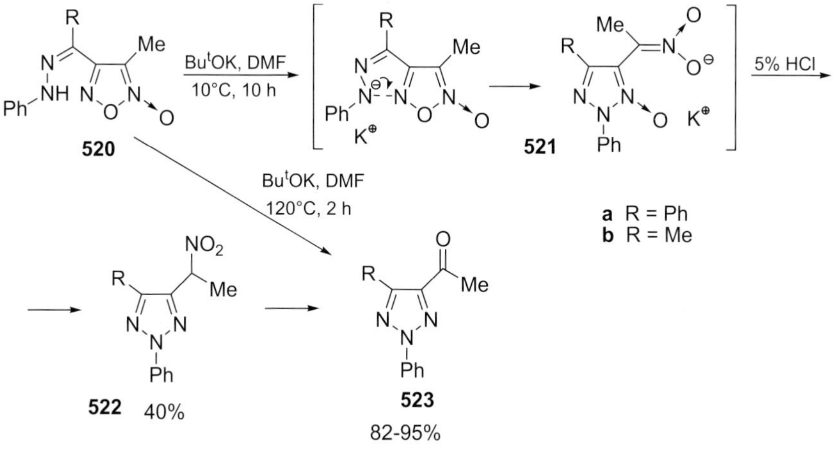

Scheme 13.180

Analogously, the rearrangement of 3-aryl(alkyl)-1-(3-R-furoxan-4-yl)amidines **526** – synthesized by reaction of aminofuroxans **524** with triethyl orthoformate or triethyl orthoacetate, followed by the action of various amines on the resulting iminoethers **525** – afforded the 1,5-disubstituted 3-[1-nitroethyl(benzyl)]1,2,4-triazoles **527** (Scheme 13.181) [455].

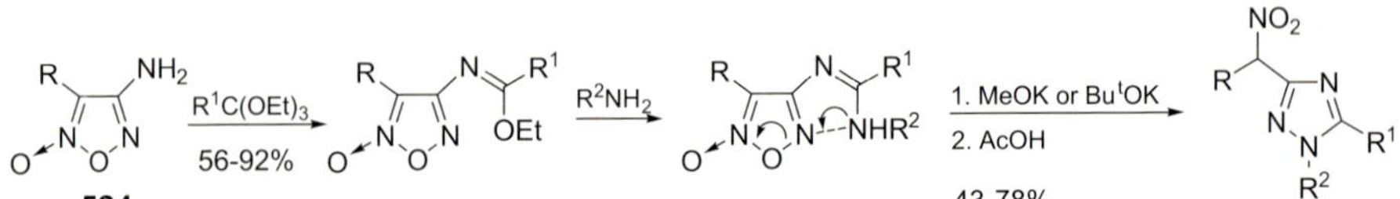

$R^1C(OEt)_3$, 56-92%; R^2NH_2; 1. MeOK or Bu^{t}OK 2. AcOH, 43-78%

524 **525** **526** **527**

R = H, Me
R^1 = Me, Et, Ph
R^2 = Me, Et, i-Pr

Scheme 13.181

5-Ethoxycarbonylamino-3-(1-nitroalkyl)-1,2,4-thiadiazole derivatives **530** have been obtained by refluxing a mixture of aminofuroxans and ethoxycarbonyl isothiocyanate in various solvents: the reaction proceeds through a not-isolated 4-(3-ethoxycarbonylthioureido)-3-substituted-furoxan intermediate (**529**) (Scheme 13.182) [456].

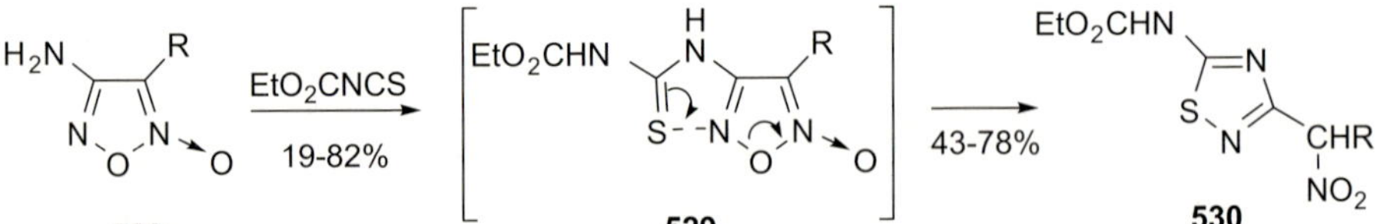

EtO_2CNCS, 19-82%; 43-78%

528 **529** **530**

a R = Ph
b R = Me
d R = COMe

a R = Ph
b R = Me
e R = H

Scheme 13.182

A different kind of rearrangement has been described that proceeds through a dinitrosoethylene intermediate. In particular *(Z)*-isomers of 4-benzoyl or 4-acetyl-3-methylfuroxan phenylhydrazones **520**, thermally or in the presence of various bases, give oximes of 5-acetyl-4-phenyl(methyl)-2-phenyl-2*H*-1,2,3-triazole 1-oxide **532** (Scheme 13.183) [457]. It has been suggested that the reaction starts with rupture of the O1–N2 bond in the furoxan ring, which results in formation of dinitrosoethylene intermediates **531**, followed by the reaction of one nitroso group with the phenylhydrazone moiety and transformation of the second nitroso group into the oxime group.

Another example of this rearrangement is represented by the thermally induced transformation of 3,3′-disubstituted-4,4′-azofuroxans **533** in an oxidizing medium into triazole 1-oxide derivatives **536** (Scheme 13.184) [458].

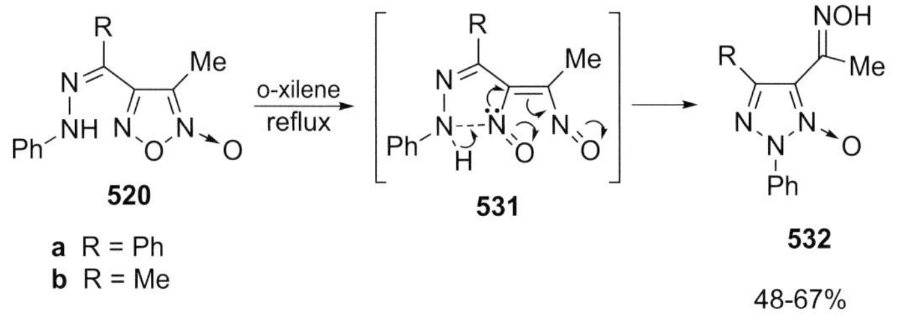
o-xilene
reflux
520
531
532
a R = Ph
b R = Me
48-67%

Scheme 13.183

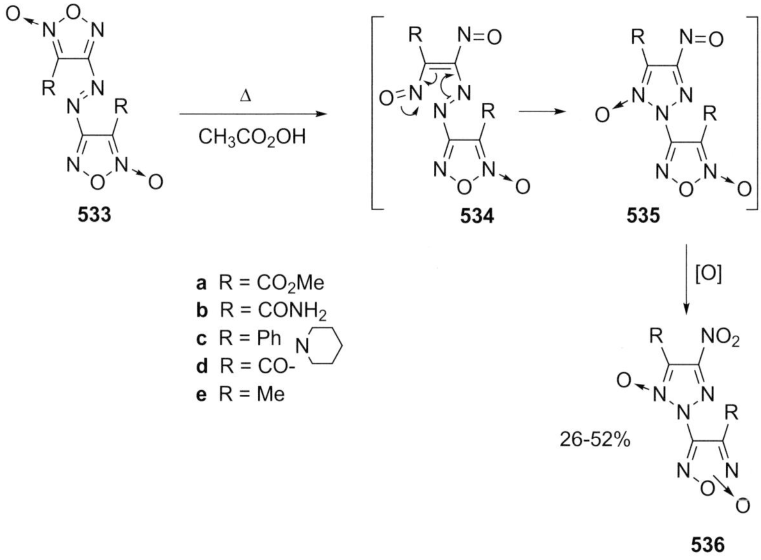
Δ
CH_3CO_2OH
533
534
535
536
[O]
a R = CO_2Me
b R = $CONH_2$
c R = Ph
d R = CO-
e R = Me
26-52%

Scheme 13.184

Furoxans **537** bearing a methyl group at C3 give hydroximino derivatives of isoxazolines **538** on treatment with alkoxides or alcoholic alkali hydroxides (the so-called isoxazoline transposition or Angeli rearrangement) (Scheme 13.185) [459].

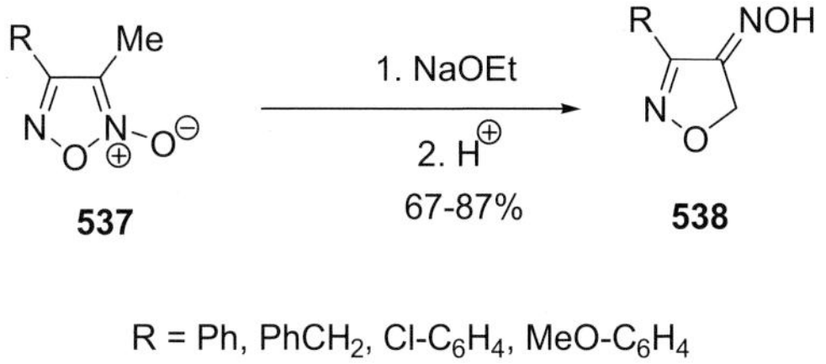
1. NaOEt
2. $H^{\oplus}$
67-87%
537
538
R = Ph, $PhCH_2$, Cl-C_6H_4, MeO-C_6H_4

Scheme 13.185

13.4.5.6 **Rearrangements of Benzofuroxans**

The Boulton–Katritzky rearrangement of benzofuroxans bearing at the 4-position a ring-conjugated side chain has been studied in detail from both a synthetic and mechanistic point of view [342, 370, 393, 460–465]. As a rule, rearrangements are initiated thermally, photochemically or in the presence of bases; the first example of acid catalysis has been published [466]. Several types of unsaturations can constitute the X = Ygroup, such as C=O, N=N, C=N, and N=N, so allowing the construction of a series of new heterocyclic systems. Thus, 4-acetylbenzofuraxan **539** rearranges spontaneously to 3-methyl-7-nitrobenzo[*c*]isoxazole **540**, while nitroindazoles **542** are formed from 4-formylbenzofurazan **541** and primary amines (Scheme 13.186).

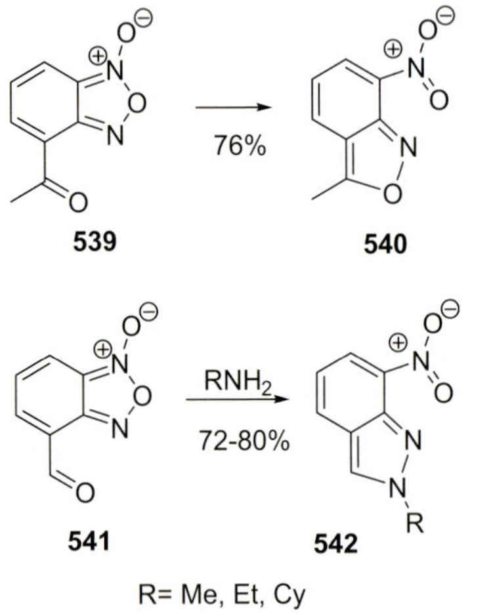

76%

539 **540**

RNH_2

72-80%

541 **542**

R= Me, Et, Cy

Scheme 13.186

Analogously, nitrosation of 5-(dimethylamino)benzofuroxan (**543**) affords 4-(dimethylamino)-7-nitrobenzofurazan (**545**) through the rearrangement of the intermediate 4-nitroso compound **544**. Similar behavior has been observed for 4-arylazobenzofuroxans **546** yielding 4-nitrobenzo-1,2,3-triazoles **547** (Scheme 13.187).

Quinoxaline-1,4-dioxides **551**, **553**, and **555** are readily obtained from the reaction of benzofurazans with enamines or carbonyl compounds in the presence of ammonia or amines (Beirut reaction) (Scheme 13.188) [467]. In the absence of the α-hydrogen required for the elimination, the 2,3-dihydroquinoxaline intermediate **550** can be isolated. Enolates derived from β-diketones react in an analogous way, affording 2-acylquinoxalines [468].

This process formally involves the insertion of a two-carbon fragment between the N–O groups of the furoxan (Scheme 13.188). Various quinoxalines are endowed with interesting biological activity: this feature has considerably extended the scope of this reaction. In the same context, the reaction also gives ready access to polycyclic compounds: for example, phenazine derivatives result from benzofuroxans and phenolates, *p*-benzoquinone, or hydroquinone [312, 469].

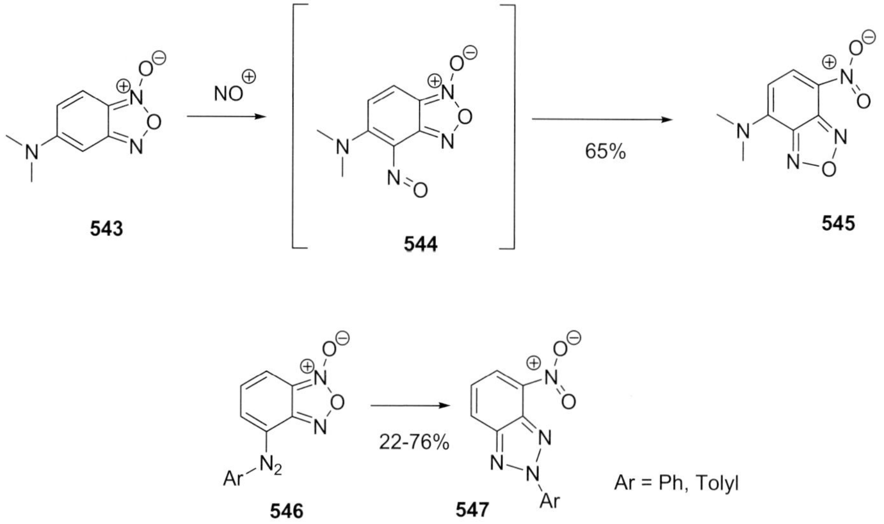

NO⊕ → [544] → 65%

543 544 545

22-76%

546 547 Ar = Ph, Tolyl

Scheme 13.187

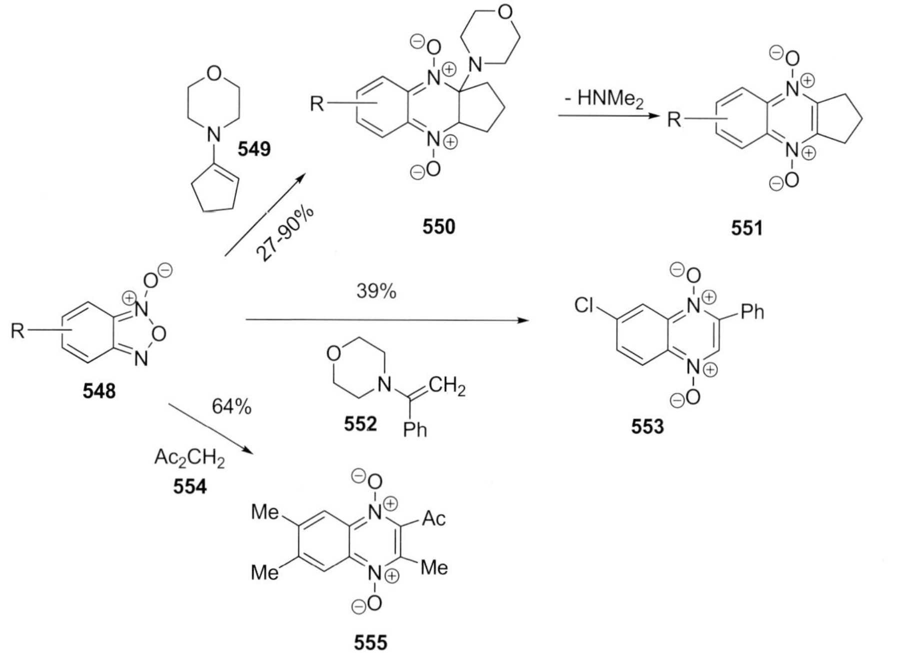

549 27-90% 550 - $HNMe_2$ 551

548 39% 552 553

64% Ac_2CH_2 554 555

R = H, 5(6)Cl, 4,7-Cl_2, 5,6-Cl_2, 4(7)-Me, 5(6)Me, 5,6-Me_2, 5(6)-CF_3, 4(7)-MeO, 5(6)-MeO, 5(6)-Ac

Scheme 13.188

Benzimidazole oxides, in 25–90% yields, are also accessible from benzofuroxans. The reaction with primary nitroalkanes leads to 2-substituted-1-hydroxybenzimidazole-3-oxides **556** via displacement of the NO_2 group; similarly, the nitrile group of α-cyanoacetamides is removed with formation of 2-amide derivatives **557** (R′ = $CONR_2$). Secondary nitroalkyl compounds afford 2,2-disubstituted-2*H*-benzimidazole-1,3-dioxides **558**.

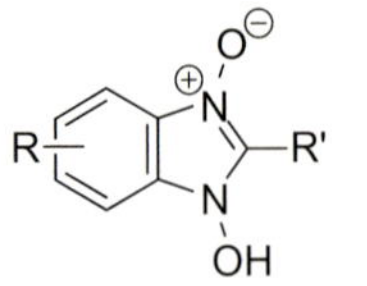

556 : R' = Alkyl

557 : R' = $CONR_2$

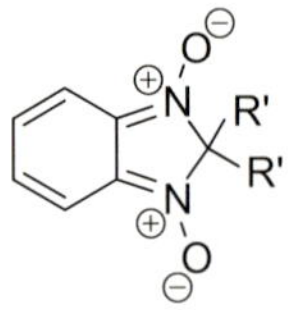

558

Benzimidazoles are also obtained from the reaction of benzofuroxans with phosphorus ylides [470], nitrones [471], and diazo compounds [472].

A mechanism has been suggested that involves the nucleophilic attack at N3 of the benzofuroxan **396** (or at one of the nitroso groups of the *o*-dinitroso tautomer), followed by cleavage of the O2–N3 bond to give the di-N-oxide **561**, with subsequent cyclization to five- or six-membered ring products, according to the nature of the nucleophile (Scheme 13.189).

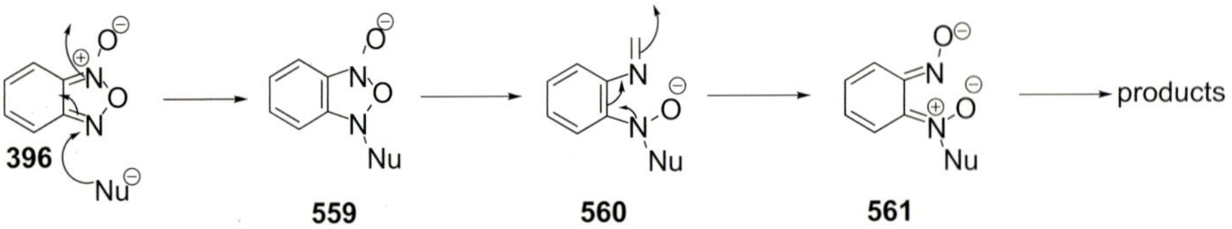

396 → **559** → **560** → **561** → products

Scheme 13.189

13.4.5.7 Cycloaddition Reactions of Benzofuroxans

The double bonds at the 4,5- and 6,7-positions in benzofuroxans are sufficiently localized and activated to undergo [3 + 2] and [4 + 2] cycloaddition reactions. Thus, isoprene (**563**) gives in 80% yield the 1:1 Diels–Alder adduct **562** with 4-nitro-6-trifluoromethansulfonylbenzofuroxan (**564**) (Scheme 13.190) [473].

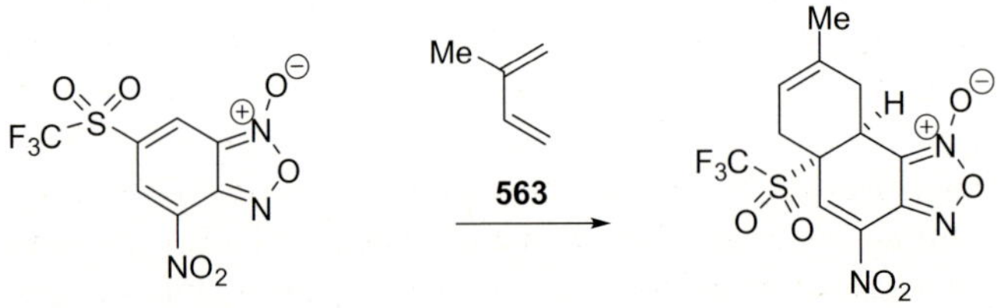

564 —**563**→ **562**

Scheme 13.190

Similarly, alkyl diazoacetates **566** afford pyrazolo derivatives **567** by reaction with 6-nitrobenzofuroxan (**565**) (Scheme 13.191) [474].

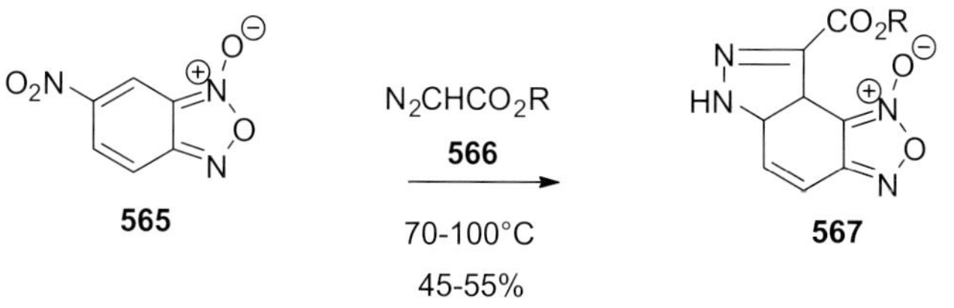

565 — N_2CHCO_2R **566**, 70-100°C, 45-55% → 567

R=Et, Me

Scheme 13.191

Mesitonitrile oxide gives mixtures of 1 : 1 and 2 : 1 adducts [475], and diazomethane reacts at C5–C6 of 4-nitrobenzofuroxan to give the 5-6-cyclopropa-fused derivative **568**, probably by loss of nitrogen from the initial pyrazoline cycloadduct [476].

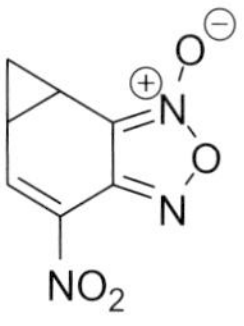

568

13.4.5.8 **Alkyl and Aryl Furazans and Furoxans**

Reactions of aryl furazans and furoxans have been studied in detail. The heterocyclic ring exerts an *ortho–para*-directing influence with the predominant formation of *para* products. For example, nitration or chlorosulfonation of phenylfurazan and phenylfuroxan take place at the phenyl group, at the 4′-position, leaving the heterocycle intact. In benzofurazans and benzofuroxans, electrophilic attack occurs, as previously reported (Section 13.4.5.1.1), at the 4-position; a second group can be sometimes inserted at the 6-position.

Alkyl groups on the furazan or furoxan ring can undergo functional transformations that depend on the electron-withdrawing properties of the rings. Treatment of alkylfurozans **569** with NBS in the presence of BPO or AIBN gives α-bromoalkyl derivatives **570** (Scheme 13.192) [477]. Similar results were obtained with 3-methylfuroxans **571** [478].

These α-haloyl derivatives are excellent starting materials for side-chain substituted furazans and furoxans through classical nucleophilic substitution reactions. The halogen atom is readily displaced by a wide range of oxygen, sulfur, nitrogen, phosphorus, and carbon nucleophiles to give the corresponding products in good yields [477–483].

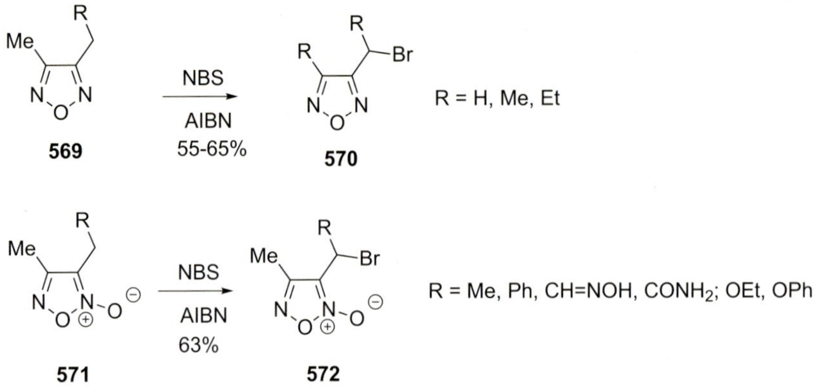

Scheme 13.192

α-Metalation offers an alternative approach to functionalization of the methyl compounds. 3,4-Dimethylfurazan readily undergoes lithiation by treatment with *n*-butyllithium: the lithiated intermediate **573** reacts with electrophiles at −55 °C to give various α-functionalized alkylfurazans (Scheme 13.193) [484–486]. The electrophile

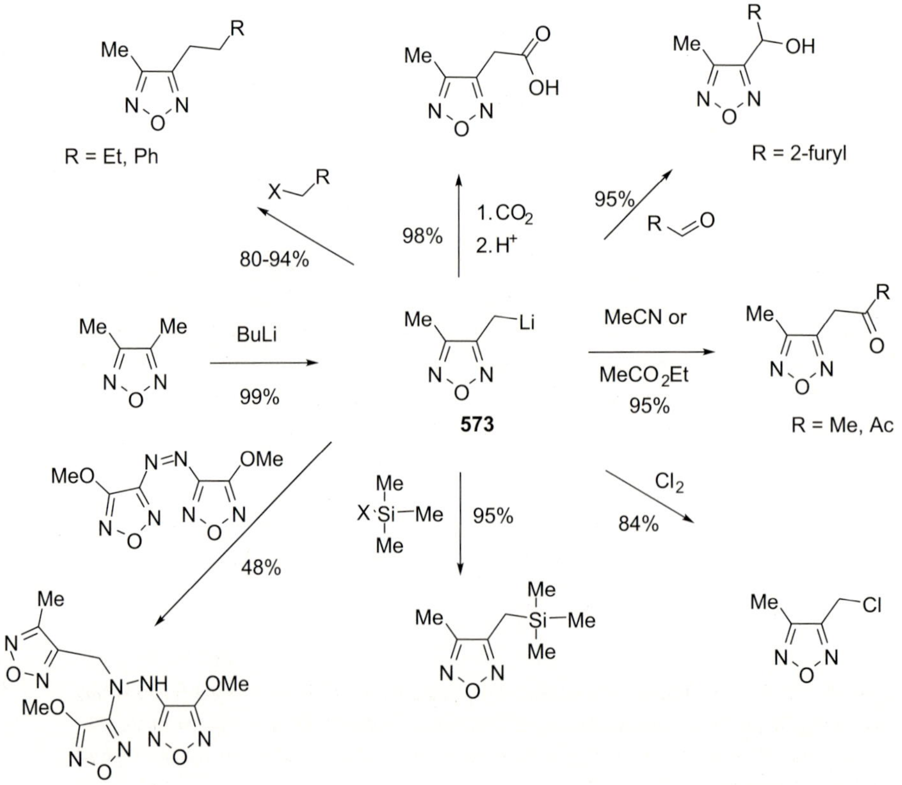

Scheme 13.193

can be an alkyl halide, a chlorosilane, a carbonyl compound, a nitrile, or an ester, an azo compound, and chlorine.

A similar procedure using two equivalents of BuLi and two equivalents of the electrophile offers access to α,α'-difunctionalized derivatives.

13.4.6 Furazans, Furoxans, and Benzo-Related Compounds in Medicine

Furazans, benzofurazans, and in particular furoxans and benzofuroxans are very important bioactive compounds. They have shown anti-microbial, anti-parasitic, anti-viral; mutagenic, anticancer and immunosuppressive, anti-aggregating, and vasorelaxant activity. Moreover, compounds containing the furoxan or benzofuroxan moiety inserted in a classical active principle have produced hybrid compounds that have been used, very recently, as new anti-ulcer drugs, calcium channel modulators, and vasodilatators. Several furoxan and furazan derivatives have been evaluated as antibacterial (Gram-negative and Gram-positive), antiprotozoal (*Trichomonas vaginalis* and *Entamoeba histolytica*), and antifungal compounds. 4,7-Dicyanobenzofurazan (**574**) presents a bacteriostatic effect in *Escherichia coli*, due to inactivation of 2,3-dihydroisovalerate [487]. 3-Nitro-4-phenylfuroxan (**575**) and its tautomer **576** displayed anti-infective properties, but with mutagenic activity; 3-bromo-4-phenylfuroxan (**577**) shows strong antimicrobial activity [488].

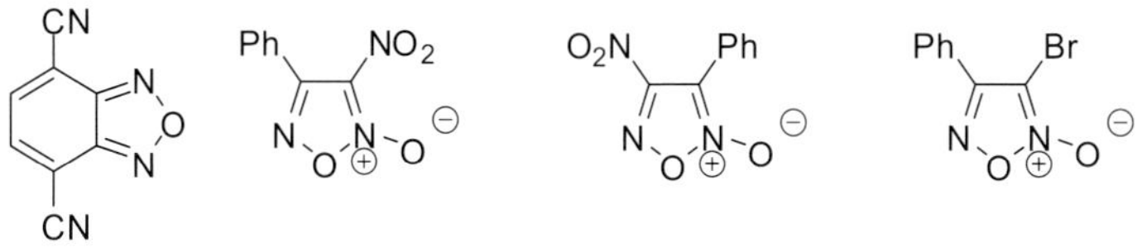

574 **575** **576** **577**

Furoxans and benzofuroxan **578–580** have been reported to inhibit *in vitro* the growth of *Trypanosoma cruzi*, the etiologic agent of Trypanosomiasis americana, the so-called Chagas' disease, and their activity is in the order **580** > **579** > **578**.

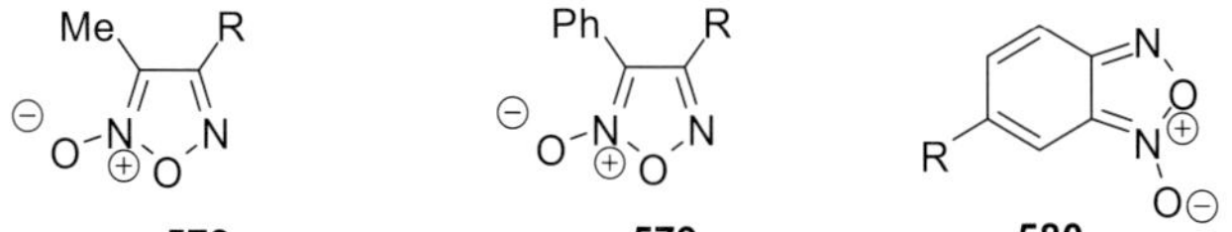

578 **579** **580**

R = CH=NNHCONH-butyl

Anti HIV-1 reverse transcriptase activity has been described for compounds **581** and **582**. In particular, compound **582** has shown the best anti-viral activity with a selectivity index (ratio of cytotoxic concentration to effective concentration) ranked in the order of **582** > **581**.

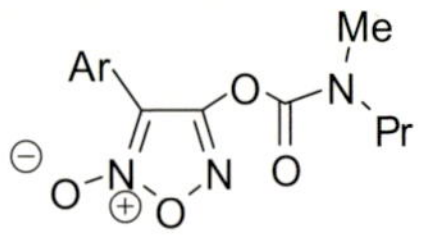

581 Ar = Ph
582 Ar = 2,6-di-Cl-Ph

4-Nitro (**583**), 4-thio (**585**) or 4-phenoxy- (**587**) benzofurazans and 4-nitro- (**584**), 7-thio (**586**) or 7-phenoxy (**588**) benzofuroxans present optimal drug activity as inhibitors of RNA synthesis in sheep lymphocytes [489].

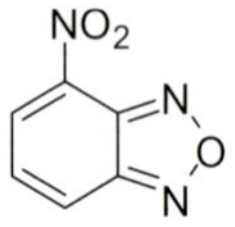

583 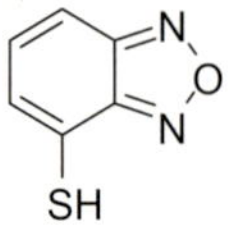 **585** 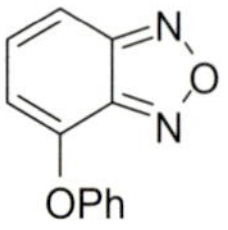 **587**

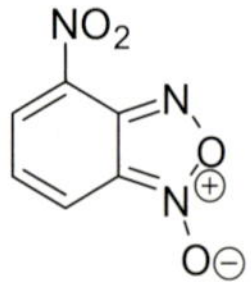

584 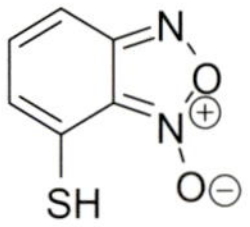 **586** 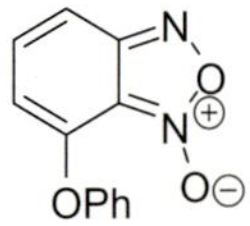 **588**

Since the first report indicating that some nitrobenzofurazans displayed anti-leukemic properties, numerous 7-nitro-2,1,3-benzoxadiazole derivatives have been evaluated as anticancer agents. In particular, 7-nitro-2,1,3-benzoxadiazoles such as **589** [R_1 = alkyl, cycloalkyl, alkenyl, cycloalkenyl, alkynyl, cycloalkynyl, aryl, etc.; X = O, S] have been prepared recently and used as agents able to inhibit glutathione S-transferase (GST). These compounds are useful in the production of pharmaceutical drugs to be used in anticancer therapy, and may be employed either alone or in combination with other chemotherapeutic agents. Thus, 4-[(7-nitro-2,1,3-benzoxadiazol-4-yl)sulfanyl]butanol **590**, prepared by reacting 4-chloro-7-nitro-2,1,3-benzoxadiazole with 4-mercapto-1-butanol in EtOH and potassium phosphate buffer, has shown a relevant activity against different cancer cell lines, such as K562 human myeloid leukemia, HepG2 human hepatic carcinoma, CEM1.3 human T-lymphoblastic leukemia, and GLC-4 human small cell lung carcinoma [490].

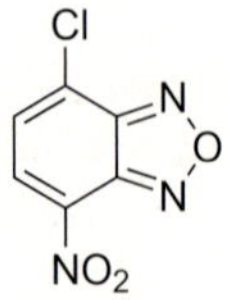

589 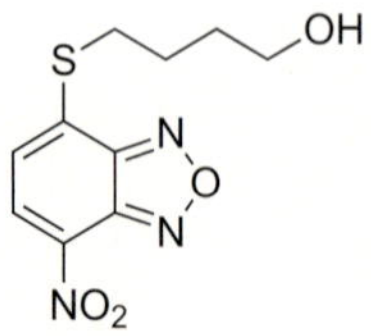 **590**

Combretafurazan (**592**), obtained from combretastatin A-4 (**591**), an antitumoral and antitubulin agent that is active only in its cis configuration, has shown to be a potent *in vitro* cytotoxic compound compared to combretastatin in neuroblastoma cells, while maintaining a similar structure–activity relationship and pharmacodynamic profiles [492].

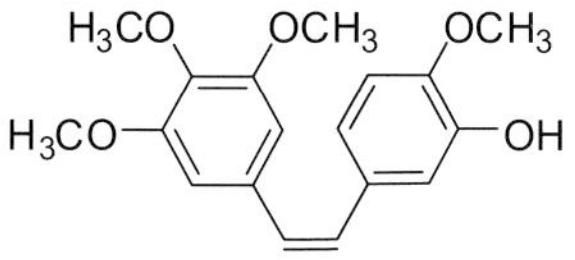

Combrestatin A-4 **(591)** 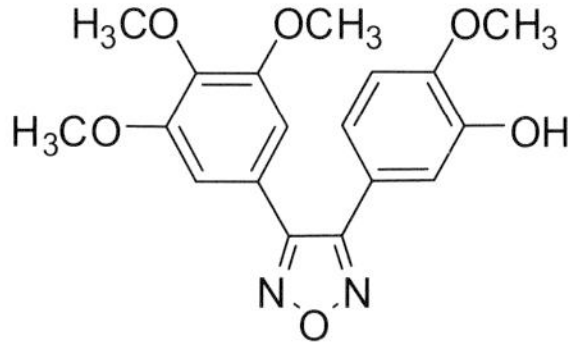 Combretafurazan **(592)**

Recently, another class of furazans, and in particular the furazano[3,4-*b*]pyrazines **593** have been prepared and used as antitumoral agents. Their activity is not limited to sarcomas, melanomas, neuroblastomas, carcinomas (including but not limited to lung, renal cell, ovarian, liver, bladder, and pancreatic carcinomas), and mesotheliomas. Moreover, specific assays, conduced for compound **594**, have demonstrated that it exhibited an IC_{50} of 0.00834 nM against sarcoma tumors [492].

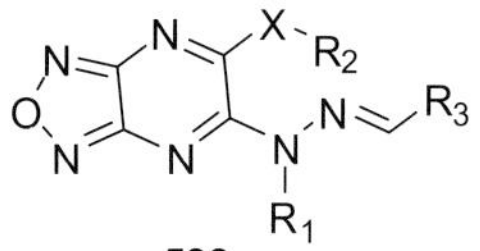

593 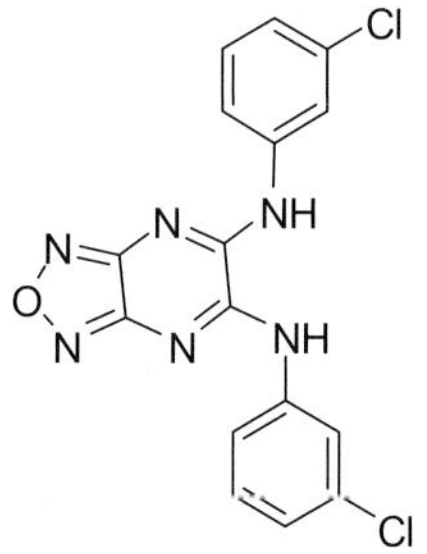 **594**

X = O, S, NH R_1 = H, Alkyl

R_2 = H, Aryl, Heteroaryl, alkyl

R_3 = H, Ary, Heteroaryl, alkyl

Some furazanobenzimidazoles **595** (R = aryl, haloaryl, etc.; R_1, R_2 = H, alkyl, cycloalkyl, etc.; R_3, R_4, R_5, R_6 = H, alkyl, haloalkyl, cycloalkyl, etc.; X = O, C:Y; Y = O, NOH, etc.) and their salts have been synthesized as apoptosis inducers for the treatment of neoplastic and autoimmune diseases. In particular, compound **596** exhibits a strong apoptotic activity in the Hoechst 33342 nuclear staining assay [493].

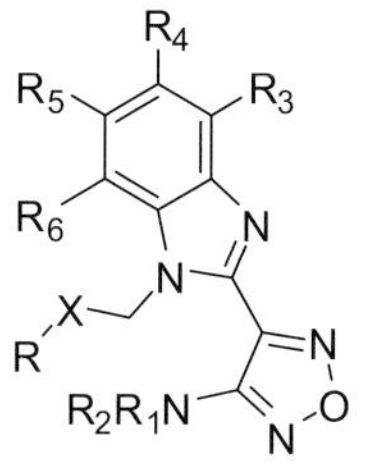

595 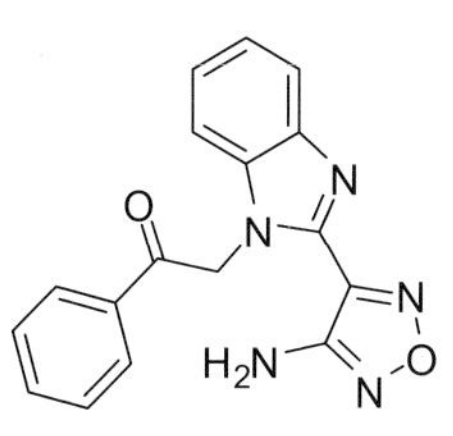 **596**

It has also been found that azabenzimidazoles **597** – in which R_1 is H or C_{1-6} alkyl; R_2 is halo or optionally substituted Ph, heteroaryl, or carboxamide; R_3 is halo, (un) substituted C_{1-6} alkoxy, (un)substituted phenoxy, heteroaryloxy, or heterocyclyloxy – are inhibitors of Rho-kinases. Rho-kinase is implicated in the phosphorylation of myosin light chain downstream of Rho, which is thought to induce smooth muscle contraction and stress fiber formation in non-muscle cells [494].

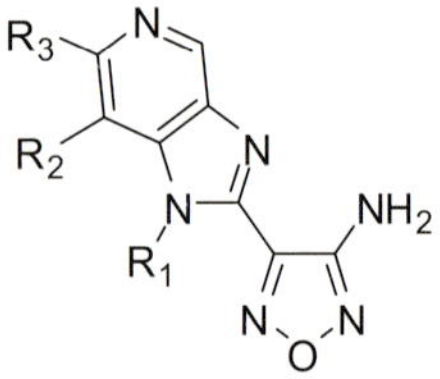

597

Moreover, compound **598** is useful for the treatment of diseases such as hypertension, heart failure, and ischemic angina [495].

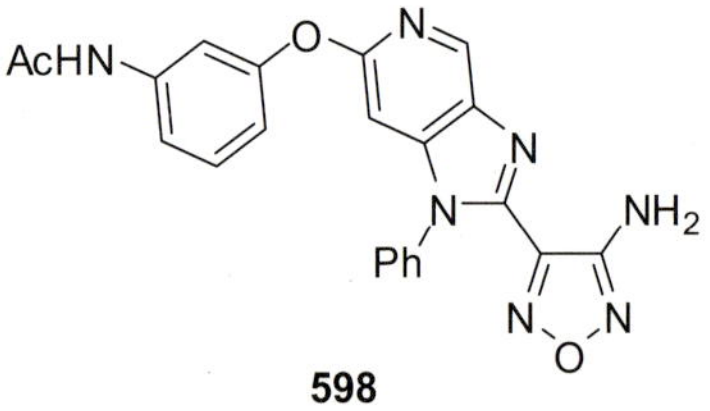

598

Many other furoxans and furazans have been synthesized to improve their cytotoxic activity; their biological assays have established that the furazan analogues are, usually, less active than the corresponding furoxans. These results indicate the relevance of N-oxide in terms of the bio-response.

One of the most interesting pharmacological properties of furoxans and benzofuroxans is the nitric oxide (NO) releasing capacity. NO displays diverse potent physiological actions. As regards the cardiovascular system, it plays a crucial role in vascular homeostasis through several mechanisms, including vasodilation, inhibition of platelet aggregation, and modulation of platelet and leukocyte adherence. In the central nervous system, it plays roles in learning and memory formation. In the peripheral nervous system, it regulates several gastrointestinal, genitourinay, and respiratory functions as neurotransmitter at the endings of nonadrenergic, noncholinergic nerves. NO is also potentially toxic and can induce genomic alterations.

Figure 13.5 gives some examples of these drugs.

13.5 1,3,4-Oxadiazoles

1,3,4-Oxadiazole (**599**) is a partially aromatic and thermally stable molecule [496]. Exocyclic-conjugated mesoionic 1,3,4-oxadiazoles (**600**), 1,3,4-oxadiazolium cations

Vasodilating agents:

Human platelet SGS activators:

Hybrid Antiulcer agents:

Figure 13.5 Examples of drugs that contain a furoxan moiety.

(**601**), and 1,3,4-oxadiazolines (**602**) are also stable molecules. The partially and fully reduced systems designated as 4,5-dihydro-(Δ^2) (**603**), 2,5-dihydro-(Δ^3) (**604**), and 2,3,4,5-tetrahydro-1,3,4-oxadiazole (**605**) are also known.

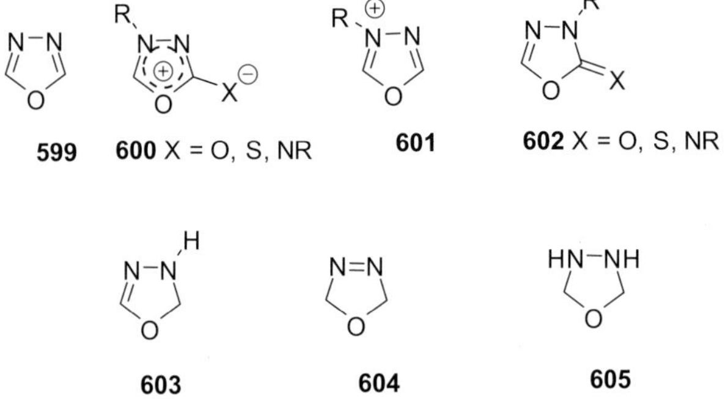

1,3,4-Oxadiazoles are of great practical importance. In particular, these compounds are used in medicine, as leprostatics, tuberculostatics, antibacteric, antiproteolytic, and anticonvulsants. They also possess analgesic, antipyretic, antiphlogistic, bactericides, insecticides, fungicidal, and several other biological activities [496]. More recently, compounds containing the 1,3,4-oxadiazole motif have been used as HIV integrase and angiogenesis inhibitors [497]. 1,3,4-Oxadiazoles have also been used in agriculture, in the production of polymers, laser dyes, photographic materials, or scintillators. Furthermore, they show a combination of interesting properties, which makes them suitable for the development of new electrical and electro-optical devices.

A good number of reviews on the chemistry of the 1,3,4-oxadiazoles are present in the literature [496]: the most recent report covers the literature up to the early part of 2007.

13.5.1 Structure

13.5.1.1 Theoretical Aspects

1,3,4 Oxadiazole is not fully aromatic; it has an aromaticity index of 50 (43 and 66 for furan and thiophene, respectively), while the bond orders for O–C, C–N, and N–N bonds are 1.3124, 1.9062, and 1.3348, respectively.

Theoretical studies on the structure and properties of 1,3,4-oxadiazoles are numerous [496]. In particular, MNDO and STO-3G *ab initio* methods have been used to calculate the proton affinities; CNDO/2 methods have been used to calculate the total energies, ionization potentials, and net atomic populations; INDO/S has been used to calculate the electron distribution of ground and excited states. Diels–Alder reactions of several 1,3,4-oxadiazoles used as dienes with alkenes have been investigated through molecular orbital calculations at the B3LYP/6-31G(d)AM1 theory level [498]. MNDO-PM3 calculations have been used to calculate the geometry and electronic structure of mesoionic 1,3,4-oxadiazolium-2-aminides **606** (R = 4-MeO-3-$O_2NC_6H_3$, R^1 = Me, Ph; R = 4-Cl-3-$O_2NC_6H_3$, R^1 = Me; R = Me, R^1 = Ph) and 1,3,4-oxadiazolium-2-olates **607** (Ar = 4-Cl-3-$O_2NC_6H_3$, 4-MeO-3-$O_2NC_6H_3$) [499].

606 **607**

2-Hydroxy, 2-amino and 2-thiol derivatives **608** are in tautomeric equilibrium with Δ^2-1,3,4-oxadiazolin-5-ones, 5-imino, and 5-thiones **609**, respectively. Usually, one of the forms distinctly predominates.

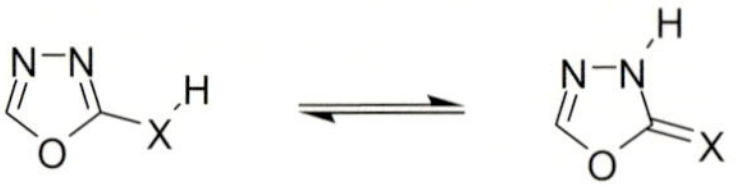

608 **609**

X = O, NH, NR, S

Most of 1,3,4-oxadiazoles are solids, apart from the parent compound (bp 150 °C) and its lower alkyl derivatives, which are liquids. Some of them are soluble in water, with a solubility that decreases with increasing molecular weight.

13.5.1.2 **Structural Aspects**

13.5.1.3 **X-Ray Diffraction**

Many papers related to the X-ray structures of 1,3,4-oxadiazoles have been reported [500]. The oxadiazole ring has a nearly flat structure: all the atoms of the ring lie in the same plane with very slight deviation from it.

Table 13.8 shows the reported bond lengths and bond angles for 2,5-di(4-pyridyl)-1,3,4-oxadiazole **610** and for 2-(4-cyanophenyl)-5-(4-dimethylaminophenyl)-1,3,4-oxadiazole **611**.

In the crystal, compound **610** has an almost planar structure. All three rings of the molecule are planar but they show a slight torsion relative to each other. The deviation of the planes of the two pyridyl rings relative to the plane of the oxadiazole ring is +3.3° for one ring and −3.4° for the other. Owing to the absence of significant steric hindrance the torsion angle between the neighboring rings is small and, therefore, the conjugation is not lost. Compound **611** is almost planar. The planarity of the three single rings is nearly perfect but again the rings show a slight torsion relative to each other. The rotation of the inter-ring bond between the oxadiazole ring and the benzonitrile is +6.5° and between the oxadiazole ring and the dimethylaniline is +4.2. In this molecule the two substituents are tilted in the same direction relative to the oxadiazole ring whereas in **610** the rings are tilted in opposite directions [501].

Table 13.8 Molecular dimensions for 2,5-di(4-pyridyl)-1,3,4-oxadiazole (**610**) and for 2-(4-cyanophenyl)-5-(4-dimethylaminophenyl)-1,3,4-oxadiazole (**611**).

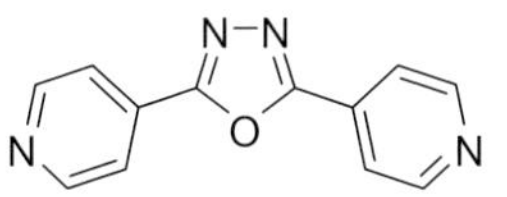

610 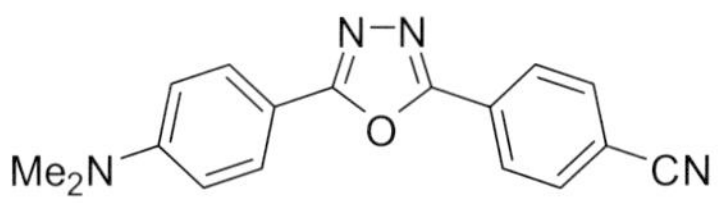 **611**

Compound 610				Compound 611			
Bond lengths (nm)		Bond angles (°)		Bond lengths (nm)		Bond angles (°)	
O1−C2	0.1365	O1−C2−N3	112.50	O1−C2	0.1368	O1−C2−N3	111.36
C2−N3	0.1292	C2−N3−N4	106.20	C2−N3	0.1292	C2−N3−N4	106.97
N3−N4	0.1409	N3−N4−C5	106.20	N3−N4	0.1407	N3−N4−C5	106.13
N4−C5	0.1292	N4−C5−O1	112.50	N4−C5	0.1294	N4−C5−O1	112.36
C5−O1	0.1365	C2−O1−C5	102.50	C5−O1	0.1374	C2−O1−C5	102.82

Table 13.9 Proton NMR data (ppm) for ring hydrogens of 1,3,4-oxadiazoles.

R	Solvent	δ
H	$CDCl_3$	8.73
Me	$CDCl_3$	8.53
Et	$CDCl_3$	8.48
$PhCH_2$	$CDCl_3$	8.26
Ph	$CDCl_3$	8.50
MeS	d_6-DMSO	9.42

13.5.1.4 **NMR Spectroscopy**

The ^{1}H NMR spectrum of the parent compound **599** shows the relative signals at 8.73 δ in $CDCl_3$ [496, 502]. The presence of alkyl groups or phenyl group moves the proton of the ring upfield, while the shift is downfield for 2-alkylthio derivatives (Table 13.9).

The ^{13}C chemical shift for C2, or C5 carbon in the parent compound is centered at 152.1 ppm. The presence of a phenyl ring at C2 moves this carbon to 164 δ [502]. The chemical shifts of the ring carbon atoms in several 1,3,4-oxadiazoles have also been reported. For example, in 2-methoxy-1,3,4-oxadiazole the C2 signal is shifted downfield in comparison with signal of C5. The same trend has been observed for the oxadiazolinone and oxadiazolinethione derivatives where the C2 carbons resonate downfield with respect to C5 carbon [496]. Recently the structure of some 2,5-disubstitued-1,3,4-oxadiazoles has been elucidated by spectral (IR, ^{1}H NMR, ^{13}C NMR) analysis. The ^{13}C NMR analysis revealed that the presence of alkyl groups attached to C2 and C5 of the ring induced a downfield shift of both carbons by at least about 20–22 ppm in comparison with the relative signal present in **599** [503].

The ^{15}N and ^{17}O data of different 1,3,4-oxadiazoles have also been used to elucidate their structures. In particular, the ^{17}O resonances registered for the 1,3,4-oxadiazolium-2-olate have demonstrated that the value centered at 181 ppm, relative to exocyclic oxygen, is that expected for an enolate form instead of that for a carbonylic function [496].

13.5.1.5 **UV and IR Spectroscopy**

The electronic spectrum of the 1,3,4-oxadiazole system is equivalent to that of benzene and the maxima are only slightly hypsochromically shifted. Substituted 2,5-diaryl- derivatives show strong fluorescence in solution on stimulation by UV or β-irradiation, and some of them are electroluminescent with irradiation of blue light [504]. The electronic effects of conjugated rings on 1,3,4-oxadiazoles are maintained and the absorption and emission spectra of these compounds have been extensively studied and reported [499, 505].

The IR absorption spectra for 1,3,4-oxadiazoles show bands at 1640–1560 (ν_{CN}), 1030–1020 (ν_{CO}), and 970 cm^{-1} [496, 503, 506]. These bands occur at longer wavelengths in the spectra of 2,5-dialkyl-1,3,4-oxadiazoles and at shorter values in 5-thione derivatives [496]. A band in the range 1785–1740 cm^{-1} is reported for C=O absorption in the case of oxazolidin-5-ones [496].

13.5.1.6 Mass Spectrometry

The electron impact mass spectra of most 1,3,4-oxadiazoles exhibit a very intense signal for the molecular ion [496, 503, 506]. Moreover, the predominant fragment is represented by R-C = $O^{+\bullet}$. In the case of 2-substituted 1,3,4-oxadiazoles, diagnostic fragments derive from loss of CO and HCO. Loss of HNCO is fundamental in the spectrum of 2-amino-5-phenyl-1,3,4-oxadiazole. Oxazolidin-5-ones easily lose CO_2 to give the corresponding ions of general formula R-C=N=NH^{+}.

A study regarding the electron-spray ionization mass spectra of 2,5-diaryl and 2-arylamino-5-aryl-1,3,4-oxadiazoles together with their complexes with copper cations has been reported [507]. In this latter case, loss of NH_3 and HNCO was observed. In some protonated 2,5-diaryl derivatives an unusual elimination of HNCO was also detected [508].

13.5.2 Synthesis of 1,3,4-Oxadiazoles

The common synthetic routes to these compounds involve:

1) cyclization of diacylhydrazines with various anhydrous reagents such as $BF_3{\cdot}OEt_2$ [503], thionyl chloride [509], phosphorous pentoxide [510], phosphorous oxychloride [511], triflic anhydride [512], triphenylphosphine [513], polyphosphoric acid [514], and sulfuric acid [515];
2) cyclization of acylhydrazones [516], semicarbazones, and thiosemicarbazides;
3) ring transformations [517].

13.5.2.1 Cyclization of Diacylhydrazines

The first synthesis of this ring, reported by Robert Stolle, exploits a condensation reaction of *N,N'*-diacid hydrazides **612** under vigorous conditions to produce in variable yields 2,5-diaryl(alkyl)-1,3,4-oxadiazoles **613** (Scheme 13.194) [518].

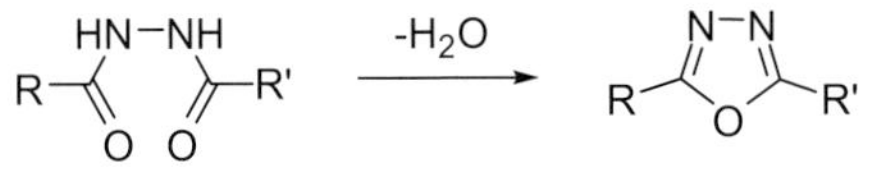

HN−NH -H_2O N−N
R R' → R O R'
O O
612 **613**

Scheme 13.194

Useful as agricultural fungicide, Boesch has reported the synthesis of 2-cyanooxadiazole **617** by cyclocondensation of ethyl 2-(2-benzoylhydrazinyl)-2-oxoacetate **614** with P_2O_5 followed by NH_3 treatment and dehydration with $POCl_3$ (Scheme 13.195) [519].

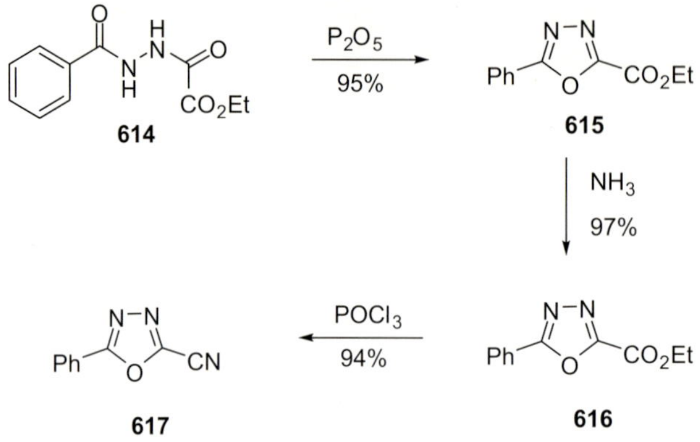

Scheme 13.195

Using hot polyphosphoric acid (PPA) 5-substituted [1,3,4]oxadiazol-2-yl compounds **619** have been obtained in good yield (75–95%) (Scheme 13.196).

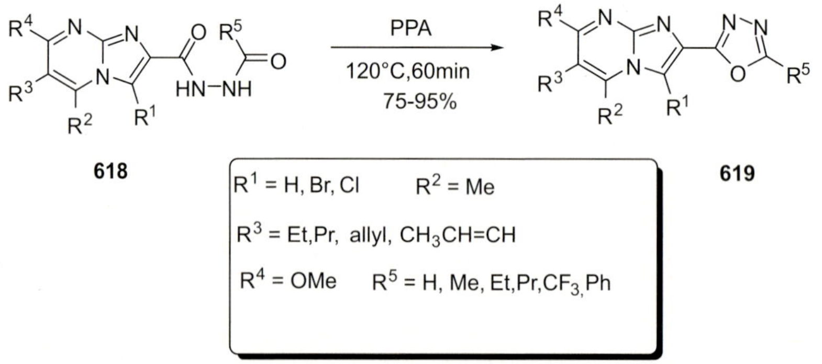

Scheme 13.196

The 2-(oxadiazolyl)imidazo[1,2-*a*]pyrimidines (**619**) thus obtained are a class of compounds that bind to benzodiazepine receptors with moderate to weak affinity, and yet display antianxiety properties of similar potency to chlordiazepoxide in animal models, while demonstrating reduced or negligible myorelaxant effects [514].

A polymer-supported Burgess reagent under microwave conditions has been efficaciously used for the cyclodehydration of 1,2-diacylhydrazines **620** to provide 1,3,4-oxadiazoles **621** in excellent yields (Scheme 13.197) [520].

A convenient, one-pot procedure has been reported by Mashraqui and coworkers for the synthesis of various 2,5-disubstituted-1,3,4-oxadiazoles **625** by condensing mono-aryl hydrazides **622** with acid chlorides **623** in HMPA solvent under microwave heating (involving as intermediates diaroylhydrazines **624**) (Scheme 13.198) [521].

The yields are good to excellent; the process is rapid and does not need any added acid catalyst or dehydrating reagent.

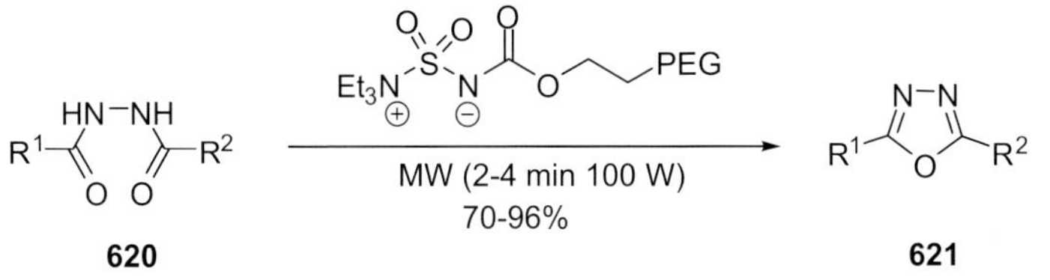

R^1	R^2	Yield	HPLC Purity (%)
Ph	Ph	96	91
2-Methoxyphenyl	Me	89	>99
2-Chlorophenyl	Me	70	97
2-Nitrophenyl	Me	95	>99
2-Tyhiophenyl	Ph	95	97
3-Pyridyl	Ph	95	>99
4-Pyridyl	NHPh	95	>99
4-Chloro-3-Nitroamminophenyl	Ph	90	92

Scheme 13.197

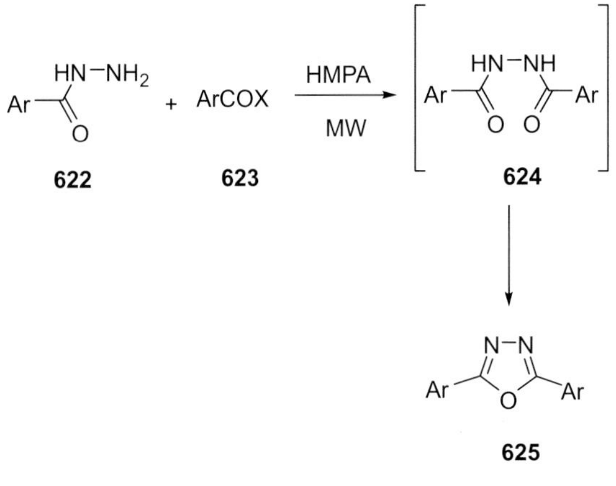

Scheme 13.198

13.5.2.2 **Cyclization of Acylhydrazones, Semicarbazones, and Thiosemicarbazides**
Mono and disubstituted 1,3,4-oxadiazoles **629** have been prepared by oxidation of acylhydrazones **628** prepared *in situ* by the condensation of aryl carboxylic acid hydrazides **626** with orthoesters **627** (Scheme 13.199). In two examples, the 1-acyl-2-ethoxymethylenehydrazine **628** intermediate was isolated [522]. This reaction has been used to prepare the parent 1,3,4-oxadiazole (**599**) [523].

The above reaction has been revised recently by Varma *et al.*, who have used a green protocol to synthesize 1,3,4-oxadiazoles **629** [502]. In particular, various

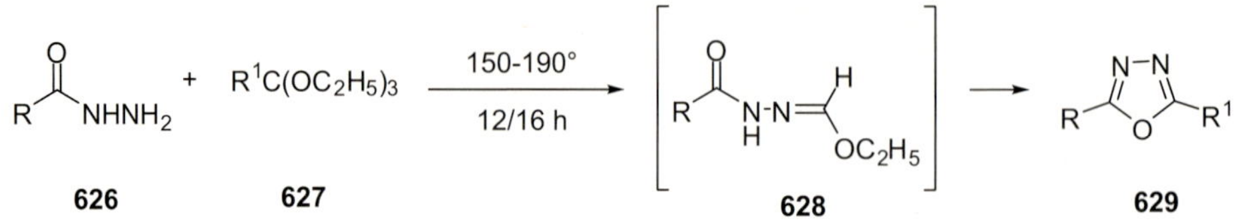

R	R^1	Yield (%)
Phenyl	H	70
o-Methoxyphenyl	H	79
p-Chlorophenyl	H	78
p-Nitrophenyl	H	79
α-Naphthyl	H	63
4-Pyridyl	H	82
3-Pyridyl	H	68
2-Quinolyl	H	70
Phenyl	Me	70
4-Pyridyl	Me	70
Phenyl	Et	80
o-Methoxyphenyl	Et	79
p-Chlorophenyl	Et	86
p-nitrophenyl	Et	92
α-Naphthyl	Et	80
4-Pyridyl	Et	84

Scheme 13.199

hydrazides **626** have been reacted with triethyl orthoalkanates or triethyl orthobenzoate (**627**), in the presence of Nafion NR50, under microwave irradiations and in the absence of any solvents to afford the desired 1,3,4-oxadiazoles **629** in good yields (68–90%) (Scheme 13.200).

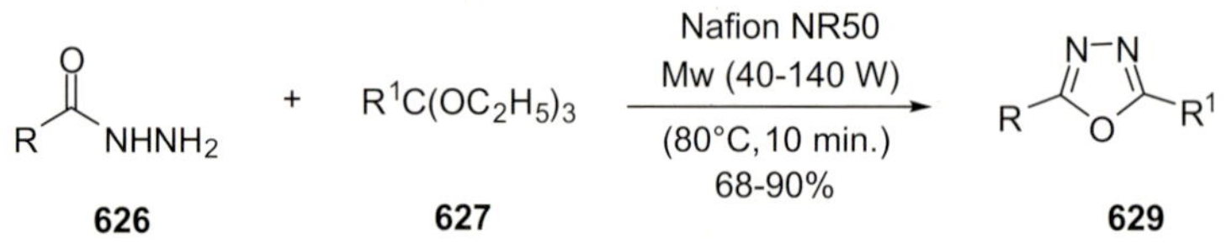

R = Ph, p-F-C_6H_4, p-OMe-C_6H_4, 2-furyl, 2-thienyl, 4-pyridyl, $PhCH_2$
R^1 = H, Et, Ph

Scheme 13.200

Owing to the selective absorption of catalyst, the reaction rate is strongly accelerated (10 min). Moreover, Nafion NR50 is easy to handle because it involves a simple addition of Nafion beads in a reaction vessel, which can be physically removed by forceps after completion of the reaction.

Electrolytic oxidation of ketone *N*-acylhydrazones **630** and aldehyde *N*-acylhydrazones **631** in methanolic sodium acetate affords, through their intramolecular cyclization, the corresponding 2-methoxy-Δ^3-1,3,4-oxadiazolines **632** and oxadiazoles **633**, respectively (Scheme 13.201) [524].

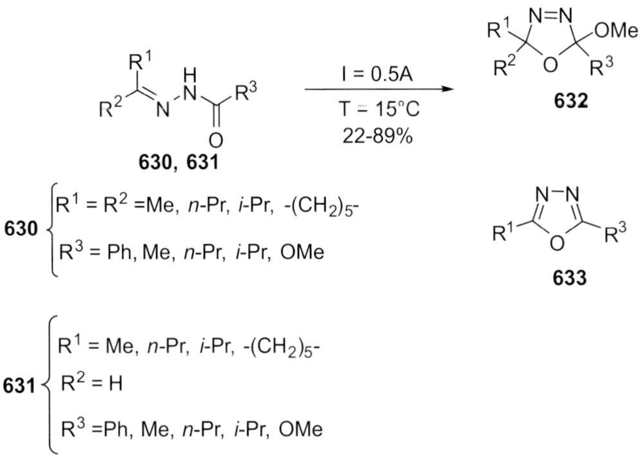

630, 631 → (I = 0.5A, T = 15°C, 22-89%) → **632**, **633**

630: $R^1 = R^2$ = Me, *n*-Pr, *i*-Pr, -$(CH_2)_5$-; R^3 = Ph, Me, *n*-Pr, *i*-Pr, OMe

631: R^1 = Me, *n*-Pr, *i*-Pr, -$(CH_2)_5$-; R^2 = H; R^3 = Ph, Me, *n*-Pr, *i*-Pr, OMe

Scheme 13.201

The formation of heterocycles **632** and **633** has been rationalized according to a three-step process. The first step involves the formation of a cationic intermediate generated from **630** or **631** by the loss of two electrons and one proton. In the second step, a 1,3,4-oxazolidinyl carbocation (**634**) is formed by an intramolecular cyclization promoted by the oxygen of the carbonyl group; and the third step consists of attack by methanol followed by expulsion of a hydrogen in the case of compound **630** or hydrogen extrusion in the case of **631** (Scheme 13.202).

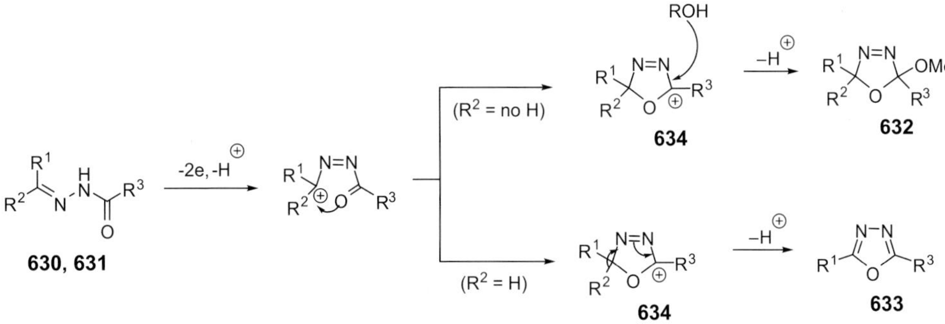

630, 631 → (-2e, -H⁺) → (R^2 = no H) **634** + ROH → ($-H^+$) **632**; (R^2 = H) **634** → ($-H^+$) **633**

Scheme 13.202

Another method of oxidation of *N*-acylhydrazones **635** and **636** involves a hypervalent iodine reagent. Thus, Dai *et al.* have reported the synthesis of 2-alkoxy-Δ^3-1,3,4-

oxadiazolines **639** and 2,5-disubstitued-1,3,4-oxadiazoles **641** in good to excellent yield by means of phenyl-iodine(III) diacetate (Scheme 13.203). The yields in 1,3,4-oxadiazoles was improved by the use of 2 mmol of NaOAc [525].

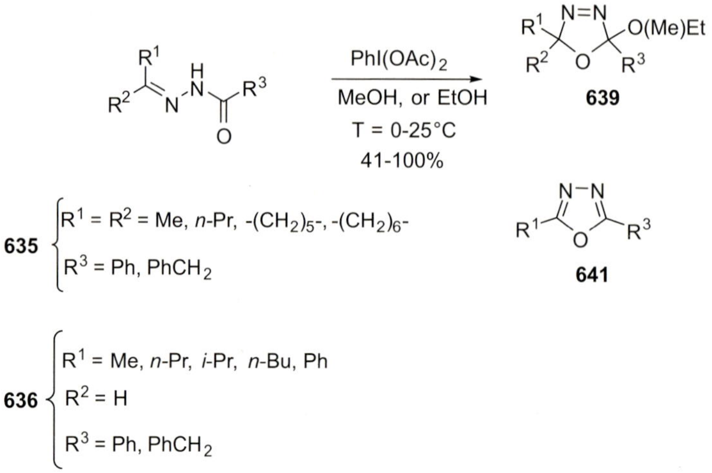

$PhI(OAc)_2$, MeOH, or EtOH, T = 0-25°C, 41-100%

635 $R^1 = R^2$ = Me, *n*-Pr, $-(CH_2)_5-$, $-(CH_2)_6-$; R^3 = Ph, $PhCH_2$

636 R^1 = Me, *n*-Pr, *i*-Pr, *n*-Bu, Ph; R^2 = H; R^3 = Ph, $PhCH_2$

639, **641**

Scheme 13.203

Scheme 13.204 explains the formation of these compounds.

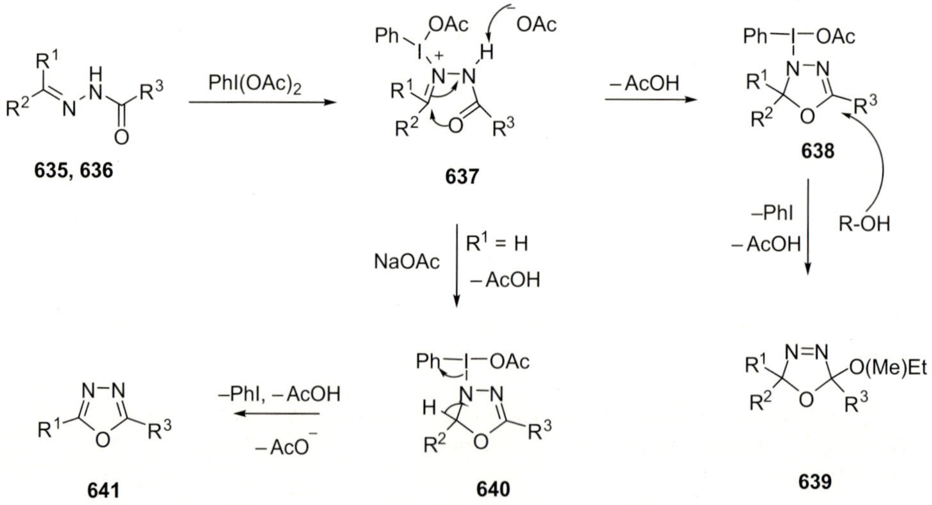

635, 636 → $PhI(OAc)_2$ → **637** → −AcOH → **638** → −PhI, −AcOH, R-OH → **639**

637 → NaOAc, R^1 = H, −AcOH → **640** → −PhI, −AcOH, $-AcO^-$ → **641**

Scheme 13.204

More recently, starting from 1-aroyl-2-arylidene hydrazines **644**, a microwave assisted synthesis of 2,5-disubstituted 1,3,4-oxadiazoles **645** has been reported using as oxidizing agent potassium permanganate supported by montmorillonite K10. The reaction was more efficient when acetone/H_2O (20 : 5) was used as solvent, and occurs in only 10 min, with a yield in the range 59–100%. Interestingly, the starting hydrazines **644** have been prepared in good yields under microwave irradiation, mixing in ethanol 6 mmol of acid hydrazide **642**, 6 mmol of aldehyde **643**, and three drops of phosphoric acid (Scheme 13.205) [526].

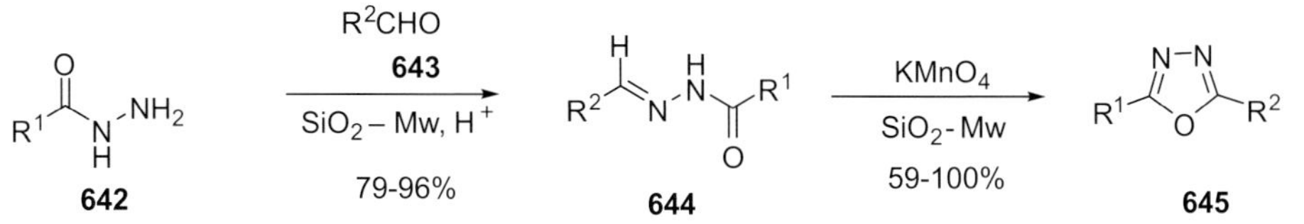

642 → (R^2CHO **643**, SiO_2 – Mw, H^+, 79-96%) → **644** → ($KMnO_4$, SiO_2 - Mw, 59-100%) → **645**

R^1 = Me, Ph, 4-Cl-C_6H_4

R^2 = Ph, 4-NO_2-C_6H_4, 3-NO_2-C_6H_4, 4-Cl-C_6H_4, 3-Cl-C_6H_4, Me, 4-I-C_6H_4, 4-Br-C_6H_4, 4-Cl-C_6H_4, 4-Me-C_6H_4, 4-MeO-C_6H_4, 4-CN-C_6H_4, 4-Cl-C_6H_4, 4-MeOOC-C_6H_4, 4-$(NMe)_2$-C_6H_4,

Scheme 13.205

Semicarbazones have also been used for the synthesis of 2-amino-1,3,4-oxadiazoles, through their cyclization performed with a mixture of sodium acetate, bromine, and glacial acetic acid. In fact, this procedure was used recently to prepare some Schiff bases of 2-amino-5-aryl-1,3,4-oxadiazoles (**649a–t**) that posses antibacterial activities. In particular, semicarbazones **646** have been reacted with a mixture of sodium acetate, bromine, and glacial acetic acid, and transformed into the corresponding 2-amino-5-aryl-1,3,4-oxadiazoles **647** that in turn have been condensed with aldehydes **648** to give the expected 1,3,4-oxadiazole derivatives **649a–t** (Scheme 13.206). The antibacterial properties of the compounds were investigated against *Proteus mirabilis*, *Pseudomonas aeruginosa*, *Bacillus subtilis*, and *Staphylococcus aureus*. The most active compounds were **649c**,**649f**,**649m**, and **649q** with a MIC in the range 62–68 $\mu g\, ml^{-1}$. Antifungal activity against *Aspergillus niger* and *Candida albicans* were also found for compounds **649g,h,i,m.** The corresponding MICs are in the range 52–60 $\mu g\, ml^{-1}$. The biocidal activities of these compounds were attributed to the toxophoric C=N linkage [527].

The mixture of sodium acetate, bromine, and glacial acetic acid has also been used to prepare several antibacterial 1,2-bis(1,3,4-oxadiazol-2-yl)ethanes **651** from the corresponding diacylhydrazones **650** (Scheme 13.207) [528].

In particular, compounds **651c–e** show a good antibacterial activity against *Pseudomonas aeruginosa*, *Bacillus subtilis*, *Staphylococcus aureus*, and *Escherichia coli*, with a MIC of 6 $\mu g\, ml^{-1}$.

Another cyclization method towards the synthesis of 1,3,4-oxadiazoles involves the cyclodesulfurization of thiosemicarbazides **652** using either dicyclohexylcarbodii-

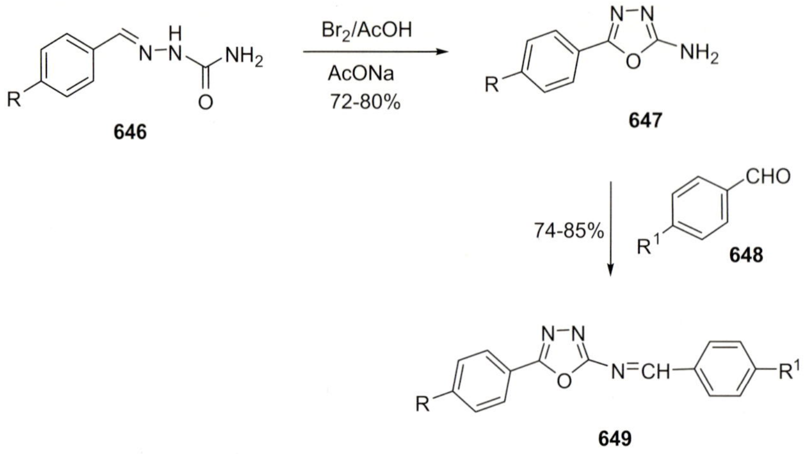

a: R =–OMe R^1 = –OMe	f: R =– NO_2 R^1 = –OMe	k: R =–Cl R^1 = –OMe	p: R =–Me R^1 = –OMe
b: R =–OMe R^1 = – NO_2	g: R =– NO_2 R^1 = – NO_2	l: R =–Cl R^1 = –NO_2	q: R =–Me R^1 = –NO_2
c: R =–OMe R^1 = –Cl	h: R =– NO_2 R^1 = –Cl	m: R =–Cl R^1 = –Cl	r: R =Me R^1 = –Cl
d: R =–OMe R^1 = –Me	i: R =– NO_2 R^1 = –Me	n: R =–Cl R^1 = –Me	s: R =–Me R^1 = –Me
e: R =–OMe R^1 = –OH	j: R =–NO_2 R^1 = –OH	o: R =–Cl R^1 = –OH	t: R =–Me R^1 = –OH

Scheme 13.206

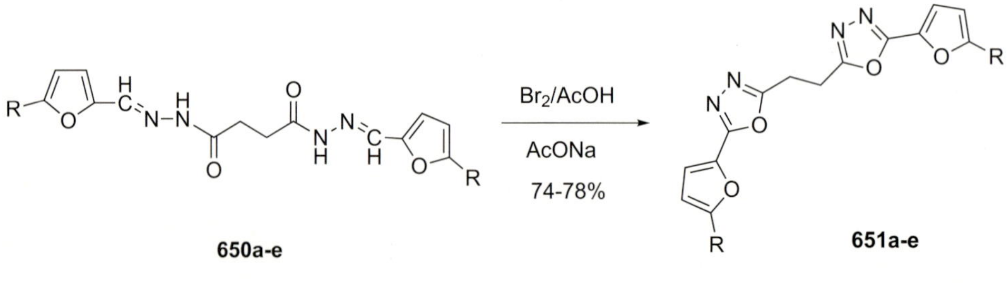

R = **a**: NO_2; **b**: *p*-nitrophenyl; **c**: *p*-chlorophenyl; **d**: *p*-bromophenyl; **e**: 2,4-dichlorophenyl

Scheme 13.207

mide (DCC), or a mixture of I_2/NaOH. By this procedure 2-amino-substituted-1,3,4-oxadiazoles **653**, having anti-inflammatory activity, have been synthesized (Scheme 13.208) [529]. The anti-inflammatory activity was investigated by determining the inhibitory effect of the oxadiazole derivatives **653a–p** on histamine-induced edema in rat abdomen. Compounds **653a,653c,653e,653j**, and **653n** proved to be more potent anti-inflammatory agents at 200 mg kg^{-1} p.o. than Ipobrufen, the standard reference drug.

2-Amino-5-aryl-1,3,4-oxadiazoles **656** have also been prepared, in good yield, by cyclodesulfurization of thiosemicarbazides **654** using 1,3-dibromo-5,5-dimethylhydantoin (**655**) as primary oxidant in the presence of potassium iodide (Scheme 13.209) [530].

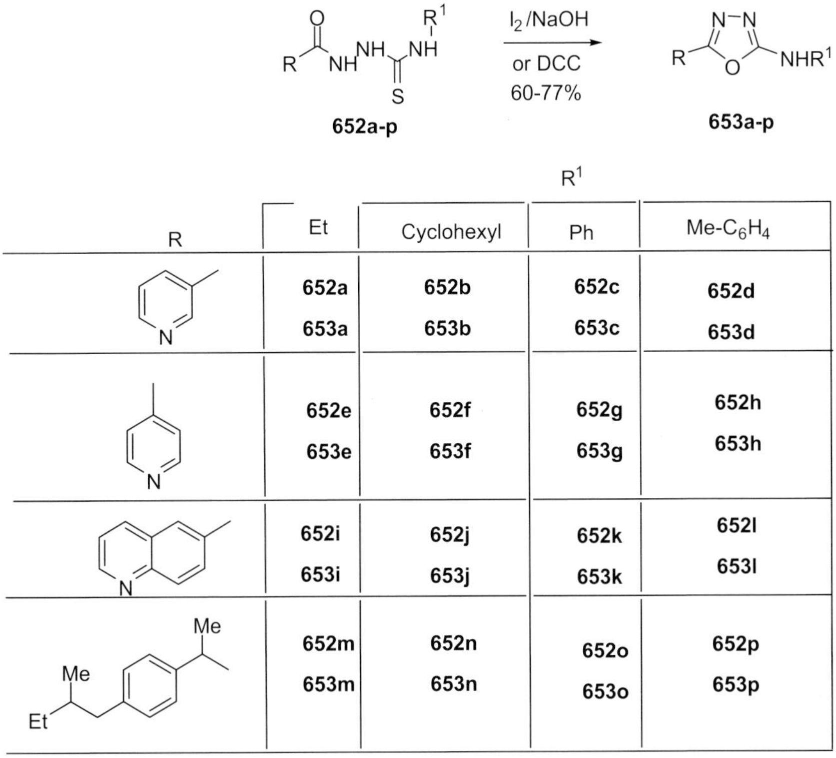

652a-p → 653a-p (I_2/NaOH or DCC, 60-77%)

R	R^1 = Et	Cyclohexyl	Ph	Me-C_6H_4
N	652a 653a	652b 653b	652c 653c	652d 653d
N	652e 653e	652f 653f	652g 653g	652h 653h
N	652i 653i	652j 653j	652k 653k	652l 653l
Me, Me, Et	652m 653m	652n 653n	652o 653o	652p 653p

Scheme 13.208

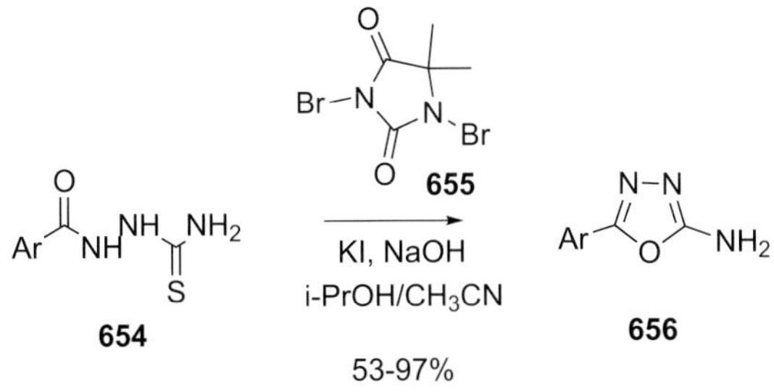

654 → 656 (655, KI, NaOH, i-PrOH/CH_3CN, 53-97%)

Ar: Ph, 4.Cl-Ph, 4-OMe-Ph, 2-Furyl, $PhCH_2CH_2$, PhCH=CH

Scheme 13.209

13.5.2.3 **Ring Transformations**

Thermal decomposition of *N*-acyl-tetrazoles **657** is a common way to obtain 1,3,4-oxadiazoles **659** [496]. This reaction has been easily explained by the loss of nitrogen, formation of the corresponding nitrilimines **658**, and an intramolecular 1,5-cycloaddition (Scheme 13.210).

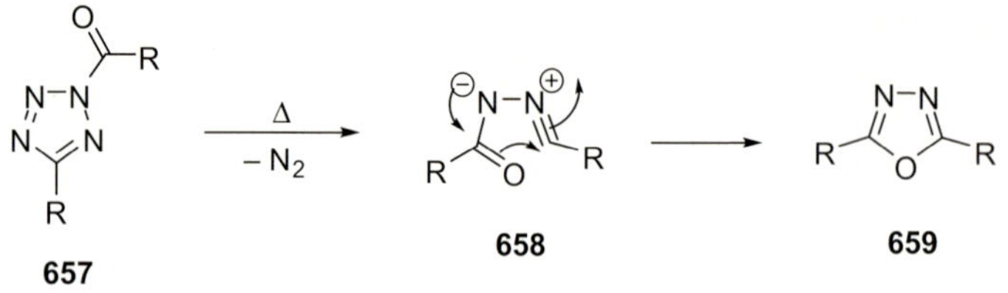

Scheme 13.210

The mechanism of the process was demonstrated by labeling with ^{15}N the atoms N1 and N4 in 5-phenyltetrazole. Half of the ^{15}N was found in the 2,5-diphenyl-1-3-4-oxadiazole, obtained by the breakdown of the *N*-benzoyl-5-phenyltetrazole. Therefore, either the atoms N1 and N2 or N3 and N4 were eliminated as N_2 [531]. Scheme 13.211 shows that the thermolysis of *N*-aroyl-5-phenyltetrazoles **660** affords oxadiazoles **661** [532].

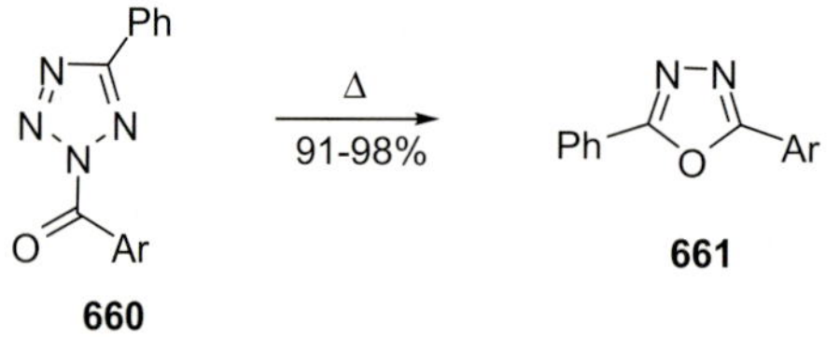

Ar = Ph, 2-Cl-C_6H_4; 2-Br-C_6H_4; 4-NO_2-C_6H_4; 4-Me-C_6H_4

Scheme 13.211

(5-Nonyl-1,3,4-oxadiazol-2-yl)benzothiazine dioxide **664**, a compound that shows anti-inflammatory properties by virtue of its inhibition of arachinodate 5-lipoxygenase [533], has been prepared starting from tetrazol-5-yl derivative **662**, which undergoes a Huisgen rearrangement on refluxing with decanoic anhydride **663** in toluene, followed by saponification (Scheme 13.212).

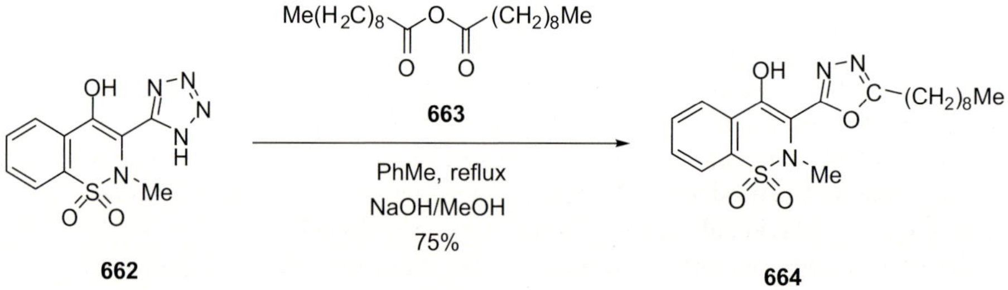

Scheme 13.212

By exploiting the thermal decomposition of alkoxycarbonyl tetrazoles, the 4-[5-(dipyrido[3,2-a:2′,3′-c]phenain-11-yl)-1,3,4-oxadiazol-2-yl]-N,N-diphenylaniline **668** has been prepared. Thus, the dipyrido[3,2-*a*,2′,3′-*c*]phenazine 8-carboxyl chloride **665** in dry pyridine was refluxed for 72 h with triphenylamine tetrazole **666** to afford **668** via the corresponding alkoxycarbonyl tetrazole intermediate **667**, in 60% yield (Scheme 13.213) [534].

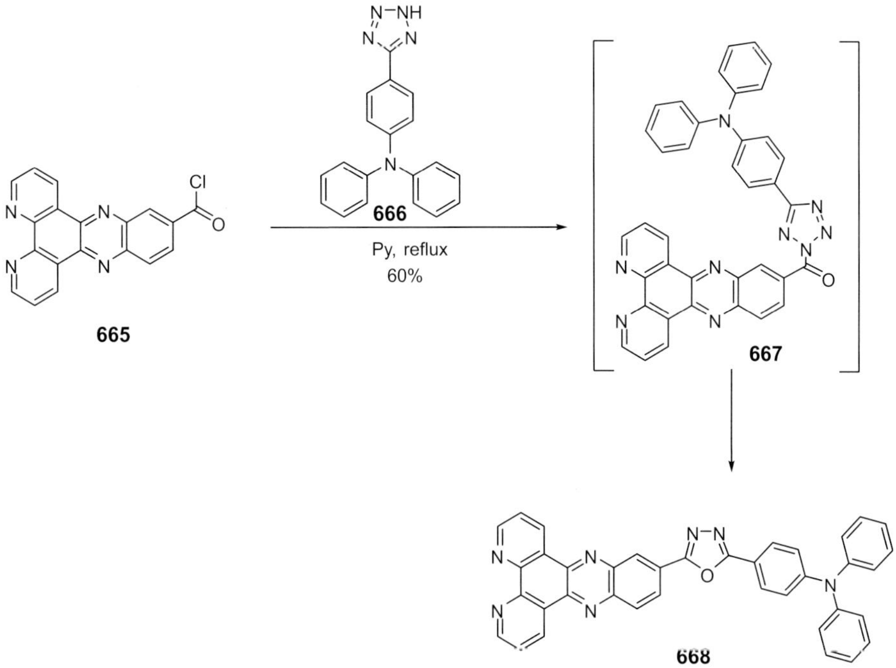

Scheme 13.213

The above compound, which contains a hole-transporting triphenylamine and an electron-transporting 1,3,4-oxadiazole unit, is an efficient light-emitting material. In particular, the absorption spectrum of **668** is extended into the visible region and shows a typical CT band around 416 nm and a broad π–π^* transition band around 347 nm. This is the result of the extended conjugation of the phenanthroline moiety. The emission maximum of **668** is at 635 nm (red) with λ_{ex} at 416 nm, and a quantum yield of 40%. Moreover, preliminary studies performed on this compound showed that it can be used as anorganic light emitting diode (OLED).

Microwave methodology has also been used to prepare various 3-(1,3,4-oxadiazol-2-yl)pyridines **671** in good to excellent yields, by reaction of 3-(5-tetrazolyl)pyridines **669** with different acid anhydrides (**670**) (Scheme 13.214) [535].

1,3,4-Oxadiazoles can also be obtained by photo-isomerization of 1,2,4-oxadiazoles. Irradiation of 5-alkyl-3-amino-1,2,4-oxadiazoles **672** at 254 nm in methanol and in presence of Et_3N, even if in moderate yields, leads to 2-amino-5-alkyl-1,3,4-

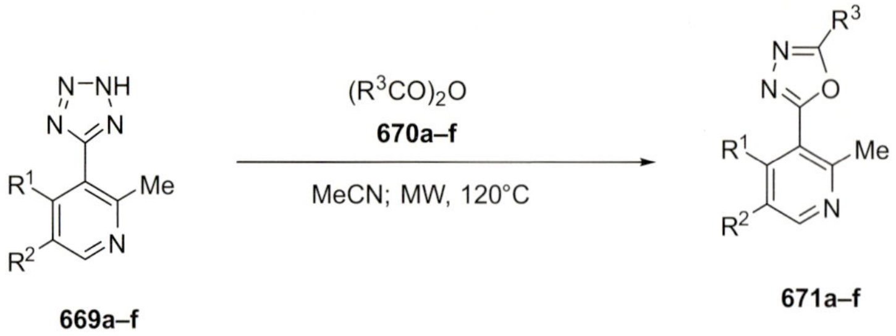

669a–f $\xrightarrow[\text{MeCN; MW, 120°C}]{(R^3CO)_2O \ \textbf{670a–f}}$ 671a–f

	R^1	R^2	R^3	Yield (%)
a:	Ph	H	Me	95
b:	Ph	H	CF_3 [a]	100
c:	$-o-C_6H_4-(CH_2)_2-$		*t*-Bu	96
d:	$-o-C_6H_4-OCH_2-$		$MeO(CH_2)_2$	80
e:	$-(CH_2)_3-$		*i*-Pr	99
f:	$-(CH_2)_5-$		4–Cl– C_6H_4	68

[a] The reaction was performed at room temperature for 2 h.

Scheme 13.214

oxadiazoles **675**. A small amount of the ring-degenerate isomers 3-alkyl-5-amino-1,2,4-oxadiazoles **678** (Scheme 13.215) was also obtained [536].

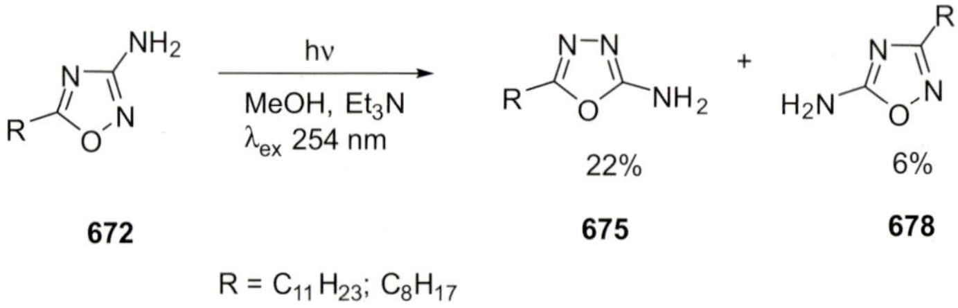

672 $\xrightarrow[\text{MeOH, Et}_3\text{N; } \lambda_{ex} \text{ 254 nm}]{h\nu}$ 675 (22%) + 678 (6%)

R = $C_{11}H_{23}$; C_8H_{17}

Scheme 13.215

The formation of **675** has been explained according to the ring contraction–ring expansion (RCRE) route, while **678** originates via a competing internal cyclization (IC)–isomerization mechanism, with an anionic species (**673**) as a common precursor (Scheme 13.216).

By a similar approach, 2-amino-5-alkylfluorinated-1,3,4-oxadiazoles **680** have been prepared utilizing the photochemical interconversion of 3-*N*-alkylamino-5-perfluoalkyl-1,2,4-oxadiazoles **679** in the presence of triethylamine. A moderate yield of 5-amino-3-alkyl-1,2,4-oxadiazoles **681** was also obtained, as a rearrangement product (Scheme 13.217) [537].

In this context, the same authors have reported that 1,2,5-oxadiazoles containing an acetylamino moiety are able to photochemically interconvert into 1,3,4-oxadiazoles. Thus, irradiation of 2,2,2-trifluoro-*N*-(4-phenyl-1,2,5-oxadiazol-3-yl)acetamide

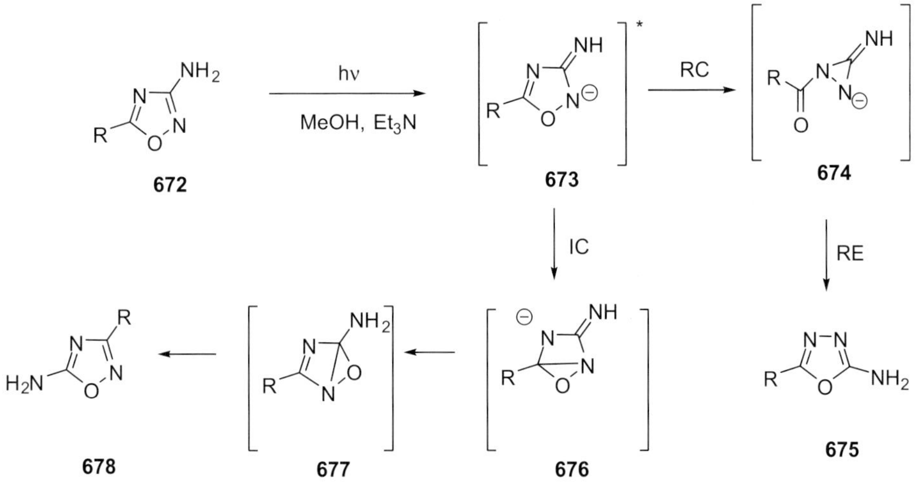

Scheme 13.216

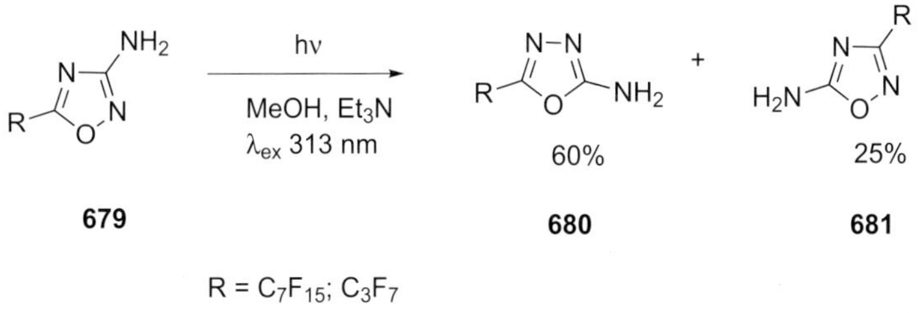

Scheme 13.217

682, in methanol in the presence of methylamine, produces, via 3-*N*-methylamino-5-trifluoromethyl-1,2,4-oxadiazole (**683**), a mixture of 2-*N*-methylamino-5-trifluoro-methyl-1,3,4-oxadiazole (**684**) and 1-methyl-3-*N*-methylamino-5-trifluoro-methyl-1,2,4-triazole (**685**) in 32% and 15% yield respectively (Scheme 13.218) [538].

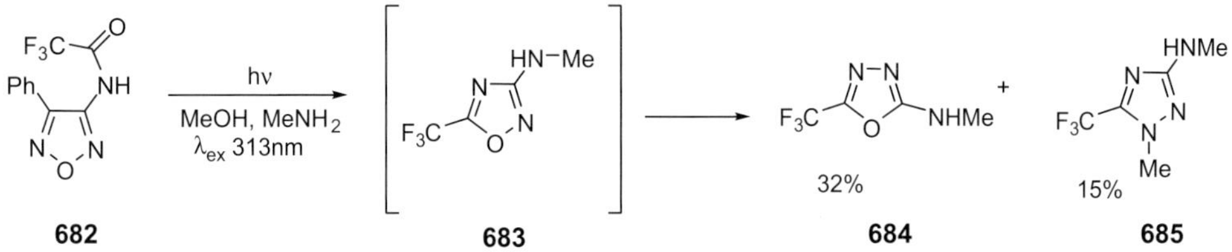

Scheme 13.218

The study of organic transformations within constrained media is a research topic that has received considerable attention in recent years. In this regard the first intrazeolite-photoinduced rearrangement of 1,2,4-oxadiazoles leading to 1,3,4-oxadiazoles has been reported. Irradiation of a perfluorohexane slurry of 3,5-diphenyl-1,2,4-oxadiazole (**686**) in zeolite (NaY) at 254 nm for 24 h furnished 2,5-diphenyl-1,3,4-oxadiazole (**689**) in 60% yield together with *N*-benzoyl-*N*′-phenylurea (**691**) (4%) and unreacted starting compound (35%). The formation of **689** and **691** is unknown in solution (Scheme 13.219) [539].

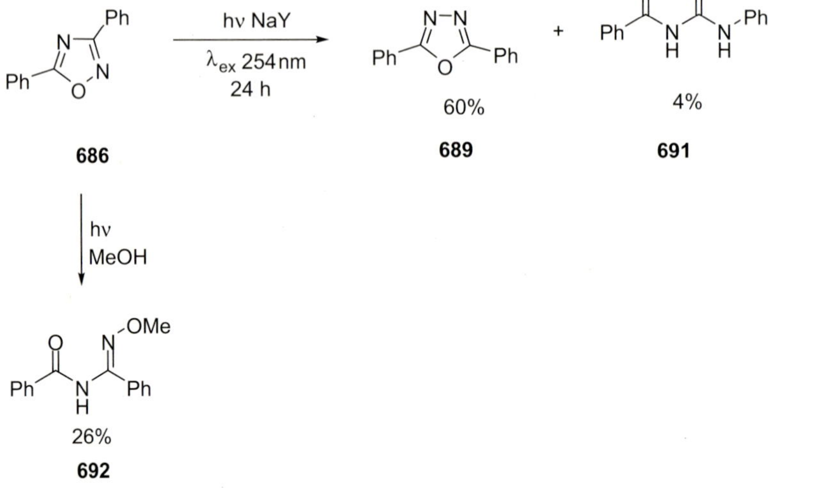

hν NaY, λ_{ex} 254 nm, 24 h; 60% **689**; 4% **691**; **686**; hν MeOH; 26% **692**

Scheme 13.219

In fact, a photochemical study performed on **686** in MeOH afforded product **692**. This dramatic difference of photo-behavior is explainable through a common intermediate (**687**), which in the zeolite cage leads to compounds **689** and **691**, via intermediates **688** and **690**, respectively, while **687** in methanol undergoes a nucleophilic addition of the solvent, giving rise to **692** (Scheme 13.220).

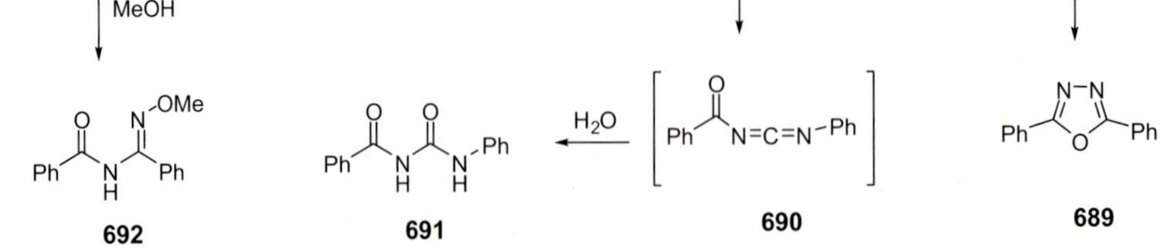

hν MeOH; hν NaY; MeOH; H_2O; **687**, **686**, **688**, **692**, **691**, **690**, **689**

Scheme 13.220

It was found recently that a thermal rearrangement promoted by base is also effective in the formation of 1,3,4-oxadiazoles **698** starting from 3-acylamino-1,2,4-oxadiazoles **693** (Scheme 13.221) [540].

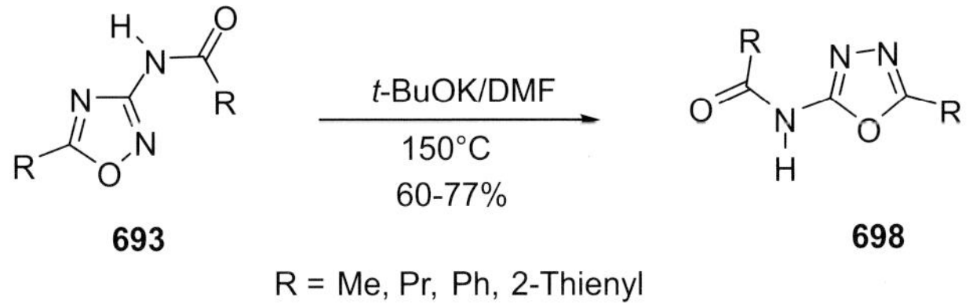

t-BuOK/DMF, 150°C, 60-77%

693 → **698**

R = Me, Pr, Ph, 2-Thienyl

Scheme 13.221

The reaction consists of a one-atom side-chain rearrangement that is base activated, occurs at higher temperature, and irreversibly leads to the corresponding 2-acylamino-1,3,4-oxadiazoles.

The reaction mechanism has been studied in depth by computational methods, using the hybrid DFT B3LYP method and the 6-31 + + G(d,p) basis set. Following these computational studies, the proposed mechanism is that reported in Scheme 13.222, where the route involving migration–nucleophilic attack–cyclization (MNAC) is the activated route, while the ring contraction–ring expansion route (RCRE) is ruled out.

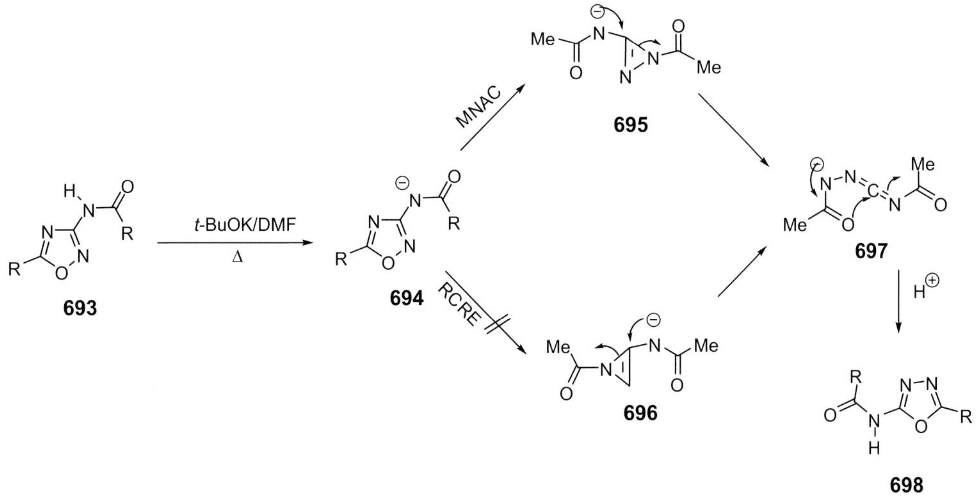

693 → (*t*-BuOK/DMF, Δ) → **694**; MNAC → **695** → **697**; RCRE (blocked) → **696**; **697** → ($H^{\oplus}$) → **698**

Scheme 13.222

13.5.2.4 **Synthesis of Mesoionic 1,3,4-Oxadiazoles (600), and 1,3,4-Oxadiazolium Cations (601)**

The simplest way to prepare mesoionic 1,3,4-oxadiazoles **600** is by the thermal cyclization of 1-carbonyl-substituted-1-substituted hydrazine hydrochloride and phosgene [541]. Thus, the 4,5-dihydro-3-methyl-5-oxo-2-phenyl-1,3,4-oxadiazolium inner salt **701** has been synthesized from 1-benzoyl-1-methylhydrazine hydrochloride (**699**) and phosgene (**700**) (Scheme 13.223) [541].

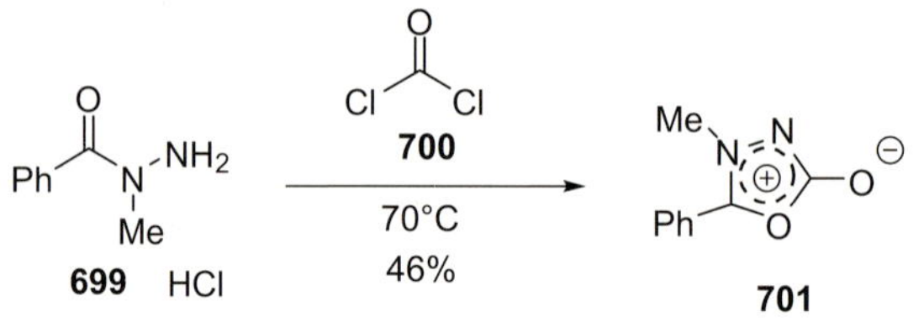
Ph
NH2
N
Me
699 HCl
Cl
Cl
O
700
70°C
46%
Me
N
N
Ph
O
O
701

Scheme 13.223

In the same way, 5-(dimethylamino)-4-methylisosydnone **705** has been prepared, via the aminoisocyanate **704** intermediate, by heating at 70 °C the 2,4,4-trimethylsemicarbazide **702** with trichloromethyl chloroformate (**703**) (Scheme 13.224) [542].

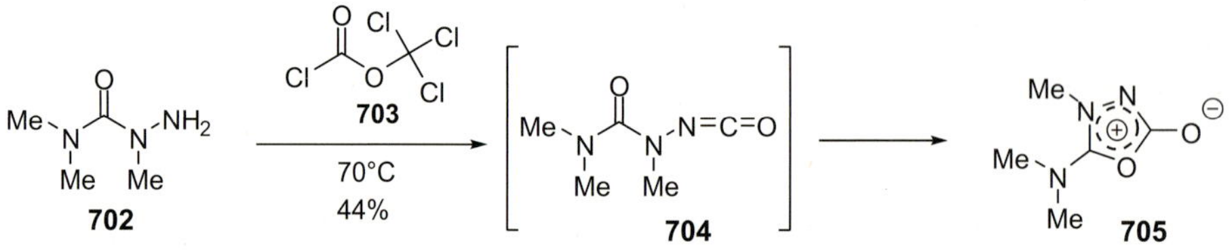
Me
NH2
Me Me
702
703
70°C
44%
N=C=O
Me Me
704
Me
Me
Me
705

Scheme 13.224

1,3,4-Oxadiazolo[3,2-*a*]pyridylium-2-aminides **708** are available in 75–91% yield by reaction, in refluxing toluene, of 1,3,4-oxadiazolo[3,2-*a*]pyridylium-2-olate **706** with *N*-aryliminotriphenyl-phosphoranes **707** (Scheme 13.225) [543].

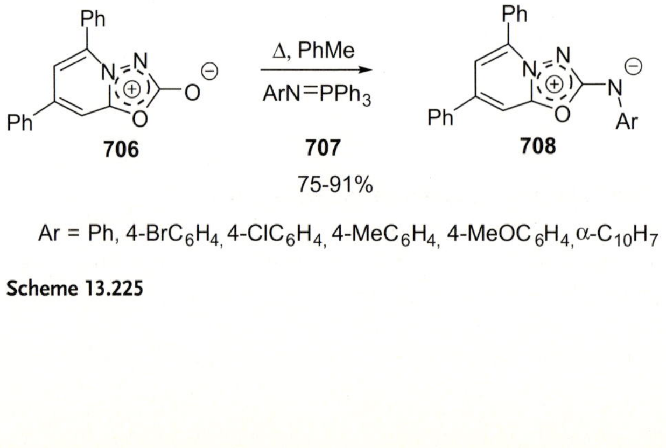
Ph
Ph
706
Δ, PhMe
ArN=PPh3
707
75-91%
Ph
Ph
Ar
708
Ar = Ph, 4-BrC6H4, 4-ClC6H4, 4-MeC6H4, 4-MeOC6H4, α-C10H7

Scheme 13.225

1,3,4-Oxadiazolium cations **601** are easily obtained by treatment of 1,3,4-oxadiazoles with several alkylating agents to give in about 100% yield the corresponding salts. A recent example is the synthesis of 2,5-diaryl-3-trimethylsilylmethyl-1,3,4-oxadiazolium trifluoromethanesulfonates (**711**) [544]. These compounds have been prepared in 99–100% yield by mixing a solution of 2,5-diaryl-1,3,4-oxadiazoles **709** with trimethylsilylmethyl trifluoromethanesulfonate **710** in dry CH_2Cl_2 at 50 °C under reflux condenser for 24 h (Scheme 13.226).

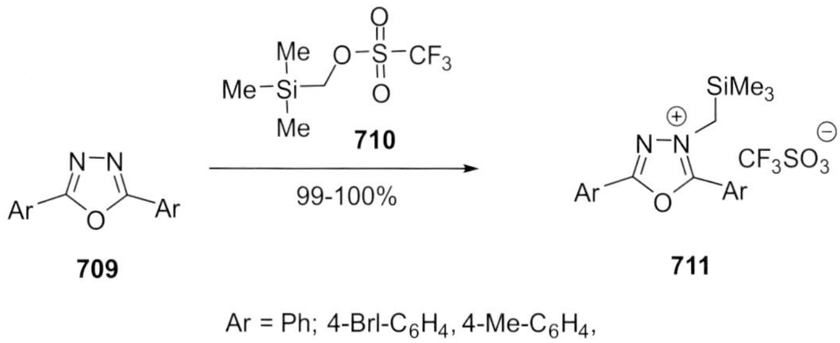
710
99-100%
709
711
Ar = Ph; 4-BrI-C6H4, 4-Me-C6H4,

Scheme 13.226

Another method involves the cyclization reaction of N-substituted diacylhydrazides with a mixture of $HClO_4$–Ac_2O. Thus, cyclization of $RCONR^1NHAc$ **712** with $HClO_4$–Ac_2O gave 47–98% yields of 1,3,4-oxadiazolium salts **713** (Scheme 13.227) [545].

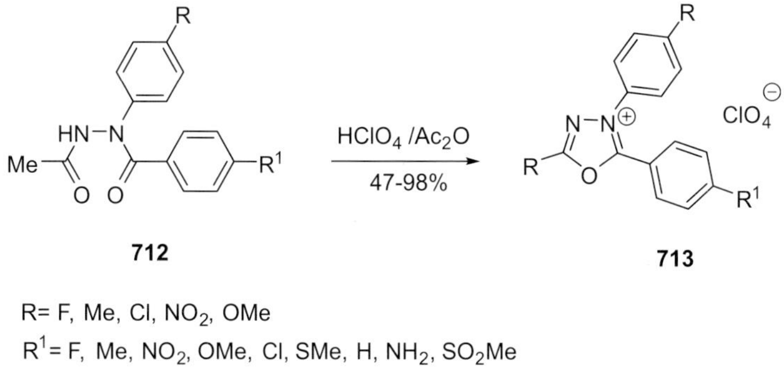
HClO4 /Ac2O
47-98%
712
713
R= F, Me, Cl, NO2, OMe
R1= F, Me, NO2, OMe, Cl, SMe, H, NH2, SO2Me

Scheme 13.227

13.5.2.5 **Synthesis of Oxadiazolinones, Oxadiazolinethiones, and Oxadiazolimines (602)**

1,3,4-Oxadiazolin5-ones are, usually, synthesized by reaction of substituted acid hydrazide with phosgene or by thermal cyclization of acylcarbazates.

5-*tert*-Butyl-3-(2,4-dichloro-5-isopropoxyphenyl)-1,3,4-oxadiazolin-2-one – a very active herbicide, that goes under the commercial name of Oxadiazon (**716**, R = CMe_3), commonly used in rice production for controlling weeds and increasing seed yield in soya beans, containing the oxadiazolinones ring – has been prepared by reaction of 1-trimethylacetyl-2-(2,4-dichloro-5-isopropoxyphenyl)hydrazide **714** with phosgene (**715**) in toluene at 100–110 °C (Scheme 13.228) [546].

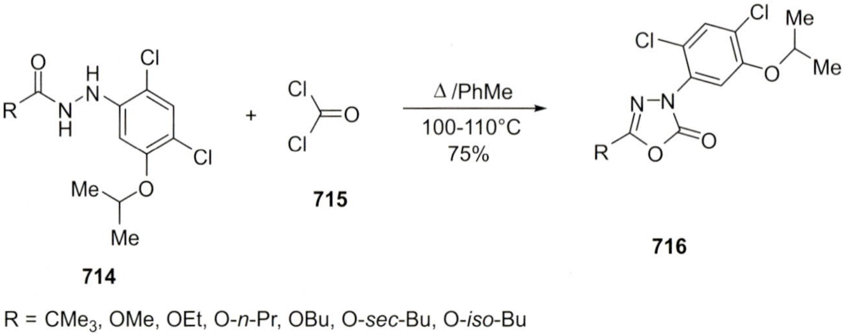

Δ /PhMe
100-110°C
75%

714 + **715** → **716**

R = CMe_3, OMe, OEt, O-*n*-Pr, OBu, O-*sec*-Bu, O-*iso*-Bu

Scheme 13.228

Similarly, 5-substituted-3-(2,4-dichloro-5-isopropoxyphenyl)-1,3,4-oxadiazolin-2-ones have been prepared (R = MeO, EtO, *n*-PrO, BuO, *sec*-BuO, *iso*-BuO).

Condensation of ethyl 2-(2-chlorophenyl)hydrazine-carboxylate **717** with phosgene has furnished 5-ethoxy-3-(2-chlorophenyl)-1,3,4-oxadiazolin-2-one **718**, which is active orally against gastrointestinal nematodes of domestic animals and man (Scheme 13.229) [547].

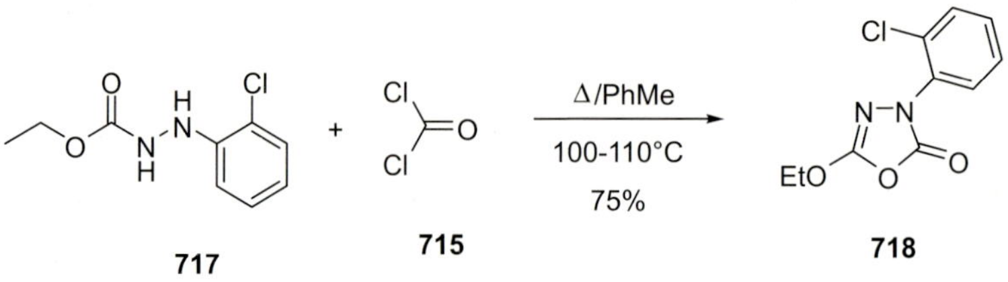

Δ/PhMe
100-110°C
75%

717 + **715** → **718**

Scheme 13.229

Cyclization of acylcarbazates has been utilized fruitfully for the synthesis of 2-(2,3-dihydro-2-oxo-1,3,4-oxadiazol-5yl)benzoxazoles, a class of compounds that are potent inhibitors of anaphylactically induced histamine release from rat peritoneal mast cells and are orally active as inhibitors of IgE-mediated passive cutaneous anaphylaxis in the rat [548]. The 2-(2,3-dihydro-2-oxo-1,3,4-oxadiazol-5-yl)benzoxazole **722a** (R = H), chosen as an example, was synthesized by heating ethyl 3-(2-benzoxazolyl)hydrazine-carboxylate (**721**) in Dowtherm at 230–240 °C for 1 h. Intermediate **721** was prepared by treating 3-chloro-1,4-benzoxazin-2-one **719** with ethyl carbazate **720** in dioxane and triethylamine at room temperature for 4 h. In a similar manner, compounds **722b–g** have been prepared (Scheme 13.230).

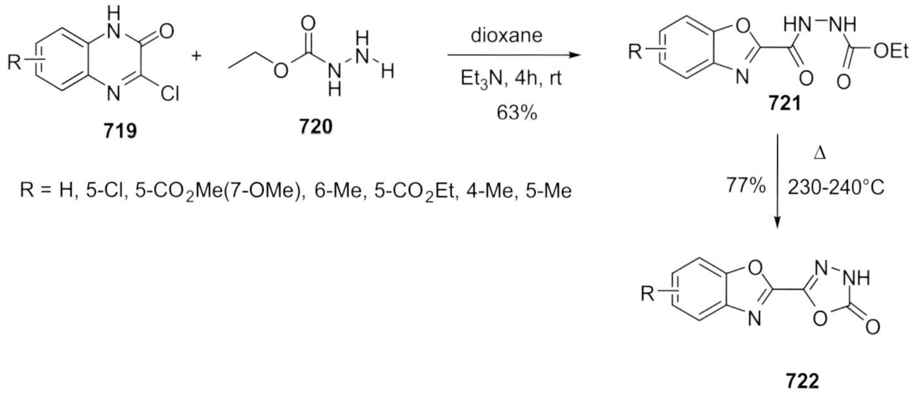
719
720
dioxane
Et_3N, 4h, rt
63%
721
Δ
77%
230-240°C
722
R = H, 5-Cl, 5-CO_2Me(7-OMe), 6-Me, 5-CO_2Et, 4-Me, 5-Me

Scheme 13.230

5-Imino-2-substituted Δ^2-1,3,4-oxadiazolines **725**, as hydrochloride salts, which are able to produce a profound flaccid paralysis in rats, have been prepared by hydrochloric reaction of 2-amino-5-aryl-1,3,4-oxadiazoles **724**, in DMF/H_2O or in ethanol–ether as solvents. These latter compounds have been synthesized by reaction of 1-acyl-3-thiosemicarbazide **723** with Pb_3O_4 (Scheme 13.231) [549].

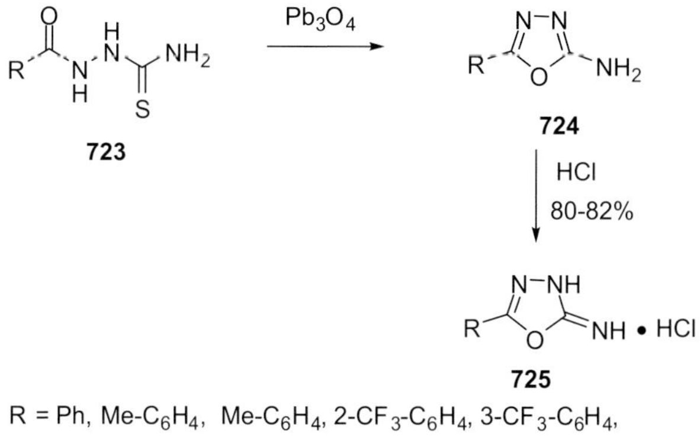
Pb_3O_4
723
724
HCl
80-82%
725
R = Ph, Me-C_6H_4, Me-C_6H_4, 2-CF_3-C_6H_4, 3-CF_3-C_6H_4,

Scheme 13.231

Diphenylnitrilimine **728**, prepared from 2,5-diphenyltetrazole (**726**) or by triethylamine treatment of diphenylchlorohydrazone **727**, adds to C=O and C=N double bond of aryl isocyanate to give 2,4-diphenyl-1,3,4-oxadiazol-5-phenylimino **729**, and 1,3-diphenyl-4-aryl-1,2,4-triazolin-5-ones (**730**) in a 2 : 1 ratio respectively (Scheme 13.232) [550].

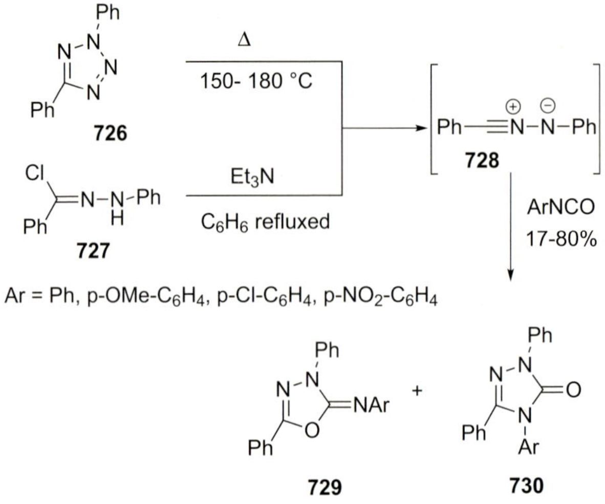

Ar = Ph, p-OMe-C_6H_4, p-Cl-C_6H_4, p-NO_2-C_6H_4

Scheme 13.232

5-Hydroxy-2-methyl-6-phenyl-7*H*-[1,3,4]oxadiazolo[3,2-*a*]pyrimidin-7-one (**733**) has been obtained in 79% yield via a solvent-free microwave cyclocondensation reaction using di(2,4,6-trichlorophenyl) 2-phenylmalonate (**732**) and 2-methyl-5-amino-1,3,4-oxadiazole (**731**) in a 1 : 2 ratio under heating at 250 °C for 15 min (Scheme 13.233) [551].

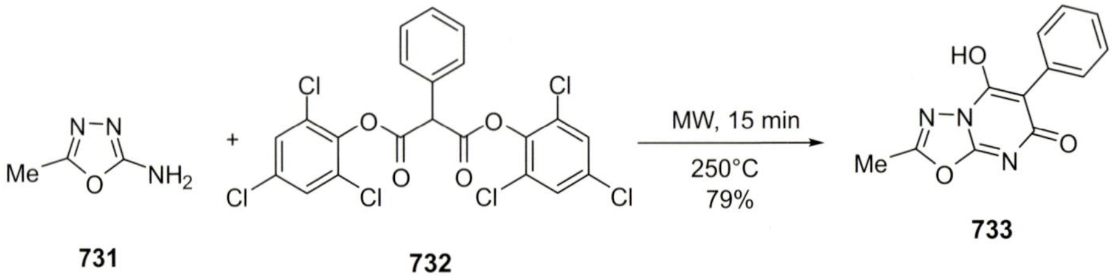

Scheme 13.233

The synthetic procedure based on the ring closure of substituted acid hydrazide **734** with carbon disulfide leads to 5-aryl-2,3-dihydro-1,3,4-oxadiazole-2-thiones **735** in excellent yield (Scheme 13.234).

The obtained compounds have been conveniently transformed into 3,5-disubstituted-2,3-dihydro-1,3,4-oxadiazole-2-thiones **738** in 72–92% yield, by reaction with dapsone (**736**) and aromatic aldehydes **737** in methanolic solution. Compound **738a** has been shown to be very active against *Mycobacterium tuberculosis* H37Rv and isoniazid (INH) resistant *M. tuberculosis* with MIC of 0.1 and 1.10 μM respectively [552].

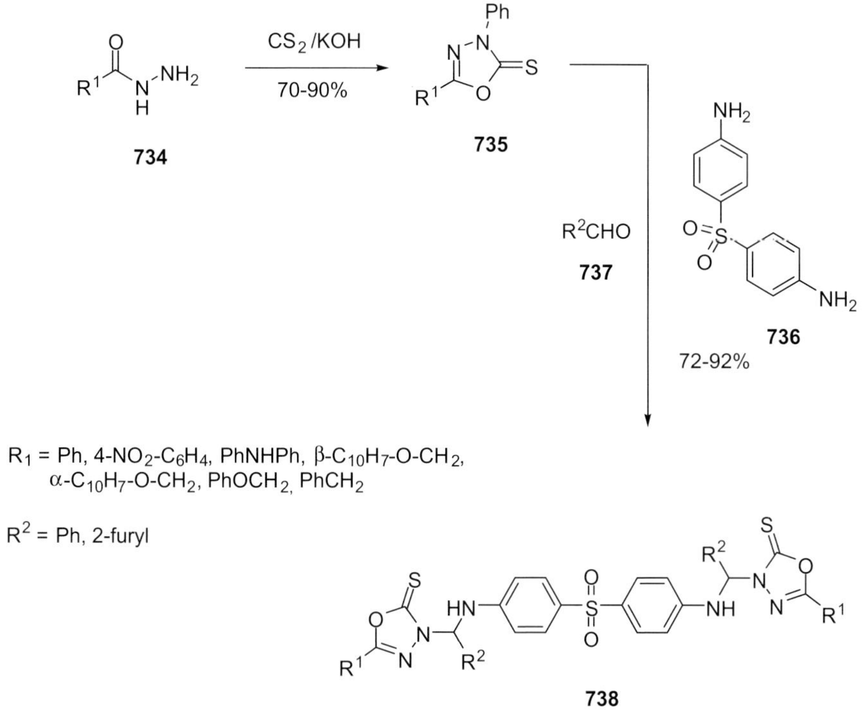
CS2 /KOH
70-90%
734
735
R2CHO
737
736
72-92%
R1 = Ph, 4-NO2-C6H4, PhNHPh, β-C10H7-O-CH2, α-C10H7-O-CH2, PhOCH2, PhCH2
R2 = Ph, 2-furyl
738

Scheme 13.234

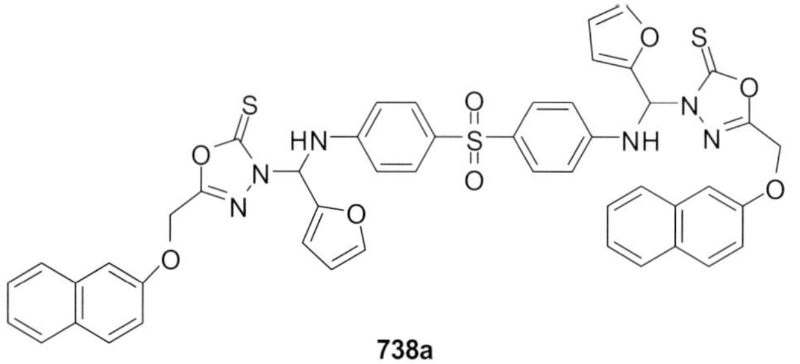

738a

Acyclic C-nucleoside 5-(1,2-dihydroxyethyl)-*3H*-[1,3,4]oxadiazole-2-thione **742**, containing an 1,3,4-oxadiazolinethione ring, as mimic of ribose unit, has been synthesized starting from (±)-2,2-dimethyl-[1,3]dioxolan-4-carboxylic acid methyl ester **739** via reaction with hydrazine to give the corresponding hydrazide **740**, followed by CS_2 treatment and subsequent deacetonation with Amberlyst 15 (Scheme 13.235) [553].

The synthesis of this optical active compound has been performed by the same common route using as chiral source the D-mannitol. Furthermore, the use of D-xylose **743**, via (tetrahydro-[1,3]dioxino[5,4-d][1,3]dioxin-4-yl)-methanol **744**, leads

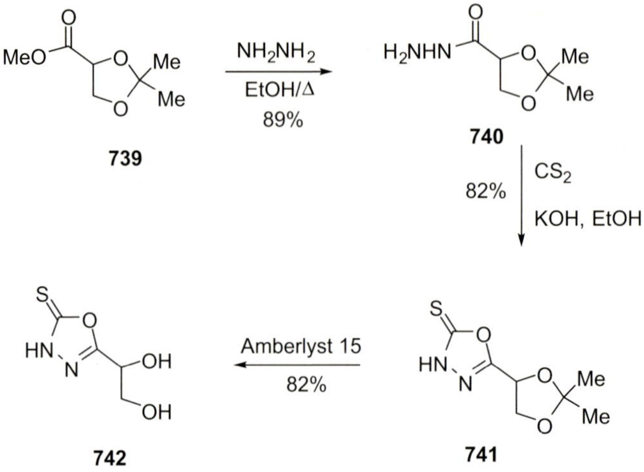

Scheme 13.235

to 5-(1,2,3,4-tetrahydroxybutyl)-3*H*-[1,3,4]oxadiazole-2-thione **745** in enantiomerically pure form, in 8.5% total yield (Scheme 13.236).

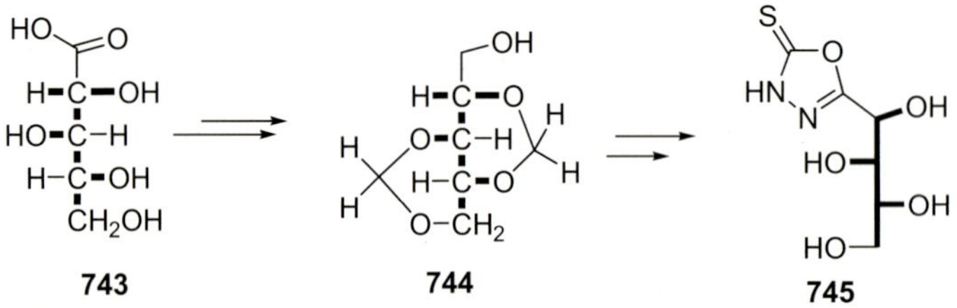

Scheme 13.236

13.5.2.6 Synthesis of (Δ^2) (603), (Δ^3) (604), and 2,3,4,5-Tetrahydro-1,3,4-Oxadiazoles (605)

The Δ^2-1,3,4-oxadiazoline **751**, which inhibits cell proliferation and binds to tubulin, has been synthesized, as reported in Scheme 13.237, according to a process that involves as the key reaction the cyclization of acylhydrazone **749** with acetic anhydride [554]. The synthesis starts from the 4-azido-3-methylbenzoic acid (**746**), which was reacted with *tert*-butylcarbazate in the presence of 1-[3-(dimethylamino) propyl]-3-ethylcarbodiimide hydrochloride (EDCI) and a small amount of 4-(*N,N*,-dimethylamino)pyridine as base, to give, after trifluoroacetic acid (TFA) treatment, 98% of 4-azido-3-methylbenzoylhydrazide (**747**). This compound was condensed with 3,4,5-trimethoxybenzaldehyde (**748**) to give in 67% yield the corresponding acylhydrazone **749**, which was then cyclized to 2-(4-azido-3-methlphenyl)-4-acetyl-5-(3,4,5-trimethoxyphenyl)-Δ^2-1-3-4-oxadiazoline (**750**). Compound **750** was converted into the target compound **751** by reduction of the azido group with a suspension of $SnCl_2$, thiophenol, and triethyl amine, in 80% yield (total overall yield 32%).

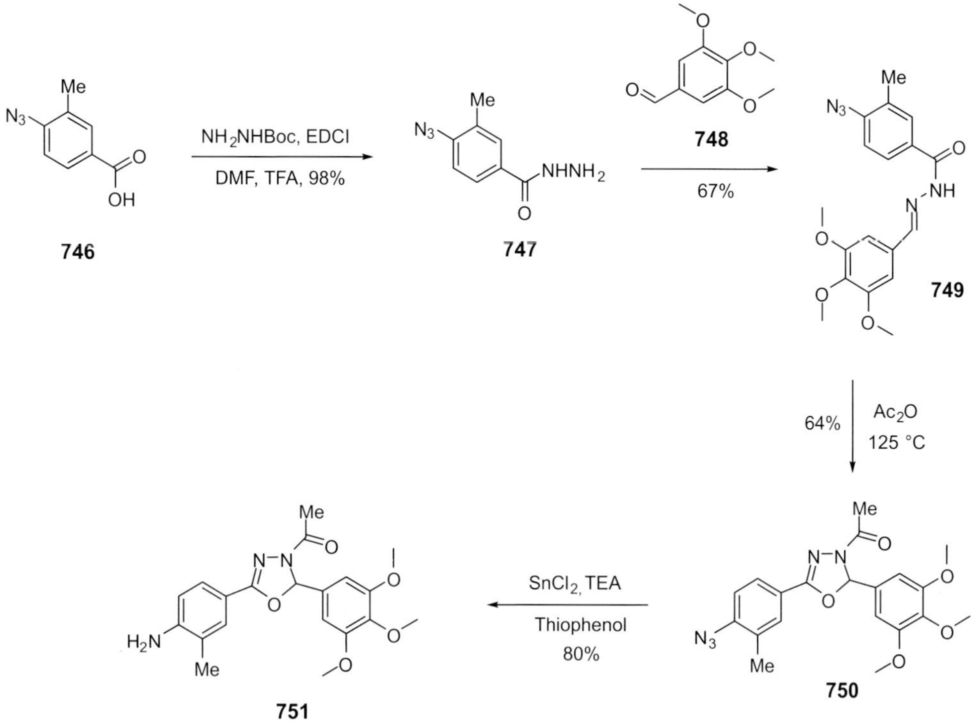

Scheme 13.237

The method has been exploited to prepare various 2-[4-(*N,N*-dimethylaminophenyl]-4-substituted-(3,4,5-trimethoxyphenyl)-Δ^2-1-3-4-oxadiazolines **753**, which present interesting antitumoral activity (Scheme 13.238)[555].

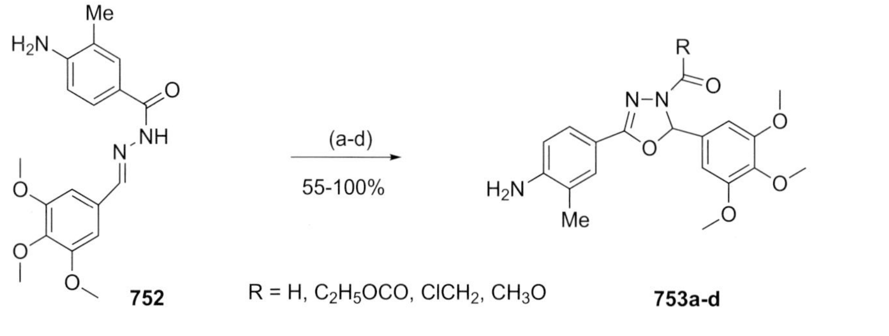

(a) Ac_2O/HCO_2H; (b) ethyl oxalyl chloride; (c) chloroacetic anhydride; (d) dimethylpyrocarbonate

Scheme 13.238

C,*N*-Diphenyl nitrilimine **728** reacts, by 1,3-dipolar cycloaddition, with aldehydes **754** to produce in 50–75% yield 5-aryl-substituted-2,4-diphenyl-1,3,4-oxadiazolines (**755**) (Scheme 13.239) [555, 556].

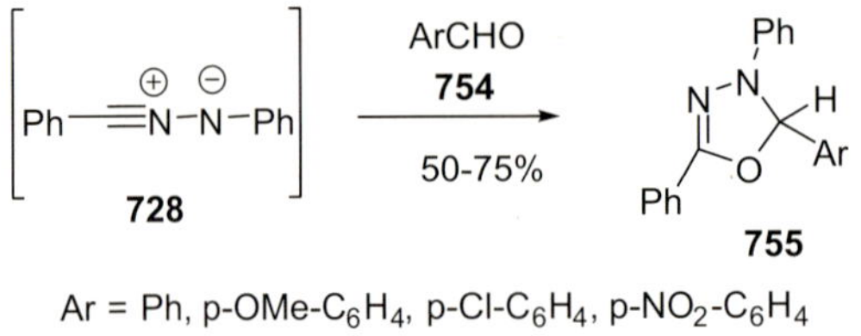

Ar = Ph, p-OMe-C_6H_4, p-Cl-C_6H_4, p-NO_2-C_6H_4

Scheme 13.239

N-Substituted 2,3-dihydrooxadiazoles have been prepared recently by intramolecular cyclization of protected amino-aldehydes (1,5-dipolar cycloaddition) [557]. Thus, the *N*,*N*-dibenzylated aldehyde **756** after heating at reflux for 72 h, in toluene, with N^1-acetyl-N^2-methylhydrazine **758**, for 16 h, gave rise to the 2,3- dihydro-1,3,4-oxadiazole **760a** (7%), through a cyclization involving the hydrazine N-acetyl group of the not-isolable intermediate **759** (Scheme 13.240). 2-Benzyloxypropanal **757** gave the analogous product **760b** (11%).

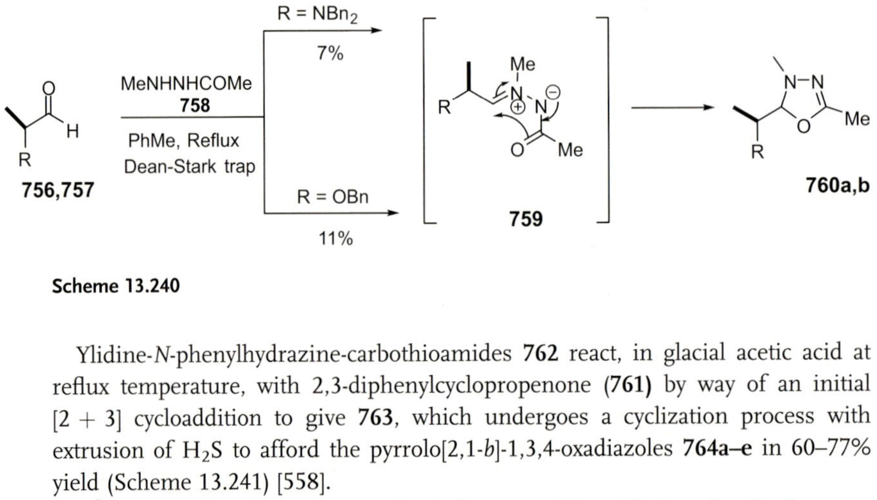

Scheme 13.240

Ylidine-*N*-phenylhydrazine-carbothioamides **762** react, in glacial acetic acid at reflux temperature, with 2,3-diphenylcyclopropenone (**761**) by way of an initial [2 + 3] cycloaddition to give **763**, which undergoes a cyclization process with extrusion of H_2S to afford the pyrrolo[2,1-*b*]-1,3,4-oxadiazoles **764a–e** in 60–77% yield (Scheme 13.241) [558].

Δ^3-1,3,4-Oxadiazolines (**604**) can be easily prepared by oxidation of acylhydrazones (see Schemes 13.201 and 13.203) [524, 525]. Thus, the oxidation of methoxycarbonyl hydrazone of acetone (**766**) with lead tetraacetate (LTA) furnishes the 2-acetoxy-2-methoxy-5,5-dimethy-Δ^3-1,3,4-oxadiazoline (**767**) in 60–72% yield (Scheme 13.242). This reaction route was further developed, transforming **767** into various Δ^3-1,3,4-oxadiazolines **768** and **769** by reaction with alcohols or phenols. Notably, these compounds have been used as a carbene source, because they fragment quite cleanly in solution at about 100 °C to give as by-products acetone and N_2 [559].

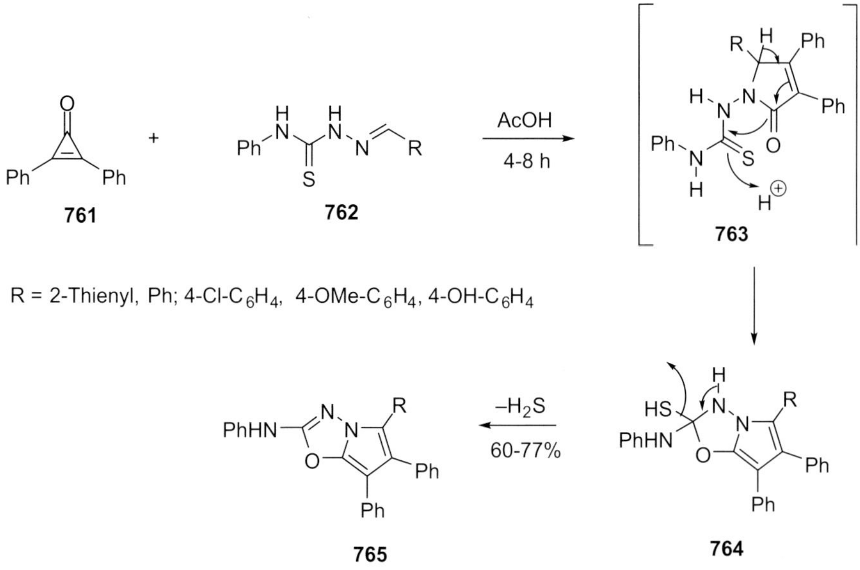

R = 2-Thienyl, Ph; 4-Cl-C_6H_4, 4-OMe-C_6H_4, 4-OH-C_6H_4

Scheme 13.241

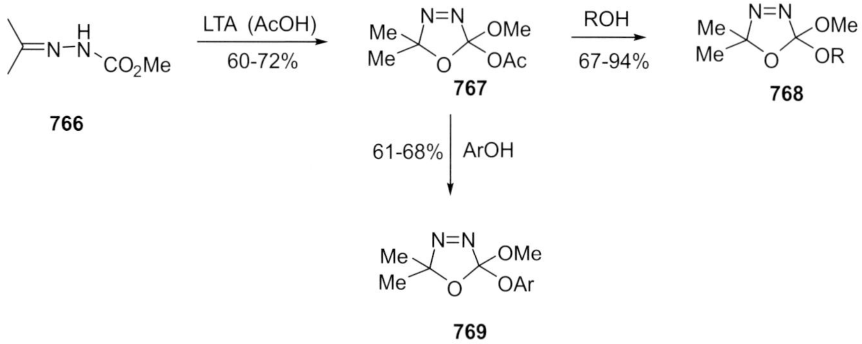

R	Me	Et	Pr	i-Pr	Bu	t-Bu	CH_2CF_3	But-3ynyl	Pent-4ynyl
Ar	Ph	4-CN-C_6H_4	4-OMe-C_6H_4	—	—	—	—	—	—

Scheme 13.242

2,3,4,5-Tetrahydro1,3,4-oxadiazoles are obtained by reaction of 1,2-disubstituted hydrazines with aldehydes. Zwanesburg *et al.* have reported the synthesis of a series of 2,3,4,5-tetraalkyl-1-3-4-oxadiazolidines (**772**) by reaction of 1,2-dimethylhydrazine (**771**) with aliphatic aldehydes **770**. Thus, the appropriate aldehyde (2 equivalents) in 50 ml of dry ether and a few grams of $MgSO_4$ treated drop-wise during 15 min, below 5 °C, with one equivalent of aldehyde gave after 1 h the corresponding 1,3,4 oxadiazolines in 60–75% yield (Scheme 13.243) [560].

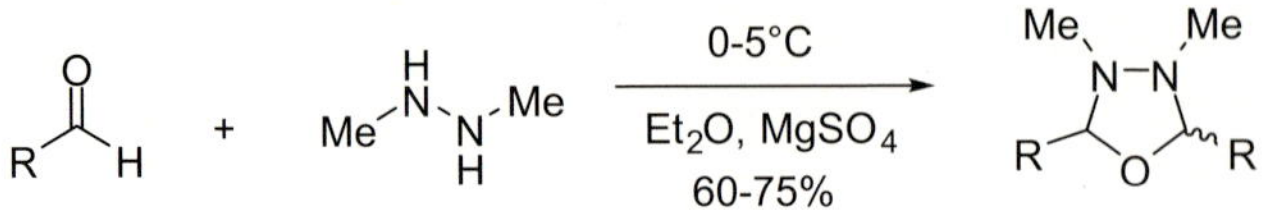

0-5°C

Et_2O, $MgSO_4$

60-75%

770 771 772

R = Me, Et, Pr, Bu

Scheme 13.243

In a similar way Zinner and Kliwing have prepared other 1,3,4-oxadiazolines (**775**) using different hydrazines (**774**) (Scheme 13.244) [561].

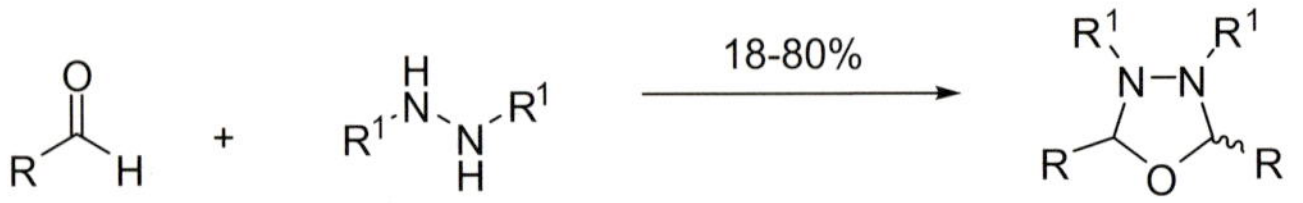

18-80%

773 774 775

R = H, Me, Et, Pr, *i*-Pr, *t*-Bu, Ph

R^1 = Me, i-Pr, Cyclohexyl, $-CH_2CHMeCH_2-$

Scheme 13.244

3,4-Diaryl substituted 1,3,4-oxadiazolidines **777** have been synthesized in moderate to good yields (16–54%) by TiO_2-photocatalyzed reaction of azobenzenes **776** in methanol through a uranium glass filter ($\lambda_{ex} > 320$ nm). The reaction probably involves a photoreduction of azobenzenes to 1,2-diarylhydrazine, an oxidation of methanol to formaldehyde, and the subsequent cyclization to 1,3,4-oxadiazolidines (Scheme 13.245) [562].

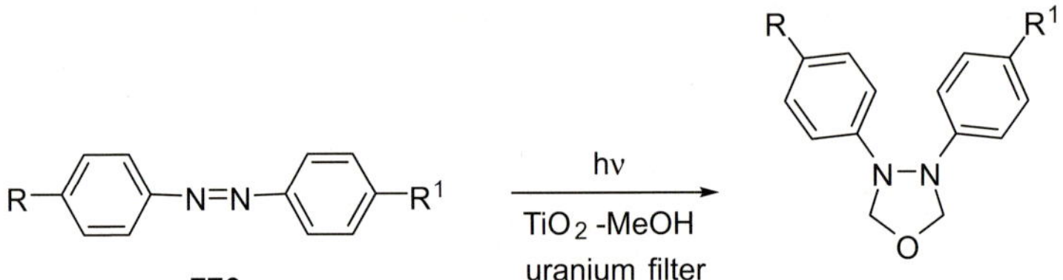

hν

TiO_2 -MeOH

uranium filter

16-54%

776 777

R = H, Ph

R^1 = 4-Cl-C_6H_4, 4-Me-C_6H_4, 3-Cl-C_6H_4

Scheme 13.245

Tricyclic fused-1,3,4-oxadiazole systems that display *in vitro* fungitoxicity comparable to that of fungicide Dithane M45 at 1000 ppm concentration against *Aspergillus niger* and *Fusarium oxysporum* have been obtained starting from 2-aryl-3-thioureido-4-

thiazolidinones **778**. These compounds were treated with a mixture of KI/I_2 under basic conditions to afford bicyclic compounds containing the Δ^2-1,3,4-oxadiazoline core (**779**) that were then converted into the target compounds by reaction with formaldehyde and various α-amino acids **780** (Scheme 13.246) [563].

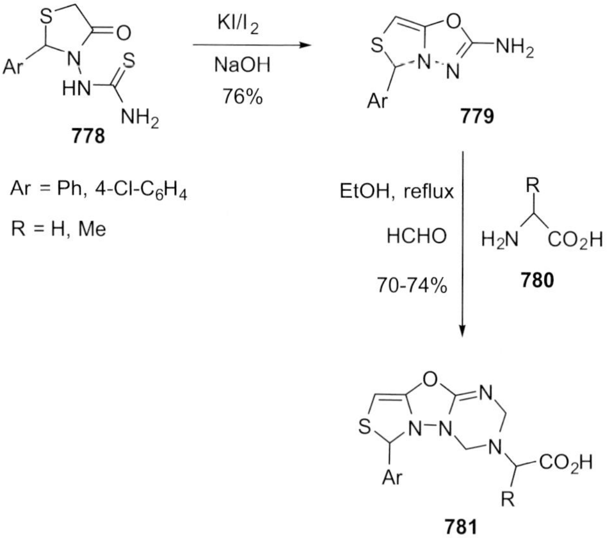

KI/I2
NaOH
76%
Ar = Ph, 4-Cl-C6H4
R = H, Me
EtOH, reflux
HCHO
70-74%
778
779
780
781

Scheme 13.246

Another versatile method for the synthesis of 1,3,4-oxadiazolidines (**605**) is the 1,3-dipolar cycloaddition of azomethinimines to carbonyl compounds. In particular, thermal decomposition of 8,8a,16,16a-tetrahydro-8,16-diphenyl-[1,2,4,5]tetrazino[6,1-*a*:3,4-*a'*]diisoquinoline (**782**) at 50–80 °C gave the 3,4-dihydroisoquinolineazomethinimine **783** that in presence of various carbonyl compounds (**784**) gave rise to substituted 1,3,4-oxadiazolidines **785** (79–100% yield) (Scheme 13.247) [564].

The obtained compounds, at high temperature, are thermolabile and decompose to azomethinimines and carbonyl compounds. This is recognizable by a color change to reddish-brown. Moreover, the azomethinimines thus obtained can be trapped with a wide range of dipolarophiles [565].

Hydrazide **786** (R = Me) and semicarbazides **787** (R = NHAr) have been used for the synthesis of azomethine imines **789** that, *in situ*, react with methyl glyoxylate hemiacetal **788** to give the corresponding 1,3,4-oxadiazolidines **790** in good yields (60–70%) (Scheme 13.248) [566].

Similarly, the cyclic hydrazine **791** reacts with ethyl glyoxylate (**792**) in the presence of $MgBr_2{\cdot}Et_2O$ in THF at 65 °C to give in 71% yield the bicyclic 1,3,4-oxadiazolidine **794** with a diastereomeric ratio of 46/38/16, via the azomethinimine intermediate **793** (Scheme 13.249) [567].

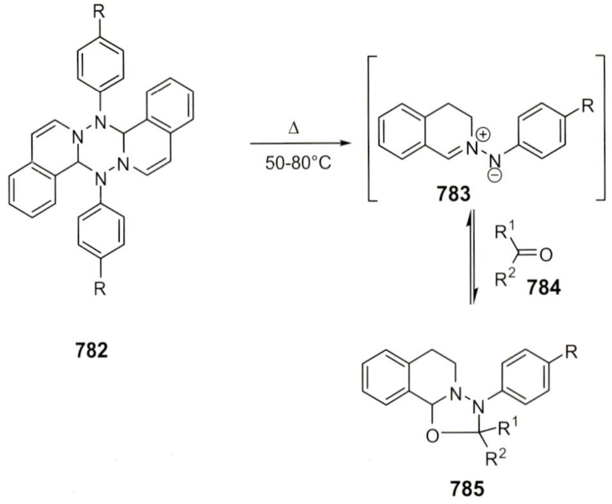

R	R^1	R^2	Yield
H	H	H	88
H	Me	H	79
H	CCl_3	H	84
NO_2	CCl_3	H	99
H	Ph	H	93
H	p-MeO-C_6H_4	H	92
H	p-Cl-C_6H_4	H	92
H	2-pyridyl	H	98
H	2-furyl	H	85
H	CO_2Et	H	93
NO_2	CO_2Et	H	100

Scheme 13.247

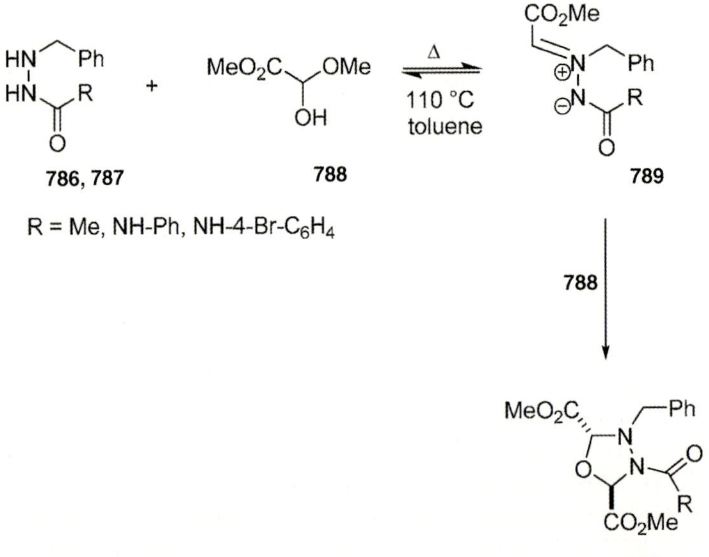

R = Me, NH-Ph, NH-4-Br-C_6H_4

Scheme 13.248

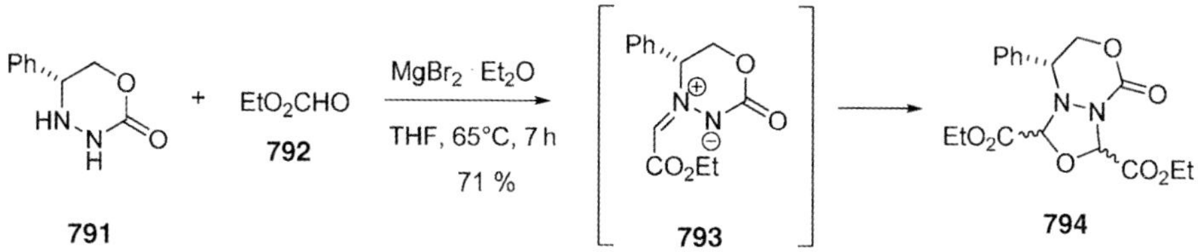

Scheme 13.249

13.5.3
Reactivity

The reactivity of the 1,3,4-oxadiazole system is expressed in a series of different chemical transformations that can be amenable to (i) ring cleavage reactions, (ii) reactions due to the reactivity of heterocycle ring, and (iii) reactions of substituents.

13.5.3.1 Ring Cleavage Reactions

Ring-opening reactions of 1,3,4-oxadiazoles can be achieved by the action of nucleophilic reagents; in some cases the ring opening occurs by thermolysis or photolysis.

2,5-Dialkyl-1,3,4-oxadiazoles are cleaved by water in basic or acid conditions to produce diacylhydrazines that suffer further hydrolysis under vigorous conditions to give carboxylic acids and hydrazine (Scheme 13.250).

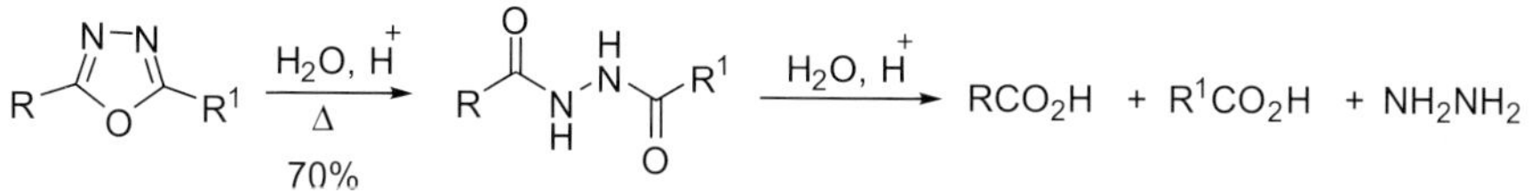

Scheme 13.250

This ring cleavage is dependent on the solubility. In fact, no hydrolysis was observed for 2,5-diphenyl-1,3,4-oxadiazole, which has a solubility in water of 0.03% [568].

The presence of an electron-withdrawing group at C2 and C5 of the ring increases the reactivity towards nucleophiles. Thus, 2,5-bis(perfluoroalkyl)-1,3,4-oxadiazoles **797a–c** are very sensitive to nucleophilic attack. They react with ammonia to give the corresponding 1-(perfluoroalkylimidoyl)-2-(perfluoroacyl)hydrazines **798a–c** (R = H). The reaction occurs by attack of nucleophiles on the electron-deficient oxadiazole ring carbon to afford the **798a–c**. The reaction with the more nucleophilic methylamine provides 1,2-bis(*N*-alkyl)perfluoroalkylimidoyl)hydrazines **799a–c** via the hydrazine intermediates **798a–c** (R = Me) (Scheme 13.251) [569].

Interestingly, thermal dehydration or deamination of these hydrazine derivatives **798** and **799** produces the corresponding 4-substituted-3,5-bis(perfluoroalkyl)-4*H*-1,2,4-triazoles **800a–c** or **801a–c** in 88–94% yield (Scheme 13.251).

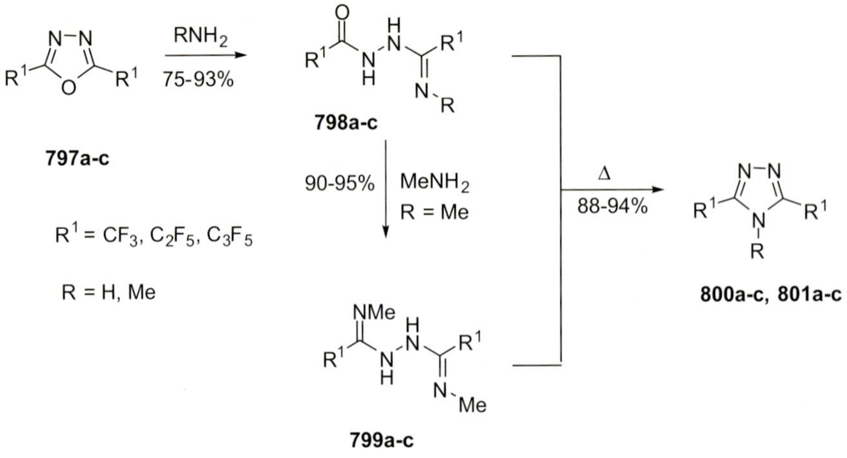

Scheme 13.251

In a similar fashion 3,5-bis(trifluoromethyl)-1,3,4-oxadiazole **802** reacts with hydrazine in methanol at −42 °C to afford the N^2-(α-hydrazonotrifluoromethyl)-N^1-(trifluoroacetyl)hydrazine **803**, which under heating is converted into the 4-amino-3,5-bis(trifluoromethyl)-4*H*-1,2,4-triazole (**804**) (85%). Dihydrotetrazine **805** in 36% yield is, instead, obtained if the reaction is performed in ethanol at 0 °C (Scheme 13.252) [570].

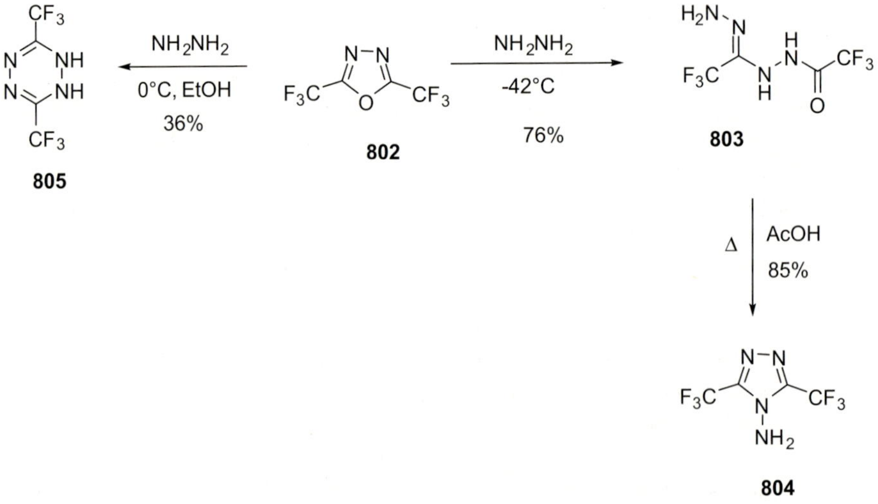

Scheme 13.252

2-Aryloxadiazolinethiones **806** also react with hydrazines, giving rise to triazoline thione derivatives **807** (Scheme 13.253) [571].

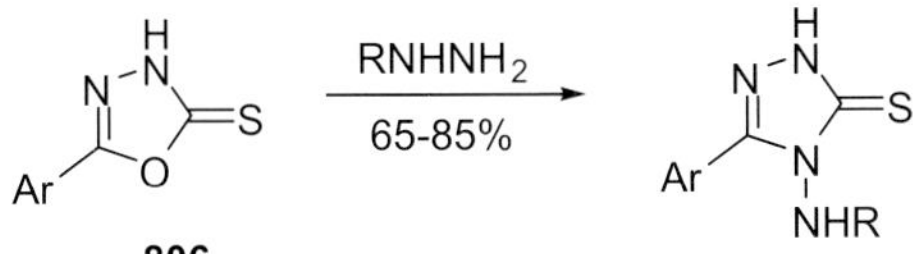

806 → 807 (RNHNH$_2$, 65-85%)

Ar = Ph, 4-MeO-C_6H_4, 4-NO_2-C_6H_4, 4-Pyridyl R = H, Ph, Me

Scheme 13.253

The ring-opening reaction promoted by hydrazine has been used by El-masry *et al.* to prepare, starting from 5-[2-(2-methylbenzimidazol-1-yl)ethyl-[1,3,4]-oxadiazole-2(3*H*)thione (**808**), 1-[(1-amino-2-mercapto-1,3,4-triazol-5-yl)ethyl]-2-methylbenzimidazole **809**, a biologically active compound, which possesses a moderate activity against *Bacillus cereus* (Scheme 13.254) [572].

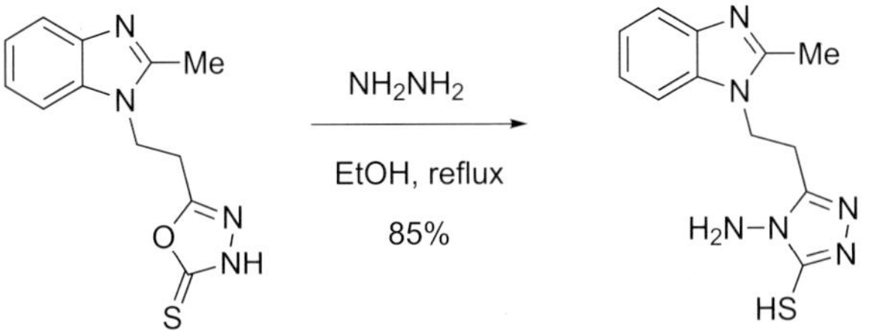

808 → 809 (NH_2NH_2, EtOH, reflux, 85%)

Scheme 13.254

The reaction of **802** with primary alkyl amines **810** in methanol at 42 °C leads to complexes **811** whose structure has been elucidated by X-ray crystal analysis. These complexes can be conveniently transformed into 4-substituted-3,5-bis(trifluoromethyl)-4*H*-1,2,4-triazoles **812** by heating in methanol. The reaction of **802** with aromatic amines **813**, performed at reflux, provides directly the triazole derivatives **814** in moderate to good yields (Scheme 13.255) [573].

1,3,4-Oxadiazol-2-ones **815** react with water to form 1,5-diacylcarbohydrazides **819**. The pathway of this reaction appears to be the hydrolytic ring opening to form the hydrazide **818**, via either an acylhydrazinoformic acid **816** or the acyl hydrazonoformic acid **817**, followed by loss of carbon dioxide to produce **818**, which, in turn, attacks the remaining oxadiazolone **815** to form the observed product **819** (Scheme 13.256) [574].

Oxadiazolones **815** also react with hydrazine and amines to give semicarbazides **820** and carbohydrazides **821**, respectively (Scheme 13.257) [575].

This reaction is quite general and similar to the above reported reactions. The reaction with NaOMe promotes the formation of **822** (Scheme 13.258) [575].

3-Substituted 5-trifluoromethyl-1,3,4-oxadiazolones **823a–d** are attacked by N and S-nucleophiles to give, as initial products, compounds deriving from the ring-opening reaction. In some cases, ring-enlargement products are formed. The

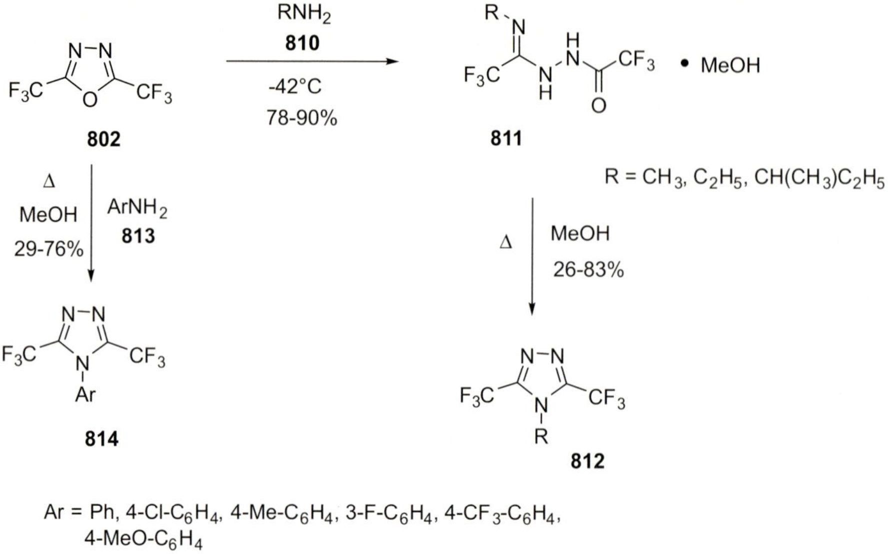

$R = CH_3, C_2H_5, CH(CH_3)C_2H_5$

$Ar = Ph, 4\text{-}Cl\text{-}C_6H_4, 4\text{-}Me\text{-}C_6H_4, 3\text{-}F\text{-}C_6H_4, 4\text{-}CF_3\text{-}C_6H_4, 4\text{-}MeO\text{-}C_6H_4$

Scheme 13.255

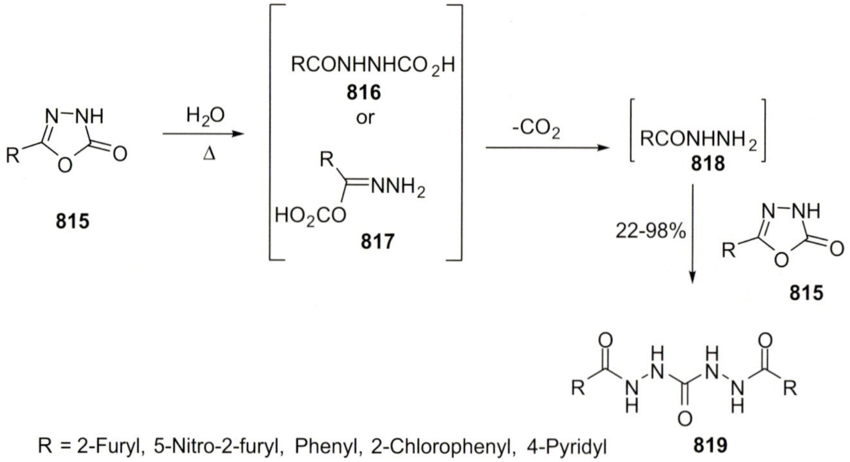

R = 2-Furyl, 5-Nitro-2-furyl, Phenyl, 2-Chlorophenyl, 4-Pyridyl

Scheme 13.256

reaction of 3-iodomethyl-5-trifluoromethyl-1,3,4-oxadiazol-2-(3*H*)-one (**823d**) with thiols (**827**) produces **830** through a Grob-type fragmentation (Scheme 13.259) [576].

2-Methyl-6-phenylimidazo[2,1-*b*]oxadiazole **831**, a cyclic oxadiazolimine, is cleaved with concentrated HCl or an 8% solution of KOH to give **832** and then **833** in quantitative yield. The reaction of **831** with 48% HBr gives **834** in 40% yield. In addition, **833** has been transformed into **834** by reaction with HBr (Scheme 13.260) [577].

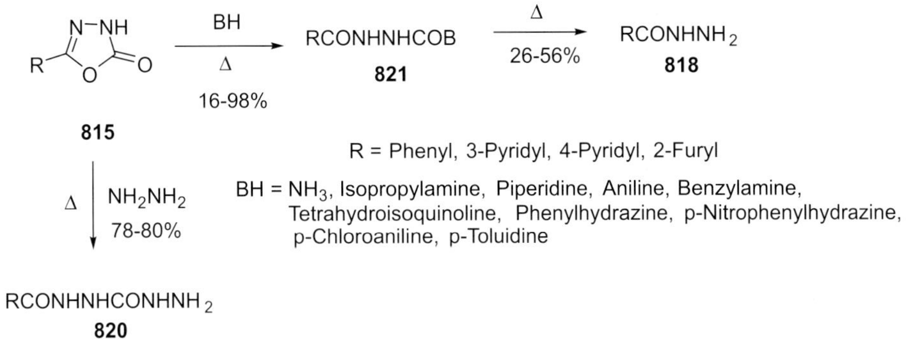

Scheme 13.257

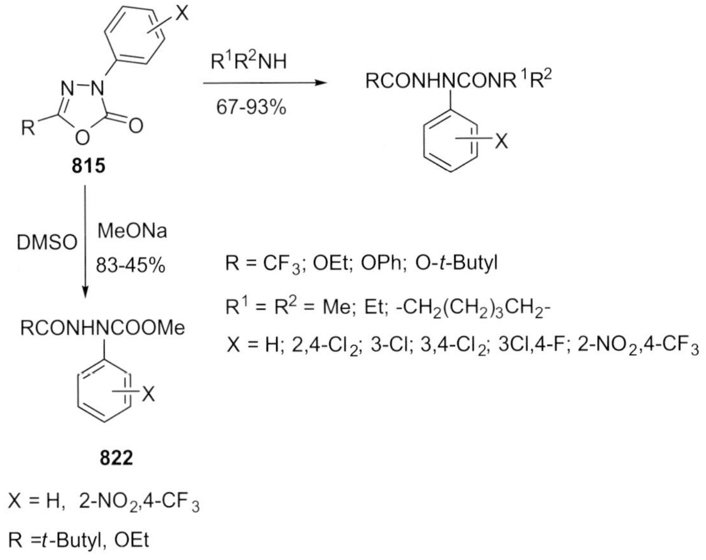

Scheme 13.258

2,5-Diphenyl-1,3,4-oxadiazole **835** is cleaved under photolytic conditions in the presence of alcohols to yield benzonitrile imine **837** and benzoic acid esters **838** in moderate yields [578]. These compounds are produced by an initial nucleophilic attack of alcohols on the C=N bond of the oxadiazole ring, followed by cycloelimination. Moreover, as expected, the benzonitrile imine **837** undergoes a 1,3-dipolar cycloaddition with the unreacted 1,3,4-oxadiazole **835** to furnish the bicycloadduct **839**. This compound is then transformed into benzamide (**841**) and 3,5-diphenyl-1,2,4-triazole (**842**), via 4-benzamido-3,5-diphenyl-1,2,4-triazole (**840**) produced by ring opening of **839** and concurrent hydrogen shift (Scheme 13.261).

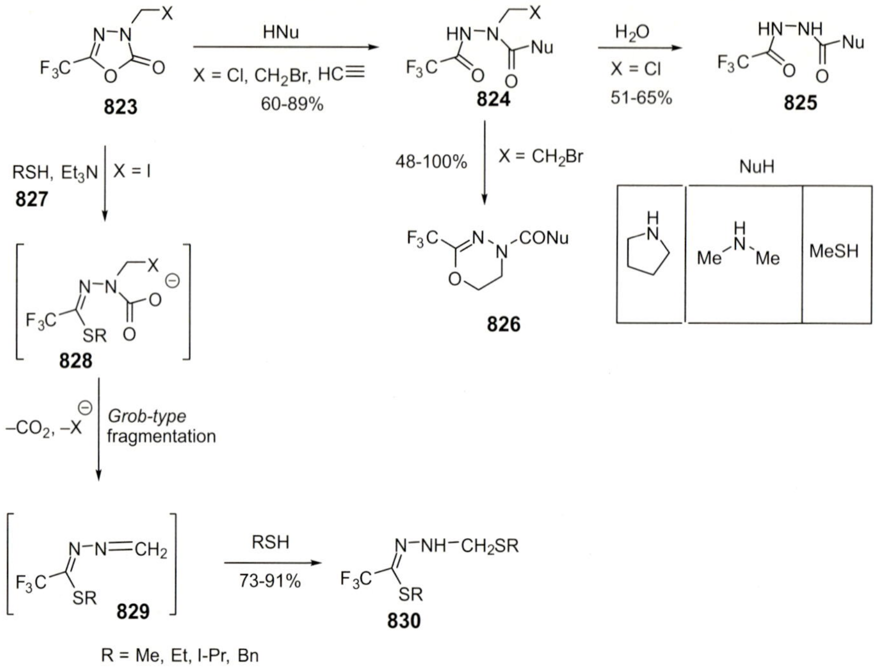

Scheme 13.259

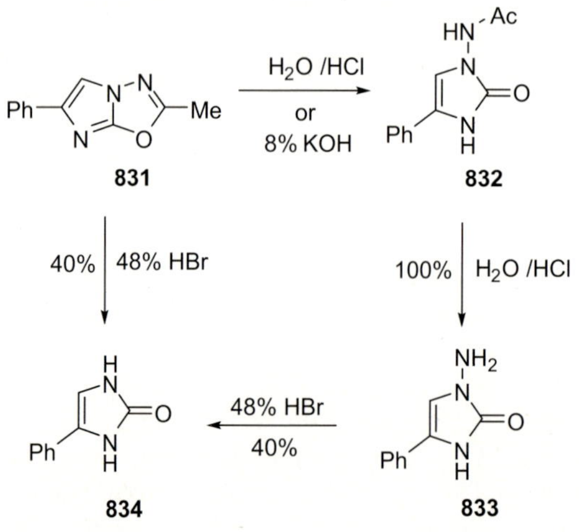

Scheme 13.260

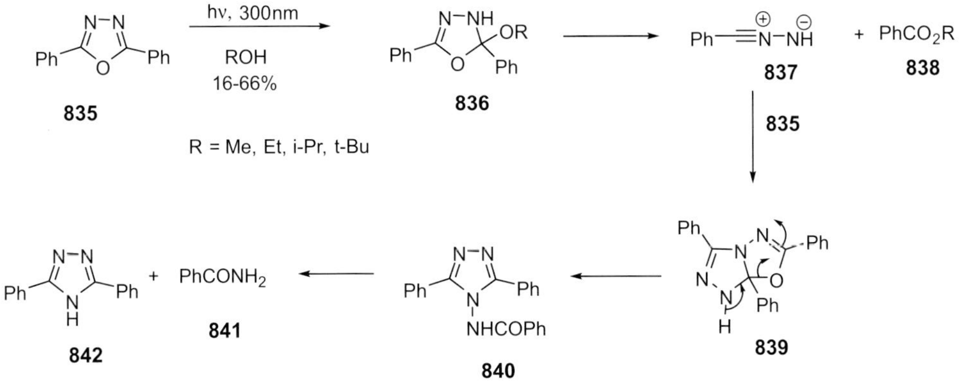

Scheme 13.261

In the case of 2-phenyl-1,3,4-oxadiazole (**843**), the regioselective addition of methoxy group at C2 of 1,3,4-oxadiazole ring affords, as the only detectable compound, 1-(α-methoxybenzylidene)-2-formylhydrazine **845** in 7% yield, produced by an initial nucleophilic attack of methanol followed by a ring opening reaction (Scheme 13.262).

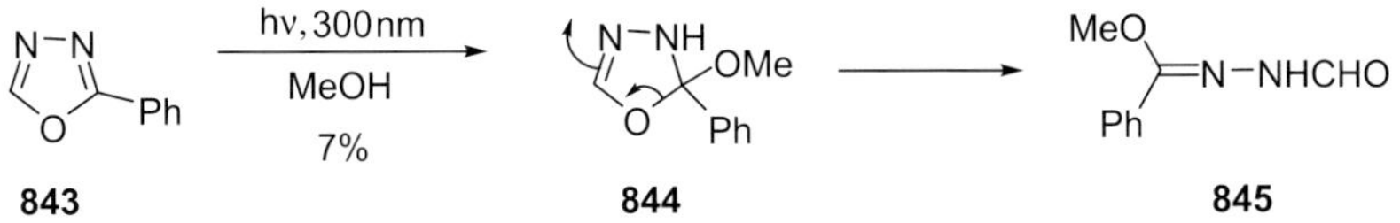

Scheme 13.262

Thione **846** forms a stable salt with *p*-toluidine (**847**), which gives rise to ring-open product **848** on heating (Scheme 13.263) [579].

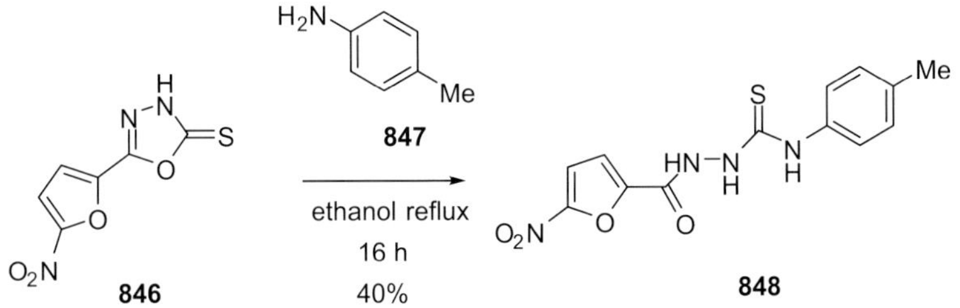

Scheme 13.263

Alkylhydrazine and arylhydrazines **850** react with 1,3,4-oxadiazolium bromides **849** to produce 2-methyl or 2-phenyl-4-acylamino-3-imino-6-aryl-2,3,4,5-tetrahydro-1,2,4-triazines **852**. These compounds are probably obtained via intermediate **851**, which rearranges to **852** through a ring opening reaction followed by a ring closure (Scheme 13.264) [580].

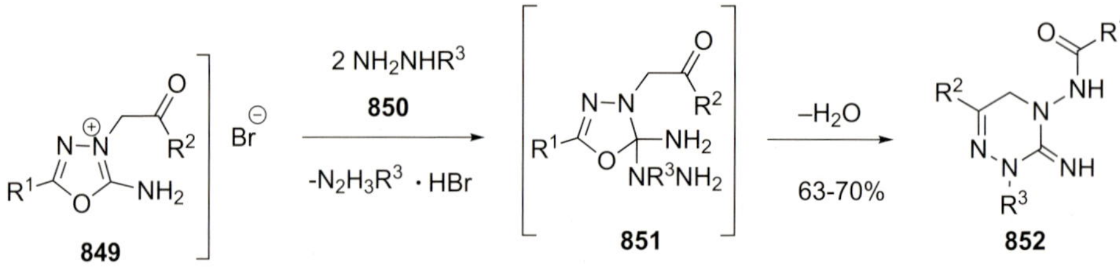

R^1 = Me, Et, Ph, Ph-CH_2 R^2 = Ph, 4-Cl-C_6H_4, 4-Me-C_6H_4, 4-MeO-C_6H_4 R^3 = Me, Ph

Scheme 13.264

1,3,4-Oxadiazolium salts **853** react with ethyl cyanoacetate (**854**) in the presence of triethylamine to yield 1,5-substituted 3-aminopyrazole-4-carboxylic esters **857**. An open chain intermediate **855** was isolated and the reaction involves an initial attack at C2 of the oxadiazole ring (Scheme 13.265) [581].

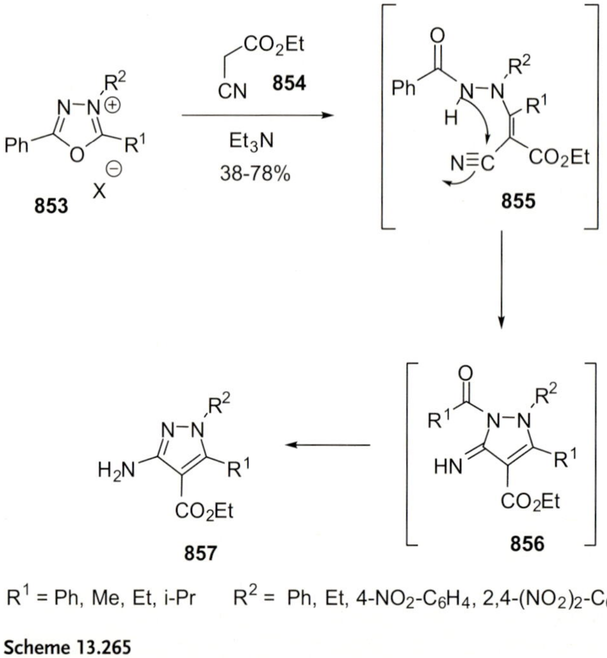

R^1 = Ph, Me, Et, i-Pr R^2 = Ph, Et, 4-NO_2-C_6H_4, 2,4-$(NO_2)_2$-C_6H_3

Scheme 13.265

Isosydnone **858** undergoes a ring-opening reaction by treatment with sodium hydroxide followed by addition of an ethanol–hydrogen chloride mixture to give 1-benzoyl-1-methylhydrazine hydrochloride (**859**) (Scheme 13.266) [541].

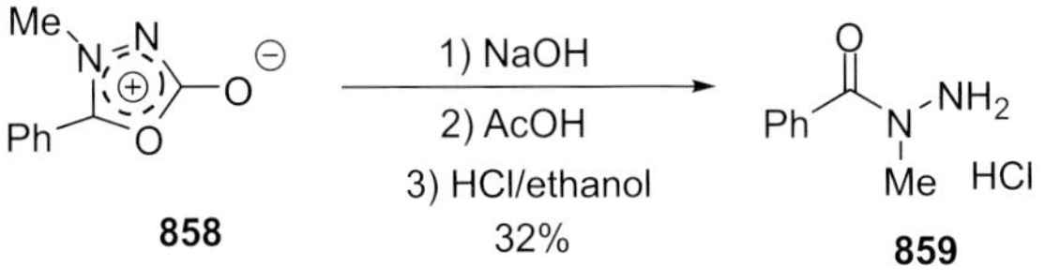

Scheme 13.266

13.5.3.2 Oxidative and Reductive Processes

1,3,4-Oxadiazoles are very stable to strong oxidizing and reducing agents. However, some oxidations or reductions involving atoms linked to the heterocycle ring have been performed. Thus, recently, sulfonyl derivatives **863**, with antifungal activity, containing trimethoxyphenyl substituted 1,3,4-oxadiazoles have been synthesized, in 67–94% yield, by hydrogen peroxide oxidation, catalyzed by ammonium molybdate in ionic liquid ([bmim]PF6), of substituted 1,3,4-oxadiazole sulfide **860** [582]. In particular, 1,3,4-oxadiazole sulfides **862** have been prepared, in 31–93% yield, by thioetherification, catalyzed by indium tribromide, of 5-(3,4,5-trimethoxyphenyl)-1,3,4-oxadiazole-2-thiol (**860**) with organic halides (Scheme 13.267).

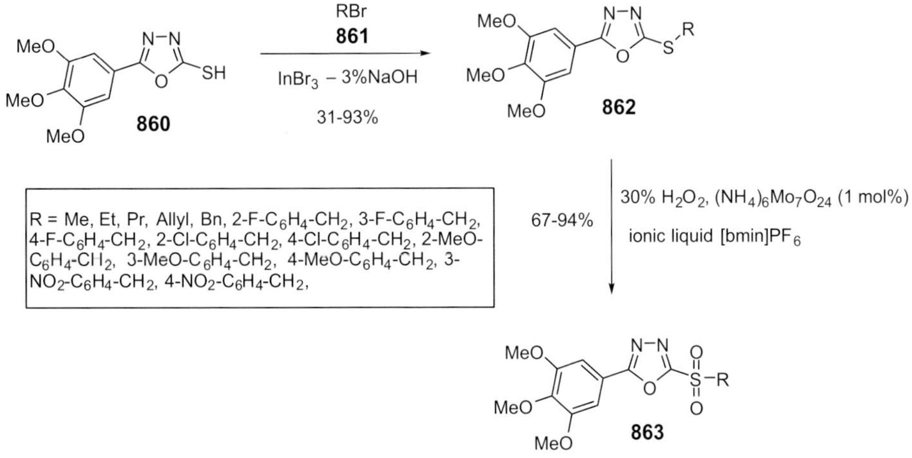

Scheme 13.267

The oxidation of sulfides **864** with m-CPBA furnishes, in contrast, the corresponding sulfoxides **865** (Scheme 13.268) [583].

It has also been reported that the oxidation of **866** with hydrogen peroxide and no catalyst affords oxadiazolone derivatives **867** (Scheme 13.269) [584].

Oxidation of 2,5-di *m*- or *p*-tolyl-1,3,4-oxadiazoles **868** with potassium permanganate/pyridine leads to the corresponding dicarboxylic acids **869** (78–94%) [585]; if the oxidation is performed with chromium trioxide/acetic anhydride, diacetoxymethyl derivatives **870** are obtained. These latter compounds can be conveniently transformed by acid hydrolysis into dialdehydes **871** (Scheme 13.270) [586].

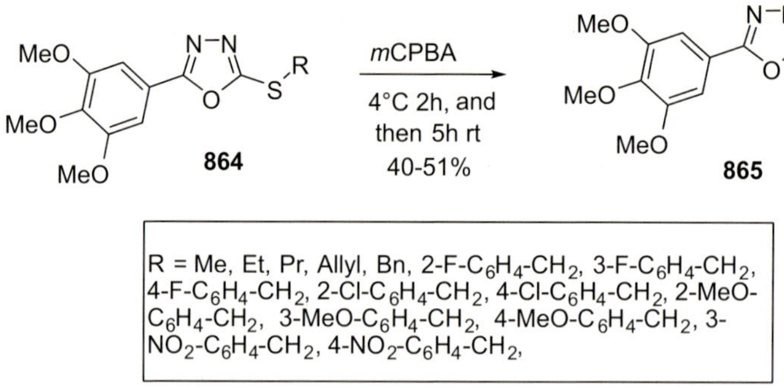

$$\textbf{864} \xrightarrow[\text{4°C 2h, and then 5h rt, 40-51\%}]{m\text{CPBA}} \textbf{865}$$

R = Me, Et, Pr, Allyl, Bn, 2-F-C_6H_4-CH_2, 3-F-C_6H_4-CH_2, 4-F-C_6H_4-CH_2, 2-Cl-C_6H_4-CH_2, 4-Cl-C_6H_4-CH_2, 2-MeO-C_6H_4-CH_2, 3-MeO-C_6H_4-CH_2, 4-MeO-C_6H_4-CH_2, 3-NO_2-C_6H_4-CH_2, 4-NO_2-C_6H_4-CH_2,

Scheme 13.268

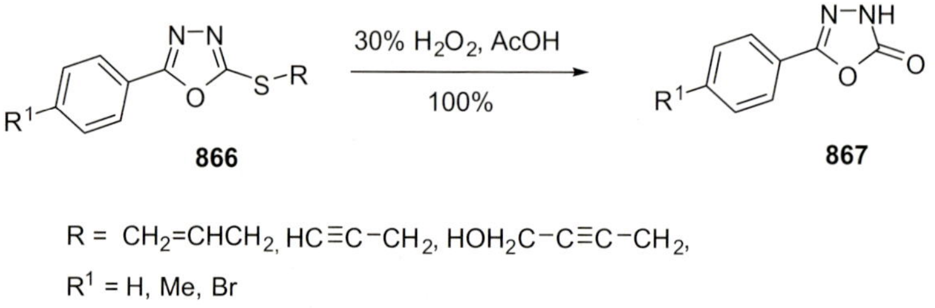

$$\textbf{866} \xrightarrow[\text{100\%}]{\text{30\% } H_2O_2\text{, AcOH}} \textbf{867}$$

R = CH_2=$CHCH_2$, HC≡C−CH_2, HOH_2C−C≡C−CH_2,

R^1 = H, Me, Br

Scheme 13.269

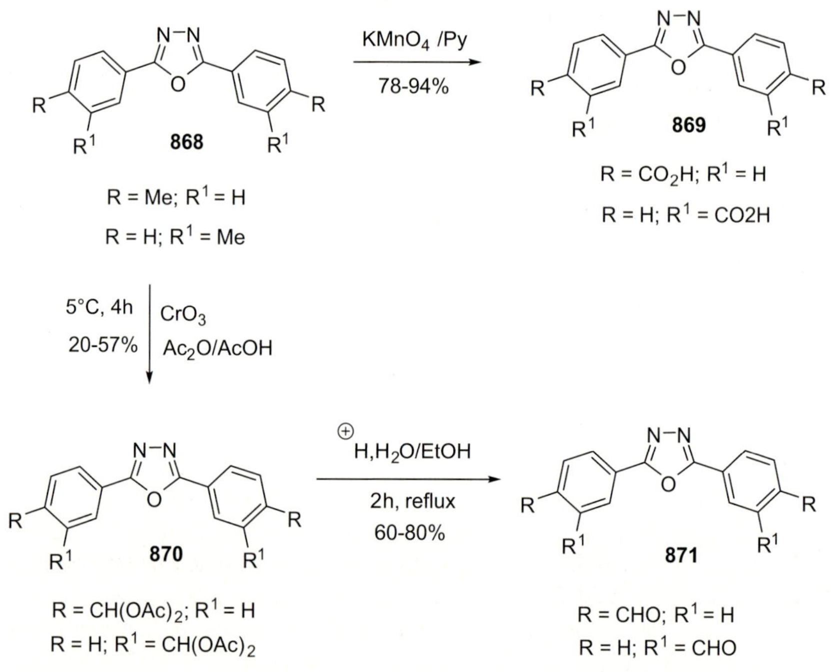

$$\textbf{868} \xrightarrow[\text{78-94\%}]{KMnO_4\text{ /Py}} \textbf{869}$$

868: R = Me; R^1 = H; R = H; R^1 = Me

869: R = CO_2H; R^1 = H; R = H; R^1 = CO2H

868 → **870**: CrO_3, Ac_2O/AcOH, 5°C, 4h, 20-57%

$$\textbf{870} \xrightarrow[\text{2h, reflux, 60-80\%}]{H^{\oplus}\text{, } H_2O\text{/EtOH}} \textbf{871}$$

870: R = $CH(OAc)_2$; R^1 = H; R = H; R^1 = $CH(OAc)_2$

871: R = CHO; R^1 = H; R = H; R^1 = CHO

Scheme 13.270

Reduction of the nitro group linked to phenyl moiety of 1,2,4-oxadiazoles with phenylhydrazine or hydrogen/palladium has been reported to give aminoaryl 1,3,4-oxadiazoles in good yield [587]. Thus, 2-phenyl-5-(*p*-nitrophenyl)-1,3,4-oxadiazole (**872**), carefully heated to 110–15 °C for 75–90 min, gives 2-phenyl-5-(*p*-aminophenyl)-1,3,4-oxadiazole **873** in 84% yield (Scheme 13.271) [588]. Similar results have been obtained with nitrophenyl derivatives such as **874**, which upon hydrogenation with Pd/C yields **875** [589].

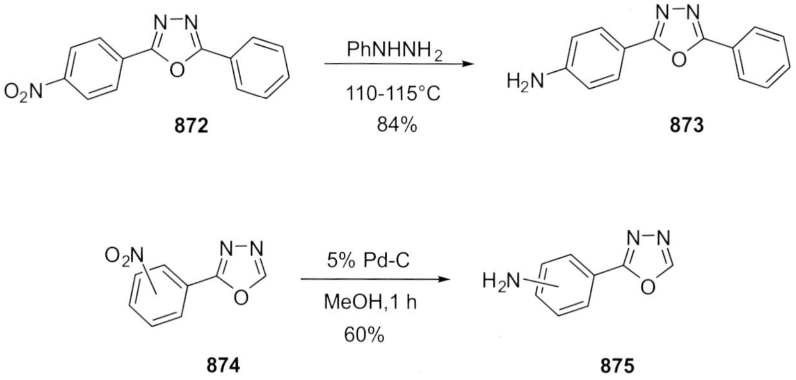

O_2N N–N O **872** → PhNHNH$_2$, 110-115°C, 84% → H_2N N–N O **873**

O_2N N–N O **874** → 5% Pd-C, MeOH,1 h, 60% → H_2N N–N O **875**

Scheme 13.271

Hydrogenation, performed on 5-*tert*-butyl-3-(4-chloro-2-nitrophenyl)-1,3,4-oxadiazolin-2-one (**876**) in AcOEt, conversely, leads to a partial reduction of nitro group with the formation of the 4-chloro-2-(hydroxyamino)phenyl derivative **877** in 44% yield (Scheme 13.272) [590].

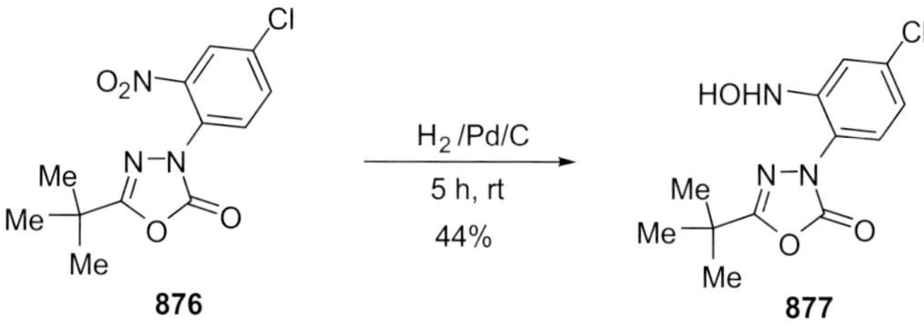

Cl O_2N Me Me Me N–N O O **876** → H_2 /Pd/C, 5 h, rt, 44% → Cl HOHN Me Me Me N–N O O **877**

Scheme 13.272

It has been reported that the hydrogenation of Δ^3-1,3,4-oxadiazoline **878**, performed in ethanol over Pd/C, gives acetic acid and *N*-cyclohexyl-*N*-benzoylhydrazine (**880**) via the corresponding 2,3,4,5-tetrahydro-1,3,4-oxadiazole derivative **879** (Scheme 13.273) [591].

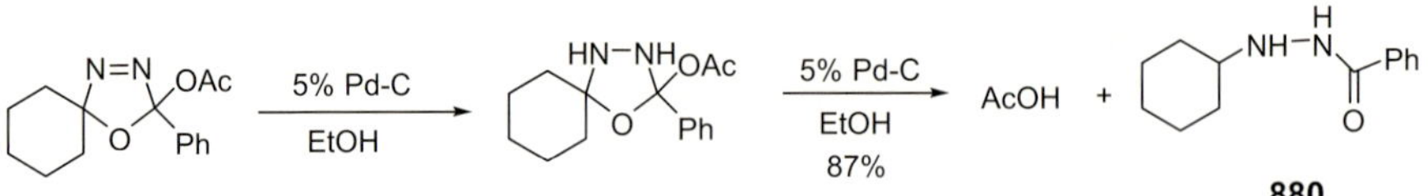

Scheme 13.273

878 → (5% Pd-C, EtOH) → 879 → (5% Pd-C, EtOH, 87%) → AcOH + 880

13.5.3.3 Reactions due to the Reactivity of the Heterocyclic Ring

1,3,4-Oxadiazoles are weak bases. The pK_a values of 2,5-diaryl-1,3,4-oxadiazoles measured by the method of Yates and MacClelland in aqueous solution of sulfuric acid are in the range of −1.15 to −2.49. 2-Amino derivatives (pK_a = 2.3−2.7) are more basic and form stable salts. As already reported, some hydrochloride salts have been obtained by hydrochloric reaction of 2-amino-5-aryl-1,3,4-oxadiazoles, in DMF/H_2O or in ethanol–ether as solvents, giving rise to compounds having muscle relaxant properties (Scheme 13.230) [549]. In the same context, 5-imino-2-phenyl-Δ^2-1,3,4-oxadiazoline-maleate, -citrate, -sulfate, and -nitrate, together with 5-imino-2-(*p*-aminophenyl)-Δ^2-1,3,4-oxadiazoline dihydrochloride, 5-imino-2-(1-ethylpropyl)-Δ^2-1,3,4-oxadiazoline hydrochloride, 5-imino-2-(1-ethylbenzyl)-Δ^2-1,3,4-oxadiazoline hydrochloride, and 1-ethyl-3-(5-phenyl-1,3,4-oxadiazol-2-yl)urea have also been prepared.

Electrophilic substitution on the C-atoms of the ring is difficult, because protonation of the nuclear nitrogen in acidic media reduces strongly the possibility of electrophilic attack. Thus, no nitrations, sulfonations, or halogenations of unsubstituted oxadiazoles are known. Mono-substituted derivatives are not able to react-with electrophiles because they are sensitive to acid conditions. In fact, for example, 2-phenyl 1,3,4-oxadiazole is easily hydrolyzed by acids at room temperature to give benzohydrazide and formic acid [588]. In addition, the mono- and 2,5-dialkyl-derivatives undergo a ring-opening reaction on treatment with acids [592].

There are several examples of reactions of alkyl halides with 1,3,4-oxadiazole derivatives. The alkylation reactions occur preferentially at the N3 ring atom, except for amino and thio derivatives, where the alkylation, essentially, occurs at the sulfur or the exo-nitrogen atom.

Thus, it has been reported that the reaction of 5-(4-chloro-3-ethyl-1-methyl-1H-pyrazole-5-yl)-1,3,4-oxadiazole-2-one (**881**) with methyl iodide in the presence of NaOH gives rise to the corresponding *N*-methyl derivative **882**, while alkylation of the 2-thioxo derivatives **883** with NaOH, tetrabutylammonium bromide (TBAB), and alkyl iodide leads to the corresponding thioalkylated derivatives **884** in good yield. In the same context, it has been noted that the 2-alkylthio derivatives so obtained are active against rice sheath blight, which is a major disease of rice in China (Scheme 13.274) [593].

Analogously, a series of bis-oxadiazolyl sulfides **887** (R = Ph, substituted Ph) have been synthesized via alkylation reaction of 5-[[2-(trifluoromethyl)-1H-benzimidazol-1-yl]methyl]-1,3,4-oxadiazole-2(3)thione(2-trifluoromethylbenzimidazol-1-ylmethyl)-5-mercapto-1,3,4-oxadiazoles **885** with 2-aryl-5-chloromethyl-1,3,4-oxadiazoles **886**. Interestingly, the relative oxidation of these sulfides performed at 0 °C

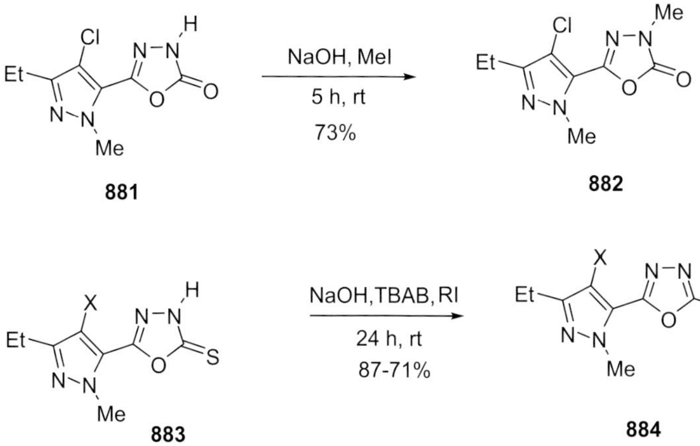

881 882

883 884

X = H, Cl
R = Me, *n*-Pr, *n*-C_5H_{11}, *n*-C_7H_{15}, *n*-C_8H_{17}, $CH(CH_3)CO_2Et$

Scheme 13.274

with HNO_3 produces the sulfoxide derivatives **888** in good yield (Scheme 13.275) [594].

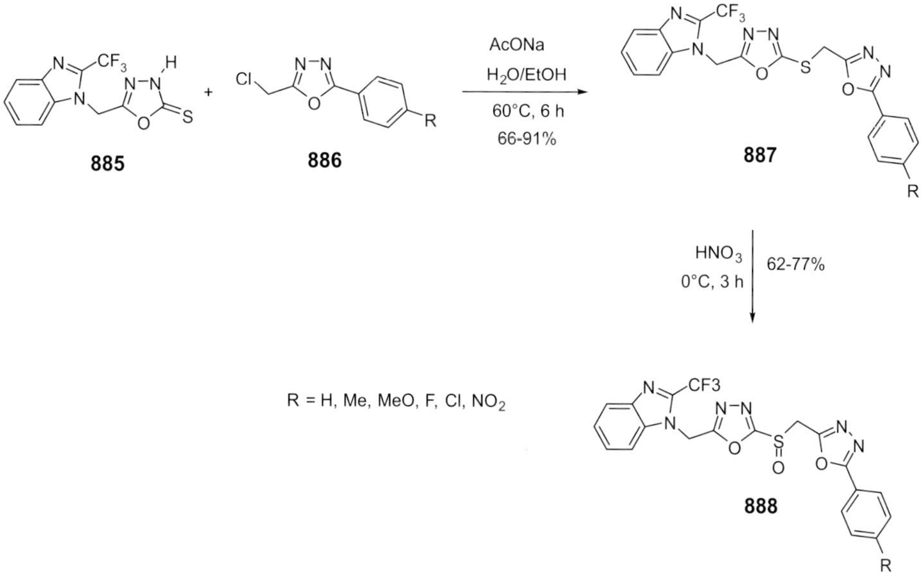

885 886 887

888

R = H, Me, MeO, F, Cl, NO_2

Scheme 13.275

Methylation of 5-methyl or 5-aryl-2-thioxo-2,3-dihydro-1,3,4-oxadiazoles **889** with trimethyloxonium tetrafluoroborate in CH_2Cl_2 at room temperature furnishes, as expected, the corresponding methyl(methylthio)oxadiazolium tetrafluoroborates **890** in 86–96% yield (Scheme 13.276) [595].

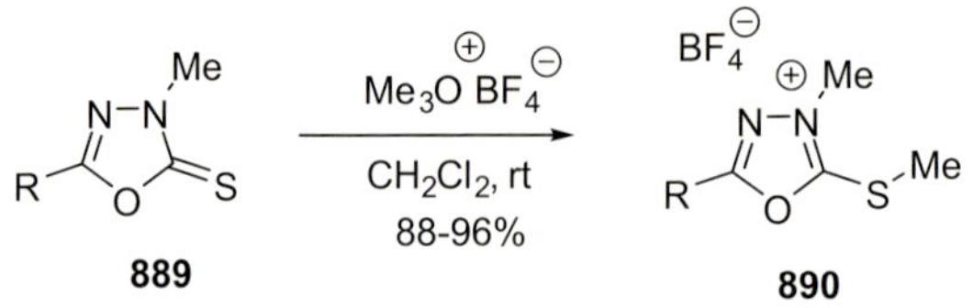

$R = Me, Ph, 4\text{-}Me\text{-}C_6H_4, 4\text{-}MeO\text{-}C_6H_4$

Scheme 13.276

The Mannich reaction of 1,3,4-oxadiazole-2-thione derivatives **891** with different secondary amines and paraformaldehyde in absolute ethanol occurs at the N3 ring atom and leads to 5-[2-(2-methylbenzimidazol-1-yl)ethyl-3-*N*-methylamino-1,3,4-oxadiazole-2-thiones (**892**) in 60–65% yield (Scheme 13.277). In particular, the diethylamino derivative **892** has shown to exhibit moderate antimicrobial activity against one strain of Gram-positive bacteria (*Bacillus cereus*) [596].

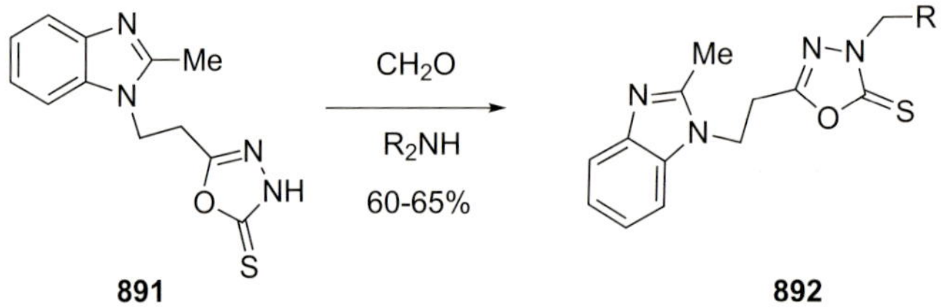

$R = N(C_2H_5)_2, N(C_4H_8O), N(C_4H_8)NMe$

Scheme 13.277

The 1,3,4-oxadiazole ring **893** has been, recently, used as 4π component in a Diels–Alder reaction. This ring is considered an electron poor diaza-diene and reacts with extremely electron rich (aminoacetylenes) or strained dienophiles in an inverse electron demand reaction. Unfortunately, the mono cycloadduct **895** thus obtained has never been isolated, but it extrudes N_2 and generates a carbonyl ylid **896**, which further reacts with olefin in a 1,3-dipolar cycloaddition to give the final product **897** (Scheme 13.278).

The first example of such a cycloaddition cascade was reported by Vasiliev *et al.* Heating at 200–220 °C of the 2,5-bis(trifluoromethyl)-1,3,4-oxadiazole **802** with ethylene (**898**) or cyclopentadiene (**899**) affords the oxabicycloheptane **900** in 41%, or oxatetracyclotridecane **901** in 33% yield respectively (Scheme 13.279 [597].

Other examples of this cycloaddition have been reported, giving rise to the formation of strained structures (Scheme 13.280) [598].

In some cases, the intramolecular version of this reaction is more productive and leads to bicyclic or monocyclic derivatives (Scheme 13.281) [599].

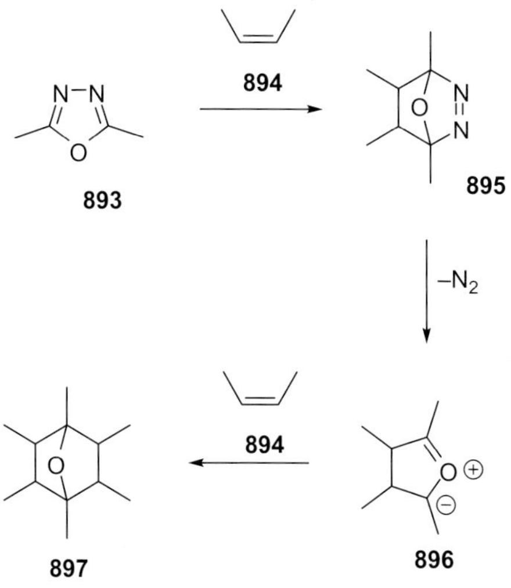

894

893

895

$-N_2$

894

897

896

Scheme 13.278

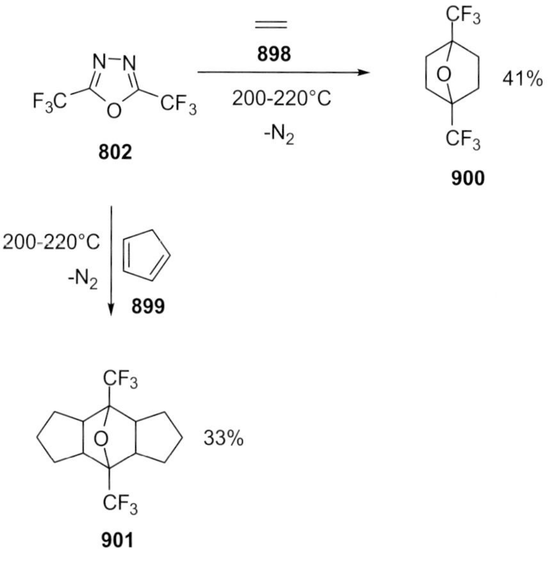

898

200-220°C

$-N_2$

802

41%

900

200-220°C

$-N_2$

899

33%

901

Scheme 13.279

The synthetic efficiency of the process can be improved through the development of domino reactions that allow the formation of complex compounds, starting from simple substrates, in a single transformation consisting of several steps. A domino reaction can be defined as a process involving two or more bond-forming transformations that take place under the same reaction conditions, without adding

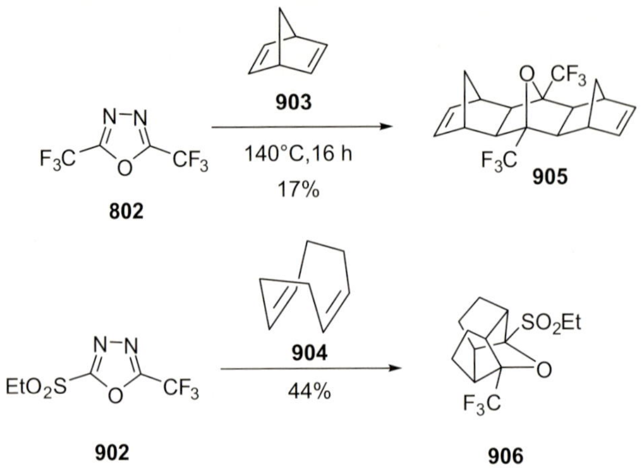

Scheme 13.280

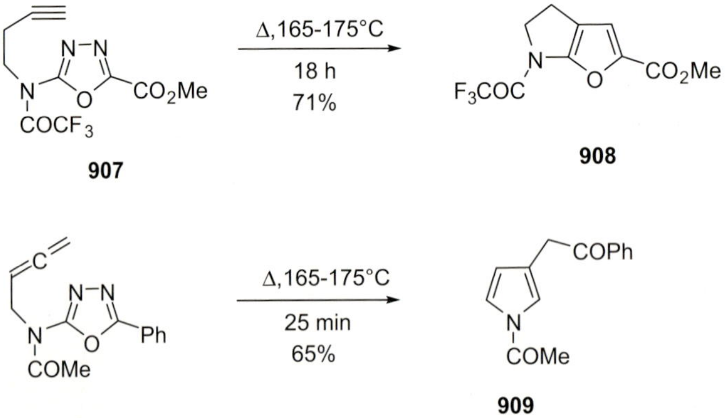

Scheme 13.281

additional reagents and catalysts and in which the subsequent reactions result as a consequence of the functionalities obtained in the previous step.

Thus, the intramolecular Diels–Alder (DA)/1,3-dipolar cycloaddition (1,3-DC) cascade of 1,3,4-oxadiazoles has became a powerful tool for the rapid generation of molecular complexity. Specifically, this methodology has been featured in the construction of the pentacyclic ring systems and ultimately in the total syntheses of vindoline and several structurally related natural products [600]. Vindoline (**913**) constitutes the most complex half of vinblastine (**914**), a member of the bisindole alkaloid family that is used as an antineoplastic drug. The method is based on a combination of a DA and 1,3-DC. The synthesis proceeds by a diastereoselective tandem [4 + 2]/[3 + 2]-cycloaddition of a substituted 1,3,4-oxadiazole. The reaction leading to vindoline is initiated by an intramolecular [4 + 2]-cycloaddition of 1,3,4-oxadiazole **910** with the tethered enol ether. Loss of N_2 from the initially formed cycloadduct **911** provides the carbonyl ylid dipole **912**, which undergoes a subsequent 1,3-dipolar cycloaddition across the proximal indole moiety (Scheme 13.282) [601].

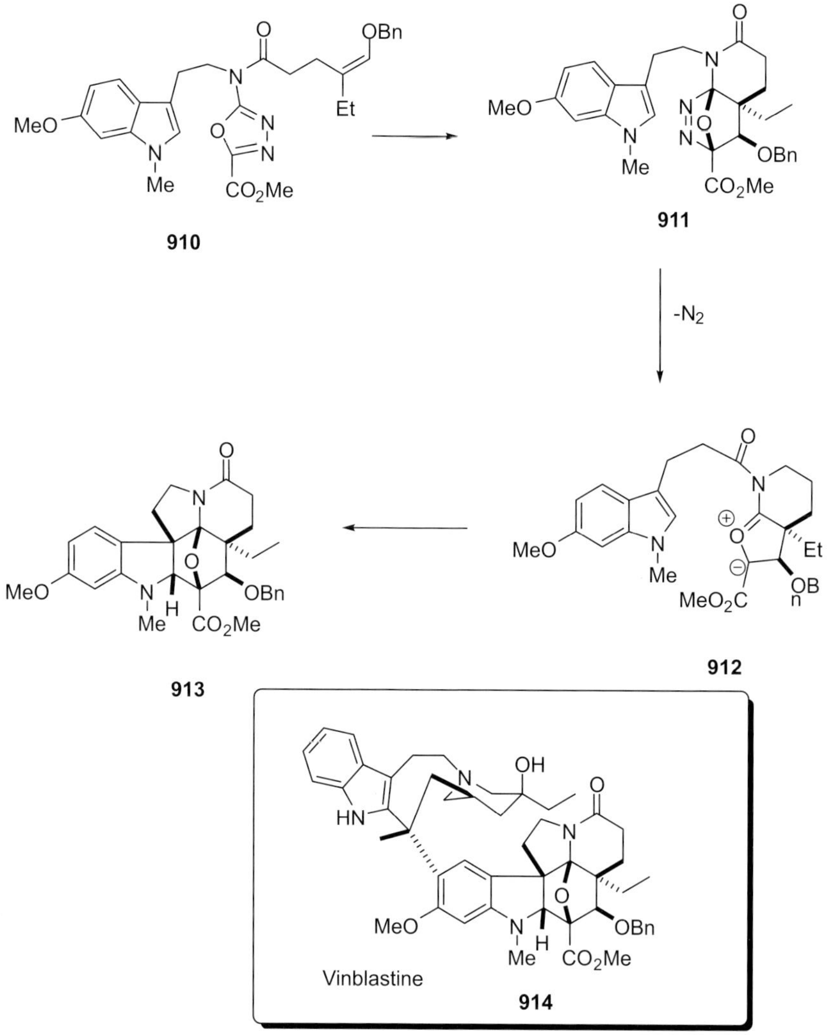

MeO N Me CO$_2$Me N O N N O OBn Et **910**

MeO N Me N N O N O OBn CO$_2$Me **911**

$-N_2$

MeO N Me O N O Et OBn MeO$_2$C **912**

MeO N H Me CO$_2$Me N O O OBn **913**

HN N OH MeO N H Me CO$_2$Me N O O OBn Vinblastine **914**

Scheme 13.282

Some other examples of this methodology (**916**, **918**) are reported in Scheme 13.283 [599].

The double bond of the 1,3,4-oxadiazole ring has also been used as 2π component in a [2 + 2] photochemical cycloaddition. Thus, it has been reported that 2,5-diphenyl-1,3,4-oxadiazole (**835**) with indene (**919**), with or without benzophenone as a sensitizer, affords the bis adduct diazetidine derivative **920**, while if the reaction is

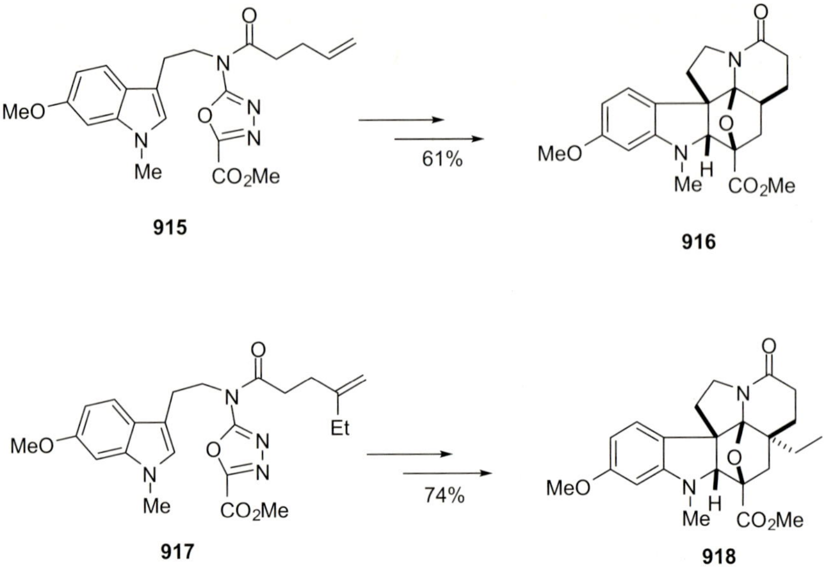

Scheme 13.283

performed in the presence of iodine the photoreaction leads to the mono-adduct **921** (Scheme 13.284) [602].

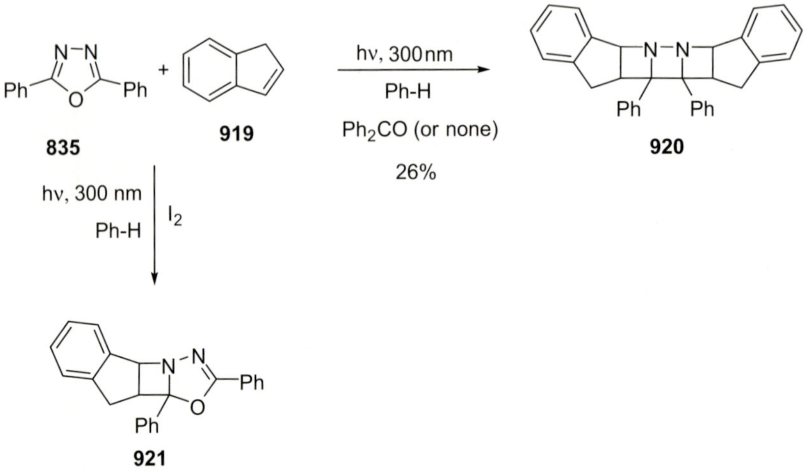

Scheme 13.284

A 1 : 1 adduct with **923** (26%) was obtained when a solution of **835** was irradiated with an excess of furan (**922**) in the presence or absence of benzophenone used as sensitizer. The reaction did not occur in the presence of a triplet quencher such as piperylene, indicating that the photoaddition takes place from an excited triplet state [603]. The presence of iodine promotes a different reaction pathway, leading to the formation of 3-benzoylfuran (**927**) (17%). This product has been rationalized by an initial valence tautomerization of furan into cyclobutadiene oxide **924**, which undergoes a [2 + 2] cycloaddition with a double bond of 1,3,4-oxadiazole ring, leading to monoadduct **925**. This compound, via ring opening and photoisomerization reaction, produces the *N′*-[3-furyl(phenyl)methylene]phenylhydrazide (**926**) that is easily hydrolyzed with trace amounts of water to give **927** (Scheme 13.285).

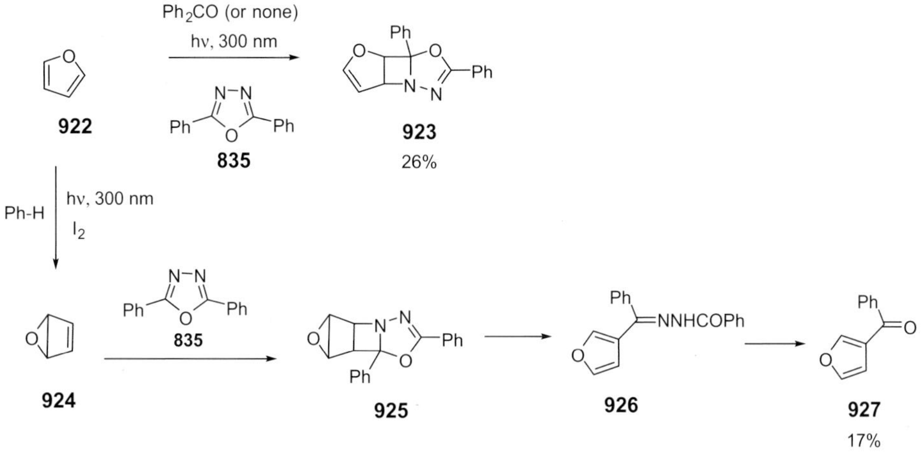

Scheme 13.285

13.5.3.4 Reactions with Nucleophiles

Direct nucleophilic substitutions of ring C-substituents in 1,3,4-oxadiazoles are seldom. These reactions occur only for compounds containing a good leaving group, via addition–elimination reaction. Thus, for example, 3-(5-phenyl-[1,3,4]oxadiazol-2-yl]pentane-2,4-dione (**930**, R = Me) and 3-(5-phenyl-[1,3,4]oxadiazol-2-yl]-dibenzoylmethane (**930**, R = Ph) have been synthesized, in 63% and 85% yield, respectively, by nucleophilic substitution of 2-methylsulfonyl-5-phenyl-1,3,4-oxadiazole (**928**) with β-diketone anions, formed by the corresponding carbonyl compounds **929** with NaH (Scheme 13.286) [604].

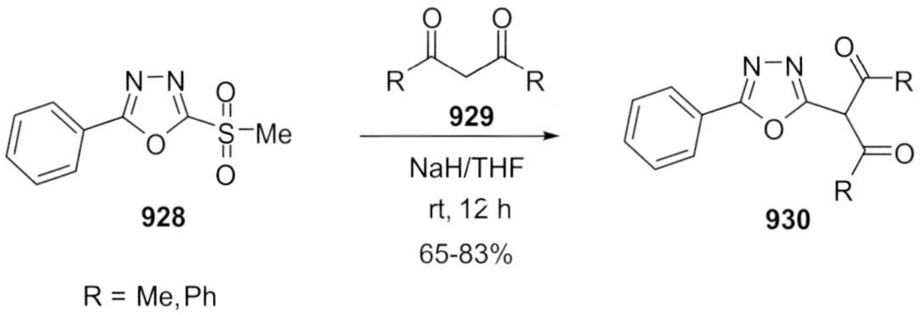

Scheme 13.286

A similar displacement reaction has been performed with 2-methylsulfonyl-5-pyrazolyl-1,3,4-oxadiazole **931** that by reaction with arylamines **932** produces bioactive 2-substituted-amino-5-pyrazolyl-1,3,4-oxadiazoles **933**, which exhibit moderate fungicidal activity (Scheme 13.287) [605].

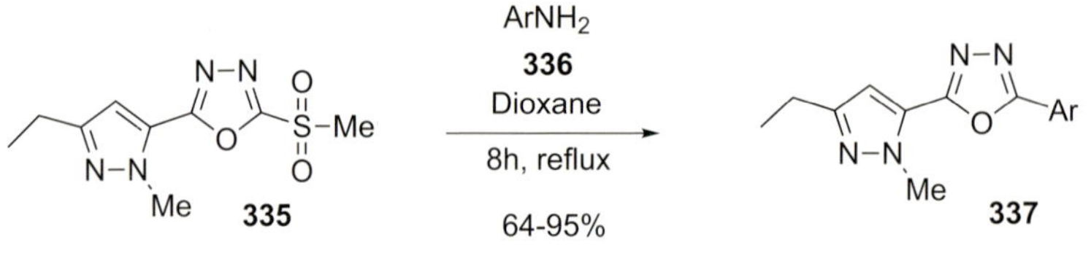

ArNH2 **336**, Dioxane, 8h, reflux, 64-95%; **335** → **337**

Ar = Ph, 4-Cl-C_6H_4, 3-Cl-C_6H_4, 4-Br-C_6H_4, 4-Me-C_6H_4, 4-MeO-C_6H_4,

Scheme 13.287

Nucleophilic substitution of 2-methylsulfonyl-1,3,4-oxadiazoles **934** has been reported to occur with other nucleophiles, such as sodium azide, amines, and acylhydrazines (Scheme 13.288). The compounds containing the acylhydrazine group have shown to posses strong antibacterial activity against *Bacillus subtilis* and *Escherichia coli* ($2.0 \times 10^{-4}\,\text{mol}\,\text{l}^{-1}$) [606].

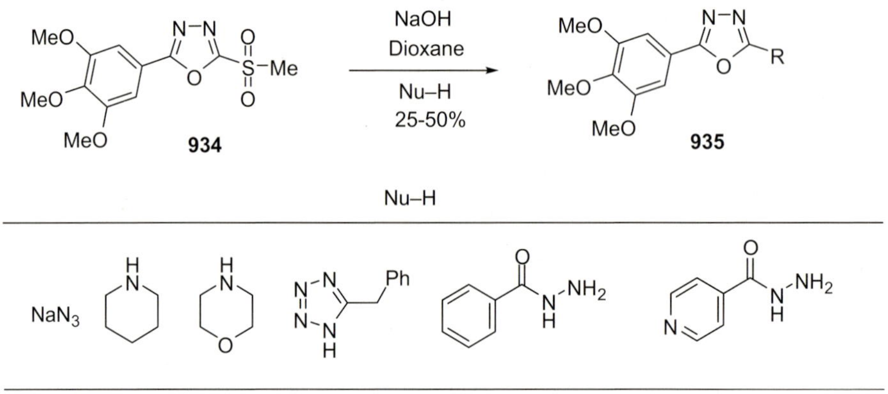

NaOH, Dioxane, Nu–H, 25-50%; **934** → **935**

Nu–H: NaN_3

Scheme 13.288

Recently an efficient conversion of 5-substituted-1,3,4-oxadiazolin-2-ones **936** into 2-amino-1,3,4-oxadiazoles **937**, via a nucleophilic aromatic substitution, appeared in the literature (Scheme 13.289) [607].

The reaction is activated from benzotriazol-1-yloxytris(dimethylamino)-phosphonium PF_6^- (**938**), and occurs according to Scheme 13.290, using as co-reagent 2 equivalents of base. The key step of the reaction is the attack of the oxygen atom of the carbonyl group on the phosphonium salt, promoted by base. The intermediate **939** thus obtained undergoes a facile reaction with the nucleophile **940** at C2 of the oxadiazole ring, followed by extrusion of HMPA (**941**).

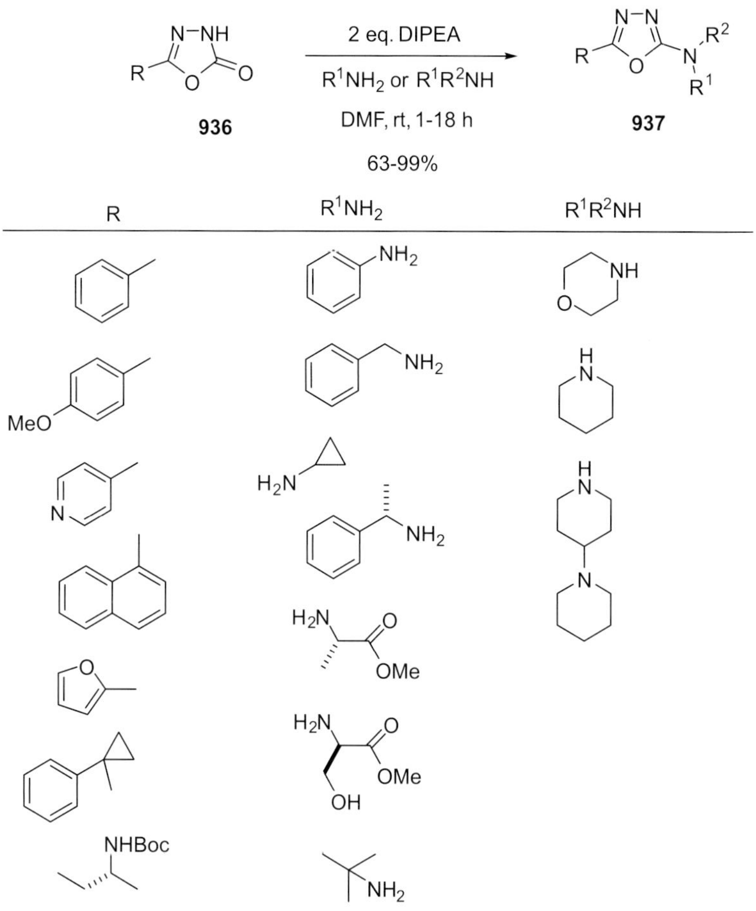

936 → 937: 2 eq. DIPEA, R^1NH_2 or R^1R^2NH, DMF, rt, 1-18 h, 63-99%

Scheme 13.289

A nucleophilic substitution has also been reported for quaternary intermediate salts obtained by reaction of alkyl iodide with phenylalkoxyoxadiazoles. The reaction produces the corresponding 3-alkyl-5-phenyl-1,3,4-oxadiazol-2-ones via a nucleophilic attack promoted by iodine ion on the R^1 group of quaternary salts (Scheme 13.291) [608].

13.5.3.5 **Reactions of Substituents**

Electrophilic reaction can occur on diphenyl derivatives. Thus, 2,5-bis(4-nitrophenyl)-, bis(3-nitrophenyl)- and bis(2-nitrophenyl)-1,3,4-oxadiazoles have been obtained in 27%, 20%, and 40% yield, respectively, by mixing 2,5-diphenyl-1,3,4-oxadiazole (**835**) with HNO_3 at 30 °C and then at 80 °C for 4 h (Scheme 13.292). The addition of HNO_3 to the oxadiazole in concentrated H_2SO_4 at 50 °C and subsequent heating for 6 h at 100 °C gave bis(3-nitrophenyl)-1,3,4-oxadiazole (38%) and 2-phenyl-5-(*m*-nitrophenyl)-1,3,4-oxadiazole (31%) [609].

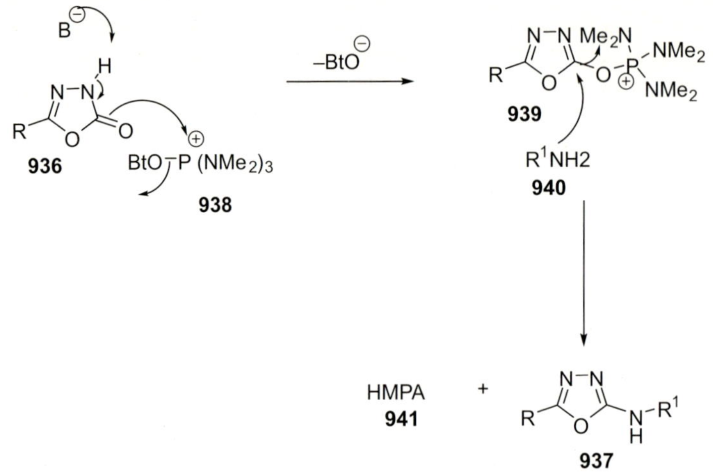

Scheme 13.290

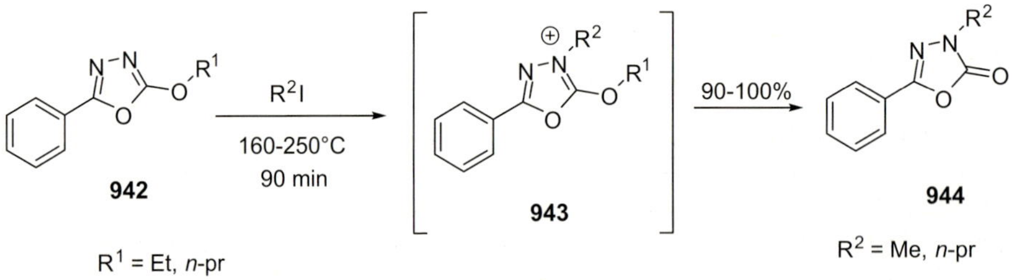

Scheme 13.291

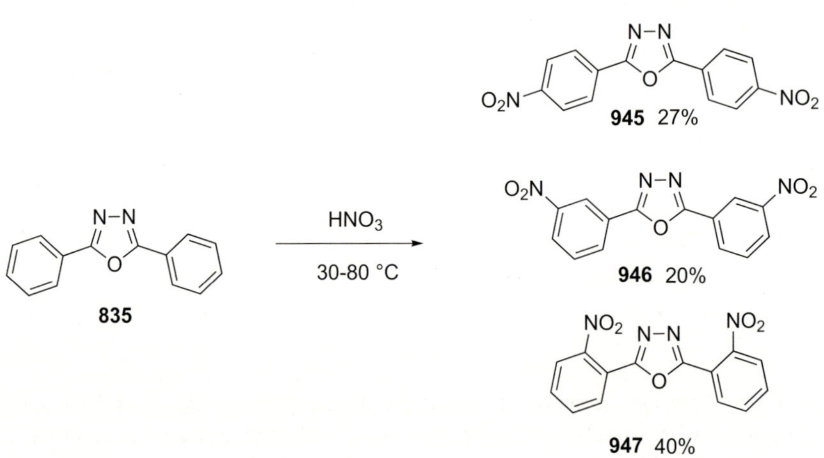

Scheme 13.292

2-(4-Nitrophenyl)-5-phenyl-1,3,4-oxadiazole undergoes selective electrophilic bromination of the phenyl ring in the presence of potassium bromate to produce *o*-, *m*-, and *p*-derivatives in 16%, 14%, and 26% yield, respectively (Scheme 13.293) [610].

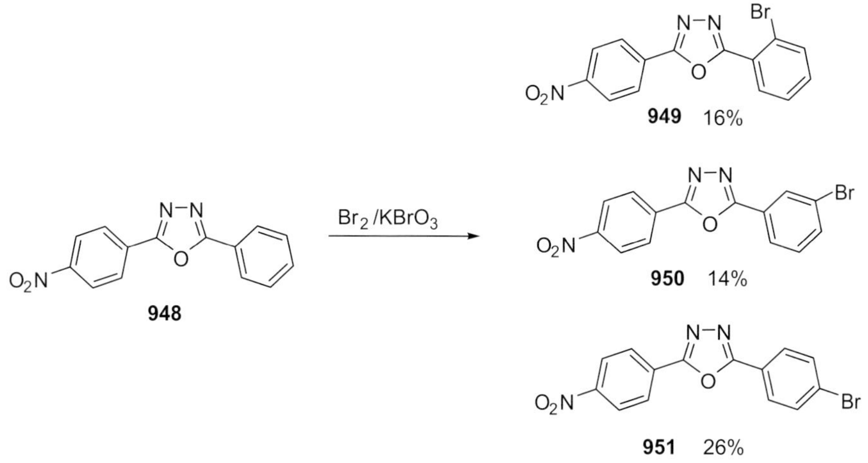

$Br_2/KBrO_3$

948

949 16%

950 14%

951 26%

Scheme 13.293

A sulfonamide group directly linked to 1,3,4-oxadiazole ring has been utilized to synthesize *N*-(anilinocarbonyl)-5-(2,4-dichlorophenyl)-1,3,4-oxadiazole-2-sulfonamide **953** (R = Cl) and *N*-(anilinocarbonyl)-5-phenyl-1,3,4-oxadiazole-2-sulfonamide **953** (R = H) with aim of preparing potential pesticides. These compounds have been obtained by reaction of phenyl isocyanate with 5-(2,4-dichlorophenyl)-1,3,4-oxadiazole-2-sulfonamide (**952**, R = Cl) or with 5-phenyl-1,3,4-oxadiazole-2-sulfonamide (**952**, R = H) respectively (Scheme 13.294). The compounds so obtained have been tested for fungicidal activity against the fungal species *Cephalosporium saccharii* and *Helminthosporium oryzae*, and have been shown to possess a good level of activity [611].

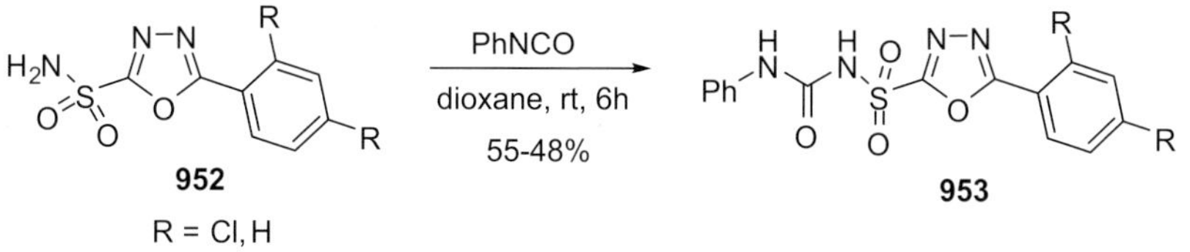

PhNCO
dioxane, rt, 6h
55-48%

952
R = Cl, H

953

Scheme 13.294

Some acetamides carrying a substituted-1,3,4-oxadiazole moiety with local anesthetic activity have synthesized by reaction of 5-aryl-2-chloroacetamido-1,3,4-oxadiazoles **956** with different secondary amines (Scheme 13.295). Compound **956** was easily prepared in 63% yield from the reaction of 5-(4-fluorophenyl)-1,3,4-oxadiazol-

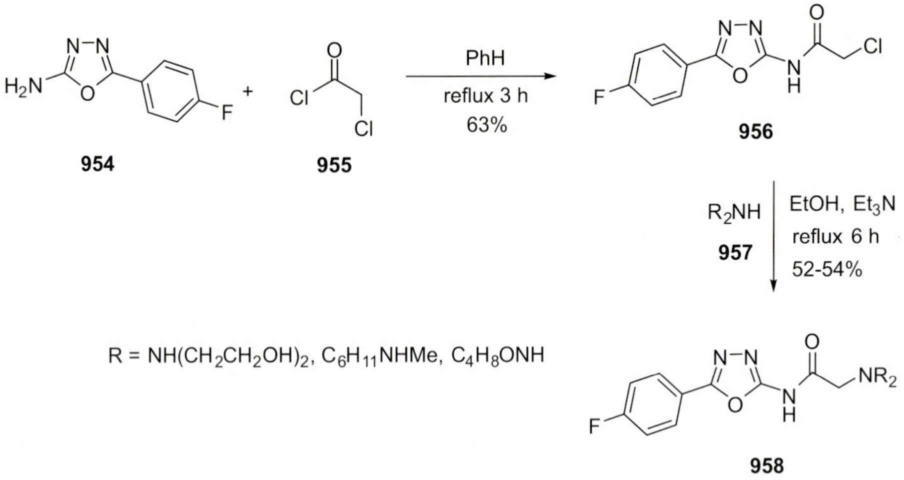

Scheme 13.295

2-amine (**954**) with chloroacetyl chloride (**955**). The local anesthetic activity was investigated using the rabbit corneal reflex method and guinea pig's wheal derm method, using lidocaine as standard drug [612].

Monosubstituted oxadiazoles are deprotonated at the ring carbon atom to give the corresponding anion, which has been subsequently alkylated with various alkylating agents. Thus, for example, 2-substituted-1,3,4-oxadiazoles **959** after treatment with butyllithium in the presence of $MgBr_2$ diethyl etherate in THF, followed by the addition of *N*-[(1*S*)-1-(methylethyl)-2-oxoethyl](*tert*-butoxy)carboxamide (*N*-Boc-L-valinal) (**961**), affords the corresponding *N*-Boc alcohols **962** in 53–79% yield. Similar treatment of (5-phenyl-1,3,4-oxadiazol-2-yl)lithium (**963**) with Boc-leucinal (**964**) produces [1-[(*R*/*S*)-hydroxy-(5-phenyl-[1,3,4]oxadiazol-2-yl]-(*S*)-methyl]-3-methylbutyl]carbamic acid *tert*-butyl ester (**965**) in 70% yield (Scheme 13.296) [613, 614].

In addition, the methyl group attached at carbons of 1,3,4-oxadiazoles shows a marked acidity when it is treated with strong bases. Thus, when 2,5-dimethyl- or 2-methyl-5-phenyl-1,3,4-oxadiazoles **966** and **971** were treated with isopropylmagnesium bromide (**967**) or NaH, followed by addition of alkyl carboxylates, 5-substituted 1,3,4-oxadiazol-2-ylmethyl ketones **969**, **970**, and **972** were obtained. The yield in ketones has been shown to depend on the nature of substituents present in the carboxylate moiety (Scheme 13.297) [615].

Interestingly, when the lithium anion of **971** was allowed to warm from −78 °C to room temperature, the *N*-benzoylated hydrazone **973** was isolated from the reaction mixture in 34% yield (Scheme 13.298) [616]. The formation of this latter compound is easily rationalized as the nucleophilic attack of the lithium anion to the not de-protonated **971** still present in solution.

1,3-4 Oxadiazoles containing a good leaving group at the methylene moiety are able to give nucleophilic substitutions. Thus, 2-aryl-5-chloromethyl-1,3,4-oxadiazoles **974**,

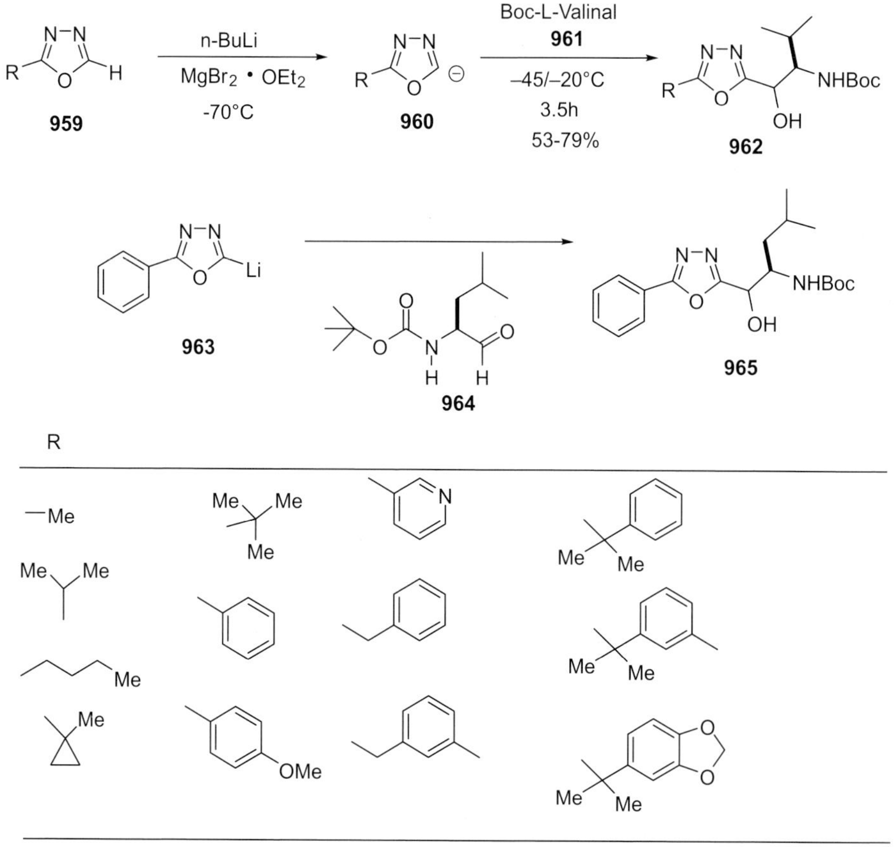

Scheme 13.296

via condensation of piperazine (**975**), give 1,4-bis[(5-aryl-1,3,4-oxadiazol-2-yl)methyl] piperazines **976** that *in vitro* displayed relatively potential antibacterial activities (Scheme 13.299) [617].

5-(*p*-Cyanomethylphenyl)-2-*n*-nonyl-1,3,4-oxadiazole **979**, useful precursor for organic light-emitting diodes (OLEDS), has been synthesized in 82% yield, by reaction of the bromide derivative **978** with tetraethylammonium cyanide. The 5-(*p*-bromomethylphenyl)-2-*n*-nonyl-1,3,4-oxadiazole (**978**) was easily prepared by the reaction of 5-(*p*-methylphenyl)-2-*n*-nonyl-1,3,4-oxadiazole (**977**) with *N*-bromosuccinimide(NBS) (Scheme 13.300) [618].

Substituents linked to 1,3,4-oxadiazoline moiety have also been involved in the synthesis of compounds having interesting properties. Thus, the herbicide 5-*tert*-butyl-3-(2,4-dichloro-5-isoprooxyphenyl)-1,3,4-oxadiazol-2-one (**981**) has been obtained starting from **877** via 5-*tert*-butyl-3-(2-amino-4-chloro-5-hydroxyphenyl)-1,3,4-oxadiazolin-2-one (**980**) (Scheme 13.301) [590].

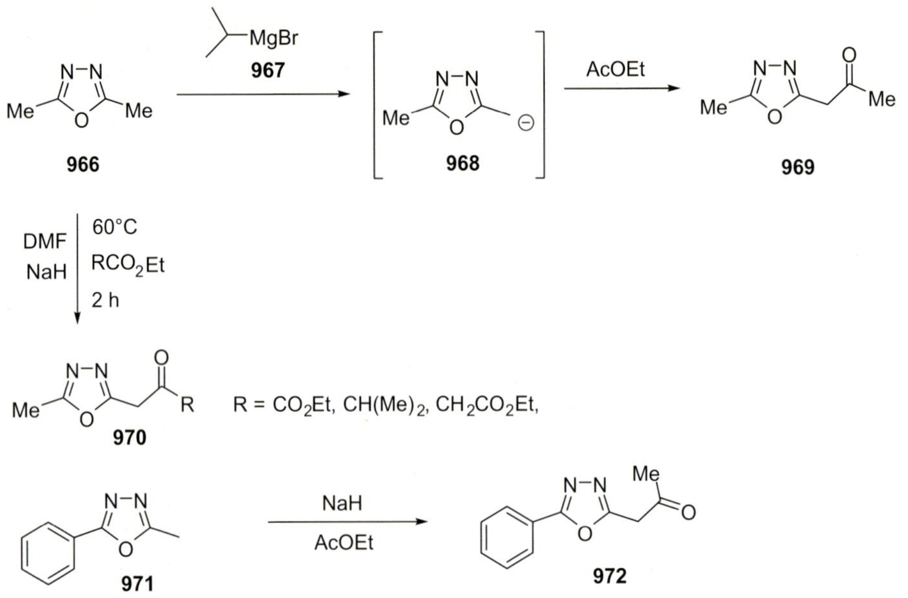

Scheme 13.297

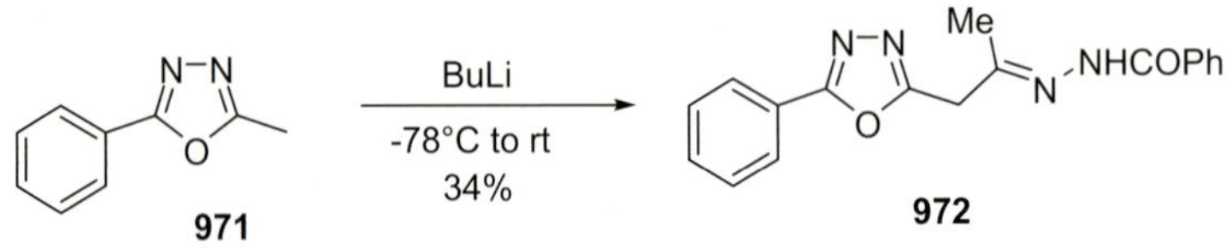

Scheme 13.298

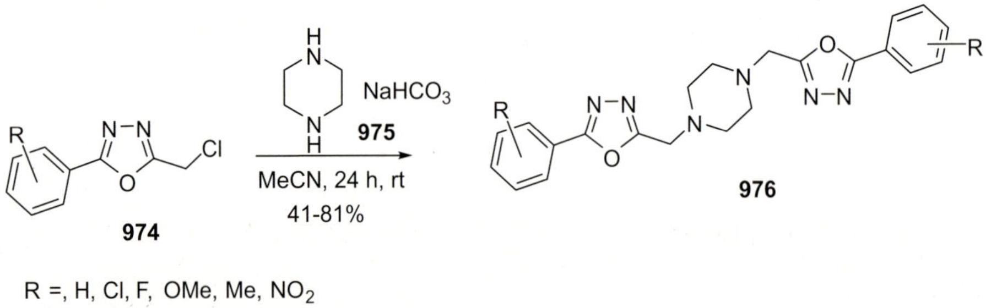

Scheme 13.299

Some 3-acyl-1,3,4-oxadiazoline derivatives **983**, having antitumoral activity, have been prepared by nucleophilic displacement of the chlorine atom in **982** (Scheme 13.302).

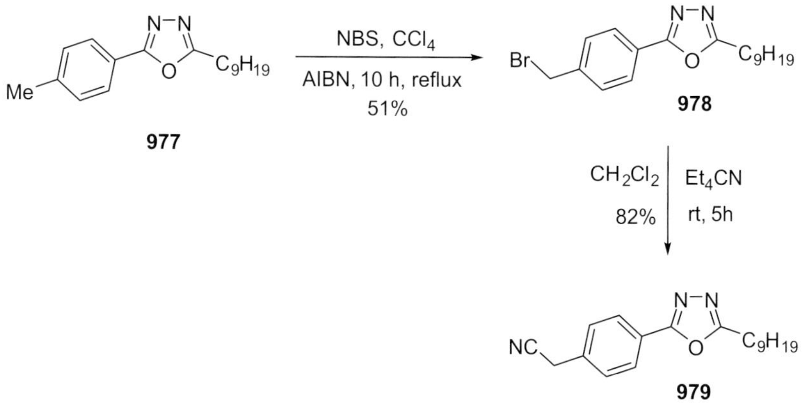

Scheme 13.300

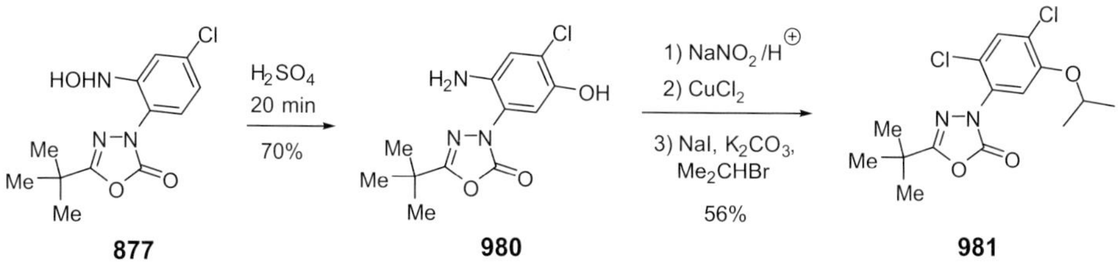

Scheme 13.301

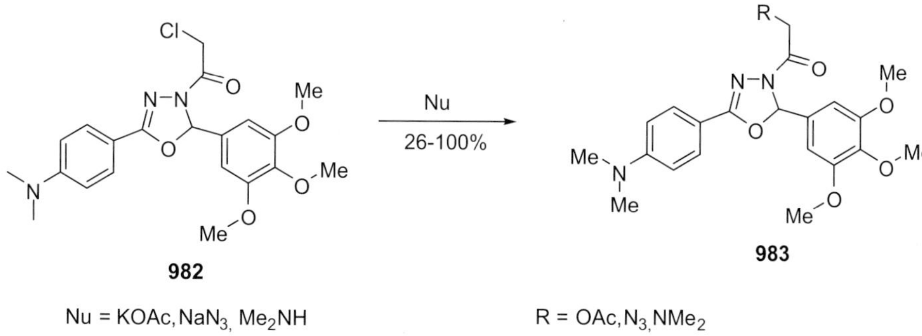

Scheme 13.302

13.5.3.6 **Metal Complexes**

1,3,4-Oxadiazole derivatives are widely used as electron-transporting groups due to their high electron deficiency and good thermal stability [619]. According to these properties, compounds containing the 1,3,4-oxadiazole core have prepared and used

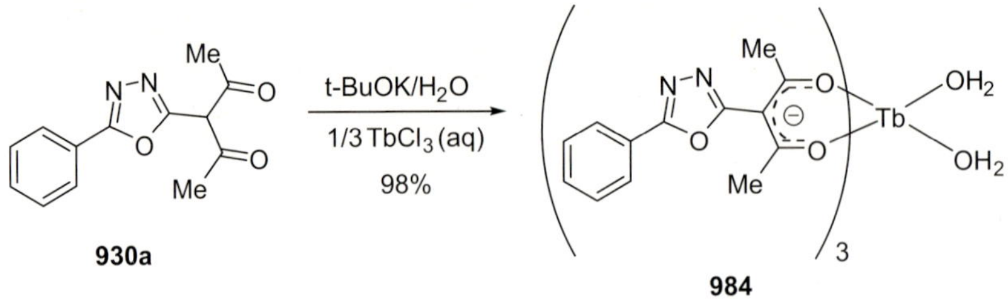

930a

t-BuOK/H_2O

1/3 $TbCl_3$ (aq)

98%

984

Scheme 13.303

in the production of organic light-emitting diodes (OLEDs). An OLED converts electrical energy into light and it is formed by an emissive chromophore, an electron-transporting group, and a hole-transporting unit. Recently, heavy metal ions have been incorporated in OLEDs as a cyclometalated ligand to increase the phosphorescence at room temperature, because these ions are able to increase the efficiency of the intersystem crossing from the singlet to triplet excited state. Based on the above consideration, here are two reported examples of 1,3,4-oxadiazole metalated complexes together with their syntheses.

The first synthesis of 1,3,4-oxadiazole-functionalized terbium (III) β-diketonate for organic electroluminescence has been reported by Zheng *et al.* The synthesis was performed by treating compound **930** with a suspension of *t*-BuOK and an aqueous solution of $TbCl_3$ (Scheme 13.303). The crystal structure of **984** was

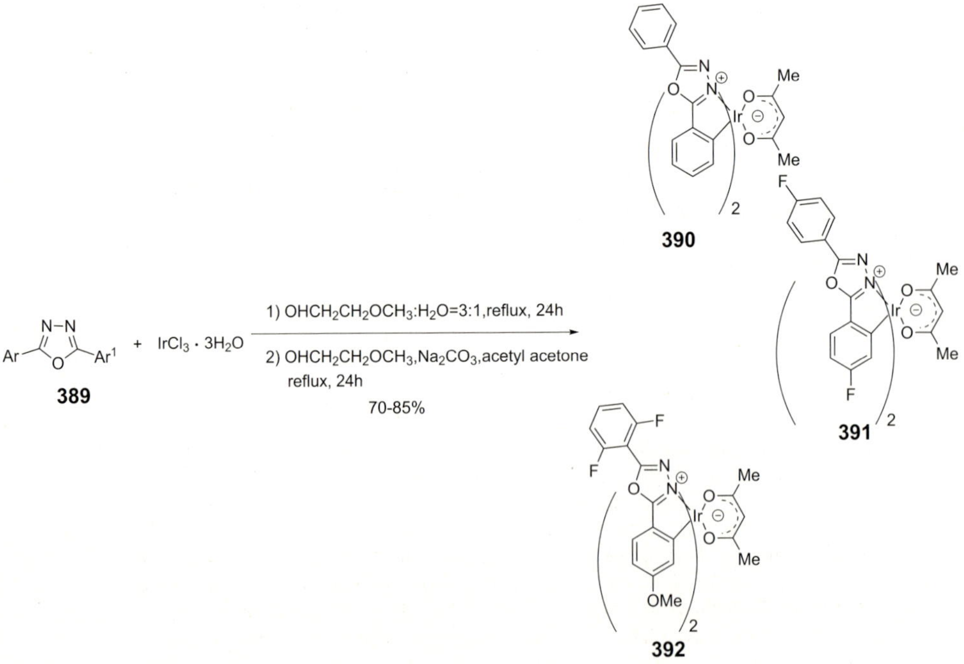

389 + $IrCl_3 \cdot 3H_2O$

1) $OHCH_2CH_2OCH_3$:H_2O=3:1,reflux, 24h

2) $OHCH_2CH_2OCH_3$,Na_2CO_3,acetyl acetone reflux, 24h

70-85%

390

391

392

Scheme 13.304

established by X-ray diffraction. The Tb(III) ion is surrounded by eight oxygen atoms, six of which are from the bidentate β-diketonate ligands and the other two from the coordinated water molecules. The coordination polyhedron is best described as square antiprismatic. This compound was used as an emitting material, and a bright and highly efficient green-emitting LED was fabricated [604].

An interesting series of iridium(III) complexes linked to 1,3,4-oxadiazole systems has been synthesized and utilized to prepare three organic light emitting diodes devices, which showed stable green-yellow luminescence. The synthetic procedure involved two steps. In the first step, $IrCl_3 \cdot 3H_2O$ was allowed to react with an excess of 2,5-diaryl-1,3,4-oxadiazoles **985** in a 2-ethoxyethanol–water mixture. In the second step, the resulting iridium compounds were treated with sodium carbonate and acetyl acetone in 2-ethoxyethanol as solvent to afford the cyclometalated derivatives **986–988** in 70–85% yield (Scheme 13.304) [620].

References

1 Thomas, E.W. (1996) *Comprehensive Heterocyclic Chemistry*, vol. 4 (ed. K.T. Potts), Pergamon Press, Oxford, p. 289.

2 (a) Meier, H. and Hanold, N. (1994) *Heteroarenes III*, Part 3, vol. E8c (ed. E. Schaumann), Goerg Thieme Verlag, Stuttgart, p. 397; (b) Newton, C.G. and Ramsden, C.A. (1982) *Tetrahedron*, **58**, 2965; (c) Rychlewska, U., Hodgson, D.J., Yeh, A., and Tien, H.-J. (1991) *Journal of the Chinese Chemical Society*, **38**, 467.

3 Clapp, L.B. (1976) *Advances in Heterocyclic Chemistry*, **20**, 65.

4 (a) Gasco, A. and Boulton, A.J. (1981) *Advances in Heterocyclic Chemistry*, **29**, 251; (b) Paton, R.M. (1984) *Comprehensive Heterocyclic Chemistry*, vol. 6 (ed. K.T. Potts), Pergamon Press, Oxford, p. 393.

5 (a) Hetzheim, A. and Mockel, K. (1966) *Advances in Heterocyclic Chemistry*, **7**, 183; (b) Hill, J. (1984) *Comprehensive Heterocyclic Chemistry*, vol. 6 (ed. K.T. Potts), Pergamon Press, Oxford, p. 427.

6 Clapp, L.B. (1984) *Comprehensive Heterocyclic Chemistry*, vol. 6 (ed. K.T. Potts), Pergamon Press, Oxford, p. 365.

7 Buckley, G.D. and Levy, W.J. (1951) *Journal of the Chemical Society*, 3016.

8 (a) Avdeev, V.I., Ruzankin, S.F., and Zhidomirov, G.M. (2005) *Kinetics and Catalysis*, **46**, 177; (b) Avdeev, V.I., Ruzankin, S.F., and Zhidomirov, G.M. (2005) *Kinetics and Catalysis*, **46**, 191.

9 Thorstad, O. and Undheim, K. (1974) *Chemica Scripta*, **6**, 222.

10 Schweig, A., Baumgartl, H., and Schultz, R. (1991) *Journal of Molecular Structure*, **247**, 135.

11 Kuchen, A., Bigler, P., and Schlunegger, U.P. (1984) *Chimia*, **38**, 387.

12 Earl, J.C. and Mackney, A.W. (1935) *Journal of the Chemical Society*, 899.

13 (a) Ollis, W.D. and Ramsden, C.A. (1976) *Advances in Heterocyclic Chemistry*, **19**, 1; (b) Newton, C.G. and Ramsden, C.A. (1982) *Tetrahedron*, **38**, 2965.

14 Stewart, F.H.C. (1964) *Chemical Reviews*, **64**, 129.

15 Thiessen, W.E. and Hope, H. (1967) *Journal of the American Chemical Society*, **89**, 5977.

16 Bridson-Jones, F.S., Buckley, G.D., Cross, L.H., and Driver, A.P. (1951) *Journal of the Chemical Society*, 2999.

17 Branchadell, V., Muray, E., Oliva, A., Ortuno, R.M., and Rodriguez-Garcia, C. (1998) *The Journal of Physical Chemistry*, **102**, 10106.

18 (a) Brundrett, R.B. (1980) *Journal of Medicinal Chemistry*, **23**, 1245; (b) Sapse, A.M., Allen, E.B., and Lowen, J.W. (1988) *Journal of the American Chemical Society*, **110**, 5671; (c) Kroeger Koepke, M.B., Schmiedekamp, A.M., and Michejda, C. (1994) *The Journal of Organic Chemistry*, **59**, 3301.

19 Loeppky, R.N., Fleischmann, E.D., Adams, J.E., Tomasik, W., Schlemper, E.O., and Wong, T.C. (1998) *Journal of the American Chemical Society*, **110**, 5946.

20 Arulsamy, N. and Bohle, D.S. (2002) *Angewandte Chemie – International Edition in English*, **41**, 2089.

21 Nelson, A.B. (1977) *Dissertation Abstract International B*, **38**, 1721.

22 Sugihara, T., Kuwahara, K., Wakabayashi, A., Takao, H., Imagawa, H., and Nishizawa, M. (2004) *Chemical Communications*, 216.

23 Wu, Y., Xue, Y., Xie, D., and Yan, G. (2005) *The Journal of Organic Chemistry*, **70**, 5045.

24 Luk'yanov, O.A., and Ternikova, T.V. (1983) *Izvestiya Akademii Nauk SSSR-Seriya Khimicheskaya*, 667.

25 Luk'yanov, O.A., Onisshchenko, A.A., Gorelik, V.P., and Tartakovskii, V.A. (1973) *Russian Chemical Bulletin*, **22**, 1251.

26 Tartakovskii, V.A., Ermekov, A.S., Strelenko, Y.A., and Vinograd, D.B. (2005) *Russian Journal of Organic Chemistry*, **41**, 120.

27 Nguyen, M.T., Egarthy, A.F., and Elguero, J. (1986) *Angewandte Chemie – International Edition in English*, **24**, 713.

28 Kroeger Kepke, M.B., Schmiedekamp, A.M., and Micheida, C.J. (1994) *The Journal of Organic Chemistry*, **59**, 3301.

29 Mais, F.-J., Dickopp, H., Middelhauve, B., Martin, H.-D., Mootz, D., and Steigel, A. (1987) *Chemische Berichte*, **120**, 27.

30 Padwa, A., Burgess, E.M., Gingrich, H.L., and Roush, D.L. (1982) *The Journal of Organic Chemistry*, **47**, 786.

31 Orvath, K., Korbonits, D., Nary-Szabo, G., and Simon, K. (1986) *Journal of Molecular Structure (Theochem)*, **136**, 215.

32 Shillady, D.D., Cutler, S., Jones, L.F., and Kier, L.B. (1990) *International Journal of Quantum Chemistry*, **24**, 153.

33 Fan, J.-M., Wang, Y., and Weng, C.-H. (1993) *The Journal of Physical Chemistry*, **97**, 8193.

34 Morley, J.O. (1995) *Journal of the Chemical Society-Perkin Transactions 2*, 253.

35 Hasek, J., Obrda, J., Huml, K., Nespurek, S., and Sorm, M. (1979) *Acta Crystallographica, Part B*, **35**, 2449.

36 King, T.J., Preston, N.P., Suffolk, J.S., and Turnbull, K. (1979) *Journal of the Chemical Society-Perkin Transactions 2*, 1751.

37 Barnighausen, H., Gellinek, F., Munnick, J., and Vos, A. (1963) *Acta Crystallographica*, **16**, 471.

38 Wheatley, P.J. (1972), in *Physical Methods in Heterocyclic Chemistry* (ed. A.R. Katritzky), vol. 5, Academic Press, New York, p. 18.

39 Semenov, S.G. and Sigolaev, Y.T. (2004) *Journal of Structural Chemistry*, **45**, 1082.

40 (a) Zhang, Z. and Duan, X. (2005) *Heterocycles*, **65**, 2649; (b) Giordano, F. (1988) *Gazzetta Chimica Italiana*, **118**, 501; (c) Ueng, C.-H., Lee, P.L., Wang, Y., and Yeh, M.-Y. (1984) *Acta Crystallographica. Section C, Crystal Structure Communications*, **49**, 1226; (d) Ueng, C.-H., Lee, P.L., Wang, Y., and Yeh, M.-Y. (1985) *Acta Crystallographica. Section C, Crystal Structure Communications*, **41**, 1776; (e) Ueng, C.-H., Wang, Y., and Yeh, M.-Y. (1987) *Acta Crystallographica. Section C, Crystal Structure Communications*, **43**, 1122; (f) Rychlewska, U., Hodgson, D.J., Yeh, A., and Tien, H.-J. (1991) *Journal of the Chinese Chemical Society*, **38**, 467; (g) Ueng, C.-H., Wang, Y., and Yeh, M.-Y. (1989) *Acta Crystallographica. Section C, Crystal Structure Communications*, **45**, 471; (h) Grossie, D.A. and Turnbull, K. (1992) *Acta Crystallographica. Section C, Crystal Structure Communications*, **48**, 377; (i) Fan, J.-M., Wang, Y., and Ueng, C.-H. (1993) *The Journal of Physical Chemistry*, **97**, 8193.

41 Blocher, A. and Zeller, K.-P. (1991) *Angewandte Chemie – International Edition in English*, **30**, 1476.

42 (a) Butkovic, K., Marinic, Z., and Sindler-Kulyk, M. (2004) *Magnetic Resonance in Chemistry*, **42**, 1053; (b) Ma, S. and Yeh, M.-Y. (1985) *Journal of the Chinese Chemical Society*, **32**, 151; (c) Araki, S., Mizuya, J., and Butsugan, Y. (1984) *Chemistry Letters*, 1045; (d) Tanaka, S. and Yokoi, M. (1983) *Bulletin of the Chemical Society of Japan*, **56**, 2198; (e) Hearn, M.T.W., and Potts, K.T. (1974) *Journal of* the Chemical Society-Perkin Transactions 2, 875; (f) Stewart, F.H.C. and Danieli, N. (1963) *Chemistry & Industry (London)*, 1926.

43 (a) Dahn, H. and Ung-Truong, M.-N. (1988) *Helvetica Chimica Acta*, **71**, 241; (b) Witanowski, M., Stefaniak, L., and Webb, G.A. (1979) *Journal of Magnetic Resonance*, **36**, 227; (c) Stefaniak, L. (1977) *Tetrahedron*, **33**, 2571.

44 (a) Araki, S., Mitsuya, J., and Butsugan, Y. (1985) *Journal of the Chemical Society-Perkin Transactions 1*, 2439; (b) Greco, C.V., Pesce, M., and Franco, J.M. (1966) *Journal of Heterocyclic Chemistry*, **3**, 391.

45 Tien, L.-L., Lin, S.-T., and Chiang, H.-J. (1989) *Heterocycles*, **29**, 185.

46 Yelamaggad, C.V., Mathews, M., Uiremath, U.S., Shankar Rao, D.S., and Prasad, S.K. (2005) *Tetrahedron Letters*, **46**, 2623.

47 Yelamaggad, C.V., Mathews, M., Uiremath, U.S., Shankar Rao, D.S., and Prasad, S.K. (2005) *Chemical Communications*, 1552.

48 Yan, H., Chan, W.L., and Szeto, Y.S. (2004) *Journal of Applied Polymer Science*, **91**, 2523.

49 Scheler, S., Buhr, G., and Bergmann, K. (1991) PCT DE 3926774 A1.

50 (a) Saulnier, M.G., Vyas, D.M., Langley, D.R., Doyle, T.W., Rose, W.C., Crosswell, A.R., and Long, B.B. (1989) *Journal of Medicinal Chemistry*, **32**, 1418; (b) Trost B. M. and Kinson, P.L. (1975) *Journal of the American Chemical Society*, **97**, 2438; (c) Horner, L. and Weber, K.H. (1962) *Chemische Berichte*, **95**, 1962; (d) Ferreira, V.F., Jorqueira, A., Leal, K.z., Pimentel, H.R.X., Seidl, P.R., da Silva, M.N., da Souza, M.C.B.V., Pinto, A.V., Wardell, J. L., and Wardell, S.M.s.v. (2006) *Magnetic Resonance in Chemistry*, **44**, 481.

51 Applegate, J. and Turnbull, K. (1988) *Synthesis*, 1011.

52 Pandeya, S.N., Kumar, A., Singh, B.N., and Mishra, D.N. (1987) *Pharmaceutical Research*, **4**, 321.

53 Tien, H.-J., Tien, M.-J., and Hung, W.J. (1994) *Huaxue*, **52**, 153.

54 Kujath, E., Schoenafinger, K., and Brendel, J. (1995) Ger. Offen DE 4.337.335.

55 Gotz, M. and Grozinger, K. (1971) *Tetrahedron*, **27**, 4449.

56 Gotz, M. and Grozinger, K. (1970) *Journal of Heterocyclic Chemistry*, **7**, 123.

57 Masuda, K., Imashiro, Y., and Kaneko, T. (1970) *Chemical & Pharmaceutical Bulletin*, **18**, 128.

58 Kroeger Kepke, M.B., Schmiedekamp, A.M., and Micheida, C.J. (1994) *The Journal of Organic Chemistry*, **59**, 3301.

59 Erb, E. (1994) *Dissertation Abstracts International B*, **54**, 4671.

60 Blocher, A. and Zeller, K.-P. (1994) *Chemische Berichte*, **127**, 551.

61 Yashunskii, V.G., Kholodov, L.E., and Peresleni, E.M. (1963) *Zhurnal Obshchei Khimii*, **33**, 3699.

62 Kopecky, K.R., Pope, P.M., and Sastre, J.A.L. (1976) *Canadian Journal of Chemistry*, **54**, 2639.

63 L-Bakoush, M.N., and Parrick, J. (1988) *Journal of Heterocyclic Chemistry*, **25**, 1055.

64 Kuo, C.-N., Wu, M.-H., Chen, S.-P., Li, T.-P., Huang, C.-Y., and Yeh, M.-Y. (1994) *Journal of the Chinese Chemical Society*, **41**, 849.

65 Nakajima, M. and Anselme, J.P. (1983) *Journal of the American Chemical Society*, **48**, 1444.

66 Ortiz de Montellano, P.R. and Grab, L.A. (1986) *Journal of the American Chemical Society*, **108**, 5584.

67 Meier, H., Heimgartner, H., and Schmid, H. (1977) *Helvetica Chimica Acta*, **60**, 1087.

68 Meier, H. and Heimgartner, H. (1986) *Helvetica Chimica Acta*, **69**, 927.

69 Angadiyavar, C.S. and George, M.J. (1971) *The Journal of Organic Chemistry*, **36**, 1589.

70 Marky, M., Meier, H., Wunderli, A., Heimgartner, H., Schmid, H., and Hansen, H.-J. (1978) *Helvetica Chimica Acta*, **61**, 1477.

71 Gottardt, H. and Reiter, F. (1979) *Chemische Berichte*, **112**, 1635.

72 Pfortner, K.-H. and Foricher, J. (1980) *Helvetica Chimica Acta*, **63**, 653.

73 Eber, G., Schneider, S., and Dorr, F. (1980) *Berichte der Bunsen-Gesellschaft-Physical Chemistry Chemical Physics*, **84**, 281.

74 Butkovic, K., Basaric, N., Lovrekovic, K., Marinic, Z., Visnjevac, A., Kojic-Prodic, B., and Syndler-Kulyk, M. (2004) *Tetrahedron Letters*, **45**, 9057.

75 Stoesser, R., Csongar, C., Lieberenz, M., and Tomaschewski, G. (1991) *Journal of Photochemistry and Photobiology (A)*, **61**, 245.

76 Huseya, Y., Chinone, A., and Ohta, M. (1972) *Bulletin of the Chemical Society of Japan*, **45**, 3202.

77 Bhat, V., Dixit, M., Ugarker, B.G., Trozzolo, A.M., and George, M.V. (1979) *The Journal of Organic Chemistry*, **44**, 2957.

78 Nespurek, S., Lucas, J., Bohm, S., and Bastl, Z. (1994) *Journal of Photochemistry and Photobiology (A)*, **84**, 257.

79 Dickopp, H. (1980) *Chemische Berichte*, **113**, 1830.

80 Greco, C.V. and Mehta, J.R. (1980) *Journal of the Chemical Society-Perkin Transactions 1*, 20.

81 Grimmett, M.R. and Iddon, B. (1995) *Heterocycles*, **41**, 1525.

82 (a) Azarifar, D., and Ghasemnejad-Bosra, H. (2006) *Synthesis*, **7**, 1123; (b) Zirngibl, L. (1983) *Prog. Drug Res.*, **27**, 253.

83 Yeh, M.-Y., Tien, H.-J., Huang, L.-Y., and Chen, M.-H. (1983) *Journal of the Chinese Chemical Society*, **30**, 29.

84 Turnbull, K., Blackburn, T.L., and Miller, J.J. (1996) *Journal of Heterocyclic Chemistry*, **33**, 485.

85 Turnbull, K., Blackburn, T.L., and McClure, D.B. (1994) *Journal of Heterocyclic Chemistry*, **31**, 1631.

86 Burson, W.C. III, Jones, D.R., Turnbull, K., and Preston, P.N. (1991) *Synthesis*, 745.

87 Chan, W.L., Waite, J.A., Lin, Y.H., and Szeto, Y.S. (1994) *Heterocycles*, **38**, 2023.

88 Fuchigami, T., Chen, C.-S., Nonaka, T., Yeh, M.-Y., and Tien, H.-J. (1986) *Bulletin of the Chemical Society of Japan*, **59**, 483.

89 Fuchigami, T., Chen, C.-S., Nonaka, T., Yeh, M.-Y., and Tien, H.-J. (1986) *Bulletin of the Chemical Society of Japan*, **59**, 487.

90 Tien, H.-J., Fang, G.-M., Lin, S.-T., and Tien, L.-L. (1992) *Journal of the Chinese Chemical Society*, **39**, **29**, 107.

91 Fleischhaker, W. and Urban, E. (1988) *Heterocycles*, **27**, 1697.

92 Kalinin, V.N. and Min, S.F. (1988) *Journal of Organometallic Chemistry*, **352**, C34.

93 Turnbull, K., Krein, D.M., and Tullis, S.A. (2003) *Synthetic Communications*, **33**, 2209.

94 Kalinin, V.N. and Min, S.F. (1989) *Journal of Organometallic Chemistry*, **379**, 195.

95 Henning, H.-G., Neumann, B.-M., and Alder, L. (1978) *Zeitschrift fur Chemie*, **18**, 262.

96 Kalluraya, B. and Rai, G. (2003) *Indian Journal of Chemistry Section B-Organic Chemistry Including Medicinal Chemistry*, **42**, 2556.

97 Shih, M. (2004) *Synthesis*, 26.

98 Kalluraya, B., Vishwanata, P., Jyothi, C.H., Priya, V.-F., and Rai, G. (2003) *Indian Journal of Heterocyclic Chemistry*, **12**, 355.

99 Yeh, M.-Y. and Tien, H.-J. (1986) *Journal of the Chinese Chemical Society*, **33**, 83.

100 Yeh, M.-Y., Pan, I.-H., Chuang, C.-P., and Tien, H.-J. (1988) *Journal of the Chinese Chemical Society*, **35**, 443.

101 Shih, H.-M., Yeh, M.-Y., Lee, M., and Su, Y. (2004) *Synthesis*, 2877.

102 Shih, H.-M. and Yeh, M.-Y. (2003) *Tetrahedron*, **59**, 4103.

103 Shih, M. and Ke, F. (2004) *Bioorganic and Medicinal Chemistry*, **12**, 4633.

104 Dumitrescu, F., Mitan, C.I., Dumitrescu, D., Barbu, L., Hrubaru, M., Caprau, D., and Vuluga, D. (2003) *Revista de Chimie*, **54**, 747.

105 Farina, F., Fernandez, P., Fraile, M.T., Martin, M.V., and Martin, M.R. (1989) *Heterocycles*, **29**, 967.

106 Takagi, K., Shiro, M., Takeda, S., and Nakamura, N. (1992) *Journal of the American Chemical Society*, **114**, 8414.

107 Gribble, G.W. and Hirth, B.H. (1996) *Journal of Heterocyclic Chemistry*, **33**, 719.

108 Combis, J.M. and Vinci, J.P. (1996) *British Journal of Clinical Pharmacology*, **41**, 409.

109 Ma, S. and Yeh, M.-Y. (1985) *Journal of the Chinese Chemical Society*, **32**, 151.

110 Bult, H., Demever, G.R.Y., and Herman, A.C. (1995) *British Journal of Clinical Pharmacology*, **114**, 1371.

111 Cai, T.B., Lu, D., Tang, X., Zhang, Y., Landerholm, M., and Wang, P.G. (2005) *The Journal of Organic Chemistry*, **70**, 3518.

112 Tang, X., Cai, T., and Wang, P.G. (2003) *Bioorganic & Medicinal Chemistry Letters*, **13**, 1687.

113 Rehse, K. and Ciborski, T. (1995) *Archiv der Pharmazie*, **328**, 71.

114 Rehse, K., Schleifer, J.K., Martens, A., and Kaempfe, M. (1994) *Archiv der Pharmazie*, **327**, 393.

115 Moustafa, M.A., Gineinah, M.M., Nasr, M.N., and Waleed, W.A.H. (2004) *Archiv der Pharmazie*, **337**, 164*ibidem*, 427.

116 Grynberg, N., Gomes, R., Shinzato, T., Echevarria, A., and Miller, J. (1992) *Anticancer Research*, **12**, 1025.

117 Dunkley, C.S. and Thoman, C. (2003) *Bioorganic & Medicinal Chemistry Letters*, **13**, 2899.

118 Satyanarayana, K. and Rao, M.N.A. (1995) *European Journal of Medicinal Chemistry*, **30**, 641.

119 Satyanarayana, K. and Rao, M.N.A. (1995) *Journal of Pharmaceutical Sciences*, **84**, 263.

120 (a) Clapp, L.B. (1984) *Comprehensive Heterocyclic Chemistry*, vol. 6 (ed. K.T. Potts), Pergamon Press, Oxford, p. 378; (b) Jochims, J.C. and Thomas, E.W. (1996) *Comprehensive Heterocyclic Chemistry*, vol. 4 (ed. K.T. Potts), Pergamon Press, Oxford, p. 179; (c) Hemming, K. (2008) *Comprehensive Heterocyclic Chemistry*, vol. 5 (ed. K.T. Potts), Pergamon Press, Oxford, p. 244.

121 Clapp, L.B. (1976) *Advances Heterocyclic Chemistry*, vol. 20 (eds A.R. Katritzky and A.J. Boulton), Academic Press, New York, p. 65.

122 De Melo, S.J., Sobral, A.D., Lopes, H.L., and Srivastava, R.M. (1998) *Journal of Brazilian Chemistry*, **9**, 465.

123 Srivastava, R.M., Oliveira, F.J.S., Machado, D.S., and Souto-Maior, R.M. (1999) *Synthetic Communications*, **29**, 1437.

124 Antunes, R.B., Srivastava, R.M., Thomas, G., and Arujo, C.C. (1988) *Bioorganic & Medicinal Chemistry Letters*, **8**, 3071.

125 Srivastava, R.M., de Morais, L.P.F., Catanho, M.T.J.A., de Souza, G.M.L., Seabra, G.M., Simas, A.M., and Rodrigues, M.A.L. (2000) *Heterocyclic Communications*, **6**, 41.

126 Andersen, K.E., Lundt, B.F., Jorgensen, A.S., and Braestrup, C. (1996) *European Journal of Medicinal Chemistry*, **31**, 417.

127 Ahn, J.-M., Boyle, N.A., MacDonald, M.T., and Janda, K.D. (2002) *Mini-Reviews in Medicinal Chemistry*, **55**, 719.

128 Kimura, S., Kawasaki, S., Watanabe, S., Fujita, R., and Sasaki, K. (2008) *Neuroscience Research*, **60**, 73.

129 Rai, M. and Kaur, B. (1982) *Journal of the Indian Chemical Society*, **59**, 1197.

130 Reddy, P.B., Reddy, S.M., Rajanarender, E., and Murthy, A.R. (1986) *National Academy Science Letters-India*, **9**, 101.

131 Bezerra, N.M., De Oliveira, S.P., Srivastava, R.M., and Da Silva, J.R. (2005) *Il Farmaco*, **60**, 955.

132 Chimirri, A., Grasso, S., Monforte, A.M., Monforte, P., Zappalà, M., and Carotti, M. (1994) *Il Farmaco*, **49**, 509.

133 Katritzky, A.R., and Barczynski, P. (1990) *Journal fur Praktische Chemie*, **332**, 885.

134 (a) Bird, C.W. (1985) *Tetrahedron*, **41**, 1409; (b) Bird, C.W. (1992) *Tetrahedron*, **48**, 335; (c) Katritzky, A.R., Jug, K., and Oniciu, D.C. (2001) *Chemical Reviews*, **101**, 1421; (d) Valavan, A.T., Oniciu;, T.D., and Katritzky, A.R. (2004) *Chemical Reviews*, **104**, 2777.

135 Buscemi, S., Pace, A., Piccionello, A.P., Macaluso, G., Vivona, N., Spinelli, D., and Giorgi, G. (2005) *The Journal of Organic Chemistry*, **70**, 3288.

136 Srivastava, R.M. and Brinn, I.M. (1977) *The Journal of Organic Chemistry*, **42**, 1555.

137 Moussebois, C. and Eloy, F. (1964) *Helvetica Chimica Acta*, **47**, 838.

138 (a) Minkin, V.I., Garnovskii, A.D., Elguero, J., Katritzky, A.R., and Denisko, O.V. (2000) *Advances in Heterocyclic Chemistry*, **44**, 157; (b) Calza, P., Mendana, C., Baiocchi, C., Hidaki, H., and Pelizzetti, E. (2006) *Chemistry - A European Journal*, **12**, 727.

139 Van Haverbeke, Y., Maquestiau, A., Muller, R.N., and Stamane, M.L. (1976) *Bulletin des Sociétés Chimiques Belges*, **85**, 35.

140 Szczepankiewiez, W., Wagner, P., Danicki, M., and Suwinski, J. (2003) *Tetrahedron Letters*, **44**, 2015.

141 Srivastava, R.M., Faustino, W.M., and Brinn, I.M. (2003) *Journal of Molecular Structure (Theochem)*, **640**, 49.

142 (a) Shokhen, M.A., Andrianov, V.G., Eremeev, A.V., and Barmina, S.V. (1987) *Khimiya Geterotsiklicheskikh Soedinenii*, **2**, 175; (b) Andrianov, V.G., Shokhen, M.A., and Eremeev, A.V. (1989) *Khimiya Geterotsiklicheskikh Soedinenii*, **4**, 508.

143 Lopez, J.P. and Rosser, R.W. (1983) *Theochem*, **11**, 203.

144 (a) Antunes, R., Batista, H., Srivastava, R.M., Thomas, G., Araujo, C.C., Longo, R.L., Magalhaes, H., Leao, M.B.C., and Pavao, A.C. (2003) *Journal of Molecular Structure*, **660**, 1. (b) Batista, H., Carpenteer, G.B., and Srivastava, R.M. (2000) *Journal of Chemical Crystallography*, **30**, 131.

145 Buscemi, S., D'Auria, M., Pace, A., Pibiri, I., and Vivona, N. (2004) *Tetrahedron*, **60**, 3243.

146 La Manna, G., Buscemi, S., and Vivona, N. (1988) *Journal of Molecular Structure (Theochem)*, **452**, 67.

147 Bottoni, A., Frenna, V., Lanza, C.Z., Macaluso, G., and Spinelli, D. (2004) *Journal of Physical Chemistry (A)*, **108**, 1731.

148 (a) Wagner, G., Danks, T.N., and Vullo, V. (2007) *Tetrahedron*, **63**, 5251; (b) Hoque, A.K.M.M., Lee, W.K., Shine, H.J., -, D., and Zhao, C. (1991) *The Journal of Organic Chemistry*, **56**, 1332.

149 Barbieux-Flammang, M., Vandevoorde, S., Flammang, R., Wong, M.H., Bibas, H., Kennard, C.H.L., and Wentrup, C. (2000) *Journal of the Chemical Society-Perkin Transactions 2*, 473.

150 Wagner, G., Haukka, M., Frausto Da Silva, J.J.R., Pombeiro, A.J.L., and Kukushkin, Yu.V. (2001) *Inorganic Chemistry*, **40**, 264.

151 Carpenter, G.P., Ventura, E., De Morais, L.P.F., Srivasta, R.M., Simas, A.M., and Faure, R. (2001) *Journal of Molecular Structure*, **561**, 29.

152 Yu, J., Zhang, S., Li, Z., Lu, W., and Cai, M. (2005) *Bioorganic and Medicinal Chemistry*, **13**, 353.

153 Srivastava, R.M., De Freitas Filho, J.R., Da Silva, M.J., De Melo Souto, S.C., Carpenter, G.B., and Faustino, W.M. (2004) *Tetrahedron*, **60**, 10761.

154 Adelfinskaya, O., Wu, W., Davisson, V.J., and Bergstrom, D.E. (2005) *Nucleosides Nucleotides and Nucleic, Acids*, **24**, 1919.

155 Meyer, E., Joussef, A.C., Gallardo, H., and Bortoluzzi, A.J. (2003) *Journal of Molecular Structure*, **655**, 361.

156 Xu, J., Li, X., Wang, Z., Yang, Q., and Yan, C. (1999) *Acta Crystallographica. Section C, Crystal Structure Communications*, **55**, 650.

157 Bruno, G., Chimirri, A., Gitto, R., Nicolò, F., and Scopelliti, R. (1999) *Acta Crystallographica. Section C, Crystal Structure Communications*, **55**, 685.

158 El Hazazi, S., Baouid, A., Hasnoui, A., and Pierrot, M. (2002) *Acta Crystallographica, Sect. E*, **58**, 548.

159 Yu, J., Zhang, S., Li, Z., Lu, W., Zhang, L., Zhon, R., Liu, Y., and Cai, M.J. (2001) *Journal of Carbohydrate Chemistry*, **20**, 877.

160 Al-Thebeit, M.S. (1991) *Journal of Carbohydrate Chemistry*, **18**, 667–674.

161 Coley, H.M., Sarju, J., and Wagner, G. (2008) *Journal of Medicinal Chemistry*, **51**, 135.

162 Srivastava, R.M., De Almeida Lima, A., Viana, O.S., Da Costa Silva, M.J., Catanho, M.T.J.A., and de Morais, J.O.F. (2003) *Bioorganic and Medicinal Chemistry*, **11**, 1821.

163 Suga, H., Shi, X., and Ibata, T. (1998) *Bulletin of the Chemical Society of Japan*, **71**, 1231.

164 Diaz-Oritz, A., Dez-Barra, E., de la Hoz, A., Moreno, A., Gomez-Escalonilla, M.J., and Loupy, A. (1996) *Heterocycles*, **43**, 1021.

165 Zhang, M., Zangh, H., Yang, Z., Ma, L., Min, J., and Zhang, L. (1999) *Carbohydrate Research*, **318**, 157.

166 Leite, L.F.C.C., Ramos, M.N., da Silva, J.B.P., Miranda, A.L.P., Fraga, C.A.M., and Barreiro, E.J. (1999) *Il Farmaco*, **54**, 747.

167 (a) Srivastava, R.M., Mendes e Silva, L.M., and Bhattacharya, J. (1989) *Quimica Nova*, **12**, 221; (b) Srivastava, R.M., da Conceicao Pereira, M., Hallwass, F., and Santana, S.R. (2002) *Journal of Molecular Structure*, **604**, 177.

168 (a) Neidlein, R., Kramer, W., and Li, S. (1998) *Journal of Heterocyclic Chemistry*, **35**, 161; (b) Outirite, M., Lebrini, M., Lagrenee, M., and Bentiss, F. (2007) *Journal of Heterocyclic Chemistry*, **44**, 152.

169 Johnson, J.E., Nwoko, D., Hotema, M., Sanchez, N., Alderman, R., and Lynch, V. (1996) *Journal of Heterocyclic Chemistry*, **33**, 1583.

170 da Silva, A.S., de Silva, M.A.A., Carvalho, C.E.M., Antunes, O.A.C., Herrera, J.O.M., Brinn, I.M., and Mangrich, A.S. (1999) *Inorganica Chimica Acta*, **292**, 1.

171 Yu, J., Zhang, S., Li, Z., Lu, W., Zhou, R., Liu, Y., and Cai, M. (2003) *Carbohydrate Research*, **338**, 257.

172 Zecchina, A., Andreoletti, G.E., and Sampietro, P. (1967) *Spectrochimica Acta. Part A, Molecular and Biomolecular Spectroscopy*, **23**, 2647.

173 Lin, X.-F., Cui, S.-L., and Wang, Y.-G. (2003) *Chemistry Letters*, **32**, 842.

174 Wagner, G., Pombeiro, A.J.L., and Kukushkin, V.Yu. (2000) *Journal of the American Chemical Society*, **122**, 3106.

175 Kaboudin, B. and Navaee, K. (2003) *Heterocycles*, **60**, 2287.

176 Leite, L.F.C.C., Barreiro, E.J., Ranmos, M.N., Silva, J.B.P., Galdino, S.L., and Pitta, I.R. (2000) *Spectroscopy*, **14**, 115.

177 Srivastava, R.M. (2005) *Mass Spectroscopy*, **24**, 328.

178 (a) Tiemann, F. and Kruger, P. (1884) *Chemische Berichte*, **17**, 1685; (b) Tiemann, F. (1885) *Chemische Berichte*, **18**, 1060.

179 (a) Beckmann, E. and Sandel, K. (1897) *Annalen der Chemie-Justus Liebig*, **296**, 279; (b) Critchley, J.P., Fear, E.J.P., and Pippett, J.S. (1964) *Chemistry & Industry*, **19**, 806; (c) Eloy, F. and Lenaers, R. (1964) *Chemical Reviews*, **62**, 155.

180 Leandri, G. (1956) *Chimica Industriale, Bologna*, **14**, 80.

181 (a) Rice, K.D. and Nuss, J.M. (2001) *Bioorganic & Medicinal Chemistry Letters*, **11**, 753; (b) Chiou, S. and Shine, H.J. (1989) *Journal of Heterocyclic Chemistry*, **26**, 125; (c) Meyer, E., Joussef, A.C., and Gallardo, H. (2003) *Synthesis*, **6**, 899.

182 Sams, C.K. and Lau, J. (1999) *Tetrahedron Letters*, **40**, 9359.

183 Buchanann, J.L., Vu, C.B., Merry, j.t., Corpuz, E.G., Pradeepan, S.G., Mani, U.N., Yang, M., Plake, H.R., Varkhedkar, V.M., Lynch, B.A., MacNeil, I.A., Loiacono, K.A., Tiong, C.L., and Holt, D.A. (1999) *Bioorganic & Medicinal Chemistry Letters*, **9**, 2359.

184 (a) Liang, G.B. and Feng, D.D. (1996) *Tetrahedron Letters*, **37**, 6627; (b) Borg, S., Estenne Bouhtou, G., Luthman, K., Csoregh, I., Hesselink, W., and Hacksell, U. (1995) *The Journal of Organic Chemistry*, **60**, 3112; (c) Borg, S., Vollinga, R.C., Labarre, M., Payza, K., Terenius, L., and Luthman, K. (1999) *Journal of Medicinal Chemistry*, **42**, 4331; (d) Buchanan, J.L., Vu, C.B., Merry, T.J., Corpuz, E.G., Pradeepan, S.G., Mani, U.N., Yang, M., Plake, H.R., Varkhedkar, V.M., and Lynch, B.A. (1999) *Bioorganic and Medicinal Chemistry*, **9**, 2359; (e) Braga, A.L., Ludtke, D.S., Alberto, E.E., Dornelles, L., Severo Filho, W.A., Corbellini, V.A., Rosa, D.M., and Schwab, R.S. (2004) *Synthesis*, 1589; (f) Ispidouki, M., Litinas, K.E., and Fylaktakidou, K.C. (2008) *Heterocycles*, **75**, 1321.

185 Buscemi, S., Pace, A., Calabrese, R., Vivona, N., and Metrangolo, P. (2001) *Tetrahedron*, **57**, 5865.

186 Poulain, R.F., Tartar, A.L., and Deprez, B.P. (2001) *Tetrahedron Letters*, **42**, 1495.

187 Elzein, E., Kalla, R., Li, X., Perry, T., Parkhill, E., Palle, V-., Varkhedkar, V., Gimbel, A., Zeng, D., Lustig, D., Leung, K., and Zablocki, J. (2006) *Bioorganic & Medicinal Chemistry Letters*, **16**, 302.

188 Vu, C.B., Corpuz, E.G., Merry, J.T., Pradeepan, S.G., Bartlett, C., Bohacek, R.S., Botfield, M.C., Eyermann, C.J., Lynch, B.A., MacNeil, I.A., Ram, M.K., van Schravendijk, M.R., Violette, S., and Sawyer, T.K. (1999) *Journal of Medicinal Chemistry*, **42**, 4088.

189 Katritzky, A.R., Shestopalov, A.A., and Suzuki, K. (2005) *Arkivoc*, 36.

190 Wang, Y., Miller, R.L., Sauer, D.R., and Djuric, S.W. (2005) *Organic Letters*, **7**, 925.

191 Chesnyuk, A.A., Mikhalichenko, S.N., Firgang, L.D., and Zaplishnyi, V.N. (2005) *Russian Chemical Bulletin*, **54**, 1900.

192 Petukhov, P.A., Zhang, M., Johnson, K.J., Tella, S.R., and Kozikowiski, A.P. (2001)

Bioorganic & Medicinal Chemistry Letters, **11**, 2079.

193 (a) Kaboudin, B. and Saadati, F. (2007) *Tetrahedron Letters*, **48**, 2829; (b) Grant, D., Dahl, R., and Cosford, N.D.P. (2008) *The Journal of Organic Chemistry*, **73**, 7219.

194 Bailey, N., Cooper, A.W.J., Deal, M.J., Dean, A.W., Gore, A.L., Hawes, M.C., Judd, D.B., Merritt, A.T., Storer, R., Travers, S., and Watson, S.P. (1997) *Chimia*, **51**, 832.

195 (a) Senzik, M. and Hui, H.C. (2003) *Tetrahedron Letters*, **44**, 8697; (b) Gangloff, A.R., Litvak, J., Shelton, E.J., Sperandio, D., Wang, W.R., and Rice, K.D. (2001) *Tetrahedron Letters*, **42**, 1441.

196 (a) Holsen, P.H., Tonder, J.E., Hansen, J.B., Hansen, H.C., and Rimvall, K. (2000) *Bioorganic and Medicinal Chemistry*, **8**, 1433; (b) Manfredini, S., Lampronti, I., Vertuani, S., Salaroli, N., Recanatini, M., Bryan, D., and McKinney, M. (2000) *Bioorganic and Medicinal Chemistry*, **8**, 1559; (c) Vieira, E., Huwyler, J., Jolidon, S., Knoflach, F., Mutel, V., and Wichmann, J. (2005) *Bioorganic & Medicinal Chemistry Letters*, **15**, 4628; (d) Amarasinghe, K.K.D., Maier, M.B., Srivastava, A., and Gray, J.L. (2006) *Tetrahedron Letters*, **47**, 3629; (e) Du, W., Hagmann, W.K., and Hale, J.J. (2006) *Tetrahedron Letters*, **47**, 4721.

197 (a) Feng, D.D., Biftu, T., Candelore, M.R., Cascieri, M.A., Colwell, L.F. Jr., Deng, L., Feeney, W.P., Forrest, M.J., Hom, G.J., MacIntyre, D.E., Miller, R.R., Stearns, R.A., Strader, C.D., Tota, L., Wyvratt, M.J., Fisher, M.H., and Weber, A.E. (2000) *Bioorganic & Medicinal Chemistry Letters*, **10**, 1427; (b) Biftu, T., Feng, D.D., Liang, G.-B., Kuo, H., Qian, X., Naylor, E.M., Colandrea, V.J., Candelore, M.R., Cascieri, M.A., Colwell, L.F. Jr., Forrest, M.J., Hom, G.J., MacIntyre, D.E., Stearns, R.A., Strader, C.D., Wyvratt, M.J., Fisher, M.H., and Weber, A.E. (2000) *Bioorganic & Medicinal Chemistry Letters*, **10**, 1431; (c) Yarovenko, V.N., Kosarev, S.A., Zavazin, I.V., and Krayuskin, M.M. (2002) *Russian Chemical Bulletin*, **51**, 1857; (d) Santos-Filho, J.M., de Lima, J.G., Leite, L.F.C.C., Ximenes, E.A., da Silva, J.B.P., Lima, P.C., and Pitta, I.R. (2005) *Heterocyclic Communications*, **11**, 29; (e) Kaboudin, B. and Saadati, F. (2005) *Journal of Heterocyclic Chemistry*, **42**, 699.

198 Buscemi, S., Pace, A., Pibiri, I., and Vivona, N. (2002) *Heterocycles*, **57**, 1891.

199 Leite, A.C.L., Viera, R.F., Wanderley, A.G., Afiatpour, P., Ximenes, E.C.P.A., Srivastava, R.M., De Oliveira, C.F., Medeiros, M.V., Antunes, E., and Brondani, D.J. (2000) *Il Farmaco*, **55**, 719.

200 Boys, M.L., Schretzman, L.A., Chandrakumar, N.S., Tollefson, M.B., Mohler, S.B., Downs, V.L., Downs, V.L., Penning, T.D., Russell, M.A., Wendt, J.A., Chen, B.B., Stenmark, H.G., Wu, H., Spangler, D.P., Clare, M., Desai, B.N., Khanna, I.K., Nguyen, M.N., Duffin, T., Engleman, V.W., Finn, M.B., Freeman, S.K., Hanneke, M.L., Keene, J.L., Klover, J.A., Nickols, G.A., Nickols, M.A., Steininger, C.N., Westlin, M., Westlin, W., Yu, Y.X., Wang, Y., Dalton, C.R., and Norring, S.A. (2006) *Bioorganic & Medicinal Chemistry Letters*, **16**, 839.

201 (a) Evans, M.D., Ring, J., Schoen, A., Bell, A., Edwards, P., Berthelo, D., Nicewonger, R., and Baldino, C.M. (2003) *Tetrahedron Letters*, **44**, 9337; (b) Bipik, B., Ho, G.-J., Williams, J.M., and Conlon, D.A. (2004) *Synthetic Communications*, **34**, 1863; (c) Santagata, V., Frecentese, F., Perissutti, E., Cirillo, D., Terracciano, S., and Caliendo, G. (2004) *Bioorganic & Medicinal Chemistry Letters*, **14**, 4491.

202 Conlon, D.A., Drahus-Panoe, A., Ho, G.-J., Pipik, B., Helmy, R., McNamara, J.M., Shi, Y.-J., Williams, J.M., Macdonald, D., Deschenes, D., Gallant, M., Mastracchio, A., Roy, B., and Scheigetz, J. (2006) *Organic Process Research & Development*, **10**, 36.

203 Deegan, T.L., Nitz, T.J., Cebzanov, D., Pufko, D.E., and Porco, J.A. Jr. (1999) *Bioorganic & Medicinal Chemistry Letters*, **9**, 209.

204 Yarovenko, V.N., Taralashvili, V.K., Zavarsin, I.V., and Krayushkin, M.M. (1990) *Tetrahedron*, **46**, 3941.

205 Pace, A., Buscemi, S., and Vivona, N. (2005) *Organic Preparations and Procedures International*, **37**, 44.

206 Abid, M., Jahromi, A.H., Tavoosi, N., Mahdavi, M., and Bijanzadeh, H.R. (2006) *Tetrahedron Letters*, **47**, 2965.

207 Young, J.R. and DeVita, R.J. (1998) *Tetrahedron Letters*, **39**, 3931.

208 Zhou, T. and Chen, Z.-C. (2002) *Synthetic Communications*, **32**, 887.

209 (a) Lin, Y.-i, Lang, S.A. Jr., Lovell, M.F., and Perkinson, N.A. (1979) *The Journal of Organic Chemistry*, **44**, 4160; (b) Eloy, F., Lenaers, R., and Buyle, R. (1964) *Bulletin des Sociétés Chimiques Belges*, **73**, 518; (c) Costanzo, A., Guerrini, G., Ciciani, G., Bruni, F., Selleri, S., Costa, B., Martini, C., Lucacchini, A., Aiello, P.M., and Ipponi, A. (1999) *Journal of Medicinal Chemistry*, **42**, 2218.

210 (a) Bock, M.G., Smith, R.L., Blaine, E.H., and Cragoe, E.J. Jr. (1986) *Journal of Medicinal Chemistry*, **29**, 1540; (b) Unangst, P.C., Shrum, G.P., Connor, D.T., Dyer, R.D., and Schrier, D.J. (1992) *Journal of Medicinal Chemistry*, **35**, 3691.

211 Stanovki, B. and Svete, J. (2000) *Synlett*, 1077.

212 Wu, W.D., Ma, L.T., Zhang, L.H., Lu, Y., Guo, F., and Zheng, Q.T. (2000) *Tetrahedron Asymmetry*, **11**, 1527.

213 Rice, K.D. and Nuss, J.M. (2001) *Bioorganic & Medicinal Chemistry Letters*, **11**, 753.

214 Humphrey, G.R. and Wright, S.H.B. (1989) *Journal of Heterocyclic Chemistry*, **26**, 23.

215 Christi, M. and Huisgen, R. (1973) *Chemische Berichte*, **106**, 3345.

216 Mukaiyama, T. and Hoshino, T. (1960) *Journal of the American Chemical Society*, **82**, 5339.

217 Neidlein, R. and Li, S. (1996) *Journal of Heterocyclic Chemistry*, **33**, 1943.

218 Bolton, R.E., Coote, S.J., Finch, H., Lowdon, A., Pegg, N., and Vinader, M.V. (1995) *Tetrahedron Letters*, **36**, 4471.

219 Itoh, K.-i., Sakamaki, H., and Horiuchi, C.A. (2005) *Synthesis*, 1935.

220 Yu, C., Lei, M., Su, W., and Xie, Y. (2007) *Synthetic Communications*, **37**, 4439.

221 Giurg, M. and Mlochowski, J. (1997) *Polish Journal of Chemistry*, **71**, 1093.

222 Bokach, N.A., Kukushkin, V.Yu., Haukka, M., and Pompeiro, J.L. (2005) *European Journal of Inorganic Chemistry*, 845.

223 Benltifa, M., Vidal, S., Gueyrard, D., Goekjian, P.G., Msaddek, M., and Praly, J.P. (2006) *Tetrahedron Letters*, **47**, 6143.

224 Quadrelli, P., Invernizzi, A.G., Falzoni, M., and Caramella, P. (1997) *Tetrahedron*, **53**, 1787.

225 Szczepankiewiez, W., Borowiak, T., Kubicki, M., Suwinski, J., and Wagner, P. (2002) *Polish Journal of Chemistry*, **76**, 1137.

226 da Costa Leite, L.F.C., Srivastava, R.M., and Cavalcante, A.P. (1989) *Bulletin des Sociétés Chimiques Belges*, **98**, 203.

227 Hemming, K., Morgan, D.T., and Smalley, R.K. (2000) *Journal of Fluorine Chemistry*, **106**, 83.

228 Buscemi, S., Pace, A., Pibiri, I., Vivona, N., Lanza, C.Z., and Spinelli, D. (2004) *European Journal of Organic Chemistry*, 974.

229 Pace, A., Pibiri, I., Buscemi, S., and Vivona, N. (2004) *Heterocycles*, **57**, 811.

230 Buscemi, S., Pace, A., and Vivona, N. (2000) *Tetrahedron Letters*, **41**, 7977.

231 Buscemi, S., Pace, A., Calabrese, R., Vivona, N., and Metrangolo, P. (2001) *Tetrahedron*, **57**, 5865.

232 Lessel, J. and Herfs, G. (2000) *Pharmazi*, **55**, 22.

233 (a) Kim, H.T., Min, J.Y., Choi, G.J., Kim, J.-C., Kim, B.S., Chung, Y.R., Kim, B.T., Kim, Y.S., Yamaguchi, I., and Cho, K.Y. (2002) *Journal of Pesticide Science*, **27**, 229; (b) Mindl, J., Kavalek, J., Strakova, H., and Sterba, V. (1999) *Collection of Czechoslovak Chemical Communications*, **64**, 1641.

234 Nicolaides, D.M., Fylakatakidou, K.C., Litinas, K.E., and Hadjipavlou-Litina, D. (1998) *European Journal of Medicinal Chemistry*, **33**, 715.

235 Nicolaides, D.M., Fylakatakidou, K.C., Litinas, K.E., and Hadjipavlou-Litina, D. (1996) *Journal of Heterocyclic Chemistry*, **33**, 967.

236 Szczepankiewiez, W., Wagner, P., Danieki, M., and Suwinski, J. (2003) *Tetrahedron Letters*, **44**, 2015.

237 Lin, X.-F., Zhang, J., and Wang, Y.-G. (2003) *Tetrahedron Letters*, **44**, 4113.

238 Baouid, A., Elhazazi, S., Hasnaoui, A., Compain, P., Lavergne, J.-P., and Huet, F. (2001) *New Journal of Chemistry*, **25**, 1479.

239 Nabih, K., Baouid, A., Hasnaoui, A., and Kenz, A. (2004) *Synthetic Communications*, **34**, 3565.

240 Boudina, A., Baouid, A., Hasnaoui, A., and Essaber, M. (2006) *Synthetic Communications*, **36**, 573.

241 Branco, P.S., Prabhakar, S., Lobo, A.M., and Williams, D.J. (1992) *Tetrahedron*, **48**, 6335.

242 Suga, H., Shi, X., and Ibata, T. (1998) *Bulletin of the Chemical Society of Japan*, **71**, 1231.

243 Naidu, B. and Sorenson, M.E. (2005) *Organic Letters*, **7**, 1391.

244 Eberson, L., McCullough, J.J., Hartshorn, C.M., and Michael, P. (1998) *Journal of the Chemical Society-Perkin Transactions 1*, 41.

245 Diaz-Ortiz, A., Diez-Barra, E., de la Hoz, A., Moreno, A., Gomez-Escalonilla, M.J., and Loupy, A. (1996) *Heterocycles*, **43**, 1021.

246 Wagner, G. and Galland, T. (2008) *Tetrahedron Letters*, **49**, 3596.

247 Coley, H.M., Sarju, J., and Wagner, G. (2008) *Journal of Medicinal Chemistry*, **51**, 135.

248 Sarju, J., Arbour, J., Sayer, J., Rohrmoser, B., Scherer, W., and Wagner, G. (2008) *Dalton Transactions*, 5302.

249 Sarju, J., Arbour, J., Sayer, J., Rohrmoser, B., Scherer, W., and Wagner, G. (2007) *Inorganic Chemistry*, **46**, 8323.

250 Consonni, R., Dalla Croce, P., Ferraccioli, R., and La Rosa, C. (1992) *Journal of Chemical Research (S)*, 32.

251 Coskun, N. and Parlar, A. (2006) *Synthetic Communications*, **36**, 997.

252 Ritter, T. and Carreira, E.M. (2005) *Angewandte Chemie*, **44**, 936.

253 Kraiem, J., Grosvalet, L., Perrin, M., and Hassine, B.B. (2001) *Tetrahedron Letters*, **42**, 9131.

254 Quadrelli, P., Srocchi, R., Piccanello, A., and Caramella, P. (2005) *Journal of Combinatorial Chemistry*, **7**, 887.

255 Buscemi, S., Pace, A., Piccionello, A.P., Vivona, N., and Pani, M. (2006) *Tetrahedron Letters*, **62**, 1158.

256 Takacsk, K., Harsanyi, K., Kolonits, P., and Ajzert, K.I. (1975) *Chemische Berichte*, **108**, 1911. (1987) *Journal of the Chemical Society-Perkin Transactions 1*, 2163.

257 Moormann, A.E., Wang, J.L., Palmquist, K.E., Promo, M.A., Snyder, J.S., Scholten, J.A., and Massa, M.A. (2004) *Tetrahedron*, **60**, 10907.

258 Cantello, B.C.C., Connor, S.C., Dean, D.K., and Hindley, R.M. (1997) *Synlett*, 263.

259 Farthing, C.N., Baldwin, J.E., Russell, A.T., Schofield, C.J., and Spivey, A.C. (1996) *Tetrahedron Letters*, **7**, 5225.

260 Greig, D.J., Hamilton, D.G., McPherson, M., and Paton, R.M. (1987) *Journal of the Chemical Society-Perkin Transactions 1*, 607.

261 Moussebois, C. and Eloy, F. (1964) *Helvetica Chimica Acta*, **47**, 838.

262 Buscemi, S., Pace, A., Pibiri, I., Vivona, N., and Spinelli, D. (2003) *The Journal of Organic Chemistry*, **68**, 605.

263 Beltrame, P., Cadoni, E., Floris, C., Gelli, G., and Lai, A. (2000) *Heterocycles*, **53**, 191.

264 Crimmin, M.J., O'Hanlon, P.J., Rogers, N.H., and Walker, G. (1989) *Journal of the Chemical Society-Perkin Transactions 1*, 2047.

265 (a) Gante, J., Juraszyk, H., Raddatz, P., and Wurziger, H. (1996) *Bioorganic & Medicinal Chemistry Letters*, **6**, 2425; (b) Liao, Y., Bottcher, H., Harting, J., Greiner, H., van Amsterdam, C., Cremers, T., Sundell, S., Marz, J., Rautenberg, W., and Wikstrom, H. (2000) *Journal of Medicinal Chemistry*, **43**, 517; (c) Palazzo, G., Strani, G., and Tavella, M. (1961) *Gazzetta Chimica Italiana*, **91**, 1085.

266 Tavella, M. and Strani, G. (1961) *Annali di Chimica (Rome)*, **51**, 361.

267 Liang, G.-B. and Qian, X. (1999) *Bioorganic & Medicinal Chemistry Letters*, **9**, 2101.

268 Malamidou-Xenikaki, E. and Coutouli-Agryropoulou, E. (1990) *Tetrahedron*, **46**, 7865.

269 Bolton, R.E., Coote, S.J., Finch, H., Lowdon, A., Pegg, N., and Vinader, M.V. (1995) *Tetrahedron Letters*, **36**, 4471.

270 (a) Ruccia, M., Vivona, N., and Spinelli, D. (1981) *Advances in Heterocyclic Chemistry*, **29**, 141; (b) Vivona, N., Buscemi, S., Frenna, V., and Cusmano, G. (1993) *Advances in Heterocyclic Chemistry*, **56**, 49; (c) Vivona, N., Cusmano, G., and Macaluso, G. (1977) *Journal of the*

Chemical Society-Perkin Transactions 1, 1616; (d) Kim, C.-K., Zielinski, P.A., and Maggiulli, C.A. *The Journal of Organic Chemistry* (1984,) **49**, 5247.

271 van der Plas, H.C. (1977) *Journal of Heterocyclic Chemistry* (2000) **37**, 427.

272 (a) D'Anna, F., Frenna, V., Macaluso, G., Marullo, S., Morganti, S., Pace, V., Spinelli, D., Spisani, R., and Tavani, C. (2006) *The Journal of Organic Chemistry*, **71**, 5616; (b) D'Anna, F., Ferroni, F., Frenna, V., Guernelli, S., Lanza, C.Z., Macaluso, G., Pace, V., Petrillo, G., Spinelli, D., and Spisani, R. (2005) *Tetrahedron*, **61**, 167.

273 Buscemi, S. and Vivona, N. (1991) *Journal of the Chemical Society-Perkin Transactions 2*, 187.

274 Buscemi, S., Vivona, N., and Caronna, T. (1996) *The Journal of Organic Chemistry*, **61**, 8397.

275 (a) Pace, A., Pibiri, I., Buscemi, S., Vivona, N., and Malpezzi, L. (2004) *The Journal of Organic Chemistry*, **69**, 4108; (b) Buscemi, S., Pace, A., Pibiri, I., Vivona, N., and Caronna, T. (2004) *Journal of Fluorine Chemistry*, **125**, 165.

276 Vivona, N., Buscemi, S., and Asta, S. (1997) *Tetrahedron*, **53**, 12629.

277 Quadrelli, P., Mella, M., and Caramella, P. (1999) *Tetrahedron Letters*, **40**, 797.

278 Quadrelli, P., Campari, G., and Mella, M. (2000) *Tetrahedron Letters*, **41**, 2019.

279 Xu, W.-M, Huang, X., and Tang, E. (2005) *Journal of Combinatorial Chemistry*, **7**, 726.

280 (a) Weidner-Wells, M.A., Henninger, T.C., Fraga-Spano, S.A., Boggs, C.M., Matheis, M., Ritchie, D.M., Argentie, D.C., Wachtd, M.P., and Hlasta, J. (2004) *Bioorganic & Medicinal Chemistry Letters*, **14**, 4307; (b) Liao, Y., Bottcher, H., Harting, J., Greiner, H., can Amsterdam, C., Cremes, T., Sundell, S., Marz, J., Rautenberg, W., and Wikstrom, H. (2000) *Journal of Medicinal Chemistry*, **43**, 517.

281 Hebert, N., Hannah, A.L., and Sutton, S.C. (1999) *Tetrahedron Letters*, **40**, 8547.

282 Gallardo, H., Cristiano, R., Vieira, A.A., Neves, F., Ricardo, A.W., and Srivastava, R.M. (2008) *Synthesis*, **4**, 605.

283 Diana, G.D., Volkots, D.L., Nitz, T.J., Bailey, T.R., Long, M.A., Vescio, N., Aldous, S., Pevear, D.C., and Dutko, F.J. (1994) *Journal of Medicinal Chemistry*, **37**, 2421.

284 Ahn, J.-M., Boyle, N.A., MacDonald, M.T., and Janda, K.D. (2002) *Mini-Reviews in Medicinal Chemistry*, **2**, 463.

285 Watjen, F., Baker, R., Engelstoff, M., Herbert, R., Macleod, A., Knight, A., Merchant, K., Moseley, J., Saunders, J., Swain, C.J., Wong, E., and Springer, J.P. (1989) *Journal of Medicinal Chemistry*, **32**, 2282.

286 Amarasinghe, K.K.D., Evidokimov, A.G., Xu, K., Clark, C.M., Maier, M.B., Srivastava, A., Colson, A.-O., Gerwe, G.S., Stake, G.E., Howard, B.W., Pokross, M.E., Graya, J.L., and Peters, K.G. (2006) *Bioorganic & Medicinal Chemistry Letters*, **16**, 4252.

287 (a) Orlek, B.S., Blaney, F.E., Brown, F., Clark, M.S.G., Hadley, M.S., Hatcher, J., Riley, G.J., Rosenberg, H.E., Wadsorth, H.J., and Wyman, P. (1991) *Journal of Medicinal Chemistry*, **34**, 2726; (b) Street, L.J., Baker, R., Book, K., Kneen, C.O., McaLeod, A.M., Merchant, K.J., Showell, G.A., Saunders, J., Fredman, S.B., and Harley, E.A. (1990) *Journal of Medicinal Chemistry*, **33**, 2690.

288 Clitherow, J.W., Beswick, P., Irving, W.J., Scopes, D.I., Barnes, J.C., Clapham, J., Brown, J.D., Evans, D.J., and Hayes, A.G. (1999) *Bioorganic & Medicinal Chemistry Letters*, **6**, 833.

289 (a) Kohara, Y., Imamiya, E., Kubo, K., Wada, T., Inada, Y., and Naka, T. (1995) *Bioorganic & Medicinal Chemistry Letters*, **5**, 1903; (b) Meyer, E., Joussef, A.C., Gallardo, A.C., and Bortoluzzi, A.J. (2003) *Journal of Molecular Structure*, **655**, 361.

290 Medebielle, M., Ait-Mohand, S., Burkhloder, C., Dolbier, W.R. Jr., Laumond, G., and Aubertin, A.-M. (2005) *Journal of Fluorine Chemistry*, **126**, 535.

291 (a) Jakopin, Z., Roskar, R., and Dolenc, M.S. (2007) *Tetrahedron Letters*, **48**, 1465; (b) Katritzky, A.R., Shestopalov, A.A., and Suzuki, K. (2005) *Arkivoc*, **7**, 36.

292 (a) Saunders, J., Cassidy, M., Freedman, S.B., Harley, E.A., Iversen, L.L., Kneen, C., MacLeod, A.M., Merchant, K.J., Snow, R.J., and Baker, R. (1990) *Journal of*

Medicinal Chemistry, **33**, 1128; (b) Showell, G.A., Gibbons, T.L., Kneen, C.O., MacLeod, A.M., Merchant, K., Saunders, J., Freedman, S.B., Patel, S., and Baker, R. (1991) *Journal of Medicinal Chemistry*, **34**, 1086.

293 Westwood, R. (1991) *Journal of Medicinal Chemistry*, **34**, 2060.

294 Chen, C.-Y., Senanayake, C.H., Bill, T.J., Larsen, R.D., Verhoeven, T.R., and Reider, P.J. (1994) *The Journal of Organic Chemistry*, **59**, 3738.

295 Swain, C.J., Baker, R., Kneen, C., Moseley, J., Saunders, J., Seward, E.M., Stevenson, G., Beer, M., Stanton, J., and Watling, K. (1991) *Journal of Medicinal Chemistry*, **34**, 140.

296 Clitherow, J.W., Beswick, P., Irving, W.J., Scopes, D.I.C., Barnes, J.C., Clapham, J., Brown, J.D., Evans, D.J., and Hayes, A.G. (1996) *Bioorganic & Medicinal Chemistry Letters*, **6**, 833.

297 Diana, G.D., Volkots, D.L., Nitz, T.J., Bailey, T.R., Long, M.A., Vescio, N., Aldous, S., Pevear, D.C., and Dutko, F.J. (1994) *Journal of Medicinal Chemistry*, **37**, 2421.

298 Ankersen, M., Peschke, B., Hansen, B.S., and Hansen, T.K. (1997) *Bioorganic & Medicinal Chemistry Letters*, **7**, 1293.

299 Chimirri, A., Grasso, S., Monforte, A.M., and Zappalà, M. (1996) *Il Farmaco*, **51**, 125.

300 Matsumoto, J., Takahashi, T., Agata, M., Toyofuku, H., and Sasada, N. (1994) *Japanese Journal of Pharmacology*, **65**, 51.

301 Ohmoto, K., Yamamoto, T., Horiuchi, T., Imanishi, H., Odagaki, Y., Kawabata, K., Sekioka, T., Hirota, Y., Matsuoka, S., Nakai, H., and Toda, M. (2000) *Journal of Medicinal Chemistry*, **43**, 4927.

302 Rudolph, J., Theis, H., Hanke, R., Endermann, R., Johannsen, L., and Geschke, F.-U. (2001) *Journal of Medicinal Chemistry*, **44**, 619.

303 Carroll, F.I., Gray, J.L., Abraham, P., Kuzemko, M.A., Lewin, A.H., Boja, J.W., and Kuhar, M.J. (1993) *Journal of Medicinal Chemistry*, **36**, 2886.

304 Nowak, I., Cannon, J.F., and Robins, M.J. (2006) *Organic Letters*, **8**, 4565.

305 (a) Seley, K.L., Salim, S., Zhang, L., and O'Daniel, P.I. (2005) *The Journal of Organic Chemistry*, **70**, 1612; (b) Charton, J., Deprez-Poulain, R., Hennuyer, N., Tailleux, A., Staels, B., and Deprez, B. (2009) Bioorganic & *Medicinal Chemistry Letters*, **19**, 489.

306 Olofson, R.A. and Michelman, J.S. (1964) *Journal of the American Chemical Society*, **86**, 1863.

307 Paton, R.M. (1984) *Comprehensive Heterocyclic Chemistry*, vol. 6 (ed. K.T. Potts), Pergamon Press, Oxford, p. 393.

308 Paton, R.M. (1996) *Comprehensive Heterocyclic Chemistry*, vol. 4 (ed. K.T. Potts), Pergamon Press, Oxford, p. 229.

309 Sliwa, W. and Thomas, A. (1984) *Heterocycles*, **22**, 1571.

310 Sliwa, W. and Thomas, A. (1985) *Heterocycles*, **23**, 399.

311 Sliwa, W., Thomas, A., and Zelichowicz, N. (1992) *Collection of Czechoslovak Chemical Communications*, **57**, 978.

312 Gasco, A. and Boulton, A.J. (1981) *Advances in Heterocyclic Chemistry*, **29**, 251.

313 Paton, R.M. (2004) *Science of Synthesis*, **13**, 185.

314 Sliva, W. (1984) *Heterocycles*, **22**, 1571.

315 Sheremetev, A.B., Strelenko, Y.A., Novokova, T.S., and Khmel'nitskii, L.I. (1993) *Tetrahedron*, **49**, 5905.

316 Wieland, H. (1903) *Annalen der Chemie-Justus Liebig*, **329**, 225.

317 Werner, A. (1904) *Lehrbuch der Stereochemie*, Fischer, Jena, p. 260.

318 Mohr, A. (1908) *Journal fur Praktische Chemie, II*, **79**, 1.

319 Sheremetev, A.B. (1995) *Journal of Heterocyclic Chemistry*, **32**, 371.

320 Struchkov, Y.T., Batsanov, A.S., Chuiguk, V.A., Batog, L.V., Kulikov, A.S., Pivina, T.S., and Strelenko, Y.A. (1992) *Khimiya Geterotsiklicheskikh Soedinenii*, **2**, 233.

321 Sasaki, T. (1970) Jpn. Pat. 45034589.

322 (a) Vichard, D., Hallè, J.C., Huguet, B., Pouet, M.J., Riou, D., and Terrier, F. (1998) *Chemical Communications*, 791; (b) Sepulcri, P., Hallè, J.C., Goumont, R., Riou, D., and Terrier, F. (1999) *The Journal of Organic Chemistry*, **64**, 9954.

323 Sepulcri, P., Hallè, J.C., Goumont, R., Riou, D., and Terrier, F. (2000) *Journal of the Chemical Society-Perkin Transactions 2*, 51.

324 Goumont, R., Sebban, M., and Terrier, F. (2002) *Chemical Communications*, 2110.

325 Lutskii, A.E., Shepel, A.V., Shvaika, O.P., and Klimisha, G.P. (1969) *Khimiya Geterotsiklicheskikh Soedinenii*, **3**, 461.

326 (a) Hafelinger, G. (1970) *Chemische Berichte*, **103**, 3370; (b) Kovalenko, I., Furer, V.L., Anisimova, L.I., and Yagund, E.M. (1994) *Zhurnal Strukturnoi Khimii*, **35**, 54.

327 Ugliengo, P., Viterbo, D., and Calleri, M. (1988) *Journal of the Chemical Society-Perkin Transactions 2*, 661.

328 Vsetecka, V., Fruttero, R., Gasco, A., and Exner, O. (1994) *Journal of Molecular Structure*, **324**, 277.

329 Trifonov, R.E., Gaenko, A.V., Vergizov, S.N., Shcherbinin, M.B., and Ostrovskii, V.A. (2003) *Croatica Chemica Acta*, **76**, 177.

330 Andrianov, V.G., Shokhen, M.A., Eremeev, A.V., and Barmina, S.V. (1986) *Khimiya Geterotsiklicheskikh Soedinenii*, 264.

331 Seminario, J.M., Concha, M.C., and Politzer, P. (1992) *Journal of Computational Chemistry*, **13**, 177.

332 Friedrichsen, W. (1995) *Journal of Chemical Research (S)*, 120.

333 Rauhut, G. (1996) *Journal of Computational Chemistry*, **17**, 1848.

334 Rauhut, G., Jarzecki, A., and Pulay, P. (1997) *Journal of Computational Chemistry*, **18**, 489.

335 Eckert, F., Rauhut, G., and Katrizky, A.R. (1999) *Journal of the American Chemical Society*, **121**, 6700.

336 Friedrichsen, W. (1994) *The Journal of Physical Chemistry*, **98**, 12933.

337 Ponder, M., Fowler, E.J., and Schaefer, H.F. (1994) *The Journal of Organic Chemistry*, **59**, 6431.

338 Dunkin, I.R., Lynch, M.A., Boulton, A.J., and Henderson, N. (1991) *Journal of the Chemical Society, Chemical Communications*, 1178.

339 Hacker, N.P. (1991) *The Journal of Organic Chemistry*, **56**, 5216.

340 Murata, S. and Tomioka, H. (1992) *Chemistry Letters*, 57.

341 (a) Himmel, H.-J., Konrad, S., Friedrichsen, W., and Rauhut, G. (2003) *Journal of Physical Chemistry A*, **107**, 6731; (b) Stevens, J., Schweizer, M., and Rauhut, G. (2001) *Journal of the American Chemical Society*, **123**, 7326.

342 Ruccia, M. and Vivona, N. (1981) *Advances in Heterocyclic Chemistry*, **29**, 141.

343 Harris, R.K., Katritzky, A.R., Oksne, A.S., Bailey, A.S., and Paterson, W.G. (1963) *Journal of the Chemical Society*, 197.

344 Katritzky, A.R. and Gordeev, M.F. (1993) *Heterocycles*, **35**, 483.

345 Friedrichsen, W. (1995) *Theochem*, **342**, 23.

346 Klenke, B. and Friedrichsen, W. (1996) *Tetrahedron*, **52**, 743.

347 Klenke, B. and Friedrichsen, W. (1998) *Theochem*, **451**, 263.

348 Beal, R.W. and Brill, T.B. (2000) *Propellants, Explosives Pyrotechnics*, **25**, 247.

349 Acree, W.E. Jr., Pilcher, G., and Ribeiro da Silva, M.D.M.C. (2005) *Journal of Physical Chemistry Reference Data*, **34**, 553.

350 Calleri, M., Chiari, G., Chiesi Villa, A., and Guastini, C. (1976) *Crystal Struct. Commun.*, **5**, 113.

351 Sheremetev, A.B., Andrianov, V.G., Mantseva, E.V., Shatunova, E.V., Aleksandrova, N.S., Yudin, I.L., Dmitriev, D.D., Averkiev, B.B., and Antipin, M.Y. (2004) *Russian Chemical Bulletin*, **53**, 596.

352 Zelenin, A.K., Trudell, M.L., and Gilardi, R.D. (1998) *Journal of Heterocyclic Chemistry*, **35**, 151.

353 Barbieux-Flammang, M., Vandevoorde, S., Flammang, R., Wong, M.W., Bibas, H., Kennard, C.H.L., and Wentrup, C. (2000) *Journal of the Chemical Society-Perkin Transactions 2*, **3**, 473.

354 Gunasekaran, A., Trudell, M.L., and Boyer, J.H. (1994) *Heteroatom Chemistry*, **5**, 441.

355 Sheremetev, A.B., Ivanova, E.A., Spiridonova, N.P., Melnikova, S.F., Tselinsky, I.V., Suponitsky, K.Y., and Antipin, M.Y. (2005) *Journal of Heterocyclic Chemistry*, **42**, 1237.

356 Sheremetev, A.B., Shamshina, J.L., Dmitriev, D.E., Lyubetskii, D.V., and Antipin, M.Y. (2004) *Heteroatom Chemistry*, **15**, 199.

357 Sheremetev, A.B., Konkina, S.M., Yudin, I.L., Dmitriev, D.E., Averkiev, B.B., and Antipin, M.Y. (2003) *Russian Chemical Bulletin*, **52**, 1413.

358 Stiefvater, O.L. (1988) *Zeitschrift für Naturforschung Teil A: Physik, Physikalische Chemie*, 597.

359 Averkiev, B.B., Antipin, M.Y., Sheremetev, A.B., and Timofeeva, T.V. (2005) *Crystal Growth Design*, **5**, 631.

360 Yudin, I.L., Sheremetev, A.B., Averkiev, B.B., and Antipin, M.Y. (2005) *Journal of Heterocyclic Chemistry*, **42**, 691.

361 Sheremetev, A.B., Aleksandrova, N.S., Dmitriev, D.E., Averkiev, B.B., and Antipin, M.Y. (2005) *Journal of Heterocyclic Chemistry*, **42**, 519.

362 Sheremetev, A.B., Ivanova, E.A., Dmitriev, D.E., Kulagina, V.O., Averkiev, B.B., and Antipin, M.Y. (2005) *Journal of Heterocyclic Chemistry*, **42**, 803.

363 Sheremetev, A.B., Shatunova, E.V., Averkiev, B.B., Dmitriev, D.E., Petukhov, V.A., and Antipin, M.Y. (2004) *Heteroatom Chemistry*, **15**, 131.

364 Strelenko, Y.A., Sheremetev, A.B., and Khmelnitskii, L.I. (1992) *Chemistry of Heterocyclic Compounds*, **28**, 927.

365 Calvino, R., Fruttero, R., Gasco, A., and Mortarini, V. (1982) *Journal of Heterocyclic Chemistry*, **19**, 427.

366 Sheremetev, A.B., Kulagina, V.O., Aleksandrova, N.S., Dmitriev, D.E., Strelenko, Y.A., Lebedev, V.P., and Matyushin, Y.N. (1998) *Propellants Explosives Pyrotechnics*, **23**, 142.

367 Sheremetev, A.B., Kharitonova, O.V., Mantseva, E.V., Kulagina, V.O., Shatunova, E.V., Aleksandrova, N.S., Mel'nikova, T.M., Ivanova, E.A., Dmitriev, D.E., Eman, V.A., Yudin, Y.L., Kuzmin, V.S., Strelenko, Y.A., Novikova, T.S., Lebedev, O.V., and Khmelnitskii, L.I. (1999) *Russian Journal of Organic Chemistry*, **35**, 1525.

368 Strelenko, Y.A., Sheremetev, A.B., and Khmel'nickii, L.I. (1992) *Khimiya Geterotsiklicheskikh Soedinenii*, **8**, 1101.

369 Terrier, F., Halle, J.C., MacCormack, P., and Pouet, M.J. (1989) *Canadian Journal of Chemistry*, **67**, 503.

370 Boulton, A.J. and Gosh, P.B. (1969) *Advances in Heterocyclic Chemistry*, **10**, 1.

371 Deady, L.W. and Quazi, N.H. (1995) *Spectroscopy Letters*, **28**, 1033.

372 Dmitriev, D.E., Strelenkp, Y.A., and Sheremetev, A.B. (2002) *Russian Chemical Bulletin*, **51**, 290.

373 Butler, A.R., Lightfoot, P., and Short, D.M. (1999) *The Journal of Organic Chemistry*, **64**, 8748.

374 Yavari, I., Botto, R.E., and Roberts, J.D. (1978) *The Journal of Organic Chemistry*, **43**, 2542.

375 Godovikova, T.I., Golova, S.P., Strelenko, Y.A., Antipin, M.Y., Struchkov, Y.T., and Khmelnitskii, L.I. (1994) *Mendeleev Communications*, 7.

376 (a) Evgen'ev, M.I., and Levinson, F.S. (1991) *Khimiya Geterotsiklicheskikh Soedinenii*, **11**, 1565. (b) Imai, K., Uzu, S., Kanda, S., and Baeyens, W.R.G. (1994) *Analytica Chimica Acta*, **290**, 3.

377 Porter, Q.N. and Baldas, J. (1971) *Mass Spectrometry of Heterocyclic Compounds*, Wiley Interscience, New York, p. 527.

378 Westphal, J. and Schmidt, R. (1973) *Journal fur Praktische Chemie*, **315**, 791.

379 Gallos, J.K., Lianis, P.S., and Rodios, N.A. (1994) *Journal of Heterocyclic Chemistry*, **31**, 481.

380 Arshadi, M.R. (1978) *Organic Mass Spectrometry*, **13**, 379.

381 Hwang, K-.J., Jo, I., Shin, Y.A., Yoo, S., and Jae Hyun, H. (1995) *Tetrahedron Letters*, **36**, 3337.

382 Deem, M.L. (1980) *Organic Mass Spectrometry*, **15**, 573.

383 Auricchio, S., Selva, A., and Truscello, A.M. (1997) *Tetrahedron*, **51**, 17407.

384 Sheremetev, A.B. (2001) *Advances in Heterocyclic Chemistry*, **78**, 66.

385 (a) Sheremetev, A.B. (2005) *Russian Chemical Bulletin*, **54**, 1032; (b) Sheremetev, A.B., Shamshina, Y.L., and Dmitriev, D.E. (2005) *Russian Chemical Bulletin*, **54**, 1057.

386 (a) Gunasekaran, A., Jayachandran, T., Boyer, J.H., and Trudell, M.L. (1995) *Journal of Heterocyclic Chemistry*, **32**, 1405; (b) Zelenin, A.K. and Trudell, M.L. (1997) *Journal of Heterocyclic Chemistry*, **34**, 1057; (c) Kusurkar, R.S., Goswami, S.K., Talawar, M.B., Gore, G.M., and Asthana, S.N. (2005) *Journal of Chemical Research*, **4**, 245.

387 Olofson, R.A. and Michelman, J.S. (1965) *The Journal of Organic Chemistry*, **30**, 1854.

388 Kamitori, Y. (1999) *Heterocycles*, **51**, 627.

389 Polyakov, B.V., Tverdoklebov, V.P., and Tselinskii, I.V. (1990) *Zhurnal Obshchei Khimii*, **60**, 2049.

390 Lakhan, R. and Singh, O.P. (1987) *Indian Journal of Chemistry Section B-Organic Chemistry Including Medicinal Chemistry*, **26**, 690.

391 Tselinskii, I.V., Mel'nikova, S.F., Romanova, T.V., Spiridonova, N.P., and Dunkunova, E.A. (2001) *Russian Journal of Organic Chemistry*, **37**, 1353.

392 Thakore, A.N., Buchsriber, J., and Oehlschlager, A.C. (1973) *Canadian Journal of Chemistry*, **51**, 2406.

393 Emmons, W.D. and Freeman, J.P. (1957) *The Journal of Organic Chemistry*, **22**, 456.

394 Andrianov, V.G. and Eremeev, A.V. (1990) *Khimiya Geterotsiklicheskikh Soedinenii*, 1443.

395 Vivona, N., Buscemi, S., Frenna, V., Ruccia, M., and Condo, M. (1985) *Journal of Chemical Research (S)*, 190.

396 Andrianov, V.G., Semenikhina, V.G., and Eremeev, A.V. (1992) *Khimiya Geterotsiklicheskikh Soedinenii*, **28**, 969.

397 Andrianov, V.G., Semenikhina, V.G., and Eremeev, A.V. (1993) *Zhurnal Organicheskoi Khimii*, **29**, 1062.

398 (a) Boulton, A.J., Ghosh, P.B., and Katritzky, A.R. (1966) *Journal of the Chemical Society*, 1004; (b) Boulton, A.J., Ghosh, P.B., and Katritzky, A.R. (1966) *Tetrahedron Letters*, **25**, 2887.

399 (a) Merritt, C. Jr., Di Pietro, C., Hand, C.W., Cornell, J.H., and Remy, D.E. (1975) *Journal of Chromatography*, **112**, 301; (b) Cadogan, J.I.G., Scott, R.J., Gee, R.D., and Gosney, I. (1974) *Journal of the Chemical Society-Perkin Transactions 1*, 1694; (c) Ghosh, P.B., Ternai, B., and Whitehouse, M.W. (1972) *Journal of Medicinal Chemistry*, **15**, 255; (d) Ghosh, P.B. and Whitehouse, M.W. (1968) *Journal of Medicinal Chemistry*, **11**, 305.

400 Bird, K.J., Rae, I.D., and White, A.M. (1973) *Australian Journal of Chemistry*, **26**, 1683.

401 Boulton, A.J., Fletcher, I.J., and Katritzky, A.R. (1971) *Journal of the Chemical Society (C)*, 1193.

402 Pedersen, C.L. (1976) *Acta Chemica Scandinavica. Series B: Organic Chemistry and Biochemistry*, **30**, 675.

403 Maksimovic-Ivanic, D., Mijatovic, S., Harhaji, L., Miljkovic, D., Dabideen, D., Cheng, K.F., Mangano, K., Malaponte, G., Al-Abed, Y., Libra, M., Garotta, G., Nicoletti, F., and Stosic-Grujicic, S. (2008) *Molecular Cancer Therapeutics*, **7**, 510.

404 Ponzio, G. (1932) *Gazzetta Chimica Italiana*, **62**, 127.

405 Niyazimbetov, M.E., Ul'yanona, E.V., and Petrosyan, V.A. (1992) *Soviet Electrochemistry (English Translation)*, **28**, 449.

406 (a) Sorba, G., Ermondi, G., Fruttero, R., Galli, U., and Gasco, A. (1996) *Journal of Heterocyclic Chemistry*, **33**, 327; (b) Bohn, H., Brendel, J., Schoenafinger, K., and Strobel, H. (1995) Eur. Pat. Appl. EP 683159; (c) Calvino, R., Fruttero, R., Ghigo, D., Bosia, A., Pescarmona, G.P., and Gasco, A. (1992) *Journal of Medicinal Chemistry*, **35**, 3296; (d) Gagneux, A.R. and Meier, R. (1970) *Helvetica Chimica Acta*, **53**, 1883.

407 Gallos, J.K., Lianis, P.S., and Rodios, N.A. (1994) *Journal of Heterocyclic Chemistry*, **31**, 481.

408 Schonafinger, K. (1999) *Farmaco (Societa Chimica Italiana: 1989)*, **54**, 316.

409 Fruttero, R., Ferrarotti, B., Serafino, A., Di Stilo, A., and Gasco, A. (1989) *Journal of Heterocyclic Chemistry*, **26**, 1345.

410 Dubonos, V.G., Ovchinnikov, I.V., Makhova, N.N., and Khmelnickii, L.I. (1992) *Mendeleev Communications*, 120.

411 Demir, A.S. and Findik, H. (2005) *Letters in Organic Chemistry*, **2**, 602.

412 Ballini, R., Barboni, L., and Filippone, P. (1997) *Chemistry Letters*, 475.

413 Ballini, R., Bosica, G., Fiorini, D., and Palmieri, A. (2005) *Tetrahedron*, **61**, 8971.

414 Nirode, W.F., Luis, J.M., and Wachter, N.M. (2006) *Bioorganic & Medicinal Chemistry Letters*, **16**, 2299.

415 (a) Huisgen, R. (1963) *Angewandte Chemie*, **2**, 565; (b) Huisgen, R. (1963) *Angewandte Chemie*, **2**, 633.

416 (a) Dondoni, A., Mangini, A., and Ghersetti, S. (1966) *Tetrahedron Letters*, **7**, 4789; (b) Barbaro, G., Battaglia, A., and Dondoni, A. (1970) *Journal of the Chemical Society (B)*, 588.

417 Mallory, F.B., Manatt, S.L., and Wood, C.S. (1965) *Journal of the American Chemical Society*, **87**, 5433.

418 Mallory, F.B. and Cammarata, A. (1966) *Journal of the American Chemical Society*, **88**, 61.

419 Hoffmann, R., Gleiter, R., and Mallory, F.B. (1970) *Journal of the American Chemical Society*, **92**, 1460.

420 Yu, Z.-X., Caramella, P., and Houk, K.N. (2003) *Journal of the American Chemical Society*, **125**, 15420.

421 Jager, V. and Colinas, P.A. (2002) Nitrile oxides in synthetic applications of 1,3-dipolar cycloaddition chemistry towards heterocycles and natural products, in *The Chemistry of Heterocyclic Compounds*, vol. 59 (eds A. Padwa and W.H. Pearson), John Wiley & Sons, Inc., Hoboken, New Jersey.

422 Caramella, P. and Grunanger, P. (1984) Nitrile oxides and imines, in *1,3-Dipolar Cycloaddition Chemistry* (ed. A. Padwa), John Wiley & Sons, Inc., New York.

423 Ayyangar, N.R., Madan Kumar, S., and Srinivasan, K.V. (1987) *Synthesis*, 616.

424 Noto, R., Rainieri, R., and Arnone, C. (1989) *Journal of the Chemical Society-Perkin Transactions 2*, **2**, 127.

425 Forster, H.J., Niclas, H.J., and Lukyanenko, N.G. (1985) *Zeitschrift fur Chemie*, **29**, 17.

426 (a) Niclas, H.J., Forster, H.J., and Zolch, L. (1985) Ger. Pat. 226286; (b) Zhao, S., Guo, Q., and Wang, Y. (2005) *Zhongguo Yiyao Gongye Zazhi*, **36**, 457; (2006) *Chemical Abstracts*, 147, 143391.

427 Munno, A., Bertini, V., Rasero, P., Picci, N., and Bonfanti, L. (1978) *Atti dell'Accademia Nazionale dei Lincei*, **64**, 385.

428 Munno, A., Bertini, V., Menconi, A., and Denti, G. (1974) *Atti della Societa Toscana di Scienze Naturali Memorie, Serie A*, **81**, 334.

429 Calvino, R., Ferrarotti, B., Gasco, A., and Serafino, A. (1983) *Gazzetta Chimica Italiana*, **113**, 811.

430 Vasilvitskii, A.E., Sheremetev, A.B., Novikova, T.S., Khmelnitskii, L.I., and Nefedov, O.M. (1989) *Bulletin of the Academy of Sciences of the USSR, Division of Chemical Sciences (English Translation)*, **38**, 2640.

431 Sheremetev, A.B., Makova, N.N., and Friedrichsen, W. (2001) *Advances in Heterocyclic Chemistry*, **78**, 65.

432 Eremeev, A.V., Andrianov, V.G., and Piskunova, I.P. (1979) *Chemistry of Heterocyclic Compounds (English Translation)*, **15**, 261.

433 Zelenov, M.P., Frolova, G.M., Mel'nikova, S.F., and Tselinskii, I.V. (1982) *Chemistry of Heterocyclic Compounds*, **18**, 21.

434 Tokura, N., Data, R., and Yokoyama, K. (1961) *Bulletin of the Chemical Society of Japan*, **34**, 270.

435 Butler, R.N., Daly, K.M., McMahon, J.M., and Burke, L.A. (1995) *Journal of the Chemical Society-Perkin Transactions 1*, 1083.

436 (a) Sheremetev, A.B., Strelenko, Y.A., Novikova, T.S., and Khmel'nitskii, L.I. (1989) *Izvestiia Akademii Nauk Seriia Biologicheskaia*, **8**, 1932; (b) Bertinaria, M., Galli, U., Sorba, G., Fruttero, R., Gasco, A., Brenciaglia, M.I., Scaltrito, M.M., and Dubini, F. (2003) *Drug Development Research*, **60**, 225.

437 Ilyushin, M.A. and Tselinskii, I.V. (1997) *Mendeleev Chemistry Journal*, **41**, 1.

438 (a) Sheremetev, A.B. and Kharitonova, O.V. (1992) *Mendeleev Communications*, 157; (b) Churakov, A.M., Semenov, S.E., Ioffe, S.L., Strelenko, Y.A., and Tartakovsky, V.A. (1995) *Mendeleev Communications*, 102; (c) Sheremetev, A.B., Kulagina, V.O., Kryazhevskikh, I.A., Melnikova, T.M., and Aleksandrova, N.S. (2002) *Russian Chemical Bulletin*, **518**, 1533; (d) Sheremetev, A.B., Andrianov, V.G., Mantseva, E.V., Shatunova, E.V., Aleksandrova, N.S., Yudin, I.L., Dmitriev, D.E., Averkiev, B.B., and Antipin, M.Yu. (2004) *Russian Chemical Bulletin*, **53**, 596.

439 (a) Ghosh, P., Ternai, B., and Whitehouse, M. (1981) *Medicinal Research Reviews*, **1**, 159; (b) Uchiyama, S., Iwai, K., and Prasanna de Silva, A. (2008) *Angewandte Chemie*, **47**, 4667; (c) Maezaki, N., Urabe,

D., Yano, M., Tominaga, H., Morioka, T., Kojima, N., and Tanaka, T. (2007) *Heterocycles*, **73**, 159.

440 Kelley, J.L., Linn, J.A., and Selway, J.W.T. (1989) *Journal of Medicinal Chemistry*, **32**, 218.

441 Boulton, A.J. and Mathur, S.S. (1973) *The Journal of Organic Chemistry*, **38**, 1054.

442 (a) Antipin, M.Y., Struchkov, Y.T., Balitskii, Y.V., and Gololobov, Y.G. (1981) *Zhurnal Strukturnoi Khimii*, **22**, 98; (b) Prokudin, V.G. and Nazin, G.M. (1987) *Izvestiya Akademli Nauk SSSR, Seriya Khimlcheskaya*, 221.

443 Heinzelmann, W. and Gilgen, P. (1976) *Helvetica Chimica Acta*, **59**, 2727.

444 Zelenov, M.P., Frolova, G.M., Mel'nikova, S.F., and Tselinskii, I.V. (1982) *Khimiya Geterotsiklicheskikh Soedinenii*, **1**, 27.

445 Boyer, J.H. and Huang, C. (1981) *Journal of the Chemical Society, Chemical Communications*, 365.

446 Zavarzina, O.V., Rakitin, O.A., and Khmelnitskii, L.I. (1994) *Khimiya Geterotsiklicheskikh Soedinenii*, 1133.

447 Sheremetev, A.B., Mantseva, E.V., Aleksandrova, N.S., and Khmelnitskii, L.I. (1995) *Mendeleev Communications*, 25.

448 Ostrowski, S. and Wojciechowski, K. (1990) *Canadian Journal of Chemistry*, **68**, 2239.

449 Sheremetev, A.B. and Ovchinnikov, Y.V. (1997) *Heteroatom Chemistry*, **8**, 7.

450 Gasco, A.M., Medana, C., and Gasco, A. (1994) *Synthetic Communications*, **24**, 2707.

451 Gohain, S., Prajapati, D., and Sandhu, J.S. (1995) *Chemistry Letters*, 725.

452 (a) Pasinszki, T. and Westwood, N.P.C. (1995) *Journal of the Chemical Society, Chemical Communications*, 1901; (b) Barness, J.F., Barrow, M.J., Harding, M.M., Paton, R.M., Aschroft, P.L., Crosby, J., and Joyce, C.J. (1979) *Journal of Chemical Research* (S), 10, 314; (c) Kulikov, A.S., Epishina, M.A., Ovchinnikov, I.V., and Makhova, N.N. (2007) *Russian Chemical Bullettin*, **56**, 1580.

453 Terrier, F. (1995), in *Organic Reactivity: Physical and Biological Aspects* (ed. B.T. Golding, R.J. Griffin, and H. Maskill), Special Publication no 148, The Royal Society of Chemistry, London, p. 399–414.

454 Boulton, A.J., Franck, F., and Huckstep, M.R. (1982) *Gazzetta Chimica Italiana*, **112**, 181.

455 Molotov, S.I., Kulikov, A.S., Strelenko, Yu.A., Makhova, N.N., and Lyssenko, K.A. (2003) *Russian Chemical Bulletin*, **52**, 1829.

456 Makhova, N.N., Ovchinnikov, I.V., Kulikov, A.S., Molotov, S.I., and Baryshnikova, E.L. (2004) *Pure and Applied Chemistry*, **76**, 1691.

457 Baryshnikova, E.L. and Makhova, N.N. (2000) *Mendeleev Communications*, 190.

458 Ovchinnikov, I.V., Epishina, M.A., Molotov, S.I., Strelenko, Y.A., Lyssenko, K.A., and Makhova, N.N. (2003) *Mendeleev Communications*, 272.

459 (a) Boulton, A.J., Coe, D.E., and Tsoungas, P.G. (1981) *Gazzetta Chimica Italiana*, **111**, 167; (b) Dannhardt, G. and Oberscruberger, I. (1989) *Archiv der Pharmazie*, **322**, 513.

460 (a) Vivona, N., Buscemi, S., Frenna, V., and Cusumano, G. (1993) *Advances in Heterocyclic Chemistry*, **56**, 49; (b) van der Plas, H.C. (1973) *Ring Transformations of Heterocycles*, vol. 1 and 2, Academic Press, London.

461 Ruccia, M., Vivona, N., and Spinelli, D. (1981) *Advances in Heterocyclic Chemistry*, **29**, 141.

462 Frenna, V., Vivona, N., Consiglio, G., Corrao, A., and Spinelli, D. (1981) *Journal of the Chemical Society-Perkin Transactions 2*, 325.

463 Vivona, N., Macaluso, G., Frenna, V., and Ruccia, M. (1983) *Journal of Heterocyclic Chemistry*, **20**, 931.

464 Guernelli, S., Laganà, M.F., Spinelli, D., Lo Meo, P., Noto, R., and Riela, S. (2002) *The Journal of Organic Chemistry*, **67**, 2948.

465 Cosimelli, B., Guernelli, S., Spinelli, D., Buscemi, S., Frenna, V., and Macaluso, G. (2001) *The Journal of Organic Chemistry*, **66**, 6121.

466 Cosimelli, B., Frenna, V., Guernelli, S., Lanza, C.Z., Macaluso, G., Petrillo, G., and Spinelli, D. (2002) *The Journal of Organic Chemistry*, **67**, 8010.

467 Mufarrij, N.A., Haddadin, M.J., Issidorides, C.H., McFarland, J.W., and Johnston, J.D. (1972) *Journal of the Chemical Society-Perkin Transactions 1*, 965.

468 Fisher, G. (1990) *Zeitschrift fur Chemie*, **30**, 305.

469 Takabataka, T., Miyazawa, T., Kojo, M., and Hasagawa, H. (2000) *Heterocycles*, **53**, 2151.

470 Argyropoulos, N.G., Gallos, J.K., and Nicolaides, D.N. (1986) *Tetrahedron*, **42**, 3631.

471 Borah, H.N., Boruah, R.C., and Sandhu, J.S. (1985) *Heterocycles*, **23**, 1625.

472 Bulacinski, A.B., Scriven, E.F.V., and Suschitzky, H. (1975) *Tetrahedron Letters*, **16**, 3577.

473 (a) Goumont, R., Sebban, M., Sepulcri, P., Marrot, J., and Terrier, F. (2002) *Tetrahedron*, **58**, 3249. (b) Lakhdar, S., Goumont, R., Boubaker, T., Mokhtari, M., and Terrier, F. (2006) *Organic and Biomolecular Chemistry*, **4**, 1910; (c) Goumont, R., Sebban, M., Marrot, J., and Terrier, F. (2004) *Arkivoc*, **3**, 85.

474 Devi, P. and Sandhu, J.S. (1983) *Journal of the Chemical Society, Chemical Communications*, 990.

475 Argyropoulos, N.G. and Gallos, J.K. (1990) *Journal of the Chemical Society-Perkin Transactions 1*, 3277.

476 Cerè, V., Pollicino, S., Sandri, E., and Scapini, G. (1976) *Tetrahedron*, **32**, 1277.

477 Kenley, R.A., Bedford, C.D., Dailey, O.D. Jr., Howd, R.A., and Miller, A. (1984) *Journal of Medicinal Chemistry*, **27**, 1201.

478 Di Stilo, A., Visentin, S., Cena, C., Gasco, A.M., Ermondi, G., and Gasco, A. (1998) *Journal of Medicinal Chemistry*, **41**, 5393.

479 Stetter, J., Ditgens, K., Thomas, R., Eue, L., and Schmidt, R.R. (1981) Pat. DE 2919293.

480 Schoenafinger, K. and Bohn, H. (1995) German Offen. DE. 4 401 150.

481 Gasco, A.M., Boschi, D., and Gasco, A. (1995) *Journal of Heterocyclic Chemistry*, **32**, 811.

482 Gasco, A.M., Cena, C., Di Stilo, A., Ermondi, G., Medana, C., and Gasco, A. (1996) *Helvetica Chimica Acta*, **79**, 1803.

483 Haworth, K.E., Owen, S.N., and Seward, E.M. (2004) PCT Int. Appl. WO96 29328 A1.

484 Sheremetev, A.B., Ivanova, E.A., Sizov, A.Yu., Kulagina, V.O., Dmitriev, D.E., and Strelenko, Yu.A. (2003) *Russian Chemical Bulletin*, **52**, 679.

485 Sheremetev, A.B. and Ivanova, E.A. (2003) *Russian Chemical Bulletin*, **52**, 2017.

486 Sheremetev, A.B., Ivanova, E.A., Shatunova, E.V., Dmitriev, D.E., and Kuz'mina, N.E. (2004) *Russian Chemical Bulletin*, **53**, 615.

487 Takabatake, T., Hasegawa, M., Nagano, T., and Hirobe, M. (1992) *Chemical & Pharmaceutical Bulletin*, **40**, 1644.

488 Calvino, R., Serafino, A., Ferrarotti, B., Gasco, A., and Sanfilippo, A. (1984) *Archiv der Pharmazie*, **317**, 695.

489 (a) Cerecetto, H. and Porcal, W. (2005) *Mini-Reviews in Medicinal Chemistry*, **5**, 57; (b) Ghosh, P.B., Ternai, B., and Whitehouse, N.V. (1972) *Journal of Medicinal Chemistry*, **15**, 255.

490 Caccuri, A.M. and Ricci, G. (2004) PCT Int. Appl. WO 2004093874 A1 20041104.

491 Tron, G.C., Pagliai, F., Del Grosso, E., Genazzani, A.A., and Sorba, G. (2005) *Journal of Medicinal Chemistry*, **48**, 3260.

492 Baures, P.W., James, D.R., Gless, R.D., Tran, T., Verheij, H.J., and Schultz, J. (2006) PCT Int. Appl. WO 2006044402 A1 20060427.

493 Eberle, M., Bachmann, M., Strebel, A., Roy, S., Srivastava, S., and Sudhir Saha, S. (2004) PCT Int. Appl. WO 2004103994 A1 20041202.

494 Lee, D. and Stavenger, R.A. (2005) PCT Int. Appl. WO 2005034866 A2 20050421.

495 Lee, D., Stavenger, R.A., Goodman, K.B., Hilfiker, M.A., Cui, H., and Viet, A.Q. (2005) PCT Int. Appl. WO 2005037197 A2 20050428.

496 (a) Thomas, E.W. (1984) *Comprehensive Heterocyclic Chemistry*, vol. 6 (ed. K.T. Potts), Pergamon Press, Oxford, p. 427; (b) Thomas, E.W. (1996) *Comprehensive Heterocyclic Chemistry*, vol. 4 (ed K.T. Potts), Pergamon Press, Oxford, p. 289; (c) Suwinski, J., Szczepankiewicz, W., and Thomas, E.W. (2008) *Comprehensive*

Heterocyclic Chemistry III, vol. 5 (ed. K.T. Potts), Pergamon Press, Oxford, p. 397.

497 (a) Johns, B.A. (2004) PCT Int. Appl. WO 101512. (b) Piatnitski, E., Kiselyov, A., Doddy, J., Hadari, Y., Ouyang, S., and Chen, X. (2004) PCT Int. Appl. WO 0522280.

498 Jursic, B.S. (1988) *Journal of Molecular Structure*, **452**, 153.

499 Montanari, C.A., Giesbrecht, A.M., Sandall, J.P.B., Miyata, Y., and Miller, J. (1996) *Heterocyclic Communications*, **2**, 71.

500 Hughes, G., Kreher, D., Wang, C., Batsanov, A.S., and Bryce, M.R. (2004) *Organic and Biomolecular Chemistry*, **2**, 3363.

501 Stockhause, S., Wickleder, M.S., Meyer, G., Orgzall, I., and Schulz, B. (2001) *Journal of Molecular Structure*, **56**, 175.

502 Polshettiwar, V. and Varma, R.S. (2008) *Tetrahedron Letters*, **49**, 879.

503 Rauf, A., Sharma, S., and Ganga, S. (2008) *Chinese Chemistry Letters*, **19**, 5.

504 Hamada, Y., Adachi, Ch. Tsutsui, T., and Saito, S. (1992) *Optoelectronics - Devices and Technology*, **7**, 83.

505 (a) Feng, L., Wang, X., and Chen, Z. (2008) *Journal of Applied Polymer Science*, **108**, 1995; (b) Wang, H., Ryu, J.-T., Han, Y.S., Kim, D.-H., Choi, B.D., and Kwon, Y. (2007) *Molecular Crystals and Liquid Crystals*, **463**, 285; (c) Wollarz, E., Chrzumnica, E., Fischer, T., and Stumpe, J. (2007) *Dyes and Pigments*, **75**, 753; (d) Buscemi, S., Pace, A., Piccionello Palumbo;, A., and Vivona, N. (2006) *Journal of Fluorine Chemistry*, **127**, 1601; (e) Feng, L. and Chen, Z. (2006) *Spectrochimica Acta*, **63**, 15; (f) Malicka, J., Gryczynski, I., Gryczynski, Z., and Lakowicz, J.R. (2004) *The Journal of Physical Chemistry. B*, **108**, 19114.

506 (a) Kumari, N. and Sah, P. (2008) *Indian Journal of Heterocyclic Chemistry*, **17**, 331; (b) Wagle, S., Adhikari, A.V., and Kumari, N.S. (2008) *Indian Journal of Chemistry*, **47**, 439.

507 Fransky, R. (2004) *Journal of Mass Spectrometry*, **39**, 272.

508 Fransky, R., Schroeder, G., Rybachenko, V., and Szwajka, O.P. (2002) *Rapid Communications. Journal of Mass Spectrometry*, **16**, 390.

509 (a) Kerr, V.N., Ott, D.G., and Hayes, F.N. (1960) *Journal of the American Chemical Society*, **812**, 186; (b) Iqbal, R., Zareef, M., Ahmed, S., Zaidi, J.H., Khan, K.M., Arfan, M., Shafique, M., and Shahzad, S.A. (2006) *Journal of the Chemical Society of Pakistan*, **28**, 165.

510 Carlsen, P.H. and Jorgensen, K.B. (1994) *Journal of Heterocyclic Chemistry*, **31**, 805.

511 Hayes, F.N., Rogers, B.S., and Ott, D.G. (1955) *Journal of the American Chemical Society*, **77**, 1850.

512 Liras, S., Allen, M.P., and Segelstein, B.E. (2000) *Synthetic Communications*, **30**, 437.

513 (a) Brown, P., Best, D.J., Broom, N.J.P., Cassels, R., Bhanlon, P.J., Mitchell, T.J., Osborne, N.F., and Wilson, M.J. (1997) *Journal of Medicinal Chemistry*, **40**, 2563; (b) Rajapakse, H.A., Zhu, H., Young, M.B., and Mott, B.T. (2006) *Tetrahedron Letters*, **47**, 4827.

514 Tully, W.R., Gardner, C.R., and Gillespie, R.J. (1991) *Journal of Medicinal Chemistry*, **34**, 2060.

515 Short, F.W. and Long, L.N. (1969) *Journal of Heterocyclic Chemistry*, **6**, 707.

516 Balchandran, K.S. and George, M.V. (1973) *Tetrahedron*, **29**, 2119.

517 Reddy, P.S.N. and Reddy, P.R. (1987) *Indian Journal of Chemistry*, **26**, 890.

518 Stolle, R. (1899) *Chemische Berichte*, **32**, 797.

519 Boesch, R. (1978) P. A.: DE 78-2808842 19780301.

520 Brain, C.B., Paul, J.M., Loong, Y., and Oakley, P.J. (1999) *Tetrahedron Letters*, **40**, 3275.

521 Mashraqui, S.H., Ghadigaonkar, S.G., and Kenny, S.R. (2003) *Synthetic Communications*, **33**, 2541.

522 Ainsworth, C. (1955) *Journal of the American Chemical Society*, **77**, 1148.

523 Ainsworth, C. (1965) *Journal of the American Chemical Society*, **86**, 5800.

524 Chiba, T. and Okimoto, M. (1992) *The Journal of Organic Chemistry*, **57**, 1375.

525 Yang, R.-Y. and Dai, L.-X. (1993) *The Journal of Organic Chemistry*, **58**, 3381.

526 Rostamizadeh, S. and Gasem Housaini, S.A. (2004) *Tetrahedron Letters*, **45**, 8753.

527 Mishra, P., Rajak, H., and Mehta, A. (2005) *Journal of General Microbiology*, **51**, 133.

528 Holla, B.S., Gonsalves, R., and Shenoy, S. (2000) *European Journal of Medicinal Chemistry*, **35**, 267.

529 Omar, F., Mahfouz, N., and Rahman, M. (1966) *European Journal of Medicinal Chemistry*, **31**, 819.

530 Rivera, N.R., Balsells, J., and Hansen, K.B. (2006) *Tetrahedron Letters*, **47**, 4889.

531 Herbest, R.M. (1961) *The Journal of Organic Chemistry*, **26**, 2372.

532 (a) Osipova, T.F., Koldobskii, G.I., and Ostrovskii, V.A. (1984) *Zhurnal Organicheskoi Khimii*, **20**, 2468; (b) Myznikov, Y.E., Vasil'eva, G.I., and Ostrovskii, V.A. (1988) *Zhurnal Organicheskoi Khimii*, **24**, 1550.

533 Nagakura, I. and Nakanishi, S. (1986) Eur. Pat. Appl. EPXXDW EP 200408 A1 19861210.

534 Bing, Y.J., Leung, L.M., and Menglian, G. (2004) *Tetrahedron Letters*, **45**, 6361.

535 Lukyanov, S.M., Bliznets, I.V., Shorshenev, S.V., Aleksandrov, G.G., Stepanov, A.E., and Vasil'ev, A.A. (2006) *Tetrahedron*, **62**, 1849.

536 (a) Buscemi, S., Cicero, M.G., Vivona, N., and Caronna, T. (1988) *Journal of the Chemical Society-Perkin Transactions 1*, 1313; (b) Buscemi, S., Cicero, M.G., Vivona, N., and Caronna, T. (1988) *Journal of Heterocyclic Chemistry*, **25**, 931; (c) Buscemi, S., Cicero, M.G., Vivona, N., and Caronna, T. (2001) *Journal of Heterocyclic Chemistry*, **38**, 1777; (d) Buscemi, S., Pace, A., Pibiri, I., and Vivona, N. (2002) *The Journal of Organic Chemistry*, **67**, 6253.

537 Buscemi, S., Pace, A., Pibiri, I., Vivona, N., and Caronna, T. (2004) *Journal of Fluorine Chemistry*, **125**, 165.

538 Pace, A., Pibiri, I., Buscemi, S., Vivona, N., and Malpezzi, L. (2004) *The Journal of Organic Chemistry*, **69**, 4108.

539 Pace, A., Buscemi, S., and Vivona, N. (2005) *The Journal of Organic Chemistry*, **70**, 2322.

540 Pace, A., Pibiri, I., Palumbo Piccionello, A., Buscemi, S., Vivona, N., and Barone, G. (2007) *The Journal of Organic Chemistry*, **72**, 7656.

541 Ainsworth, C. (1965) *Canadian Journal of Chemistry*, **43**, 1607.

542 Gibson, H.H. Jr., Weissinger, K., Abashawl, A., Hall, G., Lawshae, T., LeBlanc, K., Moody, J., and Lwowski, W. (1986) *The Journal of Organic Chemistry*, **51**, 3858.

543 Molina, P., Alajarin, M., Arques, A., Benzal, R., and Hernandez, H. (1984) *Journal of the Chemical Society-Perkin Transactions 1*, 1891.

544 (a) Butler, R.N., Cloonan, M.O., Smyth, G.M., McArdle, P., and Cunningham, D. (2003) *Arkiv*, **7**, 244; (b) Butler, R.N. and Cloonan, M.O. (1997) *Bulletin des Sociétés Chimiques Belges*, **106**, 515.

545 Bozo, E., Szilagyi, G., and Janaky, J. (1989) *Archiv der Pharmazie*, **322**, 583.

546 Metivier, J. and Boesch, R. (1976) P. A.: DE 1795773 19760429.

547 Boesch, R. (1976) P. A.: DE 76- 2604110 19760203.

548 Musser, J.H., Brown, R.E., Love, B., Bailey, K., Jones, H., Kahen, R., Huang, F.-C., Khandwala, A., Leibowitz, M., Sonnino-Goldman, P., and Donigi-Ruzza, D. (1984) *Journal of Medicinal Chemistry*, **27**, 121.

549 Yale, H.L. and Losee, K. (1966) *Journal of Medicinal Chemistry*, **9**, 478.

550 Huisgen, R., Grashey, R., Knupfer, H., Kunz, R., and Seidel, M. (1964) *Chemische Berichte*, **97**, 1085.

551 Chichetti, S.M., Ahearn, S.P., Adams, B., and Rivkin, A. (2007) *Tetrahedron Letters*, **48**, 8250.

552 Ashraf, A.M. and Shaharyar, M. (2007) *Bioorganic & Medicinal Chemistry Letters*, **17**, 3314.

553 Belkadi, M. and Othman, A.A. (2006) *Arkivoc*, **11**, 183.

554 Szczepankiewicz, B.G., Liu, G., Jae, H.-S., Tasker, A.S., Gunawardana, I.W., von Geldern, T.W., Gwaltney, S.L., Wu-Wong, J.R., Gehrke, L., Chiou, W.J., Credo, R.B., Alder, J.D., Nukkala, M.A., Zielinski, N.A., Jarvis, K., Mollison, K.W., Frost, D.J., Bauch, J.L., Hui, Y.H., Claiborne, A.K., Li, Q., and Rosenberg, S.H. (2001) *Journal of Medicinal Chemistry*, **44**, 4416.

555 Huisgen, R., Seidel, M., Sauer, J., McFarland, J.W., and Wallbillich, G.

(1959) *The Journal of Organic Chemistry*, **24**, 892.

556 Huisgen, R., Grashey, R., Seidel, M., Knupfer, H., and Schmidt, R. (1962) *Ann*, **658**, 169.

557 Jones, R.C.F., Hollis, S.J., and Iley, J.N. (2007) *Arkivoc*, **5**, 152.

558 Ashraf, A.A., Hassan;, A.A., Ameen, M.A., and Brown, A.B. (2008) *Tetrahedron Letters*, **49**, 4060.

559 Kassam, K., Pole, D.L., El-Saidi, M., and Warkentin, J. (1994) *Journal of the American Chemical Society*, **116**, 1161.

560 Zwanenburg, B., Weening, W.E., and Strafing, J. (1964) *Recueil des Travaux Chimiques*, **83**, 877.

561 (a) Zinner, G. and Kliwing, W. (1973) *Archiv der Pharmazie*, **306**, 134; (b) Lennart, E. and Kay, P. (1964) *Acta Chemica Scandinavica*, **18**, 721.

562 Matsui, M., Furukawa, K., Funabiki, K., and Shibata, K. (2002) *Shikizai Kyokaishi*, **75**, 106.

563 Yadav, L.D.S., Vaish, A., and Sharma, S. (1994) *Journal of Agricultural and Food Chemistry*, **42**, 811.

564 Grashey, R. and Adelsberger, K. (1962) *Angewandte Chemie*, **74**, 292.

565 (a) Roussi, F., Chauveau, A., Bonin, M., Micouin, L., and Husson, H.-P. (2000) *Synthesis*, **8**, 1170; (b) Chauveau, A., Martens, T., Bonin, M., Micouin, L., and Husson, H.-P. (2002) *Synthesis*, **13**, 1885.

566 Khau, V.V. and Martinelli, M.J. (1996) *Tetrahedron Letters*, **37**, 4323.

567 Chung, F., Chauveau, A., Seltki, M., Bonin, M., and Micouin, L. (2004) *Tetrahedron Letters*, **45**, 3127.

568 Grekov, A.P. and Azen, R.S. (1961) *Zhurnal Obshchei Khimii*, **31**, 407.

569 Brown, H.C. and Cheng, M.T. (1961) *The Journal of Organic Chemistry*, **27**, 3240.

570 Reitz, D.B. and Finkes, M.J. (1989) *The Journal of Organic Chemistry*, **54**, 1760.

571 (a) Artemov, V.N. and Shvaika, O.P. (1971) *Khimiya Geterotsiklicheskikh Soedinenii*, 905; (b) Joshi, S.S. and Karnik, A.V. (2006) *Indian Journal of Chemistry*, **45**, 1057.

572 El-masry, H., Fahmy, H.H., and Ali Abdelwahed, S.H. (2000) *Molecules*, **5**, 1429.

573 Reitz, D.B. and Finkes, M.J. (1989) *Journal of Heterocyclic Chemistry*, **26**, 225.

574 Sherman, W.R. and Von Esch, A. (1961) *The Journal of Organic Chemistry*, **27**, 3472.

575 (a) Stempel, A., Zelauskas, J., and Aeschlimann, J.A. (1955) *The Journal of Organic Chemistry*, **20**, 412; (b) Saegusa, Y., Harada, S., and Kanamura, S. (1988) *Journal of Heterocyclic Chemistry*, **25**, 1337; (c) Pilgram, K.H. (1982) *Journal of Heterocyclic Chemistry*, **19**, 823.

576 Kristinsson, H., Winkler, T., Winkler, T., and Mollenkopf, M. (1985) *Helvetica Chimica Acta*, **68**, 1155.

577 Hetzheim, A. and Beyer, H. (1972) *Chemische Berichte*, **103**, 272.

578 Tsuge, O., Oe, K., and Tashiro, M. (1977) *Chemistry Letters*, 1207.

579 Sherman, W.R. and Von Esch, A. (1961) *The Journal of Organic Chemistry*, **27**, 3472.

580 Hetzheim, A. and Singelmann, J. (1971) *Annalen der Chemie-Justus Liebig*, **749**, 125.

581 Boyd, G.V. and Dando, S.R. (1971) *Journal of the Chemical Society*, **2**, 225.

582 Chen, C.-J., Song, B.-A., Yang, S., Xu, G.-F., Bhadury, P.S., Jin, L.-H., Hu, D.-Y., Li, Q.-Z., Liu, F., Xue, W., Lu, P., and Chen, Z. (2007) *Bioorganic and Medicinal Chemistry*, **15**, 3981.

583 Liu, F., Luo, X.-Q., Song, B.-A., Bhadury, P.S., Yang, S., Jin, L.-H., Xue, W., and Hu, D.-Y. (2008) *Bioorganic and Medicinal Chemistry*, **16**, 3632.

584 Ioannisyan, E., Chernitsa, B.V., and Yakovlev, V.V. (2006) *Russian Journal of Organic Chemistry*, **42**, 1089.

585 Javaid, K. and Smith, D.M. (1984) *Journal of Chemical Research (S)*, **4**, 118.

586 Saegusa, Y., Seikiba, K., and Nakamura, S. (1990) *Journal of Polish Science (A)*, **28**, 3637.

587 Grekov, A.P. and Grigor'eva, V.I. (1961) *Zhurnal Obshchei Khimii*, **31**, 4012.

588 Grekov, A.P. and Shvaika, O.P. (1960) *Soveshch*, 105.

589 Vincent, M., Maillard, J., and Benard, M. (1962) *Bulletin de la Société Chimique de France*, 1580.

590 Fort, J.F. and Giraudon, R. (1974) Patent DE 2413938 19740926.

591 Hoffmann, R.W. and Luthardt, H.J. (1966) *Tetrahedron Letters*, **4**, 411.

592 Brown, H.C., Cheng, M.T., Parcell, L.J., and Pilipovich, D. (1961) *The Journal of Organic Chemistry*, **26**, 4407.

593 Chen, H., Li, Z., and Han, Y. (2000) *Journal of Agricultural and Food Chemistry*, **48**, 5312.

594 Liu, C.-J., Shi, T.-H., and Li, Y.-P. (2007) *Youji Huaxue*, **27**, 985; (2007) *Chemical Abstracts*, **149**, 200836.

595 Molina, P., Tarraga, A., and Espinosa, A. (1988) *Synthesis*, 690.

596 (a) El-masry, A.H., Fahmy, H.H., and Ali Abdelwahed, S.H. (2000) *Molecules*, **5**, 1429; (b) Mekuskiene, G., Burbuliene, M.M., Jakubkiene, V., Udrenaite, E., Gaidelis, P., and Vainilavieius, P. (2003) *Chemistry of Heterocyclic Compounds*, **39**, 1364.

597 Vasiliev, N.V., Lyashenko, Y.E., Kolomietz, A.F., and Solkolskii, G.A. (1987) *Khimiya Geterotsiklicheskikh Soedinenii*, 562.

598 Thalhammer, F., Wallfahrer, U., and Sauer, J. (1988) *Tetrahedron Letters*, **29**, 3231;*ibidem* 1995, 36, 5275.

599 Wilkie, G.D., Elliot, G.I., Blagg, B.S.J., Wolkenberg, S.E., Soenen, D.R., Miller, M.M., Pollack, S., and Boger, D.L. (2002) *Journal of the American Chemical Society*, **124**, 11294.

600 (a) Elliott, G.I., Gregory, I., Velcicky, J., Ishikawa, H., Li, Y., and Boger, D.L. (2006) *Angewandte Chemie*, **45**, 620; (b) Ishikawa, H., Elliott, G.I., Velcicky, J., Choi, Y., and Boger, D.L. (2006) *Journal of the American Chemical Society*, **128**, 10596.

601 Choi, Y., Ishikawa, H., Velcicky, J., Elliott, G.I., Miller, M.M., and Boger, D.L. (2005) *Organic Letters*, **7**, 4539.

602 Tsuge, O., Tashiro, M., and Oe, K. (1968) *Tetrahedron Letters*, **9**, 3971.

603 Tsuge, O., Oe, K., and Tashiro, M. (1973) *Tetrahedron*, **29**, 41.

604 Wang, J., Wang, R., Yang, J., Zheng, Z., Carducci, M.D., Cayou, T., Peyghambarian, N., and Jabbour, G.E. (2001) *Journal of the American Chemical Society*, **123**, 6179.

605 Yuan, D.-K., Li, Z.-M., Zhao, W.-G., and Chen, H.-S. (2003) *Yingyong Huaxue*, **20**, 624. CAN 140:4994 AN 2003: 617213.

606 Feng, X.-M., Chen, R., and Lin, T. (1994) *Youji Huaxue*, **14**, 293; (1994) *Chemical Abstracts*, 121, 205276.

607 Levins, C.G. and Wan, Z.-K. (2008) *Organic Letters*, **10**, 1755.

608 Golfier, M. and Milcent, R. (1974) *Tetrahedron Letters*, **44**, 3871.

609 Grekov, A.P., Azen, R.S., and Kharkov (1961) *Zhurnal Obshchei Khimii*, **31**, 1919.

610 Blackhall, A., Brydon, D.L., Javaid, K., Sagar, A.J.G., and Smith, D.M. (1984) *Journal of Chemical Research (S)*, 382.

611 Srivastava, M.K. (2000) *Bollettino Chimico Farmaceutico*, **139**, 161.

612 Rajak, H., Kharya, M., and Mishra, P. (2008) *Archiv der Pharmazie*, **341**, 247.

613 Ohmoto, K., Yamamoto, T., Okuma, M., Horiuchi, T., Imanishi, H., Odagaki, Y., Kawabata, K., Sekioka, T., Hirota, Y., Matsuoka, S., Nakai, H., Toda, M., Cheronis, J.C., Spuce, L.W., Gyorkos, A., and Wieczorek, M. (2001) *Journal of Medicinal Chemistry*, **44**, 1268.

614 Rydezewski, R.M., Burrill, L., Mendonca, R., Palmer, J.T., Rice, M., Tahilramani, R., Bass, K.E., Leung, L., Gjerstad, E., Janc, J.W., and Pan, L. (2006) *Journal of Medicinal Chemistry*, **49**, 2953.

615 Kuebel, B. (1982) PCT German Patent CODEN: GWXXBX, DE 3105222, A1 19820909. Application: DE 81-3105222, 19810213.

616 Knaus, G. and Meyers, A.I. (1974) *The Journal of Organic Chemistry*, **39**, 1189.

617 Hu, G., Xu, Q., Zhang, Z., Chen, B., Xu, Q., Huang, W., and Zhang, H. (2004) *Huaxue Zazhi*, **14**, 76.

618 Ryu, H., Subramanian, L.R., and Hanack, M. (2006) *Tetrahedron*, **62**, 6236.

619 Huges, G. and Bryce, M.R. (2005) *Journal of Materials Chemistry*, **15**, 94.

620 Xu, Z., Li, Y., Ma, X., Gao, X., and Tian, H. (2008) *Tetrahedron*, **64**, 1860.